Diagnostic Microbiology

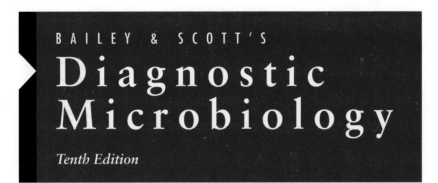

BAILEY & SCOTT'S
Diagnostic
Microbiology

Tenth Edition

Betty A. Forbes, PhD

Professor of Pathology and Medicine
Director, Clinical Microbiology Laboratories
SUNY Health Science Center
Syracuse, New York

Daniel F. Sahm, PhD

Chief Scientific Officer
MRL Pharmaceutical Services
Reston, Virginia

Alice S. Weissfeld, PhD

President, Microbiology Specialists Incorporated
Adjunct Assistant Professor
Department of Microbiology and Immunology
Baylor College of Medicine
Houston, Texas

Photography by Ernest A. Trevino, MT (ASCP)
Director of Operations
Microbiology Specialists Incorporated
Houston, Texas

with 548 illustrations

 Mosby

St. Louis Baltimore Boston Carlsbad Chicago Naples New York Philadelphia Portland
London Madrid Mexico City Singapore Sydney Tokyo Toronto Wiesbaden

Mosby
Dedicated to Publishing Excellence

A Times Mirror
Company

Publisher: Don Ladig
Senior Editor: Jennifer Roche
Developmental Editors: Sandra J. Parker, Laura MacAdam
Project Manager: Mark Spann
Senior Production Editor: Anne Salmo
Book Design Manager: Judi Lang
Manufacturing Manager: Betty Mueller
Cover Art: Michael Kilfoy

TENTH EDITION

Copyright © 1998 by Mosby, Inc.

Previous editions copyrighted 1962, 1966, 1970, 1974, 1978, 1982, 1986, 1990, 1994

Printed in the United States of America
Composition by Black Dot Group
Lithography/color film by Black Dot Group
Printing/binding by Von Hoffmann Press

Mosby, Inc.
11830 Westline Industrial Drive
St. Louis, Missouri 63146

Library of Congress Cataloging-in-Publication Data
Weissfeld, Alice S.
    Bailey & Scott's diagnostic microbiology / Alice S. Weissfeld, Daniel F. Sahm, Betty A. Forbes; photography by Ernest Trevino.— 10th ed.
        p. cm.
    Rev. ed. of: Bailey & Scott's diagnostic microbiology / Ellen Jo Baron, Lance R. Peterson, Sydney M. Finegold. 9th ed. ©1994.
    Includes bibliographical references and index.
    ISBN 0-8151-2535-6
    1. Diagnostic microbiology.   I. Sahm, Daniel F.   II. Forbes, Betty A.   III. Baron, Ellen Jo.   Bailey & Scott's diagnostic microbiology.   IV. Title.
    [DNLM: 1. Microbiological Techniques.   QW 25 W433b 1998]
QR67.B37   1998
616'.01'028—dc21
DNLM/DLC
for Library Congress   97-46902
                                                            CIP

98 99 00 01 02/9 8 7 6 5 4 3 2 1

Contributors

CHAPTER AUTHORS

Lynne S. Garcia, MS, F (AAM)
Manager, Clinical Microbiology
University of California Los Angeles Medical Center
Los Angeles, California

Gary W. Procop, MD
Instructor of Microbiology and Laboratory Medicine
Mayo Medical School
Fellow in Clinical Microbiology
Mayo Clinic and Mayo Foundation
Rochester, Minnesota

Glenn D. Roberts, PhD
Professor of Microbiology and Laboratory Medicine
Mayo Medical School
Director, Clinical Mycology and Mycobacteriology
Laboratories
Mayo Clinic and Mayo Foundation
Rochester, Minnesota

Richard B. Thomson, Jr., PhD
Director, Microbiology and Virology
Associate Professor of Pathology
Northwestern University Medical School
Department of Pathology and Laboratory
Medicine
Evanston Hospital
Evanston, Illinois

CHAPTER REVIEWER
Pamela A. Fuselier, MT (ASCP)
Technical Specialist
Microbiology Specialists Incorporated
Houston, Texas

Reviewers

Diane Berger, MHS, MT (ASCP)
Instructor, MLT Program
Kankakee Community College
Kankakee, Ill.

Louis B. Caruana, PhD, MT (ASCP)
Professor
Southwest Texas State University
Health Science Center
San Marcos, Texas

Darlene E. Clingenpeel, MS, MT (ASCP)
Assistant Professor
Illinois Central College
Peoria, Ill.

Donald Lehman, MS, MT (ASCP), SM (AAM)
Assistant Professor
Department of Medical Technology
University of Delaware
Newark, Delaware

Donna Poteet, MA, CLS (NCA)
Program Director
Medical Laboratory Technician and
Phlebotomy Programs
Central Texas College
Killeen, Texas

Foreword

Among my collection of favorite books is a slim volume published 40 years ago, entitled *Diagnostic Bacteriology, Fifth Edition,* by Isabelle G. Schaub, M. Kathleen Foley, Elvyn G. Scott, and W. Robert Bailey. Although the names of the last two authors are familiar to most everyone who has received training in clinical microbiology, the first two are little remembered as the persons whose initial works predated and inspired the classic "Bailey & Scott" series, now in its tenth edition. In the foreword of the fifth edition, Schaub and Foley wrote: "In 1940 we first published a small laboratory manual entitled *Methods for Diagnostic Bacteriology.* Our purpose at that time was to present a guide for the isolation and identification of pathogenic bacteria for use in the clinical bacteriology laboratory." Through the many iterations of this text, various authors have undertaken this same increasingly daunting task, expanding it also to include other areas of laboratory diagnosis (parasitology, mycology, and virology) now considered integral sections of the clinical microbiology laboratory.

Undoubtedly, the challenges that face clinical microbiologists as we approach the twenty-first century could not have been predicted by the authors of the original text. The mandates of the Clinical Laboratory Improvement Amendments (CLIA '88) and the demands for fiscal responsibility that affect our daily conduct in the laboratory were not of concern to these pioneers. In 1940 there was no need to perform antimicrobial susceptibility tests and the specter of antimicrobial agent-resistant bacteria, whose detection can defy the most sophisticated practitioner, had not yet appeared. Whereas determining which isolates were pathogens or nonpathogens was once relatively straightforward, the modern medical advances that sustain and prolong life require ever more cautious interpretations of the concept of normal flora. The opportunistic infections of AIDS and other immunocompromised patients, the ease of global travel, and the emerging and reemerging infectious diseases arising sometimes as the consequence of political or social upheavals or acquisition of new virulence or antimicrobial resistance determinants have further complicated today's diagnostic task. On the other hand, advances in molecular techniques have permitted a better understanding of taxonomy of and relationships among certain microorganisms so that we can better define their role in specific disease syndromes. These molecular advances have also given rise to highly specific (albeit highly expensive) methods for microbial identification and in some cases, methods for their direct detection in clinical specimens.

It is in this context that the tenth edition of *Bailey & Scott's Diagnostic Microbiology* appears, authored anew by three experienced clinical microbiologists: Betty A. Forbes, Daniel F. Sahm, and Alice S. Weissfeld. Each has contributed greatly to updating old information and providing new concepts in their respective areas of expertise. Whether a beginning or experienced practitioner, the reader will appreciate the clearly written, practical, and informative approach that has been taken in each chapter. In four parts the authors describe general issues in clinical microbiology, the scientific and laboratory basis for clinical microbiology, diagnosis by organ system, and bacteriology. The fifth through seventh parts are written by other experts in the fields of parasitology, mycology, and virology. Extensive use of tables, diagrams, and color photographs throughout the text enhances and facilitates comprehension of subject matter.

The most significant departure from previous editions is in the section on bacteriology in which the authors have approached the subject from the point of view of a bench technologist confronted with a culture plate of microbial growth. A few direct observations (e.g., Gram stain reaction and morphology, growth on certain media) and standard rapid tests (e.g., oxidase, catalase) are used to place the organisms into major groupings from which identification then proceeds. Further division of each chapter into sections describing general characteristics of the organisms, their epidemiology, pathogenesis, laboratory diagnosis, and antimicrobial susceptibility testing and therapy guide the workup of the culture. The section on susceptibility testing will be especially valuable for technologists in small clinical laboratories who often are requested to perform susceptibility tests on organisms for whom no standardized susceptibility testing method is available. Here, they will find documentation to support their refusal of such requests. In the words of the authors, "Microbiologists are going back to basics. The real challenge is how to most efficiently and effectively bring all the information together in a cost-effective and clinically relevant manner." This new edition is an admirable starting point in helping us meet that challenge.

I received my copy of Schaub, Foley, Scott, and Bailey's fifth edition of *Diagnostic Bacteriology* from a clinical microbiology laboratory supervisor who, along with the book, initiated me in what has become a fascinating and rewarding career in clinical microbiology. I expect that this new edition will inspire many students in the same way.

Josephine A. Morello, PhD
Director of Hospitals Laboratories
Co-director of Clinical Microbiology Laboratories
Professor of Pathology and Medicine
University of Chicago Medical Center
Chicago, Illinois

Preface

Undertaking the tenth edition of a best-selling microbiology textbook was an incredible challenge. The most daunting tasks were to try to reconcile extensive changes in the practice of clinical microbiology over the past several years, the need to provide students with a comprehensive blueprint for performing their future jobs, and the opportunity to write a text to bridge the gap between school and clinical practice.

To accomplish our goals, we reorganized the table of contents and expanded the scope of the book. Part 1 includes chapters covering physical design of the laboratory, organization of the laboratory, cost accounting, test validation and verification, and infection control and quality assurance. Part 2 focuses on several basic aspects of medical microbiology, such as microbial taxonomy, physiology, genetics, and host-parasite relationships that provide important background material for understanding the practices of clinical microbiology. Overviews of technologies, including microscopy, microbial cultivation, microbial identification, as well as molecular and immunologic methods used in diagnostic microbiology, are also included in Part 2. Part 3, "Diagnosis by Organ System," presents an overview of general considerations for the particular organ systems, the pathogenesis of disease, clinical manifestations, the epidemiology and etiology of disease, and laboratory diagnosis of infections. Parts 4 through 7 are presented in a format intended to be user friendly in both the classroom and at the bench. For the sake of consistency, each chapter in Part 4, "Bacteriology," is organized so that the same topic subheadings are employed from one chapter to another. Furthermore, the chapters on bacteriology have been conveniently grouped into sections, so organisms that share common characteristics with respect to key phenotypic traits (e.g., oxidase reactions, growth on MacConkey and other agars, and catalase reactions) are considered together. Care has been taken to discuss organism identification and antimicrobial susceptibility testing in light of clinical relevance and in a manner that is cost effective and efficient. Lastly, it is intended that the brand new photographs and line art for this edition will allow readers to visualize the numerous and varied facets of the microorganisms that cause infections in humans. It is hoped that by these visual means, students will more easily be able to make the transition from the classroom into the real clinical microbiology laboratory setting.

We hope that students, teachers, and clinical microbiologists will find this book a worthy successor to the previous editions.

Betty A. Forbes
Daniel F. Sahm
Alice S. Weissfeld

Acknowledgments

We acknowledge the help of individuals at Mosby who guided us through this project. These people include Jennifer Roche, our acquisitions editor; Sandra J. Parker and Laura MacAdam, our developmental editors; Anne Salmo, our production editor; and Mark Spann, our project manager. In addition, we thank Nadine Sokol, our medical illustrator, who took our scribbles and made them look professional.

We each would like to thank the individuals who helped us to complete this project. To wit, we recognize David Meliza, Debbie Mandvia, Kisha Lemelle, Cindy Peyton, and Buffy Turner, who typed the manuscripts and otherwise assisted a computer-challenged and overly committed Alice S. Weissfeld. The efforts of Dennis Vittatow that were so essential for accomplishing chapter merges are also gratefully acknowledged. In addition, we recognize the skills and patience of Judith Kelsey, who typed the initial first-draft ramblings and final manuscript versions; Deanna Kiska for her clinical microbiology expertise and moral support; Nadine Bartholoma for her critical reviews; and all the wonderful staff in the clinical microbiology laboratory who helped Betz Forbes survive this incredible process.

Finally, we acknowledge and thank all the clinical microbiologists, scientists, clinicians, and educators who, through the communication of their work, observations, and experience, made the writing of this textbook possible.

To my family—my husband, Ray, and my children, Elizabeth and Jonathan—who nurtured, loved, and faithfully supported me throughout this entire process.

BAF

To my wife, Janet, and our sons Nathan, Elias, Aaron, Zachary, and Matthew. Their unending love, support, and laughter make anything possible.

DFS

To my parents, who are always there to help and encourage me, and to everyone at MSI, who make coming to work so much fun.

ASW

Contents

Part One

General Issues in Clinical Microbiology

1 | GENERAL ISSUES AND ROLE OF LABORATORIANS

In the late 1800s, the first clinical microbiology laboratories were organized to diagnose infectious diseases such as tuberculosis, typhoid fever, malaria, intestinal parasites, syphilis, gonorrhea, and diphtheria. Between 1860 and 1900 microbiologists such as Pasteur, Koch, and Gram developed the techniques for staining and the use of solid media for isolation of microorganisms that are still used in clinical laboratories today. Microbiologists continue to look for the same organisms that these laboratorians did, as well as a whole range of others that have been uncovered in the twentieth century, for example, *Legionella*, viral infections, nontuberculosis acid-fast bacteria, and fungal infections. Microbiologists work in public health laboratories, hospital laboratories, reference or independent laboratories, and physician office laboratories (POLs). Depending on the level of service of each facility, the type of testing differs, but in general a microbiologist will perform one or more of the following functions:

- Cultivation (growth), identification, and antimicrobial susceptibility testing of microorganisms
- Direct detection of infecting organisms by microscopy
- Direct detection of specific products of infecting organisms using chemical, immunologic, or molecular techniques
- Detection of antibodies produced by the patient in response to an infecting organism (serodiagnosis)

This chapter presents an overview of issues involved in infectious disease diagnostic testing. Many of these issues are covered in detail in separate chapters, which are cited.

ROLE OF THE MICROBIOLOGIST: RESPONSIBILITY TO THE PATIENT AND CLINICIAN

The microbiologist is an integral part of the health care team. In today's world of shrinking health care dollars and potential rationing of medical services, the microbiologist can ultimately save the patient money by providing an accurate diagnosis in a timely manner

using the most cost-effective techniques. The microbiologist's primary responsibility is to communicate information about the patient quickly and accurately in both verbal and written formats. The microbiologist can also assist the patient care facility in saving money by helping to (1) identify and control nosocomial (hospital-acquired) pathogens and to quickly track organisms resistant to antimicrobial agents (a role discussed more fully in Chapter 7) and (2) coordinate the antimicrobial agents tested in the laboratory with those selected by the institution's Pharmacy and Therapeutics Committee. Finally, the microbiologist serves the patient and clinician by preparing and updating manuals regarding appropriate specimen collection and handling.

COMMUNICATION OF LABORATORY FINDINGS

To fulfill their professional obligation to the patient, microbiologists must communicate their findings to those health care professionals responsible for treating the patient. This task is not as easy as it may seem. This is nicely illustrated in a study in which a group of physicians was asked whether they would treat a patient with a sore throat given two separate laboratory reports, that is, one that stated, "many group A *Streptococcus*" and one that stated, "few group A *Streptococcus*".[1] Although group A *Streptococcus (Streptococcus pyogenes)* is considered significant in any numbers in a symptomatic individual, the physicians said that they would treat the patient with many organisms but not the one with few organisms. Thus, although a pathogen (group A *Streptococcus*) was isolated in both cases, one word on the report (either "many" or "few") made a difference in how the patient would be handled.

In communicating with the physician, the microbiologist can avoid confusion and misunderstanding by not using jargon or abbreviations and by providing reports with clear-cut conclusions. The microbiologist should not assume that the clinician is fully familiar with laboratory procedures or the latest microbial taxonomic schemes. Thus, when appropriate, interpretive statements should be included in the written report along with the specific results. One such example would be the addition of a statement, such as "suggests contamination at collection," when more than three organisms are isolated from a clean-voided midstream urine specimen.

Laboratory newsletters should be used to provide physicians with material such as details of new procedures, nomenclature changes, and changes in usual antimicrobial susceptibility patterns of frequently isolated organisms. This last information, discussed in more detail in Chapters 4 and 19, is very useful to clinicians when selecting empiric therapy. Empiric therapy is based on the physician determining the most likely organism causing a patient's clinical symptoms and then selecting an antimicrobial that, in the past, has worked against that organism in a particular hospital or geographic area. Empiric therapy is used to start patients on treatment before the results of the patient's culture are known and may be critical to the patient's well-being in cases of life-threatening illnesses.

Positive findings should be telephoned to the clinician, and all verbal reports should be followed by written confirmation of results. Results should be legibly handwritten or generated electronically using a computer.

EXPEDITING RESULTS REPORTING—COMPUTERIZATION

Before widespread computerization of clinical microbiology laboratories, result reporting was accomplished by handwriting reports and having couriers deliver hard copies that were pasted into the patient's chart. Today, microbiology computer software is available that simplifies and speeds up this task. One of the first microbiology programs was designed at Massachusetts General Hospital in Boston and was written in a language, MUMPS (Massachusetts General Hospital Utility Multi-Program System), which is still used today in some commercially available systems. FORTRAN (FORmula TRANslator), COBOL (COmmon Business-Oriented Language), Pascal, and "C" are other common microbiology computer languages.

CPUs (central processing units), disks, tape drives, controllers, printers, video terminals, communication ports, modems, and other types of hardware support running the software. The hardware and software together make up the complete Laboratory Information System (LIS). Many LIS systems are, in turn, hooked up with a Hospital Information System (HIS). Between the HIS and LIS systems, most functions involved in ordering and reporting laboratory tests can be handled electronically. Order entry, patient identification, and specimen identification can

be handled using the same type of bar coding that is commonly used in supermarkets. The LIS also takes care of result reporting and supervisory verification of results, stores quality control data, allows easy test inquiries, and assists in test management reporting by storing, for example, the number of positive, negative, and unsatisfactory specimens. Most large systems also are capable of **interfacing** (communicating) with microbiology instruments to automatically download (transfer) and store data regarding positive cultures or antimicrobial susceptibility results. Results of individual organism antibiograms (patterns) can then be retrieved monthly so hospitalwide susceptibility patterns can be studied for emergence of resistant organisms or other epidemiologic information. Many vendors of laboratory information systems are now writing software for microbiology to adapt to personal computers (PCs) so that large CPUs may no longer be needed. This should bring down the cost of microbiology systems so that even smaller laboratories will be able to afford them. Even small systems can be interfaced with teletypes or electronic facsimile machines (faxes) for quick and easy reporting and information retrieval. This further improves the quality of patient care.

GENERAL CONCEPTS FOR SPECIMEN COLLECTION AND HANDLING

Specimen collection and transportation are critical considerations, because any results the laboratory generates will be limited by the quality of the specimen and its condition on arrival in the laboratory. Specimens should be obtained to preclude or minimize the possibility of introducing extraneous (contaminating) microorganisms that are not involved in the infectious process. This is a particular problem, for example, in specimens collected from mucous membranes that are already colonized with microorganisms that are part of an individual's endogenous or "normal" flora; these organisms are usually contaminants but may also be opportunistic pathogens. For example, the throats of hospitalized patients on ventilators may frequently be colonized with *Klebsiella pneumoniae;* although *K. pneumoniae* is not usually involved in cases of community-acquired pneumonia, it can cause a hospital-acquired respiratory infection in this subset of patients. Use of special techniques that bypass areas containing normal flora when this is feasible (e.g., covered brush bronchoscopy in critically ill patients with pneumonia) will prevent many problems associated with false-positive results. Likewise, careful skin preparation before procedures such as blood cultures and spinal taps will decrease the chance that organisms normally present on the skin will contaminate the specimen.

APPROPRIATE COLLECTION TECHNIQUES

Specimens should be collected during the acute (early) phase of an illness (or within 2 to 3 days for viral infections), and *before* antibiotics are administered, if possible. Swabs generally are poor specimens if tissue or needle aspirates can be obtained. It is the microbiologist's responsibility to provide clinicians with a collection manual or instruction cards listing optimal specimen collection techniques and transport information. Instructions should be written, so specimens collected by the patient (e.g., urine, sputum, or stool) are handled properly. Most urine or stool collection kits contain instructions in several languages, but nothing substitutes for a concise set of verbal instructions. Similarly, when distributing kits for sputum collection, the microbiologist should be able to explain to the patient the difference between spitting in a cup (saliva) and producing good lower respiratory secretions from a deep cough (sputum). General collection information is shown in Table 1-1. An in-depth discussion of each type of specimen is found in Part 3.

SPECIMEN TRANSPORT

Ideally, specimens should be transported to the laboratory within 30 minutes of collection. All specimen containers should be leak-proof, and the specimens should be transported within sealable, leak-proof, plastic bags with a separate section for paperwork; resealable bags or bags with a permanent seal are common for this purpose. Bags should be marked with a biohazard label (Figure 1-1). Many microorganisms are susceptible to environmental conditions such as the presence of oxygen (anaerobic bacteria), changes in temperature *(Neisseria meningitidis),* or changes in pH *(Shigella).* Thus, use of special preservatives or holding media for transportation of specimens delayed for more than 30 minutes is important in ensuring organism viability (survival).

SPECIMEN PRESERVATION

Preservatives, such as boric acid for urine or polyvinyl alcohol (PVA) and buffered formalin for stool for ova and parasite (O&P) examination, are designed to maintain the appropriate colony counts (for urines) or the integrity of trophozoites and cysts (for O&Ps), respectively. Other transport, or **holding,** media maintain the viability of microorganisms present in a specimen without supporting the growth of any of the organisms. This maintains the organisms in a state of suspended animation so that no organism overgrows another or dies out. Stuart's medium and Amie's medium are two common holding media. Sometimes charcoal is added to these media to absorb fatty acids present in the specimen that could kill fastidious (fragile) organisms such as *Neisseria gonorrhoeae* or *Bordetella pertussis.*

Anticoagulants are used to prevent clotting of specimens such as blood, bone marrow, and synovial fluid because microorganisms will otherwise be bound up in the clot. The type and concentration of anticoagulant is very important because many organisms are inhibited by some of these chemicals. Sodium polyanethol sulfonate (SPS) at a concentration of 0.025% (w/v) is usually used, because *Neisseria* spp. and some anaerobic bacteria are particularly sensitive to higher concentrations. Because the ratio of specimen to SPS is so important, it is necessary to have both large (adult-size) and small (pediatric-size) tubes available, so organisms in small amounts of bone marrow or synovial fluid are not overwhelmed by the concentration of SPS. Heparin is also a commonly used anticoagulant, especially for viral cultures, although it may inhibit growth of gram-positive bacteria and yeast. Citrate, EDTA (ethylenediaminetetraacetic acid), or other anticoagulants should not be used for microbiology, as their efficacy has not been demonstrated for a majority of organisms. It is the microbiologist's job to make sure the appropriate anticoagulant is used for each procedure. One generally should not specify a color ("yellow-top") tube for collection without specifying the anticoagulant (SPS), as at least one popular brand of collection tube (Vacutainer, Becton Dickinson) has a yellow-top tube with either SPS or ACD (trisodium citrate/citric acid/dextrose); ACD is not appropriate for use in microbiology.

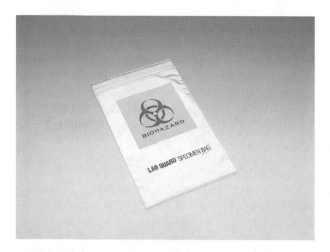

FIGURE 1-1 Specimen bag with biohazard label, separate pouch for paperwork, and zip-lok seal. (Courtesy Allegiance Healthcare Corp., McGaw Park, Ill.)

SPECIMEN STORAGE

If specimens cannot be processed as soon as they are received, they must be stored (see Table 1-1). Several storage methods are used (i.e., refrigerator temperature [4° C], ambient [room] temperature [22° C], body temperature [37° C], and freezer temperature [either −20° or −70° C]), depending on the type of transport media (if applicable) and the etiologic (infectious) agents sought. Specimens suspected of containing anaerobic bacteria, for example, should never be stored in the refrigerator, while cerebral spinal fluid (CSF) should always be kept at 37° C. Urine, stool, viral specimens, sputa, swabs, and foreign devices such as catheters should be stored at 4° C. Serum for serologic studies may be frozen for up to 1 week at −20° C, and tissues or specimens for long-term storage should be frozen at −70° C.

SPECIMEN LABELING

Specimens should be labeled at the very least with the patient's name, identifying number (hospital number) or birth date, and source. Enough information must be provided on the specimen label so that the specimen can be matched up with the requisition when it is received in the laboratory.

SPECIMEN REQUISITION

The specimen (or test) requisition is an order form that is sent to the laboratory along with a specimen. Sample requisitions can be found in Chapter 3. Often the requisition is a hard (paper) copy of the doctor's orders and the patient demographic information (such as name and hospital number). Sometimes, however, if a hospital information system offers computerized order entry, the requisition is transported to the laboratory electronically. The requisition should contain as much information as possible regarding the patient history, diagnosis, and immunization record. This information will help the microbiologist in working up the specimen and in determining which organisms are significant in the culture. A complete requisition should include the following:

- The patient's name
- Hospital number
- Age or date of birth
- Sex
- Collection date and time
- Ordering physician
- Exact nature and source of the specimen
- Diagnosis
- Immunization history
- Antimicrobial therapy

REJECTION OF UNACCEPTABLE SPECIMENS

Criteria for specimen rejection should be set up and distributed to all clinical practitioners. In general, specimens are unacceptable if any of the following conditions apply:

- The information on the label does not match the information on the requisition (patient name or source of specimen are different)
- The specimen has been transported at the improper temperature
- The specimen has not been transported in the proper medium (e.g., specimens for anaerobic bacteria submitted in aerobic transports)
- The quantity of specimen is insufficient for testing (the specimen is considered QNS [quantity not sufficient])
- The specimen is leaking
- The specimen transport time exceeds 2 hours postcollection and the specimen is not preserved
- The specimen was received in a fixative (formalin) which, in essence, kills any microorganism present
- The specimen has been received for anaerobic culture from a site known to have anaerobes as part of the normal flora (vagina, mouth)
- The specimen is dried up
- Processing the specimen would produce information of questionable medical value (e.g., Foley catheter tip)

It is an important rule to always talk to the requesting physician or another member of the health care team before discarding unacceptable specimens. In some cases, such as mislabeling of a specimen or requisition, the individual who collected the specimen and filled out the paperwork can come to the laboratory and correct the problem; identification of a mislabeled specimen or requisition should *not* be done over the telephone. Frequently, it may be necessary to do the best possible job on a less than optimal specimen, if the specimen would be impossible to collect again because the patient is taking antibiotics, the tissue was collected at surgery, or the patient would have to undergo a second invasive procedure (bone marrow or spinal tap). A notation regarding improper collection should be added to the final report in this instance, since only the primary care giver is able to determine the validity of the results.

SPECIMEN PROCESSING

Depending on the site of testing (hospital, independent lab, POL) and how the specimens are transported to the laboratory (in-house courier or driver), microbiology samples may arrive in the laboratory

Text continued on p. 14

TABLE 1-1 COLLECTION, TRANSPORT, STORAGE, AND PROCESSING OF SPECIMENS COMMONLY SUBMITTED TO A MICROBIOLOGY LABORATORY*

SPECIMEN	CONTAINER	PATIENT PREPARATION	SPECIAL INSTRUCTIONS
Aqueous/vitreous fluid	Sterile, screw-cap tube		
Abscess (also lesion, wound, pustule, ulcer) Superficial	Aerobic swab moistened with Stuart's or Amie's medium	Wipe area with sterile saline or 70% alcohol	Swab along leading edge of wound
Deep	Anaerobic transporter	Wipe area with sterile saline or 70% alcohol	Aspirate material from wall or excise tissue
Blood or Bone Marrow	Blood culture media set (aerobic and anaerobic bottle) or Vacutainer tube with SPS	Disinfect venipuncture site with 70% alcohol and disinfectant such as betadine	Draw blood at time of febrile episode; draw 2 sets from right and left arms; do *not* draw more than 3 sets in a 24-hr period
Body Fluids Amniotic, abdominal, ascites (peritoneal), bile, joint (synovial), pericardial, pleural	Sterile, screw-cap tube or anaerobic transporter	Disinfect skin before aspirating specimen	Needle aspiration
Bone	Sterile, screw-cap container	Disinfect skin before surgical procedure	Take sample from affected area for biopsy
Cerebrospinal Fluid	Sterile, screw-cap tube	Disinfect skin before aspirating specimen	Consider rapid testing, e.g., Gram stain; cryptococcal antigen
Ear Inner	Sterile, screw-cap tube or anaerobic transporter	Clean ear canal with mild soap solution before myringotomy (puncture of the ear drum)	Aspirate material behind drum with syringe if ear drum intact; use swab to collect material from ruptured ear drum

TRANSPORTATION TO LABORATORY	STORAGE PRIOR TO PROCESSING	PRIMARY PLATING MEDIA	DIRECT EXAMINATION	COMMENTS
Immediately/RT	Set up immediately on receipt	BA, CA, Mac, Sab, 7H10, thio	Gram/AO	
Within 24 hrs/RT	24 hrs/RT	BA, CA, Mac, thio	Gram	Add CNA if smear suggests mixed gram-positive and gram-negative flora
Within 24 hrs/RT	24 hrs/RT	BA, CA, Mac, Ana, thio	Gram	Wash any granules and "emulsify" in saline
Within 2 hrs/RT	Must be incubated at 37° C on receipt in laboratory	Blood culture bottles	AO or Giemsa	Other considerations: brucellosis, tularemia, cell wall-deficient bacteria, leptospirosis, or AFB
Immediately/RT	Plate as soon as received	BA, CA, Mac, Ana, thio Note: May substitute an aerobic and anaerobic blood culture bottle set for peritoneal and synovial fluids	Gram	May need to concentrate by centrifugation or filtration—stain and culture sediment
Immediately/RT	Plate as soon as received	BA, CA, Mac, thio	Gram	May need to homogenize
Immediately/RT	6 hrs/37° C except for viruses, which can be held at 4° C for up to 3 days	BA, CA, thio	Gram—best sensitivity by cytocentrifugation (may also want to do AO if cytocentrifuge not available)	
Immediately/RT	6 hrs/RT	BA, CA, Mac, thio	Gram	Add anaerobic culture plates for tympanocentesis specimens

Continued

TABLE 1-1 COLLECTION, TRANSPORT, STORAGE, AND PROCESSING
OF SPECIMENS COMMONLY SUBMITTED
TO A MICROBIOLOGY LABORATORY*—CONT'D

SPECIMEN	CONTAINER	PATIENT PREPARATION	SPECIAL INSTRUCTIONS
Ear—cont'd			
Outer	Aerobic swab moistened with Stuart's or Amie's medium	Wipe away crust with sterile saline	Firmly rotate swab in outer canal
Eye			
Conjunctiva	Aerobic swab moistened with Stuart's or Amie's medium		Sample both eyes; use swab premoistened with sterile saline
Corneal scrapings	Bedside inoculation of BA, CA, Sab, 7H10, thio	Clinician should instill local anesthetic before collection	
Foreign Bodies			
IUD	Sterile, screw-cap container	Disinfect skin before removal	
IV catheters, pins, prosthetic valves	Sterile, screw-cap container	Disinfect skin before removal	Do not culture Foley catheters; IV catheters are cultured quantitatively by rolling the segment back and forth across agar with sterile forceps 4 times; ≥15 colonies are associated with clinical significance
GI Tract			
Gastric aspirate	Sterile, screw-cap tube	Collect in early AM before patient eats or gets out of bed	Most gastric aspirates are on infants or for AFB
Gastric biopsy	Sterile, screw-cap tube		Rapid urease test or culture for *Helicobacter pylori*
Rectal swab	Swab placed in enteric transport medium		Insert swab ≈ 2.5 cm past anal sphincter
Stool culture	Clean, leak-proof container		Routine culture should include *Salmonella, Shigella,* and *Campylobacter*; specify *Vibrio, Aeromonas, Plesiomonas, Yersinia, Escherichia coli* 0157:H7, if needed

TRANSPORTATION TO LABORATORY	STORAGE PRIOR TO PROCESSING	PRIMARY PLATING MEDIA	DIRECT EXAMINATION	COMMENTS
Within 24 hrs/RT	24 hrs/RT	BA, CA, Mac	Gram	
Within 24 hrs/RT	24 hrs/RT	BA, CA, Mac, thio	Gram, AO, histologic stains, e.g., Giemsa	Other considerations: *Chlamydia trachomatis*, viruses, and fungi
Immediately/RT	Must be incubated at 28° C (Sab) or 37° C (everything else) on receipt in laboratory	BA, CA, Sab, 7H10, Ana, thio	Gram/AO	Other considerations: *Acanthamoeba* spp., herpes simplex virus and other viruses, *Chlamydia trachomatis*, and fungi
Immediately/RT	Plate as soon as received	Thio		
Immediately/RT	Plate as soon as received	BA, thio		
Immediately/RT	Must be neutralized within 1 hr of collection	BA, CA, Mac, CNA, thio	Gram/AO	Other considerations: AFB
Immediately/4° C	Must be set up immediately on receipt	Skirrow's, BA		
Within 24 hrs/4° C	72 hrs/4° C	BA, Mac, XLD or HE, Campy	Methylene blue for fecal leukocytes	Other considerations: *Vibrio*, *Yersinia enterocolitica*, *Escherichia coli* 0157:H7
Within 24 hrs/4° C	72 hrs/4° C	BA, Mac, XLD or HE, Campy	Methylene blue for fecal leukocytes	Consider *Vibrio*, *Yersinia enterocolitica*, *Escherichia coli* 0157:H7

Continued

TABLE 1-1 COLLECTION, TRANSPORT, STORAGE, AND PROCESSING OF SPECIMENS COMMONLY SUBMITTED TO A MICROBIOLOGY LABORATORY* — CONT'D

SPECIMEN	CONTAINER	PATIENT PREPARATION	SPECIAL INSTRUCTIONS
GI Tract—cont'd			
O&P	O&P transporters (e.g., 10% formalin and PVA)	Collect 3 specimens every other day at a minimum for outpatients; hospitalized patients (inpatients) should have 3 specimens collected every day; specimens from inpatients hospitalized more than 3 days should be discouraged	Wait 7-10 days if patient has received antiparasitic compounds, barium, iron, Kaopectate, metronidazole, Milk of Magnesia, Pepto-Bismol, or tetracycline
Genital Tract			
Female			
Bartholin cyst	Anaerobic transporter	Disinfect skin before collection	Aspirate fluid; consider chlamydia and GC culture
Cervix	Swab moistened with Stuart's or Amie's medium	Remove mucus before collection of specimen	Do *not* use lubricant on speculum; use viral/chlamydial transport medium, if necessary; swab deeply into endocervical canal
Cul-de-sac	Anaerobic transporter		Submit aspirate
Endometrium	Anaerobic transporter		Surgical biopsy or transcervical aspirate via sheathed catheter
Urethra	Swab moistened with Stuart's or Amie's medium	Remove exudate from urethral opening	Collect discharge by massaging urethra against pubic symphysis or insert flexible swab 2-4 cm into urethra and rotate swab for 2 seconds
Vagina	Swab moistened with Stuart's or Amie's medium *or* JEMBEC transport system	Remove exudate	Swab secretions and mucous membrane of vagina
Male			
Prostate	Swab moistened with Stuart's or Amie's medium *or* sterile, screw-cap tube	Clean glans with soap and water	Collect secretions on swab or in tube

TRANSPORTATION TO LABORATORY	STORAGE PRIOR TO PROCESSING	PRIMARY PLATING MEDIA	DIRECT EXAMINATION	COMMENTS
Within 24 hrs/RT	Indefinitely/RT			
Within 24 hrs/RT	24 hrs/RT	BA, CA, Mac, TM, Ana, thio	Gram	
Within 24 hrs/RT	24 hrs/RT	BA, CA, Mac, TM, thio	Gram	
Within 24 hrs/RT	24 hrs/RT	BA, CA, Mac, TM, Ana, thio	Gram	
Within 24 hrs/RT	24 hrs/RT	BA, CA, Mac, TM, Ana, thio	Gram	
Within 24 hrs/RT	24 hrs/RT	BA, CA, TM	Gram	Other considerations: *Chlamydia, Mycoplasma*
Within 24 hrs/RT	24 hrs/RT	BA, CA, Mac, HBT, TM	Gram	Examine Gram stain for bacterial vaginosis, especially, white blood cells, gram-positive rods indicative of *Lactobacillus,* and curved, gram-negative rods indicative of *Mobiluncus* spp.
Within 24 hrs/RT for swab; immediately if in tube/RT	Swab: 24 hrs/RT; tube: plate secretions immediately	BA, CA, Mac, TM, CNA, thio	Gram	

Continued

TABLE 1-1 COLLECTION, TRANSPORT, STORAGE, AND PROCESSING OF SPECIMENS COMMONLY SUBMITTED TO A MICROBIOLOGY LABORATORY* — CONT'D

SPECIMEN	CONTAINER	PATIENT PREPARATION	SPECIAL INSTRUCTIONS
Genital Tract—cont'd			
Male—cont'd			
Urethra	Swab moistened with Stuart's or Amie's medium *or* JEMBEC transport system		Insert flexible swab 2-4 cm into urethra and rotate for 2 seconds or collect discharge on JEMBEC transport system
Hair, Nails, or Skin Scrapings (for fungal culture)	Clean, screw-top tube	Nails or skin: wipe with 70% alcohol	Hair: collect hairs with intact shaft Nails: send clippings of affected area Skin: scrape skin at leading edge of lesion
Respiratory Tract			
Lower			
BAL, BB, BW	Sterile, screw-top container		Anaerobic culture appropriate only if sheathed (protected) catheter used
Sputum, tracheal aspirate (suction)	Sterile, screw-top container	Sputum: rinse or gargle with water before collection	Sputum: have patient collect from deep cough; specimen should be examined for suitability for culture by Gram stain; induced sputa on pediatric or unco-operative patients may be watery because of saline nebulization
Upper			
Nasopharynx	Swab moistened with Stuart's or Amie's medium		Insert flexible swab through nose into posterior nasopharynx and rotate for 5 seconds; specimen of choice for *Bordetella pertussis*
Pharynx (throat)	Swab moistened with Stuart's or Amie's medium		Swab posterior pharynx and tonsils; routine culture for group A *Streptococcus* (*S. pyogenes*) only

TRANSPORTATION TO LABORATORY	STORAGE PRIOR TO PROCESSING	PRIMARY PLATING MEDIA	DIRECT EXAMINATION	COMMENTS
Within 24 hrs/RT for swab; within 2 hrs for JEMBEC system	24 hrs/RT for swab; put JEMBEC at 37° C immediately on receipt in laboratory	BA, CA, TM	Gram	Other considerations: *Chlamydia, Mycoplasma*
Within 24 hrs/RT	Indefinitely/RT	Sab, IMAcg, Sabcg	CW	
Within 2 hrs/RT	24 hrs/4° C	BA, CA, Mac	Gram and other special stains as requested, e.g., *Legionella* DFA, acid-fast stain	Other considerations: quantitative culture for BAL, AFB, *Legionella, Nocardia, Mycoplasma, Pneumocystis,* cytomegalovirus
Within 2 hrs/RT	24 hrs/4° C	BA, CA, Mac	Gram and other special stains as requested, e.g., *Legionella* DFA, acid-fast stain	Other considerations: AFB, *Nocardia*
Within 24 hrs/RT	24 hrs/RT	BA, CA		Other considerations: add special media for *Corynebacterium diphtheriae,* pertussis, *Chlamydia,* and *Mycoplasma*
Within 24 hrs/RT	24 hrs/RT	BA		Other considerations: add special media for *C. diptheriae* and *Neisseria gonorrhoeae*

Continued

TABLE 1-1 COLLECTION, TRANSPORT, STORAGE, AND PROCESSING OF SPECIMENS COMMONLY SUBMITTED TO A MICROBIOLOGY LABORATORY* — CONT'D

SPECIMEN	CONTAINER	PATIENT PREPARATION	SPECIAL INSTRUCTIONS
Tissue	Anaerobic transporter or sterile, screw-cap tube	Disinfect skin	Do not allow specimen to dry out; moisten with sterile, distilled water if not bloody
Urine			
Clean-voided midstream (CVS)	Sterile, screw-cap container	Females: clean area with soap and water, then rinse with water; hold labia apart and begin voiding in commode; after several mL have passed, collect mid-stream Males: clean glans with soap and water, then rinse with water; retract foreskin; after several mL have passed, collect midstream	
Straight catheter (in and out)	Sterile, screw-cap container	Clean urethral area (soap and water) and rinse (water)	Insert catheter into bladder; allow first 15 mL to pass; then collect remainder
Indwelling catheter (Foley)	Sterile, screw-cap container	Disinfect catheter collection port	Aspirate 5-10 mL of urine with needle and syringe
Suprapubic aspirate	Sterile, screw-cap container or anaerobic transporter	Disinfect skin	Needle aspiration above the symphysis pubis through the abdominal wall into the full bladder

≈, approximately; *7H10*, Middlebrook 7H10 agar; *AFB*, acid-fast bacilli; *AM*, morning; *Ana*, anaerobic agars as appropriate (see Chapter 58); *AO*, acridine agar; *CNA*, Columbia agar with colistin and nalidixic acid; *CW*, calcofluor white stain; *DFA*, direct fluorescent antibody stain; *GC*, *Neisseria gonorrhoeae*; *HE*, Hektoen enteric agar; *hrs*, hours; *IMAcg*, inhibitory mold agar with chloramphenicol and gentamicin; *IUD*, intrauterine device; *Mac*, MacConkey agar; with cycloheximide and gentamicin; *SPS*, sodium polyanethol sulfate; *thio*, thioglycollate broth; *TM*, Thayer-Martin agar; *XLD*, xylose lysine deoxycholate agar.

*Specimens for viruses, chlamydia, and mycoplasma are usually submitted in appropriate transport media at 4° C to stabilize respective microorganisms.

in large numbers or as single tests. Although batch processing may be possible in large independent laboratories, most often, hospital testing is performed as specimens arrive. When multiple specimens arrive at the same time, priority should be given to those that are most critical, such as cerebrospinal fluid (CSF), tissue, blood, and sterile fluids. Urine, throat, sputa, stool, or wound drainage specimens can be saved for later. Acid-fast, viral, and fungal specimens are usually batched for processing at one time. When a specimen is received with multiple requests but the

amount of specimen is insufficient to do all of them, the microbiologist should call the clinician to prioritize the testing.

GROSS EXAMINATION OF SPECIMEN

All processing should begin with a gross examination of the specimen. Areas with blood or mucus should be located and sampled for culture and direct examination. Stool should be examined for evidence of barium (i.e., chalky white color), which would preclude O&P examination. Notations should be

TRANSPORTATION TO LABORATORY	STORAGE PRIOR TO PROCESSING	PRIMARY PLATING MEDIA	DIRECT EXAMINATION	COMMENTS
Within 24 hrs/RT	24 hrs/RT	BA, CA, Mac, Ana, thio	Gram	May need to homogenize
Within 2 hrs/4° C	24 hrs/4° C	BA, Mac	Gram or check for pyuria	Plate quantitatively at 1:1000; consider plating quantitatively at 1:100 if patient is female of child-bearing age with white blood cells and possible acute urethral syndrome
Within 2 hrs/4° C	24 hrs/4° C	BA, Mac	Gram or check for pyuria	Plate quantitatively at 1:100 and 1:1000
Within 2 hrs/4° C	24 hrs/4° C	BA, Mac	Gram or check for pyuria	Plate quantitatively at 1:1000
Immediately/RT	Plate as soon as received	BA, Mac, Ana, thio	Gram or check for pyuria	Plate quantitatively at 1:100 and 1:1000

orange stain; *BA*, blood agar; *BAL*, bronchial alveolar lavage; *BB*, bronchial brush; *BW*, bronchial wash; *CA*, chocolate agar; *Campy*, selective *Campylobacter* transport using JEMBEC system with modified Thayer-Martin medium; *GI*, gastrointestinal; *Gram*, Gram stain; *HBT*, human blood-bilayer Tween agar; *mL*, milliliters; *O&P*, ova and parasite examination; *PVA*, polyvinyl alcohol; *RT*, room temperature; *Sab*, sabouraud dextrose agar; *Sabcg*, sabouraud dextrose agar

made on the handwritten or electronic workcard regarding the status of the specimen (e.g., bloody, cloudy, clotted) so that if more than one person works on the sample, all microbiologists working it up will know the results of the gross examination.

DIRECT MICROSCOPIC EXAMINATION

All appropriate specimens should have a direct microscopic examination. The direct examination serves several purposes. First, the quality of the specimen can be assessed; for example, sputa can be rejected that represent saliva and not lower respiratory tract secretions by quantitation of white blood cells or epithelial cells present in the specimen. Second, the microbiologist and clinician can be given an early indication of what may be wrong with the patient (e.g., 4+ white blood cells [WBCs] in exudate). Third, the workup of the specimen can be guided by comparing what grows in culture to what was seen on smear. A situation in which three different morphotypes (cellular types) are seen on direct Gram stain but only two grow out in culture, for example, alerts the microbiologist to the fact that the third organism may be an anaerobic bacterium.

Direct examinations are usually not performed on throat, nasopharyngeal, or stool specimens but are indicated from most other sources.

The most common stain in bacteriology is the Gram stain, which helps to visualize rods, cocci, white blood cells, red blood cells, or epithelial cells present in the sample. The most common direct fungal stains are KOH (potassium hydroxide), PAS (periodic-acid Schiff) and calcofluor white. The most common direct acid-fast stains are AR (auramine rhodamine), ZN (Ziehl-Neelsen), and Kinyoun. Chapter 11 describes the use of microscopy in clinical diagnosis in more detail.

SELECTION OF CULTURE MEDIA

Primary culture media are divided into several categories. The first are **nutritive** media, such as blood or chocolate agars. Nutritive media support the growth of a wide range of microorganisms and are considered nonselective because, theoretically, the growth of most organisms is supported. Nutritive media can also be **differential,** in that microorganisms can be distinguished on the basis of certain growth characteristics evident on the medium. Blood agar is considered both a nutritive and differential medium because it will differentiate organisms based on whether they are alpha- (α), beta- (β), or gamma- (γ) hemolytic (Figure 1-2). **Selective** media support the growth of one group of organisms, but not another, by adding antimicrobials, dyes, or alcohol to a particular medium. MacConkey agar, for example, contains the dye crystal violet, which will inhibit gram-positive organisms. Columbia agar with colistin and nalidixic acid (CNA) is a selective medium for gram-positive organisms because the antimicrobials colistin and nalidixic acid inhibit gram-negative organisms. Selective media can also be differential media if, in addition to their inhibitory activity, they differentiate between groups of organisms. MacConkey agar, for example, differentiates between lactose-fermenting

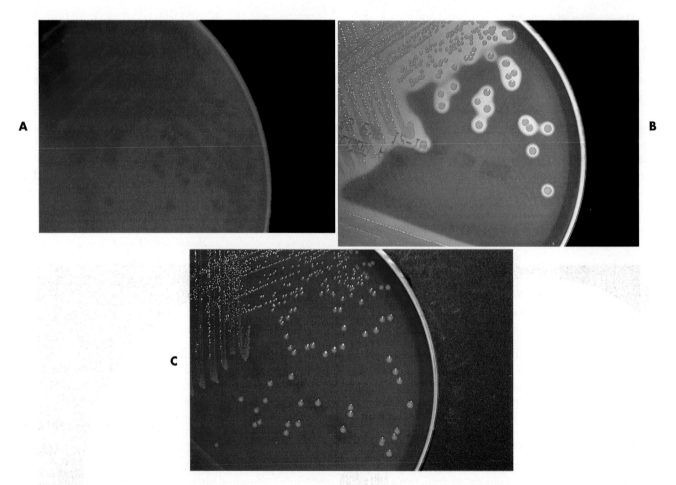

FIGURE 1-2 Examples of various types of hemolysis on blood agar. **A,** *Streptococcus pneumoniae* showing alpha (α)-hemolysis (i.e., greening around colony). **B,** *Staphylococcus aureus* showing beta (β)-hemolysis (i.e., clearing around colony). **C,** *Enterococcus faecalis* showing gamma (γ)-hemolysis (i.e., no hemolysis around colony).

and nonfermenting gram-negative rods by the color of the colonial growth (pink or clear, respectively); this is shown in Figure 1-3. In many cases, **back-up broth** (also called supplemental or enrichment broth) medium is inoculated, along with primary solid (agar) media, so small numbers of organisms present may be detected; this allows detection of anaerobes in aerobic cultures and organisms that may be damaged by either previous or concurrent antimicrobial therapy. Thioglycollate (thio) broth, brain-heart infusion broth (BHIB), and tryptic soy broth (TSB) are common back-up broths.

Selection of media to inoculate for any given specimen is usually based on the organisms most likely to be involved in the disease process. For example, in determining what to set up for a CSF, one considers the most likely pathogens that cause meningitis *(Streptococcus pneumoniae, Haemophilus influenzae, Neisseria meningitidis, Escherichia coli,* group B *Streptococcus)* and selects media that will support the growth of these organisms (blood and chocolate agar and a back-up broth at a minimum). Likewise, if a specimen is collected from a source likely to be contaminated with normal flora, for example, an anal fistula (an opening of the surface of the skin near the anus that may communicate with the rectum), one might want to add a selective medium, such as CNA, to suppress gram-negative organisms and allow gram-positive organisms and yeast to be recovered.

Routine primary plating media and direct examinations for specimens commonly submitted to the microbiology laboratory are shown in Table 1-1. Chapter 12 on bacterial cultivation reemphasizes the strategies described here for selection and use of bacterial media.

SPECIMEN PREPARATION

Many specimens require some form of initial treatment before inoculation onto primary plating media. Such procedures include **homogenization** (grinding) of tissue; **concentration** by centrifugation or filtration of large volumes of sterile fluids, such as ascites (peritoneal) or pleural (lung) fluids; or **decontamination** of specimens, such as those for legionellae or mycobacteria. Swab specimens are often vortexed (mixed) in 0.5 to 1.0 mL of saline or broth for 10 to 20 seconds to dislodge material from the fibers.

INOCULATION OF SOLID MEDIA

Specimens can be inoculated (plated) onto solid media either quantitatively by a dilution procedure or by means of a quantitative loop, or semiquantitatively using an ordinary inoculating loop. Urine cultures and tissues from burn victims are plated quantitatively; everything else is usually plated semiquantitatively. Plates inoculated for quantitation are usually streaked with a 1:100 or 1:1000 loop, as shown in Figure 1-4, *A.* The original streak line is cross-struck with an ordinary inoculating loop to produce isolated, countable colonies. Plates inoculated for semiquantitation are usually struck out in four quadrants, as shown in Figure 1-4, *B.* The inoculum is applied by swabbing a dime-sized area or placing a drop of liquid specimen on the plate. The original inoculum is then cross-struck with an ordinary nichrome inoculating loop (quadrant one). The loop is then flipped over or flamed and quadrant two is struck by pulling the loop through quadrant one a few times and streaking the rest of the area. The loop is then flamed again and quadrant three is streaked by going into quadrant two a few times and streaking the rest of the area. Finally, quadrant four is streaked

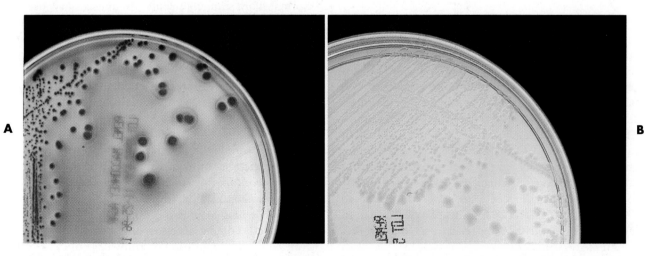

A **B**

FIGURE 1-3 MacConkey agar. **A,** *Escherichia coli,* a lactose-fermenter. **B,** *Pseudomonas aeruginosa,* a nonlactose-fermenter.

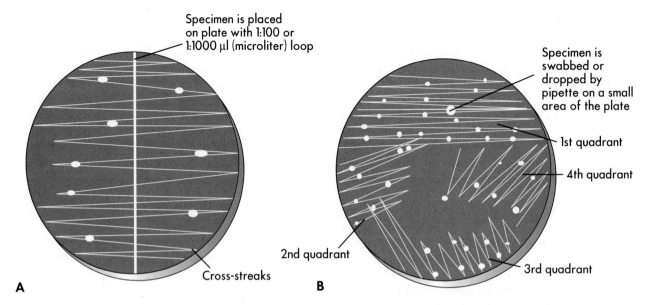

FIGURE 1-4   Methods of inoculating solid media. **A,** Streaking for quantitation. **B,** Streaking semiquantitatively.

by pulling the loop over the rest of the agar without further flaming. This process is called **streaking for isolation,** because the microorganisms present in the specimen are successively diluted out as each quadrant is streaked until finally each morphotype is present as a single colony. Numbers of organisms present can subsequently be graded as 4+ (many, heavy growth) if growth is out to the fourth quadrant, 3+ (moderate growth) if growth is out to the third quadrant, 2+ (few or light growth) if growth is in the second quadrant, and 1+ (rare) if growth is in the first quadrant. This tells the clinician the relative numbers of different organisms present in the specimen; such semiquantitative information is usually sufficient for the physician to be able to treat the patient.

INCUBATION CONDITIONS

Inoculated media are incubated under various temperatures and environmental conditions, depending on the organisms being sought, for example, 28° C for fungi and 35° to 37° C for most bacteria, viruses, and AFB. A number of different environmental conditions exist. **Aerobes** grow in ambient air, which contains 21% oxygen (O_2) and a small amount (0.03%) of carbon dioxide (CO_2). **Anaerobes** usually cannot grow in the presence of O_2 and the atmosphere in anaerobe jars, bags, or chambers is composed of 5% to 10% hydrogen (H_2), 5% to 10% CO_2, 80% to 90% nitrogen (N_2) and 0% O_2. **Capnophiles,** such as *Haemophilus influenzae* and *Neisseria gonorrhoeae,* require increased concentrations of CO_2 (5% to 10%) and approximately 15% O_2. This atmosphere can be

achieved by a candle jar (3% CO_2) or a CO_2 incubator, jar, or bag. **Microaerophiles** *(Campylobacter jejuni, Helicobacter pylori)* grow under reduced O_2 (5% to 10%) and increased CO_2 (8% to 10%). This environment can also be obtained in specially designed jars or bags.

SPECIMEN WORKUP

One of the most important functions that a microbiologist performs is to decide what is clinically relevant regarding specimen workup. Considerable judgment is required to decide what organisms to look for and report. It is essential to recognize what constitutes indigenous (normal) flora and what constitutes a potential pathogen. Indiscriminate reporting of normal flora can contribute to unnecessary use of antibiotics and potential emergence of resistant organisms. Because organisms that are clinically relevant to identify and report vary by source, the microbiologist should know which ones cause disease at various sites. Part 3 contains a detailed discussion of these issues.

EXTENT OF IDENTIFICATION REQUIRED

As health care continues to change, one of the most problematic issues for microbiologists is the extent of culture workup. Microbiologists still rely heavily on definitive identification, although shortcuts, including the use of limited identification procedures in some cases, are becoming necessary in most clinical laboratories. Careful application of knowledge of the significance of various organisms in specific situations and thoughtful use of limited approaches will keep microbiology testing cost effective and the laboratory's workload manageable, while providing for optimum patient care.

Complete identification of a blood culture isolate, such as *Clostridium septicum* as opposed to a genus identification of *Clostridium* spp., will alert the clinician to the possibility of malignancy or other disease of the colon. At the same time, a presumptive identification of *Escherichia coli* if a gram-negative, spot indole-positive rod is recovered with appropriate colony morphology on MacConkey agar (flat, lactose-fermenting colony that is precipitating bile salts) is probably permissible from an uncomplicated urinary tract infection. In the final analysis, culture results should always be compared with the suspected diagnosis. The clinician should be encouraged to supply the microbiologist with all pertinent information (e.g., recent travel history, pet exposure, pertinent radiograph findings) so that the microbiologist can use the information to interpret culture results and plan appropriate strategies for workup.

References

1. Lee, A. and McLean, S. 1977. The laboratory report: a problem in communication between clinician and microbiologist? Med. J. Aust. 2:858.

Bibliography

Aller, R.D. and Elevitch, F.R., editors. 1991. Laboratory and hospital information systems. Clinics in laboratory medicine. W.B. Saunders, Philadelphia.

Isenberg, H.D., Schoenknecht, F.D., and von Graevenitz, A. 1979. Cumitech 9: collection and processing of bacteriological specimens. S.J. Rubin, coordinating editor, American Society for Microbiology, Washington, D.C.

Miller J.M. 1985. Handbook of specimen collection and handling in microbiology. CDC Laboratory Manual. U.S. Department of Health and Human Services, Atlanta.

Miller, J.M. and Holmes, H.T. 1995. Specimen collection, transport and storage. In Murray, P.R., Baron, E.J., Pfaller, M.A., et al. editors. Manual of clinical microbiology, ed 6. ASM Press, Washington, D.C.

Pezzlo, M. 1991. Specimen collection: the role of the clinical microbiologist. Clinical Microbiology Updates 3:1, Hoechst-Roussel Pharmaceuticals Inc., Somerville, N.J.

2 | LABORATORY SAFETY

Microbiology laboratory safety practices were first published in 1913 in a textbook by Eyre.[2] They included admonitions such as the necessity to (1) wear gloves, (2) wash hands after working with infectious materials, (3) disinfect all instruments immediately after use, (4) use water to moisten specimen labels rather than the tongue, (5) disinfect all contaminated waste before discarding, and (6) report to appropriate personnel all accidents or exposures to infectious agents.

These guidelines are still incorporated into safety programs in the late twentieth-century laboratory. In addition, safety programs have been expanded to include not only the proper handling of biologic hazards encountered in processing patient specimens or handling infectious microorganisms but also fire safety; electrical safety; the safe handling, storage and disposal of chemicals and radioactive substances; and techniques for the safe lifting or moving of heavy objects. In areas of the country prone to natural disasters (e.g., earthquakes, hurricanes, snowstorms), safety programs also involve disaster preparedness plans that outline steps to take in an emergency. Although all microbiologists are responsible for their own health and safety, their institution and immediate supervisors are required to provide safety training to help familiarize microbiologists with known hazards and to avoid accidental exposure. Laboratory safety is considered such an integral part of overall laboratory services that federal law in the United States mandates preemployment safety training followed by at least quarterly safety inservices. Microbiologists should find very little reason to be afraid while performing duties if the safety regulations are internalized and followed without deviation. Investigation of the causes of accidents usually shows that accidents happen when individuals become sloppy in performing their duties or when they do not believe that they will be affected by departures from safety standards.

STERILIZATION AND DISINFECTION

Sterilization is a process whereby all forms of microbial life, including bacterial spores, are killed. Sterilization may be accomplished by physical or chemical means. **Disinfection** is a process whereby pathogenic organisms, but not necessarily all microorganisms or spores, are destroyed. As with sterilization, disinfection may be accomplished by physical or chemical methods.

METHODS OF STERILIZATION

The **physical** methods of sterilization include the following:

- Incineration
- Moist heat
- Dry heat
- Filtration
- Ionizing (gamma) radiation

Incineration is the most common method of treating infectious waste. Hazardous material is literally burned to ashes at temperatures of 870° to 980° C. Toxic air emissions and the presence of heavy metals in ash have limited the use of incineration in most large U.S. cities, however.

Moist heat (steam under pressure) is used to sterilize biohazardous trash and heat-stable objects; an autoclave is used for this purpose. An autoclave is essentially a large pressure cooker. Moist heat in the form of saturated steam under 1 atmosphere (15 psi [pounds per square inch]) of pressure causes the irreversible denaturation of enzymes and structural proteins. The most common type of steam sterilizer in the microbiology laboratory is the gravity displacement type shown in Figure 2-1. Steam enters at the top of the sterilizing chamber and, because steam is lighter than air, it displaces the air in the chamber and forces it out the bottom through the drain vent. The two common sterilization temperatures are 121° C (250° F) and 132° C (270° F). Items such as media, liquids, and instruments are usually autoclaved for 15 minutes at 121° C. Infectious medical waste, on the other hand, is often sterilized at 132° C for 30 to 60 minutes to allow penetration of the steam throughout the waste and the displacement of air trapped inside

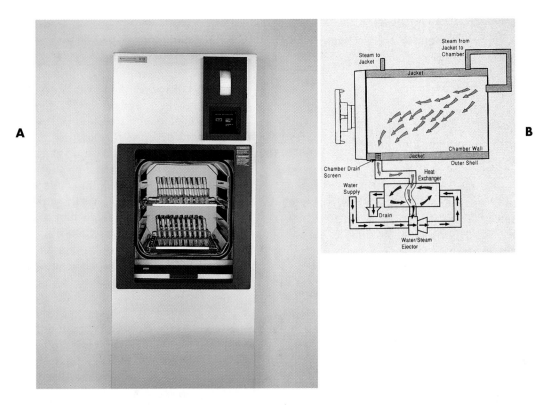

FIGURE 2-1 Gravity displacement type autoclave. **A,** Typical Eagle Century Series sterilizer for laboratory applications. **B,** Typical Eagle 3000 sterilizer piping diagram. The arrows show the entry of steam into the chamber and the displacement of air. (Courtesy AMSCO International Inc, a wholly owned subsidiary of STERIS Corp., Mentor, Ohio.)

the autoclave bag. Moist heat is the fastest and simplest physical method of sterilization.

Dry heat requires longer exposure times (1.5 to 3 hours) and higher temperatures than moist heat (160° to 180° C). Dry-heat ovens are used to sterilize items such as glassware, oil, petrolatum, or powders. Filtration is the method of choice for antibiotic solutions, toxic chemicals, radioisotopes, vaccines, and carbohydrates, which are all heat-sensitive. Filtration of liquids is accomplished by pulling the solution through a cellulose acetate or cellulose nitrate membrane with a vacuum. Filtration of air is accomplished using high-efficiency-particulate-air (HEPA) filters designed to remove organisms larger than 0.3 μm from isolation rooms, operating rooms, and biological safety cabinets (BSCs). Ionizing radiation used in microwaves and radiograph machines are short wavelength and high-energy gamma rays. Ionizing radiation is used for sterilizing disposables such as plastic syringes, catheters, or gloves before use.

The most common **chemical** sterilant is ethylene oxide (EtO), which is used in gaseous form for sterilizing heat-sensitive objects. Formaldehyde vapor and vapor-phase hydrogen peroxide (an oxidizing agent) have been used to sterilize HEPA filters in BSCs. Gluter-aldehyde, which is sporocidal (kills spores) in 3 to 10 hours, is used for medical equipment such as bronchoscopes, because it does not corrode lenses, metal, or rubber. Peracetic acid, effective in the presence of organic material, has also been used for the surface sterilization of surgical instruments. The use of gluteraldehyde or peracetic acid is called **cold sterilization.**

METHODS OF DISINFECTION

PHYSICAL METHODS OF DISINFECTION

The three **physical** methods of disinfection are:

- Boiling at 100° C for 15 minutes, which kills vegetative bacteria
- Pasteurizing at 63° C for 30 minutes or 72° C for 15 seconds, which kills food pathogens
- Using nonionizing radiation such as ultraviolet (UV) light

UV rays are long wavelength and low energy. They do not penetrate well and organisms must have direct surface exposure, such as the working surface of a BSC, for this form of disinfection to work.

CHEMICAL METHODS OF DISINFECTION

Chemical disinfectants comprise many classes, including the following:

- Alcohols
- Aldehydes
- Halogens
- Heavy metals
- Quaternary ammonium compounds
- Phenolics

When chemicals are used to destroy all life they are called chemical sterilants, or **biocides;** however, these same chemicals used for shorter periods are disinfectants. Disinfectants used on living tissue (skin) are called **antiseptics.**

A number of factors influence the activity of disinfectants such as the following:

- Types of organisms present
- Temperature and pH of process
- Number of organisms present (microbial load)
- Concentration of disinfectant
- Amount of organics (blood, mucus, pus) present
- Nature of surface to be disinfected (e.g., potential for corrosion; porous vs. nonporous surface)
- Length of contact time
- Type of water available (hard vs. soft)

Resistance to disinfectants varies with the type of microorganisms present. Bacterial spores such as *Bacillus* spp., are the most resistant, followed by mycobacteria (acid-fast bacilli); nonlipid viruses, for example, poliovirus; fungi; vegetative (nonsporulating) bacteria, for example, gram-negative rods; and, finally, lipid viruses, for example, herpes simplex virus, which are the most susceptible to the action of disinfectants. The Environmental Protection Agency (EPA) registers chemical disinfectants used in the United States and requires manufacturers to specify the activity level of each compound at the working dilution. Therefore, microbiologists who must recommend appropriate disinfectants should check manufacturers' cut sheets (product information) for the classes of microorganisms that will be killed. Generally, the time necessary for killing microorganisms increases in direct proportion with the number of organisms (microbial load). This is particularly true of instruments contaminated with organic material such as blood, pus, or mucus. The organic material should be mechanically removed before chemical sterilization to decrease the microbial load. This is analogous to removing dried food from utensils before placing them in a dishwasher and is important for cold sterilization of instruments such as bronchoscopes. The type of water and its concentration in a solution are also important. Hard water may reduce the rate of killing of microorganisms, and surprisingly, 70% ethyl alcohol is more effective as a disinfectant than 95% ethyl alcohol.

Ethyl or isopropyl alcohol is nonsporicidal (does not kill spores) and evaporates quickly. Therefore, either is best used on the skin as an antiseptic or on thermometers and injection vial rubber septa as a disinfectant. Because of their irritating fumes, the aldehydes (formaldehyde and gluteraldehyde) are generally not used as surface disinfectants. The halogens, especially chlorine and iodine, are frequently used as disinfectants. Chlorine is most often used in the form of sodium hypochlorite (NaOCl), the compound known as household bleach. The Centers for Disease Control and Prevention (CDC) recommends that tabletops be cleaned following blood spills with a 1:10 dilution of bleach. Iodine is prepared either as tincture with alcohol or as an iodophor coupled to a neutral polymer, for example, povidone-iodine. Both iodine compounds are widely employed antiseptics. In fact, 70% ethyl alcohol, followed by an iodophor, is the most common compound utilized for skin disinfection before drawing blood cultures or surgery. Because mercury is toxic to the environment, heavy metals containing mercury are no longer recommended, but an eye drop solution containing 1% silver nitrate is still instilled in the eyes of newborns to prevent infections with *Neisseria gonorrhoeae.* Quaternary ammonium compounds are used to disinfect bench-tops or other surfaces in the laboratory. However, organic materials, such as blood, may inactivate heavy metals or quaternary ammonium compounds, thus limiting their utility. Finally, phenolics, such as the common laboratory disinfectant amphyl, are derivatives of carbolic acid (phenol). The addition of detergent results in a product that cleans and disinfects at the same time, and at concentrations between 2% and 5% these products are widely used for cleaning bench-tops.

The most important point to remember when working with biocides or disinfectants is to prepare a working solution of the compound *exactly* according to the manufacturer's package insert. Many people think that if the manufacturer says to dilute 1:200, they will be getting a stronger product of they dilute it 1:10. However, the ratio of water to active ingredient may be critical, and if sufficient water is not added, the free chemical for surface disinfection may not be released.

CHEMICAL SAFETY

In 1987 the United States Occupational Safety and Health Administration (OSHA) published the Hazard Communication Standard, which provides for certain institutional educational practices to ensure all laboratory personnel have a thorough working knowledge of the hazards of chemicals with which they work. This standard has also been called the "employee right-to-know." It mandates that all hazardous chemicals in the workplace be identified and clearly marked with a National Fire Protection Association (NFPA) label stating the health risks, such as carcinogens (cause of cancer), mutagens (cause of mutations in DNA or RNA), or teratogens (cause of birth defects), and the hazard class, for example, corrosive (harmful to mucous membranes, skin, eyes, or tissues), poison, flammable, or oxidizing. Examples of these labels are shown in Figure 2-2.

Each laboratory should have a **chemical hygiene plan** that includes guidelines on proper labeling of chemical containers, manufacturer's material safety data sheets (MSDS), and the written chemical safety training and retraining programs. Hazardous chemicals must be inventoried annually. In addition, laboratories are required to maintain a file of every chemical that they use and a corresponding MSDS. These

MSDS are provided by the manufacturer for every hazardous chemical; some manufacturers also provide letters for nonhazardous chemicals, such as saline, so these can be included with the other MSDS. The MSDS include information on the nature of the chemical, the precautions to take if the chemical is spilled, and disposal recommendations. The sections in the typical MSDS include:

- Substance name
- Name, address, and phone number of manufacturer
- Hazardous ingredient(s)
- Physical and chemical properties
- Fire and explosion data
- Toxicity
- Health effects and first aid
- Stability and reactivity
- Shipping data
- Spill, leak, and disposal procedures
- Personal protective equipment
- Handling and storage

Employees should become familiar with the MSDS so that they know where to look in the event of an emergency.

Fume hoods (Figure 2-3) are provided in the laboratory to prevent inhalation of toxic fumes. Fume hoods protect against chemical odor by exhausting air to the outside but are not HEPA-filtered to trap pathogenic microorganisms. It is important to remember that a BSC is *not* a fume hood. Work with toxic or noxious chemicals should always be done in a fume hood or when wearing a fume mask. Spills should be cleaned up using a fume mask, gloves, impervious (impenetrable to moisture) apron and goggles. Acid and alkaline, flammable, and radioactive spill kits are available to assist in rendering harmless any chemical spills.

FIRE SAFETY

Fire safety is an important component of the laboratory safety program. Each laboratory is required to post fire evacuation plans that are essentially blueprints for finding the nearest exit in case of fire (Figure 2-4). Fire drills conducted quarterly or annually depending on local laws ensure that all personnel know what to do in case of fire. Exit ways should always remain clear of obstructions, and employees should be trained to use fire extinguishers. The local fire department is often an excellent resource for training in the types and use of fire extinguishers.

Type A fire extinguishers are used for trash, wood, and paper; type B are used for chemical fires, and type C are used for electrical fires. Combination type ABC extinguishers are found in most laboratories

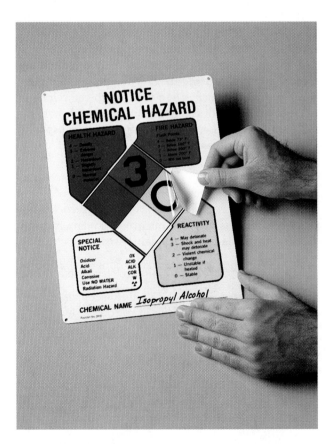

FIGURE 2-2 National Fire Protection Association diamond stating chemical hazard. This information can be customized as shown here for isopropyl alcohol by applying the appropriate self-adhesive polyester numbers to the corresponding color-coded hazard area. (Courtesy Lab Safety Supply Inc., Janesville, Wis.)

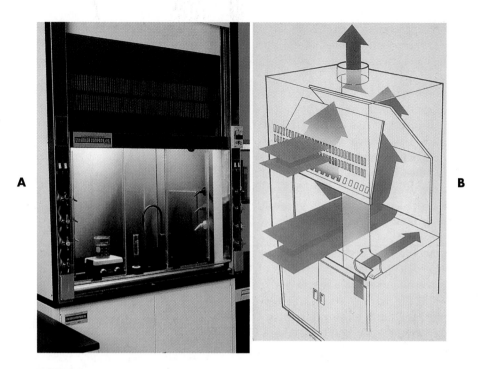

FIGURE 2-3 Fume hood. **A,** Model ChemGARD. **B,** Schematics. Arrows indicate air flow through cabinet to outside vent. (Courtesy The Baker Co., Sanford, Maine.)

so personnel need not worry about which extinguisher to reach for in the event of a fire. However, type C extinguishers, which contain carbon dioxide (CO_2) or another dry chemical to smother fire, are also used because this type of extinguisher will not damage equipment.

The important points to remember in case of fire and the order in which to perform tasks are as follows:

1. *Rescue* any injured individuals.
2. *Sound* the fire alarm.
3. *Contain* (smother) the fire, if feasible (close fire doors).
4. *Extinguish* the fire, if possible.

ELECTRICAL SAFETY

Electrical cords should be checked regularly for fraying and replaced when necessary. All plugs should be the three-prong, grounded type. All sockets should be checked for electrical grounding and leakage at least annually. No extension cords should be used in the laboratory.

HANDLING OF COMPRESSED GASES

Compressed gas cylinders (CO_2, anaerobic gas mixture) contain pressurized gas(es) and must be properly handled and secured. In cases where leaking cylinders have fallen, tanks can become missiles, resulting in loss of life and destruction of property. Therefore, gas tanks should be properly chained (Figure 2-5, *A*) and stored in well-ventilated areas. The metal cap, which is removed when the regulator is installed, should always be in place when a gas cylinder is not in use. Cylinders should be transported chained to special dollies (Figure 2-5, *B*).

BIOSAFETY

Individuals are exposed in various ways to laboratory-acquired infections in microbiology laboratories. These involve the following:

- Rubbing the eyes or nose with contaminated hands
- Inhaling aerosols produced during centrifugation, vortexing, or spills of liquid cultures
- Accidentally ingesting microorganisms by putting pens or fingers in the mouth
- Suffering percutaneous inoculation, that is, being punctured by a needlestick

Risks from a microbiology laboratory may extend to adjacent laboratories and to families of those who work in the microbiology laboratory. For example, Blaser and Feldman[1] noted that 5 of 31 individuals who contracted typhoid fever from proficiency testing specimens did not work in a microbiology laboratory. Two patients were family members

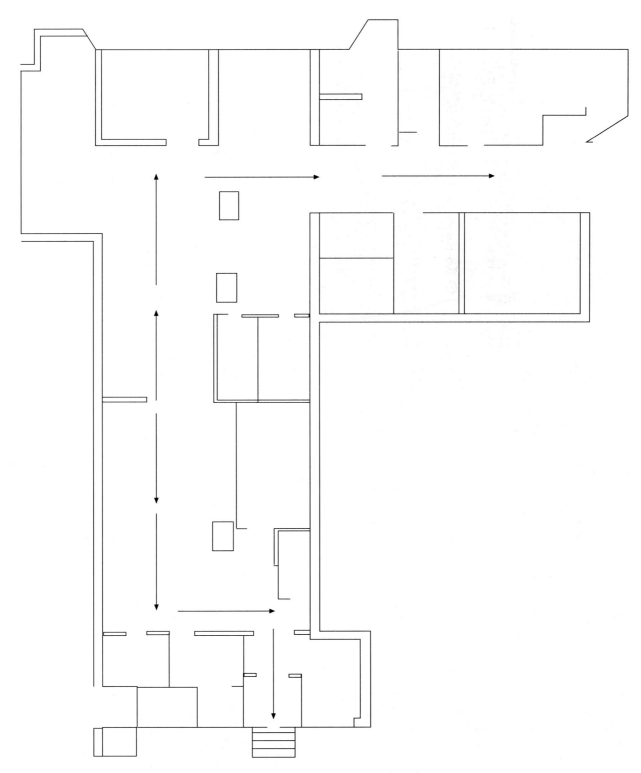

FIGURE 2-4 Fire evacuation plan for General Hospital Microbiology Laboratory. Arrows indicate quickest fire exits.

FIGURE 2-5 A, Gas cylinders chained to the wall. **B,** Gas cylinder chained to dolly during transportation. (Courtesy Lab Safety Supply Inc., Janesville, Wis.)

of a microbiologist who had worked with the *Salmonella typhi,* two were students whose afternoon class was in the laboratory where the organism had been cultured that morning, and one worked in an adjacent chemistry laboratory.

In the clinical microbiology laboratory shigellosis, salmonellosis, tuberculosis, brucellosis, and hepatitis are the five most frequently acquired laboratory infections. Viral agents transmitted through blood and body fluids cause most of the infections in nonmicrobiology laboratory workers and in health care workers in general. These include hepatitis B virus (HBV), hepatitis C virus (HCV), hepatitis D virus (HDV), and the human immunodeficiency virus (HIV). Of interest, males and younger employees (17 to 24 years old) are involved in more laboratory-acquired infections than females and older employees (45 to 64 years old).

EXPOSURE CONTROL PLAN

It is the legal responsibility of the laboratory director and supervisor to ensure that an **Exposure Control Plan** has been implemented and that the mandated safety guidelines are followed. The plan identifies

tasks that are hazardous to employees and promotes employee safety through use of the following:

- Employee education and orientation
- Appropriate disposal of hazardous waste
- Universal Precautions
- Engineering controls and safe work practices, as well as appropriate waste disposal and use of biological safety cabinets
- Personal protective equipment (PPE) such as laboratory coats, shoe covers, gowns, gloves, and eye protection (goggles, face shields)
- Postexposure plan involving the investigation of all accidents and a plan to prevent recurrences

EMPLOYEE EDUCATION AND ORIENTATION

Each institution should have a safety manual that is reviewed by all employees and a safety officer who is knowledgeable about the risks associated with laboratory-acquired infections. The safety officer should provide orientation of new employees, as well as quarterly continuing education updates for all personnel. Initial training and all retraining should be documented in writing. Hand washing should be emphasized for all laboratory personnel. The mechanical action of rubbing the hands together and soaping under the fingernails is the most important part of the process; in the laboratory, products containing antibacterial agents do not prove to be any more effective than ordinary soap, unlike the situation in areas of the hospital such as the operating room.

All employees should also be offered, at no charge, the HBV vaccine and annual skin tests for tuberculosis. For those employees whose skin tests are already positive, the employer should offer chest radiographs.

DISPOSAL OF HAZARDOUS WASTE

All materials contaminated with potentially infectious agents must be decontaminated before disposal. These include unused portions of patient specimens, patient cultures, stock cultures of microorganisms, and disposable sharp instruments, such as scalpels and syringes with needles. Infectious waste may be decontaminated by use of an autoclave, incinerator, or any one of several alternative waste-treatment methods. Some state or local municipalities permit blood, serum, urine, feces, and other body fluids to be carefully poured into a sanitary sewer. Infectious waste from microbiology laboratories is usually autoclaved on-site or sent for incineration, however.

In 1986 the EPA published a guide to hazardous waste reduction to limit the amount of hazardous waste generated and released into the environment. These regulations call for the following:

- Substituting less hazardous chemicals when possible, for example, the substitution of ethyl acetate for ether in ova and parasite concentrations and Hemo-de in place of xylene for trichrome stains
- Developing procedures that use less of a hazardous chemical, for example, the substitution of infrared technology for radioisotopes in blood culture instruments
- Segregating infectious waste from uncontaminated (paper) trash
- Substituting miniaturized systems for identification and antimicrobial susceptibility testing of potential pathogens to reduce the volume of chemical reagents and infectious waste

Recently, several alternative waste-treatment machines were developed to reduce the amount of waste buried in landfills. These systems combine mechanical shredding or compacting of the waste with either chemical (sodium hypochlorite, chlorine dioxide, peracetic acid), thermal (moist heat, dry heat) or ionizing radiation (microwaves, radio waves) decontamination. Most state regulations for these units require at least a sixfold reduction in vegetative bacte-

ria, fungi, mycobacteria, and lipid-containing viruses and at least a fourfold reduction in bacterial spores.

Infectious waste (agar plates, tubes, reagent bottles) should be placed into two leak-proof, plastic bags for sturdiness (Figure 2-6); this is known as **double-bagging** in common laboratory jargon. Pipettes, swabs, and other glass objects should be placed into rigid cardboard containers (Figure 2-7) before disposal. Broken glass is placed in thick boxes lined with plastic biohazard bags (Figure 2-8); when full, the box is incinerated or autoclaved. Sharp objects, including scalpels and needles, are placed in Sharps containers (Figure 2-9), which are autoclaved or incinerated when full.

UNIVERSAL PRECAUTIONS

In 1987 the CDC published guidelines, which have become known as **Universal Precautions,** to reduce the risk of HBV transmission in clinical laboratories and blood banks. These precautions require that blood and body fluids from every patient be treated as potentially infectious. The essentials of Universal Precautions and safe laboratory work practices are as follows:

- Do not eat, drink, smoke, or apply cosmetics (including lip balm).
- Do not insert or remove contact lenses.
- Do not bite nails or chew on pens.
- Do not mouth-pipette.
- Limit access to the laboratory to trained personnel only.
- Assume all patients are infectious for HIV or other blood-borne pathogens.
- Use appropriate barrier precautions to prevent skin and mucous membrane exposure, including wearing gloves at all times and masks, goggles, gowns, or aprons if there is a risk of splashes or droplet formation.
- Thoroughly wash hands and other skin surfaces after gloves are removed and immediately after any contamination.
- Take special care to avoid injuries with sharp objects such as needles and scalpels.

The CDC's Universal Precautions should be followed for handling blood and body fluids, including all secretions and excretions (e.g., serum, semen, all sterile body fluids, saliva from dental procedures, and vaginal secretions) submitted to the microbiology laboratory. Universal Precautions do not apply to feces, nasal secretions, saliva (except in dental procedures), sputum, sweat, tears, urine, or vomitus unless they are grossly bloody.

The consistent practice of Universal Precautions by health care workers handling all patient material will lessen the risks associated with such specimens.

FIGURE 2-6   Autoclave bags. (Courtesy Allegiance Healthcare Corp., McGaw Park, Ill.)

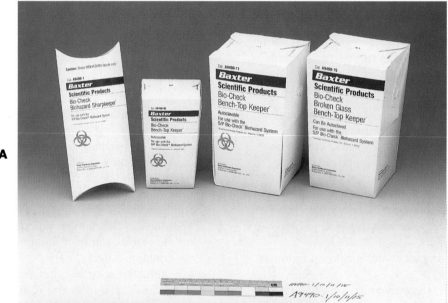

A

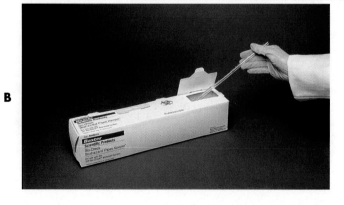

B

FIGURE 2-7   A, Various bench-top pipette discard containers. **B,** Bench-top serologic pipette discard container. (Courtesy Allegiance Healthcare Corp., McGaw Park, Ill.)

FIGURE 2-8 Cartons for broken glass. (Courtesy Lab Safety Supply Inc., Janesville, Wis.)

Mouth pipetting is strictly prohibited. Mechanical devices must be used for drawing all liquids into pipettes. Eating, drinking, smoking, and applying cosmetics are strictly forbidden in work areas. Food and drink must be stored in refrigerators in areas separate from the work area. All personnel should wash their hands with soap and water after removing gloves, handling infectious material, and before leaving the laboratory area. In addition, it is good practice to store sera collected periodically from all health care workers so that, in the event of an accident, a seroconversion (acquisition of antibodies to an infectious agent) can be documented (see Chapter 16).

All health care workers should follow Universal Precautions whether working inside or outside the laboratory. When collecting specimens outside the laboratory, individuals should follow these guidelines:

- Wear gloves and a laboratory coat.
- Deal carefully with needles and lancets.
- Discard sharps in an appropriate puncture-resistant container.
- Never recap needles by hand; if necessary, special devices should be available for resheathing needles.

ENGINEERING CONTROLS

Laboratory environment

The biohazard symbol should be prominently displayed on laboratory doors and any equipment (refrigerators, incubators, centrifuges) that contain

FIGURE 2-9 Sharps containers. (Courtesy Lab Safety Supply Inc., Janesville, Wis.)

infectious material. The air-handling system of a microbiology laboratory should move air from lower to higher risk areas, never the reverse. Ideally, the microbiology laboratory should be under negative pressure, and air should not be recirculated after passing through microbiology (see Chapter 3 for a more detailed discussion of negative pressure in microbiology laboratories). The selected use of BSCs for procedures that generate infectious aerosols is critical to laboratory safety. Many infectious diseases, such as plague, tularemia, brucellosis, tuberculosis, and legionnaires' disease, may be contracted by inhaling infectious particles, often present in a droplet of liquid. Because blood is a primary specimen that may contain infectious virus particles, subculturing blood cultures by puncturing the septum with a needle should be performed behind a barrier to protect the worker from droplets. Several other common procedures used to process specimens for culture, notably mincing, grinding, vortexing, and preparing direct smears for microscopic examination, are known to produce aerosol droplets. These procedures must be performed in a BSC.

The microbiology laboratory poses many hazards to unsuspecting and untrained people; therefore, access should be limited to employees and other necessary personnel (biomedical engineers, housekeepers). Visitors, especially young children, should be discouraged. Certain areas of high risk, such as the mycobacteriology and virology laboratories, should be closed to visitors. Custodial personnel should be

trained to discriminate among the waste containers, dealing only with those that contain noninfectious material.

Care should be taken to prevent insects from infesting any laboratory area. Mites, for example, can crawl over the surface of media, carrying microorganisms from colonies on a plate to other areas. Houseplants can also serve as a source of insects and should be carefully observed for infestation, if they are not excluded altogether from the laboratory environment. A pest control program should be in place to control rodents and insects.

Biological safety cabinet

A BSC is a device that encloses a workspace in such a way as to protect workers from aerosol exposure to infectious disease agents. Air that contains the infectious material is sterilized, either by heat, ultraviolet light, or, most commonly, by passage through a HEPA filter that removes most particles larger than 0.3 μm in diameter. These cabinets are designated by class according to the degree of biologic containment they afford. Class I cabinets allow room (unsterilized) air to pass into the cabinet and around the area and material within, sterilizing only the air to be exhausted (Figure 2-10). Class II cabinets sterilize air that flows over the infectious material, as well as air to be exhausted. The air flows in "sheets," which serve as barriers to particles from outside the cabinet, and direct the flow of contaminated air into the filters (Figure 2-11). Such cabinets are called **laminar flow**

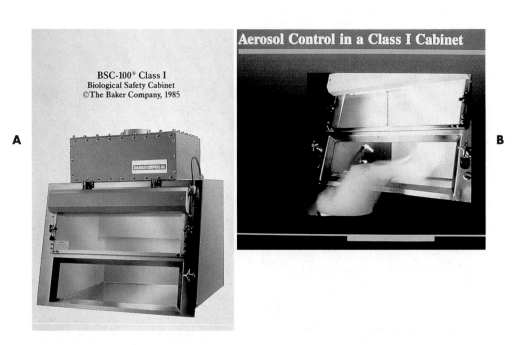

FIGURE 2-10 Class I Biological Safety Cabinet. **A,** Model BSC-100. **B,** Schematics showing airflow. (Courtesy The Baker Co., Sanford, Maine.)

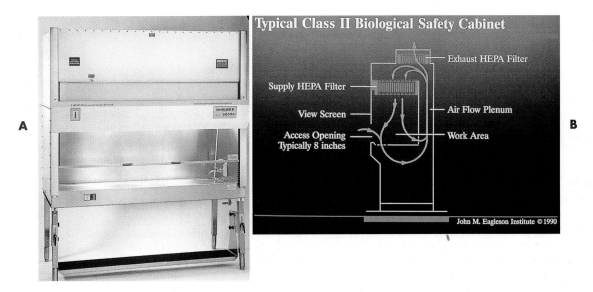

FIGURE 2-11 Class II Biological Safety Cabinet. **A,** Model SterilGARD II. **B,** Schematics showing airflow. (Courtesy The Baker Co., Sanford, Maine.)

BSCs. Type II cabinets have a variable sash opening through which the operator gains access to the work surface. The IIA cabinet is self-contained, and 70% of the air is recirculated. The class IIB cabinet is selected if radioisotopes, toxic chemicals, or carcinogens will be used; the exhaust air in class IIB cabinets is discharged outside the building. Because they are completely enclosed, with negative pressure, class III cabinets afford the most protection to the worker. Air coming into and going out of the cabinet is filter sterilized, and the infectious material within is handled with rubber gloves that are attached and sealed to the cabinet (Figure 2-12).

Most hospital clinical microbiology laboratory technologists use class II cabinets. The routine inspection and documentation of adequate function of these cabinets are critical factors in an ongoing quality assurance program. Important to proper operation of laminar flow cabinets is maintenance of an open area for 3 feet (90 cm) from the cabinet during operation of the air-circulating system to ensure that infectious material is directed through the HEPA filter. BSCs must be certified initially, whenever moved more than 18 inches, and annually, thereafter.

Personal protective equipment

OSHA regulations require that health care facilities provide employees with all **personal protective equipment** (Figure 2-13) necessary to protect them from hazards encountered during the course of work. This usually includes plastic shields or goggles to protect workers from droplets, disposal containers for sharp objects, holders for glass bottles, trays in which to carry smaller hazardous items (such as blood culture bottles), handheld pipetting devices, impervious gowns, laboratory coats, disposable gloves, masks, safety carriers for centrifuges (especially those used in the AFB laboratory), and HEPA respirators.

HEPA respirators are required for all health care workers, including phlebotomists, who enter rooms of patients with tuberculosis, and also to clean up spills of pathogenic microorganisms (see Chapter 7). All respirators should be fit-tested for each individual so that each person is assured that theirs is working properly. Males must shave their facial hair to achieve a tight fit.

Microbiologists should wear laboratory coats over their street clothes, and these coats should be removed before leaving the laboratory. Most exposures to blood-containing fluids occur on the hands or forearms, so gowns with closed wrists or forearm covers and gloves that cover all potentially exposed skin on the arms are most beneficial. If the laboratory protective clothing becomes contaminated with body fluids or potential pathogens, it should be sterilized in an autoclave immediately and cleaned before reusing. Laboratory coats should be cleaned by the institution or a uniform agency; it is no longer permissible for microbiologists to launder their own coats. Alternatively, disposable gowns may be used. Obviously, laboratory workers who plan to enter an area of the hospital in which patients at special risk of acquiring infection are present (e.g., intensive care units, the nursery, operating rooms, or areas in which immunosuppressive therapy is being administered) should take every precaution to cover their street clothes with clean or sterile protective clothing appropriate to the area being visited. Special impervious, protective

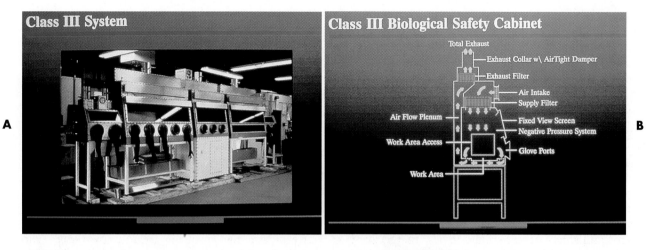

FIGURE 2-12 Class III Biological Safety Cabinet. **A**, Photograph of custom-built class III system. **B**, Schematics. Arrows show airflow through cabinet. (Courtesy The Baker Co., Sanford, Maine.)

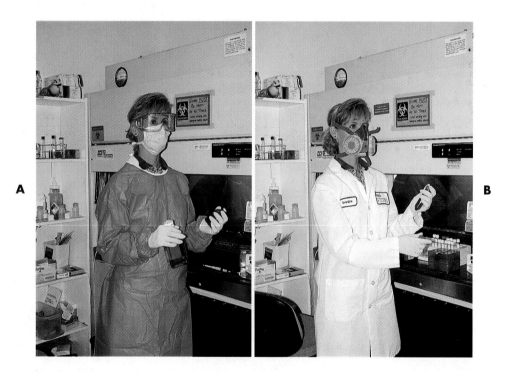

FIGURE 2-13 Personal protective equipment. **A**, Microbiologist wearing a laboratory gown, gloves, goggles, and face mask. **B**, Microbiologist wearing a laboratory coat, gloves, and respirator with HEPA filters (pink cartridges) for cleaning up spills of *Mycobacterium tuberculosis.*

clothing is advisable for certain activities, such as working with radioactive substances or caustic chemicals. Solid-front gowns are indicated for those working with specimens being cultured for mycobacteria. Unless large-volume spills of potentially infectious material are anticipated, impervious laboratory gowns are not necessary in most microbiology laboratories.

Postexposure control

All laboratory accidents and potential exposures *must* be reported to the supervisor and safety officer, who will immediately arrange to send the individual to employee health or an outside occupational health physician. Immediate medical care is of foremost importance; investigation of the accident should take place only after the employee has received appropriate care. If the accident is a needlestick injury, for example, the patient should be identified and the risk of the laboratorian acquiring a blood-borne infection should be assessed. The investigation will help the physician determine the need for prophylaxis, that is, HBIG (hepatitis B virus immunoglobulin) or an HBV booster immunization. The physician will also be able to discuss the potential for disease transmission to family members following, for example, exposure to a patient with *Neisseria meningitidis*. Follow-up treatment should also be assessed including, for example, drawing additional sera at intervals of 6 weeks, 3 months, and 6 months for HIV testing. Finally, the safety committee or, at minimum, the laboratory director and safety officer should review the accident to determine whether it could have been prevented and to delineate measures to be taken to prevent future accidents. The investigation of the accident and corrective action should be documented in writing in an incident report.

CLASSIFICATION OF BIOLOGIC AGENTS BASED ON HAZARD

A CDC booklet titled *Classification of Etiological Agents on the Basis of Hazard* served as a reference for assessing the relative risks of working with various biologic agents until an updated CDC/NIH document was published titled *Biosafety in Microbiological and Biomedical Laboratories*. In general, patient specimens pose a greater risk to laboratory workers than do microorganisms in culture, because the nature of etiologic agents in patient specimens is initially unknown.

Biosafety Level I agents include those that have no known potential for infecting healthy people. These agents are used in laboratory teaching exercises for beginning-level students of microbiology. Level 1 agents include *Bacillus subtilis* and *Mycobacterium gordonae*. Precautions for working with Level 1 agents include standard good laboratory technique as described previously.

Biosafety Level 2 agents are those most commonly being sought in clinical specimens, and they include all the common agents of infectious disease, as well as HIV and several more unusual pathogens. For handling clinical specimens suspected of harboring any of these pathogens, Biosafety Level 2 precautions are sufficient. This level of safety includes the principles outlined previously plus limiting access to the laboratory during working procedures, training laboratory personnel in handling pathogenic agents, direction by competent supervisors, and performing aerosol-generating procedures in a BSC. Employers must offer hepatitis B vaccine to all employees determined to be at risk of exposure.

Biosafety Level 3 procedures have been recommended for the handling of material suspected of harboring viruses unlikely to be encountered in a routine clinical laboratory, and for cultures growing *M. tuberculosis,* the mold stages of systemic fungi, and for some other organisms when grown in quantities greater than that found in patient specimens. These precautions, in addition to those undertaken for Level 2, consist of laboratory design and engineering features that contain potentially dangerous material by careful control of air movement and the requirement that personnel wear protective clothing and gloves, for instance. Persons working with Biosafety Level 3 agents should have baseline sera stored for comparison with acute sera that can be drawn in the event of unexplained illness.

Biosafety Level 4 agents, which include only certain viruses of the arbovirus, arenavirus, or filovirus groups, none of which are commonly found in the United States, require the use of maximum containment facilities; personnel and all materials must be decontaminated before leaving the facility, and all procedures are performed under maximum containment (special protective clothing, class III BSCs). Most of these facilities are public health or research laboratories.

MAILING BIOHAZARDOUS MATERIALS

In October 1995 the requirements for packaging and shipping biologic materials were significantly revised in response to the international community's desire to ensure safe and trouble-free shipment of infectious material. Before this revision, the United States Public Health Service (PHS), the United States Postal Service (USPS) and the United States Department of Transportation (DOT) specified the minimum requirements for packing and shipping biologic materials within the United States, and the United Nations (UN) published recommendations that were subsequently adopted for international shipments by the International Air Transport Association (IATA). Today, most U.S. carriers (Federal Express, Airborne Express, United Parcel Service) have adopted IATA standards for shipments by air or ground throughout the world, since the IATA **Dangerous Goods Regulations** incorporate both U.S. and UN regulations. Training in the proper packing and shipping of infectious material is a key feature of the regulations. Every institution that ships infectious materials, whether a hospital or POL (physician office laboratory), is required to have an appropriately trained individual; training may be obtained through carriers, package manufacturers, and special safety training organizations. These individuals are, in turn, responsible for training other people at their workplace.

The PHS categorizes regulated materials as one of the following:

- Diagnostic or clinical specimens that may or may not contain an etiologic (infectious) agent, such as urine sent for drug testing
- Material containing etiologic agents, such as blood agar slant with *Staphylococcus aureus* or a blood specimen from an HIV-positive patient
- Material containing special etiologic agents, such as multidrug-resistant organisms or organisms that potentially can be used as biologic weapons

The shipper's laboratory or medical director is responsible for defining the appropriate categorization of a specimen. This currently is the most difficult part of the procedure. If in doubt about how to classify a specimen, always follow the most stringent regulations.

In the most **basic packaging** for diagnostic specimens, the specimen should be wrapped with absorbent material and placed within a closed, watertight container, called a **primary container.** The lid of the specimen, preferably in a plastic tube or vial, should be sealed with waterproof tape. The primary container is then inserted into a **secondary container,** often a metal or plastic tube (Figure 2-14). After the secondary container is capped, an address label and an Etiologic Agents label (Figure 2-15) is affixed to the outside of the container.

In shipping material containing etiologic agents, specimens that are less than 50 mL in volume may be shipped in special containers such as those shown in Figure 2-16. The specimen is wrapped in a sponge treated with a disinfectant that will absorb and neutralize the contents in case of breakage. The specimen and sponge are then placed in a biohazard bag and rolled up inside a hard, plastic, watertight container. An address label and Infectious Substance label should be affixed to the outside of the container, which now may be shipped as an ordinary diagnostic specimen.

In instances where the specimen to be shipped is greater than 50 mL in volume, or when shipping special etiologic agents, an outer **tertiary container** constructed of fiber board must be used. This **special packaging** material is shown in Figure 2-17. A UN number on the outer box confirms that it meets all the required standards. The label shown in Figure 2-17, *A* is then affixed to the outer box. The Shippers Declaration for Dangerous Goods (Figure 2-18) must accompany the airbill or ground form.

Shippers should note that some carriers have additional requirements for coolant materials such as ice, dry ice, or liquid nitrogen. Because the shipper is liable for appropriate packaging, it is best to check with individual carriers in special circumstances.

FIGURE 2-14 Primary and secondary metal mailing tubes for basic packaging of diagnostic specimens. (Courtesy Allegiance Healthcare Corp., McGaw Park, Ill.)

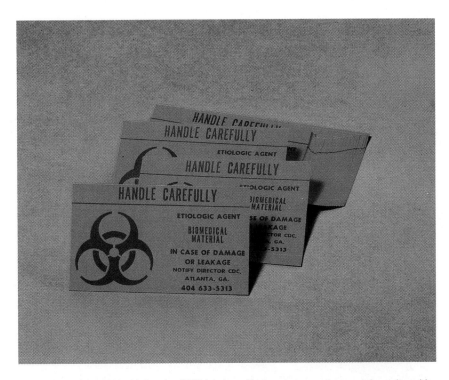

FIGURE 2-15 Public Health Service (PHS) label to affix to outer, secondary container when shipping an etiologic agent.

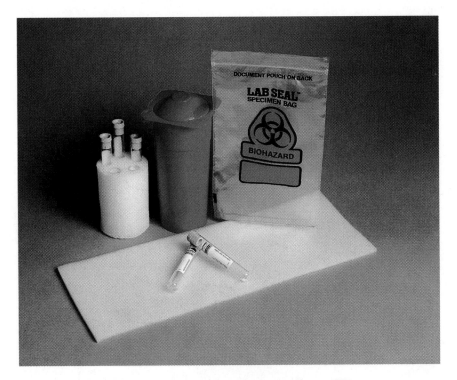

FIGURE 2-16 Device for shipping infectious substances in volumes of 50 mL or less. Tubes of blood or slants of organisms are wrapped in the treated sponge and inserted into the biohazard bag. Everything is then placed inside the bright red, plastic container. In the event of a spill, the sponge absorbs and disinfects the contents of the container and renders it noninfectious. (Courtesy Pro-Tech-Tube International, Carrollton, Texas.)

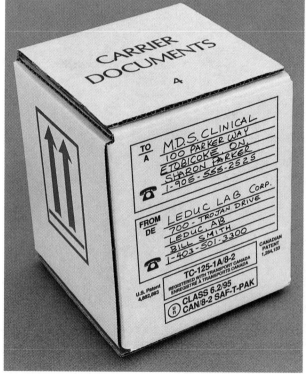

FIGURE 2-17 Special UN packaging for highly infectious substances or volumes exceeding 50 mL. **A,** All necessary components. **B,** Closed cardboard box with UN markings. (Courtesy Saf-T-Pak Inc., Edmonton, Canada.)

SHIPPER'S DECLARATION FOR DANGEROUS GOODS

Shipper	Air Waybill No.
	Page of Pages
	Shipper's Reference Number
	(optional)

Consignee	*For optional use* *for* *Company logo* *name and address*

Two completed and signed copies of this Declaration must be handed to the operator.	**WARNING**

TRANSPORT DETAILS

This shipment is within the limitations prescribed for: *(delete non-applicable)*	Airport of Departure
PASSENGER AND CARGO AIRCRAFT CARGO AIRCRAFT ONLY	

Failure to comply in all respects with the applicable Dangerous Goods Regulations may be in breach of the applicable law, subject to legal penalties. This Declaration must not, in any circumstances, be completed and/or signed by a consolidator, a forwarder or an IATA cargo agent.

Airport of Destination:

Shipment type: *(delete non-applicable)*
NON-RADIOACTIVE | RADIOACTIVE

NATURE AND QUANTITY OF DANGEROUS GOODS

Dangerous Goods Identification							
Proper Shipping Name	Class or Division	UN or ID No.	Packing Group	Subsidiary Risk	Quantity and type of packing	Packing Inst.	Authorization

Additional Handling Information

I hereby declare that the contents of this consignment are fully and accurately described above by the proper shipping name, and are classified, packaged, marked and labelled/placarded, and are in all respects in proper condition for transport according to applicable international and national governmental regulations.	Name/Title of Signatory
	Place and Date
	Signature *(see warning above)*

FIGURE 2-18 Shippers Declaration of Dangerous Goods form. (Courtesy IATA.)

References

1. Blaser, M.J. and Feldman, R.A. 1980. Acquisition of typhoid fever from proficiency testing specimens. N. Engl. J. Med. 303:1481.
2. Eyre, J.W.H. 1913. Bacteriologic technique. W.B. Saunders, Philadelphia.

Bibliography

Centers for Disease Control. 1988. Update: Universal precautions for prevention of transmission of human immunodeficiency virus, hepatitis B virus, and other blood-borne pathogens in health care settings. MMWR Morb. Mortal. Wkly. Rep. 37:377.

Centers for Disease Control. 1987. Recommendations for prevention of HIV transmission in health-care settings. MMWR Morb. Mortal. Wkly. Rep. 36:3S.

Fleming, D.O., Richardson, J.H., Tulis, J.I., and Vesley, D. 1995. Laboratory safety: principles and practices, ed 2. American Society for Microbiology, Washington, D.C.

International Air Transport Association. 1997. Dangerous Goods Regulations, ed 38. International Air Transport Association, Montreal.

Jamison, R., Noble, M.A., Proctor, E.M., and Smith, J.A. 1996. Cumitech 29, Laboratory safety in clinical microbiology. J.A. Smith, coordinating editor, American Society for Microbiology, Washington, D.C.

National Committee for Clinical Laboratory Standards. 1991. Protection of laboratory workers from infectious disease transmitted by blood and tissue, ed 2. Tentative Guideline M29-T2. National Committee for Clinical Laboratory Standards, Villanova, Pa.

Occupational Safety and Health Administration. 1991. Occupational exposure to blood-borne pathogens: final rule. Federal Register 56:64175.

Occupational Safety and Health Administration. 1991. Occupational exposure to blood-borne pathogens: correction July 1, 1992. 29 CFR Part 1910. Federal Register 57:127: 29206.

Occupational Safety and Health Administration. 1993. Draft guidelines for preventing the transmission of tuberculosis in health care facilities. Federal Register 58:52810.

Sewell, D.L. 1995. Laboratory-associated infections and biosafety. Clin. Microbiol. Rev. 8:389.

United States Department of Health and Human Services. 1993. Biosafety in microbiological and biomedical laboratories. HHS publication (CDC) 93-8395. U.S. Government Printing Office, Washington, D.C.

United States Environmental Protection Agency. 1986. EPA guide for infectious waste management. Publication EPA/530-SW-86-014. U.S. Environmental Protection Agency, Washington, D.C

3 | LABORATORY PHYSICAL DESIGN AND ORGANIZATION

Microbiology laboratories operate in various venues. During the 1990s the trend changed from centralized facilities that operate from large hospital, public health, or independent laboratories to decentralized facilities that operate outside of a main laboratory but may still be within the physical hospital facility. Decentralized testing is also known as **alternate-site testing**. Alternate-site testing is performed in physician office laboratories (POLs), in ancillary outpatient hospital clinics, or at the hospital patient's bedside. Bedside testing is also known as **point of care testing (POCT)**. POCT is the most radical departure from previous tenets involved in laboratory testing. The most common reason for POCT is enhanced turn-around time, and most POCT involves nonmicrobiologic testing (e.g., glucose, fecal occult blood, urinalysis, hemoglobin, hematocrit, arterial blood gases, and sodium or potassium levels). POCT is most often performed by nurses, perfusionists (who operate heart-lung machines during open heart surgery), respiratory therapists, and physicians themselves. The most common microbiology bedside test is the rapid group A streptococcal antigen test for sore throats; this test is also popular in POLs or outpatient pediatric clinics where support personnel can perform the test and report the results back to the physician before the patient goes home. Microbiologists are often asked to take sole or joint responsibility for test accuracy and reproducibility and quality control and proficiency test verification at alternate sites. Thus, because they may be asked to consult outside the main laboratory, microbiologists should be aware of all testing concerning infectious disease diagnosis.

SPACE REQUIREMENTS AND ORGANIZATION OF WORK FLOW

No matter where microbiology testing is performed a physical space will be set aside for this purpose. Depending on the complexity of testing and the type of pathogens sought, the laboratory space is organized based on the following considerations:

- Organization and staffing
- Functions (tests) that must be performed
- Equipment that will be used
- Special electrical, plumbing, or air handling requirements
- Safety equipment (e.g., shower/eye wash) and emergency systems (e.g., fire alarm or sprinkler system) requirements
- Waste-treatment requirements
- Containment requirements

All of these features will be used to determine the space requirements or square footage. Although architects, engineers, and contractors are hired to design and construct the space, the microbiologist should take an active role in ensuring that the design and layout optimize work flow and efficiency. In the typical hospital or independent laboratory, there is a space for each of the following:

- Specimen receiving
- Specimen accessioning and processing
- Staining and light microscopy
- Dark-field and fluorescent microscopy, if applicable
- Waste disposal
- Media preparation and glassware washing
- An isolation room for processing acid-fast bacilli (AFB) and systemic fungi, if applicable
- A separate room for specialized molecular-based tests
- Open benches for routine specimen workup

There may also be separate areas for offices, record storage, restrooms, lunchrooms, and library or conference rooms. Laboratory design consultants at the Centers for Disease Control and Prevention (CDC) recommend 200 square feet to accommodate 2 to 3 technologists; others have recommended 8 to 10 linear feet per technologist. Closely related functions (waste treatment and glassware washing) should be adjacent to each other or in close proximity. Bubble diagrams (floor plans) (Figure 3-1) allow visualization of spatial relationships and traffic flow patterns. Several rules of thumb should be considered when designing spaces. Module design is recommended for

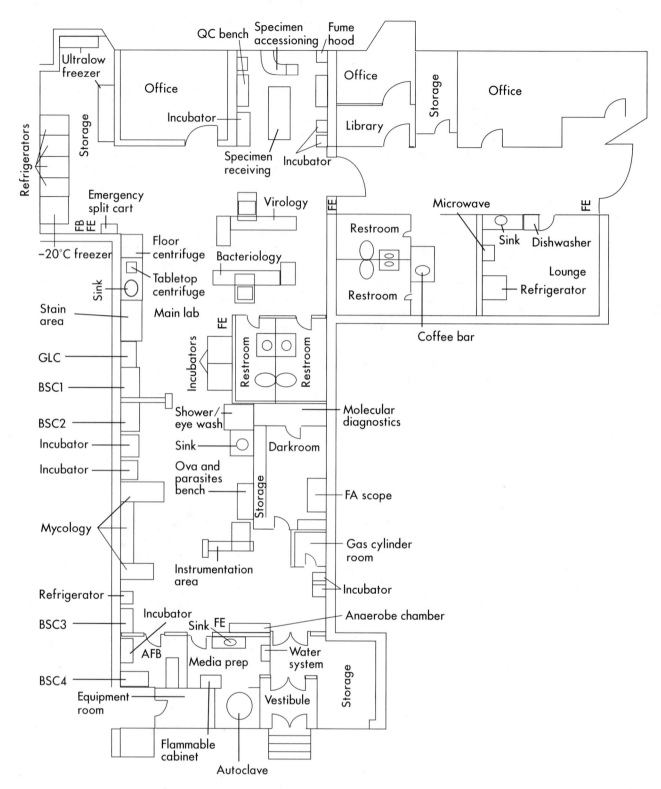

FIGURE 3-1 Bubble diagram for space layout. FE, fire extinguisher; FB, fire blanket; QC, quality control; GLC, gas liquid chromatograph; BSC, biological safety cabinet; AFB, acid-fast bacilli.

maximum flexibility. The administrative (office) area should be located near the front entrance to the laboratory away from potentially contaminated and odor-producing activities. Likewise, the areas where work with the most potentially infectious organisms is performed should be located in the most secluded space. Corridors should be at least 5 feet wide to allow people to pass each other, and total floor space should include reasonable projections for expansion.

LABORATORY DESIGN WITH RESPECT TO SAFETY

Laboratory safety is discussed in depth in Chapter 2. However, a few points are worth repeating here as they pertain to laboratory design. The emergency shower and eye wash stations should be centrally located. The standard for the location of these units is the same, that is, within 10 seconds and 100 feet of each work area. Cold water is used for emergency flushing of eyes and skin because it will do the following:

- Slow the reaction rate of the splashed chemical
- Constrict blood vessels and minimize circulation of an absorbed chemical
- Slow cellular metabolism and enzyme reaction rates
- Help reduce the pain of chemical contact

The National Fire Protection Association (NFPA) can be consulted regarding the proper storage of chemicals to prevent unintentional mixing of chemicals from leaking or broken containers. Sulfuric acid and sodium hydroxide, which are both corrosives, for example, should not be stored in the same area. Flammables should be stored in a special cabinet and alkalines, oxidizers, and carcinogens should be stored separately.

Fire extinguishers and blankets should be readily available throughout the work area. A spill cart should be centrally located and should contain first aid supplies and kits to clean up radioactive, acid, alkali, and corrosive spills. Installation of an automatic fire detection and alarm system, as well as sprinklers (to extinguish fire), is part of the building code in many locations. Moreover, two means of exit should be planned in case of fire. All electrical outlets should be grounded.

The windows in the laboratory should be closed and sealed, and laboratory doors should be self-closing. There may be special requirements for storage and disposal of radionuclides although ^{14}C (carbon-14) and ^3H (tritium), which are commonly used in the microbiology laboratory, are usually under a general license; state radiation safety offices should be consulted, if necessary. As stated in Chapter 2, the CDC and NIH (National Institutes of Health) address biosafety design criteria. Except for those facilities in which culturing of *Mycobacterium tuberculosis* (Biosafety Level 3) is conducted, clinical microbiology laboratories usually operate at Biosafety Level 2. Special or maximum containment facilities require a separate building, sealed openings into the laboratory, airlocks or liquid disinfectant barriers, dressing rooms and shower contiguous to the laboratory, double-door entry into the laboratory, biowaste treatment system, separate ventilation system, and a treatment system to decontaminate exhaust air.

DESIGN OF AIR HANDLING SYSTEM

HVAC (heating, ventilation, and air conditioning) systems must maintain a constant yearround temperature of between 68° to 72° F and a relative humidity of 40% to 60%. The large amount of heat produced by instruments must be taken into account when planning an HVAC system. The American Society of Heating, Refrigeration and Air-Conditioning Engineers (ASHRAE) publishes standards for design of HVAC systems that take these environmental conditions into account. The supply air system provides fresh air to the various locations through adjustable vane diffusers mounted in the ceiling; it may condition the air (filter out mold spores and pollens, for instance) before its delivery (e.g., in laminar flow rooms). The supply air ductwork is generally constructed of galvanized sheet metal with external insulation to facilitate cleaning of the interior of the ducts in the case of aerosolization of pathogens following an accident. The exhaust air system is also externally insulated sheet metal; activated charcoal absorption filters and HEPA filters may be used to remove contaminants before exhaust air is discharged to the atmosphere. However, except in cases of Biosafety 3 or 4 facilities, the exhaust air can be discharged to the outside without being treated (HEPA-filtered).

The laboratory areas should maintain negative pressure with respect to the administrative areas to prevent toxic or pathogenic materials used in laboratory work areas from escaping and injuring humans or contaminating the environment. To set up negative pressure in the laboratory (or positive pressure in the offices) laboratory HVAC systems are designed so that a unidirectional airflow draws supply air into the laboratory (from the administrative areas) and discharges exhaust air directly to the outside away from occupied areas and air intakes. The amount of supply air being provided to a laboratory should equal 85% of the air being exhausted from the space. The other 15% of the make-up, or supply, air is provided from adjacent corridors through doorways to maintain the negative pressure. The number of air exchanges per hour is usually 4 to 6 for office spaces and 10 to 15 for biohazard rooms.

DESIGN OF MECHANICAL SYSTEMS

Many municipal or state government agencies have established utility specifications. These include electrical, plumbing, communications, lighting, gas, and vacuum systems. Electrical wall outlets should be located every 4 to 5 feet along walls and peninsular benches and should be positioned high enough to clear the top of a stand-up bench. The sum of the electrical requirements for all equipment, lighting, and air conditioning is used in calculating the total number of amps required. Dedicated circuits (with only one outlet) should be planned for computers or other instruments sensitive to power surges. Care should be taken to provide 110- and 220-volt circuitry for instruments, as needed. Telephones should be located for ease of use as should computer terminals, facsimile machines, and teleprinters. Service outlets for natural gas, compressed air, and vacuum should be located at each workstation. Commonly used gases such as carbon dioxide (CO_2) for incubators are often piped into the laboratory work areas to prevent having to transport compressed gas tanks throughout the laboratory. Sinks should be located at the end of benches or peninsulas. Typical plumbing services include hot and cold water and deionized or distilled water. Local building codes often call for plumbing effluent (discharge) to be diluted in a neutralization tank before it joins the building's sanitary sewer system. Light fixtures should be surface-mounted on or recessed into the ceiling and centered over the front edge of laboratory benches. The number of foot-candles required for illumination of work surfaces is usually between 160 and 200, depending on local ordinances. Because they cause the least disturbance to the ceiling, surface-mounted fixtures are preferable in microbiology. Recessed lighting should be tightly sealed to make the ceiling airtight.

WALLS, FLOORS, CEILINGS, AND FURNITURE

Sheetrock walls should be painted with epoxy so that they may be easily cleaned in the event of a spill. Square-foot, vinyl composition tiles are the most economical and effective floor covering for a laboratory. Seamless flooring is difficult to repair following spills, because the spilled material may dissolve the vinyl. Acoustic tile ceilings are permissible for Biosafety Level 1 or 2 facilities. However, sealed sheetrock or plaster ceilings with three coats of epoxy

paint are recommended for Biosafety Level 3 areas; this type of ceiling will prevent airborne contamination from penetrating the space above the ceiling and can be washed with disinfectant as necessary. Flexible furniture systems that can be disassembled and moved as technologic changes occur or new equipment is purchased are preferred to permanently installed built-ins. They allow entire sections of the laboratory to be modified with minimal cost and inconvenience. Bench-tops should be impervious to water and resistant to acids, alkalis, organic solvents, and moderate heat. Tops are made of cast resin, quarried stone, particleboard or wood cores with acid-resistant plastic laminate surfaces, and stainless steel. Plastic laminates are usually acceptable for most uses and are the least expensive and easiest to replace. Standard benches have a depth of 30 inches.

INSTRUMENTATION

Instruments commonly found in the microbiology laboratory include:

- Bacterial detection devices (e.g., blood cultures)
- Bacterial identification and susceptibility testing devices
- Gas liquid chromatographs (GLC)
- Microscopes (light, fluorescent, phase dark-field)
- Centrifuges
- Autoclave
- Dry-heat oven
- Incubators (radiant heat, CO_2)
- Water baths
- Heat baths
- Vortex mixers
- Anaerobe chamber
- Refrigerators/freezers
- Ultra-low freezer
- Biological safety cabinet (BSC)
- Fume hood
- Thermal cyclers

The life expectancy of major laboratory equipment is 5 years. Some equipment, (e.g., biological safety cabinets or fume hoods) must be carefully placed in the space when designing the floor plan. This equipment, in particular, should be located away from the main traffic areas or doorways to minimize air turbulence and the potential for air spillage. Autoclaves should be in a separate, back area, if possible, to reduce the noise, odors, and heat generated by this equipment.

The decision to buy any instrument should be analyzed based on a number of factors as outlined in Chapter 4. This includes an analysis of how the instrument will assist the microbiologist in turning out quality patient results. For example, will it decrease turn-around time (e.g., same-day bacterial identification and susceptibility instruments), will it allow a single person to handle more work by automating various parts of a process, or will it help save money by allowing testing of large volumes of specimens. Manufacturers' representatives are good resources for justifying instrument purchases because they know, for example, the minimum number of tests that a laboratory should perform to make the instrument cost effective and they usually have compared their instrument to others or to conventional methodology. Microbiology still remains mainly a manual discipline although many semiautomated instruments are currently marketed.

ORGANIZATION OF THE MICROBIOLOGY LABORATORY

The organization of the microbiology laboratory is dependent on the site (public health hospital, POL, or independent laboratory) and the complexity of testing. However, some general guidelines may be applied to any situation, and they are discussed in the following section.

DIVISION OF WORK

Clinical microbiology comprises essentially seven subspecialty areas: aerobic and anaerobic bacteriology, mycology, mycobacteriology (AFB), parasitology, virology, serology, and molecular diagnostics (PCR [polymerase chain reaction] and DNA probe technology). Cross-training (educating microbiologists in more than one of these areas) is now common practice in many laboratories. Because many of the disciplines require long, specialized training to demonstrate proficiency, rotations in mycology, AFB, parasitology, virology, serology, or molecular diagnostics may span as much as 6 months to 1 year. The same may also be the case in the anaerobe laboratory if the volume is high and special identification methods are employed. On the other hand, aerobic bacteriology is usually broken down in one of two ways. Many laboratories will separate specimens by source (e.g., blood, respiratory, stool, urine, and exudate [or miscellaneous] cultures.) The advantage of this system is that microbiologists working up the cultures only have to remember rules for that particular type of culture and work usually proceeds faster and more efficiently. Other laboratories split aerobic clinical specimens using the first letter of the patient's last name. In this system, one technologist will work up all aerobic cultures on a single patient. The advantage of this system is that a single individual has an overview of the results of all specimens on one patient. For example, if Mrs. Smith has a urine, wound, and stool culture, the same microbiolo-

gist would work up all three. In large laboratories, specimen setup and accessioning is usually handled by someone other than the one who will work up the culture. In smaller laboratories, specimen setup and workup are usually handled by the same individual, although many times one technologist reads the 24-hour cultures and another reads the biochemicals and susceptibility tests. Specimens for AFB, virology, parasitology, serology, and molecular diagnostics are usually batched and set up once a day. The reason for this is that the processing protocol can be lengthy and can involve setting up positive and negative controls with each run, so it is not cost effective to set up specimens one at a time. Clinical bacteriology and mycology specimens are usually set up as they arrive unless specimens have been submitted in transport media that will allow them to be batched (see Chapter 1).

DESIGN OF LABORATORY HANDBOOK FOR CLINICAL STAFF

It is the microbiologist's responsibility to provide a laboratory handbook for the clinical practitioners or make any applicable information available electronically on a computer screen. This handbook should be present on all nursing units (in a hospital) or given out to clients (by an independent laboratory). The handbook should list all the tests included in the test menu, with an individual description of how to collect and transport all specimen types. Handbooks are usually written in tabular form and would include the following information:

- Test names
- CPT test code number (which corresponds to numbers assigned by insurance carriers, including Medicare and Medicaid) based on standardized codes developed by the American Medical Association
- Internal (billing) test code number if different from the CPT code
- The appropriate specimen to submit (e.g., thick and thin blood smears for malaria, nasopharyngeal swab for pertussis)
- The minimum specimen requirements (e.g., 1 mL serum, 0.5 mL cerebrospinal fluid)
- The methodology employed (e.g., enzyme-linked immunosorbent assay [ELISA], complement fixation [CF], direct fluorescent antibody [DFA], culture, DNA probe)
- Appropriate container or transport medium (e.g., formalin and PVA [polyvinyl alcohol] for ova and parasite [O&P] examination, modified Cary-Blair medium for stool culture)
- Special collection instructions, including patient

preparation if applicable (e.g., no O&P examination for 10 days after a barium enema, collect 3 to 5 early-morning sputum specimens for AFB)
- Reference ranges, if applicable (e.g., titer greater than 1:8 for cryptococcal antigen test is a positive test)
- A comment section indicating turn-around times or other pertinent information such as whether testing is batched or sent to a reference laboratory

Tests are usually listed in alphabetic order for easy reference and cross-referenced, if possible (e.g., *Brucella* culture may be listed under *Brucella* culture, blood culture, and bone marrow culture). The microbiologist should use the handbook as a method of communicating laboratory information.

DESIGN OF LABORATORY REQUISITION FORM

The laboratory requisition form should be designed for ease of use. A sample requisition is shown in Figure 3-2. This requisition is used by a reference laboratory, but the information required is typical except for the probable absence of the billing information on a hospital requisition. The requisition commonly includes several parts. Patient demographic information (full name, birth date or age, sex), and ordering physician is included as a minimum. Patient information is often stamped onto the requisition using a plastic card similar to a credit card; in the hospital, unique patient identification numbers are also stamped on the cards. Information about the source or type of specimen and date and time of collection is also required. Individuals setting up specimens in the laboratory should always check this information carefully to verify that the specimen has been received in a timely manner and is in an appropriate transport medium. Further, because some specimens are unacceptable for particular types of cultures (e.g., vaginal specimens for anaerobic culture) checking the source of specimen against the test information is important. Certain medical information (current antimicrobial therapy, immunization history, and clinical syndrome or suspect agent) can be critical to guiding specimen setup or workup, and a space for it should be included on the requisition. Finally, the requisition should include a list of commonly ordered tests so that clinical personnel can simply check off the tests they want to order.

With the addition of computers, many laboratories have instituted electronic ordering. In many cases, a series of video screens are used to prompt the clinicians to enter the same information that can also be collected on sheets of paper. Whatever the format, however, the basic information requested will be the same.

DESIGN OF LABORATORY WORKCARD

A workcard is the legal document that can be used to reconstruct the testing process. The record of the work performed on a specimen may be handwritten on the back of the laboratory requisition or on a workcard designed especially for this purpose (Figure 3-3). Alternatively, the technologist can record the details of the specimen workup electronically, that is, directly into the computer; this is called a "paperless workcard."

The workcard should include the following information:

- Patient name
- Specimen source
- Laboratory number
- Date and time inoculated
- Initials of microbiologist setting up or working up the culture
- All notes made by the microbiologist during specimen workup, including any test results
- Record of any phone calls or faxes to clinicians
- Instrument printouts, if applicable

DESIGN OF LABORATORY REPORT FORM

The laboratory report form is the written means of communicating patient information to the clinician. It is a confidential record about the patient. The report should contain the following information:

- Name, address, and phone number of laboratory performing test
- Laboratory accreditation numbers, as applicable (including CLIA [Clinical Laboratory Improvement Amendment], CAP [College of American Pathologists], Medicare and Medicaid, and state licensure)
- Patient name and other demographic information
- Specimen source
- Laboratory number
- Ordering physician
- Test ordered (e.g., aerobic culture, herpes culture)
- Date ordered
- Date and time inoculated
- Initials of microbiologist who set up specimen
- Type of report (direct examination, preliminary report, final report)
- Test results
- Initials of microbiologist performing the test and date of report
- Initials of supervisor reviewing report and date

The report should be legible if handwritten (Figure 3-4). Computer-generated reports should be laid out in an easy-to-read format so that the clinician can readily retrieve information (Figure 3-5).

Houston Reference Laboratory

2423 Times Boulevard, Houston, Texas 77005
Telephone: (713) 522-1762   Fax: (713) 522-7722

CLIENT INFORMATION
NAME/ADDRESS/PHONE

PATIENT INFORMATION
PATIENT NAME

BIRTH DATE   SEX

SOURCE/TYPE OF SPECIMEN

CLIENT LAB ID
PATIENT IDENTIFICATION

LAB REFERENCE

PHYSICIAN

BILLING INFORMATION
PATIENT ADDRESS   SSN

CITY, STATE, ZIP   PHONE

SUBSCRIBER   RELATION

INSURANCE INFORMATION

MEDICARE/MEDICAID   PHYSICIAN UPIN   ICD-9 DIAGNOSIS

CURRENT ANTIMICROBIAL THERAPY/IMMUNIZATION HISTORY

CLINICAL SYNDROME/SUSPECT AGENT   DATE/TIME COLLECTED

TEST REQUEST Please check the appropriate box(es).

VIROLOGY
- ❏ 071 Adenovirus culture
- ❏ 104 Chlamydia pneumoniae (strain TWAR)
- ❏ 077 Chlamydia psittaci culture
- ❏ 005 Chlamydia trachomatis culture
- ❏ 105 Chlamydia trachomatis DNA probe
- ❏ 088 Combined Chlamydia trachomatis and Neisseria gonorrhoeae DNA probe
- ❏ 064 Chlamydia trachomatis direct exam
- ❏ 003 Clostridium difficile cytotoxin assay
- ❏ 106 Clostridium difficile cytotoxin assay and rapid latex detection
- ❏ 003 Cytomegalovirus (CMV) culture (conventional culture)
- ❏ 008   Cytomegalovirus (CMV) rapid culture
  - • Blood and bone marrow specimens must be done by rapid CMV culture.
- ❏ 063 Enterovirus culture
- ❏ 130 Hemadsorbing viruses culture
- ❏ 094 Herpes family screen (includes CMV, HSV and VZ culture and/or direct exams from lesions)
- ❏ 112   With rapid CMV
- ❏ 002 Herpes simples virus (HSV) conventional culture (includes typing)
- ❏ 176   Herpes simplex virus - rapid culture (no typing)
- ❏ 177   Herpes simplex virus - typing
- ❏ 114   Herpes simplex virus - direct exam-lesions only (includes typing)
- ❏ 001   Herpes simplex virus - direct exam and culture-lesions only (includes typing)
- ❏ 032 Influenza virus culture
- ❏ 156   Influenza A & B direct exam
- ❏ 100 Mumps virus culture
- ❏ 007 Mycoplasma culture - Genital
- ❏ 119 Mycoplasma culture - Respiratory
- ❏ 125 Parainfluenza virus culture
- ❏ 127 Rapid respiratory panel
- ❏ 060 Respiratory syncytial virus (RSV) culture (includes direct exam)
- ❏ 059 RSV direct exam
- ❏ 098 Rotavirus antigen detection
- ❏ 054 Varicella-zoster virus (VZ) culture (includes direct exam)
- ❏ 073 Varicella-zoster virus (VZ) direct exam
- ❏ 085 Viral direct exam (DFA)
- ❏ 082 Viral identification
- ❏ 065 Viral screen
- ❏ 134   With rapid CMV, RSV, VZ, or Adenovirus
  - • Blood and bone marrow specimens must be done by rapid CMV culture.
- ❏ 152   With rapid respiratory panel

BACTERIOLOGY
- ❏ 080 Blood culture
- ❏ 147 Blood culture using antimicrobial removal device
- ❏ 011 GC culture (Neisseria gonorrhoeae)
- ❏ 107 GC DNA probe
- ❏ 088 GC and Chlamydia trachomatis DNA probe
- ❏ 067 Group B Strep screen
- ❏ 040 Legionella culture
- ❏ 096 Legionella culture and direct exam
- ❏ 078 Legionella direct exam
- ❏ 149 Legionella subgrouping
- ❏ 044 Pertussis culture
- ❏ 092 Pertussis culture and direct exam
- ❏ 084 Pertussis direct exam
- ❏ 014 Stool culture - Routine (includes Salmonella, Shigella and Campylobacter)
- ❏ 131   Aeromonas/Plesiomonas only
- ❏ 017   Campylobacter only
- ❏ 034   Hemorrhagic E. coli only
- ❏ 015   Vibrio only
- ❏ 016   Yersinia only
- ❏ 009 Throat culture (Strep screen only)
- ❏ 013 Urine culture
- ❏ 113 Urine culture and susceptibility
- ❏ 197 Vaginal screen

AEROBIC
- ❏ 019 Culture (includes direct exam)
- ❏ 050 Identification

ANAEROBIC
- ❏ 020 Culture (includes direct exam)
- ❏ 049 Identification
- ❏ 022 Susceptibility (per organism)
- ❏ 056 Gas Liquid Chromatography (each organism)
- ❏ 062 Clostridium difficile culture

ACID - FAST
- ❏ 038 Culture (includes direct exam)
- ❏ 170   From blood or bone marrow
- ❏ 029 DNA probe
- ❏ 047 Identification
- ❏ 048 Susceptibility - Routine
- ❏ 070 Susceptibility - Special
- ❏ 076 Stain (concentrate)
- ❏ 024 Stain (direct)

SPECIAL MICROBIOLOGY
- ❏ 045 Brucella culture
- ❏ 157 Cat Scratch Disease culture
- ❏ 043 Diphtheria culture
- ❏ 052 Francisella tularensis culture
- ❏ 012 Gardnerella vaginalis culture
- ❏ 108 Giemsa stain (modified)
- ❏ 023 Gram stain
- ❏ 058 Helicobacter culture (includes direct exam)
- ❏ 110   Rapid urea screen
- ❏ 035 Haemophilus ducreyi culture
- ❏ 083 Leptospira culture
- ❏ 116 Listeria culture
- ❏ 120 OB screen 4 (includes Group B strep screen and GC and chlamydia DNA probe)
- ❏ 087 Quantitative culture
- ❏ 074 Schlichter test (Serum cidal)
- ❏ 143 STD panel 4 (includes GC and chlamydia DNA probe, vaginal screen, and mycoplasma culture)
- ❏ 090 Tzanck study

ACTINOMYCOSIS
- ❏ 072 Aerobic Actinomycetes culture
- ❏ 190 Aerobic Actinomycetes identification
- ❏ 068 Aerobic Actinomycetes susceptibility
- ❏ 102 Anaerobic Actinomyces culture

BACTERIAL AGGLUTINATION
- ❏ 075 Haemophilus influenzae, type b
- ❏ 185 Neisseria meningitidis, type B
- ❏ 089 Salmonella grouping
- ❏ 139 Shigella grouping
- ❏ 140 Beta-hemolytic streptococcal grouping (includes groups A, B, C, D, F and G)
- ❏ 155   Group A only
- ❏ 156   Group B only
- ❏ 141 Hemorrhagic E. coli (0157:H7) grouping
- ❏ 028 BACTERIAL ANTIGEN PANEL (latex agglutination for Haemophilus influenzae type b, Streptococcus pneumoniae, Group B Strep, Neisseria meningitidis types A, C, Y and W135, and Neisseria meningitidis type B/E. coli K1)
- ❏ 041   Without Group B Strep
- ❏ 042   Individual antigens (specify below)

PARASITOLOGY
- ❏ 025 Ova & Parasite (O & P) Exam
- ❏ 093 Pinworm exam
- ❏ 079 Worm identification
- ❏ 123 Malaria exam
- ❏ 132 Trichrome stain
- ❏ 057 Fecal leukocyte stain
- ❏ 081 Acanthamoeba culture
- ❏ 111 Acanthamoeba direct exam (scrapings only)
- ❏ 053 Cryptosporidium/Giardia IFA concentrate and stain
- ❏ 189 Microsporidium stain

SUSCEPTIBILITIES
- ❏ 021 Kirby Bauer - Routine
- ❏ 037 Kirby Bauer - Fastidious
- ❏ 004 Kirby Bauer - Special drugs (2 drugs)
- ❏ 021 Oxacillin disc screen
- ❏ 086 Beta-lactamase
- ❏ 171 MIC - Routine
- ❏ 172 MIC - Fastidious
- ❏ 179 E-test - Routine (2 drugs)
- ❏ 180 E-test - Routine (5 drugs)
- ❏ 184 Synergy screen

MYCOLOGY
- ❏ 039 Fungal culture and smear (includes identification)
- ❏ 010 Fungal DNA probe
- ❏ 046 Fungal identification
- ❏ 061 Fungal smear (Calcafluor white)
- ❏ 175 Fungal smear (Periodic Acid Schiff - PAS)
- ❏ 066 Fungal susceptibility
- ❏ 018 Cryptococcal antigen - CSF
- ❏ 118 Cryptococcal antigen - Serum

OTHER/SPECIAL REQUESTS (Refer to the Test Menu for a complete listing of tests.)

FIGURE 3-2 Microbiology reference laboratory requisition.

Virology Workcard									Jane Doe	12/27/95

LD _____ LPI _____ LP2 _____ LF 12/29/95 PV FAX 12/29/95 PV | 95-38086   lesion

Time set-up: 12/27/95   23:59 | Set-up by: ET | HSV culture

DFA/IFA Specimen: _____   Result: _____   Tech: _____   Specimen FRZ No.: 12,002

	1	2	3	4	5	6	7	14			
RK	−	+							21	28	DFA/IFA Isolate: HSV/RK (12/29)
HFF	±	+									Result: HSV1 Tech: PV
HFL											
HEP-2											
A549											
PMK 1						Hemad _____					Isolate FRZ No.: 632
PMK 2											Hemadsorb: _____

Calls: 12/29/95 20:30 (F) → Dr. Jackson PV

FIGURE 3-3 Microbiology workcard.

WRITING A PROCEDURE MANUAL

The laboratory's procedure manual (also known as the SOPM [Standard Operating Procedure Manual]) is a compilation of all the tests performed by the microbiologist. It should be written in a step-by-step format. The validation of a well-written SOPM is that a microbiologist from another facility will be able to read it and perform any procedure done by your facility. SOPMs are standardized in the sense that they are all written in a format outlined by the National Committee for Clinical Laboratory Standards (NCCLS). Individual procedures should contain the following sections:

- Title (name of procedure)
- Principle (reason for performing the test)
- Preferred specimen patient preparation (if required)
- Transport container (need for anticoagulant, preservative, or holding medium)
- Transportation conditions (wet ice, room temperature)
- Specimen storage in laboratory (room temperature, 4° C, −20° C, −70° C)
- Criteria for unacceptable specimen (delay in transport, leaking container, presence of barium)
- Special safety precautions (tape plates for AFB or brucella)

- Reagents or media required and incubation conditions
- Examination of cultures
- Guidelines for identification and susceptibility testing by culture type (respiratory, urine, blood, stool)
- Required quality control
- Methods for reporting positive, negative, and unsatisfactory results
- Technical notes, including possible sources of error and helpful hints
- References, including manufacturer's package inserts, textbooks, NCCLS procedures, and research papers

The laboratory director is responsible for ensuring that a current SOPM is present, and he or she must review it at least annually. Each microbiologist should be required to review the procedures applicable to the work he or she is performing; many laboratories have their technologists and technicians initial the procedure when they review it. When procedures are updated, old procedures should be maintained for at least 2 years. Manufacturer's instrument manuals that conform to specifications of NCCLS publication GP2-A2 are acceptable to both CLIA and CAP accreditation programs. However, CAP does not accept product package inserts or manufacturer's

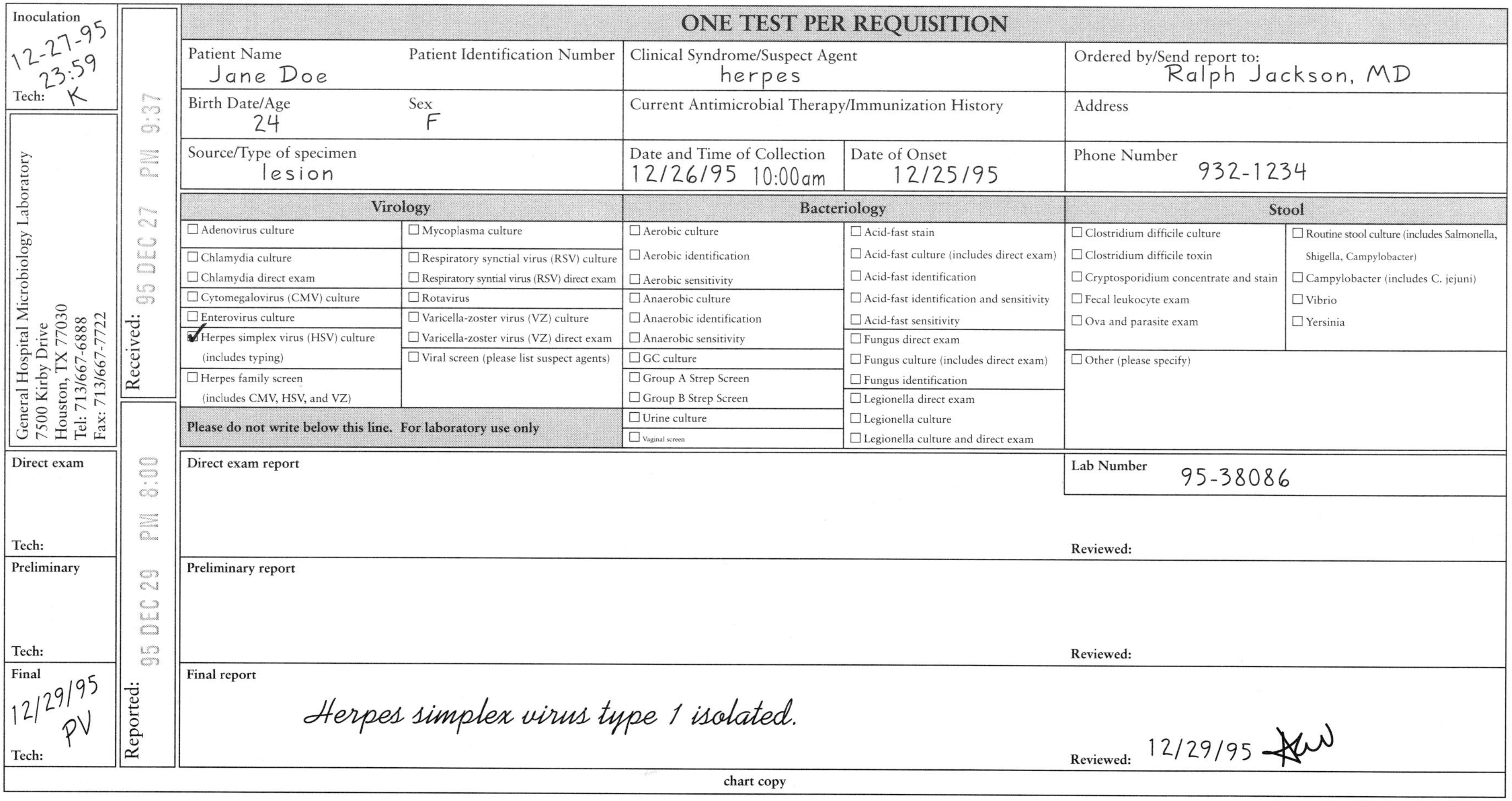

ONE TEST PER REQUISITION

Patient Name	Patient Identification Number	Clinical Syndrome/Suspect Agent	Ordered by/Send report to:
Jane Doe		herpes	Ralph Jackson, MD

Birth Date/Age	Sex	Current Antimicrobial Therapy/Immunization History	Address
24	F		

Source/Type of specimen	Date and Time of Collection	Date of Onset	Phone Number
lesion	12/26/95 10:00am	12/25/95	932-1234

Virology	Bacteriology	Stool
☐ Adenovirus culture	☐ Aerobic culture	☐ Acid-fast stain	☐ Clostridium difficile culture	☐ Routine stool culture (includes Salmonella, Shigella, Campylobacter)
☐ Chlamydia culture	☐ Aerobic identification	☐ Acid-fast culture (includes direct exam)	☐ Clostridium difficile toxin	☐ Campylobacter (includes C. jejuni)
☐ Chlamydia direct exam	☐ Aerobic sensitivity	☐ Acid-fast identification	☐ Cryptosporidium concentrate and stain	☐ Vibrio
☐ Cytomegalovirus (CMV) culture	☐ Anaerobic culture	☐ Acid-fast identification and sensitivity	☐ Fecal leukocyte exam	☐ Yersinia
☐ Enterovirus culture	☐ Anaerobic identification	☐ Acid-fast sensitivity	☐ Ova and parasite exam	
☑ Herpes simplex virus (HSV) culture (includes typing)	☐ Anaerobic sensitivity	☐ Fungus direct exam	☐ Other (please specify)	
☐ Mycoplasma culture	☐ GC culture	☐ Fungus culture (includes direct exam)		
☐ Respiratory synctial virus (RSV) culture	☐ Group A Strep Screen	☐ Fungus identification		
☐ Respiratory syntial virus (RSV) direct exam	☐ Group B Strep Screen	☐ Legionella direct exam		
☐ Rotavirus	☐ Urine culture	☐ Legionella culture		
☐ Varicella-zoster virus (VZ) culture	☐ Vaginal screen	☐ Legionella culture and direct exam		
☐ Varicella-zoster virus (VZ) direct exam				
☐ Viral screen (please list suspect agents)				
☐ Herpes family screen (includes CMV, HSV, and VZ)				

Please do not write below this line. For laboratory use only

Direct exam report	Lab Number
	95-38086

Preliminary report	Reviewed:

Final report	Reviewed:
Herpes simplex virus type 1 isolated.	Reviewed: 12/29/95 ✗

chart copy

Inoculation
12-27-95
23:59
Tech: K

Received: 95 DEC 27 PM 9:37

General Hospital Microbiology Laboratory
7500 Kirby Drive
Houston, TX 77030
Tel: 713/667-6888
Fax: 713/667-7722

Reported: 95 DEC 29 PM 8:00

Direct exam
Tech:
Preliminary
Tech:
Final 12/29/95 PN
Tech:

FIGURE 3-4 Handwritten laboratory report.

General Hospital Microbiology Laboratory
7500 Kirby Drive
Houston, TX 77030
Tel: (713) 667-6888
Fax: (713) 667-7722

CLIA 45D0000000
CAP 00000-01
Medicare CL0000

==

Confidential Laboratory Report

==

Laboratory Reference: 95-38086

Physician: Ralph Jackson, MD

Client Supplemental Identification: 95-361-1044

Patient Name: Doe, Jane   Age: 24   Sex: F
Specimen: Lesion

Test Ordered: HSV culture   Inoculated: 12/27/95 23:59   Tech: ET
Date Ordered: 12/27/95

Final Report: Reviewed: ASW   Date: 12/29/95
Herpes simplex virus type 1 isolated.
Tech: PV   Date: 12/29/95

FIGURE 3-5 Computer-generated report.

technical manuals in place of individually written laboratory procedure.

PRODUCTION OF STATISTICAL REPORTS

One of the microbiologist's most important jobs is to publish a cumulative susceptibility report that tracks susceptibility or resistance of commonly isolated organisms to commonly administered antimicrobials in a particular hospital or geographic area. This report is called an **antibiogram.** Figure 3-6 illustrates an example of such a report for a small hospital laboratory. Cumulative antibiograms should be distributed to the medical staff so that physicians can plan empiric therapy. Empiric therapy (see Chapter 1) is basically therapy begun before actual culture and susceptibility data are available; it is based on the clinician's best guess as to the most probable organism(s) causing the infection and the drug most likely to cure it. Many microbiology laboratories include the cost of drugs, provided by the pharmacy, along with the antibiogram, so clinicians will pick the most cost-effective antimicrobial. From time to time, microbiologists are also asked to provide statistical information on the percentage of positive cultures (also known as positivity rate) for comparison with other laboratories or validation of testing methodologies. For example, no more than 3% to 5% of AFB cultures should be contaminated with normal upper

General Hospital
Antimicrobial Susceptibility of Common Organisms
January 1995

Isolated organism	Urine isolates	Other isolates	Total isolates	Amox/Clav	Ampicillin	Carbenicillin*	Cefotaxime	Cefoxitin	Ceftazidime	Ceftizoxime*	Ceftriaxone	Cefuroxime	Cephalothin	Chloramphenicol	Ciprofloxacin	Clarithromycin	Clindamycin**	Erythromycin**	Gentamicin	Imipenem	Loracarbef	Nitrofurantoin*	Norfloxacin*	Oxacillin	Penicillin	Piperacillin	Sulfa*	Tetracycline	Ticarcillin	Trim/Sulfa	Vancomycin
Gram-negative Bacteria																															
Acinetobacter spp.	1	3	3			0			100						100				100	100			0				100		100		
Acinetobacter lwoffii	1	1	2			100			100	100	100				100				100	100			100				100		100	100	
Aeromonas hydrophila	0	1	1		0		100	100					0		100				100											100	
Alcaligenes spp.	1	0	1			100			100						0				100				0								
Citrobacter spp.	3	0	3	100	0	0	100	100					100		100				100			100	100				100			100	
Citrobacter freundii	7	2	9	0	0	43	67	11					0		100				100			100	100				57			89	
Citrobacter diversus	2	0	2	100	0	0	100	100					100		100				100			50	100				100			100	
Enterobacter aerogenes	3	0	3	0	0	33	33	0					0		33				67			0	33				33			33	
Enterobacter cloacae	3	1	4	0	0	33	100	0					0		100				100			33	100				67			75	
Escherichia coli	91	10	101	72	60	67	99	99					68		93				93			93	100				75			85	
Haemophilus influenzae	0	1	1		0						100			100	100	100					0							0		100	
Klebsiella oxytoca	4	0	4	75	0	0	100	100					75	0	100				100			75	100				75			100	
Klebsiella pneumoniae	37	7	44	84	0	11	84	73					75		70				95			65	70				70			77	
Moraxella (Branhamella) catarrhalis	0	1	1										100					100										100			
Morganella morganii	10	0	10	0	0	50	70	40					0		50				70			0	50				0			60	
Proteus mirabilis	23	1	24	96	46	52	100	96					88		71				100			0	83				70			71	
Proteus penneri	1	1	2	50	50	100	100	100					50		100				100			0	100				0			100	
Providencia rettgeri	3	1	4	0	50	100	75	75					0		50				75			0	33				33			50	
Providencia stuartii	6	1	7	14	0	0	100	100					14		14				57			0	17				0			14	
Pseudomonas spp.	1	4	5			0			60						80				100				100			100					
Pseudomonas aeruginosa	59	12	71			44			99	3					48				75	99		46	46				2		65	40	
Pseudomonas putida	1	0	1			0			100						0				100			0	0			100				0	
Serratia marcescens	2	0	2	0	0	0	50	0					0		50				100			0	50				0			50	
Shigella sonnei	0	9	9		0										100															100	
Gram-positive Cocci																															
Coagulase-negative Staphylococcus	9	13	22	64									64		64		77	77				100	22	64	18		56			68	100
Enterococcus spp.	54	10	64		97										9							93	7		84						100
Staphylococcus aureus	6	30	36	89									89		87		80	70				100	33	89	22		100	42		97	100
Stomatococcus spp.	0	2	2												100		100	100						100	0					0	100
Streptococcus pneumoniae	0	1	1																						0						100

Note: Values represent percentage (%) of susceptible organisms.

*Only tested with organisms recovered from urinary tract infections. Values represent percentage of susceptible organisms.

**Not routinely tested against organisms from the urinary tract.

FIGURE 3-6 General Hospital monthly antibiogram.

respiratory flora to validate the decontamination procedure used in any facility. Quality assurance testing based on analysis of positive and unsatisfactory cultures will be discussed in more detail in Chapter 6.

Bibliography

Barker, J.H. 1985. Laboratory facilities planning and design. Centers for Disease Control, Atlanta.

Barker, J.H., Blank, C.H., and Steere, N.V. 1989. Designing a laboratory. American Public Health Association, Washington, D.C.

Bickford, G. 1993. Decentralized testing in the '90s. Clin. Lab. Management Rev. 8:327.

College of American Pathologists. 1977. Manual for laboratory planning and design. College of American Pathologists, Skokie, Ill.

Elin, R.J., Robertson, E.A., and Sever, G. 1984. Workload, space, and personnel of microbiology laboratories in teaching hospitals. Am. J. Clin. Pathol. 82:78.

National Committee for Clinical Laboratory Standards. 1992. Clinical laboratory technical procedure manuals, ed 2. Approved Guideline GP2-A2. National Committee for Clinical Laboratory Standards, Villanova, Pa.

4 | LABORATORY MANAGEMENT

The age of laboratory management has come into its own at the turn of the millennium, as microbiologists are challenged to become business, as well as technical, specialists. Each laboratory must have a strategic plan that describes its long-term goals, such as a move toward more automation or molecular diagnostic techniques. Each employee's role should be clearly defined, and written job descriptions should be provided so personnel know what they are expected to do. It is a paradox of modern health care that while laboratories have never been more closely regulated, they often are reimbursed very little for their efforts. Therefore, the microbiology laboratory manager must strike a balance among microbiology laboratory regulations, fiscal responsibility, and employee competence and morale to maintain the overall quality of patient care. Although these tasks may seem intimidating, it is appropriate to remember that the two most important components of management are the use of good *common sense* and the fostering of *open communication* with laboratory staff. Finally, the term "manager" is used in a broad sense in this chapter to describe functions that may be split among various people, that is, laboratory director, supervisor, or consultant in individual institutions.

REGULATORY AGENCIES

The manager's first job is to understand the various rules and regulations that govern the practice of clinical microbiology. The regulatory environment in which microbiologists currently work has evolved in the past 30 years and began after the television news magazine show, *60 Minutes,* aired a program about sink-testing. Hidden cameras recorded a laboratorian pouring a patient's urine down the sink and then fabricating the results of the urine culture and susceptibility. The public outcry that followed this broadcast led Congress to enact the Clinical Laboratory Improvement Act of 1967, now known as **CLIA '67.** The **Health Care Financing Administration (HCFA** [pronounced hic′ va]) was created to oversee the enforcement of the guidelines published in CLIA '67; HCFA is part of the United States Department of Health and Human Services (DHHS). HCFA is also responsible for overseeing the Medicare and Medicaid programs, which were instituted to provide health care for elderly and poor persons under the Social Security Act. CLIA '67 mandated the first quality control, personnel, and proficiency testing standards for clinical laboratories but applied only to laboratories engaged in interstate commerce; this represented less than 10% of all clinical laboratories. Therefore, in 1988, following media publicity regarding deaths from uterine and ovarian cancer in women whose Pap smears had been misread, the authority of HCFA under CLIA '67 was expanded under the Clinical Laboratory Improvement Amendments of 1988, known as **CLIA '88.**

CLIA '67 and CLIA '88 are federal regulations; in addition many state health departments (e.g., New York, Florida, Maryland, and California) have established their own, sometimes more stringent, regulations. A microbiology manager who lives in a state with specific laboratory guidelines must ensure that the laboratory meets all federal and state requirements and that the technical personnel are licensed by the state. The laboratory branch of the state Public Health Department is a good source for regulatory information.

A number of other federal agencies more indirectly regulate the practice of clinical microbiology. The **Food and Drug Administration (FDA)** clears the products bought by the laboratory for use in diagnostic testing. Labels of products that the FDA has approved will state, "For Diagnostic Testing." Those that have not been cleared must be labeled "For Investigational Use." Laboratories may use devices or instruments that are not cleared by the FDA, but they must be subjected to rigorous validation and quality control testing. FDA clearance implies that the manufacturer has already verified the accuracy, precision, and reproducibility of the product, device, or instrument, and the clinical laboratory may do less rigorous in-house testing. The **Environmental Protection Agency (EPA)** regulates the disposal of toxic chemicals and biohazardous waste (see Chapter 2). The **Occupational Safety and Health Administration (OSHA)** regulates employee safety in the workplace (see Chapter 2).

ACCREDITING AGENCIES

All laboratories are required to be accredited (licensed) under CLIA '88. This means that every laboratory must register with HCFA for every analyte (test) it performs and must meet certain minimum quality standards, which are monitored by on-site inspections. The inspections include a review of a laboratory's procedures, quality control program, documentation, and patient management. Several private agencies also inspect and accredit clinical laboratories for a fee. Laboratories inspected by private agencies with **deemed status** do not have to be reinspected by HCFA because HCFA has determined that their accreditation program is at least as stringent as that specified under CLIA '88. A list of private accrediting agencies is shown in Table 4-1. Most of the private agencies have their own rules and checklists and do not accept inspections by others. Therefore, managers must be aware of their requirements, as well as CLIA and state requirements, if applicable.

CLIA '88

The CLIA '88 regulations combined and replaced former laboratory standards with a single set of requirements that apply to clinical laboratory testing wherever it is performed, that is, freestanding facility, POL (physician office laboratory), hospital; this concept has come to be known as **site neutrality.** Uniform, *minimum* standards have been published in the federal register regarding laboratory personnel, quality control, quality assurance, and proficiency testing (PT) issues. The regulations are based on a test complexity model that, in turn, is based on the potential for harm to a patient. CLIA '88 also sets up enforcement procedures and defines sanctions to be applied if laboratories fail to meet minimum standards. A Clinical Laboratory Improvement Amendments Committee (CLIAC) was established to advise the DHHS and the Centers for Disease Control and Prevention (CDC) as laws are turned into enforceable regulations. The CLIAC is composed of representatives of the laboratory industry, the manufacturing sector, health care administrators, a consumer advocate, and representatives of the CDC, FDA, and HCFA.

TEST COMPLEXITY MODEL

CLIA '88 defined four categories of testing as follows:

- Waived tests
- Practitioner-performed microscopy
- Moderate complexity
- High complexity

Waived tests are those that are so simple to perform that there is virtually no risk to the patient even if the test is not performed correctly. Inspections and PT are waived for laboratories performing only these tests. Currently, there are very few waived microbiology tests; these include some group A *Streptococcus* direct antigen tests, for example. **Practitioner-performed microscopy (PPM)** is composed of microscopy performed at the point-of-care by a physician, physician's assistant, nurse, midwife, or nurse practitioner. The PPM category used to be called physician-performed microscopy but was subsequently expanded to include the other mid-level specialists. Laboratories holding a PPM certificate are not subject to on-site inspections but must still meet quality assurance (QA) and quality control (QC) standards. Microbiology tests in this category include

TABLE 4-1 PRIVATE ACCREDITING AGENCIES

AGENCY	ACCREDITS	DEEMED STATUS	COMMENTS
College of American Pathologists (CAP)	Laboratories only	Yes	Oldest private accrediting agency; inspects by peer review (i.e., checking one another)
Joint Commission on the Accreditation of Health Care Organizations (JCAHO)	Entire hospitals or clinics, including laboratory	Yes	
Commission of Laboratory Accreditation (COLA)	Licenses laboratories in physician offices	Yes	
American Association of Bioanalysts (AAB)	Laboratories in free-standing facilities	No	

wet mounts of vaginal and cervical secretions and skin, potassium hydroxide (KOH) preparations, and pinworm examinations. The third category of testing is **moderate complexity.** Microbiology procedures in this category include urethral and cervical Gram stains and presumptive reporting of group A *Streptococcus (Streptococcus pyogenes)* using hemolysis and inhibition by bacitracin. The fourth category of testing is **high complexity.** Most microbiology cultures that require isolation, identification, and susceptibility testing of organisms are considered high complexity. Moderate and high complexity testing laboratories are both subject to on-site inspections and must comply with the same requirements for QC, QA, and PT. However, personnel requirements are different for each level.

PERSONNEL STANDARDS

Appropriately trained and experienced personnel are essential for the performance of quality laboratory testing. Therefore, a major part of CLIA '88 details the requirements for laboratory directors, supervisors, consultants, and testing personnel. Although the federal government emphasizes that the CLIA personnel standards represent *minimum* standards, education and experience requirements were lowered from those required in CLIA '67. For example, personnel in moderate testing laboratories may now be high school graduates if they have documented on-the-job training. Previously, individuals performing microbiology procedures were required to have at least an associate degree in medical laboratory technology or laboratory science and most held baccalaureate degrees in a biologic, chemical, or physical science or medical technology. Similarly, supervisory personnel who used to have baccalaureate degrees and 6 years of experience are now merely required to have an associate degree in medical technology or laboratory science plus 2 years of clinical laboratory training and 2 years of supervisory experience. Table 4-2 details the personnel requirements under CLIA '88 for moderate and high complexity laboratories.

PROFICIENCY TESTING STANDARDS

Proficiency testing (PT) was originally mandated in CLIA '67, but its scope was expanded in CLIA '88 to include sanctions (penalties) for poor performance. PT is a method whereby laboratories are sent unknown samples to test as they would test patient specimens. It is discussed in detail in Chapter 6. CLIA '88 mandates that at least three times a year each laboratory participate in a program comparable in scope to the level of testing it performs. Public and private accrediting agencies send laboratories samples to test and return critiques with the answers. Cri-

tiques are summaries of how the laboratory performed in relation to its peers, and they serve as a method by which a laboratory can compare its test systems with that of others. Laboratories that fail a particular analyte two times consecutively must submit a plan of corrective action. If the laboratory continues to fail that analyte, CLIA '88 specifies sanctions, including the withdrawal of a laboratory's permit for that particular analyte. Both moderate and high complexity laboratories must participate in PT.

QUALITY ASSURANCE STANDARDS

Quality assurance (QA) and quality control (QC) standards are discussed in depth in Chapter 6. Both represent the procedures that laboratories establish to ensure that the patient results reported are accurate and timely, and assist the clinician in treating the patient. CLIA '88 mandates auditing (studying) the preanalytic, analytic, and postanalytic phases of a laboratory test to ensure that each step led to a favorable outcome for the patient. Laboratories must also keep problem logs showing complaints from clinicians or patients and detailing the investigation and any corrective action indicated.

COST ACCOUNTING

In today's health care environment, microbiologists must understand how to prepare budgets, read balance sheets, and determine the cost effectiveness of different tests. This section will explain the basics of cost accounting and is not intended as a substitute for classic accounting texts.

In the laboratory, cost accounting is the process by which managers determine what it costs to do a test. Chapter 5 explains how a manager selects a test based on scientific accuracy. Once the test is selected, the **cost** can be determined. Cost is the amount of money spent for supplies (consumables), labor, and overhead. The amount of money related directly to test performance, supplies, and labor is also called **direct costs. Overhead,** also called **indirect costs,** applies to items necessary to run the laboratory or hospital facility, which are not test specific. Because overhead represents actual dollars spent to run the laboratory, however, a portion of it must be included in the price the laboratory charges for each test. Examples of indirect costs or overhead include the following:

- Labor to supervise performance of a test
- Quality control necessary to ensure test accuracy
- Maintenance and repairs to equipment or the physical facility
- Service contracts on equipment

— **TABLE 4-2 MICROBIOLOGY PERSONNEL REQUIREMENTS UNDER CLIA '88** —

TEST COMPLEXITY LEVEL	PERSONNEL	MINIMUM QUALIFICATIONS
Moderate	Director	Physician (MD, DO, DDS, DPM) or PhD in science; must have 1 year of supervisory experience if not a board-certified pathologist or doctoral scientist *or* Baccalaureate degree in science plus 2 years of laboratory training and experience and 2 years of supervisory experience
	Technical consultant (responsible for technical oversight if nurses or nonmedical technologists are laboratory director)	Baccalaureate degree in science plus 2 years of experience
	Clinical consultant (liaison between laboratory and clinicians)	Physician (MD, DO, DPM) or board-certified doctoral scientist
	Testing personnel	High school diploma plus documentation of on-the-job testing
High	Director	Physician (MD, DO, DDS, DPM) or board-certified doctoral scientist
	Technical supervisor	Baccalaureate degree in science plus 4 years of laboratory experience in high complexity testing
	General supervisor	Associate degree in medical laboratory science plus 2 years of laboratory experience in high complexity testing
	Clinical consultant (can be director)	Physician (MD, DO, DDS, DPM) or board-certified doctoral scientist
	Testing personnel	Associate degree in medical laboratory technology or laboratory science

- Equipment lease or rental costs
- Continuing education programs
- Travel to professional meetings
- Utilities (electricity, water, phone, vacuum, gases)
- Building security (alarm system, on-site guards)
- PT program
- Insurance, including fire, general, and professional liability; workers' compensation and medical or disability
- Property taxes
- Fees for licenses or certification
- Subscriptions to professional journals or books.

Depreciation, another indirect cost, is the amount of money the Internal Revenue Service (IRS) allows the institution to deduct over the useful life (usually 5 to 7 years) of the **capital equipment** (instruments) it buys. Depreciation takes into account that over time the equipment will need to be replaced because of wear, obsolescence, or because of procedural changes. Finally, **profit** is the amount of money made per test that exceeds the total direct and indirect costs.

In calculating how much to charge for a test, the microbiology laboratory manager must consider the cost of supplies and labor. Supplies are usually placed on **bid** to several vendors so that the best price is obtained; tax and freight costs must be included, if applicable. Labor costs are most accurately determined by performing studies, that is, timing a technologist using a stopwatch as he or she performs all parts of the testing. All the steps performed *by the* *microbiology laboratory* must be considered in pricing the labor component. This includes the three segments of each clinical test, that is, preanalytic, analytic, and postanalytic.

The preanalytic stage involves specimen collection, accessioning, and preparation of a workcard. The analytic stage involves all steps in performing the test. The postanalytic stage involves result reporting, telephoning positive results, supervisory review of the written report, and delivery of the written report. The contribution of **fringe benefits** (paid time off such as holiday, vacation, and sick leave; medical insurance; employer contribution to a pension plan) must also be considered. Typically, benefits add an additional 25% to 30% to the employee's base salary. Realistically, in today's fast-paced laboratory, it is almost impossible for the manager to tally up all labor costs. Therefore, because labor is usually shown to represent 40% to 60% of the total cost of a test, many individuals estimate the cost of labor by doubling the cost of supplies. Box 4-1 illustrates how to determine the cost of a test, in this case a chlamydia culture. In the example, the costs of a positive and negative test are the same, although in most instances positive tests cost more than negative tests because of the additional costs associated with biochemical or immunologic confirmation.

The cost of the test with a 5% profit should be $20.07. If the laboratory runs a courier service to pick up specimens, this cost should also be added into the price of each test. Courier costs per test can be estimated by averaging their salary over a 6- or 12-month period and then dividing by the total number of tests during that period.

BUDGETING

Budgeting for a business is just like budgeting personal expenses. The manager looks at each line item, such as labor, supplies, travel, and subscriptions, in the previous year's ledger and estimates expenses for the next year. The laboratory is both a **cost** and **revenue** center. A cost center spends money to perform patient testing. A revenue center brings in money as it bills for these patient tests. The fiscally sound laboratory will generate more dollars in revenue than it spends.

The institution's accounting office will usually post (write) expenses to the **general ledger** once a month. The general ledger is the accounting record of the laboratory's expenditures and is taken from actual payroll figures and expenses. This ledger also includes revenues posted from tests billed during the month. Many budget reports have a column that shows the amount the microbiology laboratory manager

BOX 4-1   **PRICING A CHLAMYDIA CULTURE**

1. Cost of Supplies

Transport and Collection

Collection kit plastic bag	$0.06
Transport medium	0.95
Dual pack Dacron swabs (male and female)	0.30
Requisition	0.06
Specimen label	0.03
Collection instructions	<u>0.03</u>
	$1.43

Processing

Workcard	$0.03
Sheet of specimen labels	0.05
Cryovial	0.30
BGMK shell vial	1.25
1 mL phosphate buffered saline (PBS)	0.20
1 mL *Chlamydia* isolation medium	0.20
1 mL 95% alcohol	0.20
2 mL PBS (wash)	0.18
75-µL anti-chlamydiae FITC monoclonal antibody	2.50
6 Pasteur pipettes	0.18
1 glass slide	<u>0.07</u>
	$5.16

Total cost of supplies	$6.59

Note: Cost of running QC is not included here but is included as part of overhead.

2. Cost of Labor

Labor cost, estimated to be equal to the total cost of supplies	**6.59**

3. Total Direct Cost of Test

Supplies + labor ($6.59 + $6.59)	$13.18

4. Overhead (see text)

Overhead typically runs between 40% and 50% of test costs.
In this example, we shall estimate overhead to be 45% of total test cost.

Total direct cost × overhead rate ($13.18 × 45%)	**5.93**
Total test cost without profit	$19.11

5. Profit

Profit is determined by the institution and usually is between 3% and 10%.
In this example, we shall calculate a 5% profit.

Total test cost × profit rate ($19.11 × 5%)	**0.96**
Total test cost with profit	**$20.07**

budgeted for a particular line item in addition to the actual amount spent. The budget report may also show a column indicating the **variance,** or difference, between the budgeted and actual costs.

Figure 4-1 is a typical monthly budget for the Microbiology Laboratory at General Hospital. The laboratory anticipates revenues of $135,000 and expenses of $129,650 for a **gross profit** (before taxes) of $5350. Figure 4-2 shows the income statement for April 1996. The laboratory shows a gross profit of

Revenue

Test income (inpatients)	$118,000
Test income (outpatients)	17,000
Total Revenue	$135,000

Expenses

Salaries	$64,000
Payroll taxes	5600
Health insurance	4000
Workers' compensation insurance	1300
Employee medical testing	200
Laboratory supplies	34,000
Office supplies	1000
Printing	1000
Shipping	800
Outside laboratory services	1500
Licenses and permits	100
Proficiency tests	200
Rent	3100
Equipment lease/rental	250
Repairs and maintenance	2500
Telephone	1500
Utilities	2500
Travel	2000
Professional and general liability insurance	2000
Depreciation	2100
Total Expenses	$129,650

FIGURE 4-1 Monthly budget for General Hospital Microbiology Laboratory.

Revenue

Test income (inpatients)	$118,562
Test income (outpatients)	17,190
Total Revenue	$135,752

Expenses

Salaries	$67,945
Payroll taxes	5959
Health insurance	4231
Workers' compensation insurance	1322
Employee medical testing	56
Laboratory supplies	37,343
Office supplies	600
Printing	2509
Shipping	967
Outside laboratory services	308
Licenses and permits	125
Proficiency tests	0
Rent	3100
Equipment lease/rental	250
Repairs and maintenance	1716
Telephone	1709
Utilities	1954
Travel	680
Professional and general liability insurance	1935
Depreciation	2108
Total Expenses	$134,817

FIGURE 4-2 Income statement for General Hospital Microbiology Laboratory, April 1996.

only $935 even though it expected to make $5350. The budget variance analysis (Figure 4-3) shows where the differences are between the estimated budget figures and the actual income and expenses. Numbers in parentheses indicate less money was spent than was budgeted. In this example, the laboratory made more money than expected ($752). It also spent more money on salaries, laboratory supplies, and printing; however, this was offset by the fact that less money was spent on outside laboratory services, repairs and maintenance, and travel. The overall effect on the "bottom line" was the small gross profit. However, the prudent manager would check out the reasons for increased salaries and laboratory supplies, for example, increased overtime may be due to the fact that someone was on maternity leave or purchase of extra supplies may have been necessary to initiate a special project.

INVENTORY CONTROL

One of the best ways to control costs is to keep inventory at a level at which supplies will neither run out nor stay on the shelf long enough to reach their expiration date. Most laboratories have lists of all reagents, media, and disposable supplies (pipettes, paper towels, slides) that they use. Past usage will help managers to establish minimum and maximum amounts to keep on hand. An inventory card system is commonly used to keep track of the status of all supplies (Figure 4-4). This card includes the following: (1) name of the item, (2) name of the vendor (supplier), (3) catalog number of the item, (4) packaging quantity (standard unit) of the item (package, gross), (5) cost per unit, (6) minimum and maximum volumes to keep on hand, (7) actual amount of the product on hand and the date the inventory was counted, (8) date and amount ordered and the **purchase order** (PO) number used (except in large facilities, where purchasing may be handled by a separate materials management department), (9) date and amount received, (10) amount **backordered,** and (11) date the backorder is finally filled. A backordered product is not currently in stock. If this product will not be released soon, the manager may choose to make arrangements to purchase a substitute product. Some microbiology computer software packages also contain an inventory component.

	Budget	Actual	Variance*
Revenue			
Test income (inpatients)	$118,000	$118,562	($562)
Test income (outpatients)	17,000	17,190	($190)
Total Revenue	$135,000	$135,752	($752)
Expenses			
Salaries	$64,000	$67,945	($3945)
Payroll taxes	5600	5959	($359)
Health insurance	4000	4231	($231)
Workers' compensation insurance	1300	1322	($22)
Employee medical testing	200	56	$144
Laboratory supplies	34,000	37,343	($3343)
Office supplies	1000	600	$400
Printing	1000	2509	($1509)
Shipping	800	967	($167)
Outside laboratory services	1500	308	$1192
Licenses and permits	100	125	($25)
Proficiency tests	200	0	$200
Rent	3100	3100	$0
Equipment lease/rental	250	250	$0
Repairs and maintenance	2500	1716	$784
Telephone	1500	1709	($209)
Utilities	2500	1954	$546
Travel	2000	680	$1320
Professional and general liability insurance	2000	1935	$65
Depreciation	2100	2108	($8)
Total Expenses	$129,650	$134,817	($5167)

*Numbers in parentheses indicate less money was spent than budgeted.

FIGURE 4-3 Budget variance analysis for General Hospital Microbiology Laboratory, April 1996.

A physical inventory is performed (i.e., stock is counted) weekly or monthly. Expired media or reagents are usually discarded at this time, and stock is rotated so that the product with the earliest expiration date will be in front and used first. Sometimes, products are put on **standing order;** standing orders are products that arrive at a predetermined interval. For example, in a virology laboratory, tissue culture cells used for growing viruses arrive once a week as prearranged with the vendor. Standing orders are usually set up for 6- or 12-month intervals before reordering is necessary.

INTERVIEWING AND HIRING EMPLOYEES

Hiring good employees is one of the hardest jobs a manager faces. Interviewing prospective candidates is an art; managers must quickly determine not only whether the individual has a working knowledge of microbiology but also whether he or she has a good work ethic and will get along with other employees. The most important points to remember during an interview is to put the applicant at ease and not ask personal questions. For example, the manager should not ask individuals about plans to marry or have children. Neither should the manager ask whether someone is divorced, separated, or cohabiting; whether a spouse is likely to be transferred; the applicant's religion or age; whether the applicant has been dishonorably discharged from the military; the applicant's sexual orientation; or whether the applicant has ever been arrested.

The interviewer should not ask about health problems unless a specific illness would preclude a candidate from safely performing their job, for example, a patient who has received a transplant and is

Item Information										
Item:			Vendor:				Catalog no.:			
							Std. unit:			
							Cost/unit:			
Usage: min.: max.:										

Inventory		Ordered		Purchase order number	Cost per unit	Received		Back ordered		
date	amount	date	amount			date	amount	amount	date complete	notes

Std., Standard; *min.*, minimum; *max.*, maximum; *no.*, number.

FIGURE 4-4 Inventory supply card.

receiving immunosuppressive therapy would not be a good candidate to work with infectious materials in a microbiology laboratory. The manager must not discriminate against people with disabilities if reasonable accommodations can mean they would be qualified for the job. An example of a reasonable accommodation would be to install a special telephone for a hearing-impaired microbiologist who would then be able to communicate with clinicians. Individuals with mental and physical disabilities are protected under the Americans with Disabilities Act.

Before starting an interview, the manager should obtain the job description of the position for which the candidate is applying. The applicant should be asked if he or she can perform all functions outlined. It is permissible to ask why the candidate is leaving (or has left) his or her previous position, and about his or her career goals. Questions should be directed toward helping to determine whether the applicant has the experience, training, and interests necessary to perform the job. The interviewer should fully describe the job's functions and how it relates to the overall functioning of the laboratory; he or she should also describe any special requirements such as rotation to the second or third shifts, phlebotomy duties, or weekend and holiday rotations. The applicant should be allowed to ask questions, and the interview should only be closed when the manager is satisfied that enough information was gathered to evaluate how the applicant will perform.

EMPLOYEE JOB PERFORMANCE STANDARDS AND APPRAISALS

All employees should know what they are supposed to do, and every laboratory should have written **job descriptions** to provide to them. Standards of performance should be developed so that the employee knows how he or she will be judged when rated on job performance (**job appraisal**). Annual merit raises are often linked to job performance, and employees should get feedback throughout the year on how they are doing so that they have an opportunity to correct poor performance.

PERSONNEL RECORDS

Personnel records should be well organized and contain all the information required by regulatory bodies. This includes the following:

- Name of employee
- Application for employment, including:
    - Colleges or universities attended
    - Specialized laboratory training
    - Laboratory certification, if applicable
    - Previous experience
- Job description, signed by employee
- Documentation of credential review by the laboratory director and level under CLIA '88 personnel standards
- Record of job training at present job
- Record of continuing education
- Competency review (which shows that the employee meets minimum standards for performing tasks for which he or she is qualified)
- Performance appraisals (evaluation)
- Record of medical testing, for example, annual PPDs
- Record of safety training

Bibliography

Federal Register. 1968. Clinical Laboratories Improvement Act of 1967. Fed. Regist. 33:15297.

Health Care Financing Administration. 1988. Medicare, Medicaid, and CLIA programs; revision of the clinical laboratory regulations for the Medicare, Medicaid, and Clinical Laboratories Improvement Act of 1967 programs. Federal Register 53:29590.

Health Care Financing Administration. 1990. Medicare, Medicaid, and CLIA programs; revision of laboratory regulations; final rule with request for comments. Federal Register 55:19538.

National Committee for Clinical Laboratory Standards. 1994. Inventory control systems for laboratory supplies; Approved guideline. GP6-A. National Committee for Clinical Laboratory Standards, Villanova, Pa.

National Committee for Clinical Laboratory Standards. 1993. Cost accounting in the clinical laboratory; Tentative guideline. GP11-T, National Committee for Clinical Laboratory Standards, Villanova, Pa.

Pollack, H.M. 1995. Guide to regulatory requirements. A special supplement for users of the Clinical Microbiology Procedures Handbook. ASM Press, Washington, D.C.

Stuart, J. and Hicks, J.M. 1991. Good laboratory management: an Anglo-American perspective. J. Clin. Pathol. 44:793.

5 | SELECTION OF DIAGNOSTIC TESTS

One of the jobs of the microbiology laboratory manager is to decide what testing should be offered on-site and to select the best method of doing any test. The decision may be as basic as deciding whether to offer a particular test or as complex as deciding between different test methodologies. Although there are several aspects to consider when deciding to add a particular test, the most important is the clinical need for the test. The test should improve the quality of care delivered to patients by offering improved turnaround time or accessibility. The more serious the disease, the more a diagnostic test is indicated, especially if treatment is available but expensive. A good example of this is the decision by many facilities to offer on-site **stat** (immediate) testing for respiratory syncytial virus (RSV). Laboratories began offering antigen tests for RSV when treatment (ribavirin) became available. Ribavirin therapy is very expensive, and infants with RSV can be very sick, so it makes sense to offer this test in order to make a rapid, specific diagnosis. Moreover, prompt therapy of infected infants may prevent spread of RSV in hospital nurseries.

Several other factors must be considered when deciding whether to add a diagnostic test:

- Projected test volume. Will it be worth the money required to verify this new procedure and perform ongoing quality control?
- Projected test cost. Will the laboratory be able to make money or save money relative to the current test?
- Medical staff needs. Is a screening test enough or should the test be confirmatory (diagnostic)?
- Should a culture be performed or is a noncultural test adequate?
- Disease prevalence. Analysis of population demographics (composition) will help to determine whether your facility will see patients with a particular illness. In the example of the RSV antigen test, you might examine the incidence of emergency room admissions of children younger than age 3 in respiratory distress.
- Availability of technically skilled personnel to

perform the test, especially if it will be offered on all shifts.
- Availability of reagents cleared by the Food and Drug Administration (FDA).
- Availability of instrumentation and separate physical space, if needed. This would be an important consideration, for example, if the laboratory was considering a molecular technique, such as polymerase chain reaction (PCR), in which a thermal cycler must be used and separate work areas are recommended.
- Extent of quality control required.
- Availability of proficiency test materials.
- Specimen requirements and storage conditions.
- Availability of a reference laboratory offering the test. Will it offer the service at a lower cost?

Once the decision has been made to offer the test, the different test methods should be studied. For example, if a laboratory is considering offering a test for chlamydia, the microbiologist would have to decide whether to do (1) a culture, (2) a direct fluorescent antibody test, (3) an enzyme-linked immunosorbent assay, (4) a DNA probe, or (5) an amplified, PCR-based DNA probe. Although not all testing decisions are as complex as this, often a decision must be made among several methods.

ANALYSIS OF TESTS

Quantitative test methods must be validated based on accuracy, precision, sensitivity, specificity, predictive value, and efficiency before a decision is made. To understand how to quantitatively analyze the performance of various tests, some basic definitions are required.

DEFINITIONS

Accuracy (efficiency) is the ability of the test under study to match the results of a standard test commonly known as the "gold standard." For example, a noncultural test may be compared with culture that is the gold standard.

Precision is the reproducibility of a test when it

is run (repeated) several times. Precision can be assessed both intralaboratory (within the same laboratory) and interlaboratory (between different laboratories). Precision does not imply accuracy.

Sensitivity is the percentage of individuals with the particular disease for which the test is used in whom positive test results are found. These are also called **true-positives**. Tests with very high sensitivity (≥99%) can, if negative, be used to exclude the presence of disease. For example, if an RSV direct antigen test is negative and the test sensitivity is 99.9%, then the infant is unlikely to have RSV. Most tests will show some **false-positives**; these represent cases in which individuals without the particular disease test positive by a certain test.

Specificity is the percentage of individuals who do *not* have the particular disease being tested and in whom negative test results are found. These are also called **true-negatives**. Tests with very high specificity (≥99%) should be negative both in healthy individuals and in individuals with symptoms similar to that of the disease being tested but not caused by the same organism. For example, in the infant with symptoms suggestive of RSV with a negative RSV direct antigen test, the physician would consider *Bordetella pertussis* or parainfluenza virus as the possible etiologic agent. Most tests will show some **false-negatives**; these represent cases in which individuals with the particular disease test negative by a certain test.

Prevalence is the frequency of disease in the population at a given time. It is based on the **incidence** of new cases of disease per year, usually calculated per 100,000 population. Prevalence is important in comparing the predictive value of a positive or negative test. The prevalence of chlamydiae in women attending a clinic for a sexually transmitted disease (STD), for example, is higher than in nuns.

Predictive value positive (PVP) is the percentage of true-positive test results based on the prevalence of the disease in the population studied. PVP measures the probability that a positive result indicates the presence of disease.

Predictive value negative (PVN) is the percentage of true-negative test results based on the prevalence of the disease in the population studied. PVN measures the probability that a negative result indicates the absence of disease.

Efficiency (accuracy) is the percentage of test results that are correctly identified by the test, that is, true positives and true negatives.

MATHEMATIC FORMULAS

To calculate sensitivity, specificity, prevalence, PVP, PVN, and efficiency the following formulas can be used. These formulas can be easily understood by

remembering a simple table that contains all the critical information:

	Gold Standard	
	POSITIVES	NEGATIVES
New test Positives	a (true positives)	b (false-positives)
Negatives	c (false-negatives)	d (true negatives)

$$Sensitivity = \frac{a}{a+c} \times 100$$

$$Specificity = \frac{d}{b+d} \times 100$$

$$PVP = \frac{a}{a+b} \times 100$$

$$PVN = \frac{d}{c+d} \times 100$$

$$Efficiency = \frac{a+d}{a+b+c+d} \times 100$$

$$Prevalence = \frac{a+c}{a+b+c+d} \times 100$$

ASSESSING THE SENSITIVITY, SPECIFICITY, PVP, PVN, AND EFFICIENCY OF A TEST

All FDA-cleared products for diagnostic use must undergo clinical trials. During the trials, laboratories from different geographic regions with high and low prevalence of the disease compare the new test to a gold standard. In most cases, culture (if available) is considered the gold standard, although this is not always true. For example, because RSV, a very fastidious virus, will not easily grow in cell culture, a new direct antigen assay might be compared with an approved direct antigen test already on the market. The **package insert** (instructions) for each FDA-cleared kit, reagent, or medium contains data from the clinical trials. This information is an indicator of the characteristics of the new test and can be used for comparison when in-house testing is performed.

An example is given in Box 5-1 to illustrate how to perform the calculations and make subsequent decisions regarding test verification.

TEST VERIFICATION AND VALIDATION

CLIA '88 (see Chapter 4) mandates that all tests used by a laboratory after September 1, 1992, be **verified**

BOX 5-1 USE OF TEST VERIFICATION DATA TO ANALYZE TWO MICROBIOLOGY TESTS

BACKGROUND INFORMATION: A laboratory that serves a pediatric outpatient clinic wants to determine if there is any difference in performance between summer and winter months for two rotavirus tests, that is, a latex test and an enzyme-linked immunosorbent assay (ELISA). The prevalence of rotavirus in winter in infants with diarrhea is high; in summer the prevalence is low.

PURPOSE: The laboratory has been using an ELISA test that is considered the gold standard. However, because the latex test is less expensive, easier to perform, and has a faster turnaround time, the laboratory wants to switch to this test.

STUDY DESIGN: One hundred diarrheal stool specimens were tested at random during January and July.

RESULTS:

JANUARY

		ELISA POSITIVES	ELISA NEGATIVES
Latex	Positives	45	1
	Negatives	45	9

$$Prevalence = 90\%$$

$$Sensitivity = \frac{45}{90} \times 100 = 50\%$$

$$Specificity = \frac{9}{10} \times 100 = 90\%$$

$$PVP = \frac{45}{46} \times 100 = 98\%$$

$$PVN = \frac{9}{54} \times 100 = 17\%$$

$$Efficiency = \frac{54}{100} \times 100 = 54\%$$

JULY

		ELISA POSITIVES	ELISA NEGATIVES
Latex	Positives	1	10
	Negatives	1	88

$$Prevalence = 2\%$$

$$Sensitivity = \frac{1}{2} \times 100 = 50\%$$

$$Specificity = \frac{88}{98} \times 100 = 90\%$$

$$PVP = \frac{1}{11} \times 100 = 9\%$$

$$PVN = \frac{88}{89} \times 100 = 99\%$$

$$Efficiency = \frac{89}{100} \times 100 = 89\%$$

CONCLUSIONS: The latex test would perform well in the summer to screen out negative patients because the prevalence of disease is low and the predictive value of a negative test is 99%. However, when the disease prevalence is high, in winter, the laboratory would want to keep the ELISA test because the sensitivity of the latex test is low (50%), the predictive value of a negative latex test is only 17%, and the accuracy (efficiency) of the latex test is only 54%.

before being used. Verification is a one-time process. Thereafter, the test must be **validated** as part of an ongoing process of quality assurance. Validation involves documentation of quality control, proficiency testing, and continued satisfactory performance of the test; it is discussed in more detail in Chapter 6. The comparability of different tests may be verified by one of several methods:

- Test patient samples in parallel with an established test using positive and negative specimens.
- Test specimens with known potency, for example, low-positive and high-positive sera.
- Determine the precision of the test by establishing the 95% confidence interval.

The manner of testing is left to the laboratory director's discretion as long as the full range of testing is evaluated. Currently, many laboratories perform concurrent testing of 20 to 30 samples using the new test and the standard test; others test 40 to 70 samples. At least 25% to 30% of the specimens tested should yield a positive result, however.

The **confidence interval** (CI) at the 95% level can be calculated as a way of determining the precision (or imprecision) of a test when a small sample size (e.g., 20 to 30 samples) is used. Wide confidence intervals point out the unreliability of a test in studies where a small number of samples are used. The **standard error** (SE) and sensitivity of the tests must be known (Box 5-2).

DETERMINATION OF TEST RELIABILITY AT THE 95% CONFIDENCE LEVEL

BOX 5-2

DATA:

Test	Number of Tests	Sensitivity
current	30	90
new	30	95

STANDARD ERROR:

Current Test: $SE = \sqrt{0.90 \times \dfrac{(1 - 0.90)}{30}} = \sqrt{0.003} = 0.05$

New Test: $SE = \sqrt{0.95 \times \dfrac{(1 - 0.95)}{30}} = \sqrt{0.0016} = 0.04$

95% CONFIDENCE INTERVAL:

Current Test	New Test
$0.90 - (1.96 \times 0.05)$ to $0.90 + (1.96 \times 0.05)$	$0.95 - (1.96 \times 0.04)$ to $0.95 + (1.96 \times 0.04)$
$0.90 - 0.10$ to $0.90 + 0.10$	$0.95 - 0.08$ to $0.95 + 0.08$
0.80 to 1.00	0.87 to 1.03

CONCLUSION: The new test would be acceptable (validated) at the 95% CI.

The standard error can be calculated using the following formula:

$$SE = \sqrt{px \frac{(1 - p)}{n}}$$

where p = sensitivity of the test
n = sample size
$\sqrt{}$ = square root

The following example shows data collected by a laboratory that is validating a new test in parallel with the test being used; the 95% CI has been calculated for each test using the following formula:

$$p - (1.96 \times SE) \text{ to } p + (1.96 \times SE)$$

where   p = sensitivity of the test
1.96 = normal distribution value for 95% CI and is found in statistical tables; this number will change if the CI changes (e.g., 99% rather than 95%).

Bibliography

Elder, B.L., Hansen, S.A., Kellogg, J.A., et al. 1977. Cumitech 31: Verification and validation of procedures in the clinical microbiology laboratory. B.W. McCurdy, coordinating editor. American Society for Microbiology, Washington, D.C.

Ilstrup, D.M. 1990. Statistical method in microbiology. Clin. Microbiol. Rev. 3:219.

McPherson, B.S. and Needham, C.A. 1987. Method evaluation and test selection. In B. B. Wentworth, editor. Diagnostic procedures for bacterial infections, ed 7. American Public Health Association, Washington, D.C.

Washington, J.A. 1990. Confidence intervals: an important component of data presentations. Clin. Microbiol. Newsletter 12:109.

6 | QUALITY IN THE CLINICAL MICROBIOLOGY LABORATORY

The issue of laboratory quality has evolved over more than three decades since the first recommendations for **quality control (QC)** were published in 1965. QC is now seen as only one part of a total laboratory quality program. Quality now also includes **total quality management (TQM), continuous quality improvement (CQI),** and **quality assurance (QA).** TQM and CQI are used as umbrella terms, encompassing the whole quality program. TQM evolved as an activity to improve patient care by having the laboratory monitor its work to detect deficiencies and subsequently correct them. CQI went a step further by seeking to improve patient care by placing the emphasis on not making mistakes in the first place; CQI advocates continuous training to guard against having to correct deficiencies. QC is now associated with the internal activities that ensure diagnostic test accuracy. QA is associated with those external activities that ensure positive patient outcomes. Positive patient outcomes in the microbiology laboratory are:

- Reduced length of stay
- Reduced cost of stay
- Reduced turnaround time for diagnosis of infection
- Change to appropriate antimicrobial therapy
- Customer (physician or patient) satisfaction

CQI through well thought out programs of QC and QA are part of the requirements for laboratory accreditation under CLIA '88 (see Chapter 4).

QC PROGRAM

The laboratory director bears the primary responsibility for the QC and QA programs. However, all laboratory personnel must actively participate in both programs. Federal guidelines (CLIA '88) are considered minimum standards and are superseded by higher standards imposed by states or private certifying agencies such as the College of American Pathologists (CAP), the Joint Commission on the Accreditation of Health Care Organizations (JCAHO), or the Commission of Laboratory Accreditation (COLA). The basic elements of a QC program are described in the following sections.

SPECIMEN COLLECTION AND TRANSPORT

The laboratory is responsible for providing instructions for the proper collection and transport of specimens. These instructions should be available to the clinical staff for use when specimens are collected. The written collection instructions should includes:

- Test purpose and limitations
- Patient selection criteria
- Timing of specimen collection (e.g., before antimicrobials are administered)
- Optimal specimen collection sites
- Approved specimen collection methods
- Specimen transport medium
- Specimen transport time and temperature
- Specimen holding instructions if it cannot be transported immediately (e.g., hold at 4° C for 24 hours)
- Availability of test (on-site or sent to reference laboratory)
- Hours test performed (daily or batched)
- Turnaround time
- Result reporting procedures

The collection instructions should include information on how the requisition should be filled "out, and the laboratory must include a statement indicating that the requisition must be filled out entirely. In addition to standard information, such as patient name, hospital or laboratory number, and ordering physician, other critical information includes (1) whether the patient is on antimicrobial therapy, (2) suspect agent or syndrome, and (3) immunization history (if applicable). The laboratory should also establish criteria for unacceptable specimens. Examples of unacceptable specimens include:

- Unlabeled or mislabeled specimens—these specimens should be identified by the collector or be recollected
- Use of improper transport medium such as

stool for ova and parasites not submitted in preservative(s)
- Use of improper swab such as use of wooden shaft or calcium alginate tip for viruses
- Excessive transport time
- Improper temperature during transport or storage
- Improper collection site for test requested such as stool for respiratory syncytial virus
- Specimen leakage out of transport container
- Sera that are excessively hemolyzed, lipemic, or contaminated with bacteria

Sometimes, even though the specimen is not acceptable, the physician asks that it be processed anyway. If this happens, a disclaimer should be put on the final report, indicating that the specimen was not collected properly and that the results should be interpreted with caution.

STANDARD OPERATING PROCEDURE MANUAL (SOPM)

The requirement for a Standard Operating Procedure Manual (SOPM) is considered part of the QC program. The SOPM, discussed in detail in Chapter 3, should define test performance, tolerance limits, reagent preparation, required quality control, result reporting, and references. The SOPM should be written in National Committee for Clinical Laboratory Standards (NCCLS) format and must be reviewed and signed annually by the microbiology laboratory supervisor and director; in addition, all changes must be approved and dated by the laboratory director. The SOPM should be available in the work areas. It is the definitive laboratory reference and is used often for questions relating to individual tests. Any obsolete procedure should be dated when removed from the SOPM and retained for at least 2 years.

PERSONNEL

It is the laboratory director's responsibility to employ sufficient qualified personnel for the volume and complexity of the work performed. For example, published studies regarding staffing of virology laboratories suggest one technologist per 500 to 1000 specimens per year. Technical on-the-job training must be documented and the employee's competency must be assessed twice each year. Figure 6-1 is an example of a personnel competency assessment. Continuing education programs must be provided and documentation of attendance should be maintained in the employee's personnel file.

REFERENCE LABORATORIES

A laboratory should use only accredited or licensed reference laboratories. The referral laboratory's name, address, and licensure numbers should be included on the patient's final report.

PATIENT REPORTS

The laboratory should establish a system for supervisory review of all laboratory reports. This review should involve checking the specimen workup to verify that the correct conclusions were drawn and no clerical errors were made in reporting results. Reports should be given only to individuals authorized by law to receive them (physicians and other mid-level practitioners). Clinicians should be notified about "panic values" immediately. Panic values are potential life-threatening results, for example, positive Gram stain for cerebrospinal fluid (CSF) or a positive blood culture. Normal ranges are included on the report where appropriate, for example, serology tests. All patient records should be maintained for at least 2 years.

PROFICIENCY TESTING (PT)

Laboratories are required to participate in an external proficiency testing (PT) program for each analyte (test) for which a program is available; the laboratory must maintain an average score of 80% to maintain licensure in any subspecialty area. The federal government no longer maintains a PT program, but some states, such as New York, as well as several private accrediting agencies, such as CAP and the American Association of Bioanalysts (AAB) send out blind unknowns. These unknowns are to be treated exactly as patient specimens, from accessioning into the laboratory computer or manual logbook through workup and reporting of results. The testing personnel and laboratory director are required to sign a statement when the PT is completed attesting to the fact that the specimen was handled exactly like a patient specimen. In this way, PT specimens serve to establish the accuracy and reproducibility of a laboratory's day-to-day performance. The laboratory's procedures, reagents, equipment, and personnel are all checked in the process. Further, errors on PT actually help to point out deficiencies, and the subsequent education of the staff can lead to overall improvements in laboratory quality. When grades (evaluations) come back, critiques (summaries) accompanying them should be discussed with the entire technical staff. Evidence of corrective action in the event of problems should be documented, including changes in procedures, retraining of personnel, or the purchase of alternative media and reagents.

Some laboratories have a system of internal PT in addition to those received from external agencies. In cases where external audit is not available for a particular test method, laboratories are required to set

Employee's Name: _____   Bench: _____

Evaluator's Name: _____   Date of Evaluation: _____

Numerical Parameters
5 = No errors for each task
4 = 1 error for each task
3 = 2 errors for each task
2 = 3 errors for each task
1 = 4 errors for each task
N/A = not applicable to individual's job description

Appearance

	Neatness					Legibility				
Workcards	1	2	3	4	5	1	2	3	4	5
Logbook	1	2	3	4	5	1	2	3	4	5
Reports	1	2	3	4	5	1	2	3	4	5

Specimen Setup and Transport

						N/A
Were specimens transported at correct temperature?	1	2	3	4	5	N/A
Were specimens preserved correctly for transport?	1	2	3	4	5	N/A
Was client inventory maintained current and at appropriate levels?	1	2	3	4	5	N/A
Were specimens received in timely manner?	1	2	3	4	5	N/A
Were specimen requests date stamped?	1	2	3	4	5	N/A
Did specimen name labels match request form?	1	2	3	4	5	N/A
Were specimens logged correctly?	1	2	3	4	5	N/A
Were shared, stat or special instructions noted on specimens label?	1	2	3	4	5	N/A
Were safety precautions for transportation and check-in of specimens followed?	1	2	3	4	5	N/A
Were all tests set up as ordered?	1	2	3	4	5	N/A
Were specimens inoculated onto appropriate media?	1	2	3	4	5	N/A
Were inoculated specimens incubated appropriately?	1	2	3	4	5	N/A
Were left over specimens saved appropriately?	1	2	3	4	5	N/A

Specimen Workup

						N/A
Were workcards easy to follow?	1	2	3	4	5	N/A
Were the correct biochemicals set up?	1	2	3	4	5	N/A
Did decision-making process yield proper results and follow accepted protocol?	1	2	3	4	5	N/A
Was each day's work dated?	1	2	3	4	5	N/A
Were all biochemical results recorded?	1	2	3	4	5	N/A
Were biochemicals inoculated properly?	1	2	3	4	5	N/A
Were isolates frozen when appropriate?	1	2	3	4	5	N/A
Was all work completed in a timely manner?	1	2	3	4	5	N/A
Were specimen stains read correctly?	1	2	3	4	5	N/A
Were specimens overworked?	1	2	3	4	5	N/A
Were safe work practices observed?	1	2	3	4	5	N/A
Were susceptibilities performed in a timely manner?	1	2	3	4	5	N/A
Were stat requests done in a timely manner?	1	2	3	4	5	N/A

FIGURE 6-1 Employee competency evaluation.

Continued

Quality Control							N/A
Were check plates done when appropriate?	1	2	3	4	5	N/A
Were check plates mixed?	1	2	3	4	5	N/A
Were necessary controls performed?	1	2	3	4	5	N/A
Were periodic QC controls performed?	1	2	3	4	5	N/A

Reporting							N/A
Were verbal reports given and documented?	1	2	3	4	5	N/A
Were final reports correct?	1	2	3	4	5	N/A
Were reports given in a timely manner?	1	2	3	4	5	N/A
Were correct susceptibilities reported?	1	2	3	4	5	N/A
Were all reports logged out?	1	2	3	4	5	N/A
Was supervisory review appropriate?	1	2	3	4	5	N/A

Conclusion							
Overall performance of employee at this bench (average of all tasks evaluated):	1	2	3	4	5	N/A
Is retraining necessary?		Yes			No		

NOTE: Overall satisfactory performance does not mean retraining is not necessary. Each task is evaluated individually for retraining purposes. Retraining is required on each task with a score less than 3.

Additional Comments:

Acknowledgement

Employee's signature: _____   Date: _____

Evaluator's signature: _____   Date: _____

Followup

Retraining completed.

Employee's signature: _____   Date: _____

Evaluator's signature: _____   Date: _____

FIGURE 6-1, cont'd Employee competency evaluation.

up an internal program to revalidate the test at least semiannually. Internal PT samples can be set up by (1) seeding a simulated specimen and labeling it as an autopsy specimen so no one will panic if a pathogen is recovered, (2) splitting a routine specimen for workup by two different technologists, or (3) sending part of a specimen to a reference laboratory to confirm the laboratory's result.

PERFORMANCE CHECKS

Instruments
Equipment logs should contain the following information:

- Instrument name, serial number, and date put in use
- Procedure and periodicity (daily, weekly, monthly, quarterly) for routine function checks
- Acceptable performance ranges
- Instrument function failures, including specific details of steps taken to correct the problems (**corrective action**)
- Date and time of service requests and response
- Date of routine preventative maintenance (PM), which should follow manufacturer's recommendations

Maintenance records should be retained in the laboratory for the life of the instrument. Specific guidelines regarding periodicity of testing for autoclaves, biological safety cabinets, centrifuges, incubators, microscopes, refrigerators, freezers, water baths, heat blocks, and other microbiology laboratory equipment can be found in a number of the references listed at the end of this chapter. An example of a QC form for checking CO_2 incubators is shown in Figure 6-2.

Commercially prepared media exempt from QC
The NCCLS Subcommittee on Media Quality Control collected data over several years regarding the incidence of QC failures of commonly used microbiology media. Based on its findings, the subcommittee published a list of media that did not require retesting in the user's laboratory if purchased from a manufacturer who follows NCCLS guidelines. The list of media is published in NCCLS standard M22-A. The laboratory must inspect each shipment for cracked media or petri dishes, hemolysis, freezing, unequal filling, excessive bubbles, clarity, and visible contamination. The manufacturer must supply written assurance that NCCLS standards were followed; this verification must be maintained along with the laboratory's QC protocol. An example of a QC sheet for an NCCLS-exempt product is shown in Figure 6-3.

User-prepared and nonexempt, commercially prepared media
QC forms for user-prepared media should contain the amount prepared, the source of each ingredient, the lot number, the sterilization method, the preparation date, the expiration date (usually 1 month for agar plates and 6 months for tubed media), and the name of the preparer. Both user-prepared and nonexempt, commercially prepared media should be checked for proper color, consistency, depth, smoothness, hemolysis, excessive bubbles, and contamination. A representative sample of the lot should be tested for sterility; 5% of any lot is tested when a batch of 100 or fewer units is received, and a maximum of 10 units are tested in larger batches. A batch is any one shipment of a product with the same lot number; if a separate shipment of the same lot number of product is received, then it is considered a different batch and needs to be tested separately.

Sterility is routinely checked by incubating the media for 48 hours at the temperature at which the medium will be used. Both user-prepared and nonexempt, commercially prepared media should also be tested with QC organisms of known physiologic and biochemical properties. Tables listing specific organisms to test for various media can be found in a number of the references listed at the end of this chapter. An example of a QC sheet for a nonexempt, commercially prepared medium (chocolate agar) is shown in Figure 6-4.

ANTIMICROBIAL SUSCEPTIBILITY TESTS
The goal of quality control testing of antimicrobial susceptibility tests (ASTs) is to ensure the precision and accuracy of the supplies and microbiologists performing the test. Criteria regarding frequency of testing are the same regardless of the methodology, such as minimum inhibitory concentration (MIC) broth dilution or Kirby-Bauer (see Chapter 18). Each new shipment of microdilution trays or Mueller-Hinton plates should be tested with NCCLS-approved American Type Culture Collection (ATCC [Rockville, Md.]) reference strains.

Reference strains for MIC testing are selected for genetic stability and give MICs that are in the midrange of each antimicrobial agent tested. Reference strains for Kirby-Bauer testing have clearly defined mean diameters of their respective zone of inhibition for each antimicrobial tested. ATCC numbers of reference strains are different for various AST methods. Moreover, quality control MICs and zone diameters are updated annually by the NCCLS Subcommittee on Antimicrobial Susceptibility Testing, so new tables should be obtained from the NCCLS regularly.

Model: _____   Month: _____

Model number: _____   Year: _____

Serial number: _____

Date	Temperature Expected results: 37°C±1.0	Humidity Expected results: 40-90%	CO_2 Content Expected results 7-10%	Tank Pressure Expected results: ≥200psi	Comments and Corrective Action	Tech
1						
2						
3						
4						
5						
6						
7						
8						
9						
10						
11						
12						
13						
14						
15						
16						
17						
18						
19						
20						
21						
22						
23						
24						
25						
26						
27						
28						
29						
30						
31						

Reviewed by: _____   Comments: _____

Date: _____   _____

FIGURE 6-2 Function check of CO_2 incubator.

IDENTIFICATION

Name	SHEEP BLOOD AGAR			
Lot number	Source	Date Received	Expiration Date	Storage conditions
				2 - 8°C

APPEARANCE (circle one in each category)

Package damage	Medium color RED	Clarity OPAQUE	Excess moisture or drying	Visible contamination	Checked Date By
NO	YES	YES	NO	NO	
YES	NO	NO	YES	YES	

STERILITY TEST

Number of units tested	Temperature and time of incubation	Results	Set Up Date By
			N/A
			Checked Date By
N/A	N/A	N/A	N/A

N/A - Commercially prepared and tested by manufacturer and conforms to NCCLS standards as defined in M22 - A.

PERFORMANCE TEST

Test Organism	Stock Number	Expected Result	Observed Result	Set Up Date By
N/A (Not required by CLIA)	N/A	N/A	N/A	N/A
				Checked Date By
				N/A

CORRECTIVE ACTION

Circle problems in red. Assign each a number and describe below.

FINAL ACTION	*REVIEWED*	*COMMENTS*
Release for use	By	
Discard	Date	

FIGURE 6-3 Quality control record for exempt commercial media.

CHOCOLATE AGAR

Name					
Lot number	Source	Date Received	Expiration Date	Storage conditions
					2 - 8°C

APPEARANCE (circle one in each category)

Package damage	Medium color	Clarity	Excess moisture or drying	Visible contamination	Checked
	BROWN	OPAQUE			Date By
NO	YES	YES	NO	NO	
YES	NO	NO	YES	YES	

STERILITY TEST

Number of units tested	Temperature and time of incubation	Results (circle one)	Set Up
			Date By
	35°C, 48 hr	GROWTH	Checked
		NO GROWTH	Date By

PERFORMANCE TEST

Test Organism	ATCC Number	Expected Result	Observed Result	Set Up
				Date By
H. influenzae		Good growth		Checked
N. gonorrhoeae		Good growth		Date By

CORRECTIVE ACTION
Circle problems in red. Assign each a number and describe below.

FINAL ACTION

Release for use	
Discard	

REVIEWED

By	
Date	

COMMENTS

FIGURE 6-4 Quality control record for commercial, nonexempt media.

Each susceptibility test system must also be tested with use (usually daily) for 30 consecutive days. If three or fewer MICs or zone of inhibition diameters per drug-reference strain combination are outside the reference range during the 30-day testing period, laboratories may switch to weekly QC testing. Thereafter, aberrant results obtained during the weekly testing must be vigorously investigated. If a source of error, such as contamination, incorrect reference strain used, or incorrect atmosphere of incubation, is found, quality control testing may simply be repeated. However, if no source of error is uncovered, 5 consecutive days of retesting must be performed. If accuracy and precision is again acceptable, weekly QC testing may resume; if the problem drug/organism combination(s) are still outside the reference ranges, 30 days of consecutive testing must be reinitiated before weekly testing can be reinstated. Under no circumstances should any drug/bug combination be reported for a patient isolate if QC testing has failed.

Stains and reagents

Containers of stains and reagents should be labeled as to contents, concentration, storage requirements, date prepared (or received), date placed in service (commonly called the date opened), expiration date, source (commercial manufacturer or user prepared), and lot number. All stains and reagents should be stored according to manufacturer's recommendations and tested with positive and negative controls before use. Tables listing specific organisms to test for various stains or reagents can be found in a number of the references at the end of this chapter. Outdated materials or reagents that fail QC even after retesting with fresh organisms should be discarded immediately. Patient specimens should not be tested using the lot number in question until the problem is solved; in the case of a repeat failure, an alternative method should be used or the patient's specimen should be sent to a reference laboratory.

Antisera

The lot number, date received, condition received, and expiration date must be recorded for all shipments of antisera. In addition, the antisera should be dated when opened. New lots must be tested concurrently with previous lots, and testing must include positive and negative controls. Periodicity of testing thereafter should follow the requirements of agencies that inspect an individual laboratory (including the Health Care Financing Administration) and may include, with use, monthly or semiannual checks.

Kits

Kits that have been approved by the U.S. Food and Drug Administration need to be tested as specified in the manufacturer's package insert. Each shipment of kits must be tested even if it is the same lot number as a previously tested lot because temperature changes during shipment may affect the performance.

MAINTENANCE OF QC RECORDS

All QC results should be recorded on an appropriate QC form. Corrective action should be noted on this form. If temperature is adjusted or a biochemical test is repeated, the new readings within the tolerance limits should be listed. The supervisor or laboratory director should review and initial all forms monthly. QC records should be maintained for at least 2 years except those on equipment, which must be saved for the life of each instrument.

MAINTENANCE OF REFERENCE QC STOCKS

Stock organisms may be obtained from the ATCC, commercial vendors, or PT programs; well-defined clinical isolates may also be used. The laboratory should have enough organisms on hand to cover the full range of testing of all necessary materials such as media, kits, and reagents.

Bacteriology

Nonfastidious (rapidly growing), aerobic bacterial organisms can be saved up to 1 year on trypticase soy agar (TSA) slants. Long-term storage (>1 year) of aerobes or anaerobes can be accomplished either by lyophilization (freeze drying) or freezing at $-70°C$. Frozen, nonfastidious organisms should be thawed, reisolated, and refrozen every 5 years; fastidious organisms should be thawed, reisolated, and refrozen every 3 years. Stock isolates may be maintained by freezing them in 10% skim milk, trypticase soy broth (TSB) with 15% glycerol, or 10% horse blood in sterile, screw-cap vials.

Mycology

Yeasts may be treated as nonfastidious bacterial organisms for maintaining stock cultures. Molds can be stored on potato dextrose agar (PDA) slants at 4° C for 6 months to 1 year. For longer-term storage, PDA slants may be overlaid with sterile mineral oil and stored at room temperature. Alternatively, sterile water can be added to an actively sporulating culture on PDA, the conidia (spores) can be teased apart to dislodge them from the agar surface and the water can then be dispensed to sterile, screw-top vials. These vials should be capped tightly and stored at room temperature.

Mycobacteriology

Acid-fast bacilli (AFB) may be kept on Lowenstein-Jenson (LJ) agar slants at 4° C for up to 1 year. They may also be frozen at −70° C in 7H9 broth with glycerol.

Virology

Viruses may be stored indefinitely at −70° C in a solution containing a cryoprotectant, such as 10% dimethyl sulfoxide (DMSO) or fetal bovine serum.

Parasitology

Slides and photographs must be available for QC purposes. Trichrome and other permanent slides may be purchased from commercial vendors. Clinical slides may be preserved indefinitely by adding a drop of permount and a coverslip.

QA PROGRAM

Because QA is the method by which the overall process of infectious disease diagnosis is reviewed, any of the steps involved in the diagnosis of an infectious disease may be studied. These include:

- Preanalytic
  - Ordering of test by the clinician
  - Processing of test request by the clerical staff
  - Collection of specimen by nursing personnel or patient
  - Transport of specimen to the laboratory
  - Initial processing of specimen in laboratory, including specimen accessioning
- Analytic
  - Examination and workup of culture by the microbiologist
  - Interpretation of specimen results by the microbiologist
- Postanalytic
  - Formulation of a written report by the microbiologist
  - Communication of the microbiologist's conclusions to the clinician in written format
  - Interpretation of report by the clinician
  - Institution of appropriate therapy by the clinician

Analytic testing (the work actually done in the microbiology laboratory) is now seen as only one part of a continuing spectrum of steps that begins when the physician orders the test and ends when they receive the results and treat the patient.

QA audits (studies) are planned and conducted by examining the three stages of testing. The goal is to look at the proficiency with which the patient is served by the whole facility, including the laboratory. The outcome is to look at the consequences to the patient of the work that has been performed. QA audits involve the analysis of how the system works and how it can be improved.

Q-PROBES

A process is selected for audit in a number of ways. One is to subscribe to the Q-Probes program, which is a national interlaboratory QA program developed and administered by CAP. CAP selects topics to be audited and provides instructions and worksheets for collection of data as well as data entry forms. Data is collected for a specified period and then returned to CAP for analysis. CAP returns a summary of the institution's performance as well as a comparison with other facilities of similar size and scope of service. That way, an individual facility can compare its results with those of its peers, a process called **benchmarking**. Q-Probes are designed for all areas of laboratory medicine so every one will not be directed toward microbiology. Since inception of the program, microbiology Q-probe audits have included areas such as (1) blood culture utilization, (2) nosocomial (hospital-acquired) infections, (3) cumulative susceptibility results, (4) antibiotic usage, (5) turnaround time of Gram stains of cerebrospinal fluid, (6) viral hepatitis test utilization, (7) laboratory diagnosis of tuberculosis, (8) blood culture contamination rates, (9) appropriateness of ordering of stools for microbiology testing, and (10) sputum quality. Several generic (laboratorywide) audits are also applicable to the microbiology laboratory, including error reporting, quality of reference laboratories, and effects of laboratory computer down time.

IN-HOUSE QA AUDITS

A facility that does not subscribe to the Q-Probe program may select topics for audits through suggestions from the medical, nursing, or pharmacy staffs; complaints from the medical or nursing staff; or deficiency or observation noted in the laboratory.

Physicians may suggest an audit to measure the transcription accuracy of their orders by nursing unit clerical personnel. Nursing administrators may suggest an audit of contaminated urine cultures to access the compliance of the nursing staff in instructing patients about proper urine culture collection techniques. Pharmacists may notice improper antibiotic utilization by the clinical staff, for example, a patient was not placed on the appropriate therapy after the pathogen was reported or the patient remains on antibiotic therapy to which their organism is resistant after the susceptibility report has been charted. Complaints from the medical or nursing staff can involve failure of the laboratory to conduct all the tests requested on the requisition, performance of the wrong test, or prolonged delay in turnaround time of test results. All complaints to the laboratory must be documented; Figure 6-5 shows a form designed for this purpose. Corrective action and followup with the laboratory, medical, and nursing staffs must also be documented.

Deficiencies or problems in the laboratory should also be documented (Figure 6-6). If, for example, the laboratory notices a dramatic rise in the number of positive respiratory syncytial virus (RSV) direct antigen tests in the summer (not RSV season) and the problem is traced back to a quality control problem that was not recognized by a new employee, a QA audit might be indicated to study the outcome of the patients, including inappropriate treatment for RSV and failure to institute treatment for the true causative agent. Alternatively, microbiologists may notice they are receiving many ova and parasite (O&P) examinations and stool cultures on patients hospitalized for more than 3 days. Because current cost containment guidelines suggest that this is inappropriate, the microbiology laboratory personnel could undertake a study to determine the percentage of positive results and the number of patients who

tested positive for *C. difficile* cytotoxin, which is the more likely cause of diarrhea in patients hospitalized for more than 3 days. If the audit showed that none of the stool cultures or O&P examinations were positive and no stools were analyzed for *C. difficile* cytotoxin, these findings would be presented to the medical staff. Some months following the medical staff in-service, the number of stool culture and O&P requests on patients hospitalized longer than 3 days would be reevaluated. It is hoped this would result in a dramatic decrease in numbers of inappropriate specimens.

CONDUCTING A QA AUDIT

Box 6-1 is an example of how an in-house QA audit may be conducted.

CONTINUOUS DAILY MONITORING

Daily activities of microbiologists and supervisory personnel ensure that patients get the best quality care. These include (1) comparing results of morphotypes seen on direct examinations with what grows on the culture to ensure that all organisms have been recovered, (2) checking antimicrobial susceptibility reports to verify that profiles match those expected from a particular species, and (3) studying culture and susceptibility reports for clusters of patients with

---

BOX 6-1   **QA AUDIT ON STAT TURNAROUND TIMES**

BACKGROUND: Following a complaint regarding turnaround time for stat RSV direct antigen tests one winter, the microbiology laboratory at General Hospital decided to audit its turnaround time. The medical staff indicated that it would like to turn the test around in 2.5 hours (150 minutes) from the time of collection to the time the physician is notified; the medical staff felt that this would ensure maximum patient benefits.

STUDY DESIGN: All RSV requests for direct antigen testing were evaluated for a 3-month period to determine if this turnaround time was being met by laboratory personnel.

RESULTS:

Month	Reports given in < 150 min		Report time exceeding 150 min		Combined averages	
	# specimens	Average time	# specimens	Average time	# specimens	Average time
December	57	114 min	15	195 min	72	130 min
January	114	108 min	14	179 min	128	116 min
February	70	114 min	3	165 min	73	116 min

ANALYSIS: Two hundred and seventy-three reports were reviewed. The average reporting time for the 3-month period was under the acceptable 150 minutes. In December and January 15 and 14 specimens, respectively, had turnaround times that exceeded 150 minutes, with an average of 195 minutes in December and 179 minutes in January; February had 3 reports exceeding 150 minutes.

CONCLUSIONS: The overall (combined) average reporting time, while remaining within 150 minutes, could be improved. There was a dramatic drop in February after the medical staff complaint. This was undoubtedly a result of in-services given to courier and clerical staff regarding the need to transport and accession the stat specimens quickly.

RECOMMENDATIONS: Because hospitalwide systems have been improved, appropriate followup would be to audit stat turnaround time for another test, for example, Gram stain of CSF, in 3 to 6 months to verify that they also meet the 150-minute turnaround time requirement.

Complaint Identification	
Date:	Facility:
Time:	Contact:
Recipient:	Phone:

Notes

_____
_____
_____
_____
_____

Corrective Action

_____
_____
_____
_____
_____

Effect on Patient Care

_____
_____
_____
_____
_____

Followup

_____
_____
_____
_____
_____

Reviewed by: _____   Date: _____

FIGURE 6-5 Client complaint form.

Problem

Steps Taken to Solve the Problem

Corrective Action Taken to Avoid this Problem in Future

Effect on Patient Care

Followup

Reviewed by: _____ Date: _____

FIGURE 6-6 Laboratory problem report.

unusual infections or multiple-drug–resistant organisms. These and many other processes result in continual improvement to all test systems that will ultimately benefit the patient.

Bibliography

August, M.J., Hindler, J.A., Huber, T.W., and Sewell, D.L. 1990. Cumitech 3A: Quality control and quality assurance practices in clinical microbiology. A.S. Weissfeld, coordinating editor. American Society for Microbiology, Washington, D.C.

Bartlett, R.C., Mazens-Sullivan, M., Tetreault, J.Z., et al. 1994. Evolving approaches to management of quality in clinical microbiology. Clin. Microbiol. Rev. 7:55.

LaRocco, M.L. 1995. Quality and productivity in the microbiology laboratory: continuous quality improvement. Clin. Microbiol. Newsletter 17:129.

Miller, J.M. 1987. Quality control in microbiology. Centers for Disease Control, Atlanta.

Miller, J.M. and Wentworth, B.B., editors. 1985. Methods for quality control in diagnostic microbiology. American Public Health Association, Washington, D.C.

National Committee for Clinical Laboratory Standards. 1984. Clinical lab procedure manuals. Approved Guideline GP2-A. National Committee for Clinical Laboratory Standards, Villanova, Pa.

National Committee for Clinical Laboratory Standards. 1990. QC assurance for commercially prepared microbiological culture media. Approved Standard M22-A. National Committee for Clinical Laboratory Standards, Villanova, Pa.

National Committee for Clinical Laboratory Standards. 1993. Methods for dilution antimicrobial susceptibility tests for bacteria that grow aerobically. Approved Standard M7-A3. National Committee for Clinical Laboratory Standards, Villanova, Pa.

National Committee for Clinical Laboratory Standards. 1993. Performance standards for antimicrobial disk susceptibility tests. Approved Standard M2-A5. National Committee for Clinical Laboratory Standards, Villanova, Pa.

National Committee for Clinical Laboratory Standards. 1994. Development of in vitro susceptibility testing criteria and quality control parameters. NCCLS document M23-A. National Committee for Clinical Laboratory Standards, Villanova, Pa.

National Committee for Clinical Laboratory Standards. 1995. Training verification for laboratory personnel. Approved Guideline GP21-A. National Committee for Clinical Laboratory Standards, Villanova, Pa.

Schifman, R.B. 1987. Quality assurance in microbiology. In Howanitz P.J. and Howanitz J.H., editors. Laboratory Quality Assurance. McGraw-Hill, N.Y.

Sewell, D.L. 1992. Quality control. In Isenberg H.D., editor. Clinical microbiology procedures handbook, vol 2. American Society for Microbiology, Washington, D.C.

Warford, A.L. 1992. QC control in clinical virology. In Specter S. and Lancz G., editors. Clinical virology manual, ed 2. Elsevier Science Publishing, N.Y.

7 INFECTION CONTROL

Every year, between 1.75 and 3 million (5% to 10%) of the 35 million patients admitted to acute-care hospitals in the United States will acquire an infection that was neither present nor in the prodromal (incubation) stage when they entered the hospital. These infections are called **nosocomial,** or hospital-acquired, infections. Treatment of nosocomial infections is estimated to add between $4.5 and $15 billion annually to the cost of health care and represents an enormous economic problem in today's environment of cost containment. In addition, many of these infections lead to the death of hospitalized patients (patient mortality) or, at minimum, lead to additional complications (patient morbidity) and antimicrobial chemotherapy.

Some of the earliest efforts to control infection followed the recognition in the nineteenth-century that women were dying in childbirth from bloodstream infections caused by group A *Streptococcus (Streptococcus pyogenes)* because physicians were spreading the organism by not washing their hands between the examination of different patients. Handwashing is still the cornerstone of modern infection control programs. Moreover, the first recommendations for isolation precautions in U.S. hospitals were published in the late 1800s when guidelines appeared advocating placement of patients with infectious diseases in separate hospital facilities. By the late 1950s, the advent of nosocomial infections caused by *Staphylococcus aureus* finally ushered in the modern age of infection control. In the past four decades, we have learned that in addition to hospitalized patients acquiring infections, health care workers are also at risk of acquiring infections from patients. Thus, present day infection control programs have evolved to prevent the acquisition of infection by patients and caregivers.

INCIDENCE OF NOSOCOMIAL INFECTIONS

The Centers for Disease Control and Prevention (CDC) has established the **National Nosocomial Infections Surveillance** (NNIS) program to monitor the incidence of nosocomial infections in the United States. Regardless of a hospital's size or medical school affiliation, the rates of infections at each body site are consistent across institutions. The most common nosocomial infections are urinary tract infections (33%), followed by pneumonia (15%), surgical site infections (15%), and bloodstream infections (13%). A companion CDC program, **Study of the Efficacy of Nosocomial Infection Control** (SENIC), keeps statistics on morbidity and mortality of hospital-acquired infections. Each nosocomial infection adds 5 to 10 days to the affected patient's hospital stay. Of individuals with hospital-acquired bloodstream or lung infections, 40% to 60% die each year. Likewise, patients with indwelling (Foley) catheters have a threefold increased chance of dying from urosepsis, a bloodstream infection that is a complication of a urinary tract infection, than those who do not have one.

Attack rates vary according to the type of hospital. Large, tertiary-care hospitals that treat the most seriously ill patients often have higher rates of nosocomial infection than do small, acute-care community hospitals; large medical school-affiliated (teaching) hospitals have higher infection rates than do small teaching hospitals. This difference in the risk of infection is probably related to several factors, including the severity of illness, the frequency of invasive diagnostic and therapeutic procedures, and variation in the effectiveness of infection control programs. Within hospitals, the surgical and medical services have the highest rates of infection; the pediatric and nursery services have the lowest. Moreover, within services, the predominant type of infections varies, that is, surgical site infections are the most common on the surgical service while urinary tract or bloodstream infections are the most common on medical services or in the nursery.

TYPES OF NOSOCOMIAL INFECTIONS

The majority of noscomial infections are endogenous in origin, that is, they involve the patient's own microbial flora. Three principal factors determine the likelihood that a given patient will acquire a nosocomial infection:

- Susceptibility of the patient to the infection
- The virulence of the infecting organism
- The nature of the patient's exposure to the infecting organism

In general, hospitalized individuals have increased susceptibility to infection. Corticosteroids, cancer chemotherapeutic agents, and antimicrobial agents all contribute to the likelihood of nosocomial infection by suppressing the immune system or altering the host's normal flora to that of resistant (hospital) microbes. Likewise, foreign objects, such as urinary or intravenous catheters, break the body's natural barriers to infection. Nonetheless, these medications or devices are necessary to cure the patient's primary medical condition. Finally, exerting influence over the virulence of the pathogens is not possible because it is not possible to immunize patients against nosocomial infections. Patients with serious community-acquired infections are frequently admitted to the hospital, and the disease may spread nosocomially by either direct contact; contact with contaminated food, water, medications, or medical devices (fomites); or by airborne transmission. Thus, nosocomial infections may never be completely eliminated, only controlled.

URINARY TRACT INFECTIONS

Gram-negative rods cause the majority of hospital-acquired urinary tract infections, and *Escherichia coli* is the number one organism involved. Gram-positive organisms, *Candida* spp., and other fungi cause the remainder of the infections. The risk factors that predispose patients to acquire a nosocomial urinary tract infection include advanced age, female gender, severe underlying disease, and the placement of indwelling urinary catheters.

LUNG INFECTIONS

The most common nosocomial pathogens causing pneumonia are gram-negative rods, *S. aureus*, and *Moraxella catarrhalis*. *Streptococcus pneumoniae* and *Haemophilus influenzae*, which cause the majority of community-acquired pneumonias, are not important etiologic agents in hospital-acquired infections except very early during the hospital course (first 2 to 5 days); these infections probably represent infections that were already incubating at the time of the hospital admission. The risk factors that predispose patients to acquire a nosocomial lung infection include (1) advanced age, (2) chronic lung disease, (3) large volume aspiration (the microorganisms in the upper respiratory tract are coughed up and lodge in the lungs instead of being spit out or swallowed), (4) chest surgery, (5) monitoring intracranial pressure (in which a catheter inserted through the skull mea-

sures the amount of fluid on the brain), (6) hospitalization in intensive care units, and (7) intubation (placement of a breathing tube down a patient's throat) or attachment to a mechanical ventilator (which controls breathing).

SURGICAL SITE INFECTIONS

Approximately 4% of surgical patients develop surgical site infections; 50% of these infections develop *after* the patient has left the hospital so this number may be an underestimate. Gram-positive organisms (*S. aureus*, coagulase-negative staphylococci and enterococci) cause the majority of these infections followed by gram-negative rods and *Candida* spp. The risk factors that predispose patients to acquire a nosocomial wound infection include (1) advanced age, (2) obesity, (3) infection at a remote site (that spreads through the bloodstream), (4) malnutrition, (5) diabetes, (6) extended preoperative hospital stay, (7) greater than 12 hours between preoperative shaving of site and surgery, (8) extended time of surgery, and (9) inappropriate timing of prophylactic antibiotics (given to prevent common infections before they seed the surgical site). Surgical wounds are classified as either clean, clean-contaminated, contaminated, or dirty depending on the number of contaminating organisms at the site. Bowel surgery is considered dirty, for example, while surgery for a total hip replacement is considered clean.

BLOODSTREAM INFECTIONS

The overall rate of nosocomial bloodstream infections increased in all NNIS hospitals between 1980 and 1989 when the incidence of infections with coagulase-negative staphylococci, enterococci, *S. aureus*, and *Candida* spp. increased. The risk factors that predispose patients to acquire a nosocomial bloodstream infection include (1) age (≤ 1 year or ≥ 60 years), (2) malnutrition, (3) immunosuppressive chemotherapy, (4) loss of skin integrity (e.g., burn or decubiti [bedsore]), (5) severe underlying illness, (6) indwelling device (e.g., catheter), (7) intensive care unit stay, and (8) prolonged hospital stay.

EMERGENCE OF ANTIBIOTIC-RESISTANT MICROORGANISMS

The organisms that cause nosocomial infections have changed over the years because of selective pressures from the use (and overuse) of antibiotics (see Chapter 17). Risk factors for the acquisition of highly resistant organisms include prolonged hospitalization and prior treatment with antibiotics. In the preantibiotic era, most hospital-acquired infections were caused by *S. pneumoniae* and group A *Streptococcus* (*Streptococcus pyogenes).* In the 1940s and '50s, with the advent of treatment of patients with penicillin and sulfonamides, resistant strains of *S. aureus* appeared. Then, in the 1970s, treatment of patients with narrow-spectrum cephalosporins and aminoglycosides led to the emergence of resistant aerobic gram-negative rods, such as *Klebsiella, Enterobacter, Serratia,* and *Pseudomonas.* During the late 1970s and early 1980s, the use of more potent cephalosporins played a role in the emergence of antibiotic-resistant, coagulase-negative staphylococci, enterococci, methicillin-resistant *S. aureus* (MRSA) and *Candida* spp. The 1990s has witnessed the emergence of beta-lactamase–producing, high-level gentamicin-resistant, and vancomycin-resistant enterococci (VRE).

Patients' normal flora changes very quickly after hospitalization from viridans streptococci, saprophytic *Neisseria* spp. and diphtheroids to potentially resistant microorganisms found in the hospital environment. Then, their colonized nares, skin, gastrointestinal tract, or genitourinary tract can serve as reservoirs for endogenously acquired infections. Moreover, if patients colonized with resistant microorganisms return to nursing homes in the community harboring these organisms, they can also transfer them to other patients. This further increases the pool of patients who harbor multidrug-resistant organisms when they, in turn, are hospitalized. These new patients recontaminate the hospital environment and serve as potential reservoirs for spread to additional patients.

HOSPITAL INFECTION CONTROL PROGRAMS

Hospital infection control programs are designed to detect and monitor hospital-acquired infections and to prevent or control their spread. The Infection Control Committee is multidisciplinary and should include the microbiologist, the infection control practitioner (often a nurse with special training), the hospital epidemiologist (usually an infectious disease physician), and a pharmacist. The infection control practitioner collects and analyzes surveillance data, monitors patient care practices, and participates in epidemiologic investigations. Daily review of charts of patients with fever or positive microbiology cultures allows the infection control practitioner to recognize problems with hospital-acquired infections and to detect outbreaks as early as possible. The infection control practitioner is also responsible for education of health care providers in techniques such as, handwashing and isolation precautions, that will minimize the acquisition of infections.

It is the infection control practitioner's job to identify all cases of an outbreak. The investigation of the cluster of cases during a particular outbreak involves its characterization in terms of commonalities, such as location in the hospital (nursery, intensive care unit), same caregiver, or prior respiratory or physical therapy. Risk factors, including underlying diseases, current or prior antimicrobial therapy, and placement of a Foley catheter, are also assessed. This information will help the infection control committee determine (1) the reservoir of the organism in the hospital, that is, the place where it exists, and (2) the means of transmission of the organism from its reservoir to the patient.

Microorganisms are spread in hospitals through several modes:

- **Direct contact,** for example, in contaminated food or intravenous solutions
- **Indirect contact,** for example, from patient to patient on the hands of health care workers (MRSA, rotavirus)
- **Droplet contact,** for example, inhalation of droplets (>5 μm in diameter) that cannot travel more than 3 feet (pertussis)
- **Airborne contact,** for example, inhalation of droplets (≤5 μm) that can travel large distances on air currents (tuberculosis)
- **Vector-borne contact,** for example, disease spread by vectors, such as mosquitoes (malaria) or rats (rat-bite fever); this mode of transmission is rare in hospitals in developed countries

Once the reservoir is known, the infection control practitioner can implement control measures, such as reeducation regarding handwashing (in the case of spread by health care workers) or hyperchlorination of cooling towers in the case of nosocomial legionellosis.

ROLE OF THE MICROBIOLOGY LABORATORY

The microbiology laboratory supplies the data on organism identification and antimicrobial susceptibility profile that the infection control practitioner reviews daily for evidence of nosocomial infection. Thus, the laboratory must be able to detect potential microbial pathogens and then accurately identify them to species level and perform susceptibility testing. The microbiology laboratory should also monitor multidrug-resistant organisms by tabulating data on antimicrobial susceptibilities of common isolates and studying trends indicating emerging resistance. Significant findings should be immediately reported to the infection control practitioner. If an outbreak is identified, the laboratory works in tandem with the Infection Control Committee by (1) saving all isolates, (2) culturing possible reservoirs (patients, personnel, or the environment), and (3) performing typing of strains to establish relatedness between isolates of the same species. Microbiology laboratories are also obligated by law to report certain isolates or syndromes to public health authorities. For example, Box 7-1 lists organisms to be reported to state health authorities in Texas. Other states have similar criteria.

CHARACTERIZING STRAINS INVOLVED IN AN OUTBREAK

The ideal system for typing microbial strains involved in outbreaks should be standardized, reproducible, sensitive, stable, readily available, inexpensive, applicable to a wide range of microorganisms, and field tested in other epidemiologic investigations. Although no such perfect system is currently available, a number of methods are used to aid in typing epidemic strains. There are two major ways to type strains using either phenotypic traits or molecular typing methods.

Classic phenotypic techniques include **biotyping** (analyzing unique biologic or biochemical characteristics), **antibiograms** (analyzing antimicrobial susceptibility patterns), and **serotyping** (serologic typing of bacterial or viral antigens, such as bacterial cell wall [O] antigens). **Bacteriocin typing,** which examines an organism's susceptibility to bacterial peptides (proteins), and **bacteriophage typing,** which examines the ability of bacteriophages (viruses capable of infecting and lysing bacterial cells) to attack certain strains, have been useful for typing *Pseudomonas aeruginosa* and *S. aureus,* respectively; these techniques, however, are not widely available.

Genotypic, or molecular, methods have largely replaced phenotypic methods as a means of confirming the relatedness of strains involved in an outbreak. *Plasmid analysis* and *restriction endonuclease analysis* of chromosomal DNA are widely used. Plasmids are extrachromosomal pieces of genetic material (nucleic acids) that self-replicate (reproduce). Plasmids may be transferred from one bacterial cell to another by conjugation or transduction (see Chapter 9). Plasmid analysis has often been used to explain the occurrence of unusual or multiple-antibiotic resistance patterns. It has been shown that plasmids or R factors (resistance genes carried on plasmids) can cause outbreaks when a specific plasmid is transmitted from one genus of bacteria to another. Plasmid profiles, patterns created when plasmids are separated based on molecular weight by agarose gel electrophoresis, can also be used to characterize the similarity of bacterial strains. Relatedness of strains is based on the number and size of plasmids, with strains from identical sources showing identical plasmid profiles. Plasmids themselves or chromosomal DNA may also be typed by means of restriction endonuclease digestion patterns. Restriction enzymes recognize specific nucleotide sequences in DNA and produce double-stranded cleavages that

---

BOX 7-1   **REPORTABLE DISEASES IN TEXAS***

Diseases to be Reported Immediately by Telephone

Botulism, foodborne	Cholera	Diphtheria	*H. influenzae,* invasive infections	Measles (Rubeola)	Meningococcal invasive infections
Pertussis	Poliomyelitis, acute paralytic	Rabies in humans	Tuberculosis	Viral hemorrhagic fevers	Yellow fever
Plague					

Other Reportable Diseases/Syndromes

Acquired immunodeficiency syndrome (AIDS)	Human immunodeficiency virus (HIV) infection
Amebiasis	Legionellosis
Anthrax	Listeriosis
Botulism, adult and infant	Lyme disease
Brucellosis	Malaria
Campylobacteriosis	Meningitis (specify type)
Chancroid	Mumps
Chickenpox	Relapsing fever
Chlamydia trachomatis infections	Rocky Mountain Spotted Fever
Cryptosporidium infections	Rubella
Dengue	Salmonellosis, including typhoid fever
Encephalitis (specify etiology)	Shigellosis
Ehrlichiosis	Streptococcal disease, invasive group A
Escherichia coli 0157:H7	Syphilis
Gonorrhea	Tetanus
Hansen's disease (leprosy)	Trichinosis
Hantavirus infection	Typhus
Hemolytic uremic syndrome (HUS)	Vibrio infections
Hepatitis, acute viral (specify type)	

*In addition to individual case reports, any outbreak, exotic disease, or unusual group expression of disease that may be of public health concern should be reported by the most expeditious means.

break the DNA into smaller fragments. The fragments of various sizes are separated using gel electrophoresis based on molecular weight. The specific recognition sequence and cleavage site have been defined for a great many of these enzymes.

Modifications of the basic restriction endonuclease technique have been developed to reduce the number of bands generated to less than 20 in an attempt to make the gels easier to interpret. These include pulsed-field gel electrophoresis (PFGE) and hybridization of ribosomal RNA with short fragments of DNA. Plasmid restriction digests have been used to type *S. aureus* and coagulase-negative staphylococci, and PFGE is the preferred method for typing enterococci, enteric gram-negative rods, and other gram-negative rods.

Other molecular methods, such as PCR (polymerase chain reaction), are used in conjunction with these methods for strain typing. All molecular methods are discussed in more detail in Chapter 14.

PREVENTING NOSOCOMIAL INFECTIONS

The CDC published guidelines in the 1970s specifying isolation precautions in hospitals. Techniques for isolation precautions included (1) health care workers washing their hands between caring for different patients, (2) segregation of infected patients in private rooms or cohorting of patients (placing patients with the same clinical syndrome in semiprivate rooms) if private rooms are not available, (3) wearing of masks, gowns, and gloves when caring for infected patients, (4) bagging of contaminated articles, such as bed linens, when removed from the room, (5) cleaning of all isolation rooms after the patient is discharged, and (6) placement of cards on the patient's door specifying the type of isolation and instructions for visitors and

nursing staff. Categories of isolation were also established and included (1) strict isolation for highly contagious diseases such as chicken pox, pneumonic plague, and Lassa fever; (2) respiratory isolation for diseases such as measles or *Haemophilus influenzae* or *Neisseria meningitidis;* (3) enteric precautions for diseases such as amebic dysentery, *Salmonella,* and *Shigella;* (4) contact isolation for patients infected with multidrug-resistant bacteria; (5) AFB (tuberculosis) isolation for persons with *M. tuberculosis;* (6) drainage and secretion precautions for persons with conjunctivitis and burns; and (7) blood and body fluid precautions for individuals with acquired immunodeficiency syndrome (AIDS). Over time, a system of disease-specific precautions was added to the category-specific ones, and hospitals were given the option of using one of the two systems. Disease-specific precautions were more cost effective, as *only* those precautions specifically necessary were utilized to interrupt transmission of that one disease.

In 1996 the CDC developed a new system of **Standard Precautions** synthesizing the features of universal precautions (described in Chapter 2) and body substance isolation. Standard precautions are used in the care of *all* patients and apply to blood; all body fluids, secretions, and excretions *except sweat,* regardless of whether they contain visible blood; nonintact skin; and mucous membranes.

In addition, **transmission-based precautions** are used for patients known (or suspected) to be infected with pathogens spread by airborne or droplet transmission or by contact with dry skin or fomites. Box 7-2 lists infection control measures for standard precautions. Table 7-1 lists the infectious agents or syndromes along with the respective infection control measures for each transmission-based precaution. Many infection control practitioners find these guide-

---

BOX 7-2 — **INFECTION CONTROL MEASURES FOR STANDARD PRECAUTIONS**

- Health care workers (HCWs) should wash hands frequently using a plain soap except in special circumstances, for example, preoperatively or after handling dressings from patients on contact isolation.
- HCWs should wear gloves when touching blood, body fluids, secretions, excretions, and contaminated items.
- HCWs should wear a mask, gown, eye protection, or face shield as appropriate.
- Each hospital should ensure that it has adequate procedures for routine care and cleaning and disinfection of environmental surfaces, beds, bed rails, and bedside equipment.
- Hospitals should handle, transport, and launder used linen soiled with blood, body fluids, secretions, and excretions in a manner that prevents skin and mucous membrane exposure and contamination of clothing, and that avoids transfer of microorganisms to other patients or the environment.
- HCWs should take care to prevent injuries when using needles, scalpels, and other sharp instruments or devices.
- HCWs should use equipment, such as mouthpieces and resuscitation bags, instead of mouth-to-mouth resuscitation.
- HCWs should refrain from handling patient care equipment if they have exudative lesions or weeping dermatitis.
- Hospitals should place incontinent or nonhygienic patients in a private room.
- Hospitals should ensure that reusable equipment is properly sterilized.
- Hospitals should ensure that single-use items are discarded properly.

Modified from Hospital Infection Control Practices Advisory Committee, 1996.

TABLE 7-1 TRANSMISSION-BASED PRECAUTIONS

TYPE OF PRECAUTION	SPECIFIC ETIOLOGIC AGENTS OR SYNDROMES	INFECTION CONTROL MEASURE TO BE UNDERTAKEN BY HOSPITAL
Airborne	Measles Varicella Tuberculosis	1. Place patient in private room that has monitored negative air pressure, 6-12 air changes per hour, and appropriate discharge of air outdoors *or* monitored HEPA filtration of room air before air is circulated to other areas of the hospital *or* cohorting of patients, that is, placing patients with the same infection in the same room, if private rooms are not available 2. Health care workers (HCWs) to wear respiratory protection when entering room of patient with known or suspected tuberculosis and, if not immune, for patients with measles or varicella as well 3. Transport patients out of their room only after placement of a surgical mask
Droplet	Invasive *Haemophilus influenzae* type b infection, including meningitis, pneumonia, epiglottitis, and sepsis Invasive *Neisseria meningitidis* infection, including meningitis, pneumonia, and sepsis Diphtheria (pharyngeal) *Mycoplasma pneumoniae* Pertussis Pneumonic plague Streptococcal pharyngitis, pneumonia, or scarlet fever in infants and young children Adenovirus, influenza virus Mumps Parvovirus B19 Rubella	1. Place patient in private room without special air handling or ventilation *or* cohort patients 2. HCW should wear mask when working within 3 feet of patient 3. Transfer patients out of their room only after placement of a surgical mask
Contact	Gastrointestinal, respiratory, skin or wound infections, or colonization with multidrug-resistant bacteria *Clostridium difficile* For diapered or incontinent patients: *Escherichia coli* 0157:H7, *Shigella,* hepatitis A virus, or rotavirus Respiratory syncytial virus, parainfluenza virus, and enterovirus infections in infants and young children Skin infections such as diphtheria (cutaneous), herpes simplex virus (neonatal or mucocutaneous), impetigo, major abscesses, cellulitis, or decubiti, pediculosis (lice infestation), scabies (mite infestation), staphylococci furunculosis (boils) in infants and young children, zoster (disseminated or in the immunocompromised	1. Place patient in private room without special air handling or ventilation *or* cohort patients 2. HCW should wear gloves when entering patient's room 3. HCWs should wash hands with a special antimicrobial agent or a waterless antiseptic agent 4. HCWs should wear a mask and eye protection during activities that are likely to generate splashes of blood, body fluids, secretions, and excretions

Continued

	TABLE 7-1 TRANSMISSION-BASED PRECAUTIONS—CONT'D	
TYPE OF PRECAUTION	**SPECIFIC ETIOLOGIC AGENTS OR SYNDROMES**	**INFECTION CONTROL MEASURE TO BE UNDERTAKEN BY HOSPITAL**
	host), viral hemorrhagic conjunctivitis, viral hemorrhagic infections (Ebola, Lassa, or Marburg)	5. HCWs should wear a gown during procedures likely to generate splashes 6. HCWs should ensure reusable equipment is properly sterilized 7. HCWs should ensure that single-use items are properly discarded

Modified from Hospital Infection Control Practices Advisory Committee, 1996.

lines a lot less cumbersome to implement than the old category- and disease-specific measures. Hospitals, however, may modify these guidelines to fit their individual situations as long as their number of nosocomial infections remains low.

SURVEILLANCE CULTURES

Routine environmental cultures in the hospital are now considered to be of little use and should *not* be performed unless there are specific epidemiologic reasons. The decision to perform these cultures should be determined by the microbiologist, infection control practitioner, and hospital epidemiologist. For example, *Legionella* spp. are frequently isolated from environmental sources without any history of causing human disease. Thus, routine cultures of cooling towers or hot water systems would not be productive. However, in the case of a cluster of respiratory infections caused by *Legionella* in patients in a certain hospital wing, the Infection Control Committee might authorize cultures of hot water faucets and showerheads to try to isolate the same strain of *Legionella* from the patient and the environment.

Routine surveillance of air handlers, food utensils, food equipment surfaces, and respiratory therapy equipment is no longer recommended; neither is monitoring infant formulas prepared in-house nor items purchased as sterile. A better approach is for the infection control team to monitor patients for the development of nosocomial infections that might be related to the use of contaminated commercial products. In the event of an outbreak or an incident related to suspected contamination, a microbiologic study would be indicated. However, most often, such infections are actually caused by in-use contamination, rather than contamination during the manufac-

turing process. Suspect lots of fluid and catheter trays should be saved, and the U.S. Food and Drug Administration should be notified if contamination of an unopened product is suspected.

Although some institutions still require preemployment stool cultures and ova and parasite examinations on food handlers, most now recognize that this is of limited value. It is much more important for food handlers to submit specimens for these tests if they develop diarrhea. Similarly, most hospitals no longer screen personnel routinely for nasal carriage of *S. aureus*. Although a significant percentage of the general population, including hospital personnel, are known to carry this organism, most individuals rarely shed enough organism to pose a hazard and there is no simple way to predict which nasal carriers will disseminate staphylococci.

All steam and dry-heat sterilizers and ethylene oxide gas sterilizers should be checked at least once each week with a liquid spore suspension. Water and dialysis fluids used for hemodialysis should be cultured once each month using a procedure similar to the one given in Procedure 7-1. If bacteria are present at levels exceeding acceptable limits, the entire system should be disinfected and retested. In addition, hemodialysis water should be tested for the presence of endotoxin (which can cause septic shock) if dialyzer membranes are reused. Physical rehabilitation centers often culture hydrotherapy equipment (whirlpools) quarterly to verify that cleaning methods are adequate; some centers culture more frequently.

Hospitals that perform bone marrow transplants or treat hematologic malignancies may also conduct surveillance cultures of severely immunocompromised patients who occupy laminar flow rooms. In these instances, isolation of specific organisms may have predictive value for subsequent systemic infection.

Procedure

7-1 CULTURING HEMODIALYSIS WATER AND DIALYSIS FLUID

The Millipore Corp., Bedford, Mass., manufactures a Total-Count Sampler, for sampling bacteria in water (Figure 7-1). The sampler is a self-contained device that consists of a paddle containing a counting grid overlaid with nutrient agar; the paddle fits into a plastic case. Before sampling, the paddle is removed from the case and placed on sterile gauze, and liquid is added to the case up to the top line marked on the case. The paddle is then replaced in the case, which is shaken several times and then held still for 30 seconds. During this time, 1 mL of fluid is absorbed onto the paddle. The paddle is then removed from the case, the case is emptied, the paddle is reinserted in the case, and the sampler is incubated, grid side up, at 35° C for 24 hours.

HEMODIALYSIS WATER

The sample should be collected at a point where water enters the proportioner or where it is placed in mixing tanks depending on the type of system used for preparing dialysis fluids. Total viable counts should not exceed 200 colonies/mL to be acceptable.

DIALYSIS FLUID

Dialysis fluid should be diluted 1:10 by filling the case with dialysate up to the lower line (1:8 mL) and filling the case to the upper line with sterile, distilled water (18 mL). Samples should be collected at the end of the dialysis treatment. In single-pass systems, samples should be collected at the point where dialysis fluid leaves the dialyzer. In recirculating systems, samples should be collected at the periphery of the recirculating canister containing the coil dialyzer. Total viable counts should not exceed 2000 colonies/mL.

Bibliography

Banerjee, S.N., Emori, T.G., Culver, D.H., et al. and the National Nosocomial Infection Surveillance System. 1991. Secular trends in nosocomial primary bloodstream infections in the United States, 1980-1989. Am. J. Med. 91(suppl. 3B):86S.

Craven, D.E., Steger, K.A., and Barber, T.W. 1991. Preventing nosocomial pneumonia: state of the art and perspectives for the 1990s. Am. J. Med. 91(suppl. 3B):44S.

Emori, T.G. and Gaynes, R.P. 1993. An overview of nosocomial infections, including the role of the microbiology laboratory. Clin. Microbiol. Rev. 6:428.

Garibaldi, R.A., Cushing, D., and Lerer, T. 1991. Risk factors for postoperative infection. Am. J. Med. 91(suppl. 3B):158S.

Garner, J.S. and Favero, M.S. 1985. Guideline for handwashing and hospital environmental control, 1985. PB85-923404. Centers for Disease Control, Atlanta.

Garner, J.S. and Simmons, B.P. 1983. CDC guideline for isolation precautions in hospitals. PB85-923401. Centers for Disease Control, Atlanta.

Garner, J.S. 1985. Guideline for prevention of surgical wound infections, 1985. PB85-923403. Centers for Disease Control, Atlanta.

Hospital Infection Control Practices Advisory Committee. 1994. Guideline for prevention of nosocomial pneumonia. PB95-176970. Centers for Disease Control and Prevention, Atlanta.

Hospital Infection Control Practices Advisory Committee. 1995. Recommendations for preventing the spread of vancomycin resistance. Infect. Control Hosp. Epidemiol. 16:105.

Hospital Infection Control Practices Advisory Committee. 1996. Guideline for isolation precaution in hospitals. Infect. Control Hosp. Epidemiol. 17:53.

Jewett, J.K., Reid, D.E., Safon, L.E., and Easterday, C.L. 1968. Childbed fever: a continuing entity. J.A.M.A. 206:344.

McGowan, J.E., Jr. and Weinstein, R.A. 1992. The role of the laboratory in control of nosocomial infection. In Bennett J.V. and Brachman, P.S., editors. Hospital infections, ed 3. Little, Brown, & Co., Boston.

Nichols, R.L. 1991. Surgical wound infection. Am. J. Med. 91(suppl. 3B):54S.

Simmons, B.P., Hooton, T.M., Wong, E.S., and Allen, J.R. 1982. Guidelines for prevention of intravascular infections. PB 84-923403, Centers for Disease Control, Atlanta.

Stamm, W.E. 1991. Catheter-associated urinary tract infections: epidemiology, pathogenesis, and prevention. Am. J. Med. 91(suppl. 3B):65S.

Wenzel, R.P. editor. 1993. Prevention and control of nosocomial infections, ed 2. Williams & Wilkins, Baltimore.

Williams, W.W. 1983. Guideline for infection control in hospital personnel. PB85-92402. Centers for Disease Control, Atlanta.

Wong, E.S. and Hooton, T.M. 1982. Guideline for prevention of catheter-associated urinary tract infections. PB84-923402. Centers for Disease Control, Atlanta.

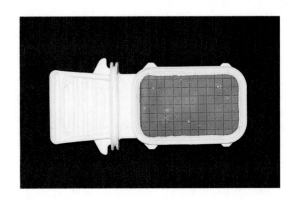

FIGURE 7-1 Millipore Total Count Water Tester, Millipore Corp., Bedford, Mass.

Part Two

Scientific and Laboratory Basis for Clinical Microbiology

CHAPTER

8 | MICROBIAL TAXONOMY

Taxonomy is defined as an area of biologic science comprising three distinct, but highly interrelated, disciplines that include classification, nomenclature, and identification. Applied to all living entities, taxonomy provides a consistent means to classify, name, and identify organisms. This consistency allows biologists worldwide to use a common label for every organism they study within their particular disciplines. The common language that taxonomy provides minimizes confusion about names and allows attention to center on more important scientific issues and phenomena. The importance of this contribution is not only realized in phylogeny (the evolutionary history of organisms) but in virtually every other discipline of biology, including microbiology.

In diagnostic microbiology, classification, nomenclature, and identification of microorganisms play a central role in providing accurate and timely diagnosis of infectious diseases. Because of taxonomy's important contribution to diagnostic microbiology, some detailed discussion of the three areas that make up taxonomy follows.

CLASSIFICATION

Classification is the organization of microorganisms that share similar morphologic, physiologic, and genetic traits into specific groups, or **taxa.** The classification system is hierarchic and consists of the following taxa designations:

- **Species**
- **Genus** (composed of similar species)
- **Family** (composed of similar genera)
- **Order** (composed of similar families)
- **Class** (composed of similar orders)
- **Division** (composed of similar classes)
- **Kingdom** (composed of similar divisions)

SPECIES

Species is the most basic taxonomic group and can be defined as a collection of bacterial strains that share many common physiologic and genetic features and as a group differ notably from other bacterial species.

Occasionally, taxonomic subgroups within a species, called **subspecies,** are recognized. Furthermore, designations such as biotype, serotype, or phagotype may be given to groups below the subspecies levels that share specific, but relatively minor, characteristics. Although these subgroups may have some taxonomic importance and occasional practical utility, their usefulness in diagnostic microbiology is usually limited.

GENUS

Genus (pl. genera) is the next higher taxon and comprises different species that have several important features in common but differ sufficiently to still maintain their status as individual species. All bacterial species belong to a genus, and relegation of a species to a particular genus is based on various genetic and phenotypic characteristics shared among the species. However, microorganisms do not possess the multitude of physical features exhibited by higher organisms such as plants and animals. For instance, they rarely leave any fossil record and they exhibit a tremendous capacity to intermix genetic material among supposedly unrelated species and genera. For these reasons the ability to confidently establish microorganism relatedness so they may be classified in higher taxa beyond the genus level is difficult. Therefore, although grouping similar genera into common families and similar families into common orders and so on is used for classification of plants and animals, these higher taxa designations (that is, division, class, order, family) are not usually useful for classifying bacteria.

OTHER BACTERIAL GROUPINGS

Instead of higher taxa designations descriptive and convenient groupings of bacteria frequently are used. This is illustrated in Table 8-1, which provides an abbreviated classification scheme of clinically relevant bacteria. The table demonstrates how bacteria are grouped into descriptive sections and how taxa beyond the genus level, such as family and order, are not assigned for several organism groups. More extensive species lists are provided in the chapters in Part 4. Fortunately, the difficulty of assigning higher

TABLE 8-1 CLASSIFICATION SCHEME FOR SELECTED CLINICALLY RELEVANT BACTERIA

GROUP DESIGNATION/ DESCRIPTION	FAMILY	GENUS	SPECIES EXAMPLES
Section 1			
The spirochetes (Order Spirochaetales)	*Spirochaetaceae*	*Treponema*	*Treponema pallidum*
		Borrelia	*Borrelia burgdorferi*
	Leptospiraceae	*Leptospira*	*Leptospira interrogans*
Section 2			
Aerobic-microaerophilic, motile, helical-vibroid, gram-negative bacteria	None assigned	*Campylobacter*	*Campylobacter jejuni*
Section 3			
Gram-negative, aerobic rods and cocci	*Pseudomonadaceae*	*Pseudomonas*	*Pseudomonas aeruginosa*
		Stenotrophomonas	*Stenotrophomonas maltophilia*
	Legionellaceae	*Legionella*	*Legionella pneumophila*
	Neisseriaceae	*Neisseria*	*Neisseria gonorrhoeae*
		Moraxella	*Moraxella catarrhalis*
		Acinetobacter	*Acinetobacter anitratus*
		Kingella	*Kingella kingae*
	None assigned	*Flavobacterium*	*Flavobacterium meningosepticum*
	None assigned	*Alcaligenes*	*Alcaligenes odorans*
	None assigned	*Brucella*	*Brucella abortus*
	None assigned	*Bordetella*	*Bordetella pertussis*
	None assigned	*Francisella*	*Francisella tularensis*
Section 5			
Facultatively anaerobic, gram-negative rods	*Enterobacteriaceae*	*Escherichia*	*Escherichia coli*
		Shigella	*Shigella sonnei*
		Salmonella	*Salmonella typhi*
		Citrobacter	*Citrobacter freundii*
		Klebsiella	*Klebsiella pneumoniae*
		Enterobacter	*Enterobacter cloacae*
		Serratia	*Serratia marcescens*
		Hafnia	*Hafnia alvei*

TABLE 8-1 CLASSIFICATION SCHEME FOR SELECTED CLINICALLY RELEVANT BACTERIA—CONT'D

GROUP DESIGNATION/ DESCRIPTION	FAMILY	GENUS	SPECIES EXAMPLES
Section 5—cont'd			
		Proteus	*Proteus mirabilis*
		Providencia	*Providencia rettgeri*
		Morganella	*Morganella morganii*
		Yersinia	*Yersinia pestis*
		Edwardsiella	*Edwardsiella tarda*
	Vibrionaceae	*Vibrio*	*Vibrio cholerae*
		Aeromonas	*Aeromonas sobria*
		Plesiomonas	*Plesiomonas shigelloides*
	Pasteurellaceae	*Pasteurella*	*Pasteurella multocida*
		Haemophilus	*Haemophilus influenzae*
		Actinobacillus	*Actinobacillus actinomycetem-comitans*
	None assigned	*Chromobacterium*	*Chromobacterium violaceum*
	None assigned	*Cardiobacterium*	*Cardiobacterium hominis*
	None assigned	*Gardnerella*	*Gardnerella vaginalis*
	None assigned	*Eikenella*	*Eikenella corrodens*
	None assigned	*Streptobacillus*	*Streptobacillus moniliformis*
Section 6			
Anaerobic, gram-negative straight, curved, and helical rods	*Bacteroidaceae*	*Bacteroides*	*Bacteroides fragilis*
		Porphyromonas	*Porphyromonas asaccharolytica*
		Prevotella	*Prevotella disiens*
		Bilophila	*Bilophila wadsworthia*
		Fusobacterium	*Fusobacterium nucleatum*
Section 8			
Anaerobic, gram-negative cocci	*Veillonellaceae*	*Veillonella*	*Veillonella parvula*
Section 9			
The rickettsias and chlamydias			
Order Rickettsiales	*Rickettsiaceae*	*Rickettsia*	*Rickettsia rickettsii*

Continued

TABLE 8-1 CLASSIFICATION SCHEME FOR SELECTED CLINICALLY RELEVANT BACTERIA—CONT'D

GROUP DESIGNATION/ DESCRIPTION	FAMILY	GENUS	SPECIES EXAMPLES
Section 9—cont'd			
		Coxiella	*Coxiella burnetii*
		Ehrlichia	*Ehrlichia chaffeensis*
	Uncertain	*Bartonella*	*Bartonella henselae*
Order Chlamydiales	*Chlamydiaceae*	*Chlamydia*	*Chlamydia trachomatis*
Section 10			
The mycoplasmas Order Mycoplasmatales	*Mycoplasmataceae*	*Mycoplasma*	*Mycoplasma pneumoniae*
		Ureaplasma	*Ureaplasma urealyticum*
Section 12			
Gram-positive cocci	*Micrococcaceae*	*Micrococcus*	*Micrococcus luteus*
		Stomatococcus	*Stomatococcus mucilaginosus*
		Staphylococcus	*Staphylococcus aureus*
	None assigned	*Streptococcus*	*Streptococcus pneumoniae*
	None assigned	*Enterococcus*	*Enterococcus faecalis*
	None assigned	*Leuconostoc*	*Leuconostoc mesenteroides*
	None assigned	*Gemella*	*Gemella haemolysans*
	None assigned	*Pediococcus*	*Pediococcus acidilactici*
	None assigned	*Aerococcus*	*Aerococcus urinae*
	None assigned	*Peptococcus*	*Peptococcus niger*
	None assigned	*Peptostreptococcus*	*Peptostreptococcus hydrogenalis*
Section 13			
Spore-forming, gram-positive rods	None assigned	*Bacillus*	*Bacillus anthracis*
	None assigned	*Clostridium*	*Clostridium botulinum*
Section 14			
Regular, nonspore-forming, gram-positive rods	None assigned	*Lactobacillus*	*Lactobacillus casei*
	None assigned	*Listeria*	*Listeria monocytogenes*
	None assigned	*Erysipelothrix*	*Erysipelothrix rhusiopathiae*

TABLE 8-1   CLASSIFICATION SCHEME FOR SELECTED CLINICALLY RELEVANT BACTERIA—CONT'D

GROUP DESIGNATION/ DESCRIPTION	FAMILY	GENUS	SPECIES EXAMPLES
Section 15			
Irregular, nonspore-forming, gram-positive rods	None assigned	*Corynebacterium*	*Corynebacterium diphtheriae*
	None assigned	*Arcanobacterium*	*Arcanobacterium haemolyticum*
	None assigned	*Propionibacterium*	*Propionibacterium acnes*
	None assigned	*Actinomyces*	*Actinomyces israelii*
	None assigned	*Eubacterium*	*Eubacterium lentum*
	None assigned	*Bifidobacterium*	*Bifidobacterium dentium*
Section 16			
The mycobacteria	None assigned	*Mycobacterium*	*Mycobacterium tuberculosis*
Section 17			
The nocardioforms	None assigned	*Nocardia*	*Nocardia asteroides*
	None assigned	*Rhodococcus*	*Rhodococcus equi*
	None assigned	*Oerskovia*	*Oerskovia turbata*

Modified from Krieg, N.R. and Holt, J.G., editors. 1984. Bergey's manual of systematic bacteriology, vol 1. Williams & Wilkins, Baltimore, and Sneath, P.H.A., Mair, N.S., Sharpe, M.E., and Holt, J.G., editors. 1986. Bergey's manual of systematic bacteriology, vol 2. Williams & Wilkins, Baltimore.

classifications to microorganism groups is not of great consequence in diagnostic microbiology. Genus and species designations usually suffice for most clinical needs.

NOMENCLATURE

Nomenclature, the naming of microorganisms according to established rules and guidelines, provides the accepted labels by which organisms are universally recognized. Because genus and species are the groups of most concern to microbiologists, the discussion of rules governing microbial nomenclature will be limited to these two taxa designations. In this binomial ("two-name") system of nomenclature, every organism is assigned a genus and species name of Latin or Greek derivation. In other words, every organism has a scientific "label" consisting of two parts: the genus designation, which is always capitalized, and the species designation, which is never capitalized. Both

components are always used simultaneously and are printed in italics, or underlined in script. For example, the streptococci include *Streptococcus pneumoniae*, *Streptococcus pyogenes*, *Streptococcus agalactiae*, and *Streptococcus bovis*. Alternatively, the name may be abbreviated by using the upper case form of the first letter of the genus designation followed by a period (.) and the full species name, which is never abbreviated, such as *S. pneumoniae*, *S. pyogenes*, *S. agalactiae*, and *S. bovis*. Frequently an informal designation (e.g., staphylococci, streptococci, enterococci) may be used to label a particular group of organisms, but such designations are not capitalized or italicized. Examples of genus and species names for several other clinically relevant bacteria are provided in Table 8-1.

As more information is gained regarding organism classification and identification, a particular species may be moved to a different genus or assigned a new genus name. The rules and criteria for these changes are beyond the scope of this chapter, but such changes are documented in the *International Journal for Systematic Bacteriology*. In the diagnostic laboratory, changes in nomenclature are phased in gradually so that physicians and laboratorians have ample opportunity to recognize that a familiar pathogen has been given a new name. This is usually accomplished by using the new genus designation while continuing to provide the previous designation in parentheses, for example, *Stenotrophomonas (Xanthomonas) maltophilia* or *Burkholderia (Pseudomonas) cepacia*.

IDENTIFICATION

Microbial **identification** is the process by which a microorganism's key features are delineated. Once those features are established, the profile is compared with those of other previously characterized microorganisms so that the organism in question can be classified within the most appropriate taxa (classification) and can be assigned an appropriate genus and species name (nomenclature); both are essential aspects of the role taxonomy plays in diagnostic microbiology and infectious diseases (Box 8-1).

IDENTIFICATION METHODS

A wide variety of methods and criteria are used to establish a microorganism's identity. These methods usually can be separated into either of two general categories: genotypic characteristics and phenotypic characteristics. **Genotypic characteristics** relate to an organism's genetic makeup, including the nature of the organism's genes and constituent nucleic acids (see Chapter 9 for more information about microbial genetics). **Phenotypic characteristics** are based on features beyond the genetic level and include readily observable characteristics and those characteristics that may require extensive analytic

procedures to be detected. Examples of characteristics used as criteria for bacterial identification and classification are provided in Table 8-2. Modern microbial taxonomy usually employs a combination of several methods so that the microorganism is characterized as completely as possible and the most appropriate classification and naming of the organism is accomplished.

Although the criteria and examples given in Table 8-2 are in the context of microbial identification for classification purposes, the principles and practices of classification exactly parallel the approaches used in diagnostic microbiology for the identification and characterization of microorganisms encountered in the clinical setting. Fortunately, because of the previous efforts and accomplishments of microbial taxonomists, microbiologists do not have to use several burdensome classification and identification schemes to identify infectious agents. Instead, microbiologists extract key organism features on which to base their identification methods so that organism identity can be made in a timely and clinically useful manner. In a sense, diagnostic microbiology uses streamlined versions of the phenotypic and genotypic approaches employed by taxonomists to identify microorganisms. This should not be taken to mean that the identification of all clinically relevant organisms is easy and straightforward, and this is also not meant to imply that microbiologists can only identify or recognize organisms that have already been characterized and named by taxonomists. Indeed, the clinical microbiology laboratory is well recognized as the place where previously unknown or uncharacterized infectious agents are initially encountered and, as such, has an ever-increasing responsibility to be the sentinel for emerging etiologies of infectious diseases.

BOX 8-1   **ROLE OF TAXONOMY IN DIAGNOSTIC MICROBIOLOGY AND INFECTIOUS DISEASES**

Establishes and maintains records of key characteristics of clinically relevant microorganisms
Facilitates communication among technologists, microbiologists, and physicians by assigning universally useful names to clinically relevant microorganisms. This is essential for:

- ■ Establishing an association of particular diseases or syndromes with specific microorganisms
- ■ Accumulating knowledge regarding the management and outcome of diseases associated with specific microorganisms
- ■ Establishing patterns of resistance to antimicrobial agents and recognition of changing microbial resistance patterns
- ■ Understanding the mechanisms of resistance and detecting new resistance mechanisms exhibited by a particular microorganism
- ■ Recognizing new and emerging pathogenic microorganisms
- ■ Recognizing changes in the types of infections or diseases caused by familiar microorganisms
- ■ Designing alterations of available technologies for the development of new methods that optimize the detection and identification of infectious agents and optimize detection of microbial resistance to antiinfective agents
- ■ Developing new antiinfective therapies

TABLE 8-2 IDENTIFICATION CRITERIA AND CHARACTERISTICS FOR MICROBIAL CLASSIFICATION

PHENOTYPIC CRITERIA EXAMPLES	PRINCIPLES
Macroscopic morphology	Characteristics of microbial growth patterns on artificial media as observed when inspected with the unaided eye. Examples of such characteristics include the size, texture, and pigmentation of bacterial colonies
Microscopic morphology	Size, shape, intracellular inclusions, cellular appendages, and arrangement of cells when observed with the aid of microscopic magnification
Staining characteristics	Ability of organism to reproducibly stain a particular color with the application of specific dyes and reagents. Staining is usually used in conjunction with microscopic morphology as part of the process of bacterial identification. For example, the Gram stain for bacteria is a critical criterion for identification
Environmental requirements	Ability of organism to grow at various temperatures, in the presence of oxygen and other gases, at various pH levels, or in the presence of other ions and salts such as NaCl
Nutritional requirements	Ability of organism to utilize various carbon and nitrogen sources as nutritional substrates when grown under specific environmental conditions
Resistance profiles	Exhibition of a characteristic inherent resistance to specific antibiotics, heavy metals, or toxins by certain microorganisms
Antigenic properties	Establishment of profiles of microorganisms by various serologic and immunologic methods that are useful for determining the relatedness among various microbial groups
Subcellular properties	Establishment of the molecular constituents of the cell that are typical of a particular taxon, or organism group, by various analytic methods. Some examples include cell wall components, components of the cell membrane, and enzymatic content of the microbial cell

GENOTYPIC CRITERIA EXAMPLES	PRINCIPLES
DNA base composition ratio	DNA comprises four bases (guanine, cytosine, adenine, and thymine). The extent to which the DNA from two organisms is made up of cytosine and guanine (i.e., G + C content) relative to their total base content can be used as an indicator of relatedness, or lack thereof. For example, an organism with a G + C content of 50% will not be closely related to an organism whose G + C content is 25%
Nucleic acid (DNA and RNA) base sequence analysis	The order of bases along a strand of DNA or RNA is known as the base sequence, and the extent to which sequences are similar (homologous) between two microorganisms can be determined directly or indirectly by various molecular methods. The degree of similarity in the sequences is a measure of the degree of organism relatedness

Bibliography

Balows, A., Turper, H.G., Dworkin, M., et al., editors. 1992. The prokaryotes. A handbook on the biology of bacteria: ecophysiology, isolation, identification, applications, vol 1-4. Springer Verlag, N.Y.

Brock, T.D., Madigan, M.T., Martinko, J.M., and Parker, J., editors. 1994. Biology of microorganisms. Prentice Hall, Englewood Cliffs, N.J.

Krieg, N.R. and Holt, J.G., editors. 1984. Bergey's manual of systematic bacteriology, vol 1. Williams & Wilkins, Baltimore.

Sneath, P.H.A., Mair, N.S., Sharpe, M.E., and Holt, J.G., editors. 1986. Bergey's manual of systematic bacteriology, vol 2. Williams & Wilkins, Baltimore.

9 | BACTERIAL GENETICS, METABOLISM, AND STRUCTURE

Microbial genetics, metabolism, and structure are the keys to microbial viability and survival. These processes involve numerous pathways that are widely varied, often complicated, and frequently interactive. Essentially, survival requires energy to fuel the synthesis of materials necessary for growth, propagation, and carrying out all other metabolic processes (Figure 9-1). Although the goal of survival is the same for all organisms, the strategies microorganisms use to accomplish this vary substantially.

Knowledge regarding genetic, metabolic, and structural characteristics of microorganisms provides the basis for understanding almost every aspect of diagnostic microbiology, including:

- The mechanism by which microorganisms cause disease
- Developing and implementing optimum techniques for microbial detection, cultivation, identification, and characterization
- Understanding antimicrobial action and resistance
- Developing and implementing tests for antimicrobial resistance detection
- Designing strategies for disease therapy and control

Microorganisms vary greatly in many genetic and physiologic aspects, and a detailed consideration of all these differences is beyond the scope of this textbook. Therefore, bacterial systems will be used as a model to discuss microbial genetics, metabolism, and structure. Information regarding characteristics of fungi, viruses, and parasites may be found in the chapters discussing these specific organism groups.

BACTERIAL GENETICS

Genetics, the process of heredity and variation, is the starting point from which all other cellular pathways, functions, and structures originate. The ability of a microorganism to maintain viability, adapt, multiply, and cause disease is founded in genetics. The three major aspects of microbial genetics that require discussion include:

- The structure and organization of genetic material
- Replication and expression of genetic information
- The mechanisms by which genetic information is changed and exchanged among bacteria

NUCLEIC ACID STRUCTURE AND ORGANIZATION

For all living entities hereditary information resides or is encoded in **nucleic acids.** The two major classes include **deoxyribonucleic acid (DNA),** which is the most common macromolecule that encodes genetic information, and **ribonucleic acid (RNA).** In some forms RNA encodes genetic information for various viruses and in other forms plays an essential role in several of the genetic processes to be discussed.

Nucleotide structure and sequence

DNA consists of deoxyribose sugars connected by phosphodiester bonds (Figure 9-2, *A*). The **bases** that are covalently linked to each deoxyribose sugar are the key to the genetic code within the DNA molecule. The four bases include two purines, adenine (A) and guanine (G), and the two pyrimidines, cytosine (C) and thymine (T) (Figure 9-3). In RNA, uracil replaces thymine. Taken together, the sugar, the phosphate, and a base form a single unit referred to as a **nucleotide.** DNA and RNA are nucleotide polymers (i.e., chains or strands), and the order of bases along a DNA or RNA strand is known as the **base sequence.** This sequence provides the information that specifies the proteins that will be synthesized by microbial cells (i.e., the sequence is the **genetic code**).

DNA molecular structure

The intact DNA molecule usually is composed of two nucleotide polymers. Each strand has a 5′ and a 3′ hydroxyl terminus (Figure 9-2, *B*). The two strands run antiparallel, with the 5′ terminus of one strand opposed to the 3′ terminal of the other. The strands are also complementary as the adenine base of one strand always binds, via two hydrogen bonds, to the thymine base of the other strand, or vice versa. Likewise, the guanine base of one strand always binds by three hydrogen bonds to the cytosine base of the

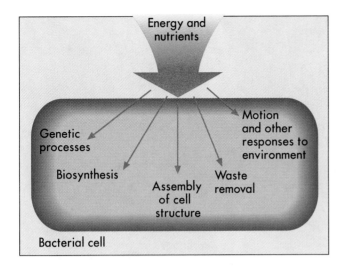

FIGURE 9-1   General overview of bacterial life processes.

other strand, or vice versa. The molecular restrictions of these base pairings, along with the conformation of the sugar-phosphate backbones oriented in antiparallel fashion, result in DNA having the unique structural conformation often referred to as a "twisted ladder" (Figure 9-2, *B*). Additionally, the dedicated base pairs provide the format essential for consistent replication and expression of the genetic code. In contrast to DNA, RNA rarely exists as a double-stranded molecule and, while DNA carries the genetic code, the three major types of RNA (**messenger RNA [mRNA]**, **transfer RNA [tRNA]**, and **ribosomal RNA [rRNA]**), play other key roles in gene expression.

Genes and the genetic code

A DNA sequence that encodes for a specific product (RNA or protein) is defined as a **gene.** Thousands of genes within an organism encode messages or blueprints for production, by gene expression, of specific protein and RNA products that play essential metabolic roles in the cell. All genes taken together within an organism comprise that organism's **genome.** The size of a gene and an entire genome is usually expressed in the number of base pairs (bp) present (e.g., kilobases [10^3 bases], megabases [10^6 bases]).

Certain genes are widely distributed among various organisms while others are limited to particular species. Also, the base pair sequence for individual genes may be highly conserved (i.e., show limited sequence differences among different organisms) or be widely variable. As discussed in Chapter 14, these similarities and differences in gene content and sequences are the basis for the development of molecular tests used to detect, identify, and characterize clinically relevant microorganisms.

Chromosomes

The genome is organized into discrete elements known as **chromosomes.** The set of genes within a given chromosome are arranged in a linear fashion, but the number of genes per chromosome is variable. Similarly, although the number of chromosomes per cell is consistent for a given species, this number varies considerably among species. For example, human cells contain 23 pairs (i.e., diploid) of chromosomes while bacteria contain a single, unpaired (i.e., haploid) chromosome.

The bacterial chromosome contains all genes essential for viability and exists as a double-stranded, closed circular, naked (i.e., not enclosed within a membrane) macromolecule. The molecule is extensively folded and twisted (i.e., supercoiled) so that it may be accommodated within the confines of the bacterial cell. The fact that the linearized, unsupercoiled chromosome of the bacterium *Escherichia coli* is about 1300 μm in length but fits within a 1 μm × 3 μm cell attests to the extreme compactness that the bacterial chromosome must achieve. For genes within the compacted chromosome to be expressed and replicated, unwinding or relaxation of the molecule is essential.

In contrast to the bacterial chromosome, the chromosomes of parasites and fungi number greater than one per cell, are linear, and are housed within a membrane structure known as the **nucleus.** This difference is a major criterion for classifying bacteria as prokaryotic organisms while fungi and parasites are classified as eukaryotes. The genome of viruses may be referred to as a chromosome, but the DNA (or RNA) is contained within a protein coat rather than within a cell.

Nonchromosomal elements of the genome

Interestingly, although the bacterial chromosome represents the majority of the genome, not all genes within a given cell are confined to the chromosome. Many genes are also located on plasmids and transposable elements. Both of these are able to replicate and encode information for the production of various cellular products. Although considered part of the bacterial genome, they are not as stable as the chromosome and may be lost during cellular replication, often without severe detrimental effects on the cell.

Plasmids exist as "miniature" chromosomes in being double-stranded, closed, circular structures with size ranges from 1 to 2 kilobases up to 1 megabase or more. The number of plasmids per bacterial cell varies extensively, and each plasmid is composed of several genes. Some genes encode products that mediate plasmid replication and transfer between bacterial cells, while others encode products that provide a

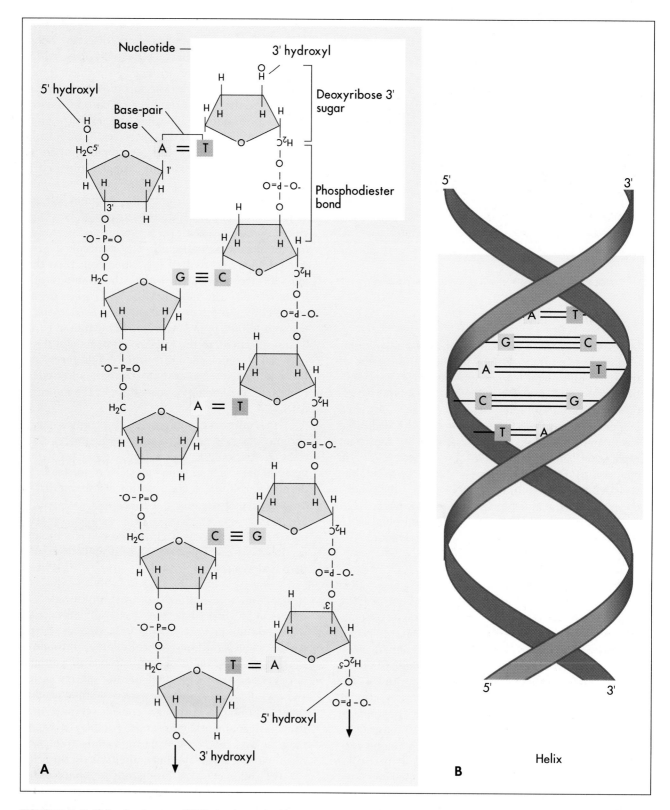

FIGURE 9-2 A, Molecular structure of DNA showing nucleotide structure, phosphodiester bond connecting nucleotides, and complementary pairing of bases (A, adenine; T, thymine; G, guanine; C, cytosine) between antiparallel nucleic acid strands. **B,** 5′ and 3′ antiparallel polarity and helical ("twisted ladder") configuration of DNA.

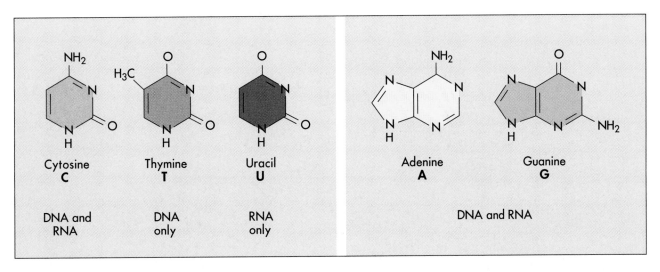

FIGURE 9-3 Molecular structure of nucleic acid bases. Pyrimidines: cytosine, thymine, and uracil. Purines: adenine and guanine.

survival edge such as determinants of antimicrobial resistance. Unlike most chromosomal genes, plasmid genes do not usually encode for products essential for viability. Plasmids, in whole or in part, may also become incorporated in the chromosome.

Transposable elements are pieces of DNA that move from one genetic element to another, from plasmid to chromosome or vice versa. Unlike plasmids, they do not exist as separate entities within the bacterial cell because they must either be incorporated into a plasmid or the chromosome. The two types of transposable elements are **insertion sequences (IS)** and **transposons.** Insertion sequences contain genes that simply encode for information required for movement among plasmids and chromosomes. Transposons contain genes for movement as well as genes that encode for other features such as drug resistance. Plasmids and transposable elements coexist with chromosomes in the cells of many bacterial species. These extra-chromosomal elements play a key role in the exchange of genetic material throughout the bacterial microbiosphere, including genetic exchange among clinically relevant bacteria.

REPLICATION AND EXPRESSION OF GENETIC INFORMATION

Replication

Bacteria multiply by cell division that results in the production of two daughter cells from one parent cell. As part of this process, the genome must be replicated so that each daughter cell receives the same complement of functional DNA. Replication is a complex process that is mediated by various enzymes such as DNA polymerase, and cofactors; replication must occur quickly and accurately. For descriptive purposes replication may be considered in four stages that are depicted together in Figure 9-4:

1. Unwinding or relaxation of the chromosome's supercoiled DNA
2. Unzipping, or disconnecting, the complementary strands of the parental DNA so that each may serve as a template (i.e., pattern) for synthesis of new DNA strands
3. Synthesis of the new DNA strands
4. Termination of replication with release of two identical chromosomes, one for each daughter cell

Relaxation of supercoiled chromosomal DNA is required so that enzymes and cofactors involved in replication can access the DNA molecule at the site where the replication process will originate (i.e., **origin of replication**). On exposure of the replication site (a specific sequence of approximately 300 base pairs that is recognized by several initiation proteins), unzipping of the complementary strands of parental DNA begins. Each parental strand serves as a template for the synthesis of a new complementary daughter strand. The site of active replication is referred to as the **replication fork;** there are two during the replication process. Each replication fork moves through the parent DNA molecule in opposite directions so that replication is a bidirectional process. Activity at each replication fork involves different cofactors and enzymes, with DNA polymerases playing a central role. Using each parental strand as a template, **DNA polymerases** add nucleotide bases to each growing daughter strand in a sequence that is complementary to the base sequence

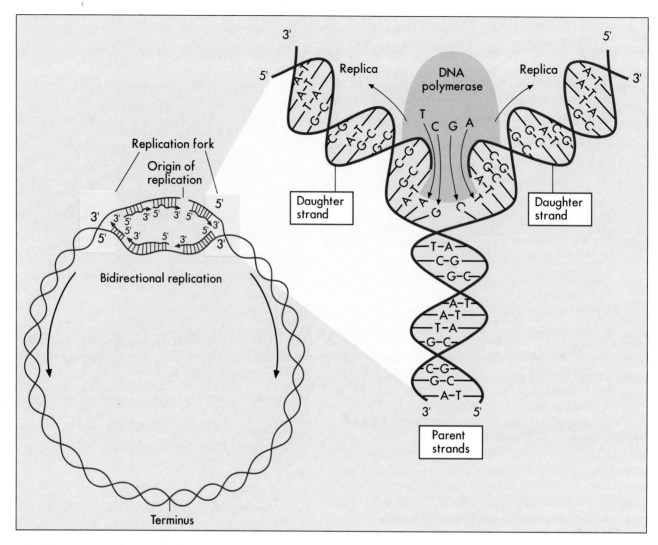

FIGURE 9-4 Bacterial DNA replication depicting bidirectional movement of two replication forks from origin of replication. Each parent strand serves as a template for production of a complementary daughter strand and, eventually, two identical chromosomes.

of the template (parent) strand. The complementary bases of each strand are then crosslinked. The new nucleotides can only be added to the 3' hydroxyl end of the growing strand so that synthesis for each daughter strand only occurs in a 5' to 3' direction.

Termination of replication occurs when the two replication forks meet, resulting in two complete chromosomes, each containing two complementary strands, one of parental origin and one newly synthesized daughter strand. Although the time required for replication can vary among bacteria, the process generally takes approximately 40 minutes in rapidly growing bacteria such as *E. coli*. However, the replication time for a particular bacterial strain can vary depending on environmental conditions such as the availability of nutrients or the presence of toxic substances (e.g., antimicrobial agents).

Expression of genetic information

Gene expression is the processing of information encoded in genetic elements (i.e., chromosomes, plasmids, and transposons) that results in the production of biochemical products. The overall process is composed of two complex steps, transcription and translation, and requires various components, including a DNA template representing a single gene or cluster of genes, various enzymes and cofactors, and RNA molecules of specific structure and function.

TRANSCRIPTION Gene expression begins with transcription, which converts the DNA base sequence of the gene (i.e., the genetic code) into a messenger RNA (mRNA) molecule that is complementary to the gene's DNA sequence (Figure 9-5). Usually only one of the two DNA strands (the sense strand) encodes

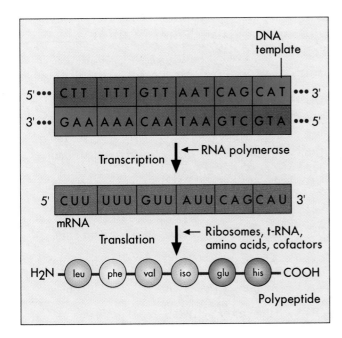

FIGURE 9-5 Overview of gene expression components; transcription for production of mRNA and translation for production of polypeptide (protein).

for a functional gene product, and this same strand is the template for mRNA synthesis.

RNA polymerase is the enzyme central to the transcription process. The enzyme is composed of four protein subunits (α [two copies], β, β′) and a sigma (σ) factor. Sigma factor is loosely affiliated with the enzyme structure and identifies the appropriate site on the DNA template where transcription of mRNA is initiated. This initiation site is also known as the **promoter sequence.** The remainder of the enzyme ($\alpha_2\beta\beta'$) functions to open double-stranded DNA at the promoter sequence and use the DNA strand as a template to sequentially add ribonucleotides (ATP, GTP, UTP, and CTP) to form the growing mRNA strand.

Transcription proceeds in a 5′ to 3′ direction. However, in mRNA the thymine triphosphate (TTP) of DNA is replaced with uracil triphosphate (UTP). Synthesis of the single-stranded mRNA product ends when specific nucleotide base sequences on the DNA template are encountered. In some instances, termination of transcription may be facilitated by a *rho* cofactor, which can disrupt the mRNA-RNA polymerase-template DNA complex.

In bacteria the mRNA molecules that result from the transcription process are **polycistronic,** that is, they encode for several gene products. Frequently, polycistronic mRNA may encode several genes whose products (proteins) are involved in a single or closely related cellular function. When a cluster of genes is under the control of a single promoter sequence, the gene group is referred to as an **operon.**

The transcription process not only produces mRNA but also tRNA and rRNA. All three types of RNA have key roles in protein synthesis.

TRANSLATION   The next phase in gene expression, translation, involves the production of proteins. By this process the genetic code within mRNA molecules is translated into specific amino acid sequences that are responsible for protein structure, and hence, function (Figure 9-5).

Before discussing translation, an understanding of the genetic code that is originally transcribed from DNA to mRNA and then translated from mRNA to protein is warranted. The code consists of triplets of nucleotide bases, referred to as **codons;** each codon encodes for a specific amino acid. Because there are 64 different codons for 20 amino acids, an amino acid can be encoded by more than one codon (Table 9-1). However, each codon specifies only one amino acid. Therefore, through translation the codon sequences in mRNA direct which amino acids are added and in what order. Translation ensures that proteins with proper structure and function are produced. Errors in the process can result in aberrant proteins that are partly or completely unable to function as needed, underscoring the need for translation to be well controlled and accurate.

To accomplish the task of translation, intricate interactions between mRNA, tRNA, and rRNA are required. Sixty different types of tRNA molecules are responsible for transferring different amino acids from intracellular reservoirs to the site of protein synthesis. These molecules, whose structure resembles an inverted t, contain one sequence recognition site for binding to specific three base sequences (codons) on the mRNA molecule (Figure 9-6). A second site binds specific amino acids, the building blocks of proteins. Each amino acid is joined to a specific tRNA molecule via the enzymatic activity of aminoacyl-tRNA synthetases. Therefore, tRNA molecules have the primary function of using the codons of the mRNA molecule as the template for precisely delivering a specific amino acid for polymerization. **Ribosomes,** composed of rRNA and proteins, also are central to translation and provide the site where translation occurs.

Translation, diagrammatically shown in Figure 9-6, involves three steps: initiation, elongation, and termination. Following termination, bacterial proteins often undergo posttranslational modifications as a final step in protein synthesis.

Initiation begins with the association of ribosomal subunits, mRNA, formylmethionine tRNA

TABLE 9-1 THE GENETIC CODE AS EXPRESSED BY TRIPLET BASE SEQUENCES OF mRNA*

CODON	AMINO ACID	CODON	AMINO ACID	CODON	AMINO ACID	CODON	AMINO ACID
UUU	Phenylalanine	CUU	Leucine	GUU	Valine	AUU	Isoleucine
UUC	Phenylalanine	CUC	Leucine	GUC	Valine	AUC	Isoleucine
UUG	Leucine	CUG	Leucine	GUG	Valine	AUG (start)†	Methionine
UUA	Leucine	CUA	Leucine	GUA	Valine	AUA	Isoleucine
UCU	Serine	CCU	Proline	GCU	Alanine	ACU	Threonine
UCC	Serine	CCC	Proline	GCC	Alanine	ACC	Threonine
UCG	Serine	CCG	Proline	GCG	Alanine	ACG	Threonine
UCA	Serine	CCA	Proline	GCA	Alanine	ACA	Threonine
UGU	Cysteine	CGU	Arginine	GGU	Glycine	AGU	Serine
UGC	Cysteine	CGC	Arginine	GGC	Glycine	AGC	Serine
UGG	Tryptophan	CGG	Arginine	GGG	Glycine	AGG	Arginine
UGA	None (stop signal)	CGA	Arginine	GGA	Glycine	AGA	Arginine
UAU	Tyrosine	CAU	Histidine	GAU	Aspartic	AAU	Asparagine
UAC	Tyrosine	CAC	Histidine	GAC	Aspartic	AAC	Asparagine
UAG	None (stop signal)	CAG	Glutamine	GAG	Glutamic	AAG	Lysine
UAA	None (stop signal)	CAA	Glutamine	GAA	Glutamic	AAA	Lysine

*The codons in DNA are complementary to those given here. Thus U here is complementary to the A in DNA, C is complementary to G, G to C, and A to T.

  The nucleotide on the left is at the 5′-end of the triplet.

†AUG codes for *N*-formylmethionine at the beginning of mRNAs of bacteria.

Modified from Brock, T.D., Madigan, M.T., Martinko, J.M., and Parker, J., editors. 1994. Biology of microorganisms. Prentice Hall, Englewood Cliffs, N.J.

(f-met; carrying the initial amino acid of the protein to be synthesized), and various initiation factors (Figure 9-6, *A*). Assembly of the complex begins at a specific 3 to 9 base sequence on the mRNA, referred to as the **ribosomal binding site, or RBS.** After the initial complex is formed, addition of individual amino acids begins.

Elongation involves tRNAs mediating the sequential addition of amino acids in a specific sequence that is dictated by the codon sequence of the mRNA molecule (Figure 9-6, *B* to *C* and Table 9-1). As the mRNA molecule threads through the ribosome in a 5′ to 3′ direction, peptide bonds are formed between adjacent amino acids still bound by their respective tRNA molecules in the P (peptide) and A (acceptor) sites of the ribosome. During the process, the forming peptide is moved to the P site and the most 5′ tRNA is released from the E (exit) site. This movement vacates the A site, which contains the codon specific for the next amino acid, so that the incoming tRNA-amino acid can join the complex (Figure 9-6, *C*).

Because multiple proteins encoded on an mRNA strand can be translated at the same time, multiple ribosomes may be simultaneously associated with one mRNA molecule. Such an arrangement is referred to as a **polysome;** its appearance resembles a string of pearls.

Termination, the final step in translation, occurs when the ribosomal A site encounters a stop or

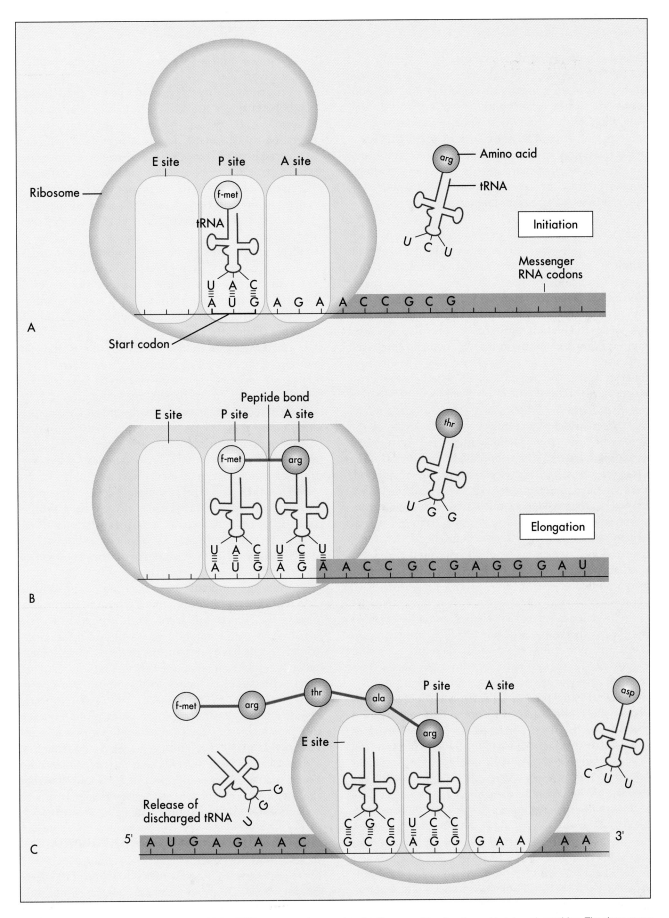

FIGURE 9-6 Overview of translation in which mRNA serves as the template for the assembly of amino acids into polypeptides. The three steps include initiation (**A**), elongation (**B** and **C**), and termination (not shown).

nonsense codon that does not specify an amino acid (i.e., a "stop signal," see Table 9-1). At this point, the protein synthesis complex disassociates and the ribosomes are available for another round of translation. Following termination most proteins must undergo some extent of modification, such as folding or enzymatic trimming, so that either protein function, transportation, or incorporation into various cellular structures can be accomplished. This process is referred to as **posttranslational modification.**

Regulation and control of gene expression

The vital role that gene expression and protein synthesis play in the survival of cells dictates that bacteria judiciously control these processes. So that a physiologic balance is maintained, the cell must regulate gene expression and control the activities of gene products. Regulation and control are also key and highly complex mechanisms by which single-cell organisms are able to respond and adapt to environmental challenges, regardless of whether the challenges occur naturally or result from medical progress (e.g., antibiotics).

Regulation occurs at one of three levels of the gene expression and protein synthesis pathways: transcriptional, translational, and posttranslational. The most common is transcriptional level regulation. Because direct interactions with genes and their ability to be transcribed to mRNA are involved, transcriptional level regulation is also referred to as **genetic level control.** Genes that encode enzymes involved in biosynthesis (**anabolic enzymes**) and genes that encode enzymes for biodegradation (**catabolic enzymes**) will be used as examples of genetic level control.

In general, genes that encode anabolic enzymes for the synthesis of particular products are **repressed** (i.e., are not transcribed and therefore are not expressed) in the presence of those products. This strategy avoids waste and overproduction of products that are already present in sufficient supply. In this system the product acts as a corepressor that forms a complex with a repressor molecule. In the absence of corepressor product (i.e., gene product) transcription occurs (Figure 9-7, A). When present in sufficient quantity the product forms a complex with the repressor. The complex then binds to a specific base region of the gene sequence known as the **operator region** (Figure 9-7, B). This binding blocks RNA polymerase progression from the promoter sequence and inhibits transcription. As the supply of product (corepressor) dwindles, an insufficient amount remains to form a complex with the repressor, the operator region is no longer bound, and transcription of the genes for the anabolic enzymes will commence

and continue until a sufficient supply of end product is again available.

In contrast to repression, genes that encode catabolic enzymes are usually **induced;** that is, transcription only occurs when the substrate to be degraded by enzymatic action is present. Production of degradative enzymes in the absence of substrates would be a waste of cellular energy and resources. When the substrate is absent in an inducible system, a repressor binds to the operator sequence of the DNA and blocks transcription of the genes for the degradative enzymes (Figure 9-7, C). In the presence of an inducer, which often is the target substrate for degradation, a complex is formed between inducer and repressor that results in the release of the repressor from the operator site, thus allowing transcription of the genes encoding for the appropriate catabolic enzymes (Figure 9-7, D).

Certain genes are not regulated, that is, they are not under the control of inducers or repressors. These genes are referred to as **constitutive.** Because they usually encode for products that are essential for viability under almost all growth and environmental conditions, these genes are continuously expressed. Also, not all regulation occurs at the genetic level (i.e., the level of transcription). For example, the production of some enzymes may be controlled at the protein synthesis (i.e., translational) level. The activities of other enzymes that have already been synthesized may be regulated at a posttranslational level, that is, certain catabolic or anabolic metabolites may directly interact with enzymes to either increase or decrease their enzymatic activity.

Among different bacteria and even among different genes within the same bacterium, the mechanisms by which inducers and corepressors are involved in gene regulation vary widely. Furthermore, bacterial cells have mechanisms to detect environmental changes and can generate signals that interact with gene expression machinery so that appropriate products are made in response to the environmental changes. Also, several complex interactions between different regulatory systems are found within a single cell so that many regulation schemes do not function independently. Such diversity and interdependence no doubt is a necessary component of metabolism so that the organism's response to the environment is timely, well coordinated, and appropriate.

GENE EXCHANGE AND GENETIC DIVERSITY

In eukaryotic organisms genetic diversity is achieved by sexual reproduction, which allows the mixing of genomes through genetic exchange. Bacteria multiply by simple cell division in which two daughter cells result by division of one parent cell and each daughter

Repression

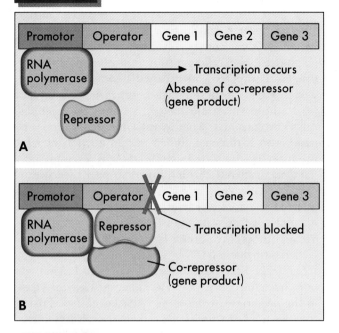

Induction

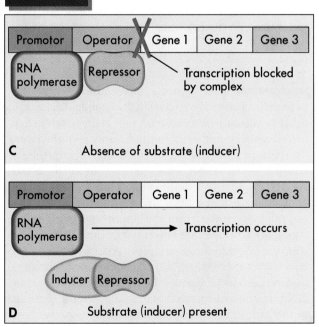

FIGURE 9-7 Transcriptional (i.e., genetic level) control of gene expression. Gene repression is depicted in **A** and **B**; induction is shown in **C** and **D**.

cell receives the full and identical genetic complement contained in the original parent cell. This process does not allow for the mixing of genes from other cells and leaves no means of achieving genetic diversity among bacterial progeny. Without genetic diversity and change, the essential ingredients for evolution

are lost. However, because bacteria have been on earth for billions of years and microbiologists are witnesses to their ability to change on an all too frequent basis, it is evident that these organisms are fully capable of evolving and changing.

Genetic change in bacteria is accomplished by three basic mechanisms: mutation, genetic recombination, and gene exchange between bacteria, with or without recombination. Throughout diagnostic microbiology and infectious diseases there are numerous examples of the impact these genetic change and exchange mechanisms have on clinically relevant bacteria and the management of the infections they cause.

Mutation

Mutation is defined as a change in the original nucleotide sequence of a gene or genes within an organism's genome, that is, a change in the organism's **genotype**. This change may involve a single DNA base within a gene, an entire gene, or several genes. Mutational changes in the sequence may arise spontaneously, perhaps by an error made during DNA replication. Alternatively, mutations may be induced by chemical or physical factors (i.e., **mutagens**) in the environment, or by biologic factors such as the introduction of foreign DNA into the cell. Changes in the DNA base sequence can result in changes in the base sequence of mRNA codons during transcription. This, in turn, can affect the types and sequences of amino acids that will be incorporated in proteins during translation.

Various outcomes may result from a mutation and are depedent on the site and extent of the mutation. For example, a mutation may be so devastating that it is lethal to the organism; the mutation therefore "dies" along with the organism. In other instances the mutation may be silent so that no changes in the organism's observable properties (i.e., the organism's **phenotype**) are detected. Alternatively, the mutation may result in a noticeable change in the organism's phenotype and the change may provide the organism with a survival advantage. This outcome, in Darwinian terms, is the basis for prolonged survival and evolution. Nonlethal mutations are considered stable if they are passed on from one generation to another as an integral part of the cell's genotype (i.e., genetic makeup). Additionally, genes that have undergone stable mutations may also be transferred to other bacteria by one of the mechanisms of gene exchange. In other instances, the mutation may be lost through repair mechanisms in the cell that restore the original phenotype and genotype, or be lost spontaneously during subsequent cycles of DNA replication.

Genetic recombination

Besides mutations, bacterial genotypes can be changed through **recombination**. In this process some segment of DNA that originated from one bacterial cell (i.e., donor) enters a second bacterial cell (i.e., recipient) and is exchanged with a DNA segment of the recipient's genome. This is also referred to as **homologous recombination** because the pieces of DNA that are exchanged usually have extensive homology or similarities in their nucleotide sequences. Recombination involves a number of binding proteins with the RecA protein playing a central role (Figure 9-8, *A*). After recombination, the recipient DNA consists of one original, unchanged strand and the second from the donor DNA fragment that has been recombined.

Recombination is a molecular event that occurs frequently in many varieties of bacteria, including most of the clinically relevant species, and may involve any portion of the organism's genome. However, the recombinational event may go unnoticed unless the exchange of DNA results in a distinct alteration in the phenotype. Nonetheless, recombination is a major means by which bacteria may achieve genetic diversity and change.

Gene exchange

As just mentioned, an organism's opportunity for undergoing recombination depends on the acquisition of "foreign" DNA from a donor cell. The three mechanisms by which bacteria physically exchange DNA include: transformation, transduction, and conjugation.

TRANSFORMATION **Transformation** involves recipient cell uptake of free DNA released into the environment when a bacterial cell (i.e., donor) dies and undergoes lysis (Figure 9-8, *B*). This DNA, which had comprised the dead cell's genome, exists as fragments in the environment. Certain bacteria are able to take up this free DNA, that is, are able to undergo transformation. Such bacteria are said to be **competent.** Among the bacteria that cause human infections, competence is a characteristic commonly associated with members of the genera *Haemophilus, Streptococcus,* and *Neisseria.*

Once the donor DNA, usually as a singular strand, gains access to the interior of the recipient cell, recombination with the recipient's homologous DNA can occur. The mixing of DNA between bacteria via transformation and recombination plays a major role in the development of antibiotic resistance and in the dissemination of genes that encode factors essential to an organism's ability to cause disease. Additionally, gene exchange by transformation is not limited to organisms of the same species, thus allowing important characteristics to be disseminated to a greater variety of medically important bacteria.

TRANSDUCTION **Transduction** is a second mechanism by which DNA from two bacteria may come together in one cell, thus allowing for recombination (Figure 9-8, *C*). This process is mediated by viruses that infect bacteria (i.e., **bacteriophages**). In their "life cycle" these viruses integrate their DNA into the bacterial cell's chromosome, where viral DNA replication and expression is directed. When the production of viral products is completed, viral DNA is excised (cut) from the bacterial chromosome and packaged within protein coats. The viruses are then released when the infected bacterial cell lyses. In transduction the virus not only packages its own DNA but may also package a portion of the donor bacterium's DNA.

The bacterial DNA may be randomly incorporated with viral DNA (**generalized transduction**) or only bacterial DNA that was adjacent to viral DNA in the bacterial chromosome is packaged (**specialized transduction**). In any case, when the viruses infect another bacterial cell they release their DNA contents, which may include bacterial donor DNA. Therefore, the newly infected cell is the recipient of donor DNA introduced by the infecting bacteriophage and recombination between DNA from two different cells may occur.

CONJUGATION The third mechanism of DNA exchange between bacteria is **conjugation**. This process occurs between two living cells, involves cell-to-cell contact, and requires mobilization of the donor bacterium's chromosome. The nature of intercellular contact is not well characterized in all bacterial species capable of conjugation. However, in *E. coli* contact is mediated by a sex pilus (Figure 9-9). The sex pilus originates from the donor and establishes a conjugative bridge that serves as the conduit for DNA transfer from donor to recipient cell. With intercellular contact established, chromosomal mobilization is undertaken and involves DNA synthesis. One new DNA strand is produced by the donor and is passed to the recipient, where a strand complementary to the donor strand is synthesized (Figure 9-8, *D*). The amount of DNA transferred depends on how long the cells are able to maintain contact, but usually only portions of the chromosome are transferred. In any case, the newly introduced DNA is then available to recombine with the recipient's chromosome.

In addition to chromosomal DNA, genes encoded in nonchromosomal genetic elements, such as plasmids and transposons, may be transferred by conjugation (Figure 9-8, *E*). Not all plasmids are

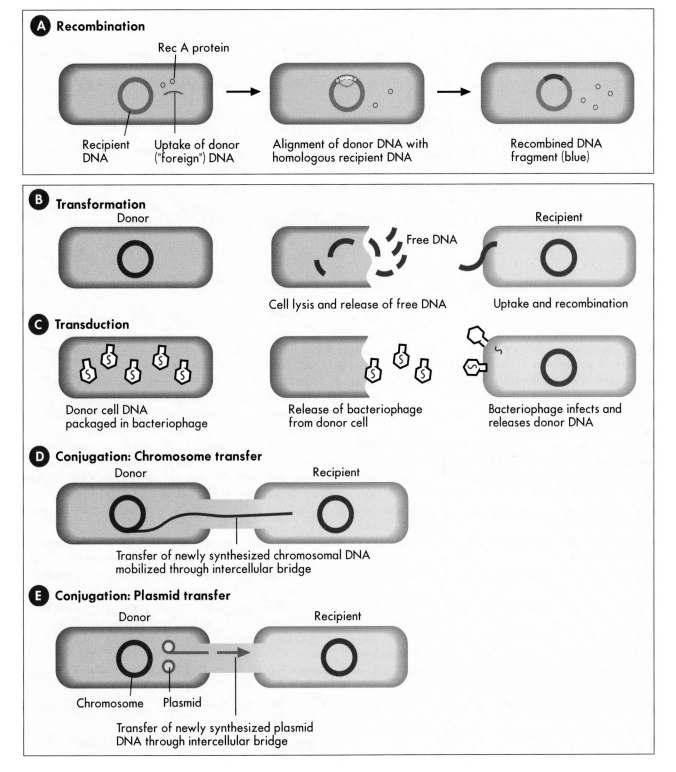

FIGURE 9-8 Genetic recombination (**A**). The mechanisms of gene exchange between bacteria: transformation (**B**), transduction (**C**), and conjugational transfer of chromosomal (**D**) and plasmid (**E**) DNA.

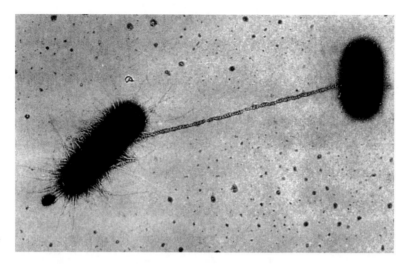

FIGURE 9-9 Photomicrograph of *E. coli* sex pilus between donor and recipient cell. (From Brock, T.D., Madigan, M.T., Martinko, J.M., and Parker, J., editors. 1994. Biology of microorganisms. Prentice Hall, Englewood Cliffs, N.J.)

capable of conjugative transfer, but for those that are, the donor plasmid usually is replicated so that the donor retains a copy of the plasmid that is being transferred to the recipient. Plasmid DNA may also become incorporated into the host cell's chromosome.

In contrast to plasmids, transposons do not exist independently within the cell. Except when they are moving from one location to another, transposons must be incorporated into the chromosome and/or into plasmids. These elements are often referred to as "jumping genes," because of their ability to change location within, and even between, the genomes of bacterial cells. **Transposition** is the process by which these genetic elements excise from one genomic location and insert into another. Transposons carry genes whose products help mediate the transpositional process as well as genes that encode for some other characteristic such as antimicrobial resistance. In the cases of both plasmids and transposons, homologous recombination between the genes of these elements and the host bacterium's chromosomal DNA can occur but is not necessary. Therefore these elements provide an alternative means for those organisms that cannot accommodate recombination to maintain foreign DNA in their genome.

Plasmids and transposons play a key role in genetic diversity and dissemination of genetic information among bacteria. Many characteristics that significantly alter the activities of clinically relevant bacteria are encoded and disseminated on these elements. Furthermore, as shown in Figure 9-10, the variety of strategies that bacteria can use to mix and match genetic elements provides them with a tremendous capacity to genetically adapt to environmental changes, including those imposed by human medical practices. A good example of this is the emergence and widespread dissemination of resistance to antimicrobial agents among clinically important bacteria. Bacteria have used their capacity for disseminating genetic information to establish resistance to most of the commonly used antibiotics (see Chapter 17 for more information regarding antimicrobial resistance mechanisms).

BACTERIAL METABOLISM

Fundamentally, bacterial metabolism involves all the cellular processes required for the organism's survival and replication. Familiarity with bacterial metabolism is essential for understanding bacterial interactions with human host cells, the mechanisms bacteria use to cause disease, and for understanding the basis of diagnostic microbiology tests and strategies used for laboratory identification of infectious etiologies. Because metabolism is an extensive and complicated topic, this section will focus on processes most typical of medically important bacteria.

For the sake of clarity, metabolism will be discussed in terms of four primary, but interdependent, processes: fueling, biosynthesis, polymerization, and assembly (Figure 9-11).

FUELING

Metabolic pathways in this category are those that are involved in acquisition of nutrients from the environment, production of precursor metabolites, and energy production.

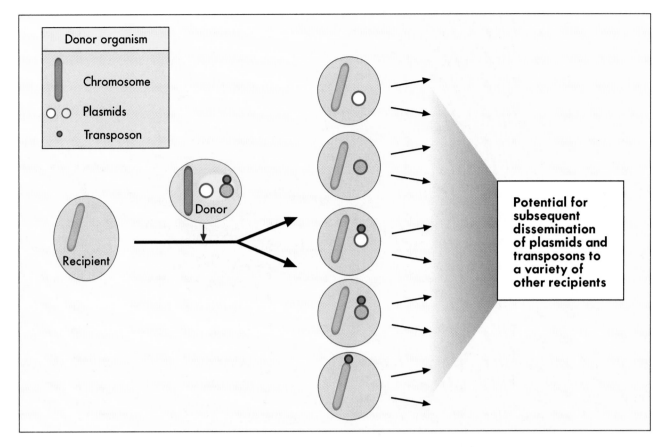

FIGURE 9-10 Pathways for bacterial dissemination of plasmids and transposons, together and independently.

Acquisition of nutrients

Bacteria employ various strategies for obtaining essential nutrients from the environment, but the common goal is to transport these substances from the external environment into the cell's interior. For nutrients to be internalized they must cross the **bacterial cell envelope,** a complex structure that helps protect the cell from environmental insults, maintains intracellular equilibrium, and transports substances into and out of the cell. Although some key nutrients such as water, oxygen, and carbon dioxide enter the cell by simple diffusion across the envelope, the uptake of other substances requires energy and selectivity by the cell's envelope.

Active transport is among the most common methods used for the uptake of nutrients such as certain sugars, most amino acids, organic acids, and many inorganic ions. The mechanism, which is driven by an energy-dependent pump, involves carrier molecules embedded in the membrane portion of the cell envelope. These carriers combine with the nutrients, transport them across the membrane, and release them within the cell. **Group translocation** is another transport mechanism that requires energy but differs from active transport in that the nutrient being transported undergoes chemical modification. Many sugars, purines, and pyrimidines, and fatty acids are transported by this mechanism.

Production of precursor metabolites

Once inside the cell, many nutrients serve as the raw materials from which precursor metabolites for subsequent biosynthetic processes are produced. These metabolites, listed in Figure 9-11, are produced by three central pathways that include the Embden-Myerhof-Parnas (EMP) pathway, the tricarboxylic acid (TCA) cycle, and the pentose phosphate shunt. These pathways and their relationship to one another are shown in Figure 9-12; not shown are the several alternative pathways (e.g., the Entner-Douderoff pathway) that play key roles in redirecting and replenishing the precursors as they are used in subsequent processes.

The production efficiency of a bacterial cell resulting from these precursor-producing pathways can vary substantially, depending on the growth conditions and availability of nutrients. This is an important consideration because the accurate identifi-

Precursor Metabolites

- Glucose 6-phosphate
- Fructose 6-phosphate
- Pentose 5-phosphate
- Erythrose 4-phosphate
- 3-Phosphoglycerate
- Phosphoenolpyruvate
- Pyruvate
- Acetyl CoA
- α–Ketoglutarate
- Succinyl CoA
- Oxaloacetate

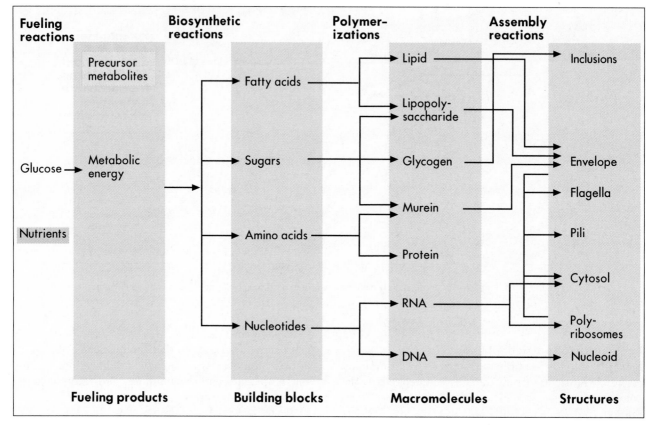

Nutrients

- Gases
  - Carbon dioxide (CO_2)
  - Oxygen (O_2)
  - Ammonia (NH_3)
- Organic compounds, including amino acids
- Water (H_2O)
- Nitrate (NO_3^-)
- Phosphate (PO_4^{3-})
- Hydrogen sulfide (H_2S)
- Sulfate (SO_4^{2-})
- Potassium (K^+)
- Magnesium (Mg^{2+})
- Calcium (Ca_2^+)
- Sodium (Na^+)
- Iron (Fe_3^+)
- Organic iron complexes

FIGURE 9-11 Overview of bacterial metabolism that includes the processes of fueling, biosynthesis, polymerization, and assembly. (Modified from Niedhardt, F.C., Ingraham, J.L., and Schaechter, M., editors. 1990. Physiology of the bacterial cell: a molecular approach. Sinauer Associates, Sunderland, Mass.).

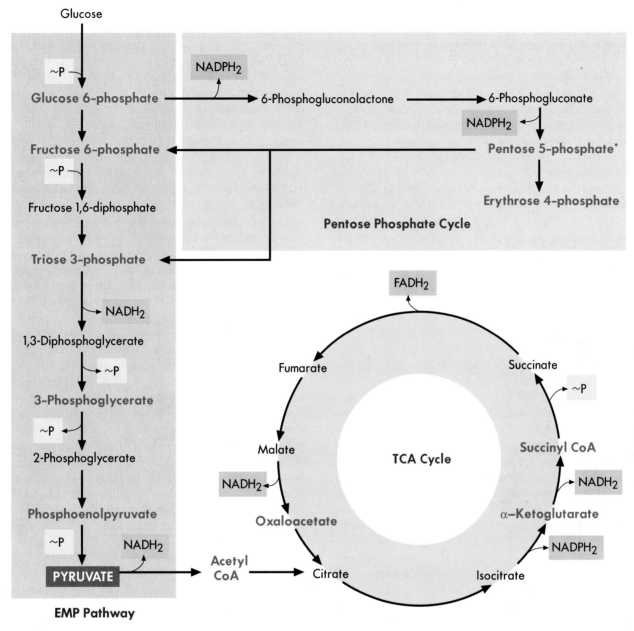

FIGURE 9-12 Overview diagram of the central metabolic pathways (Embden-Myerhof-Parnas [EMP], the tricarboxylic acid [TCA] cycle, and the pentose phosphate shunt). Precursor metabolites (see also Figure 9-11) that are produced are highlighted in red; production of energy in the form of ATP (~P) by substrate level phosphorylation is highlighted in yellow, and reduced carrier molecules for transport of electrons used in oxidative phosphorylation are highlighted in green. (Modified from Niedhardt, F.C., Ingraham, J.L., and Schaechter, M., editors. 1990. Physiology of the bacterial cell: a molecular approach. Sinauer Associates, Sunderland, Mass.).

cation of medically important bacteria often depends heavily on methods that measure the presence of products and byproducts of these metabolic pathways.

Energy production

The third type of fueling pathway is one that produces energy, which is required for nearly all cellular processes, including the two other types of fueling pathways just discussed, that is, nutrient uptake and precursor production. Energy production is accom-

plished by the breakdown of chemical substrates (i.e., chemical energy) through the degradative process of catabolism coupled with oxidation-reduction reactions. In this process the energy source molecule (i.e., **substrate**) is oxidized as it donates electrons to an electron-acceptor molecule, which is then reduced. The transfer of electrons is mediated by carrier molecules such as **nicotinamide-adenine-dinucleotide** (**NAD$^+$**) and **nicotinamide-adenine-dinucleotide-phosphate** (**NADP$^+$**). The energy released by the oxidation-reduction reaction is transferred to phosphate-containing compounds where high-energy phosphate bonds are formed. **Adenosine triphosphate (ATP)** is the most common of such molecules; the energy contained within this compound is eventually released by hydrolysis of ATP under controlled conditions. The release of this chemical energy and enzymatic activities specifically catalyze each biochemical reaction within the cell and drive nearly all cellular reactions.

The two general mechanisms for ATP production in bacterial cells are substrate level phosphorylation and electron transport, also referred to as oxidative phosphorylation. In **substrate level phosphorylation,** high-energy phosphate bonds produced by the central pathways are donated to adenosine diphosphate (ADP) to form ATP (see Figure 9-12). Additionally, pyruvate, a primary intermediate in the central pathways, serves as the initial substrate for several other pathways that also can generate ATP by substrate level phosphorylation. These other pathways constitute **fermentative metabolism,** which does not require oxygen and produces various end products, including alcohols, acids, carbon dioxide, and hydrogen. The specific fermentative pathways used, and hence end products produced, vary with different bacterial species. Detecting these products serves as an important basis for laboratory identification of bacteria (see Chapter 13 for more information on the biochemical basis for bacterial identification).

Oxidative phosphorylation involves an electron transport system that conducts a series of electron transfers from reduced carrier molecules such as $NADH_2$ and $NADPH_2$, produced in the central pathways (see Figure 9-12), to a terminal electron acceptor. The energy produced by the series of oxidation-reduction reactions is used to generate ATP from ADP. When oxidative phosphorylation uses oxygen as the terminal electron acceptor, the process is known as **aerobic respiration. Anaerobic respiration** refers to processes that use final electron acceptors other than oxygen.

Knowledge regarding which mechanisms bacteria use to generate ATP is important for designing laboratory protocols for cultivating and identifying these organisms. For example, some bacteria solely depend on aerobic respiration and cannot grow in the absence of oxygen (**strictly aerobic bacteria**). Others can use either aerobic respiration or fermentation, depending on the availability of oxygen (**facultatively anaerobic bacteria**). For still others, oxygen is absolutely toxic (**strictly anaerobic bacteria**).

BIOSYNTHESIS

The fueling reactions essentially bring together all the raw materials needed to initiate and maintain all other cellular processes. The production of precursors and energy is mostly accomplished by catabolic processes that involve the degradation of substrate molecules. The three remaining primary pathways—biosynthesis, polymerization, and assembly—depend on anabolic metabolism, in which precursor compounds are put together for the creation of larger molecules (polymers) needed for assembly of cellular structures (see Figure 9-11).

Biosynthetic processes use the 12 precursor products in dozens of pathways to produce nearly 100 different building blocks such as amino acids, fatty acids, sugars, and nucleotides (see Figure 9-11). Many of these pathways are highly complex and interdependent, while other pathways are almost completely independent. In many cases the enzymes that drive the individual pathways are encoded on a single mRNA molecule that has been cotranscribed from contiguous genes in the bacterial chromosome (i.e., an operon).

As discussed with the fueling pathways, bacterial genera and species vary extensively in their biosynthetic capabilities. Knowledge of these variations is necessary to ensure the design of optimal conditions for growing organisms under laboratory conditions. For example, some organisms may not be capable of synthesizing a particular amino acid that is a necessary building block for several essential proteins. Without the ability to synthesize the amino acid, the bacterium must obtain that building block from the environment. Similarly, if this organism is to be grown in the microbiology laboratory, this need must be fulfilled by supplying the amino acid in culture media.

POLYMERIZATION AND ASSEMBLY

Various anabolic reactions assemble (polymerize) the building blocks into macromolecules that include lipids, lipopolysaccharides, polysaccharides, proteins, and nucleic acids. This synthesis of macromolecules is driven by energy and enzymatic activity within the cell. Similarly, energy and enzymatic activities also drive the assembly of these various macromolecules into the component structures of the bacterial cell. That is, the cellular structures are the product of all the genetic and metabolic processes discussed.

STRUCTURE AND FUNCTION OF THE BACTERIAL CELL

Before discussing bacterial cellular structure, a general description of cells is useful. Based on key characteristics all cells are classified into two basic types: prokaryotic and eukaryotic. Although these two cell types share many common features, there are also many important differences in terms of structure, metabolism, and genetics.

EUKARYOTIC AND PROKARYOTIC CELLS

Among clinically relevant microorganisms, bacterial cells are prokaryotic, whereas cells of parasites and fungi are eukaryotic, as are those of plants and all higher animals. Viruses are acellular and depend on host cells for survival. Although the differences between eukaryotic and prokaryotic cells will not be detailed, they substantially influence the organisms' ability to cause disease and our ability to diagnose, treat, and control these diseases.

A notable characteristic of **eukaryotic cells,** such as those of parasites and fungi, is the presence of membrane-enclosed organelles that have specific cellular functions. Examples of these organelles include:

- Endoplasmic reticulum, which processes and transports proteins
- Golgi body, which processes substances for transport out of cell
- Mitochondria, which generate energy (ATP)
- Lysosomes, which provide environment for controlled enzymatic degradation of intracellular substances
- Nucleus, which provides membrane enclosure for chromosomes

Additionally, eukaryotic cells have an infrastructure, or cytoskeleton, that provides support for the different organelles.

Prokaryotic cells, such as those of bacteria, do not contain organelles. All functions take place in the cytoplasm or cytoplasmic membrane of the cell. Prokaryotic and eukaryotic cell types differ considerably at the macromolecular level regarding protein synthesis machinery, chromosomal organization, and gene expression. Also, one notable structure present only in bacterial cells that is absent among eukaryotes is a cell wall composed of peptidoglycan. As will be seen many times over, this structure alone has immeasurable impact on the practice of diagnostic bacteriology and the management of bacterial diseases.

BACTERIAL MORPHOLOGY

Most clinically relevant bacterial species range in size from 0.25 to 1 μm in width and 1 to 3 μm in length, thus requiring microscopy for visualization (see Chapter 11 for more information on microscopy). Just as bacterial species and genera vary in their metabolic processes, their cells also vary in size, morphology, cell-to-cell arrangements, and in the makeup of one of the most prominent cellular structures, the cell wall. In fact, cell wall differences provide the basis for the

Gram stain, the most fundamental test used in bacterial identification schemes. This staining procedure separates almost all medically important bacteria into two general types: **gram-positive bacteria,** which stain a deep blue color, and **gram-negative bacteria,** which stain a pink to red color. This simple, but important, color distinction is due to differences in the constituents of bacterial cell walls that influence the cell's ability to retain certain dyes, even after decolorization with alcohol.

Common bacterial cellular morphologies include **cocci** (round), **coccobacilli** (ovoid), and **bacillus** (rod-shaped), as well as fusiform (pointed-end), curved, or spiral shapes. Cellular arrangements are also noteworthy because cells may characteristically occur singly, in pairs, tetrads, clusters, or in chains (see Figure 11-4 for examples of bacterial staining and morphologies). The determination of Gram reaction coupled with cell size, morphology, and arrangement are essential aspects of bacterial identification.

BACTERIAL CELL COMPONENTS

Bacterial cell components can be divided into those that make up the cell envelope and its appendages and those associated with the cell's interior. Importantly, the cellular structures work together to function as a complex and integrated unit.

Cell envelope

As shown in Figure 9-13, the outermost structure, the cell envelope, comprises:

- An outer membrane (in gram-negative bacteria only)
- A cell wall composed of the peptidoglycan macromolecule (also known as the murein layer)
- Periplasm (in gram-negative bacteria only)
- The cytoplasmic membrane, interior to which is the cytoplasm

OUTER MEMBRANE   Outer membranes are only found in gram-negative bacteria, and they function as the cell's initial barrier to the environment. These membranes serve as primary permeability barriers to hydrophilic and hydrophobic compounds and also retain essential enzymes and other proteins located in the periplasmic space. The membrane is a bilayered structure composed of **lipopolysaccharide** that gives the surface of gram-negative bacteria a net negative charge and also plays a significant role in the ability of certain bacteria to cause disease.

Scattered throughout the lipopolysaccharide macromolecules are protein structures called **porins.** These water-filled structures control the passage of nutrients and other solutes, including antibiotics,

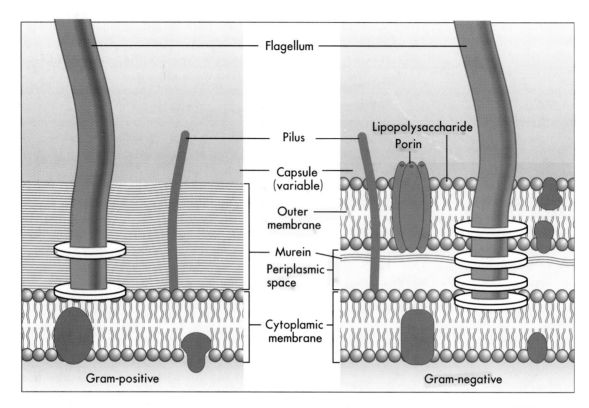

FIGURE 9-13 General structures of the gram-positive and gram-negative bacterial cell envelopes. The outer membrane and periplasmic space are only present in the gram-negative envelope. The murein layer is substantially more prominent in gram-positive envelopes. (Modified from Niedhardt, F.C., Ingraham, J.L., and Schaechter, M., editors. 1990. Physiology of the bacterial cell: a molecular approach. Sinauer Associates, Sunderland, Mass.).

through the outer membrane. The number and types of porins vary with bacterial species, and these differences can substantially influence the extent to which various substances pass through the outer membranes of different bacteria. In addition to porins, other proteins (murein lipoproteins) facilitate attachment of the outer membrane to the next deeper layer in the cell envelope, the cell wall.

CELL WALL (MUREIN LAYER) The cell wall, also referred to as the **peptidoglycan**, or murein layer, is an essential structure found in nearly all clinically important bacteria. This structure gives the bacterial cell shape and strength to withstand changes in environmental osmotic pressures that would otherwise result in cell lysis. The murein layer also protects against mechanical disruption of the cell and offers some barrier to the passage of larger substances. Because this structure is essential for the survival of bacteria, its synthesis and structure have been the primary target for the development and design of several antimicrobial agents.

The structure of the cell wall is unique in nature and is composed of **disaccharide-pentapeptide subunits.** The disaccharides *N*-acetylglucosamine and *N*-acetylmuramic acid are the alternating sugar

components (moieties), with the amino acid chain only linked to *N*-acetylmuramic acid molecules (Figure 9-14). Polymers of these subunits crosslink to one another via peptide bridges to form peptidoglycan sheets. Layers of these sheets are, in turn, crosslinked with one another to give a multilayered, crosslinked structure of considerable strength. Referred to as the murein sacculus, or sack, this peptidoglycan structure surrounds the entire cell.

A notable difference between the cell walls of gram-positive and gram-negative bacteria is that the peptidoglycan layer in gram-positive bacteria is substantially thicker. Additionally, the cell wall of gram-positive bacteria contains **teichoic acids** (i.e., glycerol or ribitol phosphate polymers that are combined with various sugars, amino acids, and amino sugars). Some teichoic acids are linked to *N*-acetylmuramic acid, and others (e.g., lipoteichoic acids) are linked to the next underlying layer, the cellular membrane. Still other gram-positive bacteria fortify their murein layer with waxy substances, such as **mycolic acids,** that make their cells even more refractory to toxic substances, including acids.

PERIPLASMIC SPACE The periplasmic space is only found in gram-negative bacteria and is bounded by

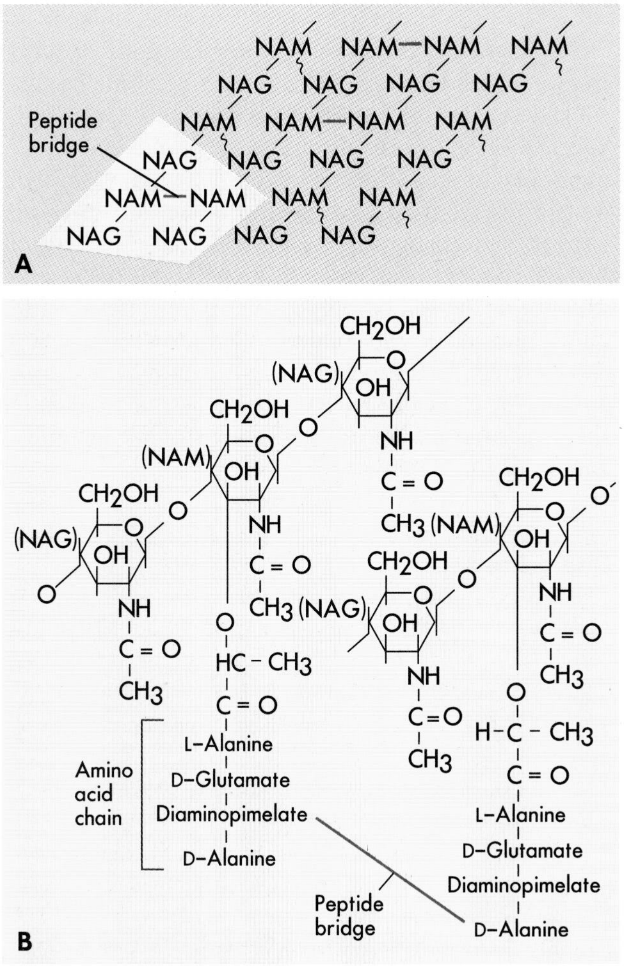

FIGURE 9-14 Peptidoglycan sheet (**A**) and subunit (**B**) structure. Multiple peptidoglycan layers compose the murein structure, and different layers are extensively crosslinked by peptide bridges (NAG, N-acetylglucosamine; NAM, N-acetylmuramic acid). Note that amino acid chains only derive from NAM. (Modified from Saylers, A.A. and Whitt, D.D. 1994. Bacterial pathogenesis: a molecular approach. American Society for Microbiology Press, Washington, D.C.)

osmotic barrier, the cytoplasmic membrane is functionally similar to several of the eukaryotic cell's organelles (e.g., mitochondria, Golgi complexes, lysosomes). The cytoplasmic membrane functions include:

- Transport of solutes into and out of the cell
- Housing enzymes involved in outer membrane synthesis, cell wall synthesis, and in the assembly and secretion of extracytoplasmic and extracellular substances
- Generation of chemical energy (i.e., ATP)
- Cell motility
- Mediation of chromosomal segregation during replication
- Housing molecular sensors that monitor chemical and physical changes in the environment

CELLULAR APPENDAGES   In addition to the components of the cell envelope proper, there are also cellular appendages (i.e., capsules, fimbriae, and flagella) that are associated with or proximal to this portion of the cell. The presence of these appendages, which can play a role in causing infections and in laboratory identification, varies among bacterial species and even among strains within the same species.

The **capsule** is immediately exterior to the murein layer of gram-positive bacteria and the outer membrane of gram-negative bacteria. Often referred to as the **slime layer,** the capsule is composed of high-molecular–weight polysaccharides whose production may depend on the environment and growth conditions surrounding the bacterial cell. The capsule does not function as an effective permeability barrier or add strength to the cell envelope but does protect bacteria from attack by cells of the human defense system. The capsule also facilitates and maintains bacterial colonization of biologic surfaces (i.e., teeth) and surfaces of inanimate objects such as implanted medical devices (e.g., prosthetic heart valves).

Fimbriae, or **pili,** are hairlike, proteinaceous structures that extend from the cell membrane into the external environment; some may be up to 2 μm in length. Two general types are known: common pili and sex pili. Common pili are adhesins that help bacteria attach to animal host cell surfaces, often as the first step in establishing infection. The sex pilus, well characterized in the gram-negative bacillus *Escherichia coli,* serves as the conduit for the passage of DNA from donor to recipient during conjugation.

Flagella are complex structures, mostly composed of the protein flagellin, intricately embedded in the cell envelope. These structures are responsible for bacterial motility. Although not all bacteria are motile, motility plays an important role in survival and the ability of certain bacteria to cause disease.

the internal surface of the outer membrane and the external surface of the cellular membrane. This area, which contains the murein layer, is not just an empty space but consists of gel-like substances that help to secure nutrients from the environment. This space also contains several enzymes that degrade macromolecules and detoxify environmental solutes, including antibiotics, that enter through the outer membrane.

CYTOPLASMIC (INNER) MEMBRANE   The cytoplasmic membrane is present in both gram-positive and gram-negative bacteria and is the deepest layer of the cell envelope. In both gram-positive and gram-negative bacteria the structure of the cell membrane is essentially the same. This lipid bilayer is heavily laced with various proteins, including a number of enzymes vital to cellular metabolism. Besides being an additional

Depending on the bacterial species, the cell may have flagella located at one end of the cell (**monotrichous flagella**), at both ends of the cell (**lophotrichous flagella**), or the entire cell surface may be covered with flagella (**peritrichous flagella**).

Cell interior

Those structures and substances that are bounded to the inside by the cytoplasmic membrane compose the cell interior and include the cytosol, polysomes, inclusions, the nucleoid, plasmids, and endospores.

The **cytosol,** where nearly all other functions not conducted by the cell membrane occur, contains thousands of enzymes and is the site of protein synthesis. The cytosol has a granular appearance caused by the presence of many polysomes (mRNA complexed with several ribosomes during translation and protein synthesis) and **inclusions** (i.e., storage reserve granules). The number and nature of the inclusions vary depending on the bacterial species and the nutritional state of the organism's environment. Two common types of granules include **glycogen,** a storage form of glucose, and **polyphosphate granules,** a storage form for inorganic phosphates that are microscopically visible in certain bacteria stained with specific dyes.

Unlike eukaryotic chromosomes, the bacterial chromosome is not enclosed within a membrane-bound nucleus. Instead the bacterial chromosome exists as a **nucleoid** in which the highly coiled DNA is intermixed with RNA, polyamines, and various proteins that lend structural support. At times, depending on the stage of cell division, there may be more than one chromosome per bacterial cell. **Plasmids** are the other genetic elements that exist independently in the cytosol, and their numbers may vary from none to several per bacterial cell.

The final bacterial structure to be considered is the **endospore.** Under adverse physical and chemical conditions, or when nutrients are scarce, some bacterial genera are able to form spores (i.e., **sporulate**). Sporulation involves substantial metabolic and structural changes in the bacterial cell. Essentially, the cell transforms from an actively metabolic and growing state to a dormant state, with a decrease in cytosol and a concomitant increase in the thickness and strength of the cell envelope. The spore state is maintained until favorable conditions for growth are again encountered. This survival tactic is demonstrated by a number of clinically relevant bacteria and frequently challenges our ability to thoroughly sterilize materials and foods for human use.

Bibliography

Brock, T.D., Madigan, M.T., Martinko, J.M., and Parker, J., editors. 1994. Biology of microorganisms. Prentice Hall, Englewood Cliffs, N.J.

Joklik, W.K., Willett, H.P., Amos, D.B., and Wilfert, C.M., editors. 1992. Zinsser microbiology. Appleton & Lange, Norwalk, Conn.

Moat, A.G. and Foster, J.W. 1995. Microbial physiology. Wiley-Liss, N.Y.

Neidhardt, F.C., Ingraham, J.L., and Schaecter, M., editors. 1990. Physiology of the bacterial cell: a molecular approach. Sinauer Associates, Sunderland, Mass.

Ryan, K.J., editor. 1994. Sherris medical microbiology: an introduction to infectious diseases. Appleton & Lange, Norwalk, Conn.

Saylers, A.A. and Whitt, D.D. 1994. Bacterial pathogenesis: a molecular approach. American Society for Microbiology Press, Washington, D.C.

10 | HOST-MICROORGANISM INTERACTIONS

Interactions between humans and microorganisms are exceedingly complex and are far from being completely understood. What is known about the interactions between these two living entities plays an important role both in the practice of diagnostic microbiology and in the management of infectious diseases. Understanding these interactions is necessary for establishing methods to reliably isolate specific microorganisms from patient specimens and for developing effective treatment strategies. This chapter provides the framework for understanding the various aspects of host-microorganism interactions (see Part 3 for information regarding specific organ systems); specific characteristics of particular microorganisms are addressed in Parts 4 through 7.

Humans and microorganisms inhabit the same planet, and their paths cross in many and varied ways so that interactions are inevitable. The most important point regarding host-microorganism interactions is that the relationships are always bidirectional in nature. The bias is to believe that humans dominate the relationship because microorganisms are used by humans in various settings, including the food and fermentation industry; as biologic insecticides for agriculture, genetically engineering a multitude of products; and even biodegrading our industrial wastes. However, as one realizes that microbial populations share the common goal of survival with humans, and that they have been successful at achieving that goal, then exactly which participant in the relationship is the user and which one is the used becomes much less clear. This is especially true when considering the microorganisms most closely associated with humans and human disease.

The complex relationships between human hosts and medically relevant microorganisms are best understood by considering the sequential steps in the development of microbial-host associations and subsequent development of infection and disease. The stages of interaction are shown in Figure 10-1 and include physical encounter between host and microorganism, colonization or survival of the microorganism on an internal (gastrointestinal, respiratory, or genitourinary tract) or external (skin) surface of the host, microbial

entry, invasion, and dissemination to deeper tissues and organs of the human body, and resolution or outcome. Each stage will be discussed from two perspectives: the human host's and the microorganism's.

THE ENCOUNTER BETWEEN HOST AND MICROORGANISM

THE HUMAN HOST'S PERSPECTIVE

Because microorganisms are found everywhere, human encounters are inevitable, but the means of encounter vary widely. Which microbial population a human is exposed to and the mechanism of exposure is often a direct consequence of a person's activity. Certain activities carry different risks for an encounter and there is a wide spectrum of activities or situations over which a person may or may not have absolute control. For example, acquiring salmonellosis because one fails to thoroughly cook the holiday turkey is an avoidable activity, whereas contracting tuberculosis as a consequence of living in conditions of extreme poverty and overcrowdedness may be less avoidable. The role that human activities play in the encounter process cannot be overstated because most of the crises associated with infectious disease could be avoided or greatly modified if human behavior and living conditions could be changed.

Microbial reservoirs and transmission

Humans encounter microorganisms when they enter or are exposed to the same environment in which the microbial agents live or when the infectious agents are brought to the human host by indirect means. The environment, or place of origin, of the infecting agent is referred to as the reservoir. As shown in Figure 10-2, microbial reservoirs include humans, animals, water, food, air, and soil. The human host may acquire microbial agents by various means referred to as the **modes of transmission.** The mode of transmission is direct when the host directly contacts the microbial reservoir and is indirect when the host encounters the microorganism by an intervening agent of transmission.

The agents of transmission that bring the microorganism from the reservoir to the host may be a living entity, in which case they are called **vectors,** or they may be a nonliving entity referred to as a **vehicle** or **fomite.** Additionally, some microorganisms may have a single mode of transmission while others may spread by various methods. From a diagnostic microbiology perspective, knowledge about an infectious agent's mode of transmission is often important for determining optimum specimens for organism isolation and for taking precautions that minimize the risk of laboratory-acquired infections (see Chapter 2 for more information regarding laboratory safety).

Humans as microbial reservoirs

Humans play a substantial role as microbial reservoirs. Indeed, the passage of a neonate from the sterile environment of the mother's womb through the birth canal, which is heavily colonized with various microbial agents, is a primary example of one human directly acquiring microorganisms from another human serving as the reservoir. This is the mechanism by which newborns first encounter microbial agents. Other examples in which humans serve as the microbial reservoir for direct transmission of microorganisms to other people include acquisition of "strep" throat through touching; hepatitis by blood transfusions; gonorrhea, syphilis, and AIDs by sexual contact; tuberculosis by coughing; and the common cold through sneezing. Indirect transfer can occur when microorganisms from one individual contaminate a vehicle of transmission such as food (e.g., salmonellosis) or water (e.g., cholera) that is then ingested by another person. In the medical setting, indirect transmission of microorganisms from one human host to another via contaminated medical devices helps disseminate infections in hospitals. Hospital-acquired infections are referred to as **nosocomial infections.**

Animals as microbial reservoirs

From animal reservoirs, infectious agents can be transmitted directly to humans such as through an animal bite (e.g., rabies) or indirectly such as through the bite of insect vectors that feed on both animals and humans (e.g., Lyme disease and Rocky Mountain Spotted Fever). Animals may also transmit infectious agents by water and food supplies. For example, beavers are often heavily colonized with parasites that cause infections of the human gastrointestinal tract. These parasites may be encountered and subsequently acquired when contaminated stream water is used by beaver and camper alike. Alternatively, animals used for human food carry numerous bacteria (e.g., salmonella and campylobacter) that, if not destroyed by appropriate cooking during preparation, can cause severe gastrointestinal illness.

Many other infectious diseases are encountered through direct or indirect animal contact, and information regarding a patient's exposure to animals is often a key component to diagnosing these infections. It should be noted that some microorganisms primarily infect animal populations and on occassion accidentally encounter and infect humans. Such infectious diseases are classified as **zoonoses;** when a human infection results from such an encounter it is referred to as a **zoonotic infection.**

Insects as vectors

The most common role of insects (arthropods) in the transmission of infectious diseases is as vectors rather than as reservoirs. A wide variety of arthropods transmit viral, parasitic, and bacterial diseases from animals to humans, while others transmit microorganisms between human hosts without an animal reservoir being involved. Malaria, a leading cause of death from infection, is a prime example of a disease that is maintained in the human population by the activities of an insect (i.e., mosquito) vector. Still other arthropods may themselves be agents of disease. These include organisms, such as lice and scabies, that are spread directly between humans and cause skin irritations but do not penetrate the body. Because they are able to survive on the skin of the host without usually gaining access to internal tissues, they are referred to as **ectoparasites.**

The environment as a microbial reservoir

The soil and natural environmental debris are reservoirs for countless types of microorganisms. There-

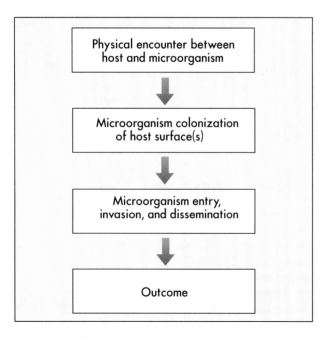

FIGURE 10-1 General stages of microbial-host interaction.

fore, it is not surprising that these also serve as reservoirs for microorganisms that can cause infections in humans. Many of the fungal agents (see Part 6 for more information regarding fungal organisms and disease) are acquired by inhalation of soil and dust particles containing the microorganisms (e.g., San Joaquin Valley Fever). Other, nonfungal, infections (e.g., tetanus) may result when microbial agents in the environment are introduced into the human body as a result of a penetrating wound.

THE MICROORGANISM'S PERSPECTIVE

Clearly, numerous activities can result in human encounters with many microorganisms. Because humans are engaged in all of life's complex activities, the tendency is to perceive the microorganism as having a passive role in the encounter process. However, this assumption is a gross oversimplification.

Microorganisms are also driven by survival, and the environments of the reservoirs they occupy must allow their metabolic and genetic needs to be fulfilled. Furthermore, most reservoirs are usually inhabited by hundreds, if not thousands, of different species. Yet human encounters with the reservoirs, either directly or indirectly, do not result in all species establishing an association with the human host. Although some species have evolved strategies that do not involve the human host to ensure survival, others have included humans to a greater or lesser extent as part of their survival tactics. Therefore the latter type of organism often has mechanisms that enhance its chances for human encounter.

Depending on factors associated with both the human host and the microorganism involved, the encounter may have a beneficial, disastrous, or inconsequential impact on each of the participants.

MICROORGANISM COLONIZATION OF HOST SURFACES

THE HOST'S PERSPECTIVE

Once brought in contact, the outcome of an encounter between a microbe and the human host depends on what happens during each step of interaction (Figure 10-1) beginning with colonization. The human host's role in microbial **colonization**, defined as the persistent survival of microorganisms on a surface of the human body, is dictated by the defenses that protect vital internal tissues and organs against microbial invasion. The first defenses are the external and internal body surfaces that are in relatively direct contact with the external environment and as such are the body areas with which microorganisms will initially associate. These surfaces include:

- Skin (including conjunctival epithelium covering the eye)
- Mucous membranes that line the mouth or oral cavity, the respiratory tract, the gastrointestinal tract, and the genitourinary tract

Because body surfaces are always present and, unless altered or damaged, their protective features generally target almost all microorganisms, skin and

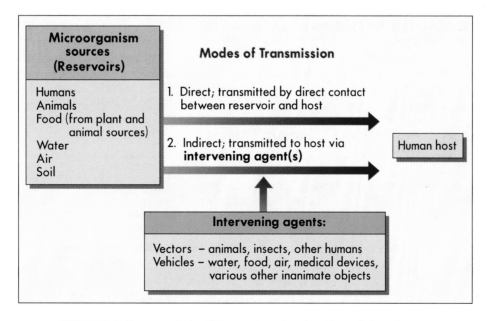

FIGURE 10-2 Summary of microbial reservoirs and modes of transmission to humans.

mucous membranes are considered constant and non-specific protection mechanisms. As will be discussed, other protective mechanisms are produced only in response to the presence of microbial agents (i.e., inducible defenses) and some are specifically directed to particular microorganisms (i.e., specific defenses).

Skin and skin structures

Skin serves as a physical and chemical barrier to microorganisms; its protective characteristics are summarized in Box 10-1 and Figure 10-3. The acellular, outermost layer of skin and tightly packed cellular layers underneath provide an unpenetrable physical barrier to all microorganisms, unless damaged. Additionally, these layers continuously shed, thus dislodging bacteria that have attached to the outer layers. The skin is also a dry and cool environment that contrasts with the warm and moist environment in which many microbial agents thrive.

The follicles and glands of the skin produce various natural antibacterial substances. However, many microorganisms can survive the conditions of the skin. These bacteria are known as skin colonizers, and they often produce substances that may be toxic and inhibit the growth of more harmful microbial agents. Beneath the outer layers of skin are various host cells that protect against organisms that breach the surface barriers. These cells, collectively known as **skin-associated lymphoid tissue,** mediate specific and nonspecific responses directed at controlling microbial invaders.

Mucous membranes

Because cells that line the respiratory tract, gastrointestinal tract, and genitourinary tract are involved in numerous functions besides protection, they are not covered with a hardened acellular layer like that found on the skin. However, the cells that compose these membranes still exhibit various protective characteristics (Box 10-2 and Figure 10-4).

GENERAL PROTECTIVE CHARACTERISTICS A major protective component of mucous membranes is the mucus itself. This substance serves to trap bacteria before they can reach the outer surface of the cells, lubricates the cells to prevent damage that may promote microbial invasion, and contains numerous specific (i.e., antibodies) and nonspecific antibacterial substances. In addition to mucous activity and flow mediated by cilia action, rapid cellular shedding and tight intercellular connections provide effective barriers. As is the case with skin, specific cell clusters, known as **mucosa-associated lymphoid tissue,** exist below the outer cell layer and mediate specific protective mechanisms against microbial invasion.

BOX 10-2 | **PROTECTIVE CHARACTERISTICS OF MUCOUS MEMBRANES**

Mucous Membrane Structures	Protective Activity
Mucosal cells	Rapid sloughing for bacterial removal Tight intercellular junctions prevent bacterial penetration
Goblet cells	Mucus production: Protective lubrication of cells Bacterial trapping Contains specific antibodies with specific activity against bacteria Provision of antibacterial substances to mucosal surface: lysozyme: degrades bacterial cell wall lactoferrin: competes for bacterial iron supply lactoperoxidase: production of substances toxic to bacteria
Mucosa-associated lymphoid tissue	Mediates specific responses against bacteria that penetrate outer layer

BOX 10-1 | **PROTECTIVE CHARACTERISTICS OF THE SKIN AND SKIN STRUCTURES**

Skin Structure	Protective Activity
Outer (dermal) layers	Physical barrier to microbial penetration Sloughing of outer layers removes attached bacteria Provide dry, acidic, and cool conditions that limit bacterial growth
Hair follicles, sweat glands, sebaceous glands	Production of acids, alcohols, and toxic lipids that limit bacterial growth
Conjunctival epithelium covering the eyes	Flushing action of tears removes microorganisms Tears contain lysozyme that destroys bacterial cell wall
Skin-associated lymphoid tissue	Mediate specific and nonspecific protection mechanisms against microorganisms that penetrate outer layers

SPECIFIC PROTECTIVE CHARACTERISTICS Besides the general protective properties of mucosal cells, the linings of the different body tracts have other characteristics specific to each anatomic site (Figure 10-5).

The mouth, or oral cavity, is protected by the flow of saliva that physically carries microorganisms away from cell surfaces and also contains antibacterial substances such as lysozyme, which destroys bacterial cell walls, and antibodies. The mouth is also heavily colonized with microorganisms that contribute to protection by producing substances that hinder successful invasion by harmful agents.

In the gastrointestinal tract, the low pH and proteolytic (protein-destroying) enzymes of the stomach help keep the numbers of microorganisms low. In the small intestine, protection is provided by the presence of bile salts that disrupt bacterial membranes and by the fast flow of intestinal contents that hinders microbial attachment to mucosal cells. Although the large intestine also produces bile salts, the movement of bowel contents is slower so that a higher concentration of microbial agents are able to attach to the mucosal cells and inhabit this portion of the gastrointestinal tract. As in the oral cavity, the high concentration of normal microbial inhabitants in the large bowel also contributes significantly to protection.

In the upper respiratory tract, nasal hairs keep out large airborne particles that may contain microorganisms and the cough-sneeze reflex also significantly contributes to the removal of potentially infective agents. The cells lining the trachea contain cilia (hairlike cellular projections) that move mucus and the microorganisms trapped within upward and away from delicate cells of the lungs (see Figure 10-4). These barriers are so effective that only inhalation of particles smaller than 2 to 3 μm have a chance of reaching the lungs.

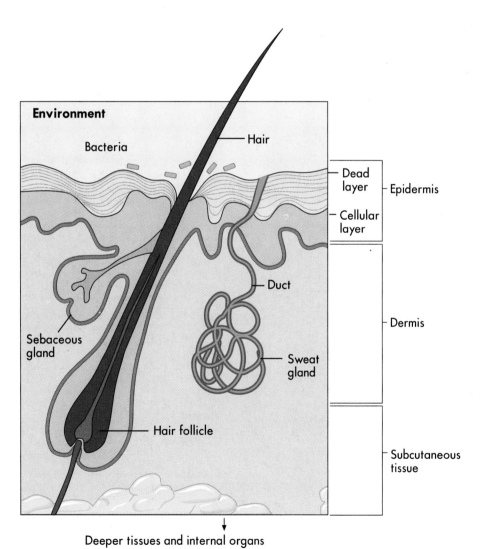

FIGURE 10-3 Skin and skin structures.

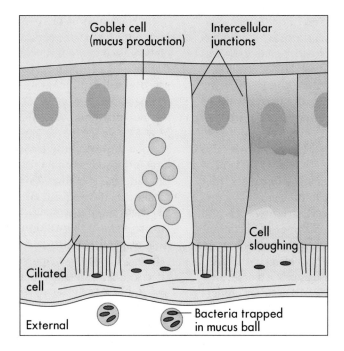

FIGURE 10-4 General features of mucous membranes highlighting protective features such as ciliated cells, mucous production, tight intercellular junctions, and cell sloughing.

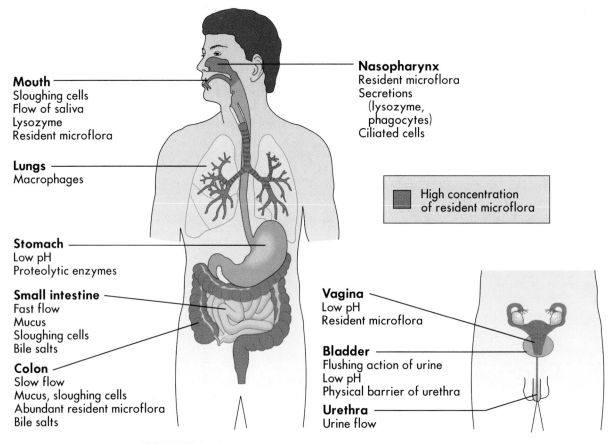

FIGURE 10-5 Protective characteristics associated with the mucosal linings of different internal body surfaces.

In the female urogential tract the vaginal lining and the ectocervix are protected by heavy colonization with normal microbial inhabitants and a low pH. A thick mucous plug in the cervical opening is a substantial barrier that keeps microorganisms from ascending and invading the more delicate tissues of the uterus, fallopian tubes, and ovaries. The anterior urethra of males and females is naturally colonized with microorganisms, but stricture at the urethral opening provides a physical barrier that, combined with low urine pH and the flushing action of urination, protects against bacterial invasion of the bladder, ureters, and kidneys.

THE MICROORGANISM'S PERSPECTIVE

As previously discussed, microorganisms that inhabit many surfaces of the human body (see Figure 10-5) are referred to as **colonizers,** or **normal flora.** Some are **transient colonizers** because they merely survive, but do not multiply, on the surface and are frequently shed with the host cells. Others, called **resident flora,** not only survive but also thrive and multiply; their presence is more permanent.

The normal flora of body surfaces vary considerably with anatomic location. For example, environmental conditions, such as temperature and oxygen availability, vary considerably between the nasal cavity and small bowel. Only microorganisms with the metabolic capacity to survive under the circumstances that each anatomic location offers will be inhabitants of those particular body surfaces.

Knowledge of the normal flora of the human body is extremely important in diagnostic microbiology, especially for determining the clinical significance of microorganisms that are isolated from patient specimens. Normal flora organisms frequently are found in clinical specimens as a result of contamination of normally sterile specimens during collection, the processing of specimens that normally contain or are in contact with colonized body surfaces, or because the colonizing organism is actually involved in the infec-

tion. Microorganisms considered as normal colonizers of the human body and the anatomic locations they colonize are addressed in Part 4.

Microbial colonization

Colonization may be the last step in the establishment of a long-lasting, mutually beneficial (i.e., **commensal**), or harmless, relationship between a colonizer and the human host. Alternatively, colonization may be the first step in the process of developing infection and disease. Whether colonization results in a harmless or harmful situation depends on the host and microorganism characteristics, but in either case successful initial colonization depends on the microorganism's ability to survive the conditions first encountered on the host surface (Box 10-3).

To avoid the dryness of the skin, organisms often seek moist areas that include hair follicles, sebaceous (oil) and sweat glands, skin folds, underarms, the anogenital region, the face, the scalp, and areas around the mouth. Microbial protection against mucous surface environments include being embedded in food particles to survive oral and gastrointestinal conditions or being contained within airborne particles to aid in surviving the respiratory tract. Microorganisms also exhibit metabolic capabilities that help survival. For example, the ability of staphylococci to thrive in relatively high salt concentrations enhances their survival in and among the sweat glands of the skin.

Besides surviving the host's physical and chemical conditions, colonization also requires that microorganisms attach and adhere to host surfaces (see Box 10-3). This can be particularly challenging in places such as the mouth and bowel, in which the surfaces are frequently washed by passing fluids. Pili, the rodlike projections of bacterial envelopes (for more information about pili see Chapter 9 and Figure 21-2), and various molecules (e.g., adherence proteins and adhesins) and biochemical complexes (e.g., biofilm) work together to enhance attachment of microorganisms to the host cell surface.

 **MICROBIAL ACTIVITIES CONTRIBUTING TO COLONIZATION OF HOST SURFACES**

Survival Against Environmental Conditions
Localization in moist areas
Protection within ingested or inhaled debris
Expression of specific metabolic characteristics (e.g., salt tolerance)
Attachment and Adherence to Host Cell Surfaces
Pili
Adherence proteins
Biofilms
Various protein adhesins

Motility

Production of Substances that Compete with Host for Acquisition of Essential Nutrients (e.g., sideophores for capture of iron)

Ability to Coexist with Other Colonizing Microorganisms

In addition, microbial motility by flagella allows organisms to move around and actively seek optimum conditions. Finally, because no single microbial species is a lone colonizer, successful colonization also requires that a microorganism be able to coexist with other microbial agents.

MICROORGANISM ENTRY, INVASION, AND DISSEMINATION

THE HOST'S PERSPECTIVE

In most instances, microorganisms must penetrate or circumvent the host's physical barriers (i.e., skin or mucosal surfaces) in order to establish infection; overcoming these defensive barriers depends on both host and microbial factors. When these barriers are broken, numerous other host defensive strategies are activated.

Disruption of surface barriers

Any situation that disrupts the physical barrier of the skin and mucosa, alters the environmental conditions (e.g., loss of stomach acidity or dryness of skin), changes the functioning of surface cells, or alters the normal flora population can facilitate the penetration of microorganisms past the barriers and into deeper host tissues. Disruptive factors may vary from accidental or intentional (medical) trauma resulting in surface destruction to the use of antibiotics that remove normal, protective, colonizing microorganisms (Box 10-4). Importantly, a number of these factors are related to medical activities and procedures.

Responses to microbial invasion of deeper tissues

Once surface barriers have been bypassed, the host responds to microbial presence in the underlying tissue in various ways. Some of these responses are nonspecific because they occur regardless of the type of invading organism, while other responses are more specific and involve the host's immune system. Both nonspecific and specific host responses are critical if the host is to survive. Without them microorganisms would multiply and invade vital tissues and organs unchecked, resulting in severe damage to the host.

NONSPECIFIC RESPONSES Some nonspecific responses are biochemical, and others are cellular. Biochemical factors remove essential nutrients, such as iron, from tissues so that they cannot be used by invading microorganisms. Cellular responses are central to tissue and organ defenses and the cells involved are known as phagocytes.

Phagocytes. **Phagocytes** are cells that ingest and destroy bacteria and other foreign particles. The two types of phagocytes are **polymorphonuclear leukocytes** (also known as **PMNs** or **neutrophils**) and **macrophages**. Phagocytes ingest bacteria by a process known as **endocytosis** and engulf them in a membrane-lined structure called a **phagosome** (Figure 10-6). The phagosome is then fused with a second structure, the lysosome. When the lysosome, which contains toxic chemicals and destructive enzymes, combines with the phagosome, the bacteria trapped within are neutralized and destroyed. This destructive process must be carried out inside membrane-lined structures; otherwise the noxious substances contained within would destroy the phagocyte itself. This is evident during the course of rampant infections when thousands of phagocytes exhibit "sloppy" ingestion of the microorganisms and toxic substances spill from the cells, thus damaging the surrounding host tissue.

Although both PMNs and macrophages are phagocytes, these cell types differ. PMNs develop in the bone marrow and spend their short lives, usually a day or less, circulating in blood and tissues. Widely dispersed in the body, PMNs usually are the first cells on the scene of bacterial invasion. Macrophages also develop in the bone marrow but first go through a cellular phase when they are called **monocytes**. During the monocyte stage, the cells circulate in the blood.

BOX 10-4   **FACTORS CONTRIBUTING TO DISRUPTION OF SKIN AND MUCOSAL SURFACE**

Trauma
Penetrating wounds
Abrasions
Burns (chemical and fire)
Surgical wounds
Needle sticks
Inhalation
Smoking
Noxious or toxic gases

Implantation of Medical Devices

Other Diseases
Malignancies
Diabetes
Previous or simultaneous infections
Alcoholism
Childbirth

Overuse of Antibiotics

.m

organs, such as the spleen, lymph nodes, liver, or lungs, where they may live for days to several weeks awaiting encounters with invading bacteria. In addition to the ingestion and destruction of bacteria, macrophages play an important role in mediating immune system defenses (see the section on Specific Response—The Immune System later in this chapter).

In addition to the inhibition of microbial proliferation by phagocytes and by biochemical substances such as lysozyme, microorganisms are "washed" from tissues by the flow of lymph fluid. The fluid carries infectious agents through the lymphatic system, where they are deposited in tissues and organs (e.g., lymph nodes and spleen) heavily populated with phagocytes. This process functions as an efficient filtration system.

Inflammation. Because microbes may survive initial encounters with phagocytes (see Figure 10-6), the inflammatory response plays an extremely important role as a reinforcement mechanism against microbial survival and proliferation in tissues and organs. Inflammation has both cellular and biochemical components that interact in various complex ways (Box 10-5).

The **complement system** is composed of a coordinated group of proteins that are activated by the immune system or by the mere presence of invading microorganisms. On activation of this system, a cascade of biochemical events occurs that attract and enhance the activities of more phagocytes. Because PMNs and macrophages are widely dispersed throughout the body, signals are needed to attract and concentrate these cells at the point of invasion. The complement system provides many of these signals.

Protective functions of the complement system also are enhanced by the **coagulation system,** which works to increase blood flow to the area of infection and can also effectively wall off the infection through the production of barrier substances.

Another key component of inflammation is a group of biochemicals known as **cytokines,** substances secreted by one type of cell that have substantial effects on the antiinfective activities of other cells.

The manifestations of inflammation are evident and familiar to most of us and include:

- **Swelling**—caused by increased flow of fluid and cells to the affected body site
- **Redness**—results from vasodilatation of blood vessels at the infection site
- **Heat**—results from increased temperature of affected tissue

On a microscopic level, ... phagocytes at the infection site is an important observation in diagnostic microbiology. Microorganisms seen associated with these host cells are frequently identified as the cause of a particular infection. An overview of inflammation is given in Figure 10-7.

SPECIFIC RESPONSES—THE IMMUNE SYSTEM   The immune system provides the human host with the ability to mount a specific protective response to the presence of a microorganism, a customized defense against the invading microorganism. In addition to this specificity, the immune system has a "memory" so that if a microorganism is encountered a second or third time, an immune-mediated defensive response is immediately available. It is important to remember that nonspecific (i.e., phagocytes, inflammation) and specific (i.e., the immune system) host defensive systems are interdependent in their efforts to limit the spread of infection.

Components of the Immune System.   The central molecule of the immune response is the **antibody.** Antibodies, also referred to as **immunoglobulins,** are specific

	COMPONENTS OF INFLAMMATION
BOX 10-5	

Component	**Functions**
Phagocytes (PMNs and macrophages)	Ingest and destroy microorganisms
Complement system (coordinated group of serum proteins)	Attracts phagocytes to site of infection
	Helps phagocytes recognize and bind to bacteria
	Directly kills gram-negative bacteria
Coagulation system (wide variety of proteins and other biologically active compounds)	Attracts phagocytes to site of infection
	Increases blood and fluid flow to site of infection
	Walls off site of infection to physically inhibit spread of microorganisms
Cytokines (proteins secreted by macrophages and other cells)	Multiple effects that enhance the activities of many different cells essential to nonspecific and specific defensive responses

of the invading microorganism's structure that are usually made of proteins or polysaccharides. Antibodies circulate in the serum portion of the host's blood and are present in secretions such as saliva. These molecules have two active areas: the **antigen binding site** and the **phagocyte binding site** (Figure 10-8).

There are five different classes of antibody (IgG, IgA, IgM, IgD, and IgE). Each class has distinctive molecular configurations. IgG, IgM, IgA, and IgE are most involved in combating infections. IgM is the first antibody produced when an invading microorganism is encountered for the first time. Then production of the most abundant antibody, IgG, follows. IgA is secreted in various body fluids and primarily protects those body surfaces lined with mucous membranes. Increased IgE is associated with various parasitic infections. As will be discussed in Chapter 16, our ability to measure specific antibody production is a valuable tool for the laboratory diagnosis of infectious diseases.

Regarding the cellular components of the immune response, there are three major types of cells:

develop into B cells in the bone marrow before being widely distributed to lymphoid tissues throughout the body. These cells primarily function as antibody producers. **T lymphocytes** also originate from bone marrow stem cells, but they mature in the thymus and either directly destroy infected cells or work with B cells to regulate antibody production. The development of **natural killer cells,** which destroy infected or malignant host cells, is uncertain. Each of the three cell types is strategically located within lymphoid tissue throughout the body to maximize the chances of encountering invading microorganisms that the lymphatic system drains from the infection site.

Two Arms of the Immune System. The immune system provides immunity that generally can be divided into two arms:

- **Antibody-mediated immunity** (or **humoral immunity**)
- **Cell-mediated immunity** (or **cellular immunity**)

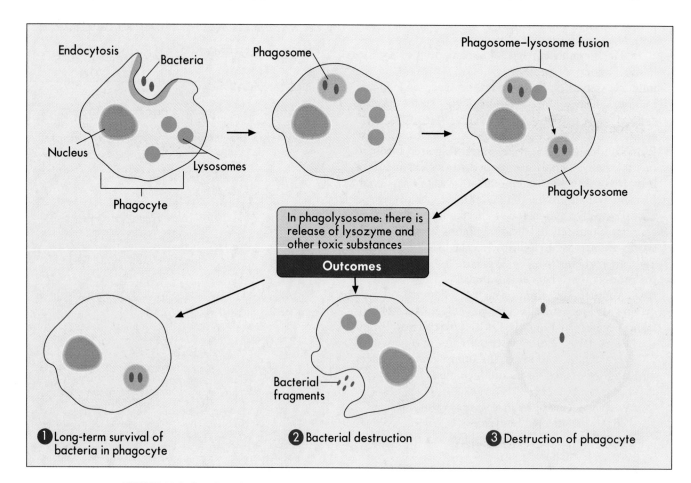

FIGURE 10-6 Overview of phagocyte activity and possible outcomes of phagocyte-bacterial interactions.

Antibody-m
ities of P
When
becom
These
cells
an explosion in the number of B cells that recognize
the antigen and the maturation of B cells into plasma
cells that produce tremendous amounts of antibodies
specific for the antigen. The process also results in the
production of B-memory cells (Figure 10-9).

Antibodies protect the host in a number of ways
that include:

■ Helping phagocytes ingest and kill microorganisms
■ Neutralizing microbial toxins detrimental to host
cells and tissues
■ Promoting bacterial clumping (agglutination)
that facilitates clearing from infection site
■ Inhibiting bacterial motility
■ Combining with microorganisms to activate the
complement system and inflammatory response

Because a population of activated specific B cells
usually is not ready for all microbial antigens, antibody production is delayed when the host is first

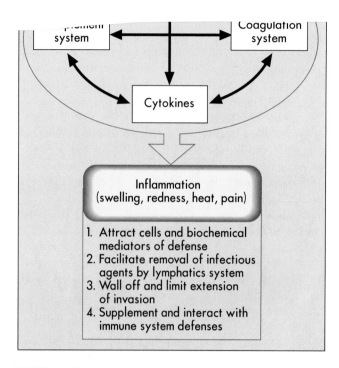

FIGURE 10-7 Overview of the components, signs, and functions of inflammation.

CELLS OF THE IMMUNE SYSTEM

BOX 10-6

B Lymphocytes (B Cells)
Residence: Lymphoid tissues (lymph nodes, spleen, gut-associated lymphiod tissure, tonsils)
Function: Antibody-producing cells
Subtypes:
B lymphocytes; cells waiting to be stimulated by an anti-gen
Plasma cells; activated B lymphocytes that are actively secreting antibody in response to an antigen
B-memory cells; long-lived cells programmed to remem-ber antigens
T Lymphocytes (T Cells)
Residence: Circulate and reside in lymphoid tissues (lymph nodes, spleen, gut-associated lymphoid tissue, tonsils)
Functions: Multiple, see different subtypes
Subtypes:
Helper T cells (T_H); interact with B cells to facilitate anti-body production
Cytotoxic T cells (T_C); recognize and destroy host cells that have been invaded by microorganisms
Suppressor T cells (T_S); shutting down immune response when no longer needed
Natural Killer Cells (NK Cells)
Function similar to cytotoxic T cells but do not require stimula-tion by presence of antigen to function

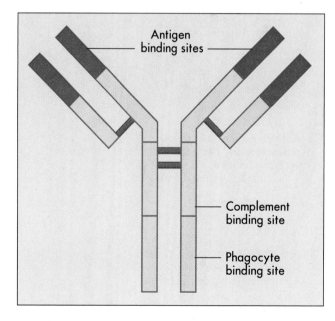

FIGURE 10-8 General structure of the IgG class antibody molecule.

tion, that work to hold the invading organisms in check while antibody production begins. This also emphasizes the importance of B-memory cell production. By virtue of this memory, any additional exposure to the same microorganism will result in a rapid, overwhelming production of protective antibodies so the body is spared the delays that are characteristic of the primary exposure.

Some antigens, such as bacterial capsules and outer membranes, activate B cells to produce antibodies without the intervention of helper T cells. How-

response on the part of the host.

The primary cells that mediate cell-mediated immunity are T lymphocytes that recognize and destroy human host cells infected with microorganisms. This function is extremely important for the destruction and elimination of infecting microorganisms (e.g., viruses, tuberculosis, some parasites, and fungi) that are able to survive within host cells where they are "hidden" from antibody action. Therefore, antibody-mediated immunity targets microorganisms outside of human cells while cell-mediated immunity targets microorganisms inside human cells. However, in many instances these two arms of the immune system overlap and work together.

Like B cells, T cells must be activated. Activation is accomplished by T-cell interactions with other cells that process microbial antigens and present them on their surface (e.g., macrophages and B cells). The responses of activated T cells are very different and depend on the subtype of T cell (Figure 10-10). Activated helper T cells work with B cells for antibody production (see Figure 10-9) and facilitate inflammation by releasing cytokines. Cytotoxic T cells directly interact with and destroy host cells that contain microorganisms. The T-cell subset, helper or cytotoxic cells, that is activated is controlled by an extremely complex series of molecular and genetic events known as the **major histocompatability complex,** or **MHC,** which is a part of cells that present antigens to the T cells.

In summary, the host presents a spectrum of challenges to invading microorganisms, from the physical barriers of skin and mucous membranes to the interactive cellular and biochemical components of inflammation and the immune system. All these systems work together to minimize microbial invasion and avoid damage to vital tissues and organs that can result from the presence of infectious agents.

THE MICROORGANISM'S PERSPECTIVE

Given the complexities of the human host's defense systems, it is no wonder that microbial strategies designed to survive these systems are equally complex. Before considering the microorganism's perspective, a number of definitions and terms must be considered.

Colonization and infections

Many of our body surfaces are colonized with a wide variety of microorganisms without apparent detriment. In contrast, an **infection** involves the growth and multiplication of microorganisms that result in damage to the host. The extent and severity of the damage depend on many factors, including the

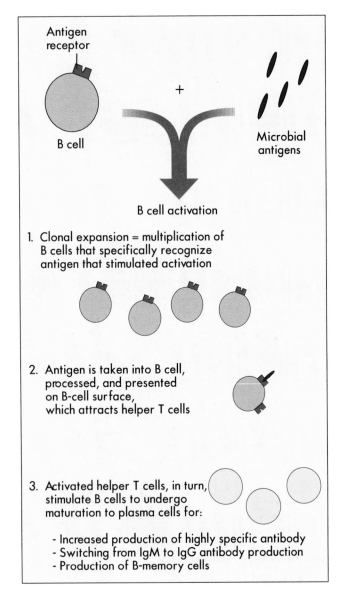

FIGURE 10-9 Overview of B-cell activation that is central to antibody-mediated immunity.

microorganisr
of the infectio.
infected. **Disease** .
notable changes in hu
associated with damages tc
organ systems.

Pathogens and virulence

Microorganisms that cause infections and/or disease
are called **pathogens,** and the characteristics that
enable them to cause disease are referred to as **virulence factors.** Most virulence factors protect the
organism against host attack or mediate damaging
effects on host cells. The terms **pathogenicity** and **virulence** reflect the degree to which a microorganism is
capable of causing disease. An organism of high pathogenicity is very likely to cause disease when encountered, whereas an organism of low pathogenicity is
much less likely to cause infection. When disease does
occur, highly virulent organisms often severely damage the human host. The degree of severity diminishes
with diminishing virulence of the microorganism.

Because host factors play a role in the development of infectious diseases, the distinction between a
pathogen and nonpathogen, or colonizer, is not
always clear. For example, many organisms that colonize our skin usually do not cause disease (i.e., exhibit
low pathogenicity) under normal circumstances.
However, when damage to the skin occurs (Box 10-4)
or this defense is disrupted in some other way, these
organisms can gain access to deeper tissues and establish an infection.

Organisms that only cause infections when one
or more of the host's defense mechanisms are disrupted or malfunctioning are known as **opportunistic
pathogens,** and the infections they cause are referred
to as **opportunistic infections.** On the other hand, several pathogens known to cause serious infections can
be part of a person's normal flora and never cause disease in that person. However, the same organism can
cause life-threatening infections when transmitted to
other persons. The reasons for these inconsistencies
are not fully understood, but such widely different
results undoubtedly involve complex interactions
between microorganism and human. Recognizing and
separating pathogens from nonpathogens presents
one of the greatest challenges in interpretng diagnostic microbiology laboratory results.

Microbial virulence factors

Virulence factors provide microorganisms with the
capacity to avoid host defenses and damage host cells,
tissues, and organs in a number of ways. Some virulence factors are specific for certain pathogenic genera
or species. Substantial differences exist in the way

and fungi cause disease
_____ vary widely. Or impor-
___ of a microorganism's capacity to
cause specific types of infections plays a major role in
developing the diagnostic microbiology procedures
used for isolating and identifying microorganisms (see
Part 3 for more information regarding diagnosis by
organ system).

ATTACHMENT Whether humans encounter microorganisms via the air, ingestion, or by direct contact, the
first step of infection and disease development (a

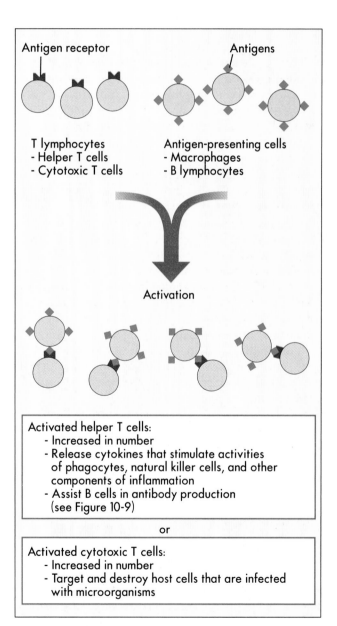

FIGURE 10-10 Overview of T-cell activation that is central to cell-mediated immunity.

process referred to as **pathogenesis**) is microbial attachment to a surface (exceptions being instances in which the organisms are directly implanted by trauma or other means into deeper tissues).

Many of the microbial factors that facilitate attachment of pathogens are the same as those used by nonpathogen colonizers (Box 10-3). The difference between pathogens and colonizers is that pathogens do not always stop at colonization. Also, many pathogens are not part of the normal microbial flora so that their successful attachment also requires that they out compete colonizers for a place on the body's surface. Often medical interventions, such as the overuse of antimicrobial agents that can destroy much of the normal flora, tilt the competition in favor of the invading organism.

INVASION   Once surface attachment has been secured, microbial invasion into subsurface tissues and organs (i.e., infection) is accomplished by traumatic factors such as those listed in Box 10-4 or by the direct action of an organism's virulence factors. Some microorganisms produce factors that force mucosal surface phagocytes (M cells) to ingest them and then release them unharmed into the tissue below the mucosal surface. Other organisms, such as staphylococci and streptococci, are not so subtle. These organisms produce an array of enzymes (e.g., hyaluronidases, nucleases, collagenases) that hydrolyze host proteins and nucleic acids that destroy host cells and tissues. This destruction allows the pathogen to "burrow" through minor openings in the outer surfaces and through deeper tissues. Once a pathogen has penetrated the body surface, strategies that allow microbial survival of the host's inflammatory and immune responses must be used. Alternatively, some pathogens cause disease from their site of attachment without further penetration. For example, in diseases such as diphtheria and whooping cough, the bacteria produce toxic substances that destroy surrounding tissues but the organisms themselves generally do not penetrate the mucosal surface they inhabit.

SURVIVAL AGAINST INFLAMMATION   If a pathogen is to survive, the action of phagocytes and the complement components of inflammation must be avoided or controlled (Box 10-7). Some organisms, such as *Streptococcus pneumoniae*, a common cause of bacterial pneumonia and meningitis, avoid phagocytosis by producing a large capsule that inhibits the phagocytic process. Other pathogens may not be able to avoid phagocytosis but are not effectively destroyed and are able to survive within phagocytes. This is the case for *Mycobacterium tuberculosis*, the bacterium that causes tuberculosis. Still other pathogens use toxins and enzymes to attack and destroy phagocytes before the phagocytes attack and destroy them.

The defenses offered by the complement system depend on a series of biochemical reactions triggered by specific microorganism molecular structures. Therefore, microbial avoidance of complement activation requires that the infecting agent either mask its activating molecules (e.g., via production of a capsule that covers bacterial surface antigens) or produce substances (e.g., enzymes) that disrupt critical biochemical components of the complement pathway.

It is important to remember that any single microorganism may produce various virulence factors and several may be expressed simultaneously. For example, while trying to avoid phagocytosis an organism may also be excreting other enzymes and toxins that help to destroy and penetrate tissue, and be producing other factors designed to interfere with the immune response. Microorganisms may also use host systems to their own advantage. For example, the lymphatic and blood circulation system used to carry pathogens away from the site of infection can also be used by surviving pathogens to become dispersed throughout the body.

SURVIVAL AGAINST THE IMMUNE SYSTEM   Microbial strategies to avoid the defenses of the immune system are outlined in Box 10-8. Again, a pathogen can use more that one strategy to avoid immune-mediated defenses, and microbial survival does not necessarily require devastation of the immune system. The pathogen may merely need to "buy" time to reach a safe area in the body or to be transferred to the next susceptible host. Also, microorganisms can avoid much of the immune response if they do not penetrate the surface layers. This strategy is the hallmark of diseases that are caused by microbial toxins.

BOX 10-7   **MICROBIAL STRATEGIES FOR SURVIVING INFLAMMATION**

Avoid Killing by Phagocytes (PMNs, Monocytes, and Macrophages)
Inhibit ability of phagocyte to ingest by producing capsule
Avoid phagocyte-mediated killing by:
  Inhibiting phagosome-lysosome fusion
  Being resistant to destructive agents (e.g., lysozyme) released by lysosomes
  Actively and rapidly multiplying within phagocyte
  Releasing toxins and enzymes that damage or kill phagocyte

Avoid Effects of the Complement System
Use capsule to hide surface molecules that would otherwise activate the complement system
Produce substances that inhibit the processes involved in complement activation
Produce substances that destroy specific complement proteins

MICROBIAL TOXINS **Toxins** are biochemically active substances that are released by microorganisms and have a particular effect on host cells. Microorganisms use toxins to help them establish infections and multiply within the host. Alternatively, a pathogen may be restricted to a particular body site from which toxins are released to cause widespread problems throughout the body. Toxins also can cause human disease in the absence of the pathogens that produced them. This common mechanism of food poisoning that involves ingestion of preformed bacterial toxins is referred to as **intoxication,** a notable example of which is botulism.

Endotoxins and exotoxins are the two general types of bacterial toxins (Box 10-9). **Endotoxins** are released by gram-negative bacteria and can have devastating effects on the body's metabolism, the most serious being endotoxic shock, which often results in death. The effects of **exotoxins** produced by gram-positive bacteria tend to be more limited and specific than the effects of gram-negative endotoxins. The activities of exotoxins range from those enzymes produced by many staphylococci and streptococci that augment bacterial invasion by damaging host tissues and cells to those that have highly specific activities (e.g., diphtheria toxin that inhibits protein synthesis or the cholera toxin that interferes with host cell signals). Examples of other highly active and specific toxins are those that cause botulism and tetanus by interfering with neuromuscular functions.

OUTCOME AND PREVENTION OF INFECTIOUS DISEASES

OUTCOME OF INFECTIOUS DISEASES

Given the complexities of host defenses and microbial virulence, it is not surprising that the factors determining outcome between these two living entities are also complicated. Basically, outcome depends on the state of the host's health, the virulence of the pathogen, and whether the host can clear the pathogen before infection and disease cause irreparable harm or death (Figure 10-11).

The time for a disease or infection to develop also depends on host and microbial factors. Infectious processes that develop quickly are referred to as **acute infections,** and those that develop and progress slowly, sometimes over a period of years, are known as **chronic infections.** Some pathogens, particularly certain viruses, can be clinically silent inside the body without any noticeable effect on the host before suddenly causing a severe and acute infection. During the silent phase, the infection is said to be **latent.** Again, depending on host and microbial factors, acute, chronic, or latent infections can result in any of the outcomes detailed in Figure 10-11.

BOX 10-8 **MICROBIAL STRATEGIES FOR SURVIVING THE IMMUNE SYSTEM**

- Pathogen multiplies and invades so quickly that damage to host is complete before immune response can be fully activated, or organism's virulence is so great that the immune response is insufficient
- Pathogen invades and destroys cells involved in the immune response
- Pathogen survives, unrecognized, in host cells and avoids detection by immune system
- Pathogen covers its antigens with a capsule so that an immune response is not activated
- Pathogen changes antigens so that immune system is constantly fighting a primary encounter (i.e., the memory of the immune system is neutralized)
- Pathogen produces enzymes (proteases) that directly destroy or inactivate antibodies

---

BOX 10-9 **SUMMARY OF BACTERIAL TOXINS**

Endotoxins
- General toxin common to almost all gram-negative bacteria
- Composed of lipopolysaccharide (LPS) portion of cell envelope
- Released when gram-negative bacterial cell is destroyed
- Effects on host include:
    Disruption of clotting, causing clots to form throughout body (i.e., disseminated intravascular coagulation [DIC])
    Fever
    Activation of complement and immune systems
    Circulatory changes that lead to hypotension, shock, and death

Exotoxins
- Most commonly associated with gram-positive bacteria
- Produced and released by living bacteria; do not require bacterial death for release
- Specific toxins target specific host cells; the type of toxin varies with the bacterial species
- Some kill host cells and help spread bacteria in tissues (e.g., enzymes that destroy key biochemical tissue components or specifically destroy host cell membranes)
- Some destroy or interfere with the specific intracellular activities (e.g., interruption of protein synthesis, interruption of internal cell signals, or interruption of neuromuscular system)

Medical intervention can help the host fight the infection but usually is not instituted until after the host is aware that an infectious process is under way. The clues that an infection is occurring are known as the **signs** and **symptoms** of disease and result from host responses (e.g., inflammatory and immune responses) to the action of microbial virulence factors (Box 10-10). The signs and symptoms reflect the stages of infection. In turn, the stages of infection generally reflect the stages in host-microorganism interactions (Figure 10-12).

Whether medical procedures contribute to controlling or clearing an infection depends on key factors, including:

- The severity of the infection, which is determined by host and microbial interactions already discussed
- Accuracy in diagnosing the pathogen(s) causing the infection, which is the primary function of diagnostic microbiology
- Whether the patient receives appropriate treatment for the infection (this heavily depends on accurate diagnosis)

PREVENTION OF INFECTIOUS DISEASES

The treatment of an infection is often difficult and not always successful. Because much of the damage is already done before appropriate medical intervention is begun, the microorganisms gain too much of a "head start." Another strategy for combating infectious diseases is to stop infections before they even start (i.e., **disease prevention**). As discussed at the beginning of this chapter, the first step in any host-microorganism relationship is the encounter. Therefore, strategies to prevent disease involve interrupting encounters or minimizing the risk of infection when encounters do occur. As outlined in Box 10-11, interruption of encounters may be accomplished by preventing transmission of the infecting agents and by controlling or destroying reservoirs of human pathogens. Of interest, most of these measures do not really involve medical practices but rather social practices and policies.

Immunization

If encounters do occur, medical strategies exist for minimizing the risk for disease development. One of the most effective methods is **immunization**. This

---

BOX 10-10   **SIGNS AND SYMPTOMS OF INFECTION AND INFECTIOUS DISEASES**

General or localized aches and pains	Cough and sneezes
Headaches	Congestion of nasal and sinus passages
Fever	Sore throat
Swollen lymph nodes	Nausea and vomiting
Rashes	Diarrhea
Redness and swelling	

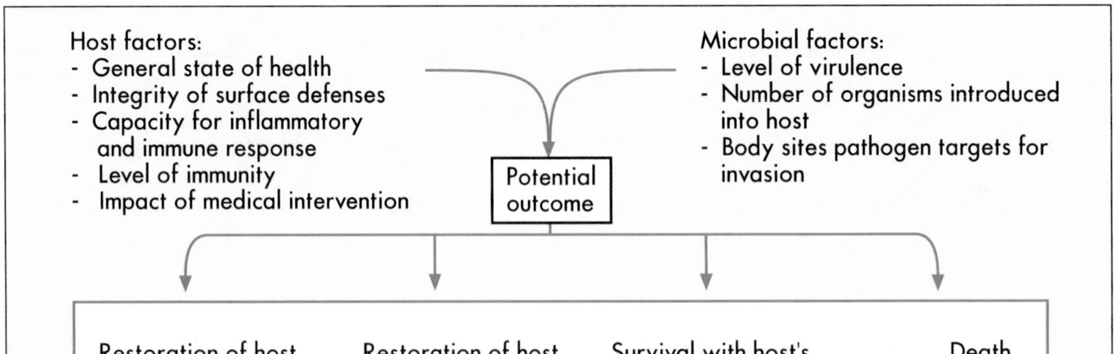

Host factors:
- General state of health
- Integrity of surface defenses
- Capacity for inflammatory and immune response
- Level of immunity
- Impact of medical intervention

Microbial factors:
- Level of virulence
- Number of organisms introduced into host
- Body sites pathogen targets for invasion

Potential outcome

Restoration of host to complete health

Restoration of host to health with residual effects

Survival with host's health severely compromised

Death

Full spectrum of outcomes

FIGURE 10-11 Possible outcomes of infections and infectious diseases.

practice takes advantage of the specificity and memory of the immune system. There are two basic approaches to immunization: active and passive. With **active immunization,** modified antigens from pathogenic microorganisms are introduced into the body and cause an immune response. If or when the host encounters the pathogen in nature, the memory of the immune system will ensure minimal delay in the immune response, thus affording strong protection. Alternatively, with **passive immunization** antibodies against a particular pathogen have been produced in one host and are transferred to a second host where they provide temporary protection. The passage of maternal antibodies to the newborn is a key example of natural passive immunization. Active immunity is generally more long lasting because the immunized host's own immune response is activated. However, for complex reasons, active immunity only has been successfully applied for relatively few infectious diseases, some of which include diphtheria, whooping cough (pertussis), tetanus, influenza, polio, small pox, measles, and hepatitis.

Prophylactic antimicrobial therapy, the administration of antibiotics when the risk of developing an infection is high, is another common medical intervention for preventing infection.

Epidemiology
To prevent infectious diseases, information is required regarding the sources of pathogens, the method(s) of transmission to and among humans, human risk factors for encountering the pathogen and developing infection, and factors that contribute to good and bad

BOX 10-11 STRATEGIES FOR PREVENTION OF INFECTIOUS DISEASES

Preventing Transmission:
Avoid direct contact with infected persons or take protective measures when direct contact is going to occur (e.g., wear gloves, wear condoms)
Block spread of airborne microorganisms by wearing masks or isolating persons with infections transmitted by air
Use of sterile medical techniques
Controlling Microbial Reservoirs:
Sanitation and disinfection
Sewage treatment
Food preservation
Water treatment
Control of pests and insect vector populations
Minimizing Risk Before or Shortly after Exposure:
Immunization
Cleansing and use of antiseptics
Prophylactic use of antimicrobial agents

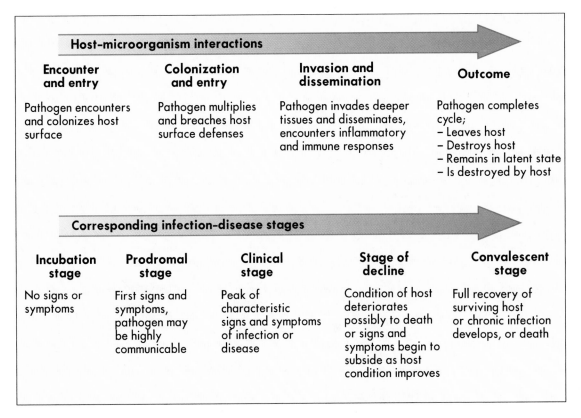

FIGURE 10-12 Host-microorganism interactions and stages of infection or disease.

outcomes resulting from the encounter. **Epidemiology** is the science that characterizes these aspects of infectious diseases and monitors the effect diseases have on public health. By fully characterizing the circumstances associated with the acquisition and dissemination of infectious diseases, there is a better chance for preventing and eliminating these diseases. Additionally, many epidemiologic strategies that have been developed for public health also apply in hospitals for the control of hospital-based (i.e., nosocomial) infec-

tions (for more information on infection control in hospitals see Chapter 7).

The field of epidemiology is broad and complex and beyond the scope of this text. However, diagnostic microbiology laboratory personnel and hospital epidemiologists often work very closely to investigate problems. Therefore, familiarity with certain epidemiologic terms and concepts is important (Box 10-12).

Because the central focus of epidemiology is on tracking and characterizing infections and infectious

---

BOX 10-12 — DEFINITIONS OF SELECTED EPIDEMIOLOGIC TERMS

Carrier:	A person who carries the etiologic agent but shows no apparent signs or symptoms of infection or disease
Common source:	The etiologic agent responsible for an epidemic or outbreak originates from a single source or reservoir
Disease incidence:	The number of diseased or infected persons in a population
Disease prevalence:	Percentage of diseased persons in a given population at a particular time
Endemic:	A disease is constantly present at some rate of occurrence in a particular location
Epidemic:	A larger than normal number of diseased or infected individuals in a particular location
Etiologic agent:	A microorganism responsible for causing infection or infectious disease
Mode of transmission:	Means by which etiologic agents are brought in contact with the human host (e.g., infected blood, contaminated water, insect bite)
Morbidity:	The state of disease and its associated effects on the host
Morbidity rate:	The incidence of a particular disease state
Mortality:	Death resulting from disease
Mortality rate:	The incidence in which a disease results in death
Nosocomial infection:	Infection in which etiologic agent was acquired in a hospital
Outbreak:	A larger than normal number of diseased or infected individuals that occurs over a relatively short period
Pandemic:	An epidemic that spans the world
Reservoir:	Origin of the etiologic agent or location from which they disseminate (e.g., water, food, insects, animals, other humans)
Strain typing:	Laboratory-based characterization of etiologic agents designed to establish their relatedness to one another during a particular outbreak or epidemic
Surveillance:	Any type of epidemiologic investigation that involves data collection for characterizing circumstances surrounding the incidence or prevalence of a particular disease or infection
Vector:	A living entity (animal, insect, or plant) that transmits the etiologic agent
Vehicle:	A nonliving entity that is contaminated with the etiologic agent and as such is the mode of transmission for that agent

diseases, this field heavily depends on diagnostic microbiology. Epidemiologic investigations cannot proceed without first knowing the etiologic agents involved. Therefore, the procedures and protocols used in diagnostic microbiology to detect, isolate, and characterize human pathogens are not only essential for patient care but also play a central role in epidemiologic studies focused on disease prevention and the general improvement of public health. In fact, microbiologists who work in clinical laboratories are often the first to recognize patterns that suggest potential outbreaks or epidemics.

Bibliography

Atlas, R.M. 1995. Principles of microbiology. Mosby, St. Louis.

Brock, T.D., Madigan, M.T., Martinko, J.M., and Parker, J., editors: 1994. Biology of microorganisms. Prentice Hall, Englewood Cliffs, N.J.

Murray, P.R., Kobayashi, G.S., Pfaller, M.A., and Rosenthal, K.S., editors: 1994. Medical microbiology, ed 2. Mosby, St. Louis.

Ryan, K.J., editor. 1994. Sherris medical microbiology: an introduction to infectious diseases. Appleton & Lange, Norwalk, Conn.

Salyers, A.A. and Whitt, D.D. 1994. Bacterial pathogenesis: a molecular approach. ASM Press, Washington, D.C.

Schaechter, M., Medoff, G., and Eisenstein, B.I., editors: 1993. Mechanisms of microbial disease, ed 2, Williams & Wilkins, Baltimore.

11 | ROLE OF MICROSCOPY IN THE DIAGNOSIS OF INFECTIOUS DISEASES

The basic flow of procedures involved in the laboratory diagnosis of infectious diseases is:

1. Direct examination of patient specimens for the presence of etiologic agents
2. Growth and cultivation of the agents from these same specimens
3. Analysis of the cultivated organisms to establish their identification and other pertinent characteristics such as susceptibility to antimicrobial agents.

For certain infectious diseases, this process may also include measuring the patient's immune response to the infectious agent.

Microscopy is the most common method used both for the detection of microorganisms directly in clinical specimens and for the characterization of organisms grown in culture (Box 11-1). **Microscopy** is defined as the use of a microscope to magnify (i.e., visually enlarge) objects too small to be visualized with the naked eye so that their characteristics are readily observable. Because the majority of infectious agents cannot be detected with the unaided eye, microscopy plays a pivotal role in the laboratory. Microscopes and microscopic methods vary, but only those of primary use in diagnostic microbiology will be discussed.

The method used to process patient specimens is dictated by the type and body source of specimen (see Part 3). Regardless of the method used, some portion of the specimen usually is reserved for microscopic examination. Specific stains or dyes applied to the specimens, combined with particular methods of microscopy, can detect etiologic agents in a rapid, relatively inexpensive, and productive way. Microscopy also plays a key role in the characterization of organisms that have been cultivated in the laboratory (for more information regarding cultivation of bacteria see Chapter 12).

The types of microorganisms to be detected, identified, and characterized determine the most appropriate types of microscopy to use. Table 11-1 outlines the four types of microscopy used in diagnostic microbiology and their relative utility for each of the four major microorganism groups. **Bright-field**

microscopy (also known as **light microscopy**) and **fluorescence microscopy** have the widest use and application. Which microorganisms can be detected or identified by each microscopic method also depends on the methods used to highlight the microorganisms and their key characteristics. This enhancement is usually achieved using various dyes or stains.

BRIGHT-FIELD (LIGHT) MICROSCOPY

PRINCIPLES OF LIGHT MICROSCOPY

For light microscopy, visible light is passed through the specimen and then through a series of lenses that reflect the light in a manner that results in magnification of the organisms present in the specimen (Figure 11-1). The total magnification achieved is the product of the lenses used.

Magnification

In most light microscopes the **objective lens,** which is closest to the specimen, magnifies objects 100× (times) and the **ocular lens,** which is nearest the eye, magnifies 10×. Using these two lenses in combination, organisms in the specimen are magnified 1000× their actual size when viewed through the ocular lens. Objective lenses of lower magnification are available so that those of 10×, 20×, and 40× magnification power can provide total magnifications of 100×, 200×, and 400×, respectively. Magnification of 1000× allows for the visualization of fungi, most parasites, and most bacteria but is not sufficient for observing viruses, which require magnification of 100,000× or more (see the section on electron microscopy in this chapter).

Resolution

To optimize visualization, other factors besides magnification must be considered. **Resolution,** defined as the extent to which detail in the magnified object is maintained, is also essential. Without it everything would be magnified as an indistinguishable blur. Therefore, **resolving power,** the closest distance between two objects that when magnified still allows the two objects to be distinguished from each other, is extremely

BOX 11-1   APPLICATIONS OF MICROSCOPY IN DIAGNOSTIC MICROBIOLOGY

- Rapid preliminary organism identification by direct visualization in patient specimens
- Rapid final identification of certain organisms by direct visualization in patient specimens
- Detection of different organisms present in the same specimen
- Detection of organisms not easily cultivated in the laboratory
- Evaluation of patient specimens for the presence of cells indicative of inflammation (i.e., phagocytes) or contamination (i.e., squamous epithelial cells)
- Determination of an organism's clinical significance. Bacterial contaminants usually are not present in patient specimens at sufficiently high numbers ($\geq 10^5$ cells/mL) to be seen by light microscopy
- Provide preculture information about which organisms might be expected to grow so that appropriate cultivation techniques are used
- Determine which tests and methods should be used for identification and characterization of cultivated organisms
- Provide a method for investigating unusual or unexpected laboratory test results

important. The resolving power of most light microscopes allows bacterial cells to be distinguished from one another but usually does not allow bacterial structures, internal or external, to be detected.

To achieve the level of resolution desired with 1000× magnification, oil immersion must be used in conjunction with light microscopy. Immersion oil is used to fill the space between the objective lens and the glass slide onto which the specimen has been affixed. The oil enhances resolution by preventing light rays from dispersing and changing wavelength after passing through the specimen. A specific objective lens, the **oil immersion lens,** is designed for use with oil; this lens provides 100× magnification on most light microscopes.

Lower magnifications (i.e., 100× or 400×) may be used to locate specimen samples in certain areas on a microscope slide, or to observe microorganisms such as some fungi and parasites. The 1000× magnification provided by the combination of ocular and oil immersion lenses usually is required for optimal detection and characterization of bacteria.

Contrast

The third key component to light microscopy is **contrast,** which is needed to make objects stand out from the background. Because microorganisms are essentially transparent, owing to their microscopic dimensions and high water content, they cannot be easily detected among the background materials and debris in patient specimens. Lack of contrast is also a problem for the microscopic examination of microorganisms grown in culture. Contrast is most commonly achieved by staining techniques that highlight organisms and allow them to be differentiated from one another and from background material and debris.

STAINING TECHNIQUES FOR LIGHT MICROSCOPY

Smear Preparation

Staining methods are either used directly with patient specimens or are applied to preparations made from microorganisms grown in culture. Details of specimen processing are presented throughout Part 3 and in most instances the preparation of every specimen

TABLE 11-1   MICROSCOPY FOR DIAGNOSTIC MICROBIOLOGY

ORGANISM GROUP	BRIGHT-FIELD MICROSCOPY	FLUORESCENCE MICROSCOPY	DARK-FIELD MICROSCOPY	ELECTRON MICROSCOPY
Bacteria	+	+	±	−
Fungi	+	+	−	−
Parasites	+	+	−	±
Viruses	−	+	−	±

+, Commonly used; ±, limited use; −, rarely used.

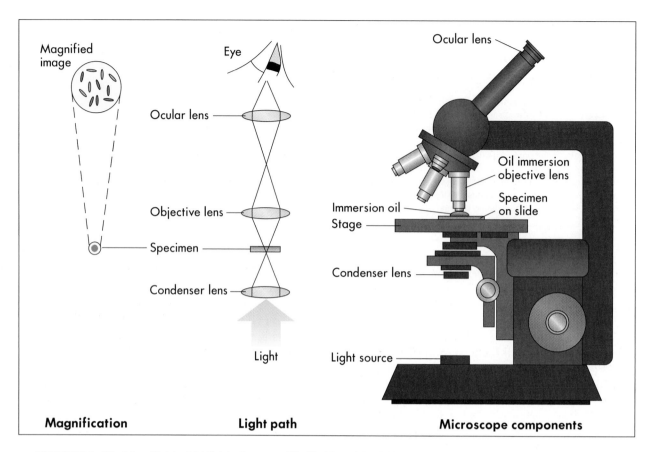

FIGURE 11-1   Principles of bright-field (light) microscopy. (Modified from Atlas, R.M. 1995. Principles of microbiology. Mosby, St. Louis.)

includes application of some portion of the specimen to a clean glass slide (i.e., **"smear" preparation**) for subsequent microscopic evaluation.

Generally, specimen samples are placed on the slide using a swab that contains patient material or by using a pipette into which liquid specimen has been aspirated (Figure 11-2). Material to be stained is dropped (if liquid) or rolled (if on a swab) onto the surface of a clean, dry, glass slide. To avoid contamination of culture media, once a swab has touched the surface of a nonsterile slide, it should not be used for subsequently inoculating media.

For staining microorganisms grown in culture, a sterile needle may be used to transfer a small amount of growth from a solid medium to the surface of the slide. This material is emulsified in a drop of sterile water or saline on the slide. For small amounts of growth that might become lost in even a drop of saline, a sterile wooden applicator stick can be used to touch the growth, this material is then rubbed directly onto the slide, where it can be easily seen. The material placed on the slide to be stained is allowed to dry and is affixed to the slide by placing

on a slide warmer (60° C) for at least 10 minutes, or by flooding with 95% methanol for 1 minute. To examine organisms grown in liquid medium an aspirated sample of the broth culture is applied to the slide, dried, and fixed before staining.

Smear preparation varies depending on the type of specimen being processed (see chapters in Part 3 that discuss specific specimen types) and on the staining methods to be used. Nonetheless, the general rule for smear preparation is that sufficient material must be applied to the slide so chances for detecting and distinguishing microorganisms are maximized. At the same time application of excessive material that could interfere with the passage of light through the specimen or that could distort the details of microorganisms must be avoided. Finally, the staining method to be used is dictated by which microorganisms are being sought.

As listed in Table 11-1, light microscopy has applications for bacteria, fungi, and parasites. However, the stains used for these microbial groups differ extensively. Those primarily designed for examination of parasites and fungi by light microscopy are discussed in

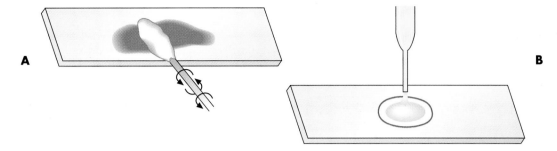

FIGURE 11-2 Smear preparations by swab roll (**A**) and pipette deposition (**B**) of patient specimen on glass microscope slide.

Chapters 5 and 6, respectively. The stains for microscopic examination of bacteria, the Gram stain and the acid-fast stains, are discussed in this chapter.

Gram stain

The Gram stain is the principle stain used for microscopic examination of bacteria. Nearly all clinically important bacteria can be detected using this method, the only exceptions being those organisms that exist almost exclusively within host cells (e.g., chlamydia), those that lack a cell wall (e.g., mycoplasma and ureaplasma), and those of insufficient dimension to be resolved by light microscopy (e.g., spirochetes). First devised by Hans Christian Gram during the late nineteenth century, the Gram stain can be used to divide most bacterial species into two large groups: those that take up the basic dye, crystal violet (i.e., gram-positive bacteria), and those that allow the crystal violet dye to wash out easily with the decolorizer alcohol or acetone (i.e., gram-negative bacteria).

PROCEDURE Although modifications of the classic Gram stain that involve changes in reagents and timing exist, the principles and results are the same for all modifications. The classic Gram stain procedure entails fixing clinical material to the surface of the microscope slide either by heating or by using methanol (Figure 11-3). Methanol fixation preserves the morphology of host cells, as well as bacteria, and is especially useful for examining bloody specimen material. Slides are overlaid with 95% methanol for 1 minute; the methanol is allowed to run off, and the slides are air dried before staining. After fixation, the first step in the Gram stain is the application of the primary stain **crystal violet.** A mordant, **Gram's iodine,** is applied after the crystal violet to chemically bond the alkaline dye to the bacterial cell wall. The decolorization step distinguishes gram-positive from gram-negative cells. After **decolorization,** organisms

that stain gram-positive retain the crystal violet and those that are gram-negative are cleared of crystal violet. Addition of the counterstain **safranin** will stain the clear gram-negative bacteria pink or red.

PRINCIPLE The difference in composition between gram-positive cell walls, which contain thick peptidoglycan with numerous teichoic acid cross-linkages, and gram-negative cell walls, which consist of a thinner layer of peptidoglycan, accounts for the Gram staining differences between these two major groups of bacteria. Presumably, the extensive teichoic acid cross-links contribute to the ability of gram-positive organisms to resist alcohol decolorization. Although the counterstain may be taken up by the gram-positive organisms, their purple appearance will not be altered.

Gram-positive organisms that have lost cell wall integrity because of antibiotic treatment, old age, or action of autolytic enzymes may allow the crystal violet to wash out with the decolorizing step and may appear gram-variable, with some cells staining pink and others staining purple. However, for identification purposes these organisms are considered to be truly gram-positive. On the other hand, gram-negative bacteria rarely, if ever, retain crystal violet (e.g., appear purple) if the staining procedure has been properly performed. Host cells, such as red and white blood cells (phagocytes), allow the crytal violet stain to wash out with decolorization and should appear pink on smears that have been correctly prepared and stained.

GRAM STAIN EXAMINATION Once stained, the smear is examined using the oil immersion (1000× magnification) lens. When clinical material is Gram-stained (e.g., the **direct smear**), the slide is evaluated for the presence of bacterial cells as well as the Gram reactions, morphologies (e.g., cocci or bacilli), and

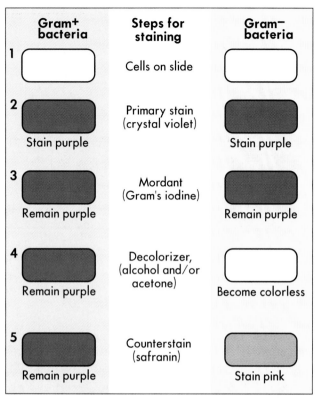

Gram+ bacteria	Steps for staining	Gram− bacteria
1	Cells on slide	1
2 Stain purple	Primary stain (crystal violet)	2 Stain purple
3 Remain purple	Mordant (Gram's iodine)	3 Remain purple
4 Remain purple	Decolorizer, (alcohol and/or acetone)	4 Become colorless
5 Remain purple	Counterstain (safranin)	5 Stain pink

1 Fix material on slide with methanol or heat. If slide is heat fixed, allow it to cool to the touch before applying stain.

2 Flood slide with crystal violet (purple) and allow it to remain on the surface without drying for 10 to 30 seconds. Rinse the slide with tap water, shaking off all excess.

3 Flood the slide with iodine to increase affinity of crystal violet and allow it to remain on the surface without drying for twice as long as the crystal violet was in contact with the slide surface (20 seconds of iodine for 10 seconds of crystal violet, for example). Rinse with tap water, shaking off all excess.

4 Flood the slide with decolorizer for 10 seconds and rinse off immediately with tap water. Repeat this procedure until the blue dye no longer runs off the slide with the decolorizer. Thicker smears require more prolonged decolorizing. Rinse with tap water and shake off excess.

5 Flood the slide with counterstain and allow it to remain on the surface without drying for 30 seconds. Rinse with tap water and gently blot the slide dry with paper towels or bibulous paper or air dry. For delicate smears, such as certain body fluids, air drying is the best method.

6 Examine microscopically under an oil immersion lens at 1000x for phagocytes, bacteria, and other cellular material.

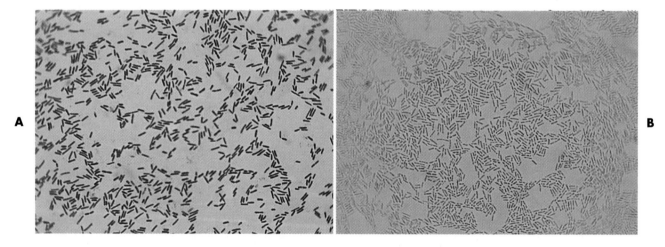

FIGURE 11-3 Gram stain procedures and principles. **A,** Gram-positive bacteria observed under oil immersion. **B,** Gram-negative bacteria observed under oil immersion. (Modified from Atlas, R.M. 1995. Principles of microbiology, Mosby, St. Louis.)

arrangements (e.g., chains, pairs, clusters) of the cells seen (Figure 11-4). This information often provides a preliminary diagnosis regarding the infectious agents and frequently is used to direct initial therapies for the patient.

The direct smears should also be examined for the presence of inflammatory cells (e.g., phagocytes)

that are key indicators of an infectious process. Noting the presence of other host cells, such as squamous epithelial cells in respiratory specimens, is also helpful because the presence of these cells may indicate contamination with organisms and cells from the mouth (for more information regarding interpretation of respiratory smears see Chapter 21). Observing

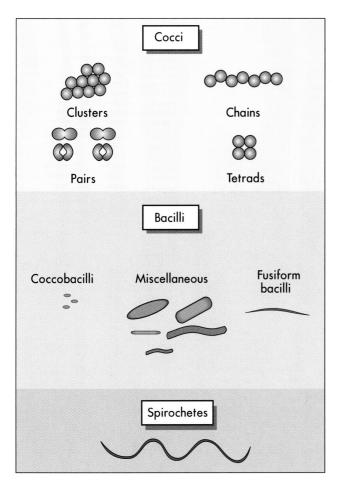

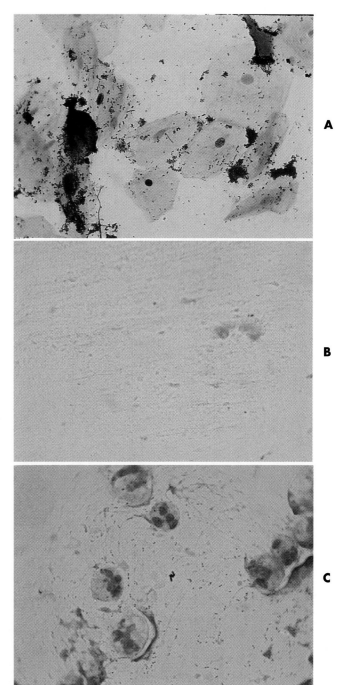

FIGURE 11-4 Examples of common bacterial cellular morphologies, Gram staining reactions, and cellular arrangements.

FIGURE 11-5 Gram stains of direct smears showing squamous cells and bacteria **(A)**, proteinaceous debris alone **(B)**, and proteinaceous debris with polymorphonuclear leukocytes and bacteria **(C)**.

background tissue debris and proteinaceous material, which generally stain gram-negative, also provides helpful information. For example, the presence of such material indicates that specimen material was adequately affixed to the slide. Therefore, the absence of bacteria or inflammatory cells on such a smear is "real" and not likely the result of loss of specimen during staining (Figure 11-5). Other ways that Gram stain evaluations of direct smears are used are discussed throughout the chapters of Part 3 that deal with infections of specific body sites.

Several examples of Gram stains of direct smears are provided in Figure 11-6. Basically, whatever is observed is also recorded and is used to produce a laboratory report for the physician. The report typically includes:

- The presence of host cells and debris
- The Gram reactions, morphologies (e.g., cocci, bacilli, coccobacilli), and arrangement of bacterial cells present. Note: reporting the absence of bacteria and host cells can be equally as important

- Optionally, the relative amounts of bacterial cells (e.g., rare, few, moderate, many) may be provided. However, it is important to remember that to visualize bacterial cells by light microscopy a minimum concentration of 10^5 cells per mL of specimen is required. This is a large number of bacteria for any normally sterile body site and to

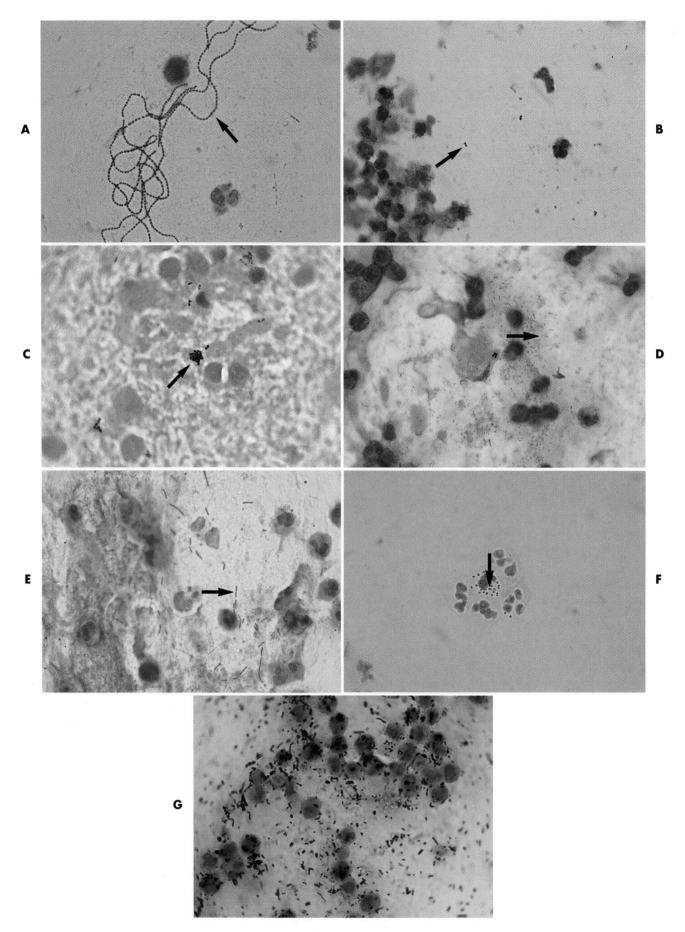

FIGURE 11-6 Gram stain of direct smears showing polymorphonuclear leukocytes, proteinaceous debris, and bacterial morphologies (*arrows*), including gram-positive cocci in chains (**A**), gram-positive diplococci (**B**), gram-positive cocci in clusters (**C**), gram-negative coccobacilli (**D**), gram-negative bacilli (**E**), gram-negative diplococci (**F**), and mixed gram-positive and gram-negative morphologies (**G**).

describe the quantity as rare or few based on microscopic observation may be understating their significance in a clinical specimen. On the other hand, noting the relative amounts seen on direct smear may be useful laboratory information to correlate smear results with the amount of growth observed subsequently from cultures

Although Gram stain evaluation of direct smears is routinely used as an aid in the diagnosis of bacterial infections, unexpected but significant findings of other infectious etiologies may be detected and cannot be ignored. For example, fungal cells and elements generally stain gram-positive, but they may take up the crystal violet poorly and appear gram-variable (e.g., both pink and purple) or gram-negative. Because infectious agents besides bacteria may be detected by Gram stain, any unusual cells or structures observed on the smear should be evaluated further before being dismissed as unimportant (Figure 11-7).

GRAM STAIN OF BACTERIA GROWN IN CULTURE   The Gram stain also plays a key role in the identification of bacteria grown in culture. Similar to direct smears, smears prepared from bacterial growth are evaluated for the bacterial cells' Gram reactions, morphologies, and arrangements (see Figure 11-4). If growth from more than one specimen is to be stained on the same slide, a wax pencil may be used to create divisions. Drawing a "map" of such a slide so that different Gram stain results can be recorded in an organized fashion is helpful (Figure 11-8). The smear results will be used to determine subsequent testing for identifying and characterizing the organisms isolated from the patient specimen.

Acid-fast stains

The acid-fast stain is the other commonly used stain for light-microscopic examination of bacteria.

PRINCIPLE   The acid-fast stain is specifically designed for a subset of bacteria whose cell walls contain long-chain fatty (mycolic) acids. Mycolic acids render the cells resistant to decolorization, even with acid alcohol decolorizers. Thus, these bacteria are referred to as being **acid-fast.** Although these organisms may Gram stain slightly or poorly as gram-positive, the acid-fast stain takes full advantage of the waxy content of the cell walls to maximize detection. Mycobacteria are the most commonly encountered acid-fast bacteria, typified by *Mycobacterium tuberculosis,* the etiologic agent of tuberculosis. Bacteria lacking cell walls fortified with mycolic acids cannot resist decolorization with acid alcohol and are categorized as being **non–acid-fast,** a trait typical of most

FIGURE 11-7 Gram stains of direct smears can reveal infectious etiologies other than bacteria, such as the yeast *Candida tropicalis.*

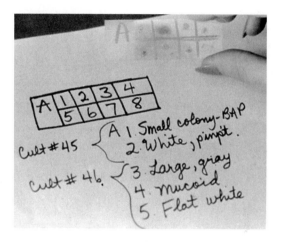

FIGURE 11-8 Example of a slide map for staining several bacterial colony samples on a single slide.

clinically relevant bacteria. However, some degree of acid-fastness is a characteristic of a few nonmycobacterial bacteria, such as *Nocardia* spp., and coccidian parasites, such as *Cryptosporidium* spp.

PROCEDURE   The classic acid-fast staining method, **Ziehl-Neelsen,** is outlined in Figure 11-9. The procedure requires heat to allow the **primary stain (carbolfuchsin)** to enter the wax-containing cell wall. A modification of this procedure, the **Kinyoun** acid-fast method, is more commonly used today. Because of a higher concentration of phenol in the primary stain solution, heat is not required for the intracellular penetration of carbolfuchsin. This modification, which is presented in more detail in Chapter 60, is referred to as the "cold" method. Another modification of the acid-fast stain that is used for identifying certain nonmycobacterial species is described and

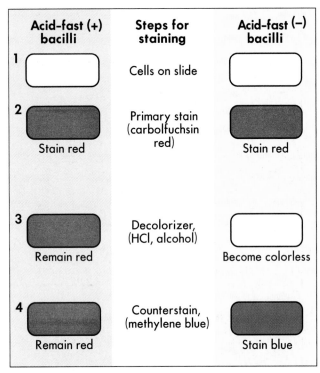

Acid-fast (+) bacilli	Steps for staining	Acid-fast (−) bacilli
1	Cells on slide	1
2 Stain red	Primary stain (carbolfuchsin red)	2 Stain red
3 Remain red	Decolorizer, (HCl, alcohol)	3 Become colorless
4 Remain red	Counterstain, (methylene blue)	4 Stain blue

1   Fix smears on heated surface (60° C for at least 10 minutes).

2   Flood smears with carbolfuchsin (Primary Stain) and heat to almost boiling by performing the procedure on an electrically heated platform or by passing the flame of a Bunsen burner underneath the slides on a metal rack. The stain on the slides should steam. Allow slides to sit for 5 minutes after heating; do not allow them to dry out. Wash the slides in distilled water (Note: tap water may contain acid-fast bacilli). Drain off excess liquid.

3   Flood slides with 3% HCl in 95% ethanol (Decolorizer) for approximately 1 minute. Check to see that no more red color runs off the surface when the slide is tipped. Add a bit more decolorizer for very thick slides or those that continue to "bleed" red dye. Wash thoroughly with water and remove the excess.

4   Flood slides with Methylene blue (Counterstain) and allow to remain on surface of slides for 1 minute. Wash with distilled water and stand slides upright on paper towels to air dry. Do not blot dry.

5   Examine microscopically, screening at 400x magnification and confirm all suspicious (i.e. red) organisms at 1000x magnification using an oil-immersion lens.

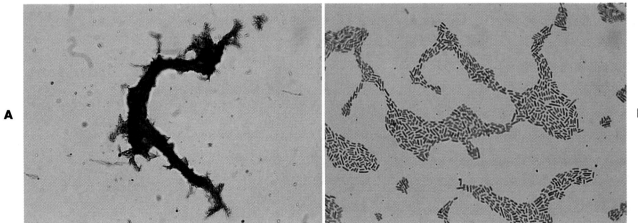

FIGURE 11-9 The Ziehl-Neelsen acid-fast stain procedures and principles. **A**, Acid-fast positive bacilli. **B**, Acid-fast negative bacilli. (Modified from Atlas, R.M. 1995. Principles of microbiology, Mosby, St. Louis.)

discussed in Part 4, Section 13. When the acid-fast–stained smear is read with 1000× magnification, acid-fast–positive organisms stain red. Depending on the type of counterstain used (e.g., **methylene blue** or **malachite green**), other microorganisms, host cells, and debris stain a blue to blue-green color (Figures 11-9 and 11-10).

As with the Gram stain, the acid-fast stain is used to detect acid-fast bacteria (e.g., mycobacteria)

directly in clinical specimens and provide preliminary identification information for suspicious bacteria grown in culture. Because mycobacterial infections are much less common than infections caused by other non-acidfast bacteria, the acid-fast stain is only performed on specimens from patients highly suspected of having a mycobacterial infection. That is, Gram staining is a routine part of most bacteriology procedures while acid-fast staining is reserved for

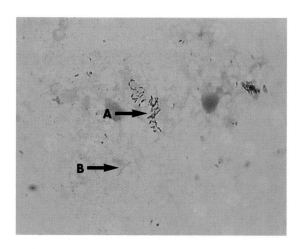

FIGURE 11-10 Acid-fast stain of direct smear to show acid-fast bacilli staining deep red (*arrow A*) and non–acid-fast bacilli and host cells staining blue with the counter stain methylene blue (*arrow B*).

specific situations. Similarly, the acid-fast stain is only applied to bacteria grown in culture when there is a suspicion based on other characteristics that mycobacteria are present (for more information regarding identification of mycobacteria see Chapter 60).

PHASE CONTRAST MICROSCOPY

Instead of using a stain to achieve the contrast necessary for observing microorganisms, altering microscopic techniques to enhance contrast offers another approach. Phase contrast microscopy is one such contrast-enhancing technique. By this method, beams of light pass through the specimen and are partially deflected by the different densities or thicknesses (i.e., refractive indices) of the microbial cells or cell structures in the specimen. The greater the refractive index of an object, the more the beam of light is slowed, which results in decreased light intensity. These differences in light intensity translate into differences that provide contrast. Therefore, phase microscopy translates differences in phases within the specimen into differences in light intensities that result in contrast among objects within the specimen being observed.

Microscopy that depends on staining microorganisms only allows dead organisms to be observed. Because staining is not part of phase contrast microscopy, this method offers the advantage of allowing observation of viable microorganisms. The method is not commonly used in most aspects of diagnostic microbiology, but it is used to identify medically important fungi grown in culture (for more information regarding the use of phase contrast microscopy for fungal identification see Part 4).

FLUORESCENT MICROSCOPY

PRINCIPLE OF FLUORESCENT MICROSCOPY

Certain dyes, called **fluors** or **fluorochromes,** can be raised to a higher energy level after absorbing ultraviolet (excitation) light. When the dye molecules return to their normal, lower energy state, they release excess energy in the form of visible (fluorescent) light. This process is called **fluorescence,** and microscopic methods have been developed to exploit the enhanced contrast and detection that this phenomenon provides.

Figure 11-11 depicts diagrammatically the principle of fluorescent microscopy in which the excitation light is emitted from above (epifluorescence). An excitation filter passes light of the desired wavelength to excite the fluorochrome that has been used to stain the specimen. A barrier filter in the objective lens prevents the excitation wavelengths from damaging the eyes of the observer. When observed through the ocular lens, fluorescing objects appear brightly lit against a dark background.

The color of the fluorescent light depends on the dye and light filters used. For example, use of the fluorescent dyes acridine orange, auramine, and fluorescein isothiocyanate (FITC) requires blue excitation light, exciter filters that select for light in the 450 to 490 λ wavelength range, and a barrier filter for 515 λ. Calcofluor white, on the other hand, requires violet excitation light, an exciter filter that selects for light in the 355 to 425 λ wavelength range, and a barrier filter for 460 λ. Which dye is used often depends on which organism is being sought and the fluorescent method used. The intensity of the contrast obtained with fluorescent microscopy is an advantage it has over the use of chromogenic dyes (e.g., crystal violet and safranin of the Gram stain) and light microscopy.

STAINING TECHNIQUES FOR FLUORESCENT MICROSCOPY

Based on the composition of the fluorescent stain reagents, fluorescent staining techniques may be divided into two general categories: **fluorochroming,** in which a fluorescent dye is used alone, and **immunofluorescence,** in which fluorescent dyes have been linked (conjugated) to specific antibodies. The principle differences between these two methods are diagrammed in Figure 11-12.

Fluorochroming
In fluorochroming a direct chemical interaction occurs between the fluorescent dye and a component of the bacterial cell; this interaction is the same as occurs with the stains used in light microscopy. The difference is that use of a fluorescent dye enhances contrast and amplifies the observer's ability to detect stained cells tenfold greater than would be observed

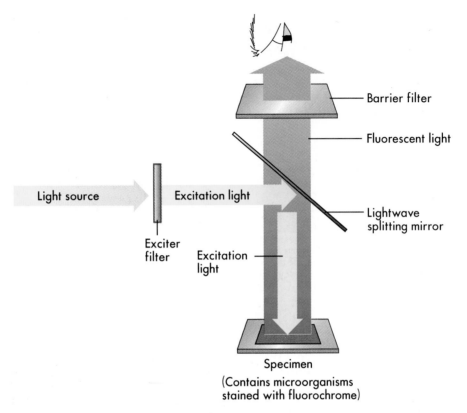

FIGURE 11-11 Principle of fluorescent microscopy. Microorganisms in a specimen are stained with a fluorescent dye. On exposure to excitation light, organisms are visually detected by the emission of fluorescent light by the dye with which they have been stained (i.e., fluorochroming) or "tagged" (i.e., immunofluorescence).

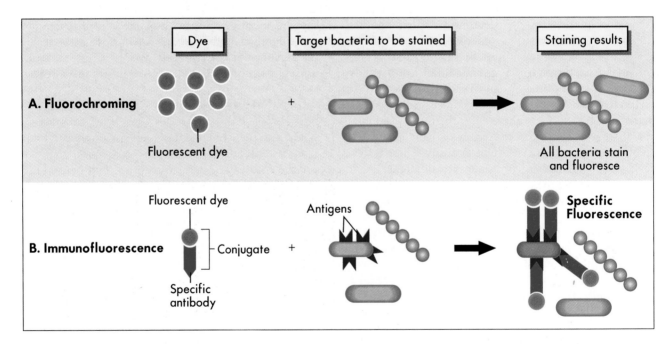

FIGURE 11-12 Principles of fluorochroming and immunofluorescence. Fluorochroming **(A)** involves nonspecific staining of any bacterial cell with a fluorescent dye. Immunofluorescence **(B)** uses antibodies labeled with fluorescent dye (i.e., a conjugate) to specifically stain a particular bacterial species.

by light microscopy. For example, a minimum concentration of at least 10^5 organisms per milliliter of specimen is required for visualization by light microscopy whereas by fluorescent microscopy that number decreases to 10^4 per milliliter. The most common fluorochroming methods used in diagnostic microbiology include acridine orange stain, auramine-rhodamine stain, and calcofluor white stain.

ACRIDINE ORANGE   The fluorochrome acridine orange binds to nucleic acid. This staining method (Procedure 11-1) can be used to confirm the presence of bacteria in blood cultures when Gram stain results are difficult to interpret or when the presence of bacteria is highly suspected but none are detected using light microscopy. Because acridine orange stains all nucleic acids, it is nonspecific. Therefore, all microorganisms and host cells will stain and give a bright orange fluorescence. Although this stain can be used to enhance detection, it does not discriminate between gram-negative and gram-positive bacteria. The stain is also used for detection of cell wall-deficient bacteria (e.g., mycoplasmas) grown in culture that are incapable of retaining the dyes used in the Gram stain (Figure 11-13).

AURAMINE-RHODAMINE   The waxy mycolic acids in the cell walls of mycobacteria have an affinity for the fluorochromes **auramine** and **rhodamine**. As shown in Figure 11-14, these dyes will nonspecifically bind to nearly all mycobacteria. The mycobacterial cells appear bright yellow or orange against a greenish background. This fluorochroming method can be used to enhance detection of mycobacteria directly in patient specimens and for initial characterization of cells grown in culture.

CALCOFLUOR WHITE   The cell walls of fungi will bind the stain calcofluor white, which greatly enhances fungal

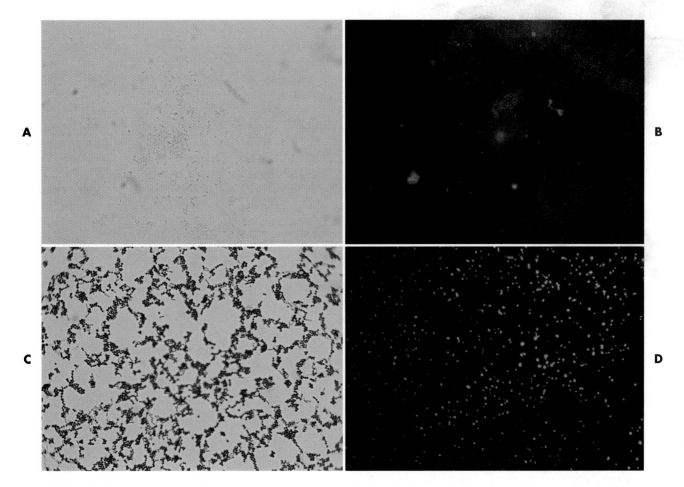

FIGURE 11-13 Comparison of acridine orange fluorochroming and Gram stain. Gram stain of mycoplasma demonstrates inability to distinguish these cell wall-deficient organisms from amorphous gram-negative debris (**A**). Staining the same specimen with acridine orange confirms the presence of nucleic acid-containing organisms (**B**). Gram stain distinguishes between gram-positive and gram-negative bacteria (**C**), but all bacteria stain the same with the nonspecific acridine orange dye (**D**).

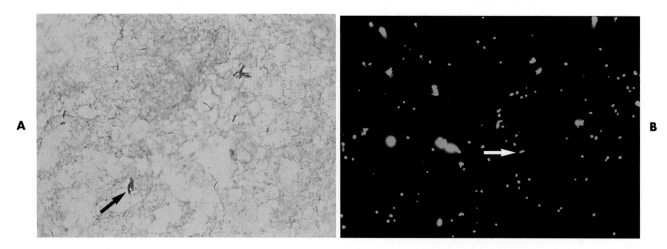

FIGURE 11-14 Comparison of the Ziehl-Neelsen–stained (**A**) and auramine-rhodamine–stained (**B**) *Mycobacterium* spp. (*arrows*).

visibility in tissue and other specimens. This fluorochrome is commonly used to directly detect fungi in clinical material and to observe subtle characteristics of fungi grown in culture (for more information regarding the use of calcofluor white for the laboratory diagnosis of fungal infections see Part 6).

Immunofluorescence

As discussed in Chapter 10, antibodies are molecules that have high specificity for interacting with microbial antigens. That is, antibodies specific for an antigen characteristic of a particular microbial species will only combine with that antigen. Therefore, if antibodies are conjugated (chemically linked) to a fluorescent dye, the resulting **dye-antibody conjugate** can be used to detect, or "tag," specific microbial agents (Figure 11-12). When "tagged," the microorganisms become readily detected by fluorescent microscopy. Thus, immunofluorescence combines the amplified contrast provided by fluorescence with the specificity of antibody-antigen binding.

This method is used to directly examine patient specimens for bacteria that are difficult or slow to grow (e.g., *Legionella* spp., *Bordetella pertussis,* and *Chlamydia trachomatis*) or to identify organisms already grown in culture. Fluorescein isothiocyanate (FITC), which emits an intense, apple green fluorescence, is the fluorochrome most commonly used for conjugation to antibodies (Figure 11-15). Immunofluorescence is also used in virology (Part 7) and to some extent in parasitology (Part 5).

Two additional types of microscopy, dark-field microscopy and electron microscopy, are not commonly used to diagnose infectious diseases. However, because of their importance in the detection and characterization of certain microorganisms, they will be discussed.

DARK-FIELD MICROSCOPY

Dark-field microscopy is similar to phase contrast microscopy in that it involves the alteration of microscopic technique rather than the use of dyes or stains to achieve contrast. By the dark-field method, the condenser does not allow light to pass directly through the specimen but directs the light to hit the specimen at an oblique angle (Figure 11-16, *A*). Only light that hits objects, such as microorganisms in the specimen, will be deflected upward into the objective lens for visualization. All other light that passes through the specimen will miss the objective, thus making the background a dark field.

This method has greatest utility for detecting certain bacteria directly in patient specimens that, because of their thin dimensions, cannot be seen by light microscopy and, because of their physiology, are difficult to grow in culture. Dark-field microscopy is used to detect spirochetes, the most notorious of which is the bacterium *Treponema pallidum,* the causative agent of syphilis (for more information regarding spirochetes see Chapter 63). As shown in Figure 11-16, *B,* spirochetes viewed using dark-field microscopy will appear extremely bright against a black field. The use of dark-field microscopy in diagnostic microbiology has substantially decreased with the advent of reliable serologic techniques for the diagnosis of syphilis.

ELECTRON MICROSCOPY

The electron microscope uses electrons instead of light to visualize small objects and, instead of lenses, the electrons are focused by electromagnetic fields and form an image on a fluorescent screen, like a television screen. Because of the substantially increased

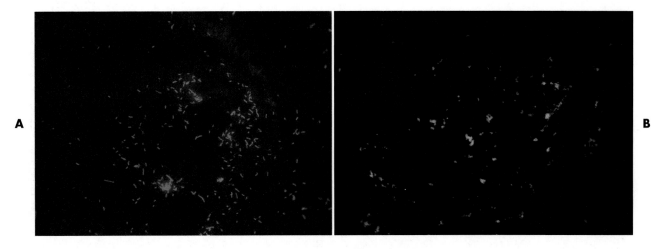

FIGURE 11-15 Immunofluoresence stains of *Legionella* spp. (**A**) and *Bordetella pertussis* (**B**) used for identification.

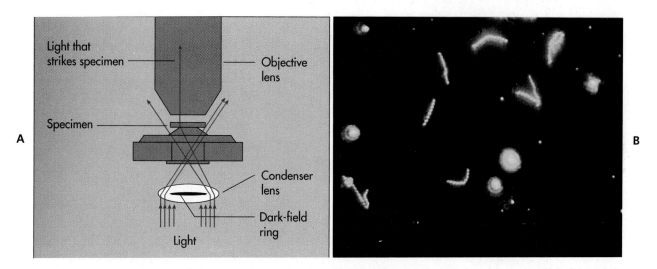

FIGURE 11-16 Dark-field microscopy. Principle (**A**) and dark-field photomicrograph showing the tightly coiled characteristics of the spirochete *Treponema pallidum* (**B**). (From Atlas, R.M. 1995. Principles of microbiology, Mosby, St. Louis.)

resolution this technology allows, magnifications in excess of 100,000× compared with the 1000× magnification provided by light microscopy are achieved.

Electron microscopes are of two general types: the **transmission electron microscope (TEM)** and the **scanning electron microscope (SEM)**. TEM passes the electron beam through objects and allows visualization of internal structures. SEM uses electron beams to scan the surface of objects and provides three-dimensional views of surface structures (Figure 11-17). These microscopes are powerful research tools, and many new morphologic features of bacteria, bacterial components, fungi, viruses, and parasites have been discovered using electron microscopy. However, because an electron microscope is a major capital investment and is not needed for the laboratory diagnosis of most infectious diseases (except for certain viruses and microsporidian parasites), few laboratories employ this method.

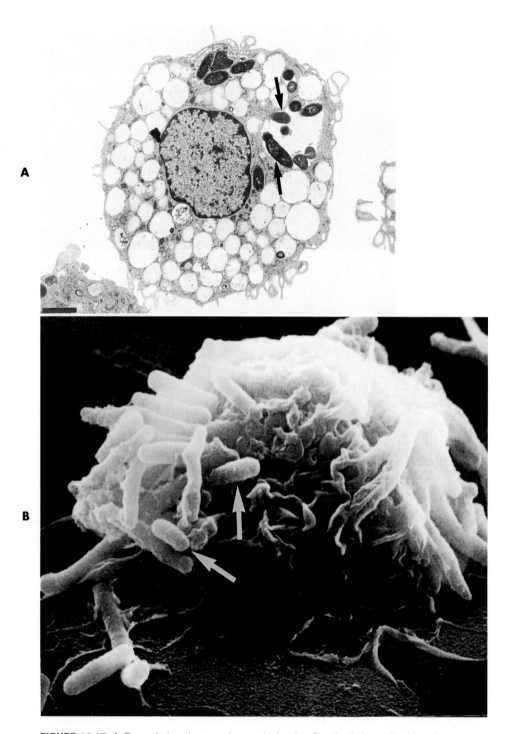

FIGURE 11-17 A, Transmission electron micrograph showing *E. coli* cells internalized by a human mast cell (*arrows*). **B,** Scanning electron micrograph of *E. coli* interacting with the surface of human mast cell (*arrows*). (Courtesy S.N. Abraham, Washington University School of Medicine).

Procedure

11-1 ACRIDINE ORANGE STAIN

PRINCIPLE

Acridine orange, a vital stain, will intercalate with nucleic acid, changing the dye's optical characteristics so that it will fluoresce bright orange under ultraviolet light. All nucleic acid-containing cells will fluoresce orange (see Figure 11-13).

METHOD

1. Fix slide, either in methanol or with heat, as described previously for Gram stain.

2. Flood slide with acridine orange stain (available from various commercial suppliers). Allow stain to remain on surface of slide for 2 minutes without drying.

3. Rinse with tap water and allow moisture to drain from slide and air dry.

4. Examine the slide using fluorescent microscopy.

EXPECTED RESULTS

Bacteria will fluoresce bright orange against a green-fluorescing or dark background. The nuclei of host cells may also fluoresce.

Bibliography

Atlas, R.M. 1995. Principles of microbiology. Mosby, St. Louis.

Murray, P.R., Baron, E.J., Pfaller, M.A., et. al., editors. 1995. Manual of clinical microbiology, ed 6. ASM Press, Washington, D.C.

12 LABORATORY CULTIVATION AND ISOLATION OF BACTERIA

Direct laboratory methods, such as microscopy, provide preliminary information about the bacteria involved in an infection, but bacterial growth is usually required for definitive identification and characterization. This chapter focuses on the principles and practices of bacterial cultivation, which has three main purposes:

- To grow and isolate all bacteria present in an infection
- To determine which of the bacteria that grow are most likely causing infection and which are likely contaminants or colonizers
- To obtain sufficient growth of clinically relevant bacteria to allow identification and characterization

PRINCIPLES OF BACTERIAL CULTIVATION

Cultivation is the process of growing microorganisms in culture by taking bacteria from the infection site (i.e., the **in vivo** environment) by some means of specimen collection and growing them in the artificial environment of the laboratory (i.e., the **in vitro** environment). Once grown in culture most bacterial populations are easily observed without microscopy and are present in sufficient quantities to allow laboratory testing to be performed.

The successful transition from the in vivo to the in vitro environment requires that the nutritional and environmental growth requirements of bacterial pathogens are met. The in vivo to in vitro transition is not necessarily easy for bacteria. In vivo they are utilizing various complex metabolic and physiologic pathways developed for survival on or within the human host. Then, relatively suddenly, they are exposed to the artificial in vitro environment of the laboratory. The bacteria must adjust to survive and multiply in vitro. Of importance, their survival depends on the availability of essential nutrients and appropriate environmental conditions.

Although growth conditions can be met for most known bacterial pathogens, the needs of certain clinically relevant bacteria are not sufficiently under-

stood to allow for the development of in vitro growth conditions. Examples include *Treponema pallidum* (the causative agent of syphilis) and *Mycobacterium leprae* (the causative agent of leprosy).

NUTRITIONAL REQUIREMENTS

As discussed in Chapter 9, bacteria have numerous nutritional needs that include different gases, water, various ions, nitrogen, sources for carbon, and energy. The latter two requirements are most commonly provided by carbohydrates (e.g., sugars and their derivatives) and proteins.

General concepts of culture media

In the laboratory, nutrients are incorporated into culture media on or in which bacteria are grown. If a culture medium meets a bacterial cell's growth requirements, that cell will multiply to sufficient numbers to allow visualization by the unaided eye. Of course, bacterial growth after inoculation also requires that the media be placed in optimal environmental conditions.

Because different pathogenic bacteria have different nutritional needs, various types of culture media have been developed for use in diagnostic microbiology. For certain bacteria the needs are relatively complex, and exceptional media components must be used for growth. Bacteria with such requirements are said to be **fastidious**. Alternatively, the nutritional needs of most clinically important bacteria are relatively basic and straightforward. These bacteria are considered **nonfastidious**.

Phases of growth media

Growth media are used in either of two phases: liquid (**broth**) or solid (**agar**). In some instances (e.g., certain blood culture methods), a biphasic medium that contains both a liquid and a solid phase is used.

In broth media nutrients are dissolved in water, and bacterial growth is indicated by a change in the broth's appearance from clear to turbid (i.e., cloudy). The turbidity, or cloudiness, of the broth is due to light deflected by bacteria present in the culture (Figure 12-1). The more bacteria growth the greater

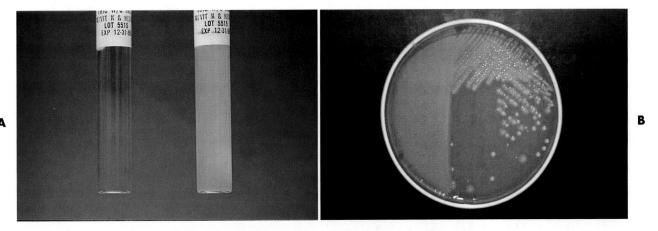

FIGURE 12-1 A, Clear broth indicating no bacterial growth (*left*) and turbid broth indicating bacterial growth (*right*). **B,** Individual bacterial colonies growing on the agar surface following incubation.

the turbidity. At least 10^6 bacteria per milliliter of broth are needed for turbidity to be detected with the unaided eye.

Solid media are made by adding a solidifying agent to the nutrients and water. **Agarose,** the most common solidifying agent, has the unique property of melting at high temperatures ($\geq 95°$ C) but resolidifying only after its temperature falls below $50°$ C. Addition of agar allows a solid medium to be prepared by heating to an extremely high temperature, which is required for sterilization, and cooling to $55°$ to $60°$ C for distribution into petri dishes. On further cooling, the agarose-containing medium forms a stable solid gel referred to as **agar.** The petri dish containing the agar is referred to as the **agar plate.** Different agar media usually are identified according to the major nutritive component of the medium (e.g., sheep blood agar, bile esculin agar, xylose-lysine-desoxycholate agar).

With appropriate incubation conditions, each bacterial cell inoculated onto the agar medium surface will proliferate to sufficiently large numbers to be observable with the unaided eye (see Figure 12-1). The resulting bacterial population is considered to be derived from a single bacterial cell and is known as a **colony.** In other words, all bacterial cells within a single colony are the same genus and species, having identical genetic and phenotypic characteristics (i.e., are a single **clone**). Bacterial cultures derived from a single colony or clone are considered "pure." Pure cultures are required for subsequent procedures used to identify and characterize bacteria. The ability to select pure (individual) colonies is one of the first and most important steps required for bacterial identification and characterization (for more information regarding bacterial identification strategies see Chapter 13).

Media classifications and functions

Media are categorized according to their function and use. In diagnostic bacteriology there are four general categories of media: **enrichment, supportive, selective, and differential.**

Enrichment media contain specific nutrients required for the growth of particular bacterial pathogens that may be present alone or with other bacterial species in a patient specimen. This media type is used to enhance the growth of a particular bacterial pathogen from a mixture of organisms by using nutrient specificity. One example of such a medium is buffered charcoal-yeast extract agar, which provides L-cysteine and other nutrients required for the growth of *Legionella pneumophila*, the causative agent of legionnaires' disease (Figure 12-2).

Supportive media contain nutrients that support growth of most nonfastidious organisms without giving any particular organism a growth advantage. **Selective media** contain one or more agents that are inhibitory to all organisms except those being sought. In other words, these media select for the growth of certain bacteria to the disadvantage of others. Inhibitory agents used for this purpose include dyes, bile salts, alcohols, acids, and antibiotics. An example of a selective medium is phenylethyl alcohol agar, which inhibits the growth of aerobic and facultatively anaerobic gram-negative rods and allows gram-positive cocci to grow (Figure 12-3).

Differential media employ some factor (or factors) that allows colonies of one bacterial species or type to exhibit certain metabolic or culture characteristics that can be used to distinguish them from other bacteria growing on the same agar plate. One commonly used differential medium is MacConkey agar, which differentiates between gram-negative

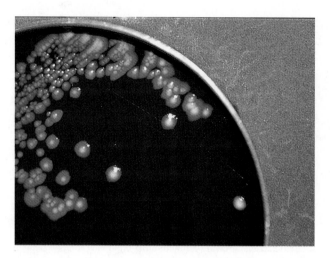

FIGURE 12-2 Growth of *Legionella pneumophila* on the enrichment medium buffered charcoal-yeast extract (BCYE) agar, used specifically to grow this bacterial genus.

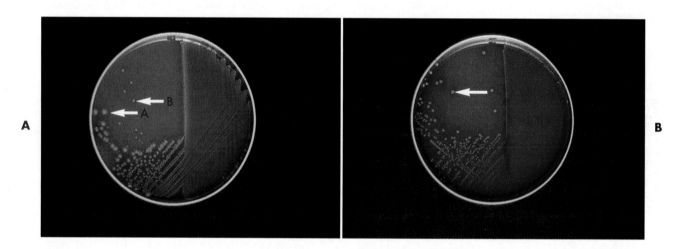

FIGURE 12-3 A, Heavy mixed growth of the gram-negative bacillus *E. coli* (*arrow A*) and the gram-positive coccus *Enterococcus* spp. (*arrow B*) on the nonselective medium sheep blood agar (SBA). **B,** The selective medium phenylethyl-alcohol agar (PEA) only allows the enterococci to grow (*arrow*).

bacteria that can and cannot ferment the sugar lactose (Figure 12-4).

Of importance, many media used in diagnostic bacteriology provide more than one function. For example, MacConkey agar is both differential and selective because it will not allow most gram-positive bacteria to grow. Another example is sheep blood agar. This is the most commonly used supportive medium for diagnostic bacteriology because it allows many organisms to grow. However, in many ways this agar is also differential because the appearance of colonies produced by certain bacterial species are readily distinguishable (Figure 12-5).

Summary of artificial media for routine bacteriology

Various broth and agar media that have enrichment, selective, and/or differential capabilities and are used frequently for routine bacteriology are listed alphabetically in Table 12-1. Anaerobic bacteriology (Chapters 58 and 59), mycobacteriology (Chapter 60), and mycology (Part 6) employ similar media strategies; details regarding these media are provided in the appropriate chapters.

Of the dozens of available media, only those most commonly used for routine diagnostic bacteriology are summarized in this discussion. Part 3

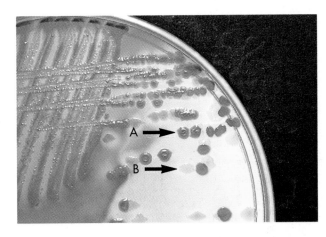

FIGURE 12-4 Differential capabilities of MacConkey agar as gram-negative bacilli capable of fermenting lactose appear deep purple (*arrow A*) while those not able to ferment lactose appear light pink or relatively colorless (*arrow B*).

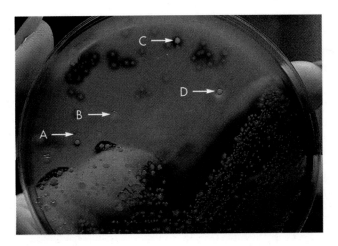

FIGURE 12-5 Different colony morphologies exhibited on sheep blood agar by various bacteria, including alpha-hemolytic streptococci (*arrow A*), gram-negative bacilli (*arrow B*), beta-hemolytic streptococci (*arrow C*), and *Staphylococcus aureus* (*arrow D*).

discusses which media should be used to culture bacteria from various clinical specimens. Similarly, Chapter 13 and other chapters throughout Part 4 discuss media used to identify and characterize specific organisms.

BRAIN-HEART INFUSION Brain-heart infusion (BHI) is a nutritionally rich medium used to grow various microorganisms, either as a broth or as an agar, with or without added blood. Key ingredients include infusion from several animal tissue sources, added peptone (protein), phosphate buffer, and a small concentration of dextrose. The carbohydrate provides

a readily accessible source of energy for many bacteria. BHI broth is often used as a major component of the media developed for culturing a patient's blood for bacteria (see Chapter 20), establishing bacterial identification, and for certain tests to determine bacterial susceptibility to antimicrobial agents (see Chapter 18).

CHOCOLATE AGAR Chocolate agar is essentially the same as blood agar except during preparation the red blood cells are lysed when added to molten agar base. This lysis releases intracellular nutrients such as hemoglobin, hemin ("X" factor), and the coenzyme nicotinamide adenine dinucleotide (NAD or "V" factor) into the agar for utilization by fastidious bacteria. Red blood cell lysis gives the medium a chocolate-brown color from which this agar gets its name. The most common bacterial pathogens that require this enriched medium for growth include *Neisseria gonorrhoeae*, the causative agent of gonorrhea, and *Haemophilus* spp., which cause infections usually involving the respiratory tract and middle ear. Neither of these species are able to grow on sheep blood agar.

COLUMBIA CNA WITH BLOOD Columbia agar base is a nutritionally rich formula containing three peptone sources and 5% defibrinated sheep blood. This supportive medium can also be used to help differentiate bacterial colonies based on the hemolytic reactions they produce. CNA refers to the antibiotics colistin (C) and nalidixic acid (NA) that are added to the medium to suppress the growth of most gram-negative organisms while allowing gram-positive bacteria to grow, thus conferring a selective property to this medium.

GRAM-NEGATIVE (GN) BROTH A selective broth, gram-negative (GN) broth is used for the cultivation of gastrointestinal pathogens (i.e., *Salmonella* spp. and *Shigella* spp.) from stool specimens and rectal swabs. The broth contains several active ingredients, including sodium citrate and sodium desoxycholate (a bile salt) that inhibit gram-positive organisms and the early multiplication of gram-negative, nonenteric pathogens. To optimize its selective nature, GN broth should be subcultured 6 to 8 hours after initial inoculation and incubation. After this time, the nonenteric pathogens begin to overgrow the pathogens.

HEKTOEN ENTERIC (HE) AGAR Hektoen Enteric (HE) agar contains bile salts and dyes (bromthymol blue and acid fuchsin) to selectively slow the growth of most nonpathogenic gram-negative bacilli found in the gastrointestinal tract and allow *Salmonella* spp.

TABLE 12-1 PLATING MEDIA FOR ROUTINE BACTERIOLOGY

MEDIUM	COMPONENTS/COMMENTS	PRIMARY PURPOSE
Bile esculin agar (BEA)	Nutrient agar base with ferric citrate. Hydrolysis of esculin by group D streptococci imparts a brown color to medium; sodium desoxycholate inhibits many bacteria	Differential isolation and presumptive identification of group D streptococci and enterococci
Bile esculin azide agar with vancomycin	Contains azide to inhibit gram-negative bacteria, vancomycin to select for resistant gram-positive bacteria, and bile esculin to differentiate enterococci from other vancomycin-resistant bacteria that may grow	Selective and differential for cultivation of vancomycin-resistant enterococci from clinical and surveillance specimens
Blood agar	Trypticase soy agar, *Brucella* agar, or beef heart infusion base with 5% sheep blood	Cultivation of fastidious microorganisms, determination of hemolytic reactions
Bordet-Gengou agar	Potato-glycerol–based medium enriched with 15%-20% defibrinated blood. Contaminants inhibited by methicillin (final concentration of 2.5 μg/mL)	Isolation of *Bordetella pertussis*
Buffered charcoal-yeast extract agar (BCYE)	Yeast extract, agar, charcoal, and salts supplemented with L-cysteine HCl, ferric pyrophosphate, ACES buffer, and α-ketoglutarate	Enrichment for *Legionella* spp.
Buffered charcoal-yeast extract (BCYE) agar with antibiotics	BCYE supplemented with polymyxin B, vancomycin, and ansamycin, to inhibit gram-negative bacteria, gram-positive bacteria, and yeast, respectively	Enrichment and selection for *Legionella* spp.
Campy-blood agar	Contains vancomycin (10 mg/L), trimethoprim (5 mg/L), polymixin B (2500 U/L), amphotericin B (2 mg/L), and cephalothin (15 mg/L) in a *Brucella* agar base with sheep blood	Selective for *Campylobacter* spp.
Campylobacter thioglycollate broth	Thioglycollate broth supplemented with increased agar concentration and antibiotics	Selective holding medium for recovery of *Campylobacter* spp.
Cefoperazone, vancomycin, amphotericin (CVA) medium	Blood-supplemented enrichment medium containing cefoperazone, vancomycin, and amphotericin to inhibit growth of most gram-negative bacteria, gram-positive bacteria, and yeast, respectively	Selective medium for isolation of *Campylobacter* spp.

TABLE 12-1 PLATING MEDIA FOR ROUTINE BACTERIOLOGY—CONT'D

MEDIUM	COMPONENTS/COMMENTS	PRIMARY PURPOSE
Cefsulodin-irgasan-novobiocin (CIN) agar	Peptone base with yeast extract, mannitol, and bile salts. Supplemented with cefsulodin, irgasan, and novobiocin; neutral red and crystal violet indicators	Selective for *Yersinia* spp.; may be useful for isolation of *Aeromonas* spp.
Chocolate agar	Peptone base, enriched with solution of 2% hemoglobin or IsoVitaleX (BBL)	Cultivation of *Haemophilus* spp. and pathogenic *Neisseria* spp.
Columbia colistin-nalidixic acid (CNA) agar	Columbia agar base with 10 mg colistin per liter, 15 mg nalidixic acid per liter, and 5% sheep blood	Selective isolation of gram-positive cocci
Cystine-tellurite blood agar	Infusion agar base with 5% sheep blood. Reduction of potassium tellurite by *Corynebacterium diphtheriae* produces black colonies	Isolation of *C. diphtheriae*
Eosin methylene blue (EMB) agar (Levine)	Peptone base with lactose and sucrose. Eosin and methylene blue as indicators	Isolation and differentation of lactose-fermenting and non–lactose-fermenting enteric bacilli
Gram-negative broth (GN)	Peptone base broth with glucose and mannitol. Sodium citrate and sodium desoxycholate act as inhibitory agents	Selective (enrichment) liquid medium for enteric pathogens
Hektoen enteric (HE) agar	Peptone base agar with bile salts, lactose, sucrose, salicin, and ferric ammonium citrate. Indicators include bromthymol blue and acid fuchsin	Differential, selective medium for the isolation and differentiation of *Salmonella* and *Shigella* spp. from other gram-negative enteric bacilli
MacConkey agar	Peptone base with lactose. Gram-positive organisms inhibited by crystal violet and bile salts. Neutral red as indicator	Isolation and differentiation of lactose fermenting and non–lactose-fermenting enteric bacilli
MacConkey sorbitol agar	A modification of MacConkey agar in which lactose has been replaced with D-sorbitol as the primary carbohydrate	For the selection and differentiation of *E. coli* O157:H7 in stool specimens
Mannitol salt agar	Peptone base, mannitol, and phenol red indicator. Salt concentration of 7.5% inhibits most bacteria	Selective isolation of staphylococci
New York City (NYC) agar	Peptone agar base with cornstarch, supplemented with yeast dialysate, 3% hemoglobin, and horse plasma. Antibiotic supplement includes vancomycin (2 μg/mL), colistin (5.5 μg/mL), amphotericin B (1.2 μg/mL), and trimethoprim (3μg/mL).	Selective for *Neisseria gonorrhoeae*

Continued

--- **TABLE 12-1** PLATING MEDIA FOR ROUTINE BACTERIOLOGY—CONT'D ---

MEDIUM	COMPONENTS/COMMENTS	PRIMARY PURPOSE
Phenylethyl alcohol (PEA) agar	Nutrient agar base. Phenylethanol inhibits growth of gram-negative organisms *(aer)*	Selective isolation of gram-positive cocci and anaerobic gram-negative bacilli
Regan Lowe	Charcoal agar supplemented with horse blood, cephalexin, and amphotericin B	Enrichment and selective medium for isolation of *Bordetella pertussis*
Salmonella-Shigella (SS) agar	Peptone base with lactose, ferric citrate, and sodium citrate. Neutral red as indicator; inhibition of coliforms by brilliant green and bile salts	Selective for *Salmonella* and *Shigella* spp.
Schaedler agar	Peptone and soy protein base agar with yeast extract, dextrose, and buffers. Addition of hemin, L-cystine, and 5% blood enriches for anaerobes	Nonselective medium for the recovery of anaerobes and aerobes
Selenite broth	Peptone base broth. Sodium selenite toxic for most *Enterobacteriaceae*	Enrichment of isolation of *Salmonella* spp.
Skirrow agar	Peptone and soy protein base agar with lysed horse blood. Vancomycin inhibits gram-positive organisms; polymyxin B and trimethoprim inhibit most gram-negative organisms	Selective for *Campylobacter* spp.
Streptococcal selective agar (SSA)	Contains crystal violet, colistin, and trimethoprim-sulfamethoxazole in 5% sheep blood agar base	Selective for *Streptococcus pyogenes* and *Streptococcus agalactiae*
Tetrathionate broth	Peptone base broth. Bile salts and sodium thiosulfate inhibit gram-positive organisms and *Enterobacteriaceae*	Selective for *Salmonella* and *Shigella* spp.
Thayer-Martin agar	Blood agar base enriched with hemoglobin and supplement B; contaminating organisms inhibited by colistin, nystatin, vancomycin, and trimethoprim	Selective for *N. gonorrhoeae* and *N. meningitidis*
Thioglycollate broth	Pancreatic digest of casein, soy broth, and glucose enrich growth of most microorganisms. Thioglycollate and agar reduce redox potential	Supports growth of anaerobes, aerobes, microaerophilic, and fastidious microorganisms
Thiosulfate citrate-bile salts (TCBS) agar	Peptone base agar with yeast extract, bile salts, citrate, sucrose, ferric citrate, and sodium thiosulfate. Bromthymol blue acts as indicator	Selective and differential for vibrios

─────── **TABLE 12-1 PLATING MEDIA FOR ROUTINE BACTERIOLOGY—CONT'D** ───────

MEDIUM	COMPONENTS/COMMENTS	PRIMARY PURPOSE
Todd-Hewitt broth supplemented with antibiotics	Todd-Hewitt, an enrichment broth for streptococci, is supplemented with nalidixic acid and gentamicin or colistin for greater selectivity	Selection and enrichment for *Streptococcus agalactiae* in female genital specimens
Trypticase soy broth (TSB)	All-purpose enrichment broth that can support the growth of many fastidious and nonfastidious bacteria	Enrichment broth used for subculturing various bacteria from primary agar plates
Xylose lysine desoxycholate (XLD) agar	Yeast extract agar with lysine, xylose, lactose, sucrose, and ferric ammonium citrate. Sodium desoxycholate inhibits gram-positive organisms; phenol red as indicator	Isolation and differentiation of *Salmonella* and *Shigella* spp. from other gram-negative enteric bacilli

and *Shigella* spp. to grow. The medium is also differential because many nonenteric pathogens that do grow will appear as orange to salmon-colored colonies. This colony appearance results from the organism's ability to ferment the lactose in the medium, resulting in the production of acid, which lowers the medium's pH and causes a change in the pH indicator bromthymol blue. *Salmonella* spp. and *Shigella* spp. do not ferment lactose so no color change occurs and their colonies maintain the original blue-green color of the medium. As an additional differential characteristic the medium contains ferric ammonium citrate, an indicator for the detection of H_2S, so that H_2S-producing organisms, such as *Salmonella* spp., can be visualized as colonies exhibiting a black precipitate (Figure 12-6).

MACCONKEY AGAR MacConkey agar is the most frequently used primary selective and differential agar. This medium contains crystal violet dye to inhibit the growth of gram-positive bacteria and fungi, and allows many types of gram-negative bacilli to grow. The pH indicator, neutral red, provides this medium with a differential capacity. Bacterial fermentation of lactose results in acid production, which decreases medium pH and causes the neutral red indicator to give bacterial colonies a pink to red color. Non-lactose fermenters, such as *Shigella* spp., remain colorless and translucent (see Figure 12-4).

PHENYLETHYL ALCOHOL (PEA) AGAR   PEA agar is essentially sheep blood agar that is supplemented with phenylethyl alcohol to inhibit the growth of gram-negative bacteria. Five percent sheep blood in PEA provides nutrients for common gram-positive cocci such as enterococci, streptococci, and staphylococci (see Figure 12-3).

SHEEP BLOOD AGAR   Most bacteriology specimens are inoculated to sheep blood agar plates because this medium supports all but the most fastidious clinically significant bacteria. Additionally, the colony morphologies that commonly encountered bacteria exhibit on this medium are familiar to most clinical microbiologists. The medium consists of a base containing a protein source (e.g., tryptones), soybean protein digest (containing a slight amount of natural carbohydrate), sodium chloride, agar, and 5% sheep blood.

Certain bacteria produce extracellular enzymes that lyse red blood cells in the agar (**hemolysis**). This activity can result in complete clearing of the red blood cells around the bacterial colony (**beta hemolysis**) or in only partial lysis of the cells to produce a greenish discoloration around the colony (**alpha hemolysis**). Other bacteria have no effect on the red blood cells, and no halo is produced around the colony (**gamma** or **nonhemolysis**). Microbiologists often use colony morphology and the degree or absence of hemolysis as criteria for determining what additional steps will be necessary for identification of a bacterial isolate. To read the hemolytic reaction on a blood agar plate accurately, the technologist must hold the plate up to the light and observe the plate with the light coming from behind (i.e., transmitted light).

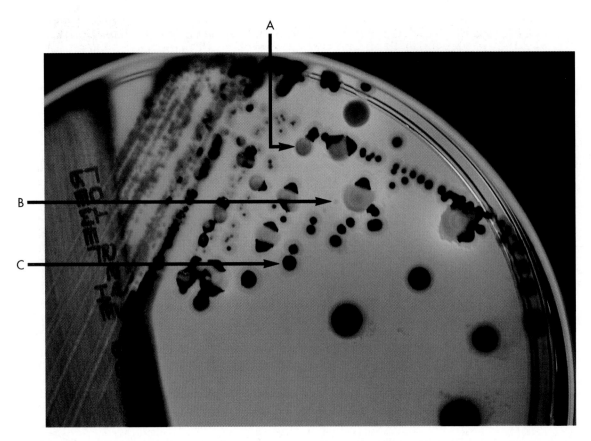

FIGURE 12-6 Differential capabilities of HE agar for lactose-fermenting, gram-negative bacilli (e.g., *E. coli, arrow A*), nonlactose fermenters (e.g., *Shigella* spp., *arrow B*) and H₂S producers (e.g., *Salmonella* spp., *arrow C*).

THAYER-MARTIN AGAR Thayer-Martin agar is an enrichment and selective medium for the isolation of *Neisseria gonorrhoeae*, the causative agent of gonorrhea, and *Neisseria meningitidis*, a life-threatening cause of meningitis. The enrichment portion of the medium is the basal components and the chocolatized blood, while the addition of antibiotics provides a selective capacity. The antibiotics include colistin to inhibit other gram-negative bacteria, vancomycin to inhibit gram-positive bacteria, and nystatin to inhibit yeast. The antimicrobial, trimethoprim, is also added to inhibit *Proteus* spp., which have a tendency to swarm over the agar surface and mask the detection of individual colonies of the two pathogenic *Neisseria* spp. A further modification, **Martin-Lewis agar,** substitutes ansamycin for nystatin and has a higher concentration of vancomycin.

THIOGLYCOLLATE BROTH Thioglycollate broth is the enrichment broth most frequently used in diagnostic bacteriology. The broth contains many nutrient factors, including casein, yeast and beef extracts, and vitamins, to enhance the growth of most medically important bacteria. Other nutrient supplements, an oxidation-reduction indicator (resazurin), dextrose, vitamin K₁, and hemin have been used to modify the basic thioglycollate formula. In addition, this medium contains 0.075% agar to prevent convection currents from carrying atmospheric oxygen throughout the broth. This agar supplement and the presence of thioglycolic acid, which acts as a reducing agent to create an anaerobic environment deeper in the tube, allows anaerobic bacteria to grow.

Gram-negative, facultatively anaerobic bacilli (i.e., those that can grow in the presence or absence of oxygen) generally produce diffuse, even growth throughout the broth, while gram-positive cocci frequently grow as discrete "puffballs." Strict aerobic bacteria (i.e., require oxygen for growth), such as *Pseudomonas* spp., tend to grow toward the surface of the broth, while strict anaerobic bacteria (i.e., those that cannot grow in the presence of oxygen) grow at the bottom of the broth (Figure 12-7).

XYLOSE LYSINE AND DESOXYCHOLATE (XLD) AGAR As with HE agar, xylose-lysine-desoxycholate (XLD) agar is selective and differential for *Shigella* spp. and *Salmonella* spp. The salt, sodium desoxycholate, inhibits

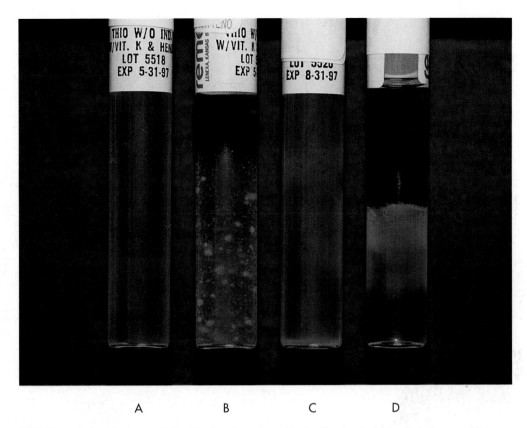

FIGURE 12-7 Growth characteristics of various bacteria in thioglycollate broth. **A,** Facultatively anaerobic gram-negative bacilli (i.e., those that grow in the presence or absence of oxygen) grow throughout broth. **B,** Gram-positive cocci grow as "puff balls." **C,** Strictly aerobic organisms (i.e., those that require oxygen for growth), such as *Pseudomonas aeruginosa,* grow toward the top of the broth. **D,** Strictly anaerobic organisms (i.e., those that do not grow in the presence of oxygen) grow in the bottom of the broth.

many gram-negative bacilli that are not enteric pathogens and inhibits gram-positive organisms. A phenol red indicator in the medium detects increased acidity from carbohydrate (i.e., lactose, xylose, and sucrose) fermentation. Enteric pathogens, such as *Shigella* spp., do not ferment these carbohydrates so their colonies remain colorless (i.e., the same approximate pink to red color of the uninoculated medium). Colonies of *Salmonella* spp. are also colorless on XLD, because of the decarboxylation of lysine, which results in a pH increase that causes the pH indicator to turn red. These colonies often exhibit a black center that results from *Salmonella* spp. producing H$_2$S. Several of the nonpathogens ferment one or more of the sugars and produce yellow colonies (Figure 12-8).

Preparation of artificial media

Nearly all media are commercially available as ready-to-use agar plates or tubes of broth. If media are not purchased, laboratory personnel can prepare agars and broths using dehydrated powders that are reconstituted in water (distilled or deionized) according to

manufacturer's recommendations. Generally, media are reconstituted by dissolving a specified amount of media powder, which usually contains all necessary components, in water. Boiling is often required to dissolve the powder, but specific manufacturer's instructions printed in media package inserts should be followed exactly. Most media require sterilization so that only bacteria from patient specimens will grow and not those that are contaminants from water or the powdered media. Broth media are distributed to individual tubes before sterilization. Agar media are usually sterilized in large flasks or bottles capped with either plastic screw caps or plugs before being placed in an autoclave.

MEDIA STERILIZATION   The timing of **autoclave sterilization** should start from the moment the temperature reaches 121° C and usually requires a minimum of 15 minutes. Once the sterilization cycle is completed, molten agar is allowed to cool to approximately 50° C before being distributed to individual petri plates (usually 25 mL of molten agar per plate). If other

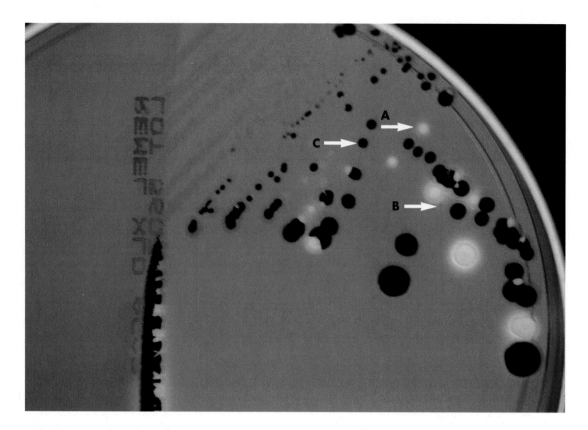

FIGURE 12-8 Differential capabilities of XLD agar for lactose-fermenting, gram-negative bacilli (e.g., *E. coli, arrow A*), nonlactose fermenters (e.g., *Shigella* spp., *arrow B*) and H₂S producers (e.g., *Salmonella* spp., *arrow C*).

ingredients are to be added (e.g., supplements such as sheep blood or specific vitamins, nutrients, or antibiotics) they should be incorporated when the molten agar has cooled, just before distribution to plates.

Delicate media components that cannot withstand steam sterilization by autoclaving (e.g., serum, certain carbohydrate solutions, certain antibiotics, and other heat-labile substances) can be sterilized by **membrane filtration**. Passage of solutions through membrane filters with pores ranging in size from 0.2 to 0.45 μm in diameter will not remove viruses but does effectively remove most bacterial and fungal contaminants. Finally, all media, whether purchased or prepared, must be subjected to stringent quality control before being used in the diagnostic setting (for more information regarding quality control see Chapter 6).

CELL CULTURES Although most bacteria grow readily on artificial media, certain pathogens require factors provided only by living cells. These bacteria are **obligate intracellular parasites** that require viable host cells for propagation. Although all viruses are obligate intracellular parasites, chlamydiae, rickettsiae, and rickettsiae-like organisms are bacterial pathogens that require living cells for cultivation.

The cultures for growth of these bacteria comprise layers of living cells growing on the surface of a solid matrix such as the inside of a glass tube or the bottom of a plastic flask. The presence of bacterial pathogens within the cultured cells is detected by specific changes in the cells' morphology. Alternatively, specific stains, composed of antibody conjugates, may be used to detect bacterial antigens within the cells. Cell cultures may also detect certain bacterial toxins (e.g., *Clostridium difficile* cytotoxin). Cell cultures are most commonly used in diagnostic virology. Cell culture maintenance and inoculation is addressed in Part 7.

ENVIRONMENTAL REQUIREMENTS

Optimizing the environmental conditions to support the most robust growth of clinically relevant bacteria is as important as meeting the organism's nutritional needs for in vitro cultivation. The four most critical environmental factors to consider include oxygen and carbon dioxide (CO_2) availability, temperature, pH, and moisture content of medium and atmosphere.

Oxygen and carbon dioxide availability

Most clinically relevant bacteria are either aerobic, facultatively anaerobic, or strictly anaerobic. **Aerobic bacteria** use oxygen as a terminal electron acceptor and grow well in room air. Most clinically significant aerobic organisms are actually **facultatively anaerobic,** being able to grow in the presence (i.e., aerobically) or absence (i.e., anaerobically) of oxygen. However, some bacteria, such as *Pseudomonas* spp., members of the *Neisseriaceae* family, *Brucella* spp., *Bordetella* spp., and *Francisella* spp., are strictly aerobic and cannot grow in the absence of oxygen. Other aerobic bacteria require only low levels of oxygen and are referred to as being **microaerophilic,** or microaerobic. **Anaerobic bacteria** are unable to use oxygen as an electron acceptor, but some aerotolerant strains will still grow slowly and poorly in the presence of oxygen. Oxygen is inhibitory or lethal for strictly anaerobic bacteria.

In addition to oxygen, the availability of CO_2 is important for growth of certain bacteria. Organisms that grow best with higher CO_2 concentrations (i.e., 5% to 10% CO_2) than is provided in room air are referred to as being **capnophilic.** For some bacteria a 5% to 10% CO_2 concentration is essential for successful cultivation from patient specimens.

Temperature

Bacterial pathogens generally multiply best at temperatures similar to those of internal human host tissues and organs (i.e., 37° C). Therefore, cultivation of most medically important bacteria is done using incubators with temperatures maintained in the 35° to 37° C range. For others, an incubation temperature of 30° C (i.e., the approximate temperature of the body's surface) may be preferable, but such bacteria are encountered relatively infrequently so that use of this incubation temperature occurs only when dictated by special circumstances.

Recovery of certain organisms can be enhanced by incubation at other temperatures. For example, the gastrointestinal pathogen *Campylobacter jejuni* grows at 42° C, while many other pathogens and nonpathogens cannot. Therefore, incubation at this temperature can be used as an enrichment procedure. Other bacteria, such as *Listeria monocytogenes* and *Yersinia enterocolitica*, can grow at 0° C, but grow best at temperatures between 20° and 40° C. Cold enrichment has been used to enhance the recovery of these organisms in the laboratory.

pH

The pH scale is a measure of the hydrogen ion concentration of an organism's environment, with a pH value of 7 being neutral. Values less than 7 indicate the environment is **acidic,** while values greater than 7 indicate **alkaline** conditions. Most clinically relevant bacteria prefer a near neutral pH range, from 6.5 to 7.5. Commercially prepared media are buffered in this range so that checking their pH is rarely necessary.

Moisture

Water is provided as a major constituent of both agar and broth media. However, when media are incubated at the temperatures used for bacterial cultivation, a large portion of water content can be lost by evaporation. Loss of water from media can be deleterious to bacterial growth in two ways: (1) less water is available for essential bacterial metabolic pathways and (2) with a loss of water there is a relative increase in the solute concentration of the media. An increased solute concentration can osmotically shock the bacterial cell and cause lysis. In addition, increased atmospheric humidity enhances the growth of certain bacterial species. For these reasons, measures, such as sealing agar plates to trap moisture or using humidified incubators, are taken to ensure appropriate moisture levels are maintained throughout the incubation period.

Methods for providing optimum incubation conditions

Although heating blocks and temperature-controlled water baths may be used occasionally, **incubators** are the primary laboratory devices used to provide the environmental conditions required for cultivating microorganisms. The conditions of incubators can be altered to accommodate the type of organisms to be grown. This section focuses on the incubation of routine bacteriology cultures. Conditions for growing anaerobic bacteria (Chapters 58 and 59), mycobacteria (Chapter 60), fungi (Part 6), and viruses (Part 7) are covered in other areas of the text.

Once inoculated with patient specimens, most media are placed in incubators with temperatures maintained between 35° and 37° C and humidified atmospheres that contain 3% to 5% CO_2. Incubators containing room air may be used for some media, but the lack of increased CO_2 may hinder the growth of certain bacteria.

Various atmosphere-generating systems are commercially available and are used instead of CO_2-generating incubators. For example, a self-contained culture medium and a compact CO_2-generating system can be used for culturing fastidious organisms such as *Neisseria gonorrhoeae*. A tablet of sodium bicarbonate is dissolved by the moisture created within an airtight plastic bag and releases sufficient CO_2 to support growth of the pathogen (see Figure 26-6). As an alternative to commercial systems, a **candle jar** can also generate a CO_2 concentration of approximately 3% and has historically been used as a common method for cultivating certain fastidious bacteria. The burning candle, which is placed in a container of inoculated agar plates that is subsequently sealed, uses up just enough oxygen before it goes out (from lack of oxygen) to lower the oxygen tension and produce CO_2 and water by combustion. Other atmosphere-generating systems are available to create conditions optimal for cultivating specific bacterial pathogens (e.g., *Campylobacter* spp. and anaerobic bacteria).

Finally, the duration of incubation required for obtaining good bacterial growth depends on the organisms being cultured. Most bacteria encountered in routine bacteriology will grow within 24 to 48 hours, if not sooner. Certain anaerobic bacteria may require longer incubation, and mycobacteria frequently take weeks before detectable growth occurs.

THE PROCESS OF BACTERIAL CULTIVATION

The process of bacterial cultivation involves the use of optimal artificial media and incubation conditions to isolate and identify the bacterial etiologies of an infection as rapidly and as accurately as possible.

ISOLATION OF BACTERIA FROM SPECIMENS

As discussed in detail throughout Part 3, the cultivation of bacteria from infections at various body sites is accomplished by inoculating processed specimens directly onto artificial media. The media summarized in Table 12-1 and incubation conditions are selected for their ability to support the growth of the bacteria most likely to be involved in the infectious process.

To enhance the growth, isolation, and selection of etiologic agents, specimen inocula are usually spread over the surface of plates in a standard pattern so that individual bacterial colonies are obtained and semiquantitative analysis can be performed. A commonly used streaking technique is illustrated in Figure 12-9. Using this method, the relative numbers of organisms in the original specimen can be estimated based on the growth of colonies past the original area of inoculation. To enhance isolation of bacterial colonies the loop should be flamed for sterilization between streaking each subsequent quadrant.

Streaking plates inoculated with a measured amount of specimen, such as when a calibrated loop is used to quantify colony-forming units (CFUs) in urine cultures, is accomplished by spreading the inoculum evenly over the entire agar surface (Figure 12-10). This facilitates counting colonies by ensuring that individual bacterial cells will be well dispersed over the agar surface.

Evaluation of colony morphologies
Initial evaluation of colony morphologies on the primary plating media is extremely important. Laboratorians can provide physicians with early preliminary information regarding the patient's culture results. This information also is important for deciding which subsequent steps to take for definitive organism identification and characterization.

TYPE OF MEDIA SUPPORTING BACTERIAL GROWTH As previously discussed, different media are used to recover particular bacterial pathogens so that determining which media support growth is a clue to the type of organism isolated (e.g., growth on MacConkey agar indicates the organism is a gram-negative bacillus). The incubation conditions that support growth may also be a preliminary indicator of which bacteria have been isolated (e.g., aerobic vs. anaerobic bacteria).

RELATIVE QUANTITIES OF EACH COLONY TYPE The predominance of a bacterial isolate is often used as one of the criteria, along with direct smear results, organism virulence, and the body site from which the culture was obtained, for establishing the organism's clinical significance.

COLONY CHARACTERISTICS Noting key features of a bacterial colony is important for any bacterial identification; success or failure of subsequent identification procedures often depends on the accuracy of these observations. Criteria frequently used to characterize bacterial growth include:

- Colony size (usually measured in millimeters or described in relative terms such as pinpoint, small, medium, large)
- Colony pigmentation
- Colony shape (includes form, elevation, and margin of the colony [Figure 12-11])
- Colony surface appearance (e.g., glistening, opaque, dull, transparent)
- Changes in agar media resulting from bacterial growth (e.g., hemolytic pattern on blood agar blood, changes in color of pH indicators, pitting of the agar surface; for examples see Figures 12-3 through 12-8)
- Odor (certain bacteria produce distinct odors that can be helpful in preliminary identification)

Many of these criteria are somewhat subjective, and the adjectives and descriptive terms used may vary among different laboratories. Regardless of the terminology used, nearly every laboratory's protocol for bacterial identification begins with some agreed upon colony description of the commonly encountered pathogens.

Although careful determination of colony appearance is important, it is unwise to place total confidence on colony morphology for preliminary identification. Bacteria of one species often exhibit colony characteristics that are nearly indistinguishable from those of many other species. Additionally, bacte-

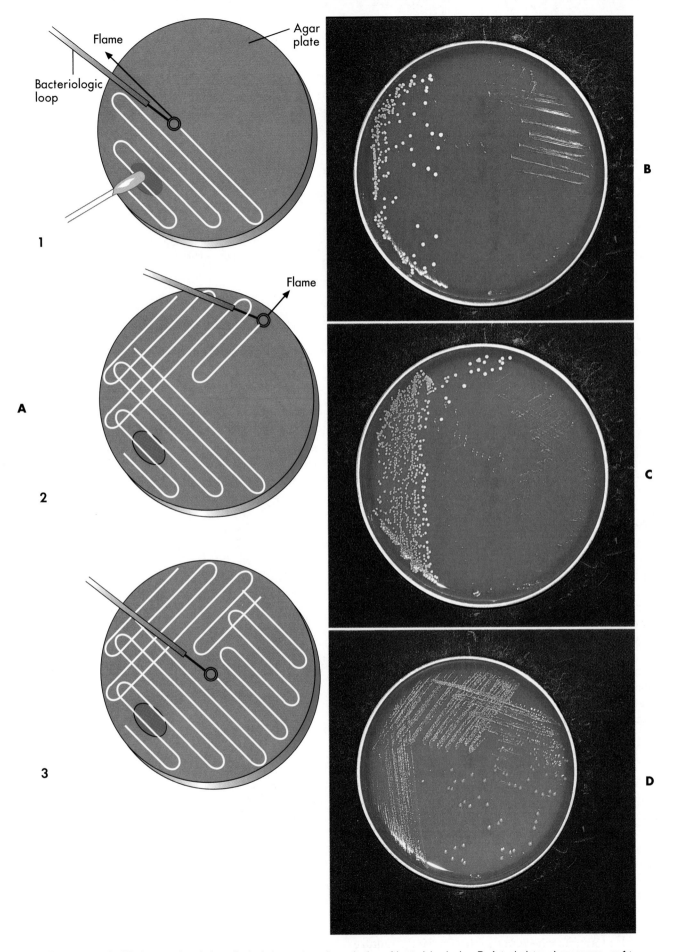

FIGURE 12-9 A, Dilution streak technique for isolation and semiquantitation of bacterial colonies. **B,** Actual plates show sparse, or 1+, bacterial growth that is limited to the first quadrant. **C,** Moderate, or 2+, bacterial growth that extends to the second quadrant. **D,** Heavy, or 3+ to 4+, bacterial growth that extends to the fourth quadrant.

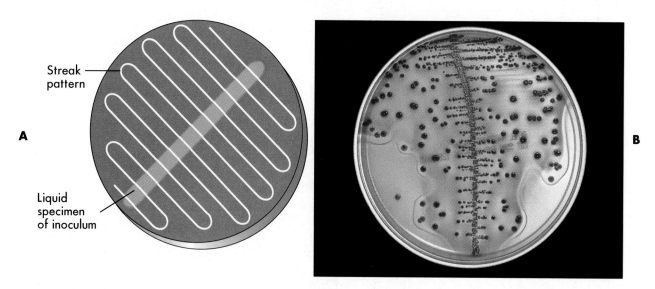

FIGURE 12-10 A, Streaking pattern using a calibrated loop for enumeration of bacterial colonies grown from a liquid specimen such as urine. **B,** Actual plate shows well-isolated and dispersed bacterial colonies for enumeration obtained with the calibrated loop streaking technique.

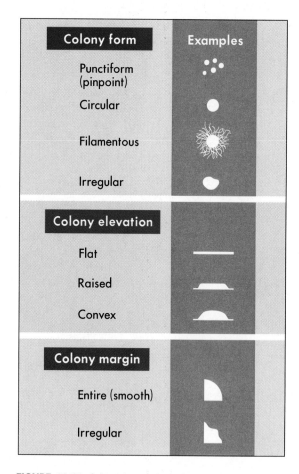

FIGURE 12-11 Colony morphologic features and descriptive terms for commonly encountered bacterial colonies.

ria of the same species exhibit morphologic diversity. For example, certain colony characteristics may be typical of a given species, but different strains of that species may have different morphologies.

GRAM STAIN AND SUBCULTURES   Isolation of individual colonies during cultivation is not only important for examining morphologies and characteristics but also is necessary for timely performance of Gram stains and subcultures.

The Gram stain and microscopic evaluation of cultured bacteria are used with colony morphology to decide which identification steps are needed. To avoid confusion, organisms from a single colony are stained. In many instances staining must be performed with each different colony morphology that is observed on the primary plate. In other cases staining may not be necessary because growth on a particular selective agar provides dependable evidence of the organism's Gram stain morphology (e.g., gram-negative bacilli essentially are the only clinically relevant bacteria that grow well on MacConkey agar).

Following characterization of growth on primary plating media, all subsequent procedures for definitive identification require the use of pure cultures (i.e., cultures containing one strain of a single species). If sufficient inocula for testing can be obtained from the primary media, then a subculture is not necessary, except as a precaution to obtain more of the etiologic agent if needed and to ensure that a pure inoculum has been used for subsequent tests (i.e., a "purity" check). However, frequently the primary media do not yield sufficient amounts of bacteria in pure culture and a subculture step is required (Figure 12-12).

Using a sterile loop, a portion of an isolated colony is taken and transferred to the surface of a suitable enrichment medium that is then incubated under conditions optimum for the organism. When making transfers for subculture it is beneficial to flame the inoculating loop between streaks to each area on the agar surface. This avoids overinoculation of the subculture media and ensures individual colonies will be obtained. Once a pure culture is available in a sufficient amount an inoculum for subsequent identification procedures can be prepared.

ALTERNATIVE APPROACHES TO BACTERIAL CULTIVATION

In addition to cultivation and visualization of bacterial growth, other methods for the detection of bacterial pathogens in human specimens have been developed and evaluated. Instrumentation for the measurement of bacterial CO_2 production to detect the presence of organisms has been most successfully applied to automated blood culture systems (for more information regarding blood culture methods see Chapter 20). Examples of other strategies include bioluminescence, colorimetry, and electrical impedance.

BIOLUMINESCENCE ASSAY FOR VIABLE ORGANISMS

In bioluminescence, the firefly enzyme luciferase catalyzes the conversion of luciferin to oxyluciferin and light. This reaction requires energy that is supplied by the dephosphorylation of adenosine triphosphate (ATP). The amount of light (measured in photons) produced by the reaction is directly proportional to the amount of ATP present in the solution. Based on this system, assays have been developed that can quantitate the amount of bacterial ATP by using a luminometer to measure emitted light. The amount of ATP detected can be used to predict the presence and even the number of bacteria in a clinical specimen. Systems have employed this principle for screening urine specimens for bacteria, but none are currently commercially available.

COLORIMETRIC FILTRATION FOR URINE SCREENING

Another technology used to rapidly determine whether a urine specimen contains significant numbers of bacteria is the Bac-T-Screen bacteriuria detection device (bioMérieux Vitek Inc., Hazelwood, Mo.). With this system, a measured amount of urine is drawn through a paper filter by vacuum suction. Particles, such as bacteria and white blood cells, adhere to the filter. A stain is then passed through the filter, imparting color, the depth of which is dependent on the number and type of stainable particles. The filter paper is then manually inserted into another section of the instrument, where a photometer compares the color with a preset standard. The filter may also be examined visually. Urines that stain at a level below the control can be considered negative and need not be processed further. Those that stain above the level (positive) are processed for cultivation and growth of bacteria.

ELECTRICAL IMPEDANCE AS AN INDICATOR OF MICROBIAL GROWTH

Bacterial metabolic products present in a growth medium can impede the flow of an electrical current through the medium (i.e., impedance). Theoretically, measuring these changes in electrical current could be used to detect the presence of multiplying organisms in a specimen. However, such systems have not been successfully developed for the detection of bacteria in clinical specimens.

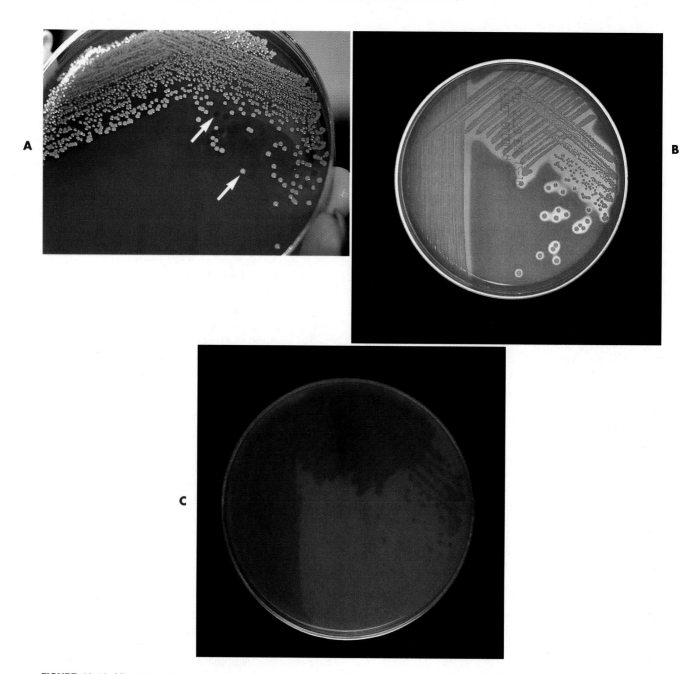

FIGURE 12-12 Mixed bacterial culture on sheep blood agar (**A**) requires subculture of individually distinct colonies (*arrows*) to obtain pure cultures of *S. aureus* (**B**) and *S. pneumoniae* (**C**).

Bibliography

Atlas, R.M. and Parks, L.C., editors. 1993. Handbook of microbiological media. CRC Press, Boca Raton, Fla.

Baldwin, M.S., Warner, M.B., and Murray. C.A. 1993. REMEL Technical manual, ed 1.

Difco manual, ed 10. 1984. Difco Laboratories, Detroit.

MacFaddin, J.F. 1985. Media for isolation-cultivation-identification-maintenance of medical bacteria, vol 1. Williams & Wilkins, Baltimore.

Oxoid: the manual. 1995. Oxoid Ltd., Basingstoke, U.K.

Power, D.A. and McCuen, P.M. 1988. Manual of BBL products and laboratory procedures, ed 6.

13 OVERVIEW OF CONVENTIONAL METHODS FOR BACTERIAL IDENTIFICATION

Microbiologists use various methods to identify organisms cultivated from patient specimens. Although many of the principles and issues about bacterial identification discussed in this chapter are generally applicable to most clinically relevant bacteria, specific information regarding particular organism groups is covered in the appropriate chapters in Part 4.

The importance of accurate bacterial identification cannot be overstated because identity is central to diagnostic bacteriology issues, including:

- Determining the clinical significance of a particular pathogen (e.g., is the isolate a pathogen or a contaminant?)
- Guiding physician care of the patient
- Determining whether laboratory testing for detection of antimicrobial resistance is warranted
- Determining the type of antimicrobial therapy that is appropriate
- Determining whether the antimicrobial susceptibility profiles are unusual or aberrant for a particular bacterial species
- Determining whether the infecting organism is a risk for other patients in the hospital, the public, or laboratory workers, that is, is the organism one that may pose problems for infection control, public health, or laboratory safety?

PRINCIPLES OF IDENTIFICATION

The identification of a bacterial isolate requires analysis of information gathered from laboratory tests that provide characteristic profiles of bacteria. The tests and the order in which they are used for organism identification are often referred to as an **identification scheme**. Identification schemes can be classified into one of two categories: those that are based on genotypic characteristics of bacteria and those based on phenotypic characteristics. Certain schemes rely on both genotypic and phenotypic characteristics. Additionally, some tests, such as the Gram stain, are an integral part of many schemes used for identifying a wide variety of bacteria, whereas other tests may only

be used in the identification scheme for a single species such as the fluorescent antibody test for identification of *Legionella pneumophila*.

ORGANISM IDENTIFICATION USING GENOTYPIC CRITERIA

Genotypic identification methods involve characterization of some portion of a bacterium's genome using molecular techniques for DNA or RNA analysis. This usually involves detecting the presence of a gene, or a part thereof, or an RNA product that is specific for a particular organism. In principle, the presence of a specific gene or a particular nucleic acid sequence is interpreted as a definitive identification of the organism. The genotypic approach is highly specific and often very sensitive. With the ever-expanding list of molecular techniques being developed, the genetic approach to organism identification will continue to grow and become more integrated into diagnostic microbiology laboratory protocols (for more information regarding molecular methods see Chapter 14).

ORGANISM IDENTIFICATION USING PHENOTYPIC CRITERIA

Phenotypic criteria are based on observable physical or metabolic characteristics of bacteria, that is, identification is through analysis of gene products rather than through the genes themselves. The phenotypic approach is the classic approach to bacterial identification, and most identification strategies are still based on bacterial phenotype. Delineation of some characteristics may require subcellular analysis involving sophisticated instrumentation (e.g., high-performance liquid chromatography [HPLC] to analyze cell wall components. For more information about these techniques see Chapter 14). Other characterizations are based on the antigenic makeup of the organisms and involve techniques based on antigen-antibody interactions (for more information regarding immunologic diagnosis of infectious diseases see Chapter 15). However, most of the phenotypic characterizations used in diagnostic bacteriology are based on tests that establish a bacterial isolate's morphology and metabolic capabilities. The most commonly used phenotypic criteria include:

- Macroscopic (colony) morphology
- Environmental requirements for growth
- Resistance or susceptibility to antimicrobial agents
- Microscopic morphology and staining characteristics
- Nutritional requirements and metabolic capabilities

Macroscopic (colony) morphology

Evaluation of colony morphology includes considering colony size, shape, color (pigment), surface appearance, and any changes that colony growth produces in the surrounding agar medium (e.g., hemolysis of blood in blood agar plates).

Although these characteristics usually are not sufficient for establishing a final or definitive identification, the information gained provides preliminary information necessary for determining what identification procedures should follow. However, it is unwise to place too much confidence on colony morphology alone for preliminary identification of isolates. Microorganisms often grow as colonies whose appearance is not that different from many other species, especially if the colonies are relatively young (i.e., less than 14 hours old). Therefore, unless colony morphology is distinctive or unless growth occurs on a particular selective medium, other characteristics must be included in the identification scheme.

Environmental requirements for growth

Environmental conditions required for growth can be used to supplement other identification criteria. However, as with colony morphologies, this information alone is not sufficient for establishing a final identification. The ability to grow in particular incubation atmospheres most frequently provides insight about the organism's potential identity. For example, organisms growing only in the bottom of a tube containing thioglycollate broth are not likely to be strictly aerobic bacteria, thus eliminating these types of bacteria from the list of identification possibilities. Similarly, anaerobic bacteria can be discounted in the identification schemes for organisms that grow on blood agar plates incubated in an ambient (room) atmosphere. An organism's requirement, or preference, for increased carbon dioxide concentrations can provide hints for the identification of other bacteria such as *Streptococcus pneumoniae*, *Haemophilus influenzae*, and *Neisseria gonorrhoeae*.

In addition to atmosphere, the ability to survive or even thrive in temperatures that exceed or are well below the normal body temperature of 37° C may be helpful for organism identification. The growth of *Campylobacter jejuni* at 42° C and the ability of *Yersinia enterocolitica* to survive at 0° C are two examples.

Resistance or susceptibility to antimicrobial agents

The ability of an organism to grow in the presence of certain antimicrobial agents or specific toxic substances is widely used to establish preliminary identification information. This is accomplished by using agar media supplemented with inhibitory substances or antibiotics (for examples see Table 12-1) or by directly measuring an organism's resistance to antimicrobial agents that may be used to treat infections (for more information regarding antimicrobial susceptibility testing see Chapter 18).

As discussed in Chapter 12, most clinical specimens are inoculated to several media, including some selective or differential agars. Therefore, the first clue to identification of an isolated colony is the nature of the media on which the organism is growing. For example, with rare exceptions, only gram-negative bacteria grow well on MacConkey agar. Alternatively, other agar plates, such as Columbia agar with colistin and nalidixic acid (CNA), support the growth of gram-positive organisms to the exclusion of most gram-negative bacilli. Certain agar media can be used to differentiate even more precisely than simply separating gram-negative and gram-positive bacteria. Whereas chocolate agar will support the growth of all *Neisseria* spp., the antibiotic-supplemented Thayer-Martin formulation will almost exclusively support the growth of the pathogenic species *N. meningitidis* and *N. gonorrhoeae*.

Directly testing a bacterial isolate's susceptibility to a particular antimicrobial agent may be a very useful part of an identification scheme. Many gram-positive bacteria (with a few exceptions, such as certain enterococci, lactobacilli, *Leuconostoc*, and *Pediococcus* spp.) are susceptible to vancomycin, an antimicrobial agent that acts on the bacterial cell wall. In contrast, most clinically important gram-negative bacteria are resistant to vancomycin. Therefore, when organisms with uncertain Gram stain results are encountered, susceptibility to vancomycin can be used to help establish the organism's Gram "status." Any zone of inhibition around a vancomycin-impregnated disk after overnight incubation is usually indicative of a gram-positive bacterium (Figure 13-1). With few exceptions (e.g., certain *Flavobacterium*, *Moraxella*, or *Acinetobacter* spp. isolates may be vancomycin susceptible) truly gram-negative bacteria are resistant to vancomycin. Conversely, most gram-negative bacteria are susceptible to the antibiotics colistin or polymyxin, whereas gram-positive bacteria are frequently resistant to these agents.

Microscopic morphology and staining characteristics

Microscopic evaluation of bacterial cellular morphology, as facilitated by the Gram stain or other enhancing methods discussed in Chapter 11, provides the most basic and important information on which final identification strategies are based. For this reason, a Gram stain of bacterial growth from isolated colonies on various media is usually the first step in any identification scheme. Based on these findings, most clinically relevant bacteria can be divided into four distinct groups: gram-positive cocci, gram-negative cocci, gram-positive bacilli, and gram-negative bacilli (see Figure 11-4). Some bacterial species are morphologically indistinct and are described as "gram-negative coccobacilli," "gram-variable bacilli," or **pleomorphic** (i.e., exhibiting various shapes). Still other morphologies include curved and/or rods and spirals.

Even without staining, examination of a wet preparation of bacterial colonies under oil immersion (1000× magnification) can provide clues as to possible identity. For example, a wet preparation prepared from a translucent, alpha-hemolytic colony on blood agar may reveal cocci in chains, a strong indication that the bacteria are probably streptococci. Also, the presence of yeast, whose colonies can closely mimic bacterial colonies but whose cells are generally much larger, can be determined (Figure 13-2).

In most instances, identification schemes for final identification are based on the cellular morphologies and staining characteristics of bacteria. To illustrate, an abbreviated identification flow chart for commonly encountered bacteria is shown in Figure 13-3. More detailed identification schemes are presented throughout Part 4; this flow chart simply illustrates how information about microorganisms is integrated into subsequent identification schemes that are usually based on the organism's nutritional requirements and metabolic capabilities. In certain cases, staining characteristics alone are used to definitively identify a bacterial species. Examples of this are mostly restricted to the use of fluorescent-labeled specific antibodies and fluorescent microscopy to identify organisms such as *Legionella pneumophila* and *Bordetella pertussis*.

Nutritional requirements and metabolic capabilities

Determining the nutritional and metabolic capabilities of a bacterial isolate is the most common approach used for determining the genus and species of an organism. The methods available for making these determinations share many commonalties but also have some important differences. In general, all methods use a combination of tests to establish the enzymatic capabilities of a given bacterial isolate as well as the isolate's ability to grow or survive the presence of certain inhibitors (e.g., salts, surfactants, toxins, and antibiotics).

ESTABLISHING ENZYMATIC CAPABILITIES As discussed in Chapter 9, enzymes are the driving force in bacterial metabolism. Because enzymes are genetically encoded,

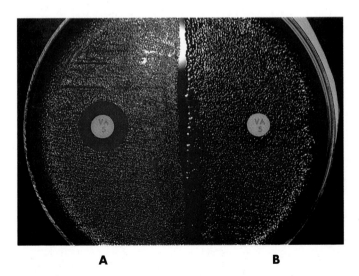

A   **B**

FIGURE 13-1 Zone of growth inhibition around the 5-μg vancomycin disk is indicative of a gram-positive bacterium (**A**). The gram-negative organism is not inhibited by this antibiotic, and growth extends to the edge of the disk (**B**).

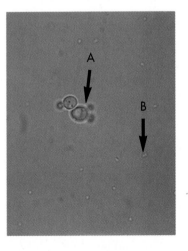

FIGURE 13-2 Microscopic examination of a wet preparation demonstrates the size difference between most yeast cells, such as those of *C. albicans* (*arrow A*), and bacteria such as *S. aureus* (*arrow B*).

the enzymatic content of an organism is a direct reflection of the organism's genetic makeup, which, in turn, is specific for individual bacterial species.

TYPES OF ENZYME-BASED TESTS In diagnostic bacteriology, enzyme-based tests are designed to measure the presence of one specific enzyme or a complete metabolic pathway that may contain several different enzymes. Although the specific tests most useful for the identification of particular bacteria are discussed in Part 4, some examples of tests commonly used to characterize a wide spectrum of bacteria are reviewed here.

SINGLE ENZYME TESTS Several tests are commonly used to determine the presence of a single enzyme. These tests usually provide rapid results because they can be performed on organisms already grown in culture. Of importance, these tests are easy to perform and interpret and often play a key role in the identification scheme. Although most single enzyme tests do not

yield sufficient information to provide species identification, they are used extensively to determine which subsequent identification steps should be followed. For example, the catalase test can provide pivotal information and is commonly used in schemes for gram-positive identifications. The oxidase test is of comparable importance in identification schemes for gram-negative bacteria (Figure 13-3).

Catalase Test. The enzyme **catalase** catalyzes the release of water and oxygen from hydrogen peroxide (H_2O_2 + catalase = H_2O + O_2); its presence is determined by direct analysis of a bacterial culture (Procedure 13-1). The rapid production of bubbles (effervescence) when bacterial growth is mixed with a hydrogen peroxide solution is interpreted as a positive test (i.e., the presence of catalase). Failure to produce effervescence or weak effervescence is interpreted as negative. If the bacterial inoculum is inadvertently contaminated with red blood cells when the test inoculum is collected from a sheep blood agar plate, weak produc-

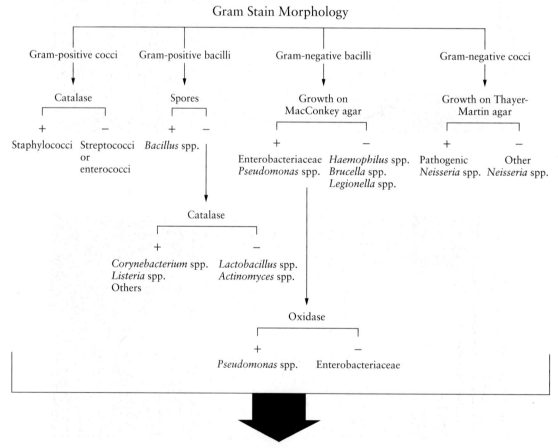

FIGURE 13-3 Flow chart example of a bacterial identification scheme.

tion of bubbles may occur, but this should not be interpreted as a positive test (Figure 13-4).

Because the catalase test is key to the identification scheme of many gram-positive organisms, interpretation must be done carefully. For example, staphylococci are catalase-positive, whereas streptococci and enterococci are negative; similarly, the catalase reaction differentiates *Listeria monocytogenes* and corynebacteria (catalase-positive) from other gram-positive, non–spore-forming bacilli (see Figure 13-3).

Oxidase Test. Cytochrome oxidase participates in electron transport and in the nitrate metabolic pathways of certain bacteria. The test for the presence of oxidase can be performed by flooding bacterial colonies on the agar surface with the reagent 1% tetramethyl-*p*-phenylenediamine dihydrochloride. Alternatively, a sample of the bacterial colony can be rubbed onto filter paper impregnated with the reagent (Kovac's method, Procedure 13-2). Positive and negative results obtained by both methods are shown in Figure 13-5. If an iron-containing wire is used to transfer growth, a false-positive reaction may result; therefore, platinum wire or wooden sticks are recommended. Certain organisms may show slight positive reactions after the initial 10 seconds have passed; such results are not considered definitive.

The test is initially used for differentiating between groups of gram-negative bacteria. Among the

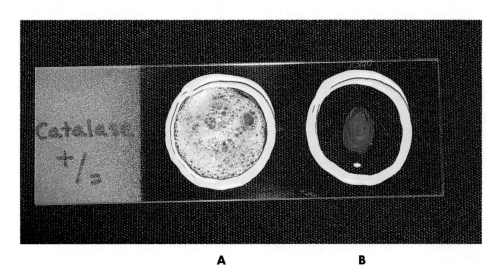

FIGURE 13-4 A positive catalase reaction is indicated by the production of copious bubbles when hydrogen peroxide is added to a bacterial inoculum (**A**). Lack of or minimal bubble production indicates a negative test (**B**).

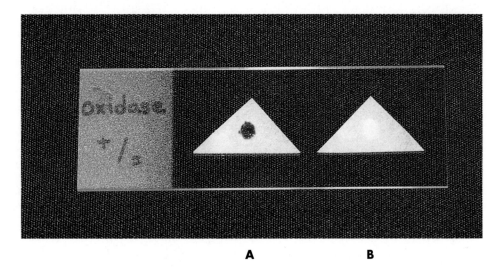

FIGURE 13-5 A positive oxidase reaction is indicated by the development of a dark purple color when the reagent substrate is inoculated with bacterial growth (**A**). No color development indicates a negative test and the absence of the enzyme oxidase (**B**).

commonly encountered gram-negative bacilli *Enterobacteriaceae, Stenotrophomonas maltophilia*, and *Acinetobacter* spp. are oxidase-negative, whereas many other bacilli, such as *Pseudomonas* spp. and *Aeromonas* spp., are positive (see Figure 13-3). The oxidase test is also a key reaction for the identification of *Neisseria* spp. (oxidase-positive).

Indole Test. Bacteria that produce the enzyme **tryptophanase** are able to degrade the amino acid tryptophan into pyruvic acid, ammonia, and indole. Indole is detected by combining with an indicator, aldehyde (1% paradimethylaminocinnamaldehyde), that results in a blue color formation (Procedure 13-3 and Figure 13-6). This test is used in numerous identification schemes, especially to presumptively identify *Escherichia coli*, the gram-negative bacillus most commonly encountered in diagnostic bacteriology.

Urease Test. **Urease** hydrolyzes the substrate urea into ammonia, water, and carbon dioxide. The presence of the enzyme is determined by inoculating an organism to broth or agar that contains urea as the primary carbon source and detecting the production of ammonia (Procedure 13-4). Ammonia increases the pH of the medium so its presence is readily detected using a pH indicator (Figure 13-7). Change in medium pH is a common indicator of metabolic process and, because **pH indicators** change color with increases (alkalinity) or decreases (acidity) in the medium's pH, they are commonly used in many identification test schemes. The urease test helps to identify certain species of *Enterobacteriaceae*, such as *Proteus* spp., and other important bacteria such as *Corynebacterium urealyticum* and *Helicobacter pylori*.

PYR Test. The enzyme L-**pyrroglutamyl-aminopeptidase** hydrolyzes the substrate L-pyrrolidonyl-β-naphthylamide (PYR) to produce a β-naphthylamine. When the β-naphthylamine combines with a cinnamaldehyde reagent, a bright red color is produced (Procedure 13-5 and Figure 13-8). The PYR test is particularly helpful in identifying gram-positive cocci such as *S. pyogenes* and *Enterococcus* spp., which are positive, while other streptococci are negative.

Hippurate Hydrolysis. **Hippuricase** is a constitutive enzyme that hydrolyzes the substrate hippurate to produce the amino acid glycine. Glycine is detected by oxidation with ninhydrin reagent that results in the production of a deep purple color (Procedure 13-6 and Figure 13-9). The hippurate test is most frequently used in the identification of *Streptococcus agalactiae, Campylobacter jejuni*, and *Listeria monocytogenes*.

TESTS FOR PRESENCE OF METABOLIC PATHWAYS Several identification schemes are based on determining what metabolic pathways an organism uses and the substrates processed by these pathways. In contrast to single enzyme tests, these pathways may involve several interactive enzymes. The presence of an end product resulting from these interactions is measured in the testing system. Assays for metabolic pathways can be classified into three general categories: carbohydrate oxidation and fermentation, amino acid degradation, and single substrate utilizations.

Oxidation and Fermentation Tests. As discussed in Chapter 9, bacteria use various metabolic pathways to produce biochemical building blocks and energy. For most

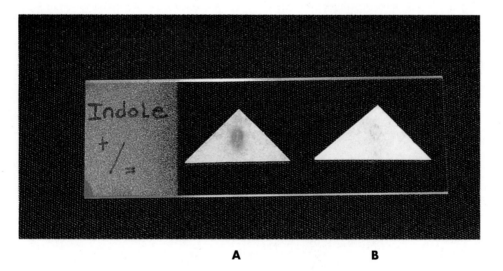

FIGURE 13-6 A positive indole test reaction is indicated by the development of a blue-green color when a filter paper disk impregnated with the test substrate is inoculated with bacterial growth (**A**). No color development indicates a negative test and the absence of the enzyme tryptophanase (**B**).

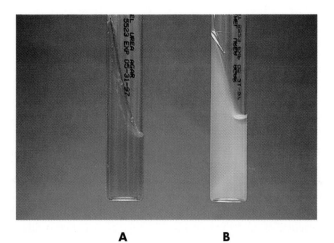

FIGURE 13-7 A positive urease test reaction is indicated by the development of a bright pink or red color when a tube containing test substrate is inoculated with bacterial growth (**A**). No color change in the tube indicates a negative test (**B**).

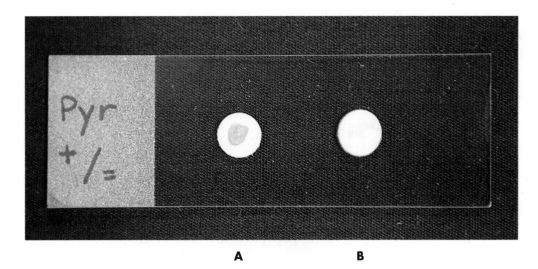

FIGURE 13-8 PYR-positive organisms produce a deep red color (**A**), whereas PYR-negative bacteria produce a slight orange color or no color change (**B**).

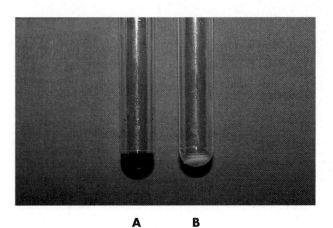

FIGURE 13-9 Hippurate-positive bacteria produce a deep purple color (**A**), whereas hippurate-negative bacteria produce a slightly yellow-pink color or fail to produce any color (**B**).

clinically relevant bacteria, this involves utilization of carbohydrates (e.g., sugar or sugar derivatives) and protein substrates. Determining whether substrate utilization is an oxidative or fermentative process is important for the identification of several different bacteria.

Oxidative processes require oxygen; fermentative ones do not. The clinical laboratory determines how an organism utilizes a substrate by observing whether acid byproducts are produced in the presence or absence of oxygen. In most instances the presence of acid byproducts is detected by a change in the pH indicator incorporated into the medium. The color changes that occur in the presence of acid depend on the type of pH indicator used.

Oxidation-fermentation determinations are usually accomplished using a special medium (**oxidative-fermentative [O-F] medium**) that contains low concentrations of peptone and a single carbohydrate substrate such as glucose. The organism to be identified is inoculated into two **glucose O-F** tubes, one of which is then overlaid with mineral oil as a barrier to oxygen. Common pH indicators used for O-F tests, and the color changes they undergo with acidic conditions, include **bromcresol purple,** which

changes from purple to yellow; **Andrade's acid fuchsin** indicator, which changes from pale yellow to pink; **phenol red,** which changes from red to yellow; and **bromthymol blue,** which changes from green to yellow.

As shown in Figure 13-10, when acid production is detected in both tubes the organism is identified as a **glucose fermenter** since fermentation can occur with or without oxygen. If acid is only detected in the open, aerobic tube, the organism is characterized as a **glucose-oxidizer.** As a third possibility, some bacteria do not use glucose as a substrate and no acid is detected in either tube (a **nonutilizer**). The glucose fermentative or oxidative capacity is generally used to separate organisms into major groups (e.g., *Enterobacteriaceae* are fermentative; *Pseudomonas* spp. are oxidative). However, the utilization pattern for several other carbohydrates (e.g., lactose, sucrose, xylose, maltose) is often needed to help identify an organism's genus and species.

Amino Acid Degradation. Determining the ability of bacteria to produce enzymes that either deaminate, dihydrolyze, or decarboxylate certain amino acids is often used in identification schemes. The amino acid

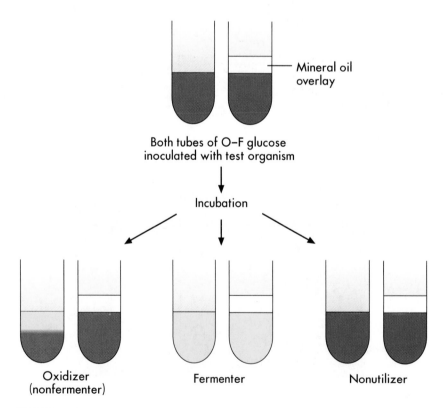

FIGURE 13-10 Principle of glucose oxidative-fermentation (O-F) test. Fermentation patterns shown in O-F tubes are examples of oxidative, fermentative, and nonutilizing bacteria.

substrates most often tested include lysine, ornithine, arginine, and phenylalanine (the indole test for tryptophan cleavage is presented earlier in this chapter).

Decarboxylases cleave the carboxyl group from amino acids so that amino acids are converted into amines; lysine is converted to cadaverine, and ornithine is converted to putrescine. Because amines increase medium pH, they are readily detected by color changes in a pH indictor indicative of alkalinity. Decarboxylation is an anaerobic process that requires an acid environment for activation. The most common medium used for this test is **Moeller decarboxylase base,** whose components include glucose, the amino acid substrate of interest (i.e., lysine, ornithine, or arginine), and a pH indicator.

Organisms are inoculated into the tube medium that is then overlaid with mineral oil to ensure anaerobic conditions. Early during incubation bacteria utilize the glucose and produce acid, resulting in a yellow coloration of the pH indicator. Organisms that can decarboxylate the amino acid then begin to attack that substrate and produce the amine product, which increases the pH and changes the indicator back from yellow to purple (if bromcresol purple is the pH indicator used; red if phenol red is the indicator). Therefore, after overnight incubation a positive test is indicated by a purple color and a negative test (i.e., lack of decarboxylase activity) is indicated by a yellow color (Figure 13-11). With each amino acid tested, a control tube of the glucose-containing broth base without amino acid is inoculated. This standard's color is compared with that of the tube containing the amino acid following incubation.

Because it is a two-step process, the breakdown of arginine is more complicated than that of lysine or ornithine. Arginine is first **dehydrolyzed** to citrulline, which is subsequently converted to ornithine. Ornithine is then decarboxylated to putrescine, which results in the same pH indicator changes as just outlined for the other amino acids.

Unlike decarboxylation, **deamination** of the amino acid phenylalanine occurs in air. The presence of the end product (phenylpyruvic acid) is detected by the addition of 10% ferric chloride, which results in the development of a green color. Agar slant medium is commercially available for this test.

Single Substrate Utilization. Whether an organism can grow in the presence of a single nutrient or carbon source provides useful identification information. Such tests entail inoculation of organisms to a medium that contains a single source of nutrition (e.g., citrate, malonate, or acetate) and, after incubation, observing the medium for growth. Growth is determined by observing the presence of bacterial colonies or by using a pH indicator to detect end products of metabolic activity.

ESTABLISHING INHIBITOR PROFILES The ability of a bacterial isolate to grow in the presence of one or more inhibitory substances can provide valuable identification information. Examples regarding the use of inhibitory substances were presented earlier in this chapter and may also be found in Chapter 12.

In addition to the information gained from using inhibitory media and/or antimicrobial susceptibility testing, other more specific tests may be incorporated into bacterial identification schemes. Because most of these tests are used to identify a particular group of bacteria, their protocols and principles are discussed in the appropriate chapters in Part 4. A few examples of such tests include:

- Growth in the presence of various NaCl concentrations (identification of enterococci and *Vibrio* spp.)
- Susceptibility to optochin and solubility in bile (identification of *Streptococcus pneumoniae*)
- Ability to hydrolyze esculin in the presence of bile (identification of enterococci)
- Ethanol survival (identification of *Bacillus* spp.)

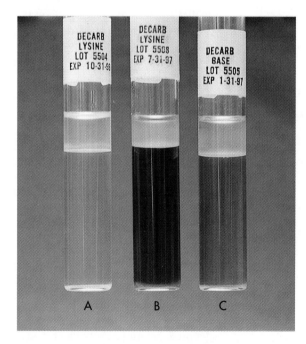

FIGURE 13-11 With the use of bromcresol purple as the pH indicator, a negative lysine decarboxylase test is indicated by development of a yellow color (**A**); a positive test is indicated by the purple color (**B**). A control tube not containing lysine is included for color comparisons (**C**).

PRINCIPLES OF PHENOTYPE-BASED IDENTIFICATION SCHEMES

As shown in Figure 13-3, growth characteristics, microscopic morphologies, and single test results are used to categorize most bacterial isolates into general groups. However, the definitive identification to species requires use of schemes designed to produce metabolic profiles of the organisms. Identification systems usually consist of four major components (Figure 13-12):

- Selection and inoculation of a set (i.e., **battery**) of specific metabolic substrates and growth inhibitors
- Incubation to allow substrate utilization to occur or to allow growth inhibitors to act
- Determination of metabolic activity that occurred during incubation
- Analysis of metabolic profiles and comparison with established profile databases for known bacterial species to establish definitive identification

SELECTION AND INOCULATION OF IDENTIFICATION TEST BATTERY

The number and types of tests that are selected for inclusion in a battery depends on various factors, including the type of bacteria to be identified, the clinical significance of the bacterial isolate, and the availability of reliable testing methods.

Type of bacteria to be identified

Certain organisms have such unique features that relatively few tests are required to establish identity. For example, *Staphylococcus aureus* is essentially the only gram-positive coccus that appears microscopically in clusters, is catalase-positive, and produces

1. Selection and inoculation of tests

- Number and type of tests selected depend on type of organism to be identified, clinical significance of isolates, and availability of reliable methods
- Identification systems must be inoculated with pure cultures

2. Incubation for substrate utilization

- Duration depends on whether bacterial multiplication is or is not required for substrate utilization (i.e., growth-based test vs. a non–growth-based test)

3. Detection of metabolic activity (substrate utilization)

- Colorimetry, fluorescence, or turbidity are used to detect products of substrate utilization
- Detection is done visually or with the aid of various photometers

4. Analysis of metabolic profiles

- Involves conversion of substrate utilization profile to a numeric code (see Figure 13-13)
- Computer-assisted comparison of numeric code with extensive taxonomic data base provides most likely identification of the bacterial isolate
- For certain organisms for which identification is based on a few tests, extensive testing and analysis are not routinely needed

FIGURE 13-12 Four basic components of bacterial identification schemes and systems.

coagulase. Therefore, identification of this common pathogen usually requires the use of only two tests coupled with colony and microscopic morphology. In contrast, identification of most clinically relevant gram-negative bacilli, such as those of the *Enterobacteriaceae* family, requires establishing metabolic profiles often involving 20 or more tests.

Clinical significance of the bacterial isolate

Although a relatively large number of tests may be required to identify a particular bacterial species, the number of tests actually inoculated may depend on the clinical significance of an isolate. For instance, if a gram-negative bacillus is mixed with five other bacterial species in a urine culture it is likely to be a contaminant. In this setting, multiple tests to establish species identity is not warranted and should not routinely be performed. However, if this same organism is isolated in pure culture from cerebrospinal fluid, the full battery of tests required for definitive identification should be performed.

Availability of reliable testing methods

With the complicated medical procedures patients are subjected to and the increasing population of immunocompromised patients, uncommon or unusual bacteria are isolated. Because of the unusual nature that some of these bacteria exhibit, reliable testing methods and identification criteria may not be established in most clinical laboratories. In these instances, only the genus of the organism may be identified (e.g., *Bacillus* spp.), or identification may not go beyond a description of the organism's microscopic morphology (e.g., gram-positive, pleomorphic bacilli, or gram-variable, branching organism). When such bacteria are encountered and are thought to be clinically significant, they should be sent to a reference laboratory whose personnel are experienced in identifying unusual organisms.

Although the number of tests included in an identification battery may vary and different identification systems may require various inoculation techniques, the one common feature of all systems is the requirement for inoculation with a pure culture. Inoculation with a mixture of bacteria produces mixed and often uninterpretable results. To expedite identification, cultivation strategies (described in Chapter 12) should focus on obtaining pure cultures as soon as possible. Furthermore, controls should be run with most identification systems as a check for purity of the culture used to inoculate the system.

INCUBATION FOR SUBSTRATE UTILIZATION

The time required to obtain bacterial identification depends heavily on the length of incubation needed before the test result is available. In turn, the duration of incubation depends on whether the test is measuring metabolic activity that requires bacterial growth or whether the assay is measuring the presence of a particular enzyme or cellular product that can be detected without the need for bacterial growth.

Conventional identification

Because the **generation time** (i.e., the time required for a bacterial population to double) for most clinically relevant bacteria is 20 to 30 minutes, growth-based tests usually require hours of incubation before the presence of an end product can be measured. Many conventional identification schemes require 18 to 24 hours of incubation, or longer, before the tests can be accurately interpreted. Although the conventional approach has been the standard for most bacterial identification schemes, the desire to produce results and identifications in a more timely fashion has resulted in the development of rapid identification strategies.

Rapid identification

In the context of diagnostic bacteriology the term *rapid* is relative. In some instances a rapid method is one that provides a result the same day that the test was inoculated. Alternatively, the definition may be more precise whereby "rapid" is only used to describe tests that provide results within 4 hours of inoculation.

Two general approaches have been developed to obtain more rapid identification results. One has been to vary the conventional testing approach by decreasing the test substrate medium volume and increasing the concentration of bacteria in the inoculum. Several conventional methods, such as carbohydrate fermentation profiles, use this strategy for more rapid results. For example, the method described in Procedure 13-7 has been used successfully for rapid identification of *Neisseria* spp., and modifications have been developed for rapid determination of carbohydrate reactions of fastidious gram-negative bacilli.

The second approach uses unique or unconventional substrates. Particular substrates are chosen, based on their ability to detect enzymatic activity at all times. That is, detection of the enzyme does not depend on multiplication of the organism (i.e., not a growth-based test) so that delays caused by depending on bacterial growth are minimized. The catalase, oxidase, and PYR tests discussed previously are examples of such tests, but many others are available as part of commercial testing batteries.

Still other rapid identification schemes are based on antigen-antibody reactions, such as latex agglutination tests, that are commonly used to quickly and easily identify certain beta-hemolytic streptococci and *S. aureus* (for more information regarding these test formats see Chapter 15).

DETECTION OF METABOLIC ACTIVITY

The accuracy of an identification scheme heavily depends on the ability to reliably detect whether a bacterial isolate has utilized the substrates composing the identification battery. The sensitivity and strength of the detection signal can also contribute to how rapidly results are available. No matter how quickly an organism may metabolize a particular substrate, if the end products are slowly or weakly detected, the ultimate production of results will still be "slow."

Detection strategies for determining the end-products of different metabolic pathways use one of the following: colorimetry, fluorescence, and turbidity.

Colorimetry

Several identification systems measure color change to detect the presence of metabolic end products. Most frequently the color change is produced by pH indicators in the media. Depending on the by-products to be measured and the testing method, additional reagents may need to be added to the reaction before result interpretation. An alternative to the use of pH indicators is the oxidation-reduction potential indicator tetrazolium violet. Organisms are inoculated into wells that contain a single, utilizable carbon source. Metabolism of that substrate generates electrons that reduce the tetrazolium violet, resulting in production of a purple color (positive reaction) that can be spectrophotometrically detected. In a third approach, the substrates themselves may be chromogenic so that when they are "broken down" by the organism the altered substrate produces a color.

Many commercial systems use a miniaturized modification of conventional biochemical batteries, with the color change being detectable with the unaided eye. Alternatively, in certain automated systems, a photoelectric cell measures the change in the wavelength of light transmitted through miniaturized growth cuvettes or wells, thus eliminating the need for direct visual interpretation by laboratory personnel. Additionally, a complex combination of dyes and filters may be used to enhance and broaden the scope of substrates and color changes that can be used in such systems. These combinations hasten identification and increase the variety of organisms that can be reliably identified.

Fluorescence

There are two basic strategies for using fluorescence to measure metabolic activity. In one approach, substrate-fluorophore complexes are used. If a bacterial isolate processes the substrate, the fluorophore is released and assumes a fluorescent configuration. Alternatively, pH changes resulting from metabolic activity can be measured by changes in fluorescence of certain fluorophore markers. In these pH-driven, fluorometric reactions, pH changes result in either the fluorophore becoming fluorescent or, in other instances, fluorescence being quenched or lost. To detect fluorescence, ultraviolet light of appropriate wavelength is focused on the reaction mixture and a special kind of photometer, a fluorometer, measures fluorescence.

Turbidity

Turbidity measurements are not commonly used for bacterial identifications but do have widespread application for determining growth in the presence of specific growth inhibitors, including antimicrobial agents, and for detecting bacteria present in certain clinical specimens.

Turbidity is the ability of particles in suspension to refract and deflect light rays passing through the suspension such that the light is reflected back into the eyes of the observer. The **optical density** (OD), a measurement of turbidity, is determined in a spectrophotometer. This instrument compares the amount of light that passes through the suspension (the **percent transmittance**) with the amount of light that passes through a control suspension without particles. A photoelectric sensor, or photometer, converts the light that impinges on its surface to an electrical impulse, which can be quantified. A second type of turbidity measurement is obtained by nephelometry, or light scatter. In this case, the photometers are placed at angles to the suspension, and the scattered light, generated by a laser or incandescent bulb, is measured. The amount of light scattered depends on the number and size of the particles in suspension.

ANALYSIS OF METABOLIC PROFILES

The metabolic profile obtained with a particular bacterial isolate is essentially the phenotypic fingerprint, or signature, of that organism. Typically, the profile is recorded as a series of pluses (+) for positive reactions and minuses (−) for negative or nonreactions (Figure 13-13). Although this profile by itself provides little information, microbiologists can compare the profile with an extensive identification database to establish the identity of that specific isolate.

Identification databases

Reference databases are available for clinical use. These databases are maintained by manufacturers of identification systems and are based on the continuously updated taxonomic status of clinically relevant bacteria. Although microbiologists typically do not establish and maintain their own databases, an overview of the general approach provides background information.

The first step in developing a database is to accumulate many bacterial strains of the same species. Each strain is inoculated to an identical battery of metabolic tests to generate a positive-negative test profile. The cumulative results of each test are expressed as a percentage of each genus or species that possesses that characteristic. For example, suppose that 100 different known *Escherichia coli* strains and 100 known *Shigella* spp. strains are tested in four biochemicals, yielding the results illustrated in Table 13-1. In reality, many more strains and tests would be performed. However, the principle—to

generate a database for each species that contains the percentage probability for a positive result with each test in the battery—is the same.

Manufacturers develop databases for each of the identification systems they produce for diagnostic use (e.g., *Enterobacteriaceae,* gram-positive cocci, nonfermentative gram-negative bacilli). Because the data are based on organism "behavior" in a particular commercial system, the databases cannot and should not be applied to interpret profiles obtained by other testing methods.

Furthermore, most databases are established

TABLE 13-1 GENERATION AND USE OF GENUS-IDENTIFICATION DATABASE PROBABILITY: PERCENT POSITIVE REACTIONS FOR 100 KNOWN STRAINS

| | BIOCHEMICAL PARAMETER | | | |
ORGANISM	LACTOSE	SUCROSE	INDOLE	ORNITHINE
Escherichia	91	49	99	63
Shigella	1	1	38	20

	Test/substrate	Test results (− or +)	Binary code conversion (0 or 1)	Octal value	Octal score	Octal triplet total	Octal profile
1	ONPG	+	1	×1	1		
2	Arginine dihydrolase	−	0	×2	0	5	
3	Lysine decarboxylase	+	1	×4	4		
4	Ornithine decarboxylase	+	1	×1	1		
5	Citrate utilization	−	0	×2	0	1	
6	H$_2$S production	−	0	×4	0		
7	Urea hydrolysis	−	0	×1	0		
8	Tryptophane deaminase	−	0	×2	0	4	
9	Indole production	+	1	×4	4		
10	VP test	−	0	×1	0		
11	Gelatin hydrolysis	−	0	×2	0	4	5144572
12	Glucose fermentation	+	1	×4	4		(*E. coli*)
13	Mannitol fermentation	+	1	×1	1		
14	Inositol fermentation	−	0	×2	0	5	
15	Sorbitol fermentation	+	1	×4	4		
16	Rhamnose fermentation	+	1	×1	1		
17	Sucrose fermentation	+	1	×2	2	7	
18	Melibiose fermentation	+	1	×4	4		
19	Amygdalin fermentation	−	0	×1	0		
20	Arabinose fermentation	+	1	×2	2	2	
21	Oxidase production	−	0	×4	0		

Column header spanning "Octal code conversion*" covers Octal value, Octal score, Octal triplet total.

*As derived from API 20E (bioMérieux Vitek, Inc.) for identification of *Enterobacteriaceae.*

FIGURE 13-13 Example of converting a metabolic profile to an octal profile for bacterial identification.

with the assumption that the isolate to be identified has been appropriately characterized using adjunctive tests. For example, if an *S. aureus* isolate is mistakenly tested using a system for identification of *Enterobacteriaceae,* the database will not identify the gram-positive coccus because the results obtained will only be compared with data available for enteric bacilli. This underscores the importance of accurately performing preliminary tests and observations, such as colony and Gram stain morphologies, before selecting a particular identification battery.

Use of the database to identify unknown isolates

Once a metabolic profile has been obtained with a bacterial isolate of unknown identity the profile must be converted to a numeric code that will facilitate comparison of the unknown's phenotypic fingerprint with the appropriate database.

A **binary code conversion** system uses the numerals 0 and 1 to represent negative and positive metabolic reactions, respectively. As shown in Figure 13-13, using binary code conversion, a 21-digit binomial number (e.g., 101100001001101111010, as read from top to bottom in the figure) is produced from the test result. This number is then used in an **octal code conversion** scheme to produce a mathematic number (**octal profile** [see Figure 13-13]). The octal profile number is easier for laboratorians to recognize and can be used to generate biotyping data for determining potential relatedness among bacterial strains.

The octal profile is established as follows: The metabolic profile results are divided into sets of three (triplets). Positive results obtained with the test in the first slot of each triplet have an **octal value** of 1, posi-

tives in the second slot have an octal value of 2, and those in the third slot have an octal value of 4. The product of the octal value and the binary code conversion result is the **octal score.** Regardless of slot position, all negative results receive an octal score of 0. The sum of the octal score for each triplet of tests is the **octal triplet total,** which provides a numeral that is part of the final octal profile.

As shown in Figure 13-13, the octal profile for the unknown organism is 5144572. This profile would then be compared with database profiles to determine the most likely identity of the organism. In this example, the octal profile indicates the unknown organism is *Escherichia coli.*

CONFIDENCE IN IDENTIFICATION Once metabolic profiles have been translated into numeric scores the probability that a correct correlation with the database has been made must be established, that is, how confident can the laboratorian be in knowing that a correct identification has been made. This is accomplished by establishing the percentage probability, which is usually provided as part of most commercially available identification database schemes.

For example, unknown organism X is tested against the four biochemicals shown in Table 13-1 and yields results as follows: lactose (+), sucrose (+), indole (−), and ornithine (+). Based on the results of each test, the percentage of known strains in the database that produced positive results are used to calculate the percentage probability that strain X is a member of one of the two genera (*Escherichia* or *Shigella*) given in the example (Table 13-2).Therefore, if 91% of *Escherichia* spp. are lactose-positive (see Table 13-1), the probability that X is a species of *Escherichia* based on lactose alone is 0.91. If 38% of

TABLE 13-2 GENERATION AND USE OF GENUS-IDENTIFICATION DATABASE PROBABILITY: PROBABILITY THAT UNKNOWN STRAIN *X* IS MEMBER OF KNOWN GENUS BASED ON RESULTS OF EACH INDIVIDUAL PARAMETER TESTED

ORGANISM	LACTOSE	BIOCHEMICAL PARAMETER SUCROSE	INDOLE	ORNITHINE
X	+	+	−	+
Escherichia	0.91	0.49	0.01	0.63
Shigella	0.01	0.01	0.62	0.20

Probability that X is *Escherichia* = 0.91 × 0.49 × 0.01 × 0.63 = 0.002809.
Probability that X is *Shigella* = 0.01 × 0.01 × 0.62 × 0.20 = 0.000012.

Shigella spp. are indole positive (see Table 13-1), then the probability that *X* is a species of *Shigella* based on indole alone is 0.62 (1.00 [all *Shigella*] − 0.38 [percent positive *Shigella*] = 0.62 [percent of all *Shigella* that are indole negative]). The probabilities of the individual tests are then multiplied to achieve a calculated likelihood that *X* is one of these two genera. In this example, *X* is more likely to be a species of *Escherichia,* with a probability of 357:1 (1 divided by 0.0028; see Table 13-2). This is still a very unlikely probability for correct identification, but only four parameters were tested, and the indole result was atypical. As more parameters are added to the formula, the importance of just one test decreases and the overall pattern prevails.

With many organisms being tested for 20 or more reactions, computer-generated databases provide the probabilities. As more organisms are included in the database, the genus and species designations and probabilities become more precise. Also, with more profiles in a database, the unusual patterns can be more readily recognized and, in some cases, new or unusual species may be discovered.

All commercial suppliers of multicomponent biochemical test systems provide their users with a computer, a computer-derived code book, or access to a telephone inquiry center to facilitate matching profile numbers with species.

COMMERCIAL IDENTIFICATION SYSTEMS

ADVANTAGES AND EXAMPLES OF COMMERCIAL SYSTEM DESIGNS

Commercially available identification systems have largely replaced compilations of conventional test media and substrates prepared in-house for bacterial identification. This replacement has mostly come about because the design of commercial systems has continuously evolved to maximize the speed and optimize the convenience with which all four identification components shown in Figure 13-12 can be achieved. Because laboratory workload has increased, conventional methodologies have had difficulty competing with the advantages of convenience and updated databases offered by commercial systems.

Some of the simplest multitest commercial systems consist of a conventional format that can be inoculated once to yield more than one result. By combining reactants, for example, one substrate can be used to determine indole and nitrate results; indole and motility results; motility, indole, and ornithine decarboxylase; or other combinations. Alternatively, conventional tests have been assembled in smaller volumes and packaged so that they can be inoculated

easily with one manipulation instead of several (e.g., microtitre tray format, Figure 13-14). When used in conjunction with a computer-generated database, species identifications are made relatively easily.

Various manufacturers produce conventional biochemicals in microtitre trays that are shipped to the user in a frozen state and maintained frozen until they are thawed, inoculated, and used. Alternatively, microdilution trays may contain lyophilized substrates that are rehydrated with a suspension of the organism during inoculation.

Another approach is to use an unsupplemented broth in microtitre trays to which various substrate-impregnated disks are added before inoculation (Figure 13-15). This approach allows the battery of tests to be designed by the user, whereas with other formats the batteries are predetermined by the manufacturer. Substrates may also be dried in plastic cupules that are arranged in series on strips into which a suspension of the test organism is placed

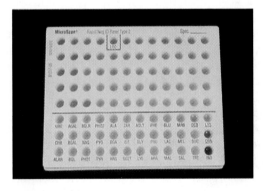

FIGURE 13-14 Microtitre tray format of a bacterial identification system (Dade International, Sacramento, Calif.). Multiple test wells are simultaneously inoculated with the same-source inoculum.

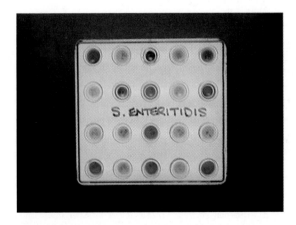

FIGURE 13-15 The Minitek system (Becton Dickinson Microbiology Systems) uses substrate-impregnated filter disks distributed to each test well before inoculation with a bacterial suspension.

(Figure 13-16). For some of these systems use of a heavy inoculum or use of substrates whose utilization is not dependent on extended bacterial multiplication allows results to be available after 4 to 6 hours of incubation.

Still other identification battery formats have been designed to more fully automate several aspects of the identification process. One example of this is use of "cards" that are substantially smaller than most microtitre trays or cupule strips (Figure 13-17). Analogous to the microtitre tray format, these cards contain dried substrates in tiny wells that are resuspended on inoculation.

Commercial systems are often categorized as either being **automated or nonautomated.** As shown in Table 13-3, various aspects of an identification system can be automated and these usually include, in whole or in part, the inoculation steps, the incubation and reading of tests, and the analysis of results. However, no strict criteria exist that state how many aspects must be automated for a whole system to be classified as automated. Therefore, whether a system is considered automated can be controversial. Furthermore, regardless of the lack or level of automation, the selection of an identification system ultimately depends on system accuracy and reliability, whether the system meets the needs of the laboratory, and limitations imposed by laboratory financial resources.

Overview of commercial systems

Various multitest bacterial identification systems are commercially available for use in diagnostic microbiology laboratories, and the four basic identification components outlined in Figure 13-12 are common to them all. However, different systems vary in their approach to each component. The most common variations involve:

- Types and formats of tests included in the test battery
- Method of inoculation (manual or automated)
- Required length of incubation for substrate utilization. This usually depends on whether utilization requires bacterial growth
- Method for detecting substrate utilization and whether detection is manual or automated
- Method of interpreting and analyzing results (manual or computer assisted), and if computer assisted, the extent to which assistance is automated

The general features of some commercial identification systems are summarized in Table 13-3. More specific information is available from the manufacturers.

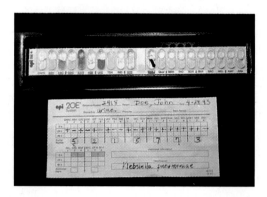

FIGURE 13-16 Biochemical test panel (API; bioMérieux Vitek, Inc., Hazelwood, Mo). The test results obtained with the substrates in each cupule are recorded, and an organism identification code is calculated by octal code conversion on the form provided. The octal profile obtained then is matched with an extensive database to establish organism identification.

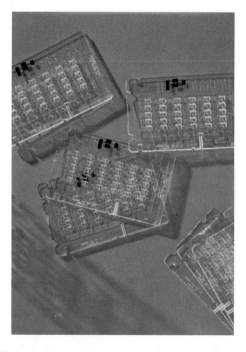

FIGURE 13-17 Plastic cards composed of multiple wells containing dried substrates that are reconstituted by inoculation with a bacterial suspension (bioMérieux Vitek, Inc., Hazelwood, Mo). Test results in the card wells are automatically read by the manufacturer's reading device.

TABLE 13-3 SUMMARY OF SELECTED COMMERCIAL BACTERIAL IDENTIFICATION SYSTEMS

PRODUCT (MANUFACTURER)	ORGANISMS IDENTIFIED	FORMAT	INOCULATION	SUBSTRATE UTILIZATION; GROWTH OR NONGROWTH (TIME TO RESULT)	DETECTION METHOD	RESULTS ANALYSIS
API Systems, several products (bioMérieux Vitek, Inc.)	Gram-negative bacilli Staphylococci Streptococci Other gram-positive cocci Corynebacteria and other gram-positive bacilli	Cupules or micro-tubes in plastic strip	Manual	Growth and nongrowth (2-72 hrs, depending on organism type)	Colorimetric, fluorescence, and turbidity (all manual)	
BBL Crystal (Becton Dickinson Microbiology Systems)	*Enterobacteriaceae* and other gram-negative bacilli Gram-positive cocci Gram-positive bacilli *Neisseria* spp. *Haemophilus* spp.	Modified microplate	Manual	Growth and nongrowth (3-20 hrs, depending on organism type)	Colorimetric or fluorometric (manual)	
BBL Minitek (Becton Dickinson Microbiology Systems)	*Enterobacteriaceae*, other gram-negative bacilli *Neisseria* spp. Miscellaneous gram-positives	Microtitre plate with substrate-impregnated disks	Manual	Growth (4-48 hrs, depending on organism type)	Colorimetric (manual)	Codebook
BBL Enterotube (Becton Dickinson Microbiology Systems)	*Enterobacteriaceae* Other gram-negative	Substrate chambers	Manual	Growth (18-48 hrs, depending on organism type)	Colorimetric	Codebook
GN-GP Microplate (Biolog)	Gram-negative bacilli and cocci Gram-positive cocci and bacilli	Microtitre plate	Manual	Growth and nongrowth (4-24 hrs)	Colorimetric (automated)	Computer-directed results analysis

Continued

TABLE 13-3 SUMMARY OF SELECTED COMMERCIAL BACTERIAL IDENTIFICATION SYSTEMS—CONT'D

PRODUCT (MANUFACTURER)	ORGANISMS IDENTIFIED	FORMAT	INOCULATION	SUBSTRATE UTILIZATION; GROWTH OR NONGROWTH (TIME TO RESULT)	DETECTION METHOD	RESULTS ANALYSIS AND INTERPRETATION
Micro-ID (Remel)	*Enterobacteriaceae* Other gram-negative bacilli	Substrate chambers in plastic strip	Manual	Nongrowth (4 hrs)	Colorimetric (manual)	Computer-assisted codebook
quad-FERM+ (bioMérieux Vitek, Inc.)	*Neisseria* and *Moraxella*	Cupules	Manual	Nongrowth (2 hrs)	Colorimetric (manual)	Computer-assisted codebook
Rapid ID, several products (Innovative Diagnostics)	*Enterobacteriaceae* and other gram-negative bacilli *Neisseria* and *Haemophilus* Streptococci, enterococci, and other gram-positive cocci	Self-inoculating tray with substrate reaction wells	Manual	Nongrowth (2-4 hrs)	Colorimetric (manual)	Computer-assisted codebook and electronic code compendium
Vitek (bioMérieux Vitek, Inc.)	*Enterobacteriaceae* and other gram-negative bacilli Staphylococci Streptococci Enterococci Miscellaneous gram-positive bacilli	Cards with miniature wells	Automated	Growth (2-18 hrs)	Colorimetric (Automated)	Computer-assisted Automated
MicroScan (Dade International)	*Enterobacteriaceae* and other gram-negative bacilli Staphylococci Streptococci Enterococci Miscellaneous gram-positive bacilli *Haemophilus* spp. *Neisseria* spp.	Microtitre plate	Manual	Growth and nongrowth (2-42 hrs, depending on organisms and substrate system used)	Colorimetric or fluorescence (manual or automated)	Computer-assisted automated and codebooks

Procedures

13-1 CATALASE TEST

PRINCIPLE

The enzyme catalase mediates the breakdown of hydrogen peroxide into oxygen and water. The presence of the enzyme in a bacterial isolate is evident when a small inoculum is introduced into hydrogen peroxide (3% solution), and the rapid elaboration of oxygen bubbles occurs. The lack of catalase is evident by a lack of or weak bubble production.

METHOD

1. Use a loop or sterile wooden stick to transfer a small amount of colony growth to the surface of a clean, dry glass slide.

2. Place a drop of 3% hydrogen peroxide (H_2O_2) onto the inoculum.

3. Observe for the evolution of oxygen bubbles (see Figure 13-4).

EXPECTED RESULTS

Catalase-positive organisms (e.g., staphylcocci, *Listeria monocytogenes*, corynebacteria) produce copious bubbles; catalase-negative organisms (e.g., streptococci and enterococci) yield no or few bubbles. Note: Some bacteria produce a peroxidase that slowly catalyzes the breakdown of H_2O_2 and the test may appear weakly positive (a few bubbles slowly elaborated). This reaction is not a truly positive test and is considered negative.

13-2 OXIDASE TEST (Kovac's Method)

PRINCIPLE

To determine the presence of bacterial cytochrome oxidase using the oxidation of the substrate tetramethyl-*p*-phenylenediamine dihydrochloride to indophenol, a dark purple-colored end product. A positive test (presence of oxidase) is indicated by the development of a dark purple color. No color development indicates a negative test and the absence of the enzyme.

METHOD

1. Moisten filter paper with the substrate (1% tetramethyl-*p*-phenylenediamine dihydrochloride) or select a commercially available paper disk that has been impregnated with the substrate.

2. Use a platinum wire or wooden stick to remove a small portion of a bacterial colony (preferably not more than 24 hours old) from the agar surface and rub the sample on the filter paper or commercial disk.

3. Observe inoculated area of paper or disk for a color change to deep blue or purple (See Figure 13-5) within 10 seconds (timing is critical).

EXPECTED RESULTS

Positive organisms, such as *Neisseria* spp. turn the filter paper dark purple within 10 seconds; negative organisms, such as *Enterobacteriaceae* (e.g., *Escherichia coli*) remain colorless or the color of the inoculum.

13-3 INDOLE TEST

PRINCIPLE

To determine the presence of the enzyme tryptophanase. Tryptophanase breaks down tryptophan to release indole, which is detected by its ability to combine with certain aldehydes to form a colored compound. For indole-positive bacteria, the blue-green compound formed by the reaction of indole with cinnamaldehyde (commercially available) is easily visualized. The absence of enzyme results in no color production (indole negative).

METHOD

1. Saturate a piece of filter paper with the 1% paradimethylaminocinnamaldehyde reagent.

2. Use a wooden stick or bacteriologic loop to remove a small portion of a bacterial colony from the agar surface and rub the sample on the filter paper. Rapid development of a blue color indicates a positive test. Most indole-positive organisms turn blue within 30 seconds.

Note: The bacterial inoculum should not be selected from MacConkey agar because the color of lactose-fermenting colonies on this medium can interfere with test interpretation.

EXPECTED RESULTS

Indole-positive organisms, such as *Escherichia coli*, display a blue-green color on the filter paper; indole-negative organisms, such as *Enterobacter cloacae*, remain colorless or turn slightly pink.

13-4 RAPID UREASE TEST

PRINCIPLE

Hydrolysis of urea by the enzyme urease releases the end product ammonia, the alkalinity of which causes a pH indicator (phenol red) to change from yellow to red. The broth method employs buffers that control the pH change and speed of the reaction.

METHOD

1. Inoculate a tube of urea broth (available commercially) with a heavy suspension of the organism to be tested.

2. Incubate the tube at 35° C and observe at 15, 30, and 60 minutes for up to 4 hours for the development of a pink or red color.

EXPECTED RESULTS

Urease-positive organisms yield a bright pink or bright red color to the broth; urease-negative organisms do not cause a change in the color of the broth (see Figure 13-7).

13-5 PYR TEST

PRINCIPLE

Predominately used in identification schemes for gram-positive cocci. Presence of the enzyme L-pyrroglutamyl-aminopeptidase that hydrolyzes the L-pyrrolidonyl-β-naphthylamide (PYR) substrate to produce a β-naphthylamine. The β-naphthylamine is detected in the presence of *N,N*-methyl-aminocinnamaldehyde reagent by the production of a bright red-colored product.

METHOD

1. Rub a small amount of a colony to be tested on the surface of a filter paper or disk impregnated with PYR (available commercially).

2. Add a drop of detector reagent, *N,N*-dimethylaminocinnamaldehyde and observe for a red color within 5 minutes (see Figure 13-8).

EXPECTED RESULTS

Positive organisms, such as enterococci and *S. pyogenes,* yield a bright-red color change within 5 minutes; negative organisms, such as most other streptococci, yield either an orange color or no change.

13-6 HIPPURATE TEST

PRINCIPLE

The end products of hydrolysis of hippuric acid by hippuricase include glycine and benzoic acid. Glycine is deaminated by the oxidizing agent ninhydrin, which is reduced during the process. The end products of the ninhydrin oxidation react to form a purple-colored product. The test medium must contain only hippurate, because ninhydrin might react with any free amino acids present in growth media or other broths.

METHOD

1. Heavily inoculate a tube containing 0.4 mL of the 1% hippurate substrate with a sample of the organism to be tested. The suspension should be milky.

2. Cap and incubate the tube for 2 hours at 35° C; use of a water bath is preferred.

3. Add 0.2 mL ninhydrin reagent and reincubate for an additional 15 minutes. Observe the solution for development of a deep purple color.

EXPECTED RESULTS

The presence of a deep purple color indicates hippurate has been hydrolyzed (positive test). A negative test appears slightly yellow-pink or colorless (see Figure 13-9).

Modified from Hwang, M. and Ederer, G.M. 1975. J. Clin. Microbiol. 1:114.

13-7 RAPID CARBOHYDRATE FERMENTATION TEST

PRINCIPLE

Bacterial fermentation of specific carbohydrate substrates results in acid production that is detected using the phenol red pH indicator. If the organism is able to ferment a carbohydrate, the phenol red indicator changes from red to yellow. The small test medium volume and the heavy inoculum of organisms allow rapid detection of fermentation by bacterial constitutive enzymes.

METHOD

1. Buffered carbohydrate broth media, or derivatives thereof, are available commercially.

2. Make an extremely heavy suspension of a pure culture of the organism to be tested in each carbohydrate tube (should be milky). One tube without carbohydrtate should be included as a negative control.

3. Cap and incubate the tubes in a water bath or heating block at 35° C for up to 4 hours.

4. Observe periodically for change in pH indicator from red to yellow.

EXPECTED RESULTS

Tubes containing carbohydrates that the organisms can ferment will change in color from red to yellow. Tubes containing carbohydrates not utilized will remain red. Many reactions are positive within 30 minutes; questionable reactions may require up to 24 hours for full development. The control tube should remain red and serves as an internal negative control for each test.

Modified from Brown, W.J. 1974. Appl. Microbiol. 27:1027; and Hollis, D.G., Sottnek, F.O., Brown, W.J., et al. 1980. J. Clin. Microbiol. 12:620.

Bibliography

Baldwin, M.S., Warner, M.B., and Murray, C.A. 1993. Remel Technical manual, ed 1.

Brown, W.J. 1974. Modification of the rapid fermentation test for *Neisseria gonorrhoeae*. Appl. Microbiol. 27:1027.

Difco manual, ed 10. 1984. Difco Laboratories, Detroit.

Hollis, D.G., Sottnek, F.O., Brown, W.J., et al. 1980. Use of the rapid fermentation test in determining carbohydrate reactions of fastidious bacteria in clinical laboratories. J. Clin. Microbiol. 12:620.

Hwang, M. and Ederer, G.M. 1975. Rapid hippurate hydrolysis method for presumptive identification of group B streptococci. J. Clin. Microbiol. 1:114.

Jorgensen, J.H., editor. 1987. Automation in clinical microbiology, CRC Press, Boca Raton, Fla.

Kellogg, D.S., Jr., Holmes, K.K., and Hill, G.A. 1976. Cumitech

4. Laboratory diagnosis of gonorrhea, American Society for Microbiology, Washington, D.C.

Koneman, E.W., Allen, S.D., Janda, M.J., et al. 1992. Color atlas and textbook of diagnostic microbiology, ed 4. J.B. Lippincott, Philadelphia.

MacFaddin, J.F. 1985. Media for isolation-cultivation-identification-maintenance of medical bacteria, vol 1. Williams & Wilkins, Baltimore.

Murray, P.R., Baron, E.J., Pfaller, M.A., et al., editors. 1995. Manual of clinical microbiology, ed 6. American Society for Microbiology, Washington, D.C.

Oxoid. 1995. The manual. Oxoid Ltd., Basingstoke, U.K.

Power, D.A. and McCuen, P.M. 1988. Manual of BBL products and laboratory procedures, ed 6.

Stagger, C.E. and Davis, J.R. 1992. Automated systems for identification of microorganisms. Clin. Microbiol. Rev. 5:302.

14 MOLECULAR METHODS FOR MICROBIAL IDENTIFICATION AND CHARACTERIZATION

The principles of bacterial cultivation and identification discussed in Chapters 12 and 13 focused on phenotypic methods. These methods analyze readily observable bacterial traits and "behavior." Although these strategies are the mainstay of diagnostic bacteriology, notable limitations are associated with the use of phenotypic methods:

- Inability to grow certain fastidious pathogens
- Inability to maintain viability of certain pathogens in specimens during transport to laboratory
- Extensive delay in cultivation and identification of slowly growing pathogens
- Lack of reliable methods to identify certain organisms grown in vitro
- Use of considerable time and resources in establishing the presence and identity of pathogens in specimens

The explosion in molecular biology over the past 20 years has provided alternatives to phenotype-based strategies used in clinical microbiology. These alternatives have the potential to avert some of the aforementioned limitations. The detection and manipulation of nucleic acids (DNA and RNA) allows microbial genes to be examined directly (i.e., **genotypic methods**) rather than by analysis of their products such as enzymes (i.e., **phenotypic methods**). Additionally, nonnucleic acid-based analytic methods that detect phenotypic traits not detectable by conventional strategies (e.g., cell wall components) have been developed to enhance bacterial detection, identification, and characterization. For the laboratory diagnosis of infectious diseases to remain timely and effective, strategies that integrate conventional, nucleic acid-based, and analytic techniques must continue to evolve.

NUCLEIC ACID-BASED METHODS

Several methods that analyze microbial DNA or RNA can detect, identify, and characterize infectious etiologies. Although technical aspects may differ, all molecular procedures involve the direct manipulation and analysis of genes, in whole or in part, rather than the analysis of gene products. Furthermore, because nucleic acids are common to all living entities, most methods are adaptable for the diagnosis of viral, fungal, parasitic, or bacterial infections. This chapter discusses the general principles and applications of molecular diagnostics.

OVERVIEW OF MOLECULAR METHODS

Because molecular diagnostic tests are based on the consistent and somewhat predictable nature of DNA and RNA, understanding these methods requires a basic understanding of nucleic acid composition and structure. Therefore, a review of the section titled, "Nucleic Acid Structure and Organization" in Chapter 9 may be helpful.

The molecular methods to be discussed are classified into one of three categories: hybridization, amplification, and sequencing and enzymatic digestion of nucleic acids.

Nucleic acid hybridization methods

Hybridization methods are based on the ability of two nucleic acid strands that have complementary base sequences (i.e., are **homologous**) to specifically bond with each other and form a double-stranded molecule, or **duplex** or **hybrid**. This duplex formation is driven by the consistent manner in which the base adenine always bonds to thymine, while the bases guanine and cytosine always form a bonding pair (see Figure 9-2). Because hybridization requires nucleic acid sequence homology, a positive hybridization reaction between two nucleic acid strands, each from a different source, indicates genetic relatedness between the two organisms that donated each of the nucleic acid strands for the hybridization reaction.

Hybridization assays require that one nucleic acid strand (the **probe**) originates from an organism of known identity and the other strand (the **target**) originates from an unknown organism to be detected or identified (Figure 14-1). Positive hybridization identifies the unknown organism as being the same as the probe-source organism. With a negative hybridization test, the organism remains undetected or unidentified.

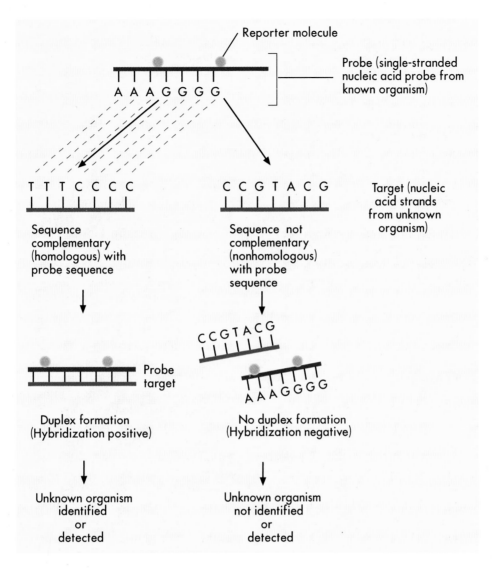

FIGURE 14-1 Principles of nucleic acid hybridization. Identification of unknown organism is established by positive hybridization (i.e., duplex formation) between a probe nucleic acid strand (from known organism) and a target nucleic acid strand from the organism to be identified. Failure to hybridize indicates lack of homology between probe and target nucleic acid.

The single-stranded nucleic acid components used in hybridization may be either RNA or DNA so that DNA-DNA, DNA-RNA, and even RNA-RNA duplexes may form, depending on the specific design of the hybridization assay.

HYBRIDIZATION STEPS AND COMPONENTS   The basic steps in a hybridization assay include:

1. Production and labeling of single-stranded probe nucleic acid
2. Preparation of single-stranded target nucleic acid
3. Mixture and hybridization of target and probe nucleic acid
4. Detection of hybridization

Production and Labeling of Probe Nucleic Acid.   In keeping with the requirement of complementation for hybridization, the probe design (i.e., probe length and its sequence of nucleic acid bases) depends on the sequence of the intended target nucleic acid. Therefore, selection and design of a probe depends on the intended use. For example, if a probe is to be used for recognizing only gram-positive bacteria, the probe's nucleic acid sequence needs to be specifically complementary to a nucleic acid sequence common to only gram-positive bacteria and not gram-negative bacteria. Even more specific probes can be designed to identify a particular bacterial genus or species, a virulence, or an antibiotic resistance gene that may only be present in certain strains within a given species.

In the past, probes were produced through a labor-intensive process that involved recombinant DNA and cloning techniques with the piece of nucleic acid of interest. More recently, probes are chemically synthesized using instrumentation, a service that is widely available commercially. This greatly facilitates probe development because the user need only supply the manufacturer with the nucleotide base sequence of the desired probe. The base sequence of potential target genes or gene fragments for probe design also is easily accessed using computer on-line services for gene sequence information (e.g., GENBANK, National Center for Biological Information). In short, the design and production of nucleic acid probes is now relatively easy. Although probes may be hundreds to thousands of bases long, oligonucleotide (i.e., 20 to 50 bases long) probes are usually sufficient for detection of most clinically relevant targets.

All hybridization tests must have a means to detect hybridization. This is accomplished with the use of a **"reporter"** molecule that chemically forms a complex with the single-stranded probe DNA. Probes may be labeled with a variety of such molecules, but most commonly radioactive (e.g., ^{32}P, ^{125}I, or ^{35}S), biotin-avidin, digoxigenin, or chemiluminescent labels are used (Figure 14-2).

With the use of radioactively labeled probes, hybridization is detected by the emission of radioactivity from the probe-target mixture (Figure 14-2, *A*). Although this is a highly sensitive method for detecting hybridization, the difficulties of working with radioactivity in a clinical microbiology laboratory have limited the use of this type of label in the diagnostic setting.

Biotinylation is a nonradioactive alternative for labeling probes and involves the chemical incorporation of biotin into probe DNA. Biotin-labeled probe-target nucleic acid duplexes are detected using avidin, a biotin-binding protein, that has been conjugated with an enzyme such as horseradish peroxidase. When a chromogenic substrate is added, the peroxidase produces a colored product that can be detected visually or spectrophotometrically (Figure 14-2, *B*).

Other nonradioactive labels are based on principles similar to biotinylation. For example, with digoxigenin-labeled probes, hybridization is detected using antidigoxigenin antibodies that have been conjugated with an enzyme. Successful duplex formation means the enzyme is present so that with the addition of a chromogenic substrate color production is interpreted as positive hybridization. Alternatively, the antibody may be conjugated with flourescent dyes that can be directly detected without the need for an enzymatic reaction to produce a colored or fluorescent end product.

Chemiluminescent reporter molecules can be directly chemically linked to the nucleic acid probe without using a conjugated antibody. These molecules (e.g., acridinium) emit light so hybridization between a chemiluminescent-labeled probe and target nucleic acid can be detected using a luminometer (Figure 14-2, *C*). The chemiluminescent approach is employed in one of the more widely available commercial hybridization systems (Gen-Probe; San Diego, Calif.).

Preparation of Target Nucleic Acid. Because hybridization is driven by complementary binding of homologous nucleic acid sequences between probe and target, target nucleic acid must be single-stranded and its base sequence integrity maintained. Failure to meet these requirements will result in negative hybridization reactions that are due to factors other than the absence of microbial target nucleic acid (i.e., false-negative results).

Because the relatively rigorous procedures for releasing nucleic acid from the target microorganism can be deleterious to the molecule's structure, obtaining target nucleic acid and maintaining its appropriate conformation and sequence can be difficult. The steps in target preparation vary depending on the organism source of the nucleic acid and the nature of the environment from which the target organism is being prepared (i.e., laboratory culture media; fresh clinical material, such as fluid, tissue, and stool; and fixed or preserved clinical material). Generally, target preparation steps involve enzymatic and/or chemical destruction of the microbial envelope to release target nucleic acid, stabilization of target nucleic acid to preserve structural integrity, and, if the target is DNA, denaturation to a single strand, which is necessary for binding to complementary probe nucleic acid.

Mixture and Hybridization of Target and Probe. Designs for mixing target and probe nucleic acids are discussed later, but some general concepts regarding the hybridization reaction require consideration.

The ability of the probe to bind the correct target depends on the extent of base sequence homology between the two nucleic acid strands and the environment in which probe and target are brought together. Environmental conditions set the **stringency** for a hybridization reaction, and the degree of stringency can determine the outcome of the reaction. Hybridization stringency is most affected by:

- Salt concentration in the hybridization buffer (stringency increases as salt concentration decreases)
- Temperature (stringency increases as temperature increases)

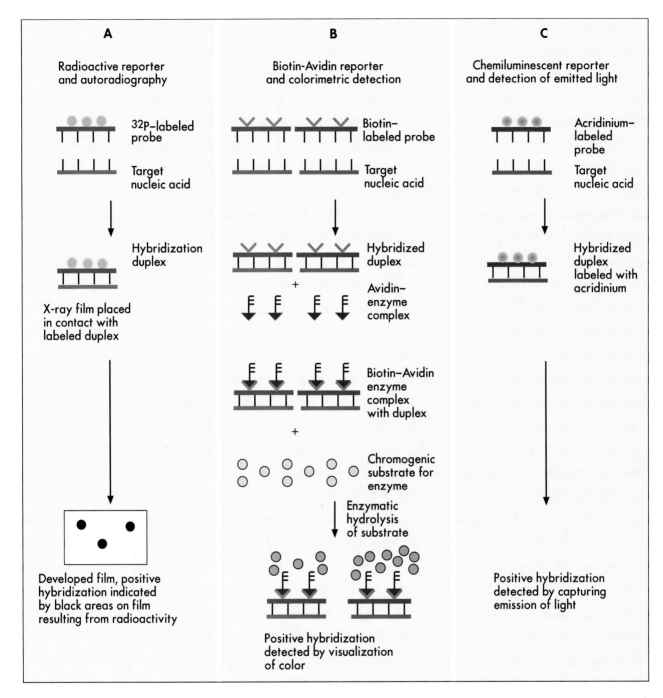

FIGURE 14-2 Reporter molecule labeling of nucleic acid probes and principles of hybridization detection. Use of probes labeled with a radioactive reporter, with hybridization detected by autoradiography (**A**); probes labeled with biotin-avidin reporter, with hybridization detected by a colorimetric assay (**B**); probes labeled with chemiluminescent reporter (i.e., acridinium), with hybridization detected by a luminometer to detect emitted light (**C**).

■ Concentration of destabilizing agents (stringency increases with increasing concentrations of formamide or urea)

With greater stringency, a higher degree of base-pair complementarity is required between probe and target to obtain successful hybridization (i.e., less tolerance for deviations in base sequence). Under less stringent conditions, strands with less base-pair complementarity (i.e., strands having a higher number of mismatched base pairs within the sequence) may still hybridize. Therefore, as stringency increases, the specificity of hybridization increases and as stringency decreases, specificity decreases. For example, under high stringency a probe specific for a target sequence in *Streptococcus pneumoniae* may only bind to target prepared from this species (high specificity), but under low stringency the same probe may bind to targets from various streptococcal species (lower specificity). Therefore, to ensure accuracy in hybridization, reaction conditions must be carefully controlled.

Detection of Hybridization. The method of detecting hybridization depends on the reporter molecule used for labeling probe nucleic acid and on the hybridization format (see Figure 14-2). Detection of hybridization using radioactively labeled probes is done by exposing the reaction mixture to radiograph film (i.e., **autoradiography**). Hybridization with nonradioactively labeled probes is detected using colorimetry, fluorescence, or chemiluminescence, and detection can be somewhat automated using spectrophotometers, fluorometers, or luminometers, respectively. The more commonly used nonradioactive detection systems (e.g., digoxigenin, chemiluminescence) are able to detect approximately 10^4 target nucleic acid sequences per hybridization reaction.

HYBRIDIZATION FORMATS Hybridization reactions can be done using either a solution format or solid support format.

Solution Format. In this format, probe and target nucleotide strands are put together in a liquid reaction mixture that facilitates duplex formation so hybridization occurs substantially faster than with the use of a solid support format. However, before detection of hybridization can be accomplished some method must be used to separate hybridized, labeled probe from nonhybridized, labeled probe (i.e., "background noise"). Separation methods include enzymatic (e.g., S1 nuclease) digestion of single-stranded probe and precipitation of hybridized duplexes, hydroxyapatite or charged magnetic microparticles that preferentially bind duplexes, or chemical destruction of the reporter molecule (e.g., acridinium dye) that is attached to unhybridized probe nucleic acid. After the duplexes have been "purified" from the reaction mixture and the background noise minimized, hybridization detection can proceed by the method appropriate for the type of reporter molecule used to label the probe (Figure 14-3).

Solid Support Format. Either probe or target nucleic acids can form a complex to a solid support and still be capable of forming duplexes with complementary strands. Various solid support materials and common solid formats exist, including: filter hybridizations, sandwich hybridizations, and in situ hybridizations.

Filter hybridizations have several variations. By one approach, the target sample, which can be previ-

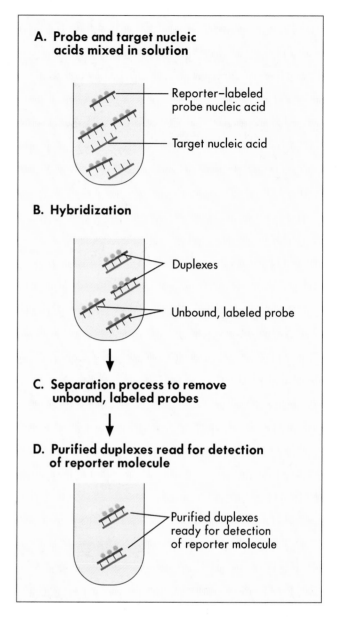

A. Probe and target nucleic acids mixed in solution

Reporter–labeled probe nucleic acid

Target nucleic acid

B. Hybridization

Duplexes

Unbound, labeled probe

C. Separation process to remove unbound, labeled probes

D. Purified duplexes read for detection of reporter molecule

Purified duplexes ready for detection of reporter molecule

FIGURE 14-3 Principle of the solution hybridization format.

ously purified DNA, the microorganism containing the target DNA, or the clinical specimen that contains the microorganism of interest, is affixed to a membrane (e.g., nitrocellulose or nylon fiber filters). The membrane is then processed to release target DNA from the microorganism and denature it to a single strand. A solution containing labeled probe nucleic acid is used to "flood" the membrane and allow for hybridization to occur. After a series of incubations and washings to remove unbound probe, the membrane is processed for detection of duplexes (Figure 14-4, *A*). An advantage of this method is that a single membrane can hold several samples for exposure to the same probe.

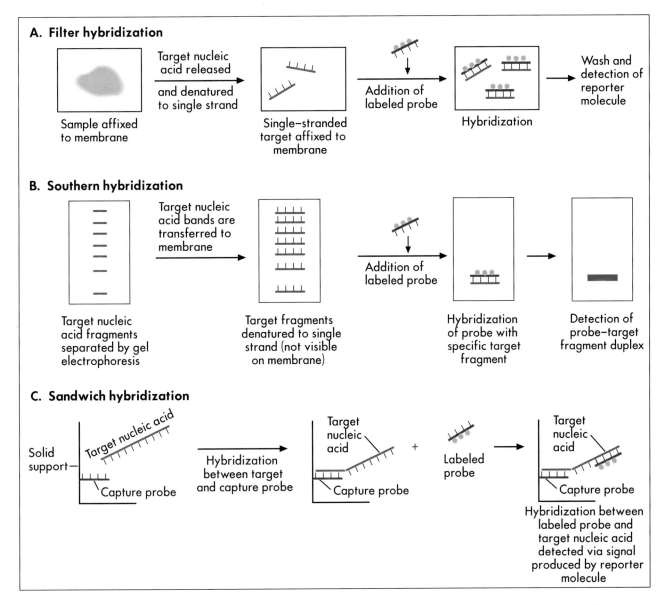

FIGURE 14-4 Principle of solid support hybridization formats. **A**, Filter hybridization. **B**, Southern hybridization. **C**, Sandwich hybridization.

Southern hybridization is another method that uses membranes as the solid support, but in this instance the nucleic acid target has been purified from the organisms and digested with specific enzymes to produce several fragments of various sizes (Figure 14-4, *B*; also see the section titled, "Enzymatic Digestion and Electrophoresis of Nucleic Acids" later in this chapter). By **gel electrophoresis,** the nucleic acid fragments, which carry a net negative charge, are subjected to an electrical field that forces them to migrate through an agarose matrix. Because fragments of different sizes migrate through the porous agarose at different rates, they can be separated over the time they are exposed to the electrical field. When electrophoresis is complete, the nucleic acid fragments are stained with the fluorescent dye **ethidium bromide** so that fragment "banding patterns" may be visualized on exposure of the gel to ultraviolet (UV) light. For Southern hybridization the target nucleic acid bands are transferred to a membrane that is then exposed to probe nucleic acid. This method allows the determination of which specific target nucleic acid fragment is carrying the base sequence of interest, but the labor intensity of the procedure precludes its common use in most diagnostic settings.

With **sandwich hybridizations** two probes are used. One probe is attached to the solid support, is not labeled, and via hybridization "captures" the target nucleic acid from the sample to be tested. The presence of this duplex is then detected using a labeled second probe that is specific for another portion of the target sequence (Figure 14-4, *C*). Sandwiching the target between two probes decreases nonspecific reactions but requires a greater number of processing and washing steps. For such formats, plastic microtitre wells coated with probes have replaced filters as the solid support material, thereby facilitating these multiple-step procedures for testing a relatively large number of specimens.

In situ hybridization allows the pathogen to be identified within the context of the pathologic lesion being produced. This method uses patient cells or tissues as the solid support phase. Tissue specimens thought to be infected with a particular pathogen are processed in a manner that maintains the structural integrity of the tissue and cells yet allows the nucleic acid of the pathogen to be released and denatured to a single strand with the base sequence intact. Although the processing steps required to obtain quality results can be technically difficult, this method is extremely useful because it combines the power of molecular diagnostics with the additional information that histopathologic examination provides.

Amplification methods

Although hybridization methods are highly specific for organism detection and identification, they are limited by their sensitivity, that is, without sufficient target nucleic acid in the reaction, false-negative results occur. Therefore, many hybridization methods require "amplifying" target nucleic acid by growing target organisms to greater numbers in culture. The requirement for cultivation detracts from the potential speed advantage that molecular methods can offer. However, the development of molecular amplification techniques not dependent on organism multiplication has contributed greatly to the circumvention of the speed problem while enhancing sensitivity and maintaining specificity. The three strategies for molecular amplification are target nucleic acid amplification, nucleic acid probe amplification, and amplification of the probe "signal."

TARGET NUCLEIC ACID AMPLIFICATION—PCR The most widely used target nucleic acid amplification method is the **polymerase chain reaction (PCR)**. This method combines the principles of complementary nucleic acid hybridization with those of nucleic acid replication that are applied repeatedly through numerous cycles. By this method, a single copy of a nucleic acid target, often undetectable by standard hybridization methods, is multiplied to 10^7 or more copies within a relatively short period. This provides ample target that can be readily detected by numerous methods.

The PCR involves 30 to 50 **repetitive cycles,** with each cycle comprising three sequential reactions: denaturation of target nucleic acid, primer annealing to single-stranded target nucleic acid, and extension of primer-target duplex.

Denaturation of Target Nucleic Acid. For PCR, target nucleic acid must be in the single-stranded conformation for the second reaction, primer annealing, to occur. Target nucleic acid is first released from the organism by heat, chemical, or enzymatic methods. Denaturation to a single strand, which is not necessary for RNA targets, is accomplished by heating to 94° C (Figure 14-5). For many PCR procedures, especially those involving commonly encountered bacterial pathogens, disruption of the organism to release DNA and the denaturation to single-stranded DNA is done in one step by heating the sample to 94° C.

Primer Annealing. **Primers** are short sequences of nucleic acid (i.e., oligonucleotides usually 20 to 30 nucleotides long) that are selected to specifically hybridize (**anneal**) to a particular nucleic acid target, essentially functioning like probes. As noted for hybridization tests, the abundance of available gene sequence data allows for the design of primers specific for a number of microbial pathogens and their virulence or antibiotic resistance genes. Thus primer nucleotide sequence design depends on the intended target, such as genus-specific genes, species-specific genes, virulence genes, or antibiotic-resistance genes.

Primers are designed to be used in pairs that flank the target sequence of interest (see Figure 14-5). When the primer pair is mixed with the denatured target DNA, one primer anneals to a specific site at one end of the target sequence of one target strand while the other primer anneals to a specific site at the opposite end of the other, complementary target strand. Usually primers are designed so that the distance between them on the target DNA is 50 to 1000 base pairs. The annealing process is conducted at 50 to 58° C, or higher. Once the duplexes are formed the last step in the cycle, which mimics the DNA replication process, begins.

Extension of Primer-Target Duplex. Annealing of primers to target sequences provides the necessary template format that allows DNA polymerase to add nucleotides to the 3' terminus of each primer and produce by extension a sequence complementary to the target template (see Figure 14-5). *Taq* polymerase is the enzyme commonly used for primer extension, which occurs at 72° C. This enzyme is used because of its ability to function efficiently at elevated temperatures and withstand the denaturing temperature of 94° C through several cycles. The ability to allow primer annealing and extension to occur at elevated temperatures without detriment to the polymerase

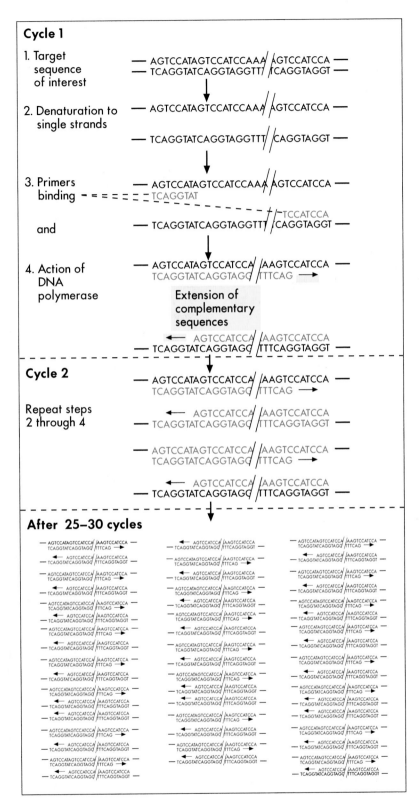

FIGURE 14-5 Overview of PCR. The target sequence is denatured to single strands, primers specific for each target strand sequence are added, and DNA polymerase catalyzes the addition of deoxynucleotides to extend and produce new strands complementary to each of the target sequence strands (*cycle 1*). The second cycle begins by both double-stranded products of cycle 1 being denatured and subsequently serving as targets for more primer annealing and extension by DNA polymerase. After 25 to 30 cycles, $\geq 10^7$ copies of target DNA may be produced. (Modified from Ryan, K.J., Champoux, J.J., Drew, W.L., et al. 1994. Medical microbiology: an introduction to infectious diseases, Appleton & Lange, Norwalk, Conn.)

increases the stringency of the reaction, thus decreasing the chance for amplification of nontarget nucleic acid (i.e., nonspecific amplification).

All three reaction components of PCR occur in the same tube that contains a mixture of target nucleic acid, primers, components to optimize polymerase activity (i.e., buffer, cation [MgCl$_2$], salt), and deoxynucleotides. To maintain continuous reaction cycles programmable **thermal cyclers** are used. These cyclers hold the reaction vessel and carry the PCR mixture through each reaction step at the precise temperature and for the optimal duration.

As shown in Figure 14-5, for each target sequence originally present in the PCR mixture, two double-stranded fragments containing the target sequence are produced after one cycle. The first step, denaturation, of the second cycle then produces four templates to which the primers will anneal. After the second cycle extension reaction there will be four double-stranded fragments containing target nucleic acid. Therefore, with completion of each cycle there is a doubling of target nucleic acid and after 30 to 40 cycles 10^7 to 10^8 target copies will be present in the reaction mixture.

Although it is possible to detect one copy of a pathogen's gene in a sample or patient specimen by PCR technology, detection is dependent on the ability of the primers to locate and anneal to the single target copy and on optimum PCR conditions. Nonetheless, PCR has proved to be a powerful amplification tool to enhance the sensitivity of molecular diagnostic techniques.

Detection of PCR Products. The specific PCR amplification product containing the target nucleic acid of interest is referred to as the **amplicon.** Because PCR produces an amplicon in substantial quantities, any of the basic methods previously described for detecting hybridization can be adopted for detecting specific amplicons. Basically this involves using a labeled probe that is specific for the target sequence within the amplicon. Therefore, solution or solid-phase formats may be used with reporter molecules that generate radioactive, colorimetric, fluorometric, or chemiluminescent signals. Probe-based detection of amplicons serves two purposes: it allows visualization of the PCR product and it provides specificity by ensuring that the amplicon is the target sequence of interest and not the result of nonspecific amplification.

When the reliability of PCR for a particular amplicon has been well established, hybridization-based detection may not always be necessary. Confirming the presence of an amplicon of the expected size may be sufficient. This is commonly accomplished by subjecting a portion of the PCR mixture, after amplification, to gel electrophoresis. After electrophoresis the gel is stained with ethidium bromide to visualize the amplicon and, using molecular–weight-size markers, the presence of amplicons of appropriate size (i.e., the size of target sequence amplified depends on the primers selected for PCR) is confirmed (Figure 14-6).

Derivations of the PCR Method. The powerful amplification capacity of PCR has prompted the development of several modifications that enhance the utility of this methodology, particularly in the diagnostic setting. Specific examples include multiplex PCR, nested PCR, quantitative PCR, RT-PCR, arbitrary primed PCR, and PCR for nucleotide sequencing.

Multiplex PCR is a method by which more than one primer pair is included in the PCR mixture. This approach offers a couple of notable advantages. First, strategies that include internal controls for PCR can be developed. For example, one primer pair can be directed at sequences present in all clinically relevant bacteria (i.e., the control or universal primers) and the second primer pair can be directed at a sequence specific for the particular gene of interest (i.e., the test primers). The control amplicon should always be detectable after PCR, and its absence would indicate that PCR conditions were not met and the test would require repeating. When the control amplicon is detected, the absence of the test amplicon can more

FIGURE 14-6 Use of ethidium-stained agarose gels to determine the size of PCR amplicons for identification. Lane A shows molecular-size markers, with the marker sizes indicated in base pairs. Lanes B, C, and D contain PCR amplicons typical of the enterococcal vancomycin-resistance genes *vanA* (783 kb), *vanB* (297 kb), and *vanC1* (822 kb), respectively.

confidently be interpreted to indicate the absence of target nucleic acid in the specimen rather than a failure of the PCR system (Figure 14-7).

Another advantage of multiplex PCR is the ability to search for different targets using one reaction. Primer pairs directed at sequences specific for different organisms or genes can be put together so that the use of multiple reaction vessels can be minimized. A limitation of multiplex PCR is that mixing different primers can cause some interference in the amplification process so that optimizing conditions can be difficult, especially as the number of different primer pairs used increases.

Nested PCR involves the sequential use of two primer sets. The first set is used to amplify a target sequence. The amplicon obtained is then used as the target sequence for a second amplification using primers internal to those of the first amplicon. Essentially this is an amplification of a sequence internal to an amplicon. The advantage of this approach is extreme sensitivity and confirmed specificity without the need for using probes. Because production of the second amplicon requires the presence of the first amplicon, production of the second amplicon automatically verifies the accuracy of the first amplicon. The problem encountered with nested PCR is that the procedure requires open manipulations of amplified DNA that is readily, albeit inadvertently, aerosolized and capable of contaminating other reaction vials.

Arbitrary primed PCR uses short primers that are not specifically complementary to a particular sequence of a target DNA. Although they are not specifically directed, their short sequence (approximately 10 nucleotides) ensures that they will randomly anneal to multiple sites in a chromosomal sequence. On cycling the multiple annealing sites result in the amplification of multiple fragments of different sizes. Theoretically, strains that have similar nucleotide sequences will have similar annealing sites and thus will produce amplified fragments (i.e., amplicons) of similar sizes. Therefore, by comparing fragment migration patterns following agarose gel electrophoresis, strains or isolates can be judged to be the same, similar, or unrelated.

Quantitative PCR is a rapidly emerging technologic approach that combines the power of PCR for the detection and identification of infectious agents with the ability to quantitate the actual number of targets originally in the clinical specimen. The ability to quantitate "infectious burden" has tremendous implications for studying and understanding the disease state (e.g., acquired immunodeficiency syndrome [AIDS]), the prognosis of certain infections, and the effectiveness of antimicrobial therapy.

The PCR methods discussed thus far have focused on amplification of a DNA target. **Reverse transcription PCR (RT-PCR)** amplifies an RNA target. Because many clinically important viruses have

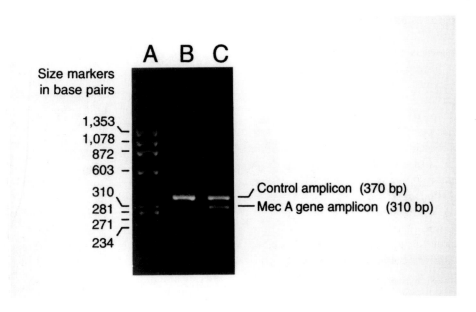

FIGURE 14-7 Ethidium-bromide–stained gels containing amplicons produced by multiplex PCR. Lane A shows molecular-size markers, with the marker sizes indicated in base pairs. Lanes B and C show amplicons obtained with multiplex PCR consisting of control primers and primers specific for the staphylococcal methicillin-resistance gene *mecA*. The presence of only the control amplicon (370 bp) in Lane B indicates that PCR was successful, but the strain on which the reaction was performed did not contain *mecA*. Lane C shows both the control and the *mecA* (310 bp) amplicons, indicating that the reaction was successful and that the strain tested carries the *mecA* resistance gene.

genomes composed of RNA rather than DNA (e.g., human immunodeficiency virus [HIV], hepatitis B virus), the ability to amplify RNA greatly facilitates laboratory-based diagnostic testing for these infectious agents. The unique step to this procedure is the use of the enzyme **reverse transcriptase** that directs synthesis of DNA from the viral RNA template. Once the DNA has been produced, relatively routine PCR technology is applied to obtain amplification.

Alternative Methods for Target Amplification. Other target nucleic acid amplification methods include transcription-based amplification systems (TAS), strand displacement amplification (SDA), and self-sustaining sequence replication (3SR). Although these methods attest to the intricate ways molecular methods can be manipulated for the amplification of target nucleic acids, their application in diagnostic microbiology has not yet achieved the same level of attention as PCR.

NUCLEIC ACID PROBE AMPLIFICATION Amplification of probe nucleic acid rather than target nucleic acid provides a second strategy for enhancing the sensitivity of molecular diagnostic techniques. By this approach an amplified probe is the final reaction product to be detected, while the target sequence is neither amplified nor incorporated into this product. One such system is the **Ligase Chain Reaction (LCR)**.

LCR uses two pairs of probes that span the target sequence of interest. Once annealed to the target sequence, a space remains between the probes that is enzymatically closed using a ligase (i.e., a ligation reaction). On heating, the joined probes are released as a single strand that is complementary to the target nucleic acid; these newly synthesized strands are then used as the template for subsequent cycles of probe annealing and ligations. Through the process, probe DNA is amplified to a level readily detectable using assays similar to those described for the biotin-avidin system. One commercial system (Abbott Laboratories, Abbott Park, Ill.) uses an LCR approach to detect *Chlamydia trachomatis* and *Neisseria gonorrhoeae* in urine or other specimens.

AMPLIFICATION OF THE PROBE SIGNAL Rather than amplifying either the target or probe, the signal used to detect hybridization between the amplicon and probe can be amplified by increasing the number of reporter molecules per probe. This is accomplished by attaching multiple signal-generating molecules to the probe that recognizes the target DNA. Thus, several reporter molecules are associated with each probe-target duplex. The potential of this approach is exemplified by the development of branched probes. These probes are complex nucleic acid constructions involv-

ing capture probes, extender probes, and multicopy DNA sequences that can accommodate thousands of reporter molecules per probe-target duplex. This method for detecting target nucleic acid is exquisitely sensitive.

Sequencing and enzymatic digestion of nucleic acids

The nucleotide sequence of a microorganism's genome is the blueprint for that organism. Therefore, it is evident that molecular methods that elucidate some part of a pathogen's genomic sequence provide a powerful tool for diagnostic microbiology. Other methods, either used independently or in conjunction with hybridization or amplification procedures, can provide nucleotide sequence information to detect, identify, and characterize clinically relevant microorganisms. These methods include nucleic acid sequencing and enzymatic digestion and electrophoresis of nucleic acids.

NUCLEIC ACID SEQUENCING Nucleic acid sequencing involves methods that determine the exact nucleotide sequence of a gene or gene fragment obtained from an organism. Although explaining the technology involved is beyond the scope of this text, nucleic acid sequencing will powerfully affect clinical microbiology for some time to come. To illustrate, nucleotide sequences obtained from a microorganism can be compared with an ever-growing gene sequence data base for:

- Detecting and classifying previously unknown human pathogens
- Identifying various known microbial pathogens and their subtypes
- Determining which specific nucleotide changes resulting from mutations are responsible for antibiotic resistance
- Establishing the relatedness between isolates of the same species

Before the development of rapid and automated methods, DNA sequencing was a laborious task only undertaken in the research setting. However, determining the sequence of nucleotides in a segment of nucleic acid from an infectious agent can be done rapidly using amplified target from the organism and an automated DNA sequencer. Because sequence information can now be rapidly produced, DNA sequencing has entered the arena of diagnostic microbiology.

ENZYMATIC DIGESTION AND ELECTROPHORESIS OF NUCLEIC ACIDS Enzymatic digestion and electrophoresis of DNA fragments are less exacting methods of identifying and characterizing microorganisms than is nucleic acid sequencing. However, enzyme digestion-electrophoresis procedures still provide valuable information for the diagnosis and control of infectious diseases.

Enzymatic digestion of DNA is accomplished using any of a number of enzymes known as **endonucleases**. Each specific endonuclease recognizes a specific nucleotide sequence (usually four to eight nucleotides in length) known as the enzyme's **recognition, or restriction, site**. Once the recognition site is located, the enzyme catalyzes the digestion of the nucleic acid strand at that site, causing a break, or **cut**, in the nucleic acid strand (Figure 14-8).

The number and size of fragments produced by enzymatic digestion depends on the length of nucleic acid being digested (the longer the strand the greater the likelihood of more recognition sites and thus more fragments), the nucleotide sequence of the strand being digested, and the particular enzyme used for digestion. For example, enzymatic digestion of a bac-

terial plasmid whose nucleotide sequence provides several recognition sites for endonuclease A, but only rare sites for endonuclease B, will produce more fragments with endonuclease A. Additionally, the size of the fragments produced will depend on the number of nucleotides between each of endonuclease A's recognition sites present on the nucleic acid being digested.

The DNA used for digestion is obtained by various methods. A target sequence may be obtained by amplification via PCR, in which case the length of the DNA to be digested is relatively short (e.g., 50 to 1000 bases). Alternatively, specific procedures may be used to cultivate the organism of interest to large numbers (e.g., 10^{10} bacterial cells) from which plasmid DNA, chromosomal DNA, or total cellular DNA may be isolated and purified for endonuclease digestion.

Following digestion, fragments are subjected to agarose gel electrophoresis, which allows them to be separated according to their size differences as previously described for Southern hybridization (see Figure 14-4, *B*). During electrophoresis all nucleic acid fragments of the same size comigrate as a single band. For many digestions, electrophoresis results in the separation of several different fragment sizes (Figure 14-8). The nucleic acid bands in the agarose gel are stained with the fluorescent dye ethidium bromide, which allows them to be visualized on exposure to UV light. Stained gels are photographed for a permanent record (see Figures 14-9 to 11).

One variation of this method, known as **ribotyping**, involves enzymatic digestion of chromosomal DNA followed by Southern hybridization using probes for genes that encode ribosomal RNA. Because all bacteria contain ribosomal genes, a hybridization pattern will be obtained with almost any isolate, but the pattern will vary depending on the arrangement of genes in a particular strain's genome.

Regardless of the method, the process by which enzyme digestion patterns are analyzed is referred to as **restriction enzyme analysis (REA)**. The patterns obtained after gel electrophoresis are referred to as **restriction patterns**, and differences between microorganism restriction patterns are known as **restriction fragment length polymorphisms (RFLPs)**. Because RFLPs reflect differences or similarities in nucleotide sequences, REA methods can be used for organism identification and for establishing strain relatedness within the same species (see Figures 14-8 to 11).

APPLICATIONS OF NUCLEIC ACID-BASED METHODS

Categories for the application of molecular diagnostic microbiology methods are the same as those for conventional, phenotype-based methods:

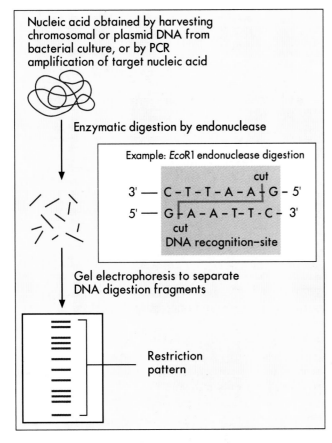

FIGURE 14-8 Principle of DNA enzymatic digestion and gel electrophoresis to separate DNA fragments resulting from the digestion. An example of a nucleic acid recognition site and enzymatic cut produced by *Eco*R1, a commonly used endonuclease, is shown in the inset.

The figure contains the following labels:

Nucleic acid obtained by harvesting chromosomal or plasmid DNA from bacterial culture, or by PCR amplification of target nucleic acid

Enzymatic digestion by endonuclease

Example: *Eco*R1 endonuclease digestion

cut
3' — C – T – T – A – A ┤ G – 5'
5' — G ├ A – A – T – T – C – 3'
cut
DNA recognition–site

Gel electrophoresis to separate DNA digestion fragments

Restriction pattern

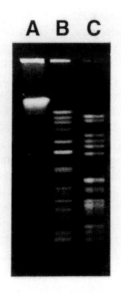

FIGURE 14-11 Restriction patterns generated by PFGE for two *Streptococcus pneumoniae* isolates, one that was susceptible to penicillin (Lane B) and one that was resistant (Lane C), from the same patient. RFLP analysis indicates that the patient was infected with different strains. Molecular-size markers are shown in Lane A.

FIGURE 14-9 Restriction fragment length polymorphisms (RFLPs) of vancomycin-resistant *E. faecalis* isolates in lanes A through G as determined by pulsed-field gel electrophoresis (PFGE). All isolates appear to be the same strain.

- Direct detection of microorganisms in patient specimens
- Identification of microorganisms grown in culture
- Characterization of microorganisms beyond identification

Direct detection of microorganisms

Nucleic acid hybridization and target or probe amplification methods are the molecular techniques most commonly used for direct organism detection in clinical specimens.

ADVANTAGES AND DISADVANTAGES When considering the advantages and disadvantages of molecular approaches to direct organism detection, comparison with the most commonly used conventional method (i.e., direct smears and microscopy) is helpful.

Specificity. Both hybridization and amplification methods are driven by the specificity of a nucleotide sequence for a particular organism. Therefore, a positive assay not only indicates the presence of an organism but also provides the organism's identity, potentially precluding the need for follow-up culture. Although molecular methods may not be faster than microscopic smear examinations, the opportunity to avoid delays associated with culture can be a substantial advantage.

However, for many infectious agents, detection and identification are only part of the diagnostic

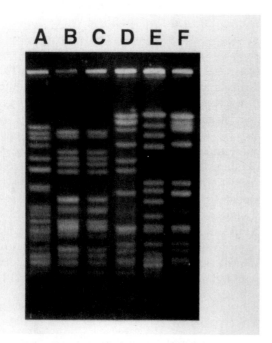

FIGURE 14-10 Although antimicrobial susceptibility profiles indicated that several methicillin-resistant *S. aureus* isolates were the same strain, RFLP analysis using PFGE (lanes A through F) demonstrates that only isolates B and C were the same.

requirement. Determination of certain characteristics, such as strain-relatedness or resistance to antimicrobial agents, is often an important diagnostic component that is not possible without the availability of culture. For this reason, most molecular direct detection methods target organisms for which antimicrobial susceptibility testing is not routinely needed (e.g., chlamydia) or for which reliable cultivation methods are not widely available (e.g., ehrlichia).

The high specificity of molecular techniques also presents a limitation in what can be detected with any one assay, that is, most molecular assays focus on detecting the presence of only one or two potential pathogens. Even if tests for those organisms are positive, the possibility of a mixed infection involving other organisms has not been ruled out. If the tests are negative, other procedures may be needed to determine if additional pathogens are present. In contrast, smear examination and cultivation procedures can detect and identify a broader selection of possible infectious etiologies. However, given the rapid development of molecular technology, protocols that widen the spectrum of detectable organisms in any particular specimen will likely be available in the future.

Sensitivity. Hybridization-based methods can have difficulty directly detecting organisms. The quantity of target nucleic acid may be insufficient, or the patient specimen may contain substances that interfere with or cross-react in the hybridization and signal-generating reactions. One approach developed by Gen-Probe (San Diego, Calif.) to enhance sensitivity has been to use DNA probes targeted for bacterial ribosomal RNA of which there are up to 10,000 copies per cell. Essentially this is amplification accomplished by the choice of a target that exists within the cell as multiple copies rather than as a single copy.

Amplification Techniques Enhance Sensitivity. As discussed with direct hybridization methods, patient specimens also may contain substances that interfere with or inhibit amplification reactions such as PCR. Nonetheless, the ability to amplify target or probe nucleic acid to readily detectable levels has provided a tremendous means for overcoming the lack of sensitivity characteristic of most direct hybridization methods.

Besides the potential for providing more reliable test results than direct hybridization (i.e., fewer false-negatives) amplification methods have other advantages that include:

- Ability to detect nonviable organisms that are not retrievable by cultivation-based methods
- Ability to detect and identify organisms that cannot be grown in culture or are extremely difficult

to grow (e.g., hepatitis B virus and the agent of Whipple's disease)
- More rapid detection and identification of organisms that grow slowly (e.g., mycobacteria, certain fungi)
- Ability to detect previously unknown agents directly in clinical specimens by using **broad-range primers** (e.g., use of primers that will anneal to a region of target DNA that is conserved among all bacteria so an amplicon can be produced without previous knowledge of the bacterial species involved in the infection)
- Ability to quantitate infectious agent burden directly in patient specimens, an application that has particular importance for managing HIV infections

Despite these significant advantages, limitations still exist, notably the ability to only find the organisms toward which the primers have been targeted. Additionally, no cultured organism is available if subsequent characterization beyond identification is necessary. As with hybridization, the first limitation may eventually be addressed using broad-range amplification methods that would be used to screen specimens for the presence of any organism (e.g., bacteria, fungi, parasite). Specimens positive by this test would then be processed further for a more specific diagnosis. The second limitation is more difficult to overcome and is one reason why culture methods will remain a major part of diagnostic microbiology for some time to come.

An interesting consequence of using highly sensitive amplification methods is the effect on clinical interpretation of results. For example, if a microbiologist detects organisms that are no longer viable, can he or she assume the organisms are or were involved in the infectious process being diagnosed? Also, amplification may detect microorganisms present in insignificant quantities as part of the patient's normal or transient flora, or as an established latent infection, that have nothing to do with the current disease state of the patient.

Finally, as previously mentioned, an underlying complication in the development and application of any direct detection method is that various substances in patient specimens can interfere with the reagents and conditions required for optimum hybridization or amplification. Specimen interference is one of the major issues that must be addressed in the design of any useful direct method for molecular diagnosis of infectious diseases.

APPLICATIONS FOR DIRECT MOLECULAR DETECTION OF MICROORGANISMS Given their inherent advantages and disadvantages, molecular direct detection methods are most useful when:

- One or two pathogens cause the majority of infections (e.g., *Chlamydia trachomatis* and *Neisseria gonorrhoeae* as common agents of genitourinary tract infections)
- Further organism characterization, such as antimicrobial susceptibility testing, is not required for management of the infection (e.g., various viral agents)
- Either no reliable diagnostic methods exist or they are notably suboptimal (e.g., various bacterial, parasitic, viral, and fungal agents)
- Reliable diagnostic methods exist but are slow (e.g., *Mycobacterium tuberculosis*)
- Quantitation of infectious agent burden that influences patient management (e.g., AIDS) is desired

Examples of commercial molecular systems for direct detection of certain infections include: Gen-Probe's Transcription-Mediated Amplification assay for *Mycobacterium tuberculosis* and nonamplification-based hybridization tests for *Chlamydia trachomatis* and *Neisseria gonorrhoeae*, Abbott Diagnostic's Ligase Chain Reaction assay for *Chlamydia trachomatis* and *Neisseria gonorrhoeae*, and Roche Diagnostic System's PCR for chlamydia and quantitation of HIV. Additionally, many direct detection assays are being developed by diagnostic manufacturers and research laboratories associated with academic medical centers. Therefore, direct molecular diagnostic methods based on amplification will continue to expand and enhance our understanding and diagnosis of infectious diseases. However, as with any laboratory method, their ultimate utility and application will depend on their accuracy, potential impact on patient care, advantages over currently available methods, and the resources required to establish and maintain their use in the diagnostic setting.

Identification of microorganisms grown in culture

Once organisms are grown in culture, hybridization, amplification, or RFLP analysis may be used to establish identity. Because the target nucleic acid is already amplified via microbial cultivation, sensitivity is not usually a problem for molecular identification methods. Additionally, extensive nucleotide sequence data are available for most clinically relevant organisms so that probes and primers that are highly specific can be readily produced. With neither specificity nor sensitivity as problems in this setting, other criteria regarding the application of molecular identification methods must be considered.

The criteria that are often considered in comparing molecular and conventional methods for microbial identification include speed, accuracy, and cost. For slow-growing organisms, such as mycobacteria and fungi, growth-based identification schemes can take weeks to months to produce a result. Molecular-based methods can identify these microorganisms almost immediately after sufficient inoculum is available, clearly demonstrating a speed advantage over conventional methods. On the other hand, phenotypic-based methods used to identify frequently encountered bacteria, such as *Staphylococcus aureus* and beta-hemolytic streptococci, can usually provide highly accurate results within minutes and are less costly and time-consuming than any currently available molecular method.

Although many of the phenotype-based identification schemes are highly accurate and reliable, in some situations phenotypic profiles may yield uncertain identifications and molecular methods may provide an alternative for establishing a definitive identification. This is especially the case when a common pathogen exhibits unusual phenotypic traits (e.g., optochin-resistant *Streptococcus pneumoniae*).

Characterization of microorganisms beyond identification

Situations exist in which characterizing a microbial pathogen beyond identification provides important information for patient management and public health. In such situations, knowledge regarding an organism's virulence, resistance to antimicrobial agents, or relatedness to other strains of the same species can be extremely important. Altough various phenotypic methods have been able to provide some of this information, the development of molecular technologies has greatly expanded our ability to generate this information in the diagnostic setting. This is especially true with regard to antimicrobial resistance and strain relatedness.

DETECTION OF ANTIMICROBIAL RESISTANCE Like all phenotypic traits, those that render microorganisms resistant to antimicrobial agents are encoded on specific genes (for more information regarding antimicrobial resistance mechanisms see Chapter 17). Therefore, molecular methods for gene amplification or hybridization can be used to detect antimicrobial resistance. In many ways, phenotypic methods for resistance detection are reliable and are the primary methods for antimicrobial susceptibility testing (see Chapter 18). However, the complexity of emerging resistance mechanisms often challenges the ability of commonly used susceptibility testing methods to detect clinically important resistance to antimicrobial agents.

Methods such as PCR play a role in the detection of certain resistance profiles that may not always readily be detected by phenotypic methods. Two such examples include detection of the *van* genes, which mediate vancomycin resistance among enterococci (see Figure 14-6), and the *mec* gene, which encodes resistance among staphylococci to all currently available drugs of the beta-lactam class (see Figure 14-7). Undoubtedly, conventional and molecular methods will both continue to play key roles in the characterization of microbial resistance to antimicrobial agents.

INVESTIGATION OF STRAIN RELATEDNESS An important component to recognizing and controlling disease outbreaks inside or outside of a hospital is identification of the reservoir and mode of transmission of the infectious agents involved. This often requires establishing relatedness among the pathogens isolated during the outbreak. For example, if all the microbial isolates thought to be associated with a nosocomial infection outbreak are shown to be identical, or at least very closely related, then a common source or reservoir for those isolates should be sought. If the etiologic agents are not the same, other explanations for the outbreak must be investigated (see Chapter 7). Because each species of a microorganism comprises an almost limitless number of strains, identification of an organism to the species level is not sufficient for establishing relatedness. **Strain typing,** the process used to establish the relatedness among organisms belonging to the same species, is required.

Although phenotypic characteristics (e.g., biotyping, serotyping, antimicrobial susceptibility profiles) historically have been used to type strains, these methods often are limited by their inability to consistently discriminate between different strains, their labor intensity, or their lack of reproducibility. In contrast, certain molecular methods do not have these limitations and have enhanced strain-typing capabilities. The molecular-typing methods either directly compare nucleotide sequences between strains or produce results that indirectly reflect similarities in nucleotide sequences among "outbreak" organisms.

Indirect methods usually involve enzymatic digestion and electrophoresis of microbial DNA to produce RFLPs for comparison and analysis.

Several molecular methods have been investigated for establishing strain relatedness. Examples of these methods include restriction enzyme analysis of plasmid or chromosomal DNA, ribotyping, restriction enzyme analysis of PCR amplicons, arbitrarily primed PCR, nucleotide sequencing, and pulsed-field gel electrophoresis (PFGE).

The method chosen primarily depends on the extent to which the following four criteria proposed by Maslow et al. are met:

- **Typeability:** the method's capacity to produce clearly interpretable results with most strains of the bacterial species to be tested
- **Reproducibility:** the method's capacity to repeatedly obtain the same typing profile result with the same bacterial strain
- **Discriminatory power:** the method's ability to produce results that clearly allow differentiation between unrelated strains of the same bacterial species
- **Practicality:** the method should be versatile, relatively rapid, inexpensive, technically simple, and provide readily interpretable results

The last criterion, practicality, is especially important for busy clinical microbiology laboratories that provide support for infection control and hospital epidemiology.

Pulsed-Field Gel Electrophoresis. Among the molecular methods used for strain typing, pulsed-field gel electrophoresis (PFGE) meets most of Maslow's criteria for a good typing system. This method is applicable to most of the commonly encountered bacterial pathogens, particularly those frequently associated with nosocomial infections and outbreaks (Box 14-1). For these reasons, PFGE has been widely accepted among microbiologists, infection control personnel,

BOX 14-1

COMMONLY ENCOUNTERED MICROORGANISMS TYPEABLE BY PULSED-FIELD GEL ELECTROPHORESIS

Gram-Positive Bacteria	Gram-Negative Bacteria	Others
Staphylococci	*E. coli*	Yeast (*Candida* spp.)
Other *Streptococcus* spp.	*Klebsiella* spp.	
Streptococcus pneumoniae	*Enterobacter* spp.	
Enterococci	*Citrobacter* spp.	
	Serratia spp.	
	Pseudomonas aeruginosa	
	Stenotrophomonas maltophilia	
	Miscellaneous other gram-negative bacteria	

and infectious disease specialists as a primary laboratory tool for epidemiology.

The principle of PFGE is to use a specialized electrophoresis device to separate chromosomal fragments produced by enzymatic digestion of intact bacterial chromosomal DNA. Bacterial suspensions are first embedded in agarose plugs, where they are carefully lysed to release intact chromosomal DNA; the DNA is then digested using restriction endonuclease enzymes. Enzymes that have relatively few restriction sites on the genomic DNA are selected so that 10 to 20 DNA fragments ranging in size from 10 to 1000 kb are produced (Figure 14-12). Because of the large DNA fragment sizes produced, resolution of the banding patterns requires the use of a **pulsed electrical field** across the agarose gel that subjects the DNA

fragments to different voltages from varying angles at different time intervals.

Although comparison and interpretation of RFLP profiles produced by PFGE can be complex, the basic premise is that strains with the same or highly similar digestion profiles share substantial similarities in their nucleotide sequences and therefore are likely to be most closely related. For example, in Figure 14-12 isolates 1 and 2 have identical RFLP patterns, whereas isolate 3 has only 7 of its 15 bands in common with either isolates 1 or 2. Therefore, isolates 1 and 2 would be considered closely related, if not identical, while isolate 3 would not be considered related to the other two isolates.

One example of PFGE application for the investigation of an outbreak is shown in Figure 14-9. Fol-

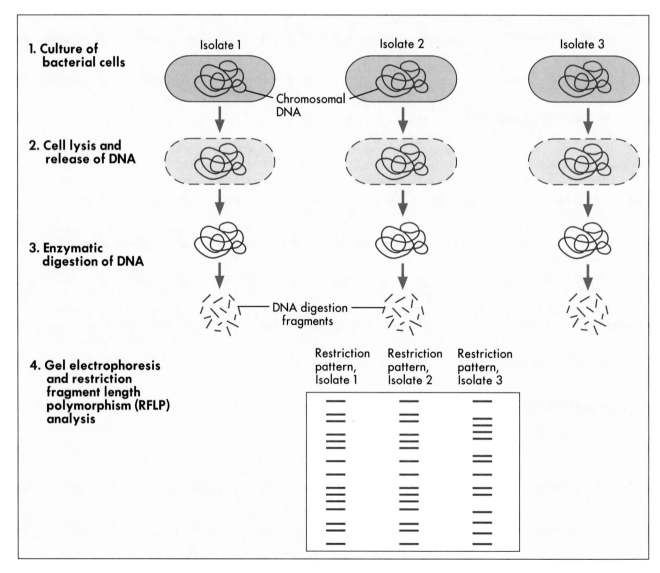

FIGURE 14-12 Procedural steps for pulsed-field gel electrophoresis (PFGE).

lowing *Sma*I endonuclease enzymatic digestion of DNA from seven vancomycin-resistant *Enterococcus faecalis* isolates, RFLP profiles show that the resistant isolates are probably the same strain. Such a finding strongly supports the probability of clonal dissemination of the same vancomycin-resistant strain among the patients from which the organisms were isolated.

The discriminatory advantage PFGE profiles have over phenotype-based typing methods is demonstrated in Figure 14-10. Because all six methicillin-resistant *Staphylococcus aureus* isolates exhibited identical antimicrobial susceptibility profiles, they were initially thought to be the same strain. However, PFGE profiling established that only isolates B and C were the same.

PFGE can also be used to determine whether a recurring infection in the same patient is due to insufficient original therapy, possibly as a result of developing antimicrobial resistance during therapy, or due to acquisition of a second, more resistant, strain of the same species. Figure 14-11 shows restriction patterns obtained by PFGE, with *Streptococcus pneumoniae* isolated from a patient with an unresolved middle ear infection. The PFGE profile of isolate A, which was fully susceptible to penicillin, differs substantially from the profile of isolate B, which was resistant to penicillin. The clear difference in PFGE profiles between the two strains indicates that the patient was most likely reinfected with a second, more resistant, strain. Alternatively, the patient's original infection may have been a mixture of both strains, with the more resistant one being lost during the original culture workup. In any case, this application of PFGE demonstrates that the method is not only useful for investigating outbreaks or strain dissemination involving several patients but also gives us the ability to investigate questions regarding reinfections, treatment failures, and mixed infections involving more than one strain of the same species.

NONNUCLEIC ACID-BASED ANALYTIC METHODS

Phenotypic traits not readily detected using the conventional methods discussed in Chapter 13 can be useful in the identification and characterization of microorganisms. Analysis of these traits requires the use of special analytic tools for detection and analysis. Unfortunately, the need for expensive and complex equipment coupled with the rapid evolution of nucleic acid-based molecular methods has limited the dissemination of this technology among diagnostic settings. However, these methods have many uses, particularly in the reference laboratory, and their principles and applications are summarized below.

CHROMATOGRAPHY

Chromatography refers to procedures used to separate and characterize substances based on their size, ionic charge, or their solubility in particular solvents. This technique has many variations that can be applied for specific purposes.

Principles

Chromatography involves two phases: the **mobile phase** and the **stationary phase**. The mobile phase contains and carries the sample to be analyzed through or across the stationary phase. The stationary phase maintains conditions necessary for separating various substances (i.e., the **analytes**) within the sample being studied. The mobile phase may be gas or liquid. The stationary phase may also be liquid or solid and is usually housed within a column of some design. Specific chromatographic methods are named according to the phases used. For example, **gas-liquid chromatography** (**GLC**) refers to any chromatographic method that uses gas as the mobile phase and liquid as the stationary phase.

Regardless of the phases used, chromatographic principles are generally the same as those shown in Figure 14-13 for gas-liquid chromatography. A sample is mixed by injection with the mobile phase, which carries the sample through a column containing the stationary phase. Depending on the nature of the solid phase, different analytes within the sample will be retained within the column based on size, ionic charge, or analyte solubility in the mobile phase. As the mobile phase continues to move through the column, analytes with different characteristics will have different **retention times** and thus will elute (come out) from the column at different times. Eluted analytes pass through a detector that generates a signal. The signal is then translated into an electronic signal that is recorded and used to produce a chromatogram.

The **chromatogram** is the plot of elution peaks produced by each analyte in the sample (Figure 14-13). With the inclusion of reliable controls and standards, retention times are used to definitively identify specific analytes. The size of each peak can be used to determine the amount of analyte present in each sample.

Applications

Many substances, including proteins, carbohydrates, organic acids, fatty acids, and mycolic acids, may be detected by chromatography. Therefore, chromatographic methods can provide analytic, phenotypic fingerprints (i.e., chromatograms) regarding the types and amounts of these substances within a bacterial cell. Comparison of chromatograms obtained with organisms of unknown identity with those of known

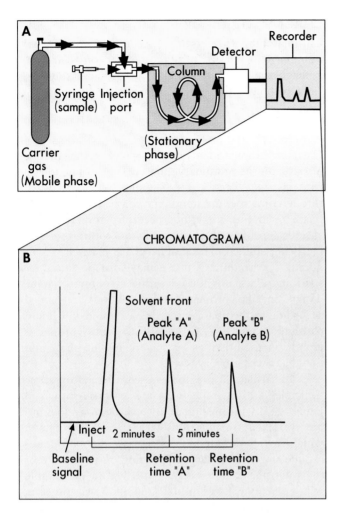

FIGURE 14-13 Principles of gas-liquid chromatography (**A**) and an example of a chromatogram (**B**). (Modified from Onderdonk, A.B. and Sasser, M. 1995. In Murray, P.R., Baron, E.J., Pfaller, M.A., et al., editors. Manual of clinical microbiology, ed 6, American Society for Microbiology, Washington, D.C.)

identity is used to identify an organism from a clinical specimen. The chromatographic approach has greatest application when conventional phenotypic-based methods for identification are poor or nonexistent.

GAS-LIQUID CHROMATOGRAPHY The application of gas-liquid chromatography (GLC) in diagnostic microbiology is most common in anaerobic bacteriology. GLC is used to separate and identify anaerobic metabolic end products (i.e., volatile fatty acids and nonvolatile organic acids) of carbohydrate fermentation and amino acid degradation. Chromatograms produced with anaerobic bacteria can greatly facilitate identification of certain genera and species that are not readily identified based on other phenotypic characteristics.

GLC application has also been expanded for the analysis of longer-chain fatty acids (i.e., 9 to 20 car-

bons in length) to produce chromatograms for identifying organisms often without the need for other phenotypic information (e.g., Gram stain morphology, oxidase profile). One such commercial system (MIDI Microbial Identification System; Newark, Del.) has more than 600 bacteria in the chromatographic data base. Although this approach may not be practical for identification of many commonly encountered bacterial species, it has great potential for use as a reference method for organisms difficult to identify by conventional methods.

HIGH-PERFORMANCE LIQUID CHROMATOGRAPHY In high-performance liquid chromatography (HPLC) a liquid mobile phase is combined with various technical advances to separate large cellular metabolites and components. HPLC is used in the research setting to separate larger precursors of bacterial cell walls from antibiotic-resistant organisms or in the medical setting to help identify mycobacteria and related organisms by characterizing their cell wall mycolic acid content. HPLC is also sufficiently versatile that levels of certain antimicrobial agents in patient serum specimens can be measured.

ELECTROPHORETIC PROTEIN ANALYSIS

Electrophoretic methods analogous to those discussed for nucleic acids can be used to generate microbial protein profiles. These profiles include analysis of proteins obtained after purification and separation from certain bacterial structures (e.g., cell membrane or cytosol) or analysis of differences in particular sets of metabolic enzymes (i.e., **multilocus enzyme electrophoresis [MLEE]**). Although these are interesting alternative approaches, the complexity of their interpretation and the widespread availability of other methods for organism identification and characterization has limited the use of protein profiling methods in most clinical laboratories.

The molecular and alternative methods for microbial detection, identification, and characterization provide opportunities for improving the laboratory diagnosis of infectious diseases. Although these are new and constantly evolving technologies that differ markedly from conventional methods, they are diagnostic tests nonetheless. As such, they must be subjected to quality assurance and quality control scrutiny, and the decision to use them must be based on accuracy, cost, and usefulness.

Bibliography

Arbeit, R.D. 1995. Laboratory procedures for the epidemiologic analysis of microorganisms. In Murray, P.R., Baron, E.J., Pfaller,

M.A., et al., editors. Manual of clinical microbiology, ed 6. American Society for Microbiology, Washington, D.C.

Maslow, J.N., Mulligan, M.E., and Arbeit, R.D. 1993. Molecular epidemiology: application of contemporary techniques to the typing of microorganisms. Clin. Infect. Dis. 17:153.

Onderdonk, A.B. and Sasser, M. 1995. Gas-liquid and high-performance liquid chromotography for the identification of microorganisms. In Murray, P.R., Baron, E.J., Pfaller, M.A., et al., editors. Manual of clinical microbiology, ed 6. American Society for Microbiology, Washington, D.C.

Persing, D.H., editor. 1996. PCR protocols for emerging infectious diseases. American Society for Microbiology, Washington, D.C.

Persing, D.H., Smith, T.F., Tenover, F.C., and White, T.J., editors. 1993. Diagnostic molecular microbiology: principles and applications. American Society for Microbiology, Washington, D.C.

Podzorski, R.P. and Persing, D.H. 1995. Molecular detection and identification of microorganisms. In Murray, P.R., Baron, E.J., Pfaller, M.A., et al., editors. Manual of clinical microbiology, ed 6. American Society for Microbiology, Washington, D.C.

Tenover, F.C., Arbeit, R.D., Goering, R.V. et al. 1995: Interpreting chromosomal DNA restriction patterns produced by pulsed-field gel electrophoresis: criteria for bacterial strain typing. J. Clin. Microbiol. 33:2233.

15 IMMUNOCHEMICAL METHODS USED FOR ORGANISM DETECTION

Certain factors can hinder the diagnosis of an infectious disease by culture and biochemical techniques. These factors include the inability to cultivate an organism on artificial media, such as with *Treponema pallidum,* the agent of syphilis, or the fragility of an organism and its subsequent failure to survive transport to the laboratory, such as with respiratory syncytial virus and varicella-zoster virus. The fastidious nature of some organisms, such as leptospira or *Bartonella,* can result in long incubation periods before growth is evident, or administration of antimicrobial therapy before a specimen is obtained, such as with a patient who has received partial treatment, also can impede diagnosis. In these cases, detecting a specific product of the infectious agent in clinical specimens is very important because this product would not be present in the specimen in the absence of the agent. This chapter discusses the direct detection of microorganisms in patient specimens using immunochemical methods and the identification of microorganisms by these methods once they have been isolated on laboratory media. Chapter 16 discusses the diagnosis of infectious diseases by using some of these same immunochemical methods to detect antibodies produced in response to the presence of an infecting agent in patient serum.

PRODUCTION OF ANTIBODIES FOR USE IN LABORATORY TESTING

Immunochemical methods use antigens and antibodies as tools to detect microorganisms. **Antigens** are foreign substances, usually high–molecular-weight proteins or carbohydrates, that elicit the production of other proteins, called **antibodies,** in a human or animal host (see Chapter 10). Antibodies attach to the antigens and aid the host in removing the infectious agent (see Chapters 10 and 16). Antigens may be part of the physical structure of the pathogen, such as the bacterial cell wall, or they may be a chemical produced and released by the pathogen, such as an enzyme or toxin; each antigen is also called an **antigenic determinant.** Figure 15-1 shows the multiple antigenic determinants of group A Streptococcus *(Streptococcus pyogenes).*

POLYCLONAL ANTIBODIES

Because one organism contains many different antigenic determinants that the host will recognize as foreign and because the host usually responds by producing antibodies to each antigen, the host's serum will contain various **polyclonal antibodies.** Polyclonal antibodies used in immunodiagnosis are prepared by immunizing animals (usually rabbits, sheep, or goats) with an infectious agent and then isolating and purifying the resulting antibodies from the animal's serum. However, because individual animals produce different antibodies, the lack of uniformity in polyclonal antibody reagents requires that different lots be continually retested for specificity and avidity (strength of binding) in any given immunochemical test system.

MONOCLONAL ANTIBODIES

The ability to create an immortal cell line producing large quantities of a completely characterized and highly specific antibody, known as a **monoclonal antibody,** has revolutionized immunologic testing. Monoclonal antibodies are produced by the offspring (clones) of a single hybrid cell, the product of fusion of an antibody-producing plasma B cell and a malignant antibody-producing myeloma cell. One technique for the production of such a clone of cells, called **hybridoma cells,** is illustrated in Figure 15-2.

The process starts by immunizing a mouse with the antigen for which an antibody is to be produced. The animal responds by producing many antibodies to the antigenic determinants injected. The mouse's spleen, which contains antibody-producing plasma cells, is removed and emulsified so that antibody-producing cells can be separated and placed into individual wells of a microdilution tray. Because cells cannot remain viable in cell culture for very long, they must be fused together with cells that are able to survive and multiply in tissue culture, that is, the continuously propagating, or immortal cells, of multiple myeloma (a malignant tumor of antibody-producing plasma cells). The special myeloma tumor cells used for hybridoma production, however, possess a very important defect—they are deficient in the enzyme hypoxanthine phosphoribosyltransferase.

This defect leads to their inability to survive in a medium containing hypoxanthine, aminopterin, and thymidine (HAT medium). Antibody-producing spleen cells, however, possess the enzyme. Thus, fused hybridoma cells survive in the selective medium and can be recognized by their ability to grow indefinitely in the medium. Unfused antibody-producing lymphoid cells die after several multiplications in vitro because they are not immortal, and unfused myeloma cells die in the presence of the toxic enzyme substrates. The only surviving cells will be true hybrids. The growth medium supernatant from the microdilution tray wells in which the hybridoma cells are growing is then tested for the presence of the desired antibody. Many such cell lines are usually examined before a suitable antibody is found because it must be specific enough to bind only the type of antigen to be tested, but not so specific that it binds only the antigen from the particular strain with which the mouse was first immunized. When a good candidate antibody-producing cell is found, the hybridoma cells are either grown in cell culture in vitro or they are reinjected into the peritoneal cavities of many mice, where the cells multiply and produce large quantities of antibody in the ascitic (peritoneal) fluid that is formed. Ascitic fluid can be removed from mice many times over the animals' lifetimes, and antibody molecules will be identical to the original clone.

Polyclonal and monoclonal antibodies are both used in commercial systems to detect infectious agents.

PRINCIPLES OF IMMUNOCHEMICAL METHODS USED FOR ORGANISM DETECTION

Numerous immunologic methods are used for the rapid detection of bacteria, fungi, parasites, and viruses in patient specimens; the same reagents can often be used to identify these organisms grown in culture. The techniques fall under the following categories: precipitin tests, particle agglutination tests, immunofluorescence assays, and enzyme immunoassays.

PRECIPITIN TESTS

The classic method of detecting soluble antigen (i.e., antigen in solution) is Ouchterlony **double immunodiffusion**.

Double immunodiffusion

In this method, small circular wells are cut out of agar or agarose, a gelatin-like matrix from seaweed through which molecules can readily diffuse, in Petri dishes. The patient specimen containing antigen is placed in one well, and antibody directed against the antigen being sought is placed in the adjacent well. The antigen and antibody will diffuse toward each other over a period of 18 to 24 hours, and they will

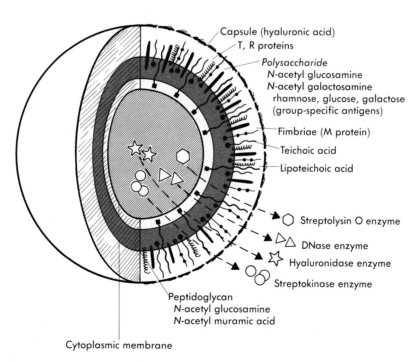

FIGURE 15-1 Group A *Streptococcus (Streptococcus pyogenes)* contains many antigenic structural components and produces various antigenic enzymes, each of which may elicit a specific antibody response from the infected host.

produce a visible precipitin band at the point that they meet, which is known as the **zone of equivalence**. This method is currently used to detect exoantigens produced by the systemic fungi to confirm their presence in culture (Figure 15-3). However, the technique is too slow to be commonly used for antigen detection directly in patient specimens.

Counterimmunoelectrophoresis

Counterimmunoelectrophoresis (CIE) is a modification of the Ouchterlony method that speeds up migration of an antigen and antibody by applying an electrical current. With some exceptions (e.g., *Streptococcus pneumoniae* serotypes 7 and 14), most bacterial antigens are negatively charged in a slightly alkaline environment, whereas antibodies are neutral.

This feature of bacterial antigens is exploited by CIE assays, in which solutions of antibody and sample fluid to be tested are placed in small wells cut into a slab of agarose on a glass surface (Figure 15-4). A paper or fiber wick is used to connect the two opposite sides of the agarose to troughs of slightly alkaline buffer, formulated for each antibody-antigen system. When an electrical current is applied through the buffer, the negatively charged antigen molecules migrate toward the positive electrode and thus toward the wells filled with antibody. The neutrally charged antibodies are carried toward the negative electrode by the flow of the buffer. At some point between the wells, a zone of equivalence occurs, and the antigen-antibody complexes form a visible precipitin band. The entire procedure usually takes about 1 hour.

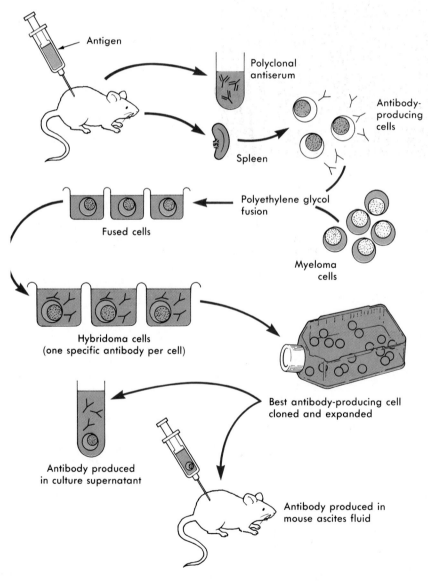

FIGURE 15-2 Production of a monoclonal antibody.

Any antigen for which antisera are available can be tested by CIE. The sensitivity appears to be less than that of particle agglutination, detecting approximately 0.01 to 0.05 mg/mL antigen, which translates to about 10^3 organisms per milliliter of fluid. Bands are often difficult to see, and the agarose gel may require overnight washing in distilled water to remove nonspecific precipitin reactions. Testing positive and negative controls is especially critical, because sera may contain nonspecifically reacting agents that form nonstable complexes in the gel. Because of the initial capital outlay for the apparatus and the large quantities of antigen and antibody that must be used, CIE is more expensive than agglutination-based tests and is, therefore, no longer widely used for immunodiagnosis.

PARTICLE AGGLUTINATION

Numerous procedures have been developed to detect antigen via the agglutination (clumping) of an artificial carrier particle such as a latex bead with antibody bound to its surface.

Latex agglutination

Antibody molecules can be bound in random alignment to the surface of latex (polystyrene) beads (Figure 15-5). Because the number of antibody molecules bound to each latex particle is large, the potential number of antigen-binding sites exposed is also large. Antigen present in a specimen being tested binds to the combining sites of the antibody exposed on the surfaces of the latex beads, forming cross-linked aggregates of latex beads and antigen. The size of the latex bead (0.8 μm or larger) enhances the ease with which the agglutination reaction is recognized. Levels of bacterial polysaccharides detected by latex agglutination have been shown to be as low as 0.1 ng/mL. Because the pH, osmolarity, and ionic concentration of the solution influence the amount of

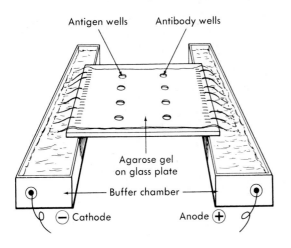

FIGURE 15-3 Exo-Antigen Identification System, Immuno-Mycologics, Inc., Norman, Okla. The center well is filled with a 50× concentrate of an unknown mold. The arrow identifies well 1; wells 2 to 6 are shown clockwise. Wells 1, 3, and 5 are filled with anti-*Histoplasma,* anti-*Blastomyces,* and anti-*Coccidioides* reference antisera, respectively. Wells 2, 4, and 6 are filled with *Histoplasma* antigen, *Blastomyces* antigen, and *Coccidioides* antigen, respectively. The unknown organism can be identified as *Histoplasma capsulatum* based on the formation of line(s) of identity (*arc*) linking the control band(s) with one or more bands formed between the unknown extract (*center well*) and the reference antiserum well (*well 1*).

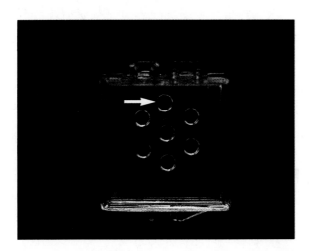

FIGURE 15-4 Apparatus for performing counterimmunoelectrophoresis.

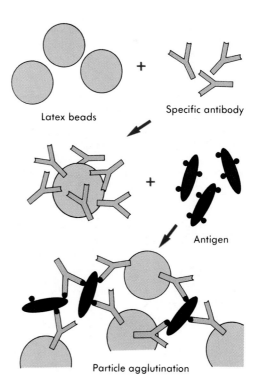

FIGURE 15-5 Alignment of antibody molecules bound to the surface of a latex particle and latex agglutination reaction.

binding that occurs, conditions under which latex agglutination procedures are carried out must be carefully standardized. Additionally, some constituents of body fluids, such as rheumatoid factor, have been found to cause false-positive reactions in the latex agglutination systems available. To counteract this problem, it is recommended that all specimens be pretreated by boiling or with ethylenediaminetetraacetic acid (EDTA) before testing. Commercial test systems are usually performed on cardboard cards or glass slides; manufacturer's recommendations should be followed precisely to ensure accurate results. Reactions are graded on a 1+ to 4+ scale, with 2+ usually the minimum amount of agglutination seen in a positive sample. Control latex (coated with antibody from the same animal species from which the specific antibody was made) is tested alongside the test latex. If the patient specimen or the culture isolate reacts with both the test and control latex, the test is considered nonspecific, and therefore, uninterpretable.

Latex tests are very popular in clinical laboratories to detect antigen to *Cryptococcus neoformans* in cerebrospinal fluid or serum (Figure 15-6) and to confirm the presence of beta-hemolytic *Streptococcus* from culture plates (Figure 15-7). Latex tests are also available to detect the agents of bacterial meningitis (*Haemophilus influenzae* type b; *Streptococcus pneumoniae*; *Neisseria meningitidis* types A, B, C, Y, and W138; and *Streptococcus agalactiae*), *Clostridium difficile* toxins A and B; and rotavirus.

Coagglutination

Similar to latex agglutination, coagglutination uses antibody bound to a particle to enhance visibility of the agglutination reaction between antigen and antibody. In this case the particles are killed and treated *S. aureus* organisms (Cowan I strain), which contain a large amount of an antibody-binding protein, protein A, in their cell walls. In contrast to latex particles, these staphylococci bind only the base of the heavy-chain portion of the antibody, leaving both antigen-binding ends free to form complexes with specific antigen (Figure 15-8). Several commercial suppliers have prepared coagglutination reagents for identification of streptococci, including Lancefield groups A, B, C, D, F, G, and N; *Streptococcus pneumoniae*; *Neisseria meningitidis*; *N. gonorrhoeae*; and *Haemophilus influenzae* types A to F grown in culture. The coagglutination reaction is highly specific but may not be as sensitive for detecting small quantities of antigen as latex agglutination. Thus, it is not usually used for direct antigen detection.

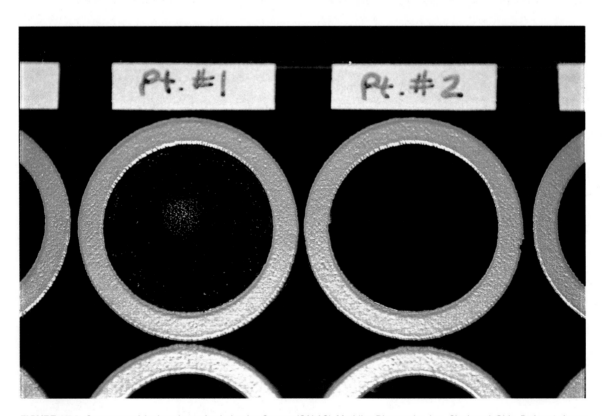

FIGURE 15-6 Cryptococcal Antigen Latex Agglutination System (CALAS), Meridian Diagnostics, Inc., Cincinnati, Ohio. Patient 1 shows positive agglutination; patient 2 is negative.

Liposome-enhanced latex agglutination

Phospholipid molecules form small, closed vesicles under certain controlled conditions. These vesicles, consisting of a single lipid bilayer, are called **liposomes**. Molecules bound to the surface of liposomes act as agglutinating particles in a reaction. By combining liposomes containing reactive molecules on their surfaces and latex particles that harbor antibody-binding sites on their surfaces, reagents are created that have the potential to transform a weak antigen-antibody particle agglutination reaction into a stronger, more easily visualized reaction (Figure 15-9).

FIGURE 15-7 Streptex, Murex Diagnostics, Ltd., Dartford, England. Colony of beta-hemolytic *Streptococcus* agglutinates with group B *Streptococcus* (*Streptococcus agalactiae*) latex suspension.

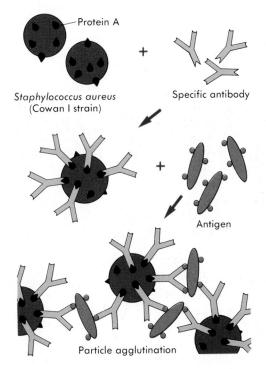

FIGURE 15-8 Coagglutination.

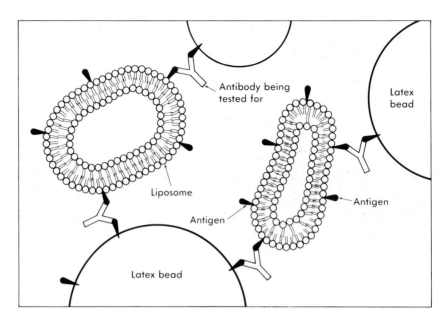

FIGURE 15-9 Diagram of liposome-latex agglutination reaction.

Liposomes have yet to reach their full potential as diagnostic reagents in clinical microbiology.

IMMUNOFLUORESCENT ASSAYS

Immunofluorescent assays are popular tests for detecting antigens in clinical laboratories. In these tests, antigenic determinants in patient specimens are immobilized and fixed onto glass slides with formalin, methanol, ethanol, or acetone. Monoclonal or polyclonal antibodies conjugated (attached) to fluorescent dyes are then applied to the specimen. After appropriate incubation, washing, and counterstaining (staining the background with a nonspecific fluorescent stain such as rhodamine or Evan's blue), the slide is viewed using a microscope equipped with a high-intensity light source (usually halogen) and filters to excite the fluorescent tag. Most kits used in clinical microbiology laboratories use fluorescein isothiocyanate (FITC) as the dye; FITC fluoresces a bright apple-green (Figure 15-10).

Fluorescent antibody tests are performed using either a direct (DFA) or indirect (IFA) technique (Figure 15-11). In the DFA, FITC is conjugated directly to the specific antibody. In the IFA, the antigen-specific antibody is unlabeled and a second antibody (usually raised against the animal species from which the antigen-specific antibody was harvested) is conjugated to the FITC. The IFA is a two-step, or sandwich, procedure. IFA is more sensitive than DFA, although DFA is faster since there is only one incubation step.

The major advantage of immunofluorescent microscopy assays is that they allow visual assessment of the adequacy of a specimen. This is a major

factor in their use in tests for chlamydial elementary bodies or respiratory syncytial virus (RSV) antigens. Microscopists can see if the specimen was taken from the columnar epithelial cells at the opening of the cervix in the case of the chlamydia DFA, or from the basal cells of the nasal epithelium in the case of RSV. Many individuals, however, consider it problematic that reading slides is completely subjective and that microbiologists must have extensive training to perform testing. Likewise, many individuals view the requirement for a fluorescent scope as an expensive luxury. Finally, fluorescence fades rapidly over time, which makes the archiving of slides difficult. Therefore, antibodies have been conjugated to other markers besides fluorescent dyes. These newer colorimetric labels use enzymes, such as horseradish peroxidase, alkaline phosphatase, and avidin-biotin, to detect the presence of antigen by converting a colorless substrate to a colored end product. Advantages of these tags are that they allow the preparation of permanent mounts because the reactions do not fade with storage and they can be detected using a simple light microscope.

Fluorescent antibody tests are commonly used to detect *Bordetella pertussis*, *Legionella pneumophila*, *Giardia*, *Cryptosporidia*, *Pneumocystis*, *Trichomonas*, herpes simplex virus, cytomegalovirus, varicella-zoster virus, RSV, adenovirus, influenza virus, and parainfluenza virus in clinical specimens.

ENZYME IMMUNOASSAYS

Enzyme immunoassay (EIA), or enzyme-linked immunosorbent assay (ELISA), systems were first

FIGURE 15-10 *Legionella* (Direct) Fluorescent Test System, Scimedx Corp., Denville, N.J. *Legionella pneumophila* serogroup 1 in sputum.

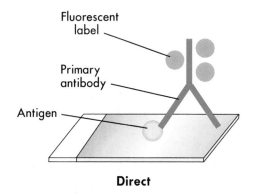

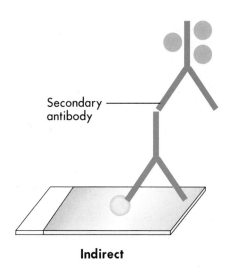

FIGURE 15-11 Direct and indirect fluorescent antibody tests for antigen detection.

developed during the 1960s. The basic test consists of antibodies bonded to enzymes; the enzymes remain able to catalyze a reaction yielding a visually discernible end product while attached to the antibodies. Furthermore, the antibody-binding sites remain free to react with their specific antigen. The use of enzymes as labels has several advantages. First, the enzyme itself is not changed during activity; it can catalyze the reaction of many substrate molecules, greatly amplifying the reaction and enhancing detection. Second, enzyme-conjugated antibodies are stable and can be stored for a relatively long time. Third, the formation of a colored end product allows direct observation of the reaction or automated spectrophotometric reading.

The use of monoclonal antibodies has helped increase the specificity of currently available ELISA systems. New ELISA systems are continually being developed for detection of etiologic agents or their products. In some instances, such as detection of RSV, HIV, and certain adenoviruses, ELISA systems may even be more sensitive than current culture methods.

Solid-phase immunoassay

Most ELISA systems developed to detect infectious agents consist of antibody directed against the agent in question firmly fixed to a solid matrix, either the inside of the wells of a microdilution tray or the outside of a spherical plastic or metal bead or some other solid matrix (Figure 15-12). Such systems are called **solid-phase immunosorbent assays** (SPIA). If antigen is present in the fluid to be tested, stable antigen-antibody complexes form when the fluid is added to the matrix. Unbound antigen is thoroughly removed by washing, and a second antibody against

the antigen being sought is then added to the system. This antibody has been complexed to an enzyme such as alkaline phosphatase or horseradish peroxidase. If the antigen is present on the solid matrix, it binds the second antibody, forming a sandwich with antigen in the middle. After washing has removed unbound, labeled antibody, the addition and hydrolysis of the enzyme substrate causes the color change and completes the reaction. The visually detectable end point appears wherever the enzyme is present (Figure 15-13). Because of the expanding nature of the reaction, even minute amounts of antigen (<1 ng/mL) can be detected. The system just described requires a specific enzyme-labeled antibody for each antigen tested. However, it is simpler to use an indirect assay in which a second, unlabeled antibody is used to bind to the antigen-antibody complex on the matrix. A third antibody, labeled with enzyme and directed against the nonvariable Fc portion of the unlabeled second antibody, can then be used as the detection marker for many different antigen-antibody complexes

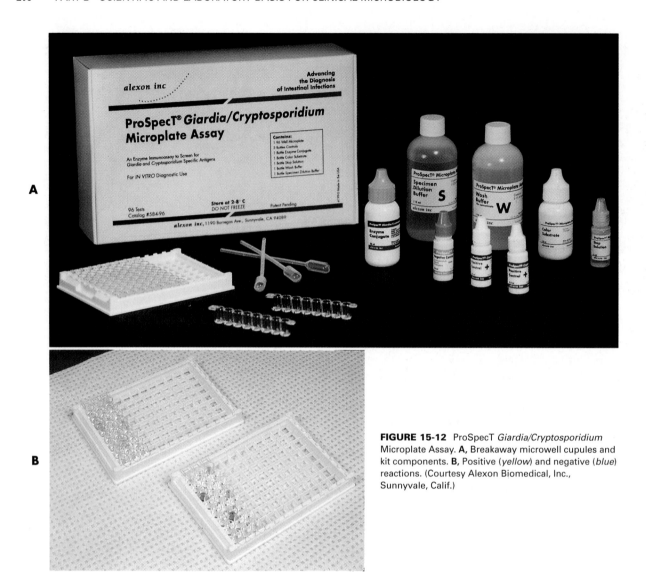

FIGURE 15-12 ProSpecT *Giardia/Cryptosporidium* Microplate Assay. **A,** Breakaway microwell cupules and kit components. **B,** Positive (*yellow*) and negative (*blue*) reactions. (Courtesy Alexon Biomedical, Inc., Sunnyvale, Calif.)

(Figure 15-14). ELISA systems are important diagnostic tools for hepatitis Bs (surface) and e antigens and HIV p24 protein, all indicators of early, active, acute infection.

Membrane-bound SPIA

The flow-through and large surface area characteristic of nitrocellulose, nylon, or other membranes have been exploited to enhance the speed and sensitivity of ELISA reactions. The presence of absorbent material below the membrane serves to pull the liquid reactants through the membrane and helps to separate nonreacted components from the antigen-antibody complexes bound to the membrane; washing steps are also simplified. Membrane-bound SPIA systems are available for several viruses (Figure 15-15), group A beta-hemolytic streptococci antigen directly from throat swabs, and group B streptococcal antigen in vaginal secretions. In addition to their use in clinical

laboratories, these assays are expected to become more prevalent for home testing systems.

OTHER IMMUNOASSAYS

Several other methods, including radioimmunoassay (RIA) and fluorescent immunoassay (FIA), are similar to ELISA except that radionucleotides (usually ^{125}I or ^{14}C) are substituted for enzymes in RIA and fluorochromes are substituted for enzymes in FIA. Although RIA was formerly the key method for antigen detection for numerous infectious agents, including hepatitis B virus, it has been largely replaced by ELISA testing, which does not require use of radioactive substances.

An optical immunoassay (OIA) has recently been introduced to detect group A streptococcal pharyngitis (Figure 15-16). It detects changes in the reflection of light occurring on the inert matrix after antigen and antibody have attached to it.

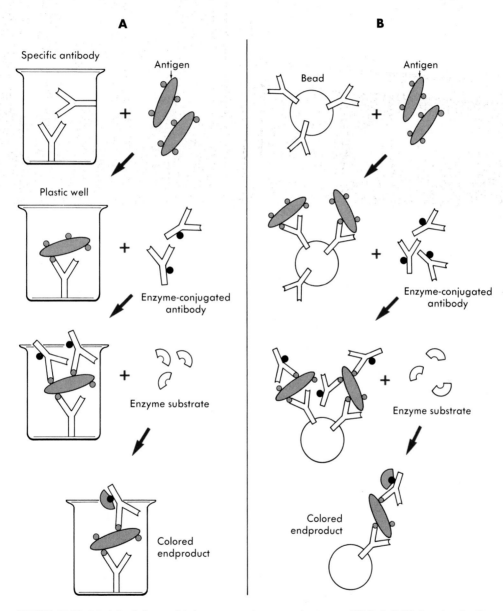

FIGURE 15-13   Principle of direct solid-phase enzyme immunosorbent assay (SPIA). **A,** Solid phase is microtiter well. **B,** Solid phase is bead.

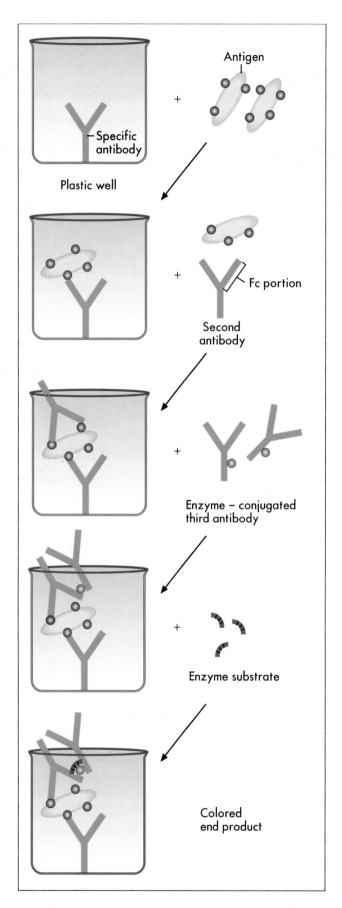

FIGURE 15-14 Principle of indirect solid-phase enzyme immunosorbent assay (SPIA).

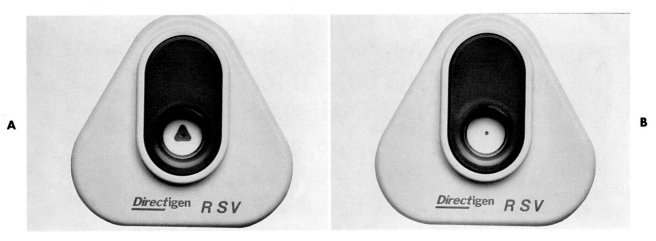

FIGURE 15-15 Directigen respiratory syncytial virus (RSV) membrane-bound cassette. **A,** Positive reaction. **B,** Negative reaction. (Courtesy Becton Dickinson Microbiology Systems, Cockeysville, Md.)

FIGURE 15-16 Optical ImmunoAssay (OIA) for group A *Streptococcus.* **A,** Positive reaction. **B,** Negative reaction. **C,** Invalid reaction. (Courtesy BioStar, Inc., Boulder, Colo.)

Bibliography

Benjamini, E., Sunshine, G., and Leskowitz, S. 1996. Immunology: a short course, ed 3. Wiley-Liss, New York.

Gaur, S., Kesarwala, H., Gavai, M., et al. 1994. Clinical immunology and infectious diseases. Pediatr. Clin. North Am. 41:745.

Kohler, G. and Milstein, C. 1975. Continuous cultures of fused cells secreting antibody of predefined specificity. Nature (London) 256:495.

Rose, N.R., deMacario, E.C., Folds, J.D., et al., editors. 1997. Manual of clinical immunology, ed 5. ASM Press, Washington, D.C.

16 | SEROLOGIC DIAGNOSIS OF INFECTIOUS DISEASES

An understanding of how immunochemical methods have been adapted as tools for serodiagnosis of infectious diseases requires a basic knowledge of the workings of the immune system. For further discussion of the host's response to foreign substances, see Chapter 10.

FEATURES OF THE IMMUNE RESPONSE

The host or patient has physical barriers, such as intact skin and ciliated epithelial cells, and chemical barriers, such as oils produced by the sebaceous glands and lysozyme found in tears and saliva, to prevent infections by foreign organisms. In addition, **natural immunity,** which is not specific, causes **chemotaxis,** the process by which phagocytes engulf organisms that enter the host and produce substances that attract white blood cells to the site of infection. **Acquired immunity** is the *specific* response of the host to the infecting organism.

The human specific immune responses are simplistically divided into the following two categories: cell-mediated and antibody-mediated.

Cell-mediated immune responses are carried out by special lymphocytes of the T (thymus-derived) class. T cells proliferate and differentiate into various effector T cells, including killer and helper cells. Killer cells, also known as **cytotoxic T lymphocytes,** specifically attack and kill microorganisms or host cells damaged or infected by these pathogens. Helper cells promote the maturation of B cells by producing activator cytokines that induce the B cells to produce antibodies and attach to and kill invading organisms. Although diagnosis of certain diseases may be aided by measuring the cell-mediated immune response to the pathogen, such tests entail skin tests performed by physicians or in vitro cell function assays performed by specially trained immunologists. These tests are usually not within the repertoire of clinical microbiology laboratories.

Antibody-mediated immune responses are those produced by specific proteins generated by lymphocytes of the B (bone marrow-derived) class. Because these proteins exhibit immunologic function and because they fold into a globular structure in the active state, they are also called **immunoglobulins.** Antibodies are either secreted into the blood or lymphatic fluid (and sometimes other body fluids) by the B lymphocytes, or they remain attached to the surface of the lymphocyte or other cells. Because the cells involved in this category of immune response chiefly circulate in the blood, this type of immunity is also called **humoral immunity.** For purposes of determining whether an antibody has been produced against a particular infectious agent by a patient, the patient's serum (or occasionally the plasma) is tested for the presence of the antibody. The study of the diagnosis of disease by measuring antibody levels in serum is called **serology.**

CHARACTERISTICS OF ANTIBODIES

By a genetically determined mechanism, immunocompetent humans are able to produce antibodies specifically directed against almost all the antigens with which they might come into contact throughout their lifetimes and which the body recognizes as foreign. Antigens may be part of the physical structure of the pathogen, or they may be a chemical produced and released by the pathogen, such as an exotoxin. One pathogen may contain or produce many different antigens that the host will recognize as foreign, so that infection with one agent may cause a number of different antibodies to be produced. In addition, some antigenic determinants of a pathogen may not be available for recognition by the host until the pathogen has undergone a physical change. For example, until a pathogenic bacterium has been digested by a human polymorphonuclear leukocyte, certain antigens deep within the cell wall are not detected by the host immune system. Once the bacterium is broken down, these new antigens are revealed and antibodies can be produced against them. For this reason, a patient may produce different antibodies at different times during the course of a single disease. The immune response to an antigen also matures with continued exposure, and the antibodies produced against it become more specific and more **avid** (able to bind more tightly).

Antibodies function by (1) attaching to the surface of pathogens and making the pathogens more amenable to ingestion by phagocytic cells (**opsonizing antibodies**), (2) binding to and blocking surface receptors for host cells, or (3) attaching to the surface of pathogens and contributing to their destruction by the lytic action of complement (**complement-fixing antibodies**). Although routine diagnostic serologic methods are usually used to measure only two antibody classes, IgM and IgG, there are five different classes of antibodies, immunoglobulin G (IgG), immunoglobulin M (IgM), immunoglobulin A (IgA), immunoglobulin D (IgD), and immunoglobulin E (IgE). IgA is the predominant class of antibody in saliva, tears, and intestinal secretions. The role of IgD in infection is unknown, and IgE rises during infections by several parasites.

The basic structure of an antibody molecule comprises two mirror images, each composed of two identical protein chains (Figure 16-1). At the terminal ends are the antigen-binding sites, which specifically attach to the antigen against which the antibody was produced. Depending on the specificity of the antibody, antigens of some similarity, but not total identity, to the inducing antigen may also be bound; this is called a **cross-reaction**. The complement-binding site is found in the center of the molecule in a structure that is similar for all antibodies of the same class. IgM is produced as a first response to many antigens, although the levels remain high only transiently. Thus, presence of IgM usually indicates recent or active exposure to antigen or infection. On the other hand, IgG antibody may persist long after infection has run its course. The IgM antibody type (Figure 16-2) consists of five identical proteins, with the basic antibody structures linked together at the bases, leaving 10 antigen-binding sites available. The second antibody class, IgG, consists of one basic antibody molecule with two binding sites (see Figure 16-1). The differences in the size and conformation between these two classes of immunoglobulins result in differences in activities and functions.

Features of the humoral immune response useful in diagnostic testing

Immunocompetent humans produce both IgM and IgG antibodies in response to most pathogens. In most cases, IgM is produced by a patient only after the first interaction with a given pathogen and is no longer detectable within a relatively short period afterward. For serologic diagnostic purposes, one important difference between IgG and IgM is that IgM cannot cross the placenta of pregnant women. Therefore, any IgM detected in the serum of a newborn baby must have been produced by the baby itself. The larger number of binding sites on IgM molecules can help to clear the offending pathogen more quickly, even though each individual antigen-binding site may not be the most efficient for attaching the antigen. Over time, the cells that were producing IgM switch to producing IgG.

IgG is often more specific for the antigen (more avid). The IgG has only two antigen-binding sites, but it can also bind complement. When IgG has bound to an antigen, the base of the molecule may be left projecting out in the environment. Structures on the IgG's base attract and bind the cell membranes of phagocytes, increasing the chances of engulfment and destruction of the pathogen by the host cells. A second encounter with the same pathogen will usually induce only an IgG response. Because the B lymphocytes retain memory of this pathogen, they can respond more quickly and with larger numbers of antibodies than at the initial interaction. This enhanced response is called the **anamnestic response**. Because the B-cell memory is not perfect, occasional clones of memory cells will be stimulated by an antigen that is similar but not identical to the original antigen; thus the anamnestic response may be polyclonal and nonspecific. For example, reinfection with cytomegalovirus may stimulate memory B cells to produce antibody against Epstein-Barr virus (another herpes family virus), which they encountered previously, in addition to antibody against cytomegalovirus. The relative humoral responses over time are diagrammed in Figure 16-3.

Interpretation of serologic tests

A central dogma of serology is the concept of rise in titer. The **titer** of antibody is the reciprocal of the highest dilution of the patient's serum in which the antibody is still detectable. Patients with large amounts of antibody have high titers, because antibody will still be detectable at very high dilutions of serum. Serum for antibody levels should be drawn during the acute phase

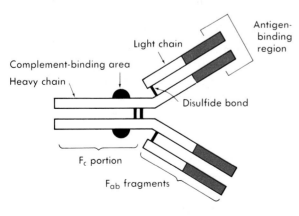

FIGURE 16-1 Structure of immunoglobulin G.

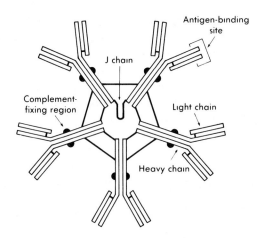

FIGURE 16-2 Structure of immunoglobulin M.

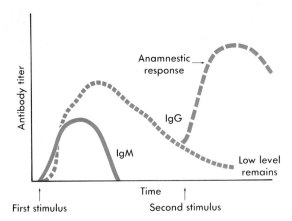

FIGURE 16-3 Relative humoral response to antigen stimulation over time.

of the disease (when it is first discovered or suspected) and again during convalescence (usually at least 2 weeks later). These specimens are called **acute** and **convalescent sera**. For some infections, such as legionnaires' disease and hepatitis, titers may not rise until months after the acute infection or may never rise.

Patients with intact humoral immunity will develop increasing amounts of antibody to a disease-causing pathogen over several weeks. If it is the patient's first encounter with the pathogen, and the specimen has been obtained early enough, no or very low titers of antibody will be detected at the onset of disease. In the case of a second encounter, the patient's serum will usually contain measurable antibody during the initial phase of the disease and the level of the antibody will quickly increase, because of

the anamnestic response. For most pathogens, an increase in the patient's titer of two doubling dilutions (e.g., from a positive result at 1:8 to a positive result at 1:32) is considered to be diagnostic of current infection. This is called a **fourfold rise in titer**.

Accurate results used for diagnosis of many infections are achieved only when *acute and convalescent sera are tested concurrently in the same test system*, because variables inherent in the procedures and laboratory error can easily result in differences of one doubling (or twofold) dilution in the results obtained from even the same sample tested at the same time. Unfortunately, a certain proportion of infected patients may never show a rise in titer, necessitating the use of other diagnostic measures. Because the delay inherent in testing paired acute and convalescent sera results in diagnostic information that arrives too late to influence initial therapy, increasing numbers of early (IgM) serologic testing assays are being commercially evaluated. Moreover, it is sometimes more realistic to see a fourfold *fall* in titer between acute and convalescent sera when they are tested concurrently in the same system because sera may be collected late in the course of an infection in many cases when antibodies have already begun to decrease.

SERODIAGNOSIS OF INFECTIOUS DISEASES

With most diseases there exists a spectrum of responses in infected humans, such that a person may develop antibody from subclinical infection or after colonization by an agent without actually having symptoms of the disease. In these cases, the presence of antibody in a single serum specimen or a similar titer of antibody in paired sera may merely indicate past contact with the agent and cannot be used to accurately diagnose recent disease. Therefore, in the vast majority of serologic procedures for diagnosis of recent infection, testing both acute and convalescent sera is the method of choice. Except for detecting the presence of IgM, testing a single serum can be recommended only in certain cases, such as diagnosis of recent infection with *Mycoplasma pneumoniae* and viral influenza B, when high titers may indicate recent infection, or if the infecting or disease-causing agent is extremely rare, and people without disease or prior immunization would have no chance of developing an immune response, such as the case with rabies virus or the toxin of botulism.

The prevalence of antibody to an etiologic agent of disease in the population relates to the number of people who have come into contact with the agent, not the number who actually develop disease. For most diseases, only a small proportion of infected individuals actually develop symptoms; others develop antibodies that are protective without experiencing actual disease. In a number of circumstances serum is tested only to determine whether a patient is immune, that is, has antibody to a particular agent either in response to a past infection or to immunization. These tests can be performed with a single serum sample. Correlation of the results of such tests with the actual immune status of individual patients must be performed to determine the level of detectable antibody present that corresponds to actual immunity to infection or reinfection in the host. For example, sensitive tests can detect the presence of very tiny amounts of antibody to the rubella virus. Certain people, however, may still be susceptible to infection with the rubella virus with such small amounts of

circulating antibody, and a higher level of antibody may be required to ensure protection from disease.

Alternatively, depending on the etiologic agent, even low levels of antibody may protect a patient from pathologic effects of disease yet not prevent reinfection. For example, a person previously immunized with killed poliovirus vaccine who becomes infected with pathogenic poliovirus will experience multiplication of the virus in the gut and virus entry into the circulation, but damage to the central nervous system will be blocked by humoral antibody in the circulation. As more sensitive testing methods are developed and these types of problems become more common, microbiologists will need to work closely with clinicians to develop guidelines for interpreting serologic test results as they relate to the immune status of individual patients. Moreover, patients may respond to an antigenic stimulus by producing antibodies that can cross-react with other antigens. These antibodies are nonspecific and thus may cause misinterpretation of serologic tests.

Table 16-1 lists the serologic tests available for immunodiagnosis of infectious diseases, the specimen required, interpretation of positive and negative test results, and examples of applications of each technique.

PRINCIPLES OF SEROLOGIC TEST METHODS

Antibodies can be detected in many ways. In some cases, antibodies to an agent may be detected in more than one way, but the different antibody detection tests may not be measuring the same antibody. For this reason, the presence of antibodies to a particular pathogen as detected by one method may not correlate with the presence of antibodies to the same agent detected by another test method. Moreover, different test methodologies have varying degrees of sensitivity in detecting antibodies even if they are present. However, because IgM is usually produced only during a patient's first exposure to an infectious agent, the detection of specific IgM can help the clinician a great deal in establishing a diagnosis. Most of the serologic test methods can be adapted for analysis of IgM.

SEPARATING IgM FROM IgG FOR SEROLOGIC TESTING

IgM testing is especially helpful for diseases that may have nonspecific clinical presentations, such as toxoplasmosis, or those for which rapid therapeutic decisions may be required. For example, rubella infection in pregnant women can lead to serious consequences to the fetus such as cataracts, glaucoma, mental retardation, and deafness. Therefore, pregnant women who are exposed to rubella virus and develop a mild febrile illness can be tested for the presence of antirubella IgM. If positive, the option of elective termination of pregnancy can be offered. An additional reason to measure for IgM alone is for diagnosis of neonatal infections. Because IgG can readily cross the placenta, newborn babies will carry titers of IgG nearly identical to those of their mothers during their first 2 to 3 months of life until they produce their own antibodies. Accurate serologic diagnosis of infection in neonates requires either demonstration of a rise in titer (which takes time to occur) or the detection of specific IgM directed against the putative agent. Because the IgM molecule does not cross the placental barrier, any IgM would have to be of fetal

	TABLE 16-1 TESTS AVAILABLE FOR SERODIAGNOSIS OF INFECTIOUS DISEASES		
TEST	**SERA NEEDED**	**INTERPRETATION**	**APPLICATION**
IgM	Single, acute (collected at onset of illness)	Newborn, positive: *in utero* (congenital) infection Adult, positive: primary or current infection Negative: no infection or past infection	Newborn: STORCH* agents; other organisms Adults: any infectious agent
IgG	Acute and convalescent (collected 2-6 weeks after onset)	Positive: fourfold rise or fall in titer between acute and convalescent sera tested at the same time in the same test system Negative: no current infection or past infection, or patient is immunocompromised and cannot mount a humoral antibody response, or convalescent specimen collected before increase in IgG (Lyme disease, *Legionella*)	Any infectious agent
IgG	Single specimen collected between onset and convalescence	Adult, positive: adult evidence of infection at some unknown time except in certain cases in which a single high titer is diagnostic (rabies, *Legionella*, *Ehrlichia*) Newborn, positive: maternal antibodies that have crossed the placenta Negative: patient has not been exposed to microorganism or patient has a congenital or acquired immune deficiency, or specimen collected before increase in IgG (Lyme disease or *Legionella*)	Any infectious agent
Immune status evaluation	Single specimen collected at any time	Positive: previous exposure Negative: no exposure	Rubella testing for women of childbearing age, syphilis testing to obtain marriage license, cytomegalovirus testing of transplant donor and recipient

*STORCH = Syphilis, toxoplasma, rubella virus, cytomegalovirus, herpes simplex virus.

origin. Agents that are difficult to culture or those that adult females would be expected to have encountered during their lifetimes, such as cytomegalovirus, herpesvirus, *Toxoplasma,* rubella virus, or *T. pallidum,* are those for which specific identification of fetal IgM is most often used. The names of some of these agents have been grouped together with the acronym **STORCH** (syphilis, *Toxoplasma,* rubella, cytomegalovirus, and herpes). These tests should be ordered separately, depending on the clinical illness of a newborn suspected of having one of these diseases. In many instances however, infected babies appear clinically well. Furthermore, in many cases serologic tests yield false-positive or false-negative results. Thus, multiple considerations, including patient history and clinical situation, must be employed in the serodiagnosis of neonatal infection, and culture in many cases is still the most reliable diagnostic method.

Several methods have been developed to measure only specific IgM in sera that may also contain specific IgG. In addition to using a labeled antibody specific for only IgM as the marker or the IgM capture sandwich assays, the immunoglobulins can be separated from each other by physical means. Centrifugation through a sucrose gradient, performed at very high speeds, has been used in the past to separate IgM, which has a greater molecular weight, from IgG. Because this method is time-consuming and requires a very expensive ultracentrifuge, many laboratories use a commercially available system that follows the principle of ion exchange chromatography (Figure 16-4).

Other available IgM separation systems operate based on the fact that certain proteins on the surface of staphylococci (protein A) and streptococci (protein G) will bind the Fc portion of IgG. A simple centrifugation step then separates the particles and their bound immunoglobulins from the remaining mixture, which contains the bulk of the IgM. Other methods use antibodies to remove IgM from sera containing both IgG and IgM. An added bonus of IgM separation systems is that **rheumatoid factor,** IgM antibodies produced by some patients against their own IgG, will often bind to the IgG molecules being removed from the serum; consequently, these IgM antibodies will be removed along with the IgG. Rheumatoid factor can cause nonspecific and interfering results with many serologic tests, and its presence should be taken into account.

METHODS FOR ANTIBODY DETECTION

Direct whole pathogen agglutination assays

The most basic tests for antibody detection are those that measure the antibody produced by a host to determinants on the surface of a bacterial agent in response to infection with that agent. Specific antibodies bind to surface antigens of the bacteria in a thick suspension and cause the bacteria to clump together in visible aggregates. Such antibodies are called **agglutinins,** and the test is called **bacterial**

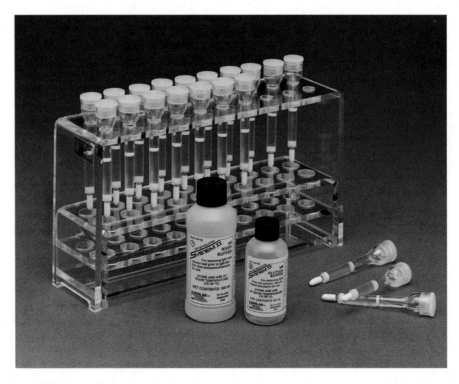

FIGURE 16-4 Quik-Sep IgM System II for separating IgM antibodies from IgG antibodies. (Courtesy Isolab, Inc., Akron, Ohio.)

agglutination. Electrostatic and other forces influence the formation of aggregates in solutions, so that certain conditions are usually necessary for good results. Because most bacterial surfaces exhibit a negative charge, they tend to repel each other. Performance of agglutination tests in sterile physiologic saline (0.85% sodium chloride in distilled water), which contains free positive ions, will enhance the ability of antibody to cause aggregation of bacteria. Although bacterial agglutination tests can be performed on the surface of both glass slides and in test tubes, tube agglutination tests are often more sensitive, because a longer incubation period (allowing more antigen and antibody to interact) can be used. The small volume of liquid used for slide tests requires a rather rapid reading of the result, before the liquid can evaporate.

Examples of bacterial agglutination tests are the tests for antibodies to *Francisella tularensis* and *Brucella* spp., which are part of a panel called **febrile agglutinin tests.** Bacterial agglutination tests are often used to diagnose diseases in which the bacterial agent is difficult to cultivate in vitro. Diseases that can be diagnosed by this technique include tetanus, yersiniosis, leptospirosis, brucellosis, and tularemia. The reagents necessary to perform many of these tests are commercially available, singly or as complete systems. Because most laboratories are able to culture and identify the causative agent, agglutination tests for certain diseases, such as typhoid fever, are seldom used today. Furthermore, the typhoid febrile agglutinin test (called the **Widal test**) is often positive in patients with infections caused by other bacteria because of cross-reacting antibodies or a previous immunization against typhoid. Appropriate specimens from patients suspected of having typhoid fever should be cultured for the presence of salmonellae instead.

Whole cells of parasites, including *Plasmodium, Leishmania,* or *Toxoplasma gondii,* have also been used for direct detection of antibody by agglutination. In addition to using the actual infecting bacteria or parasites as the agglutinating particles for the detection of antibodies, certain bacteria may be agglutinated by antibodies produced against another infectious agent. Many patients infected with one of the rickettsiae produce antibodies that can nonspecifically agglutinate bacteria of the genus *Proteus.* The **Weil-Felix test** detects these cross-reacting antibodies. As newer, more specific serologic methods of diagnosing rickettsial disease become more widely available, however, the use of the *Proteus* agglutinating test is being discontinued.

Particle agglutination tests

Numerous serologic procedures have been developed to detect antibody via the agglutination of an artificial carrier particle with antigen bound to its surface. As noted in Chapter 15, similar systems employing artificial carriers coated with antibodies are commonly used for detection of microbial antigens. Either artificial carriers, such as latex particles or treated red blood cells, or biologic carriers, such as whole bacterial cells, can carry an antigen on their surface that will bind with antibody produced in response to that antigen when it was introduced to the host. The size of the carrier enhances the visibility of the agglutination reaction, and the artificial nature of the system allows the antigen bound to the surface to be extremely specific.

Results of particle agglutination tests are dependent on several factors, including the amount and avidity of antigen conjugated to the carrier, the time of incubation together with patient's serum (or other source of antibody), and the microenvironment of the interaction (including pH and protein concentration). Commercial tests have been developed as systems, complete with their own diluents, controls, and containers. For accurate results, they should be used as units without modifications. If tests have been developed for use with cerebrospinal fluid, for example, they should not be used with serum unless the package insert or the technical representative has certified such usage.

Treated animal red blood cells have also been used as carriers of antigen for agglutination tests; these tests are called indirect **hemagglutination,** or passive hemagglutination tests, because it is not the antigens of the blood cells themselves, but rather the passively attached antigens, that are being bound by antibody. The most widely used of these tests is the microhemagglutination test for antibody to *Treponema pallidum* (MHA-TP, so called because it is performed in a microtiter plate), the hemagglutination treponemal test for syphilis (HATTS), the passive hemagglutination tests for antibody to extracellular antigens of streptococci, and the rubella indirect hemagglutination tests, all of which are available commercially. Certain reference laboratories, such as the Centers for Disease Control and Prevention, also perform indirect hemagglutination tests for antibodies to some clostridia, *Burkholderia pseudomallei, Bacillus anthracis, Corynebacterium diphtheriae, Leptospira,* and the agents of several viral and parasitic diseases.

Complete systems for the use of latex or other particle agglutination tests are commercially available for the accurate and sensitive detection of antibody to cytomegalovirus, rubella virus, varicella-zoster virus, the heterophile antibody of infectious mononucleosis, teichoic acid antibodies of staphylococci, antistreptococcal antibodies, mycoplasma antibodies, and others. Latex tests for antibodies to *Coccidioides, Sporothrix, Echinococcus,* and *Trichinella* are available, although they are not widely used because of the uncommon occurrence of the corresponding infection or its limited geographic distribution. Use of tests for *Candida* antibodies has not yet shown results reliable enough for accurate diagnosis of disease.

Flocculation tests

In contrast to the aggregates formed when particulate antigens bind to specific antibody, the interaction of soluble antigen with antibody may result in the formation of a precipitate, a concentration of fine particles, usually visible only because the precipitated product is forced to remain in a defined space within a matrix. Two variations of the **precipitin test,** flocculation and counterimmunoelectrophoresis, are widely used for serologic studies.

In **flocculation tests** the precipitin end product forms macroscopically or microscopically visible clumps. The Venereal Disease Research Laboratory test, known as the **VDRL,** is the most widely used flocculation test. Patients infected with pathogenic treponemes, most commonly *T. pallidum,* the agent of syphilis, form an antibody-like protein called **reagin** that binds to the test antigen, cardiolipin-lecithin–coated cholesterol particles, causing the particles to flocculate. Because reagin is not a specific antibody directed against *T. pallidum* antigens, the test is not highly specific but is a good screening test, detecting more than 99% of cases of secondary syphilis.

The VDRL is the single most useful test available for testing cerebrospinal fluid in cases of suspected neurosyphilis, although it may be falsely positive in the absence of this disease. The performance of the VDRL test requires scrupulously clean glassware and exacting attention to detail, including numerous daily quality control checks. In addition, the reagents must be prepared fresh each time the test is performed, patients' sera must be inactivated by heating for 30 minutes at 56° C before testing, and the reaction must be read microscopically. For all these reasons it has been supplanted in many laboratories by a qualitatively comparable test, the **rapid plasma reagin,** or **RPR,** test.

The RPR test is commercially available as a complete system containing positive and negative controls, the reaction card, and the prepared antigen suspension. The antigen, cardiolipin-lecithin–coated cholesterol with choline chloride, also contains charcoal particles to allow for macroscopically visible flocculation. Sera are tested without heating, and the reaction takes place on the surface of a specially treated cardboard card, which is then discarded (Figure 16-5). The RPR test is not recommended for testing cerebrospinal fluid. All procedures are standardized and clearly described in product inserts, and these procedures should be adhered to strictly. Over-

all, the RPR appears to be a more specific screening test for syphilis than the VDRL, and it is certainly easier to perform. Several modifications have been made, such as the use of dyes to enhance visualization of results.

Conditions and infections other than syphilis can cause a patient's serum to yield a positive result in the VDRL or RPR tests; these are called **biologic false-positive tests.** Autoimmune diseases, such as systemic lupus erythematosus and rheumatic fever; infectious mononucleosis; hepatitis; pregnancy; and old age have caused false-positive tests, so results of screening tests should be considered presumptive until confirmed with a specific treponemal test.

Counterimmunoelectrophoresis

Another variation of the classic precipitin test has been widely used to detect small amounts of antibody. This test takes advantage of the net electric charge of the antigens and antibodies being tested in a particular test buffer. Because the antigen and antibody being sought migrate toward each other in a semisolid matrix under the influence of an electrical current, the method is called **counterimmunoelectrophoresis (CIE).** The principle of this test was outlined in Chapter 15; the same methodology is used to identify specific antigen or antibody. When antigen and antibody meet in optimal proportions, a line of precipitation will appear. Because all variables, such as buffer pH, type of gel or agarose matrix, amount of electrical current, amount and concentration of antigen and antibody, size of antigen and antibody inocula, and placement of these inocula, must be carefully controlled for maximum reactivity, CIE tests are difficult to develop and perform. Other methods for

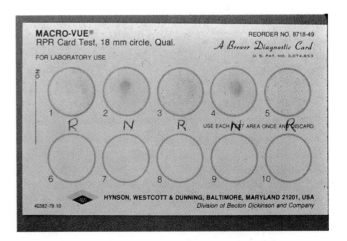

FIGURE 16-5 MACRO-VUE RPR card test. R, reactive (positive) test; N, nonreactive (negative) test. (Courtesy Becton Dickinson Microbiology Systems, Cockeysville, Md.)

detection of antibody to infectious agents are more commonly used in most diagnostic laboratories.

Immunodiffusion assays

Closely resembling the precipitin test is the Ouchterlony double immunodiffusion assay, which is widely used for detecting antibodies directed against fungal cell components. This test is described in Chapter 15. Whole-cell extracts or other antigens of the suspected fungus are placed in wells in an agarose plate, and the patient's serum and a control positive serum are placed in adjoining wells. If the patient has produced specific antibody against the fungus, precipitin lines will be visible within the agarose between the homologous (identical) antigen and antibody wells; their identity with similar lines from the control serum helps confirm the results. The type and thickness of the precipitin bands may have prognostic, as well as diagnostic, value. Antibodies against the pathogenic fungi *Histoplasma, Blastomyces, Coccidioides, Paracoccidioides,* and some opportunistic fungi are routinely detected by immunodiffusion. The test usually requires at least 48 hours, but additional time may be required to develop the bands.

Hemagglutination inhibition assays

Many human viruses can bind to surface structures on red blood cells from different species. For example, rubella virus particles can bind to human type O, goose, or chicken erythrocytes and cause agglutination of the red blood cells. Influenza and parainfluenza viruses agglutinate guinea pig, chicken, or human O erythrocytes; many arboviruses agglutinate goose red blood cells; adenoviruses agglutinate rat or rhesus monkey cells; mumps virus binds red blood cells of monkeys; and herpesvirus and cytomegalovirus agglutinate sheep red blood cells. Serologic tests for the presence of antibodies to these viruses exploit the agglutinating properties of the virus particles. Patients' sera that have been treated with kaolin or heparin-magnesium chloride (to remove nonspecific inhibitors of red cell agglutination and nonspecific agglutinins of the red cells) are added to a system that contains the virus suspected of causing disease. If antibodies to the virus are present, they will form complexes and block the binding sites on the viral surfaces. When the proper red cells are added to the solution, all of the virus particles will be bound by antibody, preventing the virus from agglutinating the red cells. Thus, the patient's serum is positive for hemagglutination-inhibiting antibodies. As for most serologic procedures, a fourfold increase in such titers is considered diagnostic. The hemagglutination inhibition tests for most agents are performed only at reference laboratories. Rubella antibodies, however, are often detected with this method in routine diagnostic laboratories. Several commercial rubella hemagglutination inhibition test systems are available.

Neutralization assays

Antibody that inhibits the infectivity of a virus by blocking its host cell receptor site is called a **neutralizing antibody**. The serum to be tested is mixed with a suspension of infectious viral particles of the same type as those with which the patient is suspected of being infected. A control suspension of viruses is mixed with normal serum. The viral suspensions are then inoculated into a cell culture system that supports growth of the virus. The control cells will display evidence of viral infection. If the patient's serum contains antibody to the virus, that antibody will bind the viral particles and prevent them from invading the cells in culture. The antibody has neutralized the "infectivity" of the virus. These tests are technically demanding and time-consuming and are performed only in reference laboratories.

Antibodies to bacterial toxins and other extracellular products that display measurable activities can be tested in the same way. The ability of a patient's serum to neutralize the erythrocyte-lysing capability of streptolysin O, an extracellular enzyme produced by *S. pyogenes* during infection, has been used for many years as a test for previous streptococcal infection. After pharyngitis with streptolysin O-producing strains, most patients show a high titer of the antibody to streptolysin O, that is, antistreptolysin O (ASO) antibody. Streptococci also produce the enzyme deoxyribonuclease B (DNase B) during infections of the throat, skin, or other tissue. A neutralization test that prevents activity of this enzyme, the anti-DNase B test, has also been used extensively as an indicator of recent or previous streptococcal disease. However, the use of particle agglutination (latex or indirect hemagglutination) tests for the presence of antibody to many of the streptococcal enzymes has replaced the use of these neutralization tests in many laboratories.

Complement fixation assays

One of the classic methods for demonstrating the presence of antibody in a patient's serum has been the **complement fixation test** (**CF**). This test consists of two separate systems. The first (the test system) consists of the antigen suspected of causing the patient's disease and the patient's serum. The second (the indicator system) consists of a combination of sheep red blood cells, complement-fixing antibody (IgG) raised against the sheep red blood cells in another animal, and an exogenous source of complement (usually guinea pig serum). When these three components are mixed together in optimum concentrations, the antisheep erythrocyte antibody will bind to the surface of the red blood cells and the complement will then bind to the antigen-antibody complex, ultimately causing lysis (bursting) of the red blood cells. For this reason the antisheep red blood cell antibody is also called **hemolysin**. For the CF test these two systems are tested in sequence (Figure 16-6). The patient's serum is first added to the putative antigen; then the limiting amount of complement is added to the solution. If the patient's serum contains antibody to the antigen, the resulting antigen-antibody complexes will bind all of the complement added. In the next step the sheep red blood cells and the hemolysin (indicator system) are added. Only if the complement has not been bound by a complex formed with antibody from the patient's serum will the complement be available to bind to the sheep cell-hemolysin complexes and cause lysis. A positive result, meaning the patient does possess complement-fixing antibodies, is revealed by failure of the red

blood cells to lyse in the final test system. Lysis of the indicator cells indicates lack of antibody and a negative CF test.

Although requiring many manipulations, at least 48 hours for both stages of the test to be completed, and often yielding nonspecific results, this test has been used over the years to detect many types of antibodies, particularly antiviral and antifungal. Many new systems provide for improved recovery of pathogens or their products and for more sensitive and less demanding procedures for detection of antibodies, including particle agglutination, indirect fluorescent antibody tests, and ELISA procedures, and have gradually been introduced to replace the CF test. At this time, CF tests are performed chiefly for diagnosis of unusual infections and are done primarily in reference laboratories. This test is still probably the most common method for diagnosing infection caused by fungi, respiratory viruses, and arboviruses, as well as to diagnose Q fever. Laboratories without experience in performing these

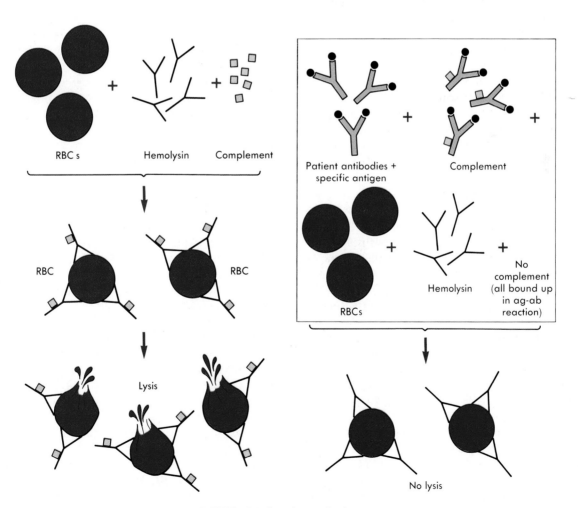

FIGURE 16-6 Complement fixation test.

tests should not adopt complement fixation tests for routine diagnostic testing when other less demanding procedures are available.

Enzyme-linked immunosorbent assays

Enzyme-linked immunosorbent assay (ELISA) technology for antibodies to infectious agents are sensitive and specific. As described more thoroughly in Chapter 15, the presence of specific antibody is detected by the ability of a second antibody conjugated to a colored or fluorescent marker to bind to the target antibody, which is bound to its homologous antigen. Various enzyme-substrate systems, including the use of avidin-biotin to bind marker substances, were also discussed in Chapter 15. The antigen to which the antibodies bind, if these antibodies are present in the patient's sera, is either attached to the inside of the wells of a microtiter plate, adherent to a filter matrix, or bound to the surface of beads or plastic paddles. Advantages of ELISA tests are that they can be performed easily on many serum samples at the same time and that the colored or fluorescent end products are easily detected by instruments, removing the element of subjectivity inherent in so many serologic procedures that rely on a technologist's interpretation of a reaction. Disadvantages include the need for special equipment, the fairly long reaction times (often hours instead of minutes for particle agglutination tests), the relative end point of the test (which relies on measuring the amount of a visible end product that is not dependent on the original antigen-antibody reaction itself but on a second enzymatic reaction as compared with a directly quantitative result), and the requirement for batch processing to ensure that performance of the test is cost effective.

Commercial microdilution or solid-phase matrix systems are available to detect antibody specific for hepatitis virus antigens, herpes simplex viruses 1 and 2, respiratory syncytial virus (RSV), cytomegalovirus, human immunodeficiency virus (HIV), rubella virus (both IgG and IgM), mycoplasmas, chlamydiae, *Borrelia burgdorferi, Entamoeba histolytica, T. gondii,* and many other agents.

The introduction of membrane-bound ELISA components has improved sensitivity and ease of use dramatically. Slot-blot and dot-blot assays force the target antigen through a membrane filter, causing it to become affixed in the shape of the hole (a dot or a slot). Several antigens can be placed on one membrane. When test (patient) serum is layered onto the membrane, specific antibodies, if present, will bind to the corresponding dot or slot of antigen. Addition of a labeled second antibody and subsequent development of the label allows visual detection of the presence of antibodies based on the pattern of antigen sites. Cassette-based membrane-bound ELISA assays, designed for testing a single serum, can be performed rapidly (often within 10 minutes). Commercial kits to detect antibodies to *Helicobacter pylori, Toxoplasma gondii,* and some other infectious agents are available. Accuracy of the results of tests using each of these formats is variable, however.

Antibody capture ELISAs are particularly valuable for detecting IgM in the presence of IgG. Anti-IgM antibodies are fixed to the solid phase, and thus only IgM antibodies, if present in the patient's serum, are bound. In a second step, specific antigen is added in a sandwich format and a second antigen-specific labeled antibody is finally added. Toxoplasmosis, rubella, and other infections are diagnosed using this technology, usually in research settings.

Indirect fluorescent antibody tests and other immunomicroscopic methods

A widely applied method to detect diverse antibodies is that of indirect fluorescent antibody determination (IFA), which is described in detail in Chapter 15. For tests of this type, the antigen against which the patient makes antibody (such as whole *Toxoplasma* organisms or virus-infected tissue culture cells) is affixed to the surface of a microscope slide. The patient's serum to be tested is diluted and placed on the slide, covering the area in which antigen was placed. If present in the serum, antibody will bind to its specific antigen. Unbound antibody is then removed by washing the slide. In the second stage of the procedure, a conjugate of antihuman globulin, which may be directed specifically against IgG or IgM, and a dye that will fluoresce when exposed to ultraviolet light (e.g., fluorescein) is placed on the slide. This labeled marker for human antibody will bind to the antibody already bound to the antigen on the slide and will serve as a detector of binding of the antibody to the antigen when viewed under a fluorescence microscope (Figure 16-7). Commercially available test kits include the slides with the antigens, positive and negative control sera, diluent for the patients' sera, and the properly diluted conjugate. As with other commercial products, IFA systems should be used as units, without modifying the manufacturer's instructions. Currently, commercially available IFA tests include those for antibodies to *Legionella* species, *B. burgdorferi, T. gondii,* varicella-zoster virus, cytomegalovirus, Epstein-Barr virus capsid antigen, early antigen and nuclear antigen, herpes simplex viruses types 1 and 2, rubella virus, *M. pneumoniae, T. pallidum* (the **fluorescent treponemal antibody absorption test [FTA-ABS]**), and several rickettsiae. Most of these tests, if performed properly, give extremely specific and sensitive results. Proper inter-

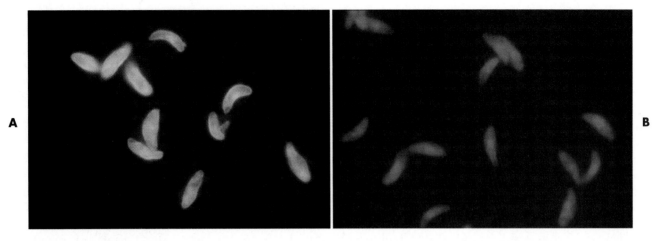

FIGURE 16-7 Indirect fluorescent antibody tests for *Toxoplasma gondii,* IgG antibodies. **A,** Positive reaction. **B,** Negative reaction. (Courtesy Gull Laboratories, Inc., Salt Lake City, Utah.)

pretation of IFA tests requires experienced and technically competent technologists. These tests can be performed rapidly and are cost effective if only a few samples are tested.

Radioimmunoassays

Radioimmunoassay (RIA) is an automated method of detecting antibodies that is usually performed in a chemistry, rather than a serology, laboratory. RIA tests were originally used to detect antibody to hepatitis B viral antigens. Radioactively labeled antibody competes with the patient's unlabeled antibody for binding sites on a known amount of antigen. A reduction in radioactivity of the antigen-patient antibody complex compared with the radioactive counts measured in a control test with no antibody is used to quantitate the amount of patient antibody bound to the antigen. The development of new marker substances, such as ELISA systems, chemiluminescence, and fluorescence, resulted in the production of diagnostic tests as sensitive as RIA without the hazards and associated disposal problems of radioactive reagents.

Fluorescent immunoassays

Because of the inconveniences associated with RIA (radioactive substances and expensive scintillation counters), fluorescent immunoassays (FIA) were developed. These tests, which use fluorescent dyes or molecules as markers instead of radioactive labels, are based on the same principle as RIA. The primary difference is that in RIA systems the competitive antibody is labeled with a radioisotope and in FIA the antigen is labeled with a compound that will fluoresce under the appropriate light rays. Binding of patient antibody to a fluorescent-labeled antigen can reduce or quench the fluorescence, or binding can cause fluo-

rescence by allowing conformational change in a fluorescent molecule. Measurement of fluorescence is thus a direct measurement of antigen-antibody binding, not dependent on a second marker system such as that in ELISA tests. Systems are commercially available to measure antibody developed against numerous infectious agents, as well as against self-antigens (autoimmune antibodies).

Western blot immunoassays

Requirements for the detection of very specific antibodies have driven the development of the Western blot immunoassay system (Figure 16-8). The method is based on the electrophoretic separation of major proteins of an infectious agent in a two-dimensional agarose gel matrix. A suspension of the organism against which antibodies are being sought is mechanically or chemically disrupted, and the solubilized antigen suspension is placed at one end of a polyacrylamide (polymer) gel. Under influence of an electrical current, the proteins migrate through the gel. Most bacteria or viruses contain several major proteins that can be recognized based on their position in the gel after electrophoresis. Smaller proteins travel faster and migrate farther in the lanes of the gel than do the larger proteins. The protein bands are transferred from the gel to a nitrocellulose or other type of thin membrane, and the membrane is treated to immobilize the proteins. The membrane is then cut into many thin strips, each carrying the pattern of protein bands. When patient serum is layered over the strip, antibodies will bind to each of the protein components represented by a band on the strip. The pattern of antibodies present can be used to determine whether the patient is infected with the agent (Figure 16-9). Antibodies against microbes with numerous cross-reacting antibodies, such as *T. pallidum, B.*

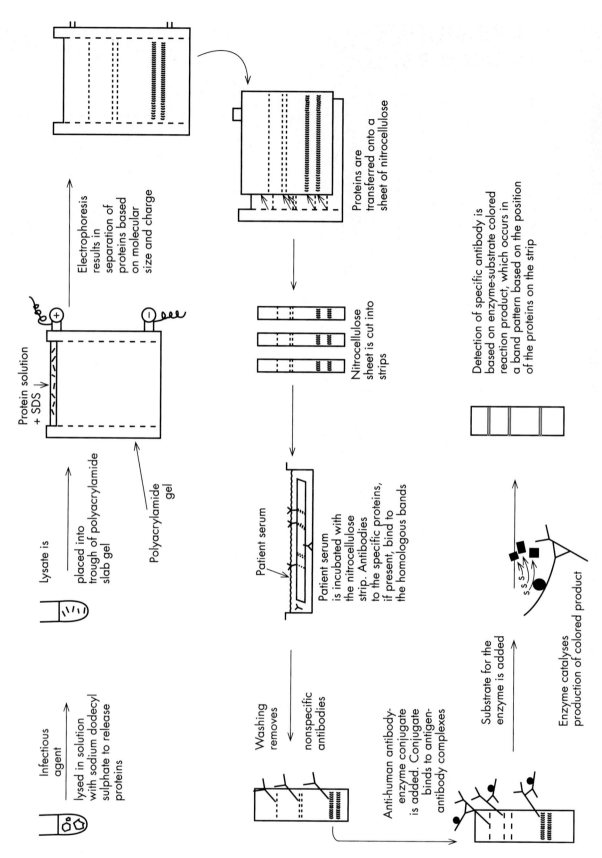

FIGURE 16-8 Diagram of Western blot immunoassay system.

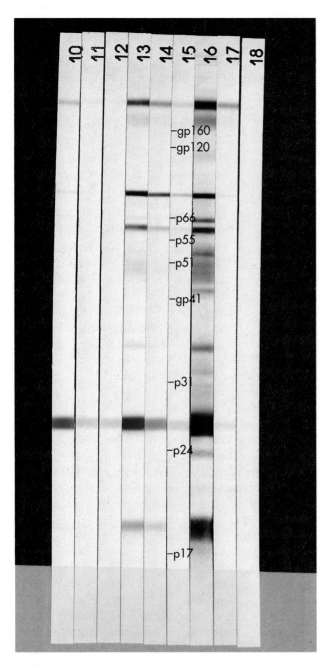

burgdorferi, herpes simplex virus types 1 and 2, and HIV, are identified more specifically using this technology than a method that tests for only one general antibody type.

Bibliography

Benjamini, E., Sunshine, G., and Leskowitz, S. 1996. Immunology: a short course, ed 3. Wiley-Liss, New York.

Gaur, S., Kesarwala, H., Gavai, M., et al. 1994. Clinical immunology and infectious diseases. Pediatr. Clin. North Am. 41:745.

James, K. 1990. Immunoserology of infectious diseases. Clin. Microbiol. Rev. 3:132.

Rose, N.R., deMacario, E.C., Folds, J.D., et al. 1997. Manual of clinical laboratory immunology, ed 5. ASM Press, Washington, D.C.

FIGURE 16-9 Human immunodeficiency virus type 1 (HIV-1) Western Blot. Samples are characterized as positive, indeterminate, or negative based on the bands found to be present in significant intensity. A positive blot has any two or more of the following bands: p24, gp41, and/or gp120/160. An indeterminate blot contains some bands but not the definitive ones. A negative blot has no bands present. Lane 13 shows antibodies from a control serum binding to the virus-specific proteins (p) and glycoproteins (gp) transferred onto the nitrocellulose paper. (Courtesy Cambridge Biotech, a wholly owned subsidiary of bioMérieux Vitek, Inc., Rockland, Mass.)

17 PRINCIPLES OF ANTIMICROBIAL ACTION AND RESISTANCE

Medical intervention in an infection primarily involves attempts to eradicate the infecting pathogen using substances that actively inhibit or kill it. Some of these substances are obtained and purified from other microbial organisms and are known as **antibiotics**. Others are chemically synthesized. Collectively, these natural and synthesized substances are referred to as **antimicrobial agents**. Depending on the type of organisms targeted, these substances are also known as antibacterial, antifungal, antiparasitic, or antiviral agents.

Because antimicrobial agents play a central role in the control and management of infectious diseases, some understanding of their mode of action and the mechanisms microorganisms deploy to circumvent antimicrobial activity is important, especially because diagnostic laboratories are expected to design and implement tests that measure a pathogen's response to antimicrobial activity (see Chapters 18 and 19). Much of what is discussed here regarding antimicrobial action and resistance is based on antibacterial agents, but the principles generally apply to almost all antiinfective agents. More information regarding antiparasitic, antifungal, and antiviral agents may be found in Parts 5, 6, and 7, respectively.

ANTIMICROBIAL ACTION

PRINCIPLES

Several key steps must be completed for an antimicrobial agent to successfully inhibit or kill the infecting microorganism (Figure 17-1). First, the agent must be in an active form. This is ensured through the pharmacodynamic design of the drug, which takes into account the route through which the patient will receive the agent (e.g., orally, intramuscularly, intravenously). Second, the antibiotic must also be able to achieve sufficient levels or concentrations at the site of infection so that it has a chance to exert an antibacterial effect (i.e., be in anatomic approximation with the infecting bacteria). The ability to achieve adequate levels depends on the pharmacokinetic properties of the agent. Table 17-1 provides examples of various anatomic limitations characteristic of a few commonly used antibacterial agents. Some agents, such as ampicillin and ceftriaxone, achieve therapeutically effective levels in several body sites, while others, such as nitrofurantoin and norfloxacin, are limited to the urinary tract. Therefore, knowledge of the site(s) of infection can substantially affect the selection of antimicrobial agent for therapeutic use.

The remaining steps in antimicrobial action relate to direct interactions between the antibacterial agent and the bacterial cell. When the antibiotic contacts the cell surface, adsorption results in the drug molecules maintaining contact with the cell surface. Next, because many targets for antibacterial agents are essentially intracellular, uptake of the antibiotic to some location within the bacterial cell is required. Once the antibiotic has achieved sufficient intracellular concentration, binding to a specific target occurs. This binding involves molecular interactions between the antimicrobial agent and one or more biochemical components that play an important role in the microorganism's cellular metabolism. Adequate binding of the target results in disruption of certain cellular processes leading to cessation of bacterial cell growth and, depending on the antimicrobial agent's mode of action, perhaps cell death. Antimicrobial agents that inhibit bacterial growth but generally do not kill the cell are known as **bacteriostatic** agents. Agents that usually kill the target organisms are said to be **bactericidal** (Box 17-1).

The primary goal in the development and design of antimicrobial agents is to optimize a drug's ability to efficiently achieve all steps outlined in Figure 17-1 while minimizing toxic effects on human cells and physiology. Different antibacterial agents exhibit substantial specificity in terms of their bacterial cell targets, that is, their **mode of action.** For this reason, antimicrobial agents are frequently categorized according to their mode of action.

MODE OF ACTION OF ANTIBACTERIAL AGENTS

Several potential antimicrobial targets exist within the bacterial cell, but those pathways or structures most frequently targeted include cell wall (peptidoglycan)

TABLE 17-1   ANATOMIC DISTRIBUTION OF SOME COMMON ANTIBACTERIAL AGENTS

	SERUM-BLOOD*	CEREBROSPINAL FLUID	URINE
Ampicillin	+	+	+
Ceftriaxone	+	+	+
Vancomycin	+	+/−	+
Ciprofloxacin	+	+/−	+
Gentamicin	+	−	+
Clindamycin	+	−	−
Norfloxacin	−	−	+
Nitrofurantoin	−	−	+

+, Therapeutic levels generally achievable at that site; +/−, therapeutic achievable levels moderate to poor;
−, therapeutic levels generally not achievable at that site.
*Serum-blood represents a general anatomic distribution.

BOX 17-1   BACTERIOSTATIC AND BACTERICIDAL ANTIBACTERIAL AGENTS*

Generally Bacteriostatic	Generally Bactericidal
Chloramphenicol	Aminoglycosides
Erythromycin and other macrolides	Beta-lactams
Clindamycin	Vancomycin
Sulfonamides	Quinolones
Trimethoprim	Rifampin
Tetracyclines	Metronidazole

*The bactericidal and bacteriostatic nature of an antimicrobial may vary depending on the concentration of the agent used and the bacterial species targeted.

synthesis, the cell membrane, protein synthesis, and DNA and RNA synthesis. The different modes of antimicrobial action are summarized in Table 17-2.

Inhibitors of cell wall synthesis

The bacterial cell wall, also known as the **peptidoglycan,** or **murein layer,** plays an essential role in the life of the bacterial cell. This fact, combined with the lack of a similar structure in human cells, has made the cell wall the focus of attention for the development of bactericidal agents that are relatively nontoxic for humans.

BETA-LACTAM ANTIMICROBIAL AGENTS   Beta-lactam antibiotics are those that contain the four-membered, nitrogen-containing, beta-lactam ring at the core of their structure. The names and basic structures of commonly used beta-lactams is shown in Figure 17-2. This drug class comprises the largest group of antibacterial agents, and dozens of derivatives are available for clinical use. The popularity of these agents results from their generally bactericidal action yet relative nontoxicity in humans and also because their molecular structures can be manipulated to achieve greater activity and wider therapeutic applications.

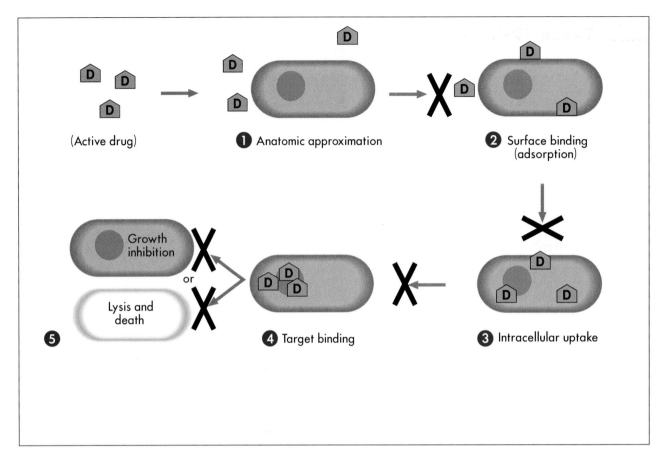

FIGURE 17-1 The basic steps required for antimicrobial activity and strategic points for bacterial circumvention or interference (marked by X) of antimicrobial action.

The beta-lactam ring is key to the mode of action of these drugs that target and inhibit cell wall synthesis by binding the enzymes involved in synthesis. Most bacterial cells cannot survive once they have lost the capacity to produce and maintain their peptidoglycan layer. The enzymes essential for this function are anchored in the cell membrane and as a group are referred to as **penicillin-binding proteins,** or **PBPs.** Bacterial species may contain between four to six different types of PBPs. The PBPs involved in cell wall cross-linking (i.e., transpeptidases) are often the most critical for survival. When beta-lactams bind to these PBPs, cell wall synthesis is essentially halted. Death results from osmotic instability caused by faulty cell wall synthesis, or the binding of the beta-lactam to PBP may trigger a series of events that leads to autolysis and death of the cell.

Because nearly all clinically relevant bacteria have cell walls, beta-lactam agents act against a broad spectrum of gram-positive and gram-negative bacteria. However, because of differences among bacteria in their PBP content, natural structural characteristics (e.g., the outer membrane present in gram-negative but not gram-positive bacteria), and their common antimicrobial resistance mechanisms, the effectiveness of beta-lactams against different types of bacteria can vary widely. Therefore, any given beta-lactam drug will have a specific group or type of bacteria against which it is considered to have the greatest activity. The types of bacteria against which a particular antimicrobial agent does and does not have activity is referred to as that drug's **spectrum of activity.** Many factors contribute to an antibiotic's spectrum of activity. Knowledge of this spectrum of activity is key to many aspects of antimicrobial use and laboratory testing.

GLYCOPEPTIDES Glycopeptides are the other major class of antibiotics that inhibit bacterial cell wall synthesis; vancomycin is the most commonly used agent in this class. Vancomycin is a large molecule and functions differently from beta-lactam antibiotics (Figure 17-3). This agent does not bind to PBPs but does bind to precursors of cell wall synthesis. The

TABLE 17-2 SUMMARY OF MECHANISMS OF ACTION FOR COMMONLY USED ANTIBACTERIAL AGENTS

ANTIMICROBIAL CLASS	MECHANISM OF ACTION	SPECTRUM OF ACTIVITY
Aminoglycosides (e.g., gentamicin, tobramycin, amikacin, netilmicin, streptomycin, kanamycin)	Inhibit protein synthesis by binding to 30S ribosomal subunit	Gram-positive and gram-negative bacteria; not anaerobic bacteria
Beta-lactams (e.g., penicillin, ampicillin, mezlocillin, piperacillin, cefazolin, cefotetan, ceftriaxone, cefotaxime, ceftazidime, aztreonam, imipenem)	Inhibit cell wall synthesis by binding enzymes involved in peptidoglycan production (i.e., penicillin-binding proteins; PBPs)	Both gram-positive and gram-negative bacteria, but spectrum may vary with the individual antibiotic
Chloramphenicol	Inhibits protein synthesis by binding 50S ribosomal subunit	Gram-positive and gram-negative bacteria
Fluoroquinolones (e.g., ciprofloxacin, ofloxacin, norfloxacin)	Inhibit DNA synthesis by binding DNA gyrases	Gram-positive and gram-negative bacteria, but spectrum may vary with individual antibiotic
Glycopeptides (e.g., vancomycin)	Inhibit cell wall synthesis by interacting with precursors and disrupting their incorporation into the growing cell wall	Gram-positive bacteria only
Macrolide-Lincosamide-Streptogramin (MLS) group (e.g., erythromycin, azithromycin, clarithromycin, clindamycin)	Inhibit protein synthesis by binding 50S ribosomal subunit	Most gram-positive bacteria; some gram-negative bacteria
Metronidazole	Exact mechanism uncertain; disruption of DNA	Gram-negative and certain genera of gram-positive (e.g., *Clostridium* spp.); anaerobic bacteria
Nitrofurantoin	Exact mechanism uncertain; may have several bacterial enzyme targets and directly damage DNA	Gram-positive and gram-negative bacteria
Polymyxins (e.g., polymyxin B and colistin)	Disruption of cell membrane	Gram-negative bacteria, poor activity against most gram-positive bacteria
Rifampin	Inhibits RNA synthesis by binding DNA-dependent, RNA polymerase	Gram-positive and certain gram-negative (e.g., *N. meningitidis*) bacteria
Sulfonamides	Interfere with folic acid pathway by binding the enzyme dihydropteroate synthase	Gram-positive and many gram-negative bacteria
Tetracycline	Inhibits protein synthesis by binding 30S ribosomal subunit	Gram-positive and gram-negative bacteria, and several intracellular bacterial pathogens (e.g., chlamydia)
Trimethoprim	Interferes with folic acid pathway by binding the enzyme dihydrofolate reductase	Gram-positive and many gram-negative bacteria

Beta-lactam class	Examples	Base molecular structure
Penicillins	Penicillin Ampicillin Piperacillin Mezlocillin	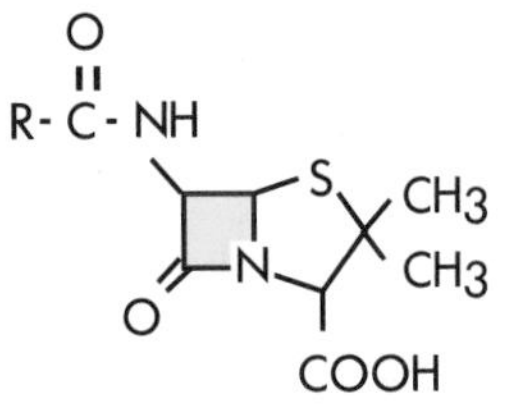
Cephalosporins	Cefazolin Cefuroxime Cefotetan Cefotaxime Ceftriaxone Ceftazidime Cefepime	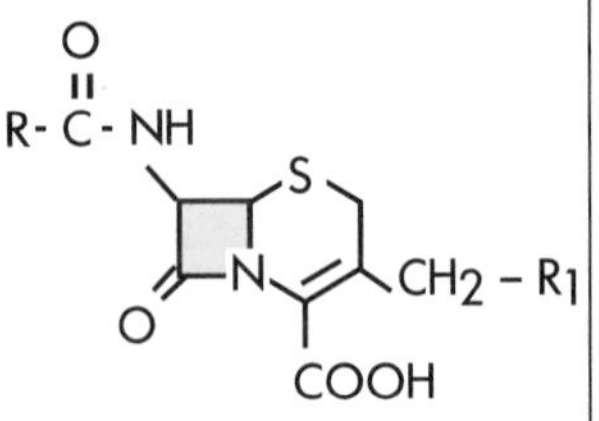
Monobactams	Aztreonam	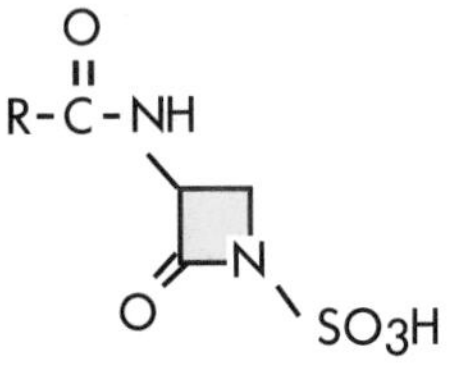
Carbapenems	Imipenem Meropenem	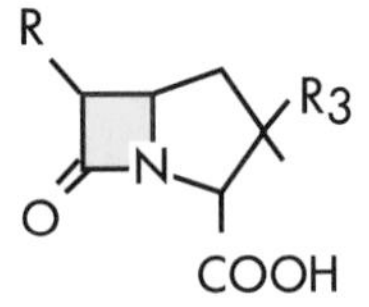

FIGURE 17-2 Basic structures and examples of commonly used beta-lactam antibiotics. The core beta-lactam ring is highlighted, in yellow, in each structure. (Modified from Salyers, A.A. and Whitt, D.D., editors. 1994. Bacterial pathogenesis: a molecular approach. ASM Press, Washington, D.C.)

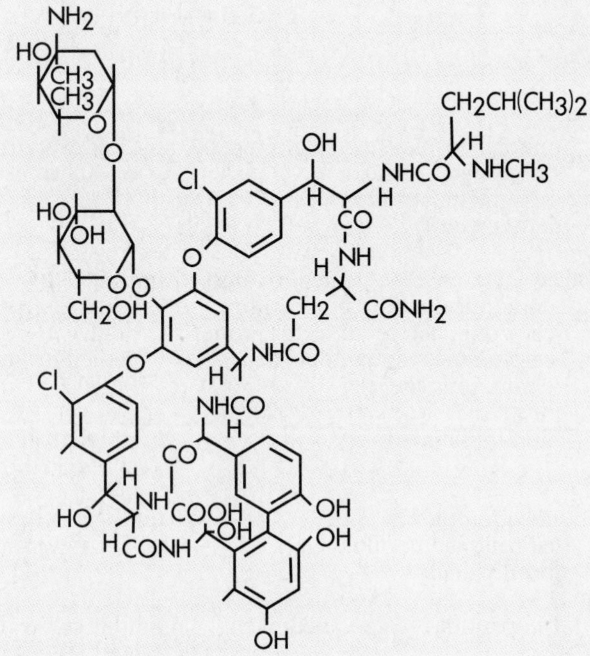

FIGURE 17-3 Structure of vancomycin, a non–beta-lactam antibiotic that inhibits cell wall synthesis. (Modified from Salyers, A.A. and Whitt, D.D., editors. 1994. Bacterial pathogenesis: a molecular approach, ASM Press, Washington, D.C.).

binding interferes with the ability of the PBP enzymes, such as transpeptidases and transglycosylases, to incorporate the precursors into the growing cell wall. With the cessation of cell wall synthesis, cell growth stops and death often follows. Because vancomycin has a different mode of action, the resistance to beta-lactam agents by gram-positive bacteria does not generally hinder vancomycin activity. However, because of its relatively large size, vancomycin cannot penetrate the outer membrane of most gram-negative bacteria to reach their cell wall precursor targets. Therefore, this agent is usually ineffective against gram-negative bacteria.

Several other cell wall-active antibiotics have been discovered and developed over the years, but toxicity to the human host has prevented their widespread clinical use. One example is bacitracin, which inhibits the recycling of certain metabolites required for maintaining peptidoglycan synthesis. Because of potential toxicity, bacitracin is usually only used as a topical antibacterial agent, and internal consumption is generally avoided.

Inhibitors of cell membrane function

Polymyxins (polymyxin B and colistin) are the agents most commonly used that disrupt bacterial cell membranes. This disruption results in leakage of macromolecules and ions essential for cell survival. Because their effectiveness varies with the molecular makeup of the bacterial cell membrane, polymyxins are not equally effective against all bacteria. Most notably they are more active against gram-negative bacteria, while activity against gram-positive bacteria tends to be poor. Furthermore, human host cells also contain membranes so that toxicity risks do exist with the use of polymyxins.

Inhibitors of protein synthesis

Several classes of antibiotics target bacterial protein synthesis and severely disrupt cellular metabolism. Antibiotic classes that act by inhibiting protein synthesis include aminoglycosides, macrolide-lincosamide-streptogramin (MLS group), chloramphenicol, and tetracyclines. Although these antibiotics are generally categorized as protein synthesis inhibitors, the specific mechanisms by which they inhibit protein synthesis differ significantly.

AMINOGLYCOSIDES Aminoglycosides inhibit bacterial protein synthesis by binding to protein receptors on the organism's 30S ribosomal subunit. This process interrupts several steps, including initial formation of the protein synthesis complex, accurate reading of the mRNA code, and disruption of the ribosomal-mRNA complex. The structure of a commonly used aminoglycoside, gentamicin, is given in Figure 17-4. Other available aminoglycosides include tobramycin, amikacin, netilmicin, streptomycin, and kanamycin. The spectrum of activity of aminoglycosides includes a wide variety of gram-negative and gram-positive bacteria. Aminoglycosides are often used in combination with cell wall-active antibiotics, such as beta-lactams or vancomycin, to achieve more rapid killing of certain bacteria. Anaerobic bacteria cannot take up these agents intracellularly so they are usually not inhibited by aminoglycosides.

MACROLIDE-LINCOSAMIDE-STREPTOGRAMIN (MLS) GROUP The most common antibiotics in the MLS group include the macrolides, such as erythromycin, azithromycin, and clarithromycin, and clindamycin (a lincosamide). Protein synthesis is inhibited by drug binding to receptors on the bacterial 50S ribosomal subunit and subsequent disruption of the growing peptide chain. Primarily because of uptake difficulties associated with gram-negative outer membranes, the macrolides and clindamycin generally are not effective against most genera of gram-negative bacteria. However, they are effective against gram-positive bacteria.

CHLORAMPHENICOL Chloramphenicol inhibits the addition of new amino acids to the growing peptide chain by binding to the 50S ribosomal subunit. This antibiotic is highly active against a wide variety of gram-negative and gram-positive bacteria; however, its use has dwindled because of serious toxicity associated with it and the development of many other effective and safer agents, mostly of the beta-lactam class.

TETRACYCLINES Tetracyclines inhibit protein synthesis by binding to the 30S ribosomal subunit so that incoming tRNA-amino acid complexes cannot bind to the ribosome, thus halting peptide chain elongation. Tetracyclines have a broad spectrum of activity that includes gram-negative bacteria, gram-positive bacteria, and several intracellular bacterial pathogens such as chlamydia, rickettsia, and rickettsia-like organisms.

Similar to chloramphenicol, the development of several effective beta-lactams has caused a marked decrease in the use of these agents.

Inhibitors of DNA and RNA synthesis

The primary antimicrobial agents that target DNA metabolism are the fluoroquinolones and metronidazole.

FLUOROQUINOLONES Fluoroquinolones, also often simply referred to as quinolones, are derivatives of nalidixic acid, an older antibacterial agent. The structures of two commonly used quinolones, ciprofloxacin and ofloxacin, are shown in Figure 17-5. These agents bind to and interfere with DNA gyrase enzymes involved in the regulation of bacterial DNA supercoiling, a process that is essential for DNA replication and transcription. The fluoroquinolones are potent bactericidal agents, and they have a broad spectrum of activity that includes gram-negative and gram-positive bacteria. However, the spectrum of activity can vary with each individual quinolone agent.

METRONIDAZOLE The exact mechanism(s) for metronidazole's antibacterial activity is uncertain, but it is believed to involve direct interactions between the activated drug and DNA that results in breakage of DNA strands. Activation of metronidazole requires reduction under conditions of low redox potential such as that found in anaerobic environments. Therefore, this agent is most potent against anaerobic bacteria, notably those that are gram-negative.

RIFAMPIN Rifampin binds to the enzyme DNA-dependent RNA polymerase and inhibits synthesis of the RNA. Because rifampin does not penetrate the outer membrane of all gram-negative bacteria effectively, activity against these organisms is usually somewhat less than against gram-positive bacteria. Also, because spontaneous mutations that result in production of rifampin-insensitive RNA polymerases occur at relatively high frequencies, resistance can develop quickly. Therefore, rifampin is usually only used in combination with other antimicrobial agents.

Inhibitors of other metabolic processes

Antimicrobial agents that target other bacterial processes other than those already discussed include sulfonamides, trimethoprim, and nitrofurantoin.

SULFONAMIDES The folic acid pathway is used by bacteria to produce precursors important for DNA synthesis (Figure 17-6). Sulfonamides target and bind to one of the enzymes, dihydropteroate synthase, and

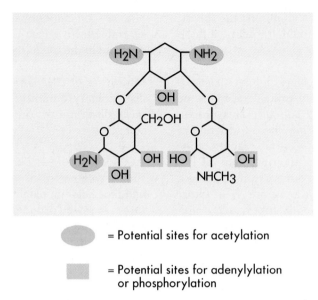

= Potential sites for acetylation

= Potential sites for adenylylation or phosphorylation

FIGURE 17-4 Structure of the commonly used aminoglycoside gentamicin. Potential sites of modification by adenylylating, phosphorylating, and acetylating enzymes produced by bacteria are highlighted. (Modified from Salyers, A.A. and Whitt, D.D., editors. 1994. Bacterial pathogenesis: a molecular approach. ASM Press, Washington, D.C.)

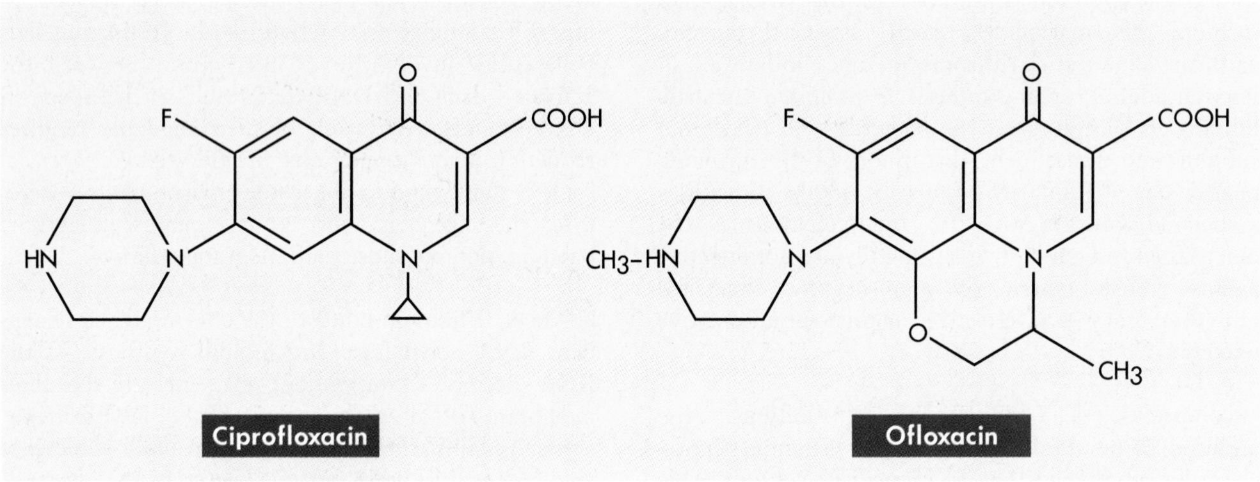

Ciprofloxacin

Ofloxacin

FIGURE 17-5 Structures of the commonly used fluoroquinolones ciprofloxacin and ofloxacin. (Modified from Katzung, B.G. 1995. Basic and clinical pharmacology, Appleton & Lange, Norwalk, Conn.)

disrupt the folic acid pathway. Several different sulfonamide derivatives are available for clinical use. These agents are active against a wide variety of gram-positive and gram-negative (except *Pseudomonas aeruginosa*) bacteria.

TRIMETHOPRIM Like sulfonamides, trimethoprim also targets the folic acid pathway. However, a different enzyme, dihydrofolate reductase, is inhibited (Figure 17-6). Trimethoprim is active against several gram-positive and gram-negative species. Frequently, tri-

methoprim is combined with a sulfonamide (usually sulfamethoxazole) into a single formulation to produce an antibacterial agent that can simultaneously attack two targets on the same folic acid metabolic pathway. This drug combination can enhance activity against various bacteria and may help avoid the emergence of bacterial resistance to a single agent.

NITROFURANTOIN The mechanism of action of nitrofurantoin is not completely known. This agent may have several targets involved in bacterial protein and

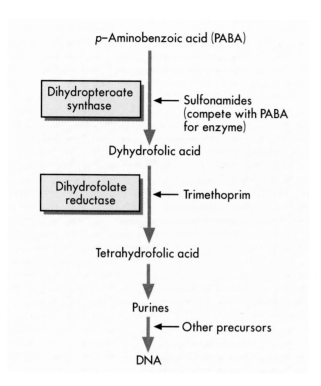

p–Aminobenzoic acid (PABA)

Dihydropteroate synthase ← Sulfonamides (compete with PABA for enzyme)

Dyhydrofolic acid

Dihydrofolate reductase ← Trimethoprim

Tetrahydrofolic acid

Purines

← Other precursors

DNA

FIGURE 17-6 Bacterial folic acid pathway indicating the target enzymes for sulfonamide and trimethoprim activity. (Modified from Katzung, B.G. 1995. Basic and clinical pharmacology. Appleton & Lange, Norwalk, Conn.)

enzyme synthesis and the drug also may directly damage DNA. Nitrofurantoin is only used to treat urinary tract infections but has good activity against most of the gram-positive and gram-negative bacteria that cause infections at that site.

MECHANISMS OF ANTIBIOTIC RESISTANCE

PRINCIPLES

Successful bacterial resistance to antimicrobial action requires interruption or disturbance of one or more of the steps essential for effective antimicrobial action (see Figure 17-1). These disturbances or resistance mechanisms can come about in various ways, but the end result is partial or complete loss of antibiotic effectiveness. Different aspects concerning antimicrobial resistance mechanisms that will be discussed include biologic vs. clinical antimicrobial resistance, environmentally mediated antimicrobial resistance, and microorganism-mediated antimicrobial resistance.

BIOLOGIC VS. CLINICAL RESISTANCE

Development of bacterial resistance to antimicrobial agents to which they were originally susceptible requires alterations in the cell's physiology or structure. **Biologic resistance** refers to changes that result

in the organism being less susceptible to a particular antimicrobial agent than has been previously observed. When antimicrobial susceptibility has been lost to such an extent that the drug is no longer effective for clinical use the organism has achieved **clinical resistance.**

Of importance, biologic resistance and clinical resistance do not necessarily coincide. In fact, because most of the laboratory methods used to detect resistance are focused on detecting clinical resistance, microorganisms may undergo substantial changes in their levels of biologic resistance without notice. For example, for some time *Streptococcus pneumoniae*, a common cause of pneumonia and meningitis, could be inhibited by penicillin at concentrations of 0.03 μg/mL or less. The clinical laboratory focused on the ability to detect strains that required 2 μg/mL of penicillin or more for inhibition. This was the threshold for resistance that was believed to be required for interference with penicillin effectiveness. However, although no isolates were being detected that required more than 2 μg/mL of penicillin for inhibition, strains were developing biologic resistance that required penicillin concentrations 10 to 50 times higher than 0.03 μg/mL for inhibition.

From a clinical laboratory and public health perspective it is important to realize that biologic development of antimicrobial resistance is an ongoing process. Our inability to reliably detect all these processes with current laboratory procedures and criteria should not be mistaken as evidence that they are not occurring.

ENVIRONMENTALLY MEDIATED ANTIMICROBIAL RESISTANCE

Antimicrobial resistance is the result of nearly inseparable interactions among the drug, the microorganism, and the environment in which they are brought together. Characteristics of the antimicrobial agents, other than the mode and spectrum of activity, are important concerns of pharmacology and are beyond the scope of this text. Microorganism characteristics are discussed in subsequent sections of this chapter. The impact that environment has on antimicrobial activity is considered here, and its importance cannot be overstated.

Environmentally mediated resistance is defined as resistance that directly results from physical or chemical characteristics of the environment that either directly alter the antimicrobial agent or alter the microorganism's normal physiologic response to the drug. Examples of environmental factors that mediate resistance include pH, anaerobic atmosphere, cation (e.g., Mg^{++} and Ca^{++}) concentrations, and thymine-thymidine content.

Several antibiotics are affected by the pH of the environment. For instance, the antibacterial activities of erythromycin and aminoglycosides diminish with decreasing pH, while the activity of tetracycline decreases with increasing pH.

Aminoglycoside-mediated shutdown of bacterial protein synthesis requires intracellular uptake across the cell membrane. Much of this aminoglycoside uptake is driven by oxidative processes, so that in the absence of oxygen, uptake, and hence activity, is substantially diminished.

Aminoglycoside activity is also affected by the concentration of cations, such as Ca^{++} and Mg^{++}, in the environment. This effect is most notable with *Pseudomonas aeruginosa*. As shown in Figure 17-1, an important step in antimicrobial activity is the adsorption of the antibiotic to the bacterial cell surface. Aminoglycoside molecules have a net positive charge and, as is true for most gram-negative bacteria, the outer membrane of *P. aeruginosa* has a net negative charge. This electrostatic attraction facilitates attachment of the drug to the surface before internalization and subsequent inhibition of protein synthesis (Figure 17-7). However, calcium and magnesium cations compete with the aminoglycosides for negatively charged binding sites on the cell surface. If the positively charged calcium and magnesium ions outcompete aminoglycoside molecules for these sites, less drug will be taken up and antimicrobial activity will be diminished. For this reason, aminoglycoside activity against *P. aeruginosa* tends to decrease as environmental cation concentrations increase.

The presence of certain metabolites or nutrients in the environment can also affect antimicrobial activity. For example, enterococci are able to use thymine and other exogenous folic acid metabolites to circumvent the activities of the sulfonamides and trimethoprim, which are folic acid pathway inhibitors (see Figure 17-6). In essence, if the environment supplies other metabolites that a microorganism can use, the activities of antibiotics that target pathways for producing those metabolites are greatly diminished, if not entirely lost. In the absence of the metabolites, full susceptibility to the antibiotics may be restored.

Information regarding environmentally mediated resistance is used to establish standardized testing methods that minimize the impact of environmental factors so that microorganism-mediated resistance (see the following discussion) is more accurately determined. Of importance, testing conditions are not established to re-create the in vivo physiology of infection but are set to optimize detection of resistance expressed by microorganisms. This is an extremely important point and a major reason why susceptibility

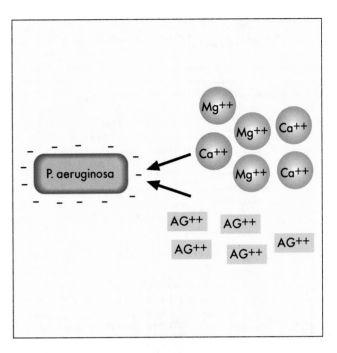

FIGURE 17-7 Cations (Mg^{++} and Ca^{++}) and aminoglycosides (AG^{++}) compete for the negatively charged binding sites on the outer membrane surface of *P. aeruginosa*. Such competition is an example of the impact that environmental factors (e.g., cation concentrations) can have on the antibacterial activity of aminoglycosides.

testing results cannot be used to predict the clinical outcome of patients undergoing antimicrobial therapy.

MICROORGANISM-MEDIATED ANTIMICROBIAL RESISTANCE

Microorganism-mediated resistance refers to antimicrobial resistance that is due to genetically encoded traits of the microorganism and is the type of resistance that in vitro susceptibility testing methods are targeted to detect (see Chapter 18). Organism-based resistance can be divided into two subcategories: intrinsic, or inherent, resistance and acquired resistance.

Intrinsic resistance

Antimicrobial resistance resulting from the normal genetic, structural, or physiologic state of a microorganism is referred to as **intrinsic resistance**. Such resistance is considered to be a natural and consistently inherited characteristic that is associated with the vast majority of strains that constitute either a particular bacterial group, genus, or species. Therefore, this is predictable resistance so that once the identity of the organism is known, so are certain aspects of its antimicrobial resistance profile. Box 17-2 lists common examples of intrinsic resistance and the underlying reason for the resistance. Intrinsic resistance profiles are useful for determining which antimicrobial agents should be included in the battery of drugs that will be tested against specific types of

| BOX 17-2 | EXAMPLES OF INTRINSIC RESISTANCE TO ANTIBACTERIAL AGENTS |

Natural Resistance	Mechanism
Anaerobic bacteria vs. aminoglycosides	Lack of oxidative metabolism to drive uptake of aminoglycosides
Gram-positive bacteria vs. aztreonam (a beta-lactam)	Lack of penicillin binding proteins (PBPs) that bind and are inhibited by this beta-lactam antibiotic
Gram-negative bacteria vs. vancomycin	Lack of uptake resulting from inability of vancomycin to penetrate outer membrane
P. aeruginosa vs. sulfonamides, trimethoprim, tetracycline, or chloramphenicol	Lack of uptake resulting from inability of antibiotics to achieve effective intracellular concentrations
Klebsiella spp. vs. ampicillin (a beta-lactam)	Production of enzymes (beta-lactamases) that destroy ampicillin before the drug can reach the PBP targets
Aerobic bacteria vs. metronidazole	Inability to anaerobically reduce drug to its active form
Enterococci vs. aminoglycosides	Lack of sufficient oxidative metabolism to drive uptake of aminoglycosides
Enterococci vs. all cephalosporins	Lack of penicillin-binding proteins (PBPs) that effectively bind and are inhibited by these beta-lactam antibiotics
Lactobacilli and *Leuconostoc* vs. vancomycin	Lack of appropriate cell wall precursor target to allow vancomycin to bind and inhibit cell wall synthesis
S. maltophilia vs. imipenem (a beta-lactam)	Production of enzymes (beta-lactamases) that destroy imipenem before the drug can reach the PBP targets

organisms. For example, referring to the information given in Box 17-2, aztreonam would not be included in antibiotic batteries tested against gram-positive cocci. Similarly, vancomycin would not be routinely tested against gram-negative bacilli. As discussed in Chapter 13, intrinsic resistance profiles are also useful markers to aid in the identification of certain bacteria or bacterial groups.

Acquired resistance

Antibiotic resistance that results from altered cellular physiology and structure caused by changes in a microorganism's usual genetic makeup is known as **acquired resistance**. Unlike intrinsic resistance, acquired resistance may be a trait associated with only some strains of a particular organism group or species, but not others. Therefore, the presence of this type of resistance in any clinical isolate is unpredictable, and this unpredictability is the primary reason why laboratory methods to detect resistance are necessary.

Because acquired resistance mechanisms are all genetically encoded, the methods for acquisition basically are those that allow for gene change or exchange. Therefore, resistance may be acquired by:

- Successful genetic mutation(s)
- Acquisition of genes from other organisms via gene transfer mechanisms
- A combination of mutational and gene transfer events

COMMON PATHWAYS FOR ANTIMICROBIAL RESISTANCE

Whether resistance is intrinsic or acquired, bacteria share similar pathways or strategies to effect resistance to antimicrobial agents. Of the pathways listed in Figure 17-8, those that involve enzymatic destruction or alteration of the antibiotic, decreased intracellular uptake or accumulation of drug, and altered antibiotic target are the most common. One or more of these pathways may be expressed by a single cell to successfully avert the action of one or more antibiotics.

Resistance to beta-lactam antibiotics

Bacterial resistance to beta-lactams may be mediated by enzymatic destruction of the antibiotics, altered antibiotic targets, or decreased intracellular uptake of the drug (Table 17-3). All three pathways play an important role in clinically relevant antibacterial resistance, but bacterial destruction of beta-lactams by producing beta-lactamases is by far the most common method of resistance. As shown in Figure 17-9, beta-lactamases open the beta-lactam ring and the altered structure of the drug prohibits subsequent effective binding to PBPs so that cell wall synthesis is able to continue.

Staphylococci are the gram-positive bacteria that most commonly produce beta-lactamase; approximately 90% or more of clinical isolates are resistant to penicillin as a result of enzyme production. Rare isolates of enterococci also produce beta-lactamase. Gram-negative bacteria, including *Enterobacteriaceae*, *P. aeruginosa*, *H. influenzae*, and *N. gonorrhoeae*, produce dozens of different beta-lactamase types that mediate resistance to one or more of the beta-lactam antibiotics.

Although the basic mechanism shown in Figure 17-9 for beta-lactamase activity is the same for all types of these enzymes, there are distinct differences. For example, beta-lactamases produced by gram-positive bacteria, such as staphylococci, are excreted into the surrounding environment where the hydrolysis of beta-lactams takes place before the drug can bind to PBPs in the cell membrane (Figure 17-10). In contrast, beta-lactamases produced by gram-negative bacteria remain intracellular in the periplasmic space where they are strategically positioned to hydrolyze beta-lactams as they transverse the outer membrane through water-filled, protein-lined porin channels (see Figure 17-10). Beta-lactamases also vary in their spectrum of substrates, that is, not all beta-lactams are susceptible to hydrolysis by every beta-lactamase. For example, staphylococcal beta-lactamase can readily hydrolyze penicillin and penicillin derivatives, such as ampicillin, mezlocillin, and piperacillin, but this enzyme can not effectively hydrolyze many cephalosporins or imipenem.

Various molecular alterations in the beta-lactam structure have been developed to protect the beta-

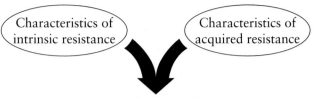

Common pathways of resistance

1. Enzymatic degradation or modification of the antimicrobial agent
2. Decreased uptake or accumulation of the antimicrobial agent
3. Altered antimicrobial target
4. Circumvention of the consequences of antimicrobial action
5. Uncoupling of antimicrobial agent-target interactions and subsequent effects on bacterial metabolism
6. Any combination of mechanisms 1 through 5

FIGURE 17-8 Overview of common pathways bacteria use to effect antimicrobial resistance.

FIGURE 17-9 Mode of beta-lactamase enzyme activity. By cleaving the beta-lactam ring the molecule can no longer bind to penicillin binding proteins (PBPs) and is no longer able to inhibit cell wall synthesis. (Modified from Salyers, A.A. and Whitt, D.D., editors. 1994. Bacterial pathogenesis: a molecular approach. ASM Press, Washington, D.C.)

lactam ring against enzymatic hydrolysis. This development has resulted in the production of more effective antibiotics in this class. For example, methicillin and closely related agents oxacillin and nafcillin are molecular derivatives of penicillin that by the nature of their structure are not susceptible to staphylococcal beta-lactamases. These agents are the mainstay of antistaphylococcal therapy. Similar strategies have been applied to develop penicillins and cephalosporins that are more resistant to the variety of beta-lactamases produced by gram-negative bacilli.

Another therapeutic strategy has been to combine two different beta-lactam drugs. One of the beta-lactams (the **beta-lactamase inhibitor**) avidly and irreversibly binds to the beta-lactamase and renders the enzyme incapable of hydrolysis, while the second

TABLE 17-3   SUMMARY OF RESISTANCE MECHANISMS FOR BETA-LACTAMS, VANCOMYCIN, AMINOGLYCOSIDES, AND FLUOROQUINOLONES

ANTIMICROBIAL CLASS	RESISTANCE PATHWAY	SPECIFIC MECHANISM	EXAMPLES
Beta-lactams (e.g., penicillin, ampicillin, mezlocillin, piperacillin, cefazolin, cefotetan, ceftriaxone, cefotaxime, ceftazidime, aztreonam, imipenem)	Enzymatic destruction	Beta-lactamase enzymes destroy beta-lactam ring so antibiotic cannot bind to PBP and interfere with cell wall synthesis (see Figure 17-9)	Staphylococcal resistance to penicillin Resistance of *Enterobacteriaceae* and *P. aeruginosa* to several penicillins, cephalosporins, and aztreonam
	Altered target	Mutational changes in original PBPs or acquisition of different PBPs that do not bind beta-lactams sufficiently to inhibit cell wall synthesis	Staphylococcal resistance to methicillin and other available beta-lactams Penicillin and cephalosporin resistance in *Streptococcus pneumoniae* and viridans streptococci
	Decreased uptake	Porin channels (through which beta-lactams cross the outer membrane to reach PBP of gram-negative bacteria) change in number or character so that beta-lactam uptake is substantially diminished	*P. aeruginosa* resistance to imipenem
Glycopeptides (e.g., vancomycin)	Altered target	Alteration in the molecular structure of cell wall precursor components decreases binding of vancomycin so that cell wall synthesis is able to continue	Enterococcal resistance to vancomycin
Aminoglycosides (e.g., gentamicin, tobramycin, amikacin, netilmicin, streptomycin, kanamycin)	Enzymatic modification	Modifying enzymes alter various sites on the aminoglycoside molecule so that the ability of drug to bind the ribosome and halt protein synthesis is greatly diminished or lost	Gram-positive and gram-negative resistance to aminoglycosides
	Decreased uptake	Porin channels (through which aminoglycosides cross the outer membrane to reach the ribosomes of gram-negative bacteria) change in number or character so that aminoglycoside uptake is substantially diminished	Aminoglycoside resistance in a variety of gram-negative bacteria

Continued

___ TABLE 17-3 SUMMARY OF RESISTANCE MECHANISMS FOR BETA-LACTAMS, ___
VANCOMYCIN, AMINOGLYCOSIDES, AND FLUOROQUINOLONES—CONT'D

ANTIMICROBIAL CLASS	RESISTANCE PATHWAY	SPECIFIC MECHANISM	EXAMPLES
Aminoglycosides—cont'd	Altered target	Mutational changes in ribosomal binding site diminishes ability of aminoglycoside to bind sufficiently and halt protein synthesis	Enterococcal resistance to streptomycin (may also be mediated by enzymatic modifications)
Quinolones (e.g., ciprofloxacin, ofloxacin, levofloxacin, norfloxacin, lomefloxacin)	Decreased uptake	Alterations in the outer membrane diminishes uptake of drug and/or activation of an "efflux" pump that removes quinolones before intracellular concentration sufficient for inhibiting DNA metabolism can be achieved	Gram-negative and staphylococcal (efflux mechanism only) resistance to various quinolones
	Altered target	Changes in the DNA gyrase subunits decrease ability of quinolones to bind this enzyme and interfere with DNA processes	Gram-negative and gram-positive resistance to various quinolones

beta-lactam, which is susceptible to beta-lactamase activity, exerts its antibacterial activity. Examples of beta-lactam/beta-lactamase inhibitor combinations include ampicillin/sulbactam, amoxicillin/clavulanic acid, and piperacillin/tazobactam.

Altered targets also play a key role in clinically relevant beta-lactam resistance (see Table 17-3). By this pathway the organism changes, or acquires from another organism, genes that encode altered cell wall-synthesizing enzymes (i.e., PBPs). These "new" PBPs continue their function even in the presence of a beta-lactam antibiotic, usually because the beta-lactam lacks sufficient affinity for the altered PBP. This is the mechanism by which staphylococci are resistant to methicillin and all other beta-lactams (e.g., cephalosporins and imipenem). Therefore, strains that exhibit this mechanism of resistance must be challenged with a non–beta-lactam agent, such as vancomycin, that acts on the cell wall. Changes in PBPs are also responsible for ampicillin resistance in *Enterococcus faecium* and in the widespread beta-lactam resistance observed in *Streptococcus pneumoniae* and viridans streptococci, organisms that to date have not been known to produce beta-lactamase.

Because gram-positive bacteria do not have outer membranes through which beta-lactams must pass before reaching their PBP targets, decreased uptake is not a pathway for beta-lactam resistance among these bacteria. However, diminished uptake can contribute significantly to beta-lactam resistance seen in gram-negative bacteria (see Figure 17-10). Changes in the number or characteristics of the outer membrane porins through which beta-lactams pass contribute to absolute resistance (e.g., *P. aeruginosa* resistance to imipenem). Additionally, porin changes combined with the presence of certain beta-lactamases in the periplasmic space may result in clinically relevant levels of resistance.

Resistance to glycopeptides

To date, acquired resistance to vancomycin has been described for enterococci but not for staphylococci or streptococci. The mechanism involves the production of altered cell wall precursors that do not bind vancomycin with sufficient avidity to allow inhibition of peptidoglycan synthesizing enzymes. The altered targets are readily incorporated into the cell wall so that synthesis progresses as usual (see Table 17-3).

Vancomycin is the only cell wall inhibiting agent for use against gram-positive organisms that are resistant to all currently available beta-lactams (e.g., methicillin-resistant staphylococci and ampicillin-

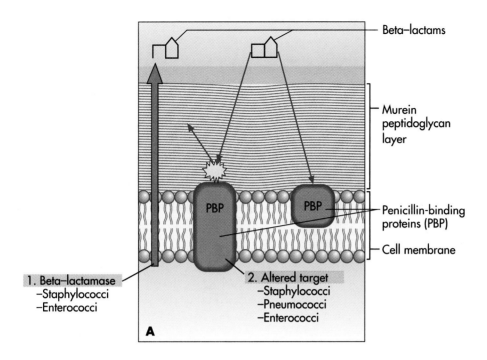

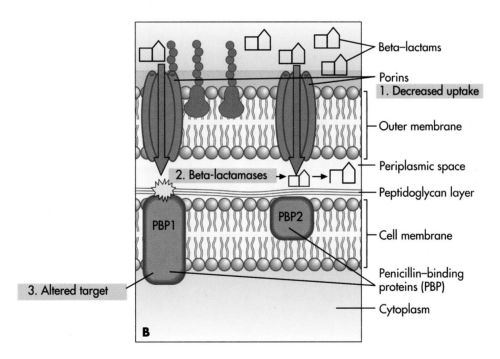

FIGURE 17-10 Diagrammatic summary of beta-lactam resistance mechanisms for gram-positive and gram-negative bacteria. Among gram-positive bacteria resistance is mediated by beta-lactamase production and altered PBP targets (**A**). In gram-negative bacteria resistance can also be mediated by decreased uptake through the outer membrane porins (**B**).

resistant enterococci). Therefore, the potential for vancomycin resistance to spread to other gram-positive genera poses a serious threat to public health. Resistance to vancomycin by enzymatic modification or destruction has not been described.

Resistance to aminoglycosides

Analogous to beta-lactam resistance, aminoglycoside resistance is accomplished by enzymatic, altered target, or decreased uptake pathways (see Table 17-3). Several different aminoglycoside-modifying enzymes are produced by gram-positive and gram-negative bacteria. Three general types of enzymes catalyze one of the following modifications of an aminoglycoside molecule (see Figure 17-4):

■ Phosphorylation of hydroxyl group(s)
■ Adenylylation of hydroxyl group(s)
■ Acetylation of amine group(s)

Once an aminoglycoside has been modified, its affinity for binding to the 30S ribosomal subunit may be sufficiently diminished or totally lost so protein synthesis is able to continue unabated.

Aminoglycosides enter the gram-negative cell by passing through outer membrane porin channels. Therefore, porin alterations may also contribute to aminoglycoside resistance among these bacteria. Although some mutations that result in altered ribosomal targets have been described, the altered target pathway is thought to be a rare means for bacteria to achieve resistance to most commonly used aminoglycosides.

Resistance to quinolones

Enzymatic degradation or alteration of quinolones has not been described as a pathway for resistance. Resistance is most frequently mediated by either decreased uptake or accumulation or by production of an altered target (see Table 17-3). Components of the gram-negative cellular envelope can limit quinolone access to the cell's interior site of DNA processing. Other bacteria, notably staphylococci, exhibit a mechanism by which the drug is "pumped"

out of the cell after having entered, thus keeping the intracellular quinolone concentration sufficiently low to allow DNA processing to continue relatively unaffected. Therefore, this "efflux" process is a pathway of diminished accumulation of drug rather than of diminished uptake.

The other quinolone resistance pathway involves mutational changes in the subunits of the DNA gyrases that are the target of quinolone activity. With a sufficient number or substantial type of changes, the gyrases no longer bind quinolones so DNA processing is able to continue.

Resistance to other antimicrobial agents

Bacterial resistance mechanisms for other antimicrobial agents involve modifications or derivations of the recurring pathway strategies of enzymatic activity, altered target, or decreased uptake. These are summarized in Box 17-3.

EMERGENCE AND DISSEMINATION OF ANTIMICROBIAL RESISTANCE

The resistance pathways that have been discussed are not necessarily new mechanisms that have recently evolved among bacteria. By definition, antibiotics originate from microorganisms. Therefore, antibiotic resistance mechanisms always have been part of the evolution of bacteria as a means of survival among antibiotic-producing competitors. However, with the introduction of antibiotics into medical practice, clinically relevant bacteria have had to adopt resistance mechanisms as part of their survival strategy. With our use and abuse of antimicrobial agents, a survival of

---

 **BOX 17-3**   **BACTERIAL RESISTANCE MECHANISMS FOR MISCELLANEOUS ANTIMICROBIAL AGENTS**

Chloramphenicol
Enzymatic modification (chloramphenicol acetyltransferase)
Decreased uptake
Tetracyclines
Diminished accumulation (efflux system)
Altered or protected ribosomal target
Enzymatic inactivation
Macrolides (i.e., erythromycin) and clindamycin
Altered ribosomal target
Diminished accumulation (efflux system)
Enzymatic modification
Sulfonamides and trimethoprim
Altered enzymatic targets (dihydropteroate synthase and dihydrofolate reductase for sulfonamides
    and trimethoprim, respectively) that no longer bind the antibiotic
Rifampin
Altered enzyme (DNA-dependent RNA polymerase) target

the fittest strategy has been used by bacteria to adapt to the pressures of antimicrobial attack (Figure 17-11).

All bacterial resistance strategies are encoded by one or more genes, and these resistance genes are readily shared between strains of the same species, between species of different genera, and even between more distantly related bacteria. When a resistance mechanism arises, either by mutation or gene transfer, in a particular bacterial strain or species, there is a propensity for this mechanism to then be passed on to other organisms using commonly described paths of genetic communication (see Figure 9-10). Therefore, resistance may spread to a wide variety of clinically important bacteria, and any single organism can acquire multiple genes and become resistant to the full spectrum of available antimicrobial agents. For example, there already exists strains of enterococci and *Pseudomonas aeruginosa* for which no effective therapy is currently available. Alternatively, multiple resistance may be mediated by a gene that encodes for a single very potent resistance mechanism. One such example is the *mecA* gene that encodes staphylococcal resistance to methicillin and to all other beta-lactams currently available for use against these organisms, leaving vancomycin as the single available and effective cell wall-inhibiting agent.

In summary, antibiotic overuse and abuse, coupled with the formidable repertoire bacteria have for thwarting antimicrobial activity, and their ability to genetically share these strategies, drives the ongoing process of resistance emergence and dissemination (see Figure 17-11). This has been manifested by the emergence of new genes of unknown origin (e.g., methicillin-resistant staphylococci and vancomycin-resistant enterococci), the movement of old genes into new bacterial hosts (e.g., penicillin-resistant *N. gonorrhoeae*), mutations in familiar resistance genes that result in greater potency (e.g., beta-lactamase–mediated resistance to cephalosporins in *E. coli*), and the emergence of new pathogens whose most evident virulence factor is intrinsic or natural resistance to many of the antimicrobial agents used in the hospital setting (e.g., *Stenotrophomonas maltophilia*).

The continuous and ongoing nature of resistance emergence and dissemination dictates that reliable laboratory procedures be used to detect resistance as an aid to managing the infected patient and as a means for monitoring changing resistance trends among clinically relevant bacteria.

Bibliography

Bryan, L.E., editor. 1989. Microbial resistance to drugs. Handb. Exp. Pharm. 91. Springer-Verlag, Berlin.

Courvalin, P. 1994. Transfer of antibiotic resistance genes between gram-positive and gram-negative bacteria. Antimicrob. Agents Chemother. 38:1447.

Davies, J. 1994. Inactivation of antibiotics and the dissemination of resistance genes. Science. 264:375.

Lorian, V., editor. 1991. Antibiotics in laboratory medicine. Williams and Wilkins, Baltimore.

Mandell, G.L., Bennett, J.E., and Dolin, R., editors. 1995. Principles and practice of infectious diseases, ed 4. Churchill Livingstone, New York.

Neu, H.C. 1992. The crisis in antibiotic resistance. Science. 257:1064.

Nikaido, H. 1994. Prevention of drug access to bacterial targets: permeability barriers and active efflux. Science 264:382.

Quintiliani, R. and Courvalin, P. 1995. Mechanisms of resistance to antimicrobial agents. In Murray, P.R., Baron, E.J., Pfaller, M.A., et. al., editors. Manual of clinical microbiology, ed 6. American Society for Microbiology, Washington, D.C.

Yao, J.D.C. and Moellering, R.C. 1995. Antibacterial agents. In Murray, P.R., Baron, E.J., Pfaller, M.A., et. al., editors. Manual of clinical microbiology, ed 6. American Society for Microbiology, Washington, D.C.

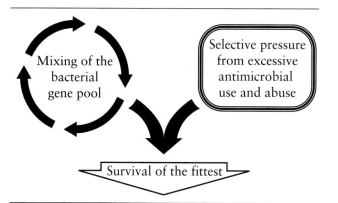

Mixing of the bacterial gene pool

Selective pressure from excessive antimicrobial use and abuse

Survival of the fittest

1. Emergence of "new" genes (e.g., methicillin-resistant staphylococci, vancomycin-resistant enterococci)

2. Spread of "old" genes to new hosts (e.g., penicillin-resistant *Neisseria gonorrhoeae*)

3. Mutations of "old" genes resulting in more potent resistance (e.g., beta-lactamase–mediated resistance to advanced cephalosporins in *Escherichia coli* and *Klebsiella* spp.)

4. Emergence of intrinsically resistant opportunistic bacteria (e.g., *Stenotrophomonas maltophilia*)

FIGURE 17-11 Factors contributing to the emergence and dissemination of antimicrobial resistance among bacteria.

18 | LABORATORY METHODS FOR DETECTION OF ANTIBACTERIAL RESISTANCE

Most clinically important bacteria are capable of acquiring and expressing resistance to antimicrobial agents commonly used to treat infections. Therefore, once an organism is isolated in the laboratory, characterization frequently includes tests to detect antimicrobial resistance. The procedures used to detect antimicrobial resistance are referred to as **antimicrobial susceptibility testing** methods. The methods used for testing aerobic and facultatively anaerobic bacteria will be the focus of this chapter. Strategies for when and how these methods should be applied are presented in Chapter 19. Antimicrobial susceptibility testing is also performed with anaerobic bacteria and mycobacteria; these procedures are covered in Chapters 58 and 60, respectively.

GOAL AND LIMITATIONS

The primary goal of antimicrobial susceptibility testing is to determine whether the bacterial etiology of concern is capable of expressing resistance to the antimicrobial agents that are potential choices as therapeutic agents for managing the infection. Because intrinsic resistance is usually well known for most organisms, testing to establish this resistance usually is not necessary and organism identification is sufficient. In essence, antimicrobial susceptibility tests are assays designed to determine the extent of an organism's acquired resistance.

STANDARDIZATION

For laboratory tests to accurately determine organism-based resistances the potential influence of environmental factors on antibiotic activity must be minimized (see Chapter 17). This is not to suggest that environmental resistance never plays a clinically relevant role, but the major focus of the in vitro tests is to measure an organism's expression of resistance. To control the impact of environmental factors, the conditions for susceptibility testing are extensively standardized. Standardization serves three important purposes:

- To optimize bacterial growth conditions so that inhibition of growth can be attributed to the antimicrobial agent against which the organism is being tested and not to limitations of nutrients, temperature, or atmosphere
- To optimize conditions for maintaining antimicrobial integrity and activity so that failure to inhibit bacterial growth can be attributed to organism-associated resistance mechanisms rather than to environmental drug inactivation
- To maintain reproducibility and consistency of results so that the same organism will produce the same resistance profile, regardless of which microbiology laboratory performs the test

Standard conditions for antimicrobial susceptibility testing methods have been established based on numerous laboratory investigations. Guidelines and recommendations for their use are published by the National Committee for Clinical Laboratory Standards (NCCLS), Subcommittee on Antimicrobial Susceptibility Testing. The NCCLS documents that describe various methods for antimicrobial susceptibility testing are continuously updated and may be obtained by contacting NCCLS, 940 W. Valley Road, Suite 1400, Wayne, Pa. 19087.

The components of antimicrobial susceptibility testing that are standardized and controlled include:

- Bacterial inoculum size
- Growth medium (most frequently a Mueller-Hinton base is used)
    pH
    cation concentration
    blood and serum supplements
    thymidine content
- Incubation atmosphere
- Incubation temperature
- Incubation duration
- Antimicrobial concentrations tested

LIMITATIONS OF STANDARDIZATION

Although standardization of in vitro conditions is essential, the use of standard conditions imparts some limitations. Most notably, the laboratory test conditions in no way mimic the in vivo environment at the infection site(s) where the antimicrobial agent and bacteria will actually interact. Factors, such as bacterial inoculum size, pH, cation concentration, and oxygen tension (i.e., atmosphere), can differ substantially depending on the site of infection. Because of the lack of correlation between in vitro test conditions and the in vivo setting, antimicrobial susceptibility testing results cannot, and should not, be used as predictors of therapeutic outcome for the use of particular antimicrobial agents. Additionally, several other important factors that play key roles in patient outcome are not taken into account by susceptibility testing. Some of these factors include:

- Antibiotic diffusion in tissues and host cells
- Serum protein binding of antimicrobial agents
- Drug interactions and interference
- Status of patient defense and immune systems
- Multiple simultaneous illnesses
- Virulence and pathogenicity of infecting bacterium
- Site and severity of infection

Despite these limitations, it is known that antimicrobial resistance substantially affects the morbidity and mortality of infected patients and the early and accurate recognition of resistant bacteria significantly aids in the optimal management of patients. Thus, the goal of in vitro susceptibility testing to detect resistance provides valuable data that is used in conjunction with other diagnostic information to optimize therapy. Additionally, as discussed in Chapter 19, in vitro susceptibility testing provides the data to track resistance trends among clinically relevant bacteria.

TESTING METHODS

PRINCIPLES

Three general methods are available to detect and evaluate antimicrobial resistance:

- Methods that directly measure the activity of one or more antimicrobial agents against a bacterial isolate
- Methods that directly detect the presence of a specific resistance mechanism in a bacterial isolate
- Special methods that measure complex antimicrobial-organism interactions

Which of these methods is used depends on factors that involve clinical need, accuracy, and convenience. Given the complexities of antimicrobial resistance patterns, a laboratory may commonly use methods from one or more of these categories.

METHODS THAT DIRECTLY MEASURE ANTIMICROBIAL ACTIVITY

Methods that directly measure antimicrobial activity involve bringing the antimicrobial agents of interest and the infecting bacterium together in the same in vitro environment to determine the impact of the drug's presence on bacterial growth or viability. The level of impact on bacterial growth is measured and interpreted so that the organism's resistance, or diminished susceptibility, to each agent tested can be reported to the clinician. Direct measures of antimicrobial activity is accomplished using:

- Conventional susceptibility testing methods such as broth dilution, agar dilution, and disk diffusion
- Commercial susceptibility testing systems
- Special screens and indicator tests

Conventional testing methods: general considerations
Some general considerations apply to all three methods regarding inoculum preparation and selection of antimicrobial agents for testing.

INOCULUM PREPARATION Properly prepared inocula are key to any of the antimicrobial susceptibility testing methods. Inconsistencies in inoculum preparation often lead to inconsistencies and inaccuracies in susceptibility test results. The two important requirements for appropriate inoculum preparation include use of a pure culture and a standardized inoculum.

Interpretation of results obtained with a mixed inoculum is not reliable, and failure to use a pure culture can substantially delay reporting of results. Pure inocula are obtained by selecting four to five colonies of the same morphology, inoculating them to a broth medium, and allowing the culture to achieve good active growth (i.e., mid-logarithmic phase), as indicated by observable turbidity in the broth. For most organisms this requires 3 to 5 hours of incubation. Alternatively, four to five colonies 16 to 24 hours of age may be selected from an agar plate and suspended in broth or 0.85% saline solution to achieve a turbid suspension.

Use of a standard inoculum size is as important as culture purity and is accomplished by comparison of the turbidity of the organism suspension with a turbidity standard. McFarland turbidity standards, prepared by mixing various volumes of 1% sulfuric acid and 1.175% barium chloride to obtain solutions with specific optical densities, are most commonly used. The 0.5 McFarland standard, which is commercially available, provides an optical density comparable to the density of a bacterial suspension of 1.5 ×

10^8 colony forming units (CFU)/mL. Pure cultures are grown or are directly prepared from agar plates to match the turbidity of the 0.5 McFarland standard (Figure 18-1). Matching turbidity using the unaided eye is facilitated by holding the bacterial suspension and McFarland tubes side by side and viewing them against a black-lined background. Alternatively, any one of various commercially available instruments that measure turbidity can be used to standardize inocula to match the 0.5 McFarland standard. If the bacterial suspension initially does not match the standard's turbidity, the suspension may be diluted, or supplemented with more organisms, as needed.

SELECTION OF ANTIMICROBIAL AGENTS FOR TESTING The antimicrobial agents that are chosen for testing against a particular bacterial isolate is referred to as the **antimicrobial battery.** A laboratory may use different testing batteries, but the content and application of each battery is based on various criteria. Although the criteria given in Box 18-1 affect decisions regarding battery content, the final decision is one that cannot and should not be made by the laboratory alone. Input from the medical staff (particularly infectious diseases specialists) and pharmacy is imperative.

The NCCLS publishes up-to-date tables that list potential antimicrobial agents to include in batteries for testing against particular organisms or organism groups. Of particular interest is Table 1, titled, "Suggested Groupings of U.S. FDA-Approved Antimicrobial Agents That Should be Considered for Routine Testing and Reporting on *Nonfastidious* Organisms by Clinical Microbiology Laboratories," and Table 1A, titled, "Suggested Groupings of U.S. FDA-Approved Antimicrobial Agents That Should

be Considered for Routine Testing and Reporting on *Fastidious* Organisms by Clinical Microbiology Laboratories." Because revisions can occur annually, up-to-date tables should be consulted when changes in testing batteries are being contemplated.

Further considerations about antibiotics that may be tested against particular organism groups are presented in Chapter 19 (see Table 19-2) and in appropriate parts of the chapters in Part 4. Most commonly, individual testing batteries are considered for each of the following organism groups:

- *Enterobacteriaceae*
- *Pseudomonas aeruginosa*
- *Staphylococcus* spp.
- *Enterococcus* spp.
- *Streptococcus* spp. (not including *S. pneumoniae*)
- *Streptococcus pneumoniae*
- *Haemophilus influenzae*
- *Neisseria gonorrhoeae*

Just as the antimicrobials selected for testing against each organism group depends on numerous factors, various criteria may be applied to determine under what circumstances results obtained with certain agents should be reported. Most frequently these decisions involve the suppression of results obtained with toxic or expensive antimicrobial agents when the organism is susceptible to other less toxic or less expensive drugs. However, because laboratory personnel usually are not familiar with a patient's full clinical picture, the application of **results suppression,** or **reporting cascades,** as they are also called, can be complex. Therefore, any decisions regarding selective reporting should be made in close conjunction with the medical staff and pharmacy.

| BOX 18-1 | **CRITERIA FOR ANTIMICROBIAL BATTERY CONTENT AND USE** |

ORGANISM IDENTIFICATION OR GROUP: Antimicrobials to which the organism is intrinsically resistant are routinely excluded from the test battery (e.g., vancomycin vs. gram-negative bacilli). Similarly, certain antimicrobials were specifically developed for use against particular organisms but not others (e.g., ceftazidime for use against *P. aeruginosa* but not against *S. aureus*), and should only be included in the appropriate battery.

ACQUIRED RESISTANCE PATTERNS COMMON TO LOCAL MICROBIAL FLORA: If resistance to a particular agent is common, the utility of the agent may be sufficiently limited so that routine testing is not warranted and only more potent antimicrobials are included in the test battery. Conversely, more potent agents may not need to be in the test battery if susceptibility to less potent agents is common.

ANTIMICROBIAL SUSCEPTIBILITY TESTING METHOD USED: Depending on the testing method, some agents do not reliably detect resistance and should not be included in the battery. For example, *S. pneumoniae* resistance to cefotaxime can be detected by dilution methods and this drug should be included in the dilution test battery. However, cefotaxime resistance cannot be detected by disk diffusion and so should not be part of a disk test battery.

SITE OF INFECTION: Antimicrobial agents, such as nitrofurantoin, that only achieve effective levels in the urinary tract should not be included in batteries tested against bacterial isolates from other body sites (i.e., must be able to achieve anatomic approximation).

AVAILABILITY OF ANTIMICROBIAL AGENTS IN FORMULARY: Antimicrobial agent test batteries are selected for their ability to detect bacterial resistance to agents that are used by the medical staff that the laboratory services.

Conventional testing methods: broth dilution

Broth dilution testing involves challenging the organism of interest with antimicrobial agents in a broth environment. Each antimicrobial agent is tested using a range of concentrations commonly expressed as µg of active drug/mL of broth (i.e., µg/mL). The concentration range tested for a particular drug depends on various criteria, including the concentration that is safely achievable in a patient's serum. Therefore, the concentration range tested will often vary from one drug to the next, depending on the pharmacologic properties of each. Additionally, the concentration range tested may be based on the level of drug that is needed to most reliably detect a particular underlying resistance mechanism. In this case, the test concentration for a drug may vary depending on the organism and its associated resistances that the test is attempting to detect. For example, to detect clinically significant resistance to cefotaxime in *S. pneumoniae* the dilution scheme need only go as high as 2 µg/mL, but to detect cefotaxime resistance in *E. coli* the scheme must go up to 16 µg/mL or beyond.

Typically, the range of concentrations tested for each antibiotic are a series of doubling dilutions (e.g., 16, 8, 4, 2, 1, 0.5, 0.25 µg/mL); the lowest antimicrobial concentration that completely inhibits visible bacterial growth is recorded as the **minimal inhibitory concentration (MIC)**.

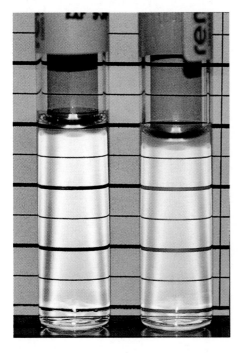

FIGURE 18-1 Bacterial suspension prepared to match the turbidity of the 0.5 McFarland turbidity standard. Matching this turbidity provides a bacterial inoculum concentration of 1 to 2 × 10⁸ CFU/mL.

PROCEDURES The key features of broth dilution testing procedures are given in Table 18-1. Because changes in procedural recommendations occur, the NCCLS-M7 series titled, "Methods for Dilution Antimicrobial Susceptibility Tests for Bacteria that Grow Aerobically" should be consulted annually.

Medium and Antimicrobial Agents. For any in vitro susceptibility testing method, it is necessary to alter certain conditions when testing particular types of organisms in order to optimize growth of some fastidious bacteria and facilitate expression of certain types of bacterial resistance. For example, Mueller-Hinton is the standard medium used for most broth dilution testing, and conditions of this medium (e.g., pH, cation concentration, thymidine content) are well controlled by commercial media manufacturers. However, media supplements or different media altogether are required to obtain good growth and reliable susceptibility profiles with relatively fastidious bacteria such as *S. pneumoniae* and *Haemophilus influenzae*. Also, although staphylococci are not fastidious organisms, media supplementation with NaCl is needed to enhance expression and detection of methicillin-resistant isolates (Table 18-1).

Broth dilution testing is divided into two categories: microdilution and macrodilution. The principles of the tests are the same; the only difference is the volume of broth in which the test is performed. For microdilution testing the total broth volume is 0.05 to 0.1 mL, and for macrodilution testing the broth volumes are usually 1.0 mL or greater. Because most susceptibility test batteries require testing several antibiotics at several different concentrations, the smaller volume used in microdilution allows this to be conveniently accomplished in a single microtitre tray format (Figure 18-2).

Use of test tubes as required by the macrodilution method becomes substantially cumbersome and labor intensive, especially because most laboratories must test several bacterial isolates daily. For this reason, macrodilution is rarely used in most clinical laboratories and subsequent comments regarding broth dilution will focus on the microdilution approach.

A key component to broth dilution testing is proper preparation and dilution of the antimicrobial agents that will be incorporated into the broth medium. Most laboratories that perform broth microdilution use commercially supplied microdilution panels in which the broth is already supplemented with appropriate antimicrobial concentrations. Therefore, antimicrobial preparation and dilution is not commonly done in most clinical laboratories, but details of this procedure are outlined in the NCCLS-M7 documents. In most instances, each antimicrobial agent is presented in the

TABLE 18-1 SUMMARY OF BROTH DILUTION SUSCEPTIBILITY TESTING CONDITIONS

ORGANISM GROUPS	TEST MEDIUM	INOCULUM SIZE (CFU/mL)	INCUBATION CONDITIONS	INCUBATION DURATION
Enterobacteriaceae	Mueller-Hinton	5×10^5	35° C; air	16-20 hrs
P. aeruginosa				
Enterococci				
Staphylococci				
To detect methicillin-resistant staphylococci	Mueller-Hinton plus 2% NaCl		30-35° C; air	24 hrs
S. pneumoniae and other streptococci	Mueller-Hinton plus 2%-5% lysed horse blood	5×10^5	35° C; air	20-24 hrs
H. influenzae	Haemophilus Test Medium	5×10^5	35° C; air	20-24 hrs
N. meningitidis	Mueller-Hinton plus 2%-5% lysed horse blood	5×10^5	35° C; 5%-7% CO_2	24 hrs

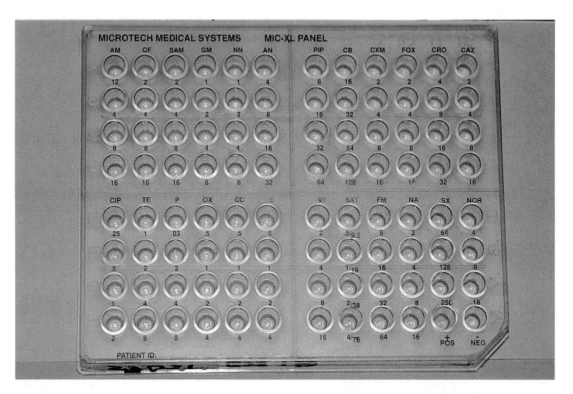

FIGURE 18-2 Microtitre tray for broth microdilution testing. Doubling dilutions of each antimicrobial agent in test broth occupies one vertical row of wells.

microtitre trays as a series of doubling twofold dilutions. To ensure against loss of antibiotic potency, the antibiotic microdilution panels are stored at $-20°$ C or lower, if possible, and are thawed just before inoculation and use. Once thawed the panels should never be refrozen because doing so can result in substantial loss of antimicrobial potency. Alternatively, the antimicrobial agents my be lyophilized or freeze-dried with the medium in each well and upon inoculation with the bacterial suspension the medium and drug are simultaneously reconstituted to the appropriate concentration.

Inoculation and Incubation. Standardized bacterial suspensions that match the turbidity of the 0.5 McFarland standard (i.e., 1.5×10^8 CFU/mL) usually serve as the starting point for dilutions that ultimately allow the final standard bacterial concentration of 5×10^5 CFU/mL in each microtitre well to be achieved. Of importance the standard inoculum should be prepared from a fresh, overnight, pure culture of the test organism. Inoculation of the microdilution panel is readily accomplished using manual or automated multiprong inoculators that are calibrated to deliver the precise volume of inoculum to each well in the panel simultaneously (see Figure 18-2).

Inoculated trays are incubated under environmental conditions that optimize bacterial growth but do not interfere with antimicrobial activity (i.e., avoiding environmentally mediated results). For the most commonly tested bacteria (e.g., *Enterobacteriaceae*, *P. aeruginosa*, staphylococci, and enterococci) the environment is air at $35°$ C (see Table 18-1). Exceptions exist for the sake of testing more fastidious bacteria (e.g., *N. meningitidis* optimally requires 5% to 7% CO_2) and for enhancing the detection of certain resistance mechanisms (e.g., for detecting methicillin resistance among staphylococci, incubation as low as $30°$ C may be used). Similarly, incubation durations for some organisms may need to be extended beyond the usual 16 to 20 hours (see Table 18-1). However, prolonged incubation times beyond recommended limits should be avoided because antimicrobial deterioration may result in false resistance interpretations. This is a primary factor that limits our ability to perform accurate testing with some slow-growing bacteria.

Reading and Interpretation of Results. Following incubation, the microdilution trays are examined for bacterial growth. Each tray should include a growth control that does not contain antimicrobial agent and a sterility control that was not inoculated. Once growth in the growth control and no growth in the sterility control wells have been confirmed, the growth profiles for each antimicrobial dilution can be established and

the MIC determined. Detecting the presence of growth in microdilution wells is often augmented through the use of light boxes and reflecting mirrors. When a panel is placed in these devices, the presence of bacterial growth, which may be manifested as light to heavy turbidity or a button of growth on the well bottom, is more reliably visualized (Figure 18-3).

When the dilution series of each antibiotic is inspected, the microdilution well containing the lowest drug concentration that completely inhibited visible bacterial growth is recorded as the MIC. Once the MICs for the antimicrobials in the test battery for a particular organism have been recorded they are usually translated into **interpretive categories** of **susceptible, intermediate,** or **resistant** (Box 18-2). The interpretive criteria for these categories are based on extensive studies that correlate MIC with serum achievable levels for each antimicrobial agent, particular resistance mechanisms, and successful therapeutic outcomes. The interpretive criteria for an array of antimicrobial agents are published in the NCCLS-M7 series document titled, "Methods for Dilution Antimicrobial Susceptibility Tests for Bacteria that Grow Aerobically (M100 supplements)." For example, using these standards, an isolate of *P. aeruginosa* with an imipenem MIC of ≤ 4 µg/mL would be classified as being susceptible, one with an MIC of 8 µg/mL would be considered intermediate, and one with an MIC of ≥ 16 µg/mL would be classified as resistant to imipenem.

After the MICs are determined and their respective and appropriate interpretive categories assigned, the laboratory may report either the MIC, the category, or both. Because the MIC alone will not provide most physicians with a meaningful interpretation of data, either the category result with or without the MIC is usually reported.

In some settings, the full range of antimicrobial dilutions are not used; only the concentrations that separate the categories of susceptible, intermediate, and resistant are used. The specific concentrations that separate or define the different categories are known as **breakpoints,** and panels that only contain these antimicrobial concentrations are referred to as **breakpoint panels.** In this case, only category results are produced; precise MICs are not available because the full range of dilutions is not tested.

Advantages and Disadvantages. Broth dilution testing allows the option of providing both quantitative (i.e., MIC) and qualitative (i.e., category interpretation) results. Whether this is an advantage or not is controversial. On one hand, an MIC can be helpful in establishing the level of resistance of a particular bacterial strain and can substantially affect the decision to use certain antimicrobial agents. For example, the penicillin MIC for *S. pneumoniae* may determine whether penicillin or alternative agents will be used to treat a case of meningitis. On the other hand, for most antimicrobial susceptibility testing methods, a category report is sufficient so that once this determination is made the actual MIC data is superfluous. This is one reason why other methods (e.g., disk dif-

fusion) that focus only on producing interpretive categories have been maintained in the clinical microbiology community.

Conventional testing methods: agar dilution

With agar dilution the antimicrobial concentrations and organisms to be tested are brought together on an agar-based medium rather than in a broth. Each dou-

BOX 18-2
DEFINITIONS OF SUSCEPTIBILITY TESTING INTERPRETIVE CATEGORIES*

SUSCEPTIBLE: This category indicates that the antimicrobial agent in question may be an appropriate choice for treating the infection caused by the bacterial isolate tested. Bacterial resistance is absent or at a clinically insignificant level.

INTERMEDIATE: This category is used to indicate a number of possibilities, including:
- The potential utility of the antimicrobial agent in body sites where it may be concentrated (e.g., the urinary tract) or if high concentrations of the drug are used
- The antimicrobial agent may still be effective against the tested isolate but possibly less so than against a susceptible isolate
- As an interpretive safety margin to prevent relatively small changes in test results from leading to major swings in interpretive category (e.g., resistant to susceptible or vice versa)

RESISTANT: This category indicates that the antimicrobial agent in question may not be an appropriate choice for treating the infection caused by the bacterial isolate tested, either because the organism is not inhibited by serum-achievable levels of the drug or because the test result highly correlates with a resistance mechanism that renders treatment success doubtful.

*Although definitions are adapted from NCCLS M7-A3 titled, "Methods for Dilution Antimicrobial Susceptibility Tests for Bacteria that Grow Aerobically," they are commonly applied to results obtained by various susceptibility testing methods.

FIGURE 18-3 Bacterial growth profiles in a broth microdilution tray. The wells containing the lowest concentration of an antibiotic that completely inhibits visible growth (*arrow*) is recorded, in μg/mL, as the minimal inhibitory concentration (MIC).

bling dilution of an antimicrobial agent is incorporated into a single agar plate so that testing a series of six dilutions of one drug would require the use of six plates, plus one positive growth control plate without antibiotic. The standard conditions for agar dilution are given in Table 18-2 and, as shown in Figure 18-4, the surface of each plate is inoculated with 1×10^4 CFU. By this method, one or more bacterial isolates is tested per plate. After incubation, the plates are examined for growth and the MIC is the lowest concentration of an antimicrobial agent in agar that completely inhibits visible growth. The same MIC breakpoints and interpretive categories used for broth dilution are applied for interpretation of agar dilution methods. Similarly, test results may be reported as MICs only, as category only, or as both.

Preparing agar dilution plates (see NCCLS-M7 series document titled, "Methods for Dilution Antimicrobial Susceptibility Tests for Bacteria that Grow Aerobically") is sufficiently labor intensive to preclude the use of this method in most clinical laboratories in which multiple antimicrobial agents must be tested, even though several isolates may be tested per plate. As with broth dilution, the standard medium is Mueller-Hinton, but supplements and substitutions are made as needed to facilitate growth of more fastidious organisms (Table 18-3). In fact, one advantage of this method is that it provides a means for determining MICs for *N. gonorrhoeae*, which does not grow sufficiently well to be tested by broth dilution methods.

Conventional testing methods: disk diffusion

As more antimicrobial agents became available for treating bacterial infections, the limitations of the macrobroth dilution method became apparent. Before microdilution technology was available, a more practical and convenient method for testing multiple antimicrobial agents against bacterial strains was needed. Out of this need the disk diffusion test was developed, spawned by the landmark study by Bauer, Kirby, Sherris, and Turck in 1966. These investigators standardized and correlated the use of antibiotic-impregnated filter paper disks (i.e., antibiotic disks) with MICs using many bacterial strains. Using the disk diffusion susceptibility test, antimicrobial resistance is detected by challenging bacterial isolates with antibiotic disks that are placed on the surface of an agar plate that has been seeded with a lawn of bacteria (Figure 18-5).

When disks containing a known concentration of antimicrobial agent are placed on the surface of a freshly inoculated plate, the agent immediately begins

TABLE 18-2 SUMMARY OF AGAR DILUTION SUSCEPTIBILITY TESTING CONDITIONS

ORGANISM GROUPS	TEST MEDIUM	INOCULUM SIZE (CFU/spot)	INCUBATION CONDITIONS	INCUBATION DURATION
Enterobacteriaceae	Mueller-Hinton	1×10^4	35° C; air	16-20 hrs
P. aeruginosa				
Enterococci				
Staphylococci				
To detect methicillin-resistant staphylococci	Mueller-Hinton plus 2% NaCl		30°-35° C; air	24 hrs
N. meningitidis	Mueller-Hinton plus 5% sheep blood	1×10^4	35° C; 5%-7% CO_2	24 hrs
S. pneumoniae	Agar dilution not recommended method for testing this organism			
Other streptococci	Mueller-Hinton plus 5% sheep blood	1×10^4	35° C; air, CO_2 may be for some isolates	20-24 hrs
N. gonorrhoeae	GC agar plus supplements	1×10^4	35° C; 5%-7% CO_2	24 hrs

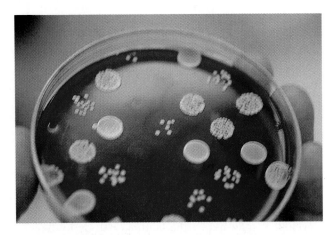

FIGURE 18-4 Growth pattern on an agar dilution plate. Each plate contains a single concentration of antibiotic, and growth is indicated by a spot on the agar surface. No spot is seen for isolates inhibited by the concentration of antibiotic incorporated into the agar of that particular plate.

to diffuse and establish a concentration gradient around the paper disk. The highest concentration is closest to the disk. Upon incubation the bacteria grow on the surface of the plate except where the antibiotic concentration in the gradient around each disk is sufficiently high to inhibit growth. Following incubation the diameter of the zone of inhibition around each disk is measured in millimeters (see Figure 18-5).

To establish reference inhibitory zone-size breakpoints that laboratories use to define susceptible, intermediate, and resistant categories for each antimicrobial agent-bacterial species combination, hundreds of strains are tested. The inhibition zone sizes obtained are then correlated with MICs obtained by broth or agar dilution, and a regression analysis (i.e., line of best fit) is performed by plotting the zone size in millimeters against the MIC (Figure 18-6). As the MIC of the bacterial strains tested increase (i.e., more resistant bacterial strains), the corresponding inhibition zone sizes (i.e., diameters) decrease. Using Figure 18-6 to illustrate, horizontal lines are drawn from the MIC resistant breakpoint and the susceptible MIC breakpoint, 8 μg/mL and 2 μg/mL, respectively. Where the horizontal lines intersect the regression line, vertical lines are drawn to delineate the corresponding inhibitory zone size breakpoints (in millimeters). Using this approach, zone size interpretive criteria have been established for most of the commonly tested antimicrobial agents and are published in the NCCLS-M2 series titled, "Performance Standards for Antimicrobial Disk Susceptibility Tests (M100 supplements)."

PROCEDURES The key features of disk diffusion testing procedures are summarized in Table 18-3 with

more details and updates available through the NCCLS.

Medium and Antimicrobial Agents. Mueller-Hinton is the standard agar base medium for testing most bacterial organisms, with certain supplements and substitutions again required for testing more fastidious organisms. In addition to factors such as pH and cation content, the depth of the agar medium can also affect test accuracy and must be carefully controlled. Because antimicrobial agents diffuse in all directions from the surface of the agar plate, the thickness of the agar affects the antimicrobial drug concentration gradient. If the agar is too thick, zone sizes would be smaller and if too thin the inhibition zones would be larger. For many laboratories that perform disk diffusion, commercial manufacturers are reliable sources for properly prepared and controlled Mueller-Hinton plates.

The appropriate concentration of drug for each disk is predetermined and set by NCCLS and the Food and Drug Administration. The disks are available from various commercial sources and should be kept at −20° C in a dessicator until used. Thawed, unused disks may be stored at 4° to 8° C for up to a week. Inappropriate storage can lead to deterioration of the antimicrobial agents and result in misleading zone size results.

To ensure equal diffusion of the drug into the agar the disks must be placed flat on the surface and be firmly applied to ensure adhesion. This is most easily accomplished by using any one of several disk dispensers that are available through commercial disk manufacturers. With these dispensers, all disks in the test battery are simultaneously delivered to the inoculated agar surface and are adequately spaced to minimize the chances for inhibition zone overlap and significant interactions between antimicrobials. In most instances, a maximum of 12 antibiotic disks may be applied to the surface of a single 150-mm Mueller-Hinton agar plate (see Figure 18-5).

Inoculation and Incubation. Before disk placement, the plate surface is inoculated using a swab that has been submerged in a bacterial suspension standardized to match the turbidity of the 0.5 McFarland turbidity standard (i.e., 1.5×10^8 CFU/mL). The surface of the plate is swabbed in three directions to ensure an even and complete distribution of the inoculum over the entire plate. Within 15 minutes of inoculation the antimicrobial agent disks are applied and the plates are inverted for incubation to avoid accumulation of moisture on the agar surface that can interfere with interpretation of test results.

For most organisms, incubation is at 35° C in air, but increased CO_2 is used when testing certain

TABLE 18-3 SUMMARY OF DISK DIFFUSION SUSCEPTIBILITY TESTING CONDITIONS

ORGANISM GROUPS	TEST MEDIUM	INOCULUM SIZE (CFU/mL)	INCUBATION CONDITIONS	INCUBATION DURATION
Enterobacteriaceae	Mueller-Hinton agar	Swab from 1.5×10^8 suspension	35° C; air	16-18 hrs
P. aeruginosa				
Enterococci				(24 hrs for vancomycin)
Staphylococci				
To detect methicillin-resistant staphylococci			30°-35° C; air	24 hrs
S. pneumoniae and other streptococci	Mueller-Hinton agar plus 5% sheep blood	Swab from 1.5×10^8 suspension	35° C; 5%-7% CO_2	20-24 hrs
H. influenzae	Haemophilus Test Medium	Swab from 1.5×10^8 suspension	35° C; 5%-7% CO_2	16-18 hrs
N. gonorrhoeae	GC agar plus supplements	Swab from 1.5×10^8 suspension	35° C; 5%-7% CO_2	20-24 hrs

fastidious bacteria (see Table 18-3). Similarly, the incubation time may be increased beyond 18 hours to enhance detection of certain resistance patterns (e.g., methicillin resistance in staphylococci and vancomycin resistance in enterococci) and to ensure accu-

rate results in general for certain fastidious organisms such as *N. gonorrhoeae.*

The dynamics and timing of antimicrobial agent diffusion to establish a concentration gradient coupled with the growth of organisms over a 16- to 24-

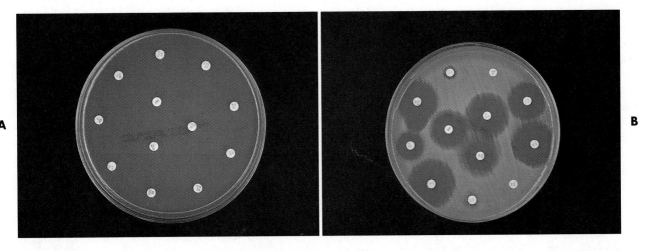

FIGURE 18-5 A, By the disk diffusion method, antibiotic disks are placed on the surface just after the agar surface was inoculated with the test organism. **B,** Zones of growth inhibition around various disks are apparent following 16 to 18 hours of incubation.

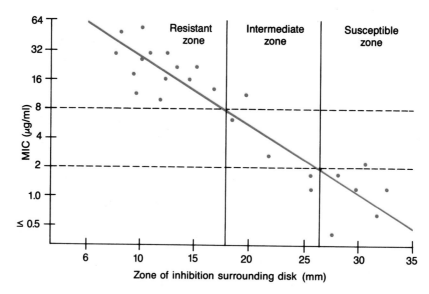

FIGURE 18-6 Example of a regression analysis plot to establish zone size breakpoints for defining the susceptible, intermediate, and resistant categories for an antimicrobial agent. In this example, the maximum achievable serum concentration of the antibiotic is 8 μg/mL; disk inhibition zones ≤18 mm in diameter indicate resistance, zones ≥26 mm indicate susceptibility, and the intermediate category is indicated by zones ranging from 19 to 25 mm.

hour duration is critical for reliable results. Therefore, incubation of disk diffusion plates beyond the allotted time should be avoided and disk diffusion generally is not an acceptable method for testing organisms that require extended incubation times to grow.

Reading and Interpretation of Results. Before results with individual antimicrobial agent disks are read, the plate is examined to confirm that a confluent lawn of good growth has been obtained (see Figure 18-5). If growth between inhibitory zones around each disk is poor and nonconfluent, then the test should not be interpreted but should be repeated. The lack of confluent growth may be due to insufficient inoculum. Alternatively, a particular isolate may have undergone mutation so that growth factors supplied by the standard susceptibility testing medium are no longer sufficient for supporting robust growth. In the latter case, medium supplemented with blood and/or incubation in CO_2 may enhance growth. However, caution in interpreting results is required when extraordinary measures are used to obtain good growth and the standard medium recommended for testing a particular type of organism is not used. Plates should also be examined for purity because mixed cultures are most evident as different colony morphologies scattered throughout the lawn of bacteria that is being tested (Figure 18-7). Mixed cultures require purification and repeat testing of the bacterial isolate of interest.

Using a dark background and reflected light (Figure 18-8), the plate is situated so that a ruler or caliper may be used to measure inhibition zone diam-

eters for each antimicrobial agent. Certain motile organisms, such as *Proteus* spp., may swarm over the surface of the plate and complicate clear interpretation of the zone boundaries. In these cases, the swarming haze is ignored and zones are measured at the point where growth is obviously inhibited. Similarly, hazes of bacterial growth may be observed when testing sulfonamides and trimethoprim as a result of the organism population going through several doubling generations before inhibition; the resulting haze of growth should be ignored for disk interpretation with these agents.

In instances not involving swarming organisms or the testing of sulfonamides and trimethoprim, hazes of growth that occur within more obvious inhibition zones should not be ignored. In many instances this is the only way in which clinically relevant resistance patterns are manifested by certain bacterial isolates when tested using the disk diffusion method. Key examples in which this can occur include cephalosporin resistance among several species of *Enterobacteriaceae*, methicillin resistance in staphylococci, and vancomycin resistance in some enterococci. In fact, detection of the haze produced by some staphylococci and enterococci can best be detected using transmitted rather than reflected light. In these cases, the disk diffusion plates are held in front of the light source when methicillin and vancomycin inhibition zones are being read (see Figure 18-8). Still other significant resistances may be subtly evident and appear as individual colonies within an obvious zone of inhibition (Figure 18-9). When such colonies are

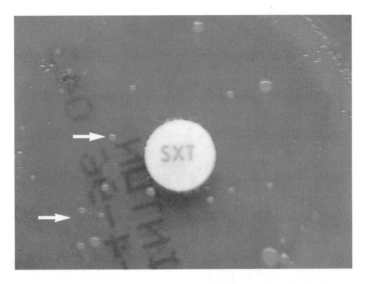

FIGURE 18-7 Disk diffusion plate that was inoculated with a mixed culture as evidenced by different colony morphologies (*arrows*) appearing throughout the lawn of growth.

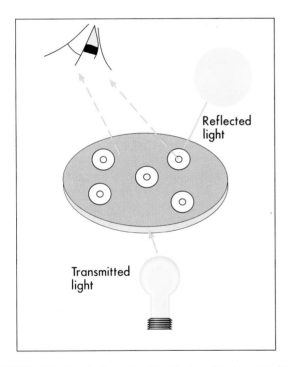

FIGURE 18-8 Examination of a disk diffusion plate by transmitted and reflected light.

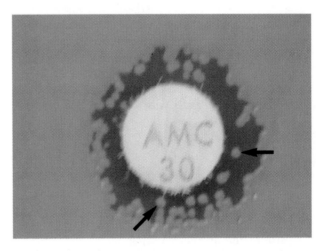

FIGURE 18-9 Individual bacterial colonies within a more obvious zone of inhibition (*arrows*). This could indicate inoculation with a mixed culture. However, emergence of resistant mutants of the test isolate is a more likely reason for this growth pattern.

seen, purity of the test isolate must be confirmed. If purity is confirmed, the individual colonies are variants or resistant mutants of the same species and the test isolate should be considered resistant.

Once zone sizes are recorded, interpretive categories are assigned. Interpretive criteria for antimicrobial agent-organism combinations that can be tested by disk diffusion are provided in the NCCLS-M2 series titled, "Performance Standards for Antimicrobial Disk Susceptibility Tests (M100 supplements)."

Definitions for susceptible, intermediate, and resistant are the same as those used for dilution methods (see Box 18-2). For example, using the NCCLS interpretive standards, an *E. coli* isolate that produces an ampicillin inhibition zone diameter of ≤13 mm is classified as resistant; if the zone is 14 to 16 mm the isolate is considered intermediate to ampicillin, and if the zone is ≥17 mm the organism is categorized as susceptible.

Unlike MICs, inhibition zone sizes are only used to produce a category interpretation and have no clinical utility in and of themselves. Therefore, when testing is performed by disk diffusion, only the category interpretation of susceptible, intermediate, or resistant is reported.

Advantages and Disadvantages. One of the greatest advantages of the disk diffusion test is convenience and user friendliness. Up to 12 antimicrobial agents can be tested against one bacterial isolate with minimal use of extra materials and devices. Because results are generally accurate and most commonly encountered bacteria are reliably tested, the disk diffusion test is still among the most commonly used methods for antimicrobial susceptibility testing. The major disadvantages of this method include the lack of interpretive criteria for organisms not included in Table 18-3 and the inability to provide more precise data regarding the level of an organism's resistance or susceptibility as can be provided by MIC methods.

Commercial susceptibility testing systems

The variety and widespread use of commercial susceptibility testing methods reflect the key role that resistance detection plays among the responsibilities of clinical microbiology laboratories. In many instances the commercial methods are variations of the conventional dilution or disk diffusion methods, and their accuracies have been evaluated by comparison of results with those obtained by conventional methods. Additionally, many of the media and environmental conditions standardized for conventional methods are maintained with the use of commercial systems. The goal of detecting resistance is the same for all commercial methods, but the principles and practices that are applied to achieve that goal vary with respect to:

- Format in which bacteria and antimicrobial agents are brought together
- Extent of automation for inoculation, incubation, interpretation, and reporting
- Method used for detection of bacterial growth inhibition
- Speed with which results are produced
- Accuracy

Accuracy is an extremely important aspect of any susceptibility testing system and is addressed in more detail in Chapter 19.

BROTH MICRODILUTION METHODS Several systems have been developed that provide microdilution panels already prepared and formatted according to the guidelines for conventional broth microdilution methods (e.g., BBL Sceptor, BD Microbiology Systems, Cockeysville, Md.; Sensititre, MicroMedia, and Just One, AccuMed, Westlake, Ohio; MicroMedia, Cleveland, Ohio; Pasco Laboratories, Wheat Ridge, Colo.; MicroTech, Aurora, Colo.). These systems enable laboratories to perform broth microdilution without having to prepare their own panels.

The systems may differ to some extent regarding volume within the test wells, how inocula are prepared and added, the availability of different supplements for the testing of fastidious bacteria, the types of antimicrobial agents and the dilution schemes used, and the format of medium and antimicrobial agents (e.g., dry-lyophilized or frozen). Furthermore, the degree of automation for inoculation of the panels and the devices available for reading results vary among the different products. In general, these commercial panels are designed to receive the standard inoculum and are incubated using conditions and durations recommended for conventional broth microdilution. They are growth-based systems that require overnight incubation, and NCCLS interpretive criteria apply for interpretation of most results.

AGAR DILUTION DERIVATIONS One commercial system (Spiral Biotech Inc., Bethesda, Md.) uses an instrument to apply antimicrobial agent to the surface of an already prepared agar plate in a concentric spiral fashion. Starting in the center of the plate the instrument deposits the highest concentration of antibiotic and from that point drug application proceeds to the periphery of the plate. Diffusion of the drug in the agar establishes a concentration gradient from high (center of plate) to low (periphery of plate). Starting at the periphery of the plate bacterial inocula are applied as a single streak perpendicular to the established gradient in a spoke wheel fashion. Following incubation, the distance from where growth is noted at the edge of the plate to where growth is inhibited toward the center of the plate is measured (Figure 18-10). This value is used to calculate the MIC of the antimicrobial agent against each of the bacterial isolates streaked on the plate.

DIFFUSION IN AGAR DERIVATIONS One test has been developed that combines the convenience of disk diffusion with the ability to generate MIC data. The E-test (AB Biodisk, Solna, Sweden) uses plastic strips; one side of the strip contains the antimicrobial agent concentration gradient and the other contains a numeric scale that indicates the drug concentration (Figure 18-11). Mueller-Hinton plates are inoculated as for disk diffusion and the strips are placed on the inoculum lawn. Several strips may be placed radially on the same plate so that multiple antimicrobials can be tested against a single isolate. Following overnight incubation, the plate is examined and the number present at the point where the border of growth inhibition intersects the E-strip is taken as the MIC (see Figure 18-11). The same MIC interpretive criteria used for dilution methods are used with the E-test value to assign an interpretive category of susceptible, interme-

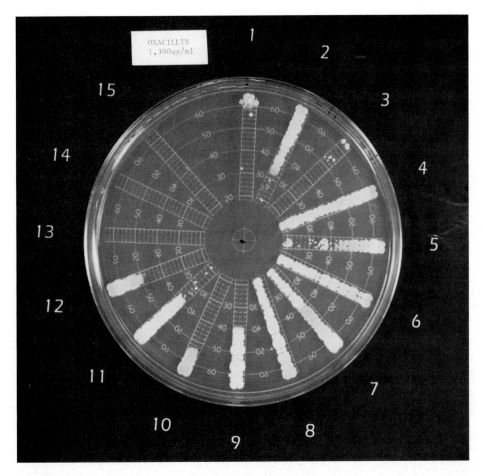

FIGURE 18-10 Growth patterns on a plate containing an antibiotic gradient (concentration decreases from center of the plate to the periphery) applied by the Spiral Gradient instrument. The distance from where growth is noted at the edge of the plate to where growth is inhibited toward the center of the plate is measured. This value is used in a formula to calculate the MIC of the antimicrobial agent against each of the bacterial isolates streaked on the plate. (Courtesy Spiral Biotech, Inc. Bethesda, Md.)

diate, or resistant. This method provides a means for producing MIC data in those situations in which the level of resistance can be clinically important (e.g., penicillin or cephalosporins against *S. pneumoniae*).

Another method (BIOMIC, Giles Scientific, Inc., New York) combines the use of conventional disk diffusion methodology with video digital analysis to automate interpretation of inhibition zone sizes. Automated zone readings and interpretations are combined with computer software to produce MIC values and to allow for data manipulations and evaluations for detecting unusual resistance profiles and producing antibiogram reports (for more information regarding antibiogram reports see Chapter 19).

AUTOMATED ANTIMICROBIAL SUSCEPTIBILITY TEST SYSTEMS
Two systems for which substantial portions of the methodology are automated and for which production of results does not necessarily depend on overnight growth of the test organisms are commonly

used in the United States. These are the Vitek System (BioMérieux Vitek, Hazelwood, Mo.) and the Walk-Away System (Dade International, Sacramento, Calif.).

In the Vitek system test organisms and the antimicrobial agents are brought together in a liquid medium that is distributed to 30 or 45 small wells contained in a small plastic card (Figure 18-12). Each well contains a specified concentration of antimicrobial agent. The inoculation of the card, which is automated through the use of a filling-sealer module, results in the distribution of inoculum to each of the antibiotic-containing wells in the card. Once inoculated, the cards are placed in an incubator-reader module. Every hour during incubation each card is automatically selected within the reader, and the kinetics of growth in each well, including a growth control, is turbidimetrically determined. Linear regression analysis and algorithmic evaluation of the kinetic data is used by the system's software to derive MICs and appropriate

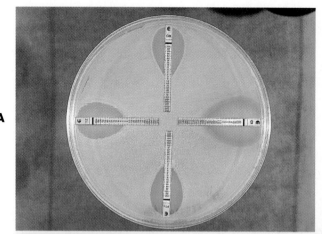

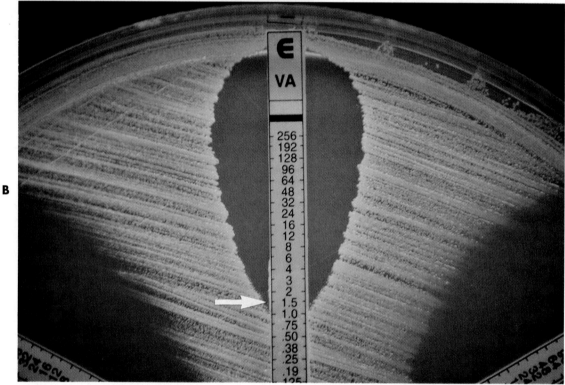

FIGURE 18-11 The E-test method uses the principle of disk diffusion to generate an MIC value. **A,** Individual antibiotic strips are placed on an inoculated agar surface. **B,** Following incubation, the MIC is determined by reading the numeric scale at the point where growth inhibition intersects the antibiotic strip (*arrow*). Several antibiotic strips can be tested per plate. (Courtesy AB Biodisk, Solna, Sweden.)

interpretive category assignment. Therefore, this system provides automated inoculation, reading, and interpretation with results for many bacteria available in 6 to 15 hours (Figure 18-13).

The WalkAway System uses the broth microdilution panel format that is manually inoculated with a multiprong device. Inoculated panels are placed in an incubator-reader unit, where they are incubated for the required length of time, and then the growth patterns are automatically read and interpreted. Depending on the microdilution tray used, bacterial growth may be detected spectrophotometrically or fluorometrically (Figure 18-14). Spectrophotometrically analyzed panels require overnight incubation, and the growth patterns may be read manually as described for routine microdilution testing. Fluorometric analysis is based on the degradation of fluorogenic substrates by viable bacteria as the means for detecting bacterial inhibition by the antimicrobial agents. The fluorogenic approach can provide susceptibility results in 3.5 to 5.5 hours. Either full dilution schemes or breakpoint panels are available for use.

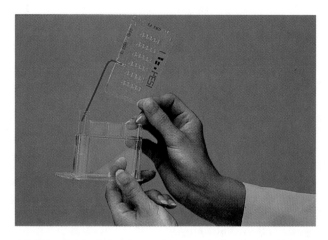

FIGURE 18-12 Vitek susceptibility cards contain 30 to 45 miniature wells supplemented with specific antibiotic concentrations. The inoculum resuspends the dried antibiotic when introduced into the card via a straw and automated filling device. (Courtesy bioMérieux Vitek, Inc., Hazelwood, Mo.)

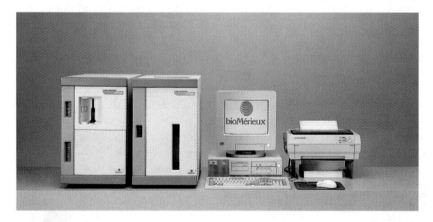

FIGURE 18-13 Vitek instrument components for automated inoculation, incubation, reading, and interpretation of antimicrobial susceptibility testing cards. The same instrument is used for bacterial identifications as discussed in Table 13-3. (Courtesy bioMérieux Vitek, Inc., Hazelwood, Mo.)

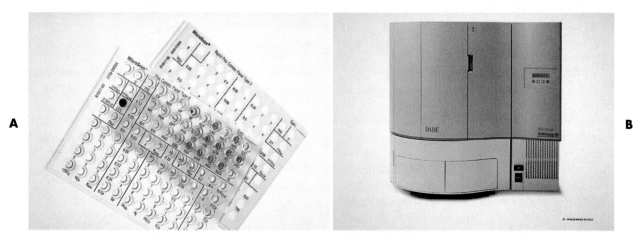

FIGURE 18-14 Microdilution tray format (**A**) used with the MicroScan WalkAway instrument (**B**) for automated incubation, reading, and interpretation of antimicrobial susceptibility tests. This same instrument is used for bacterial identifications as discussed in Table 13-3. (Courtesy Dade International, Sacramento, Calif.)

In addition to speed and facilitating work flow, the automated systems also provide increasingly powerful computer-based data management systems that can be used to evaluate the accuracy of results (for more information regarding analysis of accuracy in susceptibility testing see Chapter 19), manage larger data bases, and interface with pharmacy areas to enhance the utility of antimicrobial susceptibility testing data.

Alternative approaches for enhancing resistance detection

Although the various conventional and commercial antimicrobial susceptibility testing methods provide accurate results in most instances, certain clinically relevant resistance mechanisms can be difficult to detect. In these instances supplemental tests and alternative approaches are needed to ensure reliable detection of resistance.

SUPPLEMENTAL TESTING METHODS Table 18-4 highlights some of the features of commonly used supplemental tests that may be used to enhance resistance detection. For certain strains of staphylococci, conventional and commercial systems may have difficulty detecting resistance to oxacillin and the related drugs methicillin and nafcillin. The oxacillin agar screen provides a backup test that may be used when other methods provide equivocal or uncertain profiles. Growth on the screen correlates highly with the presence of oxacillin (or methicillin) resistance, and no growth is strong evidence that an isolate is susceptible. This is an important determination; strains that are classified as resistant are considered resistant to all other currently available beta-lactam antibiotics so that therapy must include the use of vancomycin. The agar screen plates can be made in-house, and they are available commercially (e.g., Remel, Lenexa, Kan., or BBL, Cockeysville, Md.). Additionally, other commercial tests designed to detect oxacillin resistance more rapidly (i.e., 4 hours) have been developed and may provide another approach to supplemental testing (e.g., Crystal MRSA ID System, BBL, Cockeysville, Md.).

Similarly, detection of enterococcal resistance to vancomycin can be difficult by some conventional and commercial methods and the agar screen can be helpful in confirming this resistance pattern (see Table 18-4). However, as a screen, not all enterococcal isolates that grow are resistant to vancomycin at clinically relevant levels. Therefore, strains that are detected using this method should be more fully characterized with respect to the level and nature of their resistance.

Aminoglycosides also play a key role in therapy for serious enterococcal infections, and acquired high-level resistance, which essentially destroys the therapeutic value of these drugs for combination therapy

with ampicillin or vancomycin, is not readily detected by conventional methods. Therefore, screens that use high concentrations of aminoglycosides have been developed specifically for detecting this resistance and are available commercially (e.g., Remel, Lenexa, Kan., or BBL, Cockeysville, Md.).

With the emergence of penicillin resistance among *S. pneumoniae,* the penicillin disk diffusion test was not sufficiently sensitive to detect subtle but significant changes in susceptibility to this agent. To address this issue the oxacillin disk screen described in Table 18-4 is useful but does have a notable limitation. Although organisms that give zones ≥20 mm can be accurately characterized as penicillin susceptible, the penicillin susceptibility status of those with zones <20 mm remains uncertain and use of some method that produces an MIC value is required.

Undoubtedly, as complicated resistance mechanisms that require laboratory detection continue to emerge, screening and supplemental testing methods will continue to be developed. Some of these will be maintained as the primary method for detecting a particular resistance mechanism, while others may tend to fade away as adjustments in conventional and commercial procedures enhance resistance detection and preclude the need to use a supplemental test.

PREDICTOR ANTIMICROBIAL AGENTS Another approach that may be used to ensure accuracy in resistance detection is the use of "predictor" antimicrobial agents in the test batteries. The basic premise of this approach is to use antimicrobial agents (i.e., the **predictor drugs**) that are the most sensitive indicators of certain resistance mechanisms. The profile obtained with such a battery of agents is then used to deduce the underlying resistance mechanism(s). A susceptibility report is subsequently produced based on the likely effect the resistance mechanisms would have on the antimicrobials being considered for therapeutic management of the patient. This approach avoids the need to test all agents in a hospital formulary.

Use of predictor drugs is not a new concept since this approach already is used in various situations. For example:

- Staphylococcal resistance to oxacillin is used to determine and report resistance to all currently available beta-lactams, including penicillins, cephalosporins, and imipenem
- Enterococcal high-level gentamicin resistance predicts resistance to nearly all other currently available aminoglycosides, including amikacin, tobramycin, netilmicin, and kanamycin
- Enterococcal resistance to ampicillin predicts resistance to all penicillin derivatives and imipenem

TABLE 18-4 SUPPLEMENTAL METHODS FOR DETECTION OF ANTIMICROBIAL RESISTANCE

TEST	PURPOSE	CONDITIONS		INTERPRETATION
Oxacillin Agar Screen	Detection of staphylococcal resistance to penicillinase-resistant penicillins (e.g., oxacillin, methicillin, or nafcillin)	Medium:	Mueller-Hinton agar plus 6 μg oxacillin/mL plus 4% NaCL	Growth = Resistance No Growth = Susceptible
		Inoculum:	Swab or spot from 1.5×10^8 standard suspension	
		Incubation:	30°-35° C 24 hrs, up to 48 hrs for non - *S. aureus*	
Van~~…~~	~~…~~ococcal ~~…~~comycin	Medium:	Brain heart infusion agar plus 6 μg van-comycin/mL	Growth = Resistance No Growth = Susceptible
		Inoculum:	Spot of 10^5-10^6 CFU	
		Incubation:	35° C, 24 hrs	
~~…~~	~~…~~enter-~~…~~sis-~~…~~des~~…~~se~~…~~l-~~…~~pi-	Medium:	Brain heart infusion broth: 500 μg/mL gentamicin; 1000 μg/mL streptomycin	Growth = Resistance No Growth = Susceptible
			Agar: 500 μg/mL gentamicin; 2000 μg/mL streptomycin	
		Inoculum:	Broth; 5×10^5 CFU/mL agar; 10^6 CFU/spot	
		Incubation:	35° C, 24 hrs; 48 hrs for strepto-mycin, only if no growth at 24 hrs	
Oxacillin Disk Screen	Detection of *S. pneumoniae* resistance to penicillin	Medium:	Mueller-Hinton agar plus 5% sheep blood plus 1 μg oxacillin disk	Inhibition zone ≥20 mm; penicillin susceptible
		Inoculum:	as for disk diffusion	Inhibition zone <20 mm; penicillin resistant, intermediate, or susceptible. Further testing by MIC method is needed
		Incubation:	5%-7% CO_2 35° C; 20-24 hrs	

[Handwritten note: methods for Det of Antimicrob Resistance]

As different resistance mechanisms present greater difficulties for conventional susceptibility testing protocols, the predictor drug strategy is likely to become a more widely applied approach to in vitro susceptibility testing. This may be especially true in the area of detecting beta-lactam resistance among gram-negative bacilli as the ever increasing variety and prevalence of beta-lactamases expressed by these organisms continue to challenge conventional and commercial susceptibility testing protocols.

METHODS THAT DIRECTLY DETECT SPECIFIC RESISTANCE MECHANISMS

As an alternative to detecting resistance by measuring the effect of antimicrobial presence on bacterial growth, some strategies focus on assaying for the presence of a particular mechanism. When the presence or absence of the mechanism is established, the resistance profile of the organism can be generated without having to test several different antimicrobial-organism combinations. The utility of this approach, which can involve phenotypic and genotypic methods, depends on the presence of a particular resistance mechanism as being a sensitive and specific indicator of clinical resistance.

Phenotypic methods

The most common phenotypic-based assays are those that test for the presence of beta-lactamase enzyme(s) in the clinical bacterial isolate of interest. Less commonly used are tests to detect the chloramphenicol-modifying enzyme chloramphenicol acetyltransferase.

BETA-LACTAMASE DETECTION Beta-lactamases play a key role in bacterial resistance to beta-lactam agents, and detection of their presence can provide useful information (see Chapter 17). Various assays are available to detect beta-lactamases, but the most useful one for clinical laboratories is the chromogenic cephalosporinase test. Beta-lactamases exert their effect by opening the beta-lactam ring (see Figure 17-9). With the use of a chromogenic cephalosporin as the substrate this process results in a colored product. One such assay is the cefinase disk (BD Microbiology Systems, Cockeysville, Md.) test shown in Figure 18-15.

Useful application of tests to directly detect beta-lactamase production is limited to those organisms for which the list of beta-lactams significantly affected by the enzyme are known. Further, this list must include the beta-lactams commonly considered for therapeutic eradication of the organism. Examples of useful applications include detection of:

- *N. gonorrhoeae* resistance to penicillin
- *H. influenzae* resistance to ampicillin
- Staphylococcal resistance to penicillin
- Enterococcal resistance to penicillin and ampicillin (however, resistance by altered target is substantially more common than beta-lactamase–mediated resistance)
- Penicillin resistance among anaerobic bacteria

The actual utility of this approach even for the organisms listed is decreasing. As beta-lactamase–mediated resistance has become widespread among *N. gonorrhoeae*, *H. influenzae*, and staphylococci, other agents not affected by the beta-lactamases have become the antimicrobials of choice for therapy. Therefore, the need to know the beta-lactamase status of these bacterial species has become substantially less urgent. Even so, there still may be clinical situations in which the information is useful and these methods provide a simple and rapid approach. For example, staphylococci that produce low but significant levels of beta-lactamase may not be detected by conventional methods and performance of the beta-lactamase assay on apparently penicillin-susceptible isolates is warranted.

Several *Enterobacteriaceae* and *P. aeruginosa* produce beta-lactamases. However, the effect of these

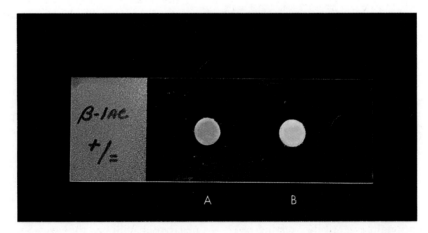

FIGURE 18-15 The chromogenic cephalosporin test allows direct detection of beta-lactamase production. When the beta-lactam ring of the cephalosporin substrate in the disk is hydrolyzed by the bacterial inoculum, a deep pink color is produced (**A**). Lack of color production indicates the absence of beta-lactamase (**B**).

enzymes on the various beta-lactams depends on which enzyme(s) are produced. Therefore, even though such organisms would frequently produce a positive beta-lactamase assay, very little, if any, information regarding which antimicrobial agents are affected would be gained. Detection of beta-lactam resistance among these organisms is best accomplished using conventional and commercial systems for directly evaluating antimicrobial agent-organism interactions.

CHLORAMPHENICOL ACETYLTRANSFERASE DETECTION
Chloramphenicol modification by chloramphenicol acetyltransferase (CAT) detection is only one mechanism by which bacteria may express resistance to this agent. This, coupled with the substantially diminished use of chloramphenicol in today's clinical settings, significantly limits the utility of this test. Commercial colorimetric assays, such as that produced by Remel (Lenexa, Kan.), do provide a convenient method for establishing the presence of this enzyme. If positive, chloramphenicol resistance can be reported, but a negative test does not rule out resistance that may be mediated by other mechanisms such as decreased uptake.

Genotypic methods
The genes that encode many of the clinically relevant acquired resistance mechanisms are known, as is all or part of their nucleotide sequences. This has allowed for the development of molecular methods involving nucleic acid hybridization and amplification for the study and detection of antimicrobial resistance (for more information regarding molecular methods for the characterization of bacteria see Chapter 14). The ability to definitively determine the presence of a particular gene that encodes antimicrobial resistance has several advantages. However, as with any laboratory procedure, certain disadvantages and limitations also exist.

From a research and development perspective, molecular methods are extremely useful for more thoroughly characterizing the resistances of bacterial collections used to establish and evaluate conventional standards recommended by the NCCLS; phenotype-based commercial (automated and nonautomated) susceptibility testing methods and systems can also be evaluated.

Molecular methods also may be directly applied in the clinical setting as an important backup resource to investigate and arbitrate equivocal results obtained by phenotypic methods. For example, the clinical importance of accurately detecting methicillin resistance among staphylococci coupled with the inconsistencies of phenotype-based methods is problematic. In

doubtful situations molecular detection of the *mec* gene that encodes methicillin resistance can be usefully applied to definitively establish an isolate's methicillin resistance (see Figure 14-7). Similarly, doubt raised by equivocal phenotypic results obtained with potentially vancomycin-resistant enterococci can be definitively resolved by establishing the presence and classification of *van* genes that mediate this resistance (see Figure 14-6).

Although molecular methods have been and will continue to be extremely important in antimicrobial resistance detection, numerous factors still complicate their use beyond supplementing phenotype-based susceptibility testing protocols. These factors include:

- Use of probes or oligonucleotides for specific resistance genes only allows those particular genes to be found. Resistance mediated by divergent genes or totally different mechanisms could be missed (i.e., the absence of one gene does not guarantee antimicrobial susceptibility).
- Phenotypic resistance to a level that is clinically significant for any one antimicrobial agent may be due to a culmination of processes that involve enzymatic modification of the antimicrobial, decreased uptake, altered affinity of the target for the drug, or some combination of these mechanisms (i.e., the presence of one gene does not guarantee resistance).
- The presence of a gene encoding resistance does not provide information regarding the status of the control genes necessary for expression of resistance. That is, although present, the gene(s) may be silent or nonfunctional and the organism may be incapable of expressing the resistance encoded by the detected gene.
- From a clinical laboratory perspective, there is practical difficulty in adopting molecular methods specific for only a few resistance mechanisms when the vast majority of the susceptibility testing would be accomplished using phenotypic-based methods.

Although there are challenges to the widespread adoption of molecular methods for routine antimicrobial susceptibility testing, the significant contributions that this approach has made to resistance detection will continue to expand.

SPECIAL METHODS FOR COMPLEX ANTIMICROBIAL-ORGANISM INTERACTIONS
Certain in vitro tests have been developed to investigate aspects of antimicrobial activity not routinely addressed by commonly used susceptibility testing procedures. Specifically, these are tests designed to

measure bactericidal activity (i.e., bacterial killing) or to measure the antibacterial effect of antimicrobial agents used in combination.

These tests are often labor intense, fraught with the potential for technical problems, frequently difficult to interpret, and of uncertain clinical utility. For these reasons, their use should be substantially limited. If performed at all, they should be done with the availability of expert microbiology and infectious disease consultation.

Bactericidal tests

Bactericidal tests are designed to determine the ability of antimicrobial agents to kill bacteria. The killing ability of most drugs is already known, and they are commonly classified as bacteriostatic or bactericidal agents. However, many variables, including the concentration of antimicrobial agent and the species of targeted organism, can influence this classification. For example, beta-lactams, such as penicillins, typically are bactericidal against most gram-positive cocci but are usually only bacteriostatic against enterococci. If bactericidal tests are clinically appropriate, they should only be applied to evaluate antimicrobials usually considered to be bactericidal (e.g., beta-lactams and vancomycin) and not to those agents known to be bacteriostatic (e.g., macrolides).

Key clinical situations in which achieving bactericidal activity is of greatest clinical importance include severe and life-threatening infections, infections in the immunocompromised host, and infections in body sites where assistance from the patient's own defenses is minimal (e.g., endocarditis or osteomyelitis). Based on research trials in animal models and clinical trials in humans, the most effective therapy for these types of infections is often already known. However, occasionally the laboratory may be asked to substantiate that bactericidal activity is being achieved or is achievable. The methods available for this include minimal bactericidal concentration (MBC) testing, time-kill studies, and serumcidal testing. Regardless of which method is used, the need to interpret the results cautiously with the understanding of uncertain clinical correlation and the potential for substantial technical artifacts cannot be overemphasized.

MINIMAL BACTERICIDAL CONCENTRATION The minimal bactericidal concentration (MBC) test involves continuation of the procedure for conventional broth dilution testing. After incubation and determination of the antimicrobial agent's MIC, an aliquot from each tube or well in the dilution series showing inhibition of visible bacterial growth is subcultured to an enriched agar medium (usually sheep blood agar).

Following overnight incubation the plates are examined and the CFUs counted. Knowing the volume of the aliquot sampled and the number of CFUs obtained, the number of viable cells per milliliter for each antimicrobial dilution well can be calculated. This number is compared with the known CFU/mL in the original inoculum. The antimicrobial concentration that resulted in a 99.9% reduction in CFU/mL compared with the organism concentration in the original inoculum is recorded as the MBC.

Although the clinical significance of MBC results is uncertain, applications of this information include considering whether treatment failure could be occurring because an organism's MBC exceeds the serum achievable level of the antimicrobial agent. Alternatively, if an antibiotic's MBC is ≥ 32 times higher than the MIC, the organism may be tolerant to that drug. **Tolerance** is a phenomenon most commonly associated with bacterial resistance to beta-lactam antibiotics and reflects an organism's ability to be only inhibited by an agent that is usually bactericidal. Although the physiologic basis of tolerance has been studied in several bacterial species, the actual clinical relevance of this phenomenon has not been well established.

TIME-KILL STUDIES Another approach to examine bactericidal activity involves exposing a bacterial isolate to a concentration of antibiotic in a broth medium and measuring the rate of killing over a specified period. By this time-kill analysis samples are taken from the antibiotic-broth solution immediately after the inoculum was added and at regular intervals afterward. Each time-sample is plated to agar plates; following incubation, CFU counts are performed as described for MBC testing. The number of viable bacteria from each sample is plotted over time so that the rate of killing can be determined. Generally, a 1000-fold decrease in the number of viable bacteria in the antibiotic-containing broth after a 24-hour period compared with the number of bacteria in the original inoculum is interpreted as bactericidal activity. Although time-kill analysis is frequently used in the research environment to study the in vitro activity of antimicrobial agents, the labor intensity and technical specifications of the procedure preclude its use in most clinical microbiology laboratories for the production of data used to manage a patient's infection.

SERUM BACTERICIDAL TEST The serum bactericidal test (SBT) is analogous to the MIC-MBC test except that the medium used is patient's serum that contains the therapeutic antimicrobial agent(s) the patient has been receiving. By using patient serum to detect bacteriostatic and bactericidal activity, the antibacterial impact of factors other than the antibiotics (e.g., antibodies and complement) also are observed.

For each test, two serum samples are required. One is collected just before the patient is to receive the next antimicrobial dose and is referred to as the **trough specimen**. The second sample is collected when the serum antimicrobial concentration is highest and is referred to as the **peak specimen**. Timing for when to collect the peak specimen varies with the pharmacokinetic properties of the antimicrobial agents and the route through which they are being administered. Peak levels for intravenously, intramuscularly, and orally administered agents are generally obtained 30 to 60 minutes, 60 minutes, and 90 minutes after administration, respectively. The trough and peak levels should be collected around the same dose and tested simultaneously.

Serial twofold dilutions of each specimen are prepared and inoculated with the bacterial isolate (final inoculum of 5×10^5 CFU/mL) that is causing the patient's infection. Dilutions are incubated overnight, and the highest dilution inhibiting visibly detectable growth is the **serumstatic titre** (e.g., 1:8, 1:16, 1:32). Aliquots of known size are then taken from each dilution at or below the serumstatic titre (i.e., those dilutions that inhibited bacterial growth) and are plated on sheep blood agar plates. Following incubation, the CFUs per plate are counted and the serum dilution that results in a 99.9% reduction in the CFU/mL as compared with the original inoculum is recorded as the **serumcidal titre**. For example, if a bacterial isolate showed a serumstatic titre of 1:32, then the tubes containing dilutions of 1:2, 1:4, 1:8, 1:16, and 1:32 would be subcultured. If the 1:8 dilution was the highest dilution to yield a 99.9% decrease in CFUs, then the serumcidal titre would be recorded as 1:8.

The SBT was originally developed to help predict the clinical efficacy of antimicrobial therapy for staphylococcal endocarditis. Peak serumcidal titres of 1:32 to 1:64 or greater have been thought to correlate best with a positive clinical outcome. However, even though the test is performed in patient serum, there are still many differences not accounted for between the in vitro test environment and the in vivo site of infection. Therefore, although the test is used to evaluate whether effective bactericidal concentrations are being achieved, the predictive clinical value for staphylococcal endocarditis or any other infection caused by other bacteria is still uncertain.

Details regarding the performance of these bactericidal tests are provided in the NCCLS document M26-T titled, "Methods for Determining Bactericidal Activity of Antimicrobial Agents."

Tests for activity of antimicrobial combinations

Therapeutic management of bacterial infections often require simultaneous use of more than one antimicrobial agent. Multiple therapies are used for reasons that include:

- Treating polymicrobial infections caused by organisms with different antimicrobial resistance profiles
- Achieving more rapid bactericidal activity than could be achieved with any single agent
- Achieving bactericidal activity against bacteria for which no single agent is lethal (e.g., enterococci)
- Minimizing the chance of resistant organisms emerging during therapy (e.g., M. *tuberculosis*)

Testing the effectiveness of antimicrobial combinations against a single bacterial isolate is referred to as **synergy testing**. When combinations are tested, three outcome categories are possible:

- **Synergy:** the activity of the antimicrobial combination is substantially greater than the activity of the single most active drug alone
- **Indifference:** the activity of the combination is no better or worse than the single most active drug alone
- **Antagonism:** the activity of the combination is substantially less than the activity of the single most active drug alone (an interaction to be avoided)

The checkerboard assay and the time-kill assay are two basic methods for synergy testing. In the checkerboard method, MIC panels are set up that contain two antimicrobial agents serially diluted alone and in combination with each other. Following inoculation and incubation, the MICs obtained with the single agents and the various combinations are recorded. By calculating the MIC ratios obtained with single and combined agents, the drug combination in question is classified as synergistic, indifferent, or antagonistic.

Using the time-kill assay, the same procedure described for testing bactericidal activity is used except the killing curve obtained with a single agent is compared with the killing curve obtained with antimicrobial combinations. Synergy is indicated when the combination exhibits 100-fold or more greater killing than the most active single agent tested alone following 24 hours of incubation. Similar killing rates between the most active agent and the combination is interpreted as indifference. Antagonism is evidenced by the combination being less active than the most active single agent.

The decision to use more than one antimicrobial agent may be based on antimicrobial resistance profiles or identifications of particular bacterial pathogens reported by the clinical microbiology laboratory. However, the decision regarding which antimicrobial agents to combine should not rely on the results of complex synergy tests performed in the clinical laboratory. Most clinically useful antimicrobial combinations have been investigated in a clinical research setting and are well described in the medical literature. These data should be used to guide the decision for combination therapy. The technical difficulties associated with performing and interpreting synergy tests, which at most would only be performed rarely in the clinical laboratory, precludes their reliable utility in the diagnostic setting.

Bibliography

Bauer, A.W., Kirby, W.M.M., Sherris, J.C., et al. 1966. Antibiotic susceptibility testing by a single disc method. Am. J. Clin. Pathol. 45:493.

Lorian, V. editor: 1991. Antibiotics in laboratory medicine. Williams & Wilkins. Baltimore.

Murray, P.R., Baron, E.J., Pfaller, M.A., et. al., editors: 1995. Manual of clinical microbiology, ed. 6, American Society for Microbiology, Washington, D.C.

National Committee for Clinical Laboratory Standards. 1997. Performance standards for antimicrobial susceptibility testing. M100-S7. NCCLS, Villanova, Pa.

National Committee for Clinical Laboratory Standards. 1997. Performance standards for antimicrobial disk susceptibility testing. M2-A6, ed. 6, NCCLS, Villanova, Pa.

National Committee for Clinical Laboratory Standards. 1997. Methods for dilution antimicrobial susceptibility testing for bacteria that grow aerobically. M7-A4, ed. 4, NCCLS, Villanova, Pa.

National Committee for Clinical Laboratory Standards. 1992. Methods for determining bactericidal activity of antimicrobial agents; tentative guideline. M26-T, NCCLS, Villanova, Pa.

19 | LABORATORY STRATEGIES FOR ANTIMICROBIAL SUSCEPTIBILITY TESTING

The clinical microbiology laboratory is responsible for maximizing the positive impact that susceptibility testing information can have on the use of antimicrobial agents to therapeutically manage infectious diseases. However, meeting this responsibility is difficult because of demands for more efficient use of laboratory resources, the increasing complexities of important bacterial resistance profiles, and the continued expectations for high-quality results. To ensure quality in the midst of dwindling resources and expanding antimicrobial resistance, strategies for antimicrobial susceptibility testing must be carefully developed. These strategies should target relevance, accuracy, and communication as their goals (Figure 19-1).

RELEVANCE

Antimicrobial susceptibility testing should only be performed when there is sufficient potential for providing clinically useful and reliable information regarding those antimicrobial agents that are appropriate for the bacterial isolate in question. Therefore, for the sake of relevance two questions must be addressed:

- When should testing be performed?
- Which antimicrobial agents should be tested?

WHEN TO PERFORM A SUSCEPTIBILITY TEST

The first issue that must be resolved is whether antimicrobial susceptibility testing is appropriate for a particular isolate. Although the answer may not always be clear, the issue must always be addressed. The decision to perform susceptibility testing depends on the following criteria:

- Clinical significance of a bacterial isolate
- Predictability of a bacterial isolate's susceptibility to the antimicrobial agents most commonly used against them, often referred to as the **drugs of choice**
- Availability of reliable standardized methods for testing the isolate

Determining clinical significance

Performing tests and reporting antimicrobial susceptibility data on clinically insignificant bacterial isolates is a waste of resources and, more importantly, can mislead physicians, who depend on laboratory information to help establish the clinical significance of a bacterial isolate. Useful criteria for establishing the clinical importance of a bacterial isolate include:

- Detection and/or the abundance of the organism on direct Gram stain of a patient's specimen, preferably in the presence of white blood cells, and growth of an organism with the same morphology in culture
- Known ability of the bacterial species isolated to cause infection at the body site from which the specimen was obtained (see Part 3)
- Whether the organism is generally considered either an epithelial or mucosal colonizer or is generally considered a pathogen
- Body site from which the organism was isolated (normally sterile vs. normally colonized site)

Although these criteria are helpful and heavily depend on the capacity of the bacterial species isolated to cause disease, the final designation of clinical significance often still requires dialogue between the laboratorian and physician.

Reporting susceptibility results on organisms of doubtful clinical significance can be falsely interpreted by clinicians as an indicator of clinical significance. Therefore, using criteria such as those just listed should be a major feature of any laboratory's antimicrobial susceptibility testing strategy.

Predictability of antimicrobial susceptibility

If the organisms are clinically significant, what are the chances they could be resistant to the antimicrobial agent(s) commonly used to eradicate them (i.e., the drugs of choice)? Unfortunately, the increasing dissemination of resistance among clinically relevant bacteria has diminished the list of bacteria whose antimicrobial susceptibility can be confidently predicted based on identification without the need to perform testing. Table 19-1 categorizes many of the

TABLE 19-1 CATEGORIZATION OF BACTERIA ACCORDING TO NEED FOR ROUTINE PERFORMANCE OF ANTIMICROBIAL SUSCEPTIBILITY TESTING*

TESTING COMMONLY REQUIRED	TESTING OCCASIONALLY REQUIRED[†]	TESTING RARELY REQUIRED
Staphylococci	*Haemophilus influenzae*	Beta-hemolytic streptococci
Streptococcus pneumoniae	*Neisseria gonorrhoeae*	(groups A, B, C, F, and G)
Viridans streptococci[‡]	*Moraxella catarrhalis*	*Neisseria meningitidis*
Enterococci	Anaerobic bacteria	*Listeria monocytogenes*
Enterobacteriaceae		
Pseudomonas aeruginosa		
Acinetobacter spp.		

*Based on the assumption that the organism is clinically significant. Table only includes those bacteria for which standardized testing procedures are available as outlined and recommended by the National Committee for Clinical Laboratory Standards.

[†]Testing only needed if an antimicrobial to which the organisms are frequently resistant is still considered for use (e.g., penicillin vs. *Neisseria gonorrhoeae*).

[‡]Viridans streptococci only require testing when implicated in endocarditis or isolated in pure culture from a normally sterile site with a strong suspicion of being clinically important.

commonly encountered bacteria according to the need to perform testing to detect resistance.

Acquired resistance to various antimicrobial agents dictates that susceptibility testing be performed on all clinically relevant isolates of several bacterial groups, genera, and species. For other organisms, such as *Haemophilus influenzae* and *Neisseria gonorrhoeae,* resistance to the original drugs of choice (ampicillin and penicillin) has become so widespread that more potent antibiotics (e.g., ceftriaxone) for which no resistance has been described to date have become the drugs of choice. Therefore, whereas testing used to be routinely indicated to detect ampicillin

and penicillin resistance, testing for resistance to currently recommended antimicrobials for these organisms is not routinely necessary.

One notable exception to the widespread emergence of resistance has been the absence of penicillin resistance among beta-hemolytic streptococci. Because susceptibility to penicillin is extremely predictable among these organisms, testing against penicillin provides little, if any, information that is not already provided by accurate organism identification. However, in instances in which the patient cannot tolerate penicillin, alternative agents, such as erythromycin, may be considered. Resistance among

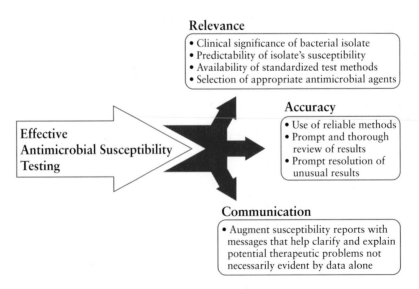

FIGURE 19-1 Goals of effective antimicrobial susceptibility testing strategies.

beta-hemolytic streptococci to this agent has been well documented, and susceptibility testing in this instance would be indicated.

The recommendations outlined in Table 19-1 are guidelines. In any clinical setting there will be exceptions that must be considered in consultation with physicians. Also, these guidelines are for providing data regarding the management of a single patient's infection. When susceptibility testing is performed as a means of gathering surveillance data for monitoring emerging resistance (see the section on accuracy and antimicrobial resistance surveillance in this chapter) the guidelines may not necessarily apply.

Availability of reliable susceptibility testing methods

If a reliable, standardized method for testing a particular bacterial genus or species does not exist, then the ability to produce accurate and meaningful data is substantially compromised. Although standard methods exist for most of the commonly encountered bacteria (see Tables 18-1 to 18-3), clinically relevant isolates of bacterial species for which standard testing methods do not exist are encountered. In these instances, the dilemma stems from the conflict between the laboratory's urge to contribute in some way by providing data and the lack of confidence in producing interpretable and accurate information.

Many organisms not listed in Table 19-1 will grow on the media and under the conditions recommended for testing commonly encountered bacteria. However, the ability to grow and the ability to detect important antimicrobial resistances are not the same thing. For example, the gram-negative bacillus *Stenotrophomonas maltophilia* grows extremely well under most susceptibility testing conditions, but the results obtained with beta-lactam antibiotics can be widely variable and seriously misleading. This organism produces potent beta-lactamases that seriously compromise the effectiveness of most beta-lactams, yet certain isolates may appear susceptible by standard in vitro testing criteria. Therefore, even though testing can provide an answer, substantial potential to obtain the wrong answer exists.

Given the uncertainty that surrounds testing of bacteria for which standardized methods are lacking, two approaches may be used. One is to not perform testing but provide physicians with information based on clinical studies published in the medical literature about the antimicrobial agents generally accepted as the drugs of choice for the bacterial species in question. This approach is best handled when the laboratory medical director and infectious disease specialists are involved. The other option is to still provide the literature information but also perform the test to the best of the laboratory's ability. In this case, results must be accompanied by a message that clearly indicates testing was performed by a nonstandardized method and results should be interpreted with caution. When such tests are undertaken, customized antimicrobial batteries that include the agents most commonly used to eradicate the bacterial species of interest need to be assembled and used.

SELECTION OF ANTIMICROBIAL AGENTS FOR TESTING

Selection of relevant antimicrobial agents is based on the criteria outlined in Box 18-1. These criteria should be carefully considered in consultation with the medical staff so that reports contain information that is most appropriate for the therapeutic management of the patient. Careful selection of antimicrobial agents avoids cluttering reports with superfluous information, minimizes the risk of confusing physicians, and substantially decreases the waste of time and resources in the clinical microbiology laboratory.

Antimicrobial agents that may be considered for inclusion in batteries to be tested against certain bacterial groups are provided in Table 19-2. The list is not exhaustive but is useful for illustrating some points regarding the development of relevant testing batteries. For example, with all the penicillins, cephalosporins, and other beta-lactam antibiotics available for testing, only penicillin and oxacillin need to be tested against staphylococci. The information gained with these two agents alone will reflect the general effectiveness of any other beta-lactam. In essence these drugs are predictor agents as discussed in Chapter 18. Similarly, ampicillin can be used alone as an indicator of enterococcal susceptibility to various penicillins and, because of intrinsic resistance, cephalosporins should never be tested against these organisms.

In contrast to the relatively few agents that may be included in testing batteries for gram-positive cocci, several potential choices exist for use against gram-negative bacilli. This is mostly due to the commercial availability of several beta-lactams with similar activities against *Enterobacteriaceae* and the general inability of one beta-lactam to serve as a reliable predictor drug for other beta-lactams. For example, an organism resistant to cefazolin may or may not be resistant to cefotetan, and an organism resistant to cefotetan may or may not be resistant to ceftazidime. With the lack of potential for selecting a predictor drug in these instances, more agents must be tested. However, in some instances overlap in activities do exist so that some duplication of effort can be avoided. For example, the spectrum of activity of ceftriaxone and cefotaxime is sufficiently similar to allow selection of only one for the testing battery.

TABLE 19-2 SELECTION OF ANTIMICROBIAL AGENTS FOR TESTING AGAINST COMMON BACTERIAL GROUPS*

ANTIMICROBIAL AGENTS	ENTEROBACTERIACEAE	P. AERUGINOSA	STAPHYLOCOCCI	ENTEROCOCCI	S. PNEUMONIAE	VIRIDANS STREPTOCOCCI
Penicillins						
Penicillin	−	−	+	−	+	+
Oxacillin	−	−	+	−	−	−
Ampicillin	+	−	−	+	−	−
Mezlocillin	+	+	−	−	−	−
Piperacillin	+	+	−	−	−	−
Cephalosporins						
Cefazolin	+	−	−	−	−	−
Cefotetan	+	−	−	−	−	−
Ceftriaxone	+	−	−	−	+	+
Cefotaxime	+	−	−	−	+	+
Ceftazidime	+	+	−	−	−	−
Other Beta-Lactams						
Aztreonam	+	+	−	−	−	−
Imipenem	+	+	−	−	+/−	−
Glycopeptides						
Vancomycin	−	−	+	+	+	+
Aminoglycosides						
Gentamicin	+	+	+/−	+†	−	−
Tobramycin	+	+	−	−	−	−
Amikacin	+	+	−	−	−	−
Quinolones						
Ciprofloxacin	+	+	+	−	−	−
Other Agents						
Erythromycin	−	−	+	−	+	+
Clindamycin	−	−	+	−	+	+
Trimethoprim/sulfamethoxazole	+	−	+	−	+	−

+, may be selected for inclusion in testing batteries (not all agents with + need be selected); +/−, may be selected in certain situations; −, selection for testing is not necessary or not recommended.

*Not all available antimicrobial agents are included. Selection recommendation is based on nonurinary tract infections only.

†Gentamicin testing against enterococci requires use of high-concentration disks or a special screen (see Table 18-4).

Many scenarios exist in which the spectrum of activity and other criteria listed in Box 18-1 are considered for the sake of designing the most relevant and useful testing batteries. Again, it must be emphasized that consideration of these criteria must be done in close consultation with the medical staff.

ACCURACY

Susceptibility testing strategies focused on production of accurate results have two key components:

- Use of methods that produce accurate results
- The application of real-time review of results before reporting

USE OF ACCURATE METHODOLOGIES

The complexities of the various resistance mechanisms have left no one method (i.e., conventional, automated, or molecular) singularly sufficient for the detection of all clinically relevant resistance patterns. Therefore, the selection of testing methods and careful consideration of how different methods are most effectively used together is necessary to ensure accurate and reliable detection of resistance.

Microbiologists must be aware of the strengths and weaknesses of their primary susceptibility testing methods (e.g., disk diffusion, commercial broth microdilution, automation) for detecting relevant resistance patterns and know when adjunct or supplemental testing is necessary. This awareness is accomplished by reviewing studies published in peer-reviewed journals that focus on the performance of antimicrobial testing systems and by periodically challenging one's own system with organisms that have been thoroughly characterized with respect to their resistance profiles (e.g., proficiency testing programs). Furthermore, accurate and relevant testing not only means using various conventional methods, or even using a mixture of automated, conventional, and screening methods, but also encompasses the potential application of molecular techniques and predictor drug panels (see Chapter 18).

Testing of *Streptococcus pneumoniae* provides one example of the need to be aware of testing limitations and the importance of implementing supplemental tests. Not long ago, susceptibility testing of *S. pneumoniae* was not considered routinely necessary. However, with the emergence of beta-lactam resistance, testing has become imperative. As the need for testing emerged, the inability of conventional tests, such as penicillin disk diffusion, to detect resistance became apparent. Fortunately, a test that utilizes the penicillin derivative oxacillin was developed and widely employed as a reliable screen for detecting nonsusceptibility to penicillin. However, this test is only a screen since the level of penicillin resistance (i.e., the minimal inhibitory concentration [MIC]) can vary greatly among nonsusceptible isolates and some strains that appear resistant by the screen may actually be susceptible. Because the level of resistance can affect therapeutic decisions, another method that allows for MIC determinations should be used for testing these organisms. Additionally, the emergence of cephalosporin (i.e., ceftriaxone or cefotaxime) resistance requires the use of tests that detect resistance to these agents.

Other important examples in which more than one method is required to obtain complete and accurate susceptibility testing data for certain organism groups or species include vancomycin-resistant enterococci, methicillin-resistant staphylococci, and extended-spectrum, beta-lactamase–producing *Enterobacteriaceae*. In addition, molecular methods also may be used in the clinical setting as an important backup resource to investigate and arbitrate equivocal results obtained by phenotypic methods. However, multiple testing protocols are not routinely necessary for every organism encountered. In most laboratories, one conventional or commercial method is likely to be the mainstay of testing, with adjunct testing used as a supplement when necessary.

REVIEW OF RESULTS

In addition to selecting one or more methods to accurately detect resistance, the strengths and weaknesses of these testing systems must be continuously monitored. This is primarily accomplished by carefully reviewing the susceptibility data that is produced daily. In the past, establishing and maintaining aggressive and effective monitoring programs often have been prohibitively labor intensive. Also, the effectiveness varied with the laboratorian's level of interest and knowledge. However, the speed and flexibility afforded by computerization of results-review and reporting greatly facilitates the administration of such quality assurance programs, even in laboratories with modest resources. Effective computer programs may be a part of the general laboratory information system, or in some cases, such programs are available through the commercial susceptibility testing system being used. Because automated expert data review greatly facilitates the review process and enhances data accuracy, this feature should be seriously considered when selecting an antimicrobial susceptibility testing system.

Susceptibility profiles must be scrutinized manually or with the aid of computers according to what profiles are likely, somewhat likely, somewhat unlikely, and nearly impossible. This awareness not only pertains to profiles exhibited by organisms within a particular institution but also to those exhibited by clinically relevant bacteria in general. The unusual resistance profiles must be discovered and evaluated expeditiously to determine whether they are due to technical or clerical errors or are truly indicative of an emerging resistance problem. The urgency of making this determination is twofold. First, if the profile results from laboratory error, it must be corrected and the physician notified so that the patient is not subjected to ineffective or inappropriate antimicrobial therapy. Second, if the profile is valid and presents a threat to the patient and to others (e.g., the emergence of vancomycin-resistant staphylococci), the immediate notification of infection control and infectious disease personnel is warranted.

Components of results review strategies

Any laboratory-based strategy for monitoring the accuracy of results and the emergence of resistance must have two components:

- Data review—a mechanism for recognizing new or unusual susceptibility profiles

- Resolution—the application of protocols for determining whether an unusual profile is due to error (technical or clerical) or accurately reflects the emergence of a new resistance mechanism

Both components must be integrated into the review process to ensure efficient and timely use of resources.

DATA REVIEW Recognition of unusual resistance profiles is primarily accomplished by carefully reviewing the susceptibility data produced daily. Examples of unusual susceptibility profiles for gram-positive and gram-negative bacteria are given in Table 19-3. The examples are a mixture of profiles that clearly demonstrate a likely error (i.e., clindamycin-resistant, erythromycin-susceptible staphylococci), profiles that have not been described but if observed require immediate attention (i.e., vancomycin resistance in staphylococci), or profiles that have been described but may not be common (i.e., imipenem resistance in *Enterobacteriaceae*). Another example is vancomycin resistance in enterococci. Although this profile has been well described, the implications of such a profile are

TABLE 19-3 EXAMPLES OF SUSCEPTIBILITY TESTING PROFILES REQUIRING FURTHER EVALUATION

ORGANISM	SUSCEPTIBILITY PROFILE
Staphylococci	Vancomycin intermediate or resistant
	Clindamycin resistant; erythromycin susceptible
Enterococci	Vancomycin intermediate or resistant (for an institution's initial cases)
Viridans streptococci	Vancomycin intermediate or resistant
Streptococcus pneumoniae	Vancomycin intermediate or resistant
Beta-hemolytic streptococci	Penicillin intermediate or resistant
Enterobacteriaceae	Imipenem resistant
Klebsiella spp./*E. coli*/*P. mirabilis*	Cefoxitin or cefotetan resistant
Enterobacter/Citrobacter/Serratia Morganella/Providencia	Susceptible to ampicillin or cefazolin
Pseudomonas aeruginosa	Amikacin resistant; gentamicin or tobramycin susceptible
Stenotrophomonas maltophilia	Imipenem susceptible; trimethoprim/sulfamethoxazole resistant
Neisseria gonorrhoeae	Ceftriaxone resistant; ciprofloxacin or ofloxacin resistant
Neisseria meningitidis	Penicillin resistant

Modified from Courvalin, P. 1992. American Society for Microbiology News 58:368.

so extensive that a particular institution's first encounter should be noted and substantiated.

The data review process for evaluation of profiles should not be the responsibility of a single person within the laboratory. Furthermore, the process requires checks and balances that do not impede the workflow or increase the time needed to get information to physicians. How this is established will vary depending on a particular laboratory's division of labor and workflow, but several key aspects to be considered include:

- The identification of the organism must be known. To evaluate the accuracy of a susceptibility profile, identification and susceptibility data must be simultaneously analyzed in a timely fashion. Without knowing the organism's identification, it is frequently difficult to determine whether the susceptibility profile is unusual.
- Susceptibility results should be analyzed and reported as early in the day as possible. The workflow should allow time for corrective action of errors found during data review so that corrected, or substantiated, results can be provided to physicians as soon as possible.
- Use of two or more tiers of data review. The first tier is at the bench level where technologists are simultaneously reading the results and evaluating an organism's susceptibility profile for appropriateness. When unusual profiles are found, the technologist should be able to initiate troubleshooting protocols (see the following section on resolution). To avoid the release of erroneous and potentially dangerous information, results should not be reported at this point. Review at this level, which is greatly facilitated by automated expert review systems, maintains proficiency among technologists regarding the nuances of susceptibility profiles and important resistance patterns. The second tier is at the level of supervisory or laboratory director. The purpose of review at this level is to track and monitor the efficiency of the first tier, to take ultimate responsibility for the accuracy of results, to provide constructive and educational feedback to the technologists performing the first-line review, and to provide guidance for resolution of the unusual profiles. Again, a computer-based review process that searches all reports for predefined unusual profiles (similar to those outlined in Table 19-3) can greatly enhance the efficiency and accuracy of the second level of review. An example of a computer-generated review report is shown in Figure 19-2.
- The review process must be flexible and updated. Because bacterial capabilities for antimicrobial

resistance profiles change, laboratory resistance detection systems can become outdated. Therefore, the list of unusual profiles requires periodic review and updating.

RESOLUTION The importance of having strategies for resolving unusual profiles cannot be overstated. However, developing detailed procedures for every possible contingency is not possible or practical. Most resolution strategies should focus on certain general approaches, with supervisory or laboratory director consultation always being among the options available to technologists. Although the steps taken to investigate and resolve an unusual profile often depend on the organism and antibiotics involved,

---

Patient name: John Doe

Patient registration number: 11111111

Laboratory accession number: 22222

Report date: Apr 7 1996 12:00:00:000AM

Collection date: Apr 5 1996 11:09:00:000AM

Collection location: X-ray

Specimen type: Aspirate—Kidney

Organism identification (code): *Klebsiella pneumoniae* (KP)

Susceptibility profile:

Antimicrobial agent	Interpretive category
Gentamicin	Intermediate
Tobramycin	Intermediate
Ampicillin	Resistant
Cefazolin	Intermediate
Cefotetan	Susceptible
Ceftazidime	Resistant
Ceftriaxone	Susceptible
Aztreonam	Resistant
Imipenem	Susceptible
Trimethoprim/ sulfamethoxazole	Resistant

Unusual profile detected:

Klebsiella pneumoniae susceptible to cefotetan but resistant to ceftazidime and resistant to aztreonam.

Significance:

Profile suggests possible extended-spectrum, beta-lactamase production; effectiveness of other cephalosporins against this bacterial isolate is uncertain.

FIGURE 19-2 Example of a laboratory report that has identified an unusual susceptibility profile.

most protocols for resolution should include one or more of the following approaches:

- Review data for possible clerical error
- Determine if susceptibility panel and identification system were inoculated with same isolate
- Reexamine test panel or plate for reading error (e.g., misreading of actual zone of inhibition)
- Confirm purity of inoculum and proper inoculum preparation
- For commercial systems, determine if manufacturer's recommended procedures were followed
- Confirm accuracy of organism identification
- Confirm resistance by using a second method or screening test

Often a quick review of the data recording and interpretation aspects, or purity of culture, will reveal the reason why an unusual profile was obtained. Other times more extensive testing, perhaps by more than one method, may be needed to establish the validity of an unusual or unexpected resistance profile.

ACCURACY AND ANTIMICROBIAL RESISTANCE SURVEILLANCE

Antimicrobial resistance surveillance involves tracking the susceptibility profiles produced by the bacteria that are encountered within a particular institution, within a specific geographic location (i.e., regionally, nationally, or internationally). For laboratories that serve a particular institution or group of institutions, periodic publication of an antibiogram report that contains susceptibility data is the extent of the surveillance offered (see Figure 3-6). These reports, which may be further organized in various ways (e.g., according to hospital location, site of infection, outpatient vs. inpatient, duration of hospital stay), provide valuable information for monitoring emerging resistance trends among the local microbial flora. Such information is also helpful for establishing **empirical therapy** guidelines (i.e., therapy that is instituted before knowledge of the infecting organism's identification or its antimicrobial susceptibility profile), detecting areas of potential inappropriate or excessive antimicrobial use, and contributing data to larger, more extensive surveillance programs.

Data that has been validated through a results-review and resolution program not only enhance the reliability of laboratory reports for patient management but also strengthen the credibility of susceptibility data used for resistance surveillance and antibiogram profiling. Therefore, the need for each institution to scrutinize susceptibility profiles daily can be accomplished by establishing a results-review and resolution format that ensures accuracy for patient management, detects emerging resistance patterns quickly, and ensures accuracy of the data that is included in summary antibiogram reports.

COMMUNICATION

Susceptibility testing profiles produced for each bacterial isolate are typically reported to the physician as a listing of the antimicrobial agents, with each agent accompanied by the category interpretation of susceptible, intermediate, or resistant (see Figure 19-2). In most instances, this reporting approach is sufficient. However, as resistance profiles and their underlying mechanisms become more varied and complex, laboratory personnel must ensure that the significance of susceptibility data is clearly and accurately communicated to clinicians in a way that optimizes both patient care and antimicrobial utilization. In many situations, passively communicating only the susceptibility data to the physician without adding comments or appropriately amending the reports is no longer sufficient.

For example, methicillin-resistant staphylococci are to be considered cross-resistant to all beta-lactams, but in vitro results occasionally may indicate susceptibility to certain cephalosporins, beta-lactam/beta-lactamase inhibitor combinations, or imipenem. Simply reporting these findings without editing such profiles to reflect probable resistance to all beta-lactams would be seriously misleading. Similarly, certain species of *Enterobacteriaceae* (e.g., *E. coli, Klebsiella pneumoniae*) capable of producing potent beta-lactamases that hydrolyze several different cephalosporins may appear susceptible to certain cephalosporins, while the clinical efficacy of these antimicrobial agents for infections caused by such organisms is doubtful. Again, simply reporting the profile obtained by the laboratory would be inadequate and somewhat misleading. Preferably, the laboratory should contact the physician, either directly or through an infectious disease consultant, so that concerns regarding the true utility of the cephalosporins may be adequately addressed (see "Significance" note at bottom of Figure 19-2). As a final example, serious enterococcal infections often require combination therapy that includes both a cell wall-active agent (ampicillin or vancomycin) and an aminoglycoside (i.e., gentamicin). This important infor-mation would not be conveyed in a report that simply lists the agents and their interpretive category results. Such an approach can leave the false impression that a "susceptible" result for any single agent indicates that one drug used alone provides appropriate therapy. Therefore, an explanatory note that clearly states the recommended use of combination therapy should accompany the enterococcal susceptibility report.

To avoid misinterpretations that may result by providing only antimicrobial susceptibility data, strategies must consider those organism-antimicrobial combinations that may require reporting supplemental messages to the physician. Consultations with infectious diseases specialists and other members of the medical staff is an important part of determining when such messages are needed and what the content should include. Finally, if a laboratory does not have the means to reliably relay these messages, either via the computer or by paper, then a policy of direct communication with the attending physician by phone or in person should be established.

Bibliography

Courvalin, P. 1992. Interpretive reading of antimicrobial susceptibility tests. American Society for Microbiology News 58:368.

Jorgensen, J.H. and Sahm, D.F. 1995. Antimicrobial susceptibility testing: general considerations. In Murray, P.R., Baron, E.J., Pfaller, M.A., et al., editors: Manual of clinical microbiology, ed 6. American Society for Microbiology, Washington, D.C.

Sahm, D.F. and O'Brien, T.F. 1994. Detection and surveillance of antimicrobial resistance. Trends in microbiology, virulence, infection, and pathogenesis. 2:366.

Sahm, D.F. 1996. The role of clinical microbiology in the control and surveillance of antimicrobial resistance. American Society for Microbiology News 62:25.

Part Three 3

 Diagnosis by Organ System

20 | BLOODSTREAM INFECTIONS

Microorganisms present in the circulating blood, whether continuously, intermittently, or transiently, are a threat to every organ in the body. Microbial invasion of the bloodstream can have serious immediate consequences, including shock, multiple organ failure, disseminated intravascular coagulation (DIC), and death. Approximately 200,000 cases of bacteremia and fungemia occur annually, with mortality rates ranging from 20% to 50%. Thus, invasion of the bloodstream by microorganisms constitutes one of the most serious situations in infectious disease, and as a result, timely detection and identification of bloodborne pathogens is one of the most important functions of the microbiology laboratory. Pathogens of all four major groups of microbes—bacteria, fungi, viruses, and parasites—may be found circulating in blood during the course of many diseases. Positive blood cultures may help provide a clinical diagnosis, as well as a specific etiologic diagnosis.

GENERAL CONSIDERATIONS

The successful recovery of microorganisms from blood by the laboratory is dependent on many, often complex, factors: the possible types of **bacteremia** (presence of bacteria in blood), specimen collection methods, blood volumes, the number and timing of blood cultures, interpretation of results, and the type of patient population being served by the laboratory. If each of these factors or issues is not addressed by the laboratory in the development of their blood culture protocols, the detection and recovery of microorganisms may be severely compromised.

CAUSES

As previously mentioned, all major groups of microbes can be present in the bloodstream during the course of many diseases.

Bacteria

The organisms most commonly isolated from blood are gram-positive cocci, including coagulase-negative staphylococci, *Staphylococcus aureus*, and *Enterococ-*

cus spp., and other organisms likely to be inhabitants of the hospital environment that colonize the skin, oropharyngeal area, and gastrointestinal tract of patients. Some of the most common, clinically significant bacteria isolated from blood cultures are listed in Box 20-1. In general, the number of fungi and coagulase-negative staphylococci has increased while the number of clinically significant anaerobic isolates has decreased over the past decade.[3,13,16]

Of importance, the laboratory isolation of certain bacterial species from blood can indicate the presence of an underlying, occult, or undiagnosed neoplasm.[2] Alterations in local conditions at the site of the neoplasm that allow bacteria to proliferate and seed the bloodstream have been suggested as a possible mechanism for this association between bacteremia and cancer. Reduced killing of bacterial cells by the host phagocytes is a second possible mechanism. Organisms associated with neoplastic disease include *Clostridium septicum* and other uncommonly isolated clostridial species, *Streptococcus bovis*, *Aeromonas hydrophila*, *Plesiomonas shigelloides*, and *Campylobacter* spp.

Fungi

Fungemia (presence of fungi in blood) is usually a serious condition, occurring primarily in immunosuppressed patients and in those with serious or

BOX 20-1 **BACTERIA COMMONLY ISOLATED FROM BLOOD CULTURES**

Coagulase-negative staphylococci
Staphylococcus aureus
Viridans streptococci
Enterococcus spp.
Beta-hemolytic streptococci
Streptococcus pneumoniae
Escherichia spp.
Klebsiella spp.
Pseudomonas spp.
Enterobacter spp.
Proteus spp.
Anaerobic bacteria—*Bacteroides* and *Clostridium* spp.

TABLE 20-1 MAYO CLINIC FUNGAL BLOOD CULTURE EXPERIENCE: 1972 THROUGH SEPTEMBER 1992*

ORGANISM	NO. POSITIVE CULTURES	NO. PATIENTS
Candida albicans	3038	889
Candida tropicalis	682	274
Candida glabrata	666	261
Candida parapsilosis	612	302
Histoplasma capsulatum	420	135
Cryptococcus neoformans	306	106
Candida lusitaniae	63	22
Candida krusei	59	25
Coccidioides immitis	42	9
Candida guilliermondi	35	14
Trichosporon cutaneum	30	13
Malassezia furfur	28	14
Rhodotorula spp.	20	14
Nocardia asteroides	20	19
Fusarium spp.	19	17
Cryptococcus albidus	14	13
Cryptococcus laurentii	12	11
Candida spp.	11	7
Candida zeylanoides	7	3
Saccharomyces spp.	5	4
Pichia burtonii	5	1
Candida pseudotropicalis	3	1
Candida famata	3	3
Blastomyces dermatitidis	3	3
Sporothrix schenckii	1	1
Blastoschizomyces capitatus	1	1
Trichophyton rubrum	1	1

*Table represents 6106 fungi recovered from 2163 patients; overall positivity rate was 6106/400,538 (1.52%) cultures.

From Geha, D.J. and Roberts, G.D. Laboratory detection of fungemia, Clin. Lab. Med. 14:83, 1994.

terminal illness. *Candida albicans* is by far the most common species, but *Malassezia furfur* can often be isolated in patients, particularly neonates, receiving lipid-supplemented parenteral nutrition. *Candida* spp. account for approximately 10% to 15% of all nosocomial bloodstream infections.[4] Table 20-1 shows the fungi recovered from blood cultures over a 20-year period at the Mayo Clinic, a major medical referral center in Minnesota.

Except for *Histoplasma*, which multiply in leukocytes (white blood cells), fungi do not invade blood cells, but their presence in the blood usually indicates a focus of infection elsewhere in the body. Fungi in the bloodstream can **disseminate** (be carried) to all organs of the host, where they may grow, invade normal tissue, and produce toxic products. Fungi gain entrance to the circulatory system via loss of integrity of the gastrointestinal or other mucosa; through damaged skin; from primary sites of infection, such as the lung or other organs; or by means of intravascular catheters.

Systemic fungal infections that begin as pneumonia disseminate from the lungs, which serve as the portal of entry. Arthroconidia of *Coccidioides immitis* and microconidia of *Histoplasma capsulatum* and *Blastomyces dermatitidis* are ingested by alveolar macrophages in the lung. These macrophages carry the fungi to nearby lymph nodes, usually the hilar nodes. The fungi multiply within the node tissue and ultimately are released into the circulating blood, from which they go on to seed other organs or are destroyed by the body's defenses. The large size and sterol-containing cell walls of molds make them particularly insensitive to the primary host defenses, that is, antibody and phagocytic cells.

Parasites

Eukaryotic parasites may be found transiently in the bloodstream as they migrate to other tissues or organs. Their presence, however, cannot be considered consistent with a state of good health. For example, tachyzoites of the parasite *Toxoplasma gondii* may be found in circulating blood. They invade cells within lymph nodes and other organs, including the lungs, liver, heart, brain, and eyes. The resulting cellular destruction accounts for the manifestations of toxoplasmosis. Also, microfilariae are seen in peripheral blood during infection with *Dipetalonema*, *Mansonella*, *Loa loa*, *Wuchereria*, or *Brugia*.

Malarial parasites invade host erythrocytes and hepatic parenchymal cells. The significant anemia and subsequent tissue hypoxia (reduction in oxygen levels) may result from destruction of red blood cells by the parasite. Vascular trapping of normal erythrocytes by the infected red cells, which are less flexible and tend to clog small capillaries, are major causes of morbid-

ity. The host's immunologic response is to remove the parasites and damaged red blood cells; the immune response may also have deleterious effects.

Parasites in the bloodstream are usually detected by direct visualization. Those parasites for which diagnosis is dependent on observation of the organism in peripheral blood smears include *Plasmodium*, *Trypanosoma*, and *Babesia*. Patients with malaria or filariasis may display a periodicity in their episodes of fever that allows the physician to time the collection of blood for microscopic examination for optimal detection.

Viruses

Although many viruses do circulate in the peripheral blood at some stage of disease, the primary pathology relates to infection of the target organ or cells. Those viruses that preferentially infect blood cells are Epstein-Barr virus (invades lymphocytes), cytomegalovirus (invades monocytes, polymorphonuclear cells, and lymphocytes), and HIV (involves only certain T lymphocytes and perhaps macrophages) and other human retroviruses that attack lymphocytes. The pathogenesis of viral diseases of the blood is the same as that of viral diseases of any organ; by diverting the cellular machinery to create new viral components or by other means, the virus may prevent the host cell from performing its normal function. The cell may be destroyed or damaged by viral replication, and immunologic responses of the host may also contribute to the pathogenesis.

Although many viral diseases have a viremic stage, recovery of virus particles or detection of circulating viruses is used in the diagnosis of only a few diseases. Chapter 66 will discuss recovery of viruses from blood in greater detail.

TYPES OF BACTEREMIA

Bacteremia may be **transient, continuous,** or **intermittent.** Transient, incidental bacteremia may occur spontaneously or with such minor events as brushing teeth or chewing food. Other conditions in which bacteria are only transiently present in the bloodstream include manipulation of infected tissues, instrumentation of contaminated mucosal surfaces, and surgery involving nonsterile sites. These circumstances may also lead to significant septicemia.

In septic shock, bacterial endocarditis, and other endovascular infections, organisms are released into the bloodstream at a fairly constant rate (continuous bacteremia). Also, during the early stages of specific infections, including typhoid fever, brucellosis, and leptospirosis, bacteria are continuously present in the bloodstream.

In most other infections, such as in patients with undrained abscesses, bacteria can be found intermittently in the bloodstream. Of note, the causative agents of meningitis, pneumonia, pyogenic arthritis, and osteomyelitis are often recovered from blood during the early course of these diseases. In the case of transient seeding of the blood from a sequestered focus of infection, such as an abscess, bacteria are released into the blood approximately 45 minutes before a febrile episode.

TYPES OF BLOODSTREAM INFECTIONS

The two major categories of bloodstream infections are: **intravascular** (those that originate within the cardiovascular system) and **extravascular** (those that result from bacteria entering the blood circulation through the lymphatic system from another site of infection). Of note, other organisms, such as fungi, may also cause intra- or extravascular infections. However, because bacteria account for the majority of significant vascular infections, these types of bloodstream infections will be addressed. Factors that contribute to the initiation of bloodstream infections are immunosuppressive agents, widespread use of broad-spectrum antibiotics that suppress the normal flora and allow the emergence of resistant strains of bacteria, invasive procedures that allow bacteria access to the interior of the host, more extensive surgical procedures, and prolonged survival of debilitated and seriously ill patients.

Intravascular infections

Intravascular infections include infective endocarditis, mycotic aneurysm, suppurative thrombophlebitis, and intravenous, catheter-associated bacteremia. Because these infections are within the vascular system, organisms are present in the bloodstream at a fairly constant rate (i.e., a continuous bacteremia). These infections in the cardiovascular system are extremely serious and are considered life-threatening.

INFECTIVE ENDOCARDITIS The development of **infective endocarditis** (infection of the endocardium most commonly caused by bacteria) is believed to involve several independent events. Cardiac abnormalities, such as congenital valvular diseases that lead to turbulence in blood flow or direct trauma from intravenous catheters, can damage cardiac endothelium. This damage to the endothelial surface results in the deposition of platelets and fibrin. If bacteria transiently gain access to the bloodstream (this can occur after an innocuous procedure such as brushing the teeth) after alteration of the capillary endothelial cells, the organisms may stick to and then colonize the damaged cardiac endothelial cell surface. After colonization, the surface will rapidly be covered with a protective layer of fibrin and platelets. This protective environment is favorable to further bacterial multiplication. This web of platelets, fibrin, inflammatory cells, and entrapped

organisms is called a **vegetation** (Figure 20-1). The resulting vegetations ultimately seed bacteria into the blood at a slow but constant rate.

The primary causes of infective endocarditis are the viridans streptococci, comprising several species (Box 20-2). These organisms are normal inhabitants primarily of the oral cavity, often gaining entrance to the bloodstream as a result of gingivitis, periodontitis, or dental manipulation. Heart valves, especially those that have been previously damaged, present convenient surfaces for attachment of these bacteria. Identification of these streptococci may be useful to the patient's clinician because certain species, such as *Streptococcus anginosus,* may be associated with increased frequency of metastatic abscess formation. *Streptococcus sanguis* and *Streptococcus mutans* are most frequently isolated in streptococcal endocarditis.

With the ever-increasing use of intravenous catheters (see facing page), intraarterial lines, and vascular prostheses, organisms found as normal or hospital-acquired inhabitants of the human skin are able to gain access to the bloodstream and find a surface on which to grow, including heart valves and vascular endothelium. In such a setting, *Staphylococcus epidermidis* and other coagulase-negative staphylococci have been increasingly implicated as causes of infection. *S. epidermidis* is the most common etiologic agent of prosthetic valve endocarditis, with *S. aureus* the second most common. *S. aureus* is an important cause of septicemia without endocarditis and is found in association with other foci, such as abscesses, wound infections, and pneumonia, as well as sepsis related to indwelling intravascular catheters.

MYCOTIC ANEURYSM AND SUPPURATIVE THROMBOPHLEBITIS Two other intravascular infections, mycotic aneurysms and suppurative thrombophlebitis, result from damage

BOX 20-2 | **AGENTS OF INFECTIVE ENDOCARDITIS**

Viridans streptococci*
Nutritionally deficient streptococci
Enterococci*
Streptococcus bovis
*Staphylococcus aureus**
Staphylococci (coagulase-negative)
Enterobacteriaceae
Pseudomonas spp. (usually in drug abusers)
Haemophilus spp., particularly *H. aphrophilus*
Unusual gram-negative bacilli (e.g., *Actinobacillus, Cardiobacterium, Eikenella*)
Yeast spp.
Other (including polymicrobial infectious endocarditis)

---

*Most common organisms associated with native valve endocarditis in nondrug-abusing adults.

FIGURE 20-1 Vegetations of bacterial endocarditis. Arrow indicates the vegetations. (Courtesy Celeste N. Powers, M.D. Ph.D., SUNY Health Science Center, Syracuse, NY.)

to the endothelial cells lining blood vessels. With respect to **mycotic aneurysm,** an infection causes inflammatory damage and weakening of an arterial wall; this weakening causes a bulging of the arterial wall (i.e., aneurysm) that can eventually rupture. The etiologic agents are similar to those that cause endocarditis.

Suppurative thrombophlebitis is an inflammation of a vein wall. The pathogenesis of this intravascular infection involves an alteration in a vein's endothelial lining that is followed by clot formation. This site is then seeded with organisms, thereby establishing a primary site of infection. Suppurative thrombophlebitis represents a frequent complication of hospitalized patients that is caused by the increasing use of intravenous catheters.

INTRAVENOUS CATHETER-ASSOCIATED BACTEREMIA Intravenous (IV) catheters are an integral part of the care of hospitalized patients; more than 3 million central venous catheters are used annually in the United States. For example, central venous catheters are used to administer fluids, blood products, medications, antibiotics, nutrition, and for hemodynamic monitoring. A short-term, triple lumen (channel opening within a tube) central venous catheter is shown in Figure 20-2. Unfortunately, one major consequence of these medical devices is colonization of the catheter by either bacteria or fungi, which can lead to catheter infection and serious bloodstream infection. This consequence is a major nosocomial source of illness and even death.

IV catheter-associated bacteremia (or fungemia)

is believed to occur primarily by two routes (Figure 20-3).[15] The first route involves the movement of organisms from the catheter skin entry site down the external surface of the catheter, to the catheter tip within the bloodstream. After arriving at the tip, the organisms multiply and may cause a bacteremia. The second way that IV catheter-associated bacteremia may occur is by migration of organisms along the inside of the catheter (the lumen) to the catheter tip. The catheter's hub, where tubing connects into the IV catheter, is considered the site where organisms can gain access to the patient's bloodstream through the catheter lumen. The most common etiologic agents for IV catheter-associated bloodstream infections, regardless of the route of infection, are organisms found on the skin (Box 20-3). Certain strains of *S. epidermidis* appear to be uniquely suited for causing catheter-related infections because of their ability to produce a "slime" that consists of complex sugars

COMMON AGENTS OF IV CATHETER-ASSOCIATED BACTEREMIA

BOX 20-3

Staphylococcus epidermidis
Other coagulase-negative staphylococci
Staphylococcus aureus
Enterobacteriaceae
Pseudomonas aeruginosa
Candida spp.
Corynebacterium spp.
Other gram-negative rods

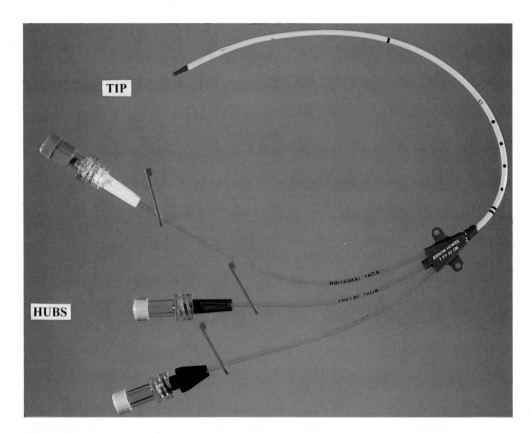

FIGURE 20-2 Short-term, triple-lumen central venous catheter. The ends from which the catheter is accessed are usually referred to as the hub(s). After the catheter is inserted, the tip will reside within the bloodstream.

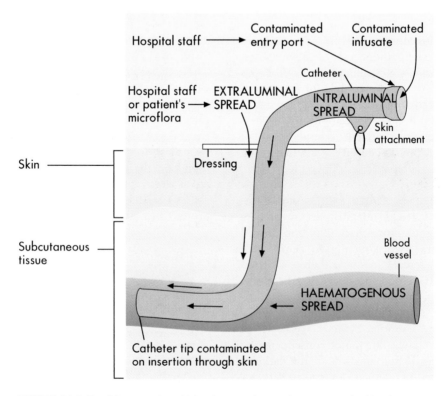

FIGURE 20-3 Possible routes by which microorganisms gain access to the bloodstream to cause intravenous catheter-associated bacteremias. (Modified from Elliott, T.S.J. 1993. PHLS Communicable disease report: line-associated bacteraemias. CDR Review 3(7):R91.)

(polysaccharides) that are believed to help the organism adhere to the catheter's surface. Uncommon routes of IV catheter-tip infection include contaminated fluids or bloodborne seeding from another infection site.[15]

Extravascular infections

Except for intravascular infections, bacteria usually enter the circulation through the lymphatic system. Most cases of clinically significant bacteremia are a result of extravascular infection. When organisms multiply at a local site of infection such as the lung, they can be drained by the lymphatics and reach the bloodstream. In most individuals, organisms in the bloodstream are effectively and rapidly removed by the reticuloendothelial system in the liver, spleen, and bone marrow and by circulating phagocytic cells. But, depending on the extent of local control of the infection, the organism may be circulated more widely, thereby causing a bacteremia or fungemia.

The most common **portals of entry** for bacteremia are the genitourinary tract (25%), respiratory tract (20%), abscesses (10%), surgical wound infections (5%), biliary tract (5%), miscellaneous sites (10%), and uncertain sites (25%). For the most part, the probability of bacteremia occurring from an extravascular site is dependent on the site of infection, its severity, and the organism. For example, any organism producing meningitis is likely to produce bacteremia at the same time. Of importance, certain organisms causing extravascular infections commonly invade the bloodstream; some of these organisms are listed in Table 20-2. In addition to these organisms, a large number of other bacteria and fungi that cause extravascular infections also invade the bloodstream but not as frequently. Whether these organisms invade the bloodstream depends on the host's ability to control the local infection and the organism's pathogenic potential. Some of the organisms that can frequently seed the bloodstream from a localized site of infection, if conditions allow, include: members of the family *Enterobacteriaceae*, *Streptococcus pneumoniae*, *Staphylococcus aureus*, *Neisseria gonorrhoeae*, anaerobic cocci, *Bacteroides*, *Clostridium*, beta-hemolytic streptococci, and *Pseudomonas*. Although lengthy, this represents only some of the organisms frequently isolated from blood and should not be considered inclusive. Almost every known bacterial species and many fungal species have been implicated in extravascular bloodstream infections.

CLINICAL MANIFESTATIONS

As previously discussed, bacteremia may indicate the presence of a focus of disease, such as intravascular infection, pneumonia, or liver abscess, or may merely represent transient release of bacteria into the bloodstream. **Septicemia** or sepsis indicates a situation in which bacteria or their products (toxins) are causing harm to the host. Unfortunately, clinicians often use the terms bacteremia and septicemia interchangeably. Signs and symptoms of septicemia may include fever or hypothermia (low body temperature), chills, hyperventilation (abnormally increased breathing that leads to excess loss of carbon dioxide from the body) and subsequent respiratory alkalosis (condition caused by the loss of acid leading to an increase in pH), skin lesions, change in mental status, and diarrhea. More serious manifestations include hypotension or shock, DIC, and major organ system failure. The syndrome, known as septic shock, characterized by fever, acute respiratory distress, shock, renal failure, intravascular coagulation, and tissue destruction, can be initiated

---

TABLE 20-2 ORGANISMS COMMONLY ASSOCIATED WITH BLOODSTREAM INVASION FROM EXTRAVASCULAR SITES OF INFECTION

ORGANISM	EXTRAVASCULAR SITE OF INFECTION
Haemophilus influenzae type b	Meninges, epiglottis, periorbital region
Streptococcus pneumoniae	Meninges, sometimes the lung
Neisseria meningitidis	Meninges
Brucella spp.	Reticuloendothelial system
Salmonella typhi	Small intestine, regional lymph nodes of the intestine, reticuloendothelial system
Listeria	Meninges

by either exotoxins or endotoxins. Septic shock is mediated by activated mononuclear cells producing cytokines, such as tumor necrosis factor and interleukins.

Shock is the gravest complication of septicemia. In **septic shock,** the presence of bacterial products and the host's responding defensive components act to shut down major host physiologic systems. Manifestations include a drop in blood pressure, increase in heart rate, impairment of function of vital organs (brain, kidney, liver, and lungs), acid-base problems, and bleeding problems. Gram-negative bacteria contain a substance in their cell walls, called endotoxin, that has a strong effect on several physiologic functions. This substance, a lipopolysaccharide (LPS) comprising part of the cell wall structure (see Chapter 9), may be released during the normal growth cycles of bacteria or after the destruction of bacteria by host defenses. Endotoxin (or the core of the LPS, lipid A) has been shown to mediate numerous systemic reactions, including a febrile (producing fever) response, and the activation of complement and certain blood-clotting factors. Although most gram-positive bacteria do not contain endotoxin, many produce exotoxins, and the effects of their presence in the bloodstream may be equally devastating to the patient.

Disseminated intravascular coagulation (DIC) is a disastrous complication of sepsis. DIC is characterized by numerous small blood vessels becoming clogged with clotted blood and bleeding as a result of the depletion of coagulation factors. DIC can occur with septicemia involving any circulating pathogen, including parasites, viruses, and fungi, although it is most often a consequence of gram-negative bacterial sepsis.

IMMUNOCOMPROMISED PATIENTS

One of the greatest challenges facing microbiologists is the handling of blood cultures from immunocompromised patients. The number of immunocompromised patients has steadily increased in recent years in large part as the result of advances in medicine. People undergoing organ transplantation, elderly persons, individuals with malignant disease (such as hematologic malignancies and cancer), and those receiving therapy for the malignancy are examples of immunosuppressed patients. Most recently, acquired immunodeficiency syndrome (AIDS) has contributed to the increase in the number of immunocompromised individuals. The marked immunosuppression brought about by infection with the human immunodeficiency virus (HIV) in patients with AIDS is a result of this virus' profound impairment of cellular immunity. Patients with AIDS have the greatest diversity of pathogens recovered from blood, including

mycobacterial species, *Bartonella henselae, Corynebacterium jeikeium, Shigella flexneri,* unusual *Salmonella* species, *Histoplasma capsulatum, Cryptococcus neoformans,* and cytomegalovirus.

As is typically observed in other hospitalized patients, organisms such as gram-positive aerobic bacteria (e.g., *Staphylococcus aureus, Enterococcus*) and gram-negative aerobic bacteria (e.g., *Enterobacteriaceae, Pseudomonas aeruginosa*) are common causes of bloodstream infections in immunocompromised patients. In addition, bloodstream infections in immunocompromised patients are frequently caused by either unusual pathogens whose recovery from blood requires special techniques or by organisms that are normally considered contaminants when isolated from blood cultures. Therefore, microbiologists must be aware of the potential pathogenicity of some organisms in immunosuppressed patients that are generally regarded as probable blood culture contaminants. Without this knowledge, aerobic gram-positive rods isolated from blood cultures may be dismissed as contaminating diphtheroids, when, in fact, the organism is *C. jeikeium,* which is known to cause bacteremia in HIV-infected patients. By the same token, microbiologists must be familiar with the unusual pathogens isolated from blood cultures obtained from immunocompromised patients and in particular, those organisms that require special techniques for isolation (some of the special considerations are covered later in this chapter).

DETECTION OF BACTEREMIA

Mortality rates associated with bloodstream infection range from 20% to 50%. Because bacteremia frequently portends life-threatening infection, the prompt detection and recovery of microorganisms from blood is of paramount importance.

To detect bloodstream infections, a patient's blood must be obtained by aseptic venipuncture and then incubated in culture media. Bacterial growth can be detected using techniques ranging from manual to totally automated methods. Once growth is detected, the organism(s) is isolated, identified, and tested for its susceptibility to various antimicrobial agents when appropriate (see Chapter 19).

SPECIMEN COLLECTION

Preparation of the site

Because blood culture media have been developed as enrichment broths to encourage the multiplication of even one bacterium, it follows that these media will enhance the growth of any stray contaminating bacterium, such as a normal inhabitant of human skin. Therefore, careful skin preparation before collecting the blood sample is of paramount importance to reduce the risk of introducing contaminants into blood culture media.

The vein from which the blood is to be drawn must be chosen before the skin is disinfected. If a patient has an existing IV line, the blood should be drawn below the existing line; blood drawn above the line will be diluted with fluid being infused. It is less desirable to draw blood through a vascular shunt or catheter, because these prosthetic devices are difficult to decontaminate completely.

ANTISEPSIS   Once a vein is selected, the skin site is defatted with 70% alcohol and an antiseptic is then applied to kill surface and subsurface bacteria. Regardless of the antiseptic used, it is critical that the manufacturer's recommendation be followed as to length of time the antiseptic is allowed to remain on the skin. The steps necessary for drawing blood for culture are given in Procedure 20-1 at the end of this chapter.

As part of ongoing quality assurance, laboratories should determine the rate of blood culture contamination by clinically evaluating patients' conditions in conjunction with the organism isolated from culture. Laboratories that recover contaminants at rates greater than 3% should suspect improper phlebotomy techniques and should institute measures to reeducate the phlebotomists in proper skin preparation methods.

PRECAUTIONS   Universal precautions require that phlebotomists wear gloves for blood drawing. Because blood for culture must be obtained aseptically, it is important that any contaminated surfaces that might come in contact with the disinfected venipuncture site be disinfected. For example, if the site must be touched after preparation, the phlebotomist must disinfect the gloved fingers used for palpation in identical fashion. Also, if the rubber stopper or septum of the container into which blood is to be inoculated (e.g., test tubes or commercial culture bottles) is potentially contaminated, the phlebotomist must disinfect the septum.

Specimen volume

ADULTS   For many years, it has been recognized that most bacteremias in adults have a low number of colony-forming units (CFU) per milliliter (mL) of blood. For example, in several studies, fewer than 30 CFU per mL of blood were commonly found in patients with clinically significant bacteremias. Therefore, a sufficient sample volume is critical for the successful detection of bacteremia.

Because there is a direct relationship between the volume of blood and the yield, it follows that the more blood that is cultured, the greater the chance of isolating the organism. Therefore collection of 10 to 20 mL of blood per culture is strongly recommended for adults; a volume of 10 mL appears to be the absolute minimum. Unfortunately, a recent study confirmed that it is common practice to underinoculate blood culture bottles; findings from this study suggested that the yield increases by 3.2% for each milliliter of blood cultured.[10]

CHILDREN   It is not safe to take large samples of blood from children, particularly infants. Fortunately, infants with more serious disease usually yield more than 10 CFU of bacteria per milliliter of blood. For infants and small children, only 1 to 5 mL of blood can usually be drawn for bacterial culture. Blood culture bottles are available that have been designed specifically for the pediatric patient. Because blood specimens from septic children may yield fewer than 5 CFU/mL of the organism, quantities less than 1 mL may not be adequate to detect pathogens. Nevertheless, smaller volumes should still be cultured because high levels of bacteremia (more than 1000 CFU/mL of blood) are detected in some infants.

If one of the many commercial methods for culturing blood (e.g., automated, instrument-based systems, biphasic broth slide, lysis centrifugation) is being used, the laboratory must adhere to the specific recommendations of the manufacturer regarding blood culture volumes.

Number of blood cultures

Because periodicity of microorganisms in the bloodstream may be characteristic for some diseases, continuous for some, and random in others, these patterns of bacteremia must be considered in establishing standards for the timing and number of blood cultures. If the volume of blood is adequate, usually two or three blood cultures are sufficient to achieve the optimum blood culture sensitivity. In patients with endocarditis who have not received antibiotics, a single blood culture will be positive in 90% to 95% of the cases, while a second blood culture will establish the diagnosis in ≥98% of patients, depending on the study. For patients who have received prior antibiotic therapy, three separate blood collections of 10 to 20 mL each, and an additional blood culture or two taken on the second day, if necessary, will detect most etiologic agents of endocarditis. This presumes use of a culture system that is adequate for growth of the organism involved, which often entails extending the incubation period. Similarly, in bacteremic patients without endocarditis, 80% to 92% are detected by the first blood culture, 90% to 99% by the first two cultures, and 99.6% by at least one of the first three cultures.[17]

Timing of collection

The timing of cultures is not as important as other factors in patients with intravascular infections because organisms are released into the bloodstream at a fairly constant rate. Because the timing of intermittent bacteremia is unpredictable, it is generally accepted that two or three blood cultures be spaced an hour apart. However, a recent study found no significant difference in the yield between multiple blood cultures obtained simultaneously or those obtained at intervals.[8] The authors concluded that the overall volume of blood cultured was more critical to increasing organism yield than was timing.

When a patient's condition requires that therapy be initiated as rapidly as possible, little time is available to collect multiple blood culture samples over a timed interval. An acceptable compromise is to collect 40 mL of blood at one time, 20 mL from each of *two separate venipuncture sites, using two separate needles and syringes* before the patient is given antimicrobial therapy. For initial evaluation of fever of unknown origin, four separate blood cultures, two drawn on each of 2 days, will detect most causative agents.

Miscellaneous matters

ANTICOAGULATION Blood drawn for culture must not be allowed to clot. If bacteria become entrapped within a clot, their presence may go undetected. Thus, blood drawn for culture may be either inoculated directly into the blood culture broth media or into a sterile blood collection tube containing an anticoagulant for transport to the laboratory for subsequent inoculation. Heparin, ethylenediaminetetraacetic acid (EDTA), and citrate inhibit numerous organisms and are not recommended for use. Sodium polyanethol sulfonate (SPS, Liquoid) in concentrations of 0.025% to 0.03% is the best anticoagulant for blood. As a result, the most commonly used preparation in blood culture media today is 0.025% to 0.05% SPS. In addition to its anticoagulant properties, SPS is also anticomplementary and antiphagocytic, and interferes with the activity of some antimicrobial agents, notably aminoglycosides. SPS, however, may inhibit the growth of a few microorganisms, such as some strains of *Neisseria* spp., *Gardnerella vaginalis, Streptobacillus moniliformis,* and all strains of *Peptostreptococcus anaerobius.* Although the addition of 1.2% gelatin has been shown to counteract this inhibitory action of SPS, the recovery of other organisms decreases.

DILUTION In addition to volume of blood cultured and type of medium chosen, the dilution factor for the blood in the medium must be considered. To conserve space and materials, it is desirable to combine the largest feasible amount of blood from the patient (usually 10 mL) with the smallest amount of medium that will still encourage the growth of bacteria and dilute out or inactivate the antibacterial components of the blood. For this purpose, a 1:5 ratio of blood to unmodified medium has been found to be adequate in conventional blood cultures. All commercial blood culture systems (discussed later in this chapter) specify the appropriate dilution.

BLOOD CULTURE MEDIA The diversity of bacteria that is recovered from blood requires an equally diverse and large number of media to enhance the growth of these bacteria. Basic blood culture media contain a nutrient broth and an anticoagulant. Several different broth formulations are available, including those that can be prepared by the laboratory in-house and those that are commercially prepared. Most blood culture bottles available commercially contain trypticase soy broth, brain-heart infusion broth, supplemented peptone, or thioglycolate broth (see Chapter 12). More specialized broth bases include Columbia or *Brucella* broth.

ADDITIVES Growth of cell wall-deficient bacteria (e.g., antibiotic-damaged organisms or *Mycoplasma pneumoniae*) may be enhanced by adding osmotic stabilizers, such as sucrose, mannitol, or sorbose, to create a hyperosmotic (hypertonic) medium. Media without osmotic additives are known as **isotonic**. The hypertonic bottles are difficult to inspect visually for evidence of bacterial growth, because red blood cells in the media become partially lysed and no longer settle as sediment to the bottom, giving a muddy appearance to the broth. Hypertonic media should be used only for specific problems, because numerous species of bacteria are inhibited in such media.[14]

The addition of penicillinase to blood culture media for inactivation of penicillin has been largely superseded in recent years by the availability of a resin-containing medium that inactivates most antibiotics nonselectively by adsorbing them to the surface of the resin particles. Resin-containing media may enhance isolation of staphylococci, particularly when patients are receiving bacteriostatic drugs. The BACTEC system (Becton Dickinson Microbiology Systems, Sparks, Md.) offers several resin-containing media. Preliminary studies have shown one medium (BACTEC Plus), which allows inoculation of a larger volume of blood (10 mL per bottle), to enhance the recovery of agents of septicemia compared with the Isolator system and the B-D Septi-Chek system.

In addition to blood culture media containing resin particles, a vial is commercially available (ARD,

Becton Dickinson Microbiology Systems, Sparks, Md.) that contains SPS, a cationic resin, and a polymeric adsorbent resin. Introduction of blood into this vial will remove antimicrobials present in the blood before the blood is inoculated into culture media. Blood must be rotated in the vial with the resins for 15 minutes and then inoculated into the media of choice. This device can remove up to 100 μg of most antibiotics. Some studies indicate no significant advantage with this system while others have found increased and faster isolation of pathogens from the blood of patients being treated with antimicrobial agents. Negative aspects include inhibition of the growth of some strains of gram-negative bacilli.

CULTURE TECHNIQUES

In general, each blood culture set includes a blood culture bottle designated for aerobic recovery and one for anaerobic recovery of bacteria. Because of the decline over the past 15 years in the proportion of positive blood cultures that yield anaerobic bacteria coupled with the increasing pressure for laboratories to be cost effective, some investigators have recommended that laboratories discard this routine practice of processing all blood samples aerobically and anaerobically.[12,13] Rather, it has been proposed that anaerobic cultures should be selectively performed and, in place of the anaerobic blood culture, a second aerobic bottle be included. Because this is a controversial proposal, laboratories must deal with conflicting recommendations as they attempt to provide clinically useful blood culture results. Also, depending on the patient population served by the laboratory, numbers of blood cultures submitted, and personnel and financial resources, the laboratory may have one or more methods available to ensure detection of the broadest range of organisms in the least possible time.

Conventional blood cultures

INCUBATION CONDITIONS The atmosphere in commercially prepared blood culture bottles is usually at a low oxidation-reduction potential, permitting the growth of most facultative and some anaerobic organisms. To encourage the growth of obligate (strict) aerobes, such as yeast and *Pseudomonas aeruginosa,* transient venting of the bottles with a sterile, cotton-plugged needle may be necessary. Constant agitation of the bottles during the first 24 hours of incubation will also enhance the growth of most aerobic bacteria.

DETECTING GROWTH After 6 to 18 hours of incubation, most bacteria responsible for clinically significant infections are present in numbers large enough to recover by blind subculture or detected by acridine

orange stain. In addition to daily visual examination, blind subcultures from conventional bottles after the first 6 to 12 hours of incubation are performed by aseptically removing a few drops of the well-mixed medium and spreading this inoculum onto a chocolate blood agar plate. The plate is incubated in 5% to 10% CO_2 at 35° C for 48 hours. Culture-negative bottles are then reincubated for 5 to 7 days unless the patient's condition requires special consideration, as is discussed later. Growth of anaerobic bacteria can be detected in stationary bottles by visual inspection with such success that blind anaerobic subcultures are not recommended. However, if growth is detected in the anaerobic bottles, subcultures are made and incubated both anaerobically and aerobically. After 48 hours of incubation, a second blind subculture or acridine orange stain may be performed.

Self-contained subculture system

Recent modifications of the biphasic blood culture medium are the BD Septi-Chek system (Figure 20-4; Becton Dickinson Microbiology Systems, Sparks, Md.) and the Vacutainer Agar Slant system (Becton Dickinson Microbiology Systems, Sparks, Md.), consisting of a conventional blood culture broth bottle with an attached chamber containing a slide coated with agar

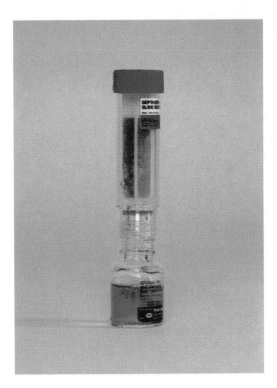

FIGURE 20-4 Becton Dickinson Septi-Chek pediatric-size biphasic blood culture bottle. The medium-containing base bottle is inoculated with blood, and the top piece containing agar paddles is added in the laboratory. The agar is inoculated by tipping the bottle to allow the blood-containing medium to flow over the agar.

or several types of agars. Special media for isolation of fungi and mycobacteria are also available. To subculture, the entire broth contents are allowed to contact the agar surface by inverting the bottle, a simple procedure that does not require opening the bottle or using needles. The large volume of broth that is subcultured and the ease of subculture allow faster detection time for many organisms than is possible with conventional systems. The Septi-Chek system appears to enhance the recovery of *Streptococcus pneumoniae*, but such biphasic systems do not efficiently recover anaerobic isolates.

Lysis centrifugation

The lysis-centrifugation system commercially available is the Isolator (Wampole Laboratories, Cranbury, N.J.). The Isolator consists of a stoppered tube containing saponin to lyse blood cells, polypropylene glycol to decrease foaming, SPS as an anticoagulant, EDTA to chelate calcium ions and thus inhibit the complement cascade and coagulation, and a small amount of an inert fluorochemical (Fluorinert, 3M Co., St. Paul, Minn.) to cushion and concentrate the microorganisms during 30-minute centrifugation at $3000 \times g$ (Figure 20-5). After centrifugation, the supernatant is discarded, the sediment containing the pathogen(s) is vigorously vor-

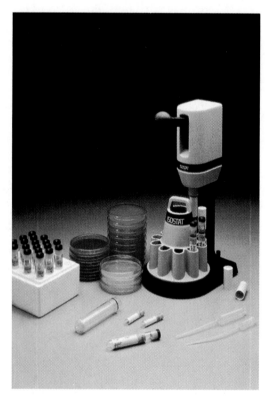

FIGURE 20-5 Lysis-centrifugation blood culture (Isolator System, Wampole Laboratories) uses vacuum-draw collection tubes with a lysing agent and special apparatus (Isostat Press) to facilitate removal of the supernatant without use of needles. (Courtesy Wampole Laboratories, Cranbury, NJ.)

texed, and the entire sediment is plated to solid agar. Benefits of this system include the more rapid and greater recovery of filamentous fungi, the presence of actual colonies for direct identification and susceptibility testing after initial incubation, the ability to quantify the colony-forming units present in the blood, rapid detection of polymicrobial bacteremia, dispensing with the need for a separate antibiotic-removal step, the ability to choose special media for initial culture setup based on clinical impression (such as direct plating onto media supportive of *Legionella* spp. or *Mycobacterium* spp.), and possible greater recovery of intracellular microorganisms caused by lysis of host cells. Possible limitations of the system seem to be a relatively high rate of plate contamination and a decreased ability to detect certain bacteria, such as *Streptococcus pneumoniae*, *Listeria monocytogenes*, *Haemophilus influenzae*, and anaerobic bacteria, compared with conventional systems.

Instrument-based systems

Conventional blood culture techniques are labor-intensive and time-consuming. During these times of cost constraints in health care and a corresponding requirement for clinically relevant care, the development of instrumentation for blood cultures has been accomplished. Instruments can rapidly and accurately detect organisms in blood specimens. By using such instrumentation, laboratories that process many blood cultures can also provide results cost effectively. The decision to purchase a blood culture instrument is difficult and must take into account such matters as volume, patient population, and cost. More than half of all hospital microbiology laboratories use an automated blood culture system.

BACTEC SYSTEMS Many laboratories use the BACTEC system (Becton Dickinson Microbiology Systems, Sparks, Md.), which measures the production of carbon dioxide (CO_2) by metabolizing organisms. Blood or sterile body fluid for routine culture is inoculated into bottles that contain the substrates.

The first BACTEC systems were semi-automated. Vials, containing ^{14}C-labeled substrates (glucose, amino acids, and alcohols), were incubated and often agitated on a rotary shaker. At predetermined time intervals thereafter, the bottles were placed into the monitoring module, where they were automatically moved to a detector. The detector inserted two needles through a rubber septum seal at the top of each bottle and withdrew the gas that had accumulated above the liquid medium and replaced it with fresh gas of the same mixture (aerobic or anaerobic). Any amount of radiolabeled CO_2, the final end product of metabolism of the ^{14}C-labeled substrates (above a preset baseline level), was considered to be suspicious for microbial growth. Microbiologists retrieved suspicious bottles and worked them up for possible microbial growth.

Subsequent modifications further automated the incubation and measuring device, and detection was accomplished by nonradioactive means. The BACTEC NR-860 is fully automated with the incubator, shaker, and detector all in one instrument and measures CO_2 produced by microbial metabolism by infrared spectrophotometry. Most recently, the BACTEC 9240 and BACTEC 9120 were introduced. These fully automated blood culture systems use fluorescence to measure CO_2 that is released; a gas-permeable fluorescent sensor is on the bottom of each vial (Figure 20-6). As CO_2 diffuses into the sensor and dissolves in water present in the sensor matrix, hydrogen (H^+) ions are generated. These H^+ ions cause a decrease in pH, which, in turn, increases the fluorescent output of the sensor. These systems differ from all other BACTEC instruments in that there is continuous monitoring of each bottle and detection is external to the bottle. Of importance, the noninvasion of the blood culture bottle eliminates the potential for cross-contamination of cultures during repeated measurements, the need for a separate gas supply, and the use of ^{14}C-labeled substrates.

BacT/ALERT MICROBIAL DETECTION SYSTEM   Other laboratories use the BacT/Alert System (Organon Teknika, Durham, N.C.), which measures CO_2-derived pH changes by a colorimetric sensor in the bottom of each bottle (see Figure 20-6). The sensor is separated from the broth medium by a membrane that is only permeable to CO_2. As organisms grow they release CO_2, which diffuses across the membrane and is dissolved in water present in the matrix of the sensor. As CO_2 is dissolved, free hydrogen ions are generated. These free hydrogen ions cause a color change in the sensor (blue to light green to yellow as the pH decreases); this color change is read by the instrument.

ESP SYSTEM   The ESP System (Accumed International, Inc., Chicago, Ill.) differs from the other previously discussed systems in that microbial growth is detected by the consumption and/or production of gases as organisms metabolize nutrients in the culture medium (see Figure 20-6). The consumption and/or production of gases is detected by monitoring changes in headspace pressure by a sensitive detector that is attached to the blood culture bottles. Similar to the other systems, this is also a continuously monitoring instrument.

VITAL   Another continuous monitoring blood culture system is the Vital (bioMérieux Vitek, Inc., Hazel-wood, Mo. [see Figure 20-6]). A fluorescent molecule that decreases its fluorescence output in the presence of CO_2, changes in pH, or modification of oxidation-reduction is incorporated in the broth solution and serves as an indicator, detecting any organism present in the culture. Vital is currently available in Europe and the Asia/Pacific regions. Table 20-3 summarizes salient characteristics of the continuous monitoring blood culture instruments.

Techniques to detect IV catheter-associated infections

The insertion of an IV catheter during hospitalization is common practice. Infection, either locally at the catheter insertion site or bacteremia, is one of the most common complications of catheter placement. Because the skin of all patients is colonized with microorganisms that are also common pathogens in catheters, techniques used to diagnose catheter-related infections attempt to quantitate bacterial growth. Diagnosis of an IV catheter-related bacteremia (or fungemia) is difficult, because there are often no signs of infection at the catheter insertion site and the typical signs and symptoms of sepsis can overlap with other clinical manifestations; even the finding of a positive blood culture does not identify the catheter as the source. To date, various methods, such as semiquantitative cultures or Gram stains of the skin entry site and culture of IV catheter tips following catheter removal, have been described to identify these infections. Many of these methods involve some type of quantitation in an attempt to differentiate colonization of the catheter from probable infection. Unfortunately, no single method has demonstrated a clear clinical benefit in large numbers of patients.

Handling positive blood cultures

Most laboratories employ a broth-based blood culture method. Bottles should be examined visually at least daily. Growth is usually indicated by hemolysis of the red blood cells, gas bubbles in the medium, turbidity, or the appearance of small aggregates of bacterial or fungal growth in the broth, on the surface of the sedimented red cell layer, or occasionally along the walls of the bottle. When macroscopic evidence of growth is apparent, a gram-stained smear of an air-dried drop of medium should be performed. Methanol fixation of the smear preserves bacterial and cellular morphology, which may be especially valuable for detecting gram-negative bacteria among red cell debris. As soon as a morphologic description can be tentatively assigned to an organism detected in blood, the physician should be contacted and given all available information. Determining the clinical significance of an isolate is the physician's responsibility. If no organisms are seen on

Text continued on p. 300.

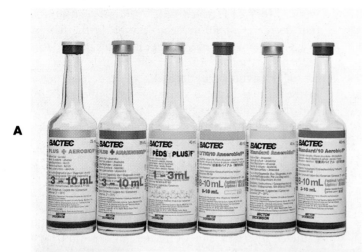

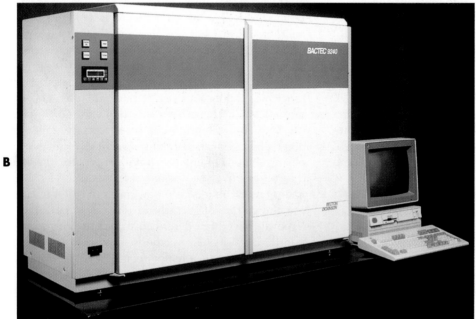

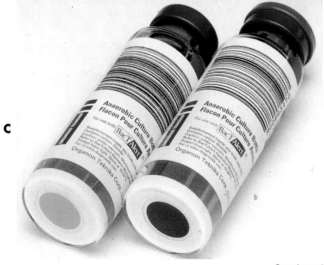

FIGURE 20-6 **A,** Blood culture bottles for the BACTEC 9240 and 9120 continuous monitoring instruments. **B,** The BACTEC 9240 continuous monitoring blood culture system. (**A** and **B** courtesy Becton Dickinson Microbiology Systems, Sparks, Md. BACTEC is a trademark of Becton Dickinson Microbiology Systems.) **C,** Blood culture bottles for the BacT/Alert continuous monitoring blood culture instruments. **D,** The BacT/Alert continuously monitoring blood culture system. (**C** and **D** courtesy Organon Teknica, Durham, NC.) **E,** Blood culture bottles for Accumed's ESP continuous monitoring instrument. **F,** ESP continuous monitoring blood culture system. (**E** and **F** courtesy Accumed International Inc., Chicago, Ill.) **G,** Blood culture bottles for the Vital. **H,** The Vital continuous monitoring blood culture system. (**G** and **H** courtesy bioMérieux Vitek, Inc., Hazelwood, Mo.)

Continued

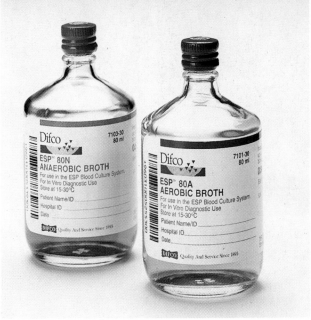

D

E

FIGURE 20-6, cont'd For legend see opposite page.

Continued

F

 G

H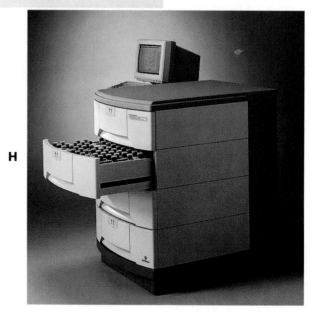

FIGURE 20-6, cont'd For legend see p. 296.

TABLE 20-3   SUMMARY CHARACTERISTICS OF CONTINUOUS-MONITORING BLOOD CULTURE SYSTEMS

SYSTEM	BOTTLES AVAILABLE	INOCULUM VOLUME (mL)	BLOOD: BROTH	VENTING REQUIRED	BOTTLE CAPACITY/ SYSTEM UNIT	DETECTION
BacT/Alert	Aerobic	5-10	1:4	Y	240 (max* = 2160 bottles)†	Colorimetric detection of CO_2
	Anaerobic	5-10	1:4	N		
	FAN (aerobic and anaerobic bottles)	5-10	1:4	Y–aerobic only		
	Pedi-Bact	1-4	1:5	Y		
BACTEC	Standard aerobic/F	3-10	1:4	N	240 (max = 1200 bottles)*	Fluorescent detection of CO_2
	Standard anaerobic/F	5-7	1:4	N		
	Plus aerobic/F	10	1:2.5	N		
	Plus anaerobic/F	10	1:2.5	N		
	Peds Plus/F	0.5-3	1:13.3	N		
ESP	80A (aerobic)	10	1:8	N	384 (max = 1920 bottles)‡	Detection of O_2 consumption and/or CO_2, H_2, and/or N_2 production
	80N (anaerobic)	10	1:8	N		
	EZ Draw 40A	5	1:8	N		
	EZ Draw 40N	5	1:8	N		
Vital	Aerobic	10	1:4	N	Max = 1200 bottles§	Fluorescent detection of growth
	Anaerobic	10	1:4	N		

Max = Maximum number of bottles that can be processed by assembling multiple system units.

†Smaller instruments available: BacT/Alert 120 or BACTEC 9120, with a total capacity of 120 bottles; BACTEC 9050, with a total capacity of 50 bottles.

‡Smaller instrument available: ESP128 with a total capacity of 128 bottles.

§Smaller instruments (units) available: Vital 400, Vital 300, and Vital 200, with a total capacity of 400, 300, and 200 bottles, respectively.

microscopic examination of a bottle that appears positive, subcultures should be performed anyway.

Subcultures from blood cultures suspected of being positive, whether proved by microscopic visualization or not, should be made to various media that would support the growth of most bacteria, including anaerobes. Initial subculture may include chocolate agar, 5% sheep blood agar, MacConkey agar (if gram-negative bacteria are seen), and supplemented anaerobic blood agar. The incidence of polymicrobial bacteremia or fungemia ranges from 3% to 20% of all positive blood cultures. For this reason, samples must be resubcultured for isolated colonies. Subculturing positive bottles a second time at the end of the incubation period is of questionable value and is not recommended.

Numerous rapid tests for identification and presumptive antimicrobial susceptibilities can be performed from the broth blood culture, if a monomicrobic infection is suspected (based on microscopic evaluation). A suspension of the organism that approximates the turbidity of a 0.5 McFarland standard, obtained directly from the broth, or by centrifuging the broth and resuspending the pelleted bacteria, can be used to perform either disk diffusion (qualitative) or broth dilution (quantitative) antimicrobial susceptibility tests. These suspensions may also be used to perform preliminary tests such as coagulase, thermostable nuclease, esculin hydrolysis, bile solubility, antigen detection by fluorescent-antibody stain or agglutination procedures for gram-positive bacteria, oxidase, and commercially available rapid identification kits for gram-negative bacteria. Presumptive results must be verified with conventional procedures using pure cultures.

All isolates from blood cultures should be stored for an indefinite period, preferably for at least 1 year, by freezing at −70° C in 10% skim milk. Storing an agar slant of the isolate under sterile mineral oil at room temperature is a good alternative to freezing. It is often necessary to compare separate isolates from the same patient or isolates of the same species from different patients, sometimes even months after the bacteria were isolated.

Interpretation of blood culture results

Because of the increasing incidence of blood/vascular infection caused by bacteria that are normally considered nonvirulent, indigenous microflora of a healthy human host, interpretation of the significance of growth of such bacteria in blood cultures has become increasingly difficult. On one hand, contaminants may lead to unnecessary antibiotic therapy, additional testing and consultation, and increased length of hospital stay.[1] Costs related to false-positive blood culture results (i.e.,

contaminants) are associated with 40% higher charges for IV antibiotics and microbiology testing. On the other hand, failure to recognize and appropriately treat indigenous microflora can have dire consequences. Guidelines that can assist in distinguishing probable pathogens from contaminants are as follows:

- Probable contaminant

    Growth of *Bacillus* spp., *Corynebacterium* spp., *Propionibacterium acnes,* or coagulase-negative staphylococci in only one of several cultures

    Growth of multiple organisms from only one of several cultures (polymicrobial bacteremia is uncommon)

    The clinical presentation and/or course is not consistent with sepsis (physician-based, not laboratory-based criteria)

    The organism causing the infection at a primary site of infection is not the same as that isolated from the blood culture

- Probable pathogen

    Growth of the same organism in repeated cultures obtained either at different times or from different anatomic sites

    Growth of certain organisms in cultures obtained from patients suspected of endocarditis, such as enterococci, or gram-negative rods in patients with clinical gram-negative sepsis

    Growth of certain organisms such as members of *Enterobacteriaceae, Streptococcus pneumoniae,* gram-negative anaerobes, and *Streptococcus pyogenes*

    Isolation of commensal microbial flora from blood cultures obtained from patients suspected to be bacteremic (e.g., immunosuppressed patients or those having prosthetic devices)

SPECIAL CONSIDERATIONS FOR OTHER RELEVANT ORGANISMS ISOLATED FROM BLOOD

The organisms discussed in this section require somewhat different conditions for their successful recovery from blood culture samples. Most of these organisms are infrequently isolated from blood. Therefore, it is important for the physician to notify the laboratory of remarkable patient history, such as travel abroad, that is leading him or her to suspect these agents.

HACEK BACTERIA

The term *HACEK* represents a group of fastidious, gram-negative bacilli that include *Haemophilus aphrophilus, Actinobacillus actinomycetemcomitans, Cardiobacterium hominis, Eikenella corrodens,* and *Kingella kingae.* Recovery of these organisms from blood cultures is usually associated with infective endocarditis. Other fastidious organisms, such as *Campylobacter* spp., *Capnocytophaga* spp., *Rothia dentocariosa, Flavobacterium* spp., and *Chromobacterium* spp. may also be isolated from blood cultures. If these less common organisms are suspected by the clinician, the laboratory should be alerted to hold the blood cultures for an extended period beyond the first week and to make blind subcultures to several enriched media, including more supportive media such as buffered charcoal-yeast extract.

FUNGI

Even though all disseminated fungal disease is preceded by fungemia, recovery of the microorganisms from blood cultures has not been accomplished until recently. One reason for this may be that cultures were not often collected during the fungemia stage, because clinical symptoms had not yet developed. However, introduction of better methods for isolating fungi from blood, including the lysis-centrifugation system, has resulted in greater recovery of fungi from peripheral blood and greater physician awareness to order fungal blood cultures.

Many fungi, particularly yeast, can be recovered in standard blood culture media, if the bottle is incubated at the appropriate temperature and has been vented and agitated to allow sufficient oxygen in the atmosphere for fungal growth. However, some fungi may grow slowly and poorly in these media, which best support bacterial growth. Until recently, the best medium for the recovery of fungi from blood had been a biphasic system. Currently, optimal isolation of fungi in blood cultures is achieved with either agitated incubation of a recently introduced commercial biphasic system, such as the Septi-Chek, or by using the lysis-centrifugation system.

Blood specimens for detecting fungemia are collected in the same manner as for bacterial culture. Manufacturers of media for automated blood culture systems have developed specific media for fungal isolation. These new formulas have dramatically increased the numbers of fungi isolated from patients with fungemia and have shortened the incubation time required for detection of the fungi.

MYCOBACTERIA

Patients with HIV infection have a high rate of disseminated infection with species of nontuberculous mycobacteria, predominantly *Mycobacterium avium* complex. Isolation of *M. tuberculosis* from the blood of these patients is also common; as many as 42% of HIV-positive patients with tuberculosis have positive blood cultures.

In the past, the use of special media was recommended, such as Middlebrook 7H9 broth with 0.05% SPS or brain-heart infusion broth with 0.5% polysorbate 80, with or without a Middlebrook 7H11 agar slant. However, newer methods of detection (Table 20-4) have increased sensitivity in their ability to detect mycobacteria present in blood specimens and have significantly shortened the time required for mycobacterial blood cultures to become positive.

BRUCELLA

Brucellosis is a common disease in many developing countries but is uncommon in developed countries. Because brucellosis may be in the differential diagnosis of many infections, microbiologists should be prepared to process blood cultures suspected of having *Brucella;* blood cultures are positive in 70% to 90% of patients with brucellosis. Septicemia occurs primarily during the first 3 weeks of illness. Special handling may be required for recovering *Brucella* spp. from blood because these organisms are fastidious, often slow-growing, intracellular parasites. Best recovery is obtained with *Brucella* or trypticase soy broth. The use of biphasic media may enhance growth, or the Isolator system may allow release of intracellular bacteria. The bottles should be continuously vented and incubated in 10% CO_2 at 37° C for at least 4 weeks. Blind subcultures should be performed at 4 days and weekly thereafter, onto *Brucella* blood agar plates incubated as described. *Brucella* spp. may grow slowly, so cultures must be incubated for a minimum of 3 weeks. Cultures should be handled in a biological safety cabinet. The use of the lysis-centrifugation method or an automated method may enhance recovery and reduce the need for prolonged incubation.

SPIROCHETES

Borrelia

Visualization in direct preparations is diagnostic for 70% of cases of relapsing fever, a febrile disease caused by *Borrelia recurrentis.* Organisms may be seen in direct wet preparations of a drop of anticoagulated blood diluted in saline as long, thin, unevenly coiled spirochetes that seem to push the red blood cells around as they move. Thick and thin smears of blood, prepared as for malaria testing (described in Chapter 64) and stained with Wright's or Giemsa stain, are also sensitive for the detection of *Borrelia.*

Leptospira

Leptospirosis can be diagnosed by isolating the causative spirochete from blood during the first 4 to 7 days of illness. Media with up to 14% (vol/vol) rabbit serum, such as Fletcher's or polysorbate 80 (Tween 80)-albumin are recommended. After adding 1 to 3 drops of fresh or SPS-anticoagulated blood to each of several tubes with 5 mL of culture medium, the cultures are incubated for 5 to 6 weeks at 28° to 30° C in air in the dark. Leptospires will grow 1 to 3 cm below the surface, usually within 2 weeks. The organisms remain viable in blood with SPS for 11 days, allowing for transport of specimens from distant locations. Direct dark-field examination of peripheral blood is not recommended because many artifacts are present that may resemble spirochetes. (Further information about *Borrelia* and *Leptospira* is provided in Chapter 63.)

VITAMIN B6-DEPENDENT STREPTOCOCCI

Abiotrophia (Streptococcus) adjacens and *A. defectivus* are unable to multiply without the addition of 0.001% pyridoxal hydrochloride (also called thiol or vitamin B_6). These streptococci are known as "nutritionally variant" or "satelliting" streptococci and have been associated with bacteremia and endocarditis. Although human blood introduced into the blood culture medium will provide enough of the pyridoxal to allow the organisms to multiply in the bottle, standard sheep blood agar plates may not support their growth. Subculturing the broth to a 5% sheep blood agar plate and either overlaying a streak of *Staphylo-coccus aureus* or dropping a pyridoxal disk to produce the supplement will generally demonstrate colonies of the streptococci growing as tiny satellites next to the streak. Some commercial media may be supplemented with enough pyridoxal (0.001%) to support growth of nutritionally variant streptococci.

MYCOPLASMA HOMINIS

Mycoplasma hominis can be recovered during postabortal or postpartum fever, following gynecologic or urologic procedures, or in patients who were immunocompromised.[7,9,11] Isolates can be recovered from both nonautomated and automated blood culture systems. However, because so few clinical isolates have been recovered to date, it has not been determined which blood culture system is optimum for recovering *M. hominis*. Although some studies report that *M. hominis* can produce sufficient CO_2 to be detected by instrumentation, the majority of isolates have been recovered only by subculture; in some cases, 7 days of incubation were required before growth was detected. It should be noted that *M. hominis* should be suspected if there are colonies on subculture yet no organisms seen on Gram stain. Thus, if *M. hominis* bacteremia is suspected, routine blind and terminal subcultures to special media to support the growth of *M. hominis* (e.g., arginine broth) and at least 7 days of incubation are recommended.

BARTONELLA (ROCHALIMAEA)

Based on phenotypic and genotypic characteristics, bacteria of the genus *Rochalimaea* were reclassified

— TABLE 20-4 NEWER METHODS USED TO DETECT MYCOBACTERIA IN BLOOD —

SYSTEM	TYPE OF SYSTEM	MANUFACTURER
BACTEC 460/BACTEC 13A vial*	Semiautomated, radiometric	Becton Dickinson
Isolator system†	Manual	Wampole Laboratories
MB/Bact system (BacT/Alert)‡	Continuous monitor, fully automated	Organon-Teknika
ESP‡	Continuous monitor, fully automated	Accumed International
Mycobacterial Growth Indicator Tube (MGIT)	Manual	Becton Dickinson
BACTEC 9000§ Myco/F Lytic Medium	Continuous monitor, fully automated	Becton Dickinson

*Media specifically for the culture of mycobacteria from blood specimens. Bottles can be directly inoculated with blood.
†Blood can be held in Isolator tubes for prolonged periods without compromising the recovery of *M. avium* complex.
‡Media specifically for the culture of mycobacteria from blood specimens. Bottles can be directly inoculated with blood (presently under development).
§Investigational product in the United States.

into the genus *Bartonella*. *Bartonella* previously contained only a single species, *B. bacilliformis*, the agent of verruga peruana and a septicemic, hemolytic disease known as Oroya fever (see Chapter 44). Two new species, *Bartonella henselae* and *B. elizabethae*, as well as *Bartonella quintana*, have been reported to cause bacteremia and endocarditis in both immunocompetent and immunocompromised patients. *B. henselae* has also been linked to cat scratch disease, a common infectious disease in the United States. Cat scratch disease is characterized by a persistent necrotizing inflammation of the lymph nodes.

Because experience in successful primary isolation of *Bartonella* from blood employing either broth-based or biphasic blood culture systems is limited to date, use of the Isolator system is recommended. Of importance, use of the Isolator overrides the inhibition of *B. henselae* growth by SPS concentrations present in broth-based systems. Once processed, blood is plated onto enriched (chocolate or blood-containing) media, incubated at 35° to 37° C under elevated CO_2 and humidity. For optimal growth, media should be freshly prepared; some manufacturers will provide media fulfilling these criteria. Plates can be sealed with either Parafilm or Shrink seals after the first 24 hours of incubation and incubated up to 30 days.

Procedure

20-1 DRAWING BLOOD FOR CULTURE

PRINCIPLE

Organisms found in circulating blood can be enriched in culture for isolation and further studies. Blood for culture must be obtained aseptically. Once removed from the circulation, unclotted blood must be diluted in growth media.

METHOD

NOTE: Universal precautions require that phlebotomists wear gloves for this procedure.

1. Choose the vein to be drawn by touching the skin before it has been disinfected.

2. Cleanse the skin over the venipuncture site in a circle approximately 5 cm in diameter with 70% alcohol, rubbing vigorously. Allow to air dry.

3. Starting in the center of the circle, apply 2% tincture of iodine (or povidone-iodine) in ever-widening circles until the entire circle has been saturated with iodine. Allow the iodine to dry on the skin for at least 1 minute. The timing is critical; a watch or timer should be used.

4. If the site must be touched by the phlebotomist after preparation, the phlebotomist must disinfect the gloved fingers used for palpation in identical fashion.

5. Insert the needle into the vein and withdraw blood. Do not change needles before injecting the blood into the culture bottle.[5,6]

6. After the needle has been removed, the site should be cleansed with 70% alcohol again, because many patients are sensitive to iodine.

References

1. Bates, D.W., Goldman, L., and Lee, T.H. 1991. Contaminant blood cultures and resource utilization. J.A.M.A. 265:365.

2. Beebe, J.L. and Koneman, E.L. 1995. Recovery of uncommon bacteria from blood: association with neoplastic disease. Clin. Micro. Rev. 8:336.

3. Dorsher, C.W., et.al. Anaerobic bacteremia: decreasing rate over a 15-year period. Rev. Infect. Dis. 13:633.

4. Geha, D.J. and Roberts, G.D. 1994. Laboratory detection of fungemia. Clin. Lab. Med. 14:83.

5. Isaacman, D.J. and Karasic, R.B. 1990. Lack of effect of changing needles on contamination of blood cultures. Pediatr. Infect. Dis. J. 9:274.

6. Krumholz, H.M., Cummings, S., and York, M. 1990. Blood culture phlebotomy: switching needles does not prevent contamination. Ann. Intern. Med. 113:290.

7. Lamey, J.R., et. al. 1982. Isolation of mycoplasmas and bacteria from the blood of postpartum women. Am. J. Obstet. Gynecol. 143:104.

8. Li, J., Plorde, J.L., and Carlson, L.G. 1994. Effects of volume and periodicity on blood cultures. J. Clin. Microbiol. 32:2829.

9. McCormack, W.M., et. al. 1975. Isolation of genital mycoplasmas from blood obtained shortly after vaginal delivery. Lancet I:596.

10. Mermel, L.A. and Maki, D.G. 1993. Detection of bacteremia in adults: consequences of culturing an inadequate volume of blood. Ann. Intern. Med. 119:270.

11. Meyer, R.D. and Clough, W. 1993. Extragenital *Mycoplasma hominis* infections in adults: emphasis on immunosuppression. Clin. Infect. Dis. 17(suppl. 1):243.

12. Morris, A.J., et. al. 1993. Rationale for selective use of anaerobic blood cultures. J. Clin. Microbiol. 31:2110.

13. Murray, P.R., Traynor, P., and Hopson, D. 1992. Critical assessment of blood culture techniques: analysis of recovery of obligate and facultative anaerobes, strict aerobic bacteria, and fungi in aerobic and anaerobic blood culture bottles. J. Clin. Microbiol. 30:1462.

14. Reimer, L.G., et. al. 1983. Controlled evaluation of hypertonic sucrose medium at a 1:5 ratio of blood to broth for detection of bacteremia and fungemia in supplemented peptone broth. J. Clin. Microbiol. 17:1045.

15. Salzman, M.B. and Rubin, L.G. 1995. Intravenous catheter-related infections. Adv. Pediatr. Infect. Dis. 10:337.

16. Scheckler, W.E., Scheibel, W., and Kresge, D. 1996. Temporal trends in septicemia in a community hospital. Am. J. Med. 91(3B):90S.

17. Washington, J.A. 1975. Blood cultures: principles and techniques. Mayo. Clin. Proc. 50:91.

Bibliography

Aronson, M.D. and Bor, D.H. 1987. Blood cultures. Ann. Intern. Med. 106:246.

Farrar, W.E. 1995. Leptospira species (leptospirosis). In Mandell, G.L., Douglas, R.G., Jr., and Bennett, J.E., editors. Principles and practice of infectious diseases, ed 4. Churchill Livingstone, New York.

Gradon, J.D., Timpone, J.G., and Schnittman, S.M. 1992. Emergence of unusual opportunistic infections in AIDS: a review. Clin. Infect. Dis. 15:134.

Scheld, W.M. and Sande, M.E. 1995. Endocarditis and intravascular infections. In Mandell, G.L., Douglas, R.G., Jr., and Bennett, J.E., editors. Principles and practice of infectious diseases, ed 4. Churchill Livingstone, New York.

Tilton, R.C. 1982. The laboratory approach to the detection of bacteremia. Ann. Rev. Microbiol. 36:467.

Washington, J.A. and Ilstrup, D.M. 1986. Blood cultures: issues and controversies. Rev. Infect. Dis. 8:792.

21 | INFECTIONS OF THE LOWER RESPIRATORY TRACT

GENERAL CONSIDERATIONS

ANATOMY

The respiratory system can be divided into upper and lower tracts (Figure 21-1). For purposes of study and discussion, the lower respiratory tract comprises structures including the trachea, bronchi, and bronchioles. The respiratory and gastrointestinal tracts are the two major connections between the interior of the body and the outside environment. The respiratory tract is the pathway through which the body acquires fresh oxygen and removes unneeded carbon dioxide. It begins with the nasal and oral passages, which serve to humidify inspired air, and extends past the nasopharynx and oropharynx to the trachea and then into the lungs. The trachea divides into bronchi, which subdivide into bronchioles, the smallest branches that terminate in the alveoli. Some 300 million alveoli are estimated to be present in the lungs; these are the primary, microscopic gas exchange structures of the respiratory tract.

One must be familiar with the anatomic structure of the thoracic cavity so specimens collected from various sites in the lower respiratory tract are appropriately processed by the laboratory. The thoracic cavity, which contains the heart and lungs, has three partitions that are separated from one another by pleura (see Figure 21-1). The lungs occupy the right and left pleural cavities while the mediastinum (space between the lungs) is occupied mainly by the esophagus, trachea, large blood vessels, and heart.

PATHOGENESIS OF THE RESPIRATORY TRACT: BASIC CONCEPTS

Microorganisms primarily cause disease by a limited number of pathogenic mechanisms (see Chapter 10). Because these mechanisms relate to respiratory tract infections, they will be discussed briefly. Encounters between the human body and microorganisms occur many times each day. However, establishment of infection after such contact tends to be the exception rather than the rule. Whether an organism is successful in establishing an infection is dependent not only on the organism's ability to cause disease (pathogenicity) but also on the human host's ability to prevent the infection.

Host factors

The human host has several mechanisms that nonspecifically protect the respiratory tract from infection: the nasal hairs, convoluted passages, and the mucous lining of the nasal turbinates; secretory IgA and nonspecific antibacterial substances (lysozyme) in respiratory secretions; the cilia and mucous lining of the trachea; and reflexes such as coughing, sneezing, and swallowing. These mechanisms prevent foreign objects or organisms from entering the bronchi and gaining access to the lungs, which remain sterile in the healthy host. Of note, aspiration of minor amounts of oropharyngeal material, as occurs often during sleep, plays an important role in the pathogenesis of many types of pneumonia. Once particles that have escaped the mucociliary sweeping activity enter the alveoli, alveolar macrophages ingest them and carry them to the lymphatics.

In addition to these nonspecific host defenses, normal flora of the nasopharynx and oropharynx help to prevent colonization of the upper respiratory tract. Some of the bacteria that can be isolated as part of the indigenous flora of healthy hosts, as well as many species that may cause disease under certain circumstances but that are often isolated from the respiratory tracts of healthy persons, are listed in Box 21-1. Under certain circumstances and for unknown reasons these colonizing organisms can cause disease—perhaps because of previous damage by a viral infection, loss of some host immunity, or physical damage to the respiratory epithelium (e.g., from smoking). (Organisms isolated from normally sterile sites in the respiratory tract by methods that avoid contamination with normal flora should be definitively identified and reported to the clinician.)

Microorganism factors

Organisms possess either certain traits and/or produce certain products that promote colonization and subsequent infection in the host.

ADHERENCE For any organism to cause disease, it must first gain a foothold within the respiratory tract to grow to sufficient numbers to produce symptoms.

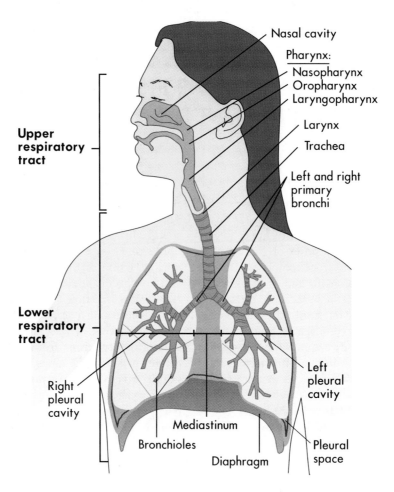

FIGURE 21-1 Anatomy of the respiratory tract, including upper and lower respiratory tract regions.

Therefore, most etiologic agents of respiratory tract disease must first adhere to the mucosa of the respiratory tract. The presence of normal flora and the overall state of the host affect the ability of microorganisms to adhere. Surviving or growing on host tissue without causing overt harmful effects is called **colonization.** Except for those microorganisms that are breathed directly into the lungs, all etiologic agents of disease must first colonize the respiratory tract to some degree before they can cause harm.

Bacteria that possess specific adherence factors include *Streptococcus pyogenes,* whose gram-positive cell wall contains lipoteichoic acids and certain proteins (M protein and others) visible as a thin layer of fuzz surrounding the bacteria. Other bacteria that possess lipoteichoic acid adherence complexes are *Staphylococcus aureus* and certain viridans streptococci. Many gram-negative bacteria (which do not have lipoteichoic acids), including *Enterobacteriaceae, Legionella* spp., *Pseudomonas* spp., *Bordetella pertussis,* and *Haemophilus* spp., adhere by means of pro-

teinaceous fingerlike surface structures called *fimbriae.*

Fimbriae are also called *pili,* although this is technically the term for a similar structure used in sexual interaction rather than just adherence (see Chapter 9). Viruses possess either a hemagglutinin (influenza and parainfluenza viruses) or other proteins that mediate their epithelial attachment.

TOXINS Certain microorganisms are almost always considered to be etiologic agents of disease if they are present in any numbers in the respiratory tract because they possess virulence factors that are expressed in every host. These organisms are listed in Box 21-2. The production of extracellular toxin was one of the first pathogenic mechanisms discovered among bacteria. *Corynebacterium diphtheriae* is a classic example of a bacterium that produces disease through the action of an extracellular toxin. Once the organism colonizes the upper respiratory epithelium, it produces a toxin that is disseminated systemically, adhering preferentially to central nervous system cells and muscle cells of the heart. Sys-

ORGANISMS PRESENT IN THE NASOPHARYNX AND OROPHARYNX OF HEALTHY HUMANS

BOX 21-1

Possible Pathogens

Acinetobacter spp.
Viridans streptococci, including *Streptococcus milleri* group
Beta-hemolytic streptococci
Streptococcus pneumoniae
Staphylococcus aureus
Neisseria meningitidis
Mycoplasma spp.
Haemophilus influenzae
Haemophilus parainfluenzae
Moraxella (Branhamella) catarrhalis
Candida albicans
Herpes simplex virus
Enterobacteriaceae
Mycobacterium spp.
Pseudomonas spp.
Burkholderia (Pseudomonas) cepacia
Filamentous fungi
Klebsiella ozaenae
Eikenella corrodens
Bacteroides spp.
Peptostreptococcus spp.
Actinomyces spp.
Capnocytophaga spp.
Actinobacillus spp., *A. actinomycetemcomitans*
Haemophilus aphrophilus
Entamocha gingivalis
Trichomonas tenax

Rarely Pathogens

Nonhemolytic streptococci
Staphylococci
Micrococci
Corynebacterium spp.
Coagulase-negative staphylococci
Neisseria spp., other than *N. gonorrhoeae* and *N. meningitidis*
Lactobacillus spp.
Veillonella spp.
Spirochetes
Rothia dentocariosa
Leptotrichia buccalis
Selenomonas
Wolinella
Stomatococcus mucilaginosus
Campylobacter spp.

---

temic disease is characterized by myocarditis, peripheral neuritis, and local disease that can lead to respiratory distress. Growth of *C. diphtheriae* causes necrosis and sloughing of the epithelial mucosa, producing a "diphtheritic (pseudo) membrane," which may extend from the anterior nasal mucosa to the bronchi, or may be limited to any area between—most often the tonsillar and peritonsillar areas. The membrane may cause sore throat and interfere with respiration and swallowing. Although nontoxic strains of *C. diphtheriae* can cause local disease, it is much milder than the version mediated by toxin.

Some strains of *Pseudomonas aeruginosa* produce a toxin very similar to diphtheria toxin. Whether this toxin actually contributes to the pathogenesis of respiratory tract infection with *P. aeruginosa* has not been established. *Bordetella pertussis,* the agent of whooping cough, also produces toxins. The role of these toxins in production of disease is not clear. They may act to inhibit the activity of phagocytic cells or to damage cells of the respiratory tract. *Staphylococcus aureus* and beta-hemolytic streptococci produce extracellular enzymes that act to damage host cells or

tissues. Extracellular products of staphylococci aid in production of tissue necrosis and destruction of phagocytic cells, contributing to the commonly seen phenomenon of abscess formation associated with infection caused by this organism. Although *S. aureus* can be recovered from throat specimens, it has not been proved to cause pharyngitis. Enzymes of streptococci, including hyaluronidase, allow rapid dissemination of the bacteria. Many other etiologic agents of respiratory tract infection also produce extracellular enzymes and toxins.

MICROORGANISM GROWTH   In addition to adherence and toxin production, pathogens cause disease by merely growing in host tissue, interfering with normal tissue function and attracting host immune effectors, such as neutrophils and macrophages. Once these cells begin to attack the invading pathogens and repair the damaged host tissue, an expanding reaction ensues with more nonspecific and immunologic factors being attracted to the area, increasing the amount of host tissue damage. Respiratory viral infections usually progress in this manner, as do many types of pneumonias, such as those caused by *Streptococcus pneumoniae, S. pyogenes,*

Staphylococcus aureus, Haemophilus influenzae, Neisseria meningitidis, Moraxella catarrhalis, Mycoplasma pneumoniae, Mycobacterium tuberculosis, and most gram-negative bacilli.

AVOIDING THE HOST RESPONSE   Another virulence mechanism that certain respiratory tract pathogens possess is the ability to evade host defense mechanisms. *S. pneumoniae, N. meningitidis, H. influenzae, Klebsiella pneumoniae,* mucoid *P. aeruginosa, Cryptococcus neoformans,* and others possess polysaccharide capsules that serve both to prevent engulfment by phagocytic host cells and to protect somatic antigens from being exposed to host immunoglobulins. The capsular material is produced in such abundance by certain bacteria, such as pneumococci, that soluble polysaccharide antigen particles can bind host antibodies, blocking them from serving as opsonins. Proof that the capsular polysaccharide is a major virulence mechanism of *H. influenzae, S. pneumoniae,* and *N. meningitidis* was established when vaccines consisting of capsular antigens alone were shown to protect individuals from disease.

Some respiratory pathogens evade the host immune system by multiplying within host cells. *Chlamydia trachomatis, C. psittaci,* and all viruses replicate within host cells. They have evolved methods for being taken in by the "nonprofessional" phagocytic cells of the host to achieve their required environment. Once within these cells, the organism is protected from host humoral immune factors and other phagocytic cells. This protection will last until the host cell becomes sufficiently damaged that the organism is then recognized as foreign by the host and is attacked. A second group of organisms that cause respiratory tract disease comprises those that are able to be taken up by phagocytic host cells (usually macrophages). Once within the phagocytic cell, these respiratory tract pathogens are able to multiply. *Legionella, Pneumocystis carinii,* and *Histoplasma capsulatum* are some of these more common intracellular pathogens.

Mycobacterium tuberculosis is the classic representative of an intracellular pathogen. In primary tuberculosis the organism is carried to an alveolus in a droplet nucleus, a tiny aerosol particle containing a few tubercle bacilli (the minimum infective dose is small). Once phagocytized by alveolar macrophages, organisms are carried to the nearest lymph node, usually in the hilar or other mediastinal chains. In the lymph node, the organisms slowly multiply within macrophages. Ultimately, *M. tuberculosis* destroys the macrophage and is subsequently taken up by other phagocytic cells. Tubercle bacilli multiply to a critical mass within the protected environment of the macrophages, which are somehow prevented from accomplishing lysosomal fusion by the bacteria. Having reached a critical mass, the organisms spill out of the destroyed macrophages, through the lymphatics, and into the bloodstream, producing mycobacteremia and carrying tubercle bacilli to many parts of the body. In most cases, the host immune system reacts sufficiently at this point to kill the bacilli; however, a small reservoir of live bacteria may be left in areas of normally high oxygen concentration, such as the apical (top) portion of the lung. These bacilli are walled off, and years later, an insult to the host, either immunologic or physical, may cause breakdown of the focus of latent tubercle bacilli, allowing active multiplication and disease (secondary tuberculosis). In certain patients with primary immune defects, the ini-

---

BOX 21-2	RESPIRATORY TRACT PATHOGENS

Definite Respiratory Tract Pathogens
Corynebacterium diphtheriae (toxin-producing)
Mycobacterium tuberculosis
Mycoplasma pneumoniae
Chlamydia trachomatis
Chlamydia (TWAR) *pneumoniae*
Bordetella pertussis
Legionella spp.
Pneumocystis carinii
Nocardia spp.
Histoplasma capsulatum
Coccidioides immitis
Cryptococcus neoformans (may also be recovered from patients without disease)
Blastomyces dermatitidis
Viruses (respiratory syncytial virus [RSV], adenoviruses, enteroviruses, herpes simplex virus [HSV], influenza and parainfluenza virus, rhinoviruses)

Rare Respiratory Tract Pathogens
Francisella tularensis
Bacillus anthracis
Yersinia pestis
Burkholderia (Pseudomonas) pseudomallei
Coxiella burnetti
Chlamydia psittaci
Brucella spp.
Salmonella spp.
Pasteurella multocida
Klebsiella rhinoscleromatis
Varicella-zoster virus (VZV)
Parasites

tial bacteremia seeds bacteria throughout a host that is unable to control them, leading to disseminated or miliary tuberculosis. Growth of the bacteria within host macrophages and histiocytes in the lung causes an influx of more effector cells, including lymphocytes, neutrophils, and histiocytes, eventually resulting in granuloma formation, then tissue destruction and cavity formation. The lesion is characteristically a semisolid, amorphous tissue mass resembling semisoft cheese, from which it received the name caseation necrosis (death of cells or tissues). The infection can extend into bronchioles and bronchi from which bacteria are disseminated via respiratory secretions by coughing. Aerosol droplets produced by coughing and containing organisms are then inhaled by the next victim. Other portions of the patient's own lungs may become infected as well through **aspiration** (inhalation of a fluid or solid).

DISEASES OF THE LOWER RESPIRATORY TRACT

BRONCHITIS

Acute

Acute bronchitis is characterized by acute inflammation of the tracheobronchial tree. This condition may be part of, or preceded by, an upper respiratory tract infection such as influenza (the "flu") or the common cold. Most infections occur during the winter when acute respiratory tract infections are common.

The pathogenesis of acute bronchitis has not been studied for all of the causative agents. But, regardless of the cause, the protective functions of the bronchial epithelium are disturbed and excessive fluid accumulates in the bronchi. Depending on the etiology, destruction of the bronchial epithelium is either extensive (e.g., influenza virus) or minimal (e.g., rhinovirus colds).

Clinically, bronchitis is characterized by cough, variable fever, and sputum production. Sputum (matter ejected from the trachea, bronchi, and lungs through the mouth) is often clear at the onset but may become purulent as the illness persists. Bronchitis may manifest as croup (a clinical condition marked by a barking cough and/or hoarseness).

Acute bronchitis is usually caused by viral agents,

but a key bacteriologic consideration in infants and preschool children is *Bordetella pertussis* (Table 21-1). The best specimen for diagnosis of pertussis is a deep nasopharyngeal swab (see Chapter 48).

Chronic

Chronic bronchitis is a common condition affecting about 10% to 25% of adults. This disease is defined by clinical symptoms in which excessive mucus production leads to coughing up sputum on most days during at least 3 consecutive months for more than 2 successive years.[16] Cigarette smoking, infection, and inhalation of dust or fumes are important contributing factors.

Patients with chronic bronchitis can suffer from acute flare-ups of infection, but determination of the cause of the infection is difficult. Potentially pathogenic bacteria, such as nonencapsulated strains of *Haemophilus influenzae, Streptococcus pneumoniae,* and *Moraxella (Branhamella) catarrhalis,* are frequently cultured from the bronchi of these patients. Because of chronic colonization, it is difficult to incriminate one of these organisms as the specific cause of an acute infection in patients with chronic bronchitis. Although the role of bacteria in acute infections in these patients is questionable, viruses are frequent causes.

PNEUMONIA

Pneumonia (inflammation of the lower respiratory tract involving the lung's airways and supporting structures) is a major cause of illness and death in both the community and hospital setting. Once the microorganism has successfully invaded the lung, disease can follow that includes the alveolar spaces and their supporting structure, the interstitium, and the terminal bronchioles.

Pathogenesis

Organisms can cause infection of the lung by four possible routes: by upper airway colonization or infection that subsequently extends into the lung; by aspiration of organisms, thereby avoiding the upper airway defenses; by inhalation of airborne droplets containing the organism; or by seeding of the lung via

---

TABLE 21-1 MAJOR CAUSES OF ACUTE BRONCHITIS

BACTERIA	VIRUSES
Bordetella pertussis, B. parapertussis, Haemophilus influenzae, Mycoplasma pneumoniae, Chlamydia pneumoniae	Parainfluenza virus, influenza virus, respiratory syncytial virus, adenovirus, measles virus

the blood from a distant site of infection. Viruses cause primary infections of the respiratory tract, as well as inhibit host defenses that, in turn, can lead to a secondary bacterial infection. For example, viruses may destroy respiratory epithelium and disrupt normal ciliary activity. Presumably, growth of viruses in host cells disrupts the function of the latter and encourages the influx of nonspecific immune effector cells that exacerbate the damage. Damage to host epithelial tissue by virus infection is known to predispose patients to secondary bacterial infection.

Aspiration of oropharyngeal contents, often not overt, is important in the pathogenesis of many types of pneumonia. Aspiration happens most often during a loss of consciousness such as might occur with anesthesia or a seizure, or after alcohol or drug abuse, but other individuals, particularly geriatric patients, may also develop aspiration pneumonia. Neurologic disease or esophageal pathology and periodontal disease or gingivitis are other important risk factors. Aided by gravity and often by loss of some host nonspecific protective mechanisms, organisms reach lung tissue, where they multiply and attract host inflammatory cells. Other mechanisms include inhalation of aerosolized material and hematogenous seeding. The buildup of cell debris and fluid contributes to the loss of lung function and thus to the pathology.

Clinical manifestations

The symptoms that suggest pneumonia are fever, chills, chest pain, and cough. In the past, pneumonias were classified into two major groups: typical or acute pneumonias (e.g., *Streptococcus pneumoniae*) and atypical pneumonias, based on whether the cough was productive or nonproductive of mucoid sputum. However, analysis of symptoms of pneumonia caused by the atypical pneumonia pathogens (*Mycoplasma pneumoniae, Legionella pneumophila,* and *Chlamydia pneumoniae*) has revealed no significant differences from those symptoms of patients with typical bacterial pneumonias.[4] Because of this overlap in symptoms, many clinicians feel that this distinction is no longer clinically useful.

It is important to keep in mind that some patients with pneumonia exhibit no signs or symptoms related to their respiratory tract (i.e., some only have fever). Therefore, physical examination of the patient, chest radiograph findings, patient history, and clinical laboratory findings are important. In addition to respiratory symptoms, 10% to 30% of patients with pneumonia will complain of headache, nausea, vomiting, abdominal pain, diarrhea, and myalgias.

Epidemiology/etiologic agents

There are two major categories of pneumonias: those that are considered community-acquired (patients are believed to have acquired their infection outside the hospital setting) and those considered hospital-acquired (patients are believed to have acquired their infection within the hospital setting, usually at least 3 days following admission). Because of the many potential etiologic agents, pneumonia in the immunocompromised patient will be addressed separately in this chapter.

COMMUNITY-ACQUIRED PNEUMONIA   In the United States, pneumonia is the sixth leading cause of death and the number one cause of death from infectious diseases. It is estimated that as many as 4 million cases of community-acquired pneumonia occur annually and roughly one fifth of these will require hospitalization.[5] The etiology of acute pneumonias is strongly dependent on age. More than 80% of pneumonias in infants and children are caused by viruses, whereas less than 10% to 20% of pneumonias in adults are viral.

Children. Among previously healthy patients 2 months to 5 years old, respiratory syncytial virus, parainfluenza, influenza, and adenoviruses are the most common etiologic agents of lower respiratory tract disease. Children suffer less commonly from bacterial pneumonia, usually caused by *H. influenzae, S. pneumoniae,* or *S. aureus.* Neonates may acquire lower respiratory tract infections with *C. trachomatis* or *P. carinii* (which likely indicates an immature immune system or an underlying immune defect).

Young Adults. The most common etiologic agent of lower respiratory tract infection among adults younger than age 30 is *Mycoplasma pneumoniae,* which is transmitted via close contact. Contact with secretions seems to be more important than inhalation of aerosols for becoming infected. After contact with respiratory mucosa, *Mycoplasma* organisms are able to adhere to and colonize respiratory mucosal cells. Both a protein adherence factor and gliding motility may be virulence determinants. Once situated in their preferred site between the cilia of respiratory mucosal cells, *Mycoplasma* organisms multiply and somehow destroy ciliary function. Cytotoxins produced by *Mycoplasma* organisms may account for the cell damage they inflict. The recently described species *C. pneumoniae* (originally called *Chlamydia* TWAR) has been found to be the third most common agent of lower respiratory tract infection in young adults, after mycoplasmas and influenza viruses; it also affects older individuals.[8,9] Chlamydiae are intracellular pathogens, thus their ability to disrupt cellular function and cause respiratory disease is similar to that of viruses.

Adults. In recent years, the epidemiology and treatment of community-acquired pneumonia has changed.[14]

Pneumonia is increasing among older patients and those with underlying diseases such as chronic obstructive lung disease and diabetes mellitus. These patients may become infected with various organisms, including newly identified or previously unrecognized pathogens. Factors such as decreased mucociliary function, decreased cough reflex, decreased level of consciousness, periodontal disease, and decreased general mobility probably contribute to a greater incidence of pneumonia in aged patients. Such patients have been found to be more frequently colonized with gram-negative bacilli than are younger people, perhaps because of poor oral hygiene, decreased saliva, or decreased epithelial cell turnover.

Although the mortality in the outpatient setting remains low (1% to 5%), mortality rates for community-acquired pneumonia approximate 25% if the patient requires hospitalization.[11,12] In addition, the epidemiology and distribution of etiologic agents in patients with severe community-acquired pneumonia who require hospitalization differs somewhat from patients with pneumonia who do not require hospitalization. For these reasons, guidelines for initial management of adults with community-acquired pneumonia take into account the severity of illness at initial presentation and the presence of either coexisting illness or advanced age.[14]

Community-acquired pneumonia in adults is most commonly due to bacterial infections. Regardless of age or coexisting illness, *Streptococcus pneumoniae* is most prevalent, causing 80% of all community-acquired bacterial pneumonia. Additional common etiologies for those patients who are not hospitalized and those who are hospitalized are listed in Tables 21-2 and 21-3, respectively.

Pneumonia secondary to aspiration of gastric or oral secretions is common and occurs in the community setting. The most common agents are primarily the oral anaerobes such as black-pigmented *Prevotella* and *Porphyromonas* spp., *Prevotella oris*, *P. buccae*, *P. disiens*, *Bacteroides gracilis*, fusobacteria, and anaerobic and microaerophilic streptococci. The anaerobic agents possess many factors, such as extracellular enzymes and capsules, that may enhance their ability to produce disease. It is their presence, however, in an abnormal site within the host producing lowered oxidation-reduction potential secondary to tissue damage that contributes most to their pathogenicity. *Staphylococcus aureus*, various *Enterobacteriaceae*, and *Pseudomonas* may also be acquired by aspiration; *Haemophilus influenzae*, *Legionella* spp., *Acinetobacter*, *Moraxella catarrhalis*, *Chlamydia pneumoniae*, meningococci, and other agents may also be implicated.

Adults also may suffer from viral pneumonia caused by influenza, adenovirus, cytomegalovirus, parainfluenza, varicella, rubeola, or respiratory syncytial virus, particularly during epidemics. After viral pneumonia, especially influenza, secondary bacterial disease caused by beta-hemolytic streptococci, pneumococci, *Staphylococcus aureus*, *Moraxella catarrhalis*, *Haemophilus influenzae*, and *Chlamydia pneumoniae* is more likely.

Unusual causes of acute lower respiratory tract infection in adults include *Actinomyces* and *Nocardia* spp. Other agents may rarely be recovered from sputum and include the agents of plague, tularemia, melioidosis (*Burkholderia* [*Pseudomonas*] *pseudomallei*), *Brucella*, *Salmonella*, *Coxiella burnetti* (Q fever), *Bacillus anthracis*, and *Pasteurella multocida*, and certain parasitic agents such as *Paragonimus westermani*, *Entamoeba histolytica*, *Ascaris lumbricoides*, and *Strongyloides* spp. (the latter may cause fatal disease in immunosuppressed patients). A high index of suspicion by the clinician is usually a prerequisite to a diagnosis of parasitic pneumonia in the United States. Psittacosis should be ruled out as a cause of acute lower respiratory tract infection in patients who have had recent contact with birds. Among the fungal etiologies, *Histoplasma capsulatum*, *Blastomyces dermatitidis*, *Paracoccidioides brasiliensis*, *Coccidioides immitis*, *Cryptococcus neoformans*, and, occasionally, *Aspergillus fumigatus* may cause acute pneumonia. Therefore, occupational history and history of

TABLE 21-2 MOST COMMON ETIOLOGIES OF COMMUNITY-ACQUIRED PNEUMONIA IN ADULTS WHO ARE NOT HOSPITALIZED

AGE	COEXISTING ILLNESS	MOST COMMON ETIOLOGIES*
<60 yrs	No	*M. pneumoniae*, respiratory viruses, *C. pneumoniae*, *H. influenzae*
>60 yrs	Yes	Respiratory viruses, *H. influenzae*, aerobic gram-negative rods, *Staphylococcus aureus*

Streptococcus pneumoniae is the most common etiology for *all* categories of adult patients with pneumonia.

exposure to animals are important in suggesting specific potential infectious agents.

HOSPITAL-ACQUIRED PNEUMONIA Pneumonia is the leading cause of death among patients with nosocomial infections (as high as 50% mortality among patients in intensive care units).[19] Some of these pneumonias are secondary to sepsis, and some are related to contaminated inhalation therapy equipment.

Nosocomial pneumonia is a risk for any hospitalized patient, particularly for intubated patients. Organisms associated with these infections can be very hospital specific, but overall the most common include *Klebsiella* spp., other *Enterobacteriaceae, S. aureus,* anaerobes, *Streptococcus pneumoniae, P. aeruginosa,* and *Legionella.* Other agents have been associated with nosocomial outbreaks, including influenza virus. Viruses, such as respiratory syncytial virus, adenovirus, and influenza A, are often implicated as causes of nosocomial pneumonia among hospitalized children.

Aspiration pneumonia with infection caused by gram-negative bacilli or staphylococci is probably the major type of hospital-acquired pneumonia, followed by pneumococcal disease. *Legionella* has been implicated in a number of hospital outbreaks, but the problem is typically specific to a given institution.

CHRONIC LOWER RESPIRATORY TRACT INFECTIONS *Mycobacterium tuberculosis* is the most likely etiologic agent of chronic lower respiratory tract infection, but fungal infection and anaerobic pleuropulmonary infection may also run a subacute or chronic course. Mycobacteria other than *M. tuberculosis* may also cause such disease, particularly *M. avium-intracellulare* and *M. kansasii.* Although possible causes of acute, community-acquired lower respiratory tract infections, fungi and parasites are more commonly isolated from patients with chronic disease. *Actinomyces* and *Nocardia* may also be associated with gradual onset of symptoms. *Actinomyces* is usually associated with an infection of the pleura or chest wall, and *Nocardia* may be isolated along with an infection caused by *M. tuberculosis.* The pathogenesis of many of the infections caused by agents of chronic lower respiratory tract disease is characterized by the requirement for some breakdown of cell-mediated immunity in the host or the ability of these agents to avoid being destroyed by host cell-mediated immune mechanisms. This may be caused by an effect on macrophages, the ability to mask foreign antigens, sheer size, or by some other factor, allowing microbes to grow within host tissues without eliciting an overwhelming local immune reaction.

Cystic fibrosis (CF) is a genetic disorder that leads to persistent bacterial infection in the lung, causing airway wall damage and chronic obstructive lung disease. Eventually, a combination of airway secretions and damage leads to poor gas exchange in the lungs, cardiac malfunction, and subsequent death. Patients with CF may present as young adults with chronic respiratory tract disease, as well as the more common presentation in children that may also include gastrointestinal problems and stunted growth. A very mucoid *Pseudomonas* sp., characterized by production of copious amounts of extracellular capsular polysaccharide, can be isolated from the sputum of almost all patients with CF older than 18 years, becoming more prevalent with increasing age after 5 years. Even if CF has not been diagnosed, isolation of a mucoid *Pseudomonas aeruginosa* from sputum should alert the clinician to the possibility of such an underlying disease. Microbiologists should always report this unusual morphologic feature if it is encountered. In addition to mucoid *Pseudomonas,* patients with CF are likely to harbor *Staphylococcus aureus, Haemophilus influenzae,* and *Burkholderia (Pseudomonas) cepacia.* Respiratory syncytial virus, influenza A, and *Aspergillus* are also important pathogens in this population.

Lung abscess is usually a complication of acute or chronic pneumonia. In these circumstances, organ-

TABLE 21-3 MOST COMMON ETIOLOGIES OF COMMUNITY-ACQUIRED PNEUMONIA IN ADULTS WHO ARE HOSPITALIZED	
GENERALLY REQUIRING INTENSIVE CARE UNIT	MOST COMMON ETIOLOGIES*
No	*H. influenzae;* polymicrobial, aerobic gram-negative bacilli; *Legionella* spp.; *Staphylococcus aureus; Chlamydia pneumoniae;* respiratory viruses
Yes	*Legionella* spp., aerobic gram-negative bacilli, *Mycoplasma pneumoniae,* respiratory viruses

Streptococcus pneumoniae is the most common etiology for *all* categories of adult patients hospitalized with pneumonia.

isms infecting the lung cause localized destruction of the lung parenchyma (functional elements of the lung). Symptoms associated with lung abscess are similar to those of acute and chronic pneumonia except symptoms fail to resolve with treatment.

IMMUNOCOMPROMISED PATIENTS

Patients with Neoplasms. Patients with cancer are at high risk to become infected because of either granulocytopenia and/or other defects in phagocytic defenses, cellular and/or humoral immune dysfunction, damage to mucosal surfaces and the skin, and various medical procedures such as blood product transfusion. In these patients the nature of the malignancy often determines the etiology (Table 21-4) and pneumonia is a frequent clinical manifestation.

Transplant Recipients. For successful organ transplantation, the recipient's immune system must be suppressed in some fashion. For this reason, these patients are predisposed to infection. Regardless of the type of organ transplant (heart, renal, bone marrow, heart/lung, liver, pancreas) most infections occur within 4 months following transplantation. Major infections can occur within the first month but are usually associated with infections carried over from

TABLE 21-4 INFECTIOUS AGENTS FREQUENTLY ASSOCIATED WITH CERTAIN MALIGNANCIES

MALIGNANCY (SITE AND TYPE OF INFECTIONS)	PATHOGENS
Acute nonlymphocytic leukemia (pneumonia, oral lesions, cutaneous lesions, urinary tract infections, hepatitis, most often sepsis without obvious focus)	*Enterobacteriaceae* *Pseudomonas* Staphylococci *Corynebacterium jeikeium* *Candida* *Aspergillus* *Mucor* Hepatitis C and other non-A, non-B
Acute lymphocytic leukemia (pneumonia, cutaneous lesions, pharyngitis, disseminated disease)	Streptococci (all types) *Pneumocystis carinii* Herpes simplex virus Cytomegalovirus Varicella zoster virus
Lymphoma (disseminated disease, pneumonia, urinary tract, sepsis, cutaneous lesions)	*Brucella* *Candida* (mucocutaneous) *Cryptococcus neoformans* Herpes simplex virus (cutaneous) Varicella zoster virus Cytomegalovirus *Pneumocystis carinii* *Toxoplasma gondii* *Listeria monocytogenes* *Mycobacteria* *Nocardia* *Salmonella* Staphylococci *Enterobacteriaceae* *Pseudomonas* *Strongyloides stercoralis*
Multiple myeloma (pneumonia, cutaneous lesions, sepsis)	*Haemophilus influenzae* *Streptococcus pneumoniae* *Neisseria meningitidis* *Enterobacteriaceae* Pseudomonas Varicella zoster virus *Candida* *Aspergillus*

the pretransplant period. Pulmonary infections are of great importance in this patient population. Some of the most common causes of pneumonia include *Streptococcus pneumoniae*, *Haemophilus influenzae*, *Pneumocystis carinii*, and cytomegalovirus. Of importance, fungi, such as *Cryptococcus neoformans*, *Aspergillus* spp., *Candida* spp., and zygomycetes, can cause life-threatening pulmonary infection.

HIV-infected Patients. Patients infected with HIV are at high risk for developing pneumonia. As discussed in the previous chapter, opportunistic infections as a result of severe immunodeficiency are a major cause of illness and death among these patients. In the United States, the most common opportunistic infection among patients with AIDS is *Pneumocystis carinii* pneumonia. Although *P. carinii* remains a major pulmonary pathogen, other organisms must be considered in this patient population, including *Mycobacterium tuberculosis* and *Mycobacterium avium* complex, as well as common bacterial pathogens such as *Streptococcus pneumoniae* and *Haemophilus influenzae*. In addition to these common pathogens, many other organisms can cause lower respiratory tract infections, including *Nocardia* spp., *Rhodococcus equi* (a gram-positive, aerobic, pleomorphic organism), and *Legionella* spp.

PLEURAL INFECTIONS

As a result of an organism infecting the lung and subsequently gaining access to the pleural space via an abnormal passage (fistula), the patient may develop an **empyema** (pus in a body cavity such as the pleural cavity). Symptoms in these patients are insidious because early on in the course of disease symptoms are related to the primary infection in the lung. Once enough purulent exudate is formed, typical physical and radiographic findings indicative of an empyema are produced.

LABORATORY DIAGNOSIS OF LOWER RESPIRATORY TRACT INFECTIONS

SPECIMEN COLLECTION AND TRANSPORT

Although the rapid determination of the etiologic agent is of paramount importance in managing pneumonia, the responsible pathogen is not determined in as many as 50% of patients, despite extensive diagnostic testing. Unfortunately, no single test is available that can identify all potential pathogens that cause lower respiratory tract infections. Refer to Table 1-1 for an overview of the collection, transport, and processing of specimens from the lower respiratory tract.

Sputum

EXPECTORATED The examination of expectorated sputum has been the primary means of determining the causes of bacterial pneumonia. However, lower respiratory tract secretions will be contaminated with upper respiratory tract secretions, especially saliva, unless they are collected using some invasive technique. For this reason, sputum is among the least clinically relevant specimens received for culture in microbiology laboratories, even though it is one of the most numerous and time-consuming specimens.

Good sputum samples depend on thorough health care worker education and patient understanding throughout all phases of the collection process. Patients should be instructed to provide a deep-coughed specimen. The material should be expelled into a sterile container, with an attempt to minimize contamination by saliva. Specimens should be transported to the laboratory immediately, because even a moderate amount of time at room temperature can result in loss of some infectious agents.

INDUCED Patients who are unable to produce sputum may be assisted by respiratory therapy technicians, who can use postural drainage and thoracic percussion to stimulate production of acceptable sputum. As an alternative, an aerosol-induced specimen may be collected that is useful for isolating the agents of mycobacterial or fungal disease. Induced sputum is now recognized for its high diagnostic yield in cases of *Pneumocystis carinii* pneumonia as well.[2] Aerosol-induced specimens are collected by allowing the patient to breathe aerosolized droplets of a solution containing 15% sodium chloride and 10% glycerin for approximately 10 minutes, or until a strong cough reflex is initiated. Lower respiratory secretions obtained in this way appear watery, resembling saliva, although they often contain material directly from alveolar spaces. These specimens are usually adequate for culture and should be accepted in the laboratory without prescreening. Obtaining such a specimen may obviate the need for a more invasive procedure, such as bronchoscopy or needle aspiration, in many cases.

The gastric aspirate is used exclusively for isolation of acid-fast bacilli and may be collected from patients who are unable to produce sputum, particularly young children. Before the patient wakes up in the morning, a nasogastric tube is inserted into the stomach and contents are withdrawn (on the assumption that acid-fast bacilli from the respiratory tract were swallowed during the night and will be present in the stomach). The relative resistance of mycobacteria to acidity allows them to remain viable for a short period. Gastric aspirate specimens must be delivered to the laboratory immediately so that the acidity can be neutralized. Specimens can be first neutralized and then transported if immediate delivery is not possible.

Endotracheal or tracheostomy suction specimens

Patients with tracheostomies are unable to produce sputum in the normal fashion, but lower respiratory tract secretions can easily be collected in a Lukens trap (Figure 21-2). Tracheostomy aspirates or tracheostomy suction specimens should be treated as sputum by the laboratory. Patients with tracheostomies rapidly become colonized with gram-negative bacilli and other nosocomial pathogens. Such colonization per se is not clinically relevant, but these organisms may be aspirated into the lungs and cause pneumonia. Thus, there can be a great deal of confusion for microbiologists and clinicians trying to ascertain the etiologic agent of pneumonia in these patients.

Bronchoscopy

The diagnosis of pneumonia, particularly in HIV-infected and other immunocompromised patients, often necessitates the use of more invasive procedures. Fiber optic bronchoscopy has dramatically affected the evaluation and management of these infections. With this method, the bronchial mucosa can be directly visualized and collected for biopsy, and the lung tissue can be sent for transbronchial biopsy to evaluate lung cancer and other lung diseases.

FIGURE 21-2 Tracheal secretions received in the laboratory in a Lukens trap.

Although transbronchial biopsy is important, the procedure is often associated with significant complications such as bleeding.

During bronchoscopy, physicians can obtain either bronchial washings or aspirates, bronchoalveolar lavage (BAL) samples, protected bronchial brush samples, or specimens for transbronchial biopsy. Bronchial washings or aspirates are obtained by instilling a small amount of sterile physiologic saline into the bronchial tree and withdrawing the fluid when purulent secretions are not visualized. Such specimens will still be contaminated with upper respiratory tract flora such as viridans streptococci and *Neisseria* spp. Recovery of potentially pathogenic organisms from bronchial washings should be attempted, because such specimens may be more diagnostically relevant than sputa.

A deeper sampling of desquamated host cells and secretions can also be obtained via bronchoscopy by BAL. Lavages are especially suitable for detecting *Pneumocystis* cysts and fungal elements.[2,15] During this procedure, a high volume of saline (100 to 300 mL) is infused into a lung segment through the bronchoscope to obtain cells and protein of the pulmonary interstitium and alveolar spaces. It is estimated that more than 1 million alveoli are sampled during this process. Recently, the value of this technique in conjunction with quantitative culture for the diagnosis of most major respiratory tract pathogens, including bacterial pneumonia, has been documented.[3,7] Scientists have found significant correlation between acute bacterial pneumonia and greater than 10^3 to 10^4 bacterial colonies per milliliter of BAL fluid. BAL has been shown to be a safe and practical method for diagnosing opportunistic pulmonary infections in immunosuppressed patients.

Another type of respiratory specimen is obtained via a protected catheter bronchial brush as part of a bronchoscopy examination.[1] Specimens obtained by this moderately invasive collection procedure are best suited for microbiologic studies, particularly in aspiration pneumonia. An overview of the collection process is shown in Figure 21-3. Upon receipt, contents of the bronchial brush may be suspended in 1 mL of broth solution with vigorous vortexing and then inoculated onto culture media using a 0.01 mL calibrated inoculating loop. Some workers have stated that specimens obtained via double-lumen–protected catheters are suitable for both anaerobic and aerobic cultures. Colony counts of greater than or equal to 1000 organisms per milliliter in the broth diluent (or 10^6/mL in the original specimen) have been considered to correlate with infection. All facets of the bronchoscopic procedure, such as order

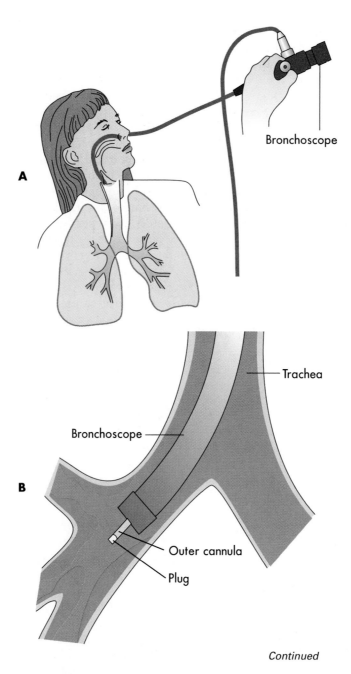

Continued

FIGURE 21-3 Overview of obtaining a protected catheter bronchial brush during a bronchoscopy examination. **A,** The bronchoscope is introduced into the nose and advanced through the nasopharyngeal passage into the trachea. The bronchoscope is then inserted into the lung area of interest. **B,** A small brush that holds 0.01 to 0.001 mL of secretions is placed within a double cannula. The end of the outermost tube or cannula is closed with a displaceable plug made of absorbable gel. The cannula is inserted to the proper area. **C,** Once in the correct area, the inner cannula is pushed out, dislodging the protective plug as it is extruded. **D,** The brush is then extended beyond the inner cannula and the specimen is collected by "brushing" the involved area. The brush is withdrawn into the inner cannula, which is withdrawn into the outer cannula to prevent contamination by upper airway organisms as it is removed.

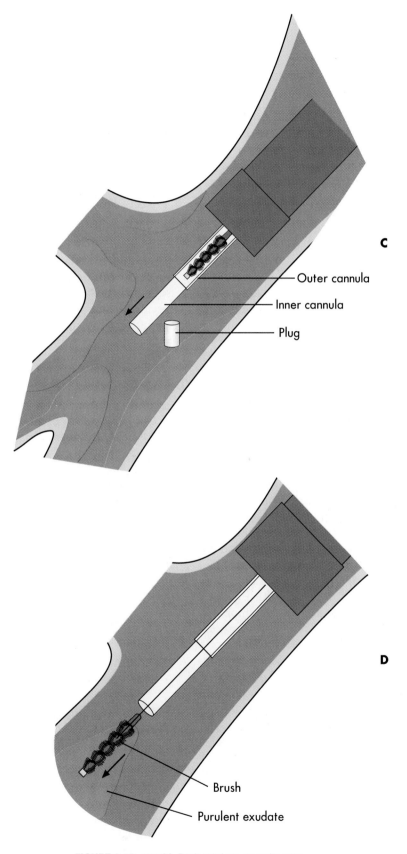

C

Outer cannula

Inner cannula

Plug

D

Brush

Purulent exudate

FIGURE 2-13, cont'd For legend see opposite page.

of sampling, use of anesthetic, and rapidity of plating, should be rigorously standardized.[1,17]

Transtracheal aspirates

Percutaneous transtracheal aspirates (TTA) are obtained by inserting a small plastic catheter into the trachea via a needle previously inserted through the skin and cricothyroid membrane. This invasive procedure, although somewhat uncomfortable for the patient and not suitable for all patients (it cannot be used in uncooperative patients, in patients with bleeding tendency, or with poor oxygenation), reduces the likelihood that a specimen will be contaminated by upper respiratory tract flora and diluted by added fluids, provided that care is taken to keep the catheter from being coughed back up into the pharynx. Although this technique is rarely used anymore, anaerobes, such as *Actinomyces* and those associated with aspiration pneumonia, can be isolated from transtracheal aspirate specimens.

Other invasive procedures

When pleural empyema is present, thoracentesis may be used to obtain infected fluid for direct examination and culture. This constitutes an excellent specimen that accurately reflects the bacteriology of an associated pneumonia. Laboratory examination of such material is discussed in Chapter 29. Blood cultures, of course, should always be obtained from patients with pneumonia, because they will be positive in about 20% of patients requiring hospitalization.

For patients with pneumonia, a thin needle aspiration of material from the involved area of the lung may be performed percutaneously. If no material is withdrawn into the syringe after the first try, approximately 3 mL of sterile saline can be injected and then withdrawn into the syringe. Patients with emphysema, uremia, thrombocytopenia, or pulmonary hypertension may be at increased risk of complication (primarily pneumothorax [air in the pleural space] or bleeding) from this procedure. The specimens obtained are very small in volume, and protection from aeration is usually impossible. This technique is more frequently used in children than in adults.

The most invasive procedure for obtaining respiratory tract specimens is the open lung biopsy. Performed by surgeons, this method is used to procure a wedge of lung tissue. Biopsy specimens are extremely helpful for diagnosing severe viral infections, such as herpes simplex pneumonia, for rapid diagnosis of *Pneumocystis* pneumonia, and for other hard-to-diagnose or life-threatening pneumonias. Ramifications of this and all other specimen collection techniques are discussed in Cumitech 7A, titled, "Laboratory Diagnosis of Lower Respiratory Tract Infections."

SPECIMEN PROCESSING

Direct visual examination

Lower respiratory tract specimens can be examined by direct wet preparation for some parasites and by special procedures for *Pneumocystis*. Fungal elements can be visualized under phase microscopy with 10% potassium hydroxide (KOH), under ultraviolet light with calcofluor white, or using periodic acid-Schiff (PAS)–stained smears.

For most other evaluations, the specimen must be fixed and stained. Bacteria and yeasts can be recognized on Gram stain. One of the most important uses of the Gram stain, however, is to evaluate the quality of expectorated sputum received for routine bacteriologic culture.[10] A portion of the specimen consisting of purulent material is chosen for the stain. Of note, the smear can be evaluated adequately even before it is stained, thus negating the need for Gram stain of specimens later judged unacceptable. An acceptable specimen will yield less than 10 squamous epithelial cells per low-power field (100×). The number of white blood cells may not be relevant, because many patients are severely neutropenic and specimens from these patients will not show white blood cells on Gram stain examination. On the other hand, the presence of 25 or more polymorphonuclear leukocytes per 100× field, together with few squamous epithelial cells, implies an excellent specimen (Figure 21-4). Until recently, only expectorated sputa were suitable for rejection based on microscopic screening. However, endotracheal aspirates (ETAs) from mechanically ventilated, adult patients can now also be screened by Gram stain. Criteria used to reject ETAs from adult patients include greater than 10 squamous epithelial cells per low-power field or no organisms seen under oil immersion (1000×).[13] It is important to realize that in *Legionella* pneumonia, sputum may be scant and watery, with few or no host cells. Such specimens may be positive by direct fluorescent antibody stain and culture, and should not be subjected to screening procedures. Conversely, sputum from patients with cystic fibrosis should be screened.[18]

Respiratory secretions may need to be concentrated before staining. The cytocentrifuge instrument has been used successfully for this purpose, concentrating the cellular material in an easily examined monolayer on a glass slide. As an alternative, specimens are centrifuged, and the sediment is used for visual examinations and cultures. For screening purposes, the presence of ciliated columnar bronchial epithelial cells, goblet cells, or pulmonary macrophages in specimens obtained by bronchoscopy or BAL indicates a specimen from the lower respiratory tract.

In addition to the Gram stain, respiratory specimens may be stained for acid-fast bacilli with either

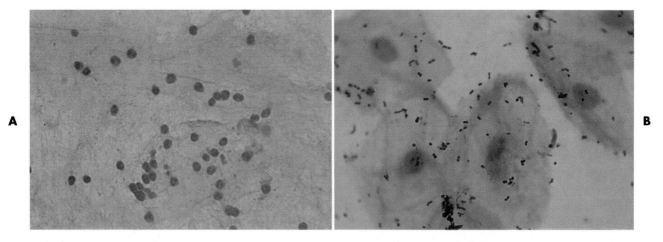

FIGURE 21-4 Gram stain of sputum specimens. **A,** This specimen contains numerous polymorphonuclear leukocytes and no visible squamous epithelial cells, indicating that the specimen is acceptable for routine bacteriologic culture. **B,** This specimen contains numerous squamous epithelial cells and rare polymorphonuclear leukocytes, indicating an inadequate specimen for routine sputum culture.

the classic Ziehl-Neelsen or the Kinyoun carbolfuchsin stain. Auramine or auramine-rhodamine is also used to detect acid-fast organisms. Because they are fluorescent, these stains are more sensitive than the carbolfuchsin formulas and are preferable for rapid screening. Slides may be restained with the classic stains directly over the fluorochrome stains as long as all of the immersion oil has been removed carefully with xylene. All of the acid-fast stains will reveal *Cryptosporidium* spp. if they are present in the respiratory tract, as may occur in immunosuppressed patients.

Immunosuppressed patients are often at risk of infection with *P. carinii.* Although the modified Gomori methenamine silver stain has been used traditionally to recognize *Nocardia, Actinomyces,* fungi, and parasites, it takes approximately 1 hour of technologist time to perform, is technically demanding, and is not suitable as an emergency procedure. A fairly rapid stain, toluidine blue O has been used in many laboratories with some success. Toluidine blue O stains not only *Pneumocystis* but also *Nocardia asteroides* and some fungi. A monoclonal antibody stain is the optimum stain for *Pneumocystis* (see Chapter 64) for less invasive specimens such as BAL and induced sputums.

Direct fluorescent antibody (DFA) staining has been used to detect *Legionella* spp. in lower respiratory tract specimens. Sputum, pleural fluid, aspirated material, and tissue are all suitable specimens. Because there are so many different serotypes of legionellae, polyclonal antibody reagents and a monoclonal antibody directed against all serotypes of *Legionella pneumophila* are used. Because of low sensitivity (50% to 75%), DFA results should not be relied on in lieu of culture. Rather, *Legionella* culture, DFA, and serology

should be performed for optimum sensitivity.

Commercially available DFA reagents are also used to detect antigens of numerous viruses, including herpes simplex, cytomegalovirus, adenovirus, influenza viruses, and respiratory syncytial virus (see Chapter 66). Commercial suppliers of reagents provide procedure information for each of these tests. Monoclonal and polyclonal fluorescent stains for *Chlamydia trachomatis* are available and may be useful for staining respiratory secretions of infants with pneumonia. A number of molecular amplification techniques (see Chapter 14) for the direct detection of respiratory pathogens have been described; however, the sensitivity and specificity of these assays vary greatly from one study to another. To date, Federal Drug Administration approved amplification assays include those for the direct detection of *Mycobacterium tuberculosis* on smear positive specimens (see Chapter 60).

Routine culture

Most of the commonly sought etiologic agents of lower respiratory tract infection will be isolated on routinely used media: 5% sheep blood agar, MacConkey agar for the isolation and differentiation of gram-negative bacilli, and chocolate agar for *Haemophilus* and *Neisseria* spp. Because of contaminating oral flora, sputum specimens, specimens obtained by bronchial washing and lavage, tracheal aspirates, and tracheostomy or endotracheal tube aspirates are not inoculated to enrichment broth or incubated anaerobically. Only specimens obtained by percutaneous aspiration (including transtracheal aspiration) and by protected bronchial brush are suitable for anaerobic culture; the latter must be done quantitatively for proper interpretation (refer to prior discussion). Transtracheal and percutaneous lung aspiration material may be inoculated to enriched thioglycollate, as well as to solid media. For suspected cases of legionnaires' disease, buffered charcoal-yeast extract (BCYE) agar and selective BCYE are inoculated. Plates should be streaked in four quadrants to provide a basis for objective semiquantitation to define the amount of growth. After 24 to 48 hours of incubation, the numbers and types of colonies are recorded. For *Legionella* cultures, colonies form on the selective agar after 3 to 5 days at 35° C.

Sputum specimens from patients known to have cystic fibrosis should be inoculated to selective agar, such as mannitol salt, for recovery of *S. aureus* and selective horse blood-bacitracin, incubated anaerobically and aerobically, for recovery of *H. influenzae* that may be obscured by the mucoid *Pseudomonas* on routine media.[6] The use of a selective medium for *B. cepacia*, such as PC or OFPBL agars, is also recommended.[6]

For interpretation of culture results on those specimens contaminated by normal oropharyngeal flora (e.g., expectorated and induced sputum, bronchial washings), growth of the predominant aerobic and facultatively anaerobic bacteria is reported. To ensure optimum culture reporting, conditions must be well defined in terms of an objective grading system for streaked plates. Finally, the clinical significance of culture findings depends not only on standardized and appropriate laboratory methods being followed but also on how specimens are collected and transported, other laboratory data, and the patient's clinical presentation.

Numerous bacterial agents that cause lower respiratory tract infections are not detected by routine bacteriologic culture. Mycobacteria, *Chlamydia*, *Nocardia*, *Bordetella pertussis*, *Legionella*, and *Mycoplasma pneumoniae* require special procedures for detection; this also applies to viruses and fungi. Refer to the appropriate chapter section dealing with these organisms.

References

1. Broughton, W.A., et al. 1987. The technique of protected brush catheter bronchoscopy. J. Crit. Illness. 18:63.
2. Caliendo, A.M. 1996. Enhanced diagnosis of *Pneumocystis carinii*: promises and problems. Clin. Microbiol. Newsletter 18:113.
3. Cantral, D.E., et al. 1993. Quantitative culture of bronchoalveolar lavage fluid for the diagnosis of bacterial pneumonia. Am. J. Med. 95:601.
4. Current topics: atypical pneumonia agents are joining the mainstream. ASM News 61:621, 1995.
5. Garibaldi, R.A. 1985. Epidemiology of community-acquired respiratory tract infections in adults: incidence, etiology, and impact. Am. J. Med. 78:325.
6. Gilligan, P. 1996. Report on the consensus document for microbiology and infectious diseases in cystic fibrosis. Clin. Microbiol. Newsletter 18:11.
7. Kahn, F.W. and Jones, J.M. 1988. Analysis of bronchoalveolar lavage specimens from immunocompromised patients with a protocol applicable in the microbiology laboratory. J. Clin. Microbiol. 26:1150.
8. Kauppinen, M. and Saikku, P. 1995. Pneumonia due to *Chlamydia pneumoniae*: prevalence, clinical features, diagnosis, and treatment. Clin. Infect. Dis. 21:244.
9. Kuo, C.C., et al. 1995. *Chlamydia pneumoniae* (TWAR). Clin. Microbiol. Rev. 8:451.
10. Lentino, J.R. 1987. The nonvalue of unscreened sputum specimens in the diagnosis of pneumonia. Clin. Microbiol. Newsletter 9:70.
11. Marrie, T.J., Durant, H., and Bates, L. 1989. Community-acquired pneumonia requiring hospitalization: a 5-year prospective study. Rev. Infect. Dis. 11:586.
12. Marrie, T.J. 1994. Community-acquired pneumonia. Clin. Infect. Dis. 18:501.
13. Morris, A.J., Tanner, D.C., and Reller, R.B. 1993. Rejection criteria for endotracheal aspirates from adults. J. Clin. Microbiol. 31:1027.
14. Niederman, M.S., et al. 1993. Guidelines for the initial management of adults with community-acquired pneumonia: diagnosis, assessment of severity, and initial antimicrobial therapy. Am. Rev. Respir. Dis. 148:1418.
15. Pisani, R.J. and Wright, A.J. 1992. Clinical utility of bronchoalveolar lavage in immunocompromised hosts. Mayo Clin. Proc. 67:221.
16. Poe, R. 1996. Management of lower respiratory tract infections. Guthrie J. 65:40.
17. Pollock, H.M., et al. 1983. Diagnosis of bacterial pulmonary infections with quantitative protected catheter cultures obtained during bronchoscopy. J. Clin. Microbiol. 17:255.
18. Sadeghi, E., et al. 1994. Utility of Gram stain in evaluation of sputa from patients with cystic fibrosis. J. Clin. Microbiol. 32:54.
19. Salemi, C., et al. 1995. Association between severity of illness and mortality from nosocomial infection. Am. J. Infect. Control. 23:188.

Bibliography

Bartlett, J.G., et al. 1987. Cumitech 7A. In Laboratory diagnosis of lower respiratory tract infections. American Society for Microbiology, Washington, D.C.

Baselski, V. 1993. Microbiologic diagnosis of ventilator associated pneumonia. Infect. Dis. Clin. North Am. 7:331.

Gradon, J.D., Timpone, J.G., and Schnittman, S.M. 1992. Emergence of unusual opportunistic pathogens in AIDS: a review. Clin. Infect. Dis. 15:134.

22 | UPPER RESPIRATORY TRACT INFECTIONS AND OTHER INFECTIONS OF THE ORAL CAVITY AND NECK

GENERAL CONSIDERATIONS

ANATOMY

The upper respiratory tract includes the epiglottis and surrounding tissues, larynx, nasal cavity, and the pharynx (throat). These anatomic structures are shown in Figure 22-1.

The pharynx is a tubelike structure that extends from the base of the skull to the esophagus (see Figure 22-1). Made of muscle, this structure is divided into three parts:

- Nasopharynx (portion of the pharynx above the soft palate)
- Oropharynx (portion of the pharynx between the soft palate and epiglottis)
- Laryngopharynx (portion of the pharynx below the epiglottis that opens into the larynx)

The tonsils are contained within the oropharynx, the larynx is located between the root of the tongue and the upper end of the trachea.

PATHOGENESIS

An overview of the pathogenesis of respiratory tract infections was presented in the previous chapter. It is important to keep in mind that upper respiratory tract infections may spread and become more serious because the mucosa (mucous membrane) of the upper tract is continuous with the mucosa lining of the sinuses, eustachian tube, middle ear, and the lower respiratory tract.

DISEASES OF THE UPPER RESPIRATORY TRACT, ORAL CAVITY, AND NECK

UPPER RESPIRATORY TRACT

Diseases of the upper respiratory tract are named according to the anatomic sites involved. Most of these infections are self-limiting and most are caused by viruses.

Laryngitis

Acute laryngitis is usually associated with the common cold or influenza syndromes. Characteristically, patients complain of hoarseness and lowering or deepening of the voice. Acute laryngitis is generally a benign illness.

Acute laryngitis is caused almost exclusively by viruses. Although numerous viruses can cause laryngitis, influenza viruses, rhinoviruses, and adenoviruses are the most common etiologic agents. If examination of the larynx reveals an exudate or membrane on the pharyngeal or laryngeal mucosa, streptococcal infection, mononucleosis, or diphtheria should be suspected (see discussion about miscellaneous infections caused by other agents later in this chapter).

Laryngotracheobronchitis

Another clinical syndrome closely related to laryngitis is acute laryngotracheobronchitis, or croup. Croup is a relatively common illness in young children, primarily those under age 3. Of significance, croup can represent a potentially more serious disease if the infection extends downward from the larynx to involve the trachea or even the bronchi. Illness is characterized by variable fever, inspiratory stridor (difficulty in moving enough air through the larynx), hoarseness, and a harsh, barking, nonproductive cough. These symptoms last for 3 to 4 days, although the cough may persist for a longer period. In young infants, severe respiratory distress and fever are common symptoms.

Similar to the etiologic agents of laryngitis, viruses are a primary cause of croup; parainfluenza viruses are the major etiologic agents. In addition to parainfluenza viruses, influenza viruses, respiratory syncytial virus, and adenoviruses can also cause croup. *Mycoplasma pneumoniae*, rhinoviruses, and enteroviruses can cause a few cases as well.

Epiglottitis

Epiglottitis is an infection of the epiglottis and other soft tissues above the vocal cords. Infection of the epiglottis can lead to significant edema (swelling) and inflammation. Most commonly, children between the ages of 2 and 6 are infected. These children typically present with fever, difficulty in swallowing because of pain, drooling, and respiratory obstruction with inspiratory stridor. Epiglottitis is a potentially

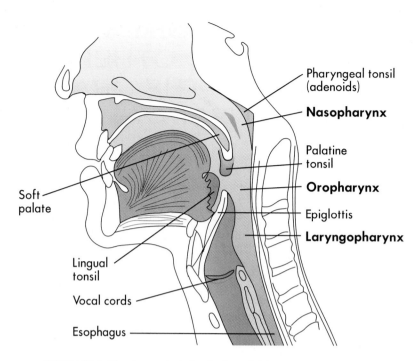

Soft palate

Lingual tonsil

Vocal cords

Esophagus

Pharyngeal tonsil (adenoids)

Nasopharynx

Palatine tonsil

Oropharynx

Epiglottis

Laryngopharynx

FIGURE 22-1 The pharynx, including its three divisions and nearby structures.

life-threatening disease because the patient's airway can become completely obstructed (blocked) if not treated.

In contrast to laryngitis, epiglottitis is usually caused by bacteria. *Haemophilus influenzae* type b is the primary cause of epiglottitis. Other organisms occasionally implicated are streptococci and staphylococci. Diagnosis is established on clinical grounds, including the visualization of the epiglottis which appears swollen and bright red in color. Bacteriologic culture of the epiglottis is contraindicated because swabbing of the epiglottis may lead to respiratory obstruction. Of importance, bacteremia with the causative agent usually occurs in children with epiglottitis.

Pharyngitis, tonsillitis, and peritonsillar abscesses

PHARYNGITIS AND TONSILLITIS Pharyngitis (sore throat) and tonsillitis are common upper respiratory tract infections affecting both children and adults.

Clinical Manifestations. Infection of the pharynx is associated with pharyngeal pain. Visualization of the pharynx reveals that affected tissues are erythematous (red) and swollen. Depending on the causative microorganism, either inflammatory exudates (fluid with protein, inflammatory cells, and cellular debris), vesicles (small blisterlike sacs containing liquid) and mucosal ulceration, or nasopharyngeal lymphoid hyperplasia (swollen lymph nodes) may be observed.

Pathogenesis. Pathogenic mechanisms will differ and are dependent on the organism causing the pharyngitis. For example, some organisms directly invade the pharyngeal mucosa (e.g., *Arcanobacterium haemolyticum*), others elaborate toxins and other virulence factors at the site (e.g., *Corynebacterium diphtheriae*), and still others invade the pharyngeal mucosa and elaborate toxins and other virulence factors (e.g., group A streptococci [*Streptococcus pyogenes*]). Therefore, pathogenic mechanisms are reviewed in Part 4 of this textbook according to various organism groups.

Epidemiology/Etiologic Agents. Most cases of pharyngitis occur during the colder months and often accompany other infections, primarily those caused by viruses. Patients with respiratory tract infections caused by influenza types A and B, adenovirus, parainfluenza, coxsackie A, rhinoviruses, or coronaviruses frequently complain of a sore throat. Pharyngitis, often with ulceration, is also commonly found in patients with infectious mononucleosis caused by either Epstein-Barr virus or cytomegalovirus.

Although different bacteria can cause pharyngitis and/or tonsillitis, the primary cause of bacterial pharyngitis is *Streptococcus pyogenes* (or group A beta-hemolytic streptococcus). Viral pharyngitis or other causes of pharyngitis/tonsillitis must be differentiated from that caused by *S. pyogenes* since pharyngitis resulting from *S. pyogenes* is treatable with penicillin whereas viral infections are not. In addition, treatment is of particular importance because infection with *S. pyogenes* can lead to complications such as acute rheumatic fever and glomerulonephritis. These complications are referred to as poststreptococcal sequelae (diseases that follow a streptococcal infection) and are primarily immunologically mediated; these sequelae will be discussed in greater detail in Chapter 53. *S. pyogenes* may also cause pyogenic infections (suppurations) of the tonsils, sinuses, and middle ear, or cellulitis as secondary pyogenic sequelae after an episode of pharyngitis. Accordingly, streptococcal pharyngitis is usually treated to prevent both the suppurative and nonsuppurative sequelae, as well as to decrease morbidity.

Cases of pharyngitis caused by groups B, C, and G and by nonhemolytic members of these groups of streptococci (including group A) have been reported.* Although bacteria other than group A streptococci may cause pharyngitis, this occurs less often, and laboratories should not routinely look for other agents except in special circumstances following discussion with the clinician involved. The agents that can cause pharyngitis or tonsillitis are listed in Table 22-1.

Although *H. influenzae, S. aureus,* and *S. pneumoniae* are frequently isolated from nasopharyngeal and throat cultures, they have not been shown to cause pharyngitis. Carriage of any of these organisms, as well as *N. meningitidis,* may have clinical importance for some patients or their contacts. Cultures of specimens obtained from the anterior nares often yield *S. aureus.* The carriage rate for this organism is especially high among health care workers, and even 20% of the general population can be colonized with this microbe.

Vincent's angina, or anaerobic tonsillitis, involves pseudomembrane formation on tonsillar surfaces. The infection is relatively rare today but is considered a very serious disease because it is often complicated by septic jugular thrombophlebitis, bacteremia, and widespread metastatic infection. Adults are more often affected than children; poor oral hygiene is a predisposing factor. Multiple anaerobes, especially *Fusobacterium necrophorum,* are implicated in this syndrome. Although Gram stain of a throat specimen is usually not predictive, in those patients with symptoms suggestive of Vincent's angina, Gram stain will reveal numerous fusiform, gram-negative bacilli and spirochetes.

PERITONSILLAR ABSCESSES   Peritonsillar abscesses are generally thought of as a complication of tonsillitis. This infection is most common in children older than age 5 and young adults. It is important to treat these infections because they can spread to adjacent tissues, as well as erode into the carotid artery to cause an acute hemorrhage. The predominant organisms in peritonsillar abscesses are non–spore-forming anaerobes, including *Fusobacterium* (especially *F. necrophorum*), *Bacteroides* (including the *B. fragilis* group), and anaerobic cocci. *Streptococcus pyogenes* and viridans streptococci may also be involved.

Rhinitis

Rhinitis (common cold) is an inflammation of the nasal mucous membrane or lining. Depending on the host response and the etiologic agent, rhinitis is characterized by variable fever, increased mucous secretions, inflammatory edema of the nasal mucosa, sneezing, and watery eyes. With rare exceptions, rhinitis is caused by viruses; some of these agents are listed in Box 22-1.

Miscellaneous infections caused by other agents

CORYNEBACTERIUM DIPHTHERIAE Although much less common than streptococcal pharyngitis, *C. diphtheriae* can still be isolated from patients with sore throat, as well as more serious systemic disease. After an incubation period of 2 to 4 days, diphtheria usually presents as pharyngitis or tonsillitis. Patients are often febrile and complain of sore throat and malaise (bodily discomfort). The hallmark for diphtheria is the presence of an exudate or membrane that is usually on the tonsils or pharyngeal wall. The gray-white membrane is a result of the action of diphtheria toxin on the epithelium at the site of infection. Complications occur frequently with diphtheria and are usually seen during the last stage of the disease (paroxysmal stage). The most feared complications are those involving the central nervous system such as seizures, coma, or blindness. Information as to how this organism causes disease is discussed in Chapter 21. Additional specifics regarding this organism are also provided in Chapter 55.

BORDETELLA PERTUSSIS Although mass immunization programs have greatly reduced the incidence of pertussis, enough cases (because of outbreaks and

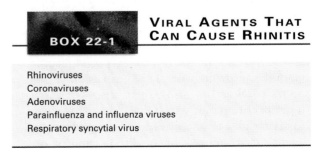

BOX 22-1   **VIRAL AGENTS THAT CAN CAUSE RHINITIS**

Rhinoviruses
Coronaviruses
Adenoviruses
Parainfluenza and influenza viruses
Respiratory syncytial virus

---

*References 1,4,7,8,10,11.

TABLE 22-1   BACTERIA THAT CAN CAUSE ACUTE PHARYNGITIS AND/OR TONSILLITIS

ORGANISM	DISEASE	RELATIVE FREQUENCY
Streptococcus pyogenes	Pharyngitis/tonsillitis/rheumatic fever/scarlet fever	15% to 35%
Group C and G beta-hemolytic streptococci	Pharyngitis/tonsillitis	3% to 11%
Arcanobacterium (Corynebacterium) haemolyticum[2,9]	Pharyngitis/tonsillitis/rash	<1% to 10%
Neisseria gonorrhoeae	Pharyngitis/disseminated disease	Rare*
Corynebacterium ulcerans	Pharyngitis	Rare
Mycoplasma pneumoniae	Pneumonia/bronchitis/pharyngitis	Rare
Yersinia enterocolitica	Pharyngitis/enterocolitis	Rare

*Less than 1%.

regional epidemics) still occur that laboratories should either be able to detect, isolate, and identify the organism or refer the specimen to a reference laboratory.

Characteristically, pertussis, or whooping cough, is a prolonged disease (lasting as long as 6 to 8 weeks) that is marked by paroxysmal (sudden and/or intense) coughing. Following an incubation period of 7 to 13 days, the patient with symptomatic infection develops upper respiratory symptoms, including a dry cough, fever, runny nose, and sneezing. This may then progress, after about 2 weeks, to spells of paroxysmal coughing. As these episodes worsen, the characteristic whoop, caused by attempted inspiration through an epiglottis undergoing spasm, begins. Vomiting may occur and usually a lymphocytosis is present. This phase of the illness may last as long as 6 weeks. Additional information about how *B. pertussis* causes this disease and how the organism is detected is provided in Chapter 48.

KLEBSIELLA SPP. Rhinoscleroma is a rare form of chronic, granulomatous infection of the nasal passages, including the sinuses and occasionally the pharynx and larynx. Associated with *Klebsiella rhinoscleromatis,* the disease is characterized by nasal obstruction appearing over a long period, caused by tumorlike growth with local extension. Another species, *Klebsiella ozaenae,* can also be recovered from upper respiratory tract infections. This organism may contribute to another infrequent condition called ozaena, characterized by a chronic, mucopurulent nasal discharge that is often foul-smelling. It is caused by secondary, low-grade anaerobic infection.

ORAL CAVITY

Stomatitis

Stomatitis is an inflammation of the mucous membranes of the oral cavity. Herpes simplex virus is the primary agent of this disease, in which multiple ulcerative lesions are seen on the oral mucosa. These lesions are painful and can be found not only in the mouth but also in the oropharynx. Herpetic infections of the oral cavity are prevalent among immunosuppressed patients.

Thrush

Candida spp. can also invade the oral mucosa. Immunosuppressed patients, including very young infants, may develop oral candidiasis, called **thrush.** Oral thrush can extend to produce pharyngitis and/or esophagitis, a common finding in patients with AIDS and other immunosuppressed patients. Thrush is suspected if whitish patches of exudate on an area of inflammation are observed on the buccal (cheek) mucosa, tongue, or oropharynx. Oral mucositis or pharyngitis in the granulocytopenic patient may be caused by *Enterobacteriaceae, S. aureus,* or *Candida* spp. and is manifested by erythema, sore throat, and possibly exudate or ulceration.

Periodontal infections

TYPES   The three dental problems for which help may be requested from clinical laboratories are root canal infections, with or without periapical abscess; orofacial odontogenic infections, with or without osteomyelitis (inflammation of a bone) of the jaw; and perimandibular space infections. Oral bacteria are clearly important in other dental processes, such as **caries** (destruction of the mineralized tissues of the tooth; a cavity), periodontal (tissues in, around, and supporting the tooth) disease, and localized juvenile periodontitis, but clinical laboratories are not involved in culturing in such cases.

ETIOLOGIC AGENTS   The bacteriology is similar in all of these infections and involves primarily anaerobic bacteria and streptococci except for perimandibular space infections, which may also involve staphylococci and *Eikenella corrodens* in about 15% of patients. The streptococci are microaerobic or facultative and are usually alpha-hemolytic, (particularly the *Streptococcus milleri* group—see Chapter 53); they are usually found in 20% to 30% of dental infections of the preceding types.

Members of the *Bacteroides fragilis* group are found in root canal infections, orofacial odontogenic infections, and bacteremia secondary to dental extraction in 5% to 10% of patients. Anaerobic cocci (both *Peptostreptococcus* and *Veillonella*), pigmented *Prevotella* and *Porphyromonas*, the *Prevotella oralis* group, and *Fusobacterium* are found in about 20% to 50% of the three conditions mentioned, as well as in postextraction bacteremia. Infection with *Actinomyces israelii* may complicate oral surgery.

Salivary gland infections

Acute suppurative parotitis (inflammation of the salivary glands located under the cheek in front of and below the external ear) is seen in very ill patients, especially those who are dehydrated, malnourished, elderly, or recovering from surgery. It is associated with painful, tender swelling of the parotid gland; purulent drainage may be evident at the opening of the duct of the gland in the mouth. *Staphylococcus aureus* is the major pathogen but on occasion viridans streptococci and oral anaerobes may play a role. A chronic bacterial parotitis has been described that usually involves *Staphylococcus aureus*. Less often, other salivary glands may be involved with a bacterial infection, usually because of ductal obstruction.

The mumps virus is traditionally the major viral agent involved in parotitis; however, since the advent of childhood vaccination, infection with mumps virus is rarely diagnosed. Influenza virus and enteroviruses may also cause this syndrome. Diagnosis of viral parotitis is usually done serologically. Infrequently, *Mycobacterium tuberculosis* may involve the parotid gland in conjunction with pulmonary tuberculosis.

NECK

Infections of the deep spaces of the neck are potentially serious because they may spread to critical structures such as major vessels of the neck or to the mediastinum, leading to mediastinitis, purulent pericarditis, and pleural empyema. The oral flora are responsible for these infections. Accordingly, the predominant organisms are anaerobes, primarily *Peptostreptococcus*, various *Bacteroides*, *Prevotella*, *Porphyromonas*, *Fusobacterium* spp., and *Actinomyces*. Streptococci, chiefly of the viridans variety, are also important. *Staphylococcus aureus* and various aerobic, gram-negative bacilli may be recovered, particularly from patients developing these problems in the hospital.

DIAGNOSIS OF UPPER RESPIRATORY TRACT INFECTIONS

COLLECTION AND TRANSPORT OF SPECIMENS

Either cotton-, Dacron-, or calcium alginate-tipped swabs are suitable for collecting most upper respiratory tract microorganisms. If the swab remains moist, no further precautions need to be taken for specimens that are cultured within 4 hours of collection. After that period, some kind of transport medium to maintain viability and prevent overgrowth of contaminating organisms should be used. Swabs for detection of group A streptococci *(Streptococcus pyogenes)* are the only exception. This organism is highly resistant to desiccation and will remain viable on a dry swab for as long as 48 to 72 hours. Throat swabs of this type can be placed in glassine paper envelopes for mailing or transport to a distant laboratory. Throat swabs are also adequate for recovery of adenoviruses and herpesviruses, *Corynebacterium diphtheriae*, *Mycoplasma*, *Chlamydia*, and *Candida* spp. Recovery of C. *diphtheriae* is enhanced by culturing both the throat and nasopharynx.

Nasopharyngeal swabs are better suited for recovery of respiratory syncytial virus, parainfluenza virus, *Bordetella pertussis*, *Neisseria* spp., and the other viruses causing rhinitis. Optimum conditions for the collection and transport of specimens for viral detection or culture are described in Chapter 66. Although swabs made of calcium alginate are commonly used to collect nasopharyngeal specimens (excluding those specimens for chlamydia or viral culture), aspirated nasopharyngeal secretions collected in a soft rubber bulb or plastic-tipped catheter are the best specimens for *Bordetella pertussis*.[6] Specimens for *B. pertussis* ideally should be inoculated directly to fresh culture media at the patient's bedside. If this is not possible, transport for less than 2 hours in 1%

casamino acid medium, charcoal horseblood transport medium, or 5% serum inositol in fetal bovine serum is acceptable.[3,5]

DIRECT VISUAL EXAMINATION

A Gram stain of material obtained from upper respiratory secretions or lesions can do very little to help with diagnosis. Yeastlike cells can be identified, helpful in identifying thrush, and the characteristic pattern of fusiforms and spirochetes of Vincent's angina may be visualized. Plain Gram's crystal violet (allowed to remain on the slide for 1 minute before rinsing with tap water) and the Gram stain can be used to identify the agents of Vincent's angina. However, if only crystal violet is used, remember to make the smear very thin because everything will be intensely gram positive, making a thick smear difficult to read.

For causes of pharyngitis, Gram stains are unreliable. Direct smears of exudate from membranelike lesions to differentiate diphtheria from other causes of such membranes are also not reliable and are not recommended.

Fungal elements, including yeast cells and pseudohyphae, may be visualized with either a 10% potassium hydroxide (KOH) preparation, calcofluor white fluorescent stain, or periodic acid-Schiff (PAS) stain. Direct examination of material obtained from the nasopharynx of suspected cases of whooping cough using a fluorescent antibody stain (see Chapter 48) has been shown to yield some early positive results for detection of *B. pertussis*.

Direct fluorescent antibody stains have been used to identify group A beta-hemolytic streptococci in throat specimens. This technology has largely been supplanted by direct detection of antigen by enzyme immunoassay, for which many commercial products are available. Direct antigen detection systems are discussed later in this chapter. Various methods, including fluorescent antibody stain reagents and enzyme immunoassays, are also commercially available to detect numerous viral agents.

CULTURE

Streptococcus pyogenes (beta-hemolytic group A streptococci)

Because the primary cause of bacterial pharyngitis in North America is *Streptococcus pyogenes*, most laboratories routinely screen throat cultures for this organism. Group A streptococci are usually beta-hemolytic, with <1% being nonhemolytic. Therefore, classically, throat swabs have been plated onto 5% sheep blood agar (trypticase soy base). Drawbacks to using a sheep blood agar plate for isolation of *S. pyogenes* are that both overnight growth for colony formation and further manipulations of the

beta-hemolytic organisms are required for definitive identification (see Chapter 53). If sufficient numbers of pure colonies are not available for identification, a subculture requiring additional incubation is necessary. By placing a 0.04 unit differential bacitracin filter paper disk (Taxo A, BBL, or Bacto Bacitracin, Difco Laboratories, Detroit, Mich.) directly on the area of initial inoculation, presumptive identification of *S. pyogenes* can be made after overnight incubation (all of group A and a very small percentage of group B streptococci are susceptible). However, use of the bacitracin disk in the primary area of inoculation reduces the sensitivity and specificity of culture and identification of *S. pyogenes*. Sometimes growth of too few beta-hemolytic colonies or overgrowth of other organisms make interpretation difficult. Therefore, using the bacitracin disk as the only method of identification of *S. pyogenes* is not recommended. New selective agars, such as streptococcal selective agar, have been developed that suppress the growth of almost all normal flora and beta-hemolytic streptococci except for groups A and B and *Arcanobacterium haemolyticum*.

Direct antigen detection tests (coagglutination or latex agglutination) or the PYR test (see Chapter 53) can also be carried out on the more isolated beta-hemolytic colonies that appear after overnight incubation on selective agar. A nucleic acid probe test was released for commercial use in 1992 (Gen-Probe, Inc., San Diego, Calif.) that provides highly sensitive and specific results following a 4-hour incubation step in selective media.

Corynebacterium diphtheriae

If diphtheria is suspected, the physician must communicate this information to the clinical laboratory. Because streptococcal pharyngitis is in the differential diagnosis of diphtheria and because dual infections do occur, cultures for *Corynebacterium diphtheriae* should be plated onto sheep blood agar or streptococcal selective agar, as well as onto special media for recovery of this agent. These special media include a Loeffler's agar slant and a cystine-tellurite agar plate. Chapter 48 discusses the identification of the organism. Recovery of this organism is enhanced by culturing specimens from both the throat and nasopharynx of potentially infected patients.

Bordetella pertussis

Freshly prepared Bordet-Gengou agar was the first medium developed for isolation of *Bordetella pertussis*. However, because it was inconvenient to use, other media were subsequently developed (see Chapter 48). Today, Regan-Lowe or charcoal horse blood agar is recommended for use in diagnostic laboratories.

Because the organisms are extremely delicate, specimens should be plated directly onto media, if possible. Technologists should use both fluorescent antibody stains to directly detect *B. pertussis* in clinical material, as well as cultures of the nasopharynx to achieve maximum detection. The yield of positive isolations from clinical cases of pertussis seems to vary from 20% to 98% depending on the stage of disease, previous treatment of the patient, and laboratory techniques.

Neisseria

Specimens received in the laboratory for isolation of *Neisseria meningitidis* (for detection of carriers) or *N. gonorrhoeae* should be plated to a selective medium, either modified Thayer-Martin or Martin-Lewis agar. After 24 to 48 hours of incubation in 5% to 10% carbon dioxide, typical colonies of *Neisseria* spp. will be visible (see Chapter 51).

Epiglottitis

Clinical specimens from cases of epiglottitis (swabs obtained by a physician) should be plated to sheep blood agar, chocolate agar (for recovery of *Haemophilus* spp.), and a streptococcal selective medium if desired. Because *Staphylococcus aureus, Streptococcus pneumoniae,* and beta-hemolytic streptococci may cause this disease, albeit rarely, their presence should be sought. Refer to Table 1-1 for an overview of collection, transport, and processing of different specimens from the upper respiratory tract.

NONCULTURE METHODS FOR DETECTION OF *STREPTOCOCCUS PYOGENES* IN THROAT SPECIMENS

One improvement in selected areas of clinical microbiology in recent years has been the development of rapid methods for antigen detection that obviate the need for culture. Identification of group A streptococcal antigen in throat specimens is a particular application that has been a part of this development. At least 40 commercial products are available, employing latex agglutination, enzyme immunoassay, and gene probe technologies, that allow detection of group A streptococcal antigen within as little as 10 minutes.

Although the specific procedures vary with the products, several generalizations can be made. Throat swabs are incubated in an acid reagent or enzyme to extract the group A-specific carbohydrate antigen (the gene probe requires preincubation in broth media followed by nucleic acid extraction). Dacron swabs seem to be most efficient at releasing antigen, although other types of swabs may yield acceptable results. Following the recommendations of the manufacturer of the extraction kit is mandatory to achieve optimal results. In most laboratory comparisons between a rapid method and conventional culture methods for detecting the presence of group A streptococci in throat swabs, the commercial kits have shown good sensitivity (>90%) and specificity. Doubts have been expressed, however, about the commercial kits' overall ability to detect all clinically important streptococcal infections (see Chapter 53). For this reason, specimens with a negative direct test for group A streptococci should be cultured.

DIAGNOSIS OF INFECTIONS IN THE ORAL CAVITY AND NECK

COLLECTION AND TRANSPORT

A problem in collecting oral and dental infection material is to avoid or minimize contamination with oral flora. For collection of material from root canal infection, the tooth is isolated by means of a rubber dam. A sterile field is established, the tooth is swabbed with 70% alcohol, and after the root canal is exposed, a sterile paper point is inserted, removed, and placed into semisolid, nonnutritive, anaerobic transport medium. Alternatively, needle aspiration can be used if sufficient purulent material is present. Completely defining the flora of such infections is beyond the scope of routine clinical microbiology laboratories.

Specimens from neck space infections can usually be obtained with a syringe and needle or by biopsy during a procedure by the surgeon. Transport must be under anaerobic conditions.

DIRECT VISUAL EXAMINATION

All material submitted for culture should be smeared and examined by Gram stain and other appropriate techniques for fungi (i.e., calcofluor white, KOH, or PAS stains), if requested.

CULTURE

Infections such as peritonsillar abscesses, oral and dental infections, and neck space infections, usually involve anaerobic bacteria. The anaerobes involved typically originate in the oral cavity and are often more delicate than anaerobes isolated from other clinical material. Very careful attention must be paid to providing optimal techniques for anaerobic cultivation, as well as collection and transport in order to recover and identify the etiologic agents.

References

1. Benjamin, J.T. and Perriello, V.A. 1976. Pharyngitis due to group C hemolytic streptococci in children. J. Pediatr. 89:254.
2. Cambier, M., Janssens, M., and Wauters, G. 1992. Isolation of *Arcanobacterium haemolyticum* from patients with pharyngitis in Belgium. Acta. Clin. Belg. 47:303.

3. Cassiday, P.K., et al. 1994. Viability in *Bordetella pertussis* in four suspending solutions at three temperatures. J. Clin. Microbiol. 32:1550.

4. Chretien, J.H., et al. 1979. Group B beta-hemolytic streptococci causing pharyngitis. J. Clin. Microbiol. 10:63.

5. Friedman, R.L. 1988. Pertussis: the disease and new diagnostic methods. Clin. Microbiol. Rev. 1:365.

6. Hallander, H.O., et al. 1993. Comparison of nasopharyngeal aspirates with swabs for culture of *Bordetella pertussis*. J. Clin. Microbiol. 31:50.

7. Hill, H.R., et al. 1969. Epidemic of pharyngitis due to streptococci of Lancefield group G. Lancet 2:371.

8. McCue, J.D. 1982. Group G streptococcal pharyngitis: analysis of an outbreak at a college. J.A.M.A. 248:1333.

9. Sellen, T.J. and Long, D.A. 1996. *Arcanobacterium haemolyticum* pharyngitis and tonsillitis. Clin. Microbiol. Newsletter 18:30.

10. Turner, J.C., et al. 1990. Association of group C β-hemolytic streptococci with endemic pharyngitis among college students. J.A.M.A. 264:2644.

11. Turner, J.C., et al. 1993. Role of group C beta-hemolytic streptococci in pharyngitis: epidemiologic study of clinical features associated with isolation of group C streptococci. J. Clin. Microbiol. 31:808.

Bibliography

Bannatyne, R.M., Clausen, C., and McCarthy, L.R. 1979. Laboratory diagnosis of upper respiratory tract infections. In Cumitech 10. American Society for Microbiology, Washington, D.C.

Brook, I., Grazier, E.H., and Gher, M.E. 1991. Aerobic and anaerobic microbiology of periapical abscess. Oral Microbiol. Immunol. 6:123.

Dymock, D., et al. 1996. Molecular analysis of microflora associated with dentoalveolar abscesses. J. Clin. Microbiol. 34:537.

Glezen, W.P., et al. 1967. Group A streptococci, mycoplasma, and viruses associated with acute pharyngitis. J.A.M.A. 202:455.

Proctor, D.F. 1977. The upper airways. II. The larynx and trachea. Am. Rev. Respir. Dis. 115:315.

Vaughan, C.W. 1982. Current concepts in otolaryngology: diagnosis and treatment of organic voice disorders. N. Engl. J. Med. 307:863.

Vukmir, R.B. 1992. Adult and pediatric pharyngitis: a review. J. Emerg. Med. 10:607.

23 MENINGITIS AND OTHER INFECTIONS OF THE CENTRAL NERVOUS SYSTEM

GENERAL CONSIDERATIONS

ANATOMY

Diagnosis of an infection involving the central nervous system (CNS) is of critical importance. Most clinicians consider infection in the CNS to be one of the medical emergencies relating to infectious diseases. Therefore, an understanding of the basic anatomy and physiology of the CNS is helpful for the microbiologist to ensure appropriate specimen processing and interpretation of laboratory results.

Coverings and spaces of the CNS

Because of the vital and essential role of the CNS in the body's regulatory processes, the brain and spinal cord have two protective coverings: an outer covering consisting of bone and an inner covering of membranes called the **meninges**. The outer bone covering encases the brain (i.e., cranial bones or skull) and spinal cord (i.e., the vertebrae). The meninges is a collective term for three distinct layers surrounding the brain and spinal column:

- Dura mater (outermost membrane layer)
- Arachnoid
- Pia mater (innermost membrane layer)

The pia mater and the arachnoid membrane are collectively called the **leptomeninges**. The portion of the arachnoid that covers the top of the brain contains arachnoid villi, which are special structures, that absorb the spinal fluid and allow it to pass into the blood.

Between and around the meninges are spaces that include the epidural, subdural, and subarachnoid spaces. The relative location of the meninges and spaces to one another in the brain is shown in Figure 23-1. The location and nature of the meninges and spaces are summarized in Table 23-1.

Cerebrospinal fluid

Cerebrospinal fluid (CSF) envelops the brain and spinal cord and has several functions. By cushioning and providing buoyancy for the bulk of the brain, the effective weight of the brain is reduced by a factor of 30. Of importance, CSF carries essential metabolites into the neural tissue and cleanses the tissues of wastes as it circulates around the entire brain, ventricles, and spinal cord. Every 3 to 4 hours, the entire volume of CSF is exchanged. In addition to these functions, CSF provides a means by which the brain monitors changes in the internal environment.

CSF is found in the subarachnoid space (see Table 23-1) and within cavities and canals of the brain and spinal cord. There are four large, fluid-filled spaces within the brain referred to as **ventricles**. Specialized secretory cells called the **choroid plexus** produce CSF. The choroid plexus is located centrally within the brain in the third and fourth ventricles; about 23 mL of CSF are contained within the ventricles. The fluid travels around the outside areas of the brain within the subarachnoid space, driven primarily by the pressure produced initially at the choroid plexus (Figure 23-2). By virtue of its circulation, chemical and cellular changes in the CSF may provide valuable information about infections within the subarachnoid space.

ROUTES OF INFECTION

Organisms may gain access to the CNS by several primary routes:

- Hematogenous spread: followed by entry into the subarachnoid space through the choroid plexus or through other blood vessels of the brain. This is the most common way that the CNS becomes infected
- Direct spread from an infected site: the extension of an infection close to or contiguous with the CNS can occasionally occur; examples of such infections include otitis media (infection of middle ear), sinusitis, and mastoiditis
- Anatomic defects in CNS structures: anatomic defects as a result of surgery, trauma, or congenital abnormalities can allow microorganisms easy and ready access to the CNS
- Travel along nerves leading to the brain (direct intraneural): the least common route of CNS infection caused by organisms such as rabies virus, which travels along peripheral sensory nerves, and herpes simplex virus

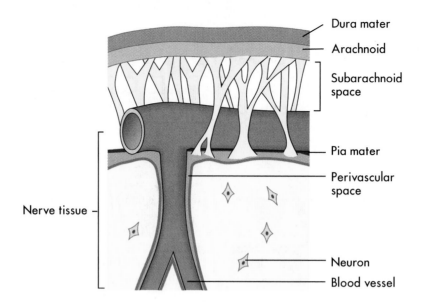

FIGURE 23-1 Cross section of the brain showing the important membrane coverings and spacings and other key structures.

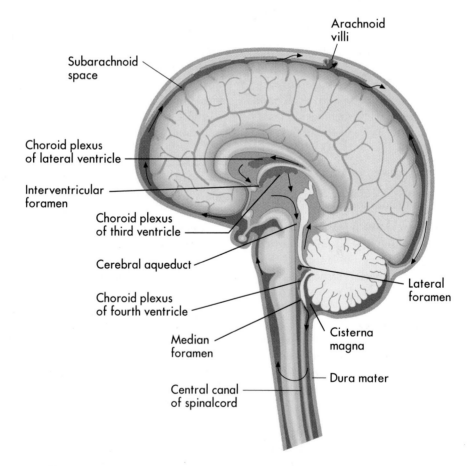

FIGURE 23-2 Flow of CSF through the brain. CSF originates in the choroid plexus, then flows through the ventricles and subarachnoid space and into the bloodstream.

TABLE 23-1   INNER COVERINGS (MENINGES) OF THE BRAIN, SPINAL CORD, AND SURROUNDING SPACES

ANATOMIC STRUCTURE	RELATIVE LOCATION	KEY FEATURES
Epidural space	Outside the dura mater yet inside the skull	Cushion of fat and connective tissues
Dura mater	Outermost membrane	Membrane that adheres to the skull; white fibrous tissue
Subdural space	Between the dura mater and the arachnoid membrane	Cushion of lubricating serous fluid
Arachnoid membrane	Between the dura mater and pia mater	Delicate, cobweblike membrane covering the brain and spinal cord
Subarachnoid space	Beneath the arachnoid membrane	Contains a significant amount of CSF in an adult (~125-150 mL)
Pia mater	Beneath the subarachnoid space	Adheres to the outer surface of the brain and spinal cord; contains blood vessels

DISEASES OF THE CENTRAL NERVOUS SYSTEM

Meningitis

Infection within the subarachnoid space or throughout the leptomeninges is called meningitis. Based on the host's response to the invading microorganism, meningitis is divided into two major categories: purulent and aseptic meningitis.

PURULENT MENINGITIS   A patient with **purulent meningitis** will typically have a marked, acute inflammatory exudate with large numbers of polymorphonuclear cells (PMNs). Frequently, the underlying CNS tissue, in particular the ventricles, may be involved. If the ventricles become involved, this process is referred to as **ventriculitis**. These infections are usually caused by bacteria.

Pathogenesis. As discussed with other organ systems, the outcome of a host-microbe interaction depends on the characteristics of both the host and microorganism. An important host defense mechanism of the CNS is the blood-brain barrier; the choroid plexus, arachnoid membrane, and the cerebral microvascular endothelium are the key structures. Because of the unique structural properties of the vascular endothelium, such as continuous intercellular tight junctions, this barrier minimizes the passage of infectious agents into CSF in addition to regulating the transport of plasma proteins, glucose, and electrolytes.

Age of the host and other underlying host factors also contribute to whether an individual will be predisposed to develop meningitis. Neonates have the highest prevalence of meningitis. This higher prevalence is probably due to the immature immune system of the newborn, the organisms present in the colonized female vaginal tract, and the increased permeability of the blood-brain barrier of newborns. Lack of demonstrable humoral antibody against *Haemophilus influenzae* type b in children has been associated with an increased incidence of meningitis. Before widespread vaccination, most children developed measurable antibody by age 5. The importance of antibody is also a factor in adults, because military recruits without antibody to *Neisseria meningitidis* are more likely to develop disease. *N. meningitidis* has been associated with epidemic meningitis among young adults in crowded conditions (e.g., military recruits and college dormitory mates).

Because the respiratory tract is the primary portal of entry for many etiologic agents of meningitis, factors that predispose adults to meningitis are often the same factors that increase the likelihood that the adult will develop pneumonia or other respiratory tract colonization or infection. Alcoholism, splenectomy, diabetes mellitus, prosthetic devices, and immunosuppression contribute to increased risk. Finally, patients with prosthetic devices, particularly central nervous system shunts, are also at increased risk of developing meningitis.

For organisms to reach the CNS (primarily by the bloodborne route), host defense mechanisms must be overcome. Most cases of meningitis caused by bacteria share a similar pathogenesis. A successful meningeal pathogen must first sequentially colonize and cross host mucosal epithelium and then enter and survive in the bloodstream. The most common causes of meningitis possess the ability to evade host defenses at each of these levels. For example, clinical isolates of *Streptococcus pneumoniae* and *N. meningitidis* secrete IgA proteases that destroy the action of the host's secretory IgA, thereby facilitating bacterial attachment to the epithelium.[8] In addition, all of the most common etiologic agents of bacterial meningitis possess an antiphagocytic capsule that helps the organisms evade destruction by the host immune system.

Organisms appear to enter the CNS by interacting and subsequently breaking down the blood-brain barrier at the level of microvascular endothelium. To date, one of the least understood processes in the pathogenesis of meningitis is how organisms cross this barrier into the subarachnoid space. Nevertheless, there appear to be specific bacterial surface components, such as pili, polysaccharide capsules, and lipoteichoic acids, that facilitate adhesion of the organisms to the microvascular endothelial cells and subsequent penetration into the CSF.[6,13-15] Organisms can enter through (1) loss of capillary integrity by disrupting tight junctions of the blood-brain barrier, (2) transport within circulating phagocytic cells, or (3) by crossing the endothelial cell lining within endothelial cell vacuoles. After gaining access, the organism multiplies within the CSF, a site initially free of antimicrobial antibodies or phagocytic cells.

Clinical Manifestations. Meningitis can be either acute or chronic in onset and progression of disease.

Acute. Cases of acute meningitis are characterized by fever, stiff neck, headache, nausea and vomiting, neurologic abnormalities, and change in mental status.[1]

With acute bacterial meningitis, CSF usually contains large numbers of inflammatory cells (>1000/mm³), primarily polymorphonuclear neutrophils. The CSF shows a decreased glucose level relative to the serum glucose level (the normal ratio of CSF to serum glucose is approximately 0.6), and shows an increased protein concentration (normal protein is 15 to 50 mg/dL in adults and as high as 170 mg/dL, with an average of 90 mg/dL, in newborns).

The sequelae of acute bacterial meningitis in children are frequent and serious. Seizures occur in 20% to 30% of patients seen at large urban hospitals, and other neurologic changes are common. Acute sequelae include cerebral edema, hydrocephalus, cerebral herniation, and focal neurologic changes. Permanent deafness can occur in 10% of children who recover from bacterial meningitis. More subtle physiologic and psychologic sequelae may follow an episode of acute bacterial meningitis. Although the morbidity associated with meningitis today is still significant, the *Haemophilus influenzae* type b conjugate vaccine has played a major role in reducing postmeningitis sequelae.[16]

Chronic. Chronic meningitis often occurs in patients who are immunocompromised, although this is not always the case. Patients experience an insidious onset of disease, with some or all of the following: fever, headache, stiff neck, nausea and vomiting, lethargy, confusion, and mental deterioration. Symptoms may persist for a month or longer before treatment is sought. The CSF usually manifests an abnormal number of cells (usually lymphocytic), elevated protein, and some decrease in glucose content (Table 23-2). The pathogenesis of chronic meningitis is similar to that of acute disease.[17]

Epidemiology/Etiologic Agents. The etiology of acute meningitis is very dependent on the age of the patient. The majority of all cases in the United States occur in children younger than age 5. Before 1985, *H. influenzae* type b (Hib) was the most common agent in children in the United States between the ages of 1 month and 6 years. Ninety-five percent of all cases were due to Hib, *Neisseria meningitidis*, and *Streptococcus pneumoniae*. In 1985, the first Hib vaccine, a polysaccharide vaccine, was licensed for use in children 18 months or older but was not efficacious in children younger than 18 months. However, the recent widespread use of conjugate vaccine, Hib polysaccharide-protein conjugate, in children as young as 2 months has significantly affected the incidence of invasive *H. influenzae* type b disease.[16] To date the risks of meningococcal and pneumococcal diseases have remained level and the incidence of Hib invasive infections has declined by at least 95% among infants and children. After age 6 patients are less likely to develop meningitis.

As previously mentioned, neonates have the highest prevalence of meningitis, with a concomitant increased mortality rate (as high as 20%). Organisms causing disease in the newborn are different from those that affect other age groups; many of them are acquired by the newborn during passage through the mother's vaginal vault. Neonates are likely to be infected with, in order of incidence, group B streptococci, *Escherichia coli*, other gram-negative bacilli, and *Listeria monocytogenes;* occasionally other organisms may be involved. For example, *Flavobacterium meningosepticum* has been associated with nursery outbreaks of meningitis. This organism is a normal inhabitant of water in the environment and is presumably acquired nosocomially.

Important causes of meningitis in the adult, in addition to the meningococcus in young adults, include pneumococci, *Listeria monocytogenes,* and,

TABLE 23-2   GUIDELINES FOR THE INTERPRETATION OF RESULTS FOLLOWING HEMATOLOGIC AND CHEMICAL ANALYSIS OF CSF FROM CHILDREN AND ADULTS (EXCLUDING NEONATES)

CLINICAL SETTING	LEUKOCYTES/MM³	PREDOMINANT CELL TYPE	PROTEIN	GLUCOSE*
Normal	0-5	None	15-50 mg/dL	45-100 mg/dL
Viral infection	2-2000 (mean of 80)	Mononuclear†	Slightly elevated (50-100 mg/dL) or normal	Normal
Purulent infection	5-20,000 (mean of 800)	PMN	Elevated (>100 mg/dL)	Low (<45 mg/dL), but may be normal early in the course of disease
Tuberculosis and fungi	5-2000 (mean of 100)	Mononuclear	Elevated (>50mg/dL)	Normal or often low (<45 mg/dL)

*Must consider CSF glucose level in relation to blood glucose level. Normally, CSF glucose serum ratio is 0.6, or 50% to 70% of the blood glucose normal value.

†About 20% to 75% of cases may have PMN leukocytosis early in the course of infection.

---

less commonly, *Staphylococcus aureus* and various gram-negative bacilli. Meningitis caused by the latter organisms results from hematogenous seeding from various sources, including urinary tract infections. The various etiologic agents of chronic meningitis are listed in Box 23-1.

ASEPTIC MENINGITIS **Aseptic meningitis** is characterized by an increase of lymphocytes and other mononuclear cells (pleocytosis) in the CSF (in contrast to purulence, the PMN response characteristic of bacterial meningitis) and negative bacterial and fungal cultures. Patients may have fever, headache, a stiff neck, and nausea and vomiting.

Aseptic meningitis is commonly associated with viral infections and is usually a self-limiting infection. Aseptic meningitis is also a component of syphilis and some other spirochetal diseases (e.g., leptospirosis). Stiff neck and CSF pleocytosis may also be produced by various processes, such as malignancy, that are not CNS infections.

ENCEPHALITIS/MENINGOENCEPHALITIS

Encephalitis is an inflammation of the brain parenchyma and is usually a result of viral infection. Concomitant meningitis that occurs with encephalitis is known as **meningoencephalitis,** and the cellular infiltrate is more likely to be lymphocytic in this situation.

The host response to these CNS infections can differ somewhat from those associated with purulent or aseptic meningitis. Early in the course of viral encephalitis, or when considerable tissue damage occurs as a part of encephalitis, the nature of the inflammatory cells found in the CSF may be no different from that associated with bacterial meningitis; cell counts, however, are typically much lower.

Viral

Viral encephalitis, which cannot always be distinguished clinically from meningitis, is common in the warmer months. The primary agents are enteroviruses (coxsackieviruses A and B, echoviruses), mumps

BOX 23-1   ETIOLOGIC AGENTS OF CHRONIC MENINGITIS

Mycobacterium tuberculosis
Cryptococcus neoformans
Coccidioides immitis
Histoplasma capsulatum
Blastomyces dermatitidis
Candida spp.
Miscellaneous other fungi
Nocardia
Actinomyces
Treponema pallidum
Brucella
Salmonella
Rare parasites—*Toxoplasma gondii,* cysticercus, *Paragonimus westermani, Trichinella spiralis*

virus, herpes simplex virus, and arboviruses (togavirus, bunyavirus, equine encephalitis, St. Louis encephalitis, and other encephalitis viruses). Other viruses, such as measles, cytomegalovirus, lymphocytic choriomeningitis, Epstein-Barr virus, hepatitis, varicella-zoster virus, rabies virus, myxoviruses, and paramyxoviruses, are less commonly encountered. Any preceding viral illness and exposure history are important considerations in establishing a cause by clinical means.

Involvement of the nervous system in patients who are infected with the human immunodeficiency virus (HIV) is common. HIV is a neurotropic (attracted to nerve cells) virus that enters the CNS by macrophage transport and causes various neurologic syndromes. As HIV-infected individuals become progressively more immunosuppressed, the CNS becomes a target for opportunistic pathogens, such as cytomegalovirus, BK virus, and JC virus, that can produce meningitis and/or encephalitis.[3]

Parasitic

Parasites can cause meningoencephalitis, brain abscess (see the following discussion), or other central nervous system infection via two routes. The free-living amebae, *Naegleria fowleri* and *Acanthamoeba* spp., invade the brain via direct extension from the nasal mucosa. These organisms are acquired by swimming or diving in natural, stagnating freshwater ponds and lakes.

Other parasites reach the brain via hematogenous spread. Toxoplasmosis, caused by a parasite that grows intracellularly and destroys brain parenchyma, is a common central nervous system affliction in HIV-infected patients with acquired immunodeficiency syndrome (AIDS). *Entamoeba histolytica* and *Strongyloides stercoralis* have been visualized in brain tissue, and the larval form of *Taenia solium* (the pork tapeworm), called a cysticercus, can travel to the brain via the bloodstream and encyst in that site. Amebic brain infection and cysticercosis cause changes in the CSF that mimic meningitis.[17]

BRAIN ABSCESS

Brain abscesses (localized collections of pus in a cavity formed by the breakdown of tissue) may occasionally cause changes in the CSF and clinical symptoms that mimic meningitis. Brain abscesses may also rupture into the subarachnoid space, producing a severe meningitis with high mortality. If anaerobic organisms or viridans streptococci are recovered from CSF cultures, the diagnosis of brain abscess must be entertained; however, CSF culture is typically negative in brain abscess. Patients who are immunosuppressed or who have diabetes with ketoacidosis may show a rapidly progressive

fungal infection (phycomycosis) of the nasal sinuses or palatal region that travels directly to the brain.

LABORATORY DIAGNOSIS OF CNS INFECTIONS

MENINGITIS

Except in unusual circumstances, a lumbar puncture (spinal tap) is one of the first steps in the workup of a patient with suspected CNS infection, in particular, meningitis. Refer to Table 1-1 to review collection, transport, and processing of specimens obtained from the central nervous system.

Specimen collection and transport

CSF is collected by aseptically inserting a needle into the subarachnoid space, usually at the level of the lumbar spine. Three or four tubes of CSF should be collected and immediately labeled with the patient's name. Tube 3 or 4 is used for cell count and differential. If a small capillary blood vessel is inadvertently broken during the spinal tap, blood cells picked up from this source will usually be absent from the last tube collected; comparison of counts between tubes 1 and 4 is occasionally needed if such bleeding is suspected. The other tubes can be used for both microbiologic and chemical studies, including protein and glucose levels. In this way, a larger proportion of the total fluid can be concentrated, facilitating detection of infectious agents present in low numbers, and the supernatant can still be used for other required studies. The volume of CSF is critical for detection of certain microorganisms, such as mycobacteria and fungi. A minimum of 5 to 10 mL is recommended for detection of these agents by centrifugation and subsequent culture. When an inadequate volume of CSF is received, the physician should be consulted regarding the order of priority of laboratory studies. Processing too little specimen lowers the sensitivity of the testing, which leads to false-negative results. This is potentially more harmful to patient care than performing an additional lumbar puncture to obtain the necessary or required amount of sample.

CSF should be hand-delivered *immediately* to the laboratory. Specimens should *never* be refrigerated. Certain agents, such as *Streptococcus pneumoniae*, may not be detectable after an hour or longer unless antigen detection methods are used. If not rapidly processed, CSF should be incubated (35° C) or left at room temperature. One exception to this rule involves CSF for viral studies. These specimens may be refrigerated for as long as 23 hours after collection or frozen at −70° C if a longer delay is anticipated until they are inoculated. CSF for viral studies should never be frozen at temperatures above −70° C.

CSF is one of the few specimens handled by the laboratory for which information promptly relayed to the clinician can directly affect therapeutic outcome. Such specimens should be processed immediately upon receipt in the laboratory and all results reported to the physician.

Initial processing

Initial processing of CSF for bacterial, fungal, or parasitic studies includes centrifugation of all specimens greater than 1 mL in volume for at least 15 minutes at 1500× *g*. Specimens in which cryptococci or mycobacteria are suspected must be handled differently. (Discussions of techniques for culturing CSF for mycobacteria and fungi are found in Chapters 60 and 65, respectively.) The supernatant is removed to a sterile tube, leaving approximately 0.5 mL of fluid in which to suspend the sediment before visual examination or culture. Mixing of the sediment after the supernatant has been removed is critical. Forcefully aspirating the sediment up and down into a sterile pipette several times will adequately disperse the organisms that remained adherent to the bottom of the tube after centrifugation. Laboratories that use a sterile pipette to remove portions of the sediment from underneath the supernatant will miss a significant number of positive specimens. The supernatant can be used to test for the presence of antigens or for chemistry evaluations (e.g., protein, glucose, lactate). As a safeguard, keep the supernatant even if it has no immediate use.

CSF findings

As previously mentioned, CSF is also removed for analysis of cells, protein, and glucose. Ideally, the glucose content of the peripheral blood is determined simultaneously for comparison to that in CSF. General guidelines for interpretation of results are shown in Table 23-2.

Because the results of hematologic and chemical tests directly relate to the probability of infection, communication between the physician and the microbiology laboratory is essential. Among 555 cerebrospinal fluids from patients older than age 4 months tested at the University of California–Los Angeles, only 2 showed normal cell count and protein in the presence of bacterial meningitis.[4] Thus the diagnosis of acute bacterial meningitis can be excluded in patients with normal fluid parameters in almost all cases, precluding further expensive and labor-intensive microbiologic processing beyond a standard smear and culture (which must be included in all cases). Similar criteria have been used to exclude performance of smear and culture for tuberculosis, as well as syphilis serology, on CSF specimens.[1,2]

Visual detection of etiologic agents

Following centrifugation, the resulting CSF sediment may be visually examined for the presence of cells and organisms.

STAINED SMEAR OF SEDIMENT   Gram stain must be performed on all CSF sediments. False-positive smears have resulted from inadvertent use of contaminated slides. Therefore, use of alcohol-dipped and flamed, or autoclaved slides is recommended. After thoroughly mixing the sediment, a heaped drop is placed on the surface of a sterile or alcohol-cleaned slide. The sediment should never be spread out on the slide surface, because this increases the difficulty of finding small numbers of microorganisms. The drop of sediment is allowed to air dry, is heat- or methanol-fixed, and is stained by either Gram or acridine orange stains. The acridine orange fluorochrome stain may allow faster examination of the slide under high-power magnification (400×) and thus a more thorough examination. The brightly fluorescing bacteria will be easily visible. All suspicious smears can be restained by the Gram stain (directly over the acridine orange stain) to confirm the presence and morphology of any organisms seen.

An alternative procedure using a cytospin centrifuge to prepare slides for staining has also been found to be beneficial.[9] Use of this method for preparing smears for staining concentrates cellular material and bacterial cells up to a 1000-fold. By centrifugation, a small amount of CSF (or other body fluid) is concentrated onto a circular area of a microscopic slide (Figure 23-3), fixed, stained, and then examined.

The presence or absence of bacteria, inflammatory cells, and erythrocytes should be reported following examination. Based on demographic and clinical patient data and Gram stain morphology, the etiology of the majority of cases of bacterial meningitis can be presumptively determined within the first 30 minutes after receiving the specimen.

WET PREPARATION   Amoebas are best observed by examining thoroughly mixed sediment as a wet preparation under phase-contrast microscopy. If a phase-contrast microscope is not available, observing under light microscopy with the condenser closed slightly is an alternative technique. Amoebas are recognized by typical slow, methodical movement in one direction by advancing pseudopodia. (The organisms may require a little time under the warm light of the microscope before they begin to move.) Organisms must be distinguished from motile macrophages, which occasionally occur in CSF. Following a suspicious wet preparation, a trichrome stain can help differentiate amoebas from somatic cells. The pathogenic

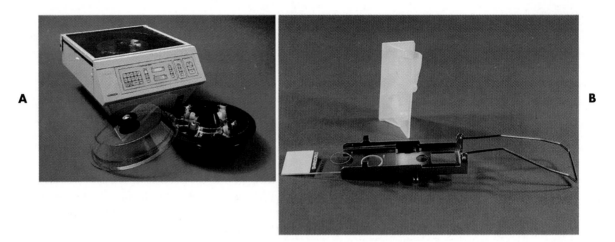

FIGURE 23-3 **A**, Cytocentrifuge. (Courtesy Cytospin 2, Shandon, Inc, Pittsburgh, Pa.) **B**, Device that is used to prepare the concentrated smears of material from body fluid specimens, such as CSF, by cytocentrifugation.

amoebas can be cultured on a lawn of *Klebsiella pneumoniae* or *Escherichia coli* (see Chapter 64).

INDIA INK STAIN The large polysaccharide capsule of *Cryptococcus neoformans* allows these organisms to be visualized by the India ink stain. However, latex agglutination testing for capsular antigen is more sensitive and extremely specific so that antigen testing is recommended for use in place of an India ink stain. Furthermore, strains of *C. neoformans* that infect patients with AIDS may not possess detectable capsules, so culture is also essential. To perform the India ink preparation, a drop of CSF sediment is mixed with one-third volume of India ink (Pelikan Drawing Ink, Block, Gunther, and Wagner; available at art supply stores). The India ink can be protected against contamination by adding 0.05 mL thimerosal (Merthiolate, Sigma Chemical Co., St. Louis, Mo.) to the bottle when first opened. After mixing the CSF and ink to make a smooth suspension, a coverslip is applied to the drop and the preparation is examined under high-power magnification (400×) for characteristic encapsulated yeast cells, which can be confirmed by examination under oil immersion. The inexperienced microscopist must be careful not to confuse white blood cells with yeast. The presence of encapsulated buds, smaller than the mother cell, is diagnostic.

Direct detection of etiologic agents

ANTIGEN Reagents and complete systems for the rapid detection of antigen in CSF have been available for several years.

Bacteria. Countercurrent immunoelectrophoresis (CIE), although the first widely accepted method for rapid antigen detection from CSF, has been largely replaced by the more sensitive and simpler techniques of latex agglutination and coagglutination (see Chapter 15).

All commercial agglutination systems use the principle of an antibody-coated particle that will bind to specific antigen, resulting in macroscopically visible agglutination. The soluble capsular polysaccharide produced by the most common etiologic agents of meningitis, including the group B streptococcal polysaccharide, are well suited to serve as bridging antigens. The systems differ in that certain antibodies are polyclonal and others are monoclonal, and not all systems detect all antigens. These reagents should only be used as an adjunct to standard procedures.

In general, the commercial systems have been developed for use with CSF, urine, or serum, although results with serum have not been as useful diagnostically as those with CSF. Soluble antigens from *Streptococcus agalactiae* and *Haemophilus influenzae* may concentrate in the urine. Urine, however, seems to produce a higher incidence of nonspecific reactions than either serum or CSF. The manufacturers' directions must be followed for performance of antigen detection test systems for different specimen types. Although some of the systems require pretreatment of samples (usually heating for 5 minutes), not all manufacturers recommend such a step. The reagents, however, may yield false-positive or cross-reactions unless specimen pretreatment is performed. Interference by rheumatoid factor and other substances, more often present in body fluids other than CSF, has also been reported. The method of Smith et al.[10] has been shown to effectively reduce a substantial portion of nonspecific and false-positive reactions, at least for tests performed with latex particle reagents. This pretreatment, called **rapid extraction of antigen procedure (REAP;** Procedure 23-1), is recommended for laboratories that use commercial body fluid antigen detection kits. Certain commercial systems have such an extraction procedure included in their protocols.

The use of these reagents is controversial. A

recent survey of laboratories regarding bacterial antigen testing (BAT) revealed a significant lack of uniformity and consensus in applying these tests.[11] Based on the findings of several studies, only a limited number of situations exist in which BAT may be clinically useful and is warranted.[5,7,12] Some of these situations include CSF specimens from previously treated patients and gram-negative CSF specimens with abnormal parameters (elevated protein, decreased glucose, and/or an abnormal white blood cell count). Of significance, these systems are not substitutes for properly performed smears and cultures because they provide less than 100% sensitivity and specificity. Nevertheless, these assays are an additional diagnostic test to aid clinicians in initial management of serious meningitis in some situations.

Cryptococcus neoformans.   Reagents for the detection of the polysaccharide capsular antigen of *Cryptococcus neoformans* are available commercially. CSF specimens that yield positive results for cryptococcal antigen should be tested with a second latex agglutination test for rheumatoid factor. The commercial test systems incorporate rheumatoid factor testing in their protocol. A positive rheumatoid factor test renders the cryptococcal latex test uninterpretable, and the results should be reported as such, unless the rheumatoid factor antibodies have been inactivated. Undiluted specimens that contain large amounts of capsular antigen may yield a false-negative reaction caused by a prozone phenomenon. Patients with AIDS may have an antigen titer in excess of 100,000, requiring many dilutions to reach an end point. Such dilutions are usually done because the test is used for following a patient's response to treatment, as well as for initial diagnosis.

MOLECULAR METHODS   With the introduction of amplification technologies, such as polymerase chain reaction, many reports in the literature detail the application of these technologies to diagnose CNS infections caused by various microorganisms. Published data indicate that many of these assays may prove to be more sensitive and specific compared with presently available techniques. As of this writing, most of these amplification assays are not commercially available; however, several tests are being developed.

Miscellaneous tests

Other tests, such as the limulus lysate test, CSF lactate determinations, C-reactive protein, mass spectrometry, and gas-liquid chromatography, have been evaluated for use in the diagnosis of CNS infections. However, the utility and value of these tests are either controversial, remain to be defined, or are impractical for routine use in the clinical laboratory.

Culture

Because the majority of cases of bacterial meningitis are usually caused by only one organism, a minimal number of media types are required.

BACTERIA AND FUNGI   Routine bacteriologic media should include a chocolate agar plate, 5% sheep blood agar plate, and an enrichment broth, usually thioglycolate without indicator. The chocolate agar plate is needed to recover fastidious organisms, most notably *H. influenzae* and isolates of *N. meningitidis* that do not grow on blood agar plates; the use of the blood agar plate also aids in the recognition of *S. pneumoniae*. After vortexing the sediment and preparing smears, several drops of the sediment should be inoculated to each medium. Plates should be incubated at 37° C in 5% to 10% carbon dioxide (CO_2) for at least 72 hours. If a CO_2 incubator is not available, a candle jar can be used. The broth should be incubated in air at 37° C for at least 5 days. The broth cap must be loose to allow free exchange of air. If organisms morphologically resembling anaerobic bacteria are seen on the Gram stain or if a brain abscess is suspected, an anaerobic blood agar plate may also be inoculated. These media will support the growth of almost all bacterial pathogens and several fungi.

The symptoms of chronic meningitis that prompt a physician to request fungal cultures are the same as those for tuberculous meningitis, which should always be sought by the laboratory if chronic or fungal meningitis is suspected. Cultures for mycobacteria are addressed in Chapter 60. For CSF fungal cultures, two drops of the well-mixed sediment should be inoculated onto Saboraud dextrose agar or other non–blood-containing medium and brain-heart infusion with 5% sheep blood. Fungal media should be incubated in air at 30° C for 4 weeks. If possible, two sets of media should be inoculated, with one set incubated at 30° C and the other at 35° C.

PARASITES AND VIRUSES   Conditions for the culture of free-living amoebae and viral agents are discussed in Chapters 64 and 66, respectively. The physician must notify the laboratory to culture these agents.

BRAIN ABSCESS/BIOPSIES

Specimen Collection, Transport and Processing.   Whenever possible, biopsy specimens or aspirates from brain abscesses should be submitted to the laboratory under anaerobic conditions. Several devices are commercially available to transport biopsy specimens under anaerobic conditions. Swabs are not considered an optimum specimen but if used to collect abscess material they should be sent in a transport device that maintains an anaerobic environment.

Biopsy specimens should be homogenized in sterile saline before plating and smear preparation. This processing should be kept to a minimum in order to reduce oxygenation.

Culture. Abscess and biopsy specimens submitted for culture should be inoculated onto 5% sheep blood and chocolate agar plates. Plates should be incubated in 5% to 10% CO_2 for 72 hours at 35° C. In addition, an anaerobic agar plate and broth with an anaerobic indicator, Vitamin K, and hemin should be inoculated and incubated in an anaerobic environment at 35° C. Anaerobic culture plates are incubated for a minimum of 72 hours but are examined after 48 hours of incubation. Anaerobic broths should be incubated for a minimum of 5 days. If a fungal etiology is suspected, fungal media, such as brain-heart infusion with blood and antibiotics or inhibitory mold agar, should be inoculated as well.

Procedure

23-1 RAPID EXTRACTION OF ANTIGEN PROCEDURE (REAP)

PRINCIPLE
Removal of nonspecific cross-reactive material can improve the specificity of direct antigen detection particle agglutination tests. Ethylenediaminetetraacetic acid (EDTA) forms complexes with cross-reactive materials, and they are removed from the reaction mixture by centrifugation.

METHOD
1. Pipette 0.05 mL fluid to be tested (CSF, serum, or urine) into a 1.5-mL plastic, conical microcentrifuge tube.

2. Add 0.15 mL of 0.1 M EDTA (Sigma Chemical Co.) to the microcentrifuge tube, close the cap tightly, and vortex the tube.

3. Heat in a dry bath (available from instrument supply companies) for 3 minutes at 100° C.

4. Centrifuge the tubes for 5 minutes at 13,000x g in a tabletop microcentrifuge. Be certain that the instrument achieves the required centrifugal force.

5. Remove the supernatant with a capillary pipette and use one drop of this solution as the test sample in the antigen detection test, following the manufacturer's instructions for performance of the test.

EXPECTED RESULTS
Nonspecific agglutination should not occur.

References

1. Albright, R.E., et al. 1991. Issues in cerebrospinal fluid management: CSF venereal disease research laboratory testing. Am. J. Clin. Pathol. 95:387.

2. Albright, R.E., et al. 1991. Issues in cerebrospinal fluid management: acid-fast bacillus smear and culture. Am. J. Clin. Pathol. 95:48.

3. Gradon, J.D., Timpone, J.G., and Schnittman, S.M. 1992. Emergence of unusual opportunistic pathogens in AIDS: a review. Clin. Infect. Dis. 15:134.

4. Hayward, R.A., Shapiro, M.F., and Oye, R.K. 1987. Laboratory testing on cerebrospinal fluid: a reappraisal. Lancet 1:1.

5. Kiska, D.L., et al. 1995. Quality assurance study of bacterial antigen testing of cerebrospinal fluid. J. Clin. Microbiol. 33:1141.

6. Parkkinen, J., et al. 1988. Binding sites in the rat brain for *Escherichia coli* S fimbriae associated with neonatal meningitis. J. Clin. Invest. 81:860.

7. Perkins, M.D., Mirrett, S., and Reller, L.B. 1995. Rapid bacterial antigen detection is not clinically useful. J. Clin. Microbiol. 33:1486.

8. Plaut, A.G. 1983. The IgA1 proteases of pathogenic bacteria. Ann. Rev. Microbiol. 37:603.

9. Shanholtzer, C.J., Schaper, P.J., and Peterson, L.R. 1982. Concentrated gram stain smears prepared with a cytospin centrifuge. J. Clin. Microbiol. 16:1052.

10. Smith, L.P., et al. 1984. Improved detection of bacterial antigens by latex agglutination after rapid extraction from body fluids. J. Clin. Microbiol. 20:981.

11. Thomas, J.G. 1994. Survey results of routine CSF antigen detection: nothing to be proud of. Clin. Microbiol. Newsletter 16:187.

12. Thomas, J.G. 1994. Routine CSF antigen detection for agents associated with bacterial meningitis: another point of view. Clin. Microbiol. Newsletter 16:89.

13. Virji, M., et al. 1992. Variations in the expression of pili: the effect on adherence of *Neisseria meningitidis* to human epithelial and endothelial cells. Mol. Microbiol. 6:1271.

14. Virji, M., et al. 1991. The role of pili in the interactions of pathogenic *Neisseria* with cultured human endothelial cells. Mol. Microbiol. 5:1831.

15. Virji, M., et al. 1991. Interactions of *Haemophilus influenzae* with cultured human endothelial cells. Microb. Pathol. 10:231.

16. Wenger, J.D. 1994. Impact of *Haemophilus influenzae* type b vaccines on the epidemiology of bacterial meningitis. Infect. Agents Dis. 2:323.

17. Wilhelm, C. and Ellner, J.J. 1986. Chronic meningitis. Neurol. Clin. 4:115.

Bibliography

Kriskern, P.J., Marburg, S., and Ellis, R.W. 1995. *Haemophilus influenzae* type b conjugate vaccines. Pharm. Biotechnol. 6:673.

Perry, J.L. 1995. Utility of cytocentrifugation for direct examination of clinical specimens. Clin. Microbiol. Newsletter 17:29.

Quagliarello, V. and Scheld, V.M. 1992. Bacterial meningitis: pathogenesis, pathophysiology, and progress. N. Engl. J. Med. 327:864.

Ray, C.G., et al. 1993. Laboratory diagnosis of central nervous system infections. In Smith, J.A., coordinating editor. Cumitech 14A. American Society for Microbiology, Washington, D.C.

Townsend, G.C. and Scheld, W.M. 1995. Microbe-endothelium interactions in blood-brain barrier permeability during bacterial meningitis. ASM News 61:294.

Tunkel, A.R. and Scheld, M. 1993. Pathogenesis and pathophysiology of bacterial meningitis. Clin. Microbiol. Rev. 6:118.

Wilson, M.L. 1996. General principles of specimen collection and transport. Clin. Infect. Dis. 22:766.

24 | INFECTIONS OF THE EYES, EARS, AND SINUSES

EYES

ANATOMY

Only a small portion of the eye is exposed to the environment; about five sixths of the eyeball is enclosed within bony orbits that are shaped like four-sided pyramids. In general, ocular (eye) infections can be divided into those that involve the exposed, or external, structures and those that involve internal sites.

The external structures of the eye—eyelids, conjunctiva, sclera, and cornea—are shown in Figure 24-1. The eyeball comprises three layers. From the outside in, these tissues are the sclera, choroid, and retina. The sclera is a tough, white, fibrous tissue (i.e., "white" of the eye). The anterior (toward the front) portion of the sclera is the cornea, which is transparent and has no blood vessels. A mucous membrane, called the **conjunctiva,** lines each eyelid and extends onto the surface of the eye itself.

The large interior space of the eyeball is divided into two sections: the anterior and posterior cavities (see Figure 24-1). The anterior cavity is filled with a clear and watery substance called **aqueous humor;** the posterior cavity is filled with a soft, gelatin-like substance called **vitreous humor.**

Infections can also occur in the eyes' lacrimal (pertaining to tears) system. The major components of the lacrimal apparatus include the lacrimal gland, lacrimal canaliculi (short channel), and lacrimal sac.

RESIDENT MICROBIAL FLORA

Rather sparse indigenous flora exist in the conjunctival sac. *Staphylococcus epidermidis* and *Lactobacillus* spp. are the most frequently encountered organisms; *Propionibacterium acnes* may also be present. *Staphylococcus aureus* is found in less than 30% of people, and *Haemophilus influenzae* colonizes 0.4% to 25%. *Moraxella catarrhalis,* various *Enterobacteriaceae,* and various streptococci (*Streptococcus pyogenes, Streptococcus pneumoniae,* other alpha-hemolytic and gamma-hemolytic forms) are found in a very small percentage of people.

DISEASES

The eye and its associated structures are uniquely predisposed to infection by various microorganisms. The major infections of the eye are listed in Table 24-1 along with a brief description of the disease.

PATHOGENESIS

The eye has a number of defense mechanisms. The eyelashes prevent entry of foreign material into the eye. The lids blink 15 to 20 times per minute, during which time secretions of the lacrimal glands and goblet cells wash away bacteria and foreign matter. Lysozyme and immunoglobulin A (IgA) are secreted locally and serve as part of the eye's natural defense mechanisms. Also, the eyes themselves are enclosed within the bony orbits. The delicate intraocular structures are enveloped in a tough collagenous coat (sclera and cornea). If these barriers are broken by a penetrating injury or ulceration, infection may occur. Infection can also reach the eye via the bloodstream from another site of infection. Finally, because three of the four walls of the orbit are contiguous with the paranasal (facial) sinuses, sinus infections may extend directly to the periocular orbital structures.

EPIDEMIOLOGY AND ETIOLOGY OF DISEASE

Blepharitis

Bacteria, viruses, and occasionally, lice can cause blepharitis. Although occasionally isolated from surfaces surrounding the healthy eye, *Staphylococcus aureus* and *S. epidermidis* are the most common infectious cause of blepharitis in developed countries. Symptoms include burning, itching, sensation of a foreign body, and crusting of the eyelids.

Viruses can also cause a vesicular (blisterlike) eruption of the eyelids. Herpes simplex virus produces vesicles on the eyelids that typically crust and heal with scarring over 2 weeks. Unfortunately, once this vesicular stage has resolved, the lesions can be confused with bacterial blepharitis.

Finally, the pubic louse *Phthirius pubis* has a predilection for eyelash hair. Presence of this organism

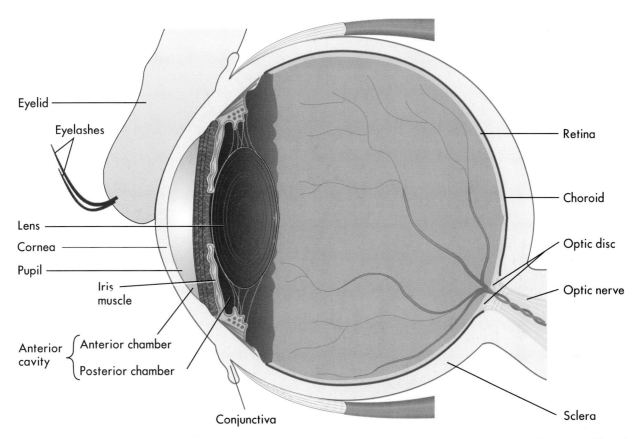

FIGURE 24-1 Key anatomic structures of the eye. (Modified from Thibodeau, G.A. and Patton, K.T. 1993. Anatomy and physiology, ed 2. Mosby, St. Louis.)

produces irritation, itch, and swelling of the lid margins (edges).

Conjunctivitis

Bacterial conjunctivitis is the most common type of ocular infection. The principle causes of acute conjunctivitis in the normal host are listed in Table 24-1. Age-related factors are key determinants of the etiologic agent. In neonates, neisserial and chlamydial infections are frequent and are acquired during passage through an infected vaginal canal. With the common practice of instilling antibiotic drops into the eyes of newborns in the United States, the incidence of gonococcal and chlamydial conjunctivitis has dropped dramatically. However, *Chlamydia trachomatis* is responsible for one of the most important types of conjunctivitis, trachoma, one of the leading causes of blindness in the world, primarily in underdeveloped countries.

In children the most common causes of bacterial conjunctivitis are *Haemophilus influenzae, S. pneumoniae,* and perhaps *S. aureus.*[3] *S. pneumoniae* and *H. influenzae* (especially *aegyptius* subsp.) have been responsible for epidemics of conjunctivitis.

Numerous other bacteria may also cause conjunctivitis. For example, diphtheritic conjunctivitis may occur in conjunction with diphtheria elsewhere in the body. *Moraxella lacunata* produces a localized conjunctivitis with little discharge from the eye. Distinctive clinical pictures may also occur with conjunctivitis caused by *Mycobacterium tuberculosis, Francisella tularensis, Treponema pallidum,* and *Yersinia enterocolitica.*

Fungi may be responsible for this type of infection as well, often in association with a foreign body in the eye or an underlying immunologic problem. However, these infections are infrequently encountered.

Viruses are an important cause of conjunctivitis; 20% of such infections in children resulted from adenoviruses in one large U.S. study and 14% of infections in adult patients in another study.[3,5] Adenoviruses types 4, 3, and 7A are common. Most viral conjunctivitis is self-limited but is highly contagious, with potential to cause major outbreaks. Worldwide, enterovirus 70 and coxsackievirus A24 are responsible for outbreaks and epidemics of acute hemorrhagic conjunctivitis.

TABLE 24-1 MAJOR INFECTIONS OF THE EYE

INFECTION	DESCRIPTION	MAJOR CAUSES			
		BACTERIA	VIRUSES	FUNGI	PARASITES
Blepharitis	Inflammation of the margins (edges) of the eyelids. Symptoms include irritation, redness, burning sensation, and occasional itching.	*Staphylococcus aureus*			
Conjunctivitis	Inflammation of the conjunctiva. Symptoms vary according to the etiologic agent but most patients have swelling of the conjunctiva and inflammatory exudates	*Streptococcus pneumoniae; Haemophilus influenzae; S. aureus; Haemophilus aegyptus; Chlamydia trachomatis; Neisseria gonorrhoeae; Streptococcus pyogenes*	Adenoviruses; herpes simplex (HSV); varicella zoster		
Keratitis	Inflammation of the cornea. Although there are no specific clinical signs to confirm infection, most patients complain of pain and usually some decrease in vision, with or without discharge from the eye	*S. aureus; S. pneumoniae; Pseudomonas aeruginosa*	HSV	*Fusarium solani; Aspergillus* spp.; *Acremonium; Curvularia*	*Acanthamoeba* spp.
Keratoconjunctivitis	Infection involving both the conjunctiva and cornea. *Ophthalmia neonatorum* is an acute conjunctivitis or keratoconjunctivitis of the newborn caused by either *N. gonorrhoeae* or *C. trachomatis*	Refer to agents for keratitis/conjunctivitis	Refer to agents for keratitis/ conjunctivitis	Refer to agents for keratitis/ conjunctivitis	

Endophthalmitis	Infection of the aqueous or vitreous humor. This infection is usually caused by bacteria or fungi, is rare, develops suddenly and progresses rapidly, often leading to blindness. Pain, especially while moving the eye, and decreased vision, are prominent features	*S. aureus; S. epidermidis; S. pneumoniae*; other streptococcal spp.; *P. aeruginosa*; other gram-negative organisms	*Candida* spp.; *Aspergillus* spp.; *Volutella* spp.; *Acremonium* spp. *Toxocara*
Lacrimal infections: Canaliculitis	A rare, chronic inflammation of the lacrimal canals in which the eyelid swells and there is a thick, mucopurulent discharge	Actinomyces; *Propionibacterium propionicum*	
Dacryocystis	Inflammation of the lacrimal sac that is accompanied by pain, swelling, and tenderness of the soft tissue in the medial canthal region	*S. pneumoniae; S. aureus; S. pyogenes; Haemophilus influenzae*	
Dacryoadenitis	Acute infection of the lacrimal gland. These infections are rare and can be accompanied by pain, redness and swelling of the upper eyelid, and conjunctival discharge	*S. pneumoniae; S. aureus; S. pyogenes*	*C. albicans; Aspergillus* spp.

Keratitis

Keratitis (corneal infection) may be caused by various infectious agents, usually only after some type of trauma produces a defect in the ocular surface. Keratitis should be regarded as an emergency, because corneal perforation and loss of the eye can occur within 24 hours when organisms such as *Pseudomonas aeruginosa, Staphylococcus aureus,* or herpes simplex virus (HSV) are involved. Bacteria account for 65% to 90% of corneal infections.

In the United States, *S. aureus, S. pneumoniae,* and *P. aeruginosa* account for more than 80% of all bacterial corneal ulcers. Many culture-positive cases are now being recognized as polymicrobial.[7] A toxic factor known as exopeptidase has been implicated in the pathogenesis of corneal ulcer produced by *S. pneumoniae.* With *P. aeruginosa,* proteolytic enzymes are responsible for the corneal destruction. The gonococcus may cause keratitis in the course of inadequately treated conjunctivitis. *Acinetobacter,* which may look identical microscopically to the gonococcus and is resistant to penicillin and many other antimicrobial agents, can cause corneal perforation. Many other bacteria, several viruses other than HSV, and many fungi, may cause keratitis. Fungal keratitis is usually a complication of trauma.

Although still unusual, a previously rare etiologic agent of corneal infections has become more common in users of soft and extended-wear contact lenses. *Acanthamoeba* spp., free living amebae, can survive in improperly sterilized cleaning fluids and be introduced into the eye with the contact lens. Other bacterial and fungal causes of infections in such patients have also been traced to inadequate cleaning of lenses.

Endophthalmitis

Surgical trauma, nonsurgical trauma (infrequently), and hematogenous spread from distant sites of infection are the background factors in endophthalmitis. The infection may be limited to specific tissues within the eye or may involve all of the intraocular contents. Bacteria are the most common infectious agents responsible for endophthalmitis.

After surgery or trauma, evidence of the disease is usually found within 24 to 48 hours but can be delayed for several days. Postoperative infection involves primarily bacteria from the ocular surface microflora. Although *Staphylococcus epidermidis* and *S. aureus* are responsible for the majority of cases of endophthalmitis after cataract removal, any bacterium, including those considered to be primarily saprophytic, may cause endophthalmitis. In hematogenous endophthalmitis, a septic focus elsewhere is usually evident before onset of the intraocular infection. *Bacillus cereus* has caused endophthalmitis in people addicted to narcotics and after transfusion with contaminated blood. Endophthalmitis associated with meningitis may involve various organisms, including *Haemophilus influenzae,* streptococci, and *Neisseria meningitidis. Nocardia* endophthalmitis may follow pulmonary infection with this organism.

Mycotic infection of the eye has increased significantly over the past 3 decades because of increased use of antibiotics, corticosteroids, antineoplastic chemotherapy, addictive drugs, and hyperalimentation. Fungi generally considered to be saprophytic are important causes of postoperative endophthalmitis (see Table 24-1). Endogenous mycotic endophthalmitis is most often caused by *Candida albicans.* Patients with diabetes and underlying disease are most at risk. Other causes of hematogenous ocular infection include *Aspergillus, Cryptococcus, Coccidioides, Sporothrix,* and *Blastomyces.*

Viral causes of endophthalmitis include HSV, varicella (herpes) zoster virus (VZV), cytomegalovirus, and measles viruses. The most common parasitic cause is *Toxocara. Toxoplasma gondii* is a well-known cause of chorioretinitis. Thirteen percent of patients with cysticercosis have ocular involvement. *Onchocerca* usually produces keratitis, but intraocular infection also occurs.

Periocular

Canaliculitis, one of three infections of the lacrimal apparatus (see Table 24-1), is an inflammation of the lacrimal canal and is usually caused by *Actinomyces* or *Propionibacterium propionicum* (formerly *Arachnia*). Infection of the lacrimal sac (dacryocystis) may involve numerous bacterial and fungal agents; the major causes are listed in Table 24-1. Dacryoadenitis is an uncommon infection of the lacrimal gland characterized by pain of the upper eyelid with erythema and often involves pyogenic bacteria such as *S. aureus* and streptococci. Chronic infections of the lacrimal gland occur in tuberculosis, syphilis, leprosy, and schistosomiasis. Acute inflammation of the gland may occur in the course of mumps and infectious mononucleosis.

Orbital cellulitis is an acute infection of the orbital contents and is most often caused by bacteria. This is a potentially serious infection because it may spread posteriorly to produce central nervous system complications.[1] Most cases involve spread from contiguous sources such as the paranasal sinuses. In children, bloodborne bacteria, notably *Haemophilus influenzae,* may lead to orbital cellulitis. *S. aureus* is the most common etiologic agent; *Streptococcus pyogenes* and *S. pneumoniae* are also common. Anaerobes may cause a cellulitis secondary to chronic sinusitis, primarily in adults. Mucormycosis of the orbit is a serious, invasive fungal infection seen particularly in patients with diabetes who have poor control of their disease, patients with acidosis from other causes, and patients with malignant disease receiving cytotoxic and immunosuppressive therapy. *Aspergillus* may produce a similar infection in the same settings but also can cause mild, chronic infections of the orbit.

Newer surgical techniques involving the ocular implantation of prosthetic or donor lenses have resulted in increasing numbers of iatrogenic (resulting from the activities of a physician) infections. Isolation of *Propionibacterium acnes* may have clinical significance in such situations, in contrast to many other sites in which it is usually considered to be a contaminant.

Other infections

Opportunistic infections in human immunodeficiency virus (HIV)-infected individuals can involve the eye. Ocular manifestations have been reported in up to 70% of HIV-infected patients.[2] Systemic infections that involve the eye include cytomegalovirus, *Pneumocystis carinii*, *Cryptococcus neoformans*, *Mycobacterium avium* complex, and *Candida* spp. Most often the retina, choroid, and optic nerve are involved with these agents, resulting in significant visual morbidity (unhealthy condition) if left untreated.

LABORATORY DIAGNOSIS

Specimen collection and transport

Purulent material from the surface of the lower conjunctival sac and inner canthus (angle) of the eye is collected on a sterile swab for cultures of conjunctivitis. Both eyes should be cultured separately. Chlamydial cultures are taken with a dry calcium alginate swab and placed in 2-SP transport medium. An additional swab may be rolled across the surface of a slide, fixed with methanol, and sent if direct fluorescent antibody (DFA) chlamydia stains are used for detection.

In the patient with keratitis, an ophthalmologist should obtain scrapings of the cornea with a heat-sterilized platinum spatula. Multiple inoculations with the spatula are made to blood agar, chocolate agar, an agar for fungi, thioglycollate broth, and an anaerobic blood agar plate. Other special media may be used if indicated. For culture of HSV and adenovirus, corneal material is transferred to viral transport media.

Cultures of endophthalmitis specimens are inoculated with material obtained by the ophthalmologist from the anterior and posterior chambers of the eye, wound abscesses, and wound dehiscences (splitting open). Lid infection material is collected on a swab in a conventional manner. For microbiologic studies of canaliculitis, material from the lacrimal canal should be transported under anaerobic conditions. Aspiration of fluid from the orbit is contraindicated in patients with orbital cellulitis. Because sinusitis is the most common background factor, an otolaryngolo-gist's assistance in obtaining material from the maxillary sinus by antral puncture is helpful. Blood cultures should also be obtained. Tissue biopsy is essential for microbiologic diagnosis of mucormycosis; because cultures are usually negative, the diagnosis is made by histologic examination.

Direct visual examination

All material submitted for culture should always be smeared and examined directly by Gram stain or other appropriate techniques. In bacterial conjunctivitis, polymorphonuclear leukocytes predominate; in viral infection, the host cells are primarily lymphocytes and monocytes. Specimens in which chlamydia is suspected can be stained immediately with monoclonal antibody conjugated to fluorescein for detection of elementary bodies or inclusions. Using histologic stains, basophilic intracytoplasmic inclusion bodies are seen in epithelial cells. Cytologists and anatomic pathologists usually perform these tests. Direct examination of conjunctivitis specimens using histologic methods (Tzanck smear) may reveal multinucleated epithelial cells typical of herpes group viral infections. However, DFA stains available for both HSV and VZV are most reliable for rapid diagnosis of these viral infections. In patients with keratitis, scrapings are examined by Gram, Giemsa, periodic acid-Schiff (PAS), and methenamine silver stains. If *Acanthamoeba* or other amebae are suspected, a direct wet preparation should be examined for motile trophozoites, and a trichrome stain should be added to the regimen. For this diagnosis, however, culture is by far the most sensitive detection method. In patients with endophthalmitis, material is also examined by Gram, Giemsa, PAS, and methenamine silver stains. When submitted in large volumes of fluid, ophthalmic specimens must be concentrated by centrifugation before additional studies are performed.

Culture

Because of the constant washing action of the tears, the number of organisms recovered from cultures of certain eye infections may be relatively low. Unless the clinical specimen is obviously purulent, using a relatively large inoculum and various media is recommended to ensure recovery of an etiologic agent. Conjunctival scrapings placed directly onto media yield the best results. At a minimum, one should use blood and chocolate agar plates incubated under increased carbon dioxide tension (5% to 10% CO_2). Because potential pathogens may be present in an eye without causing infection, it may be very helpful to the clinician, when only one eye is infected, to culture both eyes. If a potential pathogen grows in cultures of the infected and the uninfected eye, the organism may not be causing the infection; however, if the organism only grows in culture from the infected eye, it is most likely the causative agent. When *Moraxella lacunata* is suspected, Loeffler's medium may prove useful; the growth of the organism often leads to proteolysis and pitting of the medium, although nonproteolytic strains may be found. If diphtheritic conjunctivitis is suspected, Loeffler's or cystine-tellurite medium should be used. For more serious eye infections, such as keratitis, endophthalmitis, and orbital cellulitis, one should always include, in addition to the media just noted, a reduced anaerobic blood agar plate, a medium for fungi, and a liquid medium such as thioglycolate broth. Blood cultures are also important in serious eye infections.

Cultures of material for chlamydiae and viruses should be inoculated to appropriate media from transport broth. Cycloheximide-treated McCoy cells for chlamydiae isolation and human embryonic kidney, primary monkey kidney, and Hep-2 cell lines for virus isolation should be inoculated.

Nonculture methods

Although acute and convalescent serologic tests for viral agents might be used in the event of epidemic conjunctivitis, they typically are not performed because the infections are self-limited. Enzyme-linked immunosorbent assay (ELISA) tests and DFA staining are now available for detection of *Chlamydia trachomatis*. It is anticipated that the direct antigen tests should perform well, particularly because so many eyes have been partially treated before culture. An ELISA test of aqueous humor is available for diagnosis of *Toxocara* infection.

EARS

ANATOMY

The ear is divided into three anatomic parts: the external, middle, and inner ear. Important anatomic structures are shown in Figure 24-2.

The middle ear is part of a continuous system that includes the nares, nasopharynx, auditory tube, and the mastoid air spaces. These structures are lined with respiratory epithelium (e.g., ciliated cells, mucus-secreting globet cells).

RESIDENT MICROBIAL FLORA

The normal flora of the external ear canal are rather sparse, similar to flora of the conjunctival sac qualitatively except that pneumococci, *Proprionibacterium acnes*, *Staphylococcus aureus*, and *Enterobacteriaceae* are encountered somewhat more often. *Pseudomonas aeruginosa* is found on occasion. *Candida* spp. (non-*C. albicans*) are also common.

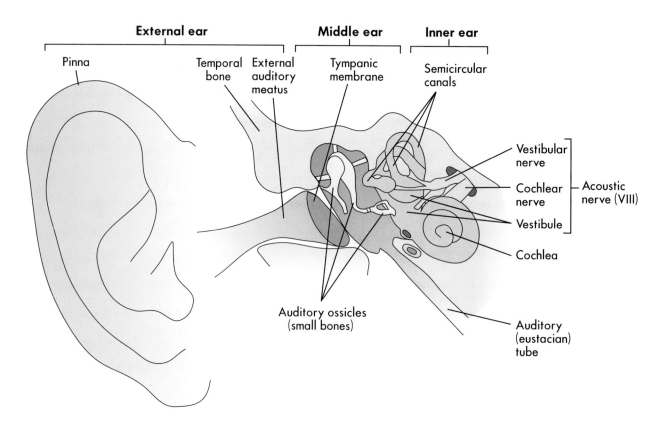

FIGURE 24-2 The ear. (Modified from Thibodeau, G.A. and Patton, K.T. 1993. Anatomy and physiology, ed 2. Mosby, St. Louis.)

DISEASES, EPIDEMIOLOGY, AND ETIOLOGY OF DISEASE

Otitis externa (external ear infections)

Otitis externa is similar to skin and soft tissue infections elsewhere. Two major types of external otitis exist: acute or chronic. Acute external otitis may be localized or diffuse. Acute localized disease occurs in the form of a pustule or furuncle and typically results from *Staphylococcus aureus*. Erysipelas caused by group A streptococci may involve the external ear canal and the soft tissue of the ear itself. Acute diffuse otitis externa (swimmer's ear) is related to maceration (softening of tissue) of the ear from swimming and/or hot, humid weather. Gram-negative bacilli, particularly *Pseudomonas aeruginosa*, play an important role. A severe, hemorrhagic external otitis caused by *P. aeruginosa* is very difficult to treat and has occasionally been related to hot tub use.

Chronic otitis externa results from the irritation of drainage from the middle ear in patients with chronic, suppurative otitis media and a perforated eardrum. Malignant otitis externa is a necrotizing infection that spreads to adjacent areas of soft tissue, cartilage, and bone. If allowed to progress and spread into the central nervous system or vascular channel, a life-threatening situation may develop. *P. aeruginosa*, in particular, and anaerobes are frequently associated with this process. Malignant otitis media is seen in patients with diabetes who have blood vessel disease of the tissues overlying the temporal bone in which the poor local perfusion of tissues results in a milieu for invasion by bacteria. On occasion, external otitis can extend into the cartilage of the ear, usually requiring surgical intervention. Certain viruses may infect the external auditory canal, the soft tissue of the ear, or the tympanic membrane; influenza A virus is a suspected, but not an established, cause. Varicella zoster virus may cause painful vesicles within the soft tissue of the ear and the ear canal. *Mycoplasma pneumoniae* is a cause of bullous myringitis (a painful infection of the eardrum with hemorrhagic bullae); the ear canal itself may be involved as well.

Otitis media (middle ear infections)

In children (in whom otitis media is most common), pneumococci (33% of cases) and *Haemophilus influenzae* (20%) are the usual etiologic agents in acute disease. Group A streptococci (*Streptococcus pyogenes*) are the third most frequently encountered agents, found in 8% of cases. Other organisms, encountered in only 1% to 3% of cases, include *Moraxella catarrhalis*, *Staphylococcus aureus*, gram-negative enteric bacilli, and anaerobes. Viruses, chiefly respiratory syncytial virus (RSV) and influenza virus, have been recovered from the middle ear fluid of 4% of children with acute or chronic otitis media. *Chlamydia trachomatis* and *Mycoplasma pneumoniae* have occasionally been isolated from middle ear aspirates.

Chronic otitis media yields a predominantly anaerobic flora, with *Peptostreptococcus* spp., *Bacteroides fragilis* group, *Prevotella melaninogenica* (pigmented, anaerobic, gram-negative rods) and *Porphyromonas*, other *Prevotella* spp., and *Fusobacterium nucleatum* as the principal pathogens; less frequently present are *S. aureus*, *Pseudomonas aeruginosa*, *Proteus* spp., and other gram-negative facultative bacilli. Table 24-2 summarizes the major causes of ear infections.

The mastoid is a portion of the temporal bone (lower sides of the skull) that contains the mastoid sinuses (cavities). Mastoiditis is a complication of chronic otitis media in which organisms find their way into the mastoid sinuses.

TABLE 24-2 MAJOR INFECTIOUS CAUSES OF EAR DISEASE

DISEASE		COMMON CAUSES
Otitis Externa		
	Acute:	*Staphylococcus aureus*; *Streptococcus pyogenes*; *Pseudomonas aeruginosa*; other gram-negative bacilli
	Chronic:	*P. aeruginosa*; anaerobes
Otitis Media		
	Acute:	*Streptococcus pneumoniae*; *Haemophilus influenzae*; *S. pyogenes*; respiratory syncytial virus; influenza virus
	Chronic:	Anaerobes

PATHOGENESIS

Local trauma, the presence of foreign bodies, or excessive moisture can lead to otitis externa (external ear infections). Infrequently, an infection from the middle ear can extend by purulent drainage to the external ear.

Anatomic or physiologic abnormalities of the auditory tube can predispose individuals to develop otitis media. The auditory tube is responsible for protecting the middle ear from nasopharyngeal secretions, draining secretions produced in the middle ear into the nasopharynx, and ventilating the middle ear so that air pressure is equilibrated with that in the external ear canal. If any of these functions becomes compromised and fluid develops in the middle ear, infection may occur. To illustrate, if a person has a viral upper respiratory infection, the auditory tube becomes inflamed and swollen. This inflammation and swelling may, in turn, compromise the auditory tube's ventilating function, thereby resulting in a negative, rather than a positive, pressure in the middle ear. This change in pressure can then allow for potentially pathogenic bacteria present in the nasopharynx to enter the middle ear.

LABORATORY DIAGNOSIS

Specimen collection and transport

For the laboratory diagnosis of external otitis, the external ear should be cleansed with a mild germicide such as 1:1000 aqueous solution of benzalkonium chloride to reduce the contaminating skin flora before obtaining the culture. Material from the ear, especially that obtained after spontaneous perforation of the eardrum or by needle aspiration of middle ear fluid (tympanocentesis), should be collected by an otolaryngologist, using sterile equipment. Cultures from the mastoid are generally taken on swabs during surgery, although actual bone is preferred. Specimens should be transported anaerobically.

Direct visual examination

Material aspirated from the middle ear or mastoid is also examined directly for bacteria and fungi. The calcofluor white or PAS stains can reveal fungal elements. Methenamine silver stains have the added efficiency of staining most bacterial, fungal, and several parasitic species.

Culture

Ear specimens submitted for culture should be inoculated to blood, MacConkey, and chocolate agars. Anaerobic cultures should also be set up on those specimens obtained by tympanocentesis or those obtained from patients with chronic otitis media or mastoiditis.

SINUSES

ANATOMY

The sinuses, like the mastoids, are unique, air-filled cavities within the head (Figure 24-3). The sinuses are normally sterile. These structures, as well as the eustachian tube, the middle ear, and the respiratory portion of the pharynx, are lined by respiratory epithelium. The clearance of secretions and contaminants depends on normal ciliary activity and mucous flow.

DISEASES

Acute sinusitis usually develops during the course of a cold or influenzal illness and tends to be self-limited, lasting 1 to 3 weeks. Acute sinusitis is often difficult to distinguish from the primary illness. Symptoms include purulent nasal and postnasal discharge, a feeling of pressure over the sinus areas of the face, cough, and a nasal quality to the voice. Fever is sometimes present.

Occasionally, acute sinusitis persists and reaches a chronic state in which bacterial colonization occurs and the condition no longer responds to antibiotic treatment. Ordinarily, surgery or drainage is required for successful management. Patients with chronic sinusitis may have acute exacerbations (flare-ups). Other complications include local extension into the orbit, skull, meninges, or brain, and development of chronic sinusitis.

PATHOGENESIS

Most cases of acute sinusitis are believed to be bacterial complications of viral colds. The exact mechanisms involved are unknown. Of significance, about

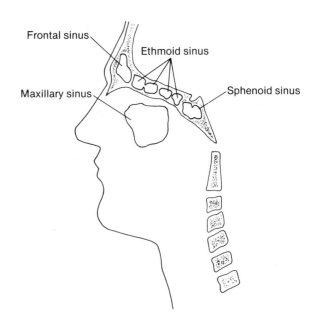

FIGURE 24-3 Location of the paranasal sinuses. (From Milliken, M.E. and Campbell, G. 1985. Essential competencies for patient care, Mosby, St. Louis.)

5% to 10% of acute maxillary sinus infections result from infection originating from a dental source. The maxillary sinuses are close to the roots of the upper teeth so that dental infections can extend into these sinuses. The primary problems believed to be associated with chronic sinusitis are inadequate drainage, impaired mucociliary clearance, and mucosal damage.

EPIDEMIOLOGY AND ETIOLOGY OF DISEASE

Although difficult to access, the actual incidence of acute sinusitis parallels that of acute upper respiratory tract infections (i.e., being most prevalent in the fall through spring).

Most studies of the microbiology of acute sinusitis have dealt with maxillary sinusitis because it is the most common type and the only one really accessible for puncture and aspiration. Bacterial cultures are positive in about three fourths of patients. In a study involving young adults, *Haemophilus influenzae* was recovered from 50% and *Streptococcus pneumoniae* from 19% of patients. *Streptococcus pyogenes* and *Moraxella catarrhalis* were also found, in addition to normal skin flora such as *Propionibacterium acnes*. Anaerobes were considered pathogens in only 2% of cases.[4]

Among children, *S. pneumoniae*, *H. influenzae*, and *M. catarrhalis* are most common. Rhinovirus is found in 15% of patients, influenza virus in 5%, parainfluenza virus in 3%, and adenovirus in less than 1%. The major causes of acute sinusitis are summarized in Table 24-3. Bacteria, particularly anaerobes, are more frequently involved in chronic sinusitis in adults. A recent report has shown *M. catarrhalis* to be an important agent in chronic sinusitis in children.[6]

LABORATORY DIAGNOSIS

In most cases, a diagnosis can be made on the basis of physical findings, history, radiograph studies, and other imaging techniques such as magnetic resonance imaging. However, if a laboratory diagnosis is needed, an otolaryngologist obtains the material from the maxillary sinus by puncture and aspiration or during surgery. Sinus drainage is unacceptable for smear or culture because this material will be contaminated with aerobic and anaerobic normal respiratory flora. Once received by the laboratory, Gram-stained smears and aerobic and anaerobic cultures should be performed. Aerobic culture media should include blood, chocolate, and MacConkey agars.

TABLE 24-3 MAJOR INFECTIOUS CAUSES OF ACUTE SINUSITIS

AGE GROUP	COMMON CAUSES
Young adults	*Haemophilus influenzae; Streptococcus pneumoniae; Streptococcus pyogenes; Moraxella catarrhalis*
Children	*S. pneumoniae; H. influenzae; M. catarrhalis;* rhinovirus

References

1. Chow, A.W. 1992. Life-threatening infections of the head and neck. Clin. Infect. Dis. 14:991.
2. Dugel, P.V. and Rao, N.A. 1993. Ocular infections in the acquired immunodeficiency syndrome. Int. Ophthalmol. Clin. 33:103.
3. Gigliotti, F. et al. 1981. Etiology of acute conjunctivitis in children. J. Pediatr. 98:531.
4. Jousimies-Somer, H.R., Savolainen, S., and Yikoski, J.S. 1988. Bacteriologic findings of acute maxillary sinusitis in young adults. J. Clin. Microbiol. 26:1919.
5. Leibowitz, H.M. et al. 1976. Human conjunctivitis: a diagnostic evaluation. Arch. Ophthalmol. 94:1747.
6. Tinkelman, D.G. and Silk, H.J. 1989. Clinical and bacteriologic features of chronic sinusitis in children. Am. J. Dis. Child. 143:938.
7. Wilhelmus, K.R. et al. 1994. Laboratory diagnosis of ocular infections. In Spector SC, editor. 1994. In Cumitech 13A. American Society for Microbiology, Washington, D.C.

Bibliography

Baum, J. 1995. Infections of the eye. Clin. Infect. Dis. 21:479.

Jones, D.B., Liesegang, T.J., and Robinson, N.M. 1981. Laboratory diagnosis of ocular infections. In Washington, J.A., editor: Cumitech 13. American Society for Microbiology, Washington, D.C.

O'Brien, T.P. and Green, W.R. 1995. Conjunctivitis, keratitis and endophthalmitis. In Mandell, G.L., Bennett, J.E., and Dolin, R., editors. Principles and practice of infectious diseases, ed 4. Churchill Livingstone, New York.

Shrader, S.K. et al. 1990. The clinical spectrum of endophthalmitis: incidence, predisposing factors, and features influencing outcome. J. Infect. Dis. 162:115.

Wilhelmus, K.R. et al. 1994. Laboratory diagnosis of ocular infections. In Spector, S.C., editor: Cumitech 13A. American Society for Microbiology, Washington, D.C.

25 | INFECTIONS OF THE URINARY TRACT

GENERAL CONSIDERATIONS

ANATOMY

The urinary tract consists of the kidneys, ureters, bladder, and urethra (Figure 25-1). Often, urinary tract infections (UTIs) are characterized as being either upper or lower based primarily on the anatomic location of the infection: the **lower urinary tract** encompasses the bladder and urethra, and the **upper urinary tract** encompasses the ureters and kidneys.

The anatomy of the female urethra is of particular importance to the pathogenesis of UTIs. The female urethra is relatively short compared with the male urethra and also lies in close proximity to the warm, moist, perirectal region, which is teeming with microorganisms. Because of the shorter urethra, bacteria can reach the bladder more easily in the female host.

RESIDENT MICROORGANISMS OF THE URINARY TRACT

The urethra has resident microflora that colonize its epithelium in the distal portion. Some of these organisms are listed in Box 25-1. Potential pathogens, including gram-negative aerobic bacilli (primarily *Enterobacteriaceae*) and occasional yeasts, are also present as transient colonizers. All areas of the urinary tract above the urethra in a healthy human are sterile. Urine is typically sterile, but noninvasive methods for collecting urine must rely on a specimen that has passed through a contaminated milieu. Therefore, quantitative cultures for diagnosis of UTIs have been used to discriminate between contamination, colonization, and infection.

INFECTIONS OF THE URINARY TRACT

EPIDEMIOLOGY

UTIs are among the most common bacterial infections that lead patients to seek medical care. It has been estimated that more than 6 million outpatient visits and 300,000 hospital stays every year are due to UTIs.[11,15,19] Approximately 10% of humans will have a UTI at some time during their lives.

The exact prevalence of UTIs is age and sex dependent. During the first year of life, UTIs are more common in males. However, the incidence of UTIs among males is low after age 1 and until approximately age 60 when enlargement of the prostate interferes with emptying of the bladder. Therefore, UTI is predominantly a disease of females. Extensive studies have shown that the incidence of **bacteriuria** (presence of bacteria in urine) among girls age 5 through 14 is 1% to 2%. This incidence increases to 5% in girls over age 10. The prevalence of bacteriuria in females increases gradually with time to as high as 10% to 20% in elderly women. In women between the ages of 20 and 40 who have had UTIs, as many as 50% may become reinfected within 1 year. The association of UTIs with sexual intercourse may also contribute to this increased incidence because sexual activity serves to increase the chances of bacterial contamination of the female urethra. Finally, as a result of anatomic and hormonal changes that favor development of UTIs, the incidence of bacteriuria increases during pregnancy. These infections can lead to serious infections in both mother and fetus.

UTIs are important complications of diabetes, renal disease, renal transplantation, and structural and neurologic abnormalities that interfere with urine flow. In addition, UTIs are a leading cause of gram-negative sepsis in hospitalized patients and are the origin for about half of all nosocomial infections caused by urinary catheters.

ETIOLOGIC AGENTS

Community-acquired

Escherichia coli is by far the most frequent cause of uncomplicated community-acquired UTIs. Other bacteria frequently isolated from patients with UTIs are *Klebsiella* spp., other *Enterobacteriaceae*, and *Staphylococcus saprophyticus*. In more complicated UTIs, particularly in recurrent infections, the relative frequency of infection caused by *Proteus, Pseudomonas, Klebsiella,* and *Enterobacter* spp. increases.

Hospital-acquired

The hospital environment plays an important role in determining the organisms involved in UTIs. Hospitalized patients are most likely to be infected by *E.*

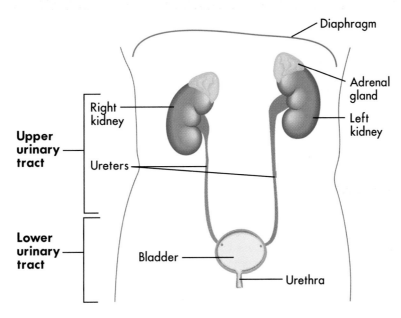

FIGURE 25-1 Overview of the anatomy of the urinary tract. (From Potter, P.H. and Perry, A.G. 1985. Fundamentals of nursing. Mosby, St. Louis.)

coli, Klebsiella spp., *Proteus mirabilis,* staphylococci, other *Enterobacteriaceae, Pseudomonas aeruginosa,* and enterococci. The introduction of a foreign body into the urinary tract, especially one that remains in place for a time (e.g., Foley catheter), carries a substantial risk of infection, particularly if obstruction is present. As many as 20% of all hospitalized patients who receive short-term catheterization develop a UTI. Consequently, UTI is the most common nosocomial infection in the United States, and the infected urinary tract is the most frequent source of bacteremia.

Miscellaneous

Other less frequently isolated agents are other gram-negative bacilli, such as *Acinetobacter* and *Alcaligenes* spp., other *Pseudomonas* spp., *Citrobacter* spp., *Gardnerella vaginalis,* and beta-hemolytic streptococci. Bacteria such as mycobacteria, *Chlamydia trachomatis, Ureaplasma urealyticum, Campylobacter* spp., *Haemophilus influenzae, Leptospira,* and certain *Corynebacterium* spp. (e.g., *C. renale*) are rarely recovered from urine. *Salmonella* spp. may be recovered during the early stages of typhoid fever; their presence should be immediately reported to the physician. If anaerobes are suspected, a percutaneous bladder tap should be done by the physician unless urine can be obtained from the upper urinary tract by another means (e.g., from a nephrostomy tube). Communication by the clinician to the laboratory that such an agent is suspected is most important for detecting such agents. However, the laboratory can exert some initiative as well. In patients with "sterile pyuria," Gram stain may reveal unusual organisms

with distinctive morphology (e.g., *H. influenzae,* anaerobes). Presence of any organisms on smear that do not grow in culture is an important clue to the cause of the infection. The laboratory can then take the action necessary to optimize chances for recovery.

In general viruses and parasites are not usually considered urinary tract pathogens. *Trichomonas vaginalis* may occasionally be observed in urinary sediment, and *Schistosoma haematobium* can lodge in the urinary tract and release eggs into the urine. Adenoviruses type 11 and 21 have been implicated as causative agents in hemorrhagic cystitis in children.

PATHOGENESIS

Routes of infection

Bacteria can invade and cause a UTI via two major routes: ascending and hematogenous pathways.[7,15] Although the **ascending route** is the most common

RESIDENT MICROFLORA OF THE URETHRA

BOX 25-1

Coagulase-negative staphylococci (excluding *S. saprophyticus*)
Viridans and nonhemolytic streptococci
Lactobacilli
Diphtheroids (*Corynebacterium* spp.)
Nonpathogenic (saprobic) *Neisseria* spp.
Anaerobic cocci
Propionibacterium spp.
Anaerobic gram-negative bacilli
Commensal *Mycobacterium* spp.
Commensal *Mycoplasma* spp.

route of infection in females, ascent in association with instrumentation (e.g., urinary catheterization, cystoscopy) is the most common cause of hospital-acquired UTIs in both sexes. For UTIs to occur by the ascending pathway, enteric gram-negative bacteria and other microorganisms that originate in the gastrointestinal tract must be able to colonize the periurethral area. Once these organisms gain access to the bladder, they may multiply and then pass up the ureters to the kidneys. UTIs occur more often in women than men, at least partially because of the short female urethra and its proximity to the anus. As previously mentioned, sexual activity can increase chances of bacterial contamination of the female urethra.

In most hospitalized patients, UTI is preceded by urinary catheterization or other manipulation of the urinary tract.[5] The pathogenesis of catheter-associated UTI is not fully understood. It is certain that soon after hospitalization, patients become colonized with bacteria endemic to the institution, often gram-negative aerobic and facultative bacilli carrying resistance markers. These bacteria colonize patients' skin, gastrointestinal tract, and mucous membranes, including the anterior urethra. With insertion of a catheter, the bacteria may be pushed along the urethra into the bladder or, with an indwelling catheter, may migrate along the track between the catheter and the urethral mucosa, gaining access to the bladder.

UTIs may also occur by the **hematogenous, or bloodborne, route.**[7] Hematogenous spread usually occurs as a result of bacteremia. Any systemic infection can lead to seeding of the kidney, but certain organisms, such as *Staphylococcus aureus* or *Salmonella* spp., are particularly invasive. Although most infections involving the kidneys are acquired by the ascending route, yeast (usually *Candida albicans*), *Mycobacterium tuberculosis, Salmonella* spp., *Leptospira* spp., or *Staphylococcus aureus* in the urine often indicate pyelonephritis acquired via hematogenous spread, or the descending route. Hematogenous spread accounts for less than 5% of UTIs.

The host-parasite relationship

Many individuals, women in particular, are colonized in the periurethral area with organisms originating from the gastrointestinal tract, yet they do not develop urinary infections. Whether an organism is able to colonize and then cause a UTI is determined in large part by a complex interplay of host and microbial factors.

In the majority of cases, the host defense mechanisms are able to eliminate the organisms. Urine itself is inhibitory to some of the urethral flora such as anaerobes. In addition, if urine has a low pH, high or low osmolality, high urea concentration, or high organic acid content, even organisms that can grow in urine may be inhibited. Of importance, if bacteria do gain access to the bladder, the constant flushing of contaminated urine from the body either eliminates bacteria or maintains their numbers at low levels. Clearly, any interference with the act of normal voiding, such as mechanical obstruction due to kidney stones or strictures, will promote the development of UTI. Also, the bladder mucosal surface has antibacterial properties. If the infection is not eradicated, the site of infection remains in the superficial mucosa; deep layers of the bladder are rarely involved.

In addition to the previously described host defenses, a valvelike mechanism at the junction of the ureter and bladder prevents the reflux (backward flow) of urine from the bladder to the upper urinary tract. Therefore, if the function of these valves is inhibited or compromised in any way, such as by obstruction or congenital abnormalities, urine reflux provides a direct route for organisms to reach the kidney. Hormonal changes associated with pregnancy and their effects on the urinary tract increase the chance for urine reflux to the upper urinary tract.

Although many microorganisms can cause UTIs, the majority are caused by only a few organisms. To illustrate, only a limited number of serogroups of *Escherichia coli* cause a significant proportion of UTIs.[15] Numerous investigations suggest that the strains of *E. coli* that cause UTIs possess certain virulence factors that enhance their ability to colonize and invade the urinary tract. Some of these virulence factors include increased adherence to vaginal and uroepithelial cells by bacterial surface structures (adhesins, in particular, pili), α-hemolysin production, and resistance to serum-killing activity.

The importance of adherence in the pathogenesis of UTIs has also been demonstrated with other species of bacteria. Once introduced into the urinary tract, *Proteus* strains appear to be uniquely suited to cause significant disease in the urinary tract. Data indicate that these strains are able to facilitate their adherence to the mucosa of kidneys. Also, *Proteus* is able to hydrolyze urea via urease production. Hydrolysis of urea results in an increase in urine pH that is directly toxic to kidney cells and also stimulates the formation of kidney stones. Similar findings have been made with *Klebsiella* spp. *Staphylococcus saprophyticus* also adheres better to uroepithelial cells than do *S. aureus* or *S. epidermidis*.

Finally, other bacterial characteristics may be important in the pathogenesis of UTIs. Motility may be important for organisms to ascend to the upper urinary tract against the flow of urine and cause pyelonephritis. Some strains demonstrate greater production of K antigen; this antigen protects bacteria from being phagocytosed.

TYPES OF INFECTION AND THEIR CLINICAL MANIFESTATIONS

UTI encompasses a broad range of clinical entities that differ in terms of clinical presentation, degree of tissue invasion, epidemiologic setting, and requirements for antibiotic therapy. There are four major types of UTIs: urethritis, cystitis, the urethral syndrome, and pyelonephritis. Sometimes UTIs are classified as uncomplicated or complicated. Uncomplicated infections occur primarily in otherwise healthy females and occasionally in male infants and adolescent and adult males. Most uncomplicated infections respond readily to antibiotic agents to which the etiologic agent is susceptible. Complicated infections occur in both sexes. In general, individuals who develop complicated infections often have certain risk factors. Some of these risk factors are listed in Box 25-2. In general, complicated infections are more difficult to treat and have greater morbidity (e.g., kidney damage, bacteremia) and mortality compared with uncomplicated infections.

The clinical presentation of UTIs may vary, ranging from asymptomatic infection to full-blown pyelonephritis (infection of the kidney and its pelvis). Some UTI symptoms may be nonspecific and frequently, symptoms overlap considerably in patients with lower UTIs and those with upper UTIs.

Urethritis

Symptoms associated with **urethritis** (infection of the urethra)—**dysuria** (painful or difficult urination) and frequency—are similar to those associated with lower UTIs. Urethritis is a common infection. Because *Chlamydia trachomatis*, *Neisseria gonorrhoeae*, and *Trichomonas vaginalis* are common causes of urethritis and are considered to be sexually transmitted, urethritis will be discussed as a sexually transmitted disease in Chapter 26.

Cystitis

Typically, patients with **cystitis** (infection of the bladder) complain of dysuria, frequency, and urgency (compelling need to urinate). These symptoms are due not only to inflammation of the bladder but also to multiplication of bacteria in the urine and urethra. Often, there is tenderness and pain over the area of the bladder. In some individuals, the urine is grossly bloody. The patient may note urine cloudiness and a bad odor. Because cystitis is a localized infection, fever and other signs of a systemic (affecting the body as a whole) illness are usually not present.

Acute urethral syndrome

Another UTI is the **acute urethral syndrome**. Patients with this syndrome are primarily young, sexually active women, who experience dysuria, frequency, and urgency but yield fewer organisms than 10^5 CFU/mL urine on culture.[6,8,13,14] (The classic criterion of greater than 10^5 colony-forming units of bacteria per milliliter [CFU/mL] of urine is highly indicative of infection in most patients with UTIs.) Almost 50% of all women who seek medical attention for complaints of symptoms of acute cystitis fall into this group. Although *Chlamydia trachomatis* and *N. gonorrhoeae* urethritis, anaerobic infection, genital herpes, and vaginitis account for some cases of acute urethral syndrome, most of these women are infected with organisms identical to those that cause cystitis but in numbers less than 10^5 CFU/mL urine. One must use a cutoff of 10^2 CFU/mL, rather than 10^5 CFU/mL, for this group of patients but must insist on concomitant **pyuria** (presence of eight or more leukocytes per cubic millimeter on microscopic examination of uncentrifuged urine). Approximately 90% of these women have pyuria, an important discriminatory feature of infection.

Pyelonephritis

Pyelonephritis refers to inflammation of the kidney parenchyma, calices (cup-shaped division of the renal pelvis), and pelvis (upper end of the ureter that is located inside the kidney), and is usually caused by bacterial infection. The typical clinical presentation of an upper urinary tract infection includes fever and flank (lower back) pain and frequently, lower tract symptoms (frequency, urgency, and dysuria). Patients can also exhibit systemic signs of infection such as vomiting, diarrhea, chills, increased heart rate, and lower abdominal pain. Of significance, 40% of patients with acute pyelonephritis will be bacteremic.

LABORATORY DIAGNOSIS OF URINARY TRACT INFECTIONS

As previously mentioned, because noninvasive methods for collecting urine must rely on a specimen that has passed through a contaminated milieu, quantitative cultures for the diagnosis of UTI are used to discriminate

RISK FACTORS ASSOCIATED WITH COMPLICATED UTIs

BOX 25-2

Underlying diseases that predispose the kidney to infection (e.g., diabetes, sickle cell anemia)
Kidney stones
Structural or functional abnormalities of the urinary tract (e.g., a tipped bladder)
Indwelling urinary catheters

between contamination, colonization, and infection. Refer to Table 1-1 for a quick reference for collecting, transporting, and processing urinary tract specimens.

SPECIMEN COLLECTION

Prevention of contamination by normal vaginal, perineal, and anterior urethral flora is the most important consideration for collection of a clinically relevant urine specimen.

Clean-catch, midstream urine

The least invasive procedure, the clean-catch, midstream urine specimen collection must be performed carefully for optimal results, especially in females. Good patient education is essential. Guidelines for proper specimen collection should be prepared on a printed card (bilingual, if necessary), with the procedure clearly described and preferably illustrated to help ensure patient compliance. The patient should be instructed to clean the periurethral area well with a mild detergent to avoid contamination. Of importance, the patient should also be instructed to rinse well because the detergent may be bacteriostatic. Once cleansing is completed, the patient should retract the labial folds or glans penis, begin to void, and then collect a midstream urine sample. Studies showed that uncleansed, first-void specimens from males were as sensitive as (but less specific than) midstream urine specimens.

Straight catheterized urine

Although slightly more invasive, urinary catheterization may allow collection of bladder urine with less urethral contamination. This procedure is performed by either a physician or other trained health professional. Risk exists, however, that urethral organisms will be introduced into the bladder with the catheter. An example of a collection device to obtain a "straight," or "in and out," catheterized urine is shown in Figure 25-2.

Suprapubic bladder aspiration

With suprapubic bladder aspiration, urine is withdrawn directly into a syringe through a percutaneously inserted needle, thereby ensuring a contamination-free specimen. The bladder must be full before performing the procedure. This collection technique may be indicated in certain clinical situations, such as pediatric practice, when urine is difficult to obtain. In brief, the full bladder is punctured using a needle and syringe and sampled following proper skin preparation (antisepsis). If good aseptic techniques are used, this procedure can be performed with little risk in premature infants, infants, small children, and pregnant women and other adults with full bladders.

Indwelling catheter

The number of patients in hospitals and nursing homes with long-term, indwelling urinary catheters continues to increase. These patients ultimately develop bacteriuria, which predisposes them to more severe infections.[5] Specimen collection from patients with indwelling catheters requires scrupulous aseptic technique. Health care workers who manipulate a urinary catheter in any way should wear gloves. The catheter tubing should be clamped off above the port to allow collection of freshly voided urine. The catheter port or wall of the tubing should then be cleaned vigorously with 70% ethanol, and urine should be aspirated via a needle and syringe; the integrity of the closed drainage system must be maintained to prevent the introduction of organisms into the bladder. Specimens obtained from the collection bag are inappropriate, because organisms can multiply there, obscuring the true relative numbers. Cultures should be obtained when patients are ill; routine monitoring does not yield clinically relevant data.

SPECIMEN TRANSPORT

Urine, being an excellent supportive medium for growth of most bacteria, must be immediately refrigerated or preserved. Bacterial counts in refrigerated (4° C) urine remain constant for as long as 24 hours. Urine transport tubes (B-D Urine Culture Kit [Becton Dickinson Vacutainer Kits, Rutherford, N.J.]) containing boric acid, glycerol, and sodium formate have been shown to preserve bacteria without refrigeration for as long as 24 hours when greater than 10^5 CFU/mL (100,000 organisms per milliliter) were present in the initial urine specimen. The system may inhibit the growth of certain organisms, and it must be used with 3 to 5 mL of urine. Another preservative system (Sage Products, Cary, Ill.) is also available. Both boric acid products preserve bacterial viability in urine for 24 hours in the absence of antibiotics. For populations of patients from whom colony counts of organisms of less than 100,000/mL might be clinically significant, plating within 2 hours of collection is recommended. None of the kits has any advantage over refrigeration, except perhaps for convenience or for transport of urine from remote areas in which refrigeration is not practical.

SCREENING PROCEDURES

As many as 60% to 80% of all urine specimens received for culture by the acute care medical center laboratory may contain no etiologic agents of infection or contain only contaminants. Procedures developed to identify quickly those urine specimens that will be negative on culture and circumvent excessive use of media, technologist time, and the overnight

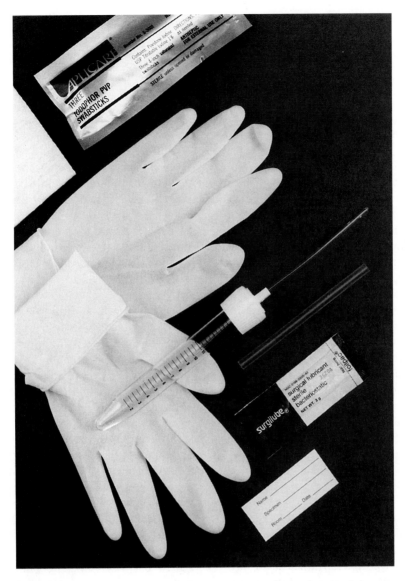

FIGURE 25-2 Collection device to obtain a urine by "in and out," or "straight," catheterization. (Courtesy Tristate Hospital Supply Corp., Howell, Mich.)

incubation period are discussed in this section. Of importance, a reliable screening test for the presence or absence of bacteriuria gives physicians important same-day information that a conventional urine culture may take a day or longer to provide. Many screening methods have been advocated for use in detecting bacteriuria and/or pyuria. These include microscopic methods, colorimetric filtration, bioluminescence, electrical impedance, enzymatic methods, photometric detection of growth, and enzyme immunoassay. Because a discussion of all available urine screening methods is beyond the scope of this chapter, only the more commonly used methods are highlighted.

Gram stain

The Gram stain is the easiest, least expensive, and probably the most sensitive and reliable screening method for identifying urine specimens that contain greater than 10^5 CFU/mL. After a drop of well-mixed urine is allowed to air dry, the smear is fixed, stained, and examined under oil immersion (1000×). Presence of at least one organism per oil immersion field (examining 20 fields) correlates with significant bacteriuria (>10^5 CFU/mL).[18] The Gram stain should not be relied on for detecting polymorphonuclear leukocytes in urine.[1] Many microbiologists are unwilling to adopt Gram stain examination as a part of the routine laboratory examination of urine specimens, probably because of the low number of positive results coupled with the large number of specimens that must be processed.

Pyuria

Pyuria is the hallmark of inflammation, and the presence of polymorphonuclear neutrophils (PMNs) can be detected and enumerated in *uncentrifuged* specimens. This method of screening urine correlates fairly well with the number of PMNs excreted per hour, the best indicator of the host's state. Patients with more than 400,000 PMNs excreted into the urine per hour are likely to be infected, and the presence of more than 8 PMNs/mm³ correlates well with this excretion rate and with infection.[2] This test can be performed using a hemocytometer, but it is not easily incorporated into the work flow of most microbiology laboratories. The standard urinalysis (usually done in hematology or chemistry sections) includes an examination of the *centrifuged* sediment of urine for enumeration of PMNs, results of which do not correlate well with either the PMN excretion rate or the presence of infection. Pyuria also can be associated with other clinical diseases, such as vaginitis, and therefore is not specific for UTIs.

Indirect indices

Frequently, screening tests detect bacteriuria or pyuria by examining for the presence of bacterial enzymes and/or PMN enzymes rather than the organisms or PMNs themselves.

NITRATE REDUCTASE (GREISS) TEST   This screening procedure looks for the presence of urinary nitrite, an indicator of UTI. Nitrate-reducing enzymes that are produced by the most common urinary tract pathogens reduce nitrate to nitrite. This test has been incorporated onto a paper strip that also tests for leukocyte esterase, an enzyme produced by PMNs (see below).

LEUKOCYTE ESTERASE TEST   As previously mentioned, evidence of a host response to infection is the presence of PMNs in the urine. Because inflammatory cells produce leukocyte esterase, a simple, inexpensive, and rapid method that measures this enzyme has been developed. Studies have shown that leukocyte esterase activity correlates with hemocytometer chamber counts. The nitrate reductase and leukocyte esterase tests have been incorporated into a paper strip. These strips are sold commercially by numerous manufacturers and is one of the most widely used enzymatic tests. Although the sensitivity of the combination strip is higher than either test alone, the sensitivity of this combination screening is not great enough to recommend its use as a stand-alone test in

most circumstances. Of note, the leukocyte esterase test is not sensitive enough for determining pyuria in patients with acute urethral syndrome.

CATALASE   The API Uriscreen (Analytab Products, Plainview, N.Y.) is another rapid urine screening system based on the detection of catalase present in somatic (pertaining to the body) cells and in most bacterial species commonly causing UTIs except for streptococci and enterococci. Approximately 1.5 to 2.0 mL of urine are added to a tube containing dehydrated substrate. Hydrogen peroxide is added to the urine, and the solution is mixed gently. The formation of bubbles above the liquid surface is interpreted as a positive test. A recent study reported that this system did not offer significant advantages over the leukocyte esterase-nitrite strip.[4]

Automated and semiautomated systems

Automated screening systems offer the promise of a large throughput with minimal labor and a rapid turnaround time compared with conventional cultures. However, these advantages may be offset by a substantial cost for the instrumentation. Often these costs can be justified only in laboratories receiving many specimens.

Several automated or semiautomated urine screening systems that are either bacterial growth independent or dependent are commercially available. By examining images of uncentrifuged urine samples using a video camera, the Yellow IRIS system (International Remote Imaging Systems, Inc., Chatsworth, Calif.), is able to recognize many cellular structures, including leukocytes and bacteria. With the semiautomated Bac-T-Screen 2000 (bioMeriéux Vitek, Hazelwood, Mo.), urine is forced through a filter paper, which retains microorganisms, somatic cells, and other particles. A dye is then added to the filter paper to visualize the particulate matter that has adhered. The intensity of color relates to the number of particles. This procedure, which takes approximately 2 minutes or less, has been shown to detect greater than 90% of all positive urine specimens, even if 100 organisms per milliliter are considered to be significant. The detection of somatic cells and particles, as well as bacteria and yeast, probably accounts for the increased sensitivity of this system at low numbers of CFUs.[1] Organisms likely to be associated with false-negative results include enterococci and *Pseudomonas aeruginosa*; the reason for this is not known.

The AutoMicrobic System (bioMeriéux Vitek Systems, Hazelwood, Mo.) is the most labor-free screening system. Once the urine is introduced into the tiny wells of a plastic substrate card by vacuum suction and the card is placed into the instrument, it is incubated and examined periodically by measuring the amount of light that passes through the individual wells. The wells contain substrates that can be utilized only by certain etiologic agents of UTI. After 6 to 8 hours, microorganisms grow in the appropriate substrate wells and are detected by increased turbidity. The internal computer issues a colony count and a preliminary identification of the agent(s). In this system as well, low counts of bacteria or yeast are considered negative urine cultures. Table 25-1 lists some of the automated screening methods.

General comments regarding screening procedures

In general, screening methods are insensitive at levels below 10^5 CFU/mL. Therefore, they are not acceptable for urine specimens collected by suprapubic aspiration, catheterization, or cystoscopy. Screening methods may also fail to detect a significant number of infections in symptomatic patients with low colony counts (10^2 to 10^3 CFU/mL) such as young, sexually active females with the acute urethral syndrome. Further complicating the laboratory's decision whether to

TABLE 25-1 SUMMARY INFORMATION ABOUT SOME AUTOMATED URINE SCREENS

INSTRUMENT	MANUFACTURER	PRINCIPLE	TIME TO RESULT	URINE VOLUME
Growth Independent				
Bac-T-Screen[1]	bioMeriéux Vitek	Colorimetric filtration	2 min	1 mL
Yellow IRIS[16]	International Remote Imaging Systems, Inc., Chatsworth, Calif.	Image processing by computer	1 to 4 min	6 mL
Autotrak[12,*]	Roche Ltd., England	Acridine orange staining	1 hr (120 specimens/hr)	1 mL
UTI screen BacterialATP test[16]	Coral Biomedical, San Diego, Calif.	Bioluminescence	10 to 45 min	10 to 25 μL
ATP bioluminescence assay using the Amerlite Analyzer[12,†]	Amersham International Ltd., Buckinghamshire, England	Bioluminescence	15-20 min	10 μL
Ramus 256[12]	Orbec Ltd., Berkshire, England	Particulate counting by electrical impedance	1 hr (45 specimens/hr)	50 μL
Growth Dependent				
Vitek System	bioMeriéux Vitek	Photometry	1 to 13 hrs	200 μL
Malthus 128H Growth Analyzer[12,†]	Malthus Ltd., Stoke-on-Trent, England	Electrical conductance	5 hrs (128 specimens at any one time)	Dipping of a ceramic strip provided by the manufacturer
Questor[17]	Difco Laboratories Ltd., East Molesey, Surrey, England	Particulate counting by electrical impedance	1 min	100 μL

*Currently not available.
† To date, industrial applications have been more successful.

adopt a screening method is whether screening results will be used to rule out infection in asymptomatic patients. Under these circumstances, testing for pyuria is essential.

Therefore, given the importance of the 10^2 CFU/mL count and the PMN count, no screening test should be used indiscriminately. Selecting a screening method largely depends on the laboratory and the patient population being served by the laboratory. For example, there will be a cost advantage in screening urines in laboratories that receive many culture-negative specimens. On the other hand, urine from patients with symptoms of UTI plus a selected group expected to have asymptomatic bacteriuria should be cultured. For example, patients in their first trimester of pregnancy should be cultured because these women might appear asymptomatic but have a covert infection and become symptomatic later; UTIs in pregnant women may lead to pyelonephritis and the likelihood of a premature birth. Other situations in which patients with no symptoms of UTI might be cultured are:

- Bacteremia of unknown source
- Urinary tract obstruction
- Follow-up after removal of an indwelling catheter
- Follow-up of previous therapy

Other factors that must be considered when selecting a rapid urine screen include accuracy, ease of test performance, reproducibility, turnaround time, and whether bacteriuria and/or pyuria are detected.

URINE CULTURE

Inoculation and incubation of urine cultures

Once it has been determined that a urine specimen should be cultured for isolation of the common agents of UTI, a measured amount of urine is inoculated to each of the appropriate media. The urine should be mixed thoroughly before plating. The plates can be inoculated using disposable sterile plastic tips with a displacement pipetting device calibrated to deliver a constant amount, but this method is somewhat cumbersome. Most often, microbiologists use a calibrated loop designed to deliver a known volume, either 0.01 or 0.001 mL of urine. These loops, made of platinum, plastic, or other material, can be obtained from laboratory supply companies.

The calibrated loop that delivers the larger volume of urine (0.01 mL) is recommended to detect lower numbers of organisms in certain specimens. For example, urines collected from catheterization, nephrostomies, ileal conduits, and suprapubic aspirates should be plated with the larger calibrated loop.

The communication of pertinent clinical history to the laboratory is essential so that appropriate processing can be performed.

The choice of which media to inoculate depends on the patient population served and the microbiologist's preference. The use of a 5% sheep blood agar plate and a MacConkey agar plate allows detection of most gram-negative bacilli, staphylococci, streptococci, and enterococci. To save cost and somewhat streamline culture processing, many laboratories employ an agar plate split in half (biplate); one side contains 5% sheep blood agar and the other half contains MacConkey agar.

In some circumstances, enterococci and other streptococci, however, may be obscured by heavy growth of *Enterobacteriaceae*. Because of this possibility, some laboratories add a selective plate for gram-positive organisms, such as Columbia colistin-nalidixic acid agar (CNA) or phenylethyl alcohol agar. Although some discriminatory capability may be added, cost is also added to the procedure. In addition to increased cost, inclusion of plated media selective for gram-positive organisms generally provides no or limited additional information. Many European laboratories use cystine-lactose electrolyte-deficient (CLED) agar.

Before inoculation, urine is mixed thoroughly and the top of the container is then removed. The calibrated loop is inserted *vertically* into the urine in a cup. Otherwise, more than the desired volume of urine will be taken up, potentially affecting the quantitative culture result (Figure 25-3). A widely used

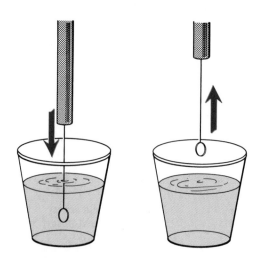

FIGURE 25-3 Method for inserting a calibrated loop into urine to ensure that the proper amount of specimen adheres to the loop.

method is described in Procedure 25-1. If the urine is in a small-diameter tube, the surface tension will alter the amount of specimen picked up by the loop. A quantitative pipette should be considered if the urine cannot be transferred to a larger container. Once inoculated, the plates are streaked to obtain isolated colonies (Figure 25-4).

Once plated, urine cultures are incubated overnight at 35° C. For the most part, incubation for a minimum of 24 hours is necessary to detect uropathogens.[10] Thus, some specimens inoculated late in the day cannot be read accurately the next morning. These cultures should either be reincubated until the next day or possibly, interpreted later in the day when a full 24-hour incubation has been completed.

Interpretation of urine cultures

As previously mentioned, UTIs may be completely asymptomatic, produce mild symptoms, or cause life-threatening infections. Of importance, the criteria most useful for microbiologic assessment of urine specimens is dependent not only on the type of urine submitted (e.g., voided, straight catheterization) but the clinical history of the patient (e.g., age, sex, symptoms, antibiotic therapy).

One major problem in interpreting urine cultures arises because urine cultures collected by the voided technique may be contaminated with normal flora, including *Enterobacteriaceae*. Determining what colony count represents true infection from contamination is of utmost importance and is related to the patient's clinical presentation. A number of studies have proposed the use of different cutoffs in colony counts based on clinical presentation; an example of one such set of guidelines is shown in Table 25-2.

Ideally, the clinician caring for the patient should provide the laboratory with enough clinical information to allow specimens from different patient populations to be identified.[3,9] These specimens could then be selectively processed using the guidelines in Table 25-2. However, because microbiology laboratories frequently receive little or no clinical information about patients, questions have been raised as to whether these cutoffs are practical and realistic for

TABLE 25-2 CRITERIA FOR CLASSIFICATION OF URINARY TRACT INFECTIONS BY CLINICAL SYNDROME

CATEGORY	CRITERIA	
	CLINICAL	LABORATORY
Acute, uncomplicated UTI in women	Dysuria, urgency, frequency, suprapubic pain No urinary symptoms in last 4 weeks before current episode No fever or flank pain	≥ 10 WBC/mm^3 $\geq 10^3$ CFU/mL uropathogens* in CCMS† urine
Acute, uncomplicated pyelonephritis	Fever, chills Flank pain on examination Other diagnoses excluded No history or clinical evidence of urologic abnormalities	≥ 10 WBC/mm^3 $\geq 10^4$ CFU/mL uropathogens in CCMS urine
Complicated UTI and UTI in men	Any combination of symptoms listed above One or more factors associated with complicated UTI‡	≥ 10 WBC/mm^3 $\geq 10^5$ CFU/mL uropathogens in CCMS urine
Asymptomatic bacteriuria	No urinary symptoms	$\pm >10$ WBC/mm^3 $\geq 10^5$ CFU/mL in two CCMS cultures >24 hrs apart

*Uropathogens: Organisms that commonly cause UTIs.

†CCMS: Clean-catch midstream urine.

‡Factors associated with complicated UTI include: any UTI in a male; indwelling or intermittent urinary catheter; >100 mL of postvoid residual urine; obstructive uropathy; urologic abnormalities; azotemia (excess urea in the blood, even without structural abnormalities) and renal transplantation.

From Stamm, W.E. 1992. Infection 20(suppl. 3):S151.

routine laboratory use. Further complicating urine culture interpretation is the increasing difficulty in distinguishing between infection and contamination as the criterion for a positive culture is lowered from 10^5 CFU/mL to 10^2 CFU/mL. Because of these issues, many laboratories establish their own interpretative criteria for urine cultures based on the type of urine submitted (e.g., clean-catch midstream, catheterized, and surgically obtained specimens such as suprapubic aspirates). Variations in interpretative guidelines occur from one laboratory to another but some generalities can be made; these are listed in Table 25-3. Some examples of urine culture results are shown in Figure 25-5 to illustrate some of these interpretations.

(For delineation of complete urine protocols, refer to sources listed in the bibliography.)

In addition to the previously described guidelines, a pure culture of *S. aureus* is considered to be significant regardless of the number of CFUs, and antimicrobial susceptibility tests are performed. The presence of yeast in any number is reported to physicians, and pure cultures of a yeast may be identified to the species level. In all urines, regardless of the extent of final workup, all isolates should be enumerated (e.g., three different organisms present at 10^3 CFU/mL), and those present in numbers greater than 10^4 CFU/mL should be described morphologically (e.g., nonlactose-fermenting gram-negative rods).

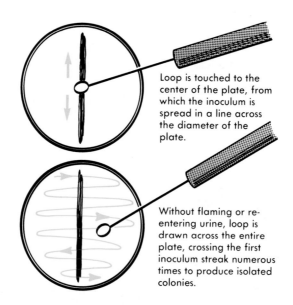

Loop is touched to the center of the plate, from which the inoculum is spread in a line across the diameter of the plate.

Without flaming or re-entering urine, loop is drawn across the entire plate, crossing the first inoculum streak numerous times to produce isolated colonies.

FIGURE 25-4 Method for streaking with calibrated urine loop to produce isolated colonies and countable colony-forming units.

FIGURE 25-5 Culture results illustrating some of the various interpretative guidelines. **A,** Growth of ≥10^5 CFU/mL of a lactose-fermenting gram-negative rod in a CCMS urine from a patient with pyelonephritis; complete workup would be done. **B,** Growth of ≥10^5 CFU/mL of a lactose-fermenting gram-negative rod (*arrow A*) and <10^4 CFU/mL of another organism type (*arrow B*) from a CCMS urine; only the organism with a colony count of ≥10^4 to 10^5 CFU/mL would be worked up completely.

--- **TABLE 25-3** GENERAL INTERPRETATIVE GUIDELINES FOR URINE CULTURES ---

RESULT	SPECIFIC SPECIMEN TYPE/ASSOCIATED CLINICAL CONDITION, IF KNOWN	WORKUP
$\geq 10^4$ CFU/mL of a single potential pathogen or for each of two potential pathogens	CCMS urine/pyelonephritis, acute cystitis, asymptomatic bacteriuria, *or* catheterized urines	Complete*
$\geq 10^3$ CFU/mL of a single potential pathogen	CCMS urine/symptomatic males *or* catheterized urines *or* acute urethral syndrome	Complete
$\geq$ Three organism types with no predominating organism	CCMS urine *or* catheterized urines	None. Because of possible contamination, ask for another specimen
Either two or three organism types with predominant growth of one organism type and $<10^4$ CFU/mL of the other organism type(s)	CCMS urine	Complete workup for the predominating† organism(s); description of the other organism(s)
$\geq 10^2$ CFU/mL of any number of organism types (set up with a 0.001- and 0.01-mL calibrated loop)	Suprapubic aspirates, any other surgically obtained urines (including ileal conduits, cystoscopy specimens)	Complete

*A complete workup includes identification of the organism and appropriate susceptibility testing.
†Predominant growth = 10^4 to $\geq 10^5$ CFU/mL.

Procedure

25-1 INOCULATING URINE WITH A CALIBRATED LOOP

PRINCIPLE

The number of microorganisms per milliliter recovered on urine culture can aid in the differential diagnosis of UTI. Plastic or wire loops, available commercially, have been calibrated to deliver a known volume of liquid when handled correctly, thus enabling the microbiologist to estimate numbers of organisms in the original specimen based on CFU of growth on cultures.

METHOD

1. Flame a calibrated wire inoculating loop and allow it to cool without touching any surface. Alternatively, aseptically remove a plastic calibrated loop from its package.

2. Mix the urine thoroughly and remove the top of the container. If the urine is in a small-diameter tube, the surface tension will alter the amount of specimen picked up by the loop. A quantitative pipette should be considered if the urine cannot be transferred to a larger container.

3. Insert the loop vertically into the urine (see Figure 25-3) to allow urine to adhere to the loop.

4. Spread the loopful of urine over the surface of the agar plate, as shown in Figure 25-4. A standard quadrant streaking technique is also acceptable.

5. Without reflaming, insert the loop vertically into the urine again for transfer of a loopful to a second plate. Repeat for each plate.

6. Incubate plates for at least 24 hours at 35° to 37° C in air. Colonies are counted on each plate. The number of CFUs is multiplied by 1000 (if a 0.001-mL loop was used) or by 100 (if a 0.01-mL loop was used) to determine the number of microorganisms per milliliter in the original specimen.

7. Because antimicrobial treatment or other factors may inhibit initial growth, reincubate plates with no growth or tiny colonies for an additional 24 hours before discarding plates.

8. To store the inoculating loop, place (handle down) in a test tube taped to the wall, rather than flat on the bench, to prevent bending, which would destroy the calibration.

References

1. Baron, E.J., et al. 1988. Visual and clinical analysis of Bac-T-Screen urine screen results. J. Clin. Microbiol. 26:2382.

2. Brumfitt, W. 1965. Urinary cell counts and their value. J. Clin. Pathol. 18:550.

3. Carroll, K.C., et al. 1994. Laboratory evaluation of urinary tract infections in an ambulatory clinic. Am. J. Clin. Pathol. 10:100.

4. Dalton, M.T., et al. 1993. A comparison of the API Uriscreen with the Vitek urine identification-3 and the leukocyte esterase or nitrite strip as a screening test for bacteriuria. Diagn. Microbiol. Infect. Dis. 16:93.

5. Falkiner, F.R. 1993. The insertion and management of indwelling urethral catheters: minimizing the risk of infection. J. Hosp. Infect. 25:79.

6. Hamilton-Miller, J.M.T. 1994. The urethral syndrome and its management. J. Antimicrob. Chemother. 33(suppl. A):63.

7. Kunin, C.M. 1994. Urinary tract infections in females. Clin. Infect. Dis. 18:1.

8. Kunin, C.M., White, L.V., and Hua, T.H. 1993. A reassessment of the importance of "low-count" bacteriuria in young women with acute urinary symptoms. Ann. Intern. Med. 119:454.

9. Morgan, M.G. and McHenzie, H. 1993. Controversies in the laboratory diagnosis of community-acquired urinary tract infection. Eur. J. Clin. Microbiol. Infect. Dis. 12:491.

10. Murray, P.R., Traynor, P., and Hopson, D. 1992. Evaluation of microbiological processing of urine specimens: comparison of overnight versus two-day incubation. J. Clin. Microbiol. 30:1600.

11. Palac, D.M. 1986. Urinary tract infections in women: a physician's perspective. Lab. Med. 17:25.

12. Smith, T.K., Hudson, A.J., and Spencer, R.C. 1988. Evaluation of six screening methods for detecting significant bacteriuria. J. Clin. Pathol. 41:904.

13. Stamm, W.E. 1992. Criteria for the diagnosis of urinary tract infection and for the assessment of therapeutic effectiveness. Infection 20(suppl. 3):S151.

14. Stamm, W.E. 1988. Protocol for the diagnosis of urinary tract infection: reconsidering the criterion for significant bacteriuria. Urology XXXII(suppl.):6.

15. Stamm, W.E., et al. 1989. Urinary tract infections: from pathogenesis to treatment. J. Infect. Dis. 159:400.

16. Stevens, M. 1989. Screening urines for bacteriuria. Med. Lab. Sci. 46:194.

17. Stevens, M. 1993. Evaluation of Questor urine screening for bacteriuria and pyuria. J. Clin. Pathol. 46:817.

18. Washington, J.A., et al. 1981. Detection of significant bacteriuria by microscopic examination of urine. Lab. Med. 12:294.

19. Wong, E.S. 1983. Guideline to prevention of catheter-associated urinary tract infections. Am. J. Infect. Control 11:28.

Bibliography

Clarridge, J.E., Pezzlo, M.T., and Vosti, K.L. 1987. Laboratory diagnosis of urinary tract infections. In Weissfeld, A.S., coordinating editor. Cumitech 2A. American Society for Microbiology, Washington, D.C.

Pezzlo, M. 1988. Detection of urinary tract infections by rapid methods. Clin. Microbiol. Rev. 1:268.

Pezzlo, M. 1992. Urine culture procedures. Section 1.17. In Isenberg, H.D., editor. Clinical Microbiology Procedures Handbook, vol 1. American Society for Microbiology, Washington, D.C.

Sobel, J.D. 1987. Pathogenesis of urinary tract infections: host defenses. Infect. Dis. Clin. North Am. 1:751.

26 | GENITAL TRACT INFECTIONS

GENERAL CONSIDERATIONS

ANATOMY

Familiarity with the anatomic structures is important for appropriate processing of specimens from genital tract sites and interpretation of microbiologic laboratory results. The key anatomic structures for the female and male genital tract in relation to other important structures are shown in Figure 26-1.

RESIDENT MICROBIAL FLORA

The lining of the normal human genital tract is a mucosal layer made up of transitional, columnar, and squamous epithelial cells. Various species of commensal bacteria colonize these surfaces, causing no harm to the host except under abnormal circumstances and helping to prevent the adherence of pathogenic organisms. Normal urethral flora include coagulase-negative staphylococci and corynebacteria, as well as various anaerobes. The vulva and penis, especially the area underneath the prepuce (foreskin) of the uncircumcised male, may harbor *Mycobacterium smegmatis* along with other gram-positive bacteria.

The flora of the female genital tract varies with the pH and estrogen concentration of the mucosa, which depends on the host's age. Prepubescent and postmenopausal women harbor primarily staphylococci and corynebacteria (the same flora present on surface epithelium), whereas women of reproductive age may harbor large numbers of facultative bacteria such as *Enterobacteriaceae*, streptococci, and staphylococci, as well as anaerobes such as lactobacilli, anaerobic non–spore-forming bacilli and cocci, and clostridia. Lactobacilli are the predominant organisms in secretions from normal, healthy vaginas. Recent studies have shown that hydrogen peroxide-producing lactobacilli are most associated with a healthy state.[12] The numbers of anaerobic organisms remain constant throughout the monthly cycle. Many women carry group B beta-hemolytic streptococci *(Streptococcus agalactiae)*, which may be transmitted to the neonate. Although yeasts (acquired from the gastrointestinal tract) may be transiently recovered from the female vaginal tract, they are not considered normal flora.

SEXUALLY TRANSMITTED DISEASES AND OTHER GENITAL TRACT INFECTIONS

Genital tract infections may be classified as endogenous or exogenous. **Exogenous infections** may be acquired as people engage in sexual activity, and these infections are referred to as **sexually transmitted diseases (STDs)**. In contrast, **endogenous infections** result from organisms that are members of the patient's normal genital flora.

In the female, genital tract infections can be divided between lower tract (vulva, vagina, and cervix) and upper tract (uterus, fallopian tubes, ovaries, and abdominal cavity) infections.[7] **Lower tract infections** are commonly acquired by sexual or direct contact. Although the organisms that cause lower tract infections are not usually part of the normal genital tract flora, some organisms that are normally present in very low numbers can increase sufficiently to cause disease. **Upper tract infections** are frequently an extension of a lower tract infection in which organisms from the vagina or cervix are believed to travel into the uterine cavity and on through the endometrium to the fallopian tubes and ovaries. Similarly, an organism can spread along contiguous mucosal surfaces in the male from a lower genital tract site of infection (i.e., urethra) and cause infection in a reproductive organ such as the epididymis.

GENITAL TRACT INFECTIONS

STDS AND OTHER LOWER GENITAL TRACT INFECTIONS

Lower genital tract infections may either be acquired through sexual contact with an infected partner or through nonsexual means. These infections are some of the most common infectious diseases.

Epidemiology/etiologic agents

STDs are major public health problems in all populations and socioeconomic groups worldwide. To illustrate, an estimated 4 million genital infections caused by *Chlamydia trachomatis* occur annually in the United States alone.[27] The incidence and spread of STDs are greatly influenced by numerous factors such as the

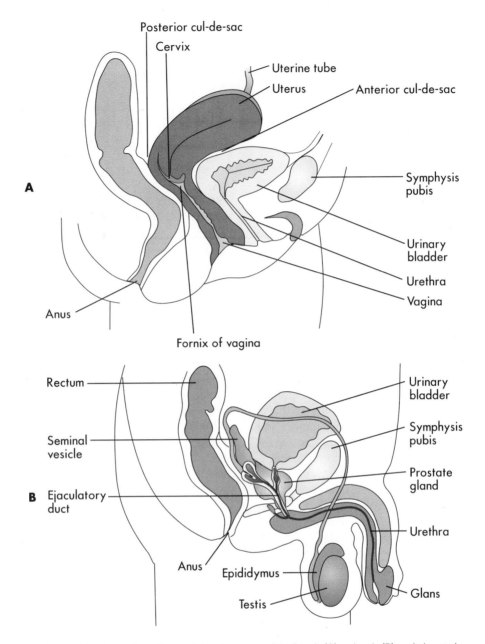

FIGURE 26-1 Location of key anatomic structures of the female (**A**) and male (**B**) genital tracts in relation to other major anatomic structures.

availability of multiple sexual partners, the presence of asymptomatic infection, the frequent movement of people within populations, and increasing affluence.[20]

The number of microorganisms that can cause genital tract infections is large. Of importance, these organisms are diverse, representing all four major groups of microorganisms (bacteria, viruses, fungi, and parasites). The major causes of genital tract infections are listed in Table 26-1.

Routes of transmission

Although genital tract infections can be caused by members of the patient's genital flora (endogenous

infections), the overwhelming majority of lower genital tract infections are sexually transmitted.

SEXUALLY TRANSMITTED *Chlamydia trachomatis, Gardnerella vaginalis, Neisseria gonorrhoeae* and *N. meningitidis, Treponema pallidum, Ureaplasma urealyticum, Mycoplasma hominis,* other mycoplasmas, herpes simplex virus (HSV), and others may be acquired as people engage in sexual activity. In addition, other agents that cause genital tract disease and may be sexually transmitted include adenovirus, coxsackievirus, molluscum contagiosum virus (a member of the poxvirus group), the human papillomaviruses

TABLE 26-1   MAJOR CAUSES OF GENITAL TRACT INFECTIONS AND SEXUALLY TRANSMITTED DISEASES

FREQUENCY	DISEASE	AGENT	ORGANISM GROUP
More common	Genital and anal warts (condyloma); cervical dysplasia; cancer	Human papillomavirus	Viruses
	Vaginitis	*Gardnerella/Mobiluncus, Trichomonas vaginalis, Candida albicans*	Bacteria, parasites, fungi
	Urethritis/cervicitis (also acute salpingitis, acute perihepatitis, urethritis, pharyngitis)	*Neisseria gonorrhoeae, Chlamydia trachomatis, Ureaplasma urealyticum*	Bacteria
	Herpes genitalis (genital/skin ulcers and vesidis papulis)	Herpes simplex virus type 2 (less commonly type 1)	Viruses
	AIDS	Human immunodeficiency virus (HIV)	Viruses
	Hepatitis (acute and chronic infection)	Hepatitis B virus	Viruses
Less common	Lymphogranuloma venereum	*C. trachomatis* (L-1, L-2, L-3 serovars)	Bacteria
	Granuloma inguinale	*Calymmatobacterium granulomatis* (Donovania)	Bacteria
	Syphilis	*Treponema pallidum*	Bacteria
	Chancroid	*Haemophilus ducreyi*	Bacteria
	Scabies, mites	*Sarcoptes scabei*	Ectoparasites
	Pediculosis pubis, "crabs" infestation	*Phthirius pubis*	Ectoparasites
	Enteritis (homosexuals/proctitis)	*Giardia lamblia, Entamoeba histolytica, Shigella* spp., *Salmonella* spp., *Enterobius vermicularis, Campylobacter* spp., *Helicobacter* spp.	Bacteria, parasites
	Molluscum contagiosum	Poxlike virus	Viruses
	Heterophile-negative mononucleosis, congenital infections	Cytomegalovirus	Viruses

of genital warts (condylomata acuminata; types 6, 11, and others) and those associated with cervical carcinoma (predominantly types 16 and 18, but numerous others are also implicated), *Calymmatobacterium granulomatis,* and ectoparasites such as scabies and lice. Some of these agents are not routinely isolated from clinical specimens. Infections with more than one agent may occur and therefore dual or concurrent infections should always be considered.

Individuals' sexual habits and practices dictate potential sites of infection. Homosexual practices and increasingly common heterosexual practices of anal-genital or oral-genital intercourse allow for transmission of a genital tract infection to other body sites such as the pharynx or anorectal region. In addition, these practices have required that other gastrointestinal and systemic pathogens also be considered etiologic agents of STDs. The intestinal protozoa *Giardia lamblia, Entamoeba histolytica,* and *Cryptosporidium* spp. are significant causes of STDs, especially among homosexual populations. In the same group of patients, fecal pathogens, such as *Salmonella, Shigella, Campylobacter,* and *Microsporidium,* are often transmitted sexually. Oral-genital practices probably allow *N. meningitidis* to colonize and infect the genital tract. Viruses shed in secretions or present in blood (cytomegalovirus [CMV]; hepatitis B, possibly C and E; other non-A, non-B hepatitis viruses; human T-cell lymphotropic virus type I [HTLV-I]; and human immunodeficiency virus [HIV] of acquired immunodeficiency syndrome [AIDS]) are increasingly spread by sexual practices.

Certain infections that are sexually transmitted occur on the surface epithelium of or near the lower genital tract. The major pathogens of these types of infections include HSV, *Haemophilus ducreyi,* and *T. pallidum.*

OTHER ROUTES Organisms may also be introduced into the genital tract by instrumentation, presence of a foreign body, or irritation and can subsequently cause infection. Infections transmitted in this way are often caused by the same organisms that cause skin/wound infections. Of great significance, infection can also be transmitted from mother to infant either in vivo (within the living body) or during delivery. For example, transplacental infection may occur with syphilis, HIV, CMV, or HSV. Infection in the newborn can also be acquired during delivery by direct contact with an infectious lesion or discharge in the mother and a susceptible area in the infant (e.g., the eye). STDs, such as HSV, *C. trachomatis,* and *N. gonorrhoeae,* may be transmitted from mother to newborn in this manner. Other organisms, such as group B streptococci, *Escherichia coli,* and *Listeria monocytogenes,* that originate from the mother may also be transmitted to the infant either before, during, or after birth. (Infections in the fetus and newborn are discussed later in this chapter.)

Clinical manifestations

Clinical manifestations of lower genital tract infections are as varied and diverse as the etiologies.

ASYMPTOMATIC Although symptoms of genital tract infections generally cause the patient to seek medical attention, a patient with an STD may be free of symptoms (i.e., asymptomatic), especially a female. For example, gonorrhea or chlamydia infection in the male is usually obvious because of a urethral discharge, yet females with either or both of these infections may have either minimal symptoms or no symptoms at all. Also, the primary lesion of syphilis (chancre) can be unremarkable and go unnoticed by the patient. Therefore, the lack of symptoms does not guarantee the absence of disease. Unfortunately, these asymptomatic individuals can then serve as reservoirs for infection and unknowingly spread the pathogen to other individuals. Also, as in the case for asymptomatic infections in the female caused by *N. gonorrhoeae* or *C. trachomatis,* untreated infections can lead to serious sequelae such as pelvic inflammatory disease or infertility.

DYSURIA Although a frequent presenting symptom associated with urinary tract infection, dysuria can commonly result from an STD caused by organisms such as *N. gonorrhoeae, C. trachomatis,* and HSV.

URETHRAL DISCHARGE The presence of an inflammatory exudate at the tip of the urethral meatus is generally only observed in males; the symptoms of urethral infection in females are infrequently localized. Most males complain of discomfort at the penile tip, as well as dysuria. Urethritis may be gonococcal, caused by *N. gonorrhoeae,* or nongonococcal. Nongonococcal urethritis can be caused by *C. trachomatis; Trichomonas vaginalis* (less frequently); and genital mycoplasmas such as *Mycoplasma hominis, M. genitalium,* and *Ureaplasma urealyticum.*

LESIONS OF THE SKIN AND MUCOUS MEMBRANES Numerous organisms can cause genital lesions that are diverse in both their appearance and their associated symptoms (Figure 26-2). The agents and their features of infection are summarized in Table 26-2. Some of these infections, such as genital herpes or warts, are common, whereas others, such as lymphogranuloma venereum and granuloma inguinale, are uncommon in the United States.

VAGINITIS Inflammation of the vaginal mucosa, called **vaginitis,** is a common clinical syndrome whose incidence appears to be increasing.[15,24] Women who present with vaginal symptoms often complain of an abnormal discharge and possibly other symptoms such as an offensive odor or itching. *T. vaginalis* and *Candida albicans* are well recognized causes of vaginitis.

C. albicans causes about 80% to 90% of cases of vaginal candidiasis, other species of *Candida* account for the remaining cases. Yeast can be carried vaginally in small numbers and produce no symptoms. However, if conditions in the vagina change so as to give the yeast an advantage over other normal vaginal flora, candidiasis can result. Most patients complain of perivaginal itching, often with little or no discharge. Frequently, candidal discharge is classically thick and "cheesy" in appearance.

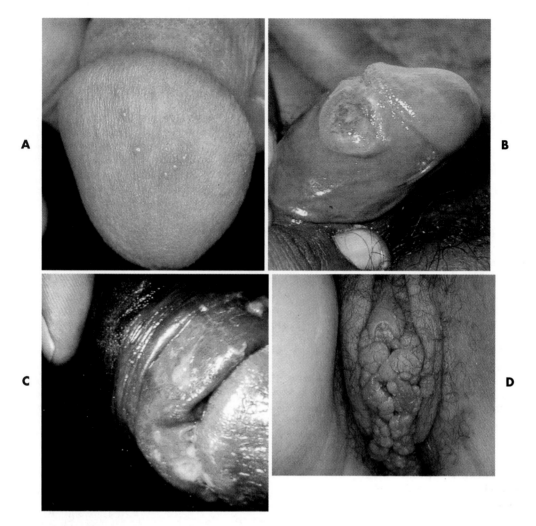

FIGURE 26-2 Genital lesions of the skin and mucous membranes that are sexually transmitted. **A,** Genital herpes showing vesicular lesions; **B,** Typical chancre of primary syphilis; **C,** Early chancroid lesion of the penis; and **D,** Condyloma acuminatum. (All photographs from Farrar W.E., Wood, M.J., Innes, J.A., and Hubbs, H. 1992. Infectious diseases text and color atlas, ed 2. Gower Medical Publishing, London.)

Vaginal infection with *T. vaginalis*, a protozoan parasite, produces a profuse, slightly offensive, yellow-green discharge; patients frequently complain of itching. About 25% of women carrying tricho-\monas are asymptomatic.

In addition to vaginitis caused by these two organisms, there is a third type of vaginitis referred to as **bacterial vaginosis** (BV). Initially, BV was believed to be caused by *Gardnerella vaginalis*, but *G. vaginalis* was isolated from 40% of women without vaginitis. Current understanding is that BV is polymicrobial in etiology, involving *G. vaginalis* and other facultative and anaerobic organisms.[17] Synergistic activity of various anaerobic organisms, including *Prevotella* spp., peptostreptococci, *Mobiluncus* spp. (curved, motile rods), and mycoplasmas, as well as *G. vaginalis*, seems to contribute to the pathology of BV.

BV is characterized by inflammation and perivaginal irritation that are considerably milder than trichomoniasis or candidiasis and is usually associated with a foul-smelling discharge that is often described as having a "fishy" odor. This odor is a result of products of bacterial metabolism (polyamines) being volatilized by vaginal fluids. Some patients also complain of abdominal discomfort. Because BV can recur in the absence of sexual reexposure and other settings (e.g., nonsexually active women, virgins), BV is not exclusively sexually transmitted. BV has been associated with serious complications such as premature labor and delivery.[11]

CERVICITIS Polymorphonuclear neutrophils (PMNs) are normally present in the endocervix; however, an abnormally increased number of PMNs may be

TABLE 26-2 SUMMARY OF COMMON CAUSES OF GENITAL LESIONS OF THE SKIN AND MUCOUS MEMBRANES

AGENT	DISEASE	LESION	MAJOR ASSOCIATED SYMPTOMS
Herpes simplex virus	Genital herpes	Papules, vesicles (blisters), pustules, and/or ulcers	Multiple lesions that are usually painful and tender, can recur (see Fig 26-2, *A*)
Treponema pallidum	Primary syphilis	Genital ulcer (chancre)	Usually a single lesion, painless; lesion has even edges, represents the first of three stages of syphilis (see Fig 26-2, *B*)
Haemophilus ducreyi	Chancroid	Papule that becomes pustular and ulcerates (chancroid); multiple ulcers may develop	Ulcer is deeply invasive, tender, painful, and purulent in appearance; edges of lesion are ragged (see Fig 26-2, *C*)
Chlamydia trachomatis serotype L1, L2, and L3	Lymphogranuloma venereum	Small ulcer or vesicle that heals spontaneously without leaving a scar	After lesion heals, painful, swollen lymph nodes (lymphadenopathy) develop 2-6 weeks later; fever and chills; severe lymphatic obstruction and lymphedema can develop
Calymmatobacterium granulomatis	Granuloma inguinale	Single or multiple subcutaneous nodules	Indolent and chronic course; nodules enlarge and erode through the skin, producing a deep red, sharply defined ulcer that is painless
Human papillomavirus (primary types 6 and 11)	Condylomata acuminata	Genital warts	Warts have a cauliflower-like appearance; usually multiple lesions that can be flat or elevated; usually asymptomatic apart from physical presence (see Figure 26-2, *D*)

associated with **cervicitis** (inflammation of the cervix). Therefore, a purulent discharge from the endocervix can be observed in some cases of cervicitis. The endocervix is the site from which *N. gonorrhoeae* is most frequently isolated in women with gonococcal infections. In patients presenting with cervicitis, *C. trachomatis* can also be isolated; chlamydia have not been associated with vaginitis. Frequently, patients are infected with both pathogens. Because most women with cervicitis caused by gonococci or chlamydia are asymptomatic and cervical abnormalities are either subtle or absent in these women, an appropriate culture must be performed.

HSV and human papillomavirus (HPV) can also infect the cervix. In women with herpes cervicitis, the cervix is friable (bleeds easily) and may have ulcers. Affected patients may also have lower abdominal pain.

ANORECTAL LESIONS As previously mentioned, because of homosexual practices and increasingly common heterosexual practices of anal-genital intercourse, other sites of infections in addition to those in the genital tract must be considered. The anorectum and pharynx are commonly infected with the classic STDs, including anal warts caused by HPV, as well as other viruses and parasites.[26] Patients with symptoms of proctitis (inflammation of the rectum) caused by *N. gonorrhoeae* or *C. trachomatis* complain of itching, mucopurulent anal discharge, anal pain, bleeding, and tenesmus (painful straining during a bowel movement). Anorectal infections caused by HSV are associated with severe anal pain, rectal discharge, tenesmus, and systemic signs and symptoms such as fever, chills, and headaches.

In HIV-infected individuals and other immunocompromised patients, these infections tend to last longer, be more severe, and are more difficult to treat compared with immunocompetent individuals. Anorectal lesions are common in HIV-infected patients and include anal condylomata, anal abscesses, and ulcers.[26] In this patient population, anal abscesses and ulcers can be due to various organisms, including CMV, *Mycobacterium avium* complex, HSV, *Campylobacter* spp., and *Shigella,* as well as traditional etiologic agents of STDs.

BARTHOLINITIS In adult women, the Bartholin's gland is a 1-cm mucus-producing gland on each side of the vaginal orifice. Each gland has a 2-cm duct that opens on the inner surface of the labia minora. If infected, this duct can become blocked and result in a Bartholin's gland abscess. Although *N. gonorrhoeae* and *C. trachomatis* can cause infection, anaerobic and polymicrobic infections originating from normal genital flora are more common.[3]

Infections of the reproductive organs and other upper tract infections

Besides the lower genital tract, infections can occur in the reproductive organs of both males and females.

FEMALES Infection of the female reproductive organs (i.e., uterus, fallopian tubes, ovaries, and even the abdominal cavity) can occur. The organisms are believed to be frequently acquired as they ascend from lower-tract sites of infection. Organisms may also be introduced to the reproductive organs by surgery, instrumentation, or during childbirth.

Pelvic Inflammatory Disease (PID). *Pelvic inflammatory disease (PID)* is an infection that results when cervical microorganisms travel upward to the endometrium, fallopian tubes, and other pelvic structures. This infection can produce one or more of the following inflammatory conditions: endometritis, salpingitis (inflammation of the salpinges), localized or generalized peritonitis, or abscesses involving the fallopian tubes or ovaries. Patients with PID often have intermittent abdominal pain and tenderness, vaginal discharge, dysuria, and possibly systemic symptoms such as fever, weight loss, and headache. Serious complications, such as permanent scarring of the fallopian tubes and infertility, can arise if PID is untreated.

Infection with *N. gonorrhoeae* and/or *C. trachomatis* in her lower genital tract can lead to PID if a woman is not adequately treated. Other organisms, such as anaerobes, gram-negative rods, streptococci, and mycoplasmas, may ascend through the cervix, particularly after parturition (childbirth), dilation of the cervix, or abortion. The presence of an intrauter-ine device (IUD) is associated with a slightly higher rate of PID. Such infections caused by *Actinomyces* have been associated with the use of IUDs.

Infections After Gynecologic Surgery. Following gynecologic surgery, such as a vaginal hysterectomy, women frequently develop postoperative infections that include pelvic cellulitis or abscesses. For the most part, these infections arise from the patient's own vaginal flora. Therefore, the major pathogens mirror the normal flora organisms: aerobic gram-positive cocci, gram-negative bacilli, anaerobes such as *Peptostreptococcus* spp., and genital mycoplasmas.

Infections Associated with Pregnancy. Infections can also occur during pregnancy (prenatal) or after birth (postpartum) in the mother. Of further significance, these infections may, in turn, be transmitted to the infant. Thus these infections not only can compromise the mother's health but also the health of the developing fetus or neonate.

While developing within the uterus, the fetus is protected from most environmental influences, including infectious agents. The human immune system does not become fully competent until several months after birth. Immunoglobulins that cross the placental barrier, primarily immunoglobulin G (IgG), serve to protect the newborn from many infections until the infant begins to produce immunoglobulins of his or her own in response to antigenic stimuli. This unique environmental niche, however, does expose the vulnerable fetus to pathogens present in the mother.

Prenatal infections (those that occur anytime before birth) may be acquired by the bloodborne or ascending routes from mother to infant. If the mother has a bloodstream infection, organisms can reach and cross the placenta, with possible spread of infection to the developing fetus. Organisms that can cross the placenta are listed in Table 26-3. Alternatively, organisms can also infect the fetus by the ascending route from the vagina through torn or ruptured fetal membranes. **Chorioamnionitis** is an infection of the uterus and its contents during pregnancy. This infection is commonly acquired by organisms ascending from the vagina or cervix after premature or prolonged rupture of the membranes or labor. Organisms that are commonly isolated from amniotic fluid are listed in Box 26-1.

MALES Infections in male reproductive organs can also occur and include epididymitis, prostatitis, and orchitis. **Epididymitis,** an inflammation of the epididymis, is commonly seen in sexually active men. Patients complain of fever and pain and swelling of the testicle. *N. gonorrhoeae* and *C. trachomatis* are

TABLE 26-3 COMMON ETIOLOGIC AGENTS OF PRENATAL AND NEONATAL INFECTIONS

TIME OF INFECTION*	ROUTE OF INFECTION	COMMON AGENTS
Prenatal	Transplacental	**Bacteria:** *Listeria monocytogenes, Treponema pallidum* **Viruses:** Cytomegalovirus (CMV), rubella, HIV, parvovirus B19, enteroviruses **Parasites:** *Toxoplasma gondii*
	Ascending	**Bacteria:** Group B streptococci, *Escherichia coli, L. monocytogenes, Chlamydia trachomatis,* genital mycoplasmas **Viruses:** CMV, Herpes simplex virus (HSV)
Natal	Passing through the birth canal	**Bacteria:** Group B streptococci, *E. coli, L. monocytogenes, N. gonorrhoeae, C. trachomatis* **Viruses:** CMV, HSV, enteroviruses, hepatitis B virus, HIV
Postnatal	All of the above routes or from the nursery environment	All agents listed above and in addition, various organisms from the nursery environment, including gram-negative bacteria and viruses such as respiratory syncytial virus

*Some newborns develop infections during the first 4 weeks of postnatal life. Infections may be delayed manifestations of earlier prenatal (before birth), natal, or postnatal (after birth) acquisition of pathogens.

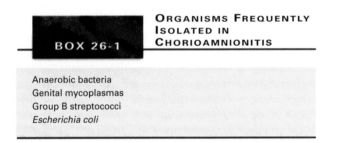

BOX 26-1 ORGANISMS FREQUENTLY ISOLATED IN CHORIOAMNIONITIS

Anaerobic bacteria
Genital mycoplasmas
Group B streptococci
Escherichia coli

common causes of epididymitis. However, enterics and coagulase-negative staphylococci can also cause infection in men over age 35 and in homosexual men; these infections are often associated with obstruction by the prostate gland.

Prostatitis is a term to clinically describe adult male patients who have perineal, lower back, or lower abdominal pain, urinary discomfort, or ejaculatory complaints. Prostatitis is caused by both infectious and noninfectious means. Bacteria can cause an acute or chronic prostatitis. Patients with acute bacterial prostatitis have dysuria and urinary frequency, symptoms that are associated with lower urinary tract infection. Frequently, these patients have systemic signs of illness such as fever. Chronic bacterial prostatitis is an important cause of persistent bacteriuria in the male that leads to recurrent bacterial urinary tract infections. The common causes of these infections are similar to the bacterial causes of lower urinary tract infections such as *Escherichia coli* and other enterics.

Finally, inflammation of the testicles, **orchitis,** is uncommon and generally acquired by bloodborne dissemination of viruses. Mumps is associated with most cases. Patients exhibit testicular pain and swelling following infection. Infections range from mild to severe.

LABORATORY DIAGNOSIS OF GENITAL TRACT INFECTIONS

LOWER GENITAL TRACT INFECTIONS

Urethritis, cervicitis, and vaginitis

SPECIMEN COLLECTION This discussion will focus only on those specimens submitted for culture and/or direct examination. Procedures for the collection and transport of specimens for detection of agents by other noncultural methods (e.g., detection of *Chlamydia trachomatis* by enzyme immunoassay) should be followed according to the respective manufacturer's instructions. Refer to Table 1-1 for a review of collection, transport, and processing of genital tract specimens.

Urethral. Urethral discharge may occur in both males and females infected with pathogens such as *Neisseria gonorrhoeae* and *Trichomonas vaginalis*. The presence of infection is more likely to be asymptomatic in females because the discharge is usually less profuse and may be masked by normal vaginal secretions. *Ureaplasma urealyticum* can also be isolated from male urethral discharge.

A urogenital swab designed expressly for collection of such specimens should be used. These swabs are made of cotton or rayon that has been treated with charcoal to adsorb material toxic to gonococci and wrapped tightly over one end of a thin wire shaft. Cotton- or rayon-tipped swabs on a thin wire may also be used to collect specimens for isolation of mycoplasmas and chlamydiae. Calcium alginate swabs are generally more toxic for HSV, gonococci, chlamydiae, and mycoplasmas than are treated cotton swabs. Because Dacron swabs are least toxic, they are recommended for viral specimens. Dacron-tipped swabs on plastic shafts are also acceptable for chlamydiae and genital mycoplasmas.

To obtain a urethral specimen, a swab is inserted approximately 2 cm into the urethra and rotated gently before withdrawing. Because chlamydiae are intracellular pathogens, it is important to remove epithelial cells (with the swab) from the urethral mucosa. Separate swabs for cultivation of gonococci, chlamydiae, and ureaplasma are required. When profuse urethral discharge is present, particularly in males, the discharge may be collected externally without inserting a sampling device into the urethra. However, a urethral swab for chlamydia must be collected on males. A few drops of first-voided urine has also been used successfully to detect gonococci in males.[16]

Because *T. vaginalis* may be present in urethral discharge, material for culture should be collected by swab as just described and another specimen collected on a swab and placed into a tube containing 0.5 mL of sterile physiologic saline. This specimen should be hand delivered to the laboratory immediately. Direct wet mounts and cultures for *T. vaginalis* can be performed from this second specimen. Commercial media for culture of *Trichomonas* are available. The first few drops of voided urine may also be a suitable specimen for recovery of *Trichomonas* from infected males, if it is inoculated into culture media immediately. Alternatively, material may be smeared onto a slide for later performance of a fluorescent antibody stain. Plastic envelopes for examination and culture have recently been introduced.[1,6]

Cervical/Vaginal. Organisms that cause purulent vaginal discharge (vaginitis) include *T. vaginalis,* gono-

cocci, and rarely, beta-hemolytic streptococci. The same organisms that cause purulent infections in the urethra may also infect the epithelial cells in the cervical opening (os), as can HSV. Mucus is removed by gently rubbing the area with a cotton ball. The urethral swab just described is inserted into the cervical canal and rotated and moved from side to side for 30 seconds before removal.

Swabs are handled as described for urethral swabs for isolation of *Trichomonas* and gonococci. Chlamydiae cause a mucopurulent cervicitis with discharge. Endocervical specimens are obtained after the cervix has been exposed with a speculum, which allows visualization of vaginal and cervical architecture, and after ectocervical mucus has been adequately removed. The speculum is moistened with warm water, because many lubricants contain antibacterial agents. Because normal vaginal secretions contain great quantities of bacteria, care must be taken to avoid or minimize contaminating swabs for culture by contact with these secretions. A small, nylon-bristled cytology brush, or cytobrush, may be used to ensure that cellular material is collected, but its use is associated with discomfort and bleeding. Some controversy exists over whether the cytobrush results in better specimens, at least for detection of *Chlamydia trachomatis*.[14]

In addition to cervical specimens, which are particularly useful for isolating herpes, gonococci, mycoplasmas, and chlamydiae, vaginal discharge specimens may be collected. Organisms likely to cause vaginal discharge include *Trichomonas,* yeast, and the agents of BV. Swabs for diagnosis of BV are dipped into the fluid that collects in the posterior fornix of the vagina.

Genital tract infections caused by sexually transmitted agents in children (preadolescents) are most often the result of sexual abuse. Because of medicolegal implications, the laboratory should treat specimens from such patients with extreme care, carefully identifying and documenting all isolates. Cultures should always be obtained, especially for *Chlamydia trachomatis*, because antigen detection methods are less sensitive in children than in adults and specificity is not 100%.[13]

Because it is impossible to exclude contamination with vaginal flora, obtaining swabs of Bartholin gland exudate are not recommended. Infected Bartholin glands should be aspirated with needle and syringe after careful skin preparation, and cultures should be evaluated for anaerobes and aerobes.

Transport. Swabs collected for isolation of gonococci may be transported to the laboratory in modified Stuart's or Amie's charcoal transport media and held at room temperature until inoculated to culture media. Good recovery of gonococci is possible if swabs are cultured within 12 hours of collection. Material that must be held longer than 12 hours should be inoculated directly to one of the commercial systems designed for recovery of gonococci, described later in this chapter.

Swabs for isolation of chlamydiae and mycoplasmas are best transported in specific transport media that have antibiotics and other essential components. Specimens for chlamydiae culture should be transported on ice. (Specimens transported at room temperature should be inoculated within 15 minutes of collection.) Specimens can be stored at 4° C for up to 24 hours. If culture inoculation will be delayed more than 24 hours, specimens should be quick-frozen in a dry ice and 95% ethanol bath and stored at −70° C until cultured. If collected and transported in specific transport media, specimens for genital mycoplasma culture may be transported on ice or at room temperature. If not in genital mycoplasma transport media, specimens should be transported on ice to suppress the growth of contaminating flora.

DIRECT MICROSCOPIC EXAMINATION   In addition to culture, urethral discharge may be examined by Gram stain for the presence of gram-negative intracellular diplococci (Figure 26-3), usually indicative of gonorrhea in males. After inoculation to culture media, the swab is rolled over the surface of a glass slide, covering an area of at least 1 cm². If the Gram stain is characteristic, cultures of urethral discharge need not be performed. Urethral smears from females may also

be examined. However, presumptive diagnosis of gonorrhea from these smears is reliable only if the microscopist is experienced, because normal vaginal flora, such as *Veillonella* or occasional gram-negative coccobacilli, may resemble gonococci. If extracellular organisms resembling *Neisseria gonorrhoeae* are seen, the microscopist should continue to examine the smear for intracellular diplococci. Presumptive diagnosis can be useful when decisions are to be made regarding immediate therapy, but confirmatory cultures or an alternative nonculture method should always be performed on specimens from females. Some strains of *N. gonorrhoeae* are sensitive to the amount of vancomycin present in selective media. If suspicious organisms seen on smear fail to grow in culture, reculture on chocolate agar without antibiotics may be warranted.

Fluorescein-conjugated monoclonal antibody reagents are sensitive and specific for visualization of the inclusions of *Chlamydia trachomatis* in cell cultures or elementary bodies in urethral and cervical specimens containing cells. Reagents for direct staining of specimens are available commercially in complete collection and test systems, but the relatively greater technologist time required for this method limits its usefulness for laboratories that receive many specimens, except as a confirmatory test for other antigen detection systems with borderline results. In some studies, the sensitivity of visual detection of chlamydiae with these newer reagents has been similar to that of culture, although such comparative results are obtained only by technologists experienced in fluorescent techniques and when the slides are

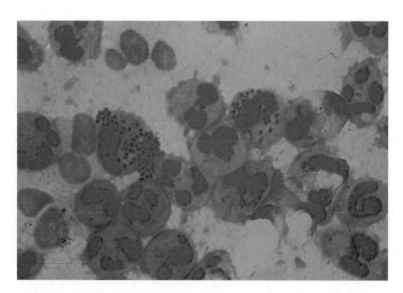

FIGURE 26-3 Gram-negative intracellular diplococci; diagnostic for gonorrhea in urethral discharge and presumptive for gonorrhea in vaginal discharge.

examined thoroughly.[23] False-positive results should not occur if at least 10 morphologically compatible fluorescing elementary bodies are seen on the entire smear. No direct visual methods exist for detection of mycoplasmas at this time, but nucleic acid probes have been evaluated.

Direct microscopic examination of a wet preparation of vaginal discharge provides the simplest rapid diagnostic test for *Trichomonas vaginalis* when such a specimen is available and can be examined immediately. The plastic envelope method combines direct visualization with culture.[6] Motile trophozoites of *Trichomonas* can be visualized in a routine wet preparation performed by a proficient technologist in two thirds of cases or a direct fluorescent antibody (DFA) stain (Meridian Diagnostics, Cincinnati, Ohio) may be used.

Budding cells and pseudohyphae of yeast can also be easily identified in wet preparations by adding 10% potassium hydroxide (KOH) to a separate preparation, thereby dissolving host cell protein and enhancing the visibility of fungal elements.[8]

Bacterial vaginosis, characterized by a foul-smelling discharge, can be diagnosed microscopically or clinically.[8,11,18] The discharge is primarily sloughed epithelial cells, many of which are completely covered by tiny, gram-variable rods and coccobacilli. These cells are called **clue cells** (Figure 26-4). The absence of inflammatory cells in the vaginal discharge is another sign of BV. Although *Gardnerella vaginalis* has been historically associated with the syndrome and can be cultured on a human blood bilayer plate, culture is not recommended for diagnosis of BV.[22] A clinical diagnosis of BV is best made using three or more of the following criteria: homogeneous, gray discharge;

clue cells seen on wet mount or Gram stain; an amine or fishy odor elicited by the addition of a drop of 10% KOH to the discharge on a slide or on the speculum; and a pH greater than 4.5.[8]

Probably the best way to differentiate BV from other vaginal infections is by Gram stain (Figure 26-5). A grading system for Gram stains of vaginal discharge has been developed by Nugent et al. (Procedure 26-1).[18] This system is based on the presence or absence of certain bacterial morphologies. Typically, in patients with BV, lactobacilli are either absent or few in number while curved, gram-variable rods (*Mobiluncus* spp.)[19] and/or *G. vaginalis* and *Bacteroides* morphotypes predominate. The Gram stain is more sensitive and specific than either the wet mount for detection of clue cells or culture for *G. vaginalis*, and the smear can be saved and reexamined later.[8]

CULTURE   Samples for isolation of gonococci may be inoculated directly to culture media, obviating the need for transport medium. Commercially produced systems have been developed for this purpose, and many clinicians inoculate standard plates directly if convenient access to an incubator is available. Modified Thayer-Martin medium is most often used, although New York City (NYC) medium has the added advantage of supporting the growth of mycoplasmas and gonococci. Excellent recovery of gonococci is the rule when specimens are inoculated directly to any of these media in self-contained incubation systems such as JEMBEC plates (Figure 26-6). The specimen swab containing material is rolled across the agar with constant turning to expose all

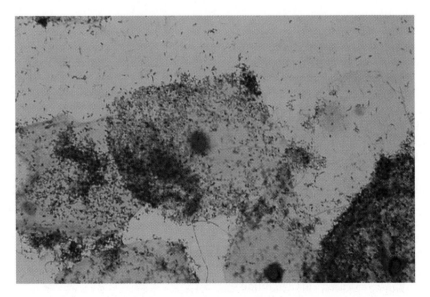

FIGURE 26-4 Clue cells in vaginal discharge suggestive of bacterial vaginosis (BV).

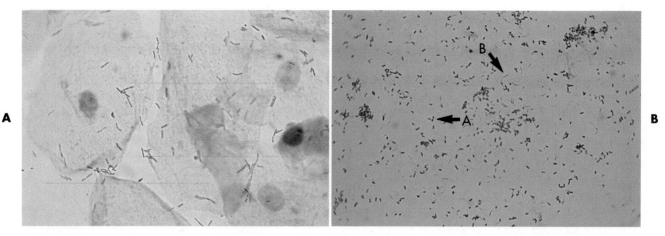

FIGURE 26-5 **A,** Predominance of lactobacilli in Gram stain from healthy vagina. **B,** Absence of lactobacilli and presence of *Gardnerella vaginalis* (*arrow A*) and *Mobiluncus* spp. (*arrow B*) morphologies.

FIGURE 26-6 JEMBEC plate containing modified Thayer-Martin medium in a plastic, snap-top box with a self-contained CO_2-generating tablet, all sealed inside a Zip-lock plastic envelope after inoculation.

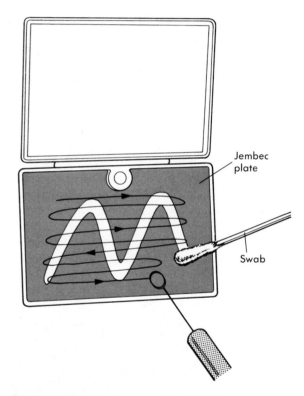

FIGURE 26-7 Method of cross-streaking JEMBEC plate after original specimen has been inoculated by rolling the swab over the surface of the agar in a W pattern.

surfaces to the medium. The JEMBEC plate, which generates its own increased carbon dioxide atmosphere by means of a sodium bicarbonate tablet, is inoculated in a **W** pattern. The plate may be cross-streaked with a sterile loop in the laboratory (Figure 26-7).

Specimens must be inoculated to additional media for isolation of yeast, streptococci, and mycoplasmas. Yeast grow well on Columbia agar base with 5% sheep blood and colistin and nalidixic acid (CNA), although more selective media are available. Most yeast and streptococci also grow on standard blood agar; thus, adding special fungal media such as Sabouraud brain-heart infusion agar (SABHI) is unwarranted.

Culture of the vaginal canal at 35 to 37 weeks' gestation can reliably predict the presence of group B streptococci at delivery.[4] For culture, group B streptococci are detected best by inoculating a swab obtained from the perianal area onto CNA agar and into Todd-Hewitt enrichment broth with blood, gentamicin, and nalidixic acid for subculture to agar the next day.

T. vaginalis may be cultured in Diamond's medium (available commercially) or plastic envelopes inoculated with discharge material. Culture techniques

are most sensitive.[1,6] A commercially available biphasic genital mycoplasma culture system (Mycotrim-GU, Irvine Scientific, Santa Ana, Calif.) can be used to culture *Mycoplasma hominis* and *Ureaplasma urealyticum,* although commercially prepared media are not as sensitive as fresh media.[27] *Mycoplasma genitalium* may not grow on commercial media because of the presence of thallium acetate.[25]

NONCULTURE METHODS Various nonculture methods may be used to diagnose genital tract diseases, includ-

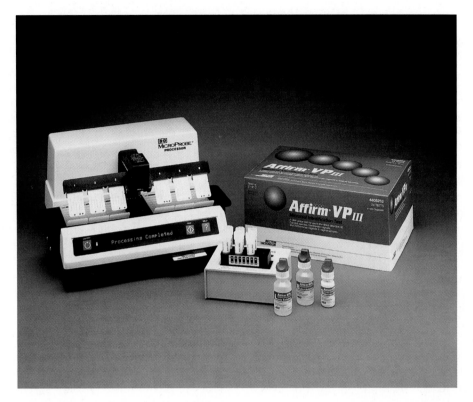

FIGURE 26-8 Affirm VP III Microbial Identification Test used to differentiate the three major causes of vaginitis/bacterial vaginosis from a single sample within 1 hour. (Courtesy Becton Dickinson Microbiology Systems. Affirm is a trademark of Becton Dickinson and Company.)

ing serology, latex agglutination, nucleic acid hybridization and amplification assays, and enzyme immunoassays. Most assays detect a single or possibly two genital tract pathogens, and the majority are commercially available. These methods are described in more detail in the chapter relating to individual pathogens in Parts 4 through 7.

As previously discussed, BV involves several organisms. Besides the Gram stain, BV can be diagnosed by an increased succinic/lactic acid ratio in vaginal secretions, as detected by direct gas liquid chromatography. The Gram stain, however, is equally sensitive and more readily available. Of note, a hybridization assay (Affirm VP III Microbial Identification Test; Becton Dickinson Microbiology Systems, Sparks, Md.) is commercially available to diagnose BV, as well as genital tract infections caused by *Candida* spp. and *Trichomonas vaginalis*. Once the appropriate reagents and specimen are added to special trays, the entire hybridization assays are then performed by instrumentation (Figure 26-8). Initial evaluations indicate this system is sensitive and specific.[2,10]

Genital skin and mucous membrane lesions
External genital lesions are usually either vesicular or ulcerative. Causes of lesions can be determined by physical examination, histologic/cytologic examina-

tion, or microscopic examination and/or culture of exudate. *Because any genital lesion may be highly contagious, clinicians should wear gloves when carrying out all manipulations of lesion material.*

Vesicles in the genital area are almost always attributable to viruses, and herpes simplex is the most common cause. Epithelial cells from the base of a vesicle may be spread onto the surface of a slide and examined for the typical multinucleated giant cells of HSV or stained by immunofluorescent antibody stains for viral antigens. Additionally or alternatively, the material may be transported for culture of the virus, as outlined in Procedure 26-2.

Several commercial fluorescein-conjugated monoclonal and polyclonal antibodies directed against herpetic antigens of either type 1 or 2 are available. When fluorescent-antibody–stained lesion material containing enough cells is viewed under ultraviolet light, the diagnosis can be made in 70% to 90% of patients. Laboratories that routinely process genital material for herpes should be using immunofluorescent staining reagents when a rapid answer is desired; otherwise, culture, which is generally positive in 2 days, is the method of choice. Nonfluorescent markers, such as biotin-avidin-horseradish peroxidase or alkaline phosphatase, have also been conjugated to these specific antibodies, often allowing earlier detection of herpes-infected cells in tissue culture monolayers.

Such reagents have been developed for use directly on clinical material, although their sensitivity is not great enough to forego culture if a definitive diagnosis is necessary.

Material from lesions suggestive of syphilis should be examined by dark-field or fluorescent microscopy. These procedures are described in Chapter 63.

All lesions suspected of infectious etiology may be Gram stained in addition to the procedures described. The smear of lesion material from a patient with chancroid may show many small, pleomorphic, gram-negative rods and coccobacilli arranged in chains and groups, characteristic of *H. ducreyi*. However, culture has been shown to be more sensitive for diagnosis of this agent. Material collected on cotton or Dacron swabs may be transported in modified Stuart's medium. Specimens should be inoculated to culture media within 1 hour of collection. A special agar, consisting of chocolate agar enriched with 1% IsoVitaleX (BBL Microbiology Systems) and vancomycin (3 mg/mL), has yielded good isolation if cultures are incubated in 5% to 7% carbon dioxide in a moist atmosphere, such as a candle jar. *H. ducreyi* grows best at 33° C.[5,21]

Granuloma inguinale is diagnosed by staining a crushed preparation of a small piece of biopsy tissue obtained from the edge of the base of the ulcer with Wright's or Giemsa stain and finding characteristic Donovan bodies (bipolar staining rods intracellularly within macrophages). Cytologists or pathologists usually examine such specimens rather than microbiologists. No acceptable media for isolation of *C. granulomatis* are available.

Bubo

Buboes, swollen lymph glands that occur in the inguinal (pelvic) region, are often evidence of a genital tract infection. Buboes are common in patients with primary syphilis, genital herpes, lymphogranuloma venereum, and chancroid. Patients with AIDS may show generalized lymphadenopathy. Other diseases that are not sexually transmitted, such as plague, tularemia, and lymphoma, can also produce buboes. Material from buboes may be aspirated for microscopic examination and culture.

INFECTIONS OF THE REPRODUCTIVE ORGANS

Pelvic inflammatory disease

Pelvic inflammatory disease is often caused by the same organisms that cause cervicitis or by organisms that make up the normal flora of the vaginal mucosa. Because of the profuse normal flora of the vaginal tract, specimens must be collected in such a way as to prevent vaginal flora contamination. Aspirated material collected by needle and syringe represents the best specimen. If this cannot be obtained at the time of surgery or laparoscopy, collection of intrauterine contents using a protected suction curetting device or double-lumen sampling device inserted through the cervix is also acceptable. Culdocentesis (aspiration of fluid in the cul-de-sac), after decontamination of the vagina by povidone-iodine, is satisfactory but rarely practiced today.

Aspirated material should be placed into an anaerobic transport container. The presence of either mixed anaerobic flora, gonococci, or both can be rapidly detected from a Gram stain. Direct examination with fluorescent monoclonal antibody stain may also detect chlamydiae. All such specimens should be inoculated to media that allow the recovery of anaerobic, facultative, and aerobic bacteria, gonococci, fungi, mycoplasmas, and chlamydiae. All material collected from normally sterile body sites in the genital tract should be inoculated to chocolate agar and placed into a suitable broth, such as chopped meat medium or thioglycollate, in addition to the other types of media noted. If only specimens obtained on routine swabs inserted through the cervix are available, cultures should be performed only for detection of gonococci and chlamydiae.

Miscellaneous infections

Infections of the male prostate, epididymis, and testes are usually bacterial. In younger men, chlamydiae predominate as the cause of epididymitis and possibly of prostatitis. Urine or discharge collected via the urethra is the specimen of choice unless an abscess is drained surgically or by needle and syringe. The first few milliliters of voided urine may be collected before and after prostatic massage to try to pinpoint the anatomic site of the infection. Cultures are inoculated to support the growth of anaerobic, facultative, and aerobic bacteria, as well as gonococci.

Infections of neonates and human products of conception

Suspected infections acquired by the fetus as a result of a maternal infection that crosses the placenta (congenital infection) can be diagnosed culturally or serologically in the newborn. Because maternal IgG crosses the placenta, serologic tests are often difficult to interpret (see Chapter 16). For culturable agents, the most definitive diagnoses involve recovery of the pathogen in culture. HSV, varicella-zoster virus (VZV), enteroviruses, and cytomegalovirus (CMV) can be cultured easily, as can most bacterial agents. Rubella and parvovirus B19 are more difficult to culture. Nasal and urine specimens offer the greatest yield for viral isolation, whereas blood, cerebrospinal fluid, and material from a lesion can also be productive. Systemic neonatal herpes without lesions may be difficult to diagnose unless tissue biopsy material is examined, because the viruses may not be present in cerebrospinal fluid or blood. Bacteria and fungi can be isolated from lesions, blood, and other normally sterile sites.

Determining the presence of fetal immunoglobulin M (IgM) directed against the agent in question establishes the serologic diagnosis of congenital infection. Until recently, ultracentrifugation was required for separation of IgM from IgG, the only definitive means of preventing false-positive results caused by maternal IgG or fetal rheumatoid factor. Ion-exchange chromatography columns, antihuman IgG, and bacterial proteins that bind to IgG specifically are now commercially available for removing cross-reactive IgG and rheumatoid factor to obtain more homogeneous IgM for differentiation of fetal antibody. Indirect fluorescent antibody and enzyme-linked immunosorbent assay (ELISA) test systems are commercially available to detect IgM against *T. gondii*, rubella, CMV, HSV, and VZV. Interference by rheumatoid factor is still a consideration in most commercial IgM test systems (see Chapter 16). Our ability to detect viral inclusions in tissue, conjunctival scrapings, and vesicular lesions, traditionally performed with Giemsa stain, has been improved as a result of monoclonal and polyclonal fluorescent antibody reagents, which are described in the chapters that discuss individual agents.

Infections that infants can acquire as they pass through an infected birth canal or are related to difficult labor, premature birth, premature rupture of the membranes, or other events include:

- HSV and CMV infections
- Gonorrhea
- Group B streptococcal sepsis
- Chlamydial conjunctivitis and pneumonia
- *Escherichia coli* or other neonatal meningitis

In the laboratory these infections are diagnosed by direct detection or culturing for the agents when possible, or by performing serologic tests. The appropriate specimens (e.g., cerebrospinal fluid, serum, pus, tracheal aspirate) should be examined and inoculated immediately. Routine body surface cultures of infants in intensive care have not been shown to be helpful for predicting subsequent disease.[9]

Finally, certain infectious agents are known to cause fetal infection and even abortion. For example, *Listeria monocytogenes*, although usually causing only mild flulike symptoms in the mother, can cause extensive disease and abortion of the fetus if infection occurs late in the pregnancy. Therefore, isolation of the organism from the placenta and from tissues of the fetus is important.

rocedures

26-1 SCORING VAGINAL GRAM STAINS FOR BACTERIAL VAGINOSIS

ORGANISM MORPHOTYPE	NUMBER/OIL IMMERSION FIELD	SCORE	ORGANISM MORPHOTYPE	NUMBER/OIL IMMERSION FIELD	SCORE
Lactobacillus-like (parallel-sided, gram-positive rods)	>30	0	*Gardnerella/Bacteroides*-like (tiny, gram-variable coccobacilli and rounded, pleomorphic, gram-negative rods with vacuoles)	>30	4
	5-30	1		5-30	3
	1-4	2		1-4	2
	<1	3		<1	1
	0	4		0	0
Mobiluncus-like (curved, gram-negative rods)	>5	2			
	<1-4	1			
	0	0			

1. Roll the swab of vaginal discharge over the surface of a slide.

2. Allow the smear to air dry, fix with methanol, and Gram stain.

3. Assign scores; refer to box above.

4. Add up total score and interpret as follows:

Score	Interpretation
0-3	Normal
4-6	Intermediate, repeat test later
7-10	Bacterial vaginosis

From Nugent, R.P., Krohn, M.A., and Hillier, S.L. 1991. J. Clin. Microbiol. 29:297.

26-2 COLLECTION OF MATERIAL FROM SUSPECTED HERPETIC LESIONS

PRINCIPLE

Herpesvirus is best recovered from the base of active lesions in the vesicular stage. The older the lesion, the less likely it will yield viable virus.

METHOD

1. Open the vesicles with a small-gauge needle or Dacron-tipped swab.

2. Rub the base of the lesion vigorously with a small cotton-tipped or Dacron-tipped swab to recover infected cells.

3. Place the swab into viral transport medium (see Chapter 66) and refrigerate until inoculated to culture media. Specimens in media may be stored at $-70°$ C for extended periods without loss of viral yield.

4. If large vesicles are present, material for culture may be aspirated directly by needle and syringe.

5. Material from another lesion can be applied directly to a glass slide for a Tzanck preparation (cytology) with Wright-Giemsa stain for detection of multinucleated giant cells, or for fluorescent antibody stain for detection of viral antigens.

References

1. Beal, C., et al. 1992. The plastic envelope method: a simplified technique for culture diagnosis of trichomoniasis. J. Clin. Microbiol. 30:2265.

2. Briselden, A.M. and Hillier, S.L. 1994. Evaluation of Affirm VP microbial identification tests for *Gardnerella vaginalis* and *Trichomonas vaginalis*. J. Clin. Microbiol. 32:148.

3. Brooks, I. 1989. Aerobic and anaerobic microbiology of Bartholin's abscess. Surg. Gynecol. Obstet. 169:32.

4. Centers for Disease Control and Prevention. 1994. Federal Register 59:64764.

5. Clarridge, J.E., Shawar, R., and Simon, B. 1990. *Haemophilus ducreyi* and chancroid: practical aspects for the clinical microbiology laboratory. Clin. Microbiol. Newsletter 12:137.

6. Draper, D., et al. 1993. Detection of *Trichomonas vaginalis* in pregnant women with the In Pouch TV culture system. J. Clin. Microbiol. 31:1016.

7. Eschenbach, D., Pollock, H.M., and Schachter, J. 1983. Laboratory diagnosis of female genital tract infections. In Rubin, S.J., coordinating editor. 1983. Cumitech 17. American Society for Microbiology, Washington, D.C.

8. Eschenbach, D.A., et al. 1988. Diagnosis and clinical manifestations of bacterial vaginosis. Am. J. Obstet. Gynecol. 158:819.

9. Evans, M.E., et al. 1988. Sensitivity, specificity, and predictive value of body surface cultures in a neonatal intensive care unit. J.A.M.A. 259:248.

10. Ferris, D.G., et al. 1995. Office laboratory diagnosis of vaginitis–clinician-performed tests compared with rapid nucleic acid hybridization test. J. Fam. Pract. 41:575.

11. Hillier, S.L., et al. 1995. The role of bacterial vaginosis and vaginal bacteria in amniotic fluid infection in women in preterm labor with intact fetal membranes. Clin. Infect. Dis. 2(suppl. 2):S276.

12. Hillier, S.L., et al. 1993. Normal vaginal flora, H_2O_2-producing lactobacilli and bacterial vaginosis in pregnant women. Clin. Infect. Dis. 16(suppl. 4):S273.

13. Janda, W.M. 1991. Sexually transmitted diseases in children: the role of the microbiology laboratory. Clin. Microbiol. Newsletter 13:9.

14. Kellogg, J.A., et al. 1992. Comparison of cytobrushes with swabs for recovery of endocervical cells and for Chlamydiazyme detection of *Chlamydia trachomatis*. J. Clin. Microbiol. 30:2988.

15. Kent, H.L. 1991. Epidemiology of vaginitis. Am. J. Obstet. Gynecol. 165:1168.

16. Luciano, A.A. and Grubin, M. 1980. Gonorrhea screening: comparison of three techniques. J.A.M.A. 243:680.

17. Majeroni, B.A. 1991. New concepts in bacterial vaginosis. Am. Fam. Pract. 44:1215.

18. Nugent, R.P., Krohn, M.A., and Hillier, S.L. 1991. Reliability of diagnosing bacterial vaginosis is improved by a standardized method of Gram stain interpretation. J. Clin. Microbiol. 29:297.

19. Roberts, M.C., Hillier, S.L., Schoenknecht, F.D., and Holmes, K.K. 1985. Comparison of Gram stain, DNA probe, and culture for the identification of species of *Mobiluncus* in female genital specimens. J. Infect. Dis. 152:74.

20. Romanowski, B. and Harris, J.R.W. 1984. Sexually transmitted diseases. Clin. Symp. 36:1.

21. Schmid, G.P., et al. Enhanced recovery of *Haemophilus ducreyi* from clinical specimens by incubation at 33° versus 35° C. J. Clin. Microbiol. 33:3257.

22. Schreckenberger, P.C. Diagnosis of bacterial vaginosis by gram-stained smears. Clin. Microbiol. Newsletter 14:126.

23. Schwebke, J.R., Stamm, W.E., and Handsfield, H.H. 1990. Use of sequential enzyme immunoassay and direct fluorescent antibody tests for detection of *Chlamydia trachomatis* infections in women. J. Clin. Microbiol. 28:2473.

24. Sobel, J.D. 1993. Candidal vulvovaginitis. Clin. Obstet. Gynecol. 36:153.

25. Tully, J.G., et al. 1983. *Mycoplasma genitalium*: a new species from the human urogenital tract. Int. J. Syst. Bacteriol. 33:387.

26. Weiss, E.G. and Wexner, S.D. 1995. Surgery for anal lesions in HIV-infected patients. Ann. Med. 27:467.

27. Wood, J.C., et al. 1985. Evaluation of Mycotrim-GU for isolation of *Mycoplasma* species and *Ureaplasma urealyticum*. J. Clin. Microbiol. 22:789.

Bibliography

Baron, E.J., et al. 1993. Laboratory diagnosis of female genital tract infections. In Baron, E.J., coordinating editor. Cumitech 17A. American Society for Microbiology, Washington, D.C.

Kreiger, J.N. 1995. Prostatitis, epididymitis, and orchitis. In Mandell, G., Douglas, R.G., Jr., and Bennett, J.E., editors. Principles and practice of infectious disease, ed 4. Churchill Livingstone, New York.

McCormack, W.M. 1994. Pelvic inflammatory disease. N. Engl. J. Med. 330:115.

Rein, M.F. 1995. Vulvovaginitis and cervicitis. In Mandell, G., Douglas, R.G., Jr., and Bennett, J.E., editors. Principles and practice of infectious disease, ed 4. Churchill Livingstone, New York.

27 GASTROINTESTINAL TRACT INFECTIONS

GENERAL CONSIDERATIONS

ANATOMY

We are all connected to the external environment through our gastrointestinal (GI) tract (Figure 27-1). What we swallow enters the GI tract and passes through the esophagus into the stomach, through the small and large intestines, and finally to the anus. During passage, fluids and other components are added to this material as secretory products of individual cells and as enzymatic secretions of glands and organs, and removed from this material by absorption through the gut epithelium.

The major components of the tract are listed in Box 27-1. The nature of the epithelial cells lining the GI tract varies with each portion. The lining of the GI tract is called the **mucosa**. Because of the differing nature of the mucosal surfaces of various segments of the bowel, specific infectious disease processes tend to occur in each segment.

The wall of the small intestine has folds that have millions of tiny, hairlike projections called **villi.** Each villus contains an arteriole, venule, and lymph vessel (Figure 27-2). The function of villi is to absorb fluids and nutrients from the intestinal contents. Epithelial cells lining the surface of villi have a surface that resembles a fine brush and is referred to as a **brush border**. The brush border is formed by nearly 2000 microvilli per epithelial cell. Intestinal digestive enzymes are produced in brush-border cells toward the top of the villi. Villi and microvilli help make the small intestine the primary site of digestion and absorption by significantly increasing the surface area; more than 90% of physiologic net fluid absorption occurs here. Mucus-secreting goblet cells are found in large numbers of villi and intestinal crypts.

Similar to the small intestine, the large intestine is composed of several segments (see Box 27-1). The wall of the large intestine consists of columnar epithelial cells, many of which are mucus-producing goblet cells. In contrast to the small intestine, there are no villous projections into the lumen. The remaining excess fluid within the GI tract is reabsorbed by the cells lining the large intestine before waste is finally discharged through the rectum.

In addition to the previously discussed components of the GI tract, numerous other organs and structures are either located in the main digestive organs or open into them. These accessory organs and structures include the salivary glands, tongue, teeth, liver, gallbladder, and pancreas. Except for the teeth and salivary glands, these organs are illustrated in Figure 27-1.

RESIDENT MICROBIAL FLORA

The GI tract contains vast, diverse normal flora. Although the acidity of the stomach prevents any significant colonization in a normal host under most circumstances, many species can survive passage through the stomach to become resident within the lower intestinal tract. Normally, the upper small intestine contains only sparse flora (bacteria, primarily streptococci; lactobacilli; and yeasts; 10^1 to 10^3/mL), but in the distal ileum, counts are about 10^6 to 10^7/mL, with *Enterobacteriaceae* and *Bacteroides* spp. predominately present.

Babies usually are colonized by normal human epithelial flora, such as staphylococci, *Corynebacterium* spp., and other gram-positive organisms (bifidobacteria, clostridia, lactobacilli, streptococci), within a few hours of birth. Over time, the content of the intestinal flora changes. The normal flora of the adult large bowel (colon) is established relatively early in life and consists predominantly of anaerobic species, including *Bacteroides, Clostridium, Peptostreptococcus, Bifidobacterium,* and *Eubacterium.*

Aerobes, including *Escherichia coli,* other *Enterobacteriaceae,* enterococci, and streptococci are outnumbered by anaerobes 1000:1. The number of bacteria per gram of stool within the bowel lumen increases steadily as material approaches the sigmoid colon (the last segment). Eighty percent of the dry weight of feces from a healthy human consists of bacteria, which can be present in numbers as high as 10^{11} to 10^{12} colony-forming units (CFU)/g of stool.

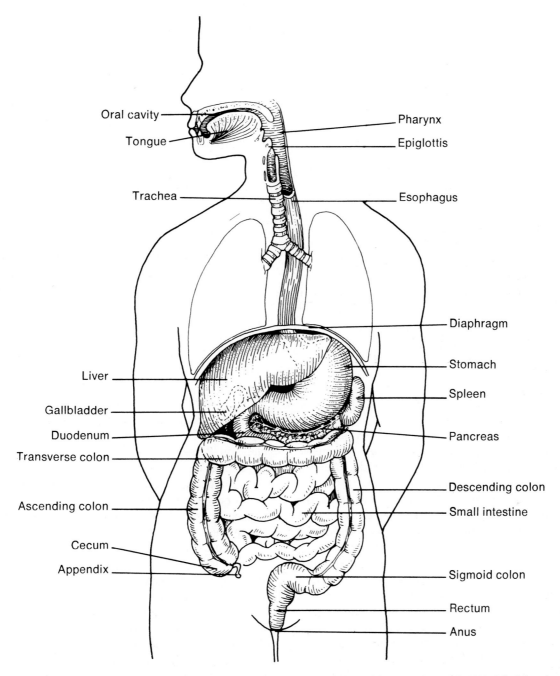

FIGURE 27-1 General anatomy of the gastrointestinal tract. (From Broadwell, D.C. and Jackson, B.S. 1982. Principles of ostomy care. Mosby, St. Louis.)

GASTROENTERITIS

SCOPE OF THE PROBLEM

Worldwide, diarrheal diseases are the second leading cause of death; about 25 million enteric infections occur each year. These infections cause significant morbidity and death, particularly in elderly people and children younger than age 5. It has been estimated that 4 to 6 million children die each year from diarrheal diseases, particularly in developing countries in Asia and Africa. Even in developed countries, sig-

nificant morbidity occurs as a result of diarrheal illness. Although acute diarrheal syndromes are usually self-limited, some persons with infectious diarrhea will require diagnostic studies and treatment.

PATHOGENESIS

Similar to the pathogenesis of urinary tract infections, the host and the invading microorganism possess key features that determine whether an enteric pathogen is able to cause microbial diarrhea.

BOX 27-1   COMPONENTS OF THE GASTROINTESTINAL TRACT

Mouth
Oropharynx
Esophagus
Stomach
  ■ Fundus—enlarged portion of the stomach to the left and above the opening of the esophagus into the stomach
  ■ Body—central part of the stomach
  ■ Pylorus—lower portion of the stomach
Small intestine
  ■ Duodenum—uppermost division; attached to pyloric end of the stomach
  ■ Jejunum—midsection of the small intestine
  ■ Ileum—lower portion of the small intestine
Large intestine
  ■ Cecum
  ■ Colon
      Ascending colon—lies on the right side of the abdomen and extends up to the lower portion of the liver; the ileum joins the large intestine at the junction of the cecum and the ascending colon
      Transverse colon—passes horizontally across the abdomen
      Descending colon—lies on the left side of the abdomen in a vertical position
      Sigmoid colon—extends downward, subsequently joining the rectum
  ■ Rectum
  ■ Anal canal

Host factors

The human host has numerous defenses that normally prevent or control disease produced by enteric pathogens. For example, the acidity of the stomach effectively restricts the number and types of organisms that enter the lower GI tract. Normal peristalsis helps to move organisms toward the rectum, interfering with their ability to adhere to the mucosa. The mucous layer coating the epithelium entraps microorganisms and helps to propel them through the gut. The normal flora prevent colonization by potential pathogens.

Mucous membranes line the GI tract, as well as the respiratory and urogenital tracts. Although technically inside the body, some of these membranes can be considered as outside in the sense that they are exposed to the external environment in the form of food, water, or air. These membranes contain multiple cell types; some are secreting or absorbing cells that perform physiologic functions of the membrane while others serve a protective function. For example, sets of specialized cells called **follicles** are part of the mucous membrane lining the GI tract and serve a protective function. Collections of follicles are called **Peyer's patches**. Follicles contain M cells, macrophages, and B and T cells. As a result of the collective action of the follicle components following uptake and processing of the bacteria or antigens, **secretory immunoglobulin A (sIgA)** is subsequently released. Phagocytic cells and sIgA within the gut help to destroy etiologic agents of disease, as do eosinophils, which are particularly active against parasites. Follicles and Peyer's patches are found in the small and large intestine.

Other factors that come into play include the host's personal hygiene and age. An initial step in the pathogenesis of enteric infections is ingestion of the pathogen. The majority of enteric pathogens, including bacteria, viruses, and parasites, are acquired by the fecal-oral route. Enteric infections can be spread by contamination of food products or drinking water and then subsequent ingestion. The age of the host also plays a role in whether disease is established. For example, diarrheal infections caused by rotavirus or enteropathogenic *Escherichia coli* tend to affect young children and not adults.

Finally, the normal intestinal flora is an important factor in the host response to the introduction of a potentially harmful microorganism. Whenever a reduction in normal flora occurs because of antibiotic treatment or some host factor, resistance to GI infection is significantly reduced. The most common example of the protective effect of normal flora is the development of the syndrome pseudomembranous colitis (PMC). This inflammatory disease of the large bowel is caused by the toxins of the anaerobic organism *Clostridium difficile* and occasionally other clostridia and perhaps even *Staphylococcus aureus*, and seldom occurs except after antimicrobial or antimetabolite treatment has altered the normal flora.

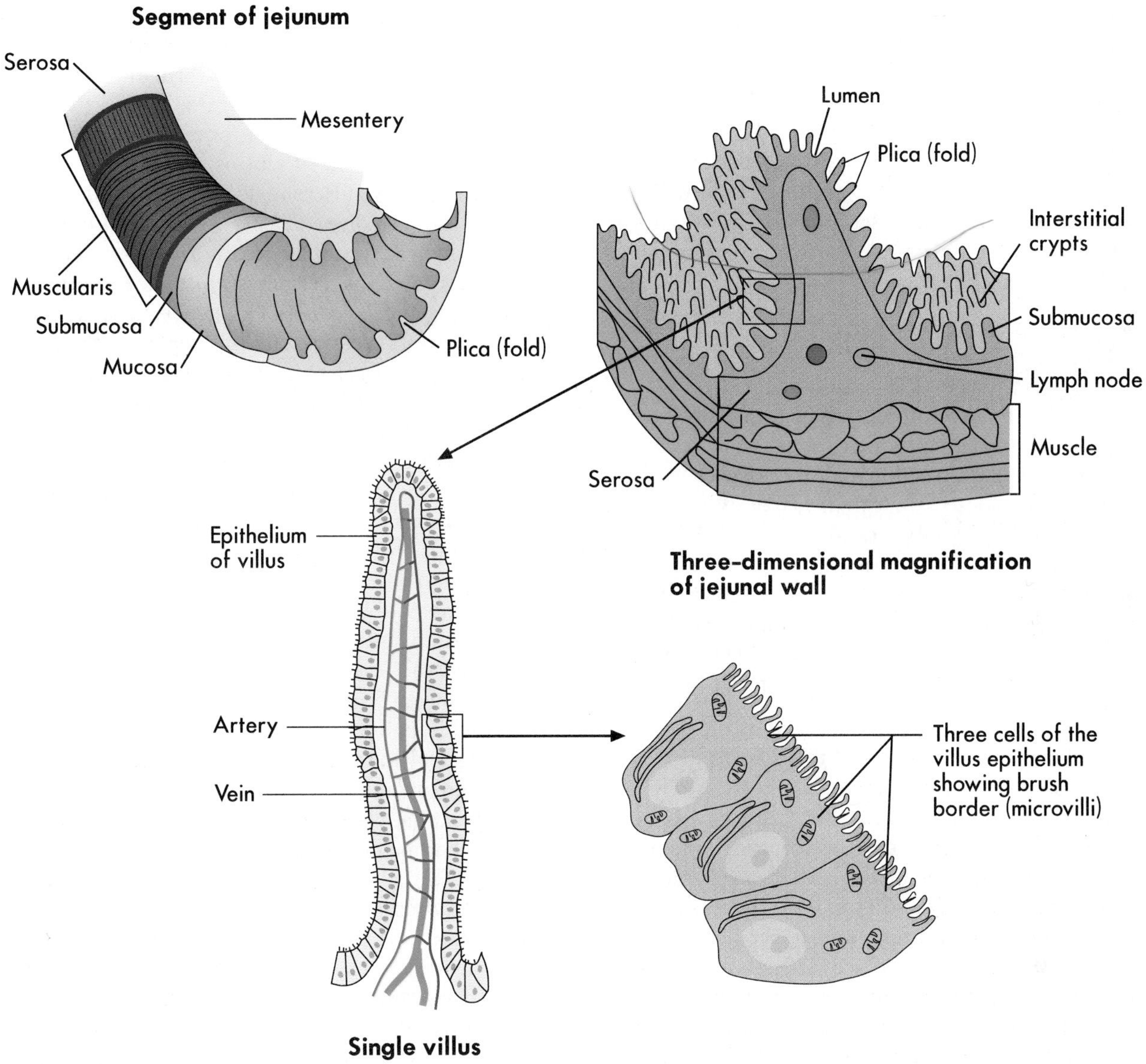

Segment of jejunum

Three-dimensional magnification of jejunal wall

Single villus

FIGURE 27-2 Wall of the small intestine. Villi cover the folds of the mucosal layer; in turn, each villus is covered with epithelial cells.

Almost every antimicrobial agent and several cancer agents have been associated with the development of PMC. *C. difficile*, usually acquired from the hospital environment, is suppressed by normal flora. When normal flora are reduced, *C. difficile* is able to multiply and produce its toxins. This syndrome is also known as antibiotic-associated colitis. Other microorganisms that may gain a foothold when released from selective pressure of normal flora include *Candida* spp., staphylococci, *Pseudomonas* spp., and various *Enterobacteriaceae*.

Microbial factors

The ability of an organism to cause GI infection depends not only on the susceptibility of the human host to the invading organism but also on the organism's virulence traits. To cause GI infection, a microorganism must possess one or more factors that allow it to overcome host defenses or it must enter the host at a time when one or more of the innate defense systems is inactive. For example, certain stool pathogens are able to survive gastric acidity only if the acidity has been reduced by bicarbonate, other buffers, or by medications for ulcers (e.g., cimetidine, ranitidine, H_2 blockers). Pathogens taken in with milk have a better chance of survival, because milk neutralizes stomach acidity. Organisms such as *Mycobacterium tuberculosis*, *Shigella*, and *C. difficile* (a spore-forming clostridium) are able to withstand exposure to gastric acids and thus require much smaller infectious inocula than do acid-sensitive organisms such as *Salmonella*.

PRIMARY PATHOGENIC MECHANISMS   Because the normal adult GI tract receives up to 8 L of fluid daily in the form of ingested liquid, plus the secretions of the various glands that contribute to digestion (salivary glands, pancreas, gallbladder, stomach), of which all but a small amount must be reabsorbed, any disruption of the normal flow or resorption of fluid will profoundly affect the host. Enteric pathogens may cause disease in one or more of three ways depending on their interaction with the human host by:

- Changing the delicate balance of water and electrolytes in the small bowel, resulting in massive fluid secretion; in many cases, this process is mediated by enterotoxin production. This is a noninflammatory process
- Causing cell destruction and/or a marked inflammatory response following invasion of host cells and possible cytotoxin production usually in the colon
- Penetrating the intestinal mucosa with subsequent spread to and multiplication in lymphatic or reticuloendothelial cells outside of the bowel; these infections are considered systemic infections

Examples of microorganisms for each of these pathogenic mechanisms are listed in Table 27-1.

Toxins.

Enterotoxins.   Enterotoxins alter the metabolic activity of intestinal epithelial cells, resulting in an outpouring of electrolytes and fluid into the lumen. They act primarily in the jejunum and upper ileum, where most fluid transport takes place. The stool of patients with enterotoxic diarrheal disease involving the small bowel is profuse and watery, and polymorphonuclear neutrophils or blood are not prominent features.

The classic example of an enterotoxin is that of *Vibrio cholerae* (Figure 27-3). This toxin consists of two subunits, A and B.[16] The A subunit is composed of one molecule of A1, the toxic moiety, and one molecule of A2, which binds an A1 subunit to five B subunits. The B subunits bind the toxin to a receptor (a ganglioside, an acidic glycolipid) on the intestinal cell membrane. Once bound, the toxin acts on adenylate cyclase enzyme, which catalyzes the transformation of adenosine triphosphate (ATP) to cyclic adenosine monophosphate (cAMP). Increased levels of cAMP stimulate the cell to actively secrete ions into the intestinal lumen. To maintain osmotic stabilization, the cells then secrete fluid into the lumen. The fluid is drawn from the intravascular fluid store of the body. Patients therefore can become dehydrated and hypotensive very rapidly. *V. cholerae* inhabit sea and stagnant water and are spread in contaminated water. They have been isolated from coastal waters of several states, and a few cases of cholera that were acquired in the United States in recent years have been reported.[7] Additional information about *V. cholerae* is provided in Chapter 35.

Other organisms also produce a cholera-like enterotoxin. A group of vibrios similar to *V. cholerae* but serologically different, known as the **noncholera vibrios,** produce disease clinically identical to cholera, effected by a very similar toxin. The heat-labile toxin (LT) elaborated by certain strains of *E. coli*, called **enterotoxigenic *E. coli* (ETEC),** is similar to cholera toxin, sharing cross-reactive antigenic determinants. The enterotoxins of some *Salmonella* spp. (including *S. arizonae*), *Vibrio parahaemolyticus*, the *Campylobacter jejuni* group, *Clostridium perfringens*, *Clostridium difficile*, *Bacillus cereus*, *Aeromonas*, *Shigella dysenteriae*, and many other *Enterobacteriaceae* also cause positive reactions in at least one of the tests for enterotoxin (discussed below). The exact contribution of these enterotoxins to the pathogenicity of most stool pathogens remains to be elucidated.

Certain strains of *E. coli*, in addition to producing a heat-labile toxin similar to cholera toxin (LT), also produce a heat-stable toxin (ST) with other properties. Although ST also promotes fluid secretion into the intestinal lumen, its effect is mediated by activation of guanylate cyclase, resulting in increased levels of cyclic guanylate monophosphate (GMP), which yields the same net effect as increased cAMP. Tests for ST include enzyme-linked immunosorbent assay (ELISA), immunodiffusion, and the classic suckling mouse assay, in which culture filtrate is placed into the stomach of a suckling mouse, with the intestinal contents later measured for fluid volume increase.

Although several tests are available for the detection of enterotoxin, they are not performed routinely in diagnostic microbiology laboratories. These tests include the ligated rabbit ileal loop test, the Chinese hamster ovary cell assay, and the Y-1 adrenal cell assay. Because many enterotoxins are antigenic, homologous antibodies can be used to identify them specifically. Immunodiffusion, ELISA, and latex agglutination tests are all available to identify specific toxins. Molecular probes for toxin detection are available for research use.

Cytotoxins.   The second category of toxins, cytotoxins, acts to disrupt the structure of individual intestinal epithelial cells. When destroyed, these cells slough from the surface of the mucosa, leaving it raw and unprotected. The secretory or absorptive functions of the cells are no longer performed. The damaged tissue evokes a strong inflammatory response from the host, further inflicting tissue damage. Numerous polymorphonuclear neutrophils and blood are often seen in

TABLE 27-1 EXAMPLES OF MICROORGANISMS THAT CAUSE GI INFECTION FOR EACH PRIMARY PATHOGENIC MECHANISM

MECHANISM	EXAMPLES OF MICROORGANISMS
Toxin Production	
Enterotoxin	*Vibrio cholerae* Noncholera vibrios *Shigella dysenteriae* type 1 Enterotoxigenic *Escherichia coli* *Salmonella* spp. *Clostridium difficile* (toxin A) *Aeromonas* *Campylobacter jejuni*
Cytotoxin	*Shigella* spp. *Clostridium difficile* (toxin B) Enterohemorrhagic *Escherichia coli*
Neurotoxin	*Clostridium botulinum* *Staphylococcus aureus* *Bacillus cereus*
Attachment Within or Close to Mucosal Cells/Adherence	Enteropathogenic *Escherichia coli* *Cryptosporidium parvum* *Isospora belli* Rotavirus Hepatitis A, B, C Norwalk virus
Invasion	*Shigella* spp. Enteroinvasive *Escherichia coli* *Entamoeba histolytica* *Balantidium coli* ? *Campylobacter jejuni* *Plesiomonas shigelloides* *Yersinia enterocolitica* *Edwardsiella tarda*

the stool, and pain, cramps, and tenesmus (painful straining during a bowel movement) are common symptoms. The term **dysentery** refers to this destructive disease of the mucosa, almost exclusively occurring in the colon. Cytotoxin has not yet been shown to be the sole virulence factor for any etiologic agent of GI disease, because most agents produce a cytotoxin in conjunction with another factor.

Escherichia coli strains seem to possess virulence mechanisms of many types.[10] Some strains produce a cytotoxin that destroys epithelial cells and blood cells. Certain strains produce a cytotoxin that affects Vero cells (African green monkey kidney cells) and resembles the cytotoxin produced by *Shigella dysenteriae* (Shiga toxin); such strains of *E. coli* are associated with hemorrhagic colitis and hemolytic uremic syndrome.[8,9]

C. difficile produces a cytotoxin, the presence of which is a most useful marker for diagnosis of PMC. *S. dysenteriae*, *Staphylococcus aureus*, *C. perfringens*, and *V. parahaemolyticus* produce cytotoxins that probably contribute to the pathogenesis of diarrhea, although they may not be essential for initiation of disease. Other vibrios, *Aeromonas hydrophila* (a relatively newly described agent of GI disease), and *Campylobacter jejuni*, the most common cause of GI disease in many areas of the United States, have been shown to produce cytotoxins. The role that these toxins actually play in the pathogenesis of the disease syndromes is not yet delineated.

Neurotoxins. **Food poisoning**, or intoxication, may occur as a result of ingesting toxins produced by microorganisms. The microorganisms usually produce

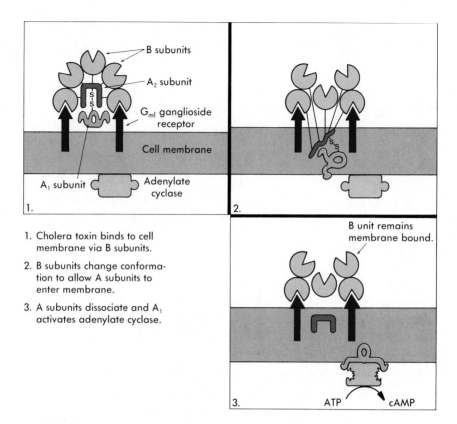

1. Cholera toxin binds to cell membrane via B subunits.

2. B subunits change conformation to allow A subunits to enter membrane.

3. A subunits dissociate and A₁ activates adenylate cyclase.

FIGURE 27-3 Diagrammatic representation of the structure and action of cholera toxin.

their toxins in foodstuffs before they are ingested; thus the patient ingests preformed toxin. Although, strictly speaking, these syndromes are not GI infections but intoxications, they are acquired by ingestion of microorganisms or their products and are considered in this chapter. Particularly in staphylococcal food poisoning and botulism, the causative organisms may not be present in the patient's bowel at all.

Bacterial agents of food poisoning that produce neurotoxins include *Staphylococcus aureus* and *Bacillus cereus*. Toxins produced by these organisms cause vomiting, independent of other actions on the gut mucosa. Staphylococcal food poisoning is one of the most frequently reported categories of foodborne disease. The organisms grow in warm food, primarily meat or dairy products, and produce the toxin. Onset of disease is usually within 2 to 6 hours of ingestion. *B. cereus* produces two toxins, one of which is preformed, called the emetic toxin, because it produces vomiting. The second type, probably involving several enterotoxins, causes diarrhea. Often acquired from eating rice, *B. cereus* has also been associated with cooked meat, poultry, vegetables, and desserts.

Perhaps the most common cause of food poisoning is from type A *Clostridium perfringens,* which produces toxin in the host after ingestion. As a result, a relatively mild, self-limited (usually 24 hours) gas-

troenteritis occurs often in outbreaks in hospitals. Meats and gravies are typical offending foods.

One of the most potent neurotoxins known is produced by the anaerobic organism *Clostridium botulinum*. This toxin prevents the release of the neurotransmitter acetylcholine at the cholinergic nerve junctions, causing flaccid paralysis. The toxin acts primarily on the peripheral nerves but also on the autonomic nervous system. Patients exhibit descending symmetric paralysis and ultimately die from respiratory paralysis unless they are mechanically ventilated. In most cases, adult patients who develop botulism have ingested the preformed toxin in food (home-canned tomato products and canned, cream-based foods are often implicated), and the disease is considered to be an intoxication, although *C. botulinum* has been recovered from the stools of many adult patients. A relatively recently recognized syndrome, **infant botulism,** is a true GI infection. Probably the flora of the normal adult bowel usually prevent colonization by *C. botulinum,* whereas the organism is able to multiply and produce toxin in the infant bowel. Infant botulism is not an infrequent condition; babies acquire the organism by ingestion, although the source of the bacterium is not always clear. Because an association has been found with honey and corn syrup, babies under 9 months of age should not be fed honey. The effect of the toxin is the same, whether ingested in food or produced by growing organisms within the bowel.

Attachment. An organism's ability to cause disease can also depend on its ability to colonize and adhere to a relevant region of the bowel. To illustrate, ETEC must be able to adhere to and colonize the small intestine, as well as produce an enterotoxin. These organisms produce an adherence antigen, called colonization factor antigen (CFA), that gives the organism this adherence capacity. Certain strains of *E. coli* referred to as the enteropathogenic *E. coli*, (EPEC), attach and then adhere to the intestinal brush border. This localized adherence is mediated by the production of pili. Subsequent to attaching, EPEC disrupts normal cell function by effacing the brush epithelium, thereby causing diarrheal disease. Genes responsible for the adherence of ETEC and EPEC reside on a transmissible plasmid.[10]

Giardia lamblia, a parasite, has increasingly become more common as an etiologic agent of GI disease in the United States. Excreted into fresh water by natural animal hosts such as the beaver, the organism can be acquired by drinking stream water or even city water in some localities, particularly in the Rocky Mountain states, as well as throughout the world. The organism, a flagellated protozoan, adheres to the intestinal mucosa of the small bowel, possibly by means of a ventral sucker, destroying the mucosal cells' ability to participate in normal secretion and absorption. No evidence indicates invasion or toxin production.

Cryptosporidium and *Isospora* spp., parasitic etiologic agents of diarrhea in animals and poultry and more recently recognized as causing human disease, probably also act by adhering to intestinal mucosa and disrupting function. Cryptosporidia are often seen in the diarrhea of patients with acquired immunodeficiency syndrome (AIDS), as well as in travelers' diarrhea, day care epidemics, and diarrhea in people with animal exposure; cryptosporidia and *Isospora* spp. may cause severe, protracted diarrhea in AIDS patients. Other coccidian parasites, such as microsporidia, produce diarrhea by destroying intestinal cell function.

Invasion. Following initial and essential adherence to GI mucosal cells, some enteric pathogens are able to gain access to the intracellular environment. Invasion allows the organism to reach deeper tissues, access nutrients for growth, and possibly avoid the host immune system.

In the case of diarrhea caused by *Shigella,* the primary mechanism of disease production consists of

(1) the triggering and directing by *Shigella* of its own entry into colonic epithelial cells by genes located on a plasmid, and once internalized, (2) the rapid multiplication of *Shigella* in the submucosa and lamina propria and its intra- and extracellular spread to other adjacent colonic epithelial cells.[15] These activities lead to extensive superficial tissue destruction. If these two steps do not occur, one does not get the clinical presentation of classic dysentery (Table 27-2). The entry process is illustrated in Figure 27-4.

Salmonellae interact with the apical (top) microvilli of colonic epithelial cells, disrupting the brush border. Similar to *Shigella, Salmonella* spp. also stimulate the host cell to internalize them through rearrangements of host actin filaments and other

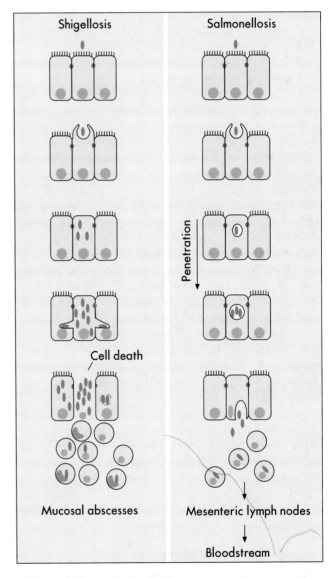

FIGURE 27-4 The invasion of *Shigella* and *Salmonella* into intestinal epithelial cells. (Modified from Sansonetti, P.J. 1991. Genetic and molecular basis of epithelial invasion by *Shigella* species. Rev. Infect. Dis. 13 (suppl. 4):S282, University of Chicago Press, Chicago.)

cytoskeleton proteins.[4-6] Once the whole bacteria are internalized within endocytic vesicles of the host epithelial cell, organisms begin to multiply within the vacuoles. In contrast to *Shigella* spp. that use the colonic mucosal epithelium as a site of multiplication, certain serotypes of *Salmonella,* such as *S. typhi* and *S. cholerasuis,* use the colonic epithelium as a route to gain access to the submucosal layers, mesenteric lymph nodes, and subsequently to the bloodstream. The entry of *Salmonella* is a very complex process that involves several essential genes, as well as particular environmental conditions of the host cell; this process is still being delineated. Invasiveness is also thought to contribute to the pathogenesis of disease associated with species of vibrios, campylobacters, *Yersinia enterocolitica, Plesiomonas shigelloides,* and *Edwardsiella tarda.*

Certain parasites, particularly *Entamoeba histolytica* and *Balantidium coli,* invade the intestinal epithelium of the colon as a primary site of infection. The ensuing amebic dysentery is characterized by blood and numerous white blood cells, and the patient experiences cramping and tenesmus. Other parasites that are acquired by ingestion, such as *Schistosoma* spp. and *Trichinella,* may cause transient bloody diarrhea and pain during migration through the intestinal mucosa to their preferred sites within the host.

Other organisms selectively destroy absorptive cells (e.g., villus tip cells) in the mucosa, disrupting their normal cell function and thereby causing diarrhea. Rotaviruses and Norwalklike viruses are both visualized by electron microscopy within the absorptive cells at the ends of the intestinal villi, where they multiply and destroy cellular function. As a result, the villi become shortened, and inflammatory cells infiltrate the mucosa, further contributing to the pathologic condition. In addition to these viral agents, hepatitis A, B, and C and occasionally enteric adenoviruses have been associated with diarrheal symptoms in infected patients.

MISCELLANEOUS VIRULENCE FACTORS Other virulence traits appear to be involved in the development of GI infections and include characteristics such as motility, chemotaxis, and mucinase production. Also, the possession of certain antigens, such as the Vi antigen of *Salmonella typhi* and certain cell wall components, are also associated with virulence.

CLINICAL MANIFESTATIONS

The clinical symptoms experienced by a patient are largely dependent on how the enteric pathogen causes disease. To illustrate, patients infected with an enteric pathogen that upsets fluid and electrolyte balance have no fecal leukocytes present in the stool and complain of watery diarrhea; fever is usually absent or mild. Although nausea, vomiting, and abdominal pain may also be present, the dominant feature is intestinal fluid loss. In contrast, patients infected with an enteric pathogen that causes significant cell destruction and inflammation have fecal leukocytes present in the stool (Figure 27-5). Their diarrhea is often characterized by the presence of mucus and possibly blood; in many of these patients, fever is a prominent component of their disease, as well as abdominal pain, cramps, and tenesmus. And finally, patients who become infected with a pathogen that is able to penetrate the intestinal mucosa of the small intestine without producing enterocolitis and then subsequently spread and multiply at other sites will present with signs and symptoms of a systemic illness such as headache, sore throat, malaise, and fever; diarrhea in these patients is not a prominent feature and is absent or mild in many cases. Features of these three types of enteric infections are summarized in Table 27-2.

EPIDEMIOLOGY

Gastrointestinal infections occur in numerous epidemiologic settings. Awareness of these different settings is important because knowledge of a particular epidemiologic setting can help provide a basis for the diagnosis and clues to possible etiologies. When this knowledge is combined with clinical findings, the etiology of the infection can often be narrowed to three or four organisms.

Institutional settings

Diarrheal illness can be a major problem in institutional settings such as day care centers, hospitals, and nursing homes. Because individual hygiene is often difficult to maintain in these settings, coupled with

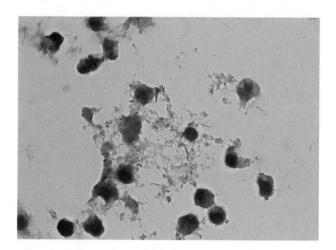

FIGURE 27-5 Wright's stain of stool from a patient with shigellosis showing moderate numbers of polymorphonuclear cells.

several organisms with relatively low infecting doses such as *Shigella* and *Giardia lamblia,* numerous outbreaks of diarrheal illness caused by various organisms have been reported. Organisms such as *Shigella, Campylobacter jejuni, Giardia lamblia, Cryptosporidium,* and rotaviruses have been reported to cause outbreaks in day care centers. Of significance, these infections can be spread to family members. Similarly, outbreaks caused by these organisms, as well as hemorrhagic *Escherichia coli* 0157:H7, have been reported in nursing homes and other extended care facilities.

Nosocomial diarrheal illness is also a problem for hospital patients and personnel. Of importance, rotaviruses, adenoviruses, and Coxsackie viruses are also nosocomially transmitted. In addition to these organisms, *Clostridium difficile* has emerged as a major nosocomial enteric pathogen in hospitals and other settings, including nursing homes and extended care facilities. This organism is a very hardy pathogen that readily survives on fomites (inanimate objects) such as floors, bed rails, call buttons, doorknobs, and on the hands of hospital personnel caring for the patient.

Travelers diarrhea

Individuals who travel into developing geographic areas that have poor sanitation are at particularly high risk for developing diarrhea if they do not pay attention to their eating and drinking habits. In areas with poor sanitation, enteric pathogens heavily contaminate the water and food. Although many types of enteric pathogens can cause diarrhea in travelers, enterotoxigenic *Escherichia coli* is a leading cause in Asia, Africa, and Latin America, accounting for about 50% of cases. Salmonellae, Shigellae, *Campylobacter* spp., vibrios, rotavirus, and Norwalk virus can also cause diarrhea in travelers, depending on the area or country they visit.

Food and waterborne outbreaks

The Centers for Disease Control and Prevention report that more than 12,000 cases of foodborne illness occur in the United States each year. Because most of these illnesses are not reported, some estimate that there are millions of cases annually. Eating raw or undercooked fish, shellfish, or meats, and drinking unpasteurized milk increases the risks of certain bacterial, parasitic, and viral infections. Many foodborne outbreaks can be traced to poor hygienic practices of food handlers such as not washing their hands after using the toilet; hepatitis A, Norwalk virus, and *Salmonella* are a few such examples in which contamination of the food during preparation by a food handler occurred and caused diarrheal disease. Since 1968 the number of cases of salmonellosis has gradually increased, with many of these infections associated with raw or undercooked eggs. Also, the potential for widespread dissemination of foodborne pathogens has increased because of factors such as the tendency to eat outside the home, the export and import of food sources worldwide, and travel.

In addition to foodborne outbreaks of GI tract infections, waterborne outbreaks of diarrheal disease caused by *Giardia lamblia* and *Cryptosporidium* have been traced to inadequately filtered surface water. Recreational waters, including swimming pools, can also become contaminated with enteric pathogens such as *Shigella* and *G. lamblia* because of poor toilet facilities or practices.

TABLE 27-2 TYPES OF ENTERIC INFECTIONS

PATHOGENIC MECHANISM	MAJOR SYMPTOMS	EXAMPLES OF ETIOLOGIC AGENTS
Upsetting of fluid and electrolyte balance/noninflammatory	Watery diarrhea No fecal leukocytes No fever	*Vibrio cholerae* Rotavirus Norwalk virus Enterotoxigenic *Escherichia coli* *Giardia lamblia* *Bacillus cereus*
Invasion and possible cytotoxin production/inflammatory (dysentery)	Dysenteric-like diarrhea (mucus, blood, white cells) Fever Fecal leukocytes	*Shigella* spp. Enteroinvasive *E. coli* *Salmonella enteritidis* *Entamoeba histolytica*
Penetration with subsequent access to the bloodstream (enteric fever)	Signs of systemic infection (headache, malaise, sore throat) Fever	*Salmonella typhi* *Yersinia enterocolitica*

Immunocompromised hosts

GI tract infections in HIV-infected individuals and other patients who are immunosuppressed, such as organ transplant recipients or individuals receiving chemotherapy, are a diagnostic challenge for the clinician and microbiologist. For example, cytotoxic chemotherapy and/or antibiotic therapy may predispose patients to develop *C. difficile* colitis.

Diarrhea is a common clinical manifestation of HIV infection, developing in about 30% to 80% of cases. Numerous pathogens and opportunistic pathogens have been identified and are believed to cause recurrent or chronic diarrhea. Commonly reported etiologic agents are:

- Species of *Salmonella*, *Shigella*, and *Campylobacter*
- Cytomegalovirus
- Cryptosporidia, *Isospora belli*
- Microsporidia
- *Entamoeba histolytica*
- Mycobacterial spp.
- *Giardia lamblia*

ETIOLOGIC AGENTS

Many microorganisms are able to cause enteric infections. A discussion of each organism is beyond the scope of this chapter. Rather, these organisms are addressed in Parts 4 through 7 of the textbook. Table 27-3 summarizes the general characteristics of the more common agents of enteric infections.

OTHER INFECTIONS OF THE GI TRACT

Besides causing disease in the small and large intestine, microorganisms can also infect other sites of the GI tract, as well as the GI tract's accessory organs.

ESOPHAGITIS

Infections of the mucosa of the esophagus (**esophagitis**) can cause painful or difficult swallowing, and/or the sensation that something is lodged in the throat while swallowing. Individuals who have esophagitis usually have local or systemic underlying illnesses such as hematologic malignancies or HIV infection, or are receiving immunosuppressive therapy. The most common etiologic agents are *Candida* spp. (primarily *C. albicans*), herpes simplex virus, and cytomegalovirus.

GASTRITIS

Gastritis refers to inflammation of the gastric mucosa. This illness is associated with nausea and upper abdominal pain; vomiting, burping, and fever may also be present. A curved organism called *Helicobacter pylori* is seen on the surface of gastric epithelial cells of patients with gastritis. The organism is recovered from gastric biopsy material obtained endoscopically but not from stool. Following acute infection, *H. pylori* can persist for years in most individuals, with most remaining asymptomatic.[2] *H. pylori* is now considered the causative agent of peptic ulcer disease and a significant risk factor for stomach cancer.[17]

PROCTITIS

Proctitis is the inflammation of the rectum (distal portion of the large intestine). Common symptoms associated with proctitis are itching and a mucous discharge from the rectum; if the infection progresses, ulcers and abscesses may form in the rectum. The majority of infections are sexually transmitted through anal intercourse. *Chlamydia trachomatis*, herpes simplex, syphilis, and gonorrhea are the most common etiologic agents.

MISCELLANEOUS

Unusual agents and those that have not been cultured, such as the mycobacteria that may be associated with Crohn's disease and the bacterium associated with Whipple's disease, identified by molecular methods as a new agent, *Trophyrema whippelii*, are also candidates as etiologic agent of GI disease. Occasionally, stool cultures from patients with diarrheal disease yield heavy growth of organisms such as enterococci, *Pseudomonas* spp. or *Klebsiella pneumoniae*, not usually found in such numbers as normal flora. Only anecdotal evidence suggests that these organisms actually contribute to the pathogenesis of the diarrhea. Agents of sexually transmitted disease may cause GI symptoms when they are introduced into the colon via sexual intercourse. *Mycobacterium avium-intracellulare* complex may be transmitted in this way, going on to cause systemic disease in patients with AIDS. The pathogenesis of infections resulting from *Blastocystis hominis* (a possible coccidian etiologic agent of human diarrheal disease) is not well documented, although these organisms are associated with GI symptoms.

LABORATORY DIAGNOSIS OF GASTROINTESTINAL TRACT INFECTIONS

SPECIMEN COLLECTION AND TRANSPORT

If enteric pathogens are to be detected by the laboratory, adherence to appropriate guidelines for specimen collection and transport is imperative (see Table 1-1 for a quick guide to specimen collection, transport, and processing). If an etiologic agent is not isolated with the first culture or visual examination, two additional specimens should be submitted to the laboratory over the next few days. Because organisms may be shed intermittently, collection of specimens at different times over several days enhances recovery. Certain infectious agents, such as *Giardia*, may be difficult to detect, requiring the processing of multiple specimens over many weeks, duodenal aspirates (in the case of *Giardia*), or additional alternative methods.

General comments

Specimens that can be delivered to the laboratory within 1 hour may be collected in a clean, waxed cardboard or plastic container. Stool for direct wet-mount examination, *Clostridium difficile* toxin assay, immunoelectron microscopy for detection of viruses, and ELISA or latex agglutination test for rotavirus must be sent to the laboratory without any added preservatives or liquids. Volume of a liquid stool at least equal to 1 teaspoon (5 mL) or a pea-sized piece of formed stool is necessary for most procedures.

Stool specimens for bacterial culture

If a delay longer than 2 hours is anticipated for stools for bacterial culture, the specimen should be placed in transport medium. Cary-Blair transport medium best preserves the viability of intestinal bacterial pathogens, including *Campylobacter* and *Vibrio* spp. However, the media produced by different manufacturers can vary.[11] Some workers recommend reducing the agar content of Cary-Blair medium from 0.5% to 0.16% for maintenance of *Campylobacter* spp. Buffered glycerol transport medium does not maintain these bacteria. Several manufacturers produce a small vial of Cary-Blair with a self-contained plastic scoop suitable for collecting samples.

Because *Shigella* spp. are delicate, a transport medium of equal parts of glycerol and 0.033 M phosphate buffer (pH 7.0) supports viability of *Shigella* better than does Cary-Blair. For this purpose, maintaining the glycerol transport medium at refrigerator or freezer temperatures yields better results.

If stool is unavailable, a rectal swab may be substituted as a specimen for bacterial or viral culture, but it is not as good, particularly for diagnosis in adults. For suspected intestinal infection with *Campylobacter*, the swab must be placed in transport medium immediately to avoid drying. Swabs are not acceptable for detection of parasites, toxins, or viral antigens. The swab should be placed in Cary-Blair transport medium.

Stool specimens for ova and parasites

For detection of ova and parasites, specimen preservation with a fixative is recommended for visual examination (see Chapter 64).

Stool specimens for viruses

Stools for virus culture must be refrigerated if they are not inoculated onto media into cell cultures within 2 hours. A rectal swab, transported in modified Stuart's transport medium or another viral transport medium, is adequate for recovery of most viruses from feces. See Chapter 66 for more information regarding collection and transport of specimens for viral culture.

Miscellaneous specimen types

Other specimens that may be obtained for diagnosis of GI tract infection include duodenal aspirates (usually for detection of *Giardia* or *Strongyloides*), which should be examined immediately by direct microscopy for the presence of motile protozoan trophozoites, cultured for bacteria, and placed into polyvinyl alcohol (PVA) fixative for subsequent parasitic examination. The laboratory should be informed in advance that such a specimen is going to be collected so that the specimen can be processed and examined efficiently.

The **string test** has proved useful for diagnosing duodenal parasites, such as *Giardia*, and for isolating *Salmonella typhi* from carriers and patients with acute typhoid fever. The patient swallows a weighted gelatin capsule containing a tightly wound length of string, which is left protruding from the mouth and taped to the cheek. After a predetermined period, during which the capsule reaches the duodenum and dissolves, the string, now covered with duodenal contents, is retracted and delivered immediately to the laboratory. There the technologist, using sterile-gloved fingers, strips the mucus and secretions attached to the string and deposits some material on slides for direct examination and some material into fixative for preparation of permanent stained mounts. The technologist also inoculates some material to appropriate media for isolation of bacteria.

DIRECT DETECTION OF AGENTS OF GASTROENTERITIS IN FECES

Wet mounts

A direct wet mount of fecal material, particularly with liquid or unformed stool, is the fastest method for detection of motile trophozoites of *Dientamoeba fragilis*, *Entamoeba*, *Giardia*, and other intestinal parasites that may not contribute to disease but may alert the microbiologist to the possibility of finding other parasites, such as *Entamoeba coli*, *Endolimax nana*, *Chilomastix mesnili*, and *Trichomonas hominis*. Occasionally the larvae or adult worms of other parasites may be visualized. Experienced observers can also see the refractile forms of cryptosporidia and many types of cysts on the direct wet mount, including *Cyclospora cayetanensis*, a recently characterized parasite that is associated with the consumption of

TABLE 27-3 GENERAL CHARACTERISTICS OF THE COMMON AGENTS OF ENTERIC INFECTIONS

ORGANISM	COMMON SOURCES OR PREDISPOSING CONDITION	DISTRIBUTION	CLINICAL PRESENTATION	PREDOMINANT PATHOGENIC MECHANISM	FECAL LEUKOCYTES
Bacillus cereus	Meats, vegetables, rice	Worldwide	Intoxication: vomiting or watery diarrhea	Ingestion of preformed toxin (food poisoning)	−
Clostridium botulinum	Improperly preserved vegetables, meat, fish	Worldwide	Neuromuscular paralysis	Ingestion of preformed toxin (food poisoning)	−
Staphylococcus aureus	Meats, salads, dairy products	Worldwide	Intoxication: vomiting	Ingestion of preformed toxin (food poisoning)	−
Clostridium perfringens	Meats, poultry	Worldwide	Watery diarrhea	Ingestion of organism followed by toxin production	−
Aeromonas	Water	Worldwide	Watery diarrhea or dysentery	?* Enterotoxin ? Cytotoxin	−
Campylobacter spp.	Water, poultry, milk	Worldwide	Dysentery	? Invasion ? Cytotoxins	+
Clostridium difficile	Antimicrobial therapy	Worldwide	Dysentery	Enterotoxin and cytotoxin	+/−
Diarrheagenic Escherichia coli: Enteropathogenic	?	Worldwide	Watery diarrhea	Adherence/? invasion without multiplication	−
Enterotoxigenic	Food, water	Worldwide—more prevalent in developing countries	Watery diarrhea	Enterotoxin	−
Enteroinvasive	Food	Worldwide	Dysentery	Invasion, enterotoxin	+
Enterohemorrhagic	Meats	Worldwide	Watery, often bloody diarrhea	Cytotoxin	−/+†
Plesiomonas shigelloides	Fresh water, shellfish	Worldwide	? Dysentery	Unknown ? Enterotoxin	+/−‡
Salmonella spp. (nontyphoidal)	Food, water	Worldwide	Dysentery	Invasion	+

Organism	Source	Distribution	Clinical features	Mechanism	White cells
Salmonella typhi	Food, water	Tropical, developing countries	Enteric fever	Penetration	+ (monocytes, *not* PMNs)
Shigella spp.	Food, water	Worldwide	Dysentery	Invasion	+
Shigella dysenteriae	Water	Tropical, developing countries	Dysentery	Invasion, cytoxin	+
Vibrio cholerae	Water, shellfish	Asia, Africa, Middle East, South and North American (along coastal areas)	Watery diarrhea	? Enterotoxin Cytotoxin	−/+
Yersinia enterocolitica	Milk, pork, water	Worldwide	Watery diarrhea and/or enteric fever	? Invasion ? Penetration	−
Giardia lamblia	Food, water	Worldwide	Watery diarrhea	Unknown—impaired absorption	−
Cryptosporidium parvum	Animals, water?	Worldwide	Watery diarrhea	? Adherence	−
Entamoeba histolytica	Food, water	Worldwide (more common in developing countries)	Dysentery	Invasion, cytotoxin	+/− (amoeba destroy the white cells)
Rotavirus	?	Worldwide	Watery diarrhea	Mucosal damage leading to impaired absorption in small intestine	−
Norwalk viruses	Shellfish, salads	Worldwide	Watery diarrhea	Mucosal damage leading to impaired absorption in small intestine	−

*?: Questionable; uncertain.
†−/+: More frequently negative.
‡+/−: More frequently positive.

contaminated food such as raspberries.[1] If present in sufficient numbers, the ova of intestinal parasites can be seen.

Examination of a direct wet mount of fecal material taken from an area with blood or mucus, with the addition of an equal portion of Loeffler's methylene blue, is helpful for detection of leukocytes, which occasionally aids in differentiating among the various types of diarrheal syndromes. Under phase-contrast and dark-field microscopy, the darting motility and curved forms of *Campylobacter* may be observed in a warm sample. Water, which will immobilize *Campylobacter*, should not be used. However, for practical reasons most laboratories do not use a wet mount. Trained observers working in endemic areas can recognize the characteristic appearance and motility of *Vibrio cholerae*.

Stains

Feces may be Gram stained for detection of certain etiologic agents. For example, many thin, comma-shaped, gram-negative bacilli may indicate *Campylobacter* infection (if vibrios have been ruled out). In addition, polymorphonuclear cells may also be detected. Of importance, an acid-fast stain can be used to detect *Cryptosporidium* spp., mycobacteria, and *Isospora* spp. Examination of fixed fecal material for parasites by trichrome or other stains is thoroughly covered in Chapter 64. A permanent stained preparation should be made from all stool specimens received for detection of parasites.

Antigen detection

An accurate, sensitive, indirect fluorescent antibody stain for giardiasis and cryptosporidiosis is now commercially available. These organisms can be visualized easily and unequivocally with a monoclonal antibody fluorescent stain (Meridian Diagnostics, Cincinnati, Ohio). Park et al. described a simple and rapid screening procedure using a DFA stain for *E. coli* 0157:H7.[13]

Enzyme immunoassays (EIAs) or latex agglutination can detect numerous microorganisms that cause GI tract infections. For example, rotavirus is detected using a solid-phase EIA procedure or a latex agglutination test. EIA methods are also available for detection of antigens of *Cryptosporidium* and *Giardia lamblia*. EIA methods have also been evaluated for detection of certain bacterial pathogens.[12]

Molecular biological techniques

The recent development of amplification techniques has led to numerous publications for the direct detection of many enteric pathogens, including all major organism groups—bacteria, viruses, and parasites. Probes for *Salmonella, Shigella,* and *Yersinia* are being evaluated. A disadvantage with probe technology is that the organism itself is not available for susceptibility testing, which is important for certain bacterial pathogens (e.g., *Shigella*) for which susceptibility patterns vary.

CULTURE OF FECAL MATERIAL FOR ISOLATION OF ETIOLOGIC AGENTS

Bacteria

Fecal specimens for culture should be inoculated to several media for maximal yield, including solid agar and broth. The choice of media is arbitrary and based on the particular requirements of the clinician and the laboratory. Recommendations for selection of media are given in this section.

ORGANISMS FOR ROUTINE CULTURE Stools received for routine culture in most clinical laboratories in the United States should be examined for the presence of *Campylobacter, Salmonella,* and *Shigella* spp. under all circumstances. Detection of *Aeromonas* spp. should be incorporated into the routine stool culture procedures. The cost of doing a stool examination on every patient for all potential enteric pathogens is prohibitive. The decision as to what other bacteria are routinely cultured should take into account the incidence of GI tract infections caused by particular etiologic agents in the area served by the laboratory. For example, if the incidence of *Yersinia enterocolitica* gastroenteritis is high enough in the area served by the laboratory, this agent should also be sought routinely. Similarly, because of the increasing prevalence of disease caused by *Vibrio* spp. in individuals living in high-risk areas of the United States (seacoast), laboratories in these localities may routinely look for these organisms. Conversely, unless a patient has a significant travel history, a laboratory located in the Midwestern part of the United States should not routinely look for these organisms except by special request. Protocols for culture of enterohemorrhagic *Escherichia coli* (e.g., *E. coli* 0157:H7) vary greatly; based on incidence of disease, some laboratories routinely culture for this organism while others carry out culture on request only. Other laboratories set up cultures routinely on bloody stool specimens from children only.

ROUTINE CULTURE METHODS An in-depth discussion regarding culture of all enteric pathogens is beyond the scope of this chapter. Because U.S. laboratories should routinely examine stools for the presence of *Salmonella, Shigella,* and *Campylobacter* spp., culture of these organisms will be addressed. Culture conditions for all other pathogens, including viruses, are covered in Parts 4, 5, and 7.

Specimens received for detection of the most frequently isolated *Enterobacteriaceae* and *Salmonella* and *Shigella* spp. should be plated to a supportive medium, a slightly selective and differential medium, and a moderately selective medium. A highly selective medium does not seem to be cost effective for most microbiology laboratories.

Blood agar (tryptic soy agar with 5% sheep blood) is an excellent general supportive medium. Blood agar medium allows growth of yeast species, staphylococci, and enterococci, in addition to gram-negative bacilli. Of importance, the absence of normal gram-negative fecal flora and/or the presence of significant quantities of organisms such as *Staphylococcus aureus*, yeasts, and *Pseudomonas aeruginosa* can be evaluated. Another benefit of blood agar is that it will allow oxidase testing of colonies. Several colonies that do not resemble *Pseudomonas* from the third or fourth quadrant should be routinely screened for production of cytochrome oxidase. If many are present, *Aeromonas*, *Vibrio*, or *Plesiomonas* spp. should be suspected.

The moderately selective agar should support growth of most *Enterobacteriaceae*, vibrios, and other possible pathogens; MacConkey agar works well. Some laboratories use eosin-methylene blue (EMB), which is slightly more inhibitory. All lactose-negative colonies should be tested further, ensuring adequate detection of most vibrios and most pathogenic *Enterobacteriaceae*. Lactose-positive vibrios *(V. vulnificus)*, pathogenic *E. coli*, some *Aeromonas* spp., and *Plesiomonas* spp. may not be distinctive on MacConkey agar.

Salmonella/Shigella. The specimen should also be inoculated to a moderately selective agar such as Hektoen enteric (HE) or xylose-lysine desoxycholate (XLD) media. These media inhibit growth of most *Enterobacteriaceae*, allowing *Salmonella* and *Shigella* spp. to be detected. Colony morphologies of lactose-negative, lactose-positive, and H_2S-producing organisms are illustrated in Figure 27-6. Other highly selective enteric media, such as salmonella-shigella, bismuth sulfite, deoxycholate, or brilliant green, may inhibit some strains of *Salmonella* or *Shigella*. All these media are incubated at 35° to 37° C in air and examined at 24 and 48 hours for suspicious colonies.

Campylobacter. Cultures for isolation of *Campylobacter jejuni* and *Campylobacter coli* should be inoculated to a selective agar containing antimicrobial agents that suppress the growth of normal flora but not *Campylobacter* spp. The introduction of a blood-free, charcoal-containing media that has more selective antibiotic components has resulted in better recovery of most enteropathogenic *Campylobacter* spp. compared with the earlier media.[3] Brucella broth base has yielded less satisfactory recovery of *Campylobacter* spp. Commercially produced agar plates for isolation of campylobacters are available from several manufacturers. These plates are incubated in a microaerophilic atmosphere at 42° C and examined at 24 and 48 hours for suspicious colonies. Culture methods for other campylobacters that are associated with GI disease, such as *C. hyointestinalis* and *C. fetus* subsp. *fetus*, are provided in Chapter 45.

Enrichment Broths. Enrichment broths are sometimes used for enhanced recovery of *Salmonella*, *Shigella*, *Campylobacter*, and *Y. enterocolitica* although *Shigella* will usually not survive enrichment. Gram-negative broth (Hajna GN) or selenite F broth yields good recovery. Enrichment broths for *Enterobacteriaceae* should be incubated in air at 35° C for 6 to 8 hours and then several drops should be subcultured to at least two selective medium plates. A commercial system that allows such broth to be tested for antigen of *Salmonella* or *Shigella* directly has been described; however, the reported sensitivity is lower than desired. Stool would be inoculated to broth only initially; those broths that tested negative could be discarded without subculturing. Campy-thioglycollate enrichment broth increases the yields of positive cultures for *Campylobacter* spp., although it is not necessary for routine use. Enrichment broth for *Campylobacter* is refrigerated overnight or for a minimum of 8 hours before a few drops are plated to *Campylobacter* agar and incubated at 42° C in a microaerophilic atmosphere.

LABORATORY DIAGNOSIS OF *CLOSTRIDIUM DIFFICILE*-ASSOCIATED DIARRHEA

The definitive diagnosis of *C. difficile*-associated diarrhea is based on clinical criteria combined with laboratory testing.[14] Visualization of a characteristic pseudomembrane or plaque on endoscopy is diagnostic for pseudomembranous colitis and, with the appropriate history of prior antibiotic use, meets the criteria for diagnosis of antibiotic-associated pseudomembranous colitis. No one laboratory test will establish the diagnosis unequivocally. Three tests are currently available for routine use: culture, detection of cytoxin by tissue culture, and antigen detection assays (e.g., enzyme immunoassay, latex agglutination) for *C. difficile* toxin(s). The laboratory diagnosis of *C. difficile*-associated diarrhea is reviewed in Chapter 59.

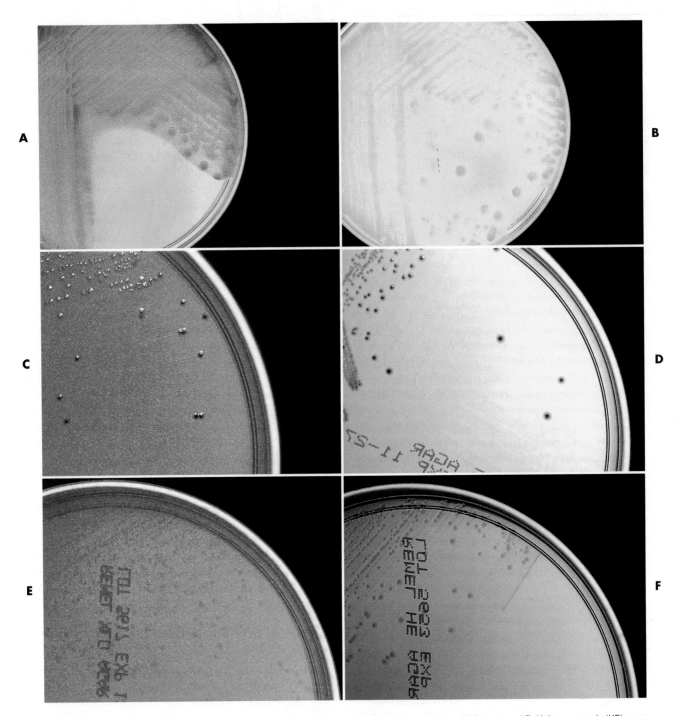

FIGURE 27-6 Colonies of a lactose-positive organism growing on **A**, xylose-lysine deoxycholate (XLD) agar and **B**, Hektoen enteric (HE) agar. Colonies of *Salmonella enteritidis* (lactose-negative) growing on **C**, XLD and **D**, HE agar. (Note how both agars detect H_2S production.) Colonies of *Shigella* (lactose-negative) growing on **E**, XLD and **F**, HE agar.

References

1. Chew, D., et al. 1996. *Cyclospora cayetanensis* infection associated with consumption of raspberries, New Jersey. Abstract LB25, 36th ICAAC, American Society for Microbiology, New Orleans, La.

2. Cover, T.L. and Blaser, M.J. 1995. *Helicobacter pylori*: a bacterial cause of gastritis, peptic ulcer disease, and gastric cancer. A.S.M. News 61:21.

3. Endtz, H.P., et al. 1991. Comparison of six media, including a semisolid agar, for the isolation of various *Campylobacter* species from stool specimens. J. Clin. Microbiol. 29:1007.

4. Finlay, B.B. and Falkow, S. 1990. *Salmonella* interactions with polarized human intestinal Caco-2 epithelial cells. J. Infect. Dis. 162:1096.

5. Galán, J.E., Ginocchio, C., and Costeas, P. 1992. Molecular and functional characterization of the *Salmonella typhimurium* invasion gene *invA*: homology of *invA* to members of a new protein family. J. Bacteriol. 17:4338.

6. Ginocchio, C., Pace, J., and Galan, J.E. 1992. Identification and molecular characterization of a *Salmonella typhimurium* gene involved in triggering the internalization of Salmonellae into cultured epithelial cells. Proc. Natl. Acad. Sci. U.S.A. 89:5976.

7. Kaper, J.B., Morris, J.G., and Levine, M.M. 1995. Cholera. Clin. Microbiol. Rev. 8:48.

8. Kaplan, B.S. 1992. Commentary on the relationships between HUS and TTP. In Kaplan, B.S. et al., editors. Hemolytic uremic syndrome and thrombotic thrombocytopenic purpura, Marcel Dekker, New York.

9. Kaye, S.A. and Obrig, T.G. 1996. Pathogenesis of *E. coli* hemolytic uremic syndrome. Clin. Microbiol. Newsletter 18:49.

10. Levine, M.M. 1987. *Escherichia coli* that cause diarrhea: enterotoxigenic, enteropathogenic, enteroinvasive, enterohemorrhagic, and enteroadherent. J. Infect. Dis. 155:377.

11. Mundy, L.S., et al. 1991. An evaluation of three commercial fecal transport systems for the recovery of enteric pathogens. Am. J. Clin. Pathol. 96:364.

12. Pal, T., et al. 1985. Modified enzyme-linked immunosorbent assay for detecting enteroinvasive *Escherichia coli* and virulent *Shigella* strains. J. Clin. Microbiol. 26:948.

13. Park, C.H., et al. 1994. Rapid diagnosis of enterohemorrhagic *Escherichia coli* 0157:H7 directly from fecal specimens using immunofluorescence stain. Am. J. Clin. Pathol. 101:91.

14. Peterson, L.R. and Kelly, P.J. 1993. The role of the clinical microbiology laboratory in the management of *Clostridium difficile*-associated disease. Infect. Dis. Clin. North Am. 7:277.

15. Sansonetti, P.J. 1991. Genetic and molecular basis of epithelial cell invasion by *Shigella* species. Rev. Infect. Dis. 13(suppl. 4):S282.

16. Sears, C.L. and Kaper, J.B. 1996. Enteric bacterial toxins: mechanisms of action and linkage to intestinal secretion. Microbiol. Rev. 60:167.

17. Solnick, J.V. and Tompkins, L.S. 1993. *Helicobacter pylori* and gastroduodenal disease: pathogenesis and host-parasite interaction. Infect. Agents Dis. 1:294.

Bibliography

Gilligan, P.H., et al. 1992. Laboratory diagnosis of bacterial diarrhea. Cumitech 12A:1. American Society for Microbiology, Washington, D.C.

Gradus, M.S. 1986. Public health criteria for the diagnosis of foodborne illness. Clin. Microbiol. Newsletter 8:85.

Guerrant, R.L. 1991. Bacterial and protozoal gastroenteritis. N. Engl. J. Med. 325:327.

Guerrant, R.L. 1995. Principles and syndromes of enteric infection. In Mandell, G.L., Bennett, J.E., and Dolin, R., editors. Principles and practice of infectious diseases, ed 4. John Wiley & Sons, New York.

Guerrant, R.L. and Hughes, J.M. 1995. Nausea, vomiting, and noninflammatory diarrhea. In Mandell, G.L., Bennett, J.E., and Dolin, R., editors. Principles and practice of infectious diseases, ed 4. John Wiley & Sons, New York.

Sears, C.L. and Kaper, J.B. 1996. Enteric bacterial toxins: mechanisms of action and linkage to intestinal secretion. Microbiol. Rev. 60:167.

28 | SKIN, SOFT TISSUE, AND WOUND INFECTIONS

GENERAL CONSIDERATIONS

The skin serves as a barrier between the internal organs and the external environment. Skin is not only subjected to frequent trauma and thereby is at frequent risk of infection, but it can also reflect internal disease.

ANATOMY OF THE SKIN

From inside out the skin is divided into three distinct layers: the subcutaneous tissue, the dermis, and the epidermis (Figure 28-1). The subcutaneous tissue lies beneath the dermis and is rich in fat. Deeper hair follicles and sweat glands originate in this layer. Below the subcutaneous layer are thin fascial membranes (sheets or bands of fibrous tissue) that cover muscles, ligaments, and other connective tissues. Of importance, the fascia serves as a barrier to infection for the deeper tissues and organs of the body.

Above the subcutaneous tissue and fascial membranes lies the dermis, which comprises dense connective tissue that is rich in blood and nerve supply. Shorter hair follicles and sebaceous (oil-producing) glands originate in the dermis. Finally, the epidermis, which is the outermost layer of skin, is made of layered squamous epithelium. Hair follicles, sebaceous glands, and sweat glands open to the skin surface, through the epidermis.

FUNCTION OF THE SKIN

The skin is the body's largest and thinnest organ. It forms a self-repairing and protective boundary between the body's internal environment and an often hostile external environment. Skin plays a crucial role in the control of body temperature, excretion of water and salts, synthesis of important chemicals and hormones, and as a sensory receptor. Of significance, the skin has an important protective function by virtue of the epidermis' outermost epithelial layer, which comprises cells containing keratin, a water-repellant protein. The skin's normal microbial flora, pH, and chemical defenses also help to prevent colonization by many pathogens. The resident microbial flora are listed in Box 28-1.

INCIDENCE, ETIOLOGIC AGENTS, AND PATHOGENESIS

Approximately 15% of all patients who seek medical attention have either some skin disease or skin lesion, many of which are infectious. Many different bacteria, fungi, and viruses may be involved. Also, these infections can result from one or several causative agents. Because of this great diversity of etiologic agents and the complexity of these infections, only the most common infections involving the skin and subcutaneous tissues will be addressed.

Skin infections can arise from the invasion of certain organisms from the external environment through breaks in the skin or from organisms that reach the skin through the blood as part of a systemic disease. In some infections, such as staphylococcal scalded-skin syndrome, skin lesions are caused by toxins produced by the bacteria. In others, lesions can also result from the host's immune response to microbial antigens.

Because of the diversity of etiologic agents, clinicians will often describe the appearance of skin lesions to microbiologists for possible input as to appropriate culture techniques. The physical characteristics of the lesions can indicate the need for smear, culture, biopsy, or surgery, for instance. Some of the terms most frequently used to describe manifestations of skin infections are provided in Table 28-1. Figure 28-2 provides some examples of these skin lesions.

SKIN AND SOFT TISSUE INFECTIONS

SKIN

Numerous infections of the skin may occur. Several of the most common are discussed on the next few pages.

BOX 28-1 RESIDENT MICROBIAL FLORA OF THE SKIN

Diphtheroids
Staphylococcus epidermidis
Other coagulase-negative staphylococci
Propionibacterium acnes

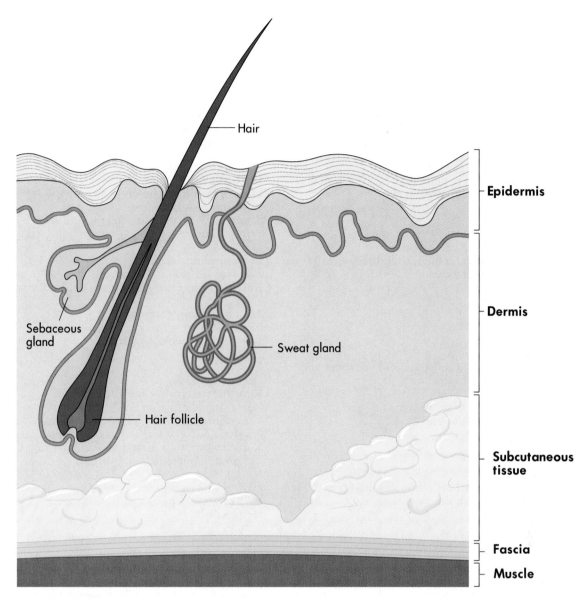

Hair

Epidermis

Dermis

Sebaceous
gland

Sweat gland

Hair follicle

Subcutaneous
tissue

Fascia

Muscle

FIGURE 28-1 Diagram of skin.

Infections in or around hair follicles

Folliculitis, furunculosis, and carbuncles are localized abscesses either in or around hair follicles. These infections are distinguishable from one another based on size and the extent of involvement in subcutaneous tissues. Table 28-2 summarizes each infection's respective clinical features. For the most part, these infections are precipitated by blockage of the hair follicle with skin oils (sebum), or minor trauma resulting from friction such as that caused by clothes rubbing the skin. *Staphylococcus aureus* is the most common etiologic agent for all three infections. Members of the family *Enterobacteriaceae* can also cause folliculitis, albeit much less commonly. Also, outbreaks of folliculitis caused by *Pseudomonas aeruginosa* and associated with the use of whirlpools, swimming pools, and hot tubs have been reported.[7]

Infections in the keratinized layer
of the epidermis

Because of their ability to utilize keratin in the cells of the epidermis, the dermatophyte fungi are significant and well-suited pathogens for this site. Unlike the previously discussed infections, dermatophytes cannot invade the deeper layers of skin. Because keratin is also present in hair and nails, these fungi may also cause superficial infections at these sites (see Chapter 65 for more information).

TABLE 28-1 MANIFESTATIONS OF SKIN INFECTIONS[8]

TERM	DESCRIPTION	POSSIBLE ETIOLOGIC AGENTS
Macule	A circumscribed (limited), flat discoloration of the skin	Dermatophytes *Treponema pallidum* (infection—secondary syphilis) Viruses such as enteroviruses (viral exanthems [rashes])
Papule	An elevated, solid lesion up to 0.5 cm in diameter. Multiple papules may become confluent and form plaques	Pox virus (infection—Molluscum contagiosum) *Sarcoptes scabei* (infection—scabies) Human papillomavirus types 3 and 10 (infection—flat warts) *S. aureus, P. aeruginosa*, etc. (infection—folliculitis)
Nodule	A circumscribed, raised, solid lesion greater than 0.5 cm in diameter	*Sporothrix schenckii* *S. aureus* (infection—furuncle) Miscellaneous fungi (infection—subcutaneous mycoses) *Mycobacterium marinum* *Nocardia* spp. *Corynebacterium diphtheriae*
Pustule	A circumscribed collection of leukocytes and fluid that varies in size	*Candida* spp. Dermatophytes *S. aureus* (infection—folliculitis) Herpes simplex virus (HSV) Varicella zoster virus (VZV) *Neisseria gonorrhoeae* (infection—gonococcemia) *S. aureus* or group A streptococci (*Streptococcus pyogenes* [infection—impetigo])
Vesicle	A circumscribed collection of fluid up to 0.5 cm in diameter (blisterlike)	HSV VZV Herpes zoster
Scales	Excess dead epidermal cells	Dermatophytes (infection—tinea) Group A streptococci (*Streptococcus pyogenes* [infection—scarlet fever])
Ulcer	A loss of epidermis and dermis	*Haemophilus ducreyi* (infection—chancroid) Bowel flora (infection—decubiti) *T. pallidum* (infection—chancre of primary syphilis) *Bacillus anthracis* (infection—anthrax)
Bulla	A circumscribed lesion of fluid more than 0.5 cm in diameter	Clostridial species (infection—necrotizing gas gangrene) *Staphylococcus aureus* (infection—staphylococcal scalded skin syndrome; bullous impetigo) HSV *Vibrio vulnificus* and other vibrios Group A streptococci (*Streptococcus pyogenes*) Other gram-negative rod

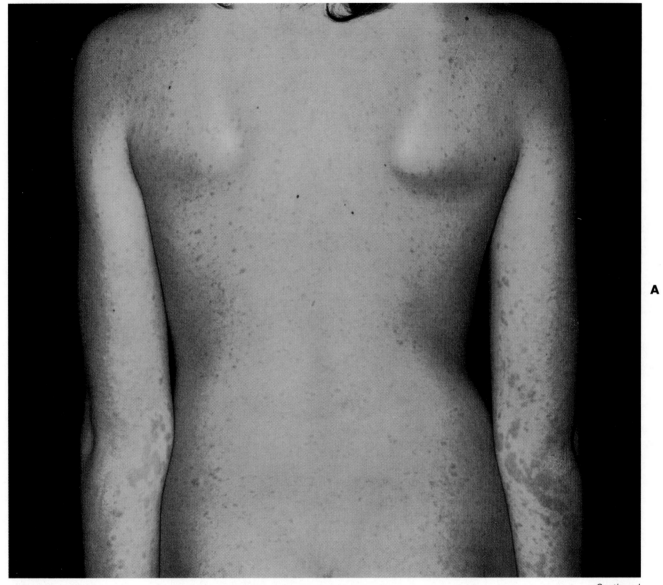

A

Continued

FIGURE 28-2 A, Viral maculopapular rash. (From Habif, T.B. 1996. Clinical dermatology: a color guide to diagnosis and therapy, ed 3. Mosby, St. Louis.)

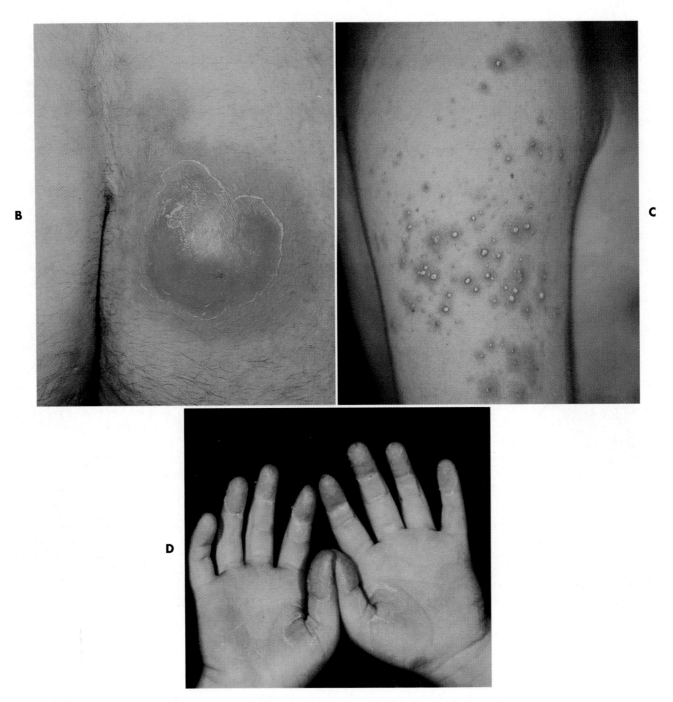

FIGURE 28-2, cont'd **B,** Furuncle. **C,** Folliculitis caused by *Staphylococcus aureus* showing numerous pustules. **D,** Desquamation (shedding or scaling) of skin resulting from scarlet fever caused by group A streptococci (*Streptococcus pyogenes*). (From Habif, T.B. 1996. Clinical dermatology: a color guide to diagnosis and therapy, ed 3. Mosby, St. Louis.)

TABLE 28-2   INFECTIONS INVOLVING HAIR FOLLICLES

INFECTION	SKIN MANIFESTATIONS
Folliculitis—minor infection of hair follicles	Papules or pustules that are pierced by a hair and surrounded with redness
Furuncle (boil)	Abscess that begins as a red nodule in a hair follicle that ultimately becomes painful and full of pus
Carbuncle	Furuncles that spread more deeply to the dermis and subcutaneous tissues; these infections may be associated with fever and malaise

Infections in the deeper layers of the epidermis and the dermis

Most infections in the deeper layers of the epidermis and dermis result from the inoculation of microorganisms by traumatic breaks in the skin. These superficial skin infections usually do not require surgical intervention. Table 28-3 summarizes these infections. In most instances, these infections resolve with local care such as heat application. Antibiotics are only occasionally required.

Cutaneous ulcers usually involve a loss of epidermal and part of the dermal tissues. In contrast, nodules are inflammatory foci in which the epidermal and dermal layers remain largely intact. Various bacteria and fungi can cause ulcerative and/or nodular skin lesions following direct inoculation. Examples of these etiologies include *Corynebacterium diphtheriae*, *Bacillus anthracis*, *Nocardia* spp., *Mycobacterium marinum*, and *Sporothrix schenckii*.

INFECTIONS OF THE SUBCUTANEOUS TISSUES

Infections of the subcutaneous tissues may manifest as abscesses, ulcers, and boils. *Staphylococcus aureus* is the most common etiologic agent of subcutaneous abscesses in healthy individuals. Many subcutaneous abscesses contain mixed bacteria. To a large degree, the organisms isolated from subcutaneous abscesses depend on the site of infection. For example, anaerobes are commonly isolated from abscesses of the perineal, inguinal, and buttock area, while nonperineal infections are caused by mixed facultative aerobic organisms.

Chronic undermining ulcer, or **Meleney's ulcer**, is a slowly progressive infection of the subcutaneous tissue with associated ulceration of portions of the overlying skin. The causative organism is classically a microaerobic (see Chapter 1) streptococcus, but anaerobic streptococci and, occasionally, other organisms may be involved.

In many instances, infections of the epidermis and dermis extend and can become subcutaneous infections and even reach the fascia and/or muscle. For example, erysipelas (Figure 28-3) can become a subcutaneous cellulitis and thereafter a streptococcal necrotizing fasciitis. Similarly, folliculitis can readily become a subcutaneous abscess or a carbuncle that can extend to the fascia. Cellulitis also can frequently extend to the subcutaneous tissues (Figure 28-4). Anaerobic cellulitis is associated with considerable amounts of gas from organisms that are usually present in the subcutaneous tissue. This infection is most often found in the extremities and is particularly common among patients with diabetes and may involve the neck, abdominal wall, perineum, or connective tissue in other areas. Anaerobic cellulitis also may occur as a postoperative problem. Although the onset and spread of this lesion are not usually rapid and patients do not show impressive systemic effects at first, it is not an illness to be taken lightly. The organisms are almost always a mixture of aerobes and anaerobes. The aerobes include *E. coli*, alpha-hemolytic or nonhemolytic streptococci, and *S. aureus* predominantly, but group A streptococci (*Streptococcus pyogenes*) and other *Enterobacteriaceae* are

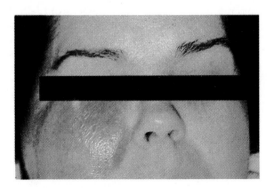

FIGURE 28-3 Erysipelas caused by group A streptococci (*Streptococcus pyogenes*).

TABLE 28-3 INFECTIONS OF THE EPIDERMAL AND DERMAL LAYERS OF THE SKIN

INFECTION	KEY FEATURES OF INFECTION	ETIOLOGIES	OTHER COMMENTS
Erysipelas	Primarily involves the dermis and most superficial parts of the subcutaneous tissue. Lesions are painful, red, swollen, and indurated. Patients are febrile, and regional lymphadenopathy (swollen glands) is often present. Lesion has a marked, well-demarcated, raised border (see Figure 28-3)	Group A streptococci (*Streptococcus pyogenes* [sometimes groups B, C, or G streptococci])	Infants, children, and elderly individuals are most affected Primarily a clinical diagnosis
Erythrasma	Chronic infection of the keratinized layer of the epidermis. Lesions are dry and scaly	*Corynebacterium minutissimum*	Common in diabetics Resembles dermatophyte infection
Erysipeloid	Purplish-red, nonvesiculated skin lesion with an irregular, raised border. The lesions itch and burn. Fever and other systemic symptoms are uncommon	*Erysipelothrix rhusiopathiae*	Uncommon. Considered an occupational disease
Impetigo	Erythematous (red) lesions that may be bullous (less common) or nonbullous	Nonbullous—group A streptococci (*Streptococcus pyogenes*) Bullous—*Staphylococcus aureus*	
Cellulitis	Diffuse, spreading infection involving the deeper layers of the dermis. Lesions are ill-defined, flat, painful, red, and swollen. Patients have fever, chills, and regional lymphadenopathy (see Figure 28-4)	Group A streptococci *Staphylococcus aureus* Less common: *Aeromonas, Vibrio* spp., and *Hemophilus influenzae*—typically affects young children	Primarily a clinical diagnosis
Dermatophytoses	Superficial infections of the skin and its appendages (i.e., ringworm, athlete's foot, jock itch, as well as infections of nails and hair)	*Microsporum, Trichophyton,* and *Epidermophyton* spp.	
Hidradenitis	Chronic infection of obstructed apocrine (sweat) glands in the axillas, genital, or perianal areas with intermittent discharge of often foul-smelling pus	*S. aureus, Streptococcus milleri* group, anaerobic streptococci, and *Bacteroides* spp.	
Infected pilonidal tuft cyst or hairs	Pain and swelling, redness	Anaerobes, including *Bacteroides fragilis* group, *Prevotella, Fusobacterium,* anaerobic gram-positive cocci, and *Clostridium* spp.	

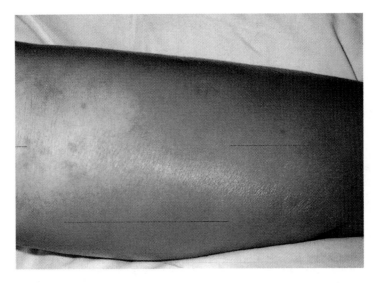

FIGURE 28-4 Cellulitis. (From Farrar et al. 1992. Infectious diseases: text and color atlas, ed 2. Mosby-Wolfe Limited, London.)

encountered as well. The anaerobes are typically found in greater numbers and in more variety than the aerobes; *Peptostreptococcus* spp., *B. fragilis* group strains, *Prevotella* and *Porphyromonas,* other anaerobic gram-negative bacilli, and clostridia are seen. Bacteremia is not usually present.

INFECTIONS OF THE MUSCLE FASCIA AND MUSCLES

There are several rare, yet serious or potentially serious, forms of deep and sometimes extensive soft tissue and skin infections.

Necrotizing fasciitis

Necrotizing fasciitis is a serious infection that occurs relatively infrequently. The basic pathology is infection of the fascia overlying muscle groups, often with involvement of the overlying soft tissue. At the fascial level, no barrier exists to spread of infection, so fasciitis may extend widely and rapidly to involve huge areas of the body in short periods. This process, once known as hospital gangrene, typically involves group A streptococci or *S. aureus.* Necrotizing fasciitis also frequently involves anaerobic bacteria, especially *Bacteroides* and *Clostridium* spp.

Progressive bacterial synergistic gangrene

Progressive bacterial synergistic gangrene is usually a chronic gangrenous condition of the skin most often encountered as a postoperative complication, particularly after abdominal or thoracic surgery. The lesions may be extensive and, with involvement of the abdominal wall, may lead to evisceration (extrusion of the internal organs). As the name suggests, this is typically a mixed infection with microaerobic streptococci and *S. aureus.* At times other organisms may be present, including anaerobic streptococci, *Proteus,* and other facultative and anaerobic bacteria. This infection occurs very infrequently. Cultures should be taken from the advancing outer edge of the lesion (not the central portion of the wound) or the microaerobic streptococcus will be missed.

Myositis

Myositis (involvement of muscle) is caused by various infectious agents. The nature of the pathologic process is variable, sometimes involving extensive necrosis of muscle, as in gas gangrene or clostridial myonecrosis, necrotizing cutaneous myositis or synergistic nonclostridial anaerobic myonecrosis, anaerobic streptococcal myonecrosis, myonecrosis caused by *Bacillus* spp., or myonecrosis caused by *Aeromonas.* Focal collections of suppuration in muscle (staphylococcal or other pyomyositis [purulent myositis]) are sometimes seen.[5,6] Abscess in the psoas muscle (muscle arising from the lumbar vertebrae and extending to a muscle, the lesser trochanter, of the leg's femur) may involve *M. tuberculosis, S. aureus,* or various facultative or anaerobic gram-negative bacilli. Serious vascular problems resulting from loss of blood supply may lead to death of muscle; such muscle may become secondarily infected (infected vascular gangrene). Organisms producing myositis or other muscle pathology are listed in Box 28-2.

WOUND INFECTIONS

Besides skin and soft tissue infections that occur primarily as a result of a break in the skin surface,

ORGANISMS PRODUCING MYOSITIS OR OTHER MUSCLE PATHOLOGY

BOX 28-2

Clostridium perfringens	*Staphylococcus aureus*
C. novyi	Group A streptococci *(Streptococcus pyogenes)*
C. septicum	*Pseudomonas mallei*
C. bifermentans	*P. pseudomallei*
C. histolyticum	*Vibrio vulnificus*
C. sordellii	*Mycobacterium tuberculosis*
C. sporogenes	*Salmonella typhi*
Bacillus spp.	*Legionella* spp.
Aeromonas spp.	*Rickettsia*
Peptostreptococcus spp.	Viruses
Microaerobic streptococci	*Trichinella*
Bacteroides spp.	*Taenia solium*
Enterobacteriaceae	*Toxoplasma*

wound infections can occur as complications of surgery, trauma, and bites or diseases that interrupt a mucosal or skin surface.

Postoperative infections

Sources of surgical wound infections can include the patient's own normal flora or organisms present in the hospital environment that are introduced to the patient by medical procedures and/or a specific underlying disease or trama (e.g., burns) that may interrupt a mucosal or skin surface. The nature of the infecting flora depends on the underlying problem and the location of the process. In the case of wound infections following appendectomy or other lower bowel surgery, indigenous flora of the lower gastrointestinal tract are involved, primarily *Escherichia coli*, streptococci, *Bacteroides fragilis* group, and other anaerobic gram-negative rods, including *Bilophila* spp., *Peptostreptococcus* spp., anaerobic non–spore-forming rods, and *Clostridium* spp.[2] Principal pathogens are listed in Box 28-3. *Mycobacterium chelonae* and *Mycobacterium fortuitum* may also cause infections following cardiac surgery, mammoplasty, and other clean surgeries.

Bites

Human bites (Figure 28-5) and clenched-fist injuries yield, in order of frequency, alpha-hemolytic streptococci, *S. aureus*, group A streptococci *(Streptococcus pyogenes)*, and *Eikenella corrodens*. Anaerobes that may be involved include, in order of frequency, *Peptostreptococcus*, pigmented *Prevotella* and *Porphyromonas*, *Prevotella oris*, *P. buccae*, and *Fusobacterium nucleatum*.

Dog bites introduce organisms commonly found in their oral and nasal fluids, including Centers for Disease Control (CDC) group EF-4, *Weeksella zoohelcum*, *Pasteurella* spp., and *Staphylococcus*

intermedius, with much smaller numbers of *S. aureus*. *Simonsiella* is found in the oral cavity of most dogs and is also found in the oral cavity of humans and cats, and other animals. The oral flora of snakes include various gram-negative bacilli, including *Pseudomonas*, *Klebsiella*, *Proteus*, and *E. coli*. Clostridia may also be recovered from snakebite wounds. In infected animal bite wounds (Figure 28-6), the most frequently encountered aerobic and facultative bacteria are alpha-hemolytic streptococci, *S. aureus*, *Pasteurella* spp., and *Enterobacter cloacae*. Predominant anaerobes in animal bites are anaerobic gram-positive cocci, *Fusobacterium* spp., and anaerobic gram-negative rods.[6]

ORGANISMS ENCOUNTERED IN POSTOPERATIVE WOUND INFECTIONS

BOX 28-3

Staphylococcus aureus
Coagulase-negative staphylococci
Streptococcus pyogenes
Streptococcus milleri group streptococci *(S. anginosus, S. constellatus, S. intermedius)*
Microaerobic streptococci
Enterococci
Proteus, Morganella, Providencia
Other *Enterobacteriaceae*
Escherichia coli
Pseudomonas spp.
Candida spp.
Bacteroides spp.
Prevotella and *Porphyromonas* spp.
Fusobacterium spp.
Clostridium spp.
Peptostreptococcus spp.
Non–spore-forming, anaerobic, gram-positive rods

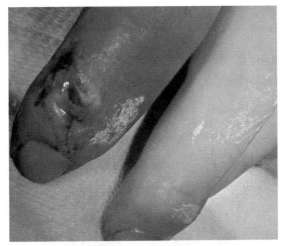

FIGURE 28-5 Human bite infection.

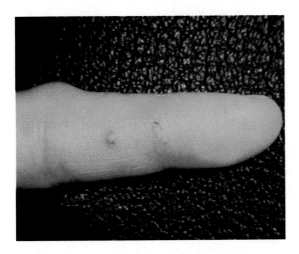

FIGURE 28-6 Animal bite infection caused by *Pasteurella* spp.

Capnocytophaga canimorsus (former CDC group DF-2) and CDC group DF-3 have been responsible for several types of serious infections, including bacteremia, endocarditis, and meningitis. Most patients had a history of dog bite, and most had underlying diseases that impair host defense mechanisms.

Burns

Infection of burn wounds may be associated with bacteremia, carry a significant mortality, and interfere with the acceptance of skin grafts. Many organisms are capable of infecting the eschar (scab) of a burn. Those most often encountered are various streptococci, *Staphylococcus aureus*, *S. epidermidis*, *Enterobacteriaceae*, *Pseudomonas* spp., other gram-negative bacilli, *Candida*, and *Aspergillus*. Anaerobes, including clostridia and *Bacteroides*, have been recov-

ered occasionally but are probably more frequently involved than has been appreciated to date.

SPECIAL CIRCUMSTANCES REGARDING SKIN AND SOFT TISSUE INFECTIONS

In addition to the infections previously discussed, other circumstances can cause the skin and underlying soft tissue to become infected. Some of these infections are associated with the host with compromised defenses; others are manifestations of systemic infection.

Infections related to vascular and neurologic problems

Classically, a patient with one of these common infections has diabetes mellitus, poor arterial circulation (often both large-vessel and small-vessel disease), and peripheral neuropathy (neurologic problems such as numbness). Because of a loss of sensation resulting from the neuropathy, these individuals traumatize their feet readily (often just by virtue of wearing a new pair of shoes) without being aware of it. The traumatized area develops an ulcer that does not heal readily because of the poor vascular supply and that often becomes infected.[1] The infections tend to be chronic and difficult to clear up, particularly because these patients may also have poor vision and therefore may not recognize the problem and may not seek medical attention until the process has gone on for some time.

These infections primarily have purulent discharge and necrotic tissue at the base of the ulcer, often with a foul odor. Extension to the underlying bone, producing a difficult-to-manage osteomyelitis (infection of the bone), occurs often. Periodically, an acute cellulitis and lymphangitis may be associated with chronic, low-grade infection, thereby making control of the patient's diabetes difficult. Peripheral vascular disease unrelated to diabetes mellitus may also predispose a patient to skin or soft tissue infections, but usually these infections are less difficult to manage because no associated neuropathy is present.

Venous insufficiency also predisposes individuals to infection, again primarily of the lower extremities (in this case, often in the area of the calf or lower leg rather than the foot). Infections related to poor blood supply often involve *S. aureus* and group A streptococci (*Streptococcus pyogenes*). Those with open ulcers often become colonized with *Enterobacteriaceae* and *P. aeruginosa*, which may or may not play a role in the infection. Although less well appreciated, anaerobes are frequently involved in infection, particularly in patients with diabetes or peripheral vascular disease. The poor blood supply contributes to anaerobic conditions. Various anaerobes may be recovered, including the *Bacteroides fragilis* group, *Prevotella* and *Porphyromonas, Peptostreptococcus* spp., and less frequently, *Clostridium* spp.

Another common type of infection in this general category, especially with the elderly or very ill, bedridden patient, is infected decubitus ulcer (pressure sore [Figure 28-7]). Anaerobic conditions are present in such lesions because of tissue necrosis. Because most of these lesions are located near the anus or on the lower extremities and because so many of these patients are relatively helpless, the ulcers become contaminated with bowel flora, which leads to chronic infection. This contributes to further death of tissue and extension of the decubitus ulcer. Bacteremia is a possible complication, with *B. fragilis* group often being involved, along with clostridia and enteric bacteria. The ulcers yield various anaerobes and aerobes characteristic of the colonic flora; nosocomial pathogens such as *S. aureus* and *P. aeruginosa* may also be recovered.

Sinus tract and fistulas

Sometimes, a deep-seated infection beneath the skin and subcutaneous soft tissue spontaneously drains itself externally by way of a sinus (channel or cavity) to the skin's surface. Draining sinus tracts are most often associated with a chronic osteomyelitis. Unfortunately, this type of drainage does not usually cure the underlying process, and such sinuses themselves tend to be chronic. The organisms most often involved in sinuses with an underlying osteomyelitis are *S. aureus*, various *Enterobacteriaceae*, *P. aeruginosa*, anaerobic gram-negative bacilli, anaerobic gram-positive cocci, and occasionally other anaerobes. In the case of actinomycosis (with or without bone involvement [Figures 28-8 and 28-9]), one would expect to recover *Actinomyces* spp., *Propionibac-*

terium propionicum, Prevotella or *Porphyromonas* and other non–spore-forming anaerobes, and *Actinobacillus actinomycetemcomitans*. With other types of draining sinuses, the organisms involved depend on the nature of the underlying process.

Chronic draining sinuses are also found in patients with tuberculosis and atypical mycobacterial infection, *Nocardia* infection, and certain infections associated with implanted foreign bodies. Curettings or biopsy from the debrided, cleansed sinus is the best specimen.

Abnormal passages or communications between two organs or leading from an internal organ to the

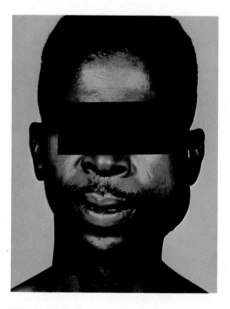

FIGURE 28-8 Actinomycosis. Note "lumpy jaw."

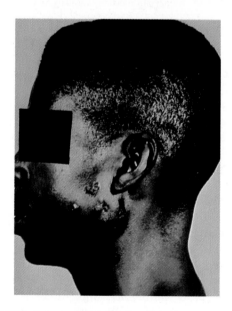

FIGURE 28-9 Actinomycosis, side view. Note sinuses in skin of face and neck.

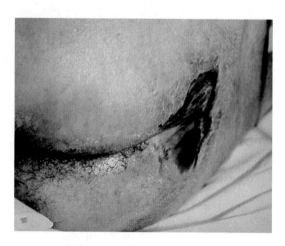

FIGURE 28-7 Sacral decubitus ulcer.

body surface, known as **fistulas,** are difficult management problems. They also often pose insurmountable problems in terms of collection of meaningful specimens, because the organ that has the abnormal communication to the skin surface often has its own profuse indigenous flora. Examples are perirectal fistulas from the small bowel to the skin in association with Crohn's disease or chronic intraabdominal infection. When the bowel is involved, only cultures for specific key organisms, such as mycobacteria or *Actinomyces,* are meaningful. Always attempt to rule out specific underlying causes such as tuberculosis, actinomycosis, and malignancy. Biopsy should be performed in such situations.

Systemic infections with skin manifestations

Cutaneous manifestations of systemic infection, such as bacteremia or endocarditis, may be important clues for the clinician, and they present an opportunity for direct or cultural demonstration of the presence of a particular organism. For example, one may be able to scrape petechiae (a tiny red spot caused by the escape of a small amount of blood) from patients with meningococcemia and demonstrate gram-negative diplococci. In other patients, the skin lesion represents a more impressive type of metastatic infection. In *Vibrio vulnificus* sepsis, dramatic-appearing cutaneous ulcers with necrotizing vasculitis or bullae may be found (Figure 28-10). In some patients, skin lesions may actually represent a noninfectious complication of a local or systemic infection such as scarlet fever or toxic shock syndrome. Various organisms that may be involved in systemic infection with cutaneous lesions are listed in Box 28-4.

LABORATORY DIAGNOSTIC PROCEDURES

INFECTIONS OF THE EPIDERMIS AND DERMIS

For many of the infections of the epidermis and dermis, such as impetigo, folliculitis, cellulitis, and erysipelas, diagnosis is generally made on a clinical basis. Table 28-3 provides the key features and etiologic agents of these infections. Of importance, awareness of the common bacterial pathogens can help guide empiric therapy.

Erysipeloid

For the most part, Gram stain or culture of superficial wound drainage is usually negative. However, culture of a full-thickness skin biopsy taken at the margin of the lesion can confirm the clinical diagnosis.

Superficial mycoses and erythrasma

If a dermatophyte infection is suspected, the lesion is cleaned and scrapings are obtained from the active border of the lesion. These scrapings should be suspended in 10% potassium hydroxide, examined for the presence of hyphae, and may also be cultured (see Chapter 65).

A Wood's lamp examination of skin lesions may reveal golden-yellow fluorescent lesions of tinea versicolor. Wood's light examination may also reveal a coral red fluorescence that is characteristic for erythrasma. *Corynebacterium minutissimum* produces porphyrin that accounts for the red fluorescence. Skin scrapings may be cultured in media containing serum, but imprint smears of the lesion should reveal gram-positive pleomorphic rods, precluding the need for culture.

Erysipelas and cellulitis

As previously mentioned, diagnosis of erysipelas and cellulitis can generally be made on a clinical basis. The value of needle aspiration for the bacteriologic diagnosis of these infections has not been clearly demonstrated, particularly in adults.[9] Of note, a higher percentage of positive cultures of soft tissue aspirates along the advancing margin of erythema using a 22- or 23-gauge needle attached to a 3- or 5-mL syringe may be achieved in children with cellulitis.[4] If received by the laboratory, aspirates should be inoculated onto blood and chocolate agars, as well as a broth such as trypticase soy broth.

Vesicles and bullae

These fluid-filled lesions characteristically involve certain organisms (see Table 28-1), so if the laboratory is

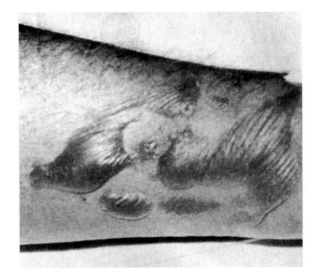

FIGURE 28-10 Bullae on the arm of a patient with *Vibrio vulnificus* sepsis. (From Pollak, S.J., Parrish, E.F. III, Barrett, T.J., et al. 1983. *Vibrio vulnificus* septicemia: isolation of organism from stool and demonstration of antibodies by indirect immunofluorescence. Arch. Intern. Med. 143:837. Copyright 1983, American Medical Association.)

BOX 28-4

ORGANISMS INVOLVED IN SYSTEMIC INFECTION WITH CUTANEOUS LESIONS

Viridans streptococci	*Mycobacterium tuberculosis*
Staphylococcus aureus	*M. leprae*
Enterococci	*Treponema pallidum*
Group A and other beta-hemolytic streptococci	*Leptospira*
Neisseria gonorrhoeae	*Streptobacillus moniliformis*
N. meningitidis	*Bartonella bacilliformis*
Haemophilus influenzae	*Bartonella (Rochalimaea) henselae*
Pseudomonas aeruginosa	*Rickettsia*
P. mallei	*Candida* spp.
P. pseudomallei	*Cryptococcus neoformans*
Listeria monocytogenes	*Blastomyces dermatitidis*
Vibrio vulnificus	*Coccidioides immitis*
Salmonella typhi	*Histoplasma capsulatum*

aware of the nature of the lesion, the flora and use of appropriate techniques to ensure recovery of the agent(s) may be anticipated. Material in the blisterlike lesion varies from serous (resembling serum) fluid to serosanguineous (composed of serum and blood) or hemorrhagic (bloody) fluid. Large bullae permit withdrawal of 0.5 to 1 mL of fluid by needle and syringe aspiration. Some vesicles are tiny, so a swab must be used for specimen collection. The clinician can usually anticipate whether the lesion is viral or bacterial in nature and may even be able to suspect a particular organism. Recovery of the agent is facilitated if the microbiologist is provided with this information. Bullous (blisterlike) lesions are caused by bacteria and are often associated with sepsis, so blood cultures are mandatory. One may see bronzed skin with bullous lesions in gas gangrene, an entity with a distinctive clinical picture; *Clostridium perfringens* and other clostridia are the key pathogens. Gram stain of the fluid from such lesions typically reveals the etiologic agent to be a gram-positive bacillus and provides the clinician with additional valuable information on which to base initial therapy.

Depending on clinical presentation, specimens would be submitted for either viral or bacterial culture. Bacteriologic diagnosis can generally be made by culturing the fluid from the lesions aerobically or in carbon dioxide on blood and MacConkey agar at 35° C.

INFECTIONS OF THE SUBCUTANEOUS TISSUES

Proper collection and transport of specimens are important factors in the laboratory diagnosis of infections of the subcutaneous tissues. Specimen collection is particularly difficult because many of these lesions are open and therefore readily colonized by nosoco-

mial pathogens that may not be involved in the infection. The most reliable specimens for determining the etiology of ulcers and nodules are those obtained from the base of the ulcer or nodule following removal of overlying debris or by surgical biopsy of deep tissues without contact with the superficial layers of the lesion. A Gram stain of the specimen should be performed and material aerobically cultured on blood agar and MacConkey agar. If fungi, *Nocardia* spp., or mycobacterial infection is considered likely, appropriate fungal media (e.g., Saboraud dextrose medium with and without chloramphenicol and cycloheximide) and mycobacterial media (e.g., Middlebrook 7H9 or 7H13, Lowenstein-Jensen) should be used. These culture methods will be addressed in greater detail in Chapters 65 and 60, respectively.

Similar problems are faced when trying to collect material for culture of sinus tracts. Material for culture should be obtained from the deepest portion of the sinus tract. If systemic symptoms such as fever are present, blood cultures should also be obtained. Again, a Gram stain should be routinely performed. Cultures should be inoculated to recover both facultative and anaerobic bacteria in the same manner as described for surgical wounds later in this chapter.

INFECTIONS OF THE MUSCLE FASCIA AND MUSCLES

Blood cultures should always be drawn from patients with significant myonecrosis. Transport of material (tissue is always better than pus, which is, in turn, better than a swab) should be under anaerobic conditions. Gram stains should be routinely performed. Cultures should be inoculated to recover both facultative and anaerobic bacteria as described for surgical wounds in the following section.

WOUND INFECTIONS

Postoperative

Because anaerobic bacteria are involved in many infections, specimen collection so as to avoid indigenous flora and specimen transport under good anaerobic conditions (which will not interfere with recovery of even obligate aerobes) are particularly important. Unusual organisms associated with postsurgical wound infections, such as *Mycoplasma hominis*, *Mycobacterium chelonae*, *Mycobacterium fortuitum*, fungi, and even *Legionella* spp. should not be overlooked. A Gram-stained smear of material submitted for culture should be examined. Exudates from superficial wounds should be inoculated to blood agar, MacConkey agar, a colistin-nalidixic acid agar plate, and a broth. Material from deep wounds should be set up for anaerobic and aerobic cultures. More detailed information regarding the processing of specimens for anaerobic cultures is presented in Chapters 58 and 59.

Bites

Bite wound infections usually involve relatively small lesions and minimal exudate, so a swab technique with anaerobic transport usually is needed. Surrounding skin should be thoroughly disinfected before the specimen is obtained. The best material for culture is pus that is aspirated from the depth of the wound or samples obtained during surgery for incision and drainage or debridement, which is the removal of all dead and necrotic tissue. Again, Gram-stained smears should be prepared and examined. For aerobic cultures blood agar, MacConkey agar, and chocolate agar should be inoculated.

Burns

For burn patients, it is important to ascertain the number of organisms present per Gram stain of tissue. Greater than 10^5 colony-forming units (CFUs) per gram of tissue is considered by some clinicians to be indicative of infection, whereas less than that number may indicate only colonization. A laboratory may be asked to perform quantitative cultures of tissue, albeit rarely. Because conventional quantitative cultures are labor-intensive and expensive, Buchanan et al.[3] cultured biopsies by a semiquantitative technique that involved the inoculation of 0.1 and 0.01 mL samples to blood agar, in duplicate (Procedure 28-1). This modified procedure, which requires fewer manipulations in specimen processing, is comparable in terms of sensitivity and specificity to more involved conventional quantitative cultures.

Procedure

28-1 SEMIQUANTITATIVE BACTERIOLOGIC CULTURE OF TISSUE

PRINCIPLE

The degree or extent of bacterial wound contamination is directly related to the risk of wound sepsis. Because of this relationship, results of a quantitative culture (the number of colony-forming units [CFUs] per gram of the eschar biopsy) are used by physicians in their management of severely burned patients.

METHOD

1. Cut a piece of tissue, measuring several cubic millimeters, aseptically onto a small, preweighed, sterile urine cup.

2. Determine the weight of the tissue by subtracting the weight of the aluminum foil from the total weight.

3. Place the specimen and 2 mL of sterile nutrient broth in a sterile tissue grinder; macerate the specimen.

4. Inoculate 0.1 mL of sample to a blood agar plate, in duplicate, and an anaerobic blood agar plate (if indicated), in duplicate. In addition, inoculate 0.01 mL of sample using a calibrated loop to a blood agar plate, in duplicate. Spread the inoculum on the plates with a sterile glass spreading rod or a loop.

5. Incubate plates in 5% to 10% carbon dioxide overnight, and count the colonies of bacteria on the plate that contains 30 to 300 CFUs. If more than 300 colonies are obtained on both plated dilutions, the factor 300 is used as N for calculations and the result is considered greater than the value.

6. Calculate the number of CFUs per gram of tissue with the following formula:

Number of CFUs counted × Reciprocal of volume of homogenate inoculated (10^{-1} or 10^{-2}) × 2 (volume of diluent used for tissue homogenization) ÷ weight of tissue

For example, for a tissue that weighed 0.002 g, 68 CFUs were observed on the plate that received the 10^{-2} dilution of suspension:

$$\frac{68 \times 10^2 \times 2}{0.002} = \frac{136 \times 10^2}{2 \times 10^{-3}}$$

$$= 6.8 \times 10^6 \text{ CFU/g}$$

Modified from a method published by Buchanan et al.[3]

References

1. Amin, N. 1988. Infected diabetic foot ulcers. Am. Fam. Physician. 37:283.

2. Baron, E.J., et al. 1992. A microbiological comparison between acute and complicated appendicitis. Clin. Infect. Dis. 14:227.

3. Buchanan, K., et al. 1986. Comparison of quantitative and semiquantitative culture techniques for burn biopsy. J. Clin. Microbiol. 23:258.

4. Fleisher, G., Ludwig, S., and Campos, J. 1980. Cellulitis: bacterial etiology, clinical features and laboratory findings. J. Pediatr. 97:591.

5. George, W.L. 1989. Other infections of skin, soft tissue, and muscle. In Finegold, S.M. and George, W.L., editors. Anaerobic infections in humans. Academic Press, San Diego.

6. Goldstein, E.J.C. 1992. Bite wounds and infection. Clin. Infect. Dis. 14:633.

7. Gustafson, T.L., et al. 1983. Pseudomonas folliculitis: an outbreak and review. Rev. Infect. Dis. 5:1.

8. Habif, T.P. 1996. Principles of diagnosis and anatomy. In Clinical dermatology: a color guide to diagnosis and therapy, ed 3. Mosby, St. Louis.

9. Hook, E.W. III, et al. 1986. Microbiologic evaluation of cutaneous cellulitis in adults. Arch. Intern. Med. 146:295.

Bibliography

Ahrenholz, D.H. and Simmons, R.L. 1995. Infections of the skin and soft tissues. In Howard, R.L. and Simmons, R.L., editors. Surgical infectious diseases, ed 3. Appleton & Lange, Norwalk, Conn.

Habif, T.P. 1996. Principles of diagnosis and anatomy. In Clinical dermatology: a color guide to diagnosis and therapy, ed 3. Mosby, St. Louis.

Simor, A.E., Roberts, F.J., and Smith, J.A. 1988. Infections of the skin and subcutaneous tissues. In Smith, J.A., editor. Cumitech 23. American Society for Microbiology, Washington, D.C.

29 | NORMALLY STERILE BODY FLUIDS, BONE AND BONE MARROW, AND SOLID TISSUES

Any body tissue or sterile body fluid site can be invaded and infected with etiologic agents of disease from all four categories of microbes: bacteria, fungi, viruses, and parasites. Although from different areas of the body, all specimens discussed in this chapter are considered normally sterile. Therefore, even one colony of a potentially pathogenic microorganism may be significant. (Refer to Table 1-1 for a quick guide regarding collection, transport, and processing of specimens from sterile body sites.)

SPECIMENS FROM STERILE BODY SITES

FLUIDS

In response to infection, fluid may accumulate in any body cavity. Infected solid tissue often presents as cellulitis or with abscess formation. Areas of the body from which fluids are typically sent for microbiologic studies (in addition to blood and cerebrospinal fluid [see Chapters 20 and 23]) include those in Table 29-1.

Pleural fluid

The parietal pleura, a serous membrane of the thoracic cavity (see Chapter 21), lines the entire thoracic cavity. The outer surface of each lung is also covered by the visceral pleura (Figure 29-1). Pleural fluid is a collection of fluid in the pleural space, normally found between the lung and the chest wall (see Figure 29-1). The fluid usually contains few or no cells and has a consistency similar to that of serum but with a lower protein content. When excess amounts of this fluid are present, it is called an **effusion**, or transudate, and is often the result of cardiac, hepatic, or renal disease. Pleural fluid that contains numerous white blood cells and other evidence of an inflammatory response (an exudate) is usually caused by infection, but malignancy, pulmonary infarction, or autoimmune diseases in which an antigen-antibody reaction initiates an inflammatory response may also be responsible. The material collected from the patient by needle aspiration (thoracentesis) is submitted to the laboratory as pleural fluid, thoracentesis fluid, or empyema fluid. Exudative pleural effusions that contain numerous polymorphonuclear neutrophils, particularly those that are grossly purulent, are called **empyema** fluids. Empyema usually occurs secondary to pneumonia, but other infections near the lung (e.g., subdiaphragmatic infection) may seed microorganisms into the pleural cavity.

Peritoneal fluid

The **peritoneum** is a large, moist, continuous sheet of serous membrane that lines the walls of the abdominal-pelvic cavity and the outer coat of the organs contained within the cavity (Figure 29-2). In the abdomen these two membrane linings are separated by a space called the **peritoneal cavity,** which contains or abuts the liver, pancreas, spleen, stomach and intestinal tract, bladder, and fallopian tubes and ovaries. The kidneys occupy a retroperitoneal (behind the peritoneum) position. Within the healthy human peritoneal cavity is a small amount of fluid that maintains moistness of the surface of the peritoneum. Normal peritoneal fluid may contain as many as 300 white blood cells per milliliter, but the protein content and specific gravity of the fluid are low. During an infectious or inflammatory process, increased amounts of fluid accumulate in the peritoneal cavity, a condition called **ascites.** The fluid, often called **ascitic fluid,** contains an increased number of inflammatory cells and an elevated protein level.

Agents of infection gain access to the peritoneum through a perforation of the bowel, through infection within abdominal viscera, by way of the bloodstream, or by external inoculation (as in surgery or trauma). On occasion, as in pelvic inflammatory disease (PID), organisms travel through the natural channels of the fallopian tubes into the peritoneal cavity.

PRIMARY PERITONITIS There are two major types of infections in the peritoneal cavity: primary and secondary peritonitis. In **primary peritonitis,** no apparent focus of infection is evident. The organisms likely to be recovered from specimens from patients with primary peritonitis vary with the patient's age. The most common etiologic agents in children are *Streptococcus pnuemoniae* and group A streptococci, *Enterobacteriaceae,* other gram-negative bacilli, and staphylococci. In adults, *Escherichia coli* is the most common

TABLE 29-1 AREAS OF THE BODY FROM WHICH FLUIDS ARE SUBMITTED TO THE MICROBIOLOGY LABORATORY

BODY AREA	FLUID NAME(S)
Thorax	Thoracentesis or pleural or empyema fluid
Abdominal cavity	Paracentesis or ascitic or peritoneal fluid
Joint	Synovial fluid
Pericardium	Pericardial fluid

bacterium, followed by *S. pneumoniae* and group A streptococci. Polymicrobic peritonitis is unusual in the absence of bowel perforation or rupture. Among sexually active young women, *Neisseria gonorrhoeae* and *Chlamydia trachomatis* are common etiologic agents of peritoneal infection, often in the form of a perihepatitis (inflammation of the surface of the liver, called Fitz-Hugh-Curtis syndrome). Tuberculous peritonitis occurs infrequently in the United States today and is more likely to be found among persons recently arrived from South America, Southeast Asia, or Africa. Fungal causes of peritonitis are not common, but *Candida* spp. may be recovered from immunosuppressed patients and patients receiving prolonged antibacterial therapy.

SECONDARY PERITONITIS **Secondary peritonitis** is a sequel to a perforated viscus (organ), surgery, traumatic injury, loss of bowel wall integrity because of destructive disease (e.g., ulcerative colitis, ruptured

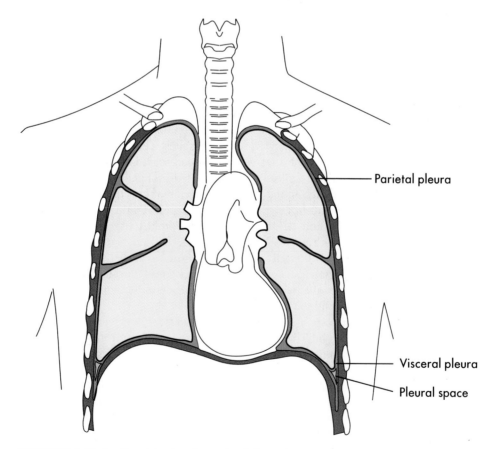

FIGURE 29-1 The location of the pleural space in relation to the parietal and visceral pleura and the rest of the respiratory tract.

appendix, carcinoma), obstruction, or a preceding infection (liver abscess, salpingitis, septicemia). The nature, location, and etiology of the underlying process govern the agents to be recovered from peritoneal fluid. With PID as the background, gonococci, anaerobes, or chlamydia are isolated. With peritonitis or intraabdominal abscess, anaerobes generally are found in peritoneal fluid, usually together with *Enterobacteriaceae* and enterococci or other streptococci. In patients whose bowel flora have been altered by antimicrobial agents, more resistant gram-negative bacilli and *Staphylococcus aureus* may be encountered. Because anaerobes outnumber aerobes in the bowel by 1000-fold, it is not surprising that anaerobic organisms play a prominent role in intraabdominal infection, perhaps acting synergistically with faculta-

tive bacteria. The organisms likely to be recovered include *E. coli*, the *Bacteroides fragilis* group, enterococci and other streptococci, *Bilophila* spp., other anaerobic gram-negative bacilli, anaerobic gram-positive cocci, and clostridia.

Peritoneal dialysis fluid

More than 5000 patients with end-stage renal disease are maintained on continuous ambulatory peritoneal dialysis (CAPD). In this treatment, fluid is injected into the peritoneal cavity and subsequently removed, which allows exchange of salts and water and removal of various wastes in the absence of kidney function. Because the dialysate fluid is injected into the peritoneal cavity via a catheter, this break in the skin barrier places the dialysis patient at significant

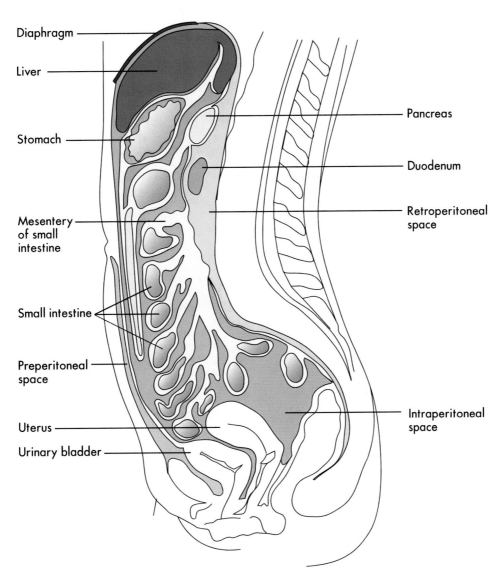

FIGURE 29-2 The abdominal cavity. The retroperitoneal and preperitoneal spaces are considered as extraperitoneal (outside) spaces. (Modified from Thibodeau, G.A. 1993. Anatomy and Physiology. Mosby, St. Louis.)

risk for infection. The average incidence of peritonitis in these patients is up to two episodes per year per patient. Peritonitis is best diagnosed clinically by the presence of cloudy dialysate with or without abdominal pain.[2,5] Although white blood cells are usually plentiful (a value of leukocytes >100/mL is usually indicative of infection), the number of organisms is usually too low for detection on Gram stain of the peritoneal fluid sediment unless a concentrating technique is used; fungi are more readily detected.

Most infections originate from the patient's own skin flora; *Staphylococcus epidermidis* and *S. aureus* are the most common etiologic agents, followed by streptococci, aerobic or facultative gram-negative bacilli, *Candida* spp., *Corynebacterium* spp., and others. The oxygen content of peritoneal dialysate is usually too high for development of anaerobic infection. Among the gram-negative bacilli isolated, *Pseudomonas* spp., *Acinetobacter* spp., and the *Enterobacteriaceae* are frequently seen.

Pericardial fluid

The heart and contiguous major blood vessels are surrounded by a protective tissue, the pericardium. The area between the epicardium which is the membrane surrounding the heart muscle, and the pericardium is called the **pericardial space** and normally contains 15 to 20 mL of clear fluid. If an infectious agent is present within the fluid, the pericardium may become distended and tight, and eventually **tamponade** (interference with cardiac function and circulation) can ensue.

Agents of **pericarditis** (inflammation of the pericardium) are usually viruses. Parasites, bacteria, certain fungi, and noninfectious causes are also associated with this disease.

Myocarditis (inflammation of the heart muscle itself) may accompany pericarditis. The pathogenesis of disease involves the host inflammatory response contributing to fluid buildup and cell and tissue damage. The most common etiologic agents of pericarditis and myocarditis are listed in Box 29-1. Other bacteria, fungi, and parasitic agents have been recovered from pericardial effusions, and therefore all agents should be sought.

Patients who develop pericarditis resulting from agents other than viruses are often compromised in some way. An example is infective endocarditis, in which a myocardial abscess develops and then ruptures into the pericardial space.

Joint fluid

Infectious arthritis may involve any joint in the body. Infection of the joint usually occurs secondary to hematogeneous spread of bacteria or, less often, fungi, or as a direct extension of infection of the bone. It

may also occur after injection of material, especially corticosteroids, into joints or after insertion of prosthetic material (e.g., total hip replacement). Although infectious arthritis usually occurs at only one site (monoarticular), a preexisting bacteremia or fungemia may seed more than one joint to establish polyarticular infection, particularly when multiple joints are diseased, such as in rheumatoid arthritis. Knees and hips are the most frequently affected joints.

In addition to active infections associated with viable microorganisms within the joint, sterile, self-limited arthritis caused by antigen-antibody interactions may follow an episode of infection, such as meningococcal meningitis. When an etiologic agent cannot be isolated from an inflamed joint fluid specimen, either the absence of viable agents or inadequate transport or culturing procedures can be blamed. For example, even under the best circumstances, *Borrelia burgdorferi* is isolated from the joints of fewer than 20% of patients with Lyme disease. Nonspecific test results, such as increased white blood cell count, decreased glucose, or elevated protein, may seem to implicate an infectious agent but are not conclusive. A role has been postulated for the persistence of bacterial L-forms (cell-wall–deficient forms) in joint fluid after systemic infection, but such theories have not been proved.

Overall, *Staphylococcus aureus* is the most common etiologic agent of septic arthritis, accounting for approximately 70% of all such infections. In adults

BOX 29-1 COMMON ETIOLOGIC AGENTS OF PERICARDITIS AND MYOCARDITIS

Viruses
Enteroviruses (primary coxsackie A and B and, less frequently, echoviruses)
Adenoviruses
Influenza viruses
Bacteria
Mycoplasma pneumoniae
Chlamydia trachomatis
Mycobacterium tuberculosis
Staphylococcus aureus
Streptococcus pneumoniae
Enterobacteriaceae and other gram-negative bacilli
Fungi
Coccidioides immitis
Aspergillus spp.
Candida spp.
Cryptococcus neoformans
Histoplasma capsulatum
Parasites
Entamoeba histolytica
Toxoplasma gondii

younger than age 30, however, *Neisseria gonorrhoeae* is isolated most frequently. *Haemophilus influenzae* has been the most common agent of bacteremia in children younger than age 2, and consequently it has been the most frequent cause of infectious arthritis in these patients, followed by *S. aureus.* The widespread use of *H. influenzae* type B vaccine should contribute to a change in this pattern. Streptococci, including groups A *(Streptococcus pyogenes)* and B *(Streptococcus agalactiae),* pneumococci, and viridans streptococci, are prominent among bacterial agents associated with infectious arthritis in patients of all ages. Among anaerobic bacteria, *Bacteroides,* including *B. fragilis,* may be recovered, as may *Fusobacterium necrophorum,* which usually involves more than one joint in the course of sepsis. Among people living in certain endemic areas of the United States and Europe, infectious arthritis is a prominent feature of Lyme disease. Some of the more frequently encountered etiologic agents of infectious arthritis are listed in Box 29-2.

These agents act to stimulate a host inflammatory response, which is initially responsible for the pathology of the infection. Arthritis is also a symptom associated with infectious diseases caused by certain agents, such as *Neisseria meningitidis,* group A streptococci (rheumatic fever), and *Streptobacillus moniliformis,* in which the agent cannot be recovered from joint fluid. Presumably, antigen-antibody complexes formed during active infection accumulate in a joint, initiating an inflammatory response that is responsible for the ensuing damage.

Infections in prosthetic joints are usually associated with somewhat different etiologic agents than those in natural joints. After insertion of the prosthesis, organisms that gained access during the surgical procedure slowly multiply until they reach a critical mass and produce a host response. This many occur long after the initial surgery; approximately half of all prosthetic joint infections occur more than 1 year after surgery. Skin flora are the most common etiologic agents, with *Staphylococcus epidermidis,* other coagulase-negative staphylococci, *Corynebacterium* spp., and *Propionibacterium* spp. predominating. However, *Staphylococcus aureus* is also a major pathogen in this infectious disease. Alternatively, organisms may reach joints during hematogeneous spread from distant, infected sites.[8]

BONE

Bone marrow aspiration or biopsy

Diagnosis of certain diseases, including brucellosis, histoplasmosis, blastomycosis, tuberculosis, and leishmaniasis, can sometimes be made only by detection of the organisms in the bone marrow. *Brucella* spp. can be isolated on culture, as can fungi, but parasitic agents must be visualized in smears or sections made from bone marrow material. Of importance, many of the etiologic agents associated with disseminated infections in HIV-infected patients may be visualized or isolated from the bone marrow. Some of these organisms include cytomegalovirus, *Cryptococcus neoformans,* and *Mycobacterium avium* complex.

Bone biopsy

A small piece of infected bone is occasionally sent to the microbiology laboratory for determination of the etiologic agent of **osteomyelitis** (infection of bone). Patients develop osteomyelitis from hematogeneous spread of an infectious agent, invasion of bone tissue from an adjacent site of infection (e.g., joint infection, dental infection), breakdown of tissue caused by trauma or surgery, or lack of adequate circulation followed by colonization of a skin ulceration with microorganisms. Once established, infections in bone

MOST FREQUENTLY ENCOUNTERED ETIOLOGIC AGENTS

BOX 29-2

Bacterial
Staphylococcus aureus
Beta-hemolytic streptococci
Streptococci (other)
Haemophilus influenzae
Haemophilus spp. (other)
Bacteroides spp.
Fusobacterium spp.
Neisseria gonorrhoeae
Pseudomonas spp.
Salmonella spp.
Pasteurella multocida
Moraxella osloensis
Kingella kingae
Moraxella catarrhalis
Capnocytophaga spp.
Corynebacterium spp.
Clostridium spp.
Peptostreptococcus spp.
Eikenella corrodens
Actinomyces spp.
Mycobacterium spp.
Mycoplasma spp.
Ureaplasma urealyticum
Borrelia burgdorferi
Fungal
Candida spp.
Cryptococcus neoformans
Coccidioides immitis
Sporothrix schenckii
Viral
Hepatitis B
Mumps
Rubella
Other viruses (rarely)

may tend to progress toward chronicity, particularly if effective blood supply to the affected area is lacking.

Staphylococcus aureus, seeded during bacteremia, is the most common etiologic agent of osteomyelitis among people of all age groups. The toxins and enzymes produced by this bacterium, as well as its ability to adhere to smooth surfaces and produce a protective glycocalyx coating, seem to contribute to its pathogenicity. Among young persons, osteomyelitis is usually associated with a single agent. Such infections are usually of hematogeneous origin. Other organisms that have been recovered from hematogeneously acquired osteomyelitis include *Salmonella* spp., *Haemophilus* spp., Enterobacteriaceae, *Pseudomonas* spp., *Fusobacterium necrophorum*, and yeasts. *S. aureus* or *Pseudomonas aeruginosa* is often recovered from drug addicts. Parasites or viruses are rarely, if ever, etiologic agents of osteomyelitis.

Bone biopsies from infections that have spread to a bone from a contiguous source or that are associated with poor circulation, especially in patients with diabetes, are likely to yield multiple isolates. Gram-negative bacilli are increasingly common among hospitalized patients; a break in the skin (surgery or intravenous line) may precede establishment of gram-negative osteomyelitis. Breaks in skin from other causes, such as a bite wound or trauma, also may be the initial event that leads to underlying bone infection. For example, a human bite may lead to infection with *Eikenella corrodens*, whereas an animal bite may lead to *Pasteurella multocida* osteomyelitis. Stepping on a nail or other sharp object while wearing tennis shoes has been associated with osteomyelitis caused by *P. aeruginosa*.[6] Poor oral hygiene may lead to osteomyelitis of the jaw with *Actinomyces* spp., *Capnocytophaga* spp., and other oral flora, particularly anaerobes. Pigmented *Prevotella* (formerly *Bacteroides melaninogenicus* group) and *Porphyromonas*, *Fusobacterium*, and *Peptostreptococcus* spp. are often involved. Pelvic infection in the female may lead to mixed aerobic and anaerobic osteomyelitis of the pubic bone.

Patients with neuropathy (pathologic changes in the peripheral nervous system) in the extremities, notably patients with diabetes, who may have poor circulation as well, are subjected to trauma that they cannot feel simply by walking. They develop ulcers on the feet that do not heal, become infected, and may eventually progress to involve underlying bone. These infections are usually polymicrobial, involving anaerobic and aerobic bacteria.[3] *Prevotella* or *Porphyromonas*, other gram-negative anaerobes, including the *Bacteroides fragilis* group, *Peptostreptococcus* spp., *Staphylococcus aureus*, and group A and other streptococci are frequently encountered.

SOLID TISSUES

Pieces of tissue are removed from patients during surgical or needle biopsy procedures or may be collected at autopsy. Any agent of infection may cause disease in tissue, and laboratory practices should be adequate to recover bacteria, fungi, and viruses and to detect the presence of parasites. Fastidious organisms (e.g., *Brucella* spp.), and agents of chronic disease (e.g., systemic fungi and mycobacteria) may require special media and long incubation periods for isolation. Some agents that require special supportive or selective media are listed in Box 29-3.

LABORATORY DIAGNOSTIC PROCEDURES

SPECIMEN COLLECTION AND TRANSPORT

Because of the numerous specimen types from normally sterile body sites that can be submitted to the laboratory, requirements for collection and transport vary.

Fluids and aspirates

Most specimens (pleural, peritoneal, pericardial, and synovial fluids) are collected by aspiration with a needle and syringe.

Collection of pericardial fluid is obviously hazardous because the sample is immediately adjacent to the beating heart. Collection is performed by needle aspiration with electrocardiographic monitoring or as a surgical procedure. Laboratory personnel should be alerted in advance so that the appropriate media, tissue culture media, and stain procedures are available immediately.

Body fluids from sterile sites should be transported to the laboratory in a sterile tube or vial that excludes oxygen. From 1 to 5 mL of specimen is adequate for isolation of most bacteria, but the larger the specimen, the better, particularly for isolation of *M. tuberculosis* and fungi; at least 5 mL should be submitted for each of these latter two cultures. Ten mL of fluid is recommended for the diagnosis of peritonitis.

BOX 29-3 **INFECTIOUS AGENTS IN TISSUE REQUIRING SPECIAL MEDIA**

Actinomyces spp.
Brucella spp.
Legionella spp.
Bartonella (Rochalimaea) henselae (cat-scratch disease bacilli)
Systemic fungi
Mycoplasma
Mycobacteria
Viruses

Anaerobic transport vials are available from several sources. These vials are prepared in an oxygen-free atmosphere and are sealed with a rubber septum or short stopper through which the fluid is injected. Fluid should never be transported in a syringe capped with a sterile rubber stopper because this method is unsafe. The use of syringes should be curtailed whenever possible because recapping and removal of needles is not permitted under universal precautions requirements for hospital safety. Most clinically significant anaerobic bacteria survive adequately in nonanaerobic transport containers (e.g., sterile, screw-capped tubes) for short periods if the specimen is frankly purulent and of adequate volume. Specimens received in anaerobic transport vials should be inoculated to routine aerobic (an enriched broth, blood, chocolate, and sometimes, MacConkey agar plates) and anaerobic media (see Chapters 58 and 59) as quickly as possible. Specimens for recovery of only fungi or mycobacteria may be transported in sterile, screw-capped tubes. At least 5 to 10 mL of fluid are required for adequate recovery of small numbers of organisms. If gonococci or chlamydiae are suspected, additional aliquots should be sent to the laboratory for smears and appropriate cultures.

With respect to synovial and peritoneal fluids, the inoculation of blood culture broth bottle(s) at the bedside or in the laboratory may prove to be beneficial.[1,9,10,12] One must always remember to send some of the specimen to the laboratory in a container other than a blood culture bottle, since putting the sample into a blood culture bottle dilutes it, making the preparation of a smear for Gram stain useless. The specimen in the blood culture bottle is processed as a blood culture, facilitating the recovery of small numbers of organisms and diluting out the effects of antibiotics. Citrate or sodium polyanetholesulfonate (SPS) may be used as an anticoagulant (see Chapter 20). Of importance, specimens collected by percutaneous needle aspiration (paracentesis) or at the time of surgery should be inoculated into aerobic and anaerobic blood culture broth bottles immediately at the bedside. Any delay results in decreased detection of true positive cultures.[10] In general, if sufficient specimen is available, cultures should be inoculated with the same volumes specified by the manufacturer for blood specimens. If the volume is insufficient to follow the manufacturer's instructions, as little as 0.1 mL can be inoculated.

Fluid from CAPD patients can be submitted to the laboratory in a sterile tube, urine cup, or in the original bag. The bag is entered only once with a sterile needle and syringe to withdraw fluid for culture. Fluid may also be directly inoculated into blood culture bottles (at least 20 mL [10 mL in each of two culture bottles] and cultured).[10] Numerous other studies indicate that in addition to blood culture bottles, an adult Isolator tube is a sensitive and specific method for culture.[5,13]

Bone

Bone marrow is typically aspirated from the interstitium of the iliac crest. Usually, this material is not processed for routine bacteria, because blood cultures are equally useful for these microbes, and false-positive cultures for skin bacteria *(Staphylococcus epidermidis)* are frequent. Some laboratories report good recovery from bone marrow material that has been injected into a Pediatric Isolator tube as a collection and transport device. The lytic agents within the Isolator tube are thought to lyse cellular components, presumably freeing intracellular bacteria for enhanced recovery. Bone removed at surgery or by percutaneous biopsy is sent to the laboratory in a sterile container.

Tissue

Tissue specimens are obtained after careful preparation of the skin site. It is critical that biopsy specimens be collected aseptically and submitted to the microbiology laboratory in a sterile container. A wide-mouthed, screw-capped bottle or plastic container is recommended. Anaerobic organisms survive within infected tissue long enough to be recovered from culture. A small amount of sterile, nonbacteriostatic saline may be added to keep the specimen moist. Because homogenizing with a tissue grinder can destroy some organisms by the shearing forces generated during grinding, it is often best to use a sterile scissors and forceps to mince larger tissue specimens into small pieces suitable for culturing (Figure 29-3). Note: *Legionella* spp. may be inhibited by saline; a section of lung should be submitted without saline for *Legionella* isolation.

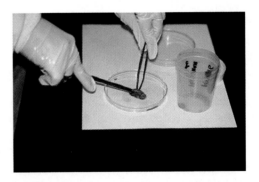

FIGURE 29-3 Mincing a piece of tissue for culture using a sterile forceps and scissors. Note: Perform this procedure in a biosafety cabinet.

If anaerobic organisms are of concern, a small amount of tissue can be placed into a loosely capped, wide-mouthed plastic tube and sealed into an anaerobic pouch system, which also seals in moisture enough for survival of organisms in tissue until the specimen is plated. The surgeon should take responsibility for seeing that a second specimen is submitted to anatomic pathology for histologic studies. Formaldehyde-fixed tissue is not useful for recovery of viable microorganisms, although some organisms can be recovered after very short periods. Therefore, an attempt may rarely be made to subculture from tissue in formalin if that is the only specimen available. Material from draining sinus tracts should include a portion of the tract's wall, obtained by deep curettage. Tissue from infective endocarditis should contain a portion of the valve and vegetation if the patient is undergoing valve replacement.

In some instances, contaminated material may be submitted for microbiologic examination. Specimens, such as tonsils or autopsy tissue, may be surface cauterized with a heated spatula or blanched by immersing in boiling water for 5 to 10 seconds to reduce surface contamination. The specimen may then be dissected with sterile instruments to permit culturing of the specimen's center, which will not be affected by the heating. Alternatively, larger tissues may be cut in half with a sterile scissors or blade and the interior portion cultured for microbes.

Because surgical specimens are obtained at great risk and expense to the patient, and because supplementary specimens cannot be obtained easily, it is important that the laboratory save a portion of the original tissue (if enough material is available) in a small amount of sterile broth in the refrigerator and at $-70°$ C (or, if necessary, at $-20°$ C) for at least 4 weeks in case additional studies are indicated. If the entire tissue must be ground up for culture, a small amount of the suspension should be placed into a sterile tube and refrigerated.

SPECIMEN PROCESSING, DIRECT EXAMINATION, AND CULTURE

Fluids and aspirates

Techniques for laboratory processing of all sterile body fluids are similar. Clear fluids may be concentrated by centrifugation or filtration, whereas purulent material can be inoculated directly to media. Any body fluid received in the laboratory that is already clotted must be homogenized to release trapped bacteria and minced or cut to release fungal cells. Either processing such specimens in a motorized tissue homogenizer or grinding them manually in a mortar and pestle or glass tissue grinder allows better recovery of bacteria. Hand grinding is often preferred, because motorized grinding can generate considerable heat and thereby kill microorganisms in the specimen. Grinding may lyse fungal elements; therefore it is not recommended with specimens processed for fungi. Small amounts of whole material from a clot should be aseptically cut with a scalpel and placed directly onto media for isolation of fungi.

All fluids should be processed for direct microscopic examination. In general, if one organism is seen per oil immersion field, at least 10^5 organisms per milliliter of specimen are present. In such cases, often only a few organisms are present in normally sterile body fluids. Therefore, organisms must be concentrated in body fluids. For microscopic examination, cytocentrifugation (see Figure 23-3) should be used to prepare Gram-stained smears because organisms can be further concentrated up to a 1000-fold.[4] Body fluids for culture should be concentrated by either filtration or high-speed centrifugation. Once the sample is concentrated, the supernatant is aseptically decanted or aspirated with a sterile pipette, leaving approximately 1 mL liquid in which to mix the sediment thoroughly. Vigorous vortexing or drawing the sediment up and down into a pipette several times adequately resuspends the sediment. This procedure should be done in a biological safety cabinet. The suspension is used to inoculate media. Direct potassium hydroxide (KOH) or calcofluor white preparations for fungi and acid-fast stain for myocbacteria can also be performed.

Specimens for fungi should be examined by direct wet preparation or by preparing a separate smear for periodic acid-Schiff (PAS) staining in addition to Gram stain. Either 10% KOH or calcofluor white is recommended for visualization of fungal elements from a wet preparation. In addition to hyphal forms, material from the thoracic cavity may contain spherules of *Coccidioides* or budding yeast cells.

Lysis of leukocytes before concentration of CAPD effluents can significantly enhance recovery of organisms.[7] Filtration of CAPD fluid through a 0.45-μm pore-size membrane filter allows a greater volume of fluid to be processed and usually yields better results. Because the numbers of infecting organisms may be low (fewer than one organism per 10 mL of fluid), a large quantity of fluid must be processed. Sediment obtained from at least 50 mL of fluid has been recommended.[1] If the specimen is filtered, the filter should be cut aseptically into three pieces, one of which is placed on chocolate agar for incubation in 5% carbon dioxide, one on MacConkey agar, and the other on a blood agar plate for anaerobic incubation.

If fluids have been concentrated by centrifugation, the resulting sediment should be inoculated to an enrichment broth and blood and chocolate agars.

Because these specimens are from normally sterile sites, selective media are inadvisable because they may inhibit the growth of some of the organisms being sought. Appropriate procedures for the isolation of anaerobes, mycobacteria, fungi, *Chlamydia* spp., and viruses should be used when such cultures are clinically indicated.

Bone

Clotted bone marrow aspirates or biopsies must be homogenized or ground to release trapped microorganisms. Specimens are inoculated to the same media as for other sterile body fluids. A special medium for enhancement of growth of *Brucella* spp. and incubation under 10% carbon dioxide may be needed. A portion of the specimens may be inoculated directly to fungal media. Sections are also made from biopsy material (bone) for fixation, staining, and examination (usually by anatomic pathologists) for the presence of mycobacterial, fungal, or parasitic agents. With respect to obtaining specimens from patients suspected of having osteomyelitis, cultures taken from open wound sites above infected bone or material taken from a draining sinus leading to an area of osteomyelitis may not reflect the actual etiologic agent of the underlying osteomyelitis.

Normal bone is very difficult to break up; however, most infected bone is soft and necrotic. Therefore, grinding the specimen in a mortar and pestle may break off some pieces. Small shavings from the most necrotic-looking areas of the bone specimen may sometimes be scraped off aseptically and inoculated to media. Pieces should be placed directly into media for recovery of fungi. Small bits of bone can be ground with sterile broth to form a suspension for bacteriologic and mycobacterial cultures. If anaerobes are to be recovered, all manipulations are best performed in an anaerobic chamber. If such an environment is unavailable, microbiologists should work quickly within a biosafety cabinet to inoculate prereduced anaerobic plates and broth with material from the bone.

Solid tissue

Tissue should be manipulated within a laminar flow biological safety cabinet by an operator wearing gloves. Processing tissue within an anaerobic chamber is even better. The microbiologist should cut through the infected area (which is often discolored) with a sterile scalpel blade. Half of the specimen can then be used for fungal cultures and the other for bacterial cultures. Both types of microbial agents should be sought in all tissue specimens. Some sample should also be sent to surgical pathology for histologic examination. Specimens should be cultured for viruses or acid-fast bacilli when such tests are requested. Material that is to be cultured for parasites should be finely minced or teased before inoculation into broth (see Chapter 64). Direct examination of stained tissue for parasites is performed by anatomic pathologists. Imprint cultures of tissues may yield bacteriologic results identical to homogenates and may help differentiate microbial infection within the tissue's center from surface colonization (growth only at the edge) when specimens are cut in half before processing. Additional media can be inoculated for incubation at lower temperatures, which may facilitate recovery of certain systemic fungi and mycobacteria.

Tissue may also be inoculated to virus tissue culture cells for isolation of viruses. Brain, lung, spinal fluid, and blood are most useful. Tissue may be examined by immunofluorescence for the presence of herpes simplex virus, varicella-zoster virus, cytomegalovirus, or rabies viral particles. Lung tissue should be examined by direct fluorescent antibody test for *Legionella* spp.

The tissues of all fetuses, premature infants, and young babies who have died from an infectious process should be cultured for *Listeria*. Specimens of the brain, spinal fluid, blood, liver, and spleen are most likely to contain the organism. The isolation procedure is given in detail by Seeliger and Cherry.[11]

References

1. Bobadilla, M., Sifuentes, J., and Garcia-Tsao, G. 1989. Improved method for the bacteriological diagnosis of spontaneous bacterial peritonitis. J. Clin. Microbiol. 27:2145.

2. Buggy, F.P. 1986. Culture methods for continuous ambulatory peritoneal dialysis-associated peritonitis. Clin. Microbiol. Newsletter 8:12.

3. Centers for Disease Control–Department of Health and Human Services, Public Health Service. 1996. The prevention and treatment of complication of diabetes: a guide for primary care practitioners. Atlanta.

4. Chapin-Robertson, K., Dahlberg, S.E., and Edberg, S.C. 1992. Clinical and laboratory analyses of Cytospin-prepared Gram stains for recovery and diagnosis of bacteria from sterile body fluids. J. Clin. Microbiol. 30:377.

5. Forbes, B.A., et al. 1988. Evaluation of the lysis-centrifugation system for culturing dialysates from continuous ambulatory peritoneal dialysis patients with peritonitis. Am. J. Kidney Dis. XI:176.

6. Jacobs, R.F., McCarthy, R.E., and Elser, J.M. 1989. *Pseudomonas* osteochondritis complicating puncture wounds of the foot in children: a 10-year evaluation. J. Infect. Dis. 160:657.

7. Ludlam, H.A., et al. 1988. Laboratory diagnosis of peritonitis in patients on continuous ambulatory peritoneal dialysis. J. Clin. Microbiol. 26:1757.

8. Maderazo, E.G., Judson, S., and Pasternak, H. 1988. Late infections of total joint prostheses. Clin. Orthop. 229:131.

9. Rheinhold, C.E., et al. 1988. Evaluation of broth media for routine culture of cerebrospinal fluid and joint fluid specimens. Am. J. Clin. Pathol. 89:671.

10. Runyon, B., et al. 1990. Bedside inoculation of blood culture bottles with ascitic fluid is superior to delayed inoculation in the detection of spontaneous bacterial peritonitis. J. Clin. Microbiol. 28:2811.

11. Seeliger, H.P.R. and Cherry, W.B. 1957. Human listeriosis: its nature and diagnosis. Washington, D.C. U.S. Government Printing Office.

12. Von Essen, R. and Holtta, A. 1986. Improved method of isolating bacteria from joint fluids by the use of blood culture bottle. Ann. Rheum. Dis. 45:454.

13. Woods, G.L. and Washington, J.A. 1987. Comparison of method for processing dialysate in suspected continuous ambulatory peritoneal dialysis-associated peritonitis. Diagn. Microbiol. Infect. Dis. 7:155.

Part Four

4

Bacteriology

30 OVERVIEW OF BACTERIAL IDENTIFICATION METHODS AND STRATEGIES

RATIONALE FOR APPROACHING ORGANISM IDENTIFICATION

Deciding how to teach microbiology in a manner that is both comprehensive and yet understandable is difficult. Most microbiology text chapters are organized by genus name and provide no obvious approach as to how to work up a clinical isolate. Some texts have flowcharts containing algorithms for organism workup; many of these, however, are either too broad to be helpful (e.g., gram-positive vs. gram-negative bacilli) or too esoteric (e.g., cellular analysis of fatty acids) to be practical in routine clinical practice. Unfortunately, the student ends up simply memorizing seemingly unrelated bits of information about various organisms.

The chapters in Part 4 have been arranged to guide the student through the workup of a microorganism. To accomplish this, chapters have been grouped into subsections using results of basic microbiology procedures, such as the Gram stain, oxidase and catalase tests, and growth on common laboratory media, such as blood, chocolate, and MacConkey agars. Each chapter begins with a short description of the organisms covered in the chapter and an assessment of what the organisms have in common. Because microbiology is ultimately the identification of organisms based on common phenotypic traits shared with known members of the same genus or family, microbiologists "play the odds" every day by finding the best biochemical fit and assigning the most probable identification. For example, the gram-negative rod known as CDC group EF-4a may be considered with either the MacConkey-positive or -negative organisms because it grows on MacConkey agar 50% of the time. Therefore although CDC group EF-4a has been arbitrarily assigned to the section on oxidase-positive, MacConkey-positive, gram-negative bacilli and coccobacilli, it is also included in the discussion of oxidase-positive, MacConkey-negative, gram-negative bacilli and coccobacilli. This approach to the identification of microorganisms is similar to that employed by clinical microbiologists. Internalization of this method will allow students entering the workforce to immediately be able to identify bacteria.

The following chapters help guide the determination of whether a clinical isolate is relevant and provide key biochemical characteristics necessary for organism identification, information on whether susceptibility testing is indicated, and information on the correct antimicrobial agents to use. Most of the procedures described in these chapters can be found either in Chapter 13 or at the end of this chapter. In this chapter, each procedure includes a photograph of positive and negative reactions so that the information discussed in the section on expected results can be easily visualized. Chapter 12 includes additional photographs of some commonly used bacteriologic tests. In addition, Table 30-1 lists several commonly used commercial identification systems for a variety of the microorganisms discussed in the following pages.

APPROACH TO USING PART 4

In most instances, the first information that a microbiologist uses in the identification process is the macroscopic description of the colony, or colony morphology. This includes the type of hemolysis (if any), pigment (if present), size, texture (opaque, translucent, or transparent), adherence to agar, pitting of agar, and many other characteristics. After careful observation of the colony, the Gram stain is used to separate the organism into one of a variety of broad categories based on Gram reaction and cellular morphology of gram-positive or gram-negative bacteria (e.g., gram-positive cocci, gram-negative rods). For gram-positive organisms, the catalase test should follow the Gram stain and testing on gram-negative organisms should begin with the oxidase test. These simple tests plus growth on MacConkey agar, if the isolate is a gram-negative rod or coccobacillus, help the microbiologist assign the organism to one of the primary categories (organized here as subsections). Subsequent testing criteria that are outlined in each chapter are then used to more definitively identify the isolate.

The clinical diagnosis and the source of the specimen also can aid in determining which group of organisms to consider. For example, if a patient has endocarditis or the specimen source is blood, and a

_____ **TABLE 30-1 EXAMPLES OF COMMERCIAL IDENTIFICATION SYSTEMS** _____
FOR VARIOUS ORGANISMS

ORGANISM GROUP	TYPE OF SYSTEM	MANUFACTURER	INCUBATION TIME
Enterobacteriaceae	**Manual**		
	API 20E	bioMérieux Vitek[a]	24-48 hrs
	API Rapid 20E	bioMérieux Vitek	4 hrs
	Crystal E/NF	Becton Dickinson Microbiology Systems[b]	18 hrs
	Enterotube II	Becton Dickinson Microbiology Systems	24 hrs
	GN Microplate	Biolog[c]	4-24 hrs
	r/b Enteric Differential System	Remel Laboratories[d]	24 hrs
	Automated:		
	GNI card	bioMérieux Vitek	4-13 hrs
	NEG ID Type 2	Dade International Dade MicroScan[e]	15-42 hrs
Enterococcus spp. and _Streptococcus_ spp.	**Manual**		
	API 20 Strep	bioMérieux Vitek	4-24 hrs
	RapID STR	Innovative Diagnostic Systems[f]	4 hrs
	Automated:		
	GPI Card	bioMérieux Vitek	4-15 hrs
	Pos ID	Dade International Dade MicroScan	24-48 hrs
Haemophilus spp.	**Manual**		
	RapID NH	Innovative Diagnostic Systems	4 hrs
	HNID	Dade International Dade MicroScan	4 hrs
	Automated:		
	NHI Card	bioMérieux Vitek	4 hrs
Neisseria spp. and _Moraxella catarrhalis_	**Manual:**		
	Gonochek II[g]	E-Y Laboratories[h]	30 min
	RIM-Neisseria	Remel Laboratories	4 hrs
	Quad-FERM+	bioMérieux Vitek	2 hrs
	RapID NH	Innovative Diagnostic Systems	4 hrs
	HNID	Dade International Dade MicroScan	4 hrs

Continued

TABLE 30-1 EXAMPLES OF COMMERCIAL IDENTIFICATION SYSTEMS FOR VARIOUS ORGANISMS—CONT'D

ORGANISM GROUP	TYPE OF SYSTEM	MANUFACTURER	INCUBATION TIME
	Automated:		
	HI Card	bioMérieux Vitek	4 hrs
Nonenteric gram-negative rods	**Manual:**		
	API NFT	bioMérieux Vitek	24-28 hrs
	API 20E[i]	bioMérieux Vitek	24-48 hrs
	Crystal E/NF	Becton Dickinson Microbiology Systems	18-24 hrs
	Automated:		
	GNI Card	bioMérieux Vitek	4-13 hrs
	Neg ID Type 2	Dade International Dade MicroScan	15-42 hrs
Staphylococcus spp.	**Manual**		
	API STAPH	bioMérieux Vitek	24 hrs
	Automated:		
	GPI Card	bioMérieux Vitek	4-15 hrs
	Pos ID	Dade International Dade MicroScan	24-48 hrs

[a] Hazelwood, Mo.

[b] Cockeysville, Md.

[c] Hayward, Calif.

[d] Lenexa, Kan.

[e] West Sacramento, Calif.

[f] Norcross, Ga.

[g] *Neisseria gonorrhoeae* only.

[h] San Mateo, Calif.

[i] Identifies commonly isolated nonfermentative bacteria, such as *Pseudomonas aeruginosa, Stenotrophomonas maltophilia,* and *Acinetobacter* spp.

small, gram-negative rod is observed on Gram stain, the microbiologist should consider a group of gram-negative bacilli known as the *HACEK* bacteria for *Haemophilus aphrophilus, Actinobacillus actinomycetemcomitans, Cardiobacterium hominis, Eikenella corrodens,* and *Kingella* spp. Similarly, if the patient has suffered an animal bite, the microbiologist should think of *Pasteurella multocida,* EF-4a, and EF-4b if the isolate is gram-negative and *Staphylococcus hyicus* and *S. intermedius* if the organism is gram-positive.

The most important thing to remember when attempting to group and identify any microorganism is that no piece of information should be overlooked. In this era of cost containment it is more important than ever to take a stepwise approach to microbial identification, beginning with the simplest procedures and proceeding to more expensive definitive testing only if the clinical situation warrants further work. The days of going from a quick glance at a colony on a plate to a rapid identification system without a Gram stain, catalase, or oxidase test are over. Microbiologists are going back to basics. The real challenge is how to most efficiently and effectively bring all the information together in a cost-effective and clinically relevant manner.

Procedures

30-1 ACETAMIDE UTILIZATION

PRINCIPLE

This test is used to determine the ability of an organism to use acetamide as the sole source of carbon. Bacteria that can grow on this medium deaminate acetamide to release ammonia. The production of ammonia results in a pH-driven color change of the medium from green to royal blue.

METHOD

1. Inoculate acetamide slant *lightly* with a needle using growth from an 18- to 24-hour culture.

2. Do not inoculate from a broth culture, because the growth will be too heavy.

EXPECTED RESULTS

Positive: Deamination of the acetamide resulting in a blue color (Figure 30-1, *A*).
   Negative: No color change (Figure 30-1, *B*).

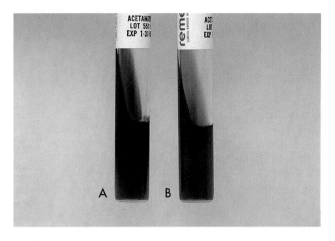

FIGURE 30-1   Acetamide utilization. **A,** Positive. **B,** Negative.

30-2 ACETATE UTILIZATION

PRINCIPLE

This test is used to determine if an organism can use acetate as the sole source of carbon. If so, breakdown of the sodium acetate causes the pH of the medium to shift toward the alkaline range, turning the indicator from green to blue.

METHOD

1. With a straight inoculating needle, inoculate acetate slant *lightly* from an 18- to 24-hour culture. Do not inoculate from a broth culture, because the growth will be too heavy.

2. Incubate at 35° C for up to 7 days.

EXPECTED RESULTS

Positive: Medium becomes alkalinized (blue) because of the growth of the organism (Figure 30-2, *A*).
   Negative: No growth or growth with no indicator change to blue (Figure 30-2, *B*).

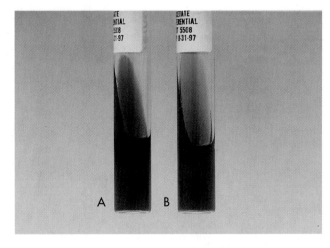

FIGURE 30-2   Acetate utilization. **A,** Positive. **B,** Negative.

30-3 BILE ESCULIN AGAR

PRINCIPLE
Gram-positive bacteria other than group D streptococci and enterococci are inhibited by the bile in this medium. Group D streptococci and enterococci can grow in the presence of 40% bile and hydrolyze esculin; they subsequently turn the indicator, ferric ammonium citrate, a dark brown color. This dark brown color results from the combination of esculetin (the end product of esculin hydrolysis) with ferric ions to form a phenolic iron complex.

METHOD
1. Inoculate one to two colonies from an 18- to 24-hour culture onto the surface of the slant.

2. Incubate at 35° C in ambient air for 48 hours.

EXPECTED RESULTS
Positive: Blackening of the agar slant (Figure 30-3, *A*).

 Negative: No blackening of medium (Figure 30-3, *B*).

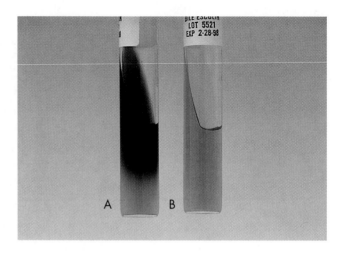

FIGURE 30-3 Bile esculin agar. **A,** Positive. **B,** Negative.

30-4 BILE SOLUBILITY TEST

PRINCIPLE
The bile solubility test differentiates *Streptococcus pneumoniae* (positive) from alpha-hemolytic streptococci (negative). Pneumococcal colonies are rapidly lysed by bile or a solution of a bile salt, such as sodium desoxycholate. Lysis depends on the presence of an intracellular autolytic enzyme. Bile salts lower the surface tension between the bacterial cell membrane and the medium, thus accelerating the organism's natural autolytic process.

METHOD
1. Place 1 to 2 drops of 10% sodium desoxycholate to the side of a young (18- to 24-hour), well-isolated colony growing on 5% sheep blood agar. Note: A tube test is performed with 2% sodium desoxycholate.

2. Gently wash liquid over colony, without dislodging colony from agar.

3. Incubate plate at 35° C in ambient air for 30 minutes.

FIGURE 30-4 Bile solubility test. **A,** Colony lysed. **B,** Intact colony.

4. Examine for lysis of colony.

EXPECTED RESULTS
Positive: Colony disintegrates; an imprint of the lysed colony may remain within the zone (Figure 30-4, *A*).

 Negative: Intact colonies (Figure 30-4, *B*).

30-5 CAMP TEST

PRINCIPLE

Certain organisms (including group B streptococci) produce a diffusible extracellular protein (CAMP factor) that acts synergistically with the beta-lysin of *Staphylococcus aureus* to cause enhanced lysis of red blood cells.

METHOD

1. Streak a beta-lysin–producing strain of *S. aureus* down the center of a sheep blood agar plate.

2. Streak test organisms across the plate perpendicular to the *S. aureus* streak. (Multiple organisms can be tested on a single plate if they are 3 to 4 mm apart.)

3. Incubate overnight at 35° C in ambient air.

EXPECTED RESULTS

Positive: Enhanced hemolysis is indicated by an arrowhead-shaped zone of beta hemolysis at the juncture of the two organisms (Figure 30-5, *A*).

Negative: No enhancement of hemolysis (Figure 30-5, *B*).

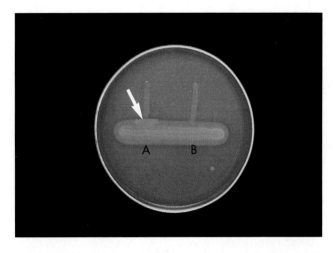

FIGURE 30-5 CAMP test. **A,** Positive, arrowhead zone of beta hemolysis (*at arrow*) typical of group B streptococci. **B,** Negative, no enhancement of hemolysis.

30-6 CETRIMIDE

PRINCIPLE

This test is used to determine the ability of an organism to grow in the presence of cetrimide, a toxic substance that inhibits the growth of many bacteria.

METHOD

1. Inoculate a cetrimide agar slant with 1 drop of an 18- to 24-hour culture in brain-heart infusion broth.

2. Incubate at 35° C for up to 7 days.

3. Examine the slant for bacterial growth.

EXPECTED RESULTS

Positive: Growth (Figure 30-6, *A*).

Negative: No growth (Figure 30-6, *B*).

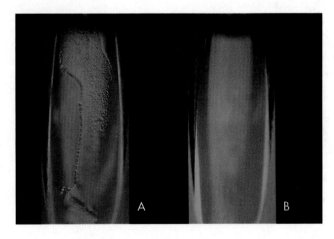

FIGURE 30-6 Cetrimide. **A,** Positive. **B,** Negative.

30-7 CITRATE UTILIZATION

PRINCIPLE

This test is used to determine the ability of an organism to utilize sodium citrate as its only carbon source and inorganic ammonium salts as its only nitrogen source. Bacteria that can grow on this medium turn the bromthymol blue indicator from green to blue.

METHOD

1. Inoculate Simmons citrate agar *lightly* on the slant by touching the tip of a needle to a colony that is 18 to 24 hours old. Note: There is no need to stab into the butt of the tube. Do not inoculate from a broth culture, because the inoculum will be too heavy.

2. Incubate at 35° C to 37° C for up to 7 days.

3. Observe for development of blue color, denoting alkalinization.

EXPECTED RESULTS

Positive: Growth on the medium, with or without a change in the color of the indicator. The color change of the indicator is due to acid or alkali production by the test organism as it grows on the medium. Growth usually results in the bromthymol blue indicator turning from green to blue (Figure 30-7, *A*).

Negative: Absence of growth (Figure 30-7, *B*).

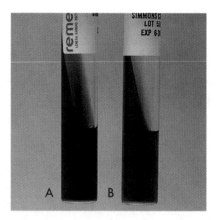

FIGURE 30-7 Citrate utilization. **A,** Positive. **B,** Negative.

30-8 COAGULASE TEST

PRINCIPLE

This test is used to differentiate *Staphylococcus aureus* (positive) from coagulase-negative staphylococci (negative). *S. aureus* produces two forms of coagulase: bound and free. Bound coagulase, or "clumping factor," is bound to the bacterial cell wall and reacts directly with fibrinogen. This results in an alteration of fibrinogen so that it precipitates on the staphylococcal cell, causing the cells to clump when a bacterial suspension is mixed with plasma. The presence of bound coagulase correlates well with free coagulase, an extracellular protein enzyme that causes the formation of a clot when *S. aureus* colonies are incubated with plasma. The clotting mechanism involves activation of a plasma coagulase-reacting factor (CRF), which is a modified or derived thrombin molecule, to form a coagulase-CRF complex. This complex in turn reacts with fibrinogen to produce the fibrin clot.

METHOD

A. Slide test
1. Place a drop of coagulase plasma (preferably rabbit plasma with EDTA) on a clean, dry glass slide.

2. Place a drop of distilled water or saline next to the drop of plasma as a control.

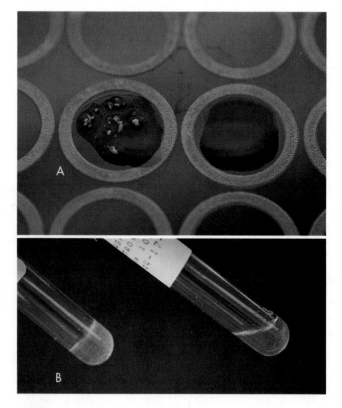

FIGURE 30-8 Coagulase test. **A,** Slide coagulase test for clumping factor. Left side is positive; right side is negative. **B,** Tube coagulase test for free coagulase. Tube on the left is positive, exhibiting clot. Tube on the right is negative.

3. With a loop, straight wire, or wooden stick, emulsify a portion of the isolated colony being tested in each drop, inoculating the water or saline first. Try to create a smooth suspension.

30-8 COAGULASE TEST—cont'd

4. Mix well with a wooden applicator stick.

5. Rock the slide gently for 5 to 10 seconds.

EXPECTED RESULTS

Positive: Macroscopic clumping in 10 seconds or less in coagulated plasma drop and no clumping in saline or water drop (Figure 30-8, *A*, left side).

Negative: No clumping in either drop. Note: *All* negative slide tests

must be confirmed using the tube test (Figure 30-8, *A*, right side).

Equivocal: Clumping in both drops indicates that the organism autoagglutinates and is unsuitable for the slide coagulase test.

B. Tube test
1. Emulsify several colonies in 0.5 mL of rabbit plasma (with EDTA) to give a milky suspension.

2. Incubate tube at 35° C in ambient air for 4 hours.

3. Check for clot formation. Note: Tests can be positive at 4 hours and then revert to negative after 24 hours.

4. If negative at 4 hours, incubate at room temperature overnight and check again for clot formation.

EXPECTED RESULTS

Positive: Clot of any size (Figure 30-8, *B*, left side).

Negative: No clot (Figure 30-8, *B*, right side).

30-9 DECARBOXYLASE TESTS (MOELLER'S METHOD)

PRINCIPLE

This test measures the enzymatic ability of an organism to decarboxylate (or hydrolyze) an amino acid to form an amine. Decarboxylation, or hydrolysis, of the amino acid results in an alkaline pH change.

METHOD

A. Glucose nonfermenting organisms
1. Prepare a very heavy suspension (≥McFarland No. 5 turbidity standard) in brain-heart infusion broth from young bacteria (18 to 24 hours old) growing on 5% sheep blood agar.

2. Inoculate each of the three decarboxylase broths (arginine, lysine, and ornithine) and the control broth (no amino acid) with 4 drops of broth.

3. Add a 4-mm layer of sterile mineral oil to each tube.

4. Incubate the cultures at 35° C in ambient air for up to 4 days.

B. Glucose-fermenting organisms
1. Inoculate tubes with 1 drop of an 18- to 24-hour brain-heart infusion broth culture.

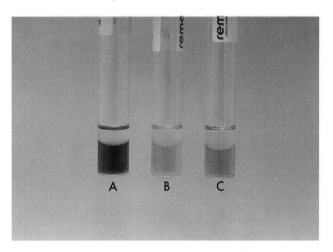

FIGURE 30-9 Decarboxylase tests (Moeller's method). **A,** Positive. **B,** Negative. **C,** Uninoculated tube.

2. Add a 4-mm layer of sterile mineral oil to each tube.

3. Incubate the cultures for 4 days at 35° C in ambient air.

EXPECTED RESULTS

Positive: Alkaline (purple) color change compared with the control tube (Figure 30-9, *A*).

Negative: No color change or acid (yellow) color in test and control tube. Growth in the control tube. Note: The fermentation of dextrose in the medium causes the acid color change. It would not, however, mask the alkaline color change brought about by a positive decarboxylation reaction (Figure 30-9, *B*). Uninoculated tube is shown in Figure 30-9, *C*.

30-10 DNA HYDROLYSIS

PRINCIPLE

This test is used to determine the ability of an organism to hydrolyze DNA. The medium is pale green because of the DNA-methyl green complex. If the organism growing on the medium hydrolyses DNA, the green color fades and the colony is surrounded by a colorless zone.

METHOD

1. Inoculate the DNase agar with the organism to be tested and streak for isolation.

2. Incubate aerobically at 35° C for 18 to 24 hours.

EXPECTED RESULTS

Positive: When DNA is hydrolyzed, methyl green is released and combines with highly polymerized DNA at a pH of 7.5, turning the medium colorless around the test organism (Figure 30-10, A and B).

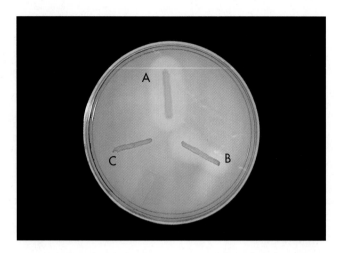

FIGURE 30-10 DNA hydrolysis. **A,** Positive, Staphylococcus *aureus.* **B,** Positive, *Serratia marcescens.* **C,** Negative.

Negative: If there is no degradation of DNA, the medium remains green (Figure 30-10, C).

30-11 ESCULIN HYDROLYSIS

PRINCIPLE

This test is used to determine whether an organism is able to hydrolyze the glycoside esculin.

METHOD

1. Inoculate the medium with 1 drop of a 24-hour broth culture.

2. Incubate at 35° C for up to 7 days.

3. Examine the slants for blackening and under the ultraviolet rays of a Wood's lamp for esculin hydrolysis.

EXPECTED RESULTS

Positive: Blackened medium (Figure 30-11, A), which would also show a loss of fluorescence under the Wood's lamp.

Negative: No blackening and no loss of fluorescence under Wood's lamp, or slight blackening with no loss of fluorescence under Wood's lamp. Uninoculated tube is shown in Figure 30-11, B.

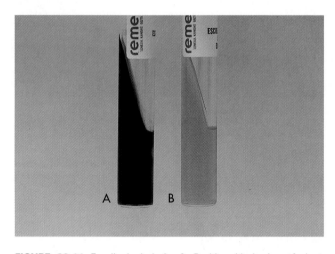

FIGURE 30-11 Esculin hydrolysis. **A,** Positive, blackening of slant. **B,** Uninoculated tube.

30-12 FERMENTATION MEDIA

PRINCIPLE

These media are used to determine the ability of an organism to ferment a specific carbohydrate that is incorporated in a basal medium, thereby producing acid with or without visible gas. A medium for *Enterobacteriaceae* and coryneforms and a medium for streptococci and enterococci are described.

METHOD

A. Peptone medium with Andrade's indicator (for enterics and coryneforms)
   1. Inoculate each tube with one drop of an 18- to 24-hour brain-heart infusion broth culture.

   2. Incubate at 35° C for up to 7 days in ambient air. Note: Tubes are held 4 days for the organisms belonging to the *Enterobacteriaceae* family.

   3. Examine the tubes for acid (indicated by a pink color) and gas production.

   4. Tubes must show growth for the test to be valid. If after 24 hours of incubation there is no growth in the fermentation tubes or control, add 1 to 2 drops of sterile rabbit serum per 5 mL of fermentation broth to each tube.

EXPECTED RESULTS

Positive: Indicator change to pink with or without gas formation in Durham tube (Figure 30-12, *A*, left and middle).

Negative: Growth, but no change in color. Medium remains clear to straw-colored (Figure 30-12, *A*, right).

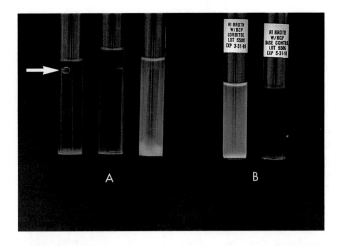

FIGURE 30-12 Fermentation media. **A,** Peptone medium with Andrade's indicator. The tube on the left ferments glucose with the production of gas (visible as a bubble [*arrow*] in the inverted [Durham] tube), the tube in the middle ferments glucose with no gas production, and the tube on the right does not ferment glucose. **B,** Heart infusion broth with bromcresol purple indicator. The tube on the left is positive; the tube on the right is negative.

B. Heart infusion broth with bromcresol purple indicator (for streptococci and enterococci)
   1. Inoculate each tube with two drops of an 18- to 24-hour brain-heart infusion broth culture.

   2. Incubate 4 days at 35° C in ambient air.

   3. Observe daily for a change of the bromcresol purple indicator from purple to yellow (acid).

EXPECTED RESULTS

Postive: Indicator change to yellow (Figure 30-12, *B*, left).

Negative: Growth, but no change in color. Medium remains purple (Figure 30-12, *B*, right).

30-13 FLAGELLA STAIN (WET-MOUNT TECHNIQUE)

PRINCIPLE
A wet-mount technique[1] for staining bacterial flagella is simple and is useful when the number and arrangement of flagella are critical in identifying species of motile bacteria.

METHOD
1. Grow the organism to be stained at room temperature on blood agar for 16 to 24 hours.

2. Add a small drop of water to a microscope slide.

3. Dip a sterile inoculating loop into sterile water.

4. Touch the loopful of water to the colony margin briefly (this allows motile cells to swim into the droplet of water).

5. Touch the loopful of motile cells to the drop of water on the slide. Note: Agitating the loop in the droplet of water on the slide causes the flagella to shear off the cell.

6. Cover the faintly turbid drop of water on the slide with a cover slip. A proper wet mount has barely enough liquid to fill the space under a cover slip. Small air spaces around the edge are preferable.

7. Examine slide immediately under 40× to 50× for motile cells. If motile cells are not seen, do not proceed with the stain.

8. If motile cells are seen, leave slide at room temperature for 5 to 10 minutes. This allows time for the bacterial cells to adhere to either the glass slide or the cover slip.

9. Apply 2 drops of RYU flagella stain (Remel Laboratories, Lenexa, Kan.) gently to the edge of the cover slip. The stain will flow by capillary action and mix with the cell suspension. Small air pockets around the edge of the wet mount are useful in aiding the capillary action.

10. After 5 to 10 minutes at room temperature, examine the cells for flagella.

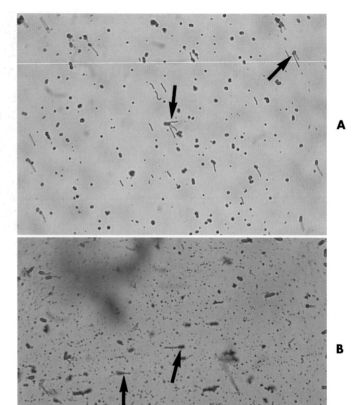

FIGURE 30-13 Flagella stain (wet mount technique). **A,** *Alcaligenes* spp., peritrichous flagella (*arrows*). **B,** *Pseudomonas aeruginosa,* polar flagella (*arrows*).

11. Cells with flagella may be observed at 100× (oil) in the zone of optimum stain concentration, about half way from the edge of the cover slip to the center of the mount.

12. Focusing the microscope on the cells attached to the coverslip rather than the cells attached to the slide facilitates visualization of the flagella. The precipitate from the stain is primarily on the slide rather than the cover slip.

EXPECTED RESULTS
Observe the slide and note the following:
1. Presence or absence of flagella

2. Number of flagella per cell

3. Location of flagella per cell
   a. Peritrichous (Figure 30-13, *A*)
   b. Lophotrichous
   c. Polar (Figure 30-13, *B*)

4. Amplitude of wavelength
   a. Short
   b. Long

5. Whether or not "tufted"

30-14 GELATIN HYDROLYSIS

PRINCIPLE
This test is used to determine the ability of an organism to produce proteolytic enzymes (gelatinases) that liquefy gelatin.

METHOD
1. Inoculate the gelatin deep with 4 to 5 drops of a 24-hour broth culture.

2. Incubate at 35° C in ambient air for up to 14 days. Note: Incubate the medium at 25° C if the organism grows better at 25° C than at 35° C.

3. Alternatively, inoculate the gelatin deep from a 24-hour–old colony by stabbing 4 to 5 times ½ inch into the medium.

4. Remove the gelatin tube daily from the incubator and place at 4° C to check for liquefaction. Do not invert or tip the tube, because sometimes the only discernible liquefaction will occur at the top of the deep where inoculation occurred.

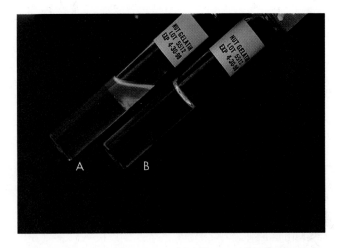

FIGURE 30-14 Gelatin hydrolysis. **A,** Positive; note liquefication at top of tube. **B,** Uninoculated tube.

5. Refrigerate an uninoculated control along with the inoculated tube. Liquefaction is determined only after the control has hardened (gelled).

EXPECTED RESULTS
Positive: Partial or total liquefaction of the inoculated tube (the control tube must be completely solidified) at 4° C within 14 days (Figure 30-14, *A*).

Negative: Complete solidification of tube at 4° C (Figure 30-14, *B*).

30-15 GROWTH AT 42° C

PRINCIPLE
The test is used to determine the ability of an organism to grow at 42° C.

PROCEDURE
1. Inoculate two tubes of trypticase soy agar (TSA) with a light inoculum by *lightly* touching a needle to the top of a single 18- to 24-hour–old colony and streaking the slant.

2. Immediately incubate one tube at 35° C and one at 42° C.

3. Record the presence of growth on each slant after 18 to 24 hours.

EXPECTED RESULTS
Positive: Good growth at both 35° and 42° C (Figure 30-15, *A*).

Negative: No growth at 42° C (Figure 30-15, *B*), but good growth at 35° C.

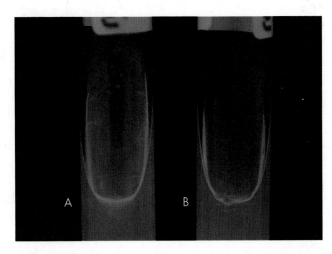

FIGURE 30-15 Growth at 42° C. **A,** Positive, good growth. **B,** Negative, no growth.

30-16 INDOLE PRODUCTION

PRINCIPLE
The test is used to determine the ability of an organism to split tryptophan to form the compound indole.

METHOD
A. *Enterobacteriaceae*
1. Inoculate tryptophane broth with one drop from a 24-hour brain-heart infusion broth culture.

2. Incubate at 35° C in ambient air for 48 hours.

3. Add 0.5 mL of Kovac's reagent to the broth culture.

B. Other gram-negative bacilli
1. Inoculate tryptophane broth with 1 drop of a 24-hour broth culture.

2. Incubate at 35° C in ambient air for 48 hours.

3. Add 1 mL of xylene to the culture.

4. Shake mixture vigorously to extract the indole and allow to stand until the xylene forms a

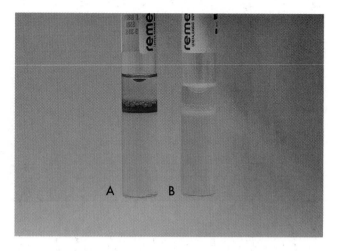

FIGURE 30-16 Indole production. **A,** Positive. **B,** Negative.

layer on top of the aqueous phase.

5. Add 0.5 mL of Ehrlich's reagent down the side of the tube.

EXPECTED RESULTS
Positive: Pink-to-wine colored ring after addition of appropriate reagent (Figure 30-16, *A*).

Negative: No color change after the addition of the appropriate reagent (Figure 30-16, *B*).

30-17 LITMUS MILK

PRINCIPLE
This test is used to determine an organism's ability to metabolize litmus milk. Fermentation of lactose is evidenced by the litmus turning pink as a result of acid production. If sufficient acid is produced, casein in the milk is coagulated, solidifying the milk. With some organisms, the curd shrinks and whey is formed at the surface. Some bacteria hydrolyze casein, causing the milk to become straw-colored and resemble turbid serum. Additionally, some organisms reduce litmus, in which case the medium becomes colorless in the bottom of the tube.

METHOD
1. Inoculate with 4 drops of a 24-hour broth culture.

2. Incubate at 35° to 37° C in ambient air.

3. Observe daily for 7 days for alkaline reaction (litmus turns blue),

FIGURE 30-17 Litmus milk. **A,** Acid reaction. **B,** Alkaline reaction. **C,** No change. **D,** Reduction of indicator. **E,** Clot. (Note separation of clear fluid from clot at arrow.) **F,** Peptonization.

acid reaction (litmus turns pink), indicator reduction, acid clot, rennet clot, and peptonization. Multiple changes can occur over the observation period.

4. Record all changes.

30-17 LITMUS MILK—cont'd

EXPECTED RESULTS

APPEARANCE OF INDICATOR (LITMUS DYE)

COLOR	pH CHANGE TO . . .	RECORD
Pink, mauve (Figure 30-17, *A*)	Acid	Acid (A)
Blue (Figure 30-17, *B*)	Alkaline	Alkaline (K)
Purple (identical to uninoculated control) (Figure 30-17, *C*)	No change	No change
White (Figure 30-17, *D*)	Independent of pH change; result of reduction of indicator	Decolorized

---

APPEARANCE OF MILK

CONSISTENCY OF MILK	OCCURS WHEN pH IS . . .	RECORD
Coagulation or clot (Figure 30-17, *E*)	Acid or alkaline	Clot
Dissolution of clot with clear, grayish, watery fluid and a shrunken, insoluble pink clot (Figure 30-17, *F*)	Acid	Digestion
Dissolution of clot with clear, grayish, watery fluid and a shrunken, insoluble blue clot	Alkaline	Peptonization

---

30-18 LYSINE IRON AGAR

PRINCIPLE

This test is used to determine whether a gram-negative rod decarboxylates or deaminates lysine and forms hydrogen sulfide (H_2S). Lysine iron agar (LIA) contains lysine, peptones, a small amount of glucose, ferric ammonium citrate, and sodium thiosulfate. When glucose is fermented, the butt of the medium becomes acidic (yellow). If the organism produces lysine decarboxylase, cadaverine is formed. Cadaverine neutralizes the organic acids formed by glucose fermentation, and the butt of the medium reverts to the alkaline state (purple). If the decarboxylase is not produced, the butt remains acidic (yellow). If oxidative deamination of lysine occurs, a compound is formed that, in the presence of ferric ammonium citrate and a coenzyme, flavin mononucleotide, forms a burgundy color on the slant. If deamination does not occur, the LIA slant remains purple.

METHOD

1. With a straight inoculating needle, inoculate LIA (Figure 30-18, *E*) by twice stabbing through the center of the medium to the bottom of the tube and then streaking the slant.

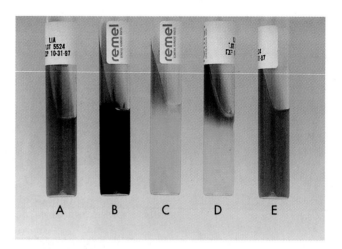

FIGURE 30-18 Lysine iron agar. **A,** Alkaline slant/alkaline butt (K/K). **B,** Alkaline slant/alkaline butt, H_2S positive (K/K H_2S^+). **C,** Alkaline slant/acid butt (K/A). **D,** Red slant/acid butt (R/A). **E,** Uninoculated tube.

2. Cap the tube tightly and incubate at 35° C in ambient air for 18 to 24 hours.

EXPECTED RESULTS

1. Alkaline slant/alkaline butt (K/K) = lysine decarboxylation and no fermentation of glucose (Figure 30-18, *A*).

2. Alkaline slant/acid butt (K/A) = glucose fermentation (Figure 30-18, *C*).

3. Note: Patterns shown in 1 and 2 above can be accompanied by a black precipitate of ferrous sulfide (FeS), which indicates production of H_2S (Figure 30-18, *B*).

4. Red slant/acid butt (R/A) = lysine deamination and glucose fermentation (Figure 30-18, *D*).

30-19 METHYL RED/VOGES-PROSKAUER (MRVP) TESTS

PRINCIPLE

This test is used to determine the ability of an organism to produce and maintain stable acid end products from glucose fermentation, to overcome the buffering capacity of the system, and to determine the ability of some organisms to produce neutral end products, (e.g., acetylmethylcarbinol or acetoin), from glucose fermentation.

METHOD

1. Inoculate MRVP broth with one drop from a 24-hour brain-heart infusion broth culture.

2. Incubate at 35° to 37° C for a minimum of 48 hours in ambient air. Tests should not be made with cultures incubated less than 48 hours, because the end products build up to detectable levels over time. If results are equivocal at 48 hours, repeat the tests with cultures incubated at 35° to 37° C for 4 to 5 days in ambient air; in such instances, duplicate tests should be incubated at 25° C.

3. Split broth into aliquots for MR test and VP test.

A. MR (methyl red) test
   1. Add 5 or 6 drops of methyl red reagent per 5 mL of broth.

   2. Read reaction immediately.

EXPECTED RESULTS

Positive: Bright red color indicative of mixed acid fermentation (Figure 30-19, *A*).
    Weakly positive: Red-orange color.
    Negative: Yellow color (Figure 30-19, *B*).

B. VP (Voges-Proskauer) test (Barritt's method) for gram-negative rods
   1. Add 0.6 mL (6 drops) of solution A (α-naphtol) and 0.2 mL (2 drops) of solution B (KOH) to 1 mL of MRVP broth.

   2. Shake well after addition of each reagent.

   3. Observe for 5 minutes.

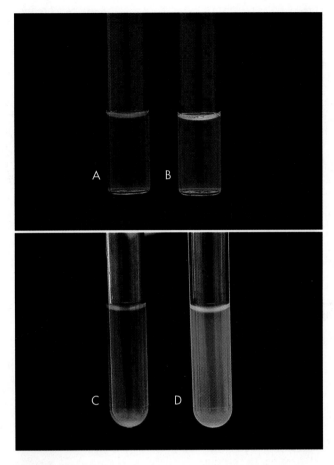

FIGURE 30-19 MRVP tests. **A,** Positive methyl red. **B,** Negative methyl red. **C,** Positive Voges-Proskauer. **D,** Negative Voges-Proskauer.

EXPECTED RESULTS

Positive: Red color indicative of acetoin production.
    Negative: Yellow color.

C. VP (Voges-Proskauer) test (Coblentz method) for streptococci
   1. Use 24-hour growth from blood agar plate to heavily inoculate 2 mL of MRVP broth.

   2. After 6 hours of incubation at 35° C in ambient air, add 1.2 mL (12 drops) of solution A (α-naphthol) and 0.4 mL (4 drops) solution B (40% KOH with creatine).

   3. Shake the tube and incubate at room temperature for 30 minutes.

EXPECTED RESULTS

Positive: Red color (Figure 30-19, *C*).
    Negative: Yellow color (Figure 30-19, *D*).

30-20 MRS BROTH

PRINCIPLE
This test is used to determine whether an organism forms gas during glucose fermentation. Gas is produced by some *Lactobacillus* spp. and *Leuconostoc* spp.

METHOD
1. Inoculate MRS broth with an 18- to 24-hour culture from agar or broth.

2. Incubate 24 to 48 hours at 35° C in ambient air.

EXPECTED RESULTS
Negative: No gas production (Figure 30-20, *A*).

 Positive: Gas production indicated by a bubble in Durham tube (Figure 30-20, *B*).

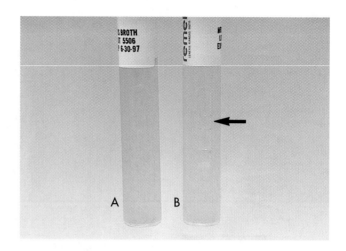

FIGURE 30-20 MRS broth. **A,** Negative, no gas production. **B,** Positive, gas production by *Leuconostoc* (*arrow*).

30-21 NITRATE REDUCTION

PRINCIPLE
This test is used to determine the ability of an organism to reduce nitrate. The reduction of nitrate to nitrite is determined by adding sulfanilic acid and alpha-naphthylamine. The sulfanilic acid and nitrite react to form a diazonium salt. The diazonium salt then couples with the alpha-naphthylamine to produce a red, water–soluble azo dye.

METHOD
1. Inoculate nitrate broth (Figure 30-21, *D*) with 1 to 2 drops from a young broth culture of the test organism.

2. Incubate for 48 hours at 35° C in ambient air (some organisms may require longer incubation for adequate growth). Test these cultures 24 hours after obvious growth is detected or after a maximum of 7 days.

3. After a suitable incubation period, test the nitrate broth culture for the presence of gas, reduction of nitrate, and reduction of nitrite according to the following steps:
   a. Observe the inverted Durham tube for the presence of gas, indicated by bubbles inside the tube.
   b. Add 5 drops each of nitrate reagent solution A (sulfanilic acid) and B (alpha-naphthylamine). Observe for at least 3 minutes for the development of a red color.
   c. If no color develops, test further with zinc powder. Dip a wooden applicator stick into zinc powder and transfer only the amount that adheres to the stick to the nitrate broth culture to which solutions A and B have been added. Observe for at least 3 minutes for the development of a red color. Breaking the stick into the tube after the addition of the zinc provides a useful marker for the stage of testing.

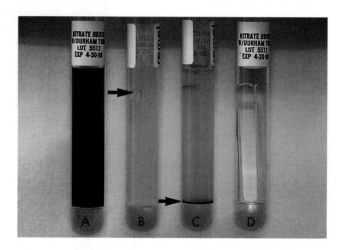

FIGURE 30-21 Nitrate reduction. **A,** Positive, no gas. **B,** Positive, gas (*arrow*). **C,** Positive, no color after addition of zinc (*arrow*). **D,** Uninoculated tube.

EXPECTED RESULTS
The Nitrate Reduction Test is read for the presence or absence of three metabolic products: gas, nitrate (NO_3), and nitrite (NO_2). The expected results can be summarized as follows:

30-21 NITRATE REDUCTION—cont'd

Reaction	Gas	Color after Addition of Solutions A and B	Color after Addition of Zinc	Interpretation
$NO_3 \rightarrow NO_2$ (Figure 30-21, A)	None	Red	—	NO_3^+, no gas
$NO_3 \rightarrow NO_2$, partial nongaseous end products	None	Red	—	NO_3^+, no gas
$NO_3 \rightarrow NO_2$, gaseous end products (Figure 30-21, B)	Yes	Red	—	NO_3^+, gas$^+$
$NO_3 \rightarrow$ gaseous end product (Figure 30-21, C)	Yes	None	None	NO_3^+, NO_2^+, gas$^+$
$NO_3 \rightarrow$ nongaseous end products	None	None	None	NO_3^+, NO_2^+, no gas
$NO_3 \rightarrow$ no reaction	None	None	Red	Negative

30-22 NITRITE REDUCTION

PRINCIPLE
This test is used to determine whether an organism can reduce nitrites to gaseous nitrogen or to other compounds containing nitrogen.

METHOD
1. Inoculate nitrite broth with 1 drop from a 24-hour broth culture.

2. Incubate for 48 hours at 35° C.

3. Examine 48-hour nitrite broth cultures for nitrogen gas in the inverted Durham tube and add 5 drops each of the nitrate reagents A and B to determine if nitrite is still present in the medium (reagents A and B are described under the nitrate reduction test).

EXPECTED RESULTS
Positive: No color change to red 2 minutes after the addition of the reagents and gas production observed in the Durham tube (Figure 30-22, A). Note: If broth does not become red and no gas production is observed, zinc dust is added to make sure the nitrite has not been oxidized to nitrate, thus invalidating the test.

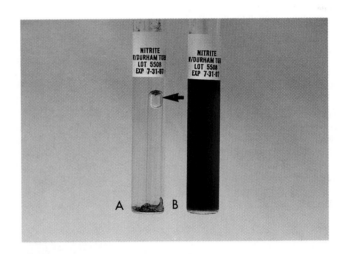

FIGURE 30-22 Nitrite reduction. **A,** Positive, no color change after addition of zinc dust and gas in Durham tube (*arrow*). **B,** Negative.

If oxidation has occurred, the mixture becomes red after the addition of zinc.

Negative: The broth becomes red after the addition of the reagents. No gas production is observed (Figure 30-22, B).

30-23 ONPG (*o*-NITROPHENYL-β-*D*-GALACTOPHYRANOSIDE) TEST

PRINCIPLE

This is used to determine the ability of an organism to produce β-galactosidase, an enzyme that hydrolyzes the substrate ONPG to form a visible (yellow) product, orthonitrophenol.

METHOD

1. Aseptically suspend a loopful of organism in 0.85% saline.

2. Place an ONPG disk in the tube.

3. Incubate for 4 hours at 37° C in ambient air.

4. Examine tubes for a color change.

EXPECTED RESULTS

Positive: Yellow (presence of β-galactosidase) (Figure 30-23, *A*).

Negative: Clear (absence of enzyme) (Figure 30-23, *B*).

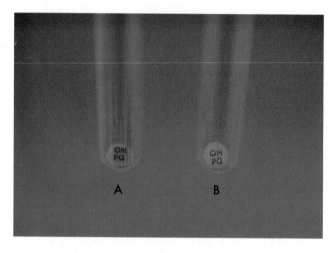

FIGURE 30-23 OPNG test. **A**, Positive. **B**, Negative.

30-24 OPTOCHIN TEST

PRINCIPLE

This test is used to determine the effect of optochin (ethylhydrocupreine hydrochloride) on an organism. Optochin lyses pneumococci (positive test), but alpha-streptococci are resistant (negative test).

METHOD

1. Using an inoculating loop, streak two or three suspect colonies of a *pure* culture onto one half of a 5% sheep blood agar plate.

2. Using heated forceps, place an optochin disk in the upper third of the streaked area. Gently tap the disk to ensure adequate contact with the agar surface.

3. Incubate plate for 18 to 24 hours at 35° C in 5% CO_2. Note: Cultures do not grow as well in ambient air and larger zones of inhibition occur.

4. Measure zone of inhibition in millimeters, including diameter of disk.

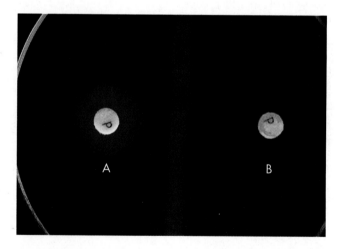

FIGURE 30-24 Optochin test. **A**, *Streptococcus pneumoniae* showing zone of inhibition >14mm. **B**, Alpha-hemolytic *Streptococcus* growing up to the disk.

EXPECTED RESULTS

Positive: Zone of inhibition is 14 mm or greater in diameter, with 6-mm disk (Figure 30-24, *A*).

Negative: No zone of inhibition (Figure 30-24, *B*).

Equivocal: Any zone of inhibition less than 14 mm is questionable for pneumococci; the strain is identified as a pneumococcus only if it is bile-soluble.

30-25 OXIDATION/FERMENTATION (OF) MEDIUM (CDC METHOD)

PRINCIPLE

This test is used to determine whether an organism uses carbohydrate substrates to produce acid byproducts. Nonfermentative bacteria are routinely tested for their ability to produce acid from six carbohydrates (glucose, xylose, mannitol, lactose, sucrose, and maltose). In addition to the six tubes containing carbohydrates, a control tube containing the OF base without carbohydrate is also inoculated. Triple sugar iron agar (TSI) (see Procedure 30-29) is also used to determine whether an organism can ferment glucose. OF glucose is used to determine whether an organism ferments (Figure 30-25, *A*) or oxidizes (Figure 30-25, *B*) glucose. If no reaction occurs in either the TSI or OF glucose, the organism is considered a nonglucose utilizer (Figure 30-25, *C*).

METHOD

1. To determine whether acid is produced from carbohydrates, inoculate agar deeps, each containing a single carbohydrate, with bacterial growth from an 18- to 24-hour culture by stabbing a needle 4 to 5 times into the medium to a depth of 1 cm. Note: Two tubes of OF dextrose are usually inoculated; one is overlaid with either sterile, melted petrolatum or sterile paraffin oil to detect fermentation.

2. Incubate the tubes at 35° C in ambient air for up to 7 days. Note: If screwcap tubes are used, loosen the caps during incubation to allow for air exchange. Other-

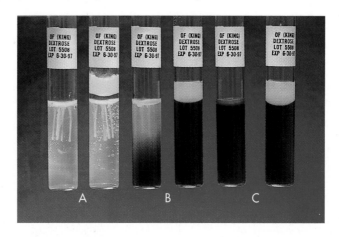

FIGURE   30-25 Oxidative-fermentative medium (CDC method). **A,** Fermenter. **B,** Oxidizer. **C,** Nonutilizer.

wise, the control tube and tubes containing carbohydrates that are not oxidized might not become alkaline.

EXPECTED RESULTS

Positive: Acid production (A) is indicated by the color indicator changing to yellow in the carbohydrate-containing deep.

Weak-positive (A^w): Weak acid formation can be detected by comparing the tube containing the medium with carbohydrate with the inoculated tube containing medium with no carbohydrate. Most bacteria that can grow in the OF base produce an alkaline reaction in the control tube. If the color of the medium in a tube containing carbohydrate remains about the same as it was before the medium was inoculated and if the inoculated medium in the control tube becomes a deeper red (i.e., becomes alkaline), the culture being tested is considered weakly positive, assuming the amount of growth is about the same in both tubes.

Negative: Red or alkaline (K) color in the deep with carbohydrate equal to the color of the inoculated control tube.

No change (NC) or neutral (N): There is growth in the media, but neither the carbohydrate-containing media or the control base turn alkaline (red).

Note: If the organism does not grow at all in the OF medium, mark the reaction as *no growth (NG)*.

30-26 PHENYLALANINE DEAMINASE

PRINCIPLE

This test is used to determine the ability of an organism to oxidatively deaminate phenylalanine to phenyl-pyruvic acid. The phenylpyruvic acid is detected by adding a few drops of 10% ferric chloride; a green-colored complex is formed between these two compounds.

METHOD

1. Inoculate phenylalanine slant with 1 drop of a 24-hour brain-heart infusion broth.

2. Incubate 18 to 24 hours (or until good growth is apparent) at 35° C in ambient air with cap loose.

3. After incubation, add 4 to 5 drops of 10% aqueous ferric chloride to the slant.

EXPECTED RESULTS

Positive: Green color develops on slant after ferric chloride is added (Figure 30-26, A).

Negative: Slant remains original color after the addition of ferric chloride (Figure 30-26, B).

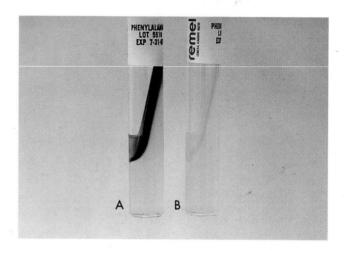

FIGURE 30-26 Phenylalanine deaminase. **A,** Positive. **B,** Negative.

30-27 PYRUVATE BROTH

PRINCIPLE

This test is used to determine the ability of an organism to utilize pyruvate. This aids in the differentiation between *Enterococcus faecalis* (positive) and *Enterococcus faecium* (negative).

METHOD

1. *Lightly* inoculate the pyruvate broth with an 18- to 24-hour culture of the organism from 5% sheep blood agar.

2. Incubate at 35° C in ambient air for 24 to 48 hours.

EXPECTED RESULTS

Positive: Indicator changes from green to yellow (Figure 30-27, A).

Negative: No color change; yellow-green indicates a weak reaction and should be regarded as negative (Figure 30-27, B).

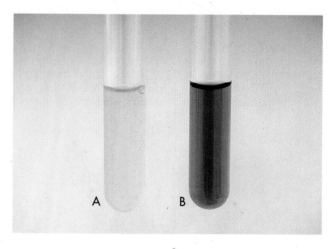

FIGURE 30-27 Pyruvate broth. **A,** Positive. **B,** Negative.

30-28 SALT TOLERANCE TEST

PRINCIPLE
This test is used to determine the ability of an organism to grow in high concentrations of salt. It is used to differentiate enterococci (positive) from nonenterococci (negative). A heart infusion broth containing 6.5% NaCl is used as the test medium. This broth also contains a small amount of glucose and bromcresol purple as the indicator for acid production.

METHOD
1. Inoculate one to two colonies from an 18- to 24-hour culture into 6.5% NaCl broth.

2. Incubate tube at 35° C in ambient air for 48 hours.

3. Check daily for growth.

EXPECTED RESULTS
Positive: Visible turbidity in broth with or without color change from purple to yellow (Figure 30-28, *A*).

   Negative: No turbidity and no color change (Figure 30-28, *B*).

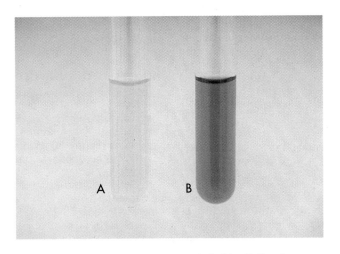

FIGURE 30-28 Salt tolerance test. **A,** Positive. **B,** Negative.

30-29 TRIPLE SUGAR IRON AGAR (TSI)

PRINCIPLE
Triple sugar iron agar (TSI) is used to determine whether a gram-negative rod utilizes glucose and lactose or sucrose fermentatively and forms hydrogen sulfide (H_2S). TSI contains 10 parts lactose: 10 parts sucrose: 1 part glucose and peptone. Phenol red and ferrous sulfate serve as indicators of acidification and H_2S formation, respectively. When glucose is utilized by a fermentative organism, the entire medium becomes acidic (yellow) in 8 to 12 hours. The butt remains acidic after the recommended 18- to 24-hour incubation period because of the presence of organic acids resulting from the fermentation of glucose under anaerobic conditions in the butt of the tube. The slant, however, reverts to the alkaline (red) state because of oxidation of the fermentation products under aerobic conditions on the slant. This change is a result of the formation of CO_2 (carbon dioxide) and H_2O and the oxidation of peptones in the medium to alkaline amines. When, in addition to glucose, lactose and/or sucrose are fer-

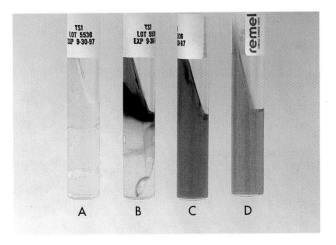

FIGURE 30-29 Triple sugar iron agar. **A,** Acid slant/acid butt with gas, no H_2S (A/Ⓐ). **B,** Alkaline slant/acid butt, no gas, H_2S-positive (K/A H_2S^+). **C,** Alkaline slant/no change butt, no gas, no H_2S (K/NC). **D,** Uninoculated tube.

mented, the large amount of fermentation products formed on the slant will more than neutralize the alkaline amines and render the slant acidic (yellow), provided the reaction is read in 18 to 24 hours. Reactions in TSI should *not* be read beyond 24 hours of incubation, because aerobic oxidation of the fermentation products from lactose and/or sucrose does proceed and the slant will eventually revert to the alkaline state. The formation of CO_2 and H_2 (hydrogen gas) is indicated

Continued

30-29 TRIPLE SUGAR IRON AGAR (TSI)—cont'd

by the presence of bubbles or cracks in the agar or by separation of the agar from the sides or bottom of the tube. The production of H_2S requires an acidic environment and is manifested by blackening of the butt of the medium in the tube.

METHOD

1. With a straight inoculation needle, touch the top of a well-isolated colony.

2. Inoculate TSI (Figure 30-29, *D*) by first stabbing through the center of the medium to the bottom of the tube and then streaking the surface of the agar slant.

3. Leave cap on loosely and incubate the tube at 35° C in ambient air for 18 to 24 hours.

EXPECTED RESULTS

Alkaline slant/no change in the butt (K/NC) = glucose, lactose, and sucrose nonutilizer; this may also be recorded as K/K (alkaline slant/alkaline butt) (Figure 30-29, *C*)

Alkaline slant/acid butt (K/A) = glucose fermentation only

Acid slant/acid butt (A/A) = glucose, sucrose, and/or lactose fermenter (Figure 30-29, *A*)

Note: A black precipitate in the butt indicates production of ferrous sulfide and H_2S gas. (H_2S^+) (Figure 30-29, *B*). Bubbles or cracks in the tube indicate the production of CO_2 or H_2. This usually is indicated by drawing a circle around the Ⓐ for the acid butt, that is, *A*/Ⓐ means the organism ferments glucose and sucrose, glucose and lactose, or glucose, sucrose, and lactose, with the production of gas.

30-30 UREA HYDROLYSIS (CHRISTENSEN'S METHOD)

PRINCIPLE

This test is used to determine the ability of an organism to produce the enzyme urease, which hydrolyzes urea. Hydrolysis of urea produces ammonia and CO_2. The formation of ammonia alkalinizes the medium, and the pH shift is detected by the color change of phenol red from light orange at pH 6.8 to magenta at pH 8.1.

PROCEDURE

1. Streak the surface of a urea agar slant with a portion of a well-isolated colony or inoculate slant with 1 to 2 drops from an overnight brain-heart infusion broth culture.

2. Leave the cap on loosely and incubate tube at 35° C in ambient air for 48 hours to 7 days.

EXPECTED RESULTS

Positive: Change in color of slant from light orange to magenta (Figure 30-30, *A*).

Negative: No color change (agar slant and butt remain light orange) (Figure 30-30, *B*).

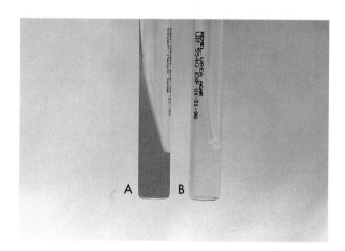

FIGURE 30-30 Urea hydrolysis (Christensen's method). **A,** Positive. **B,** Negative.

References

1. Heimbrook, M.E., Wang, W.L., and Campbell, G. 1989. Staining bacterial flagella easily. J. Clin. Microbiol. 26:2612.

2. Texas Department of Health. 1989. Laboratory Methods Workshop. King/Weaver protocol for the identification of gram-negative bacteria. Austin, Texas.

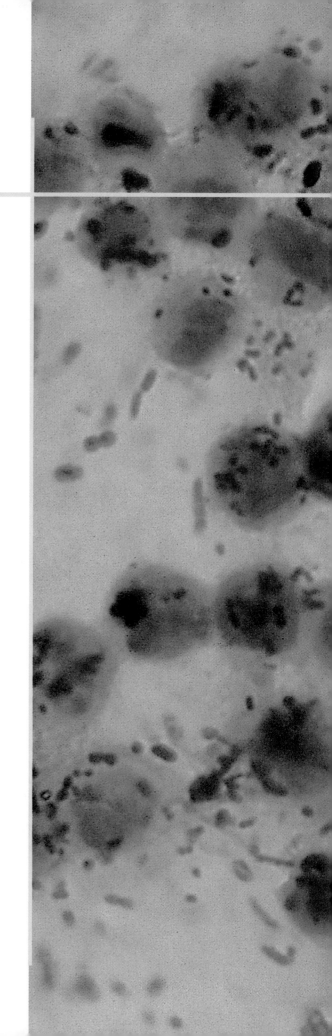

Section One

Gram-Negative Bacilli and Coccobacilli (MacConkey-Positive, Oxidase-Positive)

31 | PSEUDOMONAS, BURKHOLDERIA, AND SIMILAR ORGANISMS

Genera and Species to be Considered

Current Name	Previous Name
■ *Brevundimonas diminuta*	*Pseudomonas diminuta*
■ *Brevundimonas vesicularis*	*Pseudomonas vesicularis*
■ *Burkholderia cepacia*	*Pseudomonas cepacia*
■ *Burkholderia pseudomallei*	*Pseudomonas pseudomallei*
■ *Burkholderia mallei*	*Pseudomonas mallei*
■ *Burkholderia gladioli*	*Pseudomonas gladioli*
■ *Pseudomonas aeruginosa*	
■ *Pseudomonas alcaligenes*	
■ *Pseudomonas fluorescens*	
■ *Pseudomonas mendocina*	
■ *Pseudomonas pseudoalcaligenes*	
■ *Pseudomonas putida*	
■ *Pseudomonas* sp. CDC group 1	
■ *Pseudomonas stutzeri* (includes CDC group Vb-3)	
■ *"Pseudomonas denitrificans"**	
■ *Pseudomonas*-like group 2	
■ CDC group Ic	
■ *Ralstonia pickettii*	*Pseudomonas pickettii* and *Burkholderia pickettii*

GENERAL CHARACTERISTICS

At one time the species belonging to the genera *Brevundimonas, Burkholderia,* and *Ralstonia* were all members of the genus *Pseudomonas*. Organisms in these genera share many similar characteristics. They are aerobic, straight and slender gram-negative bacilli whose cells range from 1 to 5 μm in length and 0.5 to 1 μm in width. All species except *B. mallei* are motile. Members of these genera use a variety of carbohydrate, alcohol, and amino acid substrates as carbon and energy sources. Although they are able to survive and possibly grow at relatively low temperatures (i.e., as low as 4° C), the optimum temperature range for growth of most species is between 30° and 37° C; that is, they are mesophilic.

EPIDEMIOLOGY

BURKHOLDERIA SPP. AND RALSTONIA PICKETTII

Burkholderia spp. and *Ralstonia pickettii* are inhabitants of the environment and are not considered part of normal human flora (Table 31-1). As such, their transmission usually involves human contact with heavily contaminated medical devices or substances that are encountered in the hospital setting.

Among *Burkholderia* spp. found in the United States, *B. cepacia* is the one most commonly found in clinical specimens. Plants, soil, and water serve as reservoirs. This organism is also able to survive on or in medical devices and disinfectants. Intrinsic resistance to multiple antimicrobial agents also con-

---

*Quotation marks indicate an uncertain status.

TABLE 31-1 EPIDEMIOLOGY

SPECIES	HABITAT (RESERVOIR)	MODE OF TRANSMISSION
Burkholderia cepacia	Environmental (soil, water, plants); survives well in hospital environment; not part of normal human flora; may colonize respiratory tract of patients with cyctic fibrosis	Exposure of medical devices and solutions contaminated from the environment; Person-to-person transmission also documented
B. pseudomallei	Environmental (soil, streams, surface water such as rice paddies); limited to tropical and subtropical areas, notably Southeast Asia; not part of human flora	Inhalation or direct inoculation from environment through disrupted epithelial or mucosal surfaces
B. mallei	Causative agent of glanders in horses; mules, and donkeys; not part of human flora	Transmission to humans is extremely rare; associated with close animal contact and introduced through mucous membranes or broken skin.
B. gladioli	Environmental, pathogen of plants; occasionally found in respiratory tract of patients with cystic fibrosis but not normal part of human flora	Transmission to humans uncommon, mode of transmision is not known
Ralstonia pickettii	Environmental (multiple sources); found in variety of clinical specimens; not part of human flora	Mode of transmission is not known; likely involves exposure to contaminated medical devices and solutions
Pseudomonas aeruginosa	Environmental (soil, water, plants); survives well in domestic environments (e.g., hot tubs, whirlpools, contact lens solutions) and hospital environments (e.g., sinks, showers, respiratory equipment); rarely part of normal flora of healthy humans	Ingestion of contaminated food or water; exposure to contaminated medical devices and solutions; introduction by penetrating wounds; person-to-person transmission is assumed to occur
P. fluorescens, P. putida, P. stutzeri, (including Vb-3), and *P. mendocina*	Environmental (soil and water); not part of normal human flora	Exposure to contaminated medical devices and solutions
P. alcaligenes, P. pseudoalcaligenes, Pseudomonas sp. CDC group 1, "*P. denitrificans,*" *Pseudomonas*-like group 2, and CDC group Ic	Environmental; not part of normal human flora	Uncertain. Rarely encountered in clinical specimens
Brevundimonas vesicularis and *B. diminuta*	Environmental; not part of normal human flora	Uncertain. Rarely encountered in clinical specimens

---

tributes to the organism's survival in hospitals. Human acquisition of *B. cepacia* resulting in colonization or infection usually involves direct contact with contaminated foods; medical solutions, including disinfectants; and devices such as respiratory equip-

ment. Alternatively, person-to-person transmission also has been documented.[3]

B. pseudomallei also is an environmental inhabitant of niches similar to those described for *B cepacia,* but it is geographically restricted to tropical and subtropical areas of Australia and Southeast Asia. The organism is widely disseminated in soil, streams, ponds, and rice paddies. Human acquisition occurs through inhalation of contaminated debris or by direct inoculation through damaged skin or mucous membranes.[2]

Although *B. mallei* causes severe infections in horses and related animals, infections of humans are exceedingly rare. When transmission has occurred, it has been associated with close animal contact. *B. gladioli* is a plant pathogen that is only rarely found in sputa of patients with cystic fibrosis; the mode of transmission to humans and its clinical significance is unknown.[19]

Ralstonia pickettii is another environmental organism that is occasionally found in a variety of clinical specimens, such as blood, sputa of patients with cystic fibrosis, and urine. The mode of transmission is uncertain but is likely to involve exposure to contaminated materials.

PSEUDOMONAS SPP. AND *BREVUNDIMONAS* SPP.

These genera comprise several environmental species that rarely inhabit human skin or mucosal surfaces. In the clinical setting, *Pseudomonas aeruginosa* is the most commonly encountered gram-negative species that is not a member of the family *Enterobacteriaceae*. This organism survives in various environments in nature and in homes and hospitals (Table 31-1). Because of the ubiquitous nature of *P. aeruginosa,* there are a variety of means for transmission to humans.

P. fluorescens, P. putida, and *P. stutzeri* also are environmental inhabitants, but they are much less commonly found in clinical specimens than is *P. aeruginosa.* The other pseudomonads and *Brevundimonas* spp. listed in Table 31-1 also are environmental organisms. Because they are rarely encountered in patient specimens, the mode of transmission to humans remains uncertain.

PATHOGENESIS AND SPECTRUM OF DISEASE

BURKHOLDERIA SPP. AND *RALSTONIA PICKETTII*

Because these organisms are uncommon causes of infection in humans, very little is known about what, if any, virulence factors they exhibit.[11] Except for *B. pseudomallei,* the species listed in Table 31-2 generally are nonpathogenic for healthy human hosts.

The capacity of *B. cepacia* to survive in the hospital environment, which may be linked to the organism's intrinsic resistance to many antibiotics, provides the

opportunity for this species to occasionally colonize and infect hospitalized patients.[5,6] In patients with cystic fibrosis or chronic granulomatous disease the organism can cause fulminant lung infections and bacteremia, resulting in death.[14-16] In other types of patients, infections of the blood, urinary tract, and respiratory tract usually result from exposure to contaminated medical solutions or devices but are rarely fatal.[17]

Infections caused by *B. pseudomallei,* which can survive within human macrophages,[18] can range from being asymptomatic to severe. The disease is referred to as **melioidosis;** it has several forms, including formation of skin abscesses, sepsis and septic shock, abscess formation in several internal organs, and acute pulmonary disease.[2]

The remaining species listed in Table 31-2 are rarely encountered in human disease, and their clinical significance should be questioned when they are found in clinical specimens.[1]

PSEUDOMONAS SPP. AND *BREVUNDIMONAS* SPP.

Of the species in these two genera, *Pseudomonas aeruginosa* is the most thoroughly studied with regard to infections in humans. Although this organism is an environmental inhabitant, it is also a very successful opportunistic pathogen. Factors that contribute to pathogenicity include production of exotoxin A, which kills host cells by inhibiting protein synthesis, and production of several proteolytic enzymes and hemolysins that destroy cells and tissue. On the bacterial cell surface, pili may mediate attachment to host cells. Some strains produce alginate, a polysaccharide polymer that inhibits phagocytosis and contributes to infection potential in patients with cystic fibrosis. Additionally, *P. aeruginosa* can survive harsh environmental conditions and displays intrinsic resistance to a wide variety of antimicrobial agents that facilitates the organism's ability to survive in the hospital setting (see Table 31-2).

Even with the variety of potential virulence factors discussed, *P. aeruginosa* remains an opportunistic pathogen that requires compromised host defenses to establish infection. In normal healthy hosts infection is usually associated with events that disrupt or bypass protection provided by the epidermis (e.g., burns, puncture wounds, use of contaminated needles by IV drug abusers, eye trauma with contaminated contact lenses) and result in infections of the skin, bone, heart, or eye (see Table 31-2).

In patients with cystic fibrosis, *P. aeruginosa* has a predilection for infecting the respiratory tract. Although organisms rarely invade through respiratory tissue and into the bloodstream of these patients, the

TABLE 31-2 PATHOGENESIS AND SPECTRUM OF DISEASES

SPECIES	VIRULENCE FACTORS	SPECTRUM OF DISEASE AND INFECTIONS
Burkholderia cepacia	Unkown. Binding of mucin from patients with cystic fibrosis may be involved. Intrinsic resistance to multiple antibiotics complicates therapy and may promote organism survival in hospital	Nonpathogenic to healthy human hosts; able to colonize and cause life-threatening infections in patients with cystic fibrosis or chornic granulomatous disease; other patients may suffer nonfatal infections of the urinary tract, respiratory tract, and other sterile body sites
B. pseudomallei	Unknown. Bacilli can survive within phagocytes	Wide spectrum from asymptomatic infection to melioidosis, of which there are several forms, including infections of the skin and respiratory tract, multisystem abscess formation, and bacteremia with septic shock
B. mallei	Unknown for human infections	Human disease is extremely rare. Infections range from localized acute or chronic suppurative infections of skin at site of inoculation to acute pulmonary infections and septicemia
B. gladioli	Unknown	Role in human disease is uncertain; occasionally found in sputa of patients with cystic fibrosis, but clinical significance in this setting is unknown
Ralstonia pickettii	Unknown	Rarely encountered as cause of disease; non-pathogenic to healthy human host, but may be isolated from a variety of clinical specimens, including blood, sputum, and urine; when encountered, environmental contamination should be suspected
Pseudomonas aeruginosa	Exotoxin A, endotoxins, proteolytic enzymes, alginate, and pili; intrinsic resistance to many antimicrobial agents	Opportunistic pathogen that can cause community- or hospital- acquired infections Community-acquired infections: skin (folliculitis); external ear canal (otitis externa); eye, following trauma; bone (osteomyelitis), following trauma; heart (endocarditis) in IV drug abusers; and respiratory tract (patients with cystic fibrosis) Hospital acquired infections: respiratory tract, urinary tract, wounds, bloodstream (bacteremia), and central nervous system
P. fluorescens, P. putida, and *P. stutzeri* (includes Vb-3)	Unknown. Infection usually requires patient with underlying disease to be exposed to contaminated medical devices or solutions	Uncommon cause of infection; have been associated with bacteremia, urinary tract infections, wound infections, and respiratory tract infections; when found in clinical specimen, significance should always be questioned

Continued

—— **TABLE 31-2 PATHOGENESIS AND SPECTRUM OF DISEASES—CONT'D** ——

SPECIES	VIRULENCE FACTORS	SPECTRUM OF DISEASE AND INFECTIONS
P. mendocina, P. alcaligenes, P. pseudoalcaligenes, Pseudomonas sp. CDC group 1 "*P. denitrificans*" *Pseudomonas*-like group 2, and CDC group Ic	Unknown	Not known to cause human infections
Brevundimonas vesicularis and *B. diminuta*	Unknown	Rarely associated with human infections. *B. vesicularis* is rare cause of bacteremia in patients suffering underlying disease

consequences of respiratory involvement alone are serious and life-threatening. In other patients, *P. aeruginosa* is a notable cause of nosocomial infections of the respiratory and urinary tracts, wounds, bloodstream, and even the central nervous system. For compromised patients such infections are often severe and frequently life-threatening. In some cases of bacteremia the organism may invade and destroy walls of subcutaneous blood vessels, resulting in formation of cutaneous papules that become black and necrotic. This condition is known as **ecthyma gangrenosum**. Similarly, patients with diabetes may suffer a severe infection of the external ear canal (malignant otitis externa), which can progress to involve the underlying nerves and bones of the skull.

No known virulence factors have been associated with *P. fluorescens, P. putida,* or *P. stutzeri.* When infections do occur they usually involve a compromised patient exposed to contaminated medical materials.[12,13] Such exposure has been known to result in infections of the respiratory and urinary tracts, wounds, and bacteremia (see Table 31-2). However, because of their low virulence, whenever these species are encountered in clinical specimens their significance should be highly suspect. Similar caution should be applied whenever the other *Pseudomonas* spp. or *Brevundimonas* spp. listed in Table 31-2 are encountered.[13]

LABORATORY DIAGNOSIS

SPECIMEN COLLECTION AND TRANSPORT

No special considerations are required for specimen collection and transport of organisms discussed in this chapter. Refer to Table 1-1 for general information on specimen collection and transport.

SPECIMEN PROCESSING

No special considerations are required for processing of the organisms discussed in this chapter. Refer to Table 1-1 for general information on specimen processing.

DIRECT DETECTION METHODS

Other than Gram stain of patient specimens, there are no specific procedures for the direct detection in clinical material of the organisms discussed in this chapter. These organisms usually appear as medium-sized, straight rods by Gram stain. Exceptions are *Brevundimonas diminuta,* which is a long straight rod; *Burkholderia mallei,* which is a coccobacillus; *Pseudomonas* CDC group 1, which is a medium-size, slightly curved rod; and CDC group Ic, which is a thin, pleomorphic rod.

CULTIVATION

Media of choice

Pseudomonas spp., *Brevundimonas* spp., *Burkholderia* spp., *Ralstonia pickettii,* and CDC group Ic grow well on routine laboratory media, such as 5% sheep blood agar and chocolate agar. Except for *Brevundimonas vesicularis,* all usually grow on MacConkey agar. All four genera also grow well in broth-blood culture systems and common nutrient broths, such as thioglycollate and brain-heart infusion. Specific selective media, such as *Pseudomonas cepacia* (PC) agar or oxidative-fermentative base–polymyxin B–bactracin-lactose (OFPBL) agar may be used to isolate *Burkholderia cepacia* from respiratory secretions of patients with cystic fibrosis (see Table 31-3). Ashdown medium is used to isolate *Burkholderia pseudomallei* when infections with this species are suspected.

TABLE 31-3 COLONY APPEARANCE AND OTHER CHARACTERISTICS OF *PSEUDOMONAS, BREVUNDIMONAS, BURKHOLDERIA, RALSTONIA,* AND OTHER ORGANISMS

ORGANISM	MEDIUM	APPEARANCE
P. aeruginosa	BA	Spreading and flat, serrated edges; confluent growth; often shows metallic sheen; bluish-green, red, or brown pigmentation; colonies often beta-hemolytic; grapelike or corn taco–like odor; mucoid colonies commonly seen in patients with cystic fibrosis
	Mac	NLF
P. fluorescens	BA	No distinctive appearance
	Mac	NLF
P. putida	BA	No distinctive appearance
	Mac	NLF
P. stutzeri and CDC group Vb-3	BA	Dry, rinkled, adherent, buff to brown
	Mac	NLF
P. mendocina	BA	Smooth, nonwrinkled, flat, brownish-yellow pigment
	Mac	NLF
P. alcaligenes	BA	No distinctive appearance
	Mac	NLF
P. pseudoalcaligenes	BA	No distinctive appearance
	Mac	NLF
Pseudomonas spp. CDC group 1	BA	No distinctive appearance
	Mac	NLF
Brevundimonas diminuta	BA	Chalk white
	Mac	NLF
B. vesicularis	BA	Orange pigment
	Mac	NLF, but only 66% grow
Burkholderia pseudomallei	BA	Smooth and mucoid to dry and wrinkled (may resemble *P. stutzeri*)
	Mac	NLF
	Ashdown	Dry, wrinkled, violet-purple
B. cepacia	BA	Smooth
	Mac	NLF
	PC or OFPBL	Smooth
B. gladioli	BA, Mac, OFPBL	Bright yellow pigment

Continued

TABLE 31-3 COLONY APPEARANCE AND OTHER CHARACTERISTICS OF *PSEUDOMONAS, BREVUNDIMONAS, BURKHOLDERIA, RALSTONIA,* AND OTHER ORGANISMS—CONT'D

ORGANISM	MEDIUM	APPEARANCE
Ralstonia pickettii	BA	No distinctive appearance but may take 72 hrs to produce visible colonies
	Mac	NLF
B. mallei	BA	No distinctive appearance
	Mac	NLF
"Pseudomaonas denitrificans"	BA	No distinctive appearance
	Mac	NLF
Pseudomonas-like group 2	BA	No distinctive appearance but colonies tend to stick to agar
	Mac	NLF
CDC group Ic	BA	No distinctive appearance
	Mac	NLF

BA, 5% sheep blood agar; *Mac,* MacConkey agar; *PC, Pseudomonas cepacia* agar; *OFPBL,* Oxidative-fermentative base–polymyxin B-bacitracin–lactose; *NLF,* nonlactose fermenter.

Incubation conditions and duration
Detectable growth on 5% sheep blood and chocolate agars, incubated at 35° C in carbon dioxide or ambient air, generally occurs within 24 to 48 hours after inoculation. Growth on MacConkey agar, which should only be incubated in ambient air at 35° C, also is detectable within this same time. Selective media (PC or OFPBL) used for patients with cystic fibrosis may require incubation at 35° C in ambient air for up to 72 hours before growth is detected.

Colonial appearance
Table 31-3 describes the colonial appearance and other distinguishing characteristics (e.g., hemolysis and odor) of each genus on common laboratory media.

APPROACH TO IDENTIFICATION
Most of the commercial systems available for identification of these organisms reliably identify *Pseudomonas aeruginosa* and *Burkholderia cepacia,* but their reliability for identification of other species is less certain.

Table 31-4 provides key phenotypic characteristics for identifying the species discussed in this chapter. These tests provide useful information for presumptive organism identification, but definitive identification often requires the use of a more extensive battery of tests usually performed by reference laboratories.

Comments regarding specific organisms
A convenient and reliable identification scheme for *Pseudomonas aeruginosa* involves the following conventional tests and characteristics:

- Oxidase-positive
- TSI slant with an alkaline/no change (K/NC) reaction
- Good growth at 42° C
- Production of bright bluish-green, red, or brown diffusible pigment on Mueller-Hinton agar or trypticase soy agar (Figure 31-1)

P. aeruginosa, P. fluorescens, and *P. putida* comprise the group known as the **fluorescent pseudomonads.** *P. aeruginosa* can be distinguished from the others in this group by its ability to grow at 42° C. Mucoid strains of *P. aeruginosa* from patients with cystic fibrosis may not exhibit the characteristic pigment and may react slower in biochemical tests than nonmucoid strains. Therefore biochemicals should be held for the complete 7 days before being recorded as negative. This slow biochemical activity is often what prevents the identification of mucoid *P. aeruginosa* by commercial systems.

Burkholderia cepacia should be suspected whenever a nonfermentative organism that decarboxylates lysine is encountered. Lysine decarboxylation is positive in 80% of strains. Correct identification of the occasional strains that are lysine-negative (20%), or oxidase-negative (14%), requires full biochemical profiling.

The presumptive identification of other species in this chapter is fairly straightforward using the key characteristics given in Table 31-4. However, there are a few notable exceptions. First, unlike the other pseudomonads, *P. alcaligenes, P. pseudoalcaligenes,* and *Pseudomonas* spp. CDC group 1 do not oxidize glucose. Second, strains of *Burkholderia mallei* and *Burkholderia gladioli* may be oxidase-negative.

SERODIAGNOSIS

Serodiagnostic techniques are not generally used for the laboratory diagnosis of infections caused by the organisms discussed in this chapter.

ANTIMICROBIAL SUSCEPTIBILITY TESTING AND THERAPY

Except for *Pseudomonas aeruginosa,* validated susceptibility testing methods do not exist for the organisms discussed in this chapter. Therefore, when these organisms are isolated, the laboratory has a conflict between the urge to contribute in some way to patient management by providing data and the lack of confidence in producing interpretable and accurate information.

Although many of these organisms will grow on the media and under the conditions recommended for testing the more commonly encountered bacteria (see Chapter 18 for more information regarding validated testing methods), the ability to grow under test conditions does not guarantee reliable detection of important antimicrobial resistance. Therefore, even though testing can provide an answer, substantial potential to obtain the wrong answer exists. Chapter 19 should be reviewed for strategies that can be used to provide susceptibility information and data when validated testing methods do not exist for a clinically important bacterial isolate.

The infrequency with which *Burkholderia* spp. and *Ralstonia pickettii* are encountered in infections in humans and the lack of validated in vitro susceptibility testing methods do not allow definitive treatment and testing guidelines to be given (Table 31-5). Potential therapies for *B. cepacia* and *B. pseudomallei* are provided, but antimicrobial therapy rarely eradicates *B. cepacia,* especially from the respiratory tract of patients with cystic fibrosis, and the optimum ther-

apy for melioidosis remains controversial.[5,20,22] *Burkholderia* spp. are capable of expressing resistance to various antibiotics, so devising effective treatment options can be problematic.[4,7,19] Additionally troublesome is that valid testing methods for these organisms do not exist. Because of these complex issues, the importance of establishing the clinical significance of these species when they are isolated in clinical specimens is emphasized.

Among *Pseudomonas* spp. and *Brevundimonas* spp., *P. aeruginosa* is the only species for which valid in vitro susceptibility testing methods exist[8-10] and for which there is extensive therapeutic experience (see Table 31-5; also see Chapter 18 for a discussion of available testing methods).* Therapy usually involves the use of a beta-lactam developed for antipseudomonal activity and an aminoglycoside. The particular therapy used depends on several clinical factors and on the antimicrobial resistance profile the laboratory reports for a particular *P. aeruginosa* isolate.

P. aeruginosa is intrinsically resistant to various antimicrobial agents; only those with potential activity are given in Table 31-5. However, *P. aeruginosa* also readily acquires resistance to the potentially active agents listed, necessitating that agents selected from the list be tested against each clinically relevant isolate.

Although antimicrobial resistance is also characteristic of the other *Pseudomonas* spp. and *Brevundimonas* spp.,† the fact that they are rarely clinically significant and the lack of validated testing methods prohibits the provision of specific guidelines (see Table 31-5). Antimicrobial agents used for *P. aeruginosa* infections are often considered for use against the other species;

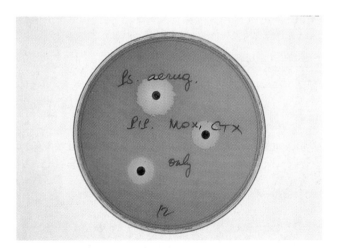

FIGURE 31-1 *Pseudomonas aeruginosa* on Mueller-Hinton agar. Note bluish-green color.

---

*Order from NCCLS, 940 W. Valley Road, Suite 1400, Wayne, Pa. 19087.

†References 4,7,12,13,16.

—— **TABLE 31-4 BIOCHEMICAL AND PHYSIOLOGIC CHARACTERISTICS** ——

ORGANISM	OXIDASE	GROWTH AT 42° C	NITRATE REDUCTION	GAS FROM NITRATE
Burkholderia pseudomallei	+	+	+	+
B. mallei	v	=	+	=
B. cepacia	v	v	v	=
B. gladioli	v	=	v	=
Ralstonia pickettii biovar 1	+	v	+	v
R. pickettii biovar 2	+	+	+	+
R. pickettii biovar 3	+	v	v	=
Pseudomonas aeruginosa	+	+	+	+
P. fluorescens	+	=	v	=
P. putida	+	=	=	=
P. stutzeri	+	v	+	+
CDC group Vb-3	+	=	+	+
P. mendocina	+	+	+	+
P. alcaligenes	+	=	v	=
P. pseudoalcaligenes	+	+	+	=
Pseudomonas spp. CDC group 1	+	v	+	+
Brevundimonas diminuta	+	v	=	=
B. vesicularis	+	v	=	=
"Pseudomonas denitrificans"	+	v	+	v
Pseudomonas-like group 2	+	v	v	=
CDC group Ic	+	+	+	=

+, >90% of strains are positive; =, >90% of strains are negative; *v*, variable.

Data compiled from Weyant, R.S., et al., editors. 1996. Identification of unusual pathogenic gram-negative aerobic and facul-
Segers, P., and Pot, B., et al. 1994. Int. J. Syst. Bacteriol. 44:499.

ARGININE DIHYDROLASE	LYSINE DECARBOXYLASE	UREA HYDROLYSIS	OXIDIZES GLUCOSE	OXIDIZES LACTOSE	OXIDIZES MANNITOL
+	=	v	+	+	+
+	=	v	+	v	v
=	v	v	+	+	+
=	=	v	+	v	+
=	=	+	+	+	=
=	=	+	+	=	=
=	=	+	+	+	+
+	=	v	+	=	v
+	=	v	+	v	v
+	=	v	+	v	v
=	=	v	+	=	+
+	=	v	+	=	v
+	=	v	+	=	=
v	=	=	=	=	=
v	=	=	=	=	=
v	=	=	=	=	=
=	=	v	v	=	=
=	=	=	+	=	=
+	=	v	+	+	=
v	=	+	+	+	+
+	=	v	+	=	=

tatively anaerobic bacteria, ed 2. Williams & Wilkins, Baltimore; Yabuuchi, E., et al. 1995. Microbiol. Immunol. 39:897;

--- **TABLE 31-5** ANTIMICROBIAL THERAPY AND SUSCEPTIBILITY TESTING ---

SPECIES	THERAPEUTIC OPTIONS	POTENTIAL RESISTANCE TO THERAPEUTIC OPTIONS	VALIDATED TESTING METHODS*	COMMENTS
Burkholderia cepacia	No definitive guidelines. Potentially active agents include piperacillin, ceftazidime, imipenem, ciprofloxacin, chloramphenicol, and trimethoprim/ sulfamethoxazole	Yes	Not available	Antimicrobial therapy rarely eradicates organism; will grow on susceptibility testing media, but standards for interpretation of results do not exist
B. pseudomallei	No definitive guidelines. Potentially active agents include ceftazidime, piperacillin/ tazobactam, ticarcillin/clavulanate, amoxicillin/clavulanate, imipenem, trimethoprim/ sulfamethoxazole, and chloramphenicol	Yes	Not available	Optimal therapy still controversial
B. mallei	No definitive guidelines. Potentially active agents may include those listed for *B. pseudomallei.*	Yes	Not available	Rarely involved in human infections, so reliable therapeutic data are limited
B. gladioli	No definitive guidelines. Potentially active agents include imipenem, piperacillin, and ciprofloxacin	Yes	Not available	Rarely involved in human infections, so reliable therapeutic data are limited
Ralstonia pickettii	No definitive guidelines. Potentially active agents include those listed for *B. cepacia*	Yes	Not available	Rarely involved in human infections, so reliable therapeutic data are limited

— **TABLE 31-5 ANTIMICROBIAL THERAPY AND SUSCEPTIBILITY TESTING—CONT'D** —

SPECIES	THERAPEUTIC OPTIONS	POTENTIAL RESISTANCE TO THERAPEUTIC OPTIONS	VALIDATED TESTING METHODS*	COMMENTS
Pseudomonas aeruginosa	An antipseudomonal beta-lactam (listed below) with or without an aminoglycoside; certain quinolones may also be used. Specific agents include ticarcillin, mezlocillin, piperacillin, ticarcillin/clavulanic acid, piperacillin/tazobactam, ceftazidime, cefoperazone, cefepime, aztreonam, imipenem, gentamicin, tobramycin, amikacin, netilmicin, ciprofloxacin, and ofloxacin (urinary tract only)	Yes	Disk diffusion, broth dilution, agar dilution, and commercial systems	In vitro susceptibility testing results important for guiding therapy
P. fluorescens, P. putida, P.stutzeri (includes Vb-3), *P.mendocina, P.alcaligenes, P. pseudoalcaligenes, Pseudomonas* sp. CDC group 1, "*P. denitrificans,*" *Pseudomonas*-like group 2, and CDC group Ic	Because rarely implicated in human infections, there are no definitive guidelines; agents used for *P. aeruginosa* may be effective for these species	Yes	Not available	Most will grow on susceptibility testing media, but standards for interpretation of results do not exist
Brevundimonas vescularis and *B. diminuta*	Because rarely implicated in human infections, there are no definitve guidelines	Unknown	Not available	Rarely involved in human infection

*Validated testing methods include those standard methods recommended by the National Committee for Clinical Laboratory Standards (NCCLS) and those commercial methods approved by the Food and Drug Administration (FDA).

however, before proceeding with the development of treatment strategies the first critical step should be to establish the clinical significance of these organisms.

PREVENTION

Because these organisms are ubiquitous in nature and many are commonly encountered without deleterious effects on healthy human hosts, there are no recommended vaccination or prophylaxis protocols. Hospital-acquired infections are best controlled by following appropriate infection control guidelines and implementing effective protocols for the sterilization and decontamination of medical supplies.

References

1. Christenson, J.C., Welch, D.F., Mukwaya, G., et al. 1989. Recovery of *Pseudomonas gladioli* from respiratory tract specimens of patients with cystic fibrosis. J. Clin. Microbiol. 27:270.
2. Dance, D.A.B. 1991. Melioidosis: the tip of the iceberg? Clin. Microbiol. Rev. 4:52.
3. Gilligan, P.H. 1991. Microbiology of airway disease in patients with cystic fibrosis. Clin. Microbiol. Rev. 4:35.
4. Godfrey, A.J., Wong, S., Dance, D.A.B., et al. 1991. *Pseudomonas pseudomallei* resistance to β-lactam antibiotics due to alterations in the chromosomally encoded β-lactamase. Antimicrob. Agent Chemother. 35:1635.
5. Gold, R., Jin, E., Levison, H., et al. 1983. Ceftazidime alone and in combination in patients with cystic fibrosis: lack of efficacy in treatment of severe respiratory infections caused by *Pseudomonas cepacia*. J. Antimicrob. Chemother. 12(suppl. A):331.
6. Lewin, C., Doherty, C., and Govan, J. 1993. In vitro activities of meropenem, PD12731, PD 131628, ceftazidime, chloramphenicol, cotrimoxazole, and ciprofloxacin against *Pseudomonas cepacia*. Antimicrob. Agents Chemother. 37:123.
7. Livermore, D. 1995. β-Lactamases in laboratory and clinical resistance. Clin. Microbiol. Rev. 8:557.
8. National Committee for Clinical Laboratory Standards. 1997. Performance standards for antimicrobial susceptibility testing. M100-S7. NCCLS, Villanova, Pa.
9. National Committee for Clinical Laboratory Standards. 1997. Performance standards for antimicrobial disk susceptibility tests. M2-A6, ed 6. NCCLS, Villanova, Pa.
10. National Committee for Clinical Laboratory Standards. 1997. Methods for dilution antimicrobial susceptibility tests for bacteria that grow aerobically. M7-A4, ed 4. NCCLS, Villanova, Pa.
11. Nelson, J.W., Butler, S.L., Krieg, D., and Govan, J.R.W. 1994. Virulence factors of *Burkholderia cepacia*. FEMS Immunol. Med. Microbiol. 8:89.
12. Noble, R.C. and Overman, S.B. 1994. *Pseudomonas stutzeri* infection: a review of hospital isolates and a review of the literature. Diagn. Microbiol. Infect. Dis. 19:51.
13. Oberhelman, R.A., Humbert, J.R., and Santorelli, F.W. 1994. *Pseudomonas vesicularis* causing bacteremia in a child with sickle cell anemia. South. Med. J. 87:821.
14. O'Neil, K.M., Herman, J.H., Modlin, et al. 1986. *Pseudomonas cepacia*: an emerging pathogen in chronic granulomatous disease. J. Pediatr. 108:940.
15. Pallent, L.J., Hugo, W.B., Grant, D.J.W., et al. 1983. *Pseudomonas cepacia* as a contaminant and infective agent. J. Hosp. Infect. 4:9.
16. Papapetropoulou, M., Iliopoulou, J., Rodopoulou, G., et al. 1994. Occurrence and antibiotic-resistance of *Pseudomonas* species isolated from drinking water in Southern Greece. J. Chemother. 6:111.
17. Pegues, D.A., Carson, L.A., Anderson, R.L., et al. 1993. Outbreak of *Pseudomonas cepacia* bacteremia in oncology patients. Clin. Infect. Dis. 16:407.
18. Pruksachartvuthi, S., Aswapokee, N., and Thankerngpol, K., 1990. Survival of *Pseudomonas pseudomallei* in human phagocytes. J. Med. Microbiol. 31:109.
19. Simpson, I.N., Finlay, J., Winstanleyet, D.J., et al. 1994. Multi-resistance isolates possessing characteristics of both *Burkholderia (Pseudomonas) cepacia* and *Burkholderia gladioli* from patients with cystic fibrosis. J. Antimicrob. Chemother. 34:353.
20. Smith, M.D., Wuthiekanum, V., Walsh, A.L., et al. 1994. Susceptibility of *Pseudomonas pseudomallei* to some newer β-lactam antibiotics and antibiotic combinations using time-kill studies. J. Antimicrob. Chemother. 33:145.
21. Sookpranee, W.P., Boonma, P., Susaengrat, M., et al. 1992. Multicenter prospective randomized trial comparing ceftazidime plus co-trimoxazole with chloramphenicol plus doxycycline and co-trimoxazole for treatment of severe melioidosis. Antimicrob. Agents Chemother. 36:158.
22. Sookpranee, T., Sookpranee, M., Mellencamp, M.A., et al. 1991. *Pseudomonas pseudomallei*, a common pathogen in Thailand that is resistant to the bactericidal effects of many antibiotics. Antimicrob. Agents Chemother. 35:484.

Bibliography

Balows, A., Truper, H.G., Dworkin, M., et al., editors. 1981. The prokaryotes. A handbook on the biology of bacteria: ecophysiology, isolation, identification, applications, ed 2. Springer-Verlag, New York.

Fick, R.B., Jr. 1993. *Pseudomonas aeruginosa*: the opportunist. CRC Press, Inc., Boca Raton, Fla.

Gilligan, P.H. 1995. *Pseudomonas* and *Burkholderia*. In Murray, P.R., Baron, E.J., Pfaller, M.A., et al., editors. Manual of clinical microbiology, ed 6. American Society for Microbiology, Washington, D.C.

Pollack, M. 1995. *Pseudomonas aeruginosa*. In Mandell, G.L., Bennett, J.E., and Dolin, R. editors. Principles and practice of infectious diseases. Churchill Livingstone, New York.

Segers, P., Vancanneyt, M., Pot, B., et al. 1994. Classification of *Pseudomonas diminuta* (Leifson and Hugh 1954) and *Pseudomonas vesicularis* (Basing, Dîll, and Freytag 1953) in *Bre-*

vundimonas gen. nov. as *Brevundimonas diminuta* comb. nov. and *Brevundimonas vesicularis* comb. nov., respectively. Int. J. Syst. Bacteriol. 44:499.

Weyant, R.S., Moss, C.W., Weaver, R.E., et al., editors. 1996. Identification of unusual pathogenic gram-negative aerobic and facultatively anaerobic bacteria, ed 2. Williams & Wilkins, Baltimore.

Yabuuchi, E., Kosako, Y., Yano, H., et al. 1995. Transfer of two *Burkholderia* and an *Alcaligenes* species to *Ralstonia* gen. nov.: proposal of *Ralstonia pickettii* (Ralston, Palleroni, and Doudoroff 1973) comb. nov., *Ralstonia solanacearum* (Smith 1896) comb. nov. and *Ralstonia eutropha* (Davis 1969) comb. nov. Microbiol. Immunol. 39:897.

32 | ACHROMOBACTER, AGROBACTERIUM, OCHROBACTRUM, AND SIMILAR ORGANISMS

Genera and Species to be Considered

Current Name	Previous Name
■ *"Achromobacter"** group	
■ *Acidovorax* spp.	*Pseudomonas delafieldii*
	Pseudomonas facilis
	Pseudomonas temperans
■ *Agrobacterium radiobacter*	CDC group Vd-3
■ CDC group EF-4b	CDC group EF-4
■ CDC group EO-2	
■ CDC group EO-3	
■ *Ochrobactrum anthropi*	CDC group Vd
■ CDC group OFBA-1	
■ *Psychrobacter immobilis*	Part of CDC group EO-2

GENERAL CHARACTERISTICS

The organisms discussed in this chapter are environmental inhabitants, although CDC group EF-4b inhabits the upper respiratory tract of certain animals and *Ochrobactrum anthropi* may occasionally inhabit the human gastrointestinal tract. All are oxidase-positive, grow on MacConkey agar and oxidize glucose, but their specific morphologic and physiologic features are somewhat diverse and are considered later in this chapter in the discussion of laboratory diagnosis.

EPIDEMIOLOGY

As environmental organisms, these bacteria are rarely encountered in human specimens or infections. When they are encountered, they are found on contaminated medical devices or are isolated from immunocompromised or otherwise debilitated patients. Of the organisms listed in Table 32-1, *Agrobacterium radiobacter* and *Ochrobactrum anthropi* are the species most commonly encountered in the clinical setting. The other bacteria have rarely been discovered in clinical material, and several have never been established as the cause of human infection.[7-9,11,12]

Agrobacterium radiobacter inhabits the soil; other species of this genus colonize and cause infections in plants. In fact, the only difference between *A. radiobacter* and *A. tumefaciens*, a plant pathogen that causes tumor formation, is that the latter species contains a plasmid that encodes for plant pathogenicity. Human infections occur by exposure to contaminated medical devices usually involving already ill patients.[5]

The specific environmental niche of *Ochrobactrum anthropi* is unknown, but this organism is capable of survival in water, including moist areas within the hospital environment. The organism may also be a transient colonizer of the human gastrointestinal tract. Similar to *A. radiobacter*, human infections caused by *Ochrobactrum anthropi* are associated with implantation of intravenous catheters or other foreign bodies in patients with a debilitating illness. Acquisition by contaminated pharmaceuticals and by puncture wound has also been documented.[1-4,10]

The epidemiology of CDC group EF-4b is unlike that of the other bacteria discussed in this chapter. Animals, rather than the environment, are the reservoir, and transmission to humans occurs by dog or cat bites and scratches.

PATHOGENESIS AND SPECTRUM OF DISEASE

Because these organisms rarely cause human infections, little is known about what, if any, virulence

---

*Quotation marks indicate a proposed organism name.

——————————— **TABLE 32-1 EPIDEMIOLOGY** ———————————

SPECIES	HABITAT (RESERVOIR)	MODE OF TRANSMISSION
"Achromobacter" group	Uncertain, probably environmental; not part of human flora	Unknown. Rarely found in humans
Acidovorax spp.	Environmental, soil; not part of human flora	Unknown. Rarely found in humans
Agrobacterium radiobacter	Environmental, soil and plants; not part of human flora	Contaminated medical devices such as intravenous and peritoneal catheters
CDC group EF-4b	Animal oral and respiratory flora; not part of human flora	Animal contact, particularly bites or scratches from dogs and cats
CDC group EO-2 and CDC group EO-3	Environmental. Not part of human flora	Infections in humans not yet described
Psychrobacter immobilis	Environmental, particularly cold climates such as the Antarctic; not part of human flora	Unknown. Rarely found in humans. Has been found in fish, poultry, and meat products
CDC group OFBA-1	Uncertain, probably environmental; not part of human flora	Unknown. Rarely found in humans
Ochrobactrum anthropi	Uncertain, probably environmental; found in water and hospital environments; may also be part of human flora	Uncertain. Most likely involves contaminated medical devices, such as catheters or other foreign bodies, or contaminated pharmaceuticals. Also can be acquired in community by puncture wounds

factors they may produce to facilitate infectivity (Table 32-2). The fact that *A. radiobacter* and *O. anthropi* infections frequently involve contaminated medical materials and compromised patients, and rarely if ever occur in healthy hosts, suggests that these bacteria have relatively low virulence. One report suggests that *A. radiobacter* is capable of capsule production[5], and the ability of *O. anthropi* to adhere to the silicone material of catheters may contribute to this organism's propensity to cause catheter-related infections.[1] No known virulence factors have been described for CDC group EF-4b. Infection appears to require traumatic introduction by a puncture wound (i.e., bite) or scratch, which indicates that the organism itself does not express any invasive properties.

For both *A. radiobacter* and *O. anthropi*, bacteremia is the most common type of infection (Table 32-2); peritonitis, endocarditis, meningitis, urinary tract, and pyogenic infections are much less commonly encountered.[2,5,6] Cellulitis and abscess formation typify the infections resulting from the traumatic introduction of CDC group EF-4b into the skin and subcutaneous tissue.

Although other species listed in Table 32-2 may be encountered in clinical specimens, their association with human infection is rare and their clinical significance in such encounters should be carefully analyzed.

LABORATORY DIAGNOSIS

SPECIMEN COLLECTION AND TRANSPORT

No special considerations are required for specimen collection and transport the organisms discussed in this chapter. Refer to Table 1-1 for general information on specimen collection and transport.

SPECIMEN PROCESSING

No special considerations are required for processing the organisms discussed in this chapter. Refer to Table 1-1 for general information on specimen processing.

DIRECT DETECTION METHODS

Other than Gram stain of patient specimens, there are no specific procedures for the direct detection of these organisms in clinical material. *Achromobacter, Acidovorax, Ochrobactrum,* and OFBA-1 are medium-sized straight rods. *Agrobacterium radiobacter* is a short, pleomorphic rod. *Psychrobacter immobilis,*

TABLE 32-2 PATHOGENESIS AND SPECTRUM OF DISEASES

SPECIES	VIRULENCE FACTORS	SPECTRUM OF DISEASE AND INFECTIONS
"Achromobacter" group	Unknown	Rarely isolated from humans. Patients with septicemia have been reported
Acidovorax spp.	Unknown	Rarely isolated from clinical specimens. Not yet implicated in human infections
Agrobacterium radiobacter	Unknown. One blood isolate described as mucoid, suggestive of exopolysaccharide capsule production	Exposure of immunocompromised or debilitated patient to contaminated medical devices resulting in bacteremia, and less commonly, peritonitis, endocarditis, or urinary tract infection
CDC group EF-4b	Unknown	Infected bite wounds of fingers, hands, or arm leading to cellulitis or abscess formation. Systemic infections are rare
CDC group EO-2 and CDC group EO-3	Unknown	No infections described in humans. Rarely encountered in clinical specimens
Psychrobacter immobilis	Unknown	Rare cause of infection in humans. Has been described in wound and catheter site infections, meningitis, and eye infections
CDC group OFBA-1	Unknown	Rarely isolated from clinical specimens; found in blood, respiratory, wound, and catheter specimens
Ochrobactrum anthropi	Unknown. Exhibits ability to adhere to silicone catheter material in a manner similar to staphylococci	Catheter and foreign body-associated bacteremia. May also cause pyogenic infections, community-acquired wound infections, and meningitis in tissue graft recipients. Patients are usually immunocompromised or otherwise debilitated

---

EF-4b, EO-2, and EO-3 are coccobacilli. EO-2 has a characteristic O appearance on Gram stain.

CULTIVATION

Media of choice
Achromobacter, Acidovorax, Agrobacterium, EO-2, EO-3, EF-4b, Ochrobactrum, OFBA-1, and Psychrobacter grow well on routine laboratory media such as 5% sheep blood, chocolate, and MacConkey agars. These organisms also grow well in the broth of blood culture systems and in common nutrient broths such as thioglycollate and brain-heart infusion.

Incubation conditions and duration
These organisms will produce detectable growth on 5% sheep blood, MacConkey, and chocolate agars when incubated at 35° C in either carbon dioxide or ambient air for a minimum of 24 hours. However, MacConkey agar should only be incubated in ambient air. Psychrobacter spp. is an exception in that it

usually grows poorly at 35° C and grows best between 20° and 25° C.

Colonial appearance
Table 32-3 describes the colonial appearance and other distinguishing characteristics (e.g., hemolysis and odor) of each genus when grown on 5% sheep blood or MacConkey agars.

APPROACH TO IDENTIFICATION

The ability of most commercial identification systems to accurately identify the organisms discussed in this chapter is limited or uncertain. Identification often requires the use of conventional biochemical profiles.

The key biochemical reactions that can be used to presumptively differentiate among the genera discussed in this chapter are provided in Table 32-4. However, definitive identification of these organisms often requires performing an extensive battery of biochemical tests not commonly available in many clini-

— TABLE 32-3 COLONY APPEARANCE AND CHARACTERISTICS —

ORGANISM	MEDIUM	APPEARANCE
"Achromobacter" group	BA	Smooth, glistening, entire
	Mac	NLF
Acidovorax spp.	BA	No distinctive appearance
	Mac	NLF
Agrobacterium radiobacter	BA	No distinctive appearance
	Mac	NLF
EO-2	BA	Growth frequently mucoid
	Mac	NLF
EO-3	BA	Yellow pigment
	Mac	NLF
EF-4b	BA	No distinctive appearance, but cultures smell like popcorn
	Mac	NLF
Ochrobactrum anthropi	BA	Resemble colonies of *Enterobacteriaceae,* only smaller
	Mac	NLF
OFBA-1	BA	Beta-hemolytic
	Mac	NLF
Psychrobacter immobilis	BA	No distinctive appearance but usually will not grow well at 35° C; grows best at 20° C; cultures smell like roses
	Mac	NLF

BA, 5% sheep blood agar; *Mac,* MacConkey agar; *NLF,* nonlactose fermenter.

cal microbiology laboratories. Therefore, full identification of clinically relevant isolates may require that they be sent to a reference laboratory.

Comments regarding specific organisms

Although the EF portion of the EF-4b designation stands for *eugonic* (an organism that grows well on common laboratory media) *f*ermenter, most EF-4b strains oxidize glucose, so the designation as a eugonic fermenter is a misnomer. EO-2 (a eugonic oxidizer) has a biochemical profile very similar to that of *Acinetobacter baumannii* (see Chapter 36 for more information regarding this genus), except the latter is oxidase negative.

The notable characteristic of OFBA-1 is that it produces an acidlike reaction in the OF medium control tube, even though no carbohydrates are present. In contrast, *Agrobacterium radiobacter* produces acid from various carbohydrates, but it does not acidify the OF control tube.[14]

O. anthropi and the *Achromobacter* group are phenotypically very similar so that cellular fatty acid analysis is often required for their differentiation. *Acidovorax* spp. are rarely isolated in the clinical laboratory. *Psychrobacter* spp. can be either saccharolytic or asaccharolytic, although all members of this genus have in common an optimal growth temperature of less than 35° C.

Because *Alcaligenes xylosoxidans* may oxidize glucose and consistently oxidizes xylose, this organism could be considered along with the other organisms of this chapter. Because other *Alcaligenes* spp.

TABLE 32-4 KEY BIOCHEMICAL AND PHYSIOLOGIC CHARACTERISTICS

ORGANISM	OXIDIZES GLUCOSE	OXIDIZES XYLOSE	OXIDIZES MANNITOL	NITRATE REDUCTION	GAS FROM NITRATE	ARGININE DIHYDROLASE	ESCULIN HYDROLYZED	GROWTH ON CETRIMIDE
"Achromobacter" group*	+	+	v	+	v	v	+	v
Acidovorax spp.	+	v	v	+	v	v	−	−
Agrobacterium radiobacter	+	+	+	−	−	−	+	−
EF-4b	+	−	−	+	−	−	−	ND
EO-2	+	+	−	+	v	−	−	−
EO-3	+	+	+	−	−	ND	v	−
Ochrobactrum anthropi	+	+	v	v	v	v	v	−
OFBA-1	+	+	+	+	+	+	−	+
Psychrobacter immobilis†	+	+	−	v	−	v	−	−

+, >90% of strains are positive; −, >90% of strains are negative; v, variable; ND, no data available.

*Includes biovars B, E, and F.

†Saccharolytic variety; prefers growth at 25° C.

Data compiled from Sawada, H., Hiroyuki, H., Oyaizu, H. et al. 1993. Int. J. Syst. Bacteriol. 43:694; and Weyant, R.S., Moss, C.W., Weaver, R.E., et al., editors. 1996. Identification of unusual pathogenic gram-negative aerobic and facultatively anaerobic bacteria, ed 2. Williams & Wilkins, Baltimore.

TABLE 32-5   ANTIMICROBIAL THERAPY AND SUSCEPTIBILITY TESTING

SPECIES	THERAPEUTIC OPTIONS	POTENTIAL RESISTANCE TO THERAPEUTIC OPTIONS	VALIDATED TESTING METHODS*	COMMENTS
"*Achromobacter*" group	No definitive guidelines. Human infections rare	Unknown	Not available	No clinical experience
Acidovorax spp.	No definitive guidelines	Unknown	Not available	No clinical experience
Agrobacterium radiobacter	Optimal therapy uncertain. Treatment involves removal of foreign body. Potentially active agents include ceftriaxone, cefotaxime, imipenem, gentamicin, and ciprofloxacin	Yes	Not available	Will grow on susceptibility testing media, but standards for interpretation of results do not exist
CDC group EF-4b	No definitive guidelines. Potentially active agents include penicillin, ampicillin, ciprofloxacin, and ofloxacin	Unknown; some cephalosporins may be less active than the penicilins	Not available	Limited clinical experience
CDC group EO-2 and CDC group EO-3	No definitive guidelines	Unknown	Not available	No clinical experience
Psychrobacter immobilis	No definitive guidelines	Unknown	Not available	Limited clinical experience
CDC group OFBA-1	No definitive guidelines	Unknown	Not available	No clinical experience
Ochrobactrum anthropi	Optimal therapy uncertain. Treatment involves removal of foreign body. Potentially active agents include trimethoprim/sulfamethoxazole, ciprofloxacin, and imipenem; aminoglycoside activity variable	Commonly resistant to all penicillins and cephalosporins	Not available	Will grow on susceptibility testing media, but standards for interpretation of results do not exist

*Validated testing methods include those standard methods recommended by the National Committee for Clinical Laboratory Standards (NCCLS) and those commercial methods approved by the Food and Drug Administration (FDA).

usually do not utilize glucose, all *Alcaligenes* spp. are considered together in Chapter 34.

SERODIAGNOSIS

Serodiagnostic techniques are not generally used in the laboratory diagnosis of infections caused by the organisms discussed in this chapter.

ANTIMICROBIAL SUSCEPTIBILITY TESTING AND THERAPY

Validated susceptibility testing methods do not exist for the organisms discussed in this chapter. Although many of these organisms will grow on the media and under the conditions recommended for testing the more commonly encountered bacteria (see Chapter 18 for more information regarding validated testing methods), the ability to grow and the ability to detect important antimicrobial resistances are not the same thing. Therefore, the lack of validated in vitro susceptibility testing methods does not allow definitive treatment and testing guidelines to be given for any of the organisms listed in Table 32-5. Although susceptibility data for some of these bacteria can be found in the literature, the lack of understanding of potential underlying resistance mechanisms prohibits the validation of such data. Chapter 19 should be reviewed for preferable strategies that can be used to provide susceptibility information and data when validated testing methods do not exist for a clinically important bacterial isolate.

Because *A. radiobacter* and *O. anthropi* infections are frequently associated with implanted medical devices, therapeutic management of the patient often involves removal of the contaminated material. Although definitive antimicrobial therapies for these infections have not been established, in vitro data suggest that certain agents could be more effective than others (see Table 32-5). For *A. radiobacter*, certain cephalosporins and aminoglycosides show in vitro activity. However, this organism is known to produce inactivating enzymes of both beta-lactams and aminoglycosides, so that resistance to potentially effective agents is a possibility.[13]

O. anthropi is commonly resistant to all currently available penicillins and cephalosporins but usually is susceptible to imipenem. This resistance profile is sufficiently consistent with the species that it may be useful in confirming the identification. The organism may also appear susceptible to trimethoprim/sulfamethoxazole and ciprofloxacin, but antimicrobial therapy without removal of the contaminated medical device may not successfully eradicate the organism.

PREVENTION

Because these organisms are ubiquitous in nature and are not generally a threat to human health, there are no recommended vaccination or prophylaxis protocols. Hospital-acquired infections are best controlled by following appropriate sterile techniques and infection control guidelines, and implementing effective protocols for the sterilization and decontamination of medical supplies.

References

1. Alnor, D., Frimodt-Moller, N., Espersen, F., et al. 1994. Infections with the unusual human pathogens *Agrobacterium* species and *Ochrobactrum anthropi*. Clin. Infect. Dis. 18:914.
2. Chang, H.J., Christenson, J.C., Pavia, A.T., et al. 1996. *Ochrobactrum anthropi* meningitis in pediatric pericardial allograft transplant recipients. J. Infect. Dis. 173:656.
3. Cieslak, T.J., Drabick, C.J., and Robb, M.L. 1996. Pyogenic infections due to *Ochrobactrum anthropi*. Clin. Infect. Dis. 22:845.
4. Cieslak, T.J., Robb, M.L., Drabick, C.J., et al. 1992. Catheter-associated sepsis caused by *Ochrobactrum anthropi*: report of a case and review of related non-fermentative bacteria. Clin. Infect. Dis. 14:902.
5. Dunne, W.M., Tillman, J., and Murray, J.C. 1993. Recovery of a strain of *Agrobacterium radiobacter* with a mucoid phenotype from an immunocompromised child with bacteremia. J. Clin. Microbiol. 31:2541.
6. Edmond, M.B., Riddler, S.A., Baxter, C.M., et al. 1993. *Agrobacterium radiobacter*: a recently recognized opportunistic pathogen. Clin. Infect. Dis. 16:388.
7. Gini, G.A. 1990. Ocular infection caused by *Psychrobacter immobilis* acquired in the hospital. J. Clin. Microbiol. 28:400.
8. Holmes, B., Lewis, R., and Trevett, A. 1992. Septicemia due to *Achromobacter* group B: a report of two cases. Med. Microbiol. Lett. 1:177.
9. Hulse, M., Johnson, S., and Ferrieri, P. 1993. *Agrobacterium* infections in humans: experience at one hospital and review. Clin. Infect. Dis. 16:112.
10. Kern, W.V., Oethinger, M., Kaufhold, A., et al. 1993. *Ochrobactrum anthropi* bacteremia: report of four cases and short review. Infection. 21:306.
11. Lloyd-Puryear, M., Wallace, D., Baldwin, T., et al. 1991. Meningitis caused by *Psychrobacter immobilis* in an infant. J. Clin. Microbiol. 29:2041.
12. Lozano, F., Florez, C., Recio, F.J., et al. 1994. Fatal *Psychrobacter immobilis* infection in a patient with AIDS. AIDS. 8:1189.
13. Martinez, J.L., Martinez-Suarez, J., Culebras, E., et al. 1989. Antibiotic inactivating enzymes from a clinical isolate of *Agrobacterium radiobacter*. J. Antimicrob. Chemother. 23:283.

14. Sawada, H., Hiroyuki, I., Oyaizu, H., et al. 1993. Proposal for rejection of *Agrobacterium tumefaciens* and revised descriptions for the genus *Agrobacterium* and for *Agrobacterium radiobacter* and *Agrobacterium rhizogenes.* Int. J. Syst. Bacteriol. 43:694.

Bibliography

Balows, A., Truper, H.G., Dworkin, M., et al., editors. 1981. The prokaryotes. A handbook on the biology of bacteria: ecophysiology, isolation, identification, applications, ed 2. Springer-Verlag, New York.

Holmes, B., Pickett, J.M., and Hollis, D.G. 1995. Unusual gram-negative bacteria, including *Capnocytophaga, Eikenella, Pasteurella,* and *Streptobacillus.* In Murray, P.R., Baron, E.J., Pfaller, M.A., et al., editors. Manual of clinical microbiology, ed 6. American Society for Microbiology, Washington, D.C.

McGowan, J.E., Jr. and Steinberg, J.P. 1995. Other gram-negative bacilli. In Mandell, G.L., Bennett, J.E., and Dolin, R., editors. Principles and practice of infectious diseases. Churchill Livingstone, New York.

von Graevenitz, A. 1995. *Acinetobacter, Alcaligenes, Moraxella,* and other nonfermentative gram-negative bacteria. In Murray, P.R., Baron, E.J., Pfaller, M.A., et al., editors. Manual of clinical microbiology, ed 6. American Society for Microbiology, Washington, D.C.

Weyant, R.S., Moss, C.W., Weaver, R.E., et al., editors. 1996. Identification of unusual pathogenic gram-negative aerobic and facultatively anaerobic bacteria, ed 2. Williams & Wilkins, Baltimore.

33 | CHRYSEOBACTERIUM, SPHINGOBACTERIUM, AND SIMILAR ORGANISMS

Genera and Species to be Considered

Current Name	Previous Name
■ *Chryseobacterium* spp.	
■ *Chryseobacterium meningosepticum*	*Flavobacterium meningosepticum* and CDC group IIa
■ *Empedobacter brevis*	*Flavobacterium breve*
■ *Sphingobacterium mizutaii*	*Flavobacterium mizutaii*
■ *Sphingobacterium multivorum*	*Flavobacterium multivorum* and CDC group IIK-2
■ *Sphingobacterium spiritivorum*	*Flavobacterium spiritivorum* and CDC group IIK-3
■ *Sphingobacterium thalpophilum*	

GENERAL CHARACTERISTICS

The organisms that constitute the genera discussed in this chapter are environmental inhabitants that are occasionally encountered in human specimens. They are considered together because they share similar physiologic and morphologic characteristics, that is, most are yellow-pigmented, oxidase-positive, glucose oxidizers. The majority of them grow on MacConkey agar. At one time, all of the species presented were members of the *Flavobacterium* genus.

EPIDEMIOLOGY

As environmental inhabitants, these organisms may be found in various niches (Table 33-1). Most notable in terms of clinical relevance is their ability to survive in hospital environments, especially in moist areas. Although they are not considered part of normal human flora, these species can colonize a patient's respiratory tract during hospitalization, probably as a result of exposure to a contaminated water source or medical devices. Alternatively, transmission may occur directly by contaminated pharmaceutical solutions and, in the case *of Chryseobacterium meningosepticum,* from person to person.

Because of their ability to survive well in hospital environments, these organisms also have the potential to contaminate laboratory culture media and blood culture systems. Whenever these species are encountered, their clinical significance and the potential for contamination should be seriously considered.

PATHOGENESIS AND SPECTRUM OF DISEASE

As environmental organisms, no specific virulence factors have been identified for these species. However, the ability to survive in chlorinated tap water may give these organisms an edge in their ability to thrive in hospital water systems.

The development of infection basically requires exposure of debilitated patients to a contaminated source that results in respiratory colonization (Table 33-2). Depending on the health of the patient, subsequent infections such as bacteremia and pneumonia may develop. These infections are most frequently caused by *C. meningosepticum* or *Myroides odoratum* (see Chapter 34). In addition, infections of several other body sites, which may or may not be preceded by respiratory colonization, have been associated with the other species.[1-10]

Meningitis caused by *C. meningosepticum* is the most notable infection associated with the organisms listed in Table 33-2. This life-threatening infection, which may be accompanied by bacteremia, originally gained attention because it occurred in neonates. However, *C. meningosepticum* meningitis can also occur in compromised adults, and the organism has been implicated in hospital-based outbreaks of both meningitis and pneumonia.[11]

TABLE 33-1 EPIDEMIOLOGY

SPECIES	HABITAT (RESERVOIR)	MODE OF TRANSMISSION
Chryseobacterium meningosepticum, Chryseobacterium spp., *Empedobacter brevis, Sphingobacterium* spp.	Soil; plants; water; foodstuffs; hospital water sources, including incubators, sinks, faucets, tap water, hemodialysis systems, saline solutions, and other pharmaceuticals. Not part of human flora	Exposure of patients to contaminated medical devices or solutions, but source is not always known. May colonize upper respiratory tract. *C. meningosepticum* occasionally may be transmitted from birth canal to neonate

TABLE 33-2 PATHOGENESIS AND SPECTRUM OF DISEASES

SPECIES	VIRULENCE FACTORS	SPECTRUM OF DISEASE AND INFECTIONS
Chryseobacterium meningosepticum, Chryseobacterium spp., *Empedobacter brevis, Sphingobacterium* spp.	Specific virulence factors are unknown. Able to survive chlorinated tap water. *C. meningosepticum,* the species most often associated with human infections, can be encapsulated and produces proteases and gelatinases that may be destructive for host cells and tissues	Bacteremia (often associated with implanted devices such as catheters or contaminated medical solutions). *C. meningosepticum,* particularly associated with meningitis in neonates and less commonly in adults. Other organisms associated with pneumonia, mixed infections of wounds, ocular and urinary tract infections, and occasionally other infections, including sinusitis, endocarditis, peritonitis, and fasciitis

LABORATORY DIAGNOSIS

SPECIMEN COLLECTION AND TRANSPORT

No special considerations are required for specimen collection and transport of the organisms discussed in this chapter. Refer to Table 1-1 for general information on specimen collection and transport.

SPECIMEN PROCESSING

No special considerations are required for processing the organisms discussed in this chapter. Refer to Table 1-1 for general specimen processing information.

DIRECT DETECTION METHODS

Other than Gram stain of patient specimens, there are no specific procedures for the direct detection of these organisms in clinical material. *Chryseobacterium* spp. are medium to long, straight rods and often appear as "II-forms" (i.e., cells that appear thin in the center and thicker at the termini). *Empedobacter brevis* varies in being short to long rods. *Sphingobacterium* spp. are short, straight rods, and certain species (i.e., *S. mizutaii* and *S. thalpophilum*) may exhibit II-forms.

CULTIVATION

Media of choice

Empedobacter, Chryseobacterium, and *Sphingobacterium* spp. grow well on routine laboratory media such as 5% sheep blood and chocolate agars. *Sphingobacterium mizutaii* is the only species that consistently fails to grow on MacConkey agar. All three genera grow well in the broth of blood culture systems and in common nutrient broths such as thioglycollate and brain-heart infusion.

Incubation conditions and duration

These organisms will produce detectable growth on blood and chocolate agars when incubated at 35° C in either carbon dioxide or ambient air for a minimum of 24 hours. Growth, when it occurs on MacConkey agar, is usually detectable within 24 hours of inoculation.

Colonial appearance

Table 33-3 describes the colonial appearance and other distinguishing characteristics of each genus on 5% sheep blood and MacConkey agars.

TABLE 33-3 COLONY APPEARANCE AND CHARACTERISTICS

ORGANISM	MEDIUM	APPEARANCE
Chryseobacterium meningosepticum	BA	Usually slight yellow, pigmented colonies; smooth, circular, large, shiny with entire edge
	Mac	NLF
Chryseobacterium spp.	BA	Circular, smooth, shiny with entire edge, light yellow to orange
	Mac	NLF
Empedobacter brevis	BA	Circular, smooth, shiny with entire edge, light yellow
	Mac	NLF
Sphingobacterium mizutaii	BA	Yellow pigment at room temperature
	Mac	No growth
S. multivorum	BA	Small, circular, convex, smooth, opaque with pale yellow pigment after overnight incubation at room temperature
	Mac	NLF
S. spiritivorum	BA	Small, circular, convex, smooth with pale yellow pigment
	Mac	NLF, if growth
S. thalpophilum	BA	Pale yellow
	Mac	NLF

BA, 5% sheep blood agar; *Mac*, MacConkey agar; *NLF*, nonlactose fermenter.

APPROACH TO IDENTIFICATION

The ability of most commercial identification systems to accurately identify the organisms discussed in this chapter is limited or uncertain. The key biochemical reactions that can be used to presumptively differentiate among the genera discussed in this chapter are provided in Table 33-4. However, definitive identification of these organisms often requires performing a battery of biochemical tests not commonly available in many clinical microbiology laboratories. Therefore, full identification of clinically relevant isolates may require that they be sent to a reference laboratory.

Comments regarding specific organisms

Sphingobacterium mizutaii does not grow on MacConkey agar, and the growth of *S. spiritivorum* and *Chryseobacterium* spp., other than *C. meningosepticum*, is variable. Therefore, these organisms often need to be differentiated from yellow-pigmented, MacConkey-negative, oxidase-positive genera considered in Chapters 38 through 42.

Indole and urea hydrolysis are key biochemical tests for the separation of *Empedobacter brevis* and *Chryseobacterium* spp. from *Sphingobacterium* spp.

SERODIAGNOSIS

Serodiagnostic techniques are not generally used for the laboratory diagnosis of infections caused by the organisms discussed in this chapter.

ANTIMICROBIAL SUSCEPTIBILITY TESTING AND THERAPY

Validated susceptibility testing methods do not exist for these organisms. Although they will grow on the media and under the conditions recommended for testing the more commonly encountered bacteria (see Chapter 18 for more information regarding validated testing methods), the ability to grow and the ability to detect important antimicrobial resistances are not the same thing. Therefore, the lack of validated in vitro susceptibility testing methods does not allow definitive treatment and testing guidelines to be given for any of the organisms listed in Table 33-5.

TABLE 33-4   KEY BIOCHEMICAL AND PHYSIOLOGIC CHARACTERISTICS

ORGANISM	GROWTH ON MACCONKEY	OXIDIZES GLUCOSE	OXIDIZES MANNITOL	INDOLE	GELATIN	UREA	NITRATE REDUCTION	DNase
Chryseobacterium meningosepticum	+	+	+	+	+	−	−	+
Chryseobacterium spp.*	v	+	−	+	v	v	v	−
Empedobacter brevis	+	+	−	+	+	−	−	+
Sphingobacterium mizutaii	−	+	−	−	−	−	−	−
S. multivorum	+	+	−	−	−	+	−	v
S. spiritivorum	v	+	+	−	v	+	−	+
S. thalpophilum	+	+	−	−	v	+	+	v

+, >90% of strains are positive; −, >90% of strains are negative; *v*, variable.

*Includes *Chryseobacterium gleum* and *C. indologenes.*

Data compiled from Pickett, M.J, Hollis, D.G., and Bottone, E.J. 1991. Miscellaneous gram-negative bacteria. In Balows, A., Hausler, W.J., Jr., Herrmann, K.L., et al., editors. Manual of clinical microbiology, ed 5. American Society for Microbiology, Washington, D.C.; Vandamme P., Bernardet, J.F., Segers, P., et al. 1994. Int. J. Syst. Bacteriol. 44:827; and Weyant, R.S., Moss, C.W., Weaver, R.E., et al. 1996. Identification of unusual pathogenic gram-negative aerobic and facultatively anaerobic bacteria, ed 2. Williams & Wilkins, Baltimore.

TABLE 33-5 ANTIMICROBIAL THERAPY AND SUSCEPTIBILITY TESTING

SPECIES	THERAPEUTIC OPTIONS	POTENTIAL RESISTANCE TO THERAPEUTIC OPTIONS	VALIDATED TESTING METHODS*	COMMENTS
Chryseobacterium indologenes, Chryseobacterium meningosepticum, Empedobacter brevis, Sphingobacterium spp.	No definitive guidelines Potentially active agents include ciprofloxacin, rifampin, clindamycin, trimethoprim/ sulfamethoxazole, and vancomycin	Yes, most produce beta-lactamases and are also frequently resistant to aminoglycosides	Not available	In vitro susceptibility results with disk diffusion may be seriously misleading

*Validated testing methods include those standard methods recommended by the National Committee for Clinical Laboratory Standards (NCCLS) and those commercial methods approved by the Food and Drug Administration (FDA).

Although susceptibility data for some of these bacteria can be found in the literature, the lack of understanding of potential underlying resistance mechanisms prohibits the validation of such data. Chapter 19 should be reviewed for preferable strategies that can be used to provide susceptibility information and data when validated testing methods do not exist for a clinically important bacterial isolate.

In general, the species considered in this chapter are frequently resistant to the penicillins (including carbapenems such as imipenem), cephalosporins, and aminoglycosides commonly used to treat infections caused by other gram-negative bacilli. However, the susceptibility data can vary substantially with the type of testing method used.[1,2,11] An unusual feature of many of these species is that they often appear susceptible, and may be treated with, antimicrobial agents that are usually considered only effective against gram-positive bacteria; clindamycin, rifampin, and vancomycin are notable examples.[11]

PREVENTION

Because these organisms are ubiquitous in nature and are not generally a threat to human health, there are no recommended vaccination or prophylaxis protocols. Hospital-acquired infections are best controlled by following appropriate sterile techniques and infection control guidelines, and implementing effective protocols for the sterilization and decontamination of medical supplies.

References

1. Blahovea, J., Hupkova, M., Krcmery, V., et al. 1994. Resistance to and hydrolysis of imipenem in nosocomial strains *of Flavobacterium meningosepticum.* Eur. J. Clin. Microbiol. Infect. Dis. 13:833.
2. Fass, R.J. and Barnishan, J. 1980. In vitro susceptibilities of nonfermentative gram-negative bacilli other than *Pseudomonas aeruginosa* to 32 antimicrobial agents. Rev. Infect. Dis. 2:841.
3. Ferrer, C., Jakob, E., Pastorino, G., et al. 1995. Right-sided bacterial endocarditis due to *Flavobacterium odoratum* in a patient on chronic hemodialysis. Am. J. Nephrol. 15:82.
4. Hsueh, P., Wu, J., Hsiue, T., et al. 1995. Bacteremic necrotizing fasciitis due to *Flavobacterium odoratum.* Clin. Infect. Dis. 21(5):1337.
5. Jorgensen, J.H., Maher, L.A., and Howell, A.W. 1991. Activity of meropenem against antibiotic-resistant or infrequently encountered gram-negative bacilli. Antimicrob. Agents Chemother. 35:2410.
6. Marnejon, T. and Watanakunakorn, C. 1992. *Flavobacterium meningosepticum* septicemia and peritonitis complicating CAPD. Clin. Nephrol. 38:176.
7. Pokrywka, M., Viazanko, K., Medvick, J., et al. 1993. A *Flavobacterium meningosepticum* outbreak among intensive care patients. Am. J. Infect. Control. 21:139.
8. Reina, J., Borrell, N., and Figuerola, J. 1992. *Sphingobacterium multivorum* isolated from a patient with cystic fibrosis. Eur. J. Clin. Microbiol. Infect. Dis. 11:81.
9. Sader, H.S., Jones, R.N., and Pfaller, M.A. 1995. Relapse of catheter-related *Flavobacterium meningosepticum* bacteremia demonstrated by DNA macrorestriction analysis. Clin. Infect. Dis. 21:997.
10. Skapek, S.X., Jones, W.S., Hoffman, K.M., et al. 1992. Sinusitis and bacteremia caused *by Flavobacterium meningosepticum* in a sixteen-year-old with Shwachman Diamond Syndrome. Pediatr. Infect. Dis. J. 11:411.
11. Tizer, K.B., Cervia, J.S., and Dunn, A., et al. 1995. Successful combination of vancomycin and rifampin therapy in a newborn with community acquired *Flavobacterium meningosepticum* neonatal meningitis. Pediatr. Infect. Dis. J. 14:916.

Bibliography

Balows, A., Truper, H.G., Dworkin, M., et al., editors. 1981. The prokaryotes. A handbook on the biology of bacteria: ecophysiology, isolation, identification, applications, ed 2. Springer-Verlag, New York.

McGowan, J.E., Jr. and Steinberg, J.P. 1995. Other gram-negative bacilli. In Mandell, G.L., Bennett, J.E., and Dolin, R., editors. Principles and practice of infectious diseases. Churchill Livingstone, New York.

Pickett, M.J., Hollis, D.G., and Bottone, E.J. 1991. Miscellaneous gram-negative bacteria. In Balows, A., Hausler, W.J., Jr., Herrmann, K.L., et al., editors. Manual of clinical microbiology, ed 5. American Society for Microbiology, Washington, D.C.

Vandamme, P., Bernardet, J.F., Segers, P., Kersters, K., and Holmes, B. 1994. New perspectives in the classification of the flavobacteria: description of *Chryseobacterium* gen. nov., *Bergeyella* gen. nov., and *Empedobacter* nov. rev. Int. J. Syst. Bacteriol. 44:827.

von Graevenitz, A. 1985. Ecology, clinical significance, and antimicrobial susceptibility of infrequently encountered glucose-nonfermenting gram-negative rods. In Gilardi, G.L., editor. Nonfermentative gram-negative rods: laboratory identification and clinical aspects. Marcel Dekker, New York.

von Graevenitz, A. 1995. *Acinetobacter, Alcaligenes, Moraxella,* and other nonfermentative gram-negative bacteria. In Murray, P.R., Baron, E.J., Pfaller, M.A., et al., editors. Manual of clinical microbiology, ed 6. American Society for Microbiology, Washington, D.C.

Weyant, R.S., Moss, C.W., Weaver, R.E., et al., editors. 1996. Identification of unusual pathogenic gram-negative aerobic and facultatively anaerobic bacteria, ed 2. Williams & Wilkins, Baltimore.

34 ALCALIGENES, BORDETELLA (NONPERTUSSIS), COMAMONAS, AND SIMILAR ORGANISMS

Genera and Species to be Considered

Current Name	Previous Name
■ *Alcaligenes denitrificans*	*Alcaligenes denitrificans*
■ *Alcaligenes xylosoxidans*	*Achromobacter xylosoxidans*
■ *Alcaligenes faecalis*	*Pseudomonas* or *Alcaligenes odorans*
■ *Alcaligenes piechaudii*	*Alcaligenes faecalis* type I
■ CDC *Alcaligenes*-like group 1	
■ *Bordetella parapertussis*	
■ *Bordetella bronchiseptica*	CDC group IVa
■ *Bordetella holmesii*	CDC group NO-2
■ *Bordetella trematum*	
■ CDC group IVc-2	
■ CDC group IIg	
■ *Comamonas acidovorans*	*Pseudomonas acidovorans*
■ *Comamonas testosteroni*	*Pseudomonas testosteroni*
■ *Comamonas* spp.	
■ Gilardi rod group 1	
■ *Oligella urethralis*	*Moraxella urethralis*
■ *Oligella urealytica*	CDC group IVe
■ *Myroides* spp.	*Flavobacterium odoratum*
■ *Pyschrobacter* spp.	*Moraxella phenylpyruvia*
■ "*Roseomonas*"* spp.	
■ *Shewanella putrefaciens*	*Pseudomonas putrefaciens*

GENERAL CHARACTERISTICS

The genera discussed in this chapter are considered together because they are all generally MacConkey-positive, oxidase-positive, nonglucose utilizers. Their specific morphologic and physiologic features are presented later in this chapter in the discussion of laboratory diagnosis.

EPIDEMIOLOGY

The habitats of the species listed in Table 34-1 vary from the soil and water environment to the upper respiratory tract of various mammals. Certain species have been exclusively found in humans, whereas the natural habitat for other organisms remains unknown.

The diversity of the organisms' habitats is reflected in the various ways they are transmitted. For example, transmission of environmental isolates such as *Alcaligenes xylosoxidans* frequently involves expo-sure of debilitated patients to contaminated fluids or medical solutions.[10] In contrast, *Bordetella bronchiseptica* transmission primarily occurs by close contact with animals, whereas *B. holmesii* has been detected only in human blood, and no niche or mode of transmission is known.[19,20]

PATHOGENESIS AND SPECTRUM OF DISEASE

Identifiable virulence factors are not known for most of the organisms listed in Table 34-2. However, because infections usually involve exposure of compromised patients to contaminated materials, most of these species are probably of low virulence.[3,13,14] Among the environmental organisms listed, *A. xylosoxidans* is the species most frequently associated with various infections. This species also has been implicated in

*Quotation marks indicate a proposed organism name.

TABLE 34-1   EPIDEMIOLOGY

SPECIES	HABITAT (RESERVOIR)	MODE OF TRANSMISSION
Alcaligenes denitrificans	Environment; not part of human flora	Unknown. Rarely found in humans
Alcaligenes xylosoxidans	Environment, including moist areas of hospital. Transient colonizer of human gastrointestinal or respiratory tract of patients with cystic fibrosis	Not often known. Usually involves exposure to contaminated fluids (e.g., intravenous fluids, hemodialysis fluids, irrigation fluids), soaps, and disinfectants
Alcaligenes faecalis	Environment; soil and water, including moist hospital environments. May transiently colonize the skin	Exposure to contaminated medical devices and solutions
Alcaligenes piechaudii	Environment	Unknown. Rarely found in humans
Bordetella bronchiseptica	Normal respiratory flora of several mammals, including dogs, cats, and rabbits. Not part of human flora	Probably by exposure to contaminated respiratory droplets during close contact with animals
Bordetella holmseii	Unknown. Not part of human flora	Unknown. Rarely found in humans
Bordetella trematum	Unknown	Unknown
CDC group IVc-2	Uncertain. Probably water sources, including those in the hospital setting. Not part of human flora	Usually involves contaminated dialysis systems or exposure of wounds to contaminated water
Comamonas acidovorans *Comamonas testosteroni* *Comamonas* spp.	Environmental, soil and water; can be found in hospital environment. Not part of human flora	Uncertain. Rarely found in humans. Probably involves exposure to contaminated solutions or devices
Gilardi rod group 1	Unknown. Probably environmental. Not part of human flora	Unknown. Rarely found in humans
Oligella urethralis *Oligella urealytica*	Unknown. May colonize distal urethra	Manipulation (e.g., catheterization) of urinary tract
Pyschrobacter spp.	Unknown	Unknown
"Roseomonas" spp.	Unknown	Unknown. Rarely found in humans
Shewanella putrefaciens	Environmental and foodstuffs. Not part of human flora	Unknown. Rarely found in humans

outbreaks of nosocomial infections.[4,6,9,10] Other organisms, such as *Oligella urethralis, O. urealytica, B. holmesii,* and *"Roseomonas"* spp., whose reservoirs are unknown, are rarely implicated in human infections and are likely to be of low virulence.[2,7,12,15-17,19]

LABORATORY DIAGNOSIS

SPECIMEN COLLECTION AND TRANSPORT

No special considerations are required for collection and transport of the organisms discussed in this chapter. Refer to Table 1-1 for general information on specimen collection and transport.

SPECIMEN PROCESSING

No special considerations are required for processing of the organisms discussed in this chapter. Refer to Table 1-1 for general information on specimen processing.

DIRECT DETECTION METHODS

Other than Gram stain of patient specimens, there are no specific procedures for the direct detection of these organisms in clinical material. The *Bordetella* spp. exhibit diverse cellular morphologies. *B. bronchiseptica* is a medium-sized, straight rod, whereas *B. parapertussis* is a short, straight rod. *B. holmesii* may

TABLE 34-2 PATHOGENESIS AND SPECTRUM OF DISEASES

SPECIES	VIRULENCE FACTORS	SPECTRUM OF DISEASE AND INFECTIONS
Alcaligenes denitrificans	Unknown	Rare cause of human infection
Alcaligenes xylosoxidans	Unknown. Survival in hospital the result of inherent resistance to disinfectants and antimicrobial agents	Infections usually involve compromised patients and include bacteremia, urinary tract infections, meningitis, wound infections, pneumonia, and peritonitis, and occur in various body sites; can be involved in nosocomial outbreaks
Alcaligenes faecalis	Unknown	Infections usually involve compromised patients. Often a contaminant; clinical significance of isolates should be interpreted with caution. Has been isolated from blood, respiratory specimens, and urine
Alcaligenes piechaudii	Unknown	Rare cause of human infection
Bordetella bronciseptica	Unknown for humans. Has several factors similar to *B. parapertusis.*	Opportunistic infection in compromised patients with history of close animal contact. Infections are uncommon and include pneumonia, bacteremia, urinary tract infections, meningitis, and endocarditis
Bordetella holmesii	Unknown	Bacteremia is only type of infection described
Bordetella trematum	Unknown	Unknown
CDC group IVc-2	Unknown	Rare cause of human infection. Infections in compromised patients include bacteremia and peritonitis
Comamonas acidovorans *Comamonas testosteroni* *Comamonas* spp.	Unknown	Isolated from respiratory tract, eye, and blood but rarely implicated as being clinically significant
Gilardi rod group 1	Unknown	Clinical significance uncertain, has been isolated from wounds, urine, and blood
Oligella urethralis	Unknown	Urinary tract infections, particularly in females
Oligella urealytica		Also isolated from kidney, joint, and peritoneal fluid
Pyschrobacter spp.	Unknown	Rare cause of human infection
"*Roseomonas*" spp.	Unknown	Clinical significance uncertain. Most isolated from blood, wounds, or genitourinary tract of immunocompromised or debilitated patients
Shewanella putrefaciens	Unknown	Clinical significance uncertain; often found in mixed cultures. Has been implicated in cellulitis, otitis media, and septicemia; also may be found in respiratory tract, urine, feces, and pleural fluid

appear as a coccobacillus, a short, straight rod; or, occasionally, a long rod. *B. trematum* is a medium to long, straight rod.

Oligella urethralis, *Pyschrobacter* spp. and "*Roseomonas*" spp. are all coccobacilli, although *P. phenylpyruvicus* may appear as a broad rod, and some "*Roseomonas*" spp. may appear as short, straight rods. *Oligella ureolytica* is a short, straight rod, and the cells of Gilardi rod group 1 tend to be short and broad; *Myroides* spp. are pleomorphic rods, and are either short or long and straight to slightly curved.

Alcaligenes spp. are medium to long, straight rods, as are CDC *Alcaligenes*-like group 1, CDC group IVc-2, *Shewanella putrefaciens,* and *Comamonas acidovorans.* The other *Comamonas* spp. are pleomorphic, and may appear as long, paired, curved rods or filaments. The cells of CDC group IIg appear as small coccoid-to-rod forms or occasionally as rods with long filaments.

CULTIVATION

Media of choice

Bordetella parapertussis does not grow well on 5% sheep blood or chocolate agar and may require 2 to 4 days to form visible colonies on these media. Although this species will grow on MacConkey agar, better growth is achieved with a selective medium, such as Regan-Lowe, that is used for isolation of *Bordetella pertussis* (see Chapter 48 for more information on the growth of these species). *B. bronchiseptica, B. holmesii,* and *B. trematum* all will grow on 5% sheep blood, chocolate, and MacConkey agars, but *B. bronchiseptica* grows the fastest, usually within 1 to 2 days after inoculation. Any of these strains should grow in thioglycollate broth.

Psychrobacter spp., Gilardi rod group 1, *Myroides* spp., *Oligella* spp., CDC group IVc-2, *Alcaligenes* spp., CDC *Alcaligenes*-like group 1, *Comamonas* spp., *Shewanella putrefaciens,* "*Roseomonas*" spp., and CDC group IIg all grow well on 5% sheep blood, chocolate, and MacConkey agars. Most of these genera should also grow well in the broth of blood culture systems, as well as in common nutrient broths such as thioglycollate and brain-heart infusion.

Incubation conditions and duration

The organisms mentioned above will produce detectable growth on media incubated at 35° C in ambient air. Incubation in carbon dioxide also is acceptable, but *Bordetella parapertussis* may be inhibited in this atmosphere. The length of time for incubation depends on the bacterial species. *B. parapertussis* may require 5 to 7 days of incubation, whereas *B. holmesii* may grow in as few as 3 days. *Pyschrobacter* spp. usually grow better at 25° C than at 35° C.

Colonial appearance

Table 34-3 describes the colonial appearance and other distinguishing characteristics (e.g., pigment and odor) of each genus on 5% sheep blood and MacConkey agars. *B. parapertussis* is more fully discussed in Chapter 48, because it grows better on selective media used for *B. pertussis.*

APPROACH TO IDENTIFICATION

The ability of most commercial identification systems to accurately identify the organisms discussed in this chapter is limited or uncertain. Strategies for identification of these genera therefore are based on the use of conven-
Text continued on p. 484

—— TABLE 34-3 COLONY APPEARANCE AND CHARACTERISTICS ——

ORGANISM	MEDIUM	APPEARANCE
Alcaligenes denitrificans	BA	Small, convex, and glistening
	Mac	NLF
Alcaligenes faecalis	BA	Feather-edged colonies usually surrounded by zone of green discoloration; produces a highly characteristic, fruity odor resembling apples or strawberries
	Mac	NLF
Alcaligenes piechaudii	BA	Nonpigmented, glistening, convex colonies surrounded by zone of greenish-brown discoloration
	Mac	NLF
Alcaligenes xylosoxidans	BA	Smooth, glistening, entire
	Mac	NLF

BA, 5% sheep blood agar; *Mac,* MacConkey agar; *NLF,* nonlactose fermenter.   *Continued*

TABLE 34-3 COLONY APPEARANCE AND CHARACTERISTICS—CONT'D

ORGANISM	MEDIUM	APPEARANCE
Bordetella bronchiseptica	BA	Small, convex, round
	Mac	NLF
Bordetella holmesii	BA	Pinpoint, semiopaque, convex, round, may take 3 days to exhibit visible growth
	Mac	NLF
Bordetella parapertussis	BA	Small, shiny, round; may take 2-4 days to exhibit visible growth
	Mac	NLF
Bordetella trematum	BA	Round, greyish-cream colonies
	Mac	NLF
CDC *Alcaligenes*-like group 1	BA	Resembles *A. xylosoxidans* subsp. *denitrificans*
	Mac	NLF
CDC group IIg	BA	No distinctive appearance
	Mac	NLF
CDC group 1Vc-2	BA	Resembles *B. bronchiseptica,* but colonies are larger
	Mac	NLF
Comamonas spp.	BA	No distinctive appearance
	Mac	NLF
Gilardi rod group 1	BA	No distinctive appearance
	Mac	NLF
Myroides spp.	BA	Most colonies are yellow; have a characteristic fruity odor, and tend to spread
	Mac	NLF
Oligella spp.	BA	Small, opaque, whitish
	Mac	NLF
Psychrobacter phenylpyruvicus	BA	Smooth, small, translucent to semiopaque
	Mac	NLF
"*Roseomonas*" spp.	BA	Pink-pigmented; some colonies may be mucoid
	Mac	NLF
Shewanella putrefaciens	BA	Convex, circular, smooth; brown-to-tan pigment; may be mucoid; green discoloration of medium
	Mac	NLF

TABLE 34-4 KEY BIOCHEMICAL AND PHYSIOLOGIC CHARACTERISTICS FOR NONMOTILE SPECIES

ORGANISM	OXIDASE	NITRATE REDUCTION	NITRITE REDUCTION	UREA HYDROLYSIS	PHENYLALANINE DEAMINASE	CITRATE UTILIZATION	INDOLE
Bordetella parapertussis	−	−	ND	+	ND	v	−
Bordetella holmesii	−	−	−	−	ND	−	−
CDC group IIg	+	−	+	−	ND	−	+
Gilardi rod group 1	+	−	−	−	+	−	−
Myroides spp.	+	−	v	+	ND	−	−
Oligella urethralis	+	−	+	−	+	v	−
Psychrobacter phenylpyruvicus	+	v	−	+	+	−	−
*Psychrobacter immobilis**	+	v	ND	v	ND	v	−

+, >90% of strains are positive; −, >90% of strains are negative; *v*, variable; *ND,* no data available.

*Assacharolytic variety; prefers growth at 25° C.

Data compiled from Bowman, J.P., Cavanagh, J.J., Sanderson, K. 1996. Int. J. Syst. Bacteriol. 46:841; Moss, C.W., Daneshvar, M.I., and Hollis, D.G. 1993. J. Clin. Microbiol. 31:689; Pickett, M.J. Hollis, D.G., and Bottone, E.J. 1991. Miscellaneous gram-negative bacteria. In Balows, A., Hausler, A., Hausler, W.J., Jr., Herrmann, K.L., et al., editors. Manual of clinical microbiology, ed 5. American Society for Microbiology, Washington, D.C.; Vancanneyt, M.P., Segers, P., Torck, U., et al. 1996. Int. J. Syst. Bacteriol. 46:926; Weyant, R.S., Hollis, D.G., Weaver, R.E., et al. 1995. J. Clin. Microbiol. 33:1; and Weyant R.S., Moss, C.W., Weaver, R.E., et al. 1996. Identification of unusual pathogenic gram-negative aerobic and facultatively anaerobic bacteria, ed 2. Williams & Wilkins. Baltimore.

TABLE 34-5 KEY BIOCHEMICAL AND PHYSIOLOGIC CHARACTERISTICS OF SPECIES WITH PERITRICHOUS FLAGELLA

ORGANISM	OXIDASE	OXIDIZES XYLOSE	NITRATE REDUCTION	NITRITE REDUCTION	UREA HYDROLYSIS	CITRATE UTILIZATION
Bordetella bronchiseptica	+	−	+	−	+	+
Bordetella trematum	−	−	v	v	−	−
CDC group IVc-2	+	−	v	−	+	+
Oligella ureolytica	+	−	+	+	+	v
Alcaligenes xylosoxidans	+	+	+	ND	−	+
Alcaligenes denitrificans	+	−	+	+	−	+
Alcaligenes faecalis	+	−	−	+	−	+
Alcaligens piechaudii	+	−	+	−	−	+
CDC *Alcaligenes*-like group 1	+	−	+	+	v	+

+, >90% of strains are positive; −, >90% of strains are negative; *v*, variable; *ND*, no data available.

Data compiled from Pickett, M.J., Hollis, D.G., and Bottone, E.J., 1991. Miscellaneous gram-negative bacteria. In Balows, A., Hausler, W.J., Jr., Herrmann, K.L., et al., editors. Manual of clinical microbiology, ed 5. American Society for Microbiology, Washington, D.C.; Vandamme, P., Heyndrickx, M., Vancanneyt, M., et al, 1996. Int. J. Syst. Bacteriol. 46:849; Weyant R.S., Moss, C.W., Weaver, R.E., et al. 1996. Identification of unusual pathogenic gram-negative aerobic and facultatively anaerobic bacteria, ed 2. Williams & Wilkins. Baltimore.

TABLE 34-6 KEY BIOCHEMICAL AND PHYSIOLOGIC CHARACTERISTICS OF SPECIES WITH POLAR OR POLAR TUFT FLAGELLA

SPECIES	OXIDIZES GLUCOSE	OXIDIZES MANNITOL	OXIDIZES SUCROSE	UREA HYDROLYSIS	CITRATE UTILIZATION	H_2S IN TSI BUTT	PINK COLONIES
Comamonas acidovorans	−	+	−	−	+	−	−
Comamonas spp.*	−	−	−	v	v	−	−
Shewanella putrefaciens (biotype 1)	v	−	+	v	−	+	−
Shewanella putrefaciens (biotype 2)	−	−	−	v	−	+	−
"*Roseomonas*" spp.[†]	v	v[‡]	−	+	v	−	+

+, >90% of strains are positive; −, >90% of strains are negative; v, variable; TSI, triple sugar iron agar.
*Includes the species *Comamonas testosteroni* and *C. terrigena.*
[†]"*Roseomonas*" genomospecies 5 is nonmotile.
[‡]"*Roseomonas gilardii*" has variable mannitol oxidation.

Data compiled from Hollis, D.G., Weaver, R.E., Moss, C.W., et al., 1992. J. Clin. Microbiol. 30:291; Rihs, J.D., Brenner, D.J., Weaver, R.E., et al. 1993. J. Clin. Microbiol. 31:3275; Weyant, R.S., Moss, C.W., Weaver, R.E., et al. 1996. Identification of unusual pathogenic gram-negative aerobic and facultatively anaerobic bacteria. ed 2. Williams & Wilkins, Baltimore.

tional biochemical tests and special staining for flagella. Although clinical microbiology laboratories do not routinely perform flagella stains, motility and flagella placement is the easiest way to separate these organisms.

Many microbiologists groan at the mere mention of having to perform a flagella stain, but the method described in Procedure 30-13 is a wet mount that is easy to perform. At the very least, a simple wet mount to observe cells for motility will help separate the motile and nonmotile genera. The pseudomonads, *Brevundimonas*, *Burkholderia*, and *Ralstonia* described in Chapter 31 are motile by means of single or multiple polar flagella, and many of the organisms described in this chapter have peritrichous flagella (e.g., *B. bronchiseptica* and *Alcaligenes* spp.).

Identification characteristics are divided into three tables, based on whether the organisms are nonmotile (Table 34-4), peritrichously flagellated (Table 34-5), or flagellated by polar tufts (Table 34-6).

Comments regarding specific organisms

Identification of *Bordetella parapertussis* is usually confirmed by using fluorescein-labeled antisera (see Chapter 48 for description of this identification method). Using other criteria, *B. parapertussis* is nonmotile, oxidase-negative, and urea-positive, which dif-

---

TABLE 34-7 ANTIMICROBIAL THERAPY AND SUSCEPTIBILITY TESTING

SPECIES	THERAPEUTIC OPTIONS
Alcaligenes denitrificans	No definitive guidelines
Alcaligenes xylosoxidans	No definitive guidelines. Potentially active agents include mezlocillin, piperacillin, ticarcillin/clavulanic acid, ceftazidime, imipenem, trimethoprim/sulfamethoxazole, and quinolones
Alcaligenes faecalis	No definitive guidelines
Alcaligenes piechaudii	No definitive guidelines
Bordetella bronchiseptica	No definitive guidelines. Potentially active agents include aminoglycosides, mezlocillin, piperacillin, ceftazidime, imipenem, and quinolones
Bordetella holmesii	No definitive guidelines. Potentially active agents are same as for *B. bronchiseptica*
CDC group IVc-2	No definitive guidelines. Potentially active agents include cefotaxime, ceftazidime, ceftriaxone, and imipenem
Comamonas acidovorans, Comamonas testosteroni, Comamonas spp.	No definitive guidelines. Potentially active agents include ceftazidime, piperacillin, imipenem, and ciprofloxacin
Gilardi rod group 1	No definitive guidelines
Oligella urethralis, Oligella urealytica	No definitive guidelines. Potentially active agents include several penicillins, cephalosporins, and quinolones
"*Roseomonas*" spp.	No definitive guidelines. Potentially active agents include aminoglycosides, imipenem, and quinolones
Shewanella putrefaciens	No definitive guidelines. Generally susceptible to various antimicrobial agents

*Validated testing methods include those standard methods recommended by the National Committee for Clinical Laboratory Standards (NCCLS) and those commercial methods approved by the Food and Drug Administration (FDA).

ferentiates it from *B. holmesii* (i.e., nonmotile, oxidase-negative, and urea-negative). *B. bronchiseptica* is oxidase-positive, motile, and urease-positive, sometimes in as little as 4 hours.

CDC group IVc-2 closely resembles *B. bronchiseptica,* but is usually nitrate-negative. Urea hydrolysis is a key test for *Myroides* spp. which is also distinguished by production of a characteristic, fruity odor. CDC group IIg is the only indole-positive, nonmotile species included in this chapter.

The genus *Oligella* includes one nonmotile species *(O. urethralis)* and one motile species *(O. ureolyticus).* Urease hydrolysis is a key test for differenti-

ating between these species; *O. ureolyticus* often turns positive within minutes. *Psychrobacter phenylpyruvicus* is nonmotile and both urea- and phenylalanine deaminase–positive.

Alcaligenes xylosoxidans can be quickly recognized by its strong oxidation of xylose, even when glucose oxidation is slow or negative. *A. denitrificans* and *A. piechaudii* reduce nitrate to nitrite, but only *A. denitrificans* reduces nitrite to gas. *Alcaligenes faecalis* has a fruity odor and also reduces nitrite to gas. CDC *Alcaligenes*-like group 1 is similar to *Alcaligenes denitrificans* but is usually urea-positive.

POTENTIAL RESISTANCE TO THERAPEUTIC OPTIONS	VALIDATED TESTING METHODS*	COMMENTS
Unknown	Not available	Susceptibility similar to *A. xylosoxidans*
Capable of beta-lactamase production	Not available	
Capable of beta-lactamase production	Not available	Susceptibility similar to *A. xylosoxidans*
Unknown	Not available	
Commonly resistant to ampicillin and several cephalosporins	Not available	
Unknown	Not available	
Often resistant to penicillins, even with beta-lactamase inhibitor, and aminoglycosides	Not available	
Unknown	Not available	*C. acidovorans* tends to be more resistant than the other two species, especially to aminoglycosides
Unknown	Not available	Generally susceptible to various antimicrobial agents
Produces beta-lactamases; may develop resistance to quinolones	Not available	
Generally resistant to cephalosporins and penicillins	Not available	
Often resistant to ampicillin and cephalothin	Not available	

Comamonas acidovorans is unique in producing an orange color when Kovac's reagent is added to tryptone broth (indole test). *Shewanella putrefaciens* produces a large amount of H_2S in the butt of triple sugar iron (TSI) agar. These organisms resemble H_2S-positive enteric bacilli but may be separated from them by their positive oxidase test.

"*Roseomonas*" must be separated from other pink-pigmented gram-negative (e.g., *Methylobacterium*) and gram-positive (e.g., certain *Rhodococcus* spp. or *Bacillus* spp.) organisms. "*Roseomonas*" differs from *Rhodococcus* and *Bacillus* spp. by being resistant to vancomycin, as determined by using a vancomycin disk on an inoculated 5% blood agar plate. Unlike *Methylobacterium*, "*Roseomonas*" will grow on MacConkey agar and at 42° C.

SERODIAGNOSIS

Serodiagnostic techniques are not generally used for the laboratory diagnosis of infections caused by the organisms discussed in this chapter.

ANTIMICROBIAL SUSCEPTIBILITY TESTING AND THERAPY

Validated susceptibility testing methods do not exist for these organisms. Although they will grow on the media and under the conditions recommended for testing the more commonly encountered bacteria (see Chapter 18 for more information regarding validated testing methods), this does not necessarily mean that interpretable and reliable results will be produced. Chapter 19 should be reviewed for preferable strategies that can be used to provide susceptibility information when validated testing methods do not exist for a clinically important bacterial isolate.

The lack of validated in vitro susceptibility testing methods does not allow definitive treatment and testing guidelines to be given for most organisms listed in Table 34-7. *B. parapertussis* is an exception; significant clinical experience indicates that erythromycin is the antimicrobial agent of choice for whooping cough caused by this organism (see Chapter 48 for more information regarding therapy for *B. pertussis* and *B. parapertussis* infections). Standardized testing methods do not exist for this species,[8] but the recent recognition of erythromycin resistance in *B. pertussis* indicates that development of such testing may be warranted for the causative agents of whooping cough.[11]

Even though standardized methods have not been established for the other species discussed in this chapter, in vitro susceptibility studies have been published and antimicrobial agents that have potential activity are noted, where appropriate, in Table 34-7.*

---

*References 1, 5, 9, 10, 12, 14-17, 21.

PREVENTION

The vaccine for *Bordetella pertussis* probably offers protection against *B. parapertussis* and is discussed in Chapter 48. Because the other organisms may be encountered throughout nature and do not generally pose a threat to human health, there are no recommended vaccination or prophylaxis protocols. For those organisms occasionally associated with nosocomial infections, prevention of infection is best accomplished by following appropriate sterile techniques and infection control guidelines.

References

1. Bizet, C., Tekaia, F., and Philippon, A. 1993. In vitro susceptibility of *Alcaligenes faecalis* compared with those of other *Alcaligenes* spp. to antimicrobial agents including seven β-lactams. J. Antimicrob. Chemother. 32:907.

2. Bowman, J. P., Cavanagh, J., Austin, J.J., et al. 1996. Novel *Psychrobacter* species from antarctic ornithogenic soils. Int. J. Syst. Bacteriol. 46:841.

3. Castagnola, E., Tasso, L., Conte, M., et al. 1994. Central venous catheter-related infection due to *Comamonas acidovorans* in a child with non-Hodgkin's lymphoma. Clin. Infect. Dis. 19:559.

4. Cheron, M., Abachin, E., Guerot, E., et al. 1994. Investigation of hospital-acquired infections due to *Alcaligenes denitrificans* subsp. *xylosoxidans* by DNA restriction fragment length polymorphism. J. Clin. Microbiol. 32:1023.

5. Decre, D., Arlet, G., Bergogne-Berezin, E., et al. 1995. Identification of a carbenicillin-hydrolyzing β-lactamase in *Alcaligenes denitrificans* subsp. *xylosoxidans*. Antimicrob. Agents Chemother. 39:771.

6. Dunne, W.M. and Maisch, S. 1995. Epidemiological investigation of infections due to *Alcaligenes* species in children and patients with cystic fibrosis: use of repetitive-element–sequence polymerase chain reaction. Clin. Infect. Dis. 20:836.

7. Hollis, D.G., Weaver, R.E., Moss, C.W., et al. 1992. Chemical and cultural characterization of CDC group WO-1, a weakly oxidative gram-negative group of organisms isolated from clinical sources. J. Clin. Microbiol. 30:291.

8. Hoppe, J.E. and Tschirner, T. 1995. Comparison of media for agar dilution susceptibility testing of *Bordetella pertussis* and *Bordetella parapertussis*. Eur. J. Clin. Microbiol. Infect. Dis. 14:775.

9. Knippschild, M., Schmid, E.N., Uppenkamp, M., et al. 1996. Infection by *Alcaligenes xylosoxidans* subsp. *xylosoxidans* in neutropenic patients. Oncology 53:258.

10. Legrand, C. and Anaissie, E. 1992. Bacteremia due to *Achromobacter xylosoxidans* in patients with cancer. Clin. Infect. Dis. 14:479.

11. Lewis, K., Saubolle, M.A., Tenover, F.C., et al.1995. Pertussis caused by an erythromycin-resistant strain of *Bordetella pertussis*. Pediatr. Infect. Dis. J. 14:388.

12. Lindquist, S.W., Weber, D.J., Mangum, M.E., et al. 1995.

Bordetella holmesii sepsis in an asplenic adolescent. Pediatr. Infect. Dis. J. 14:813.

13. Moss, C.W., Daneshvar, M.I., and Hollis, D.G. 1993. Biochemical characteristics and fatty acid composition of Gilardi rod group 1 bacteria. J. Clin. Micro. 31:689.

14. Musso, D., Drancourt, M., Bardot, J., et al. 1994. Human infection due to the CDC group IVc-2 bacterium: case report and review. Clin. Infect. Dis. 18:482.

15. Pugliese, A., Pacris, B., Schoch, P.E., et al. 1993. *Oligella urethralis* urosepsis. Clin. Infect. Dis. 17:1069.

16. Rihs, J.D., Brenner, D.J., Weaver, R.E., et al. 1993. *Roseomonas:* a new genus associated with bacteremia and other human infections. J. Clin. Microbiol. 31:3275.

17. Riley, U.B.G., Bignardi, G., Goldberg, L., et al. 1996. Quinolone resistance in *Oligella urethralis*-associated chronic ambulatory peritoneal dialysis peritonitis. J. Infect. 32:155.

18. Vandamme, P., Heyndrickx, M., Vancanneyt, M., et al. 1996. *Bordetella trematum* sp. nov., isolated from wounds and ear infections in humans, and reassessment of *Alcaligenes dentrificans* (Rüger and Tan 1983). Int. J. Syst. Bacteriol. 46:849.

19. Weyant, R.S., Hollis, D.G., Weaver, R.E., et al. 1995. *Bordetella holmesii* sp. nov.: a new gram-negative species associated with septicemia. J. Clin. Microbiol. 33:1.

20. Woolfrey, B.F. and Moody, J.A. 1991. Human infections associated with *Bordetella bronchiseptica.* Clin. Microbiol. Rev. 4:243.

Bibliography

Balows, A., Truper, H.G., Dworkin, M., et al., editors. 1981. The prokaryotes. A handbook on the biology of bacteria: ecophysiology, isolation, identification, applications ed 2. Springer-Verlag, New York.

Marcon, M.J. 1995. *Bordetella.* In Murray, P.R., Baron, E.J., Pfaller, M.A., editors. Manual of clinical microbiology, ed 6. American Society for Microbiology, Washington, D.C.

Pickett, M.J., Hollis, D.G., and Bottone, E.J. 1991. Miscellaneous gram-negative bacteria. In Balows, A., Hausler, W.J., Jr., Herrmann, K.L., et al., editors. Manual of clinical microbiology, ed 5. American Society for Microbiology, Washington, D.C.

McGowan, J.E., Jr. and Steinberg, J.P. 1995. Other gram-negative bacilli. In Mandell, G.L., Bennett, J.E., and Dolin, R., editors. Principles and practice of infectious diseases. Churchill Livingstone, New York.

Vancanneyt, M., Segers, P., Torck, U., et al. 1996. Reclassification of *Flavobacterium odoratum* (Stutzer 1929) strains to a new genus, *Myroides,* as *Myroides odoratus* comb. nov. and *Myroides odoratimimus* sp. nov. Int. J. Syst. Bacteriol. 46:926.

von Graevenitz, A. 1985. Ecology, clinical significance, and antimicrobial susceptibility of infrequently encountered glucose-nonfermenting gram-negative rods. In Gilardi, G.L., editor. Nonfermentative gram-negative rods: laboratory identification and clinical aspects. Marcel Dekker, New York.

von Graevenitz, A. 1995. *Acinetobacter, Alcaligenes, Moraxella,* and other nonfermentative gram-negative bacteria. In Murray, P.R., Baron, E.J., Pfaller, M.A., editors. Manual of clinical microbiology, ed 6. American Society for Microbiology, Washington, D.C.

Weyant, R.S., Moss, C.W., Weaver, R.E., editors. 1996. Identification of unusual pathogenic gram-negative aerobic and facultatively anaerobic bacteria, ed 2. Williams & Wilkins, Baltimore.

CHAPTER 35

VIBRIO, AEROMONAS, PLESIOMONAS SHIGELLOIDES, AND CHROMOBACTERIUM VIOLACEUM

Genera and Species to be Considered

Current Name	Previous Name
■ *Vibrio alginolyticus*	*Vibrio parahaemolyticus* biotype 2
■ *Vibrio cholerae*	
■ *Vibrio cincinnatiensis*	
■ *Vibrio damsela*	
■ *Vibrio fluvialis*	CDC group EF-6
■ *Vibrio furnissii*	
■ *Vibrio hollisae*	CDC group EF-13
■ *Vibrio metschnikovii*	CDC enteric group 16
■ *Vibrio mimicus*	*Vibrio cholerae* (sucrose-negative)
■ *Vibrio parahaemolyticus*	*Pasteurella parahaemolyticus*
■ *Vibrio vulnificus*	CDC group EF-3
■ *Aeromonas caviae*	
■ *Aeromonas hydrophila*	
■ *Aeromonas jandaei*	
■ *Aeromonas schubertii*	
■ *Aeromonas sobria*	
■ *Aeromonas veronii*	
■ *Plesiomonas shigelloides*	*Aeromonas* or *Pseudomonas shigelloides*
■ *Chromobacterium violaceum*	

GENERAL CHARACTERISTICS

The organisms discussed in this chapter are considered together because they are all oxidase-positive, glucose-fermenting, gram-negative bacilli that grow on MacConkey agar. Their individual morphologic and physiologic features are presented later in this chapter in the discussion of laboratory diagnosis.

EPIDEMIOLOGY

The epidemiology of *Vibrio* spp., *Aeromonas* spp., *Plesiomonas shigelloides,* and *Chromobacterium violaceum* are similar in many aspects (Table 35-1). The primary habitat for most of these organisms is water, generally brackish or marine water for *Vibrio* spp. and fresh water for *Aeromonas* spp., *P. shigelloides,* and *C. violaceum.* None of these organisms are considered part of the normal human flora. Transmission to humans is by ingestion of contaminated water or seafood or by exposure of disrupted skin and mucosal surfaces to contaminated water.

The epidemiology of the most notable human pathogen in this chapter, *Vibrio cholerae,* is far from being fully understood. This organism causes epidemics and pandemics (i.e., epidemics that span worldwide) of the diarrheal disease cholera. Since 1817 the world has witnessed seven cholera pandemics. During these outbreaks the organism is spread among people by the fecal-oral route, usually in environments with poor sanitation.

Of interest, the niche that *V. cholerae* inhabits between epidemics is uncertain. The form of the organism shed from infected humans is somewhat fragile and cannot survive long in the environment. However, evidence does suggest that survival, or dormant, stages of the bacillus exist to allow long-term survival in brackish water or saltwater environments during interepidemic periods.[3,8] Asymptomatic carriers

TABLE 35-1 EPIDEMIOLOGY

SPECIES	HABITAT (RESERVOIR)	MODE OF TRANSMISSION
Vibrio cholerae	Niche outside of human gastrointestinal tract between occurrence of epidemics and pandemics is uncertain; may survive in a dormant state in brackish or salt water; human carriers also are known, but are uncommon	Fecal-oral route, by ingestion of contaminated washing, swimming, cooking, or drinking water; also by ingestion of contaminated shellfish or other seafood
V. alginolyticus	Brackish or salt water	Uncertain; exposure to contaminated water
V. cincinnatiensis	Unknown	Unknown
V. damsela	Brackish or salt water	Exposure of wound to contaminated water
V. fluvialis	Brackish or salt water	Ingestion of contaminated water or seafood
V. furnissii	Brackish or salt water	Ingestion of contaminated water or seafood
V. hollisae	Brackish or salt water	Ingestion of contaminated water or seafood
V. metschnikovii	Unknown	Unknown
V. mimicus	Brackish or salt water	Ingestion of contaminated water or seafood
V. parahaemolyticus	Brackish or salt water	Ingestion of contaminated water or seafood
V. vulnificus	Brackish or salt water	Ingestion of contaminated water or seafood
Aeromonas caviae *A. hydrophila,* *A. sobria,* *A. jandaei,* *A. schubertii,* and *A. veronii*	Aquatic environments around the world, including fresh water, polluted or chlorinated water, brackish water, and, occasionally, marine water; may transiently colonize gastrointestinal tract; often infect various warm- and cold-blooded animal species	Ingestion of contaminated food (e.g., dairy, meat, produce) or water; exposure of disrupted skin or mucosal surfaces to contaminated water or soil; traumatic inoculation by fish fins or fishing hooks
Plesiomonas shigelloides	Fresh water, especially in warmer climates	Ingestion of contaminated water or seafood; exposure to cold-blooded animals, such as amphibia and reptiles
Chromobacterium violaceum	Environmental, soil and water of tropical and subtropical regions. Not part of human flora	Exposure of disrupted skin to contaminated soil or water

of *V. cholerae* have been documented, but they are not thought to be a significant reservoir for maintaining the organism between outbreaks.

PATHOGENESIS AND SPECTRUM OF DISEASE

As a notorious pathogen, *V. cholerae* elaborates several toxins and factors that play important roles in the organism's virulence.[8] Cholera toxin (CT) is primarily responsible for the key features of cholera (Table 35-2). Release of this toxin causes mucosal cells to hypersecrete water and electrolytes into the lumen of the gastrointestinal tract. The result is profuse watery diarrhea, leading to dramatic fluid loss. The fluid loss results in severe dehydration and hypotension that, without medical intervention, frequently leads to death. This toxin-mediated disease does not require the organism to penetrate the mucosal barrier. Therefore, blood and the inflammatory cells typical of dysenteric stools are notably absent in cholera (see Chapters 27 and 37 for more information regarding dysenteric infections). Instead, "rice water stools," composed of fluids and mucous flecks, are the hallmark of cholera toxin activity.

TABLE 35-2 PATHOGENESIS AND SPECTRUM OF DISEASES

SPECIES	VIRULENCE FACTORS	SPECTRUM OF DISEASE AND INFECTIONS
Vibrio cholerae	Cholera toxin, Zot toxin, Ace toxin, O1 and O139 somatic antigens, hemolysin/cytoxins, motility, chemotaxis, mucinase, and TCP pili	Cholera: profuse, watery diarrhea leading to dehydration, hypotension, and, often, death; occurs in epidemics and pandemics that span the globe. May also cause nonepidemic diarrhea and, occasionally, extraintestinal infections of wounds, respiratory tract, urinary tract, and central nervous system
V. alginolyticus	Specific virulence factors for the non-*V. cholerae* species are uncertain	Ear infections, wound infections; rare cause of septicemia; involvement in gastroenteritis is uncertain
V. cincinnatiensis		Rare cause of septicemia
V. damsela		Wound infections and rare cause of septicemia
V. fluvialis		Gastroenteritis
V. furnissii		Rarely associated with human infections
V. hollisae		Gastroenteritis; rare cause of septicemia
V. metschnikovii		Rare cause of septicemia; involvement in gastroenteritis is uncertain
V. mimicus		Gastroenteritis; rare cause of ear infection
V. vulnificus		Wound infections and septicemia; involvement in gastroenteritis is uncertain
Aeromonas caviae, A. hydrophila, A. sobia, A. jandaei, A. schubertii, and *A. veronii*	*Aeromonas* spp. produce various toxins and factors, but their specific role in virulence is uncertain	Gastroenteritis, wound infections, bacteremia, and miscellaneous other infections, including endocarditis, meningitis, pneumonia, conjunctivitis, and osteomyelitis
Plesiomonas shigelloides	Unknown	Gastroenteritis; septicemia in compromised adults and infants experiencing complicated delivery
Chromobacterium violaceum	Unknown	Rare but dangerous infection. Begins with cellulitis or lymphadenitis and can rapidly progress to systemic infections with abscess formation in various organs and septic shock

---

The somatic antigens O1 and O139 associated with the *V. cholerae* cell envelope are positive markers for strains capable of epidemic and pandemic spread of the disease. Strains carrying these markers almost always produce cholera toxin, whereas non-O1/non-O139 strains do not produce the toxin and hence do not produce cholera. Therefore, although these somatic antigens are not virulence factors per se, they are important virulence and epidemiologic markers that provide important information regarding *V.*

cholerae isolates. The non-O1/non-O139 strains are associated with nonepidemic diarrhea and extraintestinal infections.

V. cholerae produces several other toxins and factors whose exact role in disease is still uncertain (Table 35-2). For the organism to effectively release toxin, it must first infiltrate and distribute itself along the cells lining the mucosal surface of the gastrointestinal tract. Motility and chemotaxis mediate the distribution of organisms, and mucinase production allows for penetration of the mucous layer. TCP pili provide the means by which bacilli attach to mucosal cells for release of cholera toxin.

Depending on the species, other vibrios are variably involved in three types of infection: gastroenteritis, wound infections, and bacteremia. Although some of these organisms have not been definitively associated with human infections, others, such as *V. vulnificus,* are known to cause fatal septicemia, especially in patients suffering from an underlying liver disease.

Aeromonas spp. and *P. shigelloides* are similar to *Vibrio* spp. in terms of the types of infections they cause.[5-7,9] Although these organisms can cause gastroenteritis, most frequently in children, their role in intestinal infections is not always clear. Therefore the significance of their isolation in stool specimens should be interpreted with caution. Although noncholera vibrios, *Aeromonas* spp., and *P. shigelloides* produce factors that may contribute to virulence, little is known about any specific virulence factors.

C. violaceum is not associated with gastrointestinal infections, but acquisition of this organism by contamination of wounds can lead to fulminant, life-threatening, systemic infections.[14]

LABORATORY DIAGNOSIS

SPECIMEN COLLECTION AND TRANSPORT

Because there are no special considerations for isolation of these genera from extraintestinal sources, general specimen collection and transport information provided in Table 1-1 is applicable. However, stool specimens suspected of containing *Vibrio* spp. should be collected and transported only in Cary-Blair medium. Buffered glycerol saline is not acceptable, because glycerol is toxic for vibrios. Feces is preferable, but rectal swabs are acceptable during the acute phase of diarrheal illness.

SPECIMEN PROCESSING

No special considerations are required for processing the organisms discussed in this chapter. Refer to Table 1-1 for general information on specimen processing.

DIRECT DETECTION METHODS

V. cholerae toxin can be detected in stool using an enzyme-linked immunosorbent assay (ELISA) or a commercially available latex agglutination test (Oxoid Inc., Odgensburg, New York), but these tests are not widely used in the United States.

Microscopically, vibrios are gram-negative, straight or slightly curved rods. When stool specimens from cholera patients are examined using dark-field microscopy, the bacilli exhibit characteristic rapid darting or shooting-star motility. However, direct microscopic examination of stools by any method is not commonly used for the laboratory diagnosis of enteric bacterial infections.

Aeromonas spp. are gram-negative, straight rods, whereas *P. shigelloides* cells tend to be pleomorphic gram-negative rods that occur singly, in pairs, in short chains, or even as long, filamentous forms. Cells of *C. violaceum* are slightly curved, medium to long, gram-negative rods with rounded ends.

CULTIVATION

Media of choice

Stool cultures for *Vibrio* spp. are plated on the selective medium, thiosulfate citrate bile salts sucrose (TCBS) agar. Although some *Vibrio* spp. grow very poorly on this medium, those that grow well produce either yellow or green colonies, depending on whether they are able to ferment sucrose (and produce yellow colonies). Alkaline peptone water (pH 8.4) may be used as an enrichment broth for obtaining growth of vibrios from stool. After inoculation, the broth is incubated for 5 to 8 hours at 35° C and then subcultured to TCBS.

All of the genera considered in this chapter grow well on 5% sheep blood, chocolate, and MacConkey agars. They also grow well in the broth of blood culture systems and in thioglycollate or brain-heart infusion broths.

Incubation conditions and duration

These organisms produce detectable growth on 5% sheep blood, chocolate, and MacConkey agars when incubated at 35° C in carbon dioxide or ambient air for a minimum of 24 hours. However, MacConkey and TCBS agars only should be incubated at 35° C in ambient air. The typical violet pigment of *C. violaceum* colonies (Figure 35-1) is optimally produced

FIGURE 35-1 Colonies of *Chromobacterium violaceum* on DNase agar. Note violet pigment.

when cultures are incubated at room temperature (22° C).

Colonial appearance

Table 35-3 describes the colonial appearance and other distinguishing characteristics (e.g., hemolysis and odor) of each genus on 5% sheep blood and MacConkey agars. The appearance of *Vibrio* spp. on TCBS is described in Table 35-4 and shown in Figure 35-2.

APPROACH TO IDENTIFICATION

The colonies of these genera resemble those of the family *Enterobacteriaceae*, but notably differ by their positive oxidase test (except *V. metschnikovii*, which is oxidase-negative). The oxidase test must be performed from 5% sheep blood or another medium without a fermentable sugar (e.g., lactose in MacConkey agar or sucrose in TCBS). The reason for this is that fermentation of a carbohydrate results in acidification of the medium, and a false-negative oxidase test may result if the surrounding pH is below 5.1. Likewise, if the violet pigment of a suspected *C. violaceum* isolate interferes with performance of the oxidase test, the organism should be grown under anaerobic conditions (where it cannot produce pigment) and retested.

The reliability of commercial identification systems has not been widely validated for identification of these organisms, although most are listed in the databases of several systems. The API 20E system (bioMérieux Vitek, Inc., Hazelwood, MO.) is one of the best for vibrios. Because the inoculum is prepared in 0.85% saline, the amount of salt often is enough to allow growth of the halophilic (salt-loving) organism.

The ability of most commercial identification systems to accurately identify *Aeromonas* spp. to the species level is limited and uncertain, and some kits have trouble separating *Aeromonas* spp. from *Vibrio* spp. Therefore, identification of potential pathogens should be confirmed using conventional biochemicals or serotyping. Tables 35-4 and 35-5 show several characteristics that can be used to presumptively group *Vibrio* spp., *Aeromonas* spp., *P. shigelloides*, and *C. violaceum*.

Comments regarding specific organisms

V. cholerae and *V. mimicus* are the only *Vibrio* spp. that do not require salt for growth. Therefore, a key test in separating the halophilic species from *V. cholerae*, *V. mimicus*, *Aeromonas* spp., and *P. shigelloides* is growth in nutrient broth with 6% salt. Further, the addition of 1% NaCl to conventional biochemicals is recommended to allow growth of halophilic species.

The string test can be used to separate *Vibrio* spp. from *Aeromonas* spp. and *P. shigelloides*. In this test, organisms are emulsified in 0.5% sodium deoxycholate, which lyses the vibrio cells, but not those of *Aeromonas* spp. and *P. shigelloides*. With cell lysis

TABLE 35-3 COLONY APPEARANCE AND CHARACTERISTICS

ORGANISM	MEDIUM	APPEARANCE
Aeromonas spp.	BA	Large, round, raised, opaque; most colonies are beta-hemolytic, except *A. caviae*, which is usually nonhemolytic
	Mac	Both NLF and LF
Chromobacterium violaceum	BA	Round, smooth, convex, some strains beta-hemolytic; most colonies appear black or very dark purple; cultures smell of ammonium cyanide
	Mac	NLF
Plesiomonas shigelloides	BA	Shiny, opaque, smooth, nonhemolytic
	Mac	Both NLF and LF
Vibrio spp.	BA	Medium to large, smooth, opaque, iridescent with a greenish hue; *V. fluvialis*, *V. mimicus*, and *V. damsela* can be beta-hemolytic
	Mac	NLF except *V. vulnificus*, which may be LF

BA, 5% sheep blood agar; *Mac*, MacConkey agar; *NLF*, nonlactose fermenter; *LF*, lactose fermenter.

there is release of DNA, which can then be pulled up into a string using an inoculating loop (Figure 35-3).

A vibriostatic test using 0/129 (2,4-diamino-6, 7-diisopropylpteridine)-impregnated disks also has been used to separate vibrios (susceptible) from other oxidase-positive, glucose-fermenters (resistant) and to differentiate *V. cholerae* 01 and non-01 (susceptible) from other *Vibrio* spp. (resistant). However, recent strains of *V. cholerae* 01 have been resistant to 0/129, so the dependability of this test may be waning.

Serotyping and biotyping are used to further characterize *V. cholerae* isolates. There are two biotypes of *V. cholerae*, Classical and El Tor, and the species also can be divided into six serogroups on the basis of "O" (somatic) antigens. Toxigenic strains of serogroups 01 are involved in cholera epidemics and can be further subtyped as Ogawa, Inaba, or Hikojima strains. All isolates of *V. cholerae* should be sent to a reference laboratory for biotyping and serotyping.

Identification of *Aeromonas* spp., *P. shigelloides*, and *C. violaceum* can be accomplished using the characteristics shown in Table 35-5. *P. shigelloides* is unusual in being among the few species of clinically relevant bacteria that decarboxylates lysine, ornithine, and arginine.

Pigmented strains of *C. violaceum* are so distinctive that a presumptive identification can be made based on colonial appearance, oxidase, and Gram stain. Nonpigmented strains (approximately 9% of isolates) may be differentiated from *Pseudomonas, Burkholderia, Brevundimonas,* and *Ralstonia* based on glucose fermentation and a positive test for indole. Negative lysine and ornithine reactions are useful criteria for separating *C. violaceum* from *P. shigelloides.* In addition to the characteristics listed in Table 35-5, failure to ferment either maltose or mannitol also differentiates *C. violaceum* from *Aeromonas* spp.

SERODIAGNOSIS

Agglutination, vibriocidal, or antitoxin tests are available for diagnosing cholera using acute and convalescent sera. However, these methods are most commonly used for epidemiologic purposes. Serodiagnostic techniques are not generally used for the laboratory diagnosis of infections caused by the other organisms discussed in this chapter.

ANTIMICROBIAL SUSCEPTIBILITY TESTING AND THERAPY

Two components to the management of patients with cholera are rehydration of the patient and antimicrobial therapy (Table 35-6). Antimicrobials serve to decrease the severity of illness and shorten the duration of organism shedding. The drugs of choice for

cholera are tetracycline or doxycycline; however, resistance to these agents is known, and the use of other agents, such as chloramphenicol, erythromycin, or trimethoprim/sulfamethoxazole, may be necessary.[8] Although standard susceptibility testing methods for *V. cholerae* do not exist, disk agar diffusion is often used as a means for detecting resistance. If testing is done by this method, errors will occur with chloramphenicol, erythromycin, and doxycycline.[13] Furthermore, in vitro testing may indicate susceptibility to ampicillin, but this agent has been found to have minimal clinical efficacy for cholera.[8]

The need for antimicrobial intervention for gastrointestinal infections caused by other *Vibrio* spp., *Aeromonas* spp., and *P. shigelloides* is less clear. In

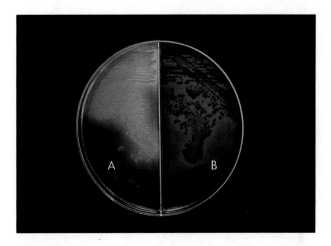

FIGURE 35-2 Colonies of *Vibrio cholerae* (**A**) and *V. parahemolyticus* (**B**) on TCBS agar.

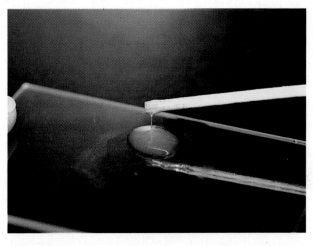

FIGURE 35-3 String test, used to separate *Vibrio* spp. (positive) from *Aeromonas* spp. and *P. shigelloides* (negative).

TABLE 35-4 KEY BIOCHEMICAL AND PHYSIOLOGIC CHARACTERISTICS OF *VIBRIO* SPP.

SPECIES	OXIDASE	D-GLUCOSE (GAS)	LACTOSE	SUCROSE	*myo*-INOSITOL	LYSINE DECARBOXYLASE[a]
		FERMENTATION				
Vibrio cholerae	+	−	−	+	−	+
V. mimicus	+	−	v	−	−	+
V. metschnikovii	−	−	v	+	v	v
V. cincinnatiensis	+	−	−	+	+	v
V. hollisae	+	−	−	−	−	−
V. damsela	+	−	−	−	−	v
V. fluvialis	+	−	−	+	−	−
V. furnisii	+	−	+	−	−	−
V. alginolyticus	+	−	−	+	−	+
V. parahaemolyticus	+	−	−	−	−	+
V. vulnificus	+	−	v	v	−	+

+, >90% of strains are positive; −, >90% of strains are negative; *v*, variable.
[a]1% NaCl added to enhance growth.
[b]Nutrient broth with 0% or 6% NaCl added.
[c]Thiosulfate citrate bile salts-sucrose agar.
[d]5% yellow.
[e]1% yellow.
[f]10% yellow.

Data compiled from McLaughlin J.C., 1995. *Vibrio*. In Murray, P.R., Baron, E.J., Pfaller, M.A., et al., editors. Manual of clini-

ARGININE DIHYDROLASE[a]	ORNITHINE DECARBOXYLASE[a]	GROWTH IN 0% NaCl[b]	GROWTH IN 6% NaCl[b]	TCBS[c] GROWTH	COLONY COLOR ON TCBS[c]
−	+	+	v	Good	Yellow
−	+	+	v	Good	Green
v	−	−	v	May be reduced	Yellow
−	−	−	+	Very poor	Yellow
−	−	−	v	Very poor	Green
+	−	−	+	Reduced at 36° C	Green[d]
+	−	−	+	Good	Yellow
+	−	−	+	Good	Yellow
−	v	−	+	Good	Yellow
−	+	−	+	Good	Green[e]
−	v	−	v	Good	Green[f]

cal microbiology, ed 6. American Society for Microbiology, Washington, D.C.

TABLE 35-5 **KEY BIOCHEMICAL AND PHYSIOLOGIC CHARACTERISTICS OF *AEROMONAS* SPP., *P. SHIGELLOIDES,* AND *C. VIOLACEUM***

SPECIES	OXIDASE	FERMENTATION GLUCOSE (GAS)	LACTOSE	SUCROSE	*myo*-INOSITOL
Aeromonas hydrophila	+	+	−	+	−
A. caviae	+	−	+	+	−
A. veronii biovar *sobria*	+	+	−	+	−
A. veronii biovar *veronii*	+	+	−	+	−
A. jandaei	+	+	−	−	−
A. schubertii	+	−	−	−	−
A. trota	+	+	−	−	−
Plesiomonas shigelloides	+	−	−	−	+
Chromobacterium violaceum‡	v	−ˢ	−	v	NT

+, >90% of strains are positive; −, >90% of strains are negative; *v*, variable; *NT,* not tested.

*Nutrient agar with 0% or 6% NaCl added.

†Thiosulfate citrate bile salts sucrose agar.

‡91% produce an insoluble violet pigment; often, nonpigmented strains are indole-positive.

ˢGas-producing strains have been described.

Data compiled from Holmes, B., Pickett, M.J., and Hollis, D.G. 1995. In Murray, P.R., Baron, E.J., Pfaller, M.A., et al., S.L., and Carnahan, A.M. 1995. In Murray, P.R., Baron, E.J., Pfaller, M.A., et al. editors. Manual of clinical microbiology, ed tion of unusual pathogenic gram-negative aerobic and facultatively anaerobic bacteria, ed 2. Williams & Wilkins, Baltimore.

LYSINE DECARBOXYLASE	ARGININE DIHYDROLASE	ORNITHINE DECARBOXYLASE	GROWTH IN 0% NaCl*	GROWTH IN 6% NaCl*	TCBS† GROWTH
+	+	−	+	−	−
−	+	−	+	−	−
+	+	−	+	−	−
+	−	+	+	−	−
+	+	−	+	−	−
+	+	−	+	−	−
+	+	−	+	−	−
+	+	+	+	−	−
−	+	−	+	−	NT

editors. Manual of clinical microbiology, ed 6. American Society for Microbiology, Washington, D.C.; Janda, J.M., Abbott, 6. American Society for Microbiology. Washington, D.C.; and Weyant, R.S., Moss, C.W., Weaver, R.E., et al. 1996. Identifica-

TABLE 35-6 ANTIMICROBIAL THERAPY AND SUSCEPTIBILITY TESTING

SPECIES	THERAPEUTIC OPTIONS	POTENTIAL RESISTANCE TO THERAPEUTIC OPTIONS	VALIDATED TESTING METHODS*	COMMENTS
Vibrio cholerae	Adequate rehydration plus antibiotics. Recommended agents include tetracycline or doxycycline; alternatives include trimethoprim/sulfamethoxazole, erythromycin, chloramphenicol, and quinolones	Yes; resistance to tetracycline, chloramphenicol, and trimethoprim/sulfamethoxazole is known	Not available	Disk diffusion is used but should be done with caution because errors are known to occur
Other *Vibrio* spp.	No definitive guidelines. For gastroenteritis therapy may not be needed; for wound infections and septicemia potentially active agents include tetracycline, chloramphenicol, nalidixic acid, most cephalosporins, and quinolones	Yes; similar to resistance reported for *V. cholerae*	Not available	Most will grow on Mueller-Hinton agar without NaCl supplement, but interpretive standards do not exist
Aeromonas spp.	No definitive guidelines. For gastroenteritis therapy may not be needed; for soft tissue infections and septicemia potentially active agents include ceftriaxone, cefotaxime, ceftazidime, imipenem, aztreonam, aminoglycosides, quinolones, and trimethoprim/sulfamethoxazole	Yes; capable of producing various beta-lactamases that mediate resistance to penicillins and certain cephalosporins	Not available	Will grow on Mueller-Hinton agar, but interpretive standards do not exist
Plesiomonas shigelloides	No definitive guidelines. For gastroenteritis therapy may not be needed; for septicemia potentially active agents include most cephalosporins, imipenem, aztreonam, beta-lactam/beta-lactamase-inhibitor combinations, and quinolones	Yes; capable of beta-lactamase production and resistance to ampicillin, mezlocillin, and piperacillin	Not available	Most will grow on Mueller-Hinton agar, but interpretive standards do not exist
Chromobacterium violaceum	No definitive guidelines. Potentially active agents include cefotaxime, ceftazidime, imipenem, and aminoglycosides	Yes; activity of penicillins is variable; activity of first- and second-generation cephalosporins is poor	Not available	Will grow on Mueller-Hinton agar, but interpretive standards do not exist

*Validated testing methods include those standard methods recommended by the National Committee for Clinical Laboratory Standards (NCCLS) and those commercial methods approved by the Food and Drug Administration (FDA).

contrast, extraintestinal infections with these organisms and with *C. violaceum* can be life-threatening, and directed therapy is required.[5-7,9,11] Unfortunately, validated susceptibility testing methods do not exist. Many of these organisms will grow on the media and under the conditions recommended for testing the more commonly encountered bacteria (see Chapter 18 for more information regarding validated testing methods), but this does not necessarily mean that interpretable and reliable results will be produced. Chapter 19 should be reviewed for preferable strategies that can be used to provide susceptibility information when validated testing methods do not exist for clinically important bacterial isolates.

Although standardized methods have not been established for these organisms, in vitro susceptibility studies have been published and antimicrobial agents that have potential activity are listed, where appropriate, in Table 35-6.* Of importance is the capacity of these species to exhibit resistance to therapeutic agents; especially noteworthy is the capacity of *Aeromonas* spp. and *P. shigelloides* to produce various beta-lactamases.[2,4,15]

PREVENTION

Two oral cholera vaccines that provide a high level of protection for several months are available worldwide. Individuals who have recently shared food and drink with a patient with cholera (e.g., household contacts) should be given chemoprophylaxis with tetracycline, doxycycline, furazolidone, or trimethoprim-sulfamethoxazole. However, mass chemoprophylaxis during epidemics is not indicated.[1] There are no approved vaccines or chemoprophylaxis for the other organisms discussed in this chapter.

References

1. Benenson, A.S., editor. 1995. Control of communicable diseases manual, ed 16. American Public Health Association, Washington, D.C.
2. Clark, R.B., Lister, P.D., Arneson-Rotert, L., et al. 1990. In vitro susceptibilities of *Plesiomonas shigelloides* to 24 antibiotics and antibiotic-β-lactamase-inhibitor combinations. Antimicrob. Agents Chemother. 34:159.
3. Colwell, R.R. 1996. Global climate and infectious disease: the cholera paradigm. Science. 274:2025.
4. Hayes, M.V., Thomson, C.J., and Amyes, S.G.B. 1996. The "hidden" carbapenemase of *Aeromonas hydrophila*. J. Antimicrob. Chemother. 37:33.
5. Janda, J.M. 1991. Recent advances in the study of the taxon-

omy, pathogenicity, and infectious syndromes associated with the genus *Aeromonas*. Clinical Microbiol. Rev. 4:397.
6. Janda, J.M., Guthertz, L.S., Kokka, R.P., et al. 1994. *Aeromonas* species in septicemia: laboratory characteristics and clinical observations. Clin. Infect. Dis. 19:77.
7. Jones, B.L. and Wilcox, M.H. 1995. *Aeromonas* infections and their treatment. J. Antimicrob. Chemother. 35:453.
8. Kaper, J.B., Morris, J.G., and Levine, M.M. 1995. Cholera. Clin. Microbiol. Rev. 8:48.
9. Ko, W.C., and Chuang, Y.C. 1994. *Aeromonas* bacteremia: review of 59 episodes. Clin. Infect. Dis. 20:1298.
10. Ko, W.C., Yu, K.W., Liu, C.Y., et al. 1996. Increasing antibiotic resistance in clinical isolates of *Aeromonas* strains in Taiwan. Antimicrob. Agents Chemother. 40:1260.
11. Lee, A.C.W., Yuen, K.Y., and Ha, S.Y. 1995. *Plesiomonas shigelloides:* case report and literature review. Ped. Hematol. and Oncology. 13:265.
12. Morita, K., Watanabe, N., Kurata, S., et al. 1994. β-Lactam resistance of motile *Aeromonas* isolates from clinical and environmental sources. Antimicrob. Agents Chemother. 38:353.
13. National Committee for Clinical Laboratory Standards. 1997. Performance standards for antimicrobial susceptibility testing. M100-S7. NCCLS, Villanova, Pa.
14. Ti, T.Y., Tan, C.W., Chong, A.P.Y., et al. 1993. Nonfatal and fatal infections caused by *Chromobacterium violaceum*. Clin. Infect. Dis. 17:505.
15. Walsh, T.R., Payne, D.J., MacGowan, A.P., et al. 1995. A clinical isolate of *Aeromonas sobria* with three chromosomally mediated inducible β-lactamases: a cephalosporinase, a penicillinase, and a third enzyme, displaying carbapenemase activity. J. Antimicrob. Chemother. 35:271.

Bibliograpy

Carpenter, C.C.J. 1995. Other pathogenic vibrios. In Mandell, G.L., Bennett, J.E., and Dolin, R., editors. Principles and practice of infectious diseases. Churchill Livingstone, New York.

Farmer, J.J., III and Hickman-Brenner, F.W. 1992. The genera *Vibrio* and *Photobacterium*, In Balows, A., Truper, H.G., Dworkin, M., et al., editors. The prokaryotes. A handbook on the biology of bacteria: ecophysiology, isolation, identification, applications, ed 2. Springer-Verlag, New York.

Farmer, J.J., III, Arduino, M.J., and Hickman-Brenner, F.W. 1992. The genera *Aeromonas* and *Plesiomonas*. In Balows, A., Truper, H.G., Dworkin, M., et al., editors. The prokaryotes. A handbook on the biology of bacteria: ecophysiology, isolation, identification, applications, ed 2. Springer-Verlag, New York.

Greenough, W.B., III. 1995. *Vibrio cholerae* and cholera. In Mandell, G.L., Bennett, J.E., and Dolin, R., editors. Principles and practice of infectious diseases. Churchill Livingstone, New York.

Holmes, B., Pickett, M.J., and Hollis, D.G. 1995. Unusual gram-negative bacteria, including *Capnocytophaga, Eikenella, Pasteurella,* and *Streptobacillus*. In Murray, P.R., Baron, E.J., Pfaller, M.A., et al., editors. Manual of clinical microbiology, ed 6. American Society for Microbiology, Washington, D.C.

---

*References 2, 3, 6, 7, 10, 12.

Janda, J.M., Abbott, S.L., and Carnahan, A.M. 1995. *Aeromonas* and *Plesiomonas*. In Murray, P.R., Baron, E.J., Pfaller, M.A., et al., editors. Manual of clinical microbiology, ed 6. American Society for Microbiology, Washington, D.C.

McGowan, J.E., Jr. and Steinberg, J.P. 1995. Other gram-negative bacilli. In Mandell, G.L., Bennett, J.E., and Dolin, R., editors. Principles and practice of infectious diseases. Churchill Livingstone, New York.

McLaughlin, J.C. 1995. *Vibrio*. In Murray, P.R., Baron, E.J., Pfaller, M.A., et al., editors. Manual of clinical microbiology, ed 6. American Society for Microbiology, Washington, D.C.

Weyant, R.S., Moss, C.W., Weaver, R.E., et al., editors. 1996. Identification of unusual pathogenic gram-negative aerobic and facultatively anaerobic bacteria, ed 2. Williams & Wilkins, Baltimore.

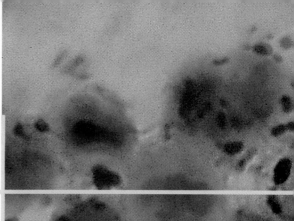

Section Two

2

Gram-Negative Bacilli and Coccobacilli (MacConkey-Positive, Oxidase-Negative)

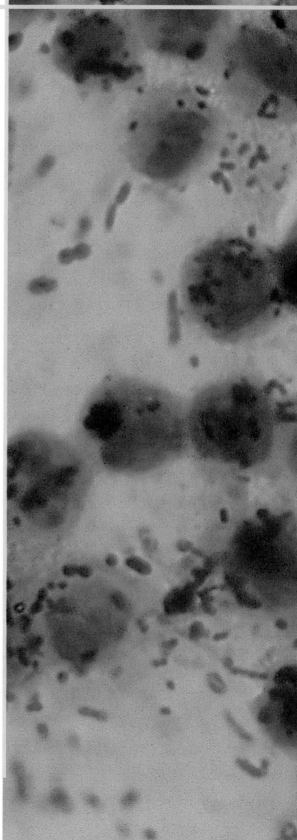

36 | ACINETOBACTER, CHRYSEOMONAS, FLAVIMONAS, AND STENOTROPHOMONAS

Genera and Species to be Considered

Current Name	Previous Name
■ *Chryseomonas luteola*	*Pseudomonas luteola*, CDC group Ve-1
■ *Flavimonas oryzihabitans*	*Pseudomonas oryzihabitans*, CDC group Ve-2
■ *Acinetobacter* spp.; saccharolytic, nonhemolytic	*Acinetobacter baumanii, Acinetobacter calcoaceticus, Acinetobacter anitratus, Acinetobacter calcoaceticus* subsp. *anitratus*
■ *Acinetobacter* spp.; saccharolytic, hemolytic	*Acinetobacter alcaligenes, Acinetobacter anitratus, Acinetobacter haemolyticus*
■ *Acinetobacter* spp.; asaccharolytic, nonhemolytic	*Acinetobacter calcoaceticus* subsp. *lwoffi, Acinetobacter johnsonii, A. junii, A. lwoffi*
■ *Acinetobacter* spp; asaccharolytic, hemolytic	
■ *Stenotrophomonas maltophilia*	*Xanthomonas maltophilia, Pseudomonas maltophilia*
■ CDC group NO-1	

GENERAL CHARACTERISTICS

The organisms discussed in this chapter are considered together because, except for CDC group NO-1, they are all oxidase-negative and grow on MacConkey agar, as do the *Enterobacteriaceae*. However, unlike the *Enterobacteriaceae*, which ferment glucose, these organisms either oxidize or do not utilize glucose. Although NO-1 is oxidase-negative and does not usually grow on MacConkey agar, it is included here because it must be distinguished from the asaccharolytic *Acinetobacter* spp. The specific morphologic and physiologic features of these organisms are considered later in this chapter in the discussion of laboratory diagnosis.

EPIDEMIOLOGY

These organisms inhabit environmental niches, with *Acinetobacter* spp. and *S. maltophilia* being widely distributed in nature and hospital environments (Table 36-1). Although none of these organisms are commonly part of the human flora, the relatively high prevalence of *Acinetobacter* spp. and *S. maltophilia* in hospitals frequently results in colonization of the skin and respiratory tract of patients.[1] The prevalence of these organisms is evidenced by the fact that, excluding *Enterobacteriaceae, Acinetobacter* spp. and *S. maltophilia* are the second and third most common gram-negative bacilli, respectively, encountered in clinical specimens. In contrast, *Chryseomonas luteola, Flavimonas oryzihabitans*, and CDC group NO-1 are not commonly encountered in the clinical setting.[4]

PATHOGENESIS AND SPECTRUM OF DISEASE

All of the organisms listed in Table 36-2 are opportunistic pathogens for which no definitive virulence factors are known. Because *Acinetobacter* spp. and *S. maltophilia* are relatively common colonizers of hospitalized patients, their clinical significance when found in patient specimens can be difficult to establish. In fact, these organisms are more frequently isolated as colonizers than as infecting agents. When infection does occur, it usually involves debilitated patients, such as those in burn or intensive care units, that have undergone medical instrumentation and/or have received multiple antimicrobial agents.[1,13] Infections caused by *Acinetobacter* spp. and *S. maltophilia*

TABLE 36-1 EPIDEMIOLOGY

SPECIES	HABITAT (RESERVOIR)	MODE OF TRANSMISSION
Acinetobacter spp.	Widely distributed in nature, including the hospital environment. May become established as part of skin and respiratory flora of patients hospitalized for prolonged periods	Colonization of hospitalized patients from environmental factors; medical instrumentation (e.g., intravenous or urinary catheters) introduces organism to normally sterile sites
Chryseomonas luteola and *Flavimonas oryzihabitans*	Environmental, including moist hospital environments (e.g., respiratory therapy equipment). Not part of normal human flora	Uncertain; probably involves exposure of debilitated, hospitalized patients to contaminated fluids and medical equipment
Stenotrophomonas maltophilia	Widely distributed in nature, including moist hospital environments. May become established as part of respiratory flora of patients hospitalized for prolonged periods	Colonization of hospitalized patients from environmental factors; medical instrumentation introduces organism to normally sterile sites (similar to transmission of *Acinetobacter* spp.)
CDC group NO-1	Oropharynx of animals. Not part of human flora	Animal bite or scratch

TABLE 36-2 PATHOGENESIS AND SPECTRUM OF DISEASES

SPECIES	VIRULENCE FACTORS	SPECTRUM OF DISEASE AND INFECTIONS
Acinetobacter spp.	Unknown	Clinical isolates are often colonizers. True infections are usually nosocomial, occur during warm seasons, and most commonly involve the genitourinary tract, respiratory tract, wounds, soft tissues, and bacteremia
Chryseomonas luteola and *Flavimonas oryzihabitans*	Unknown	Catheter-related infections, septicemia, and peritonitis usually associated with continuous ambulatory peritoneal dialysis, and miscellaneous mixed infections of other body sites
Stenotrophomonas maltophilia	Unknown. Intrinsic resistance to almost every commonly used antibacterial agent supports the survival of this organism in hospital environment	Most infections are nosocomial and include catheter-related infections, bacteremia, wound infections, pneumonia, urinary tract infections, and miscellaneous infections of other body sites
CDC group NO-1	Unknown	Animal bite wound infections

usually involve the respiratory or genitourinary tract, bacteremia, and, occasionally, wound infections, but infections involving several other body sites have been described. Community-acquired infections with these organisms can occur, but the vast majority of infections are nosocomial.[1,2,11-13]

LABORATORY DIAGNOSIS

SPECIMEN COLLECTION AND TRANSPORT

No special considerations are required for specimen collection and transport of the organisms discussed in this chapter. Refer to Table 1-1 for general information on specimen collection and transport.

SPECIMEN PROCESSING

No special considerations are required for processing of the organisms discussed in this chapter. Refer to Table 1-1 for general information on specimen processing.

DIRECT DETECTION METHODS

Other than Gram stain of patient specimens, there are no specific procedures for the direct detection of these organisms in clinical material. *Acinetobacter* spp. are plump coccobacilli that tend to resist alcohol decolorization; they may be mistaken for *Neisseria* spp.

Chryseomonas luteola and *Flavimonas oryzihabitans* are short to medium–size, slightly thick, sometimes paired, gram-negative rods. *S. maltophilia* are short to medium–size, straight rods, and CDC group NO-1 are coccoid to medium–size bacilli.

CULTIVATION

Media of choice

In addition to their ability to grow on MacConkey agar, all of the genera described in this chapter grow well on 5% sheep blood and chocolate agars. These organisms also grow well in the broth of blood culture systems and in common nutrient broths such as thioglycollate and brain-heart infusion.

Incubation conditions and duration

These organisms generally produce detectable growth on 5% sheep blood and chocolate agars when incubated at 35° C in carbon dioxide or ambient air for a minimum of 24 hours. MacConkey agar should be incubated only in ambient air.

Colonial appearance

Table 36-3 describes the colonial appearance and other distinguishing characteristics (e.g., hemolysis and odor) of each genus when grown on 5% sheep blood and MacConkey agars.

TABLE 36-3 COLONY APPEARANCE AND CHARACTERISTICS

ORGANISM	MEDIUM	APPEARANCE
Acinetobacter spp.	BA	Smooth, opaque, raised, creamy, and smaller than *Enterobacteriaceae*; some genospecies are beta-hemolytic
	Mac	NLF, but colonies exhibit a purplish hue that may cause the organism to be mistaken for an LF
Chryseomonas luteola and *Flavimonas oryzihabitans*	BA	Wrinkled and adherent colonies; rare smooth colonies; yellow pigment
	Mac	NLF
Stenotrophomonas maltophilia	BA	Large, smooth, glistening colonies with uneven edges and lavender-green to light purple pigment; greenish discoloration underneath growth; ammonia smell
	Mac	NLF
CDC group NO-1	BA	Small colonies that can be transferred intact by an inoculating needle
	Mac	NLF, but only 20% of strains grow

BA, 5% sheep blood agar; *Mac,* MacConkey agar; *NLF,* nonlactose fermenter; *LF,* lactose fermenter.

APPROACH TO IDENTIFICATION

Acinetobacter spp. and *S. maltophilia* are reliably identified by the API 20E system (bioMérieux Vitek, Inc., Hazelwood, Mo.), although other commercial systems may not perform as well. *Chryseomonas luteola, Flavimonas oryzihabitans,* and CDC group NO-1 are most reliably identified using conventional biochemical and physiologic characteristics, such as those outlined in Table 36-4.

Comments regarding specific organisms

There are 17 genospecies or genomospecies in the genus *Acinetobacter*. Each genospecies comprises a distinct DNA hybridization group and is given a numeric designation, which has replaced previous species names. The genus is also divided into two groups; one contains the saccharolytic (glucose-oxidizing) species and the other contains the asaccharolytic (nonglucose-utilizing) species.

Most glucose-oxidizing, nonhemolytic strains were previously identified as *A. baumanii,* and most glucose-nonutilizing, nonhemolytic strains were designated as *A. lwoffi,* The majority of beta-hemolytic organisms were previously called *A. haemolyticus.* Nitrate-reducing strains of asaccharolytic *Acinetobacter* spp. are difficult to differentiate from CDC group NO-1. The *Acinetobacter* transformation test provides the most dependable criterion for this purpose, but this test is not commonly performed in clinical microbiology laboratories.

S. maltophilia can produce biochemical profiles similar to those of *Burkholderia cepacia,* but a negative oxidase test most often rules out the latter. *S. maltophilia* also oxidizes maltose faster than glucose (hence the species name, *maltophilia,* "maltose loving") and produces a brown pigment on heart infusion agar that contains tyrosine.

Chryseomonas luteola can be differentiated from *Flavimonas oryzihabitans* by producing positive tests for gelatin hydrolysis, ONPG, and esculin hydrolysis.

SERODIAGNOSIS

Serodiagnostic techniques are not generally used for the laboratory diagnosis of infections caused by the organisms discussed in this chapter.

ANTIMICROBIAL SUSCEPTIBILITY TESTING AND THERAPY

Acinetobacter spp. and *S. maltophilia,* can exhibit resistance to a wide array of antimicrobial agents, making the selection of agents for optimal therapy difficult (Table 36-5). This underscores the importance of establishing clinical significance of individual isolates before antimicrobial testing is performed and

results are reported (see Chapter 19 for a discussion of criteria used to establish significance). Not doing so could lead to the inappropriate treatment of patients with expensive and potentially toxic agents.

For urinary tract infections caused by *Acinetobacter* spp., single-drug therapy is usually sufficient. In contrast, more serious infections, such as pneumonia or bacteremia, may require the use of a beta-lactam agent in combination with an aminoglycoside. Because this genus is able to acquire and express resistance to most antimicrobial agents, including imipenem, in vitro testing is recommended for clinically relevant isolates.[1,5] Methods outlined by the National Committee for Clinical Laboratory Standards appear to be suitable for testing *Acinetobacter* spp.[7-9]

S. maltophilia is notoriously resistant to most currently available antimicrobial agents, leaving trimethoprim/sulfamethoxazole as the primary drug of choice for infections caused by this species.[3,5,10] Although a few other agents, such as minocycline, ticarcillin/clavulanic acid, and chloramphenicol, often exhibit in vitro activity, clinical experience with these agents is not extensive. Therefore, trimethoprim/sulfamethoxazole remains the drug of choice.

The other agents should only be considered when trimethoprim/sulfamethoxazole-resistant strains are encountered. Even then, the potential efficacy of these other agents is suspect, because of the ability of *S. maltophilia* to rapidly develop resistance.[3] For these reasons, in vitro susceptibility testing with these either should not be performed or, if performed, should be interpreted with extreme caution.[3,6]

Even though *Chryseomonas luteola* and *Flavimonas oryzihabitans* may grow on the media and under the conditions recommended for testing other more commonly encountered bacteria (see Chapter 18 for more information regarding validated testing methods), this does not necessarily mean that interpretable and reliable results will be produced. Therefore, as with *S. maltophilia,* validated susceptibility testing methods do not exist for these two species. Chapter 19 should be reviewed for preferable strategies that can be used to provide susceptibility information when validated testing methods do not exist for clinically important bacterial isolates. Antimicrobial agents that have potential activity against these organisms are listed in Table 36-5.[2,11,12]

PREVENTION

Because these organisms are ubiquitous in nature and are not generally a threat to human health, there are no recommended vaccination or prophylaxis protocols. Hospital-acquired infections are best controlled by following appropriate sterile techniques and infection

TABLE 36-4 Key Biochemical and Physiologic Characteristics

ORGANISM	GROWTH ON MACCONKEY	OXIDIZES GLUCOSE	ESCULIN HYDROLYSIS	LYSINE DECARBOXYLASE	ARGININE DIHYDROLASE	NITRATE REDUCTION	PEPTONIZATION OF LITMUS MILK
Chryseomonas luteola	+	+	+	−	+	v	v
Flavimonas oryzihabitans	+	+	−	−	v	−	−
Stenotrophomonas maltophilia	+	+	v	+	−	v	+
Saccharolytic *Acinetobacter* spp.	+	+	−	−	v	−	−
Asaccharolytic *Acinetobacter* spp.	+	−	−	−	v	−	−
CDC group NO-1	v	−	−	−	−	+	v

+, >90% of strains are positive; −, >90% of strains are negative; *v*, variable.

Compiled from data in Weyant, R.S., Moss, C.W., Weaver, R.E., et al. 1996. Identification of unusual pathogenic gram-negative aerobic and facultatively anaerobic bacteria. ed. 2. Williams & Wilkins, Baltimore.

TABLE 36-5   ANTIMICROBIAL THERAPY AND SUSCEPTIBILITY TESTING

SPECIES	THERAPEUTIC OPTIONS	POTENTIAL RESISTANCE TO THERAPEUTIC OPTIONS	VALIDATED TESTING METHODS*	COMMENTS
Acinetobacter spp.	No definitive guidelines. Potentially active agents include beta-lactam/beta-lactamase inhibitor combinations, ceftazidime, imipenem, ciprofloxacin, and aminoglycosides	Yes; resistance to beta-lactams, aminoglycosides, and quinolones	Disk diffusion, broth dilution, and agar dilution	In vitro susceptibility testing results important for guiding therapy
Chryseomonas luteola and *Flavimonas oryzihabitans*	No definitive guidelines. Potentially active agents include cefotaxime, ceftriaxone, ceftazidime, imipenem, quinolones, and aminoglycosides	Yes, activity of penicillins is variable; commonly resistant to first- and second-generation cephalosporins	Not available	
Stenotrophomonas maltophilia	Multiple resistance leaves few therapeutic choices. Therapy of choice is trimethoprim/sulfamethoxazole. Potential alternatives include minocycline, ticarcillin/clavulanic acid, and chloramphenicol	Yes, intrinsically resistant to most beta-lactams and aminoglycosides; frequently resistant to quinolones	Not available	May be tested by various methods, but profiles obtained with beta-lactams can be seriously misleading
CDC group NO-1	No definitive guidelines. Appear susceptible to beta-lactam antibiotics	Unknown	Not available	

*Validated testing methods include those standard methods recommended by the National Committee for Clinical Laboratory Standards (NCCLS) and those commercial methods approved by the Food and Drug Administration (FDA).

control guidelines and implementing effective protocols for the sterilization and decontamination of medical supplies.

References

1. Bergogne-Berezin, E. and Towner, K.J. 1996. *Acinetobacter* spp. as nosocomial pathogens: microbiological, clinical, and epidemiological features. Clin. Microbiol. Rev. 9:148.

2. Esteban, J., Valero-Moratalla, M.L., Alcazar, R., et al. 1993. Infections due to *Flavimonas oryzihabitans*: case report and literature review. Eur. J. Clin. Microbiol. Infect. Dis. 12:797.

3. Garrison, M.W., Anderson, D.E., Campbell, D.M., et al. 1996. *Stenotrophomonas maltophilia*: emergence of multidrug-resistant strains during therapy and in an in vitro pharmacodynamic chamber model. Antimicrob. Agents Chemother. 40:2859.

4. Hollis, D.G., Moss, C.W., Daneshvar, M.I., et al. 1993. Characterization of Centers for Disease Control group NO-1, a fastidious, nonoxidative, gram-negative organism associated with dog and cat bites. J. Clin. Microbiol. 31:746.

5. Livermore, D.M. 1995. β-Lactamases in laboratory and clinical resistance. Clin. Microbiol. Rev. 8:557.

6. Metchock, B. and Thornsberry, C. 1989. Susceptibility of *Xanthomonas (Pseudomonas) maltophilia* to antimicrobial agents. Antimicob. Newsletter 6:35.

7. National Committee for Clinical Laboratory Standards. 1997. Methods for dilution antimicrobial susceptibility tests for bacteria that grow aerobically. M7-A4, ed 4. NCCLS, Villanova, Pa.

8. National Committee for Clinical Laboratory Standards. 1997. Performance standards for antimicrobial susceptibility testing. M100-S7. NCCLS, Villanova, Pa.

9. National Committee for Clinical Laboratory Standards. 1997. Performance standards for antimicrobial disk susceptibility tests. M2-A6. ed 6. NCCLS, Villanova, Pa.

10. Pankuch, G.A., Jacobs, M.R., Rittenhouse, S.F., et al. 1994. Susceptibilities of 123 strains of *Xanthomonas maltophilia* to eight β-lactams (including β-lactam-β-lactamase inhibititor combinations) and ciprofloxacin tested by five methods. Antimicrob. Agents Chemother. 38:2317.

11. Rahav, G., Simhon, A., Mattan, Y., et al. 1995. Infections with *Chryseomonas luteola* (CDC group Ve-1) and *Flavimonas oryzihabitans* (CDC group Ve-2). Medicine. 74:83.

12. Reed, R.P. 1996. *Flavimonas oryzihabitans* sepsis in children. Clin. Infect. Dis. 22:733.

13. Seifert, H., Strate, A., and Pulverer, G. 1995. Nosocomial bacteremia due *to Acinetobacter baumanii*: clinical features, epidemiology, and predictors of mortality. Medicine. 74:340.

Bibliography

Allen, D.M. and Hartman, B.J. 1995. *Acinetobacter* species. In Mandell, G.L., Bennett, J.E., and Dolin, R., editors. Principles and practice of infectious diseases. Churchill Livingstone, New York.

Balows, A., Truper, H.G., Dworkin, M., et al., editors. 1992. The prokaryotes. A handbook on the biology of bacteria: ecophysiology, isolation, identification, applications. ed 2. Springer-Verlag, New York.

Lyons, R.W. 1985. Ecology, clinical significance, and antimicrobial susceptibility of *Acinetobacter* and *Moraxella*. In Gilardi, G.L., editor. Nonfermentative gram-negative rods: laboratory identification and clinical aspects. Marcel Dekker, New York.

McGowan, J.E., Jr. and Steinberg, J.P. 1995. Other gram-negative bacilli. In Mandell, G.L., Bennett, J.E., and Dolin, R., editors. Principles and practice of infectious diseases. Churchill Livingstone, New York.

Towner, K.J. 1992. The genus *Acinetobacter*. In Balows, A., Truper, H.G., Dworkin, M., et al., editors. The prokaryotes. A handbook on the biology of bacteria: isolation, identification, applications. ed 2. Springer-Verlag, New York.

von Graevenitz, A. 1995. *Acinetobacter, Alcaligenes, Moraxella*, and other nonfermentative gram-negative bacteria. In Murray, P.R., Baron, E.J., Pfaller, M.A., et al., editors. Manual of clinical microbiology, ed 6. American Society for Microbiology, Washington, D.C.

von Graevenitz, A. 1985. Ecology, clinical significance, and antimicrobial susceptibility of infrequently encountered glucose-nonfermenting gram-negative rods. In Gilardi, G.L., editor. Nonfermentative gram-negative rods: laboratory identification and clinical aspects. Marcel Dekker, New York.

Weyant, R.S., Moss, C.W., Weaver, R.E., et al., editors 1996. Identification of unusual pathogenic gram-negative aerobic and facultatively anaerobic bacteria, ed 2. Williams & Wilkins, Baltimore.

37 | *ENTEROBACTERIACEAE*

GENERA AND SPECIES TO BE CONSIDERED

Because of the large number and diversity of genera included in the *Enterobacteriaceae*, it is helpful to consider the bacteria of this family as belonging to one of two major groups. The first group comprises those species that either commonly colonize the human gastrointestinal tract or are most notably associated with human infections (Box 37-1). Although many *Enterobacteriaceae* that cause human infections are part of our normal gastrointestinal flora, there are exceptions, such as the plague bacillus, *Yersinia pestis*. The second group consists of genera that may colonize humans but are rarely associated with human infections or are most commonly recognized as environmental inhabitants or colonizers of other animals (Box 37-2). For this reason, the discovery of these species in clinical specimens should alert laboratorians to possible identification errors, and careful confirmation of both the laboratory results and the clinical significance of such isolates is warranted.

Although the organisms that compose this second group may have substantial significance outside diagnostic microbiology and infectious diseases, this chapter focuses on organisms of the first group, whose clinical significance is well recognized.

GENERAL CHARACTERISTICS

Members of the *Enterobacteriaceae* are all oxidase-negative, glucose-fermenting organisms that grow on MacConkey agar; most also reduce nitrate. More detailed characteristics of the commonly encountered organisms are described later in this chapter in the discussion of laboratory diagnosis.

EPIDEMIOLOGY

Enterobacteriaceae inhabit a wide variety of niches that include the human gastrointestinal tract, the gastrointestinal tract of other animals, and various environmental sites. Some are agents of zoonoses, causing infections in animal populations (Table 37-1). Just as the reservoirs for these organisms vary, so do their modes of transmission to humans.

For those species that normally colonize humans, infections may result when a patient's own bacterial strains (i.e., endogenous strains) establish infections in a normally sterile body site. These organisms can also be passed from one patient to another. Such infections often depend on the debilitated state of a hospitalized patient and are nosocomially acquired. However, this is not always the case. For example, although *Escherichia*

BOX 37-1

GENERA AND SPECIES OF THE FAMILY *ENTEROBACTERIACEAE* THAT COMMONLY COLONIZE HUMANS OR ARE ASSOCIATED WITH HUMAN INFECTIONS

Citrobacter freundii	*Hafnia alvei*	*Serratia liquefaciens* group
Citrobacter (diversus) koseri	*Klebsiella pneumoniae*	*Shigella dysenteriae* (group A)
Citrobacter amalonaticus	*Klebsiella oxytoca*	*Shigella flexneri* (group B)
Edwardsiella tarda	*Klebsiella ozaenae*	*Shigella boydii* (group C)
Enterobacter aerogenes	*Morganella morganii* subsp. *morganii*	*Shigella sonnei* (group D)
Enterobacter cloacae	*Proteus mirabilis*	*Yersinia pestis*
Enterobacter agglomerans group (*Pantoea agglomerans*)	*Proteus vulgaris*	*Yersinia enterocolitica*
Enterobacter gergoviae	*Proteus penneri*	*Yersinia frederiksenii*
Enterobacter sakazakii	*Providencia rettgeri*	*Yersinia intermedia*
Enterobacter amnigenus	*Providencia stuartii*	*Yersinia pseudotuberculosis*
Enterobacter taylorae	*Salmonella*, all serotypes	
Escherichia coli	*Serratia marcescens*	

GENERA AND SPECIES OF THE FAMILY *ENTEROBACTERIACEAE* NOT COMMONLY ASSOCIATED WITH HUMAN INFECTIONS*

BOX 37-2

Budvicia aquatica	*Escherichia hermannii*	*Providencia rustigianii*
Buttiauxella agrestis	*Escherichia vulneris*	*Providencia heimbachae*
Cedecea davisae	*Escherichia blattae*	*Rahnella aquatilis*
Cedecea lapagei	*Ewingella americana*	*Serratia rubidaea*
Cedecea neteri	*Klebsiella ornithinolytica*	*Serratia odorifera*
Citrobacter farmeri	*Klebsiella planticola*	*Serratia plymuthica*
Citrobacter youngae	*Klebsiella rhinoscleromatis*	*Serratia ficaria*
Citrobacter braakii	*Klebsiella terrigena*	*Serratia entomophila*
Citrobacter werkmanii	*Kluyvera ascorbata*	*Serratia proteamaculans* subsp. *quinovora*
Citrobacter sedlakii	*Kluyvera cryocrescens*	*Tatumella pytseos*
Edwardsiella hoshinae	*Leclercia adecarboxylata*	*Trabulsiella* spp.
Edwardsiella ictaluri	*Leminorella grimontii*	*Xenorhabdus* spp.
Enterobacter asburiae	*Leminorella richardii*	*Yersinia kristensenii*
Enterobactaer hormaechei	*Moellerella wisconsensis*	*Yersinia rohdei*
Enterobacter intermedius	*Morganella morganii* subsp. *sibonii*	*Yersinia aldovae*
Enterobacter cancerogenus	*Obesumbacterium* spp.	*Yersinia bercovieri*
Enterobacter dissolvens	*Pantoea dispersa*	*Yersinia mollaretii*
Enterobacter nimipressuralis	*Pragia fontium*	*Yokenella regensburgei*
Erwinia spp.	*Proteus myxofaciens*	
Escherichia fergusonii	*Providencia alcalifaciens*	

*Does not include various CDC enteric groups or DNA groups for which genus and species names have not yet been assigned.

coli is the most common cause of nosocomial infections, it is also the leading cause of urinary tract infections in nonhospitalized patients.

Other species, such as *Salmonella* spp., *Shigella* spp., and *Yersinia enterocolitica,* only inhabit the bowel at the time they are causing infection and are acquired by ingestion of contaminated foods or water. This is also the mode of transmission for the various types of *E. coli* that are known to cause gastrointestinal infections. In contrast, *Yersinia pestis* is unique among the *Enterobacteriaceae* that infect humans. This is the only species that is transmitted from animals by the bite of an insect (i.e., flea) vector.

PATHOGENESIS AND SPECTRUM OF DISEASES

The clinically relevant members of the *Enterobacteriaceae* can be considered as two groups: the opportunistic pathogens and the overt pathogens. *Salmonella typhi,* *Shigella* spp., and *Y. pestis* are among the latter and are the causative agents of typhoid fever, dysentery, and the "black" plague, respectively. Therefore, their discovery in clinical material should always be considered significant. These organisms, as well as other *Salmonella* spp., produce various potent virulence factors and are capable of producing life-threatening infections (Table 37-2).

The opportunistic pathogens most commonly include *Citrobacter* spp., *Enterobacter* spp., *Klebsiella* spp., *Proteus* spp., and *Serratia* spp. Although consid-

ered opportunistic pathogens, these organisms produce significant virulence factors, such as endotoxins, that can mediate fatal infections. However, because they generally do not initiate disease in healthy, uncompromised human hosts, they are considered opportunistic.

Although *E. coli* is a normal bowel inhabitant, its pathogenic classification is somewhere between that of the overt pathogens and opportunistic organisms. Strains of this species, such as enterotoxigenic *E. coli* (ETEC), enteroinvasive *E. coli* (EIEC), and enteroaggregative *E. coli* (EAEC) express potent toxins and cause serious gastrointestinal infections. Additionally, in the case of enterohemorrhagic *E. coli* (EHEC), life-threatening systemic disease can result from infection. Furthermore, as the leading cause of nosocomial infections among *Enterobacteriaceae,* *E. coli* is likely to have greater virulence capabilities than the other species categorized as "opportunistic" *Enterobacteriaceae.*

LABORATORY DIAGNOSIS

SPECIMEN COLLECTION AND TRANSPORT

No special considerations are required for specimen collection and transport of the organisms discussed in this chapter. Refer to Table 1-1 for general information on specimen collection and transport.

SPECIMEN PROCESSING

No special considerations are required for processing of the organisms discussed in this chapter. Refer to

TABLE 37-1   EPIDEMIOLOGY OF CLINICALLY RELEVANT *ENTEROBACTERIACEAE*

ORGANISM	HABITAT (RESERVOIR)	MODE OF TRANSMISSION
Escherichia coli	Normal bowel flora of humans and other animals; may also inhabit female genital tract	Varies with the type of infection. For nongastrointestinal infections, organisms may be endogenous or spread person to person, especially in the hospital setting; for gastrointestinal infections, transmission mode varies with the type of *E. coli* (see Table 37-2), and may involve fecal-oral spread between humans via contaminated food or water or consumption of undercooked beef or milk from colonized cattle
Shigella spp.	Only found in humans at times of infection; not part of normal bowel flora	Person-to-person spread by fecal-oral route, especially in overcrowded areas and areas with poor sanitary conditions
Salmonella typhi and *paratyphi*	Only found in humans but not part of normal bowel flora	Person-to-person spread by fecal-oral route by ingestion of food or water contaminated with human excreta
Other *Salmonella* spp.	Widely disseminated in nature and associated with various animals	Ingestion of contaminated food products processed from animals, frequently of poultry or dairy origin. Direct person-to-person transmission by fecal-oral route can occur in health care settings when hand-washing guidelines are not followed
Edwardsiella tarda	Gastrointestinal tract of cold-blooded animals, such as reptiles	Uncertain; probably by ingestion of contaminated water or close contact with carrier animal
Yersinia pestis	Carried by urban and domestic rats and wild rodents, such as the ground squirrel, rock squirrel, and prairie dog	From rodents to humans by the bite of flea vectors, or by ingestion of contaminated animal tissues; during human epidemics of pneumonic (i.e., respiratory) disease the organism can be spread directly from human to human by inhalation of contaminated airborne droplets; rarely transmitted by handling or inhalation of infected animal tissues or fluids
Yersinia enterocolitica	Dogs, cats, rodents, rabbits, pigs, sheep, and cattle. Not part of normal human flora	Consumption of incompletely cooked food products (especially pork), dairy products such as milk, and, less commonly, by ingestion of contaminated water or by contact with infected animals
Yersinia pseudotuberculosis	Rodents, rabbits, deer, and birds. Not part of normal human flora	Ingestion of organism during contact with infected animal or by contaminated food or water
Citrobacter spp., *Enterobacter* spp., *Klebsiella* spp., *Morganella* spp., *Proteus* spp., *Providencia* spp., and *Serratia* spp.	Normal human gastrointestinal flora	Endogenous, or person-to-person spread, especially in hospitalized patients

TABLE 37-2 PATHOGENESIS AND SPECTRUM OF DISEASES FOR CLINICALLY RELEVANT *ENTEROBACTERIAEACEAE*

ORGANISM	VIRULENCE FACTORS	SPECTRUM OF DISEASE AND INFECTIONS
Escherichia coli (as a cause of extraintestinal infections)	Several, including endotoxin, capsule production, and pili that mediate attachment to host cells	Urinary tract infections, bacteremia, nosocomial infections of various body sites. Most common cause of gram-negative nosocomial infections
Enterotoxigenic *E. coli* (ETEC)	Pili that permit gastrointestinal colonization. Heat-labile (LT) and heat-stable (ST) enterotoxins that mediate secretion of water and electrolytes into the bowel lumen	Traveler's and childhood diarrhea, characterized by profuse, watery stools. Transmitted by contaminated food and water
Enteroinvasive *E. coli* (EIEC)	Virulence factors uncertain, but organism invades enterocytes lining the large intestine in a manner nearly identical to *Shigella* spp.	Dysentery (i.e., necrosis, ulceration, and inflammation of large bowel); usually in young children living in areas of poor sanitation
Enteropathogenic *E. coli* (EPEC)	Bundle-forming pilus, intimin, and other factors that mediate organism attachment to mucosal cells of the small bowel, resulting in changes in cell surface (i.e., loss of microvilli)	Diarrhea in infants in developing, low-income nations; can cause a chronic diarrhea
Enterohemorrhagic *E. coli* (EHEC)	Toxin similar to Shiga toxin produced by *Shigella dysenteriae*. Most frequently associated with certain serotypes, such as *E. coli* O157:H7	Inflammation and bleeding of the mucosa of the large intestine (i.e., hemorrhagic colitis); can also lead to hemolytic uremic syndrome resulting from toxin-mediated damage to kidneys. Transmitted by ingestion of undercooked ground beef or raw milk
Enteroaggregative *E. coli* (EAEC)	Probably involves binding by pili, ST-like, and hemolysin-like toxins; actual pathogenic mechanism not known	Watery diarrhea that, in some cases, can be prolonged. Mode of transmission is not well understood
Shigella spp.	Several factors involved to mediate adherence and invasion of mucosal cells, escape from phagocytic vesicles, intercellular spread, and inflammation. Shiga toxin role in disease is uncertain, but it does have various effects on host cells	Dysentery defined as acute inflammatory colitis and bloody diarrhea characterized by cramps, tenemus, and bloody, mucoid stools. Infections with *S. sonnei* may produce only watery diarrhea

Table 1-1 for general information on specimen processing.

DIRECT DETECTION METHODS

Other than Gram stain of patient specimens, there are no specific procedures for the direct detection of

Enterobacteriaceae. Microscopically the cells of these organisms generally appear as coccobacilli, or straight rods with rounded ends. *Yersinia pestis* resembles a closed safety-pin when it is stained with methylene blue or Wayson stain[3]; this is a key characteristic for rapid diagnosis of plague.

TABLE 37-2 PATHOGENESIS AND SPECTRUM OF DISEASES FOR CLINICALLY RELEVANT *ENTEROBACTERIAECEAE*—CONT'D

ORGANISM	VIRULENCE FACTORS	SPECTRUM OF DISEASE AND INFECTIONS
Salmonella spp.	Several factors serve to protect organisms from stomach acids, promote attachment and phagocytosis by intestinal mucosal cells, allow survival in and destruction of phagocytes, and facilitate dissemination to other tissues	Three general categories of infection: ■ Gastroenteritis and diarrhea caused by a wide variety of serotypes that produce infections limited to the mucosa and submucosa of the gastrointestinal tract. *S. typhimurium* and *S. enteritidis* are the serotypes most commonly associated with *Salmonella* gastroenteritis in the United States ■ Bacteremia and extraintestinal infections occur by spread from the gastrointestinal tract. These infections usually involve *S. choleraesuis* or *S. dublin*, although any serotype may cause bacteremia ■ Enteric fever (typhoid fever, or typhoid) is characterized by prolonged fever and multisystem involvement, including lymph nodes, liver, and spleen. This life-threatening infection is most frequently caused by *S. typhi* or *S. paratyphi* strains
Yersinia pestis	Multiple factors play a role in the pathogenesis of this highly virulent organism. These include ability to adapt for intracellular survival and production of antiphagocytic capsule, exotoxins, endotoxins, coagulase, and fibrinolysin	Two major forms of infection are bubonic plague and pneumonic plague. Bubonic plague is characterized by high fever and painful inflammatory swelling of axilla and groin lymph nodes (i.e., the characteristic bubos); infection rapidly progresses to fulminant bacteremia that is frequently fatal if untreated. Pneumonic plague involves the lungs and is characterized by malaise and pulmonary signs; the respiratory infection can occur as a consequence of bacteremic spread associated with bubonic plague or can be acquired by the airborne route during close contact with other pneumonic plague victims; this form of plague is also rapidly fatal
Yersinia enterocolitica	Various factors allow the organism to attach to and invade the intestinal mucosa and spread to lymphatic tissue	Enterocolitis characterized by fever, diarrhea, and abdominal pain; also can cause acute mesenteric lymphadenitis, which may present clinically as appendicitis (i.e., pseudoappendicular syndrome). Bacteremia can occur with this organism but is uncommon
Y. pseudotuberculosis	Similar to those of *Y. enterocolitica*	Causes similar infections as described for *Y. enterocolitica* but much less common
Citrobacter spp., *Enterobacter* spp., *Klebsiella* spp., *Morganella* spp., *Proteus* spp., *Providencia* spp., and *Serratia* spp.	Several factors, including endotoxins, capsules, adhesion proteins, and resistance to multiple antimicrobial agents	Wide variety of nosocomial infections of respiratory tract, urinary tract, blood, and several other normally sterile sites; most frequently infect hospitalized and seriously debilitated patients

CULTIVATION

Media of choice

All *Enterobacteriaceae* grow well on routine laboratory media, such as 5% sheep blood, chocolate, and MacConkey agars. In addition to these media, selective agars, such as Hektoen enteric (HE) agar, xylose-lysine-deoxycholate (XLD) agar, and *Salmonella-Shigella* (SS) agar are commonly used to cultivate enteric pathogens from gastrointestinal specimens (see Chapter 12 and Table 12-1 for more information on the characteristics and appearance of these media; also see Chapter 27 for more information about laboratory procedures for the diagnosis of bacterial gastrointestinal infections). The broths used in blood culture systems, as well as thioglycollate and brain-heart infusion broths, all support the growth of *Enterobacteriaceae*.

Cefsulodin-irgasan-novobiocin (CIN) agar is a selective medium specifically used for the isolation of *Y. enterocolitica* from gastrointestinal specimens. Similarly, MacConkey-sorbitol agar (described in Table 12-1), is used to differentiate sorbitol-negative *E. coli* 0157:H7 from other types of *E. coli* that are capable of fermenting this sugar alcohol.

Incubation conditions and duration

Under normal circumstances most *Enterobacteriaceae* produce detectable growth on commonly used broth and agar media within 24 hours of inoculation. For isolation, 5% sheep blood and chocolate agars may be incubated at 35° C in carbon dioxide or ambient air. However, MacConkey agar and other selective agars (e.g., SS, HE, XLD) should be incubated only in ambient air. Unlike most other *Enterobacteriaceae*, *Y. pestis* grows best at 25° to 30° C. Colonies of *Y. pestis* are pinpoint at 24 hours but resemble those of other *Enterobacteriaceae* after 48 hours. CIN agar, used for the isolation of *Y. enterocolitica*, should be incubated 48 hours to allow for the development of typical "bulls-eye" colonies (Figure 37-1).

Colonial appearance

Table 37-3 describes the colonial appearance and other distinguishing characteristics (pigment and odor) of the most commonly isolated *Enterobacteriaceae* on MacConkey, HE, and XLD agars (see Figures 12-4, 12-6, and 12-9 for examples). All *Enterobacteriaceae* produce similar growth on blood and chocolate agars; colonies are large, gray, and smooth. Colonies of *Klebsiella* or *Enterobacter* may be mucoid because of their polysaccharide capsule. *Escherichia coli* is often beta-hemolytic on blood agar, but most other genera are nonhemolytic. As a result of motility, *Proteus mirabilis*, *P. penneri*, and *P. vulgaris* "swarm" on blood and chocolate agars. Swarming results in the production of a thin film of growth on the agar surface (Figure 37-2) as the motile organisms spread from the original site of inoculation.

Colonies of *Y. pestis* on 5% sheep blood agar are pinpoint at 24 hours but exhibit a rough, cauliflower appearance at 48 hours. Broth cultures of *Y. pestis* exhibit a characteristic "stalactite pattern" in which clumps of cells adhere to one side of the tube.

Y. enterocolitica produces bull's-eye colonies (dark red or burgundy centers surrounded by a translucent border; see Figure 37-1) on CIN at 48 hours. However, because most *Aeromonas* spp. produce similar colonies on CIN, it is important to perform an oxidase test to verify that the organisms are *Yersinia* spp. (oxidase-negative).

APPROACH TO IDENTIFICATION

All *Enterobacteriaceae* ferment glucose, are oxidase-negative, and, with rare exception, reduce nitrates to nitrites. Further, except for *Shigella dysenteriae* type 1, all commonly isolated *Enterobacteriaceae* are catalase-positive. Practically any commercial identification system can be used to reliably identify the commonly isolated *Enterobacteriaceae*. Depending on the system, results are available within 4 hours or after overnight incubation. The extensive computer databases employed by these systems also include information on unusual biotypes. Therefore conventional biochemical

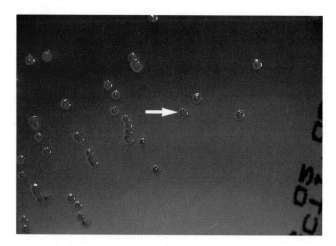

FIGURE 37-1 "Bulls-eye" (*arrow*) colony of *Yersinia enterocolitica* on cefsoludin-irgasan-novobiocin (CIN) agar.

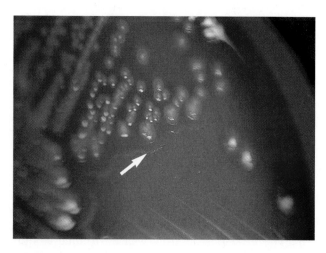

FIGURE 37-2 *Proteus mirabilis* swarming on blood agar (*arrow* at swarming edge).

─────── **TABLE 37-3   COLONY APPEARANCE AND CHARACTERISTICS*** ───────

ORGANISM	MEDIUM	APPEARANCE
Citrobacter spp.	Mac	Late lactose fermenter; therefore, NLF after 24 hours; LF after 48 hours; colonies are light pink after 48 hours
	HE	Colorless
	XLD	Red, yellow, or colorless colonies, with or without black centers (H$_2$S)
Ewardsiella spp.	Mac	NLF
	HE	Colorless
	XLD	Red, yellow, or colorless colonies, with or without black centers (H$_2$S)
Enterobacter spp.	Mac	LF; may be mucoid
	HE	Yellow
	XLD	Yellow
Escherichia coli	Mac	LF; flat, dry, pink colonies with a surrounding darker pink area of precipitated bile salts†
	HE	Yellow
	XLD	Yellow
Hafnia alvei	Mac	NLF
	HE	Colorless
	XLD	Red or yellow
Klebsiella spp.	Mac	LF; mucoid
	HE	Yellow
	XLD	Yellow
Morganella spp.	Mac	NLF
	HE	Colorless
	XLD	Red or colorless
Proteus spp.	Mac	NLF; may swarm depending on the amount of agar in the medium; characteristic foul smell
	HE	Colorless
	XLD	Yellow or colorless, with or without black centers
Providencia spp.	Mac	NLF
	HE	Colorless
	XLD	Yellow or colorless

Continued

TABLE 37-3 COLONY APPEARANCE AND CHARACTERISTICS*—CONT'D

ORGANISM	MEDIUM	APPEARANCE
Salmonella spp.	Mac	NLF
	HE	Green
	XLD	Red with black center
Serratia spp.	Mac	Late LF; *S. marcescens* may be red pigmented, especially if plate is left at 25° C (Figure 37-3)
	HE	Colorless
	XLD	Yellow or colorless
Shigella spp.	Mac	NLF; *S. sonnei* produces flat colonies with jagged edges
	HE	Green
	XLD	Colorless
Yersinia spp.	Mac	NLF; may be colorless to peach
	HE	Salmon
	XLD	Yellow or colorless

Mac, MacConkey agar; *HE,* Hektoen enteric agar; *XLD,* xylose-lysine-deoxycholate agar; *LF,* lactose fermenter; pink colony; *NLF,* nonlactose fermenter; colorless colony.

*Most *Enterobacteriaceae* are indistinguishable on blood agar; see text for colonial description.

†Pink colonies on MacConkey agar with sorbitol are sorbitol fermenters; colorless colonies are nonsorbitol fermenters.

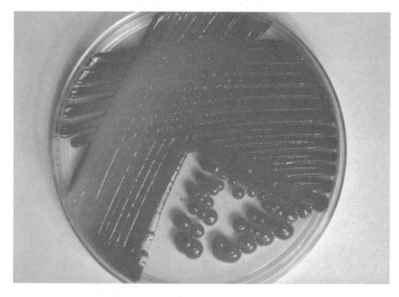

FIGURE 37-3 Red-pigmented *Serratia marcescens* on MacConkey agar.

identification of enterics using multiple tests has become a historical footnote in most clinical laboratories. However, the classical IMViC reactions can still be used to presumptively identify many of the commonly isolated genera.

IMViC profiles

IMViC is an acronym for four tests; *Indole*, *Methyl red*, *Voges-Proskauer*, and *Citrate*. Table 37-4 shows the IMViC reactions for several *Enterobacteriaceae*, and characteristics given in Tables 37-5 through 37-11 can be used to further differentiate between species that share similar IMViC profiles.

In the interests of cost containment, some clinical laboratories may use a positive spot indole test (see Procedure 13-3) to presumptively identify an organism as *E. coli*, provided that the characteristic colonial appearance on MacConkey agar, as described in Table 37-3, is present. A spot indole test can also be used to quickly separate swarming *Proteae*, such as *P. mirabilis* and *P. penneri*, which are negative, from the indole-positive *P. vulgaris*.

Specific considerations for identifying enteric pathogens

Table 37-12 illustrates the use of biochemical profiles obtained with triple sugar iron (TSI) agar and lysine iron agar (LIA) to presumptively identify enteric pathogens (see Chapter 30 for information on the principles, performance, and interpretation of these tests). Organisms exhibiting the profiles shown in Table 37-12 require further biochemical profiling and, in the case of *Salmonella* spp. and *Shigella* spp., serotyping to establish a definitive identification. Bacterial species not considered capable of causing gastrointestinal infections give profiles other than those shown.

In most clinical laboratories, serotyping of *Enterobacteriaceae* is limited to the preliminary grouping of *Salmonella* spp., *Shigella* spp., and *E. coli* 0157:H7. Typing should be performed from a non–sugar-containing medium, such as 5% sheep blood agar or LIA. Use of sugar-containing media, such as MacConkey or TSI agars, can cause the organisms to autoagglutinate.

Commercially available polyvalent antisera designated A, B, C_1, C_2, D, E, F, G, H, I, and Vi are commonly used to preliminarily group *Salmonella* spp. The antisera A through I contain antibodies against somatic ("O") antigens, and the Vi antiserum is prepared against the capsular ("K") antigen of *S. typhi*. Typing is performed using a slide agglutination test. If an isolate agglutinates with the Vi antiserum and does not react with any of the "O" groups, a saline suspension of the organism should be prepared and heated

to 100° C for 10 minutes to inactivate the Vi antigen. The organism should then be retested. *S. typhi* is positive with Vi and group D. Complete typing of *Salmonella* spp., including the use of antisera against the flagellar ("H") antigens, is performed at reference laboratories.

Preliminary serologic grouping of *Shigella* spp. is also performed using commercially available polyvalent somatic ("O") antisera designated *A*, *B*, *C*, and *D*. As with *Salmonella* spp., *Shigella* spp. may produce a capsule, and therefore heating may be required before typing is successful. Subtyping of *Shigella* spp. beyond the groups A, B, and C (*Shigella* group D only has one serotype) usually is performed by reference laboratories.

Plesiomonas shigelloides, a nonmember of the *Enterobacteriaceae* that can cause gastrointestinal infections (see Chapter 35), might cross-react with *Shigella* grouping antisera, particularly group D, and lead to misidentification. This mistake can be readily avoided by performing an oxidase test.

Sorbitol-negative *E. coli* can be serotyped using commercially available antisera to determine whether the somatic "O" antigen 157 and the flagellar "H" antigen 7 are present. Laboratory tests to identify other *E. coli* that cause gastrointestinal infections usually involve animal or tissue culture studies performed in reference laboratories.

SERODIAGNOSIS

Serodiagnostic techniques are used for only two members of the family *Enterobacteriaceae*, that is, *S. typhi* and *Y. pestis*. Agglutinating antibodies can be measured in the diagnosis of typhoid fever; a serologic test for *S. typhi* is part of the "febrile agglutinins" panel and is individually known as the **Widal test**. Because results obtained by using the Widal test are somewhat unreliable, this method is no longer widely used.

Serologic diagnosis of plague is possible using either a passive hemagglutination or ELISA test; these tests are usually performed in reference laboratories.

ANTIMICROBIAL SUSCEPTIBILITY TESTING AND THERAPY

For many of the gastrointestinal infections caused by *Enterobacteriaceae*, the inclusion of antimicrobial agents as part of the therapeutic strategy is controversial or at least uncertain (Table 37-13).

For extraintestinal infections, antimicrobial therapy is a vital component of patient management (Table 37-14). Although a broad spectrum of agents may be used for therapy against *Enterobacteriaceae*

Text continued on p. 526

TABLE 37-4 IMViC REACTIONS OF MOST COMMONLY ISOLATED MEMBERS OF THE FAMILY *ENTEROBACTERIACEAE*

ORGANISM	INDOLE	METHYL RED	VOGES-PROSKAUER	CITRATE (SIMMONS)
Citrobacte freundii	v	+	−	v
Citrobacter (diversus) koseri	+	+	−	+
Edwardsiella tarda	+	+	−	−
Enterobacter aerogenes	−	−	+	+
Enterobacter cloacae	−	−	+	+
Enterobacter agglomerans group	v	v	v	v
Escherichia coli	+	+	−	−
Hafnia alvei	−	v	v	−
Klebsiella pneumoniae	−	−	+	+
Klebsiella oxytoca	+	v	+	+
Klebsiella ozaenae	−	+	−	v
Morganella morganii subsp. *morganii*	+	+	−	−
Proteus mirabilis	−	+	v	v
Proteus vulgaris	+	+	−	v
Proteus penneri	−	+	−	−
Providencia rettgeri	+	+	−	+
Providencia stuartii	+	+	−	+
Salmonella spp. (most serotypes)	−	+	−	+
Serratia marcescens	−	v	+	+
Serratia liquefaciens group	−	+	+	+
Shigella groups A, B, and C	v	+	−	−
Shigella group D	−	+	−	−
Yersinia enterocolitica	v	+	−	−
Yersinia frederiksenii	+	+	−	v
Yersinia intermedia	+	+	−	−

+, >90% of strains positive; −, > 90% of strains negative; *v*, variable; *IMViC*, indole, methyl red, Voges-Proskauer, and citrate.

TABLE 37-5   SUPPLEMENTAL TESTS FOR IDENTIFICATION OF ORGANISMS WITH SPECIFIC IMViC PROFILES

Organisms with I M V C + + − −	H₂S PRODUCTION (IN TSI)	UREASE	PHENYLALANINE DEAMINASE	SUCROSE FERMENTATION	LYSINE DECARBOXYLASE
Citrobacter freundii	v	v	−	v	−
Edwardsiella tarda	+	−	−	−	+
Enterobacter agglomerans group*	−	v	v	v	−
Escherichia coli	−	−	−	v	+
Morganella morganii subsp. *morganii*	v	+	+	−	−
Proteus vulgaris	+	+	+	+	−
Shigella groups A, B, and C	−	−	−	−	−
Yersinia enterocolitica, frederiksenii, and *intermedia*	−	v	−	+	−

+, >90% of strains positive; −, > 90% of strains negative; *v*, variable; *I*, indole; *M*, methyl red; *V*, Voges-Proskauer; *C*, citrate; *TSI*, triple sugar iron agar.
*Because of the biochemical variability of this organism, all possible combinations of IMViC recations can occur.

TABLE 37-6   SUPPLEMENTAL TESTS FOR IDENTIFICATION OF ORGANISMS WITH SPECIFIC IMViC PROFILES

Organisms with I M V C − − + +	LYSINE DECARBOXYLASE	ARGININE DIHYDROLASE	ORNITHINE DECARBOXYLASE	LACTOSE FERMENTATION
Enterobacter aerogenes	+	−	+	+
Enterobacter agglomerans group	−	−	−	v
Enterobacter cloacae	−	+	+	+
Klebsiella pneumoniae	+	−	−	+
Serratia marcescens	+	−	+	−

+, >90% of strains positive; −, >90% of strains negative; *v*, variable; *I*, indole; *M*, methyl red; *V*, Voges-Proskauer; *C*, citrate.

	TABLE 37-7 SUPPLEMENTAL TESTS FOR IDENTIFICATION OF ORGANISMS WITH SPECIFIC IMViC PROFILES			

Organisms with I M V C − + − +	H₂S PRODUCTION (IN TSI)	PHENYLALANINE DEAMINASE	ORNITHINE DECARBOXYLASE	SALACIN FERMENTATION
Citrobacter freundii	v	−	−	−
Enterobacter agglomerans group	−	v	−	v
Klebsiella ozaenae	−	−	−	+
Proteus mirabilis	+	+	+	−
Salmonella spp. (most serotypes)	+	−	+	−

+, >90% of strains positive; −, >90% of strains negative; *v*, variable; *I*, indole; *M*, methyl red; *V*, Voges-Proskauer; *C*, citrate; *TSI*, triple sugar iron agar.

TABLE 37-8   SUPPLEMENTAL TESTS FOR IDENTIFICATION OF ORGANISMS WITH SPECIFIC IMVIC PROFILES

Organisms with IMViC	PHENYLALANINE DEAMINASE	LYSINE DECARBOXYLASE	ORNITHINE DECARBOXYLASE	SALACIN FERMENTATION	L-RHAMNOSE FERMENTATION
Citrobacter freundii (− + − −)	−	−	−	−	+
Enterobacter agglomerans group*	v	−	−	v	v
*Hafnia alvei**	−	+	+	v	+
Klebsiella ozaenae	−	v	−	+	v
*Proteus mirabilis**	+	−	+	−	−
Proteus penneri	+	−	−	−	−
Shigella groups A, B, and C	−	−	−	−	−
Shigella group D	−	−	+	−	v
Yersinia enterocolitica	−	−	+	v	−

+, >90% of strains positive; −, >90% of strains negative; *v*, variable; *I*, indole; *M*, methyl red; *V*, Voges-Proskauer; *C*, citrate.

*These three organisms may also give − − + + IMViC reactions.

TABLE 37-9 SUPPLEMENTAL TESTS FOR IDENTIFICATION OF ORGANISMS WITH SPECIFIC IMViC PROFILES

Organisms with I M V C + + − +	H₂S PRODUCTION (IN TSI)	PHENYLALANINE DEAMINASE	ORNITHINE DECARBOXYLASE	D-MANNITOL FERMENTATION
Citrobacter freundii	v	−	−	+
Citrobacter (diversus) koseri	−	−	+	+
Enterobacter agglomerans group	−	v	−	+
Proteus vulgaris	+	+	−	−
Providencia rettgeri	−	+	−	+
Providencia stuartii	−	+	−	−
Yersinia fredericksenii	−	−	+	+

+, >90% of strains positive; −, >90% of strains negative; *v*, variable; *I*, indole; *M*, methyl red; *V*, Voges-Proskauer; *C*, citrate; *TSI*, Triple sugar iron agar.

TABLE 37-10 SUPPLEMENTAL TESTS FOR IDENTIFICATION OF ORGANISMS WITH SPECIFIC IMViC PROFILES

Organisms with I M V C − + + +	H₂S PRODUCTION (IN TSI)	UREASE	PHENYLALANINE DEAMINASE	LYSINE DECARBOXYLASE	L-ARABINOSE FERMENTATION
Enterobacter agglomerans group	−	v	v	−	+
Proteus mirabilis	+	+	+	−	−
Serratia liquefaciens group	−	−	−	+	+
Serratia marcescens	−	v	−	+	−

+, >90% of strains positive; −, >90% of strains negative; *v*, variable; *I*, indole; *M*, methyl red; *V*, Voges-Proskauer; *C*, citrate; *TSI*, triple sugar iron agar.

TABLE 37-11 SUPPLEMENTAL TESTS FOR IDENTIFICATION OF ORGANISMS WITH SPECIFIC IMViC PROFILES

Organisms with I M V C + + + + *or* + − + +	LYSINE DECARBOXYLASE
Enterobacter agglomerans group	−
Klebsiella oxytoca	+

+, >90% of strains positive; −, >90% of strains negative; *v*, variable; *I*, indole; *M*, methyl red; *V*, Voges-Proskauer; *C*, citrate.

TABLE 37-12 TSI and LIA Reactions Used to Screen for Enteropathogenic *Enterobacteriaceae**†

TSI REACTIONS‡	LIA REACTIONS‡	POSSIBLE IDENTIFICATION
K/Ⓐ *or* K/A H₂S +	K/K *or* K/NC H₂S +	*Salmonella* spp. *Edwardsiella* spp.
K/A H₂S +	K/K *or* K/NC	*Salmonella* spp. (rare)
K/Ⓐ	K/K *or* K/NC H₂S +	*Salmonella* spp. (rare)
K/A	K/K *or* K/NC H₂S +	*Salmonella typhi* (rare)
K/Ⓐ	K/K *or* K/NC	*Salmonella* spp. (rare)
K/Ⓐ	K/A H₂S +	*Salmonella paratyphi* A (usually H₂S−)
K/Ⓐ	K/A *or* A/A	*Escherichia coli* *Salmonella paratyphi* A *Shigella flexneri* 6 (uncommon) *Aeromonas* spp. (oxidase-positive)
K/A	K/K *or* K/NC	*Plesiomonas* sp. (oxidase-positive) *Salmonella typhi* (rare) *Vibrio* spp. (oxidase-positive)
K/A	K/A *or* A/A	*Escherichia coli* *Shigella* groups A-D *Yersinia* spp.
A/Ⓐ H₂S +	K/K *or* K/NC H₂S +	*Salmonella* spp. (rare)
A/Ⓐ	K/A *or* A/A	*Escherichia coli* (rare)
A/A	K/A *or* A/A	*Escherichia coli* *Yersinia* spp. *Aeromonas* spp. (oxidase-positive) *Vibrio cholerae* (rare, oxidase-positive)
A/A	K/K *or* K/NC	*Vibrio* spp. (oxidase-positive)

K, alkaline; *A*, acid; *H₂S*, hydrogen sulfide; Ⓐ, acid and gas production; *NC*, no change; *TSI*, triple sugar iron agar; *LIA*, lysine iron agar.

* *Vibrio* spp., *Aeromonas* spp., and *Pleisiomonas* spp. are included in this table because they grow on the same media as the *Enterobacteriaceae* and may be enteric pathogens; identification of these organisms is discussed in Chapter 35.

†TSI and LIA reactions described in this table are only screening tests. The identity of possible enteric pathogens must be confirmed by specific biochemical and serologic testing.

‡Details regarding the TSI and LIA procedures can be found in Chapter 30.

TABLE 37-13 THERAPY FOR GASTROINTESTINAL INFECTIONS CAUSED BY *ENTEROBACTERIACEAE*

ORGANISMS	THERAPEUTIC STRATEGIES
Enterotoxigenic *E. coli* (ETEC) Enteroinvasive *E. coli* (EIEC) Enteropathogenic *E. coli* (EPEC) Enterohemorrhagic *E. coli* (EHEC) Enteroaggregative *E. coli* (EAEC)	Supportive therapy, such as oral rehydration, is indicated in cases of severe diarrhea; for life-threatening infections, such as hemolytic uremic syndrome associated with EHEC, transfusion and hemodialysis may be necessary. Antimicrobial therapy may shorten duration of gastrointestinal illness, but many of these infections will resolve without such therapy. Because these organisms may develop resistance (see Table 37-14), antimicrobial drug therapy for non–life-threatening infections may be contraindicated
Shigella spp.	Oral rehydration; antimicrobial drug therapy may be used to shorten the period of fecal excretion and perhaps limit the clinical course of the infection. However, because of the risk of resistance, using antimicrobial drug therapy for less serious infections may be questioned
Salmonella spp.	For enteric fevers (e.g., typhoid fever) and extraintestinal infections (e.g., bacteremia, etc.) antimicrobial agents play an important role in therapy. Potentially effective agents for typhoid include quinolones, chloramphenicol, trimethoprim/sulfamethoxazole, and advance-generation cephalosporins, such as ceftriaxone; however, first- and second-generation cephalosporins and aminoglycosides are not effective. For nontyphoidal *Salmonella* bacteremia, a third-generation cephalosporin (e.g., ceftriaxone) is frequently used. For gastroenteritis, replacement of fluids is most important. Antimicrobial therapy generally is not recommended for either treatment of the clinical infection or decreasing the time that a patient excretes the organism
Yersinia enterocolitica and *Yersinia pseudotuberculosis*	The need for antimicrobial therapy for enterocolitis and mesenteric lymphadenitis is not clear. In cases of bacteremia, piperacillin, third-generation cephalosporins, aminoglycosides, and trimethoprim/sulfamethoxazole are potentially effective agents. *Y. enterocolitica* is frequently resistant to ampicillin and first-generation cephalosporins, whereas *Y. pseudotuberculosis* isolates are generally susceptible

TABLE 37-14   ANTIMICROBIAL THERAPY AND SUSCEPTIBILITY TESTING
OF CLINICALLY RELEVANT *ENTEROBACTERIACEAE*

ORGANISM	THERAPEUTIC OPTIONS	POTENTIAL RESISTANCE TO THERAPEUTIC OPTIONS	TESTING METHODS*	COMMENTS
E. coli, Citrobacter spp., *Enterobacter* spp., *Morganella* spp., *Proteus* spp., *Providencia* spp., and *Serratia* spp.	Several agents from each major class of antimicrobials, including aminoglycosides, beta-lactams, and quinolones have activity. See Table 19-2 for listing of specific agents that should be selected for in vitro testing. For urinary tract infections, single agents may be used; for systemic infections, potent beta-lactams are used, frequently in combination with an aminoglycoside	Yes; every species is capable of expressing resistance to one or more antimicrobials belonging to each drug class	As documented in Chapter 18; disk diffusion, broth dilution, agar dilution, and commercial systems	In vitro susceptibility testing results are important for guiding therapy
Yersinia pestis	Streptomycin is the therapy of choice; tetracycline or chloramphenicol are effective alternatives	Yes, but rare	Not available	Manipulation of cultures for susceptibility testing is dangerous for laboratory personnel and is not necessary

*Validated testing methods include those standard methods recommended by the National Committee for Clinical Laboratory Standards (NCCLS) and those commercial methods approved by the Food and Drug Administration (FDA).

(see Table 19-1 for a detailed listing), every clinically relevant species is capable of acquiring and using one or more of the resistance mechanisms presented in Table 17-3 and Figure 17-10. The unpredictable nature of any clinical isolate's antimicrobial susceptibility requires that testing be done as a guide to therapy. As discussed in Chapter 18, several standard methods and commercial systems have been developed for this purpose.

PREVENTION

Vaccines are available for typhoid fever and bubonic plague; however, neither is routinely recommended in the United States. An oral, multiple-dose vaccine prepared against *S. typhi* strain Ty2la or a parenteral single-dose vaccine containing Vi antigen is available for people traveling to an endemic area or for household contacts of a documented *S. typhi* carrier.[2]

An inactivated multiple-dose, whole-cell bacterial vaccine is available for bubonic plague for people traveling to an endemic area. However, this vaccine will not provide protection against pneumonic plague.[2] Individuals exposed to pneumonic plague should be given chemoprophylaxis with tetracycline, sulfonamide, or chloramphenicol.[1]

References

1. Benenson, A.S., editor. 1995. Control of communicable diseases manual, ed 16. American Public Health Association, Washington, D.C.
2. Committee on Infectious Diseases. 1997. 1997 Red book: report of the committee on infectious diseases, ed 24. American Academy of Pediatrics, Elk Grove Village, Ill.
3. Quan, T.J. 1987. Plague. In Wentworth, B.B., editor. Diagnostic procedures for bacteria infections, ed 7. American Public Health Association, Washington, D.C.

Bibliography

Balows, A., Truper, H.G., Dworkin, M., et al., editors. 1992. The prokaryotes. A handbook on the biology of bacteria: ecophysiology, isolation, identification, applications, ed 2. Springer-Verlag, New York.

Butler, T. 1995. *Yersinia* species (including plague). In Mandell, G.L., Bennett, J.E., and Dolin, R., editors. Principles and practice of infectious diseases. Churchill Livingstone, New York.

DuPont, H.L. 1995. *Shigella* species (bacillary dysentery). In Mandell, G.L., Bennett, J.E., and Dolin, R., editors. Principles and practice of infectious diseases. Churchill Livingstone, New York.

Eisenstein, B.I. 1995. *Enterobacteriaceae*. In Mandell, G.L., Bennett, J.E., and Dolin, R., editors. Principles and practice of infectious diseases. Churchill Livingstone, New York.

Farmer, J.J., III. 1995. *Enterobacteriaceae*: introduction and identification. In Murray, P.R., Baron, E.J., Pfaller, M.A., et al., editors. Manual of clinical microbiology, ed 6. American Society for Microbiology, Washington, D.C.

Farmer, J.J., III. 1995. *Enterobacteriaceae*: introduction and identification. In Murray, P.R., Baron, E.J., Pfaller, M.A., et al. editors. Manual of clinical microbiology, ed 6. American Society for Microbiology, Washington, D.C.

Gilchrist, M.J.R. 1995. *Enterobacteriaceae*: opportunistic pathogens and other genera. In Murray, P.R., Baron, E.J., Pfaller, M.A., et al., editors. Manual of clinical microbiology, ed 6. American Society for Microbiology, Washington, D.C.

Gray, L.D. 1995. *Escherichia, Salmonella, Shigella,* and *Yersinia.* In Murray, P.R., Baron, E.J., Pfaller, M.A., et al., editors. Manual of clinical microbiology, ed 6. American Society for Microbiology, Washington, D.C.

Miller, S.I., Hohmann, E.L, Pegues, D.A. 1995. *Salmonella* (including *Salmonella typhi*). In Mandell, G.L., Bennett, J.E., and Dolin, R., editors. Principles and practice of infectious diseases. Churchill Livingstone, New York.

Ryan, K.J., editor 1994. Sherris medical microbiology: an introduction to infectious diseases. Appleton and Lange, Norwalk, Conn.

Salyers, A.A. and Whitt, D.D. 1994. Bacterial pathogenesis: a molecular approach. ASM Press. Washington, D.C.

Schaechter, M., Medoff, G., and Eisenstein, B.I., editors. 1993. Mechanisms of microbial disease, ed 2. Williams & Wilkins, Baltimore.

Section Three

Gram-Negative Bacilli and Coccobacilli (MacConkey-Negative, Oxidase-Positive)

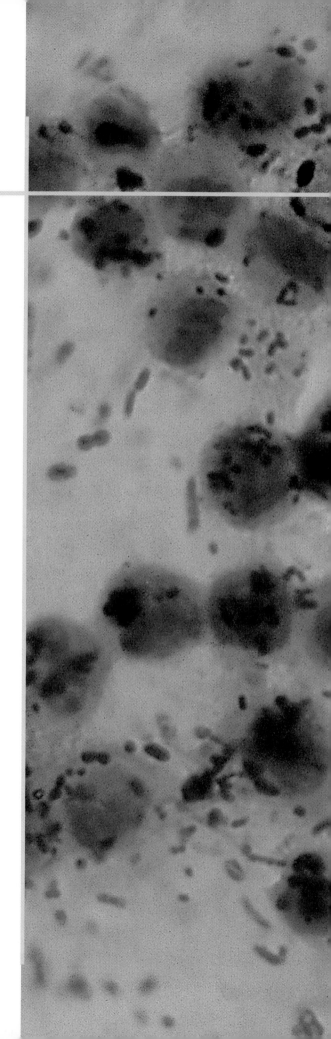

38 | SPHINGOMONAS PAUCIMOBILIS AND SIMILAR ORGANISMS

Organisms to be Considered

Current Name	Previous Name
■ *Sphingomonas paucimobilis*	*Pseudomonas paucimobilis*, CDC group IIK-1
■ CDC group IIe	
■ CDC group IIh	
■ CDC group IIi	
■ CDC group O-1	
■ CDC group O-2	

GENERAL CHARACTERISTICS

The organisms discussed in this chapter are considered together because they usually fail to grow on MacConkey agar, are oxidase-positive, and oxidatively utilize glucose.

EPIDEMIOLOGY, SPECTRUM OF DISEASE, AND ANTIMICROBIAL THERAPY

As demonstrated in Table 38-1, these organisms are rarely isolated from human materials and they have an extremely limited role as agents of infection. Because they are rarely encountered in the clinical setting, little information exists regarding their epidemiology, capacity to cause human infections, and potential for antimicrobial resistance.[1-3] For example, even though the O-1 and O-2 organisms have been submitted to CDC after being isolated from clinical materials such as blood, cerebrospinal fluid, wounds, and pleural fluid, their natural habitat is unknown. When these organisms are encountered in clinical specimens, their clinical significance and potential as contaminants should be seriously considered.

LABORATORY DIAGNOSIS

SPECIMEN COLLECTION AND TRANSPORT

No special considerations are required for specimen collection and transport of the organisms discussed in this chapter. Refer to Table 1-1 for general information on specimen collection and transport.

SPECIMEN PROCESSING

No special considerations are required for processing of the organisms discussed in this chapter. Refer to Table 1-1 for general information on specimen processing.

DIRECT DETECTION METHODS

There are no specific procedures for the direct detection of these organisms in clinical material. Microscopically, CDC groups IIe, IIh, IIi, and O-2 are all short, straight rods that may appear as "II forms." The "II forms" are bacteria with thickened ends and thin centers. As discussed in Chapter 33, *Sphingobacterium mizutaii, S. thalpophilum,* and *Chryseobacterium* spp. exhibit "II forms," as well. *Sphingomonas paucimobilis* is a medium-size, straight rod.

CULTIVATION

Media of choice

Sphingomonas spp. and all CDC groups considered in this chapter grow well on routine laboratory media, such as 5% sheep blood and chocolate agars, but most fail to grow on MacConkey agar. They grow well in thioglycollate and brain-heart infusion broths and in broths used in blood culture systems.

Incubation conditions and duration

Within 24 hours of inoculation these organisms produce detectable growth on media incubated at 35° C in carbon dioxide or in ambient air.

Colonial appearance

Table 38-2 describes the colonial appearance and other distinguishing characteristics (e.g., pigment) of each organism on 5% sheep blood agar. When these organisms (i.e., *Sphingomonas paucimobilis* and CDC groups O-1, O-2, and IIe) do grow on MacConkey agar, they appear as nonlactose fermenters.

APPROACH TO IDENTIFICATION

The ability of most commercial identification systems to accurately identify the organisms discussed in this chapter is limited or uncertain. Table 38-3 shows some conventional biochemical tests that can be used to presumptively differentiate between the various organisms.

Comments regarding specific organisms

Strains of CDC groups IIe and IIh are similar to *Empedobacter brevis* (see Chapter 33) in that they oxidize glucose and maltose and produce indole. CDC group IIi resembles *Sphingobacterium multivorum*, but IIi produces indole. Similarly, *Sphingomonas*

TABLE 38-1 EPIDEMIOLOGY, DISEASE SPECTRUM, AND ANTIMICROBIAL THERAPY

ORGANISM	EPIDEMIOLOGY	DISEASE SPECTRUM	ANTIMICROBIAL THERAPY
Sphingomonas paucimobilis	Exists in environmental niches, such as water, including hospital water systems. Not part of human flora. Mode of transmission is uncertain, probably involves patient exposure to contaminated medical devices or solutions	Virulence factors are unknown. Infections include catheter-related bacteremia, wound and urinary tract infections	No definitive guidelines; potentially active agents include; trimethoprim/sulfamethoxazole, chloramphenicol, ciprofloxacin, and aminoglycosides; resistance to beta-lactams is known, but validated susceptibility testing methods do not exist
CDC group IIe, CDC group IIh, and CDC group IIi	Soil, plants, foodstuffs, and water, including moist areas in hospitals. Not part of human flora	Rarely found in clinical material and not substantiated as cause of human infections	No guidelines; little is known about antimicrobial resistance potential
CDC group O-1 and O-2	Unknown	Rarely found in clinical material and not implicated as cause of human infections	No guidelines; nothing is known about antimicrobial resistance potential

TABLE 38-2 COLONY APPEARANCE AND CHARACTERITICS

ORGANISM	MEDIUM	APPEARANCE
CDC group IIe	BA	No distinctive appearance
CDC group IIh	BA	No distinctive appearance
CDC group IIi	BA	No distinctive appearance
CDC group O-1	BA	Yellow pigment
CDC group O-2	BA	Yellow pigment
Sphingomonas spp.	BA	Small, circular, smooth, convex, bright yellow growth pigment

BA, 5% sheep blood agar.

───── **TABLE 38-3 KEY BIOCHEMICAL AND PHYSIOLOGIC CHARACTERISTICS** ─────

ORGANISM	OXIDIZES XYLOSE	OXIDIZES SUCROSE	ESCULIN HYDROLYSIS	MOTILITY	YELLOW INSOLUBLE PIGMENT
CDC group IIe	−	−	−	−	−
CDC group IIh	−	−	+	−	−
CDC group IIi	+	+	+	−	v
Sphingomonas spp.*	+	+	+	+†	+
CDC group O-1	−	−	+	+	+
CDC group O-2	−	+	v	v‡	+

+, >90% strains positive; −, >90% strains negative; *v*, variable.

*Includes *Sphingomonas paucimobilis* and *S. parapaucimobilis*.

†Usually nonmotile in motility medium, but motility is present in wet mount.

‡Only 20% are motile. Motility is only apparent upon wet mount or flagellar staining.

From Weyant, R.S., Moss, C.W., Weaver, R.E., et al., editors. 1996. Identification of unusual pathogenic gram-negative aerobic and facultatively anaerobic bacteria, ed 2. Williams & Wilkins, Baltimore.

paucimobilis resembles CDC group O-1 but oxidizes more carbohydrates. CDC group O-2 will not oxidize xylose, mannitol, or lactose; this can help separate it from the other bright yellow–pigmented organisms discussed in this chapter.

SERODIAGNOSIS

Serodiagnostic techniques are not generally used for the laboratory diagnosis of infections caused by the organisms discussed in this chapter.

PREVENTION

Because these organisms are rarely implicated in human infections, vaccines or prophylactic measures are not necessary.

References

1. Lemaitre, D., Elaichouni, A., Hundhausen, M., et al. 1996. Tracheal colonization with *Sphingomonas paucimobilis* in mechanically ventilated neonates due to contaminated ventilator temperature probes. J. Hosp. Infect. 32:199.
2. Reina, J., Bassa, A., Llompart, I., et al. 1990. Infections with *Pseudomonas paucimobilis*: report of four cases and review. Rev. Infect. Dis. 13:1072.
3. Salazar, R., Martino, R., Suredo, A., et al. 1995. Catheter-related bacteremia due to *Pseudomonas paucimobilis* in neutropenic cancer patients: report of two cases. Clin. Infect. Dis. 20:1573.

Bibliography

von Graevenitz, A. 1995. *Acinetobacter, Alcaligenes, Moraxella,* and other nonfermentative gram-negative bacteria. In Murray, P.R., Baron, E.J., Pfaller, M.A., et al., editors. Manual of clinical microbiology, ed 6. American Society for Microbiology, Washington, D.C.

Weyant, R.S., Moss, C.W., Weaver, R.E., et al., editors. 1996. Identification of unusual pathogenic gram-negative aerobic and facultatively anaerobic bacteria. ed 2. Williams & Wilkins, Baltimore.

39 | *MORAXELLA* AND ELONGATED *NEISSERIA*

Genera and Species to be Considered

Current Name	Previous Name
■ *Moraxella nonliquefaciens*	
■ *Moraxella osloensis*	
■ *Moraxella lacunata*	
■ *Moraxella canis*	
■ *Moraxella atlantae*	
■ *Moraxella lincolnii*	
■ *Neisseria elongata*	CDC group M6
■ *Neisseria weaveri*	CDC group M5

GENERAL CHARACTERISTICS

The organisms discussed in this chapter are either coccobacilli or short to medium-sized, gram-negative rods. Subinhibitory concentrations of penicillin, such as occurs in the presence of a 10-unit penicillin disk, cause the coccoid forms of these bacteria to elongate to a bacillus morphology. In contrast, true cocci, such as most *Neisseria* spp. and *Moraxella (Branhamella) catarrhalis*, with which these organisms may be confused, maintain their original coccus shape in the presence of penicillin. In addition, the organisms discussed in this chapter do not utilize glucose. Specific morphologic and physiologic features are presented later in this chapter in the discussion of laboratory diagnosis.

EPIDEMIOLOGY, SPECTRUM OF DISEASE, AND ANTIMICROBIAL THERAPY

As normal inhabitants of mucosal surfaces, infections caused by *Moraxella* spp. and *Neisseria elongata* probably are not acquired by transfer of the organisms between persons. Infection most likely results when a breakdown of the patient's mucosal or epidermal defensive barriers allows subsequent invasion of sterile sites by an organism that is part of the patient's normal flora (i.e., an endogenous strain; Table 39-1). The fact that these organisms rarely cause infection indicates that they have low virulence.[1-3] Whenever they are encountered in clinical specimens, the possibility that they are contaminants should be seriously considered (Table 39-2). This is especially the case when the specimen source may have come in contact with a mucosal surface.

Moraxella catarrhalis is the species most commonly associated with human infections, primarily of the respiratory tract. However, because the cellular morphology of this species is more similar to that of *Neisseria* spp. than that of the other *Moraxella* spp., details of this organism's characteristics are discussed in Chapter 51.

The rarity with which these organisms are encountered as the cause of infection, and the lack of validated in vitro susceptibility testing methods, does not allow definitive treatment guidelines to be given (Table 39-3). Although many of these organisms may grow on the media and under the conditions recommended for testing other bacteria (see Chapter 18 for more information regarding validated testing methods), this does not necessarily mean that interpretable and reliable results will be produced. Chapter 19 should be reviewed for preferable strategies that can be used to provide susceptibility information when validated testing methods do not exist for a clinically important bacterial isolate.

In general, beta-lactam antibiotics are thought to be effective against these species. However, some evidence suggests that beta-lactamase–mediated resistance may be capable of spreading among *Moraxella* spp.[4]

TABLE 39-1 EPIDEMIOLOGY

ORGANISMS	HABITAT (RESERVOIR)	MODE OF TRANSMISSION
Moraxella nonliquefaciens, *Moraxella lacunata,* *Moraxella osloensis,* *Moraxella lincolnii,* *Moraxella canis,* and *Moraxella atlantae*	Normal human flora that inhabit mucous membranes covering the nose, throat, other parts of the upper respiratory tract, conjunctiva, and, for some species (i.e., *M. osloensis*), the urogenital tract. May also colonize the skin	Infections are rare. When they occur, they are probably caused by the patient's endogenous strains. Person-to-person transmission may be possible, but this has not been documented
Neisseria elongata	Normal flora of upper respiratory tract	When infections occur, they are probably caused by the patient's endogenous strains
Neisseria weaveri	Oral flora of dogs	Dog bite

TABLE 39-2 PATHOGENESIS AND SPECTRUM OF DISEASES

ORGANISMS	VIRULENCE FACTORS	SPECTRUM OF DISEASE AND INFECTIONS
Moraxella nonliquefaciens, *Moraxella lacunata,* *Moraxella osloensis,* *Moraxella lincolnii,* *Moraxella canis,* and *Moraxella atlantae*	Unknown. Because they are rarely associated with infections, they are considered opportunistic organisms of low virulence	*M. lacunata* has historically been associated with eye infections, but these infections also may be caused by other *Moraxella* spp. Other infections include bacteremia, endocarditis, septic arthritis, and, possibly, respiratory infections
Neisseria elongata	Unknown. An opportunistic organism of low virulence	Rarely implicated in infections. Has been documented as a cause of bacteremia, endocarditis, and osteomyelitis
Neisseria weaveri	Unknown	Infections of dog bite wounds

LABORATORY DIAGNOSIS

SPECIMEN COLLECTION AND TRANSPORT

No special considerations are required for specimen collection and transport of these organisms discussed in this chapter. Refer to Table 1-1 for general information on specimen collection and transport.

SPECIMEN PROCESSING

No special considerations are required for processing of the organisms discussed in this chapter. Refer to Table 1-1 for general information on specimen processing.

DIRECT DETECTION METHODS

Other than Gram stain of patient specimens, there are no specific procedures for the direct detection of these organisms in clinical material. *M. atlantae*, *M. nonliquefaciens,* and *M. osloensis* may appear as either coccobacilli or as short, broad rods that tend to resist decolorization and may appear gram-variable. This is also true for *M. canis,* which appears as cocci in pairs or short chains. *M. lacunata* is a coccobacillus or medium-sized rod, and *M. lincolnii* is a coccobacilli that may appear in chains. All subspecies of *Neisseria elongata* are either coccobacilli or short, straight rods and *N. weaveri* is a medium-length, straight bacillus.

CULTIVATION

Media of choice

Moraxella spp. and the elongated *Neisseria* spp. grow well on 5% sheep blood and chocolate agars.

TABLE 39-3 ANTIMICROBIAL THERAPY AND SUSCEPTIBILITY TESTING

ORGANISMS	THERAPEUTIC OPTIONS	POTENTIAL RESISTANCE TO THERAPEUTIC OPTIONS	VALIDATED TESTING METHODS*
Moraxella spp.	No definitive guidelines. Generally susceptible to penicillins and cephalosporins	May acquire beta-lactamase–mediated resistance to penicillins	Not available
Neisseria elongata and *Neisseria weaveri*	No definitive guidelines. Generally susceptible to penicillins and cephalosporins	None known	Not available

*Validated testing methods include those standard methods recommended by the National Committee for Clinical Laboratory Standards (NCCLS) and those commercial methods approved by the Food and Drug Administration (FDA).

Most strains grow slowly on MacConkey agar and resemble the nonlactose-fermenting *Enterobacteriaceae*. Both genera also grow well in the broth of commercial blood culture systems and in common nutrient broths, such as thioglycolate and brain-heart infusion.

Incubation conditions and duration
5% sheep blood and chocolate agars should be incubated at 35° C in carbon dioxide or ambient air for a minimum of 48 hours. For those species that may grow on MacConkey agar, the medium should be incubated at 35° C in ambient air.

Colonial appearance
Table 39-4 describes the colonial appearance and other distinguishing characteristics (e.g., pitting) of each species on 5% sheep blood and MacConkey agars.

APPROACH TO IDENTIFICATION
The ability of most commercial identification systems to accurately identify the organisms discussed in this chapter is limited or uncertain. Table 39-5 shows some conventional biochemical tests that can be used to presumptively differentiate the species in this chapter. This is a simplified scheme; clinically important isolates should be sent to a reference laboratory for definitive identification.

As just mentioned, these organisms can be difficult to differentiate from gram-negative diplococci (see Chapter 51 for more information about gram-negative diplococci). In addition, these organisms are relatively biochemically inert. Elongation in the presence of penicillin is a useful criterion for differentiating them from true cocci. The effect of penicillin is determined by streaking a blood agar plate and adding a 10-unit penicillin disk to the first quadrant before overnight incubation at 35° C. A Gram stain of the growth taken from around the edge of the zone of inhibition readily demonstrates whether the isolate in question is a true coccus or has elongated.

Comments regarding specific organisms
M. nonliquefaciens and *M. osloensis,* the two most frequently isolated species, can be differentiated by the ability of *M. osloensis* to utilize acetate. *M. lacunata* is able to liquefy serum, so depressions are formed on the surface of Loeffler's serum agar slants. Most of the species considered in this chapter do not utilize glucose; *Neisseria elongata* subsp. *glycolytica,* which produces acid from glucose in the rapid sugar test used for *Neisseria* spp., is the only exception. Unlike *Oligella* spp. (see Chapter 34 for more information regarding this genus) none of the organisms considered here is motile.

SERODIAGNOSIS
Serodiagnostic techniques are not generally used for the laboratory diagnosis of infections caused by the organisms discussed in this chapter.

PREVENTION
Because these organisms do not generally pose a threat to human health, there are no recommended vaccination or prophylaxis protocols.

TABLE 39-4 COLONY APPEARANCE AND CHARACTERISTICS

ORGANISM	MEDIUM	APPEARANCE
Moraxella atlantae	BA	Small, pitting and spreading
	Mac	NLF
M. lacunata	BA	Small colonies that pit the agar
	Mac	No growth
M. lincolnii	BA	Smooth, translucent to semiopaque
	Mac	No growth
M. nonliquefaciens	BA	Smooth, translucent to semiopaque; occasionally, colonies spread and pit agar
	Mac	NLF, if growth
M. osloensis	BA	Smooth, translucent to semiopaque
	Mac	NLF, if growth
M. canis	BA	Resemble colonies of *Enterobacteriaceae*
	Mac	NLF
Neisseria elongata (all subspecies)	BA	Gray-brown, translucent, smooth, glistening; may have dry, clay-like consistency
	Mac	NLF, if growth
N. weaveri	BA	Small, smooth, semiopaque
	Mac	NLF, if growth

BA, 5% sheep blood agar; *Mac,* MacConkey agar; *NLF,* nonlactose fermenter.

TABLE 39-5 KEY BIOCHEMICAL AND PHYSIOLOGIC CHARACTERISTICS

ORGANISM	GROWTH ON MACCONKEY	CATALASE	NITRATE REDUCTION	NITRITE REDUCTION	DNase	DIGESTS LOEFFLER'S SLANT	ACETATE UTILIZATION	GROWTH IN NUTRIENT BROTH
Moraxella atlantae	+	+	−	−	−	−	v	−
M. lacunata	−	+	+	−	−	+	−	−
M. lincolnii	−	+	−	−*	−	−	−	−
M. nonliquefaciens	−	+	+	−	−	−	−	v
M. osloensis	v	+	v	−	−	−	+	+
M. canis	−	+	+	v	+	−	+	NT
Neisseria elongata subsp. *elongata*	v	−	−	+	NT	NT	v	+
Neisseria elongata subsp. *glycolytica*	+	+	−	v	NT	NT	+	+
Neisseria elongata subsp. *nitroreducens*	v	−	+	+	NT	NT	v	v
N. weaveri	v	+	−	+	NT	NT	−	v

NT, not tested; +, >90% of strains positive; −, >90% of strains negative; *v*, variable.

*Nitrite-positive strains have been reported.

Data compiled from Grant, P.E., Brenner, D.J., Steigerwalt, A.G., et al. 1990. J. Clin. Microbiol. 28:2591; Jannes, G., Vaneechoutte, M., Lannoo, M., et al. 1993. Int. J. Syst. Bacteriol. 43:438; Kodjo, A., Richard, Y., and Tonjum, T. 1997. Int. J. Syst. Bacteriol. 47:115; Vandamme, P., Gillis, M., Vancanneyt, M., et al. 1993. Int. J. Syst. Bacteriol. 43:474; and Weyant, R.S., Moss, C.W., Weaver, R.E., et al., editors. 1996. Identification of unusual pathogenic gram-negative aerobic and facultatively anaerobic bacteria, ed 2. Williams & Wilkins, Baltimore.

References

1. Montejo, M., Ruiz-Irastorza, G., Aguirrebengoa, K., et al. 1995. Endocarditis due to *Neisseria elongata* subspecies *nitroreducens*. Clin. Infect. Dis. 20:1431.

2. Mueleman, P., Erard, K., Herregods, M.C., et al. 1996. Bioprosthetic valve endocardititis caused by *Neisseria elongata* subspecies *nitroreducens*. Infection 24:258.

3. Struuillou, L., Raffi, F., and Barrier, J.H. 1993. Endocarditis caused by *Neisseria elongata* subspecies *nitroreducens*: case report and literature review. Eur. J. Clin. Microbiol. Infect. Dis. 12:625.

4. Wallace, R. J., Steingrube, D.R., Nash, D.R. et al. 1989. BRO β-lactamases of *Branhamella catarrhalis* and *Moraxella* subgenus *Moraxella,* including evidence for chromosomal β-lactamase transfer by conjugation in *B. catarrhalis, M. nonliquefaciens,* and *M. lacunata*. Antimicrob. Agents Chemother. 33:1845.

Bibliography

Doern, G.V. 1992. The *Moraxella* and *Branhamella* subgenera of the genus *Moraxella*. In Balows, A., Truper, H.G., Dworkin, M., et al., editors. 1992. The prokaryotes. A handbook on the biology of bacteria: ecophysiology, isolation, identification, applications, ed 2. Springer-Verlag, New York.

Grant, P.E., Brenner, D.J., Steigerwalt, A.G., et al. 1990. *Neisseria elongata* subsp. *nitroreducens* subsp. nov., formerly CDC group M-6, a gram-negative bacterium associated with endocarditis. J. Clin. Microbiol. 28:2591.

Groschel, D.H.M. 1995. *Moraxella catarrhalis* and other gram-negative cocci. In Mandell, G.L., Bennett, J.E., and Dolin, R., editors. Principles and practice of infectious diseases. Churchill Livingstone, New York.

Jannes, G., Vaneechoutte, M., Lannoo, M., et al. 1993. Polyphasic taxonomy leading to the proposal of *Moraxella canis* sp. nov. for *Moraxella catarrhalis*-like strains. Int. J. Syst. Bacteriol. 43:438.

Knapp, J.S. and Rice, R.J. 1995. *Neisseria* and *Branhamella*. In Murray, P.R., Baron, E.J., Pfaller, M.A., et al., editors. Manual of clinical microbiology, ed 6. American Society for Microbiology, Washington, D.C.

Kodjo, A., Richard, Y., and Tønjum, T. 1997. *Moraxella boevrei* sp. nov., a new *Moraxella* species found in goats. Int. J. Syst. Bacteriol. 47:115.

Lyons, R.W. 1985. Ecology, clinical significance, and antimicrobial susceptibility of *Acinetobacter* and *Moraxella*. In Gilardi, G.L., editor. Nonfermentative gram-negative rods: laboratory identification and clinical aspects. Marcel Dekker, New York.

Vandamme, P., Gillis, M., Vancanneyt, M., et al. 1993. *Moraxella lincolnii* sp. nov., isolated from the human respiratory tract, and reevaluation of the taxonomic position of *Moraxella osloensis*. Int. J. Syst. Bacteriol. 43:474.

von Graevenitz, A. 1995. *Acinetobacter, Alcaligenes, Moraxella,* and other nonfermentative gram-negative bacteria. In Murray, P.R., Baron, E.J., Pfaller, M.A., et al., editors. Manual of clinical microbiology, ed 6. American Society for Microbiology, Washington, D.C.

Weyant, R.S., Moss, C.W., Weaver, R.E., et al., editors. 1996. Identification of unusual pathogenic gram-negative aerobic and facultatively anaerobic bacteria, ed 2. Williams & Wilkins, Baltimore.

40 | *EIKENELLA CORRODENS* AND SIMILAR ORGANISMS

Genera and Species to be Considered

Current Name	Previous Name
■ *Eikenella corrodens*	
■ *Methylobacterium spp.*	*Pseudomonas mesophilica, Pseudomonas extorquens*
■ *Weeksella virosa*	CDC group IIf
■ *Bergeyella zoohelcum*	*Weeksella zoohelcum,* CDC group IIj

GENERAL CHARACTERISTICS

The organisms discussed in this chapter are considered together because they are all asaccharolytic, oxidase-positive bacilli that do not grow on MacConkey agar. Their individual morphologic and physiologic features are presented later in this chapter in the discussion of laboratory diagnosis.

EPIDEMIOLOGY, SPECTRUM OF DISEASE, AND ANTIMICROBIAL THERAPY

The organisms listed in Table 40-1 are not commonly associated with human infections, but they are occasionally encountered in clinical specimens.[1,3,5,6] Of those listed, *Eikenella corrodens* is the one most commonly encountered and is usually found in mixed infections resulting from bites or clenched-fist wounds (Table 40-2). This organism also is the "E," for *Eikenella,* among the HACEK group of bacteria known to cause subacute bacterial endocarditis (see Chapter 20 for more information regarding endocarditis and bloodstream infections).

The rarity with which these organisms are encountered as the cause of infection and the lack of validated in vitro susceptibility testing methods do not allow definitive treatment guidelines to be given (Table 40-3). Although beta-lactamase production has been described in *E. corrodens,* this species is usually susceptible to penicillins and many other beta-lactam antimicrobials so that therapy with these agents is frequently used.[2,4]

LABORATORY DIAGNOSIS

SPECIMEN COLLECTION AND TRANSPORT

No special considerations are required for specimen collection and transport for the organisms discussed in this chapter. Refer to Table 1-1 for general information on specimen collection and transport.

SPECIMEN PROCESSING

No special considerations are required for processing of the organisms discussed in this chapter. Refer to Table 1-1 for general information on specimen processing.

DIRECT DETECTION METHODS

Other than Gram stain and microscopic examination, there are no specific procedures for the direct detection of these organisms in clinical material. *E. corrodens* is a medium-length, straight rod, and *Methylobacterium* is a vacuolated, pale-staining, short to medium-length bacillus that may resist decolorization. *Weeksella virosa* and *Bergeyella zoohelcum* are short, straight rods that may form "II-forms" similar to the *Sphingobacterium* (see Chapter 33 for more information regarding this genus).

CULTIVATION

Media of choice

E. corrodens and *Methylobacterium* do not grow well on either 5% sheep blood or chocolate agars, and *Weeksella* and *Bergeyella* only grow slowly on these media. None of these genera grow to a detectable

TABLE 40-1 EPIDEMIOLOGY

ORGANISM	HABITAT (RESERVOIR)	MODE OF TRANSMISSION
Eikenella corrodens	Normal human flora of mouth and gastrointestinal tract	Person to person involving trauma by human teeth incurred by bites or by clenched-fist wounds incurred by facial punches. Also, certain infections may be caused by patient's endogenous strains (e.g., endocarditis)
Methylobacterium spp.	Found on vegetation and occasionally in hospital environment. Not part of normal human flora	Uncertain; probably involves contaminated medical devices such as catheters
Weeksella virosa	Uncertain; probably environmental. Not part of normal human flora	Uncertain, rarely found in clinical material
Bergeyella zoohelcum	Oral flora of dogs and other animals. Not part of normal human flora	Bite or scratch of dog or cat

TABLE 40-2 PATHOGENESIS AND SPECTRUM OF DISEASES

ORGANISM	VIRULENCE FACTORS	SPECTRUM OF DISEASE AND INFECTIONS
Eikenella corrodens	Unknown; opportunistic organism usually requires trauma for introduction into normally sterile sites. Also may enter bloodstream to cause transient bacteremia or be introduced by intravenous drug abuse	Human bite wound infections, head and neck infections, and aspiration pneumonia as part of mixed infection. Can also cause endocarditis that is slow to develop and indolent (i.e., subacute). Less commonly associated with brain and intraabdominal abscesses
Methylobacterium spp.	Unknown; an opportunistic organism probably of low virulence that is an uncommon cause of infection	Bacteremia and peritonitis in patients undergoing chronic ambulatory peritoneal dialysis (CAPD)
Weeksella virosa	Unknown; role in human disease is uncertain	Asymptomatic bacteriuria; also isolated from female genital tract
Bergeyella zoohelcum	Unknown; an opportunistic organism that requires traumatic introduction to normally sterile site	Dog and cat bite wound infections

extent on MacConkey agar, and all grow slowly in the broth media used in blood culture systems, thioglycollate broth, and in brain-heart infusion broth.

Incubation conditions and duration
To detect growth on 5% sheep blood and chocolate agars, incubation at 35° to 37° C in carbon dioxide for a minimum of 48 hours is required. In contrast to the other genera, *Methylobacterium* grows best at 25° C.

Colonial appearance
Table 40-4 describes the colonial appearance and other distinguishing characteristics (e.g., odor and pigment) of each genus on 5% sheep blood agar.

APPROACH TO IDENTIFICATION
The ability of most commercial identification systems to accurately identify the organisms discussed in this chapter is limited, or at best, uncertain. Therefore, strategies for identification of these genera are based

TABLE 40-3   ANTIMICROBIAL THERAPY AND SUSCEPTIBILITY TESTING

ORGANISM	THERAPEUTIC OPTIONS	POTENTIAL RESISTANCE TO THERAPEUTIC OPTIONS	VALIDATED TESTING METHODS*
Eikenella corrodens	Often susceptible to penicillins, quinolones, and cephalosporins	May produce beta-lactamases. Usually resistant to clindamycin, metronidazole, and aminoglycosides	Not available
Methylobacterium spp.	No guidelines	Unknown	Not available
Weeksella virosa and *Bergeyella zoohelcum*	No guidelines; potentially active agents include beta-lactams and quinolones	Susceptibility to tetracycline, aminoglycosides, and trimethoprim/ sulfamethoxazole	Not available

*Validated testing methods include those standard methods recommended by the National Committee for Clinical Laboratory Standards (NCCLS) and those commercial methods approved by the Food and Drug Administration (FDA).

TABLE 40-4   COLONY APPEARANCE AND CHARACTERISTICS

ORGANISM	MEDIUM*	APPEARANCE
Bergeyella zoohelcum	BA	Colonies may be sticky; tan to yellow in color
Eikenella corrodens	BA	Colonies are tiny at 24 hours; mature colonies have moist, clear centers surrounded by flat, spreading growth; colonies may pit or corrode the agar surface; slight yellow pigmentation in older cultures; sharp odor of bleach
Methylobacterium spp.	BA	Pink to coral pigment; does not grow well on blood agar
Weeksella virosa	BA	Small colonies at 24 hours; mature colonies mucoid and adherent with a tan to brown pigment

BA, 5% sheep blood agar.

*These organisms usually do not grow on MacConkey agar; if breakthrough growth occurs, the organisms appear as nonlactose fermenters.

on the use of conventional biochemical tests. Table 40-5 outlines basic criteria that are useful for differentiating among the genera in this chapter.

Comments regarding specific organisms

Methylobacterium can be differentiated from other pink-pigmented, gram-negative rods (e.g., *Roseomonas* spp., discussed in Chapter 34) based on the criterion that *Methylobacterium* will utilize acetate while *Roseomonas* will not. Additionally, *Roseomonas* will grow at 42° C and *Methylobacterium* will not. Some strains of *Methylobacterium* weakly oxidize glucose, and many oxidize xylose.

    E. corrodens can be recognized in culture by its bleachlike odor. The organism does not utilize glucose or other carbohydrates and is catalase-negative. However, it reduces nitrate to nitrite and hydrolyzes both ornithine and lysine.

    Weeksella and *Bergeyella* are biochemically similar to each other, although *W. virosa* is urease-negative and *B. zoohelcum* is urease-positive. Both organisms are indole-positive, which is an unusual characteristic for most nonfermentative bacteria. *W. virosa* will grow on selective media for *Neisseria gonorrhoeae* but can be separated from the gonococci by its positive test for indole and its Gram stain morphology.

SERODIAGNOSIS

Serodiagnostic techniques are not generally used for the laboratory diagnosis of infections caused by the organisms discussed in this chapter.

TABLE 40-5   KEY BIOCHEMICAL AND PHYSIOLOGIC CHARACTERISTICS

ORGANISM	CATALASE	OXIDIZES XYLOSE	INDOLE	ARGININE DIHYDROLASE
Eikenella corrodens	−	−	−	−
Methylobacterium spp.*	+	+	−	NT
Weeksella virosa	+	−	+	−
Bergeyella zoohelcum	+	−	+	+

+, >90% of strains positive; −, >90% of strains negative; *NT*, not tested.

*Colonies are pigmented pink and must be differentiated from *Roseomonas* spp. (see Chapter 34); *Roseomonas* spp. usually grows on MacConkey agar and will grow at 42° C.

From Weyant, R.S., Moss, C.W., Weaver, R.E., et al., editors. 1996. Identification of unusual gram-negative aerobic and facultatively anaerobic bacteria, ed 2. Williams & Wilkins, Baltimore.

PREVENTION

Because these organisms do not generally pose a threat to human health, there are no recommended vaccination or prophylaxis protocols.

References

1. Chen, C.K.C. and Wilson, M.E. 1992. *Eikenella corrodens* in human oral and non-oral infections: a review. J. Periodontol. 63:941.
2. Fass, R.J., Barnishan, J., Solomon, M.C., et al.1996. In vitro activities of quinolones, β-lactams, tobramycin, and trimethoprim-sulfamethoxazole, against nonfermentative gram-negative bacilli. Antimicrob. Agents Chemother. 40:1412.
3. Kay, K.M., Macone, S., and Kazanjian, P.H.1992. Catheter infections caused by *Methylobacterium* in immunocompromised hosts: report of three cases and review of the literature. Clin. Infect. Dis. 14:1010.
4. Lacroix, J-M. and Walker, C.B. 1992. Identification of a streptomycin resistance gene and a partial Tn3 transposon coding for a β-lactamase in a periodontal strain of *Eikenella corrodens*. Antimicrob. Agents Chemother. 36:740.
5. Reina, J. and Borell, N. 1992. Leg abscess caused by *Weeksella zoohelcum* following a dog bite. Clin. Infect. Dis. 14:1162.
6. Reina, J., Gil, J., and Alomar, P. 1989. Isolation of *Weeksella virosa* (formally CDC group II f) from a vaginal sample. Eur. J. Clin. Microbiol. Infect. Dis. 8:569.

Bibliography

Holmes, B.J, Pickett, M.J., and Hollis, D.G. 1995. Unusual gram-negative bacteria, including *Capnocytophaga, Eikenella, Pasteurella,* and *Streptobacillus*. In Murray, P.R., Baron, E.J., Pfaller, M.A., et al., editors. Manual of clinical microbiology, ed 6. American Society for Microbiology, Washington, D.C.

McGowan, J.E. and Steinberg, J.P. 1995. Other gram-negative bacilli. In Mandell, G.L., Bennett, J.E., and Dolin, R., editors. Principles and practice of infectious diseases. Churchill Livingstone, New York.

von Graevenitz, A. 1995. *Acinetobacter, Alcaligenes, Moraxella,* and other non-fermentive gram-negative bacteria. In Murray, P.R., Baron, E.J., Pfaller, M.A., et al., editors. Manual of clinical microbiology, ed 6. American Society for Microbiology, Washington, D.C.

Weyant, R.S., Moss, C.W., Weaver, R.E., et. al., editors. 1996. Identification of unusual pathogenic gram-negative aerobic and facultatively anaerobic bacteria, ed 2. Williams & Wilkins, Baltimore.

41 | *PASTEURELLA* AND SIMILAR ORGANISMS

Genera and Species to be Considered

Current Name	Previous Name
■ *Pasteurella multocida*	
■ *Pasteurella bettyae*	CDC group HB-5
■ *Pasteurella aerogenes*	
■ *Pasteurella canis*	
■ *Pasteurella dagmatis*	
■ *Pasteurella haemolytica*	
■ *Pasteurella pneumotropica*	
■ *Pasteurella stomatis*	
■ *Suttonella indologenes*	*Kingella indologenes*
■ CDC group EF-4a	CDC group EF-4
■ Bisgaard's taxon 16	

GENERAL CHARACTERISTICS

The organisms discussed in this chapter are small, gram-negative, nonmotile, oxidase-positive bacilli that ferment glucose; most will not grow on Mac-Conkey agar. Their individual morphologic and physiologic features are presented later in this chapter in the discussion of laboratory diagnosis.

EPIDEMIOLOGY, SPECTRUM OF DISEASE, AND ANTIMICROBIAL THERAPY

The majority of the organisms presented in this chapter are part of animal flora and are transmitted to humans during close animal contact, including bites. For most, virulence factors are not recognized and the organisms may be considered opportunistic pathogens that require mechanical disruption of the host's anatomical barriers, such as occur with bite-induced wounds (Table 41-1). Of the organisms listed in Table 41-2, *Pasteurella multocida* is the species most commonly encountered in clinical specimens.

Because the mode of transmission often involves traumatic inoculation of the organism through human skin, it is expected that the majority of infections caused by these bacteria involve wounds and soft tissues (Table 41-2). However, in some settings involving compromised patients, infections of the respiratory tract and other body sites can occur.[1-3]

An unusual feature of the organisms considered in this chapter is that most are susceptible to penicillin. Although most other clinically relevant gram-negative bacilli are intrinsically resistant to penicillin, it is the drug of choice for infections involving *P. multocida* and several other species listed in Table 41-3.

The general therapeutic effectiveness of penicillin, the lack of resistance to this agent among *Pasteurella* spp., and the unavailability of validated in vitro susceptibility testing methods are important contraindications for performing susceptibility testing with the organisms discussed in this chapter. Even though many of these organisms may grow on the media, and under the conditions recommended for testing other bacteria (see Chapter 18 for more information regarding validated testing methods), this does not necessarily mean that interpretable and reliable results will be produced.

LABORATORY DIAGNOSIS

SPECIMEN COLLECTION AND TRANSPORT

No special considerations are required for specimen collection and transport of the organisms discussed in

—————————————— **TABLE 41-1** EPIDEMIOLOGY ——————————————

ORGANISM	HABITAT (RESERVOIR)	MODE OF TRANSMISSION
Pasteurella multocida and other *Pasteurella* spp.	Nasopharynx and gastrointestinal tract of wild and domestic animals. Part of an animal's normal flora; also associated with infections in various animals. May be found as part of upper respiratory flora of humans who have extensive exposure to animals (e.g., animal handlers)	Animal (usually cat or dog) bite or scratch. Infections may also be associated with non-bite exposure to animals. Less commonly, infections may occur without a history of animal exposure
Suttonella indologenes	Uncertain; rarely encountered in clinical specimens but may be part of human flora	Uncertain
Bisgaard's taxon 16	Uncertain; probably similar to that of *Pasteurella* spp.	Dog bite or close exposure to other animals
CDC group EF-4a	Animal oral and respiratory flora. Not part of human flora	Animal contact, particularly bites or scratches from dogs and cats

—————————— **TABLE 41-2** PATHOGENESIS AND SPECTRUM OF DISEASES ——————————

ORGANISM	VIRULENCE FACTORS	SPECTRUM OF DISEASE AND INFECTIONS
Pasteurella multocida and other *Pasteurella* spp.	Endotoxin and antiphagocytic capsule associated with *P. multocida*	Focal soft tissue infections following bite or scratch. Chronic respiratory infection, usually in patients with preexisting chronic lung disease and heavy exposure to animals. Bacteremia with metastatic abscess formation may also occur, occasionally in patients with no history of animal exposure
Suttonella indologenes	Unknown	Associated with eye infections but rare
Bisgaard's taxon 16	Unknown	Associated with animal bite wounds
CDC group EF-4a	Unknown	Infected bite wounds of fingers, hands, or arm leading to cellulitis or abscess formation. Systemic infections are rare

this chapter. Refer to Table 1-1 for general information on specimen collection and transport.

SPECIMEN PROCESSING

No special considerations are required for processing of the organisms discussed in this chapter. Refer to Table 1-1 for general information on specimen processing.

DIRECT DETECTION METHODS

Other than Gram stain of patient specimens, there are no specific procedures for the direct detection of these organisms in clinical material. *Pasteurella* spp. are typically short, straight bacilli, although *P. aerogenes* may also appear as coccobacilli. The cells of *P. bettyae* are usually thinner rods than those of the other species. *Suttonella indologenes* is a variable length, broad rod,

TABLE 41-3   ANTIMICROBIAL THERAPY AND SUSCEPTIBILITY TESTING

ORGANISM	THERAPEUTIC OPTIONS	POTENTIAL RESISTANCE TO THERAPEUTIC OPTIONS	VALIDATED TESTING METHODS*	COMMENTS
Pasteurella multocida and other *Pasteurella* spp.	Cleaning, irrigation, and debridement of bite wound. Antimicrobial of choice is penicillin. Other active agents include mezlocillin, piperacillin, cefuroxime, cefotaxime, ciprofloxacin, tetracycline, and trimethoprim/sulfamethoxazole	None known to penicillin	Not available	Unusual for gram-negative bacilli to be susceptible to penicillin
Suttonella indologenes	No guidelines. Susceptible to penicillins, similar to *Pasteurella* spp.	Unknown	Not available	
Bisgaard's taxon 16	No guidelines	Unknown	Not available	
CDC group EF-4a	No definitive guidelines. Potentially active agents include penicillin, ampicillin, ciprofloxacin, and ofloxacin	Unknown; some cephalosporins may be less active than the penicillins	Not available	Limited clinical experience

*Validated testing methods include those standard methods recommended by the National Committee for Clinical Laboratory Standards (NCCLS) and those commercial methods approved by the Food and Drug Administration (FDA).

while Bisgaard's taxon 16 is a short, straight bacillus. EF-4a appears as a coccobacillus or short, straight rod.

CULTIVATION

Media of choice

The bacteria described in this chapter grow well on routine laboratory media such as 5% sheep blood and chocolate agars; most strains do not grow on Mac-Conkey agar. All four genera also grow well in broth blood culture systems and common nutrient broths such as thioglycollate and brain-heart infusion.

Incubation conditions and duration

5% sheep blood and chocolate agars should be incubated at 35° C in carbon dioxide or ambient air for a minimum of 24 hours.

Colonial appearance

Table 41-4 describes the colonial appearance and other distinguishing characteristics (e.g., hemolysis and odor) of each genus on blood agar.

APPROACH TO IDENTIFICATION

Some commercial biochemical identification systems are able to identify strains of *P. multocida*. However, the ability of most commercial identification systems to accurately identify the other species is limited or uncertain. Table 41-5 shows some conventional biochemical tests that can be used to presumptively differentiate among the organisms discussed in this chapter. These organisms closely resemble those described in Chapter 42. Therefore, organisms discussed in Chapters 41 and 42 all should be considered when evaluating an isolate in the clinical laboratory. Definitive identification of these fermentative organisms will usually require the isolate to be sent to a reference laboratory for a full battery of biochemical tests.

Comments regarding specific organisms

The pasteurellae should all be oxidase-positive, based on the use of the tetramethyl-*p*-phenylenediamine dihydrochloride reagent; however, several subcultures may be necessary to obtain a positive reaction. Except for *P. bettyae*, these organisms are all catalase-positive, and all species reduce nitrates to nitrites. *P. aerogenes* and some strains of *P. dagmatis* ferment glucose with the production of gas. *P. multocida*, the most frequently isolated species, can be separated from the other species based on positive reactions for

TABLE 41-4 COLONY APPEARANCE AND CHARACTERISTICS ON 5% SHEEP BLOOD AGAR

ORGANISM	APPEARANCE
Bisgaard's taxon 16	No distinctive appearance
CDC group EF-4a	Some strains have a yellow to tan pigment; smells like popcorn
*Pasteurella aerogenes**	Convex, smooth, translucent, nonhemolytic†
P. bettyae‡	Convex, smooth, nonhemolytic
P. canis	Convex, smooth, nonhemolytic
P. dagmatis	Convex, smooth, nonhemolytic
*P. haemolytica**	Convex, smooth, beta-hemolytic (this feature may be lost on subculture)
P. multocida	Convex, smooth, gray, nonhemolytic; rough and mucoid variants can occur; may have a musty or mushroom smell
*P. pneumotropica**	Smooth, convex, nonhemolytic
P. stomatis	Smooth, convex, nonhemolytic
Suttonella indologenes	Resembles *Kingella* spp. (see Chapter 42)

*Breakthrough growth may occur on MacConkey agar; will appear as lactose fermenter.

†After 48 hours, colonies may be surrounded by a narrow green to brown halo.

‡Breakthrough growth may occur on MacConkey agar; will appear as nonlactose fermenter.

TABLE 41-5 KEY BIOCHEMICAL AND PHYSIOLOGIC CHARACTERISTICS

ORGANISM	INDOLE	UREA	NITRATE REDUCTION	CATALASE	ORNITHINE DECARBOXYLASE	MANNITOL	SUCROSE	MALTOSE
Pasteurella aerogenes	−	(+)	(+)	+	v	−	+	+
P. bettyae	(+)	−	(+)	−	−	−	−	−
P. canis	+	−	+	+	(+)	−	(+)	−
P. dagmatis	(+)	(+)	(+)	+	−	−	+	(+)
P. haemolytica	−	−	+	+	−	(+)	+	+
P. multocida	(+)	−	(+)	+	(+)	+	+	−
P. pneumotropica	(+)	(+)*	(+)	+	(+)	−	+	+
P. stomatis	(+)	−	+	+	−	−	(+)	−
Suttonella indologenes	(+)	−	−	v	−	−	(+)	+ or (+)
CDC group EF-4a	−	−	(+)	+	−	−	−	−
Bisgaard's taxon 16	(+)	−	(+)	+	−	−	+	(+)

+, >90% of strains positive; (+), >90% of strains positive but reaction may be delayed (i.e., 2-7 days); −, >90% of strains negative; v, variable.

*May require a drop of rabbit serum on the slant or a heavy inoculum.

Data compiled from Weyant, R.S., Moss, C.W., Weaver, R.E., et al., editors. 1996. Identification of unusual pathogenic gram-negative aerobic and facultatively anaerobic bacteria, ed 2. Williams & Wilkins, Baltimore.

ornithine decarboxylase and indole, and a negative reaction for urease.

S. indologenes can be separated from the pasteurellae by its negative nitrate test and differs from *Kingella* spp. (discussed in Chapter 42) in being both indole- and sucrose-positive. Bisgaard's taxon 16 can be separated from the pasteurellae by its fermentation of sucrose and maltose.

CDC group EF-4a is an eugonic-fermenter (EF), meaning that it grows well on routine laboratory media and ferments glucose. This distinguishes it from the dysgonic-fermenters (DF), which grow poorly on blood and chocolate agars (see Chapter 42). EF-4a cannot ferment any sugar except glucose, and is indole-negative and arginine dihydrolase-positive.

SERODIAGNOSIS

Serodiagnostic techniques are not generally used for the laboratory diagnosis of infections caused by the organisms discussed in this chapter.

PREVENTION

Because these organisms do not generally pose a threat to human health, there are no recommended vaccination or prophylaxis protocols.

References

1. Cuadrado-Gomez, L.M., Arranz-Caso, J.A., Cuadros-Gonzalez, J., et al. 1995. *Pasteurella pneumotropica* pneumonia in a patient with AIDS. Clin. Infect. Dis. 21:445.
2. Holst, E., Roloff, J., Larsson, L., et al. 1992. Characterization and distribution of *Pasteurella* species recovered from infected humans. J. Clin. Microbiol. 30:2984.
3. Shapiro, D.S., Brooks, P.E., Coffey, D.M., et al. 1996. Peripartum bacteremia with CDC group HB-5 (*Pasteurella bettyae*). Clin. Infect. Dis. 22:1125.

Bibliography

Boyce, J. 1995. *Pasteurella* species. In Mandell, G.L., Bennett, J.E., and Dolin, R., editors. Principles and practice of infectious diseases. Churchill Livingstone, New York.

Holmes, B., Pickett, J.M., and Hollis, D.G. 1995. Unusual gram-negative bacteria, including *Capnocytophaga*, *Eikenella*, *Pasteurella*, and *Streptobacillus*. In Murray, P.R., Baron, E.J., Pfaller, M.A., et al., editors. Manual of clinical microbiology, ed 6. American Society for Microbiology, Washington, D.C.

Weyant, R.S., Moss, C.W., Weaver, R.E., et al., editors. 1996. Identification of unusual pathogenic gram-negative aerobic and facultatively anaerobic bacteria, ed 2. Williams & Wilkins, Baltimore.

42 | ACTINOBACILLUS, KINGELLA, CARDIOBACTERIUM, CAPNOCYTOPHAGA, AND SIMILAR ORGANISMS

Genera and Species to be Considered

Current Name	Previous Name
■ *Actinobacillus actinomycetemcomitans*	
■ Other *Actinobacillus* spp., including	
A. ureae	
A. suis	
A. lignieresii	
A. hominis	
A. equuli	
■ *Kingella denitrificans*	
■ *Kingella kingae*	
■ *Cardiobacterium hominis*	
■ *Capnocytophaga gingivalis*	CDC group DF-1
■ *Capnocytophaga ochracea*	CDC group DF-1
■ *Capnocytophaga sputigena*	CDC group DF-1
■ *Capnocytophaga canimorsus*	CDC group DF-2
■ *Capnocytophaga cynodegmi*	CDC group DF-2
■ CDC group DF-3	

GENERAL CHARACTERISTICS

The organisms discussed in this chapter are slow or poorly growing (i.e., dysgonic). Although they all ferment glucose, their fastidious nature requires that serum be added to the basal fermentation medium to enhance growth and detection of the fermentation reactions. These bacteria are capnophiles, that is, they require additional carbon dioxide (CO_2) for growth, and most species will not grow on MacConkey agar. Their individual morphologic and physiologic features are presented later in this chapter in the discussion of laboratory diagnosis.

EPIDEMIOLOGY, SPECTRUM OF DISEASE, AND ANTIMICROBIAL THERAPY

The organisms listed in Table 42-1 are part of the normal flora of humans and other animals. As such, they generally are of low virulence and, except for those species associated with periodontal infections, usually only cause infections in humans after introduction into sterile sites following trauma such as bites or manipulations in the oral cavity.[1,4-7]

The types of infections caused by these bacteria vary from periodontitis to endocarditis (Table 42-2). Three of these organisms, *Actinobacillus actinomycetemcomitans*, *Cardiobacterium hominis*, and *Kingella* spp., are the A, C, and K, respectively, of the HACEK group of organisms that cause slowly progressive (i.e., subacute) bacterial endocarditis. *Kingella* spp. can also be involved in other serious infections involving children.[2,9]

Among the species acquired from animals, *Capnocytophaga canimorsus* can cause fulminant, life-threatening infections following dog bites.[7]

The relative rarity with which these organisms are encountered as the cause of infection, the lack of validated in vitro susceptibility testing methods, and the inability of susceptibility testing media to support good growth of many of these organisms precludes the use of in vitro testing for guiding therapy (Table 42-3). Infections are frequently treated using beta-lactam antibiotics, occasionally in combination with an aminoglycoside. Beta-lactamase production has been described in *Kingella* spp., but the impact of this resistance mechanism on clinical efficacy of beta-lactams is uncertain.[3,8]

TABLE 42-1 EPIDEMIOLOGY

ORGANISM	HABITAT (RESERVOIR)	MODE OF TRANSMISSION
Actinobacillus actinomycetemcomitans	Normal flora of human oral cavity	Endogenous; enters deeper tissues by minor trauma to mouth, such as during dental procedures
Other *Actinobacillus* spp.	Normal oral flora of animals such as cows, sheep, and pigs; not part of human flora	Rarely associated with human infection. Transmitted by bite wounds or contamination of preexisting wounds during exposure to animals
Kingella denitrificans and *Kingella kingae*	Normal flora of human upper respiratory and genitourinary tracts	Infections probably caused by patient's endogenous strains
Cardiobacterium hominis	Normal flora of human upper respiratory tract	Infections probably caused by patient's endogenous strains
Capnocytophaga gingivalis, *Capnocytophaga ochracea*, and *Capnocytophaga sputigena*	Subgingival surfaces and other areas of human oral cavity	Infections probably caused by patient's endogenous strains
Capnocytophaga canimorsus and *Capnocytophaga cynodegmi*	Oral flora of dogs	Dog bite or nonbite, long exposure to dogs
CDC group DF-3	Uncertain; possibly part of human gastrointestinal flora	Uncertain; possibly endogenous

LABORATORY DIAGNOSIS

SPECIMEN COLLECTION AND TRANSPORT

No special considerations are required for specimen collection and transport of the organisms discussed in this chapter. Refer to Table 1-1 for general information on specimen collection and transport.

SPECIMEN PROCESSING

No special considerations are required for processing of the organisms discussed in this chapter. Refer to Table 1-1 for general information on specimen processing.

DIRECT DETECTION METHODS

Other than Gram stain of patient specimens, there are no specific procedures for the direct detection of these organisms in clinical material. *Actinobacillus* spp. are short to very short bacilli. They occur singly, in pairs, and in chains and tend to exhibit bipolar staining. This staining morphology gives the overall appearance of the dots and dashes of Morse code.

Kingella spp. stain as short, plump coccobacilli with squared-off ends that may form chains. *Cardiobacterium hominis* is a pleomorphic, gram-negative rod with one rounded end and one tapered end, giving the cells a teardrop appearance. *C. hominis* tends to form clusters, or rosettes, when Gram stains are prepared from 5% sheep blood agar.

Capnocytophaga spp. are gram-negative, fusiform-shaped bacilli with one rounded end and one tapered end and occasional filamentous forms; *C. cynodegmi* and *C. canimorsus* may be curved. CDC group DF-3 stain as short gram-negative rods or coccobacilli.

CULTIVATION

Media of choice

All genera described in this chapter grow on 5% sheep blood and chocolate agars. Most actinobacilli (except *A. actinomycetemcomitans* and *A. ureae*) show light growth on MacConkey agar, but the other genera will not grow on this medium. CDC group DF-3 can be recovered from stool on CVA (cefoperazone-vancomycin-amphotericin B) agar. For recovery of CDC group DF-3, this medium, a *Campylobacter* selective agar, is incubated at 35° C instead of 42° C (preferred for *Campylobacter*).

These genera grow in the broths of commercial blood culture systems and in common nutrient broths such as thioglycollate and brain-heart infusion. Growth of *Actinobacillus* in broth media is often

TABLE 42-2 PATHOGENESIS AND SPECTRUM OF DISEASES

ORGANISM	VIRULENCE FACTORS	SPECTRUM OF DISEASES AND INFECTIONS
Actinobacillus actinomycetemcomitans	Unknown; probably of low virulence; an opportunistic pathogen	Destructive periodontitis, including bone loss; endocarditis, often following dental manipulations; soft tissue and human bite infections, often mixed with anaerobic bacteria and *Actinomyces* spp.
Other *Actinobacillus* spp.	Unknown for human disease; probably of low virulence	Rarely cause infection in humans but may be found in animal bite wounds; association with other infections, such as meningitis or bacteremia, is extremely rare and involves compromised patients
Kingella denitrificans and *Kingella kingae*	Unknown; probably of low virulence; opportunistic pathogens	Endocarditis and infections in various other sites, especially in immunocompromised patients; *K. kingae* associated with blood, bone, and joint infections of young children
Cardiobacterium hominis	Unknown; probably of low virulence	Infections in humans are rare; most commonly associated with endocarditis, especially in persons with anatomic heart defects
Capnocytophaga gingivalis, Capnocytophaga ochracea, and *Capnocytophaga sputigena*	Unknown. Produce wide variety of enzymes that may mediate tissue destruction	Most commonly associated with periodontitis and other types of periodontal disease; less commonly associated with bacteremia in immunocompromised patients
Capnocytophaga canimorsus and *Capnocytophaga cynodegmi*	Unknown	Range from mild, local infection at bite site to bacteremia culminating in shock and disseminated intravascular coagulation. Most severe in splenectomized or otherwise debilitated (e.g., alcoholism) patients but can occur in healthy people. Miscellaneous other infections such as pneumonia, endocarditis, and meningitis may also occur
CDC group DF-3	Unknown; probably of low virulence	Role in disease is uncertain. May be associated with diarrheal disease in immunocompromised patients. Rarely isolated from other clinical specimens, such as urine, blood, and wounds

barely visible, with no turbidity produced. Microcolonies may be seen as tiny puffballs growing on the blood cell layer in blood culture bottles or as a film or tiny granules on the sides of a tube.

Incubation conditions and duration
The growth of all genera discussed in this chapter occurs best at 35° C and in the presence of increased CO_2. Therefore 5% sheep blood and chocolate agars should be incubated in a CO_2 incubator or candle jar. In addition, *Actinobacillus* and *Cardiobacterium* grow best in conditions of elevated moisture; a candle jar with a sterile gauze pad moistened with sterile water is ideal for this purpose. *Capnocytophaga* will grow anaerobically but will not grow in ambient air.

Even when optimum growth conditions are met, the organisms discussed here are all slow-growing; therefore inoculated plates should be held 2 to 7 days for colonies to achieve maximal growth.

Colonial appearance
Table 42-4 describes the colonial appearance and other distinguishing characteristics (e.g., hemolysis and pigment) of each genus on 5% sheep blood agar.

—— **TABLE 42-3** **ANTIMICROBIAL THERAPY AND SUSCEPTIBILITY TESTING** ——

ORGANISM	THERAPEUTIC OPTIONS	POTENTIAL RESISTANCE TO THERAPEUTIC OPTIONS	VALIDATED TESTING METHODS*
Actinobacillus actinomycetemcomitans	No definitive guidelines. For periodontitis, debridement of affected area; potential agents include tetracycline, metronidazole, and ampicillin. For endocarditis penicillin, ampicillin, or a cephalosporin (perhaps with an aminoglycoside) may be used	Some strains appear resistant to penicillin and ampicillin, but clinical relevance of resistance is unclear	Not available
Other *Actinobacillus* spp.	No guidelines	Unknown	Not available
Kingella denitrificans, Kingella kingae	A beta-lactam with or without an aminoglycoside; other active agents include erythromycin, trimethoprim/sulfamethoxazole, and ciprofloxacin	Some strains produce beta-lactamase that mediates resistance to penicillin, ampicillin, ticarcillin, and cefazolin	Not available
Cardiobacterium hominis	For endocarditis, penicillin with or without an aminoglycoside; usually susceptible to other beta-lactams, chloramphenicol, and tetracycline	Unknown	Not available
Capnocytophaga gingivalis, Capnocytophaga ochanacea, Capnocytophaga sputigena	No definitive guidelines. Generally susceptible to clindamycin, erythromycin, tetracyclines, chloramphenicol, imipenem, and other beta-lactams	Beta-lactamase–mediated resistance to penicillin	Not available
Capnocytophaga canimorsus, Capnocytophaga cynodegmi	Penicillin is drug of choice; also susceptible to penicillin derivatives, imipenem, and third-generation cephalosporins	Unknown	Not available
CDC group DF-3	No guidelines. Potentially effective agents include chloramphenicol, trimethoprim/sulfamethoxazole, tetracycline, and clindamycin	Often resistant to beta-lactams and ciprofloxacin	Not available

*Validated testing methods include those standard methods recommended by the National Committee for Clinical Laboratory Standards (NCCLS) and those commercial methods approved by the Food and Drug Administration (FDA).

TABLE 42-4 COLONIAL APPEARANCE AND CHARACTERISTICS ON 5% SHEEP BLOOD AGAR

ORGANISM	APPEARANCE
Actinobacillus actinomycetemcomitans	Pinpoint colonies after 24 hrs; rough, sticky, adherent colonies surrounded by a slight greenish tinge after 48 hrs; characteristic finding is presence of a four- to six-pointed starlike configuration in the center of a mature colony growing on a clear medium (e.g., brain-heart infusion agar), which can be visualized by examining the colony under low power (100×) of a standard light microscope
*A. equuli**	Small colonies at 24 hrs that are sticky, adherent, smooth or rough, and nonhemolytic
*A. lignieresii**	Resembles *A. equuli*
*A. suis**	Beta-hemolytic but otherwise resembles *A. equuli* and *A. lignieresii*
A. ureae	Resembles the pasteurellae (see Chapter 41)
Cardiobacterium hominis	After 48 hrs, colonies are small, slightly alpha-hemolytic, smooth, round, glistening and opaque; pitting may be produced
Capnocytophaga spp.	After 48-72 hrs, colonies are small- to medium-size, opaque, shiny; nonhemolytic; pale beige or yellowish color may not be apparent unless growth is scraped from the surface with a cotton swab; gliding motility may be observed as outgrowths from the colonies or as a haze on the surface of the agar, similar to swarming of *Proteus*
CDC group DF-3	Pinpoint colonies after 24 hrs; small, wet, gray-white colonies at 48-72 hrs; usually nonhemolytic, although some strains may produce a small zone of beta hemolysis; characteristic odor alternately described as fruity or bitter
Kingella denitrificans	Small, nonhemolytic; frequently pits agar; can grow on *Neisseria gonorrhoeae* selective agar (e.g., Thayer-Martin agar)
K. kingae	Small, with a small zone of beta hemolysis; may pit agar

*May grow on MacConkey agar as tiny lactose fermenters.

Most species will not grow on MacConkey agar; exceptions are noted in Table 42-4.

APPROACH TO IDENTIFICATION

None of the commercial identification systems will reliably identify any of the genera in this chapter. Table 42-5 outlines some conventional biochemical tests that are useful for differentiating among *Actinobacillus, Cardiobacterium, Kingella,* and *Haemophilus aphrophilus;* these are four of the five HACEK bacteria that cause subacute bacterial endocarditis. Table 42-6 shows key conventional biochemicals that can be used to differentiate *Capnocytophaga* spp., CDC group DF-3, and aerotolerant *Leptotrichia buccalis.*

Comments regarding specific organisms

The genus *Actinobacillus* is very similar to *Pasteurella* (see Chapter 41), which must also be considered when a fastidious gram-negative rod requiring rabbit serum is isolated. *A. actinomycetemcomitans,* the most frequently isolated of the actinobacilli, can be distinguished from *H. aphrophilus* (see Chapter 43) by its positive test for catalase and negative test for lactose fermentation.

A. actinomycetemcomitans differs from *C. hominis* in being indole-negative and catalase-positive; catalase is also an important test for differentiating *Kingella* spp., which are catalase-negative, from *A. actinomycetemcomitans. C. hominis* is indole-positive following extraction with xylene and addition of Ehrlich's reagent; this is a key feature in differentiating it from *H. aphrophilus, A. actinomycetemcomitans,* and CDC group EF-4a. *C. hominis* is similar to *Suttonella indologenes* but can be distinguished by its ability to ferment mannitol and sorbitol.

Kingella spp. are catalase-negative, which helps to separate them from *Neisseria* spp. (see Chapter 51), with which they are sometimes confused. *K. den-*

TABLE 42-5   BIOCHEMICAL AND PHYSIOLOGIC CHARACTERISTICS OF ACTINOBACILLUS SPP. AND RELATED ORGANISMS

ORGANISM	CATALASE	NITRATE REDUCTION	INDOLE	UREA	ESCLULIN HYDROLYSIS	FERMENTATION OF†: XYLOSE	LACTOSE	TREHALOSE
Actinobacillus actinomycetemcomitans	+	+	−	−	−	v	−	NT
A. equuli	v	+	−	(+)*	−	+	+	(+)
A. lignieresii	v	+	−	(+)*	−	+ or (+)	v	−
A. suis	v	+	−	(+)*	+	+	+ or (+)	+
A. ureae	v	+	−	(+)*	−	−	−	NT
Cardiobacterium hominis	−	−	+	−	−	−	−	NT
Haemophilus aphrophilus	−	+	−	−	−	−	+	NT
Kingella denitrificans	−	(+)	−	−	−	−	−	NT
K. kingae	−	−	−	−	−	−	−	NT

+, >90% of strains positive; (+), >90% of strains positive but reaction may be delayed (i.e., 2 to 7 days); −, >90% of strains negative; *v*, variable; *NT*, not tested.

*May require a drop of rabbit serum on the slant or a heavy inoculum.

†May require the addition of 1 to 2 drops rabbit serum per 3 mL of fermentation broth to stimulate growth.

Data compiled from Weyant, R.S., Moss, C.W., Weaver, R. E., et al., editors. 1996. Identification of unusual pathogenic gram-negative aerobic and facultatively anaerobic bacteria, ed 2. Williams & Wilkins, New York.

TABLE 42-6 BIOCHEMICAL AND PHYSIOLOGIC CHARACTERISTICS OF *CAPNOCYTOPHAGA* SPP., CDC GROUP DF-3, AND SIMILAR ORGANISMS

ORGANISM	OXIDASE	CATALASE	ESCULIN HYDROLYSIS	INDOLE	NITRATE REDUCTION	XYLOSE FERMENTATION
Capnocytophaga spp. (CDC group DF-1)*	–	–	(v)	–	v	–
C. *canimorsus* (CDC group DF-2)†	(+)	(+)	v	–	–	–‡
C. *cynodegmi* (CDC group DF-2–like)†	(+)	(+)	+ or (+)	–	–	–
*Leptotrichia buccalis**	–	–	v	–	–	–‡
CDC group DF-3	–	–	(+)	(v)	–	+ or (+)‡
CDC group DF-3–like	v	v	v	(+)	–	–‡

+, >90% of strains positive; (+), >90% of strains positive, but reaction may be delayed (i.e., 2 to 7 days); –, >90% of strains negative; *v*, variable; (*v*), positive reactions may be delayed.

*Lactic acid is the major fermentation end product of glucose fermentation for *Leptotrichia buccalis,* and succinic acid is the major fermentation end product of glucose fermentation for *Capnocytophaga* spp. (CDC group DF-1).

†C. *canimorsus* does not ferment the sugars inulin, sucrose, or raffinose; C. *cynodegmi* will usually ferment one or all of these sugars.

‡May require the addition of 1 to 2 drops of rabbit serum per 3 mL of fermentation broth to stimulate growth.

Data compiled from Weyant, R.S., Moss, C.W., Weaver, R.E., et al., editors. 1996. Identification of unusual pathogenic gram-negative aerobic and facultatively anaerobic bacteria, ed 2. Williams & Wilkins, New York.

itrificans may be mistaken for *Neisseria gonorrhoeae* when isolated from modified Thayer-Martin agar. Nitrate reduction is a key test in differentiating *K. denitrificans* from *N. gonorrhoeae,* which is nitrate-negative.

The species in the former CDC group DF-1, that is, *C. ochracea, C. sputigena,* and *C. gingivalis,* are catalase- and oxidase-negative; however, members of CDC group DF-1 cannot be separated by conventional biochemical tests. *C. canimorsus* and *C. cynodegmi,* formerly part of CDC group DF-2, are catalase- and oxidase-positive; these species are also difficult to differentiate from each other. However, for most clinical purposes, a presumptive identification to genus, that is, *Capnocytophaga,* is sufficiently informative and precludes the need to identify an isolate to the species level. Presumptive identification of an organism as *Capnocytophaga* spp. can be made if a yellow-pigmented, thin, gram-negative rod with tapered ends that exhibits gliding motility (see Table 42-4) and will not grow in ambient air is isolated.

CDC group DF-3, although similar to the other organisms in this chapter, are oxidase-negative. They are nonmotile, unlike the *Capnocytophaga,* which exhibit gliding motility. Gas-liquid chromatography is useful in separating CDC group DF-3 and *Capnocytophaga,* but this technology is not commonly available in most clinical laboratories. CDC group DF-3 produces propionic acid, whereas *Capnocytophaga* produces succinic acid. Cellular fatty acid analysis can provide information necessary to distinguish *Capnocytophaga,* CDC group DF-3, and the aerotolerant strains of *Leptotrichia buccalis.*

SERODIAGNOSIS

Serodiagnostic techniques are not generally used for the laboratory diagnosis of infections caused by the organisms discussed in this chapter.

PREVENTION

Because the organisms discussed in this chapter do not generally pose a threat to human health, there are no recommended vaccination or prophylaxis protocols.

References

1. Gill, V.J., Travis, L.B., and Williams, D.Y. 1991. Clinical and microbiological observations on CDC group DF-3, a gram-negative coccobacillus. J. Clin. Microbiol. 29:1589.

2. Hassan, I.J. and Hayek, L. 1993. Endocarditis caused by *Kingella denitrificans.* J. Infection. 27:291.

3. Jensen, K.T., Schonheyder, H., and Thomsen, V.F. 1994. In-vitro activity of β-lactam and other antimicrobial agents against *Kingella kingae.* J. Antimicrob. Chemother. 33:635.

4. Kaplan, A.H., Weber, D.J., Oddone, E.Z., et al. 1989. Infection due to *Actinobacillus actinomycetemcomitans:* fifteen cases and review. Rev. Infect. Dis. 11:46.

5. Minamoto, G.Y. and Sordillo, E.M. 1992. *Kingella denitrificans* as a cause of granulomatous disease in a patient with AIDS. Clin. Infect. Dis. 15:1052.

6. Peel, M.M., Hornridge, K.A., Luppino, M., et al. 1991. *Actinobacillus* spp. and related bacteria in infected wounds of humans bitten by horses and sheep. J. Clin. Microbiol. 29:2535.

7. Pers, C., Gahrn-Hansen, B., and Frederiksen, W. 1996. *Capnocytophaga canimorsus* septicemia in Denmark, 1982-1995: review of 39 cases. Clin. Infect. Dis. 23:71.

8. Sordillo, E.M., Rendel, M., Sood, R., et al. 1993. Septicemia due to β-lactamase-positive *Kingella kingae.* Clin. Infect. Dis. 17:818.

9. Yagupsky, P. and Dagan, R. 1994. *Kingella kingae* bacteremia in children. Ped. Infect. Dis. J. 13:1148.

Bibliography

Gill, V.J. 1995. *Capnocytophaga.* In Mandell, G.L., Bennett, J.E., and Dolin, R., editors. Principles and practice of infectious diseases. Churchill Livingstone, New York.

Groschel, D.H.M. 1995. *Moraxella catarrhalis* and other gram-negative cocci. In Mandell, G.L., Bennett, J.E., and Dolin, R., editors. Principles and practice of infectious diseases. Churchill Livingstone, New York.

Holmes, B.J, Pickett, M.J., and Hollis, D.G. 1995. Unusual gram-negative bacteria, including *Capnocytophaga, Eikenella, Pasteurella,* and *Streptobacillus.* In Murray, P.R., Baron, E.J., Pfaller, M.A., et al., editors. Manual of clinical microbiology, ed 6. American Society for Microbiology, Washington, D.C.

McGowan, J.E. and Steinberg, J.P. 1995. Other gram-negative bacilli. In Mandell, G.L., Bennett, J.E., and Dolin, R., editors. Principles and practice of infectious diseases. Churchill Livingstone, New York.

Weyant, R.S., Moss, C.W., Weaver, R.E., et al, editors. 1996. Identification of unusual pathogenic gram-negative aerobic and facultatively anaerobic bacteria, ed 2. Williams & Wilkins, Baltimore.

Section Four

Gram-Negative Bacilli and Coccobacilli (MacConkey-Negative, Oxidase-Variable)

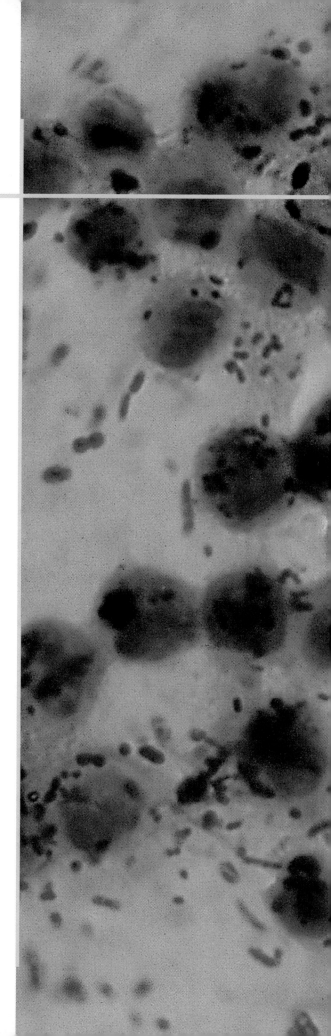

43 | *HAEMOPHILUS*

Genera and Species to be Considered

Current Name	Previous Name
■ *Haemophilus influenzae*	
■ *Haemophilus influenzae* biogroup *aegyptius*	*Haemophilus aegyptius*
■ *Haemophilus ducreyi*	
■ Other *Haemophilus* spp.	
H. parainfluenzae	
H. parahaemolyticus	
H. aphrophilus	
H. haemolyticus	
H. segnis	
H. paraphrophilus	

GENERAL CHARACTERISTICS

Species of the genus *Haemophilus,* except for *H. aphrophilus,* require hemin and nicotine adenine dinucleotide (NAD) for in vitro growth. The morphologic and physiologic features of individual species are presented later in this chapter in the discussion of laboratory diagnosis.

EPIDEMIOLOGY

As presented in Table 43-1, except for *Haemophilus ducreyi, Haemophilus* spp. normally inhabit the upper respiratory tract of humans.[5] Although *H. ducreyi* is only found in humans, the organism is not part of our normal flora and its presence in clinical specimens is indicative of infection.

Among *H. influenzae* strains, there are two broad categories: typeable and nontypeable. Strains are typed based on capsular characteristics. The capsule is composed of a sugar-alcohol phosphate (i.e., polyribitol phosphate) complex. Differences in this complex are the basis for separating encapsulated strains into one of six groups: type a, b, c, d, e, or f. Type b *H. influenzae* is most commonly encountered in serious infections in humans. Nontypeable strains do not produce a capsule[13] and are most commonly encountered as normal inhabitants of the upper respiratory tract.

Although person-to-person transmission plays a key role in infections caused by type b *H. influenzae* and *H. ducreyi,* infections caused by other *Haemophilus* strains and species likely arise endogenously as a person's own flora gains access to a normally sterile site. *H. ducreyi* is unique among the species considered here in that it is the agent of a sexually transmitted disease.

PATHOGENESIS AND SPECTRUM OF DISEASES

Production of a capsule and factors that mediate bacterial attachment to human epithelial cells are the primary virulence factors associated with *Haemophilus* spp.[7,13] In general, infections caused by type b *H. influenzae* are often systemic and life-threatening, whereas infections caused by nontypeable strains are usually localized (Table 43-2). Infections caused by other *Haemophilus* spp. occur much less frequently.[1,2,8]

Chancroid is the sexually transmitted disease caused by *H. ducreyi* (Table 43-2). Although small outbreaks of this disease have occurred in the United States, this disease is more commonly seen among socioeconomically disadvantaged populations that inhabit tropical environments.

───────────────── **TABLE 43-1** **EPIDEMIOLOGY** ─────────────────

ORGANISM	HABITAT (RESERVOIR)	MODE OF TRANSMISSION
Haemophilus influenzae (including biogroup *aegyptius*)	Normal flora of human upper respiratory tract	Person-to-person spread by contaminated respiratory droplets. Certain infections may be caused by person's endogenous strains
Haemophilus ducreyi	Not part of normal human flora; only found in humans during infection	Person-to-person spread by sexual contact (i.e., causative agent of a sexually transmitted disease known as **chancroid**)
Other *Haemophilus* spp.; species most commonly associated with infection include *H. parainfluenzae*, *H. parahaemolyticus*, and *H. paraphrophilus*	Normal flora of human upper respiratory tract	Spread of a patient's endogenous strain to sites outside the respiratory tract. Much less commonly involved than *H. influenzae* in infections in humans

LABORATORY DIAGNOSIS

SPECIMEN COLLECTION AND TRANSPORT

Haemophilus spp. can be isolated from most clinical specimens. The collection and transport of these specimens is outlined in Table 1-1, with emphasis on the following points. First, *Haemophilus* spp. are very susceptible to drying and temperature extremes. Therefore, specimens suspected of containing these organisms must be plated immediately, especially if they have not been submitted in a suitable transport medium.

Second, for recovery of *H. ducreyi* from genital ulcers, special measures are necessary because of the bacterium's fastidious nature. The ulcer should be cleaned with sterile gauze premoistened with sterile saline. A cotton swab premoistened with phosphate-buffered saline is then used to collect material from the base of the ulcer. To maximize the chance for recovering the organism, the swab must be plated to special selective media within 10 minutes of collection.

SPECIMEN PROCESSING

Other than the precautions discussed above for *H. ducreyi*, no special considerations are required for specimen processing of *Haemophilus* spp. Refer to Table 1-1 for general information on specimen processing.

DIRECT DETECTION METHODS

Direct observation

Gram stain is generally used for the direct detection of *Haemophilus* in clinical material. However, in some instances the acridine orange stain (AO; see Chapter 11 for more information on this technique) is used to detect smaller numbers of organisms than may be detected by Gram stain.

To increase the sensitivity of direct Gram stain examination of body fluid specimens, especially cerebrospinal fluid (CSF), the specimens usually are centrifuged (2000 rpm for 10 minutes) and the smear is prepared from the resulting pellet. This concentration step can increase the sensitivity of direct microscopic examination by five- to ten-fold. Moreover, cytocentrifugation of the specimen, in which clinical material is concentrated by centrifugation directly onto microscope slides, reportedly increases sensitivity of the Gram stain by as much as 100-fold (see Chapter 23 for information on infections of the central nervous system).[12]

Gram stains of the smears made with patient specimens must be examined carefully. *Haemophilus* spp. stain a pale pink and may be difficult to detect in the pink background of proteinaceous material that is often found in patient specimens.

H. influenzae appear as coccobacilli or small rods, whereas the cells of *H. influenzae* biotype *aegypticus* usually appear as long, slender rods. *H. haemolyticus* are small coccobacilli or short rods with occasional cells appearing as tangled filaments.

H. parainfluenzae produce either small pleomorphic rods or long filamentous forms, whereas *H. parahaemolyticus* usually are short to medium-length bacilli. *H. aphrophilus* and *H. paraphrophilus* are very short bacilli but occasionally are seen as filamentous forms. *H. segnis* are pleomorphic rods, and the cells of *H. ducreyi* may be either slender or coccobacillary. Traditionally, *H. ducreyi* cells are described as appearing as "schools-of-fish." However, this morphology is rarely seen in clinical laboratory practice.

Antigen detection

Detection of *H. influenzae* type b capsular polysaccharide directly in clinical specimens, such as cerebrospinal fluid and urine, can be performed using commercially available particle agglutination assays (see Chapter 15). Because organisms in clinical infections are usually present at a sufficiently high concentration to be visualized by Gram stain, most clinical labooratories no longer perform the latex test. Histor-

ically, the latex tests have been very sensitive and specific for detection of *H. influenzae* type b, especially in patients who have been on antimicrobial therapy before collection of the specimen. Unfortunately, these tests have been reported as positive in CSF and urine of patients who have recently received an *H. influenzae* type b vaccination,[4,6] so positive results of agglutination assays must be interpreted with caution.

CULTIVATION

Media of choice

Chocolate agar provides the factors, that is, hemin (X factor) and NAD (V factor), necessary for the growth of *Haemophilus* spp. Most strains will not

---

TABLE 43-2 PATHOGENESIS AND SPECTRUM OF DISEASES

ORGANISM	VIRULENCE FACTORS	SPECTRUM OF DISEASE AND INFECTIONS
Haemophilus influenzae	For encapsulated strains (of which type b is most common), the capsule is antiphagocytic and highly associated with virulence; other cell envelope factors also facilitate attachment to host cells	Infections most commonly associated with encapsulated strains can be life-threatening and include meningitis, epiglottitis, cellulitis with bacteremia, septic arthritis, and pneumonia
	For nonencapsulated strains, pili and other cell surface factors, not fully understood, play a role in attachment to host cells	Nonencapsulated strains usually cause localized infections such as otitis media, sinusitis, conjunctivitis, and exacerbations of chronic bronchitis; pneumonia and bacteremia in adults with underlying medical conditions, such as malignancy or alcoholism, also can occur
Haemophilus influenzae biogroup *aegyptius*	Uncertain; probably similar to those of other *H. influenzae*	Purulent conjunctivitis; also one strain is the cause of Brazilian purpuric fever, which is a severe infection of high mortality in children between ages 1 and 4; infection includes purulent meningitis, bacteremia, high fever, vomiting, purpura (i.e., rash), and vascular collapse
Haemophilus ducreyi	Uncertain, but capsular factors, pili, and certain toxins are probably involved in attachment and penetration of host epithelial cells	Chancroid; genital lesions that progress from tender papules (i.e., small bumps) to painful ulcers with several satellite lesions. Regional lymphadenitis is common
Other *Haemophilus* spp.	Uncertain; probably of low virulence. Opportunistic pathogens	Associated with wide variety of infections similar to those caused by *H. influenzae* but much less commonly encountered in a clinically significant setting. *H. aphrophilus* is an uncommon cause of endocarditis and is the *H* member of the HACEK group of bacteria associated with slowly progressive (subacute) bacterial endocarditis

grow on 5% sheep blood agar, which contains hemin but not NAD. Several bacterial species, including *Staphylococcus aureus,* produce NAD as a metabolic by-product. Therefore, tiny colonies of *Haemophilus* spp. may be seen growing on sheep blood agar very close to colonies of bacteria that can produce V factor; this phenomenon is known as **satelliting** (Figure 43-1). A selective medium, such as horse blood–bacitracin agar, may be occasionally used for isolation of *H. influenzae* from respiratory secretions of patients with cystic fibrosis. This medium is designed to prevent overgrowth of *H. influenzae* by mucoid *Pseudomonas aeruginosa. Haemophilus* spp. will not grow on MacConkey agar.

H. ducreyi requires special media to grow. Mueller-Hinton–based chocolate agar, supplemented with 1% IsoVitaleX and 3 µg/mL vancomycin, or heart infusion–based agar supplemented with 10% fetal bovine serum and 3 µg/mL vancomycin, are two such media. The vancomycin serves to inhibit gram-positive organisms normally colonizing the genital tract.

Haemophilus spp. will grow in the broths of commercial blood culture systems and in common nutrient broths such as thioglycollate and brain-heart infusion. However, they often produce only weakly turbid suspensions and may not be readily visible in broth cultures. For this reason, blind subcultures to chocolate agar or examination of smears by AO or Gram stain have been used to enhance detection. However, the productivity of this extra effort for detecting *Haemophilus* spp. in patient cultures has never been established.

Rabbit or horse blood agars are commonly used for detecting hemolysis by hemolysin-producing strains of *Haemophilus* strains that will not grow on 5% sheep blood.

Incubation conditions and duration

Most strains of *Haemophilus* spp. are able to grow aerobically and anaerobically. Growth is stimulated by 5% to 10% carbon dioxide (CO_2), so that incubation in a candle extinction jar, CO_2 pouch, or CO_2 incubator is recommended. These organisms usually grow within 24 hours, but cultures are routinely held 72 hours before being discarded as negative. An exception is *H. ducreyi,* which may require as long as 7 days to grow.

Optimal growth of all *Haemophilus* spp., except *H. ducreyi,* occurs at 35° to 37° C. Cultures for *H. ducreyi* should be incubated at 33° to 35° C. In addition, *H. ducreyi* requires high humidity, which may be established by placing a sterile gauze pad moistened with sterile water inside the candle jar or CO_2 pouch.

Colonial appearance

Table 43-3 describes the colonial appearance and other distinguishing characteristics (e.g., odor and hemolysis) of each species.

APPROACH TO IDENTIFICATION

Commercial identification systems for *Haemophilus* spp. are summarized in Chapter 30. All of the systems incorporate several rapid enzymatic tests and generally work well for identifying these organisms.

Traditional identification criteria include hemolysis on horse or rabbit blood and the requirement for X and/or V factors for growth. To establish X and V factor requirements, disks impregnated with each factor are placed on unsupplemented media, usually Mueller-Hinton agar or trypticase soy agar, that has been inoculated with a light suspension of the organism. After overnight incubation at 35° C in ambient air, the plate is examined for growth around each disk. Many X factor–requiring organisms are able to carry over enough factor from the primary medium to give false-negative results (i.e., growth occurs at such a distance from the X disk as to falsely indicate that the organism does not require the X factor).

The porphyrin test is another means for establishing an organism's X-factor requirements and eliminates the potential problem of carry over. This test detects the presence of enzymes that convert α-aminolevulinic acid (ALA) into porphyrins or protoporphyrins. The porphyrin test may be performed in broth, in agar, or on a disk.

Haemophilus isolates may be identified to species using rapid sugar fermentation tests; an abbreviated identification scheme for the X- and/or V-requiring organisms is shown in Table 43-4.

FIGURE   43-1 *Haemophilus influenzae* satelliting (*arrow*) around colonies of *Staphylococcus aureus.*

TABLE 43-3 COLONY APPEARANCE AND CHARACTERISTICS

ORGANISM	MEDIUM	APPEARANCE
H. aphrophilus	CA	Round; convex with opaque zone near center
H. ducreyi	Selective medium*	Small, flat, smooth, and translucent to opaque at 48-72 hrs; colonies can be pushed intact across agar surface
H. haemolyticus	CA	Resembles H. influenzae except beta-hemolytic on rabbit or horse blood agar
H. influenzae	CA	Unencapsulated strains are small, smooth, and translucent at 24 hrs; encapsulated strains form larger, more mucoid colonies; mouse nest odor; nonhemolytic on rabbit or horse blood agar
H. influenzae biotype aegyptius	CA	Resembles H. influenzae except colonies are smaller at 48 hrs
H. parahaemolyticus	CA	Resembles H. parainfluenzae except beta-hemolytic on rabbit or horse blood agar
H. parainfluenzae	CA	Medium to large, smooth, and translucent; non-hemolytic on rabbit or horse blood agar
H. paraphrophilus	CA	Resembles H. aphrophilus
H. segnis	CA	Convex, grayish white, smooth or granular at 48 hrs

CA, chocolate agar.

*See p. 559 for some formulations.

Differentiation of *H. aphrophilus* from similar organisms (e.g., *A. actinomycetemcomitans,* Chapter 42) is shown in Table 42-5. *H. aphrophilus* does not require either X or V factors for growth. However, it is catalase-negative and ferments lactose or sucrose. *A. actinomycetemcomitans* yields the opposite reactions in these tests.

SEROTYPING

Although serologic typing of *H. influenzae* may be used to establish an isolate as being any one of the six serotypes (i.e., a, b, c, d, e, and f), it is usually only used to identify type b strains. Testing can be performed using a slide agglutination test (see Chapter 15); a saline control should always be run to detect autoagglutination (i.e., the nonspecific agglutination of the test organism without homologous antiserum).

SERODIAGNOSIS

An enzyme-linked immunosorbent assay (ELISA) has been developed to detect antibodies to *H. ducreyi.* ELISA and radioimmunoassay (RIA) have been used to show seroconversion following *H. influenzae* type b vaccination. None of these assays are used commonly for diagnostic purposes.

ANTIMICROBIAL SUSCEPTIBILITY TESTING AND THERAPY

Standard methods have been established for performing in vitro susceptibility testing with clinically relevant isolates of *Haemophilus* spp. (see Chapter 18 for details on these methods). In addition, various agents may be considered for testing and therapeutic use (Table 43-5).[9–11]* Although resistance to ampicillin by production of beta-lactamase has become widespread among *H. influenzae,* resistance to cephalosporins that are not notably affected by the enzyme (i.e., ceftriaxone and cefotaxime) is rare. Therefore, routine susceptibility testing of clinical isolates as a guide to therapy may not be necessary.

In addition to beta-lactamase production, beta-lactam resistance mediated by altered penicillin-binding

*Order from NCCLS, 940 W. Valley Road, Suite 1400, Wayne, Pa. 19087.

TABLE 43-4 KEY BIOCHEMICAL AND PHYSIOLOGIC CHARACTERISTICS OF *HAEMOPHILUS* SPP. THAT REQUIRE X AND/OR V FACTORS*

ORGANISM	X FACTOR	V FACTOR	BETA-HEMOLYTIC ON RABBIT BLOOD AGAR	LACTOSE FERMENTATION†	MANNOSE FERMENTATION†
Haemophilus influenzae	+	+	-	-	-
H. influenzae biotype *aegyptius*	+	+	-	-	-
H. haemolyticus	+	+	+	-	-
H. parahaemolyticus	-	+	+	-	-
H. parainfluenzae	-	+	-	-	+
H. paraphrophilus	-	+	-	+	+
H. segnis	-	+	-	-	-
H. ducreyi	+	-	-	-	-

+, >90% of strains positive; -, >90% of strains negative.

H. aphrophilus reactions are included in Chapter 42 (Table 42-5).

†Test performed in rapid sugar fermentation medium.

Data compiled from Campos, J.M. 1995. In Murray, P.R., Baron, E.J., Pfaller, M.A., et al., editors. Manual of clincial microbiology, ed 6. American Society for Microbiology, Washington D.C.; Weyant R.S., Moss, C.W., Weaver, R.E., et al., editors. 1996. Identification of unusual pathogenic gram-negative aerobic and facultatively anaerobic bacteria, ed 2. Williams & Wilkins, Baltimore.

TABLE 43-5 ANTIMICROBIAL THERAPY AND SUSCEPTIBILITY TESTING

ORGANISM	THERAPEUTIC OPTIONS	POTENTIAL RESISTANCE TO THERAPEUTIC OPTIONS	VALIDATED TESTING METHODS*	COMMENTS
Haemophilus influenzae	Usually ceftriaxone or cefotaxime for life-threatening infections; for localized infections several cephalosporins, beta-lactam/beta-lactamase inhibitor combinations, macrolides, trimethoprim/sulfamethoxazole, and certain quinolones are effective	Beta-lactamase–mediated resistance to ampicillin is common; beta-lactam resistance by altered PBP target is rare (≤1% of strains)	As documented in Chapter 18: disk diffusion, broth dilution, and certain commercial systems	Resistance to third-generation cephalosporins has not been documented. Testing to guide therapy is not routinely needed
Haemophilus ducreyi	Erythromycin is the drug of choice; other potentially active agents include ceftriaxone and ciprofloxacin	Resistance to trimethoprim/sulfamethoxazole and tetracycline has emerged; beta-lactamase–mediated resistance to ampicillin and amoxicillin is also known	Not available	
Other *Haemophilus* spp.	Guidelines the same as for *H. influenzae*	Beta-lactamase mediated resistance to ampicillin is known	As documented in Chapter 18: disk diffusion, broth dilution, and certain commercial systems	Resistance to third-generation cephalosporins has not been documented. Testing to guide therapy is not routinely needed

*Validated testing methods include those standard methods recommended by the National Committee for Clinical Laboratory Standards (NCCLS) and those commercial methods approved by the Food and Drug Administration (FDA).

proteins in *H. influenzae* has been well described and can lead to decreased activity of both beta-lactam/beta-lactamase inhibitor combinations and first- and second-generation cephalosporins. However, this mechanism of resistance is not common and its impact on the clinical efficacy of the antimicrobial agents affected is not fully known.

Although standardized testing guidelines for *H. ducreyi* have not been established by the NCCLS, erythromycin is widely accepted as an effective drug for treating chancroid. However, this organism can exhibit resistance to other therapeutic agents, so the potential exists for the emergence of significant antimicrobial resistance.[14]

PREVENTION

Several multiple-dose protein-polysaccharide conjugate vaccines are licensed in the United States for *H. influenzae* type b. These vaccines have substantially reduced the incidence of severe infections caused by type b organisms, and vaccination of children starting at 2 months of age is strongly recommended.

Rifampin chemoprophylaxis is recommended for all household contacts of index cases of *H. influenzae* type b meningitis in which there is at least one unvaccinated contact younger than age 4. Children and staff of daycare centers should also receive rifampin prophylaxis if at least two cases have occurred among the children.[3]

References

1. Chadwick, P.R., Malnick, H., and Ebizie, A.O. 1995. *Haemophilus paraphrophilus* infection: a pitfall in laboratory diagnosis. J. Infect. 30:67.
2. Coll-Vinent, B., Suris, X., Lopez-Soto, A., et al. 1995. *Haemophilus paraphrophilus* endocarditis: case report and review. Clin. Infect. Dis. 20:1381.
3. Committe on Infectious Diseases. 1997. 1997 Red Book pp. 222-231. American Academy of Pediatrics, Elk Grove Village, Ill.
4. Darville, T., Jacobs, R.F., Lucas, R.A., et al. 1992. Detection of *Haemophilus influenzae* type b antigen in cerebrospinal fluid after immunization. Pediatr. Infect. Dis. J. 11:243.
5. Foweraker, J.E., Cooke, N.J., and Hawkey, P.M. 1993. Ecology of *Haemophilus influenzae* and *Haemophilus parainfluenzae* in sputum and saliva and effects of antibiotics on their distribution in patients with lower respiratory tract infections. Antimicrob. Agents Chemother. 37:804.

6. Jones, R.G., Bass, J.W., Weisse, M.E., et al. 1991. Antigenuria after immunization with *Haemophilus influenzae* oligosaccharide CRM197 conjugate (H6OC) vaccine. Pediatr. Infect. Dis. J. 10:557.
7. Lageragard, T. 1995. *Haemophilus ducreyi*: pathogenesis and protective immunity. Trends Microbiol. 3:87.
8. Merino, D., Saavedra, J., Pujol, E., et al. 1994. *Haemophilus aphrophilus* as a rare cause of arthritis. Clin. Infect. Dis. 19:320.
9. National Committee for Clinical Laboratory Standards. 1995. Performance standards for antimicrobial susceptibility testing. M100-S6. NCCLS, Villanova, Pa.
10. National Committee for Clinical Laboratory Standards. 1997. Methods for dilution antimicrobial susceptibility tests for bacteria that grow aerobically. M7-A4, ed 4. NCCLS, Villanova, Pa.
11. National Committee for Clinical Laboratory Standards. 1997. Performance standards for antimicrobial disk susceptibility tests. M2-A6, ed 6. NCCLS, Villanova, Pa.
12. Shanholtzer, C.J., Schaper, P.J., and Peterson, L.R. 1982. Concentrated Gram-stained smears prepared with a cytospin centrifuge. J. Clin. Microbiol. 16:1052.
13. St. Geme, J.W., III. 1993. Nontypeable *Haemophilus influenzae* disease: epidemiology, pathogenesis, and prospects for prevention. Infect. Agents Dis. 2:1.
14. Van Dyck, E., Bogaerts, J., Smet, H., et al. 1994. Emergence of *Haemophilus ducreyi* resistance to trimethoprim-sulfamethoxazole in Rwanda. Antimicrob. Agents Chemother. 38:1647.

Bibliography

Campos, J.M. 1995. *Haemophilus*. In Murray, P.R., Baron, E.J., Pfaller, M.A., et al., editors. Manual of clinical microbiology, ed 6. American Society for Microbiology, Washington, D.C.

Doern, G.V. 1995. Susceptibility testing of fastidious bacteria. In Murray, P.R., Baron, E.J., Pfaller, M.A., et al., editors. Manual of clinical microbiology, ed 6. American Society for Microbiology, Washington, D.C.

Hand, W.L. 1995. *Haemophilus* species (including chancroid). In Mandell, G.L., Bennett, J.E., and Dolin, R., editors. Principles and practice of infectious diseases. Churchill Livingstone, New York.

Moxon, E.R. 1995. *Haemophilus influenzae*. In Mandell, G.L., Bennett, J.E., and Dolin, R., editors. Principles and practice of infectious diseases. Churchill Livingstone, New York.

Weyant, R.S., Moss, C.W., Weaver, R.E., et al., editors. 1996. Identification of unusual pathogenic gram-negative aerobic and facultatively anaerobic bacteria, ed 2. Williams & Wilkins, Baltimore.

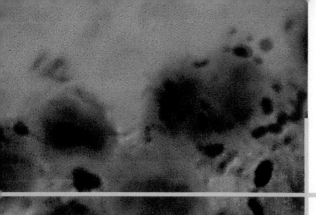

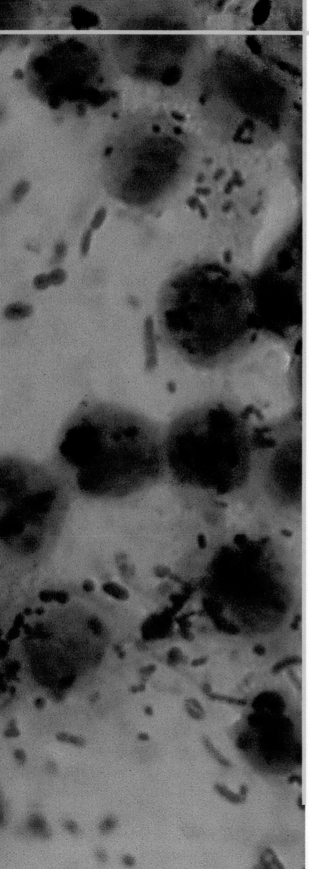

Section Five

5

Gram-Negative Bacilli that Are Optimally Recovered on Special Media

44 | BARTONELLA AND *AFIPIA*

Genera and Species to be Considered

Current Name	Previous Name
■ *Bartonella bacilliformis*	
■ Other *Bartonella* spp., including	
B. quintana	*Rochalimaea quintana*
B. henselae	*Rochalimaea henselae*
B. elizabethae	*Rochalimaea elizabethae*
B. clarridgeiae	
■ *Afipia felis*	

The two genera, *Bartonella* and *Afipia*, are able to grow on chocolate agar but not on routine blood (trypticase soy agar with 5% sheep blood) and Mac-Conkey agars. Despite these growth characteristics, the primary isolation of these organisms from clinical specimens is best accomplished using other cultivation conditions that better fulfill the growth requirements of these fastidious organisms. In light of this and the fact that two organisms, *Bartonella henselae* and *Afipia felis*, were each believed to cause cat-scratch disease, these genera are addressed together in this chapter.

BARTONELLA

GENERAL CHARACTERISTICS

The taxonomy for organisms belonging to the genus *Bartonella* has undergone extensive changes during the past several years. *Bartonella* spp. originally were grouped with members of the order *Rickettsiales*. However, because of extensive differences, the removal of the genus *Bartonella* from this order has been proposed. Another genus, *Rochalimaea*, also belonged to the same order. Extensive differences with other members of the *Rickettsiales* yet marked relatedness with *Bartonella* spp. prompted the taxonomic unification of the genera *Bartonella* and *Rochalimaea*. As a result of phylogenetic studies using molecular biologic techniques, the genus *Bartonella* now includes 10 species[7]; however, only four species are currently recognized as causes of disease in humans

(Table 44-1). *Bartonella* spp. are most closely related to *Brucella abortus* and *Agrobacterium tumefaciens*[11] and are short, gram-negative, rod-shaped, fastidious organisms that are oxidase-negative and grow best on blood-enriched media.

EPIDEMIOLOGY AND PATHOGENESIS

Organisms belonging to the genus *Bartonella* cause numerous infections in humans; most of these infections are thought to be zoonoses. During the past several years, interest in these organisms has increased because of their recognition as causes of an expanding array of clinical syndromes in immunocompromised and immunocompetent patients. Many questions remain regarding the epidemiology of these infections; some epidemiologic information is summarized in Table 44-1. Although little is known about how these agents cause disease, pathologic similarities among the different diseases caused by bartonellae exist, including bacteremia and cutaneous angioprolific (proliferation of small blood vessels) lesions.

SPECTRUM OF DISEASE

The diseases caused by *Bartonella* species are listed in Table 44-1. Because *B. quintana* and *B. henselae* are more common causes of infections in humans, these agents will be addressed in greater depth.

Trench fever, caused by *B. quintana*, was largely considered a disease of the past that afflicted about 1 million soldiers. Clinical manifestations of trench

TABLE 44-1 ORGANISMS BELONGING TO THE GENUS *BARTONELLA* AND RECOGNIZED TO CAUSE DISEASE IN HUMANS

ORGANISM	HABITAT (RESERVOIR)	MODE OF TRANSMISSION	CLINICAL MANIFESTATION
Bartonella bacilliformis	Uncertain; humans	Sand flies (?)*	Carrión's disease†
B. quintana	Uncertain; possibly small rodents	Human body louse (?)	Trench fever; relapsing fever; bacteremia; endocarditis; bacillary angiomatosis; chronic lymphadenopathy
B. henselae	Domestic cats	Domestic cat by bites and/or scratches; cat fleas (?)	Bacteremia; endocarditis; cat-scratch disease; bacillary angiomatosis; peliosis hepatitis
B. clarridgeiae	Domestic cats	Domestic cat by bites and/or scratches	Bacteremia; cat-scratch disease
B. elizabethae	Unknown	Unknown	Endocarditis

*(?), Implicated.

†Disease confined to a small endemic area in South America; characterized by a septicemic phase with anemia, malaise, fever, and enlarged lymph nodes, liver and spleen followed by a cutaneous phase with bright red cutaneous nodules that are usually self-limited.[3]

fever range from a mild influenza-like headache, bone pain, splenomegaly (enlarged spleen), and a short-lived maculopapular rash. However, recently, *B. quintana* has reemerged and has been reported to cause bacteremia, endocarditis, chronic lymphadenopathy, and bacillary angiomatosis primarily in homeless people in Europe and the United States, as well as in patients infected with the human immunodeficiency virus (HIV). **Bacillary angiomatosis** is a vascular proliferative disease most often involving the skin (other organs such as the liver, spleen, and lymph nodes may also be involved) and occurs in immunocompromised individuals such as organ transplant recipients and people who are HIV-positive.

B. *henselae* can also cause bacteremia, endocarditis, and bacillary angiomatosis. In addition, *B. henselae* causes **cat-scratch disease (CSD)** and peliosis hepatitis. About 24,000 cases of CSD occur annually in the United States, about 80% of these occur in children.[8] Usually this infection begins as a papule or pustule at the primary inoculation site; regional tender lymphadenopathy develops in 1 to 7 weeks. The spectrum of disease ranges from chronic, self-limited adenopathy to a severe systemic illness affecting multiple body organs. Although complications such as a suppurative (draining) lymph node or encephalitis are reported, fatalities are rare. Diagnosis of CSD tradi-

tionally required that a patient fulfill three of the four following criteria:

- History of animal contact plus site of primary inoculation (e.g., a scratch)
- Negative laboratory studies for other causes of lymphadenopathy
- Characteristic histopathology of the lesion
- A positive skin test using antigen prepared from heat-treated pus taken from another patient's lesion

Bartonella clarridgeiae is a newly described species that also causes cat-scratch disease and bacteremia.[5,6]

Peliosis hepatitis caused by *B. henselae* can occur alone or develop with cutaneous bacillary angiomatosis or bacteremia. Patients with peliosis hepatitis present with gastrointestinal symptoms, fever, chills, and an enlarged liver and spleen containing blood-filled cavities. This systemic disease is seen in patients infected with HIV and other immunocompromised individuals.

LABORATORY DIAGNOSIS

Specimen collection, transport, and processing
Clinical specimens frequently submitted to the laboratory for direct examination and culture include blood, collected in a lysis-centrifugation blood culture tube

(Isolator; Wampole Laboratories, Cranbury, N.J.), as well as aspirates and/or tissue specimens (e.g., lymph node, spleen, or cutaneous biopsies). There are no special requirements for specimen collection, transport, or processing of the organisms discussed in this chapter. Refer to Table 1-1 for general information on specimen collection, transport, and processing.

Direct detection methods

The presence of *Bartonella* spp. can be detected on histopathologic examination of tissue biopsies stained with the Warthin-Starry silver stain; immunofluorescence and immunohistochemical techniques have also been used. The organisms may also be directly identified in tissue and body fluids by polymerase chain reaction (PCR) amplification of *Bartonella*-specific sequences.[1,10]

Cultivation

Although still being debated as the optimum conditions required for recovery of bartonellae from clinical specimens, the two most widely used methods are direct plating onto fresh chocolate agar plates (less than 2 weeks old) and co-cultivation in cell culture.[7,12] Lysed, centrifuged sediment of blood collected in an Isolator tube or minced tissue is directly inoculated to fresh chocolate agar plates and incubated at 35° C in a very humid atmosphere with 5% to 10% carbon dioxide (CO_2), examined daily for 3 days, and again after 2 weeks of incubation. Fresh agar helps to supply moisture necessary for growth. Biopsy material is co-cultivated with an endothelial cell culture system; co-cultures are incubated at 35° C in 5% to 10% CO_2 for 15 to 20 days. Blood-enriched agar, such as Columbia or heart infusion agar base with 5% sheep blood, can also be employed, but horse or rabbit blood is reported to be more effective supplements than sheep blood.[7] Of importance, *B. henselae* appears more fastidious than *B. quintana* in vitro, and recovery of *Bartonella* spp. from patients is more successful from blood than tissue specimens.[4]

Approach to identification

Bartonella spp. should be suspected when colonies of small, gram-negative bacilli are recovered after prolonged incubation (Figure 44-1). Organisms are all oxidase- and catalase-negative. Various means can be used to confirm the identification of *Bartonella*. Species can be delineated using biochemical profiles when 100 μg/mL of hemin is added to the test medium, the MicroScan rapid anaerobe panel, and polyvalent antisera.[2,7,12]

Serodiagnosis

Several serologic methods for detecting antibodies to *Bartonella* spp. have been developed. An indirect fluorescent antibody using antigen prepared from *Bartonella* spp. co-cultivated with Vero cells[8] and enzyme-linked immunoassays have been developed and evaluated. However, the sensitivity and specificity of these assays have been questioned. Cross-reactivity between *Bartonella* and *Chlamydia* spp. and *Coxiella burnettii* have been reported.[7]

ANTIMICROBIAL SUSCEPTIBILITY TESTING AND THERAPY

Because so few strains have been isolated and bartonellae are so fastidious, antibiotic susceptibility testing is currently limited to only a few laboratories. Several antibiotics are effective against infection caused by bartonellae and include the aminoglycosides, erythromycin, and doxycycline; numerous other

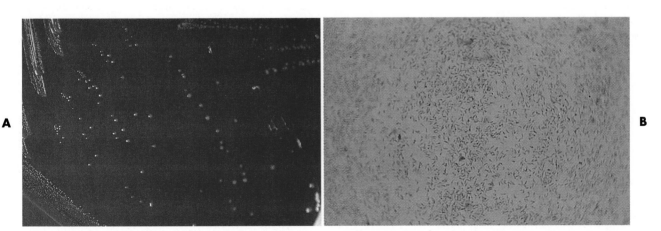

FIGURE 44-1 **A,** Colonies of *Bartonella henselae* on blood agar. **B,** Gram stain of a colony of *B. henselae* from blood agar.

antibiotics have also been successfully used in a limited number of patients.

PREVENTION

There are no vaccines available to prevent infections caused by *Bartonella* spp. Because little is known about the epidemiology of these agents, strategies to prevent infection are lacking. Because of a probable link between cats and/or cat fleas and transmission of *B. henselae* to humans, it has been recommended that immunocompromised individuals avoid contact with cats, especially kittens, and control flea infestation.

AFIPIA FELIS

Cat-scratch disease (CSD) was first reported in 1931; however, the causative agent was unknown for several decades. Finally a bacterial agent was isolated and characterized and given the name *Afipia felis*. However, the role of *A. felis* in the etiology of CSD was subsequently questioned because patients with CSD failed to mount an immune response to *A. felis* antigen and the organism was not detected from patients with CSD by either culture or PCR. Also, data was provided that patients with CSD mounted an immune response to *B. henselae*; that *B. henselae* was detected by culture, PCR, and immunocytochemistry; and that *B. henselae* was detected in CSD skin test antigens, cats, and cat fleas. In light of all the data, *B. henselae* is now recognized as the causative agent of CSD.[8]

A. felis appears to be in the environment, having been isolated from surface water. This organism grows best on buffered charcoal-yeast extract agar, like *Legionella* spp. (see Chapter 46).[9] The exact role of this organism in human disease remains unclear.

References

1. Anderson, B. et al. 1994. Detection of *Rochalimaea henselae* DNA in specimens from cat scratch disease patients by PCR. J. Clin. Microbiol. 32:942.
2. Drancourt, M. and Raoult, D. 1993. Proposed tests for the routine identification of *Rochalimaea* species. Eur. J. Clin. Microbiol. Infect. Dis. 12:710.
3. Garcia-Caceres, U. and Garcia, F.U. 1991. Bartonellosis: an immunosuppressive disease and the life of Daniel Alcides Carrión. Am. J. Clin. Pathol. 95(suppl. 1):S58.
4. Hensel, D.M. and Slater, L.N. 1995. The genus *Bartonella*. Clin. Microbiol. Newsletter 17:9.
5. Kordick, D.L., et al. 1997. *Bartonella clarridgeiae*, a newly recognized zoonotic pathogen causing inoculation papules, fever, and lymphadenopathy (cat-scratch disease). J. Clin. Microbiol. 35:1813.
6. Lawson, P.A. and Collins, M.D. 1996. Description of *Bartonella clarridgeiae* sp. nov. isolated from the cat of a patient with *Bartonella henselae* septicemia. Med. Microbiol. Lett. 5:640.
7. Maurin, M. and Raoult, D. 1996. *Bartonella (Rochalimaea) quintana* infections. Clin. Microbiol. Rev. 9:273.
8. Midani, S., Ayoub, E.M., and Anderson, B. 1996. Cat scratch disease. Adv. Pediatr. 43:397.
9. Müller, H. 1995. Investigations of culture and properties of *Afipia* spp. Zbl. Bakt. 282:18.
10. Relman, D.A. et al. 1990. The agent of bacillary angiomatosis: an approach to the identification of uncultured pathogens. N. Engl. J. Med. 323:1573.
11. Weisburg, W.G. et al. 1985. A common origin of rickettsiae and certain plant pathogens. Science 230:556.
12. Welch, R.F. et al. 1993. Bacteremia due to *Rochalimaea henselae* in a child: practical identification of isolates in the clinical laboratory. J. Clin. Microbiol. 31:2381.

45 | CAMPYLOBACTER, ARCOBACTER, AND HELICOBACTER

Genera and Species to be Considered

- *Campylobacter coli*
- *Campylobacter concisus*
- *Campylobacter curvus*
- *Campylobacter fetus* subsp. *fetus*
- *Campylobacter fetus* subsp. *venerealis*
- *Campylobacter gracilis*
- *Campylobacter helveticus*
- *Campylobacter hyoilei*
- *Campylobacter hyointestinalis* subsp. *hyointestinalis*
- *Campylobacter hyointestinalis* subsp. *lawsonii*
- *Campylobacter jejuni* subsp. *doylei*
- *Campylobacter jejuni* subsp. *jejuni*
- *Campylobacter lari*

- *Campylobacter mucosalis*
- *Campylobacter rectus*
- *Campylobacter showae*
- *Campylobacter sputorum* bv. *fecalis*
- *Campylobacter sputorum* bv. *sputorum*
- *Campylobacter upsaliensis*
- *Arcobacter cryaerophilus*
- *Arcobacter butzleri*
- *Arcobacter skirrowii*
- *Arcobacter nitrofigilis*
- *Helicobacter pylori*
- *Helicobacter cinaedi*
- *Helicobacter fennelliae*

Because of morphologic similarities and an inability to recover these organisms using routine laboratory media for primary isolation, the genera *Campylobacter*, *Arcobacter*, and *Helicobacter* are considered in this chapter. All organisms belonging to these genera are small, curved, motile, gram-negative bacilli. With few exceptions, the majority of these bacteria also have a requirement for a microaerophilic atmosphere.

CAMPYLOBACTER AND ARCOBACTER

GENERAL CHARACTERISTICS

The taxonomy of the genus *Campylobacter* has been extensively revised, particularly over the past 10 years. Being somewhat closely related to *Campylobacter* spp., *Arcobacter cryaerophilus* and *Arcobacter butzleri* belonged to this genus until 1991. Of significance, the number of campylobacters implicated in human disease has increased as methods of detection have improved. Currently, 16 species and 6 subspecies are recognized in the genus *Campylobacter*; these organisms are listed in Table 45-1. *Campylobacter* and *Arcobacter* spp. are relatively slow-growing, fastidious, and in general, asaccharolytic.

EPIDEMIOLOGY AND PATHOGENESIS

The vast majority of campylobacters appear to be pathogenic and are associated with a wide variety of diseases in humans and other animals. These organisms also demonstrate considerable ecologic diversity. *Campylobacter* spp. are microaerobic inhabitants of the gastrointestinal tracts of various animals, including poultry, dogs, cats, sheep, and cattle, as well as the reproductive organs of several animal species (Table 45-1). When fecal samples from chicken carcasses chosen at random from butcher shops in the New York City area were tested for *Campylobacter*, 83% of the samples yielded more than 10^6 colony-forming units per gram of feces. In general, *Campylobacter* spp. produce three syndromes in humans: febrile systemic disease, periodontal disease, and, most commonly, gastroenteritis. *Arcobacter* species appear to be associated with gastroenteritis as well.

Within the genus *Campylobacter*, *C. jejuni* and *C. coli* are most often associated with infections in humans and are usually transmitted via contaminated food, milk, or water. Outbreaks have been associated with contaminated drinking water and improperly pasteurized milk, among other sources. In contrast to other agents of foodborne gastroenteritis, including *Salmonella* and staphylococci, *Campylobacter* does

TABLE 45-1 *CAMPYLOBACTER* AND *ARCOBACTER* SPP.,
THEIR SOURCE, AND SPECTRUM
OF DISEASE

ORGANISM	SOURCE(S)	SPECTRUM OF DISEASE IN HUMANS
C. concisus, C. curvus, C. rectus, C. showae	Humans	Periodontal disease
C. gracilis[14]	Humans	Deep-tissue infections of head, neck, and viscera; gingival crevices
C. coli	Pigs, poultry, sheep, bulls, birds	Gastroenteritis (1°*); septicemia
C. jejuni subsp. jejuni	Poultry, pigs, bulls, dogs, cats, birds, and other animals	Gastroenteritis (1°); septicemia; meningitis; proctitis
C. jejuni subsp. doylei	Humans	Gastroenteritis (1°); gastritis; septicemia
C. lari	Birds, poultry, other animals; river and seawater	Gastroenteritis (1°); septicemia
C. hyointestinalis subsp. hyointestinalis	Pigs, cattle, hamsters, deer	Gastroenteritis
C. upsaliensis	Dogs, cats	Gastroenteritis; septicemia; abscesses
C. fetus subsp. fetus	Cattle, sheep	Septicemia; gastroenteritis; abortion; meningitis
C. fetus subsp. venerealis	Cattle	Septicemia
C. sputorum bv.† sputorum	Humans, cattle, pigs	Abscesses; gastroenteritis
C. mucosalis	Pigs	None at present
C. sputorum bv. fecalis	Sheep, bulls	None at present
C. hyoilei	Pigs	None at present
C. helveticus	Cats, dogs	None at present
C. hyointestinalis subsp. lawsonii	Pigs	None at present
Arcobacter cryaerophilus	Pigs, bulls, and other animals	Gastroenteritis (1°); septicemia
A. butzleri	Pigs, bulls, humans, other animals; water	Gastroenteritis (1°); septicemia
A. skirrowii	Sheep, bulls, pigs	None at present
A. nitrofigilis	Plant roots	None at present

*1°, Most common clinical presentation.
†bv, biovariant.

not multiply in food.[2] Other campylobacters have been isolated from patients who drank untreated water, were compromised in some way, or were returning from international travel. *C. jejuni* subsp. *doylei* has been isolated from children with diarrhea and from gastric biopsies from adults. In developed countries, the majority of *C. jejuni* infections in humans are acquired during the preparation and eating of chicken.[1] Of note, person-to-person transmission of *Campylobacter* infections plays only a minor role in the transmission of disease. There is a marked seasonality with the rates of *C. jejuni* infection in the United States; the highest rates of infection occur in late summer and early fall. *Campylobacter* spp., usually *C. jejuni,* have been recognized as the most common etiologic agent of gastroenteritis in the United States. However, this predominance may be a reflection of the laboratory methods used to detect campylobacters. Ratios of recovery of *Campylobacter* vs. *Salmonella* range from 2:1 to 46:1.[2]

Motility contributes to campylobacters' ability to colonize and infect intestinal mucosa. Although infection with *C. jejuni* results in an acute inflammatory enteritis (see Chapter 27) that affects the small intestine and colon, the pathogenesis of infection remains unclear. Blood and polymorphonuclear neutrophils are often observed in stool specimens from infected patients. Most strains of *C. jejuni* are susceptible to the nonspecific bactericidal activity of normal human serum; this susceptibility probably explains why *C. jejuni* bacteremia is uncommon.[1] Humoral immune responses are important in controlling *C. jejuni* infections and cell-mediated immunity probably plays some role as well.

SPECTRUM OF DISEASE

As previously mentioned, campylobacters can cause either gastrointestinal or extraintestinal infections. Extraintestinal disease, including meningitis, endocarditis, and septic arthritis, is being recognized increasingly, particularly in patients with acquired immunodeficiency syndrome (AIDS) and other immunocompromised individuals. The different campylobacters and the types of infections they cause are summarized in Table 45-1. Gastroenteritis caused by *Campylobacter* spp. is usually a self-limiting illness and does not require antibiotic therapy. Most recently, postinfectious complications following infection with *C. jejuni* have been recognized and include reactive arthritis and most notably, Guillain-Barré syndrome, an acute demyelination (removal of the myelin sheath from a nerve) of the peripheral nerves. Studies indicate that 20% to 40% of patients with this syndrome are infected with *C. jejuni* in the 1 to 3 weeks before the onset of neurologic symptoms.[8]

LABORATORY DIAGNOSIS

Specimen collection, transport, and processing

There are no special requirements for the collection, transport, and processing of clinical specimens for the detection of campylobacters; the two most common specimens submitted to the laboratory are feces (rectal swabs are also acceptable for culture) and blood. If a delay of more than 2 hours is anticipated, stool should be placed either in Cary-Blair transport medium or in campy thio, a thioglycollate broth base with 0.16% agar and vancomycin (10 mg/L), trimethoprim (5 mg/L), cephalothin (15 mg/L), polymyxin B (2500 U/L), and amphotericin B (2 mg/L). Cary-Blair transport medium is also suitable for other enteric pathogens; specimens received in this transport medium should be processed immediately or stored at 4° C until processed.

Direct detection

Because of their characteristic microscopic morphology, that is, small, curved or seagull-winged, faintly staining, gram-negative rods (Figure 45-1), *Campylobacter* spp. can sometimes be detected by direct Gram stain examination of stool (Figure 45-2). Although polymerase chain reaction (PCR) assays have detected *Campylobacter* spp. directly in stool, its practical utility is questionable.

Cultivation

STOOL To successfully isolate *Campylobacter* spp. from stool, selective media and optimum incubation conditions are critical. For optimum recovery, the inoculation of two selective agars is recommended.[4] Because *Campylobacter* and *Arcobacter* spp. have different optimum temperatures, two sets of selective

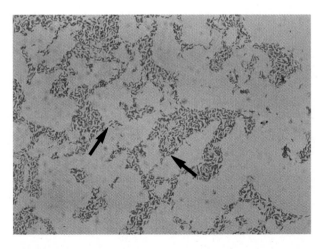

FIGURE 45-1 Gram stain appearance of *Campylobacter jejuni* subsp. *jejuni* from a colony on a primary isolation plate. Note seagull and curved forms (*arrows*).

plates should be incubated at 42° C and 37° C. Table 45-2 describes the selective plating media and incubation conditions required to recover *Campylobacter* spp. from stool.

Although not as sensitive as direct culture with selective media, a filtration method can also be employed in conjunction with a nonselective medium to recover *Campylobacter* spp. A filter (0.65-μm pore-size cellulose acetate) is placed on the agar surface, and a drop of stool is placed on the filter. The plate is incubated upright. After 60 minutes at 37° C, the filter is removed and the plates are reincubated in a microaerobic atmosphere. Because *Campylobacter* spp. are able to move through the filter, the organism is effectively removed from contaminating stool flora and colonies are produced on the agar surface.

BLOOD *Campylobacter* spp. that cause septicemia (see Table 45-1) grow in most blood culture media, although they may require as long as 2 weeks for growth to be detected. Subcultures from broths must be incubated in a microaerobic atmosphere or the organisms will not multiply. Turbidity is often not visible in blood culture media; therefore, blind subcultures or microscopic examination using acridine orange stain may be necessary. The presence of *Campylobacter* spp. in blood cultures is detected effectively by carbon dioxide (CO_2) monitoring. Isolation from sources other than blood or feces is extremely rare but is ideally accomplished by inoculating the material (minced tissue, wound exudate) to a nonselective blood or chocolate agar plate and incubating the plate at 37° C in a CO_2-enriched, microaerobic atmosphere. (Selective agars containing a cephalosporin, rifampin, and polymyxin B may inhibit growth of some strains and should not be used for isolation from normally sterile sites.)

Approach to identification

Plates should be examined for characteristic colonies, which are gray to pinkish or yellowish gray and slightly mucoid-looking; some colonies may exhibit a tailing effect along the streak line (Figure 45-3). However, other colony morphologies are also frequently seen, depending on the media used. Suspicious-looking colonies seen on selective media incubated at 42° C may be presumptively identified as *Campylobacter* spp., usually *C. jejuni* or *C. coli*, with a few simple tests. A wet preparation of the organism in broth may be examined for characteristic darting motility and curved forms on Gram stain. Almost all the pathogenic *Campylobacter* spp. are oxidase- and catalase-positive. For most laboratories, reporting of such isolates from feces as "*Campylobacter* spp." should suffice.

Most *Campylobacter* spp. are asaccharolytic, unable to grow in 3.5% NaCl, although strains of *Arcobacter* appear more resistant to salt, and except for *Arcobacter cryaerophilus*, are unable to grow in air. Growth in 1% glycine is variable. Susceptibility to nalidixic acid and cephalothin, an important differential characteristic among species (Table 45-3), is determined by inoculating a 5% sheep blood or Mueller-Hinton agar plate with a McFarland 0.5 turbidity suspension of the organism as for agar disk diffusion susceptibility testing, placing 30 μg disks on the agar surface, and incubating microaerobically at 37° C. Other tests useful for identifying these species are the rapid hippurate hydrolysis test, production of hydrogen sulfide (H_2S) in triple sugar iron agar butts, nitrate reduction, and hydrolysis of indoxyl acetate.[9] Indoxyl acetate disks are available commercially. Cellular fatty acid analysis can also help differentiate among species. This method is not available to most clinical microbiology laboratories. Several commercial products are available for species identification, including particle agglutination methods and nucleic acid probes.

Serodiagnosis

Serodiagnosis is not widely applicable for the diagnosis of infections caused by these organisms.

ANTIMICROBIAL SUSCEPTIBILITY TESTING AND THERAPY

Because susceptibility tests for *Campylobacter* spp. are not standardized, susceptibility testing of isolates is not routinely performed. *C. jejuni* and *C. coli* are susceptible to many antimicrobial agents, including macrolides, tetracyclines, aminoglycosides, and

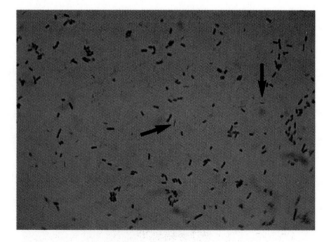

FIGURE 45-2 Appearance of *Campylobacter jejuni* subsp. *jejuni* in a direct Gram stain of stool obtained from a patient with campylobacteriosis. Arrows point to the "seagull" form.

TABLE 45-2   SELECTIVE MEDIA AND INCUBATION CONDITIONS TO RECOVER *CAMPYLOBACTER* AND *ARCOBACTER* SPP. FROM STOOL SPECIMENS

ORGANISM	PRIMARY PLATING MEDIA	INCUBATION CONDITIONS
C. jejuni C. coli	Modified Skirrow's media: Columbia blood agar base, 7% horse-lysed blood, and antibiotics (vancomycin, trimethoprim, and polymyxin B) Campy-BAP: *Brucella* agar base with antibiotics (trimethoprim, polymyxin B, cephalothin, vancomycin, and amphotericin B) and 10% sheep blood Blood-free, charcoal-based selective medium: Columbia base with charcoal, hemin, sodium pyruvate, and antibiotics (vancomycin, cefoperazone, and cyclohexamide) Modified charcoal cefoperazone deoxycholate agar (CCDA) Semisolid motility agar: Mueller-Hinton broth II, agar, cefoperazone, and trimethoprim lactate Campy-CVA: *Brucella* agar base with antibiotics (cefoperazone, vancomycin, and amphotericin B) and 5% sheep blood	42° C under microaerophilic conditions* for 72 hrs
C. fetus subsp. fetus† C. jejuni subsp. doylei C. upsaliensis C. lari C. hyointestinalis	Modified Skirrow's media Blood-free charcoal-based selective media Campy-CVA CCDA Semisolid motility agar	37° C under microaerophilic conditions for at least 72 hr up to 7 days‡
A. cryaerophilus, A. butzleri	Campy-CVA	37° C under microaerophilic conditions§ for 72 hrs

*Atmosphere can be generated in several ways, including commercially produced, gas-generating envelopes meant to be used with plastic bags or jars. Evacuation and replacement in plastic bags or anaerobic jars with an atmosphere of 10% CO_2, 5% O_2, and the balance of nitrogen (N_2) is the most cost-effective method, although it is labor intensive.

†All these organisms are susceptible to cephalothin.

‡*C. upsaliensis* will grow at 42° C but not on cephalothin-containing selective agar.

§*A. cryaerophilus* does not require microaerophilic conditions.

quinolones.[13] Erythromycin is the drug of choice for patients with more severe gastroenteritis (severe dehydration, bacteremia), with ciprofloxacin as an alternative drug. Parenteral therapy (not taken through the alimentary canal but by some other route such as intravenous) is used to treat systemic infections.

PREVENTION

No vaccines are available for *Campylobacter* spp. Because many infections caused by *Campylobacter* spp. are acquired by ingesting contaminated foodstuffs or water, all foods derived from animal sources, particularly poultry, should be thoroughly cooked. All milk should be pasteurized and drinking water chlorinated. Care must be taken during food preparation to prevent cross-contamination from raw poultry to other food items.

FIGURE 45-3 Colonies of *Campylobacter jejuni* after 48 hours' incubation on selective medium in a microaerophilic atmosphere.

TABLE 45-3   DIFFERENTIAL CHARACTERISTICS OF CLINICALLY RELEVANT CAMPYLOBACTER, ARCOBACTER, AND HELICOBACTER SPP.

GENUS AND SPECIES	GROWTH AT 25°C	GROWTH AT 42°C	HIPPURATE HYDROLYSIS	CATALASE	H₂S IN TRIPLE SUGAR IRON AGAR	INDOXYL ACETATE HYDROLYSIS	NITRATE TO NITRITE	SUSCEPTIBLE TO 30-µg DISK CEPHALOTHIN	SUSCEPTIBLE TO 30-µg DISK NALIDIXIC ACID
C. coli	-	+	-	+	-	+	+	-	+
C. concisus	-	+	-	-	+	-	+	-	-
C. curvus*	-	+	-	-	+	+	+	ND	+
C. fetus subsp. fetus	+	-/+	-	+	-	-	+	+	-
C. hyointestinalis	+/-	+	-	+	+	-	+	+	+
C. jejuni	-	+	+	+	-	+	+	-	+
C. jejuni subsp. doylei	-	+/-	+	+/- or weak +	-	+	-	+	+
C. lari	-	+	-	+	-	-	+	-	-
C. rectus*	-	Slight +	-	-	+	+	+	ND	+
C. sputorum	-	+	-	-/+	+	-	+	+	-/+
C. upsaliensis	-	+	+	-/weak+	-	+	+	+	+
A. butzleri†	+	-	-	-/weak+	-	+	+	-	+/-
A. cryaerophilus‡	+	-	-	+/-	-	+	+	-	+/-
H. cinaedi	-	-/+	-	+	-	-/+	+	+/-	+
H. fennelliae	-	-	-	+	-	+	-	+	+
H. pylori§	-	+	-	+	-	-	+/-	+	-

+, most strains positive; -, most strains negative; +/-, variable (more often positive); -/+, variable (more often negative); ND, test not done.

*Anaerobic, not microaerobic.

†Grows at 40°C.

‡Aerotolerant, not microaerobic; except for a few strains, A. cryaerophilus cannot grow on MacConkey agar, whereas A. butzleri grows on MacConkey agar.

§Strong and rapid positive urease.

HELICOBACTER

GENERAL CHARACTERISTICS

In 1983 spiral-shaped organisms resembling campylobacteria were isolated from the human stomach; these organisms were named *Campylobacter pylori.* Based on many studies, the genus *Helicobacter* was established in 1989 and *C. pylori* was renamed *Helicobacter pylori.* At least 14 species are now included in this genus, the majority of which colonize mammalian stomachs or intestines. The genus *Helicobacter* consists of curved, microaerophilic, gram-negative rods that have strong urease activity. *Helicobacter* spp. isolated from humans include *H. pylori, H. cinaedi, H. fennelliae, H. heilmannii,* and *H. rappini.* Because *H. pylori, H. cinaedi,* and *H. fennelliae* are significant human pathogens, only these species are addressed.

EPIDEMIOLOGY AND PATHOGENESIS

Helicobacter pylori's primary habitat is the human gastric mucosa. The organism is found worldwide. Although acquired early in life in underdeveloped countries, its exact mode of transmission is unknown. An oral-oral, fecal-oral or a common environmental source have been proposed as possible routes of transmission.[5] Of note, the housefly may be a significant means of transmission from person to person.[5] In industrialized nations, antibody surveys show that approximately 50% of adults older than age 60 are infected. Gastritis incidence increases with age. *H. pylori* has occasionally been cultured from feces and dental plaque, thereby suggesting a fecal-oral or oral-oral transmission.[3]

The habitat for *H. cinaedi* and *H. fennelliae* appears to be the gastrointestinal tract of humans and these organisms may be part of the resident flora; hamsters have also been proposed as a reservoir for *H. cinaedi.*[10] Although the epidemiology of these organisms is not clearly delineated, these two bacterial agents are sexually transmitted among homosexual men.

H. pylori colonizes the mucous layer of the antrum and fundus of the stomach but does not invade the epithelium. By virtue of this ability to colonize the gastric mucosa, persist despite the host immune defense, and cause host tissue damage, *H. pylori* is an effective and significant bacterial pathogen. For example, motility allows *H. pylori* to escape the acidity of the stomach and burrow through and colonize the gastric mucosa in close association with the epithelium. Although the study of specific virulence determinants of *H. pylori* has just begun, urease enzyme likely plays a significant role in the survival and growth of *H. pylori* in the stomach by creating an alkaline microenvironment. Other possible virulence determinants include adhesins for colonization of mucosal surfaces, mediators of inflammation, and a vacuolating cytotoxin that causes damage to host cells.[9] Although *H. pylori* is noninvasive, untreated infection lasts for the life of the host and persists despite a significant host immune response.

SPECTRUM OF DISEASE

H. cinaedi and *H. fennelliae* cause proctitis, enteritis, and sepsis in homosexual men. *H. cinaedi* has also been reported to cause septicemia and meningitis in a neonate.[10] *H. pylori* causes gastritis, peptic ulcer disease, and gastric cancer.[3] However, most individuals tolerate the presence of *H. pylori* for decades with few, if any, symptoms.

LABORATORY DIAGNOSIS

Specimen collection, transport, and processing
There are no special requirements for the collection, transport, or processing of stool or blood specimens for *H. cinaedi* and *H. fennelliae.* Tissue biopsy material of the stomach for detection of *H. pylori* should be placed directly into transport media such as Stuart's transport medium, to prevent drying. Specimens for biopsy may be refrigerated up to 24 hours before processing; tissues should be minced and gently homogenized.

Direct detection
The Warthin-Starry, or other silver stain, and Giemsa stains are used by pathologists for examination of biopsy specimens. One potential problem is that of sampling error. Squash preparations of biopsy material can be Gram-stained with good results; the 0.1% basic fuchsin counterstain enhances recognition of the bacteria's typical morphology.

Presumptive evidence of the presence of *H. pylori* in biopsy material may be obtained by placing a portion of crushed tissue biopsy material directly into urease broth or onto commercially available urease agar kits. A positive test is considered indicative of the organism's presence. Another noninvasive indirect test to detect *H. pylori* is the urea breath test. This test relies on the presence of *H. pylori* urease. The patient ingests radioactively labeled (^{14}C) urea, and if infection is present, the urease produced by *H. pylori* hydrolyzes the urea to form ammonia and labeled bicarbonate that is exhaled as CO_2; the labeled CO_2 is detected by either a scintillation counter or a special spectrometer. This test has excellent sensitivity and specificity.

Cultivation
Stool specimens submitted for culture of *H. cinaedi* and *H. fennelliae* are plated to selective media used

for campylobacter isolation but without cephalothin such as Campy-CVA. For the recovery of *H. pylori* from tissue biopsy specimens including gastric antral biopsies, nonselective agar media, including chocolate agar and *Brucella* agar with 5% sheep blood, have been useful. Selective agar, such as Skirrow's and modified Thayer-Martin agars, also support growth. Recently the combination of a selective agar (Columbia agar with an egg yolk emulsion, supplements, and antibiotics) and a nonselective agar (modified chocolate agar with Columbia agar, 1% Vitox and 5% sheep blood) was reported as the optimal combination for recovering *H. pylori* from antral biopsies.[12] Incubation up to 1 week in a humidified, microaerobic atmosphere at 35° to 37° C may be required before growth of this human pathogen is visible.

Approach to identification

Colonies of *Helicobacter* spp. may require 4 to 7 days of incubation before small, translucent, circular colonies are observed. Organisms are identified presumptively as *Helicobacter pylori* by the typical cellular morphology and positive results for oxidase, catalase, and rapid urease tests. Definitive identification of *H. pylori*, *H. cinaedi*, and *H. fennelliae* is accomplished using a similar approach to that for *Campylobacter* spp. (Table 45-3).

Serodiagnosis

Another approach to *H. pylori* diagnosis is serologic testing. Numerous serologic enzyme-linked immunoassays (EIA) designed to detect immunoglobulin G (IgG) antibodies to *H. pylori* are commercially available. Reported performance of these assays varies as a result of the reference method used to confirm *H. pylori* infection, antigen source for the assay, and the population studied.[7] In addition to variability in assay performance, the clinical utility of these assays has not been defined and may not differentiate between active vs. past *H. pylori* infections.

ANTIMICROBIAL SUSCEPTIBILITY TESTING AND THERAPY

Except for metronidazole, most laboratory susceptibility assays do not predict clinical outcome. Routine testing of *H. pylori* isolates' susceptibility to metronidazole is recommended using agar or broth dilution methods or possibly the E test.[6,11]

Therapy for *H. pylori* infection is problematic. Current regimens usually use triple-drug therapy that usually includes metronidazole, a bismuth salt, and either amoxicillin or tetracycline. An alternative and more simple regimen for patients with metronidazole-resistant strains is omeprazole or lansoprozole (proton pump inhibitors that cause rapid symptom relief while working synergistically with the antibiotics) and amoxicillin or clarithromycin. Relapses occur often. *Helicobacter* spp. associated with enteritis and proctitis may respond to quinolones, but ideal therapy has not been established.

PREVENTION

No vaccines are currently available that are directed against *H. pylori*. However, several vaccines are being developed that use numerous strategies.

References

1. Allos, B.M. and Blaser, M.J. 1995. *Campylobacter jejuni* and the expanding spectrum of related infections. Clin. Infect. Dis. 20:1092.
2. Centers for Disease Control. 1988. *Campylobacter* isolates in the United States. Morbid. Mortal. Weekly Rep. 37(SS-2):1.
3. Cover, T.L. and Blaser, M.J. 1995. *Helicobacter pylori*: a bacterial cause of gastritis, peptic ulcer disease and gastric cancer. ASM News 61:21.
4. Endtz, H.P. et al. 1991. Comparison of six media, including a semisolid agar, for the isolation of various *Campylobacter* species from stool specimens. J. Clin. Microbiol. 29:1007.
5. Grübel, P. et al. 1997. Vector potential of houseflies *(Musca domestica)* for *Helicobacter pylori*. J. Clin. Microbiol. 35:1300.
6. Henriken, T.H. et al. 1997. A simple method for determining metronidazole resistance of *Helicobacter pylori*. J. Clin. Microbiol. 35:1424.
7. Marchildon, P.A. et al. 1996. Evaluation of three commercial enzyme immunoassays compared with the ^{13}C urea breath test for detection of *Helicobacter pylori* infection. J. Clin. Microbiol. 34:1147.
8. Mishu, B. et al. 1993. Serologic evidence of previous *Campylobacter jejuni* infection in patients with Guillain-Barré syndrome. Ann. Intern. Med. 118:947.
9. On, S.L. 1996. Identification methods for campylobacters, helicobacters, and related organisms. Clin. Microbiol. 9:405.
10. Orlicek, S.L., Welch, D.L., and Kuhls, T.L. 1993. Septicemia and meningitis caused by *Helicobacter cinaedi* in a neonate. J. Clin. Microbiol. 31:569.
11. Pavicic, M.J.A.M.P. et al. 1993. *In vitro* susceptibility of *Helicobacter pylori* to several antimicrobial combinations. Antimicrob. Agents Chemother. 37:1184.
12. Piccolomine, R. et al. 1997. Optimal combination of media for primary isolation of *Helicobacter pylori* from gastric biopsy specimens. J. Clin. Microbiol. 35:1541.
13. Tagada, P. et al. 1996. Antimicrobial susceptibilities of *Campylobacter jejuni* and *Campylobacter coli* to 12 β-lactam agents and combinations with β-lactamase inhibitors. Antimicrob. Agents Chemother. 40:1924.
14. Vandamme, P. et al. 1995. Chemotaxonomic analysis of *Bacteroides gracilis* and *Bacteroides ureolyticus* and reclassification of *B. gracilis* as *Campylobacter gracilis* comb. nov. Int. J. Syst. Bacteriol. 45:145.

46 | *LEGIONELLA*

Genus and Species to be Considered

■ *Legionella pneumophila,*
serotypes 1 to 14

This chapter addresses organisms that will not grow on routine primary plating media and belong to the genus *Legionella. Legionella* is the causative agent of **legionnaires' disease,** a febrile and pneumonic illness with numerous clinical presentations. *Legionella* was discovered in 1976 by scientists at the Centers for Disease Control and Prevention (CDC) who were investigating an epidemic of pneumonia among Pennsylvania state American Legion members attending a convention in Philadelphia. There is retrospective serologic evidence of *Legionella* infection as far back as 1947.

GENERAL CHARACTERISTICS

There is only one genus, *Legionella,* within the family *Legionnellaceae.* All members of this genus are faintly-staining, thin, gram-negative bacilli that require a medium supplemented with L-cysteine and buffered to pH 6.9 for optimum growth. The overwhelming majority of *Legionella* spp. are motile. As of this writing, nearly 40 species belong to this genus. Nevertheless, the organism *Legionella pneumophila* predominates as a human pathogen within the genus and consists of 14 serotypes. In approximately decreasing order of clinical importance are *L. pneumophila* serotype 1 (about 50% of the cases of legionnaires' disease), *L. pneumophila* serotype 6, *L. micdadei,* and *L. dumoffii.*[12] Of note, many species of *Legionella* have only been isolated from the environment or recorded as individual cases. (See Box 46-1 for an abbreviated list of some of the species of *Legionella.*)

EPIDEMIOLOGY AND PATHOGENESIS

EPIDEMIOLOGY

Legionellae are ubiquitous and widely distributed in the environment. As a result, most individuals are exposed to *Legionella* spp.; however, few develop symptoms. In nature, legionellae are found primarily in aquatic habitats and thrive at warmer temperatures; these bacteria are capable of surviving extreme ranges of environmental conditions for long periods. *Legionella* spp. have been isolated from the majority of natural water sources investigated, including lakes, rivers, and marine waters, as well as moist soil.[10] Organisms are also widely distributed in man-made facilities, including air-conditioning ducts and cooling towers; potable water; large, warm-water plumbing systems; humidifiers; whirlpools; and technical-medical equipment in hospitals.

Legionella infections are acquired exclusively from environmental sources; no person-to-person spread has been documented.[3] Inhalation of infectious aerosols (1 to 5 μm diameter) is considered the primary means of transmission. Exposure to these aerosols can occur in the workplace or in industrial or nosocomial settings; for example, nebulizers filled with tap water and showers have been implicated. Legionnaires' disease occurs in sporadic, endemic, and epidemic forms. The incidence of disease varies greatly and appears to depend on the geographic area, but it is estimated that *Legionella* spp. cause from less than 1% to 5% of cases of pneumonia.

PATHOGENESIS

Legionella spp. can infect and multiply within some species of free-living amoebae (*Hartmannella,*

Acanthamoeba, and *Naegleria* spp.), as well as within *Tetrahymena* spp., a ciliated protozoa. This ability of legionellae contributes to its survival in the environment. Although the exact mechanism(s) by which *L. pneumophila* causes disease is not totally delineated, its ability to avoid destruction by the host's phagocytic cells plays a significant role in the disease process. *L. pneumophila* is considered a facultative intracellular pathogen. Following infection, organisms are taken up by phagocytosis into alveolar macrophages, where they survive and replicate within a specialized phagosome. This sequestering of legionellae also makes it difficult to deliver and accumulate antimicrobials within macrophages. Of significance, studies have shown that although certain antimicrobials can penetrate the macrophage and inhibit bacterial multiplication, *L. pneumophila* is not killed and, when drugs are removed the organism resumes replicating.[2] Therefore a competent cell-mediated immune response is also important in recovery from legionella infections. Humoral immunity appears to play an insignificant role in the defense against this organism. Several cellular components and extracellular products of *L. pneumophila*, such as an extracellular cytotoxin that impairs the ability of phagocytic cells to use oxygen and various enzymes (e.g., phospholipase C), have been purified and proposed as virulence factors.[10] However, their exact role in the pathogenesis of legionella infections is not completely clear.

SPECTRUM OF DISEASE

Legionella spp. are associated with a spectrum of clinical presentations, ranging from asymptomatic infection to severe, life-threatening diseases. Serologic evidence exists for the presence of asymptomatic disease, because many healthy people surveyed possess antibodies to *Legionella* spp. The following are the three primary clinical manifestations.[7]

- Pneumonia with a case fatality of 10% to 20% (referred to as *legionnaires' disease*)
- Pontiac fever, which is a self-limited, nonfatal respiratory infection
- Other infection sites, such as wound abscesses, encephalitis, or endocarditis

Individuals at risk for pneumonia are those who are immunocompromised, older than age 60, or heavy smokers. The clinical manifestations following infection with a particular species are primarily caused by differences in the host's immune response and perhaps by inoculum size; the same *Legionella* spp. gives rise to different expressions of disease in different individuals.

Of interest, a group of gram-negative bacilli has been reported that infect and multiply in the cytoplasm of amoeba. These organisms, called **Legionella-like amoeba pathogens,** are currently noncultivatable yet have been documented to induce a serologic response in humans.[1,2] It will be of interest if these organisms are subsequently documented as human respiratory tract pathogens.

LABORATORY DIAGNOSIS

SPECIMEN COLLECTION AND TRANSPORT

Specimens from which *Legionella* can be isolated include respiratory tract secretions of all types, including sputum and pleural fluid; other sterile body fluids, such as blood; and lung, transbronchial, or other biopsy material. Because sputum from patients with legionnaires' disease is usually nonpurulent and may appear bloody or watery, the grading system used for screening sputum for routine cultures is not applicable.[6] Patients with legionnaires' disease usually have detectable numbers of organisms in their respira-

BOX 46-1 SOME *LEGIONELLA* SPP. ISOLATED FROM HUMANS AND ENVIRONMENTAL SOURCES

Species Isolated from Humans
L. pneumophila, serotypes 1-14
L. micdadei
L. bozemanii
L. dumoffii
L. feelei
L. gormanii
L. hackeliae
L. longbeachae
L. oakridgensis
L. wadsworthii

Species Isolated from Environment Only
L. cherrii
L. erythra
L. gratiana
L. jamestowniensis
L. brunensis
L. fairfieldensis
L. santicrucis

tory secretions, even for some time after antibiotic therapy has been initiated. If the disease is present, the initial specimen is often likely to be positive. However, additional specimens should be processed if the first specimen is negative and suspicion of the disease persists. Pleural fluid has not yielded many positive cultures in studies performed in several laboratories, but it may contain organisms. Specimens should be transported without holding media, buffers, or saline, which may inhibit the growth of *Legionella*. The organisms are actually very hardy and are best preserved by maintaining specimens in a small, tightly closed container to prevent desiccation and transporting them to the laboratory within 30 minutes of collection. If a longer delay is anticipated, specimens should be refrigerated. If one cannot ensure that specimens will remain moist, a small amount (1 mL) of sterile broth may be added.

SPECIMEN PROCESSING

All specimens for *Legionella* culture should be handled and processed in a class II biological safety cabinet (BSC). When specimens from nonsterile body sites are submitted for culture, selective media and/or treatment of the specimen to reduce the numbers of contaminating organisms is recommended. Brief treatment of sputum specimens with hydrochloric acid before culture has been shown to enhance the recovery of legionellae.[4]

Tissues are homogenized before smears and cultures are performed, and clear, sterile body fluids are centrifuged for 30 minutes at $4000\times g$. The sediment is then vortexed and used for culture and smear preparation. Blood for culture of *Legionella* should probably be processed with the lysis-centrifugation tube system (Isolator; Wampole Laboratories, Cranbury, N.J.) and plated directly to buffered charcoal-yeast extract (BCYE) agar.

DIRECT DETECTION METHODS

Several laboratory methods are used to detect *Legionella* spp. directly in clinical specimens.

Stains

Because of their faint staining, *Legionella* spp. are not usually detectable directly in clinical material by Gram stain. Organisms can be observed on histologic examination of tissue sections using silver or Giemsa stains.

Antigens

One approach to direct detection of legionellae in clinical specimens is the direct immunofluorescent antibody (DFA) test of respiratory secretions. Polyclonal and monoclonal antisera conjugated with fluo-

rescein are available from several commercial suppliers. Specimens are first tested with pools of antisera containing antibodies to several serotypes of *L. pneumophila* or several *Legionella* spp. Those that exhibit positive results are then reexamined with specific conjugated antisera. One reagent made by Genetic Systems (Seattle, Wash.) is a monoclonal antibody directed against a cell wall protein common to *L. pneumophila*. The manufacturer's directions should be followed explicitly, and material from commercial systems should never be divided and used separately. Laboratories should decide which serotypes to test for routinely, based on the prevalence of isolates in their geographic area. The sensitivity of the DFA test ranges from 25% to 75%, and its specificity is greater than 95%.[11] If positive, organisms appear as brightly fluorescent rods (Figure 46-1). Of importance, cultures always must be performed, because *Legionella* spp. or serotypes not included in the antisera pool can be recovered.

Rapid detection of *Legionella* antigen in urine and other body fluids has been accomplished by radioimmunoassay, enzyme immunoassay, and latex agglutination. A major drawback with the urine antigen assay is that it only detects the presence of antigen of *L. pneumophila* serogroup 1. These assays have a sensitivity of 80% in their ability to detect infection caused by *L. pneumophila* serogroup 1 and are highly specific. Of importance, because bacterial antigen may persist in urine for days to weeks after initiation of antibiotic therapy, these assays may be positive when other diagnostic tests are negative.

Nucleic acid

Although not yet commercially available, direct detection of *Legionella* nucleic acid by polymerase chain

FIGURE 46-1 Fluorescent antibody-stained *Legionella pneumophila.*

reaction (PCR) has the potential to offer rapid results and increased sensitivity on respiratory samples over current methods[11]; of significance, PCR assays can detect all *Legionella* spp., not just *L. pneumophila*.

CULTIVATION

Specimens for culture should be inoculated to two agar plates for recovery of *Legionella*, at least one of which is BCYE without inhibitory agents. This medium contains charcoal to detoxify the medium, remove carbon dioxide (CO_2), and modify the surface tension to allow the organisms to proliferate more easily. BCYE is also prepared with ACES buffer and the growth supplements cysteine (required by *Legionella*), yeast extract, α-ketoglutarate, and iron. A second medium, BCYE base with polymyxin B, anisomycin (to inhibit fungi), and cefamandole, is recommended for specimens, such as sputum, that are likely to be contaminated with other flora. These media are commercially available. Several other media, including a selective agar containing vancomycin and a differential agar containing bromthymol blue and bromcresol purple, are also available from Remel (Lenexa, Kan.) and others. Specimens obtained from sterile body sites may be plated to two media without selective agents and perhaps also inoculated into the special blood culture broth without SPS. (Specimens should always be plated to standard media for recovery of pathogens other than *Legionella* that may be responsible for the disease.)

Plates are incubated in a candle jar at 35° to 37° C in a humid atmosphere. Only growth of *L. gormanii* is stimulated by increased CO_2, so incubation in air is preferable to 5% to 10% CO_2, which may inhibit some legionellae. Within 3 to 4 days, colonies should be visible. Plates are held for a maximum of 2 weeks before discarding. Blood cultures in biphasic media should be held for 1 month. At 5 days, colonies are 3 to 4 mm in diameter, gray-white to blue-green, glistening, convex, and circular and may exhibit a cut-glass type of internal granular speckling (Figure 46-2). A Gram stain yields thin, gram-negative bacilli (Figure 46-3).

APPROACH TO IDENTIFICATION

Because *Legionella* spp. are biochemically inert and many tests produce equivocal results, extensive biochemical testing is of little use.[12] Definitive identification requires the facilities of a specialized reference laboratory. Identification of *L. pneumophila* spp. can be achieved, however, by a monoclonal immunofluorescent stain (Genetic Systems). Emulsions of organisms from isolated colonies are made in 10% neutral formalin, diluted 1:100 (to produce a very thin suspension), and placed on slides for fluorescent antibody staining. Clinical laboratories probably perform sufficient service to clinicians by indicating the presence of *Legionella* spp. in a specimen. If further identification is necessary, the isolate should be forwarded to an appropriate reference laboratory.

SERODIAGNOSIS

Most patients with legionellosis have been diagnosed retrospectively by detection of a fourfold rise in anti-*Legionella* antibody with an indirect fluorescent antibody (IFA) test. Serum specimens no closer than 2 weeks apart should be tested. Confirmation of disease is accomplished by a fourfold rise in titer to more than 128. A single serum with a titer of more than 256 and a characteristic clinical picture may be presumptive for legionellosis; however, because as many as 12% of healthy persons yield titers as high as 1:256, this practice is strongly discouraged.[8] Unfortunately, individuals with legionnaires' disease may not

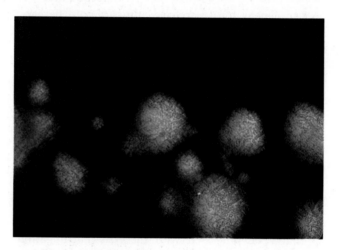

FIGURE 46-2 Colonies of *Legionella pneumophila* on buffered charcoal-yeast extract agar.

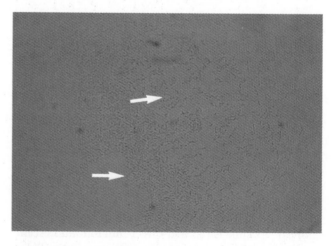

FIGURE 46-3 Gram stain of a colony of *Legionella pneumophila* showing thin, gram-negative bacilli (*arrows*).

exhibit a rise in serologic titers until as long as 8 weeks after the primary illness or may never display significant antibody titer rises. Commercially prepared antigen-impregnated slides for IFA testing are available from numerous suppliers.

ANTIMICROBIAL SUSCEPTIBILITY TESTING AND THERAPY

In vitro susceptibility studies are not predictive of clinical response and should not be performed for individual isolates of legionellae. Erythromycin is the drug of choice for treatment of disease caused by *Legionella*. High-dose trimethoprim/sulfamethoxazole and tetracyclines have been effective in some patients. The administration of rifampin along with erythromycin may prove beneficial, especially in severe cases. Clinical response usually follows within 48 hours after the introduction of effective therapy. In severely ill patients, one of the fluoroquinolones, rather than erythromycin, is recommended.[5] Penicillins, cephalosporins of all generations, and aminoglycosides are not effective and should not be used.

PREVENTION

Although under development, a vaccine against *Legionella* infections is not currently available. The effectiveness of other approaches to the prevention of legionella infections, such as the elimination of its presence from cooling towers and potable water, is uncertain.[7,9]

References

1. Adeleke, A., et al. 1996. Legionella-like amoebal pathogens: phylogenetic status and possible role in respiratory disease. Emerging Infect. Dis. 2:225.

2. Barker, J. and Brown, M.R.W. 1995. Speculations on the influence of infecting phenotype on virulence and antibiotic susceptibility of *Legionella pneumophila*. J. Antimicrob. Chemother. 36:7.

3. Breiman, R.F. 1993. Modes of transmission in epidemic and non-epidemic *Legionella* infections: directions of further study. In Barbaree, J.M., Breiman, R.F., and Dufour, A.P., editors. *Legionella*: current status and emerging perspectives. American Society for Microbiology, Washington, D.C.

4. Buesching, W.J., Brust, R.A., and Ayers, L.W. 1983. Enhanced primary isolation of *Legionella pneumophila* from clinical specimens by low pH treatment. J. Clin. Microbiol. 17;1153.

5. Edelstein, P.H. 1995. Antimicrobial chemotherapy for legionnaires' disease: a review. Clin. Infect. Dis. 21 (suppl. 3):S265.

6. Ingram, J.G. and Plouffe, J.F. 1994. Danger of sputum purulence screens in culture of *Legionella* species. J. Clin. Microbiol. 32:209.

7. Kramer, M.H.J. and Ford, T.E. 1994. Legionellosis: ecological factors of an environmentally "new" disease. Zentralbl. Hyg. Umwettmed. 195:470.

8. Plouffe, J.F., et al. 1995. Reevaluation of the definition of legionnaires' disease: use of the urinary antigen assay. Community Based Pneumonia Incidence Study Group. Clin. Infect. Dis. 20:1286.

9. Redd, S.C. and Cohen, M.L. 1987. Legionella in water: what should be done? J.A.M.A. 257:1221.

10. Reichnitzer, C. 1994. Pathogenetic aspects of Legionnaire's disease: interaction of *Legionella pneumophila* with cellular host defenses. APMIS (suppl.) 102:5.

11. Shelhamer, J.H., et al. 1996. The laboratory evaluation of opportunistic pulmonary infections, Ann. Intern. Med. 124:585.

12. Winn, W.C. 1993. Legionella and the clinical microbiologist. Infect. Dis. Clin. North Am. 7:377.

47 | *BRUCELLA*

Genera and Species to be Considered

- *Brucella abortus*
- *Brucella melitensis*
- *Brucella suis*
- *Brucella canis*

Although most isolates of *Brucella* grow on blood and chocolate agars (some isolates are also able to grow on MacConkey agar), more enriched agars and special incubation conditions are generally needed to achieve optimal recovery of these very fastidious organisms from clinical specimens.

GENERAL CHARACTERISTICS

Organisms belonging to the genus *Brucella* are small, nonmotile, aerobic, gram-negative coccobacilli or short rods that stain poorly by conventional Gram stain. Many isolates require supplementary carbon dioxide (CO_2) for growth, especially on primary isolation. *Brucella* spp. are closely related to *Bartonella*, *Rhizobium*, and *Agrobacterium*.[5] There are six known species, but only *Brucella abortus*, *B. melitensis*, *B. suis*, and *B. canis* are pathogenic for humans.

EPIDEMIOLOGY AND PATHOGENESIS

The disease **brucellosis** occurs worldwide, especially in Mediterranean and Arabian Gulf countries, India, and parts of Mexico and Central and South America.[9] Brucellosis is a zoonosis and is recognized as a cause of devastating economic losses among domestic livestock. In the United States, 105 cases were reported to the Centers for Disease Control and Prevention (CDC), but it is estimated that these represent only 4% to 10% of all cases.[7] Each of the four *Brucella* spp. that are pathogenic for humans has a limited number of preferred animal hosts (Table 47-1). In their hosts, brucella organisms tend to localize in tissues rich in erythritol, a four-carbon alcohol (e.g., placental tissue) that enhances their growth.[4] Humans become infected by four primary routes[10]:

- Ingestion of unpasteurized and contaminated milk or cheese
- Inhalation
- Penetration of ocular or oral mucosa
- Direct inoculation into the bloodstream through abrasions in the skin or vaccination

Persons considered at greatest risk for contracting brucellosis are dairy farmers, livestock handlers, slaughterhouse employees, veterinarians, and laboratory personnel.

Brucella spp. are facultative, intracellular parasites (able to exist in both intracellular and extracellular environments); only the following bacteria are classified as facultatively intracellular: *Salmonella*, *Shigella*, *Brucella*, *Yersinia*, *Listeria*, and *Francisella*.[1] Following infection, brucellae are ingested by neutrophils within which they replicate, resulting in cell lysis. Neutrophils containing viable organisms circulate in the bloodstream and are subsequently phagocytized by reticuloendothelial cells in the spleen, liver, and bone marrow. If left untreated, granulomas subsequently develop in these organs, with organisms surviving in monocytes and macrophages. Resolution of infection depends on the host's nutritional and immune status, the size of inoculum and route of infection, and the species of *Brucella* causing the infection; in general, *B. melitensis* and *B. suis* are more virulent for humans.[10] The mechanism(s) by which brucellae avoid intracellular killing are poorly understood. *Brucella* spp. can change from smooth to rough colonial morphology based on the composition of their cell wall lipopolysaccharide (LPS); those with smooth LPS are more resistant to intracellular killing by neutrophils than those with rough LPS.[6]

TABLE 47-1 *BRUCELLA* SPP. PATHOGENIC FOR HUMANS AND THEIR RESPECTIVE NATURAL HOST ANIMALS

ORGANISM	PREFERRED ANIMAL HOST
B. abortus	Cattle
B. melitensis	Sheep or goats
B. suis	Swine
B. canis	Dogs

SPECTRUM OF DISEASE

The clinical manifestations of brucellosis vary greatly, ranging from asymptomatic infection to serious, debilitating disease. For the most part, brucellosis is a systemic infection that can involve any organ of the body. Symptoms are nonspecific and include fever, chills, weight loss, sweats, headache, muscle aches, fatigue, and depression. Lymphadenopathy and splenomegaly are common physical findings. After an incubation period of about 2 to 3 weeks, the onset of disease is commonly insidious. Complications can occur, such as arthritis, spondylitis (inflammation of the spinal cord), and endocarditis.

LABORATORY DIAGNOSIS

SPECIMEN COLLECTION, TRANSPORT, AND PROCESSING

Definitive diagnosis of brucellosis requires isolation of the organisms in cultures of blood, bone marrow, or other tissues. For the following reasons it is essential that the clinical microbiology laboratory is notified whenever brucellosis is suspected:

■ To ensure that specimens are cultivated in an appropriate manner for optimum recovery from clinical specimens

■ To avoid accidental exposure of laboratory personnel handling the specimens, because *Brucella* spp. are considered class III pathogens

Blood for culture can be routinely collected (see Chapter 20) into most commercially available blood culture bottles and the lysis-centrifugation system (Isolator; Wampole Laboratories, Cranbury, N.J.) There are no special requirements for specimen collection, transport, or processing for other clinical specimens.

DIRECT DETECTION METHODS

Direct stains of clinical specimens are not particularly useful for the diagnosis of brucellosis. Preliminary studies employing polymerase chain reaction assays indicate that these assays may prove to be a reliable, sensitive, and specific means to directly detect *Brucella* spp. in clinical specimens.[3,5]

CULTIVATION

Commercial blood culture systems, such as the BacT/Alert, BACTEC, and lysis-centrifugation systems, have all successfully detected *Brucella* in blood. A recent comparison of the BACTEC 9240 with the Isolator blood culture systems found the BACTEC system more sensitive and faster in its detection of *B. melitensis*.[9] Other blood culture bottles, such as those with brain-heart infusion and trypticase soy broth, also support the growth of brucellae if bottles are continuously vented and placed in a CO_2 incubator. Although the majority of isolates can be detected within 7 days using commercial systems, prolonged incubation of 30 days and periodic blind subcultures to blood and chocolate agar plates at 7, 14, and 30 days is recommended to maximize recovery of brucellae.[8] Culture bottles may not become turbid. All subculture plates should be held for a minimum of 7 days.

Although *Brucella* spp. grow on blood and chocolate agars, supplemented media such as *Brucella* agar or some type of infusion base, is recommended for specimen types other than blood. The addition of 5% heated horse or rabbit serum enhances growth on all media. Cultures should be incubated in 5% to 10% CO_2 in a humidified atmosphere; inoculated plates are incubated for 3 weeks before being discarded as negative.

On culture, colonies appear small, convex, smooth, translucent, nonhemolytic, and slightly yellow and opalescent after at least 48 hours of incubation. The colonies may become brownish with age.

APPROACH TO IDENTIFICATION

Presumptive identification of *Brucella* can be made, but appropriate biohazard facilities must be used. The organisms are catalase-positive, and most strains are oxidase-positive. Other nonfermentative gram-negative coccobacilli that may be confused with *Brucella* are *Bordetella, Moraxella, Kingella,* and *Acinetobacter* spp. *Brucella* spp., however, are nonmotile, urease- and nitrate-positive, and strictly aerobic. The most rapid test for presumptive identification of *Brucella* is the particle agglutination test with antismooth *Brucella* serum (Difco Laboratories, Detroit, Mich.). *Brucella* spp. are differentiated by the rapidity with which they hydrolyze urea, relative ability to produce H_2S, requirements for CO_2, and susceptibility to the aniline dyes thionine and basic fuchsin. Differential characteristics are listed in Table 47-2. For determination of CO_2 requirement, identical plates of *Brucella* agar or brain-heart infusion agar should be given equal inocula (e.g., with a calibrated loop) of a broth suspension of the organism to be tested. One plate should be incubated in a candle jar and the other plate in air within the same incubator. Most strains of *B. abortus* do not grow in air but show growth in the candle jar. Isolates of *Brucella* should be sent to state or other reference laboratories for confirmation or definitive identification, because most clinical laboratories lack the necessary media and containment facilities.

SERODIAGNOSIS

Because of the difficulty of isolating the organism, a serologic test, the serum agglutination test (SAT) is widely used and detects antibodies to *Brucella abortus, B. melitensis,* and *B. suis;* SAT will not detect *B. canis* antibodies. A titer of 1:160 or greater in the SAT is considered diagnostic if this result fits the clinical and epidemiologic findings. ELISA assays also have been developed, but further evaluation must be done before these assays replace the SAT.

ANTIMICROBIAL SUSCEPTIBILITY TESTING AND THERAPY

Because of the fastidious nature of these organisms coupled with their intracellular localization, in vitro susceptibility testing is not reliable.[2] Prolonged treatment (6 weeks) is given to patients with brucellosis to prevent relapse of infection. For initial therapy, doxycycline in combination with streptomycin, gentamicin, or possibly rifampin is recommended. Sometimes surgical drainage is also required to treat localized foci of infection.

PREVENTION

Prevention of brucellosis in humans is dependent on the elimination of disease in domestic livestock. Vaccines directed against *B. abortus* or *B. melitensis* have been used in the appropriate animal host (i.e., cattle and goats/sheep, respectively) and are very successful in eradicating disease in the animals. An effective vaccine against *B. suis* and various human vaccines using killed brucellae fractions are under development.

References

1. Fortier, A.H., et al. 1994. Life and death of an intracellular pathogen: *Francisella tularensis* and the macrophage. Immunol. Series 60:349.
2. Hall, W.H. 1990. Modern chemotherapy for brucellosis in humans. Rev. Infect. Dis. 12:1060.

TABLE 47-2   CHARACTERISTICS OF *BRUCELLA* SPP. PATHOGENIC FOR HUMANS

SPECIES	CO_2 REQUIRED FOR GROWTH	TIME TO POSITIVE UREASE	H_2S PRODUCED	INHIBITION BY DYE THIONINE*	FUCHSIN*
B. abortus	+/−	2 hrs (rare 24 hrs)	+ (most strains)	+	−
B. melitensis	−	2 hrs (rare 24 hrs)	+	−	−
B. suis	−	15 min	+/−	−	+ (most)
B. canis	−	15 min	−	−	+

+, >90% of strains positive; −, >90% of strains negative; +/−, variable results.
*Dye tablets (Key Scientific Products, Inc., Round Rock, Texas).

3. Leal-Klevezas, D.S., et al. 1995. Single-step PCR for detection of *Brucella* spp. from blood and milk of infected animals. J. Clin. Microbiol. 33:3087.

4. Radolf, J.D. 1994. Brucellosis: don't let it get your goat! Am. J. Med. Sci. 307:64.

5. Romero, C., et al. 1995. Specific detection of *Brucella* DNA by PCR. J. Clin. Microbiol. 33:615.

6. Smith, L.D. and Ficht, T.A. 1990. Pathogenesis of *Brucella*. Crit. Rev. Microbiol. 17:209.

7. Wise, R.I. 1980. Brucellosis in the United States: past, present and future. J.A.M.A. 244:2318.

8. Yagupsky, P. 1994. Detection of *Brucella melitensis* by BACTEC NR660 blood culture system. J. Clin. Microbiol. 32:1899.

9. Yagupsky, P., et al. 1997. Comparison of BACTEC 9240 Peds Plus medium and Isolator 1.5 microbial tube for detection of *Brucella melitensis* from blood cultures. J. Clin. Microbiol. 35:1382.

10. Young, E.J. 1995. An overview of human brucellosis. Clin. Infect. Dis. 21:283.

48 | *BORDETELLA PERTUSSIS* AND *BORDETELLA PARAPERTUSSIS*

Genus and Species to be Considered

■ *Bordetella pertussis*
■ *Bordetella parapertussis*

The genus *Bordetella* contains three human pathogens: *Bordetella bronchiseptica, B. pertussis,* and *B. parapertussis. B. bronchiseptica* is reviewed in Chapter 34 because it grows on Mac-Conkey agar. Although *B. parapertussis* can also grow on MacConkey agar, it is addressed with *B. pertussis* in this chapter for two reasons. First, *B. pertussis* and *B. parapertussis* both cause human upper respiratory tract infections with almost identical symptoms, epidemiology, and therapeutic management. Second, optimal recovery of both organisms from respiratory specimens requires the addition of blood and/or other suitable factors to culture media.

GENERAL CHARACTERISTICS

General features of *Bordetella* spp. other than *B. pertussis* and *B. parapertussis* are summarized in Chapter 34. In contrast to *B. bronchiseptica, B. pertussis* and *B. parapertussis* are nonmotile and only infect humans. In the evolutionary process these exclusive human pathogens are believed to have arisen from a single clone and remain a very homogeneous clonal population.[9]

EPIDEMIOLOGY AND PATHOGENESIS

EPIDEMIOLOGY

Before the introduction of vaccines and in populations in which immunization is not performed, **pertussis** (whooping cough) is an epidemic disease with cycles every 2 to 5 years. Pertussis is a highly contagious, acute infection of the upper respiratory tract caused primarily by *B. pertussis* and less commonly by *B. parapertussis.*[8] Pertussis was first described in the sixteenth century and occurs worldwide. Although the incidence has decreased significantly since the widespread use of vaccination, outbreaks of pertussis occur periodically. It appears that *B. pertussis* infections are endemic in adults and adolescents; therefore these infections might serve as the source of the epidemic cycles involving unvaccinated children.[1,9] Infection is transmitted from person to person presumably by airborne transmission from the cough of an infected person; humans are the only reservoir.

PATHOGENESIS

The mechanism(s) by which *B. pertussis,* the primary pathogen of whooping cough, overcomes the immune defenses of healthy individuals is complex, involving the interplay of several virulence factors (Table 48-1). Some factors help to establish infection, others are toxigenic to the host, and still others override specific components of the host's mucosal defense system. For example, once reaching the host's respiratory tract, *B. pertussis* attaches to respiratory ciliated epithelial cells by means of adhesins and paralyzes the beating cilia by elaborating a tracheal cytotoxin.[3,9] A major virulence factor, pertussis toxin (PT) is also produced by the attached organism. PT enters the bloodstream, subsequently binding to specific receptors on host cells. After binding, PT disrupts several host cell functions, such as initiation of host cell translation; the inability of host cells to receive signals from the environment causes a generalized toxicity. The center membrane of *B. pertussis* also blocks the access of the host's lysozyme to the bacterial cell wall via its outer membrane.[11]

SPECTRUM OF DISEASE

Pertussis is usually a disease of children and can be divided into three symptomatic stages: catarrhal, paroxysmal, and convalescent. During the **catarrhal**

— **TABLE 48-1** MAJOR VIRULENCE DETERMINANTS OF *BORDETELLA PERTUSSIS* —

FUNCTION	FACTOR/STRUCTURE
Adhesion	Fimbriae
	Filamentous hemagglutinin (FHA)
	Pertactin*
	Tracheal colonization factor
	Brk A[†]
Toxicity	Pertussis toxin (an A/B toxin related to cholera toxin)
	Adenylate cyclase toxin (hemolyzes red cells and activates cyclic AMP, thereby inactivating several types of host immune cells)
	Dermonecrotic toxin (exact role unknown)
	Tracheal cytotoxin (ciliary dysfunction and damage)
	Endotoxin
Overcome host defenses	Outer membrane (host lysozyme is inhibited)
	Siderophore production (host lactoferrin and transferrin are unable to limit iron)

*Exact role in attachment is controversial.
†Might also play a role in pathogenesis by conferring serum resistance.[5]

stage, symptoms are the same as a mild cold with a runny nose and mild cough; this stage may last for several weeks.[6] Episodes of severe and violent coughing increase in number, marking the beginning of the **paroxysmal stage.** As many as 15 to 25 paroxysmal coughing episodes can occur in 24 hours and are associated with vomiting and "whooping" as air is rapidly inspired into the lungs past the swollen glottis. Lymphocytosis occurs, although typically there is no fever and there are no signs and symptoms of systemic illness. This stage may last from 1 to 4 weeks. *B. pertussis* infection in adults is common and associated with fever and far milder symptoms; these infections in adults are usually unrecognized and are incorrectly diagnosed as bronchitis.[4] The **convalescent stage** during which symptoms slowly decrease can last as long as 6 months after infection.[6]

LABORATORY DIAGNOSIS

SPECIMEN COLLECTION, TRANSPORT, AND PROCESSING

Diagnosis of pertussis is confirmed by culture, which is most sensitive early in the illness. Organisms may become undetectable by culture 2 weeks after the start of paroxysms. Nasopharyngeal aspirates or a nasopharyngeal swab (calcium alginate or Dacron on a wire handle) are acceptable specimens. For collection, the swab is bent to conform to the nasal passage and held against the posterior aspect of the nasopharynx. If coughing does not occur, another swab is inserted into the other nostril to initiate the cough. The swab is left in place during the entire cough, removed, and immediately inoculated onto a selective medium at the bedside (Table 48-2).

A fluid transport medium may be used for swabs but must be held for less than 2 hours. Half-strength Regan-Lowe agar enhances recovery when used as a transport and enrichment medium. Cold casein hydrolysate medium and casamino acid broth (available commercially) have been found to be effective transport media, particularly for preparation of slides for direct fluorescent antibody staining.

DIRECT DETECTION METHODS

A direct fluorescent antibody (DFA) stain using polyclonal antibodies against *B. pertussis* and *B. parapertussis* is commercially available for detection of *B. pertussis* in smears made from nasopharyngeal (NP) material (Difco Laboratories, Detroit, Mich.); an NP specimen DFA-positive for *B. pertussis* is shown in Figure 11-15, *B.* Although rapid, this DFA stain has

TABLE 48-2 EXAMPLES OF SELECTIVE MEDIA FOR PRIMARY ISOLATION OF *B. PERTUSSIS* AND *B. PARAPERTUSSIS*

AGAR MEDIA	DESCRIPTION
Bordet-Gengou	Potato infusion agar with glycerol and sheep blood with methicillin or cephalexin* (short shelf-life)
Modified Jones-Kendrick charcoal[10]	Charcoal agar with yeast extract, starch, and 40 μg cephalexin (2-3 mo shelf life but inferior to Regan-Lowe[6])
Regan-Lowe[†]	Charcoal agar with 10% horse blood and cephalexin (8-wk shelf-life[6])

*Cephalexin is superior to methicillin or penicillin for inhibiting normal respiratory flora.
[†]Regan-Lowe media was evaluated and found best for recovery of *B. pertussis* from nasopharyngeal swabs.[7]

limited sensitivity and variable specificity; therefore, the DFA should always be used in conjunction with culture.[6] Because of the limitations with currently available diagnostic methods, significant effort is being put into developing nucleic acid amplification methods.

CULTIVATION

Plates are incubated at 35°C in a humidified atmosphere without elevated carbon dioxide for up to 12 days[8]; most isolates are detected in 3 to 5 days. Young colonies of *B. pertussis* and *B. parapertussis* are small and shiny, resembling mercury drops; colonies become whitish-gray with age (Figure 48-1). Sensitivity of culture approaches 60% in the best of hands.

APPROACH TO IDENTIFICATION

A Gram stain of the organism reveals minute, faintly staining coccobacilli singly or in pairs (Figure 48-2). The use of a 2-minute safranin "O" counterstain or a 0.2% aqueous basic fuchsin counterstain enhances their visibility. The DFA reagent is used to presumptively identify organisms. Whole-cell agglutination reactions in specific antiserum can be used for species identification.

SERODIAGNOSIS

Although there are several serologic tests available for the diagnosis of pertussis, including agglutination, complement fixation, and enzyme immunoassay, no single method can be recommended for serologic diagnosis at this time.

ANTIMICROBIAL SUSCEPTIBILITY TESTING AND THERAPY

Laboratories currently do not perform routine susceptibility testing of *B. pertussis* and *B. parapertussis* because erythromycin remains active and is the antibiotic of choice. However, because the first case of an

FIGURE 48-1 Growth of *Bordetella pertussis* on Regan-Lowe media.

FIGURE 48-2 Typical Gram stain appearance of *Bordetella pertussis*.

erythromycin-resistant isolate of *B. pertussis* was reported in 1994,[2] this practice perhaps should be altered in the future.

PREVENTION

Whole-cell vaccines prepared from various *B. pertussis* preparations to prevent pertussis are manufactured in many countries and are efficacious in controlling epidemic pertussis. However, because of reactions to these vaccines and an apparent lack of long-term immunity, new acellular vaccines that include booster doses in older children and adults have been developed to prevent pertussis.[3] Prompt recognition of clinical cases and treatment of contacts and cases are also very important in preventing the transmission of *B. pertussis* and *B. parapertussis*.

References

1. Cattaneo, L.A., et al. 1996. The seroepidemiology of *Bordetella pertussis* infections: a study of persons ages 1-65 years. J. Infect. Dis. 173:1257.
2. Centers for Disease Control and Prevention. 1994. Erythromycin-resistant *Bordetella pertussis:* Yuma County, Arizona, May-October 1994. Morb. Mortal. Wkly. Rep. 43:807.
3. Cherry, J.D. 1996. Historical review of pertussis and the classical vaccine. J. Infect. Dis. 174(suppl. 3):S259.
4. Deville, J.G., et al. 1995. Frequency of unrecognized *Bordetella pertussis* infections in adults. Clin. Infect. Dis. 21:639.
5. Fernandez, R.C. and Weiss, A.A. 1994. Cloning and sequencing of a *Bordetella pertussiss* serum resistance locus. Infect. Immun. 62:4727.
6. Friedman, R.L. 1988. Pertussis: the disease and new diagnostic methods. Clin. Microbiol. Rev. 1:365.
7. Hoppe, J.E. and Vogl, R. 1986. Comparison of three media for cultures of *Bordetella pertussis*. Eur. J. Clin. Microbiol. 5:361.
8. Katzko, C., Hofmeister, M., and Church, D. 1996. Extended incubation of culture plates improves recovery of *Bordetella* spp. J. Clin. Microbiol. 34:1563.
9. Rappuoli, R. 1994. Pathogenicity mechanisms of *Bordetella*. Curr. Top. Microbiol. Immunol. 192:319.
10. Stauffer, L.R., Brown, D.R., and Sandstrom, R.E. 1983. Cephalexin-supplemented Jones-Kendrick charcoal agar for selective isolation of *Bordetella pertussis:* comparison with previously described media. J. Clin. Microbiol. 17:60.
11. Weiss, A. 1997. Mucosal immune defenses and the response of *Bordetella pertussis*. ASM News. 63:22.

49 | FRANCISELLA

Genera and Species to be Considered

Proposed Name	Previous Name
■ *Francisella tularensis*	*Francisella tularensis*
biogroup *tularensis*	biovar *tularensis*
biogroup *novicida*	
biogroup *palaerctica*	biovar *palaerctica*
	Francisella novicida
■ *Francisella philomiragia*	*Yersinia philomiragia*

Blood, chocolate, and MacConkey agars cannot be used for the primary isolation of organisms belonging to the genus *Francisella*. *Francisella* is a facultative, intracellular pathogen that requires cysteine and a source of iron for growth.[4] Because of this requirement for a complex medium for isolation and growth, these organisms are discussed in this chapter.

GENERAL CHARACTERISTICS

Organisms belonging to the genus *Francisella* are faintly staining, gram-negative rods that are nonmotile and obligately aerobic. Currently, the taxonomy of this genera is in flux. In Bergey's *Manual of Determinative Bacteriology,*[7] two species, *F. novicida* and *F. tularensis,* are included in the genus *Francisella*. Because *F. novicida* appears to be virtually identical to *F. tularensis,* reclassification of this organism as a biogroup of *F. tularensis* has been proposed.[6] In addition, based on the analysis of 14 human isolates initially classified as *Yersinia philomiragia*, these organisms were transferred to the genus *Francisella* as *F. philomiragia*.[6] Recently, these findings regarding *F. novicida* and *F. philomiragia,* were confirmed.[3] The taxonomy is summarized in Table 49-1. Three subspecies of *F. tularensis* have been proposed.

EPIDEMIOLOGY AND PATHOGENESIS

Francisella tularensis is the agent of human and animal tularemia. Worldwide in distribution, *F. tularensis* is carried by many species of wild rodents, rabbits, beavers, and muskrats in North America. Humans become infected by handling the carcasses or skin of infected animals, through insect vectors (primarily deerflies and ticks in the United States), by being bitten by carnivores that have themselves eaten infected animals, or by inhalation.[1]

The capsule of *F. tularensis* may be a factor in virulence, allowing the organism to avoid immediate destruction by polymorphonuclear neutrophils. The organism is extremely invasive, one of only a few infectious agents purportedly able to penetrate intact skin. In addition to invasiveness, *F. tularensis* is an intracellular parasite that is able to survive in the cells of the reticuloendothelial system, where it resides after a bacteremic phase. Granulomatous lesions may develop in various organs. Humans are infected by less than 50 organisms by either aerosol or cutaneous routes.[8] *F. philomiragia* has been isolated from several patients, many of whom were immunocompromised or were victims of near-drowning incidents. The organism is present in animals and ground water.

SPECTRUM OF DISEASE

Following inoculation of *F. tularensis* through abrasions in the skin or arthropod bites, a lesion appears at the site and progresses to an ulcer; lymph nodes adjacent to the site of inoculation become enlarged and often necrotic. Once the organism enters the bloodstream, patients become systemically ill with high

TABLE 49-1   TAXONOMY OF THE GENUS *FRANCISELLA*

OLD TAXONOMY	PROPOSED NEW TAXONOMY
F. tularensis	*F. tularensis*
biovar *tularensis*	biogroup *tularensis*
	biogroup *novicida*
biovar *palaerctica*	biogroup *palaerctica*
*F. novicida**	
Yersinia philomiragia	*F. philomiragia*

*Confirmed that *F. novicida* and *F. tularensis* are the same species.

From Hollis, D.G., et al. 1989. J. Clin. Microbiol. 27:1601.

fever, chills, headache, and generalized aching. Clinical manifestations of infection with *F. tularensis* can be glandular, ulceroglandular, oculoglandular, oropharyngeal, systemic, and pneumonic. These clinical presentations are briefly summarized in Table 49-2.

LABORATORY DIAGNOSIS

F. tularensis is a Biosafety Level 2 pathogen, a designation that requires technologists to wear gloves and to work within a biological safety cabinet (BSC) when handling clinical material that potentially harbors this agent. For cultures, the organism is designated Biosafety Level 3; a mask, recommended for handling all clinical specimens, is very important for preventing aerosol acquisition with *F. tularensis*. Because tularemia is one of the most common laboratory-acquired infections, most microbiologists do not attempt to work with infectious material from sus-

pected patients. It is recommended that specimens be sent to reference laboratories or state or other public health laboratories that are equipped to handle *Francisella*.

SPECIMEN COLLECTION, TRANSPORT, AND PROCESSING

The most common specimens submitted to the laboratory are scrapings from infected ulcers, lymph node biopsies, and sputum. Because there are no special requirements for specimen collection, transport, or processing, except for specimen handling as just described, refer to Table 1-1 for general information.

DIRECT DETECTION METHODS

Gram stain of clinical material is of little use. Fluorescent antibody stains are commercially available for direct detection of the organism in lesion smears, but such procedures are best performed by reference laboratories. Polymerase chain reaction assays are being developed to detect *F. tularensis* directly in clinical specimens.[5,9]

CULTIVATION

Isolation of *F. tularensis* is difficult. *F. tularensis* is strictly aerobic and requires enriched media (containing cysteine and cystine) for primary isolation. Commercial media for cultivation of the organism are available (glucose cystine agar, BBL Microbiology Systems, Sparks, Md.; cystine-heart agar, Difco Laboratories, Detroit, Mich.); both require the addition of 5% sheep or rabbit blood. *F. tularensis* also may grow on chocolate agar supplemented with IsoVitaleX, the nonselective buffered charcoal-yeast extract agar used for isolation of legionellae, or modified Mueller-Hinton broth.[11] Growth is not enhanced by carbon dioxide. These slow-growing organisms require 2 to 4

TABLE 49-2   CLINICAL MANIFESTATIONS OF *FRANCISELLA TULARENSIS* INFECTION

TYPES OF INFECTION	CLINICAL MANIFESTATIONS AND DESCRIPTION
Ulceroglandular	Most common; ulcer and lymphadenopathy
Glandular	Lymphadenopathy
Oculoglandular	Conjunctivitis, lymphadenopathy
Oropharyngeal	Ulceration in the oropharynx
Systemic tularemia	Acute illness with septicemia; 30%-60% mortality rate; no ulcer or lymphadenopathy
Pneumonic tularemia	Acquired by inhalation of infectious aerosols or by dissemination from the bloodstream; pneumonia

days for maximal colony formation and are weakly catalase-positive and oxidase-negative. Some strains may require up to 2 weeks to develop visible colonies. *F. philomiragia* is less fastidious than *F. tularensis*. Although *F. philomiragia* does not require cysteine or cystine for isolation, it resembles *F. tularensis* in that it is a small, coccobacillary rod that grows poorly or not at all on MacConkey agar. This organism will grow well on heart infusion agar with 5% rabbit blood or buffered charcoal-yeast extract agar with or without cysteine.[6]

APPROACH TO IDENTIFICATION

Colonies are transparent, mucoid, and easily emulsified. Although carbohydrates are fermented, isolates should be identified serologically (by agglutination) or by a fluorescent antibody stain. Ideally, isolates should be sent to a reference laboratory for characterization.

F. philomiragia differs from *F. tularensis* biochemically; *F. philomiragia* is oxidase-positive by Kovac's modification, and most strains produce hydrogen sulfide in triple sugar iron agar medium, hydrolyze gelatin, and grow in 6% sodium chloride (no strains of *F. tularensis* share these characteristics).

SERODIAGNOSIS

Because of the risk of infection to laboratory personnel and other inherent difficulties with culture, diagnosis of tularemia is usually accomplished serologically by whole-cell agglutination (febrile agglutinins or newer enzyme-linked immunosorbent assay techniques).

ANTIMICROBIAL SUSCEPTIBILITY TESTING AND THERAPY

There is no standardized antimicrobial susceptibility test for *Francisella* spp. The organism is susceptible to aminoglycosides, and streptomycin is the drug of choice. Gentamicin is a possible alternative[2]; tetracycline and chloramphenicol also have been used, although these two agents have been associated with a higher rate of relapse after treatment.

PREVENTION

The primary way to prevent tularemia is by reducing the possibility for exposure to the etiologic agent in nature, such as wearing protective clothing to prevent insect bites and not handling dead animals. An investigative live-attenuated vaccine is available.[10]

References

1. Craven, R., and Barnes, A.M. 1991. Plague and tularemia. Infect. Dis. Clin. North Am. 5:165.
2. Enderlin, G., et al. 1994. Streptomycin and alternative agents for the treatment of tularemia: review of the literature. Clin. Infect. Dis. 19:42.
3. Forsman, M., Sandstrom, G., and Sjostedt, A. 1994. Analysis of 16S ribosomal DNA sequences of *Francisella* strains and utilization for determination of the phylogeny of the genus and for identification of strains by PCR. Int. J. Syst. Bacteriol. 44:38.
4. Fortier, A.H., et al. 1994. Life and death of an intracellular pathogen: *Francisella tularensis* and the macrophage. Immunol. Series 60:349.
5. Fulop, M., Leslie, D., and Titball, R. 1996. A rapid, highly sensitive method for the detection of *Francisella tularensis* in clinical samples using the polymerase chain reaction. Am. J. Trop. Med. Hyg. 54:364.
6. Hollis, D.G., et al. 1989. *Francisella philomiragia* comb. nov. (formerly *Yersinia philomiragia*) and *Francisella tularensis* biogroup *novicida* (formerly *Francisella novicida*) associated with human disease. J. Clin. Microbiol. 27:1601.
7. Holt, J.G., Krieg, N.F., Sneath, P.H.A., et al. 1994. Bergey's manual of determinative bacteriology, ed 9. Williams & Wilkins, Baltimore.
8. Hornick, R. 1983. Tick-borne diseases. N.Y. State J. Med. 83:1036.
9. Junhui, Z., et al. 1996. Detection of *Francisella tularensis* by the polymerase chain reaction. J. Med. Microbiol. 45:477.
10. Sandstrom, G. 1994. The tularemia vaccine. J. Chem. Technol. Biotech. 59:315.
11. Stewart, S.J. 1995. *Francisella*. In Murray, P.R., Baron, E.J., Pfaller, M.A., et al., editors: Manual of clinical microbiology, ed 6. American Society for Microbiology, Washington D.C.

50 STREPTOBACILLUS MONILIFORMIS AND SPIRILLUM MINUS

Genera and Species to be Considered

- *Streptobacillus moniliformis*
- *Spirillum minus*

Streptobacillus moniliformis is a gram-negative bacillus that requires media containing blood, serum, or ascites fluid as well as incubation under carbon dioxide (CO_2) for isolation from clinical specimens. This organism causes rat-bite fever and Haverhill fever in humans. *Spirillum minus* has never been grown in culture. Because both are causative agents of **rat-bite fever**, these organisms are considered in this chapter.

STREPTOBACILLUS MONILIFORMIS

GENERAL CHARACTERISTICS

There is only one species in the genus *Streptobacillus*: *S. moniliformis*. This facultative anaerobe is non-motile and tends to be highly pleomorphic. With respect to taxonomy, it is somewhat related to members of the order *Mycoplasmatales* (see Chapter 62); however, the exact phylogenetic origin of *S. moniliformis* remains to be delineated.[8]

EPIDEMIOLOGY AND PATHOGENESIS

The natural habitat of *S. moniliformis* is the upper respiratory tract (nasopharynx, larynx, upper trachea, and middle ear) of wild and laboratory rats; only rarely has this organism been isolated from other animals, such as mice, guinea pigs, gerbils, and turkeys.[8] *S. moniliformis* is pathogenic for humans and is transmitted by two routes:

- By rat bite or possibly by direct contact with rats[8]
- By ingestion of contaminated food such as unpasteurized milk or milk products and, less frequently, water[4]

The incidence of *S. moniliformis* infections is unknown, but human infections appear to occur worldwide.

The pathogenic mechanisms of *S. moniliformis* are unknown. The organism is known to spontaneously develop L forms (bacteria without cell walls) that may allow its persistence in some sites.[2]

SPECTRUM OF DISEASE

Despite the different modes of transmission, the clinical manifestations of *S. moniliformis* infection are similar. When *S. moniliformis* is acquired by ingestion, the disease is called **Haverhill fever.**

Patients with rat-bite, or Haverhill, fever develop acute onset of chills, fever, headache, vomiting, and often, severe joint pains. Febrile episodes may persist for weeks or months.[5] Within the first few days of illness, patients develop a rash on the palms, soles of the feet, and other extremities. Complications can occur and include endocarditis, septic arthritis, pneumonia, pericarditis, brain abscess, amniotitis, prostatitis, and pancreatitis.[6,8]

LABORATORY DIAGNOSIS

Specimen collection, transport, and processing
Unfortunately, the diagnosis of rat-bite fever caused by *S. moniliformis* is often delayed owing to lack of exposure history, atypical clinical presentation, and the unusual microbiologic characteristics of the organism. Organisms may be cultured from blood or aspirates from infected joints, lymph nodes, or lesions. There are no special requirements for the collection, transport, and processing of these specimens except for blood. Because recovering *S. moniliformis* from blood cultures is impeded by concentrations of sodium polyanethol sulphonate (SPS) used in blood culture bottles, an alternative to most commercially available bottles must be employed.[3,7] After collection using routine procedures (described in Chapter 20), blood

and joint fluids are mixed with equal volumes of 2.5% citrate to prevent clotting and then inoculated to brain-heart infusion cysteine broth supplemented with Panmede (a papain digest of ox liver), commercially available fastidious anaerobe broth without SPS, or thiol broth.[3,5,7]

Direct detection methods

Pus or exudates should be smeared, stained with Gram or Giemsa stain, and examined microscopically (Figure 50-1). Direct detection of *S. moniliformis* using polymerase chain reaction is under development.[8]

Cultivation

As previously mentioned, *S. moniliformis* requires the presence of blood, ascitic fluid, or serum for growth. Growth occurs on blood agar, incubated in a very moist environment with 5% to 10% CO_2, usually after 48 hours of incubation at 37° C. Colonies are nonhemolytic. Addition of 10% to 30% ascitic fluid (available commercially from some media suppliers, such as Difco Laboratories) or 20% horse serum should facilitate recovery of the organism. In broth cultures, the organism grows as "fluff balls" or "bread crumbs" near the bottom of the tube of broth or on the surface of the sedimented red blood cell layer in blood culture media. Colonies grown on brain-heart infusion agar supplemented with 20% horse serum are small, smooth, glistening, colorless or grayish, with irregular edges.

Colonies are embedded in the agar and may also exhibit a fried egg appearance, with a dark center and a flattened, lacy edge. These colonies have undergone the spontaneous transformation to the L form. Stains of L-form colonies yield coccobacillary or bipolar

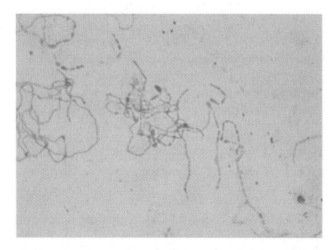

FIGURE 50-1 Gram stain of *Streptobacillus moniliformis* from growth in thioglycollate broth with 20% serum. (Courtesy Robert E. Weaver, Centers for Disease Control and Prevention, Atlanta, Ga.)

staining coccoid forms; usually a special stain, such as the Dienes stain (performed by pathologists), is required. Acridine orange stain also reveals the bacteria when Gram stain fails because of lack of cell wall constituents.

Gram-stained organisms from standard colonies show extreme pleomorphism, with long, looped, filamentous forms, chains, and swollen cells. The club-shaped cells can be 2 to 5 times the diameter of the filament. The carbolfuchsin counterstain or the Giemsa stain may be necessary for visualization (Figure 50-1).

Approach to identification

S. moniliformis does not produce indole and is catalase, oxidase, and nitrate negative, in contrast to organisms with which *Streptobacillus* may be confused, including *Actinobacillus*, *Haemophilus aphrophilus*, and *Cardiobacterium*. In addition, *S. moniliformis* is nonmotile and urea and lysine decarboxylase negative; H_2S is not produced in triple sugar iron agar but can be detected using lead acetate paper.[3]

Serodiagnosis

Serologic diagnosis of rat-bite fever is also useful; most patients develop agglutinating titers to the causative organism. The specialized serologic tests are performed only at national reference laboratories, since the disease is extremely rare in the United States. A titer of 1:80 is considered diagnostic, unless a four-fold rise in titer is demonstrated.[5]

ANTIMICROBIAL SUSCEPTIBILITY TESTING AND THERAPY

There are no standardized methods to determine *S. moniliformis* susceptibility to various antimicrobials. Different in vitro techniques, such as agar dilution and disk diffusion, have had similar results;[8] although *S. moniliformis* is susceptible to a broad spectrum of antibiotics, penicillin is regarded as the drug of choice for human rat-bite fever.

PREVENTION

There are no vaccines available to prevent rat-bite fever. Disease is best prevented by avoiding contact with animals known to harbor the organism.

SPIRILLUM MINUS

GENERAL CHARACTERISTICS

Spirillum minus is a gram-negative, helical, strictly aerobic organism.

EPIDEMIOLOGY AND PATHOGENESIS

Little information is available regarding the epidemiology or pathogenesis of *S. minus,* but it is supposed to be similar in some regards to that of *S. monili-formis*. The mode of transmission of infection is by a rat bite.

SPECTRUM OF DISEASE

S. minus also causes rat-bite fever in humans and is referred to as **Sodoku.** The clinical signs and symptoms are similar to those caused by *S. moniliformis* except that arthritis is rarely seen in patients with Sodoku and swollen lymph nodes are prominent;[1] febrile episodes are also more predictable in Sodoku. Following the bite, the wound heals spontaneously but 1 to 4 weeks later reulcerates to form a granulomatous lesion at the same time that the patient develops constitutional symptoms of fever, headache, and a generalized, blotchy, purplish, maculopapular rash. Differentiation between rat-bite fever caused by *S. minus* and that caused by *Streptobacillus moniliformis* is usually accomplished based on clinical presentations of the two infections and the isolation of the latter organism in culture. The incubation period for *S. minus* is much longer than that of streptobacillary rat-bite fever, which has occurred within 12 hours of the initial bite.

LABORATORY DIAGNOSIS

Specimen collection, transport, and processing
Specimens commonly submitted for diagnosis of Sodoku include blood, exudate, or lymph node tissues. There are no requirements for specimen collection, transport, or processing of the organisms discussed in this chapter. Refer to Table 1-1 for general information on this subject.

Direction detection methods
Because *S. minus* cannot be grown on synthetic media, diagnosis relies on direct visualization of characteristic spirochetes in clinical specimens using Giemsa or Wright stains, or dark-field microscopy. *S. minus* appears as a thick, spiral, gram-negative organism with two or three coils and polytrichous polar flagella. Diagnosis is definitively made by injection of lesion material or blood into experimental white mice or guinea pigs and subsequent recovery 1 to 3 weeks after inoculation.

Serodiagnosis
There is no specific serologic test available for *S. minus* infection.

ANTIMICROBIAL SUSCEPTIBILITY TESTING AND THERAPY

Because this spirochete is nonculturable, routine antimicrobial susceptibility testing is not performed.

PREVENTION

No vaccines are available to prevent rat-bite fever. Disease is best prevented by avoiding contact with animals known to harbor the organism.

References

1. Buranakitjaroen, P., Nilganuwong, S., and Gherunpong, V. 1994. Rat bite fever caused by *Streptobacillus moniliformis*. Southeast Asian J. Trop. Med. Public Health 25:778.
2. Freundt, E.A. 1956. Experimental investigations into the pathogenicity of the L-phase variant of *Streptobacillus moniliformis*. Acta. Pathol. Microbiol. Scand. 38:246.
3. Lambe, D.W. et al. 1973. *Streptobacillus moniliformis* isolated from a case of Haverhill fever: biochemical characterization and inhibitory effect of sodium polyanethol sulfonate. Am. J. Clin. Pathol. 60:854.
4. McEvoy, M.B., Noah, N.D., and Pilsworth, R. 1987. Outbreak of fever caused by *Streptobacillus moniliformis*. Lancet ii:1361.
5. Rogosa, M. 1985. *Streptobacillus moniliformis* and *Spirillum minus*. In Lennette, E.H., Balows, A., Hausler, W.J., Jr. and Shadomy, H.J., editors. Manual of clinical microbiology, ed 4, American Society for Microbiology, Washington D.C.
6. Rupp, M.E. 1992. *Streptobacillus moniliformis* endocarditis: case report and review. Clin. Infect. Dis. 14:769.
7. Shanson, D., Pratt, J. and Green, P. 1985. Comparison of media with and without 'Panmede' for the isolation of *Streptobacillus moniliformis* from blood cultures and observations on the inhibitory effect of sodium polyanethol sulfonate. J. Med. Microbiol. 19:181.
8. Wullenweber, M. 1995. *Streptobacillus moniliformis*–a zoonotic pathogen: taxonomic considerations, host species, diagnosis, therapy, geographical distribution. Lab. Anim. 29:1.

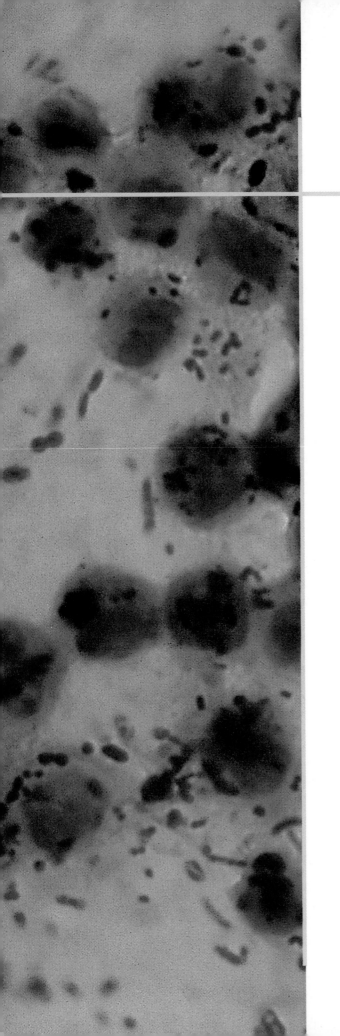

Section Six

6

Gram-Negative Cocci

51 | *NEISSERIA* AND *MORAXELLA CATARRHALIS*

Genera and Species to be Considered

Current Name	Previous Name
■ *Moraxella catarrhalis*	*Branhamella catarrhalis,* *Neisseria catarrhalis*
■ *Neisseria gonorrhoeae*	
■ *Neisseria meningitidis*	
■ Other *Neisseria* spp.	
N. cinerea	
N. lactamica	
N. polysaccharea	
N. subflava	*N. subflava, N. flava,* and *N. perflava*
N. sicca	
N. mucosa	
N. flavescens	

GENERAL CONSIDERATIONS

Species of the genus *Neisseria* discussed in this chapter and *Moraxella catarrhalis* are all oxidase-positive, gram-negative diplococci that do not elongate when exposed to subinhibitory concentrations of penicillin. The rodlike *Neisseria* spp. are described in Chapter 39.

EPIDEMIOLOGY

Except for *Neisseria gonorrhoeae,* the organisms considered in Table 51-1 are normal inhabitants of the upper respiratory tract of humans. *N. gonorrhoeae,* primarily a pathogen found in the urogenital tract, is never considered normal flora and is always considered clinically significant.

Transmission of the two pathogenic species of *Neisseria, N. gonorrhoeae* and *N. meningitidis,* is person to person. *N. gonorrhoeae* is sexually transmitted and *N. meningitidis* is spread by contaminated respiratory droplets. Infections caused by *M. catarrhalis* and the other *Neisseria* spp. usually involve a patient's endogenous strain.

PATHOGENESIS AND SPECTRUM OF DISEASE

As presented in Table 51-2, infections caused by *M. catarrhalis* are usually localized to the respiratory tract and rarely disseminate.[5,12]

N. gonorrhoeae is a leading cause of sexually transmitted disease, and infections caused by this organism usually are localized to the mucosal surfaces in the area of initial exposure to the organism (e.g., cervix, conjunctiva, pharyngeal surface, anorectal area, or urethra of males). Localized infections can be acute with a pronounced purulent response, or they may be asymptomatic. Not all infections remain localized, and dissemination from the initial infection site can lead to severe disseminated disease (Table 51-2).

N. meningitidis is a leading cause of fatal bacterial meningitis.[9] However, the virulence factors responsible for the spread of this organism from a patient's upper respiratory tract to the bloodstream and meninges to cause life-threatening infections are not fully understood (Table 51-2).

The other *Neisseria* spp. are not considered pathogens and are often referred to as the **saprophytic** *Neisseria.* Although they are most commonly

TABLE 51-1 EPIDEMIOLOGY

ORGANISM	HABITAT (RESERVOIR)	MODE OF TRANSMISSION
Moraxella catarrhalis	Normal human flora of upper respiratory tract; occasionally colonizes female genital tract	Spread of patient's endogenous strain to normally sterile sites. Person-to-person nosocomial spread by contaminated respiratory droplets also can occur
Neisseria gonorrhoeae	Not part of normal human flora. Only found on mucous membranes of genitalia, anorectal area, oropharynx, or conjunctiva at time of infection	Person-to-person spread by sexual contact, including rectal intercourse and orogenital sex. May also be spread from infected mother to newborn during birth
Neisseria meningitidis	Colonizes oro- and nasopharyngeal mucous membranes of humans. Human carriage of the organism without symptoms is common	Person-to-person spread by contaminated respiratory droplets, usually in settings of close contact
Other *Neisseria* spp.	Normal human flora of the upper respiratory tract	Spread of patient's endogenous strain to normally sterile sites. Person-to-person spread may also be possible, but these species are not common causes of human infections

encountered as contaminants in clinical specimens, they can occasionally be involved in bacteremia and endocarditis.[3]

LABORATORY DIAGNOSIS

SPECIMEN COLLECTION AND TRANSPORT

The pathogenic *Neisseria* spp. described in this chapter are very sensitive to drying and temperature extremes. In addition to general information on specimen collection and transport provided in Table 1-1, there are some special requirements for isolation of *N. gonorrhoeae* and *N. meningitidis*.

Swabs are acceptable for *N. gonorrhoeae* if the specimen will be plated within 6 hours. If cotton swabs are used, the transport medium should contain charcoal to inhibit toxic fatty acids present in the cotton fibers. Calcium alginate or rayon fibers are preferred, however. The best method for culture and transport of *N. gonorrhoeae* is to inoculate the agar immediately after specimen collection and place the medium in an atmosphere of increased carbon dioxide (CO_2) for transport. Specially packaged media consisting of selective agar in plastic trays that contain a CO_2-generating system are commercially available (JEMBEC plates) and widely used (Figure 51-1). The JEMBEC system is transported to the laboratory at room temperature. Upon receipt in the laboratory, the agar surface is cross-streaked to obtain isolated

colonies, and the plate is incubated at 35° C in 3% to 5% CO_2.

The recovery of *N. gonorrhoeae* or *N. meningitidis* from normally sterile body fluids requires no special methods, blood cultures being a notable exception. Both organisms are sensitive to sodium polyanetholsulfonate (SPS) so the content of SPS in blood culture broths should not exceed 0.025%. In addition, if blood is first collected in Vacutainer tubes (Becton Dickinson and Co., Franklin Lakes, N.J.),

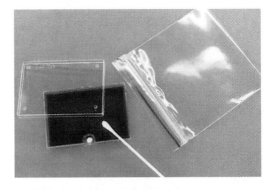

FIGURE 51-1 JEMBEC system. Plate contains modified Thayer-Martin medium. The CO_2-generating tablet is composed of sodium bicarbonate and citric acid. After inoculation the tablet is placed in the well and the plate is closed and placed in the zip-lok plastic pouch. The moisture in the agar will activate the tablet and generate a CO_2 atmosphere in the pouch.

—————— TABLE 51-2   PATHOGENESIS AND SPECTRUM OF DISEASES ——————

ORGANISM	VIRULENCE FACTORS	SPECTRUM OF DISEASE AND INFECTIONS
Moraxella catarrhalis	Uncertain; factors associated with cell envelope probably facilitate attachment to respiratory epithelial cells	Most infections are localized to sites associated with the respiratory tract and include otitis media, sinusitis, and pneumonia. Lower respiratory tract infections often target elderly patients and those with chronic obstructive pulmonary disease. Rarely causes disseminated infections such as bacteremia or meningitis
Neisseria gonorrhoeae	Several surface factors, such as pili, mediate attachment to human mucosal cell surface, invasion of host cells, and survival in the presence of inflammatory cells	A leading cause of sexually transmitted diseases. Genital infections include purulent urethritis in males and cervicitis in females. These infections also may be asymptomatic. Other localized infections include pharyngitis, anorectal infections, and conjunctivitis (e.g., ophthalmia neonatorum of newborns acquired during birth from an infected mother). Disseminated infections result when the organism spreads from a local infection to cause pelvic inflammatory disease or disseminated gonococcal infection that includes bacteremia, arthritis, and metastatic infection at other body sites
Neisseria meningitidis	Surface structures, perhaps pili, facilitate attachment to mucosal epithelial cells and invasion to the submucosa. Once in the blood, survival is mediated by production of a polysaccharide capsule. Endotoxin release mediates many of the systemic manifestations of infection such as shock	Life-threatening, acute, purulent, meningitis. Meningitis may be accompanied by appearance of petechiae (i.e., rash) that is associated with meningococcal bacteremia (i.e., meningococcemia). Bacteremia leads to thrombocytopenia, disseminated intravascular coagulation, and shock. Disseminated disease is often fatal. Less common infections include conjunctivitis, pneumonia, and sinusitis
Other *Neisseria* spp.	Unknown; probably of low virulence	Rarely involved in human infections. When infections occur they can include bacteremia, endocarditis, and meningitis

which may contain concentrations of SPS toxic to gonococci and meningococci, the blood specimen must be transferred to the broth culture system within 1 hour of collection.

Nasopharyngeal swabs collected to detect *N. meningitidis* carriers should be plated immediately to the JEMBEC system, or submitted on swabs placed in charcoal transport media.

SPECIMEN PROCESSING

The JEMBEC system should be incubated as soon as the plate is received in the laboratory. Body fluids (e.g., joint or cerebrospinal fluid [CSF]) should be kept at room temperature or placed at 37° C before culturing because both the gonococci and meningococci are sensitive to cold.

Any volume of clear fluid greater than 1 mL suspected of containing either of these pathogens should be centrifuged at room temperature at $1500 \times g$ for 15 minutes. The supernatant fluid should then be removed to a sterile tube. This fluid may be used for direct detection of soluble antigens (described below). The sediment is vortexed and inoculated onto the appropriate media (described below).

DIRECT DETECTION METHODS

Gram stain

Members of the genus *Neisseria* discussed in this chapter and *Moraxella catarrhalis* appear as gram-negative diplococci with adjacent sides flattened. The direct Gram stain of urethral discharge from symptomatic males with urethritis is an important test for gonococcal disease. The appearance of gram-negative diplococci inside polymorphonuclear leukocytes is diagnostic in this situation. However, because the normal vaginal and rectal flora is composed of gram-negative coccobacilli, which can resemble *Neisseria* spp., direct examination of endocervical secretions in symptomatic women is still only presumptive evidence of gonorrhea and the diagnosis must be confirmed by culture.

The direct Gram stain of body fluids for either *N. gonorrhoeae* or *N. meningitidis* is best accomplished using a cytocentrifuge, which can concentrate small numbers of organisms 100-fold.

Antigen detection

A commercial enzyme-linked immunosorbent assay (ELISA) system (Gonozyme, Abbott Laboratories, North Chicago, Ill.) is available to detect gonococcal antigen in the urethral and endocervical discharge from suspected cases of gonorrhea. The test is much more rapid than culture and is suitable for large-scale screening programs among sexually active individuals. However, it may cross-react with saprophytic *Neisseria* spp. and therefore is only a presumptive test.

The ELISA assay has been largely replaced in many laboratories by a newer DNA probe assay (Gen-Probe, San Diego, Calif.) with a chemiluminescent detection system for direct detection of gonococcal ribosomal RNA (rRNA) in genital and conjunctival specimens. This test performs well in high-risk patients, is rapid (results are available in 2 hours), and is suitable for screening many patients at one time.

The detection of capsular polysaccharide antigen in body fluids (e.g., urine, serum, CSF) before an organism is isolated is available for *N. meningitidis* (see Chapter 15). Particle agglutination techniques, such as latex and coagglutination, are used for this purpose. The latex particle assays have been found to be the most sensitive assays for detecting antigens of the serotypes most commonly seen in the United States, that is, *N. meningitidis* types A, B, C, Y, and W135.

CULTIVATION

Media of choice

N. meningitidis, M. catarrhalis, and saprophytic *Neisseria* spp. grow well on 5% sheep blood and chocolate agars; *N. gonorrhoeae* is more fastidious and requires an enriched chocolate agar for growth on primary culture. Because gonococci, and sometimes meningococci, must be isolated from sites that contain large numbers of normal flora (e.g., genital or upper respiratory tracts), selective media have been developed to facilitate their recovery. The first of these was Thayer-Martin medium, a chocolate agar with an enrichment supplement (IsoVitaLex) and the antimicrobials colistin (to inhibit gram-negative bacilli), nystatin (to inhibit yeast), and vancomycin (to inhibit gram-positive bacteria). This original medium was subsequently modified to include trimethoprim (to inhibit swarming *Proteus*), and its name was changed to MTM (modified Thayer-Martin) medium. Martin Lewis (ML) medium is similar to MTM except anisomycin, an antifungal agent, is substituted for nystatin and the concentration of vancomycin is increased.

A transparent medium containing lysed horse blood, horse plasma, yeast dialysate, and the same antibiotics as MTM, called **New York City (NYC) medium,** also has been used. The advantage of NYC medium is that genital mycoplasma (*Mycoplasma hominis* and *Ureaplasma urealyticum;* see Chapter 62 for more information regarding these organisms) will also grow on this agar. Some strains of *N. gonorrhoeae* are inhibited by the concentration of vancomycin in the selective media, so the addition of nonselective chocolate agar is recommended, especially in suspect cases that are culture negative or for sterile specimens (e.g., joint fluid).

Unlike the pathogenic species, some of the saprophytic *Neisseria* spp. (*N. flavescens, N. mucosa, N. sicca,* and *N. subflava*) may grow on MacConkey agar, although poorly. *N. gonorrhoeae* and *N. meningitidis* will grow in most broth blood culture media but grow poorly in common nutrient broths such as thioglycollate and brain-heart infusion. *M. catarrhalis* and the other *Neisseria* spp. grow well in almost any broth medium.

Incubation conditions and duration

Agar plates should be incubated at 35° C to 37° C for 72 hours in a CO_2-enriched, humid atmosphere. The pathogenic neisseriae and *M. catarrhalis* grow best under conditions of increased CO_2 (3% to 7%). This atmosphere can be achieved using a candle jar, CO_2-generating pouch, or CO_2 incubator. Only white, unscented candles should be used in candle jars because other types may be toxic to *N. gonorrhoeae* and *N. meningitidis*.

Humidity can be provided by placing a pan with water in the bottom of a CO_2 incubator, or by placing a sterile gauze pad soaked with sterile water in the bottom of a candle jar (Figure 51-2).

Colonial appearance

Table 51-3 describes the colonial appearance and other distinguishing characteristics (e.g., pigment) on chocolate agar.

APPROACH TO IDENTIFICATION

Various commercial systems are available for the rapid identification of the coccoid *Neisseria* spp. and *M. catarrhalis*. Some of these systems are described briefly in Table 30-1. These systems employ biochemical or enzymatic substrates and work very well for the

FIGURE 51-2 Candle jar.

pathogenic species (*N. gonorrhoeae, N. meningitidis,* and *M. catarrhalis*). A heavy inoculum of the organism is required, but because these systems detect the activity of preformed enzymes, viability of the organisms in the inoculum is not essential. Manufacturers' instructions should be followed exactly; several sys-

tems have been developed only for strains isolated on selective media and should not be used to test other gram-negative diplococci.

Biochemical identification

Table 51-4 shows some conventional biochemical tests that traditionally have been used to definitively identify these organisms. The extent to which identification of isolates is carried out depends on the source of the specimen and the suspected species of the organism involved.

An isolate from a child or a case of sexual abuse must be identified unequivocally, because of the medicolegal ramifications of these results. It is recommended that these organisms be identified using at least two different tests, that is, biochemical tests, serologic reagents, or DNA probe. Isolates from normally sterile body fluids should also be completely identified. However, isolates from genital sites of adults at risk of sexually transmitted disease can be identified presumptively, that is, oxidase-positive, gram-negative diplococci growing on gonococcal selective agar. Likewise, an oxidase-positive, gram-negative diplococcus that hydrolyzes tributyrin from an eye or ear culture can be presumptively identified as *M. catarrhalis* (Figure 51-3).

TABLE 51-3 COLONIAL APPEARANCE AND OTHER CHARACTERISTICS ON CHOCOLATE AGAR*

ORGANISM	APPEARANCE
Moraxella catarrhalis	Large, nonpigmented or gray, opaque, smooth; friable "hockey puck" consistency; colony may be moved intact over surface of agar
Neisseria gonorrhoeae	Small, grayish white, convex, translucent, shiny colonies with either smooth or irregular margins; may be up to five different colony types on primary plates
N. meningitidis	Medium, smooth, round, moist, gray to white; encapsulated strains are mucoid; may be greenish cast in agar underneath colonies
N. cinerea	Small, grayish white; translucent; slightly granular
N. flavescens	Medium, yellow, opaque, smooth
N. lactamica	Small, nonpigmented or yellowish, smooth, transparent
N. mucosa	Large, grayish white to light yellow, translucent; mucoid because of capsule
N. polysaccharea	Small, grayish white to light yellow, translucent, raised
N. sicca	Large, nonpigmented, wrinkled, coarse and dry, adherent
N. subflava	Medium, greenish yellow to yellow, smooth, entire edge

*Appearance on blood agar is the same as on chocolate agar except for pigmentation; colonies are less opaque on blood agar.

TABLE 51-4 BIOCHEMICAL AND PHYSIOLOGIC CHARACTERISTICS OF *MORAXELLA CATARRHALIS* AND COCCOID *NEISSERIA* SPP.

ORGANISM	GROWTH ON:			RAPID FERMENTATION SUGARS			NITRATE REDUCTION	GAS FROM NITRATE REDUCTION	0.1% NITRITE REDUCTION
	MODIFIED THAYER-MARTIN*	NUTRIENT AGAR AT 35°C	NUTRIENT AGAR AT 25°C	GLUCOSE	MALTOSE	LACTOSE			
Moraxella catarrhalis	v	+	v	−	−	−	+	−	v
Neisseria cinerea†	v	+	v	−‡	−	−	−	−	+
N. flavescens	−	+§	v	−	−	−	−	−	+§
N. gonorrhoeae‖	+	−	−	+	−	−	−	−	−
N. lactamica	+	v	v	+	+	+	−	−	+
N. meningitidis	+	−	−	+	+	−	−	−	v
N. mucosa	−	+	+	+ or (+)	+	−	+	+	+
N. sicca¶	−	+	+	+ or (+)	+	−	−	−	+
N. subflava¶	−	+	+	v	+	−	−	−	+

+, > 90% of strains positive; (+), > 90% of strains positive but reaction may be delayed (i.e., 2 to 7 days); −, > 90% of strains negative; *v*, variable.

*Growth defined as > 10 colonies.

†*Neisseria cinerea* may be differentiated from *N. flavescens* by a positive reaction with the amylosucrase test.

‡Some strains of *N. cinerea* may appear glucose-positive in some rapid systems and be mistaken for *N. gonorrhoeae*. However, *N. cinerea* grows on nutrient agar at 35° C and reduces nitrite, unlike the gonococcus.

§Only 2 of 10 strains were tested.

‖*Kingella denitrificans* may grow on modified Thayer-Martin agar and be mistaken for *N. gonorrhoeae* on microscopic examination. However, *K. denitrificans* can reduce nitrate and is catalase-negative, unlike the gonococcus.

¶*Neisseria subflava* produces a yellow pigment on Loeffler's agar; *N. sicca* does not.

Compiled from data in Knapp, J.S. and Rice, R.J. 1995. *Neisseria* and *Branhamella*. In Murray, P.R., Baron, E.J., Pfaller, M.A., et al., editors. Manual of clinical microbiology, ed 6. American Society for Microbiology, Washington, D.C.; and Weyant, R.S., Moss, C.W., Weaver, R.E., et al., editors. 1996. Identification of unusual pathogenic gram-negative aerobic and facultatively anaerobic bacteria, ed 2. Williams & Wilkins, Baltimore.

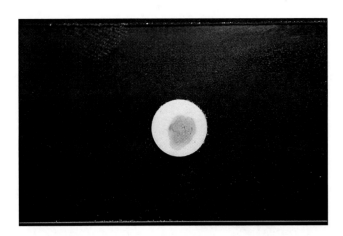

FIGURE 51-3 Butyrate disk. Positive (*Moraxella catarrhalis*) indicated by blue color.

Comments regarding specific organisms

Determination of carbohydrate utilization patterns has traditionally been performed in cystine trypticase soy agar (CTA) with 1% dextrose, maltose, lactose, and sucrose. This medium is no longer widely used because it does not work well for oxidative *Neisseria* spp., specifically *N. gonorrhoeae* and *N. meningitidis*. Therefore, carbohydrate utilization patterns are currently determined by inoculating an extremely heavy suspension of the organism to be tested in a small volume of buffered, low-peptone substrate with the appropriate carbohydrate (see Chapter 13 for a detailed description of this test).

The saprophytic *Neisseria* spp. are not routinely identified in the clinical laboratory. *M. catarrhalis* may be differentiated from the gonococci and meningococci based on its growth on blood agar at 22° C and nutrient agar at 35° C, the reduction of nitrate to nitrite, its inability to utilize carbohydrates, and its production of DNase. *M. catarrhalis* is the only member of this group of organisms that hydrolyzes DNA. *N. cinerea* may be misidentified as *N. gonorrhoeae* if the isolate produces a weak positive glucose reaction. However, it will grow on nutrient agar at 35° C while the gonococcus will not.

Immunoserologic identification

Particle agglutination methods are available for immunoserologic identification of *N. gonorrhoeae*. They include the Phadebact Monoclonal GC OMNI test (Pharmacia Diagnostics, Piscataway, N.J.) and the GonoGen II test (New Horizons Diagnostics, Columbia, Md.). These tests can be performed from colonies growing on primary plates; isolates are typed with specific monoclonal antibodies. A fluorescent mono-clonal antibody test for culture confirmation of *N. gonorrhoeae* is also available (Behring Diagnostics, Cupertino, Calif.).

Serotyping

Antisera are commercially available for identifying *N. meningitidis* serogroups A, B, C, Y, and W135, the types that most frequently cause systemic disease in the United States.

SERODIAGNOSIS

Serodiagnostic techniques are not generally used for the laboratory diagnosis of infections caused by the organisms discussed in this chapter.

ANTIMICROBIAL SUSCEPTIBILITY TESTING AND THERAPY

Although beta-lactamase production is common among *M. catarrhalis* isolates, many beta-lactam antibiotics maintain activity. Several other agents are also effective so that susceptibility testing to guide therapy is not routinely required (Table 51-5).

Standard methods have been established for performing in vitro susceptibility testing with *N. gonorrhoeae* and *N. meningitidis* (see Chapter 18 for details regarding these methods). In addition, various agents are available that may be considered for testing and therapeutic use.[6-8] Although resistance to penicillin by production of beta-lactamase has become widespread among *N. gonorrhoeae*, resistance to ceftriaxone, which is not notably affected by the enzyme, has not been described.[10] Therefore, routine testing of isolates as a guide for using this therapeutic agent does not appear to be necessary. On the other hand, quinolones also are widely used to treat gonorrhea, but resistance to these agents is emerging and testing methods to detect and track this resistance are needed.[4,11]

Beta-lactamase production in *N. meningitidis* is extremely rare, but decreased susceptibility to penicillin mediated by altered penicillin-binding proteins is emerging.[1,2,13] However, optimum laboratory methods for detecting this relatively low level of resistance have not been established, and the impact of this resistance on the clinical efficacy of penicillin is not known.

PREVENTION

A single-dose vaccine to the polysaccharide capsular antigens of *N. meningitidis* groups A, C, Y, and W135 is available in the United States. It is given to military recruits and patients without spleens who are older than age 2. Chemoprophylaxis with rifampin is indicated for close contacts of patients with meningococcal

TABLE 51-5 ANTIMICROBIAL THERAPY AND SUSCEPTIBILITY TESTING

SPECIES	THERAPEUTIC OPTIONS	POTENTIAL RESISTANCE TO THERAPEUTIC OPTIONS	VALIDATED TESTING METHODS*	COMMENTS
Moraxella catarrhalis	Several beta-lactams are effective, including: beta-lactam/beta-lactamase inhibitor combinations, cephalosporins, macrolides, quinolones, and trimethoprim/sulfamethoxazole	Commonly produce beta-lactamases that mediate resistance to ampicillin. Although not common, resistance to erythromycin, and trimethoprim/sulfamethoxazole may occur	Methods outlined in Chapter 18 for non-fastidious bacteria may be used	Testing to guide therapy is not routinely needed
Neisseria gonorrhoeae	Recommended therapy includes ceftriaxone or quinolones (e.g., ciprofloxacin, ofloxacin, levofloxacin). Macrolides also may be used	Penicillin resistance by beta-lactamase production is common. Resistance to quinolones is emerging	As documented in Chapter 18: disk diffusion, agar dilution, limited commercial methods	Testing by disk diffusion may not detect decrease in quinolone activity. No ceftriaxone resistance has been documented
Neisseria meningitidis	Supportive therapy for shock and antimicrobial therapy using penicillin, ceftriaxone, cefotaxime, or chloramphenicol	Subtle increases in beta-lactam resistance have been described, but clinical relevance is uncertain. Beta-lactamase production is extremely rare	As documented in Chapter 18: broth dilution, agar dilution	Testing to guide therapy is not routinely needed
Other *Neisseria* spp.	Usually susceptible to penicillin and other beta-lactams	Uncertain; potential for beta-lactamase production	Not available	

*Validated testing methods include those standard methods recommended by the National Committee for Clinical Laboratory Standards (NCCLS) and those commercial methods approved by the Food and Drug Administration (FDA).

meningitis. Household contacts, day care contacts, and health care workers who have given mouth-to-mouth resuscitation are at risk and should be treated within 24 hours.

A single application of either a 2.5% solution of povidone-iodine, 1% tetracycline eye ointment, 0.5% erythromycin eye ointment, or 1% silver nitrate eye drops is instilled within 1 hour of delivery to prevent gonococcal ophthalmia neonatorum.

References

1. Abadi, F.J.R., Yakubu, D.E., and Pennington, T.H. 1995. Antimicrobial susceptibility of penicillin-sensitive and penicillin-resistant meningococci. J. Antimicrob. Chemother. 35:687.

2. Blondeau, J.M., Ashton, F.E., Isaacson, M., et al. 1995. *Neisseria meningitidis* with decreased susceptibility to penicillin in Saskatachewan, Canada. J. Clin. Microbiol. 33:1784.

3. Heiddal, S., Sverrisson, J.T., Yngvason, F.E., et al. 1993. Native valve endocarditis due to *Neisseria sicca:* case report and review. Clin. Infect. Dis. 16:667.

4. Kam, K.M., Wong, P.W., Cheung, M.M., et al. 1996. Detection of quinolone-resistant *Neisseria gonorrhoeae.* J. Clin. Microbiol. 34:1462.

5. Myer, G.A., Shope, T.R., Waeker, N.J., et al. 1995. *Moraxella (Branhamella) catarrhalis* bacteremia in children. Clin. Pediatr. 34:146.

6. National Committee for Clinical Laboratory Standards. 1995. Performance Standards for Antimicrobial Susceptibility Testing. M100-S6. NCCLS, Villanova, Pa. Order from NCCLS, 940 W. Valley Road, Suite 1400, Wayne, Pa. 19087.

7. National Committee for Clinical Laboratory Standards. 1997. Performance Standards for Antimicrobial Disk Susceptibility Tests. M2-A6, ed 6. NCCLS, Villanova, Pa. Order from NCCLS, 940 W. Valley Road, Suite 1400, Wayne, Pa. 19087.

8. National Committee for Clinical Laboratory Standards. 1997. Methods for Dilution Antimicrobial Susceptibility Tests for Bacteria that Grow Aerobically. M7-A4, ed 4. NCCLS, Villanova, Pa. Order from NCCLS, 940 W. Valley Road, Suite 1400, Wayne, Pa. 19087.

9. Riedo, F.X., Plikaytis, B.D., and Broome, C.V. 1995. Epidemiology and prevention of meningococcal disease. Pediatr. Infect. Dis. J. 14:643.

10. Schwebke, J.R., Whittington, W., Rice, R.J., et al. 1995. Trends in susceptibility of *Neisseria gonorrhoeae* to ceftriaxone from 1985 through 1991. Antimicrob. Agents Chemother. 39:917.

11. Tanaka, M., Matsumoto, T., Kobayashi, I., et al. 1995. Emergence of in vitro resistance to fluoroquinolones in *Neisseria gonorrhoeae* isolated in Japan. Antimicrob. Agents Chemother. 39:2367.

12. Verghese, A. and Berk, S.L. 1991. *Moraxella (Branhamella) catarrhalis.* Infect. Dis. Clin. North Am. 5:523.

13. Woods, C.R., Smith, A.L., Wasilauskas, B.L., et al. 1994. Invasive disease caused by *Neisseria meningitidis* relatively resistant to pennicillin in North Carolina. J. Infect. Dis. 170:453.

Bibiliography

Apicella, M.A. 1995. *Neisseria meningitidis.* In Mandell, G.L., Bennett, J.E., and Dolin, R., editors. Principles and practice of infectious diseases. Churchill Livingstone, New York.

Benson, A.S. 1995. Control of communicable diseases manual, ed 16. American Public Health Association, Washington, D.C.

Doern, G.V. 1995. Susceptibility testing of fastidious bacteria. In Murray, P.R., Baron, E.J., Pfaller, M.A., et al., editors. Manual of clinical microbiology, ed 6. American Society for Microbiology, Washington, D.C.

Groschel, D.H.M. 1995. *Moraxella catarrhalis* and other gram-negative cocci. In Mandell, G.L., Bennett, J.E., and Dolin, R., editors. Principles and practice of infectious diseases. Churchill Livingstone, New York.

Handsfield, H.H. and Sparling, P.F. 1995. *Neisseria gonorrhoeae.* In Mandell, G.L., Bennett, J.E., and Dolin, R., editors. Principles and practice of infectious diseases. Churchill Livingstone, New York.

Knapp, J.S. and Rice, R.J. 1995. *Neisseria* and *Branhamella.* In Murray, P.R., Baron, E.J., Pfaller, M.A., et al., editors. Manual of clinical microbiology, ed 6. American Society for Microbiology, Washington, D.C.

Weyant, R.S., Moss, C.W., Weaver, R.E., et al., editors. 1996. Identification of unusual pathogenic gram-negative aerobic and facultatively anaerobic bacteria, ed 2. Williams & Wilkins, Baltimore.

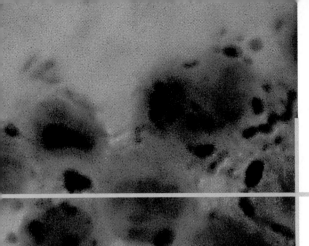

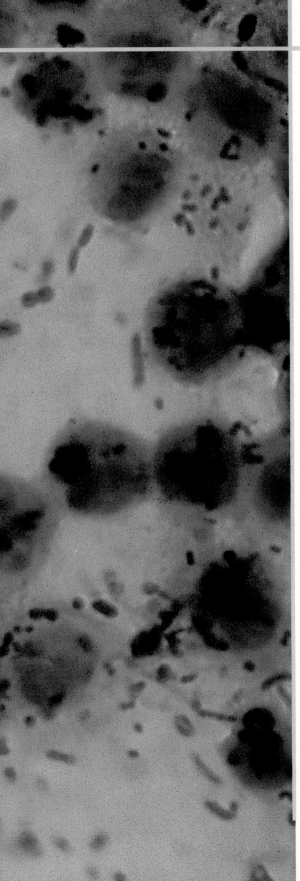

Section Seven 7

Catalase-Positive, Gram-Positive Cocci

52 | *STAPHYLOCOCCUS, MICROCOCCUS,* AND SIMILAR ORGANISMS

Genera and Species to be Considered

- *Staphylococcus aureus*
- Coagulase-negative staphylococci
  (most commonly encountered)
    Staphylococcus epidermidis
    Staphylococcus haemolyticus
    Staphylococcus saprophyticus
    Staphylococcus lugdunensis
  (less commonly encountered)
    Staphylococcus capitis
    Staphylococcus caprae

Staphylococcus warneri
Staphylococcus schleiferi
Staphylococcus hominis
Staphylococcus auricularis
Staphylococcus cohnii
Staphylococcus xylosus
Staphylococcus simulans
- *Micrococcus* spp. and related genera
- *Stomatococcus mucilaginosus*

GENERAL CHARACTERISTICS

The organisms described in this chapter are all cata-lase-positive, gram-positive cocci. However, only those belonging to the *Staphylococcus* genus are of primary clinical significance. Several of the coagulase-negative staphylococci (i.e., non-*Staphylococcus aureus*) species listed may be encountered in clinical speci-mens, but only those most prominently associated with infections in humans (i.e., *S. epidermidis*, *S. haemolyticus*, *S. saprophyticus*, and *S. lugdunensis*) are discussed in detail.

EPIDEMIOLOGY

As outlined in Table 52-1, the staphylococci that are associated with infections in humans are colonizers of various skin and mucosal surfaces. Because the carrier state is common among the human population, infec-tions are frequently acquired when the colonizing strain gains entrance to a normally sterile site as a result of trauma or abrasion to the skin or mucosal surface. However, the traumatic event that allows entry of the organism often may be so minor that it goes unnoticed.

Staphylococci are also transmitted from person to person. Upon transmission, the organisms may become established as part of the recipient's normal flora and later be introduced to sterile sites by trauma or invasive procedures. Alternatively, the organism may be directly introduced into normally sterile sites, such as by a surgeon or nurse during surgery. Person-to-person spread of staphylococci, particularly those that have acquired antimicrobial resistance, most notably occurs in hospitals and presents substantial infection control problems.[3]

PATHOGENESIS AND SPECTRUM OF DISEASE

Without question, *S. aureus* is the most virulent species of staphylococci encountered. A wide spec-trum of factors, not all of which are completely understood, contribute to this organism's ability to cause infections and disease. Several toxins and enzymes mediate tissue invasion and survival at the infection site (Table 52-2). Elaboration of these fac-tors is chiefly responsible for the various skin, wound, and deep tissue infections commonly caused by *S. aureus*. Many of these infections can rapidly become life-threatening if not treated and managed appropri-ately.

Localized skin infections may involve hair folli-cles (i.e., folliculitis) and spread deeper to cause boils (i.e., furuncles). More serious, deeper infections result when the furuncles coalesce to form carbuncles. Impetigo, the *S. aureus* skin infection that involves the epidermis, is typified by the production of vesicles that rupture and crust over. Regardless of the initial site of infection, the invasive nature of this organism

TABLE 52-1 EPIDEMIOLOGY

ORGANISM	HABITAT (RESERVOIR)	MODE OF TRANSMISSION
Staphylococcus aureus	Normal flora of human anterior nares, nasopharynx, perineal area, and skin; can colonize various epithelial or mucosal surfaces	Spread of patient's endogenous strain to normally sterile site by traumatic introduction (e.g., surgical wound or microabrasions). Also may be transmitted person to person by fomites, air, or unwashed hands of health care workers, especially in the nosocomial setting. May be transmitted from infected skin lesion of health care worker to patient
Staphylococcus epidermidis	Normal flora of human skin and mucous membranes; distributed widely, often in large numbers, over body surface	Spread of patient's endogenous strain to normally sterile site, usually as a result of implantation of medical devices (e.g., shunts, prosthetic devices) during hospitalization. Person-to-person spread in hospitals can lead to patients becoming colonized and potentially infected with antibiotic-resistant strains
Staphylococcus haemolyticus and *Staphylococcus lugdunensis*	Normal human flora similar to *S. epidermidis* but present in fewer numbers	Probably the same as for *S. epidermidis*
Staphylococcus saprophyticus	Normal flora of human skin and mucosa of genitourinary tract	Introduction of endogenous flora into sterile urinary tract, notably in young, sexually active females. A community-acquired infection, not considered an agent of nosocomial infections
Micrococcus spp., and *Stomatococcus mucilaginosus*	Normal flora of human skin, mucosa, and oropharynx	Uncertain; rarely implicated in infections. When infections occur, they likely involve endogenous strains

always presents a threat for deeper tissue invasion, bacteremia, and spread to one or more internal organs.

S. aureus can also produce toxin-mediated diseases, such as **scalded skin syndrome** and **toxic shock syndrome**. In these cases the organisms may remain relatively localized, but production of potent toxins causes systemic or widespread effects. With scalded skin syndrome, which usually afflicts neonates, the exfoliatin toxins cause extensive sloughing of epidermis to produce a burnlike effect on the patient. The toxic shock syndrome toxin (TSST-1) has several systemic effects including fever, desquamation, and hypotension potentially leading to shock and death.

The coagulase-negative staphylococci, among which S. epidermidis is the most commonly encountered, are substantially less virulent than S. aureus and are opportunistic pathogens. Their prevalence as

nosocomial pathogens is as much, if not more, related to medical procedures and practices than to the organism's capacity to establish an infection. Infections with S. epidermidis and, less commonly, S. haemolyticus and S. lugdunensis usually involve implantation of medical devices (see Table 52-2).[9] This kind of medical intervention allows invasion by these normally noninvasive organisms. Two organism characteristics that do enhance the likelihood of infection include production of a slime layer that facilitates attachment to implanted medical devices and the ability to acquire resistance to most of the antimicrobial agents used in hospital environments.[1,2,4]

Although most coagulase-negative staphylococci are primarily associated with nosocomial infections, urinary tract infections caused by S. saprophyticus are clear exceptions. This organism is most frequently associated with community-acquired urinary tract infections in young, sexually active females but is not

─────── **TABLE 52-2   PATHOGENESIS AND SPECTRUM OF DISEASES** ───────

ORGANISM	VIRULENCE FACTORS	SPECTRUM OF DISEASES AND INFECTIONS
Staphylococcus aureus	Produces and secretes toxins and enzymes that have a role in virulence; alpha, beta, gamma, and delta toxins act on host cell membranes and mediate cell destruction. Leucocidin mediates destruction of phagocytes. Clumping factor, coagulase, and hyaluronidase enhance invasion and survival in tissues. Potent exotoxins include exfoliatins, toxic shock syndrome toxin (TSST-1), and enterotoxins	Infections generally involve intense suppuration and destruction (necrosis) of tissue. Infections can be generally grouped as localized skin infections such as folliculitis, furuncles (boils), carbuncles, and impetigo; various wound infections; deep infections that spread from skin to cause bacteremia (with and without endocarditis) and to involve bones, joints, deep organs, and tissues; scalded skin syndrome in neonates; toxic shock syndrome; and food poisoning
Staphylococcus epidermidis	Certain factors facilitate initial attachment to implanted medical devices. Production of exopolysaccharide "slime" or biofilm enhances organism adhesion and provides mechanical barrier to antibiotics and host defense mechanisms. Propensity to acquire and disseminate antimicrobial resistance allows for survival in hospital setting	Ubiquitous member of normal flora makes this species the most commonly encountered in clinical specimens, usually as a contaminant. Can be difficult to establish clinical significance. Most common infections include nosocomial bacteremia associated with indwelling vascular catheters; endocarditis involving prosthetic cardiac valves (rarely involves native valves); infection at intravascular catheter sites, frequently leading to bacteremia; and other infections associated with CSF shunts, prosthetic joints, vascular grafts, postsurgical ocular infections, and bacteremia in neonates under intensive care
S. haemolyticus and *S. lugdunensis*	Uncertain; probably similar to those described for *S. epidermidis*	Similar to those described for *S. epidermidis*
S. saprophyticus	Uncertain	Urinary tract infections in sexually active, young females; infections in sites outside urinary tract are not common
Micrococcus spp. and *Stomatococcus mucilaginosus*	Unknown; probably of extremely low virulence	Usually considered contaminants of clinical specimens; rarely implicated as cause of infections in humans

commonly associated with hospital-acquired infections.

Because coagulase-negative staphylococci are ubiquitous colonizers, they are frequently found as contaminants in clinical specimens. This fact, coupled with the emergence of these organisms as nosocomial pathogens, complicates laboratory interpretation of their clinical significance. When these organisms are isolated from clinical specimens, every effort should be made to substantiate their clinical relevance in a particular patient so that unnecessary work and production of misleading information can be avoided.

What, if any, virulence factors are produced by *Micrococcus* and *Stomatococcus* is not known. Because these organisms are rarely associated with infection, they are probably of low virulence.

LABORATORY DIAGNOSIS

SPECIMEN COLLECTION AND TRANSPORT

No special considerations are required for specimen collection and transport of the organisms discussed in this chapter. Refer to Table 1-1 for general information on specimen collection and transport.

SPECIMEN PROCESSING

No special considerations are required for processing of the organisms discussed in this chapter. Refer to Table 1-1 for general information on specimen processing.

DIRECT DETECTION METHODS

Other than Gram stain of patient specimens, there are no specific procedures for the direct detection of these organisms in clinical material. All the *Micrococcaceae* produce spherical, gram-positive cells. During cell division, *Micrococcaceae* divide along both longitudinal and horizontal planes, forming pairs, tetrads, and, ultimately, irregular clusters (Figure 52-1). Gram stains should be performed on young cultures, because very old cells may lose their ability to retain crystal violet and may appear gram-variable or gram-negative. Staphylococci appear as gram-positive cocci, usually in clusters. *Micrococcus* and related genera (i.e., *Kytococcus, Nesterenkonia, Dermacoccus, Arthrobacter,* and *Kocuria*[5,9]) and *Stomatococcus* resemble the staphylococci microscopically.

CULTIVATION

Media of choice

Micrococcaceae will grow on 5% sheep blood and chocolate agars but not MacConkey agar. They will also grow well in broth-blood culture systems and common nutrient broths, such as thioglycollate and brain-heart infusion.

Selective media can also be used to isolate staphylococci from clinical material. Mannitol salt agar is commonly used for this purpose. This agar contains a high concentration of salt (10%), the sugar mannitol, and phenol red as the pH indicator. On this medium, organisms such as *S. aureus* that can grow in the presence of salt and ferment mannitol produce colonies surrounded by a yellow halo.

Incubation conditions and duration

Visible growth on 5% sheep blood and chocolate agars incubated at 35° C in carbon dioxide (CO_2) or ambient air usually occurs within 24 hours of inoculation. Mannitol salt agar and other selective media may require incubation for at least 48 to 72 hours before growth is detected.

Colonial appearance

Table 52-3 describes the colonial appearance and other distinguishing characteristics (e.g., hemolysis) of each genus and various staphylococcal species on 5% sheep blood agar. Growth on chocolate agar is similar. *S. aureus* yields colonies surrounded by a yellow halo on mannitol salt agar. However, other staphylococci (particularly *S. saprophyticus*) may also ferment mannitol and thus resemble *S. aureus* on this medium.

APPROACH TO IDENTIFICATION

The commercial systems for identification of *Staphylococcus* and *Micrococcus* are discussed in Chapter 30. Their accuracy in the identification of various staphylococcal species is highly variable and uncertain. Therefore, the ability of any commercial method to identify the different species of coagulase-negative staphylococci should be validated before use. However, in general, commercially available methods do perform well in the identification of *S. epidermidis,* the most commonly encountered coagulase-negative staphylococcal species.

Table 52-4 shows how the catalase-positive, gram-positive cocci can be differentiated. Because they may show a pseudocatalase reaction, that is, they may appear to be catalase-positive, *Aerococcus* and *Enterococcus* are included in Table 52-4. Once an organism has been characterized as a gram-positive, catalase-positive, coccoid bacterium, complete identification may involve a series of tests, including (1) atmospheric requirements, (2) resistance to 0.04 U bacitracin (Taxo A disk) and furazolidone, and (3) possession of cytochrome C as determined by the microdase (modified oxidase) test. However, in the busy setting of many clinical laboratories microbiologists proceed immediately to a coagulase test based on recognition of a staphylococcal-like colony and a positive catalase test.

Microdase disks are available commercially (Remel, Lenexa, Kan.). A visible amount of growth from an 18 to 24- hour-old culture is smeared on the disk; *Micrococcus* spp. turn blue within 2 minutes (Figure 52-2).

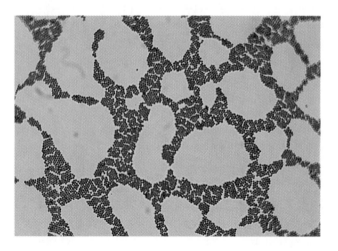

FIGURE 52-1 Gram stain of *Staphylococcus aureus* from blood agar.

TABLE 52-3   COLONIAL APPEARANCE AND CHARACTERISTICS ON 5% SHEEP BLOOD AGAR

ORGANISM	APPEARANCE
Micrococcus spp. and related organisms*	Small to medium; opaque, convex; nonhemolytic; wide variety of pigments (white, tan, yellow, orange, pink)
Staphylococcus aureus	Medium to large; smooth, entire, slightly raised, translucent; most colonies pigmented creamy yellow; most colonies beta-hemolytic
S. epidermidis	Small to medium; translucent, gray-white colonies; most colonies nonhemolytic; slime-producing strains are extremely sticky and adhere to the agar surface
S. haemolyticus	Medium; smooth, butyrous and opaque; beta-hemolytic
S. hominis	Medium to large; smooth, butyrous and opaque; may be unpigmented or cream-yellow-orange
S. lugdunensis	Medium to large; smooth, glossy, entire edge with slightly domed center; unpigmented or cream to yellow-orange
S. warneri	Resembles *S. lugdunensis*
S. saprophyticus	Large; entire, very glossy, smooth, opaque, butyrous, convex; usually white but colonies can be yellow or orange
S. schleiferi	Medium to large; smooth, glossy, slightly convex with entire edges; unpigmented
S. intermedius	Large; slightly convex, entire, smooth, glossy, translucent; usually nonpigmented
S. hyicus	Large; slightly convex, entire, smooth, glossy, opaque; usually nonpigmented
S. capitis	Small to medium; smooth, slightly convex, glistening, entire, opaque; *S. capitis* subsp. *urealyticus* usually pigmented (yellow or yellow orange); *S. capitis* subsp. *capitis* is nonpigmented
S. cohnii	Medium to large; convex, entire, circular, smooth, glistening, opaque, *S. cohnii* subsp. *urealyticum* usually pigmented (yellow or yellow orange); *S. cohnii* subsp. *cohnii* is nonpigmented
S. simulans	Large; raised, circular, nonpigmented, entire, smooth, slightly glistening
S. auricularis	Small to medium; smooth, butyrous, convex, opaque, entire, slightly glistening; non-pigmented
S. xylosus	Large; raised to slightly convex, circular, smooth to rough, opaque, dull to glistening; some colonies pigmented yellow or yellow-orange
S. sciuri	Medium to large; raised, smooth, glistening, circular, opaque; most strains pigmented yellow in center of colonies
S. caprae	Small to medium; circular, entire, convex, opaque, glistening; nonpigmented
Stomatococcus mucilaginosus	Medium white, opaque, sticky; strong adherence on agar; nonhemolytic

*Includes *Kytococcus, Nesterenkonia, Dermacoccus, Kocuria,* and *Arthrobacter.*[10]

TABLE 52-4 DIFFERENTIATION AMONG GRAM-POSITIVE, CATALASE-POSITIVE COCCI

ORGANISM	CATALASE	MODIFIED OXIDASE	AEROTOLERANCE	RESISTANCE TO: BACITRACIN (0.04 U)[a]	FURAZOLIDONE (100 µg)[a]	LYSOSTAPHIN (200 µg/mL)
Staphylococcus	+[b]	−[c]	FA	R	S	S
Micrococcus (and related organisms)	+	+	A[d]	S	R	R[e]
Stomatococcus	+/−	−	FA	R or S	R or S	R
Aerococcus	−[f]	−	FA[g]	S	S	R
Enterococcus	−[f]	−	FA	R	S	R

+, ≥90% of species or strains positive; +/−, ≥90% of species or strains weakly positive; −, ≥90% of species or strains negative; *S*, sensitive; *R*, resistant; *FA*, facultative anaerobe or microaerophile; *A*, strict aerobe.

[a]For bacitracin, susceptible ≥ 10 mm; for furazolidone, susceptible ≥ 15 mm.

[b]*S. aureus* subsp. *anaerobius* and *S. saccharolyticus* are catalase-negative and only grow anaerobically.

[c]*S. sciuri, S. caseolyticus, S. lentus,* and *S. vitulus* are modified oxidase-positive.

[d]*Kocuria (Micrococcus) kristinae* is facultatively anaerobic.

[e]Some strains of *M. luteus, Arthrobacter (Micrococcus) agilis,* and *Kytococcus (Micrococcus) sedentarius* are susceptible to lysostaphin.

[f]Some strains may show a pseudocatalase reaction.

[g]Grows best at reduced oxygen tension and may not grow anaerobically.

Compiled from Kloos, W.E. and Bannerman, T.L. 1995. In Murray, P.R., Baron, E.J., Pfaller, M.A., et al., editors. Manual of clinical microbiology, ed 6. American Society for Microbiology, Washington, D.C.

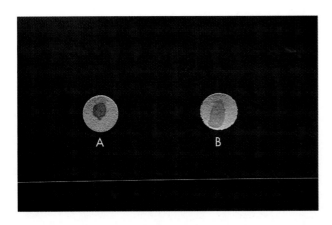

FIGURE 52-2 Microdase test. **A,** Positive *(Micrococcus luteus* [note purple color]). **B,** Negative *(Staphylococcus aureus).*

Comments regarding specific organisms

Stomatococcus sp. is weakly catalase-positive, microdase-negative, and lysostaphin-resistant and will grow in ambient air and CO_2 and under anaerobic conditions. *Micrococcus* spp. and related genera are (1) not lysed with lysostaphin, (2) resistant to the antibiotic furazolidone, (3) susceptible to 0.04 U of bacitracin, and (4) microdase-positive; they usually will only grow aerobically. In contrast, staphylococci are (1) lysed with lysostaphin, (2) resistant to 0.04 U of bacitracin, (3) susceptible to furazolidone, (4) microdase-negative, and (5) facultatively anaerobic.

For both bacitracin and furazolidone resistance, disk tests are used. A 0.04-U bacitracin-impregnated disk and a 100-µg furazolidone-impregnated disk, both available from Becton Dickinson Microbiology Systems, are placed on the surface of a 5% sheep blood agar plate that has been previously streaked in three directions with a cotton-tipped swab that has been dipped in a bacterial suspension prepared to match the turbidity of the 0.5 McFarland standard (i.e., the same as is used in preparing inocula for disk diffusion susceptibility tests as described in Chapter 18).

Once an isolate is identified as, or strongly suspected to be, a species of staphylococci, a test for coagulase production is performed to separate *S. aureus* from the other species that are collectively referred to as *coagulase-negative staphylococci* (Figure 52-3).

The enzyme coagulase produced by *S. aureus* binds plasma fibrinogen and activates a cascade of reactions that cause plasma to clot. An organism can produce two types of coagulase, referred to as *bound* and *free* (see Procedure 30-8 for further information on coagulase tests). Detection of bound coagulase, or clumping factor, is accomplished using a rapid slide test (i.e., the slide coagulase test), in which a positive test is indicated when the organisms agglutinate on a glass slide when mixed with plasma (see Figure 30-8, *A*). Most, but not all, strains of *S. aureus* produce clumping factor and thus are readily detected by this slide test.

Isolates suspected of being *S. aureus,* but failing to produce bound coagulase, must be tested for production of extracellular (i.e., free) coagulase. This test, referred to as the **tube coagulase test,** is performed by inoculating a tube containing plasma and incubating at 35° C. Production of the enzyme results in a clot formation within 1 to 4 hours of inoculation (Figure 30-8, *B*). Because citrate-utilizing organisms may yield false-positive results, plasma containing EDTA rather than citrate should be used.

Various commercial systems are available that substitute for the conventional coagulase tests just described. Latex agglutination procedures that detect clumping factor and protein A include Staphaurex Plus (Murex Diagnostics, Norcross, Ga.) and Staph Latex Kit (Remel, Lenexa, Kan.). Passive hemagglutination tests detect clumping factor and include Staphyloslide (Becton Dickinson Microbiology Systems, Cockeysville, Md.) and Hemastaph (Remel, Lenexa, Kan.).

Table 52-5 shows tests for differentiating among the coagulase-positive staphylococci; *S. intermedius* is an important agent of dog-bite wound infections and may be misidentified as *S. aureus* if only coagulase testing is performed. Microbiologists may want to consider performing the additional tests in cases in which coagulase-postive staphylococci are isolated from dog-bite infections.

Most laboratories do not identify the coagulase-negative staphylococci to species. However, exceptions may include isolates from normally sterile sites (blood, joint fluid, or CSF), isolates from prosthetic devices, catheters and shunts, and isolates from urinary tract infections that may be *S. saprophyticus.* The coagulase-negative staphylococci may be identified based on the criteria shown in Figure 52-3 and Tables 52-5 through 52-9. Isolates not identified to species usually are simply reported as "coagulase-negative staphylococci."

SERODIAGNOSIS

Antibodies to teichoic acid, a major cell wall component of gram-positive bacteria, are usually produced in long-standing or deep-seated staphylococcal infections, such as osteomyelitis. This procedure, if required, is usually performed in reference laboratories. However, the clinical utility of performing this assay is, at best, uncertain.

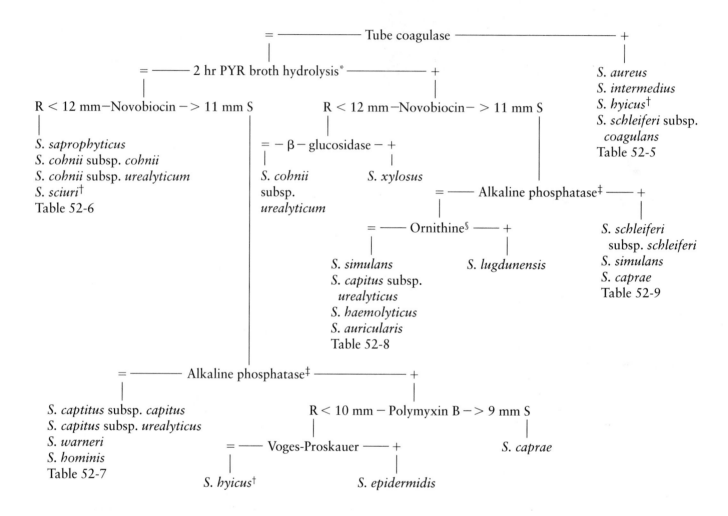

=, Signifies a negative result.

*Available commercially from Remel, Inc., Lenexa, Kan.

†Rarely involved in infections in humans.

‡Alkaline phosphatase available as a disk (Becton Dickinson Microbiology Systems, Cockeysville, Md.).

§Moeller's decarboxylase medium.

FIGURE 52-3 Staphylococcal identification to species. (Based on the methods in Hébert, G.A., Crowder, C.G., Hancock, G.A., et al. 1988. J. Clin. Microbiol. 26:1939.)

TABLE 52-5   DIFFERENTIATION AMONG COAGULASE-POSITIVE STAPHYLOCOCCI

ORGANISM	TUBE COAGULASE	ACID FROM D-TREHALOSE*	VOGES-PROSKAUER TEST	β-GALACTOSIDASE†
S. aureus	+	+	+	−
S. intermedius‡	+	+	−	+
S. hyicus‡§	v	+	−	−
S. schleiferi subsp. *coagulans*	+	−	+	ND

+, >90% of strains positive; −, >90% of strains negative; *ND*, not determined; *v*, variable.

*Performed in Andrade's peptone broth with 1% sugar.

†Performed from disk (Becton Dickinson Microbiology Systems, Cockeysville, Md.) or tablet (KEY Scientific Products, Round Rock, Texas).

‡Primarily isolated from animals.

§Rarely a cause of infections in humans.

Compiled from Kloos, W.E. and Bannerman, T.L. 1995. In Murray, P.R., Baron, E.J., Pfaller, M.A., et al., editors. Manual of clinical microbiology, ed 6. American Society for Microbiology, Washington, D.C.; Roberson, J.R., Fox, L.K., Hancock, D., et al. 1992. J. Clin. Microbiol. 30:3217; and Kloos, W.E. and Wolfshohl, J.F. 1982. J. Clin. Microbiol. 16:509.

ANTIMICROBIAL SUSCEPTIBILITY TESTING AND THERAPY

Antimicrobial therapy is vital to management of patients suffering from staphylococcal infections (Table 52-10). Although a broad spectrum of agents may be used for therapy (see Table 19-1 for a detailed listing), most staphylococci are capable of acquiring and using one or more of the resistance mechanisms presented in Table 17-3 and Figure 17-10. The unpredictable nature of any clinical isolate's antimicrobial susceptibility requires that testing be done as a guide to therapy. As discussed in Chapter 18, several standard methods and commercial systems have been developed for testing staphylococci.

Although a pencillinase-resistant penicillin, such as methicillin, nafcillin, or oxacillin, is the mainstay of antistaphylococcal therapy, resistance is common.[7] The primary mechanism for this resistance is production of an altered penicillin-binding protein (i.e., PBP 2a), which renders all currently available beta-lactams

TABLE 52-6   DIFFERENTIATION AMONG COAGULASE-NEGATIVE, PYR-NEGATIVE, NOVOBIOCIN-RESISTANT STAPHYLOCOCCI

ORGANISM	UREASE	OXIDASE	ALKALINE PHOSPHATASE*
S. saprophyticus	+	−	−
S. cohnii subsp. *cohnii*	−	−	−
S. cohnii subsp. *urealyticum*	+	−	+
S. sciuri†	−	+	+

+, >90% of strains positive; −, >90% of strains negative.

*Performed from disk (Becton Dickinson Microbiology Systems, Cockeysville, Md.) or tablet (KEY Scientific Products, Round Rock, Texas).

†Primarily isolated from animals; rarely a cause of infections in humans.

Compiled from Kloos, W.E. and Bannerman, T.L. 1995. In Murray, P.R., Baron, E.J., Pfaller, M.A., et al., editors. Manual of clinical microbiology, ed 6. American Society for Microbiology, Washington, D.C.

TABLE 52-7   DIFFERENTIATION AMONG COAGULASE-NEGATIVE, PYR-NEGATIVE, NOVOBIOCIN-SUSCEPTIBLE, ALKALINE PHOSPHATASE-NEGATIVE STAPHYLOCOCCI

ORGANISM	UREASE	β-GLUCOSIDASE*	ANAEROBIC GROWTH
S. capitis subsp. capitis	−	−	(+)
S. capitis subsp. urealyticus	+	−	(+)
S. warneri	+	+	+
S. hominis	+	−	−

+, >90% of strains positive; (+), >90% of strains delayed positive; −, >90% of strains negative.

*Performed from disk (Becton Dickinson Microbiology Systems, Cockeysville, Md.) or tablet (KEY Scientific Products, Round Rock, Texas).

Compiled from Kloos, W.E. and Bannerman, T.L. 1995. In Murray, P.R., Baron, E.J., Pfaller, M.A., et al., editors. Manual of clinical microbiology, ed 6. American Society for Microbiology, Washington, D.C.

TABLE 52-8   DIFFERENTIATION OF COAGULASE-NEGATIVE, PYR-POSITIVE, NOVOBIOCIN-SUSCEPTIBLE, ALKALINE PHOSPHATASE-NEGATIVE STAPHYLOCOCCI

ORGANISM	UREASE	β-GLUCURONIDASE*	β-GALACTOSIDASE*	ACID FROM MANNITOL†
S. simulans	+	v	+	+
S. capitis subsp. urealyticus	+	−	−	+
S. haemolyticus‡	−	v	−	v
S. auricularis‡	−	−	(v)	−

+, >90% of strains positive; −, >90% of strains negative; v, variable; (v), variable, positive reactions may be delayed.

*Performed from disk (Becton Dickinson Microbiology Systems, Cockeysville, Md.) or tablet (KEY Scientific Products, Round Rock, Texas.).

†Performed in Andrade's peptone broth with 1% sugar.

‡S. haemolyticus and S. auricularis are very difficult to separate; even fatty acid analysis does not work well.

Compiled from Kloos, W.E. and Bannerman, T.L. 1995. In Murray, P.R., Baron, E.J., Pfaller, M.A., et al., editors. Manual of clinical microbiology, ed 6. American Society for Microbiology, Washington, D.C.

TABLE 52-9   DIFFERENTIATION OF COAGULASE-NEGATIVE, PYR-POSITIVE, NOVOBIOCIN-SUSCEPTIBLE, ALKALINE PHOSPHATASE–POSITIVE STAPHYLOCOCCI

ORGANISM	β-GALACTOSIDASE*	UREASE
S. schleiferi subsp. schleiferi	(+)	−
S. simulans	+	+
S caprae	−	+

+, >90% of strains positive; (+), >90% of strains delayed positive; −, >90% of strains negative.

*Performed from disk (Becton Dickinson Microbiology Systems, Cockeysville, Md.) or tablet (KEY Scientific Products, Round Rock, Texas).

Compiled from Kloos, W.E. and Bannerman, T.L. 1995. In Murray, P.R., Baron, E.J., Pfaller, M.A., et al., editors. Manual of clinical microbiology, ed 6. American Society for Microbiology, Washington, D.C.

TABLE 52-10 ANTIMICROBIAL THERAPY AND SUSCEPTIBILITY TESTING

ORGANISM	THERAPEUTIC OPTIONS	POTENTIAL RESISTANCE TO THERAPEUTIC OPTIONS	VALIDATED TESTING METHODS*	COMMENTS
Staphylococci	Several agents from each major class of antimicrobials, including aminoglycosides, beta-lactams, quinolones, and vancomycin. See Table 19-2 for listing of specific agents that could be selected for testing and use. For many isolates a penicillinase-resistant penicillin (e.g., nafcillin, oxacillin, methicillin) is used; vancomycin is used when isolates resistant to these penicillin derivatives are encountered	Yes; resistance to every therapeutically useful antimicrobial has been described	As documented in Chapter 18: disk diffusion, broth dilution, agar dilution, and commercial systems	In vitro susceptibility testing results are important for guiding therapy. For species other than *S. aureus* clinical significance should be established before testing is done
Micrococcus spp. and *Stomatococcus mucilaginosus*	No specific guidelines, because these species are rarely implicated in infections	Unknown	Not Available	

*Validated testing methods include those standard methods recommended by the National Committee for Clinical Laboratory Standards (NCCLS) and those commercial methods approved by the Food and Drug Administration (FDA).

essentially ineffective. Therefore, vancomycin is the only cell wall–active agent that retains activity and is the alternative drug of choice for resistant strains. High-level resistance to vancomycin has not been described in clinical staphylococcal isolates, but strains with minimal inhibitory concentrations in the intermediate range (i.e., 8 to 16 µg/mL) have been encountered.[6,11] Because of the substantial clinical and public health impact of vancomycin resistance emerging among staphylococci, laboratories should have a heightened awareness of this resistance pattern.

Because *Micrococcus* spp. and *S. mucilaginosus* are rarely encountered in infections in humans, therapeutic guidelines and standardized testing methods do not exist (Table 52-10). However, in vitro results indicate that these organisms generally appear to be susceptible to most beta-lactam antimicrobials.[10]

PREVENTION

There are no approved antistaphylococcal vaccines. Health care workers identified as intranasal carriers of an epidemic strain of *S. aureus* are treated with topical mupirocin and, in some cases, with rifampin. Some physicians advocate the use of antibacterial substances such as gentian violet, acriflavine, chlorhexidine, or bacitracin to the umbical cord stump to prevent staphylococcal disease in hospital nurseries. Further, during epidemics, it is recommended that all full-term infants be bathed with 3% hexachlorophene as soon after birth as possible and daily thereafter until discharge.

References

1. Hébert, G.A., Crowder, C.G., Hancock, G.A., et al. 1988. Characteristics of coagulase-negative staphylococci that help differentiate these species and other members of the family *Micrococcacae*. J. Clin. Microbiol. 26:1939.
2. Isaac, D.W., Pearson, T.A., Hurwitz, C.A., et al. 1993. Clinical and microbiologic aspects of *Staphylococcus haemolyticus* infections. Pediatr. Infect. J. 12:1018.
3. Kloos, W.E. and Wolfshohl, J.F. 1982. Identification of *Staphylococcus* species with API STAPH-IDENT System. J. Clin. Microbiol. 16:509.
4. Kloos, W.E. and Bannerman, T.L. 1994. Update on clinical significance of coagulase-negative staphylococci. Clin. Microbiol. Rev. 7:117.
5. Koch, C., Schumann, P., and Stackerbrandt, E. 1995. Reclassification of *Micrococcus agilis* (Ali-Cohen, 1889) to the genus *Arthrobacter* as *Arthrobacter agilis* comb. nov. and emendation of the genus *Arthrobacter*. Int. J. Syst. Bacteriol. 45:837.
6. Lyytikainen, O., Vaara, M., Jaarviluoma, E., et al. 1996. Increased resistance among *Staphylococcus epidermidis* isolates in a large teaching hospital over a 12 year period. Eur. J. Clin. Microbiol. Infect. Dis. 15:133.
7. Mulligan, M.E., Murray-Leisure, K.A., Ribner, B.S., et al. 1993. Methicillin-resistant *Staphylococcus aureus*: a consensus review of the microbiology, pathogenesis, and epidemiology with implications for prevention and management. Am. J. Med. 94:313.
8. Stackerbrandt, E., Koch, C., Gvozdiak, O., et al. 1995. Taxonomic dissection of the genus *Micrococcus*: *Kocuria* gen. nov. *Nesterenkonia* gen. nov., *Kytococcus* gen. nov., *Dermacoccus* gen. nov., and *Micrococcus* (Cohn, 1872) gen. emend. Int. J. Syst. Bacteriol. 45:682.
9. Vandenesch, F., Etienne, J., Reverdy, M.E., et al. 1993. Endocarditis due to *Staphylococcus lugdunensis*: report of 11 cases and review. Clin. Infect. Dis. 17:871.
10. von Eiff, C., Herrmann, M., and Peters, G. 1995. Antimicrobial susceptibilities of *Stomatococcus mucilaginosus* and of *Micrococcus* spp. Antimicrob. Agents Chemother. 39:268.
11. Woolford, N., Johnson, A.P., Morrison, D., et al. 1995. Current perspectives on glycopeptide resistance. Clin. Microbiol. Rev. 8:585.

Bibliography

Archer, G.L. 1995. *Staphylococcus epidermidis* and other coagulase-negative staphylococci. In Mandell, G.L., Bennett, J.E., and Dolin, R., editors. Principles and practice of infectious diseases. Churchill Livingstone, New York.

Benenson, A.S. 1995. Control of communicable diseases manual, ed 16. American Public Health Association, Washington, D.C.

Kloos, W.E. and Bannerman, T.L. 1995. *Staphylococcus* and *Micrococcus*. In Murray, P.R., Baron, E.J., Pfaller, M.A., et al., editors. Manual of clinical microbiology, ed 6. American Society for Microbiology, Washington, D.C.

Roberson, J.R., Fox, L.K., Hancock, D.D., et al. 1992. Evaluation of methods for differentiation of coagulase-positive staphylococci. J. Clin. Microbiol. 30:3217.

Ruoff, K.L. 1995. *Leuconostoc, Pediococcus, Stomatococcus,* and miscellaneous gram-positive cocci that grow aerobically. In Murray, P.R., Baron, E.J., Pfaller, M.A., et al., editors. Manual of clinical microbiology, ed 6. American Society for Microbiology, Washington, D.C.

Ryan, K.J. 1994. Staphylococci. In Ryan, K.J., editor. Sherris medical microbiology: an introduction to infectious diseases. Appleton & Lange, Norwalk, Conn.

Waldvogel, F.A. 1995. *Staphylococcus aureus* (including toxic shock syndrome). In Mandell, G.L., Bennett, J.E., and Dolin, R., editors. Principles and practice of infectious diseases. Churchill Livingstone, New York.

Section Eight

8

Catalase-Negative, Gram-Positive Cocci

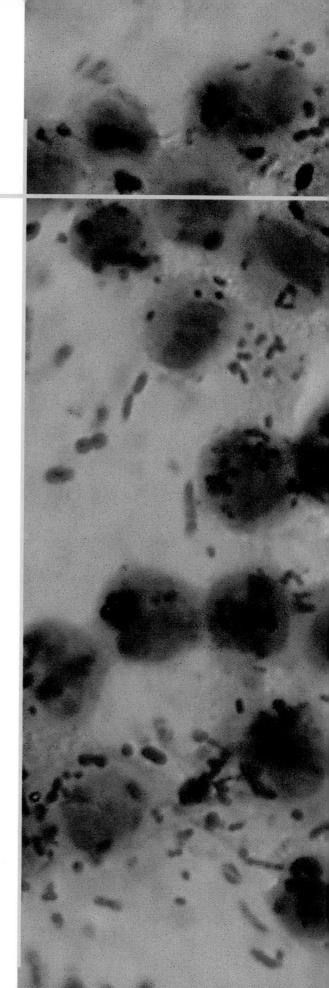

53 STREPTOCOCCUS, ENTEROCOCCUS, AND SIMILAR ORGANISMS

Genera and Species to be Considered

- Beta-hemolytic streptococci
  Streptococcus pyogenes (group A beta-hemolytic streptococci)
  Streptococcus agalactiae (group B beta-hemolytic streptococci)
  Groups C, F, and G beta-hemolytic streptococci
- *Streptococcus pneumoniae*
- Viridans (alpha-hemolytic) streptococci
  Streptococcus mutans group
  Streptococcus sanguis group
  "*Streptococcus milleri*"* group
  Streptococcus salivarius group
  Streptococcus mitis
  Streptococcus bovis
- Nutritionally variant streptococci
  Abiotrophia adjacens
  Abiotrophia defectivus
- Enterococci
  (most commonly isolated)
  Enterococcus faecalis
  Enterococcus faecium

Other *Enterococcus* spp.

E. durans	*E. avium*	*E. malodoratus*
E. mundtii	*E. hirae*	*E. raffinosus*
E. dispar	*E. sulfureus*	*E. pseudoavium*
E. flavescens	*E. gallinarum*	*E. casseliflavus*

- *Leuconostoc* spp.
- *Lactococcus* spp.
- *Globicatella* sp.
- *Pediococcus* spp.
- *Aerococcus* spp.
- *Gemella* spp.
- *Helcococcus* sp.
- *Alloiococcus otitidis*

GENERAL CHARACTERISTICS

The organisms discussed in this chapter are all catalase-negative, gram-positive cocci. *Alloiococcus*, which is catalase-negative only when tested on media devoid of whole blood (e.g., chocolate agar), is included here because it morphologically resembles the viridans streptococci. Of the organisms considered in this chapter, those that are most commonly encountered in infections in humans include *Streptococcus pyogenes, S. agalactiae, S. pneumoniae,* viridans streptococci, and enterococci, usually *E. faecalis* or *E. faecium.* The other species listed in the tables are either rarely found in clinically relevant settings or are usually considered contaminants that can be mistaken for viridans streptococci or enterococci.

EPIDEMIOLOGY

Many of these organisms are commonly found as part of normal human flora and are encountered in clinical specimens as contaminants or as components of mixed cultures with minimal or unknown clinical significance (Table 53-1). However, when these organisms gain access to normally sterile sites they can cause life-threatening infections. Other organisms, most notably *Streptococcus pneumoniae* and *Streptococcus pyogenes,* are notorious pathogens. Although *S. pneumoniae* can be found as part of the normal upper respiratory flora, this organism is also the leading cause of bacterial pneumonia and meningitis. Similarly, although *S. pyogenes* may be carried in the upper respiratory tract of humans, it is rarely considered as normal flora and should be deemed clinically important whenever it is encountered. At the other extreme, organisms such as *Leuconostoc* spp. and *Pediococcus* spp. usually are only capable of causing infections in severely compromised patients.

Many of the organisms listed in Table 53-1 are spread person to person by various means and subse-

---

*Quotation marks indicate an uncertain taxonomic status.

TABLE 53-1 EPIDEMIOLOGY

ORGANISM	HABITAT (RESERVOIR)	MODE OF TRANSMISSION
Streptococcus pyogenes	Inhabits skin and upper respiratory tract of humans. Not considered part of normal flora but may be carried on nasal, pharyngeal, and, sometimes, anal mucosa. Presence in specimens is almost always considered clinically significant	Person to person by direct contact with mucosa or secretions, or by contaminated droplets produced by coughs or sneezes. Once exposed, recipient may become colonized, with subsequent development of infection
Streptococcus agalactiae	Normal flora of female genital tract and lower gastrointestinal tract; may occasionally colonize upper respiratory tract	Infections in fetuses and infants are acquired by person-to-person transmission from mother in utero or during delivery; may also be nosocomially transmitted by unwashed hands of mother or health care personnel. Mode of transmission for infections in adults is uncertain, but probably involves endogenous isolates gaining access to sterile site(s)
Groups C, F, and G beta-hemolytic streptococci	Normal flora of human skin, nasopharynx, gastrointestinal tract, and genital tract	Endogenous isolates gain access to sterile site, or by person-to-person transmission
Streptococcus pneumoniae	Colonizes nasopharynx of humans	Person-to-person spread by contact with contaminated respiratory secretions. Once exposed, recipient may become colonized, with potential for subsequent development of infection. For pneumonia, this occurs by aspiration of the organism into the lungs
Viridans streptococci	Normal flora of human oral cavity, gastrointestinal tract, and female genital tract	Usually by endogenous strains gaining access to normally sterile site; most notably results from dental manipulations with subsequent transient bacteremia

Continued

quently establish a state of colonization or carriage; infections may then develop when colonizing strains gain entrance to normally sterile sites. In some instances this may involve trauma (medically or nonmedically induced) to skin or mucosal surfaces or, as in the case of *S. pneumoniae* pneumonia, may result from aspiration into the lungs of organisms colonizing the upper respiratory tract.

PATHOGENESIS AND SPECTRUM OF DISEASES

The capacity of the organisms listed in Table 53-2 to produce disease and the spectrum of infections they cause vary widely with the different genera and species.

BETA-HEMOLYTIC STREPTOCOCCI

S. pyogenes produces several factors that contribute to its virulence; it is one of the most aggressive pathogens encountered in clinical microbiology laboratories. Among these factors are streptolysin O and S, which not only contribute to virulence but are also responsible for the beta-hemolytic pattern on blood agar plates that is used as a guide to identification of this species. The infections caused by *S. pyogenes* may be localized or systemic; other problems may arise as a result of the host's antibody response to the infections caused by these organisms. Localized infections include acute pharyngitis, for which *S. pyogenes* is the most common bacterial etiology, and skin infections, such as impetigo and erysipelas (see Chapter 28 for more information on skin and soft tissue infections).

—— **TABLE 53-1 EPIDEMIOLOGY—CONT'D** ——

ORGANISM	HABITAT (RESERVOIR)	MODE OF TRANSMISSION
Abiotrophia spp. (Nutritionally variant streptococci)	Normal flora of human oral cavity	Uncertain; probably by endogenous strains gaining access to normally sterile sites
Enterococcus spp.	Found in soil, food, water, and as normal flora of animals, birds, and humans. Species most commonly associated with infections (i.e., *E. faecalis* and *E. faecium*) are normal flora of the human gastrointestinal tract and female genitourinary tract	Frequently by endogenous strains gaining access to sterile sites. Person-to-person transmission, directly or by contaminated medical equipment, allows nosocomial spread and colonization with multidrug-resistant strains. Once colonized, compromised patients are at risk of developing infections with resistant strains
Leuconostoc spp.	Plants, vegetables, dairy products	Mode of transmission for the miscellaneous gram-positive cocci listed is unknown. Most are likely to transiently colonize the gastrointestinal tract after ingestion. From that site they gain access to sterile sites, usually in compromised patients. All are rarely associated with human infections
Lactococcus spp.	Foods and vegetation	
Globicatella sp.	Uncertain	
Pediococcus spp.	Foods and vegetation	
Aerococcus spp.	Environmental; occasionally found on skin	
Gemella spp. and	Normal flora of human oral cavity and upper respiratory tract	
Helcococcus sp.	Uncertain	
Alloiococcus otitidis	Occasionally isolated from human sources, but natural habitat is unknown	Uncertain; rarely implicated in infections

S. pyogenes infections are prone to progression with involvement of deeper tissues and organs, a characteristic that has earned them the designation in lay publications of the "flesh-eating bacteria." Such systemic infections are life-threatening.[13] Additionally, even when infections remain localized, streptococcal pyrogenic exotoxins (SPE) may be released and produce scarlet fever, which occurs in association with streptococcal pharyngitis and is manifested by a rash of the face and upper trunk. Streptococcal toxic shock syndrome, typified by multisystem involvement that includes renal and respiratory failure, rash, and diarrhea, is another disease mediated by production of potent SPE.

Other complications that result from *S. pyogenes* infections are the poststreptococcal diseases rheumatic fever and acute glomerulonephritis. **Rheumatic fever,** which is manifested by fever, carditis (inflammation of heart muscle), subcutaneous nodules, and polyarthritis, usually follows respiratory tract infections and is thought to be mediated by antibodies produced against *S. pyogenes* that cross react with human heart tissue. **Acute glomerulonephritis,** characterized by edema, hypertension, hematuria, and proteinuria, can follow respiratory or cutaneous infections and is mediated by antigen-antibody complexes that deposit in glomeruli, where they initiate damage.

S. agalactiae infections usually are associated with neonates and are acquired before or during the birthing process (Table 53-2). Although the virulence factors associated with the other beta-hemolytic streptococci have not been definitively identified, groups C, G, and F streptococci cause infections similar to those associated with *S. pyogenes* (i.e., skin and soft tissue infections and bacteremia) but are less commonly encountered, often involve compromised patients, and do not produce postinfection sequelae.[14]

TABLE 53-2 PATHOGENESIS AND SPECTRUM OF DISEASES

ORGANISM	VIRULENCE FACTORS	SPECTRUM OF DISEASES AND INFECTION
Streptococcus pyogenes	Protein F mediates epithelial cell attachment, and M protein is antiphagocytic; produces several enzymes and hemolysins that contribute to tissue invasion and destruction, including streptolysin O, streptolysin S, streptokinase, DNase, and hyaluronidase. Streptococcal pyrogenic exotoxins mediate production of rash (i.e., scarlet fever) or multisystem effects that may result in death	Acute pharyngitis, impetigo, erysipelas, necrotizing fasciitis and myositis, bacteremia with potential for infection in any of several organs, pneumonia, scarlet fever, streptococcal toxic shock syndrome
	Cross-reactions of antibodies produced against streptococcal antigens and human heart tissue	Rheumatic fever
	Deposition of antibody-streptococcal antigen complexes in kidney results in damage to glomeruli	Acute, poststreptococcal glomerulonephritis
Streptococcus agalactiae	Uncertain. Capsular material interferes with phagocytic activity and complement cascade activation	Infections most commonly involve neonates and infants, often preceded by premature rupture of mother's membranes. Infections often present as multisystem problems, including sepsis, fever, meningitis, respiratory distress, lethargy, and hypotension. Infections may be classified as early onset (occur within first 5 days of life) or late onset (occur 7 days to 3 months after birth). Infections in adults usually involve postpartum infections, such as endometritis, which can lead to pelvic abscesses and septic shock. Infections in other adults usually reflect compromised state of the patient and include bacteremia, pneumonia, endocarditis, arthritis, osteomyelitis, and skin and soft tissue infections
Groups C, F, and G beta-hemolytic streptococci	None have been definitively identified, but likely include factors similar to those produced by *S. pyogenes* and *S. agalactiae*	Cause similar types of acute infections in adults as described for *S. pyogenes* and *S. agalactiae*, but usually involve compromised patients. A notable proportion of infections caused by group G streptococci occur in patients with underlying malignancies. Group C organisms occasionally have been associated with acute pharyngitis
Streptococcus pneumoniae	Polysaccharide capsule that inhibits phagocytosis is primary virulence factor. Pneumolysin has various effects on host cells, and several other factors likely are involved in eliciting a strong cellular response by the host	A leading cause of meningitis and pneumonia with or without bacteremia; also causes sinusitis, and otitis media

Continued

TABLE 53-2 PATHOGENESIS AND SPECTRUM OF DISEASES—CONT'D

ORGANISM	VIRULENCE FACTORS	SPECTRUM OF DISEASES AND INFECTION
Viridans strepto-cocci	Generally considered to be of low virulence. Production of extracellular complex polysaccharides (e.g., glucans and dextrans) enhance attachment to host cell surfaces, such as cardiac endothelial cells or tooth surfaces in the case of dental caries	Slowly evolving (subacute) endocarditis, particularly in patients with previously damaged heart valves. Bacteremia and infections of other sterile sites do occur in immunocompromised patients. Meningitis can develop in patients suffering trauma or defects that allow upper respiratory flora to gain access to the central nervous system. *S. mutans* plays a key role in the development of dental caries
Abiotrophia spp. (Nutritionally variant strepto-cocci)	Unknown	Endocarditis; rarely encountered in infections of other sterile sites
Enterococcus spp.	Little is know about virulence. Adhesins, cytolysins, and other metabolic capabilities may allow these organisms to proliferate as nosocomial pathogens. Multidrug resistance also contributes to proliferation	Most infections are nosocomial and include urinary tract infections, bacteremia, endocarditis, mixed infections of abdomen and pelvis, wounds, and, occasionally, ocular infections. CNS and respiratory infections are rare
Leuconostoc spp., *Lactococcus* spp., *Globicatella* sp., *Pediococcus* spp., *Aerococcus* spp., *Gemella* spp., and *Helcococcus* sp.	Unknown; probably of low virulence. Opportunistic organisms that require impaired host defenses to establish infection. Intrinsic resistance to certain antimicrobial agents (e.g., *Leuconostoc* spp. and *Pediococcus* spp. resistant to vancomycin) may enhance survival of some species in the hospital setting	Whenever encountered in clinical specimens these organisms should first be considered as probable contaminants. *Aerococcus urinae* is notably associated with urinary tract infections
Alloiococcus sp.	Unknown	Chronic otitis media in children

STREPTOCOCCUS PNEUMONIAE AND VIRIDANS STREPTOCOCCI

S. pneumoniae is a primary cause of bacterial pneumonia, meningitis, and otitis media, and the antiphagocytic properties of the polysaccharide capsule is the key to the organism's virulence. The organism may harmlessly inhabit the upper respiratory tract but may also gain access to the lungs by aspiration, where it may establish an acute suppurative pneumonia. In addition, this organism also accesses the bloodstream and the meninges to cause acute, purulent, and often life-threatening infections.

The viridans (greening) streptococci and *Abiotrophia* spp. (formally known as nutritionally variant streptococci) are generally considered to be opportunistic pathogens of low virulence. These organisms are not known to produce any factors that facilitate invasion of the host. However, when access is gained, a transient bacteremia occurs and endocarditis and infections at other sites in compromised patients may result.[1,3,4]

ENTEROCOCCI

Although virulence factors associated with enterococci are a topic of increasing research interest, very little is known about the characteristics that have allowed these organisms to become a prominent cause of nosocomial infections.[8,9,12] Compared with other clinically important gram-positive cocci, this genus is intrinsically more resistant to the antimicrobial agents commonly used in hospitals and is especially resistant to all currently available cephalosporins and aminoglycosides.[10] In addition, these organisms are capable of acquiring and exchanging genes that encode resistance to antimicrobial agents. This genus is the first clinically relevant group of gram-positive cocci to acquire and disseminate resistance to vancomycin, the single cell–wall active agent available for use against gram-positive organisms resistant to beta-lactams (e.g., methicillin-resistant staphylococci). Spread of this troublesome resistance marker from enterococci to other clinically relevant organisms is a serious public health concern.

A wide variety of enterococcal species have been isolated from human infections, but *E. faecalis* and *E. faecium* still clearly predominate as the species that are most commonly encountered. Between these two species, *E. faecalis* is the most commonly encountered, but the incidence *of E. faecium* infections is on the rise in many hospitals, which is probably related in some way to the acquisition of vancomycin resistance.

MISCELLANEOUS OTHER GRAM-POSITIVE COCCI

The other genera listed in Table 53-2 are of low virulence and are almost exclusively associated with infections involving compromised hosts.[6] A possible exception is the association of *Alloiococcus otitidis* with chronic otitis media in children.[2] Certain intrinsic features, such as resistance to vancomycin among *Leuconostoc* spp. and pediococci, may contribute to the ability of these organisms to survive in the hospital environment. However, whenever they are encountered, strong consideration must be given to their clinical relevance and potential as contaminants. These organisms can also challenge many identification schemes used for gram-positive cocci, and they may be readily misidentified as viridans streptococci.

LABORATORY DIAGNOSIS

SPECIMEN COLLECTION AND TRANSPORT

No special considerations are required for specimen collection and transport of the organisms discussed in this chapter. Refer to Table 1-1 for general information on specimen collection and transport.

SPECIMEN PROCESSING

No special considerations are required for processing of the organisms discussed in this chapter. Refer to Table 1-1 for general information on specimen processing.

DIRECT DETECTION METHODS

Antigen detection

Antigen detection screening methods are available for several streptococcal species. Detection of *S. pyogenes* antigen in throat specimens is possible using latex agglutination, coagglutination, or enzyme-linked immunosorbent assay (ELISA) technologies. These commercial kits have been reported to be very specific, but false-negative results may occur if specimens

contain low numbers of *S. pyogenes.* Sensitivity has ranged from approximately 60% to greater than 95% depending on the methodology and several other variables. Therefore, many microbiologists recommend collecting two throat swabs from each patient. If the first swab yields a positive result by a direct antigen method, the second swab can be discarded. However, for those specimens in which the rapid antigen test yielded a negative result, a blood agar plate or selective streptococcal blood agar plate should be inoculated with the second swab.

Several commercial antigen detection kits are available for diagnosis of neonatal sepsis and meningitis caused by group B streptococci. Developed for use with serum, urine, or cerebrospinal fluid (CSF), the best results have been achieved with CSF; false-positive results have been a problem using urine. Latex agglutination procedures appear to be the most sensitive and specific. Because neonates acquire *S. agalactiae* infection during passage through the colonized birth canal, direct detection of group B streptococcal antigen from vaginal swabs has also been attempted. Direct extraction and latex particle agglutination have not been sensitive enough for use alone as a screening test. However, solid-phase immunoassays (ICON Strep B, SmithKline Diagnostics, Inc., Palo Alto, Calif.; QUIDEL Group B Strep test, Quidel Corp., San Diego, Calif.) for direct detection from vaginal swab specimens have shown more promise.

Latex agglutination kits to detect the capsular polysaccharide antigen of the pneumococcus have also been developed for use with urine, serum, and CSF, although they are no longer commonly used in clinical microbiology laboratories.

Gram stain

All the genera described in this chapter are gram-positive cocci. Microscopically, streptococci are round or oval-shaped, occasionally forming elongated cells that resemble pleomorphic corynebacteria or lactobacilli. They may appear gram-negative if cultures are old or if the patient has been treated with antibiotics. *S. pneumoniae* is typically lancet-shaped and occurs singly, in pairs, or in short chains (see Figure 11-6, *B*).

Growth in broth should be used for determination of cellular morphology if there is a question regarding staining characteristics from solid media. *Streptococcus* and *Abiotrophia* growing in broth form long chains of cocci (Figure 53-1), whereas *Aerococcus, Gemella,* and *Pediococcus* grow as large, spherical cocci arranged in tetrads. or pairs or as individual cells. *Leuconostoc* may elongate to form coccobacilli, although cocci are the primary morphology. The cellular arrangements of *Lactococcus, Helcococcus, Globicatella,* and *Alloiococcus* are noted in Table 53-3.

TABLE 53-3 DIFFERENTIATION AMONG CATALASE-NEGATIVE, GRAM-POSITIVE COCCOID ORGANISMS

ORGANISMS	CELLULAR ARRANGEMENT	Hemolysis α, β, or non[a]	CYTOCHROME[b]/ CATALASE	VAN S
Leuconostoc	cb, pr, ch	α, non	−/−	−
Pediococcus[c]	c, pr, tet, d	α, non	−/−	−
Enterococcus Vancomycin S	c, ch	α, β, or non	−/+[w]	+
Enterococcus Vancomycin R	c, ch	α, β, or non	−/+[w]	−
Streptococcus (all)	c, ch	α, β, or non	−/−	+
S. agalactiae	c, ch	β, non	−/−	+
S. bovis	c, ch	α, non	−/−	+
Viridans streptococci	c, ch	α, non	−/−	+
Abiotrophia	c, ch	α, non	−/−	+
Lactococcus	cb, ch	α, non	−/−	+
Gemella	c, pr, ch, cl, tet[j]	α, non	−/−	+
Stomatococcus	c, pr, cl	non	+/− or +[w]	+
Aerococcus urinae	c, pr, tet, cl	α	−/−	+
A. viridans	c, pr, tet, cl	α	−/+[w]	+
Helcococcus kunzii	c, pr, ch, cl	non	−/−	+
Globicatella sanguis	c, ch, pr	α	−/−	+
Alloiococcus[o]	c, pr, tet	non	−/+[w]	+
Lactobacillus[q]	cb, ch	α, non	−/−	v

+, ≥90% of species or strains positive; +[w], strains or species may be weakly positive; −, ≥90% of species or strains negative; *v*, variable reactions; *Van S*, vancomycin (30 μg)-susceptible; *LAP*, leucine aminopeptidase; *PYR*, pyrrolidonyl arylamidase; *BE*, bile-esculin; *c*, cocci; *cb*, coccobacilli; *tet*, tetrads; *pr*, pairs; *ch*, chains; *cl*, clusters; *α*, alpha-hemolytic; *β*, beta-hemolytic; *non*, nonhemolytic; *S*, susceptible; *R*, resistant; *NT*, not tested.

[a]Hemolysis tested on 5% sheep blood agar.

[b]Possesses cytochrome enzymes.

[c]The most commonly isolated pediococci are arginine deamininase–positive.

[d]S. pyogenes and some isolates of *Abiotrophia* and S. pneumoniae are PYR-positive.

[e]5% to 10% of viridans streptococci and S. bovis are bile-esculin–positive.

[f]Some beta-streptococci grow in 6.5% salt broth.

[g]Occasional isolates are positive or give weakly positive reactions that are difficult to interpret.

Compiled from Christensen, J.J., Vibits, H., Ursing, J., et al. 1991. J. Clin. Microbiol. 29:1049; Facklam, R. and Elliott, J.A. M.A., et al., editors. Manual of clinical microbiology, ed 6. American Society for Microbiology, Washington, D.C.; Ruoff, Murray, P.R., Baron, E.J., Pfaller, M.A., et al., editors. Manual of clinical microbiology, ed 6. American Society for Microbiol-ual of clinical microbiology, ed 6. American Society for Microbiology, Washington, D.C.

LAP	PYR	GAS IN MRS BROTH	GROWTH			
			ON BE	IN 6.5% NaCl BROTH	AT 10° C	AT 45° C
−	−	+	v	v	v	v
+	−	−	+	v	−	v
+	+	−	+	+	+	+
+	+	−	+	+	+	+
+	v[d]	−	−[e]	−[f]	−	v
NT	−	NT	−	v	NT	NT
+	−	−	+	−	−	+
+	−	−	v[g]	−	−	v
v[h]	v	−	−	−	−	−
+	v	−	+	v	+	v[i]
+[k]	+[l]	−	−	−	−	−
+	+[m]	−	NT	−	−	−
+	−	−	−	+	−	−[n]
−	+	−	v	+	−	−
−	+	−	+	+	−	−
−	+	−	v	v	−	v
+	+	−	−	+[p]	−	−
v	−	v	v	v	+	v

[h]Some *Abiotrophia* are LAP-negative.

[i]Majority of strains will not grow at 45° C in ≤ 48 hours.

[j]*G. haemolysans* easily decolorizes when gram-stained. They resemble *Neisseria* with adjacent flattened sides of pairs of cells.

[k]*G. haemolysans* is LAP-negative.

[l]Weakly positive. Use a large innoculum.

[m]Most are positive.

[n]Conflicting results in the literature. Some isolates may grow at 45° C.

[o]No growth anaerobically.

[p]May take 2 to 7 days.

[q]May resemble catalase-negative, gram-positive coccoid organisms morphologically but is discussed in Chapter 56 with non–spore-forming catalase-negative gram-positive rods.

1995. Clin. Microbiol. Rev. 8:479; Facklam, R.R. and Sahm, D.F. 1995. *Enterococcus.* In Murray, P.R., Baron, E.J., Pfaller, K.L. 1995. *Leuconostoc, Pediococcus, Stomatococcus,* and miscellaneous gram-positive cocci that grow aerobically. In ogy, Washington, D.C.; and Ruoff, K.L. 1995. *Streptococcus.* In Murray, P.R., Baron, E.J., Pfaller, M.A., et al., editors. Man-

CULTIVATION

Media of choice

Except for *Abiotrophia*, the organisms discussed in this chapter will grow on standard laboratory media such as 5% sheep blood and chocolate agars. They will not grow on MacConkey agar but will grow on gram-positive selective media such as CNA (Columbia agar with colistin and nalidixic acid) and PEA (phenylethyl alcohol agar).

Abiotrophia will not grow on blood or chocolate agars unless pyridoxal (vitamin B$_6$) is supplied either by placement of a pyridoxal disk, by cross-streaking with *Staphylococcus*, or by inoculation of vitamin B$_6$-supplemented culture media.

Blood culture media support the growth of all of these organisms, as do common nutrient broths, such as thioglycollate or brain-heart infusion. Blood cultures that appear positive and show chaining gram-positive cocci on Gram stain but do not grow on subculture should be resubcultured with a pyridoxal disk to cover for the possibility of *Abiotrophia* spp. bacteremia.

Other selective media are available for isolating certain species from clinical specimens. For isolating group A streptococci from throat swabs the most common medium is 5% sheep blood agar supplemented with trimethoprim/sulfamethoxazole to suppress the growth of normal flora. However, this medium also inhibits growth of groups C, F, and G beta-hemolytic streptococci.

To detect genital carriage of group B streptococci during pregnancy, Todd-Hewitt broth with antimicrobials (gentamicin, nalidixic acid, or colistin and nalidixic acid) is used to suppress the growth of vaginal flora and allow growth of *S. agalactiae* following subculture to blood agar. LIM broth is one media formulation used for this purpose (see Chapter 12).

Incubation conditions and duration

Most laboratories incubate blood or chocolate agar plates in 5% to 10% carbon dioxide. This is the preferred atmosphere for *S. pneumoniae* and acceptable for all other genera discussed in this chapter. However, visualization of beta hemolysis is enhanced by anaerobic conditions. Therefore, the blood agar plates should be inoculated by stabbing the inoculating loop into the agar several times (Figure 53-2, *A*). Colonies can then grow throughout the depth of the agar, producing subsurface oxygen-sensitive hemolysins (i.e., streptolysin O) (Figure 53-2, *B*). Most organisms will

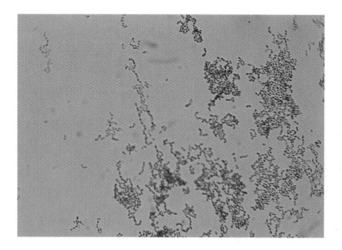

FIGURE 53-1 Chains of streptococci seen in Gram stain prepared from broth culture.

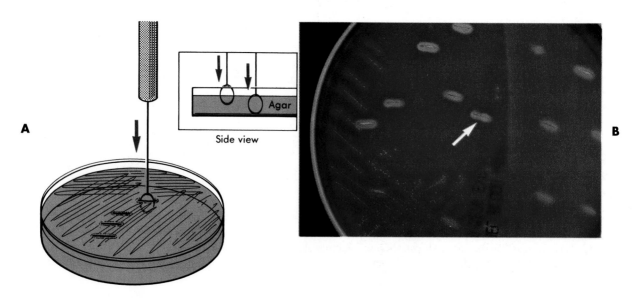

FIGURE 53-2 Stabbing the inoculating loop vertically into the agar after streaking the blood agar plate (**A**) allows subsurface colonies to display hemolysis caused by streptolysin O (**B**).

grow on the agar media within 48 hours of inoculation.

Colonial appearance

Table 53-4 describes the colonial appearance and other distinguishing characteristics (e.g., hemolysis) of each genus on 5% sheep blood agar. The beta-hemolytic streptococci may have a distinctive buttery odor.

APPROACH TO IDENTIFICATION

The commercially available biochemical test systems for streptococci and enterococci are discussed in Chapter 30. None of them has been found to accurately identify all species of viridans streptococci or enterococci.

Comments regarding specific organisms

Useful characteristics for differentiation among catalase-negative, gram-positive cocci are shown in Table 53-3. Organisms that may be weakly catalase-positive, such as *Stomatococcus,* or coccobacillary, such as *Lactobacillus,* are included in Table 53-3.

The cellular arrangement on Gram stain and the type of hemolysis are important considerations. If the presence of hemolysis is uncertain, the colony should be moved aside with a loop and the medium directly beneath the original colony should be examined by holding the plate in front of a light source.

A screening test for vancomycin susceptibility is often useful for differentiating among many alpha-hemolytic cocci. All streptococci, aerococci, gemellas, and lactococci and most enterococci are susceptible to

TABLE 53-4 COLONIAL APPEARANCE AND CHARACTERISTICS ON 5% SHEEP BLOOD AGAR

ORGANISM	APPEARANCE
Group A beta-hemolytic streptococci[a]	Grayish white, transparent to translucent, matte or glossy; large zone of beta hemolysis
Group B beta-hemolytic streptococci[b]	Larger than group A streptococci; translucent to opaque; flat, glossy; narrow zone of beta hemolysis; some strains nonhemolytic
Group C beta-hemolytic streptococci[c]	Grayish white, glistening; wide zone of beta hemolysis
Group F beta-hemolytic streptococci[d]	Grayish white, small, matte; narrow to wide zone of beta hemolysis
Group G beta-hemolytic streptococci[e]	Grayish white, matte; wide zone of beta hemolysis
S. pneumoniae	Small, gray, glistening; colonies tend to dip down in the center and resemble a doughnut (umbilicated) as they age; if organism has a polysaccharide capsule, colony may be mucoid; alpha-hemolytic
Viridans streptococci[f]	Minute to small, gray, domed, smooth or matte; alpha-hemolytic or nonhemolytic
Abiotrophia spp.[g]	Resemble viridans streptococci
Enterococcus spp.	Small, cream or white, smooth, entire; alpha-, beta-, or non-hemolytic
Leuconostoc, Aerococcus, Pediococcus, Gemella, Lactococcus, Globicatella, Helcococcus, and *Alloiococcus*	Resemble viridans streptococci; see Table 53-3 for hemolytic reactions

[a]Two colony sizes, that is, small (called *large-colony* and named *S. pyogenes*) and minute (called *small-colony* and named "*S. milleri*" group).

[b]*S. agalactiae.*

[c]Two colony sizes, that is, small (called *large-colony* and named either *S. equi, S. equisimilis,* or *S. zooepidemicus*) and minute (called *small-colony* and named "*S. milleri*" group).

[d]"*S. milleri*" group.

[e]Two colony sizes, that is, small (called *large-colony*) and minute (called *small-colony*); both named "*S. milleri*" group.

[f]Includes *S. mutans* group, *S. salivarius* group, *S. bovis,* "*S. milleri*" group, *S. sanguis* group, and *S. mitis.*

[g]May satellite around staphylococcal colonies on 5% sheep blood or chocolate agars.

vancomycin (any zone of inhibition), whereas pedio-cocci, leuconostocs, and many lactobacilli are typically resistant (growth up to the disk). Other useful tests listed in Table 53-3 include LAP (leucine aminopeptidase) and PYR (pyrrolidonyl arylamidase), which are commercially available as disks (see Chapter 13).

Leuconostoc produces gas from glucose in MRS broth; this distinguishes it from all other genera, except the lactobacilli. However, unlike *Leuconostoc* spp., lactobacilli appear as elongated bacilli when Gram stained from thioglycollate broth. Several organisms (e.g., *Leuconostoc, Pediococcus, Lactococcus, Helcococcus, Globicatella,* and *Aerococcus viridans*) will show growth on bile-esculin agar and in 6.5% salt broth; this is the reason these two tests no longer solely can be used to identify enterococci.

Serologic grouping of cell wall carbohydrates has classically been used to identify species of beta-hemolytic streptococci. The original Lancefield precipitin test is now rarely performed in clinical laboratories. It has been replaced by either latex agglutination or coagglutination procedures that are available as commercial kits. Serologic tests have the advantage of being rapid, confirmatory, and easily performed on one or two colonies. However, they are more expensive than biochemical screening tests.

The PYR, hippurate, and CAMP tests can be used to presumptively identify group A and B streptococci, respectively. However, use of the 0.04-U bacitracin disk is no longer recommended for *S. pyogenes,* because groups C and G streptococci are also susceptible to this agent. *S. pyogenes* is the only species of beta-hemolytic streptococci that will give a positive PYR reaction.

S. agalactiae is able to hydrolyze hippurate and is positive in the CAMP test. The CAMP test detects production of a diffusible, extracellular protein that enhances the hemolysis of sheep erythrocytes by *Staphylococcus aureus.* A positive test is recognized by the appearance of an arrowhead shape at the juncture of the *S. agalactiae* and *S. aureus* streaks. Occasionally, non–beta-hemolytic strains of *Streptococcus agalactiae* may be encountered, but identification of such isolates can be accomplished using the serologic agglutination approach.

Table 53-5 shows the differentiation of the clinically relevant beta-hemolytic streptococci. Minute beta-hemolytic streptococci are all likely to be "*S. milleri*" group; a positive Voges-Proskauer test and negative PYR test identify a beta-hemolytic streptococcal isolate as such.

Suspicious colonies (see Table 53-3) thought to be *S. pneumoniae* must be tested for either bile solubility or susceptibility to optochin (ethylhydrocupreine hydrochloride). The bile solubility test is based on the ability of bile salts to lyse *S. pneumoniae.* In the optochin test, a filter paper disk ("P" disk) impregnated with optochin is placed on a blood agar plate previously streaked with a suspect organism.

TABLE 53-5 DIFFERENTIATION AMONG CLINICALLY RELEVANT BETA-HEMOLYTIC STREPTOCOCCI

SPECIES	COLONY SIZE	LANCEFIELD GROUP	PYR	VP	HIPP	CAMP TEST
S. pyogenes	Large	A	+	–	–	–
"*S. milleri*" group	Small	A	–	+	–	–
S. agalactiae		B	–	–	+	+
S. dysgalactiae subsp. *equisimilis*	Large	C and G	–	–	–	–
"*S. milleri*" group	Small	C and G	–	+	–	–
"*S. milleri*" group	Small	F	–	+	–	–
"*S. milleri*" group	Small	Non-groupable	–	+	–	–

PYR, pyrrolidonyl arylamidase; *VP,* Voges-Proskauer test; *Hipp,* hydrolysis of hippurate; +, >90% of strains positive; –, >90% of strains negative.

Compiled from Ruoff, K.L. 1995. In Murray, P.R., Baron, E.J., Pfaller, M.A., et al., editors. Manual of clinical microbiology, ed 6. American Society for Microbiology, Washington D.C.; Vandamme, P., Pot, B., Falsen, E., et al. 1996. Int. J. Syst. Bacteriol. 46:774.

The plate is incubated at 35° C for 18 to 24 hours and read for inhibition. *S. pneumoniae* produce a zone of inhibition, whereas viridans streptococci grow up to the disk. Serologic identification of *S. pneumoniae* is also possible using coagglutination or latex agglutination test kits.

Once *S. pneumoniae* has been ruled out as a possibility for an alpha-hemolytic isolate, viridans streptococci and enterococci must be considered. Figure 53-3 outlines the key tests for differentiating among the viridans streptococci. Carbohydrate fermentation tests are performed in heart infusion broth with bromcresol purple indicator. Although alpha-hemolytic streptococci are not often identified to species, there are cases (i.e., endocarditis, isolation from multiple blood cultures) in which full identification is indicated. This is particularly true for blood culture isolates of *S. bovis* that have been associated with gastrointestinal malignancy and may be an early indicator of gastrointestinal cancer. *S. bovis* possesses group D antigen that may be detected using commercially available typing sera. However, this is not a definitive test, because other organisms (e.g., *Leuconostoc*) may also produce a positive result.

Except for three species not usually isolated from humans (*E. saccharolyticus, E. cecorum,* and *E. columbae*), all enterococci hydrolyze PYR and possess group D antigen. A flowchart that may be used to identify enterococcal species is shown in Figure 53-4. Identifying the species of enterococcal isolates is important for understanding the epidemiology of antimicrobial resistance among isolates of this genus and for management of patients with enterococcal infections.

SERODIAGNOSIS

Individuals with disease caused by *S. pyogenes* produce antibodies against various antigens. The most common are antistreptolysin O (ASO), Anti-DNase B, antistreptokinase, and antihyaluronidase. Pharyngitis seems to be followed by rises in antibody titers against all antigens, whereas patients with pyoderma, an infection of the skin, only show a significant response to anti-DNase B. Use of serodiagnostic tests is most useful to demonstrate prior streptococcal infection in patients from whom group A *Streptococcus* has not been cultured but who present with sequelae suggestive of rheumatic fever or acute glomerulonephritis. Serum obtained as long as 2 months after infection usually demonstrates increased antibodies. As with other serologic tests, an increasing titer over time is most useful for diagnosing previous streptococcal infection.

Commercial products are available for detection of antistreptococcal antibodies. Streptozyme (Carter

Wallace Inc., Wampole Laboratories, Cranbury, N.J.), which detects a mixture of antibodies, is a commonly used test. Unfortunately, no commercial system has been shown to accurately detect all streptococcal antibodies.

ANTIMICROBIAL SUSCEPTIBILITY TESTING AND THERAPY

For *S. pyogenes* and the other beta-hemolytic streptococci, penicillin is the drug of choice (Table 53-6). Because penicillin resistance has not been encountered among these organisms, susceptibility testing of clinical isolates for reasons other than resistance surveillance is not necessary.[7] However, if a macrolide such as erythromycin, is being considered for use, as is the case with patients who are allergic to penicillin, testing is needed to detect resistance that has emerged among these organisms. For serious infections caused by *S. agalactiae,* an aminoglycoside may be added to supplement beta-lactam therapy and enhance bacterial killing.

In contrast to beta-hemolytic streptococci, the emergence of beta-lactam resistance in *S. pneumoniae* and viridans streptococci dictates that clinically relevant isolates be subjected to in vitro susceptibility testing.[11] When testing is performed, methods that

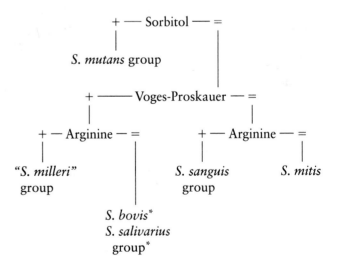

=, Signifies a negative result.

*Mannitol–positive isolates are *S. bovis.* Mannitol–negative isolates of *S. bovis* are hard to distinguish from *S. salivarius.* Of *S. salivarius,* 70% are urea–positive. Most *S. salivarius* will be negative with the bile esculin test.

FIGURE 53-3 Differentiation of viridans streptococcal groups. (Compiled from Ruoff, K.L. 1995. *Streptococcus.* In Murray, P.R., Baron, E.J., Pfaller, M.A., et al., editors. Manual of clinical microbiology, ed 6. American Society for Microbiology, Washington, D.C.)

TABLE 53-6 ANTIMICROBIAL THERAPY AND SUSCEPTIBILITY TESTING

ORGANISM	THERAPEUTIC OPTIONS
Streptococcus pyogenes	Penicillin is drug of choice; alternatives may include macrolides (e.g., erythromycin) and certain cephalosporins; vancomycin for penicillin-allergic patients with serious infections
Streptococcus agalactiae	Penicillin, with or without an aminoglycoside; ceftriaxone or cefotaxime may be used instead of penicillin; vancomycin is used for penicillin-allergic patients
Group C, F, and G beta-hemolytic streptococci	Penicillin; vancomycin for penicillin-allergic patients
Streptococcus pneumoniae	Penicillin, ceftriaxone, or cefotaxime; vancomycin for resistant strains; macrolides, trimethoprim/sulfamethoxazole, and certain quinolones may also be used for some infections
Viridans streptococci	Penicillin or ceftriaxone, with or without an aminoglycoside; vancomycin is used in cases of penicillin allergies and beta-lactam resistance
Abiotrophia spp. (nutritionally variant streptococci)	Penicillin, or vancomycin, plus an aminoglycoside
Enterococcus spp.	For systemic, life-threatening infections a cell wall–active agent (i.e., penicillin, ampicillin, or vancomycin) plus an aminoglycoside (gentamicin or streptomycin); occasionally, other agents such as chloramphenicol may be used when multidrug-resistant strains are encountered. For urinary tract isolates, ampicillin, nitrofurantoin, tetracycline, or quinolones may be effective
Leuconostoc spp., *Lactococcus* spp., *Globicatella* sp., *Pediococcus* spp., *Aerococcus* spp., *Gemella* spp., *Helcococcus* sp., and *Alloiococcus otitidis*	No definitive guidelines. Frequently susceptible to penicillins; effectiveness of cephalosporins is uncertain

*Validated testing methods include those standard methods recommended by the National Committee for Clinical Laboratory Standards (NCCLS) and those

RESISTANCE TO THERAPEUTIC OPTIONS	VALIDATED TESTING METHODS*	COMMENTS
No resistance to penicillin, cephalosporins, vancomycin known; resistance to macrolides does occur	As documented in Chapter 18: disk diffusion, broth dilution, and agar dilution	Testing to guide therapy is not routinely needed, unless a macrolide is being considered
No resistance to penicillins, cephalosporins, or vancomycin known	As documented in Chapter 18: disk diffusion, broth dilution, agar dilution, and some commercial methods	Testing to guide therapy is not routinely needed
No resistance known to penicillin or vancomycin	Same as used for *S. pyogenes* and *S. agalactiae*	Testing to guide therapy is not routinely needed
Yes. Resistance to penicillin, cephalosporins, and macrolides is frequently encountered; vancomycin resistance has not been encountered	As documented in Chapter 18: disk diffusion, broth dilution, and certain commercial methods	In vitro susceptibility testing results are important for guiding therapy
Resistance to penicillin and cephalosporins is frequently encountered; vancomycin resistance has not been encountered	As documented in Chapter 18: disk diffusion, broth dilution, agar dilution, and some commercial methods	In vitro susceptibility testing results are important for guiding therapy
Resistance to penicillin is known, but impact on efficacy of combined penicillin and aminoglycoside therapy is not known	Not available	Testing to guide therapy is not necessary
Resistance to every therapeutically useful antimicrobial agent, including vancomycin, has been described	As documented in Chapter 18: disk diffusion, broth dilution, various screens, agar dilution, and commercial systems	In vitro susceptibility testing results are important for guiding therapy
Unknown. *Leuconostoc* and pediococci are intrinsically resistant to vancomycin	Not available	Whenever isolated from clinical specimens, the potential of the isolate being a contaminant should be strongly considered

commercial methods approved by the Food and Drug Administration (FDA).

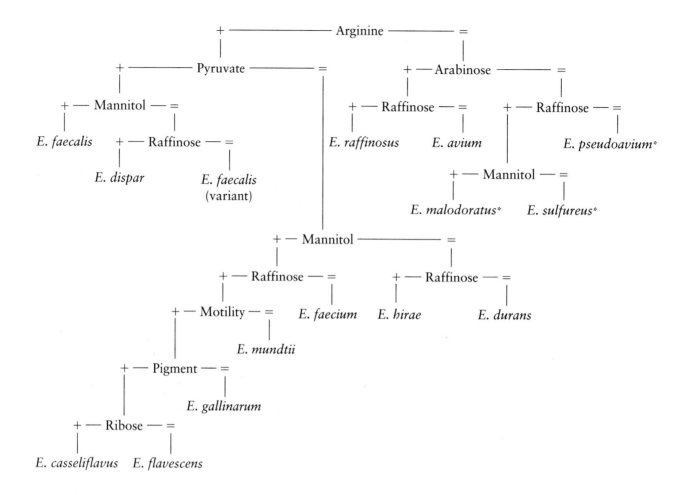

=, Signifies a negative result.

*Rarely, if ever, isolated from human sources.

FIGURE 53-4 Species identification of enterococcal isolates. (Compiled from Facklam, R.R. and Sahm, D.F. 1995. *Enterococcus.* In Murray, P.R., Baron, E.J., Pfaller, M.A., et al., editors. Manual of clinical microbiology, ed 6. American Society for Microbiology, Washington, D.C.)

produce minimal inhibitory concentration data for beta-lactams are preferred. The level of resistance (i.e., MIC in μg/mL) can provide important information regarding therapeutic management of the patient, particularly in cases of pneumococcal meningitis in which relatively slight increases in MIC can have substantial impact on clinical efficacy of penicillins and cephalosporins. Vancomycin resistance has not been described in either *S. pneumoniae* or in viridans streptococci.

Enterococci are intrinsically resistant to a wide array of antimicrobial agents, and they generally are resistant to killing by any of the single agents (e.g., ampicillin or vancomycin) that are bactericidal for most other gram-positive cocci. Therefore effective bactericidal activity can only be achieved with the combination of a cell wall–active agent, such as ampicillin or vancomycin, and an aminoglycoside, such as gentamicin or streptomycin.

Unfortunately, many *E. faecalis* and *E. faecium* isolates have acquired resistance to one or more of these components of combination therapy. This resistance generally eliminates any contribution that the target antimicrobial agent could make to the synergistic killing of the organism. Therefore performance of in vitro susceptibility testing with clinical isolates from systemic infections is critical for determining which combination of agents may still be effective therapeutic choices.

For uncomplicated urinary tract infections, bactericidal activity is usually not required for clinical efficacy, so that single agents such as ampicillin, nitrofurantoin, or a quinolone are often sufficient.

PREVENTION

A single-dose, 23-valent vaccine to prevent infection by the most common serotypes of *S. pneumoniae* is available in the United States. Vaccination is recommended for individuals over age 65 and for patients with (1) chronic pulmonary, cardiac, liver, or renal disease; (2) no spleens (asplenic); (3) sickle cell disease; (4) diabetes; (5) HIV infection; or (6) any other diseases that compromise the immune system. The vaccine is not effective in children younger than age 2.

Lifetime chemoprophylaxis with penicillin, given either monthly (intramuscular administration), or daily (oral administration), is recommended for patients with rheumatic heart disease to prevent development of bacterial endocarditis on a damaged heart valve. Likewise, penicillin may be indicated to control outbreaks of *S. pyogenes* in individuals in close physical contact, such as in households, military populations, or newborn nurseries.

References

1. Bochud, P.Y., Calandra, T., Francioli, P., et al. 1994. Bacteremia due to viridans streptococci in neutropenic patients: a review. Am. J. Med. 97:256.
2. Bosley, G.S., Whitney, A.M., Prucker, J.M., et al. 1995. Characterization of ear fluid isolates of *Alloiococccus otitidis* from patients with recurrent otitis media. J. Clin. Microbiol. 33:2876.
3. Bouvet, A. 1995. Human endocarditis due to nutritionally variant streptococci: *Streptococcus adjacens* and *Streptococcus defectivus*. Eur. Heart J. 16 (suppl. B):24.
4. Carratala, J., Alcaide, F., Fernandez-Sevilla, A., et al. 1995. Bacteremia due to viridans streptococci that are highly resistant to penicillin: increase among neutropenic patients with cancer. Clin. Infect. Dis. 20:1169.
5. Christensen, J.J., Vibits, H., Ursing, J., et al. 1991. *Aerococcus*-like organism: a newly recognized potential urinary tract pathogen. J. Clin. Microbiol. 29:1049.
6. Facklam, R. and Elliott, J.A. 1995. Identification, classification, and clinical relevance of catalase-negative, gram-positive cocci, excluding the streptococci and enterococci. Clin. Microbiol. Rev. 8:479.
7. Gerber, M.A. 1995. Antibiotic resistance in group A streptococci. Ped. Clin. North Am. 42:539.
8. Jett, B.D., Huycke, M.M., and Gilmore, M.S. 1994. Virulence of enterococci. Clin. Microbiol. Rev. 7:462.
9. Johnson, A.P. 1994. The pathogenicity of enterococci. J. Antimicrob. Chemother. 33:1083.
10. Leclercq, R. 1996. Epidemiology and control of multiresistant enterococci. Drugs 2:47.
11. McCracken, G.H. 1995. Emergence of resistant *Streptococcus pneumoniae*: a problem in pediatrics. Pediatr. Infect. Dis. J. 14:424.
12. Murray, B.E. 1990. The life and times of the enterococcus. Clin. Microbiol. Rev. 3:46.
13. Stevens, D.L. 1996. Invasive group A streptococcal disease. Infect. Agents Dis. 5:157.
14. Turner, J.C., Fox, A., Fox, K., et al. 1993. Role of group C beta-hemolytic streptococci in pharyngitis: epidemiologic study of clinical features associated with isolation of group C streptococci. J. Clin. Microbiol. 31:808.
15. Vandamme, P., Pot, B., Falsen, E., et al. 1996. Taxonomic study of Lancefield streptococcal groups C, G, and L (*Streptococcus dysgalactiae*) and proposal of *S. dysgalactiae* subsp. *equisimilis* subsp. nov. Int. J. Syst. Bacteriol. 46:774.

Bibliography

Benenson, A.S. 1995. Control of communicable diseases manual, ed 16. American Public Health Association, Washington, D.C.

Bisno, A.L. 1995. Nonsuppurative poststretococcal sequelae: rheumatic fever and glomerulonephritis. In Mandell, G.L., Bennett, J.E., and Dolin, R., editors. Principles and practice of infectious diseases. Churchill Livingstone, New York.

Bisno, A.L. 1995. *Streptococcus pyogenes*. In Mandell, G.L., Bennett, J.E., and Dolin, R., editors. Principles and practice of infectious diseases. Churchill Livingstone, New York.

Committee on Infectious Diseases. 1997. 1997 Red Book. American Academy of Pediatrics, Elk Grove Village, Ill.

Edwards, M.S. and Baker, C.J. 1995. *Streptococcus agalactiae* (group B streptococcus). In Mandell, G.L., Bennett, J.E., and Dolin, R., editors. Principles and practice of infectious diseases. Churchill Livingstone, New York.

Facklam, R.R. and Sahm, D.F. 1995. *Enterococcus*. In Murray, P.R., Baron, E.J., Pfaller, M.A., et al., editors. Manual of clinical microbiology, ed 6. American Society for Microbiology, Washington, D.C.

Johnson, C.C. and Tunkel, A.R. 1995. Viridans streptococci and groups C and G streptococci. In Mandell, G.L., Bennett, J.E., and Dolin, R., editors. Principles and practice of infectious diseases. Churchill Livingstone, New York.

Moellering, R.C., Jr. 1995. *Enterococcus* species, *Streptococcus bovis,* and *Leuconostoc* species. In Mandell, G.L., Bennett, J.E., and Dolin, R., editors. Principles and practice of infectious diseases. Churchill Livingstone, New York.

Musher, D.M. 1995. *Streptococcus pneumoniae*. In Mandell, G.L., Bennett, J.E., and Dolin, R., editors. Principles and practice of infectious diseases. Churchill Livingstone, New York.

Ruoff, K.L. 1995. *Leuconostoc, Pediococcus, Stomatococcus,* and miscellaneous gram-positive cocci that grow aerobically. In Murray, P.R., Baron, E.J., Pfaller, M.A., et al., editors. Manual of clinical microbiology, ed 6. American Society for Microbiology, Washington, D.C.

Ruoff, K.L. 1995. *Streptococcus*. In Murray, P.R., Baron, E.J., Pfaller, M.A., et al., editors. Manual of clinical microbiology, ed 6. American Society for Microbiology, Washington, D.C.

Ryan, K.J. and Falkow, S. 1994. Streptococci and enterococci. In Ryan, K.J., editor. Sherris medical microbiology: an introduction to infectious diseases. Appleton & Lange, Norwalk, Conn.

Stratton, C.W. 1995. *Streptococcus intermedius* group. In Mandell, G.L., Bennett, J.E., and Dolin, R., editors. Principles and practice of infectious diseases. Churchill Livingstone, New York.

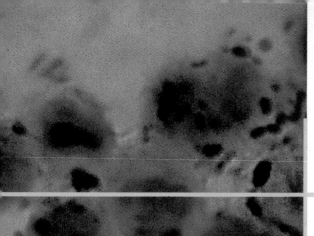

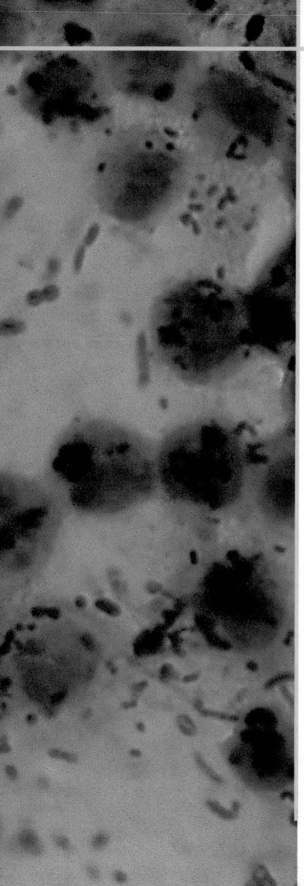

Section Nine

9

Nonbranching, Catalase-Positive, Gram-Positive Bacilli

54 | *BACILLUS* AND SIMILAR ORGANISMS

Genera and Species to be Considered

- *Bacillus anthracis*
- *Bacillus cereus*
- *Bacillus mycoides*
- *Bacillus circulans*
- *Bacillus licheniformis*
- *Bacillus subtilis*
- *Bacillus megaterium*
- Other *Bacillus* spp.
- *Brevibacillus brevis*
- *Paenibacillus* spp.

GENERAL CHARACTERISTICS

Bacillus spp. and related genera *Brevibacillus* and *Paenibacillus* all are aerobic, gram-positive, spore-forming rods.[1,3] Although these genera contain many species, only those that are most commonly associated with human infections are considered here.

EPIDEMIOLOGY

Bacillus anthracis, the most notorious pathogen of this genus, inhabits the soil. Humans acquire infections when they are inoculated with the spores, either by inhalation during exposure to contaminated animal products, such as hides, or by traumatic introduction (Table 54-1). All other *Bacillus* spp. are generally considered to be opportunistic pathogens of low virulence and are usually only associated with compromised patients exposed to contaminated materials.[2]

PATHOGENESIS AND SPECTRUM OF DISEASES

B. anthracis is the most highly virulent species for humans and is the causative agent of anthrax, of which there are three forms (Table 54-2). Anthrax is a

---

TABLE 54-1 EPIDEMIOLOGY

SPECIES	HABITAT (RESERVOIR)	MODE OF TRANSMISSION
Bacillus anthracis	Soil, from where it is contracted by various herbivores. Causes disease in these animals. Not part of normal human flora	Contact with infected animals or animal products. Infections occur by inhalation of spores or by their introduction through breaks in the skin or mucous membranes. Occasionally acquired by ingestion. Person-to-person transmission has not been documented
Bacillus cereus, Bacillus circulans, Bacillus licheniformis, Bacillus subtilis, other *Bacillus* spp., *Brevibacillus* sp., and *Paenibacillus* spp.	Vegetative cells and spores of many species are widely distributed in nature. Not commonly considered part of normal flora but may transiently colonize skin or the gastrointestinal or respiratory tracts	Traumatic introduction into normally sterile sites or by exposure to contaminated medical equipment or supplies. Infections often involve immunosuppressed patients. In cases of food poisoning, ingestion of food contaminated with *B. cereus* or toxins formed by this organism

TABLE 54-2 PATHOGENESIS AND SPECTRUM OF DISEASES

SPECIES	VIRULENCE FACTORS	SPECTRUM OF DISEASES AND INFECTIONS
Bacillus anthracis	Production of antiphagocytic capsule and potent exotoxins (i.e., edema toxin and lethal toxin) that mediate cell and tissue destruction	Causative agent of anthrax, of which there are three forms: ■ Cutaneous anthrax occurs at site of spore penetration 2 to 5 days after exposure and is manifested by progressive stages from an erythematous papule to ulceration and finally to formation of a black scar (i.e., eschar). May progress to toxemia and death ■ Pulmonary anthrax, also known as "wool sorter's" disease, follows inhalation of spores and progresses from malaise with mild fever and nonproductive cough to respiratory distress, massive chest edema, cyanosis, and death ■ Gastrointestinal anthrax may follow ingestion of spores and affects either the oropharyngeal or the abdominal area. Most patients die from toxemia and overwhelming sepsis
Bacillus cereus	Produces enterotoxins and pyogenic toxin	Food poisoning of two types: diarrheal type, which is characterized by abdominal pain and watery diarrhea; and emetic type, which is manifested by profuse vomiting. *B. cereus* is the most commonly encountered species of *Bacillus* in opportunistic infections that include post-traumatic eye infections, endocarditis, and bacteremia. Infections of other sites are rare and usually involve intravenous drug abusers or immunocompromised patients
Bacillus circulans, *Bacillus licheniformis,* *Bacillus subtilis,* other *Bacillus* spp., *Brevibacillus* sp., and *Paenibacillus* spp.	Virulence factors not known	Food poisoning has been associated with some species but is not common. These organisms may also be involved in opportunistic infections similar to those described for *B. cereus*

devastating disease that is rarely encountered in developed countries.

Although other species, such as *B. cereus,* can cause serious infections, the relative virulence of this and other *Bacillus* spp. is trivial compared with that of *B. anthracis. B. cereus* can cause food poisoning and serious eye infections may result from trauma, use of contaminated needles during intravenous drug abuse, or inoculation of the eye with contaminated dust or dirt particles. Infections at other body sites also are known to occur in compromised hosts. However, because the spores of *Bacillus* spp. are ubiquitous in nature, contamination of various clinical specimens is common. Therefore, whenever *Bacillus* spp. are encountered, the clinical significance of the isolate should be carefully established before performing extensive identification procedures.

LABORATORY DIAGNOSIS

SPECIMEN COLLECTION AND TRANSPORT

No special considerations are required for specimen collection and transport of the organisms discussed in this chapter. Refer to Table 1-1 for general information on specimen collection and transport.

SPECIMEN PROCESSING

With few exceptions, special processing considerations are not required for these organisms. Refer to Table 1-1 for general information on specimen processing. The exceptions are processing procedures for foods implicated in *B. cereus* food poisoning outbreaks and animal hides or products, and environmental samples, for the isolation of *B. anthracis.* The organisms will be present as spores in these specimens so initial processing will involve either heat or alcohol shock before plating on solid media. The shock will allow only the spore-forming bacilli to survive; thus this is an enrichment and selection technique designed to increase the chance for laboratory isolation of these species. Despite its publicity as a potential agent of biologic warfare, *Bacillus anthracis* is not highly contagious so standard safety precautions are appropriate.[4]

DIRECT DETECTION METHODS

Other than Gram stain of patient specimens, there are no specific procedures for the direct detection of *Bacillus* spp. in clinical material. The organisms are gram-positive when stained from young cultures but become gram-variable or gram-negative with age (Figure 54-1).

A feature of *Bacillus* spp. that is unique from all other clinically relevant aerobic organisms is the ability to produce spores. Although spores are not readily evident on all smears containing *Bacillus* spp., their presence confirms the genus identification. On Gram-stained smears, spores appear clear because they do not retain the crystal violet or safranin. However, spores will stain with specific dyes such as malachite green, which is forced into the spore using heat; the vegetative cell is then counterstained with safranin (Figure 54-2).

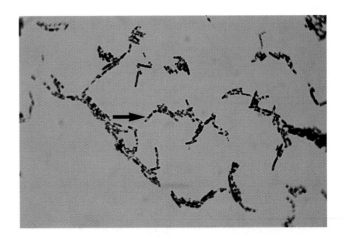

FIGURE 54-1 Gram stain of *Bacillus cereus.* The arrow is pointed at a spore, which is clear inside the gram-positive vegetative cell.

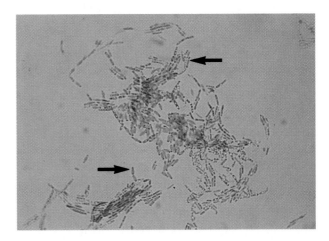

FIGURE 54-2 Spore stain of *Bacillus cereus.* The arrows are pointed at green spores in a pink vegetative cell.

CULTIVATION

Media of choice

All *Bacillus* and related genera grow well on 5% sheep blood agar, chocolate agar, routine blood culture media, and commonly used nutrient broths. They will not grow on MacConkey agar, and those that are susceptible to nalidixic acid will not grow on CNA (Columbia agar with nalidixic acid and colistin). PEA (phenylethyl alcohol agar) is useful for the isolation of *Bacillus* spp. from contaminated specimens.

Two special media are used for isolation and identification of *B. anthracis*. PLET (*polymyxin-lysozyme*-EDTA-*thallous* acetate) can be used for selection and isolation of this species from contaminated specimens. As a means of identification, bicarbonate agar is used to induce *B. anthracis* capsule formation.

Incubation conditions and duration

Most species will produce detectable growth within 24 hours following incubation on media incubated at 35° C, in ambient air, or in 5% carbon dioxide (CO_2). Bicarbonate agar must be incubated in CO_2.

Colonial appearance

Table 54-3 describes the colonial appearance and other distinguishing characteristics (e.g., hemolysis) of each species of *Bacillus* or related genera on blood agar. Colonies of *B. anthracis* growing on bicarbonate agar are large and mucoid.

APPROACH TO IDENTIFICATION

There are no commercial biochemical identification systems in routine use in clinical laboratories for *Bacillus* spp. Species differentiation within the genera *Bacillus*, *Brevibacillus*, and *Paenibacillus* is based on the size of the vegetative cell, whether the spore swells the vegetative cell, and biochemical tests (Table 54-4), including elaboration of the enzyme lecithinase (Figure 54-3).

Comments regardings specific organisms

The vegetative cell width of *B. anthracis*, *B. cereus*, *B. mycoides*, *B. thuringiensis*, and *B. megaterium* is usually greater than 1 μm, and the spores do not swell the cell. The vegetative cell width of *B. subtilis*, *B. pumilus*, and *B. licheniformis* is less than 1 μm, and the spores do not swell the cell. The cell width of *B. circulans*, *B. coagulans*, *B. sphaericus*, *B. brevis*, *P. macerans*, *P. alvei*, and *P. polymyxa* is less than 1 μm and the spores swell the cell. When determining cell width, only cells staining gram-positive should be measured. Organisms that do not retain the crystal violet appear narrower because the cell wall is not apparent.

It is important to be able to identify *B. anthracis*, even though this pathogen is rarely encountered. *B. anthracis* should be suspected if typical nonhemolytic "Medusa head" colonies are observed on 5% sheep blood agar, and the organism is subsequently found to be nonmotile and penicillin-susceptible, and produces a wide zone of lecithinase on egg yolk agar. This appearance on egg yolk agar is similar to *B. cereus*, but the latter is penicllin-resistant, beta-hemolytic, and motile. Suspected isolates of *B. anthracis* should be forwarded to a reference laboratory for gamma bacteriophage testing or animal pathogenicity testing.

The other organisms discussed in this chapter are common environmental contaminants, so most laboratories only identify isolates from sterile sites (e.g., blood) or those found in large numbers in pure culture.

SERODIAGNOSIS

Indirect hemagglutination and enzyme-linked immunosorbent assays are available to detect antibodies to *B. anthracis*, but serodiagnostic methods are not used to diagnose infections caused by other *Bacillus* spp.

ANTIMICROBIAL SUSCEPTIBILITY TESTING AND THERAPY

Although penicillin has been established as the preferred therapy for anthrax, the infrequent nature with which other species are encountered limits recommendations concerning therapy (Table 54-5). Although most *Bacillus* spp. will grow on the media and under the conditions recommended for testing the more commonly encountered bacteria (see Chapter 18 for more information regarding validated testing methods), the ability to grow does not guarantee the ability to detect important antimicrobial resistances. Therefore, the lack of validated in vitro susceptibility testing methods does not allow testing guidelines to be given. Chapter 19 should be reviewed for preferable strategies that can be used to provide susceptibility information and data when validated testing methods do not exist for a clinically important bacterial isolate.

VACCINES AND PROPHYLAXIS

A single-dose vaccine with annual boosters is available for immunizing high-risk adults (i.e., laboratory workers or workers handling potentially contaminated industrial raw materials) against anthrax.

TABLE 54-3   COLONIAL APPEARANCE AND OTHER CHARACTERISTICS

ORGANISM	APPEARANCE ON 5% SHEEP BLOOD AGAR
B. anthracis	Medium-large, gray, flat, irregular with swirling projections ("Medusa head"); nonhemolytic
B. cereus and B. thuringiensis	Large, feathery, spreading; beta-hemolytic
B. mycoides	Rhizoid colony that resembles a fungus; weakly beta-hemloytic
B. megaterium	Large, convex, entire, moist; nonhemolytic
B. licheniformis	Large blister colony; becomes opaque with dull to rough surface with age; beta-hemolytic
B. pumilus	Large, moist, blister colony; may be beta-hemolytic
B. subtilis	Large, flat, dull, with ground-glass appearance; may be pigmented (pink, yellow, orange, or brown); may be beta-hemolytic
B. circulans	Large; entire; convex; butyrous; smooth, translucent surface; may be beta-hemolytic
B. coagulans	Medium-large, entire, raised, butyrous, creamy-buff; may be beta-hemolytic
B. sphaericus	Large, convex, smooth, opaque, butyrous; nonhemolytic
Brevibacillus brevis	Medium-large, convex, circular, granular; may be beta-hemolytic
Paenibacillus macerans	Large, convex, fine granular surface; nonhemolytic
P. alvei	Swarms over agar surface; discrete colonies are large, circular, convex, smooth, glistening, translucent or opaque; may be beta-hemolytic
P. polymyxa	Large, moist blister colony with "ameboid spreading" in young cultures; older colonies wrinkled; nonhemolytic

TABLE 54-4   DIFFERENTIATION OF CLINICALLY RELEVANT
***BACILLUS* SPP., *BREVIBACILLUS*, AND**
PAENIBACILLUS

ORGANISM	BACILLARY BODY WIDTH >1.0 μm	WIDE ZONE LECITHINASE	SPORES SWELL SPORANGIUM	VOGES-PROSKAUER
Bacillus anthracis	+	+	−	+
B. cereus	+	+	−	+
B. thuringiensis	+	+	−	+
B. mycoides	+	+	−	+
B. megaterium	+	−	−	−
B. licheniformis	−	−	−	+
B. pumilus	−	−	−	+
B. subtilis	−	−	−	+
B. circulans	−	−	+	−
B. coagulans	−	−	v	v
B. sphaericus	−	−	+	−
Brevibacillus brevis	−	−	+	−
Paenibacillus macerans	−	−*	+	−
P. alvei	−	−	+	+
P. polymyxa	−	−*	+	+

+, 90% or more of species or strains are positive; −, 90% or more of species or strains are negative; *v*, variable reactions; ( ), reactions may be delayed.
*Weak lecithinase production only seen under the colonies.

Compiled from Turnbull, P.C.B. and Kramer, J.M. 1995. *Bacillus.* In Murray, P.R., Baron, E.J., Pfaller, M.A., et al., editors. and Gibson, J.R. 1983. A colour atlas of *Bacillus* species. Wolf Medical Publications Ltd., London; Hollis, D.G. and Weaver, Berkeley, R.C.W. 1986. Genus *Bacillus.* In Sneath, P.H.A., Mair, N.S., Sharpe, M.E., Holt, J.G., editors. Bergey's manual of

GLUCOSE WITH GAS	FERMENTATION OF: MANNITOL	XYLOSE	ANAEROBIC GROWTH	CITRATE	INDOLE	MOTILITY	PENICILLIN SENSITIVE	PARASPORAL CRYSTALS
−	−	−	+	v	−	−	+	−
−	−	−	+	+	−	+	−	−
−	−	−	+	+	−	+	−	+
−	−	−	+	+	−	−	−	−
−	+or(+)	v	−	+	−			
−	+	+	+	+	−			
−	+	+	−	+	−			
−	+	+	−	+	−			
−	+	+	v	−	−			
−	−	v	+	v	−			
−	−	−	−	v	−			
−	+	−	−	v	−			
+	+	+	+	−	−			
−	−	−	+	−	+			
+	v	+	+	−	−			

Manual of clinical microbiology, ed 6. American Society for Microbiology, Washington, D.C.; Parry, J.M., Turnbull, P.C.B., R.E. 1981. Gram-positive organisms: a guide to identification. Centers for Disease Control, Atlanta, Ga.; and Claus, D. and systematic bacteriology, vol 2. Williams & Wilkins, Baltimore.

TABLE 54-5 ANTIMICROBIAL THERAPY AND SUSCEPTIBILITY TESTING

SPECIES	THERAPEUTIC OPTIONS	RESISTANCE TO THERAPEUTIC OPTIONS	VALIDATED TESTING METHODS*	COMMENTS
Bacillus anthracis	Penicillin is drug of choice. Alternatives include erythromycin, tetracycline, and chloramphenicol	None known	Not available	
Other *Bacillus* spp., *Brevibacillus* sp., *Paenibacillus* spp.	No definitive guidelines. Vancomycin, ciprofloxacin, imipenem, and aminoglycosides may be effective	*B. cereus* frequently produces beta-lactamase	Not available	Whenever isolated from clinical specimens, the potential for the isolate to be a contaminant must be strongly considered

*Validated testing methods include those standard methods recommended by the National Committee for Clinical Laboratory Standards (NCCLS) and those commercial methods approved by the Food and Drug Administration (FDA).

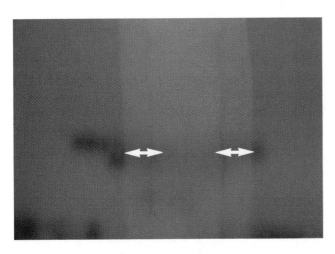

FIGURE 54-3 Lecithinase production by *Bacillus cereus* on egg yolk agar. The organism has been streaked down the center of the plate. The positive test for lecithinase is indicated by the opaque zone of precipitation around the bacterial growth (arrows).

References

1. Ash, C.F., Priest, G., and Collins, M.D. 1993. Molecular identification of rRNA group 3 bacilli (Ash, Farrow, Wallbanks and Collins) using a PCR probe test. Antonie van Leeuwenhoek 64:253.

2. Drobniewski, F.A. 1993. *Bacillus cereus* and related species. Clin. Microbiol. Rev. 6:324.

3. Shida, O., Takagi, H., Kadowaki, K., and Komagata, K. 1996. Proposal for two new genera, *Brevibacillus* gen. nov. and *Aneurinibacillus* gen. nov. Int. J. Syst. Bacteriol. 46:939.

4. Turnbull, P.C.B. and Kramer, J.M. 1995. *Bacillus*. In Murray, P.R., Baron, E.J., Pfaller, M.A., et al., editors. Manual of clinical microbiology, ed 6. American Society for Microbiology, Washington, D.C.

Bibliography

Claus, D. and Berkeley, R.C.W. 1986. Genus *Bacillus*. In Sneath, P.H.A., Mair, N.S., Sharpe, M.E., and Holt J.G., editors. Bergey's manual of systematic bacteriology, vol 2. Williams & Wilkins, Baltimore.

Committee on Infectious Diseases. 1997. 1997 Red Book. American Academy of Pediatrics, Elk Grove Village, Ill.

Hollis, D.G. and Weaver, R.E. 1981. Gram-positive organisms: a guide to identification. Centers for Disease Control, Atlanta, Ga.

Lew, D. 1995. *Bacillus anthracis* (anthrax). In Mandell, G.L., Bennett, J.E., and Dolin, R., editors. Principles and practice of infectious diseases. Churchill Livingstone, New York.

Parry, J.M., Turnbull, P.C.B., and Gibson, J.R., 1983. A colour atlas of *Bacillus* species. Wolf Medical Publications Ltd., London.

Ryan, K.J. and Falkow, S. 1994. Corynebacteria, *Listeria*, *Bacillus*, and other aerobic and facultative gram-positive rods. In Ryan, K.J., editor. Sherris medical microbiology: an introduction to infectious diseases. Appleton & Lange, Norwalk, Conn.

Tuazon, C.U. 1995. Other *Bacillus* species. In Mandell, G.L., Bennett, J.E., and Dolin, R., editors. Principles and practice of infectious diseases. Churchill Livingstone, New York.

55 | *LISTERIA, CORYNEBACTERIUM, AND SIMILAR ORGANISMS*

Genera and Species to be Considered

- *Listeria monocytogenes*
- *Corynebacterium diphtheriae*
- *Corynebacterium jeikeium*
- *Corynebacterium ulcerans*
- *Corynebacterium pseudotuberculosis*
- *Corynebacterium pseudodiphtheriticum*
- *Corynebacterium amycolatum*
- *Corynebacterium auris*

- *"Corynebacterium aquaticum"**
- *Corynebacterium minutissimum*
- *Corynebacterium urealyticum*
- *Corynebacterium xerosis*
- *Corynebacterium striatum*
- Other *Corynebacterium* spp. and CDC Coryneform groups
- *Kurthia* spp.

- *Brevibacterium* spp.
- *Dermabacter hominis*
- *Turicella otitidis*
- *Arthrobacter* spp.
- *Microbacterium* spp.
- *Aureobacterium* spp.
- *Cellulomonas* spp.
- *Exiguobacterium acetylicum*

---

*Quotation marks indicate an uncertain taxonomic status.

GENERAL CONSIDERATIONS

The genera described in this chapter are all catalase-positive, gram-positive rods. They are not acid-fast, they do not branch, and they do not form spores.

EPIDEMIOLOGY

Most of the organisms listed in Table 55-1 either are part of the normal human flora, are found in the environment, or are associated with various ani-

---

TABLE 55-1 EPIDEMIOLOGY

ORGANISM	HABITAT (RESERVOIR)	MODE OF TRANSMISSION
Listeria monocytogenes	Colonizes a wide variety of animals, soil, and vegetable matter; widespread in these environments; may colonize human gastrointestinal tract	Usually by ingestion of contaminated food, such as meat and dairy products. Colonized mothers may pass organism on to fetus. Portal of entry is probably from gastrointestinal tract to blood, and, in some instances, from blood to meninges
Corynebacterium diphtheriae	Inhabits human nasopharynx but only in carrier state; not considered part of normal flora. Isolation from healthy humans is not common	Person to person by exposure to contaminated respiratory droplets or direct contact with infected cutaneous lesions; may also be transferred by exposure to contaminated objects

TABLE 55-1 EPIDEMIOLOGY—CONT'D

ORGANISM	HABITAT (RESERVOIR)	MODE OF TRANSMISSION
Corynebacterium jeikeium	Skin flora of hospitalized patients, most commonly in the inguinal, axillary, and rectal sites	Uncertain; may be person to person or selection of endogenous resistant strains during antimicrobial therapy; introduced during placement or improper care of intravenous catheters
Corynebacterium ulcerans	Normal flora of humans and cattle	Uncertain; associated with close animal contact, especially during summer
Corynebacterium pseudotuberculosis	Associated with infections in animals such as sheep, goats, and horses	Uncertain; associated with close animal contact, but infections in humans are rare
Corynebacterium pseudodiphtheriticum	Normal human pharyngeal and, occasionally, skin flora	Uncertain; probably by access of patient's endogenous strain to normally sterile site
Corynebacterium minutissimum	Normal human skin flora	Uncertain; probably by access of patient's endogenous strain to normally sterile site
Corynebacterium urealyticum	Normal human skin flora	Uncertain; probably by access of patient's endogenous strain to normally sterile site
"Corynebacterium aquaticum"	Fresh water	Uncertain
Corynebacterium xerosis	Normal flora of human conjunctiva, skin, and nasopharynx	Uncertain; probably by access of patient's endogenous strain to normally sterile site
Corynebacterium striatum	Normal human skin flora	Uncertain; probably by access of patient's endogenous strain to normally sterile site
Corynebacterium amycolatum	Normal flora of human conjunctiva, skin, and nasopharynx	Uncertain; probably by access of patient's endogenous strain to normally sterile site
Corynebacterium auris	Uncertain; probably part of normal human flora	Uncertain; rarely implicated in human infections
Kurthia spp.	Environmental; not part of normal human flora	Uncertain; rarely implicated in human infections
Brevibacterium spp.	Normal human flora and various foods	Uncertain; rarely implicated in human infections
Dermabacter hominis	Normal flora of human skin	Uncertain; rarely implicated in human infections
Turicella otitidis	Uncertain; probably part of normal human flora	Uncertain; rarely implicated in human infections
Arthrobacter spp., *Microbacterium* spp., *Aureobacterium* spp., *Cellulomonas* spp., and *Exiguobacterium* sp.	Uncertain; probably environmental; not part of normal human flora	Uncertain; rarely implicated in human infections

mals. The two most potent pathogens are *Listeria monocytogenes* and *Corynebacterium diphtheriae*.[23,25] However, these two species differ markedly in epidemiology. *L. moncytogenes* is widely distributed in nature and occasionally colonizes the human gastrointestinal tract. *C. diphtheriae* is only carried by humans but, rarely, is isolated from healthy individuals.

In contrast to these two organisms, *C. jeikeium* is commonly encountered in clinical specimens, mostly because it tends to proliferate as skin flora of hospitalized individuals. However, *C. jeikeium* is not considered to be highly virulent. The penetration of the patient's skin by intravascular devices is usually required for this organism to cause infection.

PATHOGENESIS AND SPECTRUM OF DISEASES

L. monocytogenes, by virtue of its ability to survive within phagocytes, and *C. diphtheriae,* by production of an extremely potent cytotoxic exotoxin, are the most virulent species listed in Table 55-2. Most of the other organisms are opportunistic and infection with them requires that patients be compromised.[17] For this reason, whenever *Corynebacteriuim* spp. or the other genera of gram-positive rods are encountered,

TABLE 55-2 PATHOGENESIS AND SPECTRUM OF DISEASES

ORGANISM	VIRULENCE FACTORS	SPECTRUM OF DISEASES AND INFECTIONS
Listeria monocytogenes	Listeriolysin, a hemolytic and cytotoxic toxin that may allow for survival within phagocytes	Bacteremia, without any other known site of infection, and meningitis. Infections usually occur in neonates or immunosuppressed patients. Granulomatosis infantiseptica is an in utero infection that is disseminated systemically and causes stillbirth
Corynebacterium diphtheriae	Diphtheria toxin, a potent exotoxin that destroys host cells	Diphtheria, of which there are two forms. Respiratory diphtheria is a pharyngitis characterized by the development of an exudative membrane that covers the tonsils, uvula, palate, and pharyngeal wall; life-threatening if untreated, because respiratory obstruction develops and release of toxin in blood can damage various organs, including the heart. Cutaneous diphtheria is characterized by nonhealing ulcers and membrane formation
Corynebacterium jeikeium	Unknown; probably of low virulence. Multiple antibiotic resistance allows survival in hospital setting	Septicemia, wound infections and, rarely, endocarditis in compromised patients, especially those with intravenous catheters
Corynebacterium ulcerans	Unknown; probably of low virulence	Has been associated with diphtheria-like sore throat
Corynebacterium pseudotuberculosis	Unknown; probably of low virulence	Suppurative granulomatous lymphadenitis
Corynebacterium pseudodiphtheriticum	Unknown; probably of low virulence	Endocarditis, pneumonia, and lung abscesses
Corynebacterium minutissimum	Unknown; probably of low virulence	Superficial, pruritic skin infections known as *erythrasma*; rarely, causes septicemia and abscess formation

careful consideration must be given to their role as infecting or contaminating agents.

LABORATORY DIAGNOSIS

SPECIMEN COLLECTION AND TRANSPORT

No special considerations are required for specimen collection and transport of the organisms discussed in this chapter. Refer to Table 1-1 for general information on specimen collection and transport.

SPECIMEN PROCESSING

No special considerations are required for processing of most of the organisms discussed in this chapter.

Refer to Table 1-1 for general information on specimen processing. One exception is the isolation of *Listeria monocytogenes* from placental and other tissue. Because it may be difficult to isolate *Listeria* from these sources, the specimen may be placed in a nutrient broth at 4° C for several weeks to months. The broth is subcultured at frequent intervals to enhance recovery; this procedure is called **cold enrichment.**

DIRECT DETECTION METHODS

Other than Gram stain of patient specimens, there are no specific procedures for the direct detection of these organisms in clinical material. Most of the genera in this chapter (except *Listeria*) are classified as

---

TABLE 55-2 PATHOGENESIS AND SPECTRUM OF DISEASES—CONT'D

ORGANISM	VIRULENCE FACTORS	SPECTRUM OF DISEASES AND INFECTIONS
Corynebacterium urealyticum	Unknown; probably of low virulence. Multiple antibiotic resistance allows survival in hospital setting	Urinary tract infections and wound infections of compromised patients
"*Corynebacterium aquaticum*"	Unknown; probably of low virulence	Bacteremia
Corynebacterium xerosis	Unknown; probably of low virulence	Endocarditis or septicemia in immunocompromised patients
Corynebacterium striatum	Unknown; probably of low virulence	Bacteremia, pneumonia, and lung abscesses in compromised patients
Corynebacterium amycolatum	Unknown; probably of low virulence	Endocarditis, septicemia, and pneumonia in immunocompromised patients
Corynebacterium auris	Unknown; probably of low virulence	Uncertain disease association but has been linked to otitis media
Kurthia spp., *Brevibacterium* spp., and *Dermabacter* sp.	Unknown; probably of low virulence	Rarely cause infections in humans; when infections do occur they usually involve bacteremia in compromised patients with indwelling catheters or penetrating injuries
Turicella sp.	Unknown; probably of low virulence	Uncertain disease association but has been linked to otitis media
Arthrobacter spp., *Microbacterium* spp., *Aureobacterium* spp., *Cellulomonas* spp., and *Exiguobacterium* sp.	Uncertain; probably of low virulence	Uncertain disease association

coryneform bacteria, that is, they are gram-positive, short or slightly curved rods with rounded ends; some have rudimentary branching. Cells are arranged singly, in "palisades" of parallel cells, or in pairs that remained connected after cell division to form V or L shapes. Groups of these morphologies seen together resemble and are often referred to as *Chinese letters*. The Gram stain morphologies of clinically important species are described in Table 55-3.

L. monocytogenes is a short gram-positive rod that may occur singly or in short chains resembling streptococci.

CULTIVATION

Media of choice
All the genera described in this chapter usually will grow on 5% sheep blood and chocolate agars. Some coryneform bacteria will not grow on chocolate agar, and the lipophilic species (e.g., *C. jeikeium, C. urealyticum, C. afermentans* subsp. *lipophilum, C. accolens,* or *C. macginleyi*) produce much larger colonies if grown on 5% sheep blood agar supplemented with 1% Tween 80.

Selective and differential media for *C. diphtheriae* should be used if diphtheria is suspected. The two media commonly used for this purpose are cystine-tellurite blood agar and modified Tinsdale's agar. In addition, Loeffler medium, containing serum and egg, stimulates the growth of *C. diphtheriae* and the production of metachromatic granules within the cells. Primary inoculation of throat swabs to a Loeffler slant, overnight incubation, and subculture of any growth to cystine-tellurite blood agar may improve recovery of *C. diphtheriae*.

None of these organisms grows on MacConkey agar. They all should grow in routine blood culture broth and nutrient broths, such as thioglycollate or brain-heart infusion. The lipophilic coryneform bacteria will show better growth if broths are supplemented with rabbit serum.

Incubation conditions and duration
Detectable growth on 5% sheep blood and chocolate agars incubated at 35° C in either ambient air or in 5% to 10% carbon dioxide, should occur within 48 hours of inoculation. The lipophilic organisms grow more slowly, requiring 3 or more days to become visible on routine media. For growth of *C. diphtheriae*, cystine-tellurite blood agar and modified Tinsdale's agar should be incubated for at least 48 hours in ambient air.

Colonial appearance
Table 55-3 describes the colonial appearance and other distinguishing characteristics (e.g., hemolysis and odor) of each clinically important genus or species on blood agar. Colonies of *C. diphtheriae* on cystine-tellurite blood agar appear black or gray, whereas those on modified Tinsdale's agar are black with dark brown halos.

APPROACH TO IDENTIFICATION
Except for *Listeria monocytogenes* and a few *Corynebacterium* spp., the identification of these organisms generally is complex and problematic. A multiphasic approach is required for definitive identification. This often requires biochemical testing, whole-cell fatty acid analysis, and cell wall diamino acid analysis. The last two methods are usually not available in routine clinical laboratories, so identification of isolates requires expertise available in reference laboratories. Further complicating the situation is the fact that coryneforms are present as normal flora throughout the body. Thus, only clinically relevant isolates should be identified fully. Indicators of clinical relevance include multiple isolations from normally sterile sites (e.g., blood) or isolation in pure culture from symptomatic patients who have not yielded any other known etiologic agent.

The API Coryne strip (bioMérieux Vitek, Hazelwood, Mo.) is commercially available for rapid identification of this group of organisms, but the database may not be current with recent taxonomic changes. Therefore misidentifications can occur if the code generated using the strip is the exclusive criteria used for identification.

Table 55-4 shows the key tests needed to separate the genera discussed in this chapter. In addition to the features shown, the Gram stain and colonial morphology should be carefully noted.

Comments regarding specific organisms
Two tests (halo on Tinsdale's agar and urea hydrolysis) can be used to separate *C. diphtheriae* from other corynebacteria. Definitive identification of a *C. diphtheriae* isolate as a true pathogen requires demonstration of toxin production by the isolate in question. A patient may be infected with several strains at once, so testing is performed using a pooled inoculum of at least 10 colonies. There are several methods by which toxin testing can be performed:

- Guinea pig lethality test to ascertain if diphtheria antitoxin neutralizes the lethal effect of a cell-free suspension of the suspect organism
- Immunodiffusion test originally described by Elek (Figure 55-1)
- Tissue culture cell test to demonstrate toxicity of a cell-free suspension of the suspect organism in

TABLE 55-3 GRAM STAIN MORPHOLOGY, COLONIAL APPEARANCE, AND OTHER DISTINGUISHING CHARACTERISTICS

ORGANISM	GRAM STAIN	APPEARANCE ON 5% SHEEP BLOOD AGAR
Arthrobacter spp.	Typical coryneform gram-positive rods after 24 hrs, with "jointed ends" giving L and V forms, and coccoid cells after 72 hrs (i.e., rod-coccus cycle*)	Large colony; resembles *Brevibacterium* spp.
Aureobacterium spp.	Irregular, gram-positive coccoid and coryneform rods in V arrangements	Yellow
Brevibacterium spp.	Gram-positive rods; produce typical coryneform arrangements in young cultures (<24 hrs) and coccoid-to-coccobacillary forms that decolorize easily in older cultures (i.e., rod-coccus cycle*)	Medium to large; gray to white, convex, opaque, smooth, shiny; nonhemolytic; cheeselike odor
Cellulomonas spp.	Irregular, short, thin, branching gram-positive rods	Small to medium; two colony types, one starts out white and turns yellow within 3 days and the other starts out yellow
CDC coryneform group G[†]	Typical coryneform gram-positive rods	Small, gray to white; nonhemolytic
Corynebacterium accolens	Resembles *C. jeikeium*	Resembles *C. jeikeium*
C. afermentens subsp. *afermentens*	Typical coryneform gram-positive rods	Medium; white; nonhemolytic; non-adherent
C. afermentens subsp. *lipophilum*	Typical coryneform gram-positive rods	Small; gray, glassy
C. amycolatum	Pleomorphic gram-positive rods with single cells, V forms, or Chinese letters	Small; white to gray, dry
"*C. aquaticum*"[‡]	Irregular, slender, short gram-positive rods	Yellow
C. argentoratense	Typical coryneform gram-positive rods	Medium; cream-colored; non-hemolytic
C. auris	Typical coryneform gram-positive rods	Small to medium; dry, slightly adherent, become yellowish with time; nonhemolytic
C. diphtheriae group[§]	Irregularly staining, pleomorphic gram-positive rods	Various biotypes of *C. diphtheriae* produce colonies ranging from small, gray, and translucent (biotype *intermedius*) to medium, white, and opaque (biotypes *mitis* and *gravis*); *C. diphtheriae* biotype *mitis* may be beta-hemolytic; *C. ulcerans* and *C. pseudotuberculosis* resemble *C. diphtheriae*

Continued

TABLE 55-3 GRAM STAIN MORPHOLOGY, COLONIAL APPEARANCE, AND OTHER DISTINGUISHING CHARACTERISTICS—CONT'D		
ORGANISM	GRAM STAIN	APPEARANCE ON 5% SHEEP BLOOD AGAR
C. glucuronolyticum	Typical coryneform gram-positive rods	Small; white to yellow, convex; nonhemolytic
C. jeikeium	Pleomorphic; occasionally club-shaped gram-positive rods arranged in V forms or palisades	Small; gray to white, entire, convex; Nonhemolytic
C. macginleyi	Typical coryneform gram-positive rods	Tiny colonies after 48 hrs; non-hemolytic
C. matruchotii	Gram-positive rods with whip-handle shape and branching filaments	Small; opaque, adherent
C. minutissimum	Typical coryneform gram-positive rods with single cells, V forms, palisading and Chinese letters	Small; convex, circular, shiny, and moist
C. propinquum	Typical coryneform gram-positive rods	Small to medium with matted surface; nonhemolytic
C. pseudodiphtheriticum	Typical coryneform gram-positive rods	Small to medium; slightly dry
C. striatum	Regular medium to large gram-positive rods; can show banding	Small to medium; white, moist and smooth (resembles colonies of coagulase-negative staphylococci)
C. urealyticum	Gram-positive coccobacilli arranged in V forms and palisades	Pinpoint (after 48 hrs); white, smooth, convex; nonhemolytic
C. xerosis	Regular medium to large gram-positive rods; can show banding	Small to medium; dry, yellowish, granular
Dermabacter hominis	Coccoid to short gram-positive rods	Small; gray to white, convex; distinctive pungent odor
Exiguobacterium acetylicum	Irregular, short gram-positive rods arranged singly, in pairs, or short chains; (i.e., rod-coccus cycle*)	Golden yellow
Kurthia spp.	Regular gram-positive rods with parallel sides; coccoid cells in cultures >3 days old	Large, creamy or tan-yellow; non-hemolytic
Listeria monocytogenes	Regular, short gram-positive rods or coccobacilli occurring in pairs (resembles streptococci)	Small; white, smooth, translucent, moist; beta-hemolytic
Microbacterium spp.	Irregular, short, thin, gram-positive rods	Small to medium; yellow
Turicella otitidis	Irregular, long, gram-positive rods	Small to medium; white to cream, circular, convex

*Rod-coccus cycle means rods are apparent in young cultures; cocci are apparent in cultures >3 days old.
†Includes strains G-1 and G-2.
‡The name in quotation marks indicates an uncertain taxonomic status.
§Includes C. diphtheriae, C. ulcerans, and C. pseudotuberculosis.

Data compiled from references 3, 4, 12, 27, 30.

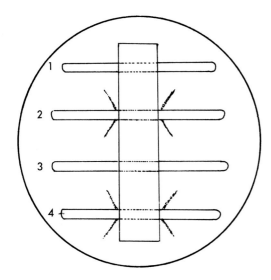

FIGURE 55-1 Diagram of an Elek plate for demonstration of toxin production by *Corynebacterium diphtheriae*. A filter paper strip impregnated with diphtheria antitoxin is buried just beneath the surface of a special agar plate before the agar hardens. Strains to be tested and known positive and negative toxigenic strains are streaked on the agar's surface in a line across the plate and at a right angle to the antitoxin paper strip. After 24 hours of incubation at 37° C, plates are examined with transmitted light for the presence of fine precipitin lines at a 45° angle to the streaks. The presence of precipitin lines indicates that the strain produced toxin that reacted with the homologous antitoxin. Line 1 is the negative control, and Line 2 is the positive control. Line 3 is an unknown organism that is a nontoxigenic strain; Line 4 is an unknown organism that is a toxigenic strain.

tissue culture cells and the neutralization of the cytopathic effect by diphtheria antitoxin
■ Polymerase chain reaction (PCR) test to detect the toxin gene

Because the incidence of diphtheria in the United States is so low, it is not practical to perform these tests in routine clinical laboratories. Toxin testing is usually performed in reference laboratories.

Identification criteria for *Corynebacterium* spp. (including *C. diphtheriae*) are shown in Tables 55-5 through 55-9. Most clinically relevant strains are catalase-positive, nonmotile, nonpigmented, or esculin- or gelatin-negative. Therefore isolation of an organism failing to demonstrate any of these characteristics provides a significant clue that another genus shown in Table 55-4 should be considered. In addition, an isolate that is an irregular gram-positive rod that is strictly aerobic and nonlipophilic and oxidizes or does not utilize glucose, will likely be "*C. aquaticum,*" *Arthrobacter, Brevibacterium,* or *Aureobacterium.*

The enhancement of growth by lipids (e.g., Tween 80 or serum) by certain coryneform bacteria, such as *C.*

jeikeium and *C. urealyticum,* is useful for preliminary identification. These two species are also resistant to several antibiotics commonly tested against gram-positive bacteria, with vancomycin frequently being the only drug demonstrating inhibition of growth.

Listeria monocytogenes can be presumptively identified by observation of motility by direct wet mount. The organism exhibits characteristic end-over-end tumbling motility when incubated in nutrient broth at room temperature for 1 to 2 hours. Alternatively, characteristic motility can be seen by an umbrella-shaped pattern (Figure 55-2) that develops after overnight incubation at room temperature of a culture stabbed into a tube of semisolid agar. *L. monocytogenes* ferments glucose and is Voges-Proskauer and esculin-positive. The isolation of a small gram-positive, catalase-positive rod with a narrow zone of beta hemolysis isolated from blood or CSF should be used as strong presumptive evidence for listeriosis.

SERODIAGNOSIS

Serodiagnostic techniques are not generally used for the laboratory diagnosis of infections caused by the organisms discussed in this chapter. Antilisteriolysin O IgG antibodies can be detected in cases of listeriosis, although IgM antibodies cannot. However, these tests are not commonly used for the clinical diagnosis.

ANTIMICROBIAL SUSCEPTIBILITY TESTING AND THERAPY

Definitive guidelines have been established regarding antimicrobial therapy for *L. monocytogenes* against certain antimicrobial agents, but because there is no resistance to the therapeutic agents of choice, antimicrobial susceptibility testing is not routinely necessary (Table 55-10).[21]

Although many of the other organisms listed in Table 55-10 will grow on the media and under the conditions recommended for testing the more commonly encountered bacteria (see Chapter 18 for more information regarding validated testing methods), the ability to grow and the ability to detect important antimicrobial resistances are not the same thing.[29] Therefore the lack of validated in vitro susceptibility testing methods does not allow testing guidelines to be given for these organisms. Chapter 19 should be reviewed for strategies that can be used to provide susceptibility information and data when warranted.

PREVENTION

The only effective control of diphtheria is through immunization with a multidose diphtheria toxoid prepared by inactivation of the toxin with formaldehyde.

Text continued on p. 660

TABLE 55-4 CATALASE-POSITIVE, NON–ACID-FAST, GRAM-POSITIVE RODS[a]

ORGANISM	AEROTOLERANCE[b]	MOTILITY	PIGMENT[c]	NITRATE REDUCTION
Corynebacterium	F and A	−	n, w, y	v
Arthrobacter	A	v[i]	w, g	v
Brevibacterium	A	−	w, g, sl. y, t	v
Microbacterium	F[k]	v[l]	y, o	v
Aureobacterium	A	v	y	v
Turicella otitidis	A	−	w	−
Dermabacter	F	−	n, w	−
Cellulomonas	F[k]	v	sl. y	+
"C. aquaticum"	A	+	y	v
Listeria monocytogenes	F	+[p]	w	−
Kurthia	A	+[p]	n, c	−
Exiguobacterium acetylicum	F	+	Golden	v

+, ≥90% of species or strains positive; −, ≥90% of species or strains negative; *v*, variable reactions; *TSI*, triple sugar iron agar; *NT*, not tested.

[a]The aerotolerent catalase-positive *Propionibacterium spp.* and *Actinomyces spp.* included in Table 56-4 should also be considered in the differential with the organisms in this table.

[b]*F*, facultative; *A*, strict aerobe.

[c]*y*, yellow; *sl.*, slightly; *t*, tan; *w*, white; *o*, orange; *g*, gray; *c*, cream; *n*, nonpigmented.

[d]CAMP test using a beta-lysin–producing strain of *Staphylococcus aureus.*

[e]Mycolic acids of various lengths are also present in the partially acid-fast *Nocardia, Gordona, Rhodococcus,* and *Tsukamurella* and the completely acid-fast *Mycobacterium* genera.

[f]*meso-DAP, meso*-diaminopimelic acid; *L-Lys,* L-lysine; *D-orn,* d-ornithine; *L-orn,* L-ornithine; *DAB,* diaminobutyric acid.

[g]Of the significant clinical *Corynebacterium* isolates, only *C. matruchotii* and *C. glucoronolyticum* are esculin-positive.

[h]Of the significant *Corynebacterium* isolates, *Corynebacterium amycolatum* does not have mycolic acid as a lipid in the cell wall, as determined by high-performance liquid chromatography (HPLC) profiling methods.

[i]Rod forms of some species are motile.

[j]Glucose may be variably oxidized, but is not fermented.

[k]Some grow poorly anaerobically

[l]Only the orange-pigmented species, *M. imperiale* and *M. arborescens,* are motile at 28° C.

[m]Reaction may be delayed.

[n]Some strains of *M. arborescens* are CAMP-positive.

[o]Glucose usually oxidized, but is not fermented.

[p]Motile at 20° to 25° C.

Data compiled from references 1-8, 10, 12-14, 16-20, 22, 24, 30, 31, 33, 34.

ESCULIN	GLUCOSE FERMENTATION	CAMP[d]	MYCOLIC ACIDS[e]	CELL WALL DIAMINO ACIDS[f]	OTHER COMMENTS:
−[g]	v	v	+[h]	*meso*-DAP	
v	−[j]	−	−	L-lys	Gelatin-positive
−	−[j]	−	−	*meso*-DAP	Gelatin-and casein-positive; cheese odor
+[m]	+	−[n]	−	L-lys	
v	−[o]	−	−	D-orn	Gelatin-and casein-variable
−	−	+	−	*meso*-DAP	Isolated from ears
+	+	−	−	*meso*-DAP	Pungent odor; decarboxylates lysine and ornithine
+	+	−	−	L-orn	Gelatin-positive, casein-negative
v	−[o]	−	−	DAB	Gelatin-and casein-negative
+	+	+	−	*meso*-DAP	Narrow zone of beta hemolysis on sheep blood agar; hippurate-positive
−	−	NT	−	L-lys	Large, "Medusa-head" colony with rhizoid growth on yeast nutrient agar; may be H$_2$S-positive in TSI butt; gelatin-negative
+	+	NT	−	L-lys	Most are oxidase-positive; casein- and gelatin-positive

TABLE 55-5   FERMENTATIVE, NONLIPOPHILIC, TINSDALE-POSITIVE CORYNEBACTERIUM SPP.*

ORGANISM	UREASE†	NITRATE REDUCTION†	ESCULIN HYDROLYSIS†	FERMENTATION OF GLYCOGEN†	REQUIRES SERUM
C. diphtheriae subsp. gravis	−	+	−	+	−
C. diphtheriae subsp. mitis	−	+	−	−	−
C. diphtheriae subsp. belfanti	−	−	−	+	−
C. diphtheriae subsp. intermedius	−	+	−	−	+
C. ulcerans	+	−	−	+	−
C. pseudotuberculosis	+	v	−	−	−

+, ≥90% of species or strains positive; −, ≥90% of species or strains are negative; v, variable reactions.
*Separation of lipophilic and nonlipophilic species can be determined by comparing growth on sheep blood agar and sheep blood agar with 1% Tween 80 or growth in brain-heart infusion broth with and without 1 drop of Tween 80 or rabbit serum.
†Reactions from API Coryne.

Data compiled from references 3, 4, 12, 19.

TABLE 55-6   FERMENTATIVE, NONLIPOPHILIC, TINSDALE-NEGATIVE CLINICALLY RELEVANT *CORYNEBACTERIUM* SPP.*,†

| ORGANISM | NITRATE REDUCTION‡ | PROPIONIC ACID§ | MOTILITY | ESCULIN HYDROLYSIS‡ | FERMENTATION OF | | | | CAMP‖ |
					MALTOSE‡	SUCROSE‡	XYLOSE‡		
C. xerosis	v	–	–	–	+	+	–		–
C striatum	+	–	–	–	–	v	–		v
C minutissimum	–	–	–	–	+	v	–		–
C. amycolatum	v	+	–	–	v	v	–		–
C. glucuronolyticum	v	+	–	v	v	+	v		+
C argentoratense	–	NT	–	–	–	–	–		–
C. matruchotii	+	NT	–	+	+	+	–		–
C. coyleae	–	NT	–	–	–	–	–		+

+, ≥90% of species or strains are positive; –, ≥90% of species or strains are negative; *v*, variable reactions; *NT*, not tested.

*Consider also *Dermabacter, Cellulomonas, Exiguobacterium,* and *Microbacterium* from Table 55-4 if the isolate is pigmented, motile, or esculin- or gelatin-positive. The aerotolerent catalase-positive *Propionibacterium* spp. and *Actinomyces* spp. included in Table 56-4 should also be considered in the differential with the organisms in this table.

†Separation of lipophilic and nonlipophilic species can be determined by comparing growth on sheep blood agar and sheep blood agar with 1% Tween 80 or growth in brain-heart infusion broth with and without one drop of Tween 80 or rabbit serum.

‡Reactions from API Coryne.

§Propionic acid as an end-product of glucose metabolism.

‖CAMP test using a beta-lysin–producing strain of *Staphylococcus aureus*.

Data compiled from references 3, 4, 11, 12, 15, 19, 27, 32.

TABLE 55-7 STRICTLY AEROBIC, NONLIPOPHILIC, NONFERMENTATIVE CLINICALLY RELEVANT *CORYNEBACTERIUM* SPP.[a,b]

ORGANISM	NITRATE REDUCTION[c]	UREASE[c]	ESCULIN HYDROLYSIS[c,d]	GELATIN[c,d]	CAMP[e]	OTHER COMMENTS
C. afermentens subsp. afermentens	−	−	−	−	v	Isolated from blood; nonadherent colony
C. auris[f]	−	−	−	−	+	Isolated from ears; dry, usually adherent colony
C. pseudodiphtheriticum	+	+	−	−	−	
C. propinquum	+	−	−	−	−	

+, ≥90% of species or strains positive; −, ≥90% of species or strains negative; v, variable reactions.

[a] *Kurthia* is also a stricly aerobic, nonlipophilic, nonfermentative organism. However, as described in Table 55-3, the colonial and cellular morphology of *Kurthia* should easily distinguish it from the organisms in this table.

[b] Separation of lipophilic and nonlipophilic species can be determined by comparing growth on sheep blood agar and sheep blood agar with 1% Tween 80 or growth in brain-heart infusion broth with and without one drop of Tween 80 or rabbit serum.

[c] Reactions from API Coryne.

[d] Consider also *Brevibacterium* and *Arthobacter* from Table 55-4 in the differential if the isolate is gelatin- or esculin-positive.

[e] CAMP test using a beta-lysin-producing strain of *Staphylococcus aureus*.

[f] For isolates from the ear, consider also *Turicella otitidis* from Table 55-4 in the differential.

Data compiled from references 3, 4, 9, 12, 16, 19, 30.

TABLE 55-8   STRICTLY AEROBIC, LIPOPHILIC, NONFERMENTATIVE CLINICALLY RELEVANT *CORYNEBACTERIUM* SPP.*

ORGANISM	NITRATE REDUCTION†	UREASE†	ESCULIN HYDROLYSIS†	OXIDATION OF	
				GLUCOSE	MALTOSE
C. jeikeium‡	−	−	−	+	v
C. afermentens subsp. lipophilum	−	−	−	−	−
C. urealyticum‡	−	+	−	−	−

+, ≥90% of species or strains positive; −, ≥90% of species or strains negative; *v*, variable reactions.

*Separation of lipophilic and nonlipophilic species can be determined by comparing growth on sheep blood agar and sheep blood agar with 1% Tween 80 or growth in brain-heart infusion broth with and without one drop of Tween 80 or rabbit serum.

†Reactions from API Coryne.

‡Isolates are usually multiply antimicrobial resistant.

Data compiled from references 3, 4, 12, 19, 28.

TABLE 55-9   LIPOPHILIC, FERMENTATIVE CLINICALLY RELEVANT *CORYNEBACTERIUM* SPP.*

ORGANISM	UREASE†	ESCULIN HYDROLYSIS†	ALKALINE PHOSPHATASE†	PYRAZINAMIDASE†
C. accolens	−	−	−	v
C. macginleyi	−	−	+	−
CDC coryneform group F-1	+	−	−	+
CDC coryneform group G	−	−	+	+

+, ≥90% of species or strains positive; −, ≥90% of species or strains negative; *v*, variable.

*Separation of lipophilic and nonlipophilic species can be determined by comparing growth on sheep blood agar and sheep blood agar with 1% Tween 80 or growth in brain-heart infusion broth with and without one drop of Tween 80 or rabbit serum.

†Reactions from API Coryne.

Data compiled from references 3, 4, 12, 19, 27.

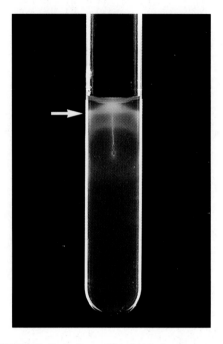

FIGURE 55-2 Umbrella motility of *Listeria monocytogenes*.

Immunization is usually initiated in infancy as part of a triple antigen vaccine (DPT) containing diphtheria toxoid, pertussis, and tetanus toxoid. Boosters are recommended every 10 years to maintain active pro-tection and are given as part of a double-antigen vaccine with tetanus toxoid.

A single dose of intramuscular penicillin or a 7- to 10-day course of oral erythromycin is recom-

TABLE 55-10 ANTIMICROBIAL THERAPY AND SUSCEPTIBILITY TESTING

ORGANISM	THERAPEUTIC OPTIONS	RESISTANCE TO THERAPEUTIC OPTIONS	VALIDATED TESTING METHODS*
Listeria monocytogenes	Ampicillin, or penicillin, with or without an aminoglycoside	Not to commonly used agents	Yes, but testing is rarely, if ever, needed to guide therapy
Corynebacterium diph-theriae	Antitoxin to neutralize diphtheria toxin plus penicillin or ery-thromycin to eradicate organism	Not to recommended agents	Not available; testing is not needed to guide therapy
Other *Corynebacterium* spp.	No definitive guidelines. All are susceptible to vancomycin	Resistance to penicillins and cephalosporins does occur	Not available
Kurthia spp., *Brevibacterium* spp., *Dermabacter* sp., *Arthrobacter* spp., *Microbacterium* spp., *Aureobacterium* spp., *Cellulomonas* spp., and *Exiguobacterium* sp.	No definitive guidelines	Unknown	Not available

*Validated testing methods include those standard methods recommended by the National Committee for Clinical Laboratory Standards (NCCLS) and those commercial methods approved by the Food and Drug Administration (FDA).

mended for all individuals exposed to diphtheria, regardless of their immunization status. Follow-up throat cultures from individuals taking prophylaxis should be obtained at least 2 weeks after therapy; if they still harbor *C. diphtheriae,* an additional 10-day course of oral erythromycin should be given. Previously immunized contacts should receive a booster dose of diphtheria toxoid; nonimmunized contacts should begin the primary series of immunizations.

References

1. Barreau, C., Bimet, F., Kiredjian, M., et al. 1993. Comparative chemotaxonomic studies of mycolic acid-free coryneform bacteria of human origin. J. Clin. Microbiol. 31:2085.

2. Bernard, K., Bellefeuille, M., Hollis, D.G., et al. 1994. Cellular fatty acid composition and phenotypic and cultural characterization of CDC fermentative coryneform groups 3 and 5. J. Clin. Microbiol. 32:1217.

3. Clarridge, J.E. and Spiegel, C.A. 1995. *Corynebacterium* and miscellaneous gram-positive rods, *Erysipelothrix* and *Gardnerella.* In Murray, P.R., Baron, E.J., Pfaller, M.A., et al., editors. Manual of clinical microbiology, ed 6. American Society for Microbiology, Washington, D.C.

4. Coyle, M.B. and Lipsky, B.A. 1990. Coryneform bacteria in infectious diseases: clinical and laboratory aspects. Clin. Microbiol. Rev. 3:227.

5. de Briel, D., Coudere, F., Riegel, P., et al. 1992. High-performance liquid chromatography of corynomycolic acids as a tool in identification of *Corynebacterium* species and related organisms. J. Clin. Microbiol. 30:1407.

6. Farrow, J.A.E., Wallbanks, S., and Collins, M.D. 1994. Phylogenetic interrelationships of round-spore-forming bacilli containing cell walls based on lysine and the non-spore-forming genera *Caryophanon, Exiguobacterium, Kurthia,* and *Planococcus.* Int. J. Syst. Bacteriol. 44:74. (Erratum, 44:377).

7. Funke, G. and Carlotti, A., 1994. Differentiation of *Brevibacterium* spp. encountered in clinical specimens. J. Clin. Microbiol. 32:1729.

8. Funke, G., Falsen, E., and Barreau, C., 1995. Primary identification of *Microbacterium* spp. encountered in clinical specimens as CDC coryneform group A-4 and A-5 bacteria. J. Clin. Microbiol. 33:188.

9. Funke, G., Lawson, P.A., and Collins, M.D. 1995. Heterogeneity within Centers for Disease Control and Prevention coryneform group ANF-1-like bacteria and description of *Corynebacterium auris* sp. nov. Int. J. Syst. Bacteriol. 45:735.

10. Funke, G., Pascual Ramos, C., and Collins, M.D. 1995. Identification of some clinical strains of CDC coryneform group A-3 and group A-4 bacteria as *Cellulomonas* species and proposal of *Cellulomonas hominis* sp. nov. for some group A-3 strains. J. Clin. Microbiol. 33:2091.

11. Funke, G., Pascual Ramos, C., and Collins, M.D. 1997. *Corynebacterium coyleae* sp. nov., isolated from human clinical specimens. Int. J. Syst. Bacteriol. 47:92.

12. Funke, G., von Graevenitz, A., Clarridge, J.E., et al. 1997. Clinical microbiology of coryneform bacteria. Clin. Microbiol. Rev. 10:125.

13. Funke, G., von Graevenitz, A., and Weiss, N., 1994. Primary identification of *Aureobacterium* spp. isolated from clinical specimens as "*Corynebacterium aquaticum.*" J. Clin. Microbiol. 32:2686.

14. Funke, G., Hutson, R.A., Bernard, K.A., et al. 1996. Isolation of *Arthrobacter* spp. from clinical specimens and description of *Arthrobacter cumminsii* sp. nov. and *Arthrobacter woluwensis* sp. nov. J. Clin. Microbiol. 34:2356.

15. Funke, G., Lawson, P.A., Bernard, K.A., et al. 1996. Most *Corynebacterium xerosis* strains identified in the routine clinical laboratory correspond to *Corynebacterium amycolatum.* J. Clin. Microbiol. 34:1124.

16. Funke, G., Stubbs, S., Altwegg, M., et al. 1994. *Turicella otitidis* gen. nov., sp. nov., a coryneform bacterium isolated from patients with otitis media. Int. J. Syst. Bacteriol. 44:270.

17. Funke, G., Stubbs, S., Pfyffer, G.E., et al. 1994. Characteristics of CDC group 3 and group 5 coryneform bacteria isolated from clinical specimens and assignment to the genus *Dermabacter.* J. Clin. Microbiol. 32:1223.

18. Gruner, E., Steigerwalt, A.G., Hollis, D.G., et al. 1994. Human infections caused by *Brevibacterium casei,* formerly CDC groups B-1 and B-3. J. Clin. Microbiol. 32:1511.

19. Hollis, D.G. and Weaver, R.E. 1981. Gram-positive organisms: a guide to identification. Special Bacteriology Section, Centers for Disease Control, Atlanta.

20. Jones, D. and Collins, M.D. 1986. Irregular, nonsporing gram-positive rods. In Sneath, P.H.A., Mair, N.S., and Sharpe, M.E., editors. Bergey's manual of systematic bacteriology, vol 2. Williams & Wilkins, Baltimore.

21. Jones, E.M. and MacGowan, A.P. 1995. Antimicrobial chemotherapy of human infection due to *Listeria monocytogenes.* Eur. J. Clin. Microbiol. Infect. Dis. 14:165.

22. Kandler, D. and Weiss, N. 1996. Regular, nonsporing, gram-positive rods. In Sneath, P.H.A., Mair, N.S., and Sharpe, M.E., editors. Bergey's manual of systematic bacteriology, vol 2. Williams & Wilkins, Baltimore.

23. Lorber, B. 1997. Listeriosis. Clin. Infect. Dis. 24:1.

24. McNeil, M.M. and Brown, J.M. 1994. The medically important aerobic actinomycetes: epidemiology and microbiology. Clin. Microbiol. Rev. 7:357.

25. Popovic, T., Wharton, M., Wenger, J.D., et al. 1995. Are we ready for diphtheria? A report from the diphtheria diagnostic workshop, Atlanta, July 11-12, 1994. J. Infect. Dis. 171:765.

26. Riegel, P., Ruimy, R., de Briel, D., et al. 1995. *Corynebacterium argentoratense* sp. nov., from the human throat. Int. J. Syst. Bacteriol. 45:533.

27. Riegel, P., Ruimy, R., deBriel, D., et al. 1995. Genomic diversity and phylogenetic relationships among lipid-requiring diphtheroids from humans and characterization of *Corynebacterium macginleyi* sp. nov. Int. J. Syst. Bacteriol. 45:128.

28. Riegel, P., de Briel, D., Prévost, G., et al. 1994. Genomic diversity among *Corynebacterium jeikeium* strains and com-

parison with biochemical characteristics. J. Clin. Microbiol. 32:1860.

29. Soriano, F., Zabardiel, J., and Nieto, E. 1995. Antimicrobial susceptibilities of *Corynebacterium* species and other non-spore-forming gram-positive bacilli to 18 antimicrobial agents. Antimicrob. Agents Chemother. 39:208.

30. Simonet, M., deBriel, D., Boucot, I., et al. 1993. Coryneform bacteria isolated from middle ear fluid. J. Clin. Microbiol. 31:1667.

31. Swaminathan, B., Rocourt, J., and Billie, J. 1995. *Listeria.* In Murray, P.R., Baron, E.J., Pfaller, M.A., et al., editors. Manual of clinical microbiology, ed 6. American Society for Microbiology, Washington, D.C.

32. Wauters, G., Driessen, A., Ageron, E., et al. 1996. Propionic acid-producing strains previously designated as *Corynebacterium xerosis, C. minutissimum, C. striatum,* and CDC group I2 and group F2 coryneforms belonging to the species *Corynebacterium amycolatum.* Int. J. Syst. Bacteriol. 46:653.

33. Yokota, A., Takeuchi, M., Sakane, T., and Weiss, N. 1993. Proposal of six new species in the genus *Aureobacterium* and transfer of *Flavobacterium esteraromaticum* Omelianski to the genus *Aureobacterium* as *Aureobacterium esteraromaticum* comb. nov. Int. J. Syst. Bacteriol. 43:555.

34. Yokota, A., Takeuchi, M., and Weiss, N. 1993. Proposal of two new species in the genus *Microbacterium: Microbacterium dextranolyticum* sp. nov. and *Microbacterium aurum* sp. nov. Int. J. Syst. Bacteriol. 43:549.

Bibliography

Armstrong, D. 1995. *Listeria monocytogenes.* In Mandell, G.L., Bennett, J.E., and Dolin, R., editors. Principles and practice of infectious diseases. Churchill Livingstone, New York.

Benenson, A.S. 1995. Control of communicable diseases manual. American Public Health Association, Washington, D.C.

Brown, A.E. 1995. Other Corynebacteria and *Rhodococcus.* In Mandell, G.L., Bennett, J.E., and Dolin, R., editors. Principles and practice of infectious diseases. Churchill Livingstone, New York.

Committee on Infectious Diseases. 1997. 1997 Red Book: report of the Committee on Infectious Diseases, ed 24. American Academy of Pediatrics, Elk Grove Village, Ill.

MacGregor, R.R. 1995. *Corynebacterium diphtheriae.* In Mandell, G.L., Bennett, J.E., and Dolin, R., editors. Principles and practice of infectious diseases. Churchill Livingstone, New York.

Ryan, K.J. and Falkow, S. 1994. Corynebacteria, *Listeria, Bacillus,* and other aerobic and faculative gram-positive rods. In Ryan, K.J., editor. Sherris medical microbiology: an introduction to infectious diseases. Appleton and Lange, Norwalk, Conn.

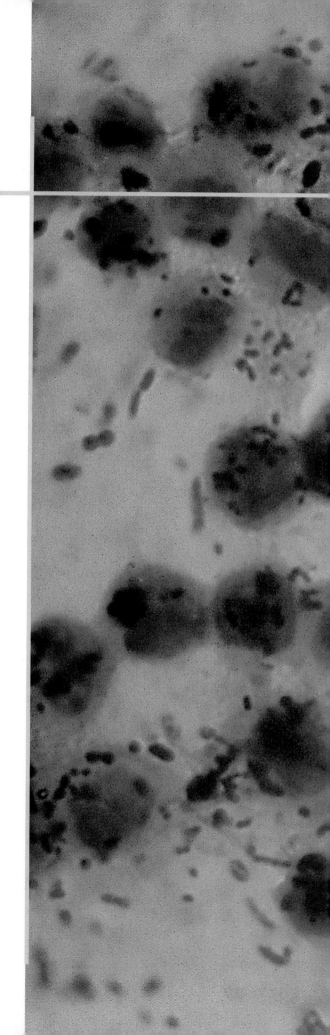

Section Ten

10

Nonbranching, Catalase-Negative, Gram-Positive Bacilli

CHAPTER

56 | ERYSIPELOTHRIX, LACTOBACILLUS, AND SIMILAR ORGANISMS

Genera and Species to be Considered

- *Erysipelothrix rhusiopathiae*
- *Arcanobacterium* spp.
- *Gardnerella vaginalis*
- *Lactobacillus* spp.

GENERAL CONSIDERATIONS

The genera described in this chapter are all catalase-negative, non–spore-forming, gram-positive rods; some may exhibit rudimentary branching.

EPIDEMIOLOGY, PATHOGENESIS, AND SPECTRUM OF DISEASES

The organisms listed in Table 56-1 include those that are closely associated with animals and are contracted by humans through animal exposure (e.g., *Ery-sipelothrix rhusiopathiae* and *Arcanobacterium pyogenes*) and those that are part of the normal human flora (e.g., *Lactobacillus* spp. and *Gardnerella vaginalis*).

These species are not commonly encountered in infections in humans, with the possible exception of *G. vaginalis*, and when they are encountered the infections are rarely life-threatening (Table 56-2). Often the primary challenge is to determine the clinical relevance of these organisms when they are found in specimens from normally sterile sites.[4,12,13,15-17]

TABLE 56-1 EPIDEMIOLOGY

SPECIES	HABITAT (RESERVOIR)	MODE OF TRANSMISSION
Erysipelothrix rhusiopathiae	Carried by and causes disease in animals. Not part of normal human flora	Abrasion or puncture wound of skin with animal exposure. Most often associated with persons who handle animals or animal products
Arcanobacterium haemolyticum	Normal flora of human skin and pharynx	Uncertain; infections probably caused by person's endogenous strains
Arcanobacterium pyogenes	Carried by and causes disease in animals. Not part of normal human flora	Uncertain; probably by abrasion or undetected wound during exposure to animals
Gardnerella vaginalis	Normal vaginal flora of humans; may also colonize distal urethra of males	Infections probably caused by person's endogenous strain
Lactobacillus spp.	Widely distributed in foods and nature. Normal flora of human mouth, gastrointestinal tract, and female genital tract	Infections are rare; when they do occur they probably are caused by person's endogenous strain

TABLE 56-2   PATHOGENESIS AND SPECTRUM OF DISEASES

ORGANISMS	VIRULENCE FACTORS	SPECTRUM OF DISEASES AND INFECTIONS
Erysipelothrix rhusiopathiae	Uncertain; certain enzymes may promote virulence	Erysipeloid; a localized skin infection that is painful and may spread slowly; may cause diffuse skin infection with systemic symptoms; bacteremia may occur, but endocarditis is rare
Arcanobacterium haemolyticum	Unknown; probably of low virulence	Pharyngitis, cellulitis, and other skin infections
Arcanobacterium pyogenes	Unknown; probably of low virulence to humans	Rarely associated with human infection; when infections do occur, they generally are cutaneous and may be complicated by bacteremia
Gardnerella vaginalis	Uncertain; produces cell adherence factors and cytotoxin	Bacterial vaginosis; less commonly associated with urinary tract infections; bacteremia is extremely rare
Lactobacillus spp.	Uncertain; probably of low virulence	Most frequently encountered as contaminant. May cause bacteremia in immunocompromised patients

LABORATORY DIAGNOSIS

SPECIMEN COLLECTION AND TRANSPORT

Generally, no special considerations are required for specimen collection and transport of the organisms discussed in this chapter. Of note, skin lesions for *Erysipelothrix* should be collected by biopsy of the full thickness of skin at the leading edge of the discolored area. Refer to Table 1-1 for other general information on specimen collection and transport.

SPECIMEN PROCESSING

No special considerations are required for processing of the organisms discussed in this chapter. Refer to Table 1-1 for general information on specimen processing.

DIRECT DETECTION METHODS

Other than Gram stain of patient specimens, there are no specific procedures for the direct detection of these organisms in clinical material. Gram stain of *Arcanobacterium* spp. shows delicate, curved, gram-positive rods with pointed ends and occasional rudimentary branching. This branching is more pronounced after these organisms have been cultured anaerobically. *Acanobacterium* spp. stain unevenly after 48 hours of growth on solid media and also exhibit coccal forms.

Lactobacillus is highly pleomorphic, occurring in long chaining rods, as well as coccobacilli and spiral forms (Figure 56-1).

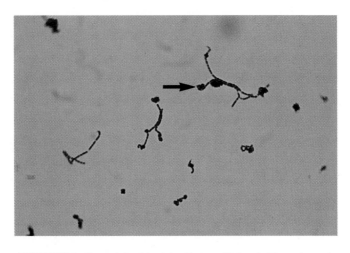

FIGURE 56-1 Gram stain of *Lactobacillus* spp. Note spiral forms (*arrow*).

E. rhusiopathiae stains as both short rods and long filaments. These morphologies correspond to two colonial types, that is, rough colonies that contain slender, filamentous gram-positive rods with a tendency to overdecolorize and become gram-negative and smooth colonies that contain small, slender rods. This variability in staining and colonial morphology may be mistaken for a polymicrobic infection on both direct examination or culture.

Gardnerella are small, pleomorphic gram-variable or gram-negative coccobacilli and short rods. The wet mount and Gram stain of vaginal secretions are key tests for diagnosing bacterial vaginosis caused by *G. vaginalis*. A wet mount prepared in saline reveals the characteristic "clue cells," which are large, squamous epithelial cells with numerous attached small rods. A Gram-stained smear of the discharge shows the attached organisms to be gram-variable coccobacilli. Clue cells typically are present, and large numbers of gram-positive rods (i.e., lactobacilli) representing normal vaginal flora are absent or few in number.

CULTIVATION

Media of choice

All the genera described in this chapter grow on 5% sheep blood and chocolate agars. They will not grow on MacConkey agar but will grow on Columbia colistin-nalidixic acid (CNA) agar. All genera, except *Gardnerella*, will grow in commercially available blood culture broths. *Gardnerella* is inhibited by sodium polyanetholsulfonate (SPS), which currently is used as an anticoagulant in most commercial blood culture media.

Isolation of *G. vaginalis* from female genital tract specimens is best accomplished using the selective medium, human blood bilayer Tween agar (HBT).

Incubation conditions and duration

Detectable growth of these organisms should occur on 5% sheep blood and chocolate agars, CNA, and HBT incubated at 35° C in 5% to 10% carbon dioxide (CO_2) within 48 hours of inoculation.

Colonial appearance

Table 56-3 describes the colonial appearance and other distinguishing characteristics (e.g., hemolysis) of each genus on sheep blood agar. *G. vaginalis* produces small, gray, opaque colonies surrounded by a diffuse zone of beta hemolysis on HBT agar (Figure 56-2).

APPROACH TO IDENTIFICATION

The identification of the four genera described in this chapter must be considered along with that of *Actinomyces*, *Bifidobacterium*, and *Propionibacterium*, which are discussed in Chapter 59. Although the latter genera are usually considered with the anaerobic bacteria, they will grow on routine laboratory media in 5% to 10% CO_2 and some are also catalase negative. Therefore, as shown in Table 56-4, these organisms must be considered together when a laboratory encounters catalase-negative, gram-positive, non–spore-forming rods.

Several commercial systems for fastidious gram-negative bacterial identifications will adequately identify *Gardnerella*. The HNID panel (*Haemophilus-Neisseria* identification panel, Dade Behring Micro-Scan, West Sacramento, Calif.) works particularly well. However, rapid identification panels usually only are used for isolates from extragenital sources (e.g., blood).

TABLE 56-3 COLONY APPEARANCE ON 5% SHEEP BLOOD AGAR AND OTHER CHARACTERISTICS

ORGANISM	APPEARANCE
Arcanobacterium spp.	Small colonies with various appearances, including smooth, mucoid, and white and dry, friable, and gray; may be surrounded by narrow zone of beta-hemolysis
Erysipelothrix rhusiopathiae	Two colony types, that is, large and rough or small, smooth, and translucent; shows alpha hemolysis after prolonged incubation
Gardnerella vaginalis	Pinpoint; nonhemolytic
Lactobacillus spp.	Multiple colonial morphologies, ranging from pinpoint, alpha-hemolytic colonies resembling streptococci to rough, gray colonies

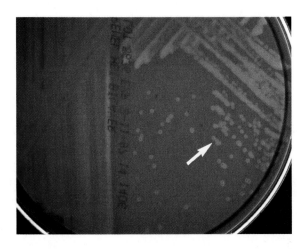

FIGURE 56-2 *Gardnerella vaginalis* on human blood bilayer Tween (HBT) agar. Note small colonies with diffuse zone of beta hemolysis (*arrow*).

Comments regarding specific organisms

A presumptive identification of *G. vaginalis* is sufficient for genital isolates, based on typical appearance on Gram stain, beta hemolysis on HBT agar, and negative tests for oxidase and catalase.

The beta-hemolytic *Arcanobacterium* resemble the beta-hemolytic streptococci but can be differentiated from them by Gram stain morphology. *A. haemolyticum* and *A. pyogenes* can be differentiated based on liquefaction of gelatin; *A. pyogenes* is positive and *A. haemolyticum* is negative. *A. bernardiae* is nonhemolytic.

Erysipelothrix is the only catalase-negative, gram-positive non–spore-forming rod that produces hydrogen sulfide (H_2S) when inoculated into triple sugar iron (TSI) agar. Some *Bacillus* spp. also blacken the butt of TSI, but they are catalase-positive and produce spores.

Lactobacillus spp. are usually identified based on colony and Gram stain morphologies and catalase reaction (negative). Differentiation from viridans streptococci may be difficult, but the formation of chains of rods rather than cocci in thioglycollate broth is helpful. Alternatively, a Gram stain of growth just outside the zone of inhibition surrounding a 10-U penicillin disk placed on a blood agar plate inoculated with a lawn of the organism should show long bacilli rather than coccoid forms if the organism is *Lactobacillus* spp.

SERODIAGNOSIS

Serodiagnostic techniques are not generally used for the laboratory diagnosis of infections caused by the organisms discussed in this chapter.

ANTIMICROBIAL SUSCEPTIBILITY TESTING AND THERAPY

The rarity with which most of these organisms are encountered as the cause of infection and the lack of validated in vitro susceptibility testing methods do not allow definitive testing guidelines to be given (Table 56-5). However, most of the organisms are susceptible to the agents used to eradicate them so that in vitro testing is not necessary for guiding therapy.[1,11] *Lactobacillus* spp. can be resistant to various antimicrobial agents. Fortunately, these organisms are rarely implicated in infections. When they are encountered in specimens from normally sterile sites, careful evaluation of their clinical significance is warranted before any attempt is made at performing a nonstandardized susceptibility test.

Although some of these organisms may grow on the media and under the conditions recommended for testing other bacteria (see Chapter 18 for more information regarding validated testing methods), this does not necessarily mean that interpretable and reliable results will be produced. Chapter 19 should be reviewed for preferable strategies that can be used to provide susceptibility information when validated testing methods do not exist for a clinically important bacterial isolate.

PREVENTION

Because these organisms are ubiquitous in nature and many are part of normal human flora that are commonly encountered without deleterious effects on healthy human hosts, there are no recommended vaccination or prophylaxis protocols.

TABLE 56-4 **BIOCHEMICAL AND PHYSIOLOGIC CHARACTERISTICS OF CATALASE-NEGATIVE, GRAM-POSITIVE, AEROTOLERANT, NON–SPORE-FORMING RODS**

	CATALASE	UREASE	NITRATE REDUCTION	β-HEMOLYSIS[b]	FERMENTATION[a] OF: GLUCOSE	MALTOSE
Actinomyces israelii	−	−	v	−	+	+
A. odontolyticus	−	−	+	−[e]	+	v
A. naeslundii	v	+	+	−	+	+
A. viscosus	+	−	+	−	+	+
A. neuii subsp. *neuii*	+	−	+	−	+	+
A. neuii subsp. *anitratus*	+	−	−	−	+	+
Arcanobacterium haemolyticum	−	−	−	+	+	+
A. pyogenes	−	−	−	+[g]	+	v
A. bernardiae	−	−	−	−	+	+
Bifidobacterium adolescentis	−	−	−	−	+	+
Erysipelothrix sp.	−	−	−	−	+[h]	−
Lactobacillus spp.	−	−	−	−	+	+
Propionibacterium acnes	+[i]	−	+	−	+	−
P. propionicum[j]	−	−	+	−	+	+
P. granulosum	+	−	−	−	+	+
P. avidum	+	−	−	+	+	+
Gardnerella vaginalis	−	−	−	−	+	+

+, ≥ 90% of strains positive; −, ≥ 90% of strains negative; *v*, variable; *NT*, not tested; *SBA*, 5% sheep blood agar; *TSI*, triple sugar iron agar; *HBT*, human blood bilayer Tween agar.

[a]Fermentation is detected in peptone base with Andrade's indicator.

[b]On sheep blood agar.

[c]CAMP test using a beta-lysin–producing strain of *Staphylococcus aureus*.

[d]End products of glucose metabolism: *A*, acetic acid; *L*, lactic acid; *S*, succinic acid; *P*, propionic acid; *( )*, may or may not produce acid end product.

[e]May show beta hemolysis on brain-heart infusion agar with sheep or human blood.

[f]Reverse CAMP test; *S. aureus* beta-lysins are inhibited by a diffusible substance produced by *A. haemolyticum*.

[g]May also show beta hemolysis on brain-heart infusion agar with human blood.

[h]Reaction may be weak or delayed.

[i]Some strains are catalase-negative.

[j]Formerly *Arachnia propionica*.

[k]*Gardenerella vaginalis*-like organisms ferment xylose.

Data compiled from references 2, 3, 5-10, 14, 18.

FERMENTATION[a] OF:					
MANNITOL	SUCROSE	XYLOSE	CAMP[c]	GLC[d]	OTHER COMMENTS:
v	+	+	−	ALS	
−	+	v	−	AS	Red pigment produced after 1 week on SBA
v	+	v	−	ALS	
−	+	v	−	ALS	
+	+	+	+	ALS	
+	+	+	+	ALS	
−	v	−	Reverse +[f]	ALS	Gelatin-negative at 48 hrs; beta hemolysis is stronger on agar containing human or rabbit blood
v	v	+	−	ALS	Gelatin-positive at 48 hrs; casein-positive
−	−	−	−	ALS	
−	+	+	NT	A>L (s)	
−	−	−		ALS	H$_2$S-positive in TSI butt
v	+	NT	NT	L (a s)	
−	−	−	+	AP (iv L s)	Indole-positive; may show beta hemolysis on rabbit blood agar
+	+	−	NT	AP S(L)	Colony may show red fluorescence under long-wavelength UV light
+	+	−	v	AP (iv s)	
−	+	v	v	AP (iv s)	
−	v	−[k]	NT	A (l s)	Beta hemolysis on HBT; usually hydrolyses hippurate

TABLE 56-5 ANTIMICROBIAL THERAPY AND SUSCEPTIBILITY TESTING

ORGANISM	THERAPEUTIC OPTIONS	RESISTANCE TO THERAPEUTIC OPTIONS	VALIDATED TESTING METHODS*	COMMENTS
Erysipelothrix rhusiopathiae	Susceptible to penicillins, cephalosporins, erythromycin, clindamycin, tetracycline, and ciprofloxacin	Not common	Not available	Susceptibility testing not needed to guide therapy
Arcanobacterium haemolyticum	No definitive guidelines. Usually susceptible to penicillin, erythromycin, and clindamycin	Not known	Not available	Susceptibility testing not needed to guide therapy
Arcanobacterium pyogenes	No definitive guidelines. Usually susceptible to cephalosporins, penicillins, ciprofloxacin, and chloramphenicol	Not known	Not available	Susceptibility testing not needed to guide therapy
Gardnerella vaginalis	Metronidazole is the drug of choice; also susceptible to ampicillin	Not known	Not available	Susceptibility testing not needed to guide therapy
Lactobacillus spp.	No definitive guidelines. Systemic infections may require use of a penicillin with an aminoglycoside	Frequently resistant to cephalosporins; not killed by penicillin alone; frequently highly resistant to vancomycin	Not available	Confirm that isolate is clinically relevant and not a contaminant

*Validated testing methods include those standard methods recommended by the National Committee for Clinical Laboratory Standards (NCCLS) and those commercial methods approved by the Food and Drug Administration (FDA).

References

1. Carlson, P., Kontiainen, S., and Renkonen, O. 1994. Antimicrobial susceptibility of *Arcanobacterium haemolyticum.* Antimicrob. Agents Chemother. 38:142.

2. Clarridge, J.E. and Spiegel, C.A. 1995. *Corynebacterium* and miscellaneous gram-positive rods, *Erysipelothrix* and *Gardnerella.* In Murray, P.R., Baron, E.J., Pfaller, M.A., et al., editors. Manual of clinical microbiology, ed 6. American Society for Microbiology, Washington, D.C.

3. Coyle, M.B. and Lipsky, B.A. 1990. Coryneform bacteria in infectious disease: clinical and laboratory aspects. Clin. Microbiol. Rev. 3:227.

4. Drancourt, M., Oules, O., Bouche, V., et al. 1993. Two cases of *Actinomyces pyogenes* infections in humans. Eur. J. Clin. Microbiol. Infect. Dis.12:55.

5. Funke, G., von Graevenitz, A., Clarridge, J.E., et al. 1997. Clinical microbiology of coryneform bacteria. Clin. Microbiol. Rev. 10:125.

6. Funke, G., Pascual Ramos, C., Fernandez-Garayzabal, J., et al. 1995. Description of human-derived Centers for Disease Control coryneform group 2 bacteria as *Actinomyces bernardiae* sp. nov. Int. J. Syst. Bacteriol. 45:57.

7. Funke, G., Martinetti Lucchini, G., Pfyffer, G.E., et al. 1993. Characteristics of CDC group 1 and group 1-like coryneform bacteria isolated from clinical specimens. J. Clin. Microbiol. 31:2907.

8. Hollis, D.G. and Weaver, R.E. 1981. Gram-positive organisms: a guide to identification. Special Bacteriology Section. Centers for Disease Control, Atlanta.

9. Jones, D. and Collins, M.D. 1986. Irregular, non-sporeforming, gram-positive rods. In Sneath, P.H.A., Mair, N.S., and Sharpe, E., editors. Bergey's manual of systematic bacteriology, vol 2. Williams & Wilkins, Baltimore.

10. Kandler, O. and Weiss, N. 1986. Regular, nonsporing gram-positive rods. In Sneath, P.H.A., Mair, N.S., and Sharpe, E., editors. Bergey's manual of systemic bacteriology, vol 2. Williams & Wilkins, Baltimore.

11. Kharsany, A.B.M., Hoosen, A.A., and Ende, J.V.D. 1993. Antimicrobial susceptibilities of *Gardnerella vaginalis.* Antimicrob. Agents Chemother. 37:2733.

12. Lidbeck, A. and Nord, C.E. 1993. Lactobacilli and the normal human anaerobic microflora. Clin. Infect. Dis. 16(suppl. 4):S181.

13. Mackenzie, A., Fuite, L.A., Chan, T.H., et al. 1995. Incidence and pathogenicity of *Arcanobacterium haemolyticum* during a 2-year study in Ottawa Clin. Infect. Dis. 21:177.

14. Pascual Ramos, C., Foster, G., and Collins, M.D. 1997. Phylogenetic analysis of the genus *Actinomyces* based on 16S rRNA gene sequences: description of *Arcanobacterium phocae* sp. nov., *Arcanobacterium bernardiae* comb. nov., and *Arcanobacterium pyogenes* comb. nov. Int. J. Syst. Bacteriol. 47:46.

15. Patel, R., Cockerill, F.R., Porayko, M.K., et al. 1994. Lactobacillemia in liver transplant patients. Clin. Infect. Dis. 18:207.

16. Schuster, M.G., Brennan, P.J., and Edelstein, P. 1993. Persistant bacteremia with *Erysipelothrix rhusiopathiae* in a hospitalized patient. Clin. Infect. Dis. 17:783.

17. Spiegel, C.A. 1991. Bacterial vaginosis. Clin. Microbiol. Rev. 4:485.

18. van Esbroeck, M., Vandamme, P., Falsen, E., et al. 1996. Polyphasic approach to the classification and identification of *Gardnerella vaginalis* and unidentified *Gardnerella*-vaginalis-like coryneforms present in bacterial vaginosis. Int. J. Syst. Bacteriol. 46:675.

Bibliography

Reboli, A.C. and Farrar, W.E. 1995. *Erysipelothrix rhusiopathiae.* In Mandell, G.L., Bennett, J.E., and Dolin, R., editors. Principles and practice of infectious diseases. Churchill Livingstone, New York.

Spiegel, C.A. 1995. *Gardnerella vaginalis* and *Mobiluncus.* In Mandell, G.L., Bennett, J.E., and Dolin, R., editors. Principles and practice of infectious diseases. Churchill Livingstone, New York.

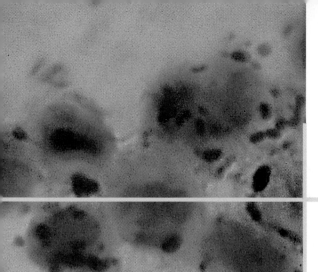

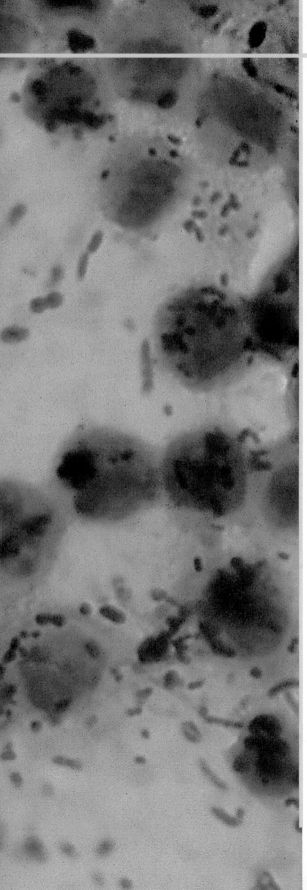

Section Eleven

11

Branching or Partially Acid-Fast, Gram-Positive Bacilli

NOCARDIA, STREPTOMYCES, RHODOCOCCUS, OERSKOVIA, AND SIMILAR ORGANISMS

Genera and Species to be Considered

- *Nocardia asteroides*
- *Nocardia nova*
- *Nocardia farcinica*
- *Nocardia brasiliensis*
- *Nocardia otitidiscaviarum*
- *Nocardia transvalensis*
- *Nocardia brevicatena*
- *Nocardia carnea*

- *Rhodococcus* spp.
- *Gordona* spp.
- *Tsukamurella* spp.
- *Streptomyces somaliensis*
- *Streptomyces paraguayensis*
- *Streptomyces anulatus*
- *Actinomadura madurae*
- *Actinomadura pelletieri*

- *Dermatophilus congolensis*
- *Nocardiopsis dassonvillei*
- *Oerskovia turbata*
- *Oerskovia xanthineolytica*
- *Rothia dentocariosa*
- Thermophilic actinomycetes

The actinomycetes are a large and diverse group of gram-positive bacilli. The cells of all actinomycetes elongate to form branching, filamentous forms. The rate and extent of filament (hyphae) elongation with lateral branching is dependent on the strain of actinomycetes, the growth medium, and the temperature of incubation.[3] Some organisms form filaments, or hyphae, on the agar surface or into the agar, whereas others extend into the air.

These organisms are either aerobic, facultatively anaerobic, or obligately anaerobic; only the aerobic actinomycetes and two facultatively anaerobic actinomycetes, *Oerskovia* and *Rothia,* are discussed in this chapter. Aerobic actinomycetes belong to the order *Actinomycetales.* There are more than 40 genera but only the clinically relevant aerobic actinomycetes genera are considered here (Table 57-1). In this chapter only those aerobic actinomycetes that can exhibit branching and/or partial acid-fastness are addressed. Although both belong to the order *Actinomycetales, Corynebacterium* spp. do not usually exhibit branching filaments nor partial acid-fastness and *Mycobacterium* spp. do not exhibit branching and are strongly (acid-alcohol) acid-fast; for these reasons, the *Corynebacteriaceae* and *Mycobacteriaceae* are addressed in Chapters 55 and 60, respectively. For purposes of discussion, the remaining genera of aerobic actinomycetes are divided into the following two large groups: those whose cell walls contain mycolic acid and are therefore partially acid-fast and those whose cell envelopes do not contain mycolic acid and therefore are non–acid-fast.

In general, the aerobic actinomycetes are not frequently isolated in the clinical laboratory; nevertheless, these organisms are causes of serious human disease. Not only are infections caused by these organisms difficult to recognize on clinical grounds, but also they are difficult to isolate. Further complicating matters from the laboratory's perspective is the difficulty in classifying, identifying, and performing antibiotic susceptibilities on aerobic actinomycetes isolated from clinical specimens. At the time of this writing, the taxonomy of the aerobic actinomycetes is complex and evolving. New and reliable methods that

TABLE 57-1 CLINICALLY RELEVANT AEROBIC ACTINOMYCETES

CELL WALL CONTAINING MYCOLIC ACID	ORGANISM
Present	*Nocardia* *Rhodococcus* *Gordona* *Tsukamurella* *Corynebacterium*
Absent	*Streptomyces* *Actinomadura* *Dermatophilus* *Nocardiopsis* *Oerskovia* *Rothia*

can identify cell wall amino acids and sugars and characterize mycolic acid, menaquinones, and phospholipids in conjunction with nucleic acid phylogenetic studies are proving extremely useful for resolving the taxonomy of the actinomycetes.

GENERAL CHARACTERISTICS

The genera *Nocardia, Rhodococcus, Gordona,* and *Tsukamurella* are partially acid-fast aerobic actinomycetes that belong to the family *Nocardiaciae.* However, the presence of this property is variable, being dependent on the particular strain and culture conditions.

PARTIALLY ACID-FAST AEROBIC ACTINOMYCETES

Nocardia

Organisms belonging to the genus *Nocardia* are gram-positive (often with a beaded appearance), variably acid-fast, catalase-positive, and strictly aerobic. As they grow, *Nocardia* spp. form branched filaments that extend along the agar surface, that is **substrate hyphae,** and into the air, that is, **aerial hyphae.** As the organisms age, nocardiae fragment into pleomorphic rods or coccoid elements.[14] Nocardiae also are characterized by the presence of mesodiaminopimelic acid (DAP) and the sugars arabinose and galactose in their cell wall peptidoglycan.

Currently, 11 validly described species are included in the genus *Nocardia;*[6] the species that are considered human pathogens or have been implicated as human pathogens are listed in Box 57-1. *N. asteroides, N. nova, N. farcinica, N. brasiliensis, N. otitidiscaviarum* (formerly *N. caviae*), and *N. trans-*

BOX 57-1

NOCARDIA SPP. THAT ARE CONSIDERED HUMAN PATHOGENS OR HAVE BEEN IMPLICATED IN HUMAN DISEASE

Nocardia asteroides
N. nova
N. farcinica
N. brasiliensis
N. otitidiscaviarum
N. transvalensis
N. brevicatena
N. carnea

valensis account for most of the diseases in humans that are caused by *Nocardia* spp.

Rhodococcus, Gordona, Tsukamurella

Organisms belonging to *Rhodococcus, Gordona,* and *Tsukamurella* are similar to *Nocardia* spp. in that they are gram-positive, aerobic, catalase-positive, partially acid-fast, branching, filamentous bacteria that can fragment into rods and cocci. The extent of acid-fastness is dependent on the amount and complexity of mycolic acids in the organism's cell envelope and on culture conditions. The differentiation of these three genera, as well as species identification, is difficult.[3] In particular, the genus *Rhodococcus* consists of a very diverse group of organisms in terms of morphology, biochemical characteristics, and ability to cause disease. As previously mentioned, the taxonomy of these organisms continues to evolve; species included in these three genera, as of this writing, are summarized in Table 57-2.

TABLE 57-2 **SPECIES INCLUDED IN THE GENERA *RHODOCOCCUS, GORDONA,* AND *TSUKAMURELLA***

GENUS	SPECIES
Rhodococcus	*chlorophenicolus, chubuensis, coprophilus, equi, erythropolis, fascians, globerulus, luteus, marinonascens, maris, obuensis, rhodnii, rhodochrous, roseus, ruber*
Gordona	*amarae, bronchialis, rubropertincta, sputi, terrae*
Tsukamurella	*paurometabolum, wratislaviensis*

Data compiled from Beaman, B.L., Saubolle, M.A., and Wallace, R.J. 1995. In Murray, P.R., Baron, E.J., Pfaller, M.A., et al., editors. Manual of clinical microbiology, ed 6. American Society for Microbiology, Washington, D.C.; Goodfellow, M., et al. 1994. Lett. Appl. Microbiol. 19:401; Klatte, S., Rainey, F.A., and Kroppenstedt, R.M. 1994. Int. J. Syst. Bacteriol. 44:769; Lasker, B.A., Brown, J.M., McNeil, M.M. 1992. Clin. Infect. Dis. 15:233.

NON–ACID-FAST AEROBIC ACTINOMYCETES: *STREPTOMYCES, ACTINOMADURA, DERMATOPHILUS, NOCARDIOPSIS, OERSKOVIA, ROTHIA,* AND THE THERMOPHILIC ACTINOMYCETES

These organisms are gram-positive, branching filaments that do not have mycolic acids present in their cell envelopes and are therefore non–acid-fast. This group of actinomycetes are heterogeneous and are encountered infrequently in the clinical laboratory. Only those non–acid-fast actinomycetes associated with human disease are addressed (Table 57-3). Another group of non–acid-fast actinomycetes, the thermophilic actinomycetes, are associated with infections in humans and include the genera *Thermoactinomyces, Saccharomonospora,* and *Saccharopolyspora,* which are medically important.

EPIDEMIOLOGY AND PATHOGENESIS

PARTIALLY ACID-FAST AEROBIC ACTINOMYCETES

Nocardia

Nocardia are inhabitants of soil and water, and are primarily responsible for the decomposition of plant material. Infections caused by *Nocardia* spp. are found worldwide. Because they are ubiquitous, the isolation of these organisms from clinical specimens does not always indicate infection; it possibly may indicate colonization of the skin and upper respiratory tract or laboratory contamination. *Nocardia* infections can be acquired by either traumatic inoculation or inhalation.

Nocardia, particularly *N. asteroides,* are facultative intracellular pathogens that can grow in various human cells. The mechanisms of pathogenesis are complex and not completely understood. However, the virulence of *N. asteroides* appears to be associated with several factors, such as its stage of growth, resistance to intracellular killing, tropism for neuronal tissue, and ability to inhibit phagosome-lysosome fusion; other characteristics, such as production of large amounts of catalase and hemolysins, may also be associated with virulence.[2]

Rhodococcus, Gordona, Tsukamurella

These organisms can be isolated from several environmental sources, especially soil and farm animals, as well as fresh and salt waters. Organisms are believed to be acquired primarily by inhalation.[3] For the most part, these aerobic actinomycetes are infrequently isolated from clinical specimens.

To date, *Rhodococcus equi* has been the organism most commonly associated with human disease, particularly in immunocompromised patients, such as those infected with human immunodeficiency virus (HIV). *R. equi* is a facultative intracellular organism that can persist and replicate within macrophages.[16] Other possible determinants of virulence for *R. equi,* such as beta-lactam resistance and the presence of mycolic acid with longer carbon chains in their cell envelope, are under investigation. Although *Gordona* spp. and *Tsukamurella* are able to cause opportunistic infections in humans, little is known regarding the pathogenic mechanisms.

TABLE 57-3 NON–ACID-FAST AEROBIC ACTINOMYCETES ASSOCIATED WITH HUMAN DISEASE

GENUS	NUMBER OF SPECIES	SPECIES ASSOCIATED WITH HUMAN DISEASE
Streptomyces	>3000	*S. somaliensis* *S. paraguayensis* *S. anulatus*
Actinomadura	>25	*A. madurae* *A. pelletieri*
Dermatophilus	1	*D. congolensis*
Nocardiopsis	1	*N. dassonvillei*
Oerskovia	2	*O. turbata* *O. xanthineolytica*
Rothia	1	*R. dentocariosa*

NON–ACID-FAST AEROBIC ACTINOMYCETES: *STREPTOMYCES, ACTINOMADURA, DERMATOPHILUS, NOCARDIOPSIS, OERSKOVIA, ROTHIA,* AND THE THERMOPHILIC ACTINOMYCETES

Aspects regarding the epidemiology of the non–acid-fast aerobic actinomycetes are summarized in Table 57-4. Little is known about how these agents cause infection.

SPECTRUM OF DISEASE

PARTIALLY ACID-FAST AEROBIC ACTINOMYCETES

The partially acid-fast actinomycetes cause various infections in humans.

Nocardia

Infections caused by *Nocardia* spp. can occur in immunocompetent and immunocompromised individuals. *N. asteroides, N. brasiliensis,* and *N. otitidis-caviarum* are the major causes of these infections, with *N. asteroides* causing greater than 80% of infections.

Nocardia spp. cause three types of skin infections in immunocompetent individuals[14]:

- Mycetoma, a chronic, localized, painless, subcutaneous infection
- Lymphocutaneous infections
- Skin abscesses or cellulitis

Of note, *N. brasiliensis* is the predominant cause of these skin infections.

TABLE 57-4 EPIDEMIOLOGY OF THE NON–ACID-FAST AEROBIC ACTINOMYCETES

ORGANISM	HABITAT (RESERVOIR)	DISTRIBUTION	ROUTES OF PRIMARY TRANSMISSION
Streptomyces somaliensis	Sandy soil	Africa, Arabia, Mexico, South America	Penetrating wound/abrasions in the skin
S. paraguayensis	Soil, decaying vegetation, thorns	Mexico, Central America, South America	Penetrating wound/abrasions in the skin
S. anulatus	Soil	Most common isolate in United States	Penetrating wound/abrasions in the skin
Actinomadura madurae	Soil	Tropical and subtropical countries	Penetrating wound/abrasions in the skin
A. pelletieri	Unknown	Tropical and subtropical countries	Penetrating wound/abrasions in the skin
Dermatophilus congolensis	Unknown; skin commensal or saprophyte in soil (?)[19]	Worldwide, but more prevalent in humid, tropical, and subtropical regions	Trauma to the epidermis caused by insect bites and thorns; contact with tissues of infected animals through abrasions in the skin
Nocardiopsis dassonvillei*	Unknown	Unknown	Unknown
Oerskovia spp.	Soil, decaying plant material	Worldwide	Unknown
Rothia dentocariosa	Human oral cavity	Worldwide	Unknown
Thermophilic actinomycetes	Ubiquitous; water, air, soil, compost piles, dust, hay	Worldwide	Inhalation

*Only a few cases of infection in the world literature.

In individuals who are immunocompromised, *Nocardia* spp. can cause invasive pulmonary infections and disseminated infections. Patients receiving systemic immunosuppression, such as transplant recipients, individuals with impaired pulmonary immune defenses, and intravenous drug abusers, are examples of immunosuppressed patients at risk for these infections. Patients with pulmonary infections caused by *Nocardia* can exhibit a wide range of symptoms, from an acute to a more chronic presentation. Unfortunately, there are no specific signs to indicate pulmonary nocardiosis. Patients usually appear systemically ill, with fever, night sweats, weight loss, and a productive cough that may be bloody. Pulmonary infection can lead to complications such as pleural effusions, empyema, and mediastinitis.

Nocardia can often spread hematogenously throughout the body from a primary pulmonary infection. Disseminated infection can result in lesions in the brain and skin; hematogenous dissemination involving the central nervous system is particularly common, occurring in about 30% of patients. Disseminated nocardiosis has a very poor prognosis.

Rhodococcus, Gordona, Tsukamurella

The types of infections caused by *Rhodococcus, Gordona,* and *Tsukamurella* are listed in Table 57-5. For the most part, these organisms are considered opportunistic pathogens because the majority of infections are in immunocompromised individuals.

NON–ACID-FAST AEROBIC ACTINOMYCETES: *STREPTOMYCES, ACTINOMADURA, DERMATOPHILUS, NOCARDIOPSIS, OERSKOVIA, ROTHIA,* AND THE THERMOPHILIC ACTINOMYCETES

Infection caused by most of these non–acid-fast aerobic actinomycetes is usually chronic, granulomatous lesions of the skin referred to as **mycetoma.** Mycetoma is an infection of subcutaneous tissues resulting in tissue swelling and draining sinus tracts. These infections are acquired by traumatic inoculation of organisms (usually in the lower limbs) and are usually caused by fungi. If a mycetoma is caused by an actinomycete, the infection is then called **actinomycetoma.**

Except for the thermophilic actinomycetes, only rarely have most of these agents been associated with other types of infections (Table 57-6). These nonmycetomic infections have occurred in immunosuppressed patients, such as those infected with HIV.[15]

The thermophilic actinomycetes are responsible for **hypersensitivity pneumonitis,** an allergic reaction to these agents. This occupational disease occurs in farmers, factory workers, and others who are repeatedly exposed to these agents. There are acute and chronic forms of this disease. With respect to acute hypersensitivity pneumonitis, patients experience malaise, sweats, chills, loss of appetite, chest tightness, cough and fever within 4 to 6 hours after exposure; typically symptoms resolve within a day.[14] Under

---

TABLE 57-5 **INFECTIONS CAUSED BY *RHODOCOCCUS, GORDONA,* AND *TSUKAMURELLA***

ORGANISM	CLINICAL MANIFESTATIONS
Rhodococcus	Pulmonary infections (pneumonia) Bacteremia Skin infection Endophthalmitis Peritonitis Catheter-associated sepsis Prostatic abscess
Gordona	Skin infections Chronic pulmonary disease Catheter-associated sepsis Wound infection
Tsukamurella	Peritonitis Catheter-associated sepsis Skin infection

TABLE 57-6 CLINICAL MANIFESTATIONS OF INFECTIONS CAUSED BY NON–ACID-FAST AEROBIC ACTINOMYCETES

ORGANISM	CLINICAL MANIFESTATIONS
Streptomyces spp. (*S. somaliensis, S. anulatus, S. paraguayensis*)	Actinomycetoma Other (rare): pericarditis, bacteremia, and brain abscess
Actinomadura spp. (*A. madurae, A. pelletieri*)	Actinomycetoma Other (rare): peritonitis, wound infection, pneumonia, and bacteremia
Dermatophilus congolensis	Exudative dermatitis with scab formation (dermatophilosis)
Nocardiopsis dassonvillei	Actinomycetoma and other skin infections
Oerskovia spp.	Catheter-associated bacteremia, meningitis, endocarditis, and endophthalmitis
Rothia dentocariosa	Endocarditis, abscess, periodontal infections, and bacteremia

some circumstances involving continued exposure to the organisms, patients suffer from a chronic form of disease in which symptoms progressively worsen with subsequent development of irreversible lung fibrosis.

LABORATORY DIAGNOSIS

SPECIMEN COLLECTION, TRANSPORT, AND PROCESSING

Appropriate specimens should be collected aseptically from affected areas. For the most part, there are no special requirements for specimen collection, transport, or processing of the organisms discussed in this chapter; refer to Table 1-1 for general information. When nocardiosis is clinically suspected, multiple specimens should be submitted for culture, because smears and cultures are simultaneously positive in only a third of the cases. Also, the random isolation of *Nocardia* spp. from the respiratory tract is of questionable significance because these organisms are so widely distributed in nature.[13] Some of the actinomycetes tend to grow as a microcolony in tissues, leading to the formation of granules. Most commonly, these granules are formed in actinomycetomas, such as those caused by *Nocardia, Streptomyces, Nocardiopsis,* and *Actinomadura* spp. Therefore material from draining sinus tracts is an excellent specimen for direct examination and culture.

DIRECT DETECTION METHODS

Direct microscopic examination of Gram-stained preparations of clinical specimens is of utmost importance in the diagnosis of infections caused by the aerobic actinomycetes. Often, the demonstration of gram-positive, branching or partially branching beaded filaments provides the first clue to the presence of an aerobic actinomycete (Figure 57-1). Unfortunately, the actinomycetes do not always exhibit such characteristic morphology; many times these organisms are not seen at all or appear as gram-positive cocci, rods, or short filaments. Nevertheless, if gram-positive, branching or partially branching organisms are observed, a modified acid-fast stain should be performed (i.e., 1% sulfuric acid rather than 3% hydrochloric acid as the decolorizing agent). Histopathologic examination of tissue specimens using various histologic stains, such as Gomori's methenamine-silver (GMS) stain, can also detect the presence of actinomycetes.

Of importance, any biopsy or drainage material from actinomycetomas should be examined for the presence of granules. If observed, the granules are washed in saline, emulsified in 10% potassium hydroxide or crushed between two slides, Gram-stained, and examined microscopically for the presence of filaments.[3]

CULTIVATION

The majority of aerobic actinomycetes do not have complex growth requirements and are able to grow on routine laboratory media, such as sheep blood, chocolate, Sabouraud dextrose, and brain-heart infusion agar. However, because many of the aerobic actinomycetes grow slowly, they may be overgrown by other normal flora present in contaminated specimens. This is particularly true for the nocardiae that

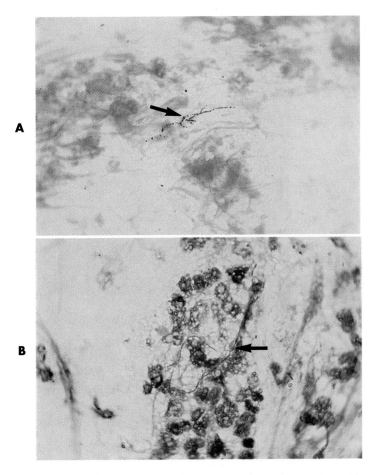

FIGURE 57-1 A, Gram stain of sputum obtained from patient with pulmonary nocardiosis caused by *Nocardia asteroides.* **B,** The same sputum stained using modified acid-fast stain. The arrows indicate the organism.

require a minimum of 48 to 72 hours of incubation before colonies become visible. Because of their slow growth and the possibility of being overgrown because of contaminating flora, various selective media have been proposed to recover nocardiae. A solid medium that uses paraffin as the sole source of carbon has been effective for isolating *Nocardia* spp. and rapidly growing mycobacteria from contaminated clinical specimens.[17] Selective media formulated for the isolation of *Legionella* spp. from contaminated specimens, such as buffered charcoal-yeast extract medium with polymyxin, anisomycin, and vancomycin, have also allowed recovery of nocardiae from contaminated specimens; Martin Lewis and colistin-nalidixic acid media have also been employed.[8,18] *Nocardia* spp. also grow well on Saboraud dextrose agar but are inhibited by chloramphenicol. Occasional isolates are first seen on mycobacterial culture media, because *Nocardia* can survive the usual decontamination procedures.

If other aerobic actinomycetes are considered, a selective medium, such as brain-heart infusion agar

with chloramphenicol and cycloheximide, is recommended in addition to routine agar to enhance their isolation from contaminated specimens.[3] Although most aerobic actinomycetes grow at 35° C, more isolates can be recovered at 30° C. Therefore, selective and nonselective agars should be incubated at 35° C and 30° C. Plates should be incubated for 2 to 3 weeks. The typical Gram-stain morphology and colonial appearance of the aerobic actinomycetes are summarized in Table 57-7. Examples of Gram stains and cultures of different aerobic actinomycetes are shown in Figures 57-2 and 57-3. A good method to demonstrate the microscopic morphology of the aerobic actinomycetes is by direct observation of a slide culture containing undisturbed colonies on tap water agar.[1,4]

Clinical laboratories are rarely asked to diagnose hypersensitivity pneumonitis that is caused by the thermophilic actinomycetes. These organisms grow rapidly on trypticase soy agar with 1% yeast extract. The ability to grow at temperatures of 50° C or greater is a characteristic of all thermophilic actinomycetes. Differentiation of the various agents is

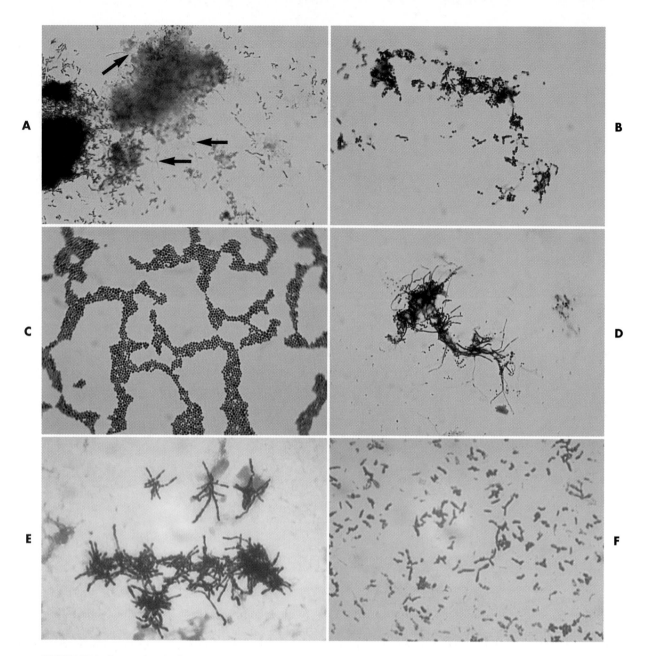

FIGURE 57-2 Gram stains of different aerobic actinomycetes. **A,** *Nocardia asteroides* grown on Löwenstein-Jensen media. The arrows indicate branching rods. **B,** *Rhodococcus equi* from broth. **C,** *R. equi* grown on chocolate agar. **D,** *Streptomyces* spp. on Saboraud dextrose agar. **E,** *Rothia denticariosa* from broth. **F,** *R. denticariosa* from solid media. (**E** and **F,** Courtesy Deanna Kiska, SUNY Science Center at Syracuse, NY.)

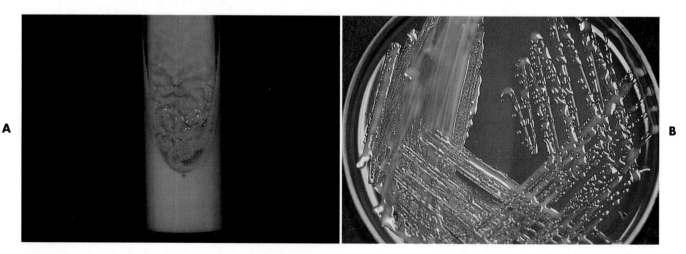

FIGURE 57-3 Aerobic actinomycetes grown on solid media. **A,** *Nocardia asteroides* grown on Löwenstein-Jensen media **B,** *Rhodococcus equi* grown on chocolate agar.

TABLE 57-7   TYPICAL GRAM-STAIN MORPHOLOGY AND COLONIAL APPEARANCE

ORGANISM	GRAM STAIN*	COLONIAL APPEARANCE ON ROUTINE AGAR
Nocardia spp.	Branching, fine, delicate filaments with fragmentation	Extremely variable; adherent; some isolates are beta-hemolytic on sheep blood agar; wrinkled; often dry, chalky-white appearance to orange-tan pigment; crumbly
Rhodococcus, Gordona, Tsukamurella	Diphtheroid-like with minimal branching or coccobacillary; colonial growth appears as coccobacilli in zigzag configuration	Nonhemolytic; round; often mucoid with salmon-pink pigment developing within 4 to 7 days (pigment may vary widely)
Streptomyces spp.	Extensive branching with chains and spores; does not fragment easily	Glabrous or waxy heaped colonies; variable morphology
Actinomadura spp.	Moderate, fine, intertwining branching with short chains of spores; fragmentation	White-to-pink pigment, mucoid, molar tooth appearance after 2 weeks' incubation
Dermatophilus sp.	Branched filaments divided in transverse and longitudinal planes; fine, tapered filaments	Round, adherent, gray-white colonies that later develop orange pigments; often beta-hemolytic
Nocardiopsis sp.	Branching with internal spores	Coarsely wrinkled and folded with well-developed aerial mycelium
Oerskovia spp.	Extensive branching; hyphae break up into motile, rod-shaped elements	Yellow-pigmented, extensive branching that grows on the surface and into the agar; dense centers
Rothia sp.	Extremely pleomorphic; predominately coccoid and bacillary (in broth) to branched filaments (solid media)	Small, smooth to rough colonies; dry

*Aerobic actinomycetes are gram-positive organisms that are often beaded in appearance.

Data compiled from Beaman, B.L., Saubolle, M.A., and Wallace, R.J. 1995. In Murray, P.R., Baron, E.J., Pfaller, M.A., et al., editors. Manual of clinical microbiology, ed 6. American Society for Microbiology, Washington, D.C.; McNeil, M.M. and Brown, J.M. 1994. Clin. Microbiol. Rev. 7:357.

based on microscopic and macroscopic morphologies; Hollick[10] and McNeil and Brown[14] provide practical identification schemes for this group of actinomycetes.

APPROACH TO IDENTIFICATION

If Gram-stain morphology or colonial morphology is suggestive of a possible actinomycetes (Table 57-7), a Ziehl-Neelsen acid-fast stain should first be performed to rule out rapidly growing mycobacteria (see Chapter 60), followed by a modified acid-fast stain (Procedure 57-1). If the modified acid-fast stain results are positive, the isolate is a probable partially acid-fast aerobic actinomycete, that is, *Nocardia,* *Rhodococcus, Tsukamurella,* or *Gordona.* (If acid-fast stain-negative, these organisms are still not completely ruled out because of the variability of acid-fastness among isolates belonging to this group.) Aerobic actinomycetes can be initially placed into major groupings by taking into consideration the following:

- Gram-stain morphology (see Figure 57-1, Figure 57-2)
- Modified acid-fast stain results
- Presence or absence of aerial hyphae when grown on tap water agar
- Growth or no growth in nutrient broth containing lysozyme (250 μg/mL; Procedure 57-2)

TABLE 57-8 PRELIMINARY GROUPING OF THE CLINICALLY RELEVANT AEROBIC ACTINOMYCETES

CHARACTERISTIC	NOCARDIA	RHODOCOCCUS	GORDONA	TSUKAMURELLA
Partially acid-fast	+	±	±	±
Appearance on tap water agar*: branching/aerial hyphae	Extensive/+	Minimal/−	Minimal/−	Minimal/−
Lysozyme resistance	+	±	−	+
Urea hydrolysis	+	±	+	+
Nitrate reduction	±	±	+	−
Growth anaerobically	−	−	−	−

+, predominantly positive; −, predominantly negative; ±, mostly positive with some negative isolates; ∓, mostly negative with some positive isolates.

*Tap water agar: Bacto agar (Difco Laboratories, Detroit, Mich.) is added to 100 mL of tap water, sterilized, and then poured into plates.[4] Plates are lightly inoculated using a single streak and incubated at 30° C for up to 7 days and examined daily.

Modified from Beaman B.L., Saubolle, M.A., and Wallace, R.J. 1995. In Murray, P.R., Baron, E.J., Pfaller, M.A., et al. editors,

■ Other tests: urea hydrolysis, nitrate reduction, and ability to grow anaerobically

Table 57-8 summarizes these key characteristics for aerobic actinomycetes.

Identification of the pathogenic *Nocardia* to the species level can be accomplished by the use of casein, xanthine, and tyrosine hydrolysis; growth at 45° C, acid production from rhamnose; gelatin hydrolysis; and opacification of Middlebrook agar.[5,7,10] Some of these reactions with the nocardial pathogens are summarized in Table 57-9. To confirm the identification of the other actinomycetes and speciate them, many tests are needed; these are beyond the capabilities of most routine clinical microbiology laboratories and should therefore be referred to a reference laboratory.

SERODIAGNOSIS

Currently, no serodiagnostic tests are available to help identify patients with active nocardiosis with certainty and can only be used to augment culture results. Infections caused by other aerobic actinomycetes cannot presently be diagnosed on a serologic basis.

ANTIMICROBIAL SUSCEPTIBILITY TESTING AND THERAPY

Although various methods, including modified disk-diffusion, agar dilution, broth microdilution, E-test, and radiometric growth index, are available, the antimicrobial susceptibility testing of *Nocardia* spp.

remains problematic.[13] Some of the problems associated with susceptibility testing include the lack of standardized, validated methods, lack of correlation of in vitro susceptibility testing results with clinical outcome, and the inability to achieve a uniform suspension of organisms for testing for all strains. Nevertheless, antimicrobial susceptibility testing should be performed on clinically significant isolates of *Nocardia*. However, this usually requires sending the isolate to a reference laboratory. A similar situation exists for all other actinomycetes in that there are no standardized methods currently available. However, in some instances susceptibility studies of *Rhodococcus*, *Oerskovia*, *Gordona*, and *Rothia* can be used as a guide for directing therapy. The primary drugs of choice against the aerobic actinomycetes are shown in Table 57-10; there is no effective antimicrobial therapy for hypersensitivity pneumonitis caused by the thermophilic actinomycetes.

PREVENTION

There are no vaccines available that are directed against any aerobic actinomycete, although some have been developed but with little success. With respect to hypersensitivity pneumonitis caused by the thermophilic actinomycetes, patients must prevent disease by avoiding exposure to these sensitizing microorganisms.

STREPTOMYCES	OERSKOVIA	ROTHIA	ACTINOMADURA	DERMATOPHILUS	NOCARDIOPSIS
−	−	−	−	−	−
Extensive/+	Extensive/−	Minimal/−	Variable/sparse	Branching	Extensive/+
−	−		−	−	−
±	∓	−	−	+	+
±	+	+	+	−	+
−	+	+	−	−	−

Manual of clinical microbiology, ed 6. American Society for Microbiology, Washington, D.C.

TABLE 57-9   KEY TESTS FOR DIFFERENTIATION OF THE PATHOGENIC NOCARDIA

TEST	N. ASTEROIDES	N. FARCINICA*	N. NOVA	N. TRANSVALENSIS	N. BRASILIENSIS	N. OTITIDISCAVIARUM
Hydrolysis of: Casein	−	−	−	∓	+	−
Xanthine	−	−	−	±	−	−
Tyrosine	−	−	−	∓	+	+
Growth at 42° C after 3 days	±	+	−	−	−	±
14-day arylsulfatase	−	−	+	−	−	−
Acid from rhamnose	∓	±	−	−	−	−
Gelatin hydrolysis	−	−	−	−	+	−
Opacification of Middlebrook agar	−	+	−	−	−	∓

+, predominantly positive; −, predominantly negative; ±, mostly positive with some negative isolates; ∓, mostly negative with some positive isolates.

*Cefotaxime resistant.

_____ **TABLE 57-10 PRIMARY DRUGS OF CHOICE FOR INFECTIONS** _____
CAUSED BY AEROBIC ACTINOMYCETES

ORGANISMS	PRIMARY DRUGS OF CHOICE
Nocardia spp.	Sulfonamides
	Trimethoprim/sulfamethoxazole
	Supplemental agents: amikacin, ceftriaxone, cefotaxime, or imipenem
	Minocycline
Rhodococcus, Gordona, Tsukamurella	Erythromycin and rifampin
	Gentamicin, tobramycin, or ciprofloxacin
	Vancomycin and imipenem
Streptomyces spp.	Streptomycin and trimethoprim/sulfamethoxazole or dapsone
Rothia dentocariosa	Penicillin G
Oerskovia spp.	Penicillin or ampicillin
	Rifampin
	Vancomycin
Actinomadura spp.	Streptomycin and trimethoprim/sulfamethoxazole or dapsone
	Amikacin and imipenem
Nocardiopsis dassonvillei	Trimethoprim/sulfamethoxazole
Dermatophilus congolensis	Highly variable susceptibilities; no specific drugs of choice

Procedures

57-1 PARTIALLY ACID-FAST STAIN FOR IDENTIFICATION OF *NOCARDIA*

PRINCIPLE

The nocardiae, because of unusual long-chain fatty acids in their cell walls, can retain carbolfuchsin dye during mild acid decolorization, whereas other aerobic branching bacilli cannot.

METHOD

1. Emulsify a very small amount of the organisms to be stained in a drop of distilled water on the slide. A known positive control and a negative control should be stained along with the unknown strain.

2. Allow to air dry and heat fix.

3. Flood the smear with Kinyoun carbolfuchsin and allow the stain to remain on the slide 3 minutes.

4. Rinse with tap water, shake off excess water, and decolorize briefly with 1% sulfuric acid alcohol until no more red color rinses off the slides.

5. Counterstain with Kinyoun methylene blue for 30 seconds.

6. Rinse again with tap water. Allow the slide to air dry, and examine the unknown strain compared with the controls. Partially acid-fast organisms show reddish to purple filaments, whereas non–acid-fast organisms are blue only.

57-2 LYSOZYME RESISTANCE FOR DIFFERENTIATING *NOCARDIA* FROM *STREPTOMYCES*

PRINCIPLE

The enzyme lysozyme, present in human tears and other secretions, can break down cell walls of certain microorganisms. Susceptibility to the action of lysozyme can differentiate certain morphologically similar genera and species.

METHOD

1. Prepare basal broth as follows:
   Peptone (Difco Laboratories), 5 g
   Beef extract (Difco), 3 g
   Glycerol (Difco), 70 mL
   Distilled water, 1000 mL
   Dispense 500 mL of this solution into 16 × 125-mm screw-cap glass test tubes, 5 mL per tube. Autoclave the test tubes and the remaining solution for 15 minutes at 120° C. Tighten caps and store the tubes in the refrigerator for a maximum of 2 months.

2. Prepare lysozyme solution as follows:
   Lysozyme (Sigma Chemical Co.), 100 mg
   HCl (0.01 N), 100 mL
   Sterilize through a 0.45-μm membrane filter.

3. Add 5 mL lysozyme solution to 95 mL basal broth; mix gently, avoiding bubbles, and aseptically dispense in 5-mL amounts to sterile, screw-cap tubes as in step 1. Store refrigerated for a maximum of 2 weeks.

4. Place several bits of the colony to be tested into a tube of the basal glycerol broth without lysozyme (control) and into a tube of broth containing lysozyme.

5. Incubate at room temperature for up to 7 days. Observe for growth in the control tube. An organism that grows well in the control tube but not in the lysozyme tube is considered to be susceptible to lysozyme.

References

1. Barksdale, L. and Kim, K.S. 1977. *Mycobacterium.* Bacteriol. Rev. 41:217.

2. Beaman, B.L. and Beaman, L. 1994. *Nocardia* species: host parasite relationships. Clin. Microbiol. Rev. 7:213.

3. Beaman, B.L., Saubolle, M.A., and Wallace, R.J. 1995. *Nocardia, Rhodococcus, Streptomyces, Oerskovia,* and other aerobic actinomycetes of medical importance. In Murray, P.R., Baron, E.J., Pfaller, M.A., et al., editors. Manual of clinical microbiology, ed 6. American Society for Microbiology, Washington, D.C.

4. Berd, D. 1973. Laboratory identification of clinically important aerobic actinomycetes. Appl. Microbiol. 25:665.

5. Carson, M. and Hellyar, A. Opacification of Middlebrook agar as an aid in distinguishing *Nocardia farcinica* within the *Nocardia asteroides* complex. J. Clin. Microbiol. 32:2270.

6. Chun, J. and Goodfellow, M. 1995. A phylogenetic analysis of the genus *Nocardia* with 16S rRNA gene sequences. Int. J. Syst. Bacteriol. 45:240.

7. Flores, M. and Desmond, E. Opacification of Middlebrook agar as an aid in identification of *Nocardia farcinica.* J. Clin. Microbiol. 31:3040.

8. Garrett, M.A., Holmes, H.T. and Nolte, F.S. 1992. Selective buffered charcoal-yeast extract medium for isolation of nocardiae from mixed cultures. J. Clin. Microbiol. 30:1891.

9. Goodfellow, M. et al. 1994. Transfer of *Nocardia amarae* (Lechevalier and Lechevalier 1974) to the genus *Gordona* as *Gordona amarae* comb. nov. Letters Appl. Microbiol. 19:401.

10. Hollick, G. 1995. Isolation and identification of aerobic actinomycetes. Clin. Microbiol. Newsletter 17:25.

11. Klatte, S., Rainey, F.A. and Kroppenstedt, R.M. 1994. Transfer of *Rhodococcus aichensis* (Tsukamura 1982) and *Nocardia amarae* (Lechevalier and Lechevalier 1974) to the genus *Gordona* as *Gordona aichiensis* comb. nov. and *Gordona amarae* comb. nov. Int. J. Syst. Bacteriol. 44:769.

12. Lasker, B.A., Brown, J.M., and McNeil, M.M. 1992. Identification and epidemiological typing of clinical and environmental isolates of the genus *Rhodococcus* with use of a digoxigenin-labeled rDNA gene probe. Clin. Infect. Dis. 15:233.

13. Lerner, P.I. Nocardiosis. Clin. Infect. Dis. 22:891.

14. McNeil, M.M. and Brown, J.M. 1994. The medically important aerobic actinomycetes: epidemiology and microbiology. Clin. Microbiol. Rev. 7:357.

15. McNeil, M.M., et al. 1992. Nonmycetomic *Actinomadura madulae* infection in a patient with AIDS. J. Clin. Microbiol. 30:1008.

16. Mosser, D.M. and Hondalus, M.K. 1996. *Rhodococcus equi:* an emerging opportunistic pathogen. Trends Microbiol. 4:29.

17. Shawar, R.M., Moore, D.G., and LaRocco, M.T. 1990. Cultivation of *Nocardia* spp. on chemically defined media for selective recovery of isolates from clinical specimens. J. Clin. Microbiol. 28:508.

18. Vickers, R.M., Rihs, J.D., and Yu, V.L. 1992. Clinical demonstration of isolation of *Nocardia asteroides* on buffered charcoal-yeast extract media. J. Clin. Microbiol. 30:227.

19. Zaria, L.T. 1993. *Dermatophilus congolensis* infection in animals and man: an update. Comp. Immun. Microbiol. Infect. Dis. 16:179.

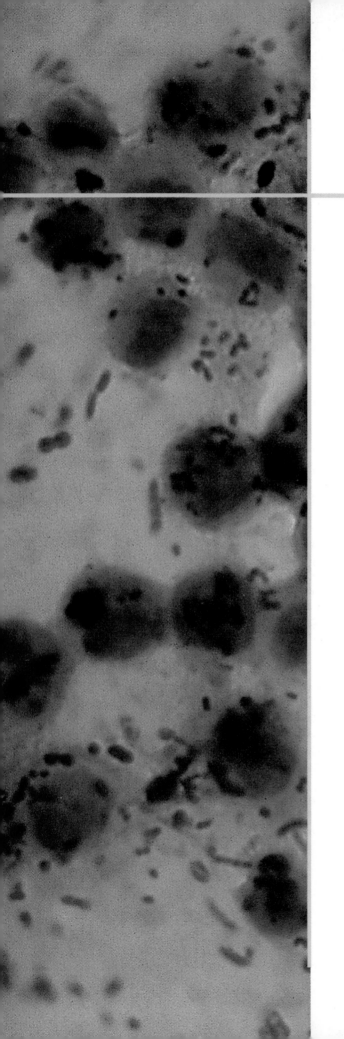

Section Twelve

12

Anaerobic Bacteriology

58 OVERVIEW AND GENERAL CONSIDERATIONS

Genera and Species to be Considered

- *Clostridium perfringens*
- *Clostridium tetani*
- *Clostridium botulinum*
- *Clostridium difficile*
- Other *Clostridium* spp.
- *Actinomyces israelii*
- *Actinomyces naeslundii*
- *Actinomyces odontolyticus*

- Other *Actinomyces* spp.
- *Bacteroides fragilis* group
- *Bacteroides ureolyticus*
- Other *Bacteroides* spp.
- *Eubacterium* spp.
- *Bifidobacterium* spp.
- *Bilophila wadsworthii*
- *Prevotella* spp.

- *Porphyromonas* spp.
- *Fusobacterium* spp.
- *Peptostreptococcus anaerobius*
- Other *Peptostreptococcus* spp.
- *Veillonella* spp.
- *Peptococcus niger*
- *Lactobacillus* spp.
- *Propionibacterium* spp.

GENERAL CHARACTERISTICS

The organisms described in this chapter and in Chapter 59 usually will not grow in the presence of oxygen. These chapters also include some microaerophilic organisms, such as *Actinomyces* spp. and *Bifidobacterium* spp., that can grow in the presence of either reduced or atmospheric oxygen but grow best under anaerobic conditions.

EPIDEMIOLOGY

Most of the anaerobic bacteria that cause infections in humans are also part of our normal flora. The ecology of these organisms is such that various species and genera exhibit preferences for the body sites that they inhabit (Table 58-1). Other pathogenic anaerobes (e.g., *Clostridium botulinum* and *Clostridium tetani*) are soil and environmental inhabitants and are not considered part of human flora.

The ways in which anaerobic infections are acquired are summarized in Table 58-2. Although person-to-person nosocomial spread of *Clostridium difficile* among hospitalized patients presents an enormous clinical and infection control dilemma, the majority of anaerobic infections occur when a patient's normal flora gains access to a sterile site as a result of disruption of some anatomic barrier.[1]

PATHOGENESIS AND SPECTRUM OF DISEASES

The types of infections and diseases in humans that are caused by anaerobic bacteria span a wide spectrum. Certain species, such as *C. botulinum* and *C. tetani*, produce some of the most potent toxins known. In contrast, specific virulence factors for the organisms most commonly encountered in infections (e.g., *B. fragilis* group, *C. difficile*) are not well understood (Table 58-3).

Most anaerobic infections involve a mixture of anaerobic and facultatively anaerobic organisms (e.g., *Enterobacteriaceae*), so that it is problematic to establish the extent to which a particular anaerobic species is contributing to infection. In addition, as ubiquitous members of our normal flora, anaerobic organisms frequently contaminate clinical materials. For these reasons, assigning clinical significance to anaerobic bacteria isolated in the laboratory is important, although often difficult.

SPECIMEN COLLECTION AND TRANSPORT

The proper collection and transport of specimens for anaerobic culture cannot be overemphasized. Because indigenous anaerobes are often present in large numbers as normal flora on mucosal surfaces, even minimal

TABLE 58-1 Incidence of Anaerobes as Normal Flora of Humans

GENUS	SKIN	UPPER RESPIRATORY TRACT*	INTESTINE	EXTERNAL GENITALIA	URETHRA	VAGINA
Gram-negative bacteria						
Bacteroides	0	±	2	±	±	±
Prevotella	0	2	2	1	±	1
Porphyromonas	0	1	1	U	U	±
Fusobacterium	0	2	1	U	U	±
Veillonella	0	2	1	0	U	1
Gram-positive bacteria						
Peptostreptococcus	1	2	2	1	±	1
Clostridium	±	±	2	±	±	±
Actinomyces	0	1	1	0	0	±
Bifidobacterium	0	1	2	0	0	±
Eubacterium	0	1	2	U	U	1
Lactobacillus	0	1	1	0	±	2
Propionibacterium	2	1	±	±	±	1

U, unknown; *0*, not found or rare; ±, irregular; *1*, usually present; *2*, present in large numbers.

*Includes nasal passages, nasopharynx, oropharynx, and tonsils.

Modified from Summanen, P.E., Baron, E.J., Citron, D.M., et al. 1993. Wadsworth anaerobic bacteriology manual, ed 5. Star, Belmont, Calif.

TABLE 58-2 Acquisition of Anaerobic Infections and Diseases

MODE OF ACQUISITION	EXAMPLES
Endogenous strains of normal flora gain access to normally sterile sites, usually as result of one or more predisposing factors that compromise normal anatomic barriers (e.g., surgery or accidental trauma) or alter other host defense mechanisms (e.g., malignancy, diabetes, burns, immunosuppressive therapy, aspiration, etc.)	Wide variety of infections involving several anatomic locations, including bacteremia, head and neck infections, dental and orofacial infections, pneumonia and other infections of the thoracic cavity, intraabdominal and obstetric and gynecologic infections, bite wound and other soft tissue infections, and gangrene (i.e., clostridial myonecrosis). Organisms most commonly encountered in these infections include *Bacteroides fragilis* group, *Prevotella* spp., *Porphyromonas* spp., *Fusobacterium nucleatum*, *Peptostreptococcus*, and *Clostridium perfringens*
Contamination of existing wound or puncture by objects contaminated with toxigenic *Clostridium* spp.	Tetanus (*Clostridium tetani*), gas gangrene (*Clostridium perfringens*, and, less commonly, *C. septicum*, *C. novyi*, and others)
Ingestion of preformed toxins in vegetable or meat-based foods	Botulism (*Clostridium botulinum*), and other clostridial food poisonings (*C. perfringens*)
Colonization of gastrointestinal tract with potent toxin-producing organism	Infant botulism (*C. botulinum*)
Person-to-person spread	Nosocomial spread of *Clostridium difficile*–induced diarrhea and pseudomembranous colitis; bite-wound infections caused by variety of anaerobic species

TABLE 58-3 PATHOGENESIS AND SPECTRUM OF DISEASES FOR ANAEROBIC BACTERIA

ORGANISM	VIRULENCE FACTORS	SPECTRUM OF DISEASE AND INFECTIONS
Clostridium perfringens	Produces several exotoxins; α-toxin is most important and mediates destruction of host cell membranes; enterotoxin inserts and disrupts membranes of mucosal cells	Gas gangrene, a life-threatening, toxin-mediated destruction of muscle and other tissues following traumatic introduction of the organism. Food poisoning caused by release of the toxin after ingestion of large quantities of organism. Disease is usually self-limiting and benign and is manifested by abdominal cramps, diarrhea, and vomiting
Clostridium tetani	Produces tetanospasmin, a neurotoxic exotoxin that disrupts nerve impulses to muscles	Tetanus (also commonly known as *lockjaw*). Organism establishes a wound infection and elaborates the potent toxin that mediates generalized muscle spasms. If untreated, spasms continue to be triggered by even minor stimuli, leading to exhaustion and, eventually, respiratory failure
Clostridium botulinum	Produces extremely potent neurotoxins	The disease botulism results from ingestion of preformed toxin in nonacidic vegetable or mushroom foodstuffs. Absorption of the toxin leads to nearly complete paralysis of respiratory and other essential muscle groups. Other forms of botulism can occur when the organism elaborates the toxin after it has colonized the gastrointestinal tract of infants (i.e., infant botulism). Wound botulism is more rare than the other forms and occurs when *C. botulinum* produces the toxin from an infected wound site
Clostridium difficile	Produces toxin A, which is an enterotoxin that is thought to be primarily responsible for the gastrointestinal disease caused by this organism. Toxin B, a cytotoxin, has a less clear role in *C. difficile* infections	Organism requires diminution of normal gut flora by the activity of various antimicrobial agents to become established in the gut of hospitalized patients. Once established, elaboration of toxin(s) results in diarrhea (i.e., antibiotic-associated diarrhea) or potentially life-threatening inflammation of the colon. When the surface of the inflamed bowel is overlaid with a "pseudomembrane" composed of necrotic debris, white blood cells, and fibrin, the disease is referred to as *pseudomembranous colitis*
Actinomyces spp., including *A. israelii* *A. meyeri* *A. naeslundii* *A. odontolyticus*	No well-characterized virulence factors. Infections usually require disruption of protective mucosal surface of the oral cavity, respiratory tract, gastrointestinal tract, and/or female genitourinary tract	Usually involved in mixed oral or cervicofacial, thoracic, pelvic, and abdominal infections caused by patient's endogenous strains; certain species (*A. viscosus* and *A. naeslundii*) also involved in periodontal disease and dental caries
Propionibacterium spp.	No definitive virulence factors known	Associated with inflammatory process in acne but only rarely implicated in infections of other body sites. As part of normal skin flora they are the most common anaerobic contaminants of blood cultures
Bifidobacterium spp.	No definitive virulence factors known	Not commonly found in clinical specimens. Usually encountered in mixed infections of pelvis or abdomen

Continued

TABLE 58-3 PATHOGENESIS AND SPECTRUM OF DISEASES FOR ANAEROBIC BACTERIA—CONT'D

ORGANISM	VIRULENCE FACTORS	SPECTRUM OF DISEASE AND INFECTIONS
Eubacterium spp.	No definitive virulence factors known	Usually associated with mixed infections of abdomen, pelvis, or genitourinary tract
Mobiluncus spp.	No definitive virulence factors known	Organisms are found in the vagina and have been associated with bacterial vaginosis, but their precise role in gynecologic infections is unclear. Rarely encountered in infections outside the female genital tract
Bacteroides fragilis group, other *Bacteroides* spp., *Bacteroides gracilis*, *Bacteroides ureolyticus*, *Prevotella* spp., *Porphyromonas* spp., *Fusobacterium nucleatum*, and other *Fusobacterium* spp.	These anaerobic gram-negative bacilli produce capsules, endotoxin, and succinic acid, which inhibit phagocytosis, and various enzymes that mediate tissue damage. Most infections still require some breach of of mucosal integrity that allows the organisms to gain access to deeper tissues	Organisms most commonly encountered in anaerobic infections. Infections are often mixed with other anaerobic and facultatively anaerobic organisms. Infections occur throughout the body, usually as localized or enclosed abscesses, and may involve the cranium, periodontium, thorax, peritoneum, liver, and female genital tract. May also cause bacteremia, aspiration pneumonia, septic arthritis, chronic sinusitis, decubitus ulcers, and other soft tissue infections. The hallmark of most, but not all, infections is the production of a foul odor. In general, infections caused by *B. fragilis* group occur below the diaphragm; pigmented *Prevotella* spp., *Porphyromonas* spp., and *F. nucleatum* generally are involved in head and neck and pleuropulmonary infections
Peptostreptococcus spp.	No definitive virulence factors known	Most often found mixed with other anaerobic and facultatively anaerobic bacteria in cutaneous, respiratory, oral, or female pelvic infections
Veillonella spp.	No definitive virulence factors known	May be involved in mixed infections but rarely play a significant role

contamination of a specimen can give misleading results. Box 58-1 shows the specimens acceptable for anaerobic culture; Box 58-2 shows the specimens that are likely to be contaminated and are unacceptable for anaerobic culture. In general, material for anaerobic culture is best obtained by tissue biopsy or by aspiration using a needle and syringe. Use of swabs is a poor alternative because of excessive exposure of the specimen to the deleterious effects of drying, the possibility of contamination during collection, and the easy retention of microorganisms within the fibers of the swab. If a swab must be used, it should be from an oxygen-free transport system.

There are special collection instructions for some clostridial illnesses, that is, foodborne *C. perfringens* and *C. botulinum*, *C. difficile* pseudomembranous enterocolitis, and *C. septicum* neutropenic enterocolitis. Food and fecal specimens must be sent to a public health laboratory for confirmation of *C. perfringens* food poisoning; these should be trans-ported at 4° C. The clinical diagnosis of botulism is confirmed by demonstration of botulinal toxin in serum, feces, vomitus, or gastric contents, as well as the recovery of the organism from the stool of patients. The Centers for Disease Control and Prevention maintains a 24 hour/day, 365 day/year hotline to provide emergency assistance in cases of botulism. Specimens for infant botulism should include serum and stool; those for wound botulism should include serum, stool, and tissue biopsy. Feces for *C. difficile* culture and toxin assay should be liquid or unformed; solid, formed stool or rectal swabs are adequate to detect carriers but not to detect active cases of enterocolitis. Stools should be placed in an anaerobic transport container if culture is to be performed. Specimens for toxin assay may be collected in leak-proof containers and may be stored for up to 3 days at 4° C or frozen at −70° C. The specimens of choice for neutropenic enterocolitis involving *C. septicum* are three different blood cultures, stool, and lumen contents or

CLINICAL SPECIMENS SUITABLE FOR ANAEROBIC CULTURE

BOX 58-1

Bile

Biopsy of endometrial tissue obtained with an endometrial suction curette (Pipelle, Unimar, Inc., Wilton, Conn.)

Blood

Bone marrow

Bronchial washings obtained with double-lumen plugged catheter

Cerebrospinal fluid

Culdocentesis aspirate

Decubitus ulcer, if obtained from base of lesion after thorough debridement of surface debris

Fluid from normally sterile site (e.g., joint)

Material aspirated from abscesses (the best specimens are from loculated or walled-off lesions)

Percutaneous (direct) lung aspirate or biopsy

Peritoneal (ascitic) fluid

Sulfur granules from draining fistula

Suprapubic bladder aspirate

Thoracentesis (pleural) fluid

Tissue obtained at biopsy or autopsy

Transtracheal aspirate

Uterine contents, if collected using a protected swab

CLINICAL SPECIMENS UNSUITABLE FOR ANAEROBIC CULTURE

BOX 58-2

Bronchial washing or brush (except if collected with double-lumen plugged catheter)

Coughed (expectorated) sputum

Feces (except for *Clostridium difficile*)

Gastric or small-bowel contents (except in blind loop syndrome)

Ileostomy or colostomy drainage

Nasopharyngeal swab

Rectal swab

Secretions obtained by nasotracheal or orotracheal suction

Swab of superficial (open) skin lesion

Throat swab

Urethral swab

Vaginal or cervical swab

Voided or catheterized urine

tissue from the involved ileocecal area; a muscle biopsy should also be collected if myonecrosis (death of muscle tissue) is suspected.

A crucial factor in the final success of anaerobic cultures is the transport of the specimen; the lethal effect of atmospheric oxygen must be nullified until the specimen can be processed in the laboratory. Recapping a syringe and transporting the needle and syringe to the laboratory is no longer acceptable because of safety concerns involving needle stick injuries. Therefore, even aspirates must be injected into some type of oxygen-free transport tube or vial. Three different kinds of anaerobic transport systems are shown in Figures 58-1 to 58-3. Figure 58-1 is a rubber-stoppered collection vial containing an agar indicator system. The vial is gassed out with oxygen-free carbon dioxide (CO_2) or nitrogen. The specimen (pus, body fluid, or other liquid material) is injected through the rubber stopper after all air is expelled from the syringe and needle. If only a swab specimen can be obtained, a special collection device with an oxygen-free atmosphere is required (Figure 58-2). When reinserting the swab, care must be taken not to tip the container, which would cause the oxygen-free CO_2 or nitrogen to spill out and be displaced by ambient air. Tissue can be placed in a small amount of liquid to keep it from drying out and then placed in an anaerobic pouch (Figure 58-3). All specimens should be held at room temperature pending processing in the laboratory, because refrigeration can oxygenate the specimen.

SPECIMEN PROCESSING

Specimens for anaerobic culture may be processed on the open bench-top with incubation in anaerobic jars or pouches or in an anaerobic chamber. The roll tube method developed at the Virginia Polytechnic Institute is no longer widely used and is not discussed here.

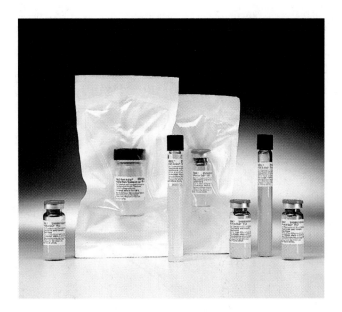

FIGURE 58-1 Anaerobic transport system for liquid specimens. Specimen is injected into tube through the rubber septum. Agar at the bottom contains oxygen tension indicator. (Courtesy Becton Dickinson Microbiology Systems, Cockeysville, Md. Port-A-Cul is a trademark of Becton Dickinson and Co.)

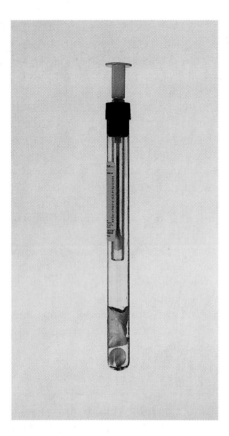

FIGURE 58-2 Anaerobic transport system for swab specimens. Vacutainer Anaerobic Specimen Collector, Becton Dickinson Microbiology Systems (Cockeysville, Md.). Sterile pack contains a sterile swab and an oxygen-free inner tube. When the specimen is collected, the swab is inserted back into the inner tube. Agar on the bottom of the outer tube contains an oxygen tension indicator. (Courtesy Fisher Scientific, Pittsburgh, Pa.)

Anaerobe jars or pouches

The most frequently used system for creating an anaerobic atmosphere is the anaerobe jar. Anaerobe jars are available commercially from several companies, including Becton Dickinson Microbiology Systems (Cockeysville, Md.) (Gas-Pak, Figure 58-4), EM Diagnostic Systems (Gibbstown, N.J.), and Oxoid U.S.A. (Columbia, Md.). These systems all use a clear, heavy plastic jar with a lid that is clamped down to make it airtight. Anaerobic conditions can be setup by two different methods. The easiest method uses a commercially available hydrogen and CO_2 generator envelope that is activated by either adding water (Gas-Pak; Becton Dickinson Microbiology Systems) or by the moisture from the agar plates (EM Diagnostic Systems; Oxoid). Production of heat within a few minutes (detected by touching the top of the jar) and subsequent development of moisture on the walls of the jar are indications that the catalyst and generator envelope are functioning properly. Reduced conditions are achieved in 1 to 2 hours, although the methylene blue or resazurin indicators take longer to decolorize. Alternatively, the "evacuation-replacement" method can be used. Air is removed from the sealed

FIGURE 58-3 Anaerobic transport system for tissue specimens. Tissue is placed in a small amount of saline to keep it moist. It is inserted into a self-contained atmosphere-generating anaerobic bag for transportation. This system is called the *Gas-Pak Pouch* (Becton Dickinson Microbiology Systems, Cockeysville, Md.).

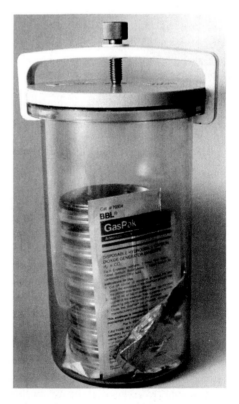

FIGURE 58-4 Gas-Pak anaerobe jar (Becton Dickinson Microbiology Systems, Cockeysville, Md.) containing inoculated plates, activated gas-generating envelope, and indicator strip. Palladium-coated alumina pellets that catalyze the reaction to remove oxygen are in a wire-mesh basket attached to the lid of the jar.

jar by drawing a vacuum of 25 inches (62.5 cm) of mercury. This process is repeated 2 times, filling the jar with an oxygen-free gas, such as nitrogen, between evacuations. The final fill of the jar is made with a gas mixture containing 80% to 90% nitrogen, 5% to 10% hydrogen, and 5% to 10% CO_2. Many anaerobes require CO_2 for maximal growth. The atmosphere in the jars is monitored by including an indicator to check anaerobiosis. Anaerobe bags or pouches are useful for laboratories processing small numbers of anaerobic specimens. A widely used anaerobic pouch is shown in Figure 58-3. Besides specimen transport, the pouch also can be used to incubate one or two agar plates.

Holding jars

If anaerobic jars or pouches are used for incubation, the use of holding jars is recommended during specimen processing and examination of cultures. Holding jars are anaerobic jars with loosely fitted lids that are attached by rubber tubing to nitrogen gas. Uninoculated plates are kept in holding jars pending use for culture setup, and inoculated plates are kept in holding jars pending incubation or examination; this minimizes exposure to oxygen.

Anaerobe chamber

Anaerobic chambers, or glove boxes, are made of molded or flexible clear plastic. Specimens and other materials are placed in the chamber through an air lock. The technologist uses gloves (Coy Corp., Grass Lake, Mich.; Forma Scientific, Marietta, Ohio) or sleeves (Anaerobe Systems, San Jose, Calif.) that form airtight seals around the arms to handle items inside the chamber (Figure 58-5). Media stored in the chamber are kept oxygen-free, and all work on a specimen from inoculation through workup is performed under anaerobic conditions. A gas mixture of 5% CO_2, 10% hydrogen, and 85% nitrogen and a palladium catalyst maintain the anaerobic environment inside the chamber.

ANAEROBIC MEDIA

Initial processing of anaerobic specimens involves inoculation of appropriate media. Table 58-4 lists commonly used anaerobic media. Primary plates should be freshly prepared or used within 2 weeks of preparation. Plates stored for longer periods accumulate peroxides and become dehydrated; this results in growth inhibition. Reduction of media in an anaerobic environment eliminates dissolved oxygen but has no effect on the peroxides. Prereduced, anaerobically sterilized (PRAS) media are produced, packaged, shipped, and stored under anaerobic conditions. They

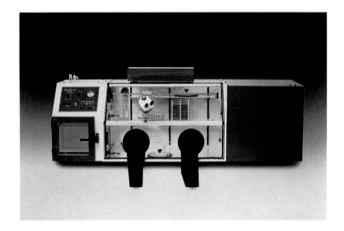

FIGURE 58-5 Gloveless anaerobe chamber. (Courtesy Anaerobe Systems, San Jose, Calif.)

are commercially available from Anaerobe Systems (San Jose, Calif.) (Figure 58-6) and have an extended shelf life of up to 6 months.

PREVENTION

A multiple-dose vaccine is available for prevention of tetanus. The immunogen is adsorbed tetanus toxid and is generally administered with diphtheria toxoid and pertussis vaccine as a triple antigen called *DTP*. Single boosters of diphtheria and tetanus (DT) or tetanus alone are recommended every 10 years. Immunoprophylaxis in wound management is based on the type of wound. Completely immunized individuals with minor and/or uncontaminated wounds do not need any specific treatment. However, completely immunized individuals with major and/or contaminated wounds should get a booster of tetanus

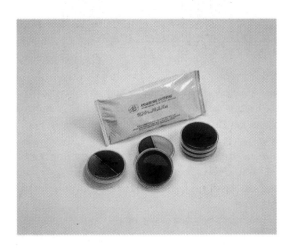

FIGURE 58-6 Prereduced, anaerobically sterilized (PRAS) plated media. (Courtesy Anaerobe Systems, San Jose, Calif.)

TABLE 58-4 COMMON ANAEROBIC MEDIA

MEDIUM	COMPONENTS/COMMENTS	PRIMARY PURPOSE
Anaerobic blood agar	May be prepared with Columbia, Schaedler, CDC, *Brucella,* or brain-heart infusion base supplemented with 5% sheep blood, 0.5% yeast extract, hemin, L-cystine, and vitamin K_1	Nonselective medium for isolation of anaerobes and facultative anaerobes
Bacteroides bile esculin agar (BBE)	Trypticase soy agar base with ferric ammonium citrate and hemin; bile salts and gentamicin act as inhibitors	Selective and differential for *Bacteroides fragilis* group; good for presumptive identification
Laked kanamycin-vancomycin blood agar (LKV)	*Brucella* agar base with kanamycin (75 μg/mL), vancomycin (7.5 μg/mL), vitamin K_1, (10 μg/mL), and 5% laked blood	Selective for isolation of *Prevotella* and *Bacteroides* spp.
Anaerobic phenylethyl alcohol agar (PEA)	Nutrient agar base, 5% blood, phenylethyl alcohol	Selective for inhibition of enteric gram-negative rods and swarming by some clostridia
Egg-yolk agar (EYA)	Egg yolk base	Nonselective for determination of lecithinase and lipase production by clostridia and fusobacteria
Cycloserine cefoxitin fructose agar (CCFA)	Egg yolk base with fructose, cycloserine (500 mg/L), and cefoxitin (16 mg/L); neutral red indicator	Selective for *Clostridium difficile*
Cooked meat (also called *chopped meat*) broth	Solid meat particles initiate growth of bacteria; reducing substances lower oxidation-reduction potential (Eh)	Nonselective for cultivation of anaerobic organisms; with addition of glucose, can be used for gas-liquid chromatography
Peptone-yeast extract glucose broth (PYG)	Peptone base, yeast extract, glucose, cysteine (reducing agent), resazurin (oxygen tension indicator), salts	Nonselective for cultivation of anaerobic bacteria for gas-liquid chromatography
Thioglycollate broth	Pancreatic digest of casein, soy broth, and glucose enrich growth of most bacteria. Thioglycollate and agar reduce Eh. May be supplemented with hemin and vitamin K_1	Nonselective for cultivation of anaerobes, as well as facultative anaerobes and aerobes

toxoid if they have not had one in the previous 5 years. Finally, a partially immunized individual or one who has never been immunized should receive a dose of tetanus toxoid immediately. In addition, passive immunization with human tetanus immune globulin (TIG) should be given if the individual has a major wound or a wound contaminated with soil containing animal feces.

Individuals who have eaten food suspected of containing botulinum toxin should be purged with cathartics (laxatives), have their stomach pumped, and be given high enemas.

References

1. Knoop, F.C., Owens, M., and Crocker, I.C. 1993. *Clostridium difficile:* clinical disease and diagnosis. Clin. Microbiol. Rev. 6:251.

Bibliography

Benenson, A.S. 1995. Control of communicable diseases manual, ed 16. American Public Health Association, Washington, D.C.
Bleck, T.P. 1995. *Clostridium botulinum.* In Mandell, G.L., Bennett, J.E., and Dolin, R., editors. Principles and practice of infectious diseases. Churchill Livingstone, New York.

Bleck, T.P. 1995. *Clostridium tetani.* In Mandell, G.L., Bennett, J.E., and Dolin, R., editors. Principles and practice of infectious diseases. Churchill Livingstone, New York.

Brook, I. 1995. Anaerobic cocci. In Mandell, G.L., Bennett, J.E., and Dolin, R., editors. Principles and practice of infectious diseases. Churchill Livingstone, New York.

Brook, I. 1995. Anaerobic gram-positive cocci. In Mandell, G.L., Bennett, J.E., and Dolin, R., editors. Principles and practice of infectious diseases. Churchill Livingstone, New York.

Finegold, S.M. 1995. Anaerobic bacteria: general concepts. In Mandell, G.L., Bennett, J.E., and Dolin, R., editors. Principles and practice of infectious diseases. Churchill Livingstone, New York.

Hillier, S.L, and Moncla, B.J. 1995. *Peptostreptococcus, Propionibacterium, Eubacterium,* and other nonsporeforming anaerobic gram-positive bacteria. In Murray, P.R., Baron, E.J., Pfaller, M.A., et al., editors. Manual of clinical microbiology, ed 6. American Society for Microbiology, Washington, D.C.

Jousimies-Somer, H.R., Summanen, P.H., and Finegold, S.M. 1995. *Bacteroides, Porphyromonas, Prevotella, Fusobacterium,* and other anaerobic gram-negative bacteria. In Murray, P.R., Baron, E.J., Pfaller, M.A., et al., editors. Manual of clinical microbiology, ed 6. American Society for Microbiology, Washington, D.C.

Lorber, B. 1995. *Bacteroides, Prevotella,* and *Fusobacterium* species (and other medically important anaeraobic gram-negative bacilli). In Mandell, G.L., Bennett, J.E., and Dolin, R., editors. Principles and practice of infectious diseases. Churchill Livingstone, New York.

Lorber, B. 1995. Gas gangrene and other clostridium-associated diseases. In Mandell, G.L., Bennett, J.E., and Dolin, R., editors. Principles and practice of infectious diseases. Churchill Livingstone, New York.

Onderdonk, A.B. and Allen, S.D. 1995. *Clostridium.* In Murray, P.R., Baron, E.J., Pfaller, et al., editors. Manual of clinical microbiology, ed 6. American Society for Microbiology, Washington, D.C.

Summanen, P.E., Baron, E.J., Citron, D.M., et al. 1993. Wadsworth anaerobic bacteriology manual, ed 5. Star, Belmont, Calif.

59 LABORATORY CONSIDERATIONS

MACROSCOPIC EXAMINATION OF SPECIMENS

Upon receipt in the laboratory, specimens should be inspected for characteristics that strongly indicate the presence of anaerobes, such as (1) foul odor; (2) sulfur granules associated with *Actinomyces* spp., *Propionibacterium* spp., or *Eubacterium nodatum;* or (3) brick-red fluoresence under long-wavelength ultraviolet (UV) light associated with pigmented *Prevotella* or *Porphyromonas.*

DIRECT DETECTION METHODS

ANTIGEN DETECTION

The cytotoxin (toxin B) of *Clostridium difficile* can be detected using a tissue culture assay. This assay, performed in various cell lines, is based on the neutralization of cytopathic effect if a cell-free fecal extract is adsorbed by either *Clostridium sordellii* or *C. difficile* antitoxins. A latex particle agglutination test or an enzyme-linked immunosorbent assay (ELISA) to detect toxins A and/or B are also available.

GRAM STAIN

The Gram stain is an important rapid tool for anaerobic bacteriology. Not only does it reveal the types and relative numbers of microorganisms and host cells present but it also serves as a quality control measure for the adequacy of anaerobic techniques. The absence of leukocytes does not rule out the presence of a serious anaerobic infection, however, because certain organisms, such as clostridia, produce necrotizing toxins that destroy white blood cells. A positive Gram stain with a negative culture may indicate (1) poor transport methods, (2) excessive exposure to air during specimen processing, (3) failure of the system (jar, pouch, or chamber) to achieve an anaerobic atmosphere, (4) inadequate types of media or old media, or (5) that microorganisms have been killed by antimicrobial therapy.

Standard Gram stain procedures and reagents are used, except that the safranin counterstain is left on for 3 to 5 minutes. Alternatively, 0.5% aqueous basic fuchsin can be used as the counterstain.

Table 59-1 indicates the cellular morphology on Gram stain of common anaerobes.

CULTIVATION

MEDIA OF CHOICE

The primary plating media for inoculating anaerobic specimens can be picked from the media listed in Table 58-4 on p. 694. In general, they include a nonselective anaerobic blood agar and one or all of the following selective media: *Bacteroides* bile esculin agar (BBE), laked kanamycin-vancomycin blood agar (LKV), or anaerobic phenylethyl alcohol agar (PEA). In addition, aerobic 5% sheep blood agar, chocolate agar, and MacConkey agar are setup because most anaerobic infections are polymicrobic and may include aerobic or facultative anaerobic bacteria. In addition, a backup broth, usually thioglycollate, to enrich for small numbers of anaerobes is included. Most anaerobes grow well on any of the foregoing media.

Cultures for *C. difficile* are plated on a special selective medium, cycloserine cefoxitin fructose agar (CCFA). There are also selective media for certain groups of anaerobes, such as *Actinomyces,* although these are rarely used in the clinical laboratory.

Special anaerobic blood bottles containing various media, including thioglycollate broth, thiol broth, and Schaedler's broth are commercially available. Although many anaerobes will grow in the aerobic blood culture bottle, it is better to use an unvented anaerobic broth when attempting to isolate these organisms from blood or bone marrow.

INCUBATION CONDITIONS AND DURATION

Inoculated plates should be immediately incubated under anaerobic conditions at 35° to 37° C for 48 hours. In general, cultures should not be exposed to oxygen until after 48 hours' incubation, because anaerobes are most sensitive to oxygen during their log phase of growth. Furthermore, colony morphology often changes dramatically between 24 and 48 hours. Plates incubated in an anaerobe chamber or

Text continued on p. 704

TABLE 59-1 GRAM-STAIN MORPHOLOGY, COLONIAL APPEARANCE, AND OTHER DISTINGUISHING FEATURES OF COMMON ANAEROBIC BACTERIA

ORGANISM	GRAM STAIN*	MEDIUM	APPEARANCE
Actinomyces spp.	Gram-positive, branching, beaded or banded, thin, filamentous rods	Ana BA	Colonies of most species are small, smooth, flat, convex, gray-white, translucent, with entire margins; colonies of *A. israelii* and *A. gerencseriae* are white, opaque, and may resemble a "molar tooth"; *A. odontolyticus* turns red after several days in ambient air and may be beta-hemolytic
Bacteroides distasonis	Gram-negative, straight rods with rounded ends; occur singly or in pairs	Ana BA	Gray-white, circular, entire, convex, smooth, translucent to opaque; nonhemolytic
		BBE	At 48 hrs, colonies are >1 mm, circular, entire, raised, and either (1) low convex, dark gray, friable, and surrounded by a dark gray zone (esculin hydrolysis) and sometimes a precipitate (bile) or (2) glistening, convex, light to dark gray, and surrounded by a gray zone
B. fragilis	Gram-negative, pale-staining, pleomorphic rods with rounded ends; occur singly or in pairs; cells often described as resembling a safety pin (Figure 59-1)	Ana BA	White to gray, circular, entire, convex, translucent to semiopaque; nonhemolytic (Figure 59-2)
		BBE	At 48 hrs, colonies are >1 mm, circular, entire, raised, and either (1) low convex, dark gray, friable, and surrounded by a dark gray zone (esculin hydrolysis) and sometimes a precipitate (bile) or (2) glistening, convex, light to dark gray, and surrounded by a gray zone (Figure 59-3)
B. ovatus	Gram-negative, ovoid rods with rounded ends; occur singly or in pairs	Ana BA	Pale buff, circular, entire, convex, semiopaque; often mucoid; nonhemolytic
		BBE	At 48 hrs, colonies are >1 mm, circular, entire, raised, and either (1) low convex, dark gray, friable, and surrounded by a dark gray zone (esculin hydrolysis) and sometimes a precipitate (bile) or (2) glistening, convex, light to dark gray, and surrounded by a gray zone
B. thetaiotaomicron	Gram-negative, irregularly staining, pleomorphic rods with rounded ends; occur singly or in pairs	Ana BA	White, circular, entire, convex, semiopaque, shiny, punctiform; nonhemolytic

Continued

TABLE 59-1 GRAM-STAIN MORPHOLOGY, COLONIAL APPEARANCE, AND OTHER DISTINGUISHING FEATURES OF COMMON ANAEROBIC BACTERIA—CONT'D

ORGANISM	GRAM STAIN*	MEDIUM	APPEARANCE
		BBE	At 48 hrs, colonies are >1 mm, circular, entire, raised, and either (1) low convex, dark gray, friable, and surrounded by a dark gray zone (esculin hydrolysis) and sometimes a precipitate (bile) or (2) glistening, convex, light to dark gray, and surrounded by a gray zone
B. ureolyticus	Gram-negative, pale-staining, thin delicate rods with rounded ends; some curved	Ana BA	Small, translucent or transparent; may produce greening of agar on exposure to air; colonies corrode (pit) the agar (Figure 59-4); or may be smooth and convex or spreading
B. vulgatus	Gram-negative, pleomorphic rods with rounded ends; occur singly, in pairs, or in short chains; swellings or vacuoles may be seen	Ana BA	Gray, circular, entire, convex, semiopaque; nonhemolytic
		BBE	At 48 hrs, colonies are >1 mm, circular, entire, raised, glistening, convex, light to dark gray but with no gray zone (esculin not hydrolyzed)
Bifidobacterium spp.	Gram-positive diphtheroid; coccoid or thin, pointed shape; or larger, highly irregular, curved rods with branching; rods terminate in clubs or thick, bifurcated (forked) ends ("dog bones")	Ana BA	Small, white, convex, shiny, with irregular edge
Bilophila wadsworthia	Gram-negative, pale-staining, delicate rods	Ana BA	Small, translucent
		BBE	Grows at 3 to 5 days; colonies are usually gray with a black center because of production of hydrogen sulfide (H_2S); black center may disappear after exposure to air
Clostridium botulinum	Gram-positive, straight rods occurring singly or in pairs; spores usually subterminal and resemble a tennis racket	Ana BA	Gray-white; circular to irregular; usually beta-hemolytic
C. clostridioforme	Gram-positive rod that stains gram-negative; long, thin rods; spores usually not seen; elongated football shape with cells often in pairs	Ana BA	Small, convex, entire edge; nonhemolytic

TABLE 59-1 GRAM-STAIN MORPHOLOGY, COLONIAL APPEARANCE, AND OTHER DISTINGUISHING FEATURES OF COMMON ANAEROBIC BACTERIA — CONT'D

ORGANISM	GRAM STAIN*	MEDIUM	APPEARANCE
C. difficile	Gram-positive straight rods; may produce chains of up to 6 cells aligned end to end; spores oval and subterminal	Ana BA	Large, white, circular, matte to glossy, convex, opaque; nonhemolytic; horse stable odor; fluoresces yellow-green
		CCFA	Yellow, ground-glass colony (Figure 59-5)
C. perfringens	Gram-variable straight rods with blunt ends occurring singly or in pairs; spores seldom seen but if present are large and central to subterminal, oval, and swell cell; large boxcar shapes (Figure 59-6)	Ana BA	Gray to grayish yellow; circular, glossy, dome-shaped, entire, translucent; double zone of beta hemolysis (Figure 59-7)
C. ramosum	Gram-variable straight or curved rods; spores rarely seen but are round and terminal; more slender and longer than C. perfringens	Ana BA	Small, gray-white to colorless; circular to slightly irregular, smooth, translucent or semiopaque; nonhemolytic
C. septicum	Gram-positive in young cultures but becomes gram-negative with age; stains unevenly; straight or curved rods occurring singly or in pairs; spores subterminal, and oval, and swell cells	Ana BA	Gray; circular, glossy, translucent; markedly irregular to rhizoid margins resembling a "Medusa head"; beta-hemolytic; swarms over entire agar surface in less than 24 hrs
C. sordellii	Gram-positive rods; subterminal spores	Ana BA	Large colony with irregular edge
C. sporogenes	Gram-positive rods; subterminal spores	Ana BA	Colonies firmly adhere to agar; may swarm over agar surface
C. tertium	Gram-variable rods; terminal spores	Ana BA	Resembles *Lactobacillus* spp.
C. tetani	Gram-positive becoming gram-negative after 24 hrs' incubation; occur singly or in pairs; spores oval and terminal or subterminal with drumstick or tennis racket appearance	Ana BA	Gray; matte surface, irregular to rhizoid margin, translucent, flat; narrow zone of beta hemolysis; may swarm over agar surface
Eubacterium spp.	Gram-positive pleomorphic rods or coccobacilli occurring in pairs or short chains; *E. alactolyticum* has a seagull-wing shape similar to *Campylobacter*; *E. nodatum* is similar to *Actinomyces*, with beading, filaments, and branching; *E. lentum* is a small, straight rod with rounded ends	Ana BA	Small, gray, transparent to translucent, raised to convex; colonies of *E. nodatum* may resemble *A. israelii*

Continued

TABLE 59-1 GRAM-STAIN MORPHOLOGY, COLONIAL APPEARANCE, AND OTHER DISTINGUISHING FEATURES OF COMMON ANAEROBIC BACTERIA—CONT'D

ORGANISM	GRAM STAIN*	MEDIUM	APPEARANCE
Fusubacterium mortiferum	Gram-negative, pale-staining, irregularly stained, highly pleomorphic rods with swollen areas, filaments, and large, bizarre, round bodies	Ana BA	Circular; entire or irregular edge, convex or slightly umbonate, smooth, translucent; nonhemolytic
		BBE	>1 mm in diameter, flat and irregular
F. necrophorum	Gram-negative, pleomorphic rods with round to tapered ends; may be filamentous or contain round bodies; becomes more pleomorphic with age	Ana BA	Circular, umbonate, ridged surface, translucent to opaque; fluoresces chartreuse; greening of agar on exposure to air; some strains beta-hemolytic
F. nucleatum	Gram-negative, pale-staining, long, slender, spindle-shaped with sharply pointed or tapered ends; occasionally the cells occur in pairs end to end; resembles *Capnocytophaga* (Figure 59-8)	Ana BA	Three colony types: bread crumblike (white; Figure 59-9), speckled, or smooth (gray to gray-white); greening of the agar on exposure to air; fluoresces chartreuse; usually nonhemolytic
F. varium	Gram-negative, unevenly staining, pleomorphic, coccoid, and rod shapes occuring singly or in pairs	Ana BA	Gray-white center with colorless edge resembling a fried egg; circular, entire, convex, translucent; nonhemolytic
		BBE	>1 mm in diameter, flat and irregular
Lactobacillus spp.	Gram-variable pleomorphic rods or coccobacilli; straight, uniform rods have rounded ends; short coccobacilli resemble streptococci	Ana BA	Resemble *Lactobacillus* spp. colonies on aerobic blood or chocolate agar, except colonies are usually larger when incubated anaerobically
Leptotrichia spp.	Gram-negative, large fusiform rods with one pointed end and one blunt end	Ana BA	Large, raspberry-like colonies
Mobiluncus spp.	Gram-variable, small, thin, curved rods; the two species can be divided based on length of the cell	Ana BA	Tiny colonies after 48 hrs' incubation; after 3 to 5 days colonies are small, low convex, and translucent
Peptococcus niger	Gram-positive, spherical cells occurring singly and in pairs, tetrads, and irregular masses	Ana BA	Tiny, black, convex, shiny, smooth, circular, entire edge; becomes light gray when exposed to air

TABLE 59-1 GRAM-STAIN MORPHOLOGY, COLONIAL APPEARANCE, AND OTHER DISTINGUISHING FEATURES OF COMMON ANAEROBIC BACTERIA—CONT'D

ORGANISM	GRAM STAIN*	MEDIUM	APPEARANCE
Peptostreptococcus anaerobius	Gram-positive large coccobacillus; often in chains	Ana BA	Medium, gray-white, opaque; sweet, fetid odor; colonies usually larger than most anaerobic cocci (Figure 59-10)
P. asaccharolyticus	Gram-positive cocci occurring in pairs, tetrads, or irregular clusters	Ana BA	Small, gray, translucent
P. magnus	Gram-positive cocci with cells >0.6 μm in diameter; in pairs and clusters; resemble staphylococci	Ana BA	Tiny, gray, translucent; nonhemolytic
P. micros	Gram-positive cocci with cells <0.7μm in diameter; occur in packets and short chains	Ana BA	Tiny, white, opaque; nonhemolytic
Porphyromonas spp.	Gram-negative coccobacilli	Ana BA	Dark brown to black; more mucoid than *Prevotella*; except for *P. gingivalis*, fluoresces brick red (Figure 59-11)
Prevotella disiens	Gram-negative rods; occur in pairs or short chains	Ana BA	White, circular, entire, convex, translucent to opaque, smooth, shiny; nonhemolytic; fluoresces brick red
		LKV	Black pigment (Figure 59-12)
P. melaninogenica	Gram-negative coccobacilli	Ana BA	Dark center with gray to light brown edges, circular, entire, convex, smooth, shiny; nonhemolytic; fluoresces brick red
		LKV	Black pigment
Propionibacterium spp.	Gram-positive pleomorphic diphtheroid-like rod; club-shaped to palisade arrangements; called *anaerobic diphtheroids*	Ana BA	Young colonies are small and white to gray-white and become larger and more yellowish tan with age; *P. avidum* is beta-hemolytic
Veillonella parvula	Gram-negative, tiny diplococci in clusters, pairs, and short chains; unusually large cocci, especially in clusters, suggests *Megasphaera* or *Acidaminococcus*	Ana BA	Small, almost transparent; grayish white; smooth, entire, opaque, butyrous; may show red fluorescence under UV light (360 nm)

Ana BA, anaerobic blood agar; *LKV*, laked kanamycin-vancomycin blood agar; *CCFA*, cycloserine cefoxitin fructose agar; *BBE*, *Bacteroides* bile esculin agar.

*Typical Gram stain appearance will be seen from broth (thioglycollate or peptone-yeast-glucose).

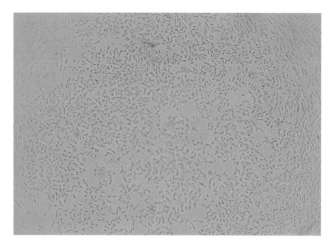

FIGURE 59-1 Gram stain of *Bacteroides fragilis*.

FIGURE 59-2 *Bacteroides fragilis* on anaerobic blood agar.

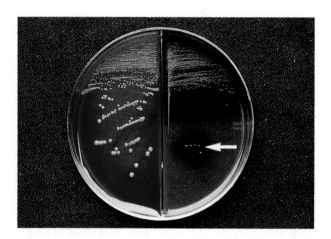

FIGURE 59-3 *Bacteroides fragilis* on *Bacteroides* bile esculin agar (BBE) (*arrow*). (Courtesy Anaerobe Systems, San Jose, Calif.)

FIGURE 59-4 *Bacteroides ureolyticus* on anaerobic blood agar. Note pitting of agar (*arrow*). (Courtesy Anaerobe Systems, San Jose, Calif.)

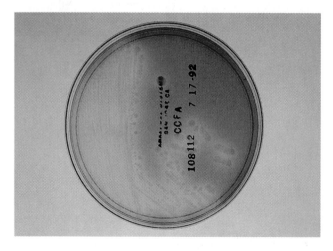

FIGURE 59-5 *Clostridium difficile* on cycloserine cefoxitin fructose agar (CCFA). (Courtesy Anaerobe Systems, San Jose, Calif.)

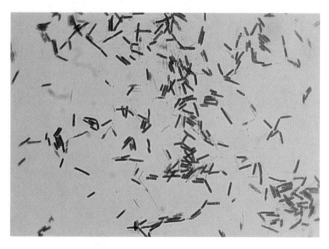

FIGURE 59-6 Gram stain of *Clostridium perfringens*.

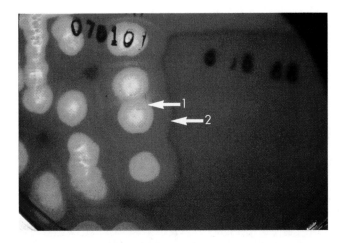

FIGURE 59-7 *Clostridium perfringens* on anaerobic blood agar. Note double zone of beta hemolysis. *1,* First zone; *2,* Second zone. (Courtesy Anaerobe Systems, San Jose, Calif.)

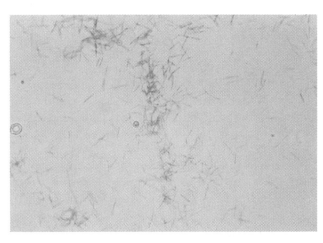

FIGURE 59-8 Gram stain of *Fusobacterium nucleatum.* Note pointed ends.

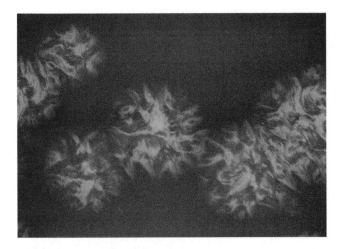

FIGURE 59-9 *Fusobacterium nucleatum* on anaerobic blood agar. Note bread crumblike colonies and greening of agar.

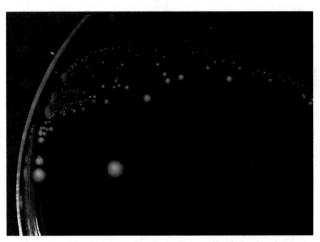

FIGURE 59-10 *Peptostreptococcus anaerobius* on anaerobic blood agar.

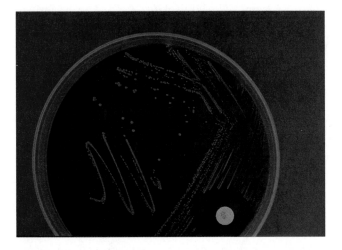

FIGURE 59-11 *Porphyromonas* spp. on anaerobic blood agar. Red fluorescence under ultraviolet light (365 nm). (Courtesy Anaerobe Systems, San Jose, Calif.)

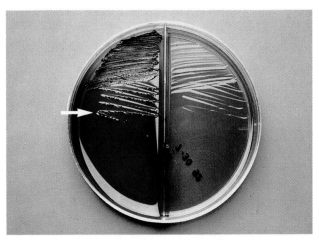

FIGURE 59-12 *Prevotella intermedia* on laked kanamycin-vancomycin blood agar. Note black pigment (*arrow*).

bag can be examined at 24 hours without oxygen exposure for typical colonies of *B. fragilis* group or *C. perfringens*. Plates showing no growth at 48 hours should be incubated for at least 5 days before discarding.

The thioglycollate broth can be incubated anaerobically with the cap loose or aerobically with the cap tight. Broths should be inspected daily for 7 days.

COLONIAL APPEARANCE

The colonial appearance and other characteristics (e.g., hemolysis and odor) of common anaerobes are shown in Table 59-1.

APPROACH TO IDENTIFICATION

Complete identification of anaerobes can be very costly, often requiring various biochemical tests, gas-liquid chromatography to analyze the metabolic end products of glucose fermentation, and/or gas-liquid chromatography to analyze cellular fatty acids. Most clinical laboratories no longer perform complete identification of anaerobes, because presumptive identification is just as useful in assisting the physician in determining appropriate therapy. Therefore, the approach to identification taken in this chapter emphasizes simple, rapid methods to identify commonly isolated anaerobic bacteria. Identification should proceed in a stepwise fashion, beginning with examination of the primary plates.

EXAMINATION OF PRIMARY PLATES

Anaerobes are usually present in mixed culture with other anaerobes and facultative bacteria. The combination of selective and differential agar plates yields information that suggests the presence and perhaps the types of anaerobe(s). Primary anaerobic plates should be examined with a hand lens ($\times$ 8) or, preferably, a stereoscopic microscope. Colonies should be described from the various media and semiquantitated.

All colony morphotypes from the nonselective anaerobic blood agar should be characterized and subcultured to purity plates, because facultative and obligately anaerobic bacteria frequently have similar colony appearances. Colonies on the PEA (phenylethyl alcohol agar) are processed further only if they are different from colonies growing on the anaerobic blood agar or if colonies on the anaerobic blood agar are impossible to subculture because of overgrowth by swarming clostridia, *Proteus,* or other organisms.

The back-up broth (e.g., thioglycollate) should be Gram-stained; if cellular types are seen that were not present on the primary plates, the broth should be subcultured.

SUBCULTURE OF ISOLATES

A single colony of each distinct morphotype is examined microscopically using a Gram stain and is subcultured for **aerotolerance testing.** A sterile wooden stick or platinum loop should be used to subculture colonies to:

- An anaerobic blood agar plate to be incubated anaerobically (purity plate)
- A chocolate agar plate to be incubated in carbon dioxide (CO_2)

The chocolate agar plate should be inoculated first, so that if only the anaerobic blood agar plate grows there is no question of not having enough organisms to initiate growth. The following antibiotic identification disks are placed on the first quadrant of the purity plate (Procedure 59-1):

- Kanamycin, 1 mg
- Colistin, 10 μg
- Vancomycin, 5 μg

These disks aid in preliminary grouping of anaerobes and serve to verify the Gram stain, but they *do not* imply susceptibility of an organism for antibiotic therapy.

Three other disks may be added to the anaerobic blood agar plate at this time. A nitrate disk may be placed on the second quadrant for subsequent determination of nitrate reduction. A sodium polyanetholsulfonate (SPS) disk can be placed near the colistin disk for rapid presumptive identification of *Peptostreptococcus anaerobius* if gram-positive cocci are seen on Gram stain. A bile disk may be added to the second quadrant to detect bile inhibition if gram-negative rods are seen on Gram stain.

If processing is performed on the open bench, all plates should promptly be incubated anaerobically, because some clinical isolates (e.g., *Fusobacterium necrophorum* and some *Prevotella* spp.) may die after relatively short exposure to oxygen. The primary plates are reincubated along with the purity plates for an additional 48 to 72 hours and are again inspected for slowly growing or pigmenting strains.

PRESUMPTIVE IDENTIFICATION OF ISOLATES

Information from the primary plates in conjunction with the atmospheric requirements, Gram stain, and colony morphology of a pure isolate provides preliminary separation of many anaerobic organisms. Table 59-2 summarizes the extent to which isolates can be identified using this information. It is also useful to consider the specimen source and expected organisms from that site to aid in this process.

TABLE 59-2   PRELIMINARY GROUPING OF ANAEROBIC BACTERIA BASED ON MINIMAL CRITERIA

ORGANISM	GRAM REACTION	CELL SHAPE	GRAM-STAIN MORPHOLOGY	AEROTOLERANCE	DISTINGUISHING CHARACTERISTICS
Bacteroides fragilis group	–	B	Can be pleomorphic with safety pin appearance	–	Grows on BBE; > 1mm in diameter; some strains hydrolyze esculin
Pigmented gram-negative bacilli	–	B, CB	Can be very coccoid or *Haemophilus*-like	–	Foul odor; black or brown pigment; some fluoresce brick red
Bacteroides ureolyticus	–	B	Thin; some curved	–	May pit agar or spread; transparent colony
Fusobacterium nucleatum	–	B	Slender cells with pointed ends	–	Foul odor; three colony types; bread crumblike, speckled, or smooth
Gram-negative bacillus	–	B		–	
Gram-negative coccus	–	C	*Veillonella* cells are tiny	–	
Gram-positive coccus	+	C, CB	Variable size	–	
Clostridium perfringens (presumptive)	+	B	Large; boxcar shape; no spores observed; may appear gram-negative	–	Double-zone beta hemolysis
Clostridium spp.	+	B	Spores usually observed; may appear gram-negative	–*	
Gram-positive bacillus	+	B, CB	No spores observed, no boxcar-shaped cells	–*	
Actinomyces-like	+	B	Branching cells	–*	Sulfur granules on direct examination; "molar tooth" colony

–, Negative; +, positive; B, bacillus; C, coccus; CB, coccobacillus; BBE, *Bacteroides* bile esculin agar.

*Some strains are aerotolerant; these include *Clostridium tertium*, *C. histolyticum*, some bifidobacteria, some propionibacteria, and most *Actinomyces* spp.

Presumptive identification of many clinically relevant anaerobic bacteria can be accomplished using a few simple tests. These tests are shown in Tables 59-3 and 59-4 and are discussed in detail in another text.[6]

The gram-negative organisms are divided into the following major categories, based on microscopic morphology, antibiotic identification disks pattern, or a few other simple tests.

Bacteroides fragilis group

The gram-negative *Bacteroides fragilis* group grow in 20% bile and are almost always resistant to all three special-potency antibiotic disks. Rare strains of *B. fragilis* are susceptible to colistin.

Nonpigmented *Prevotella* spp.

Most bile-sensitive, kanamycin-resistant, gram-negative rods belong in the genus *Prevotella*. Colistin susceptibility is variable, and almost all strains are catalase- and indole-negtive.

Pigmented *Prevotella* and *Porphyromonas* spp.

Colonies that flouresce brick red or produce brown to black pigment are placed among the pigmented *Pre-*

— TABLE 59-3 PRESUMPTIVE IDENTIFICATION OF GRAM-NEGATIVE ANAEROBES —

	CELL SHAPE	SLENDER CELLS WITH POINTED ENDS	KANAMYCIN (1 mg)	VANCOMYCIN (5 μg)	COLISTIN (10 μg)	GROWTH IN BILE	INDOLE
Gram-negative rods							
Bacteroides fragilis group	B	–	R	R	R	+	V
Pigmented spp.	B, CB	–	R	R	V	–	V
Prevotella intermedia	B, CB	–	R	R	S	–	+
Prevotella loescheii	B, CB	–	R	R	R[S]	–	–
Other *Prevotella* spp.	B, CB	–	R	R	V	–	V
Porphyromonas spp.	B, CB	–	R	S	R	–	+
Bilophila sp.	B	–	S	R	S	+	–
B. ureolyticus	B	–	S	R	S	–	–
Fusobacterium spp.	B	V	S	R	S	V	V
F. nucleatum	B	+	S	R	S	–	+
F. necrophorum	B	–	S	R	S	–[+]	+
F. mortiferum-varium	B	–	S	R	S	+	V
Gram-negative cocci							
Veillonella	C	–	S	R	S	–	–

Reactions in **bold type** are key tests; *B*, bacillus; *CB*, coccobacillus; *C*, coccus; *R*, resistant; *S*, sensitive; *V*, variable; +, positive; −, negative; superscripts indicate

*P. melaninogenica group often requires prolonged incubation before pigment is observed.

†*P. gingivalis* does not fluoresce.

‡*P. bivia* produces pigment on prolonged incubation.

Modified from Summanen, P.E., Baron, E.J., Citron, D.M., et al. 1993. Wadsworth anaerobic bacteriology manual, ed 5. Star, Belmont, Calif.

votella and *Porphyromonas* spp. Some species are coccobacillary.

Bacteroides ureolyticus

Bacteroides ureolyticus reduces nitrate and requires formate and fumarate for growth in broth culture. Its disk pattern is the same as the fusobacteria; however, colony morphology is different. *B. ureolyticus* forms small, translucent to transparent colonies that may corrode the agar, whereas the *Fusobacterium* colony is generally larger and more opaque. *B. ureolyticus* formerly was grouped with organisms that have been transferred to

the genus *Campylobacter recta,* and *C. curva).* The ⁕ new genus, *Sutterella,* are anaerobes, and are discussed in motile organisms that grow anae⁕ 5% CO_2 should be retested in a micr⁕ phere with approximately 6% oxygen.

Other Bacteroides or Prevotella spp.

The term **other *Bacteroides* or *Prevotella*** ⁕ applied to gram-negative bacilli that do not fu⁕ preceding categories or the *Fusobacterium* spp.

CATALASE	PIGMENTED COLONY	BRICK RED FLUORESCENCE	LIPASE	PITS THE AGAR	REQUIRES FORMATE- FUMARATE	NITRATE REDUCTION	UREASE	MOTILE
V	−	−	−	−	−	−	−	−
−	+*	V	V	−	−	−	−	−
−	+	+	+	−	−	−	−	−
−	+*	+	−+	−	−	−	−	−
−	−‡	−	−	−	−	−	−	−
−	+	+†	−	−	−	−	−	−
+	−	−	−	−	−	+	+−	−
−	−	−	−	+−	+	+	+	−
−	−	−	V	−	−	−	−	−
−	−	−	−	−	−	−	−	−
−	−	−	+−	−	−	−	−	−
−	−	−	−	−	−	−	−	−
V	−	V	−	−	−	V	−	−
V	−	−+	−	−	−	+	−	−

reactions of occasional strains.

TABLE 59-4 PRESUMPTIVE IDENTIFICATION OF GRAM-POSITIVE ANAEROBES

	CELL SHAPE	SPORES OBSERVED	BOXCAR-SHAPED CELLS	DOUBLE-ZONE BETA HEMOLYSIS	KANAMYCIN (1 mg)	VANCOMYCIN (5 µg)	COLISTIN (10 µg)	INDOLE	SODIUM POLYANETHOL SULFONATE
Gram-positive cocci	C, CB	−	−	−	V	S	R	V	V
Peptostreptococcus anaerobius	C, CB	−	−	−	**R**^{s}	S	R	−	S
Peptostreptococcus asaccharolyticus	C	−	−	−	S	S	R	**+**	R
Gram-positive rods **Spore-forming**									
Clostridium spp.	B	**+**$^-$	−$^+$	−	V	S	R	V	
Nagler-positive C. perfringens	B	−	**+**	**+**$^-$	S	S	R	−	
C. baratii	B	+	−	−	S	S	R	−	
C. sordellii	B	+	−	−	S	S	R	**+**	
C. bifermentans	B	+	−	−	S	S	R	**+**	
Nagler-negative Presumptive C. difficile	B	+$^-$	−	−	S	S	R	−	
Non–spore-forming	B, CB	−	−	−	V	S	R	V	
Propionibacterium acnes	B, CB	−	−	−	S	S	R	**+**$^-$	
Eubacterium lentum	B	−	−	−	S	S	R	−	

Reactions in **bold type** are key tests; *B*, bacillus; *CB*, coccobacillus; *C*, coccus; *R*, resistant; *S*, sensitive; *V*, variable; +, positive; −, negative; *w*, weak; superscripts
*Cycloserine cefoxitin fructose agar.

Modified from Summanen, P.E., Baron, E.J., Citron, D.M., et al. 1993. Wadsworth anaerobic bacteriology manual, ed 5. Star, Belmont, Calif.

CATALASE	SURVIVES ETHANOL SPORE TEST	LECITHINASE	NAGLER TEST	STRONG REVERSE-CAMP TEST	ARGININE STIMULATION	UREASE	NITRATE REDUCTION	GROUND-GLASS, YELLOW COLONIES ON CCFA* MEDIUM	COMMENTS
V	−	−	−		−		$-^+$	−	
−	−	−	−		−		−	−	Sweet, putrid odor; may chain
V	−	−	−		−		−	−	Indole odor
−	$+^-$	V	−		−	V	$-^+$	V	
−	$-^+$	+	+	+	−	−	$+^-$	−	
−	+	+	$+^w$	−	−	−	V	−	
−	+	+	$+^w$	−	−	$+^-$	−	−	
−	+	+	$+^w$	−	−	−	−	−	
−	+	−	−	−	−		−	+	Horse stable odor; fluoresces charteuse
V	−	V	−		V	V	V	−	
+	−	−	−	−	−	−	+	−	May branch; diphtheroid
$-^+$	−	−	−	−	+	−	+	−	Small rod

indicate reactions of occasional strains.

Fusobacterium spp.

The gram-negative *Fusobacterium* spp. are sensitive to kanamycin, and most strains flouresce a chartreuse color. Different species have characteristic cell and colony morphology.

Anaerobic gram-negative cocci

The category of anaerobic gram-negative cocci is based on Gram stain morphology and includes *Veillonella*, *Megasphaera*, and *Acidaminococcus*.

The gram-positive organisms are divided into three major categories based on microscopic morphology and the presence of spores.

Anaerobic gram-positive cocci

Among anaerobic gram-positive cocci, the genera of clinical importance are *Peptostreptococcus* and *Streptococcus*. If a coccus is resistant to metronidazole, it probably is an anaerobic *Streptococcus* or *Staphylococcus*.

Anaerobic, gram-positive, spore-forming bacilli

The clostridia are the endospore-forming, anaerobic, gram-positive bacilli. If spores are not present on Gram stain, the ethanol spore or heat spore test will separate this group from the nonspore-forming anaerobic bacilli. Some strains of *C. perfringens*, *C. ramosum*, and *C. clostridioforme* may not produce spores or survive a spore test, so it is important to recognize these organisms using other characteristics. Some clostridia typically stain gram-negative, although they are susceptible to vancomycin on the disk test. Several species of clostridia grow aerobically, but they produce spores only under anaerobic conditions.

Anaerobic, gram-positive, non–spore-forming bacilli

The genera *Actinomyces*, *Bifidobacterium*, *Eubacterium*, anaerobic *Lactobacillus*, and *Propionibacterium* are among the anaerobic, gram-positive, non-spore-forming bacilli. It is difficult to differentiate this group accurately to the genus level without end-product analysis, except for *P. acnes* and *E. lentum*.

DEFINITIVE IDENTIFICATION

Various techniques are available for definitive identification of anaerobic bacteria. These methods may include:

- Prereduced anaerobically sterilized (PRAS) biochemicals
- Miniaturized biochemical systems (e.g., API 20A, bioMérieux Vitek, Hazelwood, Mo. and Minitek, Becton Dickinson Microbiology Systems, Cockeysville, Md.)

- Rapid, preformed enzyme detection panels (e.g., An-Ident, bioMérieux Vitek; RapID-ANA II, Innovative Diagnostic Systems, Norcross, Ga.; BBL Brand Crystal Anaerobe ID, Becton Dickinson Microbiology Systems; Rapid Anaerobe Identification Panel, Dade Behring MicroScan, West Sacramento, Calif.; Vitek ANI card, bioMérieux Vitek)
- Gas-liquid chromatography (GLC) for end products of glucose fermentation (Dodeca, Fremont, Calif.)
- High-resolution gas-liquid chromatography (GLC) for cellular fatty acid analysis (Microbial Identification System, MIDI, Newark, Del.)

None of the commercial identification systems reliably identify all anaerobic bacteria. Therefore, their high cost alone probably does not justify their use in most clinical laboratories. To ensure accurate identification reference laboratories use a combination of PRAS biochemicals and GLC or high-resolution GLC. Several approaches to complete anaerobic identification are discussed in other texts.[1-3,6] Identification criteria for the aerotolerant *Actinomyces* spp. are presented in Table 56-4 and in Tables 59-2 and 59-4.

ANTIMICROBIAL SUSCEPTIBILITY TESTING AND THERAPY

When mixed infections are encountered, definitive information regarding the identification of each species present usually will not affect therapeutic management. Because most clinically relevant anaerobes are susceptible to first-line antimicrobials (Table 59-5), knowledge of their presence and Gram-stain morphologies in mixed cultures is usually sufficient for guiding therapy. Therefore, definitive identification methods that follow the schemes just outlined should be judiciously applied to clinical situations in which an anaerobic organism is isolated in pure culture from a normally sterile site (e.g., clostridial myonecrosis), when there is solid evidence for clinical relevance, or when antimicrobial susceptibility testing is being considered for the reasons listed in Box 59-1.

The therapeutic options listed for each of the major groups of anaerobic bacteria in Table 59-5 are the antimicrobial agents known to be effective against 95% or more of the organisms that constitute a particular organism group.[4] Therefore, therapeutic use of the antimicrobial agents listed generally precludes the need to routinely perform antimicrobial susceptibility testing with anaerobic isolates.

Although standard susceptibility testing methods have been established for testing anaerobic bacteria against various antimicrobial agents[5] (Table 59-6),

TABLE 59-5   ANTIMICROBIAL THERAPY AND SUSCEPTIBILITY TESTING OF ANAEROBIC BACTERIA

ORGANISM GROUP	THERAPEUTIC OPTIONS	POTENTIAL RESISTANCE TO THERAPEUTIC OPTIONS	VALIDATED TESTING METHODS*
Bacteroides fragilis group, other *Bacteroides* spp., *Porphyromonas* spp., *Prevotella* spp., and *Fusobacterium* spp.	Highly effective agents include most beta-lactam/beta-lactamase–inhibitor combinations, imipenem, metronidazole, and chloramphenicol	Beta-lactamase production does occur, but generally does not significantly affect imipenem or beta-lactamase–inhibitor combinations. However, isolates of *B. fragilis* are known to produce beta-lactamases capable of hydrolyzing imipenem. Metronidazole resistance is extremely rare; resistance to various cephalosporins or clindamycin does occur, and susceptibility to these agents cannot be assumed	Yes; see Table 59-6
Clostridium spp.	Penicillins, with or without beta-lactamase–inhibitor combinations and imipenem; metronidazole or vancomycin for *C. difficile*-induced gastrointestinal disease; for botulism and *C. perfringens* food poisoning antimicrobial therapy is not indicated	Resistance to therapeutic options is not common, but cephalosporins and clindamycin exhibit uncertain clinical efficacy	Yes; see Table 59-6
Actinomyces spp., *Propionibacterium* spp., *Bifidobacterium* spp., *Eubacterium* spp.	Penicillins, with or without beta-lactamase–inhibitor combinations, imipenem, cefotaxime, and ceftizoxime	Resistance to therapeutic options not common; generally resistant to many cephalosporins and metronidazole	Yes; see Table 59-6
Peptostreptococcus spp., and *Peptococcus niger*	Penicillins, most cephalosporins, imipenem, vancomycin, clindamycin, and chloramphenicol	Resistance to therapeutic options is not common	Yes; see Table 59-6

*Validated testing methods include those standard methods recommended by the National Committee for Clinical Laboratory Standards (NCCLS) and those commercial methods approved by the Food and Drug Administration (FDA).

TABLE 59-6 SUMMARY OF ANTIMICROBIAL SUSCEPTIBILITY TESTING METHODS FOR ANAEROBIC BACTERIA

| | TEST METHODS | |
TEST CONDITIONS	AGAR DILUTION	BROTH MICRODILUTION AND MACRODILUTION
Medium	*Brucella* blood agar	Various broths may be used, including Schaedler's, brain-heart infusion, or a Wilkens-Chalgren formulation without agarose. For some fastidious anaerobes, supplements such as horse serum may be required for growth. For microdilution the total volume for each antimicrobial test well is usually 0.1 mL; for macrodilution the final volume generally is 5 mL
Inoculum size	1×10^5 CFU/spot	1×10^6 CFU/mL
Incubation conditions	Anaerobic, 35°-37° C	Anaerobic, 35°-37° C
Incubation duration	48 hrs	48 hrs

From National Committee for Clinical Laboratory Standards. 1993. Methods for antimicrobial susceptibility testing of anaerobic bacteria, ed 3. M11-A3. NCCLS, Villanova, Pa.

BOX 59-1 INDICATIONS FOR PERFORMING ANTIMICROBIAL SUSCEPTIBILITY TESTING WITH ANAEROBIC BACTERIA

To establish patterns of susceptibility of anaerobes to new antimicrobial agents

To periodically monitor susceptibility patterns of anaerobic bacteria collected within and among specific geographic areas or particular health care institutions

To assist in the therapeutic management of patients, but only when such information may be critical because of the following:

Known resistance of a particular species to most commonly used agents

Therapeutic failures and/or persistence of organism at site of infection

Lack of a precedence for therapeutic management of a particular infection

Severity of the infection (e.g., brain abscess, osteomyelitis, infections of prosthetic devices, and refractory or recurrent bacteremia)

From National Committee for Clinical Laboratory Standards. 1993. Methods for antimicrobial susceptibility testing of anaerobic bacteria, ed 3. M11-A3. NCCLS, Villanova, Pa.

the fastidious nature of many species and the labor intensity of using these methods indicate that testing should only be done under special circumstances (Box 59-1).

Although certain commercial methods (e.g., E-test and Spiral Biotech, Inc., described in Chapter 18) may facilitate anaerobic susceptibility testing in some way, the difficulty in assigning clinical significance to many anaerobic isolates and the availability of several highly effective empiric therapeutic choices significantly challenges a laboratory policy of routinely performing susceptibility testing with these organisms.

Procedure

59-1 ANTIBIOTIC IDENTIFICATION DISKS

PRINCIPLE

Most anaerobes have a characteristic susceptibility pattern to colistin (10 μg), vancomycin (5 μg), and kanamycin (1 mg) disks. The pattern generated will usually confirm a dubious Gram stain reaction (with few exceptions, most gram-positive anaerobes are susceptible to vancomycin) and aid in subdividing the anaerobic gram-negative bacilli into groups.

METHOD

1. Allow the three cartridges of disks to equilibrate to room temperature.

2. Transfer a portion of one colony to an anaerobic blood agar plate. Streak the first quadrant several times to produce a heavy lawn of growth, and then steak the other quadrants for isolation.

3. Place the colistin, kanamycin, and vancomycin disks in the first quadrant, well separated from each other (Figure 59-13).

4. Incubate the plates anaerobically for 48 hours at 35° C.

EXPECTED RESULTS

Observe for a zone of inhibition of growth. A zone of 10 mm or less indicates resistance, and a zone greater than 10 mm indicates susceptibility.

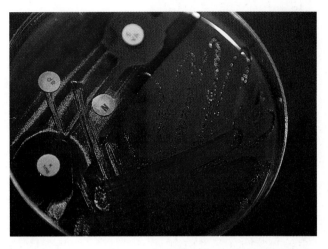

FIGURE 59-13 Special potency antibiotic and other disks. *Actinomyces ondontolyticus.* Note red pigment. (Courtesy Anaerobe Systems, San Jose, Calif.)

References

1. Dowell, V.R. and Hawkins, T.M. 1981. Labortory methods in anaerobic bacteriology: CDC laboratory manual. Centers for Disease Control, US Department of Health and Human Services, DHHS Publication No. (CDC) 81-8272, Atlanta, Ga.

2. Holdeman, L.V., Cato, E.P., and Moore, W.E.C., editors. 1987. Anaerobic laboratory manual: update, ed 4. Virginia Polytechnic Institute and State University, Blacksburg, Va.

3. Holdeman, L.V., Cato, E.P., and Moore, W.E.C., editors. 1977. Anaerobic laboratory manual, ed 4. Virginia Polytechnic Institute and State University, Blacksburg, Va.

4. Johnson, C.C. 1993. Susceptibility of anaerobic bacteria to β-lactam antibiotics in the United States. Clin. Infect. Dis. 16(suppl. 4):S371.

5. National Committee for Clinical Laboratory Standards. 1993. Methods for antimicrobial susceptibility testing of anaerobic bacteria, ed 3. M11-A3, NCCLS, Villanova, Pa. Order from NCCLS, 940 W. Valley Road, Suite 1400, Wayne, Pa. 19087.

6. Summanen, P., Baron, E.J., Citron, D.M., et al. 1993. Wadsworth anaerobic bacteriology manual, ed 5. Star, Belmont, Calif.

Bibliography

Finegold, S.M. 1995. Anaerobic bacteria: general concepts. In Mandell, G.L., Bennett, J.E., and Dolin, R., editors. Principles and practice of infectious diseases. Churchill Livingstone, New York.

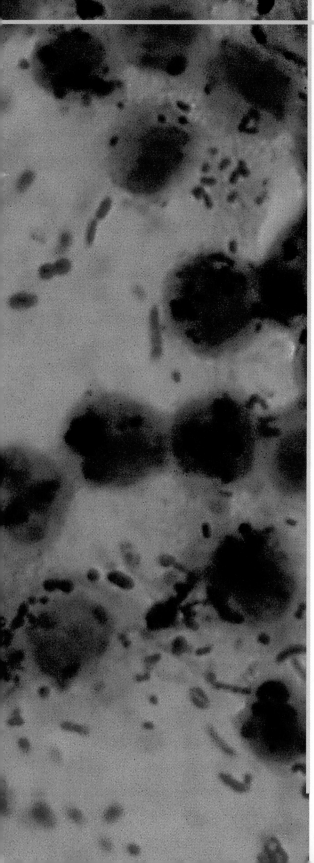

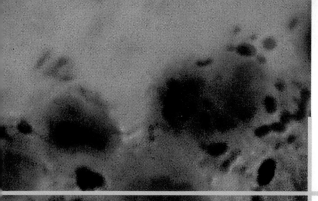

Section Thirteen 13

Mycobacteria and Other Bacteria with Unusual Growth Requirements

60 | MYCOBACTERIA

Genera and Species to be Considered

- *Mycobacterium tuberculosis* complex:
  Mycobacterium tuberculosis
  Mycobacterium bovis
  Mycobacterium bovis BCG
  Mycobacterium africanum

- Nontuberculous mycobacteria

Organisms belonging to the genus *Mycobacterium* are very thin, rod-shaped (0.2 to 0.4 × 2 to 10 μm), and nonmotile. This genus belongs in the *Mycobacteriaceae* family (the only genus in this family), *Actinomycetales* order, and *Actinomycetes* class. Genera that are closely related to members of the genus *Mycobacterium* include *Nocardia, Rhodococcus,* and *Corynebacterium. Mycobacterium* spp. have an unusual cell wall structure that contains N-glycolylmuramic acid in lieu of N-acetylmuramic acid and has a very high lipid content. Because of this cell wall structure, mycobacteria are difficult to stain with commonly used basic aniline dyes (see Chapter 11), such as those used in the Gram stain. However, these organisms resist decolorization by acidified alcohol (3% hydrochloric acid) after prolonged application of a basic fuchsin dye or with heating of this dye following its application. This important property of mycobacteria that is dependent on its cell wall is referred to as **acid-fastness.** By virtue of this acid-fast characteristic, mycobacteria can be distinguished from other genera. Another important feature of these organisms is that they grow more slowly than most other human pathogenic bacteria because of their hydrophobic cell surface. Because of this hydrophobicity, organisms tend to clump, so that nutrients are not easily allowed into the cell. Growth is slow or very slow, with colonies becoming visible in 2 to 60 days at optimum temperature.

Currently, there are 71 recognized or proposed species in the genus *Mycobacterium.*[45] These species produce a spectrum of infections in humans and animals ranging from localized lesions to disseminated disease. Although some species cause only human infections, others have been isolated from a wide variety of animals. Many species are also found in water and soil.

For the most part, mycobacteria can be divided into two major groups based on fundamental differences in epidemiology and association with disease: those belonging to the *Mycobacterium tuberculosis* complex (*M. tuberculosis, M. bovis, M. africanum*) and those referred to as **nontuberculous mycobacteria** (NTMs).[19] For the purposes of discussion, the mycobacteria are addressed using these two major groups (Box 60-1).

MYCOBACTERIUM TUBERCULOSIS COMPLEX

Tuberculosis was endemic in animals in the paleolithic period, long before it ever affected man.[50] This disease, also known as *consumption,* has been known in all ages and climates.[7] For example, tuberculosis was the subject of a hymn in a sacred text of India dating from 2500 BC. Most recently, DNA unique to *Mycobacterium tuberculosis* was identified in lesions from the lung of 1000-year-old human remains found in Peru.[44]

GENERAL CHARACTERISTICS

The term **complex** is frequently used in the clinical microbiology laboratory setting to describe two or more species whose distinction is complicated and of little or no medical importance.[45] The mycobacterial species that occur in humans and belong to the ***M. tuberculosis*** complex include *M. tuberculosis, M.*

MAJOR GROUPINGS OF ORGANISMS BELONGING TO THE GENUS *MYCOBACTERIUM*

BOX 60-1

Mycobacterium tuberculosis* complex

M. tuberculosis	M. bovis
M. bovis BCG	M. africanum

Nontuberculous Mycobacteria

Slow-Growing

   Nonphotochromogens:
     M. avium complex
     M. celatum
     M. gastri
     M. genavense
     M. haemophilum
     M. malmoense
     M. shimoidei
     M. simiae
     M. ulcerans
     M. xenopi
     M. terrae complex

   Photochromogens:
     M. kansasii
     M. asiaticum
     M. marinum
     M. intermedium

   Scotochromogens:
     M. szulgai
     M. scrofulaceum
     M. interjectum
     M. gordonae
     M. cookii
     M. hiberniae

Rapid-Growing

   Potentially pathogenic:
   M. fortuitum
   M. chelonae
   M. abscessus
   M. smegmatis
   M. peregrinum
   M. mucogenicum (former name: M. chelonae-like organism)
   M. fortuitum third biovariant complex, sorbitol-positive
   M. fortuitum third biovariant complex, sorbitol-negative

   Rarely pathogenic or not yet associated with infection:
   M. agri, M. aichiense, M. austroafricanum, M. aurum, M. chitae, M. chubuense, M. diernhoferi, M. duvalii, M. fallax, M. flavescens, M. gadium, M. gilvum, M. komossense, M. moriokaense, M. neoaurum, M. parafortuitum, M. obuense, M. phlei, M. pulveris, M. rhodesiae, M. senegalense, M. sphagni, M. thermoresistible, M. tokaiense, M. vaccae

Noncultivatable
   M. leprae

*This box is not all inclusive but rather lists only the prominent mycobacteria isolated from humans.

---

bovis, *M. bovis* BCG, and *M. africanum*; all species are capable of causing tuberculosis. It should be noted that species identification may be required for epidemiologic and public health reasons. Organisms belonging to the *M. tuberculosis* complex are considered slow-growers, and colonies are nonpigmented.

EPIDEMIOLOGY AND PATHOGENESIS

Epidemiology

All members belonging to *M. tuberculosis* complex cause tuberculous infections. *M. tuberculosis* is the cause of most cases of human tuberculosis, particularly in developed countries. An estimated 1.7 billion persons, one third of the world's population, are infected with *M. tuberculosis*. This reservoir of infected individuals results in 8 million new cases of tuberculosis and 2.9 million deaths annually.[3] Tuberculosis continues to be a public health problem in the United States, with more than 20,000 cases reported annually; an estimated 10 million persons in this country are already infected.[12] Of great concern was

the recent emergence of epidemic multidrug-resistant strains of *M. tuberculosis*[13] in conjunction with the ending of the downward trend of reported cases of tuberculosis in the United States.[14] This observed increase was largely attributed to the HIV epidemic, along with other complicating factors such as[47]:

- Noncompliance with antituberculosis treatment regimens
- Complications from infection
- Longer antituberculosis treatment regimens
- Multidrug-resistant isolates of *M. tuberculosis* and delays in identifying patients infected with these organisms
- Social factors (e.g., overcrowding and homelessness)

Although the organisms belonging to *M. tuberculosis* complex have numerous characteristics in common with one another, they do differ in certain aspects regarding their epidemiology, which is summarized in Table 60-1.

TABLE 60-1 EPIDEMIOLOGY OF ORGANISMS BELONGING TO THE *M. TUBERCULOSIS* COMPLEX THAT CAUSE HUMAN INFECTIONS

ORGANISM	HABITAT	PRIMARY ROUTE(S) OF TRANSMISSION	DISTRIBUTION
M. tuberculosis	Patients with cavitary disease are primary reservoir	Person to person by inhalation of droplet nuclei: droplet nuclei containing the organism (infectious aerosols, 1-5 μm) are produced when people with pulmonary tuberculosis cough, sneeze, speak, or sing; infectious aerosols may also be produced by manipulations of lesions or processing clinical specimens in the laboratory.[6] Droplets are so small that air currents keep them airborne for long periods; once inhaled they are small enough to reach the lung's alveoli*	Worldwide
M. bovis	Humans and wide host range of animals, cattle, nonhuman primates, goats, cats, buffalo, badgers, possums, dogs, pigs, deer, etc.	Ingestion of contaminated milk from infected cows†; airborne transmission‡	Worldwide
M. africanum	Humans§	Inhalation of droplet nuclei	East and West Africa

*Infection can occasionally occur through the gastrointestinal tract or skin.

†Incidence is significantly decreased in developed countries since introducing universal pasteurization of milk and milk products, as well as effective control programs for cattle.

‡Can be transmitted human to human, animal to human, and human to animal.

§Infections in animals have not yet been totally excluded.[39]

Pathogenesis

The pathogenesis of tuberculosis caused by organisms belonging to *M. tuberculosis* complex is discussed in Chapter 21. Inhalation of a single viable organism has been shown to lead to infection, although close contact is usually necessary for acquisition of infection. With regard to *M. tuberculosis,* 15% to 20% of persons who become infected develop disease. Disease usually occurs some years after the initial infection, when the patient's immune system breaks down for some reason other than the presence of tuberculosis bacilli within the lung. In a small percentage of infected hosts, the disease becomes systemic, affecting a variety of organs.

Following ingestion of milk from infected cows, *M. bovis* may penetrate the gastrointestinal mucosa or invade the lymphatic tissue of the oropharynx.[6] An attenuated strain of *M. bovis,* **bacille Calmette-Guérin (BCG),** has been used extensively in many parts of the world to immunize susceptible individuals against tuberculosis. BCG has recently been used as part of a controversial protocol to boost the nonspecific cellular immune response of certain immunologically deficient patients, particularly those with malignancies.

Because mycobacteria are the classic examples of intracellular pathogens and the body's response to BCG hinges on cell-mediated immunoreactivity, immunized individuals are expected to react more aggressively against all antigens that elicit cell-mediated immunity. Rarely, the unfortunate individual's immune system will be so compromised that it cannot handle the BCG, and systemic BCG infection may develop.

SPECTRUM OF DISEASE

Tuberculosis may mimic other diseases such as pneumonia, neoplasm, or fungal infections. In addition, clinical manifestations of patients infected with *M. tuberculosis* complex may range from asymptomatic to acutely symptomatic. Patients who are symptomatic can have systemic symptoms, pulmonary signs and symptoms, signs and symptoms related to other organ involvement (e.g., kidney), or a combination of these features. Of note, cases of pulmonary disease caused by *M. tuberculosis* complex organisms are clinically, radiologically, and pathologically indistinguishable.[39] Common presenting symptoms include low-grade fevers, night sweats, fatigue, anorexia (loss of appetite), and weight loss. If a patient presents with pulmonary tuberculosis, a productive cough is usually present, along with fevers, chills, myalgias (aches), and sweating, which are signs and symptoms similar for not only influenza but also acute bronchitis or pneumonia. As previously mentioned, other organs besides the lung can be involved following infection with *M. tuberculosis* complex organisms in a small percentage of patients and include the following:

- Genitourinary tract
- Lymph nodes
- Central nervous system (meningitis)
- Bone and joint (arthritis and osteomyelitis)
- Peritoneum
- Pericardium
- Larynx

It should be noted that individuals infected with HIV are particularly susceptible to develop active tuberculosis. These patients are likely to have rapidly progressive primary disease instead of a subclinical infection.[47] Of further concern, the diagnosis of tuberculosis is more difficult in persons infected with HIV, because chest radiographs of the pulmonary disease often lack specificity and frequently patients are anergic to **tuberculin skin testing**, a primary means to identify individuals infected with *M. tuberculosis*. The tuberculin skin test, or **PPD (purified protein derivative)** test, is based on the premise that following infection with *M. tuberculosis,* a patient will develop a delayed hypersensitivity cell-mediated immunity to certain antigenic components of the organism. To determine whether a person has been infected with *M. tuberculosis,* a culture extract of *M. tuberculosis* (i.e., PPD of tuberculin) is intracutaneously injected. After 48 to 72 hours a person who has been infected will exhibit a delayed hypersensitivity reaction to the PPD; this reaction is characterized by erythema (redness) and, most important, induration (firmness as a result of influx of immune cells). The diameter of induration is measured and then interpreted as to whether the patient has been infected with *M. tuberculosis;* different interpretative criteria exist for different patient populations (e.g., imunosuppressed persons, such as those infected with HIV).[6] It is important to bear in mind that this test is not 100% sensitive or specific, and a positive reaction to the skin test does not necessarily signify the presence of disease.

NONTUBERCULOUS MYCOBACTERIA

The term **nontuberculous mycobacteria (NTM)** includes all other mycobacterial species that do not belong to *M. tuberculosis* complex. This large group of mycobacteria have been known by several names; these are listed in Box 60-2. There is significant geographic variability both in the prevalence and species responsible for NTM disease.[8] As previously mentioned, NTM are present everywhere in the environment and sometimes colonize healthy individuals in the skin and respiratory and gastrointestinal tracts.[42] Little is known about how infection is acquired, but some NTM diseases appear to be acquired by either inhalation of infectious aerosols or ingestion; a few diseases are nosocomially or iatrogenically acquired. In contrast to *M. tuberculosis* complex organisms, NTM are not usually transmitted from person to person nor does their isolation necessarily mean that they are associated with a disease process. Therefore, interpretation of a positive NTM culture is complicated because these organisms are widely distributed in nature, their pathogenic potential varies greatly from one another, and humans can be colonized by these mycobacteria without necessarily developing infection or disease.[19] With few exceptions, little is known

BOX 60-2

OTHER NAMES THAT HAVE BEEN USED TO DESIGNATE THE NONTUBERCULOUS MYCOBACTERIA

Anonymous
Atypical
Unclassified
Unknown
Tuberculoid
Environmental
Opportunistic
MOTT (mycobacteria other than tubercle bacilli)

From Debrunner, M., et al. 1992. Clin. Infect. Dis. 15:330.

about the pathogenesis of infections caused by these bacterial agents.

In 1959 Runyon[43] classified NTM into four groups (Runyon groups I through IV) based on phenotypic characteristics of the various species, most notably growth rate and colonial pigmentation (Table 60-2). This large group of organisms are addressed by first discussing the slow-growing NTM (Runyon groups I to III) and then the rapid-growers (Runyon group IV); one other NTM, *M. leprae,* which is noncultivable, will also be reviewed. (It should be noted, as with many classification schemes, the Runyon classification does not always hold true. For example, there are some NTM that can be either a photochromogen or a nonphotochromogen.)

Because of the difficulty in determining the clinical significance of isolating an NTM from a clinical sample, several clinical classification schemes for the NTM have also been proposed. For example, one such scheme classifies NTM recovered from humans into four major groupings (pulmonary, lymphadenitis, cutaneous, disseminated) based on the clinical disease that they cause.[56] Other NTM classifications have been made based on the organism's pathogenic potential.[60,62]

SLOW-GROWING NTM

The slow-growing NTM can be further divided into three groups based on phenotypic characteristics of the various species. Species of mycobacteria synthesize **carotenoids** (a group of pigments that are yellow to red) in varying amounts and are categorized into three groups based on the production of these pigments. Some of these NTMs are considered potentially pathogenic for humans, whereas others are rarely associated with disease.

Photochromogens
The photochromogens are slow-growing NTM whose colonies become pigmented when exposed to light. Salient features of the photochromogens are summarized in Table 60-3.

Scotochromogens
The scotochromogens are slow-growing NTM whose colonies are pigmented when grown in the dark or the light. Salient features of the scotochromogens are summarized in Table 60-4. Of note, the epidemiology of the potentially pathogenic scotochromogens has not been definitively described. In contrast to potentially pathogenic nonphotochromogens (see below), these agents are rarely recovered in clinical laboratories.

Nonphotochromogens
The nonphotochromogens are slow-growing NTM whose colonies produce no pigment whether they are grown in the dark or the light. Of the organisms classified in this group, those belonging to *M. terrae* complex (*M. terrae, M. triviale,* and *M. nonchromogenicum*) and *M. gastri* are considered nonpathogenic for humans. The other nonphotochromogens are considered potentially pathogenic (Table 60-5), and many are frequently encountered in the clinical laboratory. Because organisms belonging to *Mycobacterium avium* complex are frequently isolated in the clinical laboratory and are able to cause infection in the human host, these nonphotochromogens are discussed in greater detail.

MYCOBACTERIUM AVIUM COMPLEX   In large part as a result of increasing immunosuppressed patient populations,

---

TABLE 60-2 RUNYON CLASSIFICATION OF NTM

RUNYON GROUP NUMBER	GROUP NAME	DESCRIPTION
I	Photochromogens	Colonies of NTM that develop pigment following exposure to light after being grown in the dark and take more than 7 days to appear on solid media
II	Scotochromogens	Colonies of NTM that develop pigment in the dark or light and take more than 7 days to appear on solid media
III	Nonphotochromogens	Colonies of NTM that are nonpigmented regardless of whether grown in the dark or light and take more than 7 days to appear on solid media
IV	Rapid-growers	Colonies of NTM that appear on solid media in less than 7 days

---

TABLE 60-3   CHARACTERISTICS OF THE NTM CLASSIFIED AS PHOTOCHROMOGENS

ORGANISM	EPIDEMIOLOGY	PATHOGENICITY	TYPES OF INFECTION
M. kansasii	Infection more common in white males. Natural reservoir is tap water[41]	Potentially pathogenic	Chronic pulmonary disease; extrapulmonary diseases, such as cervical lymphadenitis and cutaneous disease
M. asiaticum	Not commonly encountered (primarily Australia)	Potentially pathogenic	Pulmonary disease
M. marinum	Natural reservoir is fresh water and saltwater as a result of contamination from infected fish and other marine life Transmission by contact with contaminated water and organism entry by trauma or small breaks in the skin; associated with aquatic activity usually involving fish	Potentially pathogenic	Cutaneous disease
M. intermedium	Unknown	Potentially pathogenic	Pulmonary disease

---

TABLE 60-4   CHARACTERISTICS OF THE NTM CLASSIFIED AS SCOTOCHROMOGENS

ORGANISM	EPIDEMIOLOGY/HABITAT	PATHOGENICITY	TYPES OF INFECTION
M. szulgai	Water and soil	Potentially pathogenic	Pulmonary disease, predominantly in middle-age men; cervical adenitis; bursitis
M. scrofulaceum	Raw milk, soil, water, dairy products	Potentially pathogenic	Cervical adenitis in children
M. interjectum	Unknown	Potentially pathogenic	Chronic lymphadenitis (one case)
M. gordonae	Tap water, water, soil	Nonpathogenic*	NA†
M. cookii	Sphagnum, surface waters in New Zealand	Nonpathogenic	NA
M. hiberniae	Sphagnum, soil in Ireland	Nonpathogenic	NA

*Rarely, if ever, causes disease.
†NA, Nonapplicable.

---

such as individuals infected with HIV, the incidence of infections caused by these organisms and their corresponding clinical significance has changed significantly since they were first recognized as human pathogens in the 1950s.

General Characteristics. Taxonomically, the *M. avium* complex comprises *M. avium, M. intracellulare, M. paratuberculosis, M. lepraemurium,* and the "wood pigeon" bacillus. Recently, it has been proposed that *M. avium, M. paratuberculosis,* and the wood pigeon bacillus be placed in one species with three subspecies (*M. avium* subsp. *avium*, *M. avium* subsp. *paratuberculosis*, and *M. avium* subsp. *silvaticum* [wood pigeon bacillus])[52]; it has been suggested that *M. lepraemurium* be reduced to a subspecies of *M. avium*, as well.[45] Only the *M. avium* complex organisms that cause human infections are discussed and include *M. avium*, *M. intracellulare*, and *M. paratuberculosis*. Unfortunately, the nomenclature is somewhat confusing. Although *M. avium* and *M. intracellulare* are clearly different organisms, these organisms so closely resemble each other that the distinction cannot be made by routine laboratory determinations or on clinical grounds. As a result, sometimes these organisms are referred to as *M. avium-intracellulare*. Furthermore, because the isolation of *M. paratuberculosis* in a routine laboratory setting is exceedingly rare, the term *M. avium complex* is most commonly used to report the isolation of *M. avium-intracellulare*. For the purposes of this discussion, *M. avium* complex (MAC) is used and includes only *M. avium* and *M. intracellulare*.

Epidemiology and Pathogenesis.   MAC has emerged as an important pathogen in immunocompromised and immunocompetent populations. This agent is the most commonly isolated mycobacterial species in the United States today because of its high prevalence in individuals infected with HIV. Approximately 90% of mycobacterial infections in patients with AIDS involve either MAC or *M. tuberculosis;* various other NTM cause the remaining 10% of infections.[22] Autopsy series suggest that 30% to 50% of patients with AIDS will have disseminated MAC disease at the time of death.[54] The organisms are ubiquitous in the environment and have been isolated from natural water, soil, dairy products, pigs, chickens, cats, and dogs[35]; through extensive studies it is generally accepted that natural waters serve as the major reservoir for most human infections. Infections caused by MAC are acquired by inhalation or ingestion. The pathogenesis of MAC infections is not clearly understood. Cultures of MAC can exhibit opaque and translucent or transparent colony morphology. Studies suggest that transparent colonies are more virulent by virtue of being more drug resistant, isolated more frequently from blood of patients with AIDS, and appearing more virulent in macrophage and animal models.[45]

    M. paratuberculosis (*M. avium* subsp. *paratuberculosis*) is known to cause an inflammatory bowel disease referred to as Johne's disease in cattle, sheep, and goats. Recently, this organism was isolated from the bowel mucosa of patents with Crohn's disease, a chronic inflammatory bowel disease of humans.[58] This organism is extremely fastidious, seems to require a growth factor (mycobactin, produced by other species of mycobacteria, such as *M. phlei*, a saprophytic strain) and may take as long as 6 to 18 months for primary isolation. Whether these and other mycobacteria actually contribute to development of Crohn's disease or are simply colonizing an environmental niche in the bowel of these patients remains to be elucidated.

Clinical Spectrum of Disease.   The clinical manifestations of *M. avium* complex infections are summarized in Table 60-5.

OTHER NONPHOTOCHROMOGENS   There are several other mycobacterial species that are considered nonphotochromogens that are potentially pathogenic in humans. The epidemiology and spectrum of disease for these organisms are summarized in Table 60-5. In addition to the species in this table, other new species of mycobacteria that are nonphotochromogens have been recently described, such as *M. celatum*[10,17,45] and *M. conspicuum*.[48] Although not clearly established, these new agents appear to be potentially pathogenic in humans.

RAPIDLY GROWING NTM

Mycobacteria whose colonies appear on solid media in 7 days or less constitute the second major group of NTM.

General characteristics

Although a large group of organisms, only eight taxonomic groups of potentially pathogenic, rapidly growing mycobacteria are currently recognized.[55] In contrast to the majority of other mycobacteria, most rapid-growers can grow on routine bacteriologic media and on media specific for cultivation of mycobacteria. On Gram stain, organisms appear as weakly gram-positive rods resembling diphtheroids. Only the potentially pathogenic, rapid-growing mycobacteria are considered.

Epidemiology and pathogenesis

The rapidly growing mycobacteria considered as potentially pathogenic can cause disease in either healthy or immunocompromised patients. Like many other NTM, these organisms are ubiquitous in the environment and present worldwide. They have been found in soil, marshes, rivers, municipal water supplies, and in marine and terrestrial life forms.[61] Of importance, infections caused by rapidly growing mycobacteria may be acquired in the community from environmental sources or nosocomially as a result of medical intervention. They may be commensals on the skin.[33] Organisms gain entry into the host by inoculation into the skin and subcutaneous tissues during

TABLE 60-5 CHARACTERISTICS OF THE NTM CLASSIFIED AS NONPHOTOCHROMOGENS AND CONSIDERED AS POTENTIAL PATHOGENS

ORGANISM	EPIDEMIOLOGY	TYPES OF INFECTION
M. *avium* complex	Environmental sources are natural waters	Patients without AIDS: Pulmonary infections in patients with preexisting pulmonary disease; cervical lymphadenitis; disseminated disease* in immunocompromised, HIV-negative patients Patients with AIDS: Disseminated disease
M. *xenopi*†	Water, especially hot water taps in hospitals; believed to be transmitted in aerosols	Primarily pulmonary infections in adults. Less common: extrapulmonary infections (bone, lymph node, sinus tract) and disseminated disease
M. *ulcerans*	To date, has not been isolated from the environment. Little known about epidemiology, but infections occur in tropical or temperate climates	Indolent cutaneous and subcutaneous infections
M. *malmoense*	Majority of cases from England, Wales, Sweden. Rarely isolated from patients infected with HIV. Little known about epidemiology; to date, isolated from only humans and captured armadillos	Chronic pulmonary infections primarily in patients with preexisting disease; cervical lymphadenitis in children; less common, infections of the skin or bursa
M. *genavense*	Isolated from pet birds and dogs.[26,32] Mode of acquisition unknown	Disseminated disease in patients with AIDS (wasting disease characterized by fever, weight loss, hepatosplenomegaly, anemia)
M. *haemophilum*	Unknown	Disseminated disease; cutaneous infections in immunosuppressed adults. (?) mild and limited skin infections in pre-adolescence or early adolescence;[58] cervical lymphadenitis in children
M. *shimoidei*	To date, has not been isolated from environmental sources[32] Few case reports, but are widely spread geographically	Tuberculosis-like pulmonary infection, disseminated disease
M. *simiae*	Not well delineated; rarely isolated	Tuberculosis-like pulmonary infections

*Disseminated disease: can involve multiple sites, such as bone marrow, lung, liver, lymph nodes.

†Can be either nonphotochromogenic or scotochromogenic.

trauma, injections, or surgery or through animal contact; organisms can also cause disseminated cutaneous infections. The description of chronic pulmonary infections caused by rapidly growing mycobacteria suggests a possible respiratory route for acquisition of organisms present in the environment.[23] Of the potentially pathogenic, rapidly growing NTM, M. *fortuitum*, M. *chelonae*, and M. *abscessus* are commonly encountered; these three species account for approximately 90% of clinical disease.[55] To date, little is known about the pathogenesis of these organisms.

Spectrum of disease

The spectrum of disease caused by the most commonly encountered rapid-growers is summarized in Table 60-6.

TABLE 60-6   COMMON TYPES OF INFECTIONS CAUSED BY RAPIDLY GROWING MYCOBACTERIA

ORGANISM	COMMON TYPES OF INFECTION
M. abscessus	Disseminated disease primarily in immunocompromised individuals, skin and soft tissue infections, pulmonary infections, postoperative infections
M. fortuitum	Postoperative infections in breast augmentation and median sternotomy; skin and soft tissue infections
M. chelonae	Skin and soft tissue infections, postoperative wound infections, keratitis
M. fortuitum third biovariant complex, sorbitol-positive or sorbitol-negative	Skin and soft tissue infections
M. peregrinum	Skin and soft tissue infections
M. mucogenicum	Posttraumatic wound infections, catheter-related sepsis
M. smegmatis	Skin or soft tissue infections

NONCULTIVATABLE NTM-*MYCOBACTERIUM LEPRAE*

Mycobacterium leprae is an NTM that is a close relative of *M. tuberculosis*. This organism causes **leprosy** (also called **Hansen's disease**). Leprosy is a chronic disease of the skin, mucous membranes, and nerve tissue.

General characteristics

M. leprae has not yet been cultivated in vitro, although it can be cultivated in the armadillo and the footpads of mice. Largely through the application of molecular biologic techniques, information has been gained regarding this organism's genomic structure and its various genes and their products; polymerase chain reaction has been used to detect and identify *M. leprae* in infected tissues.[16]

Epidemiology and pathogenesis

Understanding of the epidemiology and pathogenesis of the disease is hampered by our inability to grow the organism in culture. In tropical countries, where the disease is most prevalent, it may be acquired from infected humans; however, infectivity is very low. Prolonged close contact and host immunologic status play a role in infectivity.

EPIDEMIOLOGY   Although leprosy is rare in the United States and most Western countries, there are about 5 million cases worldwide. The primary reservoir is infected humans. Transmission of leprosy occurs person to person through inhalation or contact with infected skin. It appears that inhalation of *M. leprae* discharged in nasal secretions of an infected individual is the more important mode of transmission.[38]

PATHOGENESIS   Although the host's immune response to *M. leprae* plays a key role in control of infection, the immune response is also responsible for the damage to skin and nerves. After acquisition of *M. leprae*, infection passes through many stages in the host that are characterized by various clinical and histopathologic features. Although there are many intermediate stages, the primary stages include a silent phase, during which leprosy bacilli multiply in the skin within macrophages, and an intermediate phase, in which the bacilli multiply in peripheral nerves and begin to cause sensory impairment. More severe disease states may follow. A patient may recover spontaneously at any stage.

Spectrum of disease

Based on the host's response, the spectrum of disease caused by *M. leprae* ranges from subclinical infection to intermediate stages of disease to full-blown and serious clinical manifestations involving the skin, upper respiratory system, testes, and peripheral nerves. The two major forms of the disease are a localized form called *tuberculoid leprosy* and a more disseminated form called *lepromatous leprosy*. Patients with lepromatous leprosy are anergic to *M. leprae* because of a defect in their cell-mediated immunity. Because of unimpeded growth, individuals display extensive skin lesions containing numerous acid-fast bacilli; organisms can spillover into blood and disseminate. In contrast, individuals with tuberculoid leprosy do not have an immune defect, so the disease is localized to the skin and nerves; few organisms are observed in skin lesions. Most of the serious sequelae associated with leprosy are a result of this organism's tropism for peripheral nerves.

LABORATORY DIAGNOSIS OF MYCOBACTERIAL INFECTIONS

Specimens received by the laboratory for mycobacterial smear and culture must be handled in a safe manner. Tuberculosis ranks high among laboratory-acquired infections. Therefore, laboratory and hospital administrators must provide laboratory personnel with facilities, equipment, and supplies that will reduce this risk to a minimum. All tuberculin-negative personnel should be skin-tested at least every year. Tuberculin-positive persons should have a chest radiograph annually. Biosafety Level 2 practices, containment equipment, and facilities for preparing acid-fast smears and culture are strongly recommended. Of great significance, all aerosol-generating procedures must be performed in a class II A or B or III biological safety cabinet (BSC) (see Chapter 2). If *M. tuberculosis* is grown and then propagated and manipulated, Biosafety Level 3 practices are recommended.

SPECIMEN COLLECTION AND TRANSPORT

Acid-fast bacilli may infect almost any tissue or organ of the body. The successful isolation of the organism depends on the quality of the specimen obtained and the appropriate processing and culture techniques employed by the mycobacteriology laboratory. In suspected mycobacterial disease, as in all other infectious diseases, the diagnostic procedure begins at the patient's bedside. Collection of proper clinical specimens requires careful attention to detail by the health care professional. Specimens should be collected in sterile, leak-proof, disposable, and appropriately labeled containers and placed into bags to contain leakage should it occur.

Pulmonary specimens

Pulmonary secretions may be obtained by any of the following methods: spontaneously produced or induced sputum, gastric lavage, transtracheal aspiration, bronchoscopy, and laryngeal swabbing. Sputum, aerosol-induced sputum, bronchoscopic aspirations, and gastric lavage constitute the majority of specimens submitted for examination. Spontaneously pro-

duced sputum is the specimen of choice. To raise sputum, patients must be instructed to take a deep breath, hold it momentarily, and then cough deeply and vigorously. Patients must also be instructed to cover their mouths carefully while coughing and to discard tissues in an appropriate receptacle. Saliva and nasal secretions are not to be collected nor is the patient to use oral antiseptics during the period of collection. Sputum specimens must be free of food particles, residues, and other extraneous matter.

The aerosol (saline) induction procedure can best be done on ambulatory patients who are able to follow instructions. Aerosol-induced sputums have been collected from children as young as age 5. This procedure should be performed only in an enclosed area with appropriate airflow by operators wearing particulate respirators and taking all appropriate safety measures to avoid exposure. The patient is told that the procedure is being performed to induce coughing to raise sputum that the patient cannot raise spontaneously and that the salt solution is irritating. The patient is instructed to inhale slowly and deeply through the mouth and to cough at will, vigorously and deeply, coughing and expectorating into a collection tube. The procedure is discontinued if the patient fails to raise sputum after 10 minutes or feels any discomfort. Ten mL of sputum should be collected; if the patient continues to raise sputum, a second specimen should be collected and submitted. Specimens should be delivered promptly to the laboratory and refrigerated if processing is delayed.

Gastric lavage specimens

Gastric lavage is used to collect sputum from patients who may have swallowed sputum during the night. The procedure is limited to senile, nonambulatory patients, children younger than age 3, and patients who fail to produce sputum by aerosol induction. The most desirable gastric lavage is collected at the patient's bedside before the patient arises and before exertion empties the stomach. Gastric lavage cannot be performed as an office or clinic procedure. A series of three specimens are collected within 3 days.

The collector should wear a cap, gown, and particulate respirator mask and stand beside (not in front of) the patient, who should sit up on the edge of the bed or in a chair, if possible. The Levine collection tube is inserted through a nostril, and the patient is instructed to swallow the tube. When the tube is fully inserted, a syringe is attached to the end of the tube and filtered distilled water is inserted through the tube. The syringe is then used to withdraw 20 to 25 mL gastric secretions that are expelled slowly down the sides of the 50-mL conical collecting tube. The top of the collection tube is screwed on tightly, and the tube is held upright during prompt delivery to the laboratory.

Bronchial lavages, washings, and brushings are collected and submitted by medical personnel. These are the specimens of choice for detecting nontuberculous mycobacteria and other opportunistic pathogens in patients with immune dysfunctions.

Urine specimens

Urogenital infections show little evidence of decreasing. Whereas 2% to 3% of patients with pulmonary tuberculosis exhibit urinary tract involvement, 30% to 40% of patients with genitourinary disease have tuberculosis at some other site. The clinical manifestations of urinary tuberculosis are variable, including frequency of urination (most common), dysuria, hematuria, and flank pain. Definitive diagnosis requires recovery of acid-fast bacilli from the urine.

Early-morning voided urine specimens in sterile containers should be submitted daily for at least 3 days. The procedure for collection is that used to collect a clean-catch midstream urine specimen, previously described in Chapter 25. Twenty-four–hour urine specimens are undesirable because of excessive dilution, higher contamination, and difficulty in concentrating.

Fecal specimens

Acid-fast stain and/or culture of stool from patients with AIDS has been used to identify patients who may be at risk for developing disseminated *M. avium* complex disease. The clinical utility of this practice remains controversial[25,36]; however, if screening stains and/or cultures are positive, dissemination often follows.[27] Feces should be submitted in a clean, dry, wax-free container without preservative or diluent. Contamination with urine should be avoided.

Tissue and body fluid specimens

Tuberculous meningitis is uncommon but still occurs in both immunocompetent and immunosuppressed patients. Sufficient quantity of specimen is most critical for isolation of acid-fast bacilli from cerebrospinal fluid. There may be very few organisms in the spinal fluid, which makes their detection difficult. At least 10 mL of cerebrospinal fluid is recommended for recovery of mycobacteria. Similarly, as much as possible (10 to 15 mL minimum) of other body fluids, such as pleural, peritoneal, and pericardial fluids, should be collected in a sterile container or syringe with a Luer tip cap.

Blood specimens

Immunocompromised patients, particularly those who are infected with HIV, can have disseminated mycobacterial infection; the majority of these infections are caused by *M. avium* complex. A blood culture positive for *M. avium* complex is always associated with clinical evidence of disease.[24] Best recovery of mycobacteria is achieved by collecting the blood in either a broth such as the radiometric BACTEC 13A vial or the Isolator lysis-centrifugation system (see Chapter 20). Some studies have indicated that the lysis-centrifugation system is advantageous, because quantitative data can be obtained with each blood culture; in patients with AIDS, quantitation of such organisms can be used to monitor therapy and determine prognosis. However, the necessity for doing quantitative blood cultures remains unclear. Blood for culture of mycobacteria should be collected in a manner as for routine blood cultures (see Chapter 20).

Wounds, skin lesions, and aspirates

If attempting to culture a skin lesion or wound, an aspirate is the best type of specimen to collect. The skin should be cleansed with alcohol before aspiration of the material into a syringe. If the volume is insufficient for aspiration, pus and exudates may be obtained on a swab and then placed in transport medium, such as Amie's or Stuart's (dry swabs are unacceptable). However, a negative culture of a specimen obtained on a swab is not considered reliable, and this should be noted in the culture report.

SPECIMEN PROCESSING

Specimen processing for the recovery of acid-fast bacilli from clinical specimens involves several complex steps, each of which must be carried out with precision. Specimens from sterile sites can be inoculated directly to media (small volume) or concentrated to reduce volume. Other specimens require decontamination and concentration. A scheme for processing is depicted in Figure 60-1. The procedures are explored in detail in the following sections.

Contaminated specimens

The majority of specimens submitted for mycobacterial culture consist of organic debris, such as mucin, tissue, serum, and other proteinaceous material that is contaminated with organisms. A typical example of such a specimen is sputum. Laboratories must process these specimens so that contaminating bacteria that can rapidly outgrow mycobacteria are either killed or reduced in numbers, and mycobacteria are released from mucin and/or cells. After decontamination, mycobacteria are concentrated usually by centrifugation to enhance their detection by acid-fast stain and culture. Unfortunately, there is not one ideal method for decontaminating and digesting clinical specimens. Although continuously faced with the inherent limitations of

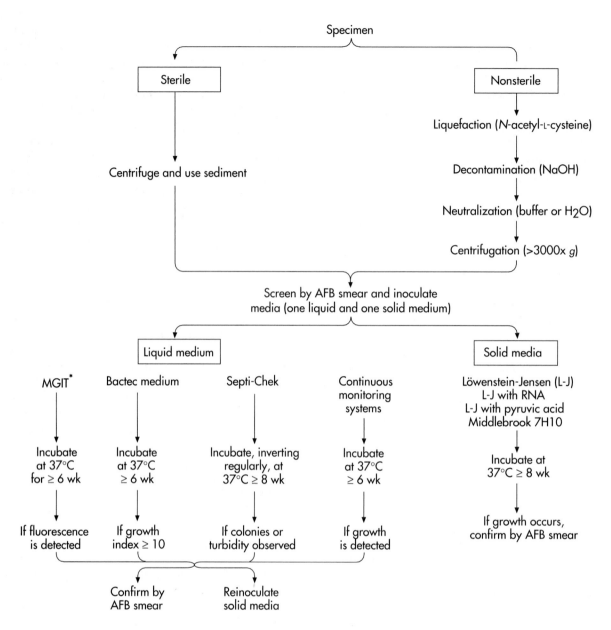

FIGURE 60-1 A flowchart for specimen processing for isolation of mycobacteria.

various methods, laboratories must strive to maximize the survival and detection of mycobacteria on the one hand, while maximizing the elimination of contaminating organisms on the other. Digestion-decontamination procedures should be as gentle as possible, with no more than an overall contamination rate of 5%.[11]

OVERVIEW Commonly used digestion-decontamination methods are the sodium hydroxide (NaOH) method, the Zephiran-trisodium phosphate method, and the N-acetyl-L-cysteine (NALC)-NaOH method; the NALC-NaOH method is detailed in Procedure 60-1. Of note, another decontaminating procedure using oxalic acid is very useful for treating specimens known to harbor gram-negative rods, particularly *Pseudomonas* and *Proteus,* which are extremely troublesome contaminants. NaOH is a commonly used decontaminant that is also mucolytic. Several agents can be used to liquefy a clinical specimen, including NALC, dithiothreitol (sputolysin), and enzymes. None of these agents are inhibitory to bacterial cells. In most procedures, liquification (release of the organisms from mucin or cells) is enhanced by vigorous mixing with a vortex type of mixer in a closed container. Following mixing, the container should be allowed to stand for 15 minutes before opening, to prevent the dispersion of fine aerosols generated during mixing. Of utmost importance during processing is strict adherence to processing and laboratory safety protocols. All of these procedures should be carried out in a BSC. Following digestion and decontamination, specimens are concentrated by centrifugation at $\geq 3000 \times g$.

SPECIAL CONSIDERATIONS   Many specimen types besides respiratory samples contain normal flora and require decontamination and concentration; handling procedures for such specimens are described briefly.

Aerosol-induced sputums should be treated as sputums. Gastric lavages should be processed within 4 hours of collection or neutralized with 10% sodium carbonate (check with pH paper to determine that the specimen is at neutral pH) and refrigerated until processed as for sputum. If more than 10 mL of watery-appearing aspirate was obtained, the specimen can be centrifuged at $3600 \times g$ for 30 minutes, the supernatant decanted, and the sediment processed as for sputum.

Urine specimens should be divided into a maximum of four 50-mL centrifuge tubes and centrifuged at $3600 \times g$ for 30 minutes. The supernatant should be decanted, leaving approximately 2 mL of sediment in each tube. The tubes are vortexed to suspend the sediments, and sediments are combined. If necessary, distilled water can be added to a total volume of 10 mL. This urine concentrate is then treated as for sputum or with the sputolysin-oxalic acid method.

For fecal specimens, approximately 0.2 g of stool (a portion about the size of a pea) is emulsified in 11 mL of sterile, filtered, distilled water.[63] The suspension is vortexed thoroughly, and particulate matter is allowed to settle for 15 minutes. Ten milliliters of the supernatant is then transferred to a 50-mL conical centrifuge tube and decontaminated using the oxalic acid or NALC-NaOH method.

Swabs and wound aspirates should be transferred to a sterile 50-mL conical centrifuge tube containing a liquid medium (Middlebrook 7H9, Dubos Tween albumin broth) at a ratio of 1 part specimen to 5 to 10 parts liquid medium. The specimen is vortexed vigorously and allowed to stand for 20 minutes. The swab is removed, and the resulting suspension is processed as for sputum.

Large pieces of tissue thought to be contaminated should be finely minced using a sterile scalpel and scissors. This material is then homogenized in a sterile tissue grinder with a small amount of sterile saline or sterile 0.2% bovine albumin; this suspension is then processed as for a sputum. If not known to be sterile, tissue is homogenized and one half is directly inoculated to solid and liquid media and the other half processed as a sputum. If the tissue is collected aseptically and not thought to be contaminated, it may be processed without treatment with NALC-NaOH.

Specimens not requiring decontamination

Tissues or body fluids collected aseptically usually do not require the digestion and decontamination methods used with contaminated specimens. Processing clinical specimens that do not routinely require decontamination for acid-fast culture is described here. If such a specimen appears contaminated because of color, cloudiness, or foul odor, perform a Gram stain to detect bacteria other than acid-fast bacilli. Specimens found to be contaminated should be processed as described in the previous section.

Cerebrospinal fluid should be handled aseptically and centrifuged for 30 minutes at $3600 \times g$ to concentrate the bacteria. The supernatant is decanted, and the sediment is vortexed thoroughly before preparing the smear and inoculating media. If insufficient quantity of spinal fluid is received, the specimen should be used directly for smear and culture. Because recovery of acid-fast bacilli from cerebrospinal fluid is difficult, additional solid and liquid media should be inoculated if material is available.

Pleural fluid should be collected in sterile anticoagulant (1 mg/mL ethylenediaminetetraacetic acid [EDTA] or 0.1 mg/mL heparin). If the fluid becomes clotted, it should be liquefied with an equal volume of sputolysin and vigorous mixing. To lower the specific gravity and density of pleural fluid, transfer 20 mL to a sterile 50-mL centrifuge tube and dilute the specimen by filling the tube with distilled water. Invert several times to mix the suspension, and centrifuge at $3600 \times g$ for 30 minutes. The supernatant should be removed, and the sediment should be resuspended for smear and culture.

Joint fluid and other sterile exudates can be handled aseptically and inoculated directly to media. Bone marrow aspirates may be injected into Wampole Pediatric Isolator tubes, which help to prevent clotting; the specimen can be removed with a needle and syringe for preparation of smears and cultures. As an alternative, these specimens are either inoculated directly to media or, if clotted, treated with sputolysin or glass beads and distilled water before concentration.

DIRECT DETECTION METHODS

Acid-fast stains

Mycobacteria possess cell walls that contain mycolic acids, which are long-chain, multiply cross-linked fatty acids. These long-chain mycolic acids probably serve to complex basic dyes, contributing to the characteristic of acid-fastness that distinguishes them from other bacteria. Mycobacteria are not the only group with this unique feature. Species of *Nocardia* and *Rhodococcus* are also partially acid-fast; *Legionella micdadei,* a causative agent of pneumonia, is partially acid-fast in tissue. Cysts of the genera *Cryptosporidium* and *Isospora* are distinctly acid-fast. The mycolic acids and lipids in the mycobacterial cell wall probably account for the unusual resistance of these organisms to the effects of drying and harsh decontaminating agents, as well as for acid-fastness.

When Gram-stained, mycobacteria usually appear as slender, poorly stained, beaded gram-positive bacilli; sometimes they appear as "gram-neutral" or "gram-ghosts" by failing to take up either crystal violet or safranin. Acid-fastness is affected by age of colonies, medium on which growth occurs, and ultraviolet light. Rapidly growing species appear to be acid-fast-variable. Three types of staining procedures are used in the laboratory for rapid detection and confirmation of acid-fast bacilli: fluorochrome, Ziehl-Neelsen, and Kinyoun. Smears for all methods are prepared in the same way (Procedure 60-2).

The visualization of acid-fast bacilli in sputum or other clinical material should be considered only presumptive evidence of tuberculosis, because stain does not specifically identify *M. tuberculosis.* The report form should indicate this. For example, *M. gordonae,* a nonpathogenic scotochromogen commonly found in tap water, has been a problem when tap water or deionized water has been used in the preparation of smears or even when patients have rinsed their mouths with tap water before the use of aerosolized saline solution for inducing sputum. However, the incidence of false-positive smears is very low when good quality control is maintained. Conversely, acid-fast stained smears of clinical specimens require at least 10^4 acid-fast bacilli per milliliter for detection from concentrated specimens.[11]

METHODS

Fluorochrome Stain. This is the screening procedure recommended for those laboratories that possess a fluorescent (ultraviolet) microscope (Procedure 60-3). This stain is more sensitive than the conventional carbolfuchsin stains because the fluorescent bacilli stand out brightly against the background (Figure 60-2); the smear can be initially examined at lower magnifications ($250\times$ to $400\times$), and therefore more fields can be visualized in a short period. In addition, a positive fluorescent smear may be restained by the conventional Ziehl-Neelsen or Kinyoun procedure, thereby saving the time needed to make a fresh smear. Screening of specimens with rhodamine or rhodamine-auramine will result in a higher yield of positive smears and will substantially reduce the amount of time needed for examining smears. One drawback associated with the fluorochrome stains is that many rapid-growers may not appear fluorescent with these reagents. It is recommended that all positive fluorescent smears be confirmed with a Ziehl-Neelsen stain or examination by another technologist. It is important to wipe the immersion oil from the objective lens after examining a positive smear, because stained bacilli can float off the slide into the oil and may possibly contribute to a false-positive reading for the next smear examined.

Fuchsin Acid-Fast Stains. The classic carbolfuchsin (Ziehl-Neelsen) stain requires heating the slide for better penetration of stain into the mycobacterial cell

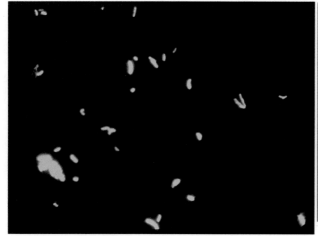

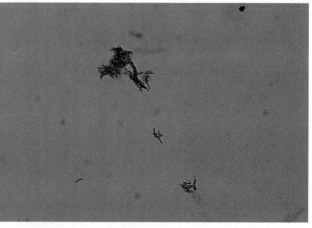

A   **B**

FIGURE 60-2 *M. tuberculosis* stained with **(A)** fluorochrome stain ($400\times$ magnification) and **(B)** Kinyoun acid-fast stain ($1000\times$ magnification).

wall; hence it is also known as the *hot stain* procedure (Procedure 60-4). Procedure 60-5 describes the Kinyoun acid-fast stain. The method is similar to the Ziehl-Neelsen stain but without heat (Figure 60-2); hence the term *cold stain*. If present, typical acid-fast bacilli appear as purple to red, slightly curved, short or long rods (2 to 8 μm); they may also appear beaded or banded (*M. kansasii*). For some nontuberculous species, such as *M. avium* complex, they appear pleomorphic, usually coccoid.

EXAMINATION, INTERPRETATION, AND REPORTING OF SMEARS
Smears should be examined carefully by scanning at least 300 oil immersion fields (equivalent to three full horizontal sweeps of a smear that is 2 cm long and 1 cm wide) before reporting a smear as negative. Because the fluorescent stain can be examined using a lower magnification than for the fuchsin-stained smear, the equivalent number of fields can be examined in less time, thus making the fluorochrome stain the preferred method.

When acid-fast organisms are observed on a smear, results must be quantified to be meaningful. Because this quantitation estimates the number of bacilli being excreted, the extent of a patent's infectiousness can be assessed for clinical and epidemiologic purposes. The positive smear is reported by the laboratory, as well as the staining method and the quantity of organisms. The recommended interpretations and ways to report smear results are given in Table 60-7.

The overall sensitivity of an acid-fast smear ranges from 20% to 80%.[6,9,11,34] Factors such as specimen type, staining method, and culture method can influence the acid-fast smear sensitivity. In general, specificity of acid-fast smear examination is very high. However, cross-contamination of slides during the staining process and use of water contaminated with saprophytic mycobacteria can lead to false-positive results. Staining receptacles should not be employed; acid-fast bacilli can also be transferred from one slide to another via immersion oil. For these reasons, it is best to confirm a positive result. Although not without some limitations, because of its simplicity and speed, the stained smear is an important and useful test, particularly for the detection of smear-positive patients ("infectious reservoirs"), who are the greatest risk to others in their environment.

Antigen-protein detection
The detection of microbial products or components has been used in recent years to diagnose infections caused by *M. tuberculosis*. For example, tuberculostearic acid is a fatty acid that can be extracted from the cell wall of mycobacteria and then detected by gas chromatography/mass spectrometry in clinical samples containing few mycobacteria.[40] Because of the limited number of species that can cause meningitis and the fact that *M. tuberculosis* appears to be the only one of these species that releases tuberculostearic acid into the surrounding environment, the presence of this substance in cerebrospinal fluid (CSF) is

---

TABLE 60-7 ACID-FAST SMEAR REPORTING

NUMBER OF AFB* SEEN FUCHSIN STAIN (1000× MAGNIFICATION)	NUMBER OF AFB SEEN FLUOROCHROME STAIN (450× MAGNIFICATION)	REPORT
0	0	No AFB seen
1-2/300 fields	1-2/70 fields	Doubtful; request another specimen
1-9/100 fields	2-18/50 fields	1+
1-9/10 fields	4-36/10 fields	2+
1-9/field	4-36/field	3+
>9/field	>36/field	4+

*AFB, acid-fast bacilli.

Modified from Kent, P.T. and Kubica, G.P. 1985. Public health mycobacteriology: a guide for the level III laboratory. U.S. Department of Health and Human Service, Public Health Service, Centers for Disease Control, Atlanta.

thought to be diagnostic of tuberculous meningitis. Performance of this assay is limited to only a few laboratories. In addition, various immunoassays for antigen detection directly in clinical specimens, including sputum and CSF, have been evaluated and show some promise.[18]

Adenosine deaminase is a host enzyme whose production is increased in certain infections caused by *M. tuberculosis*. For example, elevated levels of this enzyme were found in the majority of patients with tuberculous pleural effusions (98% sensitive); the test for the enzyme was determined to be highly specific, as well (96 specificity).[5]

Nucleic acid amplification

Recently, molecular techniques, such as polymerase chain reaction (PCR; see Chapter 14), have been used to detect *M. tuberculosis* directly in clinical specimens; two kits are available commercially that are FDA approved. The Amplicor *Mycobacterium tuberculosis* test (Roche Diagnostic Systems, Branchburg, N.J.) employs PCR to detect *M. tuberculosis* directly in respiratory specimens; the Amplified Mycobacterium Tuberculosis Direct Test (Gen-Probe, San Diego, Calif.) is based on ribosomal RNA amplification. Both of these assays currently are approved for use on acid-fast smear-positive specimens only because numerous studies have demonstrated less than optimum sensitivity on smear-negative specimens. Several clinical laboratories have developed their own PCR assays to detect *M. tuberculosis* directly in clinical specimens.[20] Manufacturers are also at different stages of development of amplification assays using various formats. The ultimate usefulness of these assays in the clinical laboratory is still being sorted out because of issues associated with amplification assays, such as expense and their clinical utility.

CULTIVATION

A combination of different culture media is required to optimize recovery of mycobacteria from culture; at least one solid medium in addition to a liquid medium should be used. The ideal media combination should support the most rapid and abundant growth of mycobacteria, allow for the study of colony morphology and pigment production, inhibit the growth of contaminants, and be economical.

Solid media

Solid media, such as those listed in Box 60-3, are recommended because of the development of characteristic, reproducible colonial morphology, good growth from small inocula, and a low rate of contamination. Optimally, at least two solid media (a serum [albumin] agar base medium, e.g., Middlebrook 7H10, and an egg-potato base medium, e.g., L-J) should be used for each specimen (these media are available from commercial sources). All specimens must be processed appropriately before inoculation. It is imperative to inoculate test organisms to commercially available products for quality control (Procedure 60-6). Wallenstein's medium, composed of egg yolk, 2.5% glycerin, malachite green, and water, is an excellent

---

BOX 60-3

SUGGESTED MEDIA FOR CULTIVATION OF MYCOBACTERIA FROM CLINICAL SPECIMENS*

Solid
  Agar-based
    Middlebrook 7H10 and Middlebrook 7H10 selective
    Middlebrook 7H11 and Middlebrook 7H11 selective
    Middlebrook biplate (7H10/7H11S agar)
  Egg-based
    Löwenstein-Jensen (L-J)
    L-J Gruft
    L-J with pyruvic acid
    L-J with iron
Liquid
BACTEC 12B medium
Middlebrook 7H9 broth
Septi-Chek AFB
Commercially supplied broths for continuously monitoring systems for mycobacteria

*For optimal recovery of mycobacteria, a minimum combination of liquid medium and solid media is recommended.

medium for the recovery of NTM, particularly *M. avium* complex.

Cultures are incubated at 35° C in the dark in an atmosphere of 5% to 10% CO_2 and high humidity. Tubed media is incubated in a slanted position with screwcaps loose for at least 1 week to allow for evaporation of excess fluid and the entry of CO_2; plated medium are either placed in a CO_2-permeable plastic bag or wrapped with CO_2-permeable tape. If specimens are obtained from skin or superficial lesions suspected to contain *M. marinum* or *M. ulcerans,* an additional set of solid media should be inoculated and incubated at 25° to 30° C. In addition, a chocolate agar plate (or the placement of an X-factor [hemin] disk on conventional media) and incubation at 25° to 33° C is needed for recovery of *M. haemophilum* from these specimens.

Cultures are examined weekly for growth. Contaminated cultures are discarded and reported as "contaminated, unable to detect presence of mycobacteria"; additional specimens are also requested. If available, sediment may be recultured after enhanced decontamination or by inoculating the sediment to a more selective medium. Most isolates will appear between 3 and 6 weeks; a few isolates will appear after 7 or 8 weeks of incubation. When growth appears, the rate of growth, pigmentation, and colony morphology are recorded. Typical colonial appearance of *M. tuberculosis* and other mycobacteria are shown in Figure 60-3. After 8 weeks of incubation, negative cultures (those showing no growth) are reported, and the cultures are discarded.

Because of the resurgence of tuberculosis in the United States in the late 1980s and early 1990s, significant effort has been put into developing methods to provide more rapid diagnosis of tuberculosis. Welch and others[59] refined a method that decreased the time to detection of mycobacterial growth by half or more compared with conventional culture methods by using a thinly poured Middlebrook 7H11 plate. These plates are inoculated in a routine manner, sealed, incubated, and examined microscopically (40× magnification) at regular intervals for the appearance of microcolonies. Of note, presumptive identification of *M. tuberculosis* or *M. avium* complex could be made for about 83% of the isolates within 10 and 11 days after inoculation, respectively.

Liquid media

In general, the use of a liquid media system reduces the turnaround time for isolation of acid-fast bacilli to approximately 10 days, compared with 17 days or longer for conventional solid media.[2,4,29,49,53] There are several different systems to culture and detect the growth of mycobacteria in liquid media. The most commonly employed systems are summarized in Table 60-8. Growth of mycobacteria in liquid media, regardless of the type, requires 5% to 10% CO_2; CO_2 is either already provided in the culture vials or is added according to manufacturer's instructions. Once growth is detected in liquid medium, an acid-fast stain of a culture aliquot is performed to confirm the presence of acid-fast bacilli and subcultured to solid agar. A Gram stain can also be performed if contamination is suspected.

Interpretation

Although isolation of *M. tuberculosis* complex organisms represents infection, the clinician must determine the clinical significance of isolating an NTM in most cases; in other words, does the organism represent mere colonization or significant infection? Because these organisms vary greatly in their pathogenic potential, can colonize an individual without causing infection, and are ubiquitous in the environment, interpretation of a positive NTM culture is complicated. Therefore, the American Thoracic Society recommends diagnostic criteria for NTM disease to help physicians interpret culture results.[56]

APPROACH TO IDENTIFICATION

The first test that is always performed on organisms growing on solid or liquid mycobacterial media is an acid-fast stain to confirm that the organisms are indeed mycobacteria. If the organism is growing on solid media only, several colonies are inoculated to Middlebrook 7H9 broth (5 mL) and incubated at 35° C for 5 to 7 days, with daily agitation to enhance growth. Either this broth subculture or mycobacterial growth in the primary liquid culture can then be used to inoculate all test media, including biochemical tests and pigmentation and growth rate determinations. In some instances, additional cultures may be inoculated and then incubated at different temperatures when more definitive identification is needed.

Growth characteristics

The preliminary identification of mycobacterial isolates depends on their rate of growth, colony morphology (Figure 60-3), colony texture, pigmentation, and, in some instances, the permissive incubation temperatures of mycobacteria. To perform identification procedures, quality control organisms should be tested along with unknowns, as listed in Table 60-9. The commonly used quality-control organisms can be maintained in broth at room temperature and transferred monthly. In this way they will always be available for inoculation to test media along with suspensions of the unknown mycobacteria being tested.

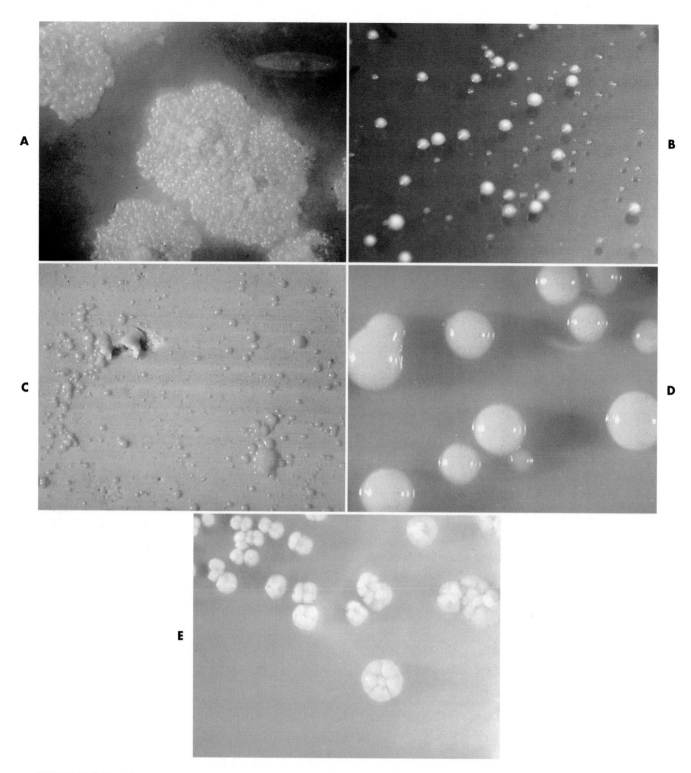

FIGURE 60-3 Typical appearance of some mycobacteria on solid agar medium. **A,** *M. tuberculosis* colonies on Löwenstein-Jensen agar after 8 weeks of incubation. **B,** Different colony morphology seen on culture of one strain of *M. avium* complex. **C,** *M. kansasii* colonies exposed to light. **D,** Scotochromogen *M. gordonae* with yellow colonies. **E,** Smooth, multilobate colonies of *M. fortuitum* on Löwenstein-Jensen medium.

TABLE 60-8   COMMONLY USED LIQUID MEDIA SYSTEMS TO CULTURE AND DETECT THE GROWTH OF MYCOBACTERIA

SYSTEM	BASIC PRINCIPLE(S) OF DETECTION
BACTEC 460 TB (Becton Dickinson Diagnostic Systems, Cockeysville, Md.)	Culture media contains ^{14}C-labeled palmitic acid. If present in the broth, mycobacteria metabolize the ^{14}C-labeled substrates and release radioactively labeled $^{14}CO_2$ in the atmosphere, which collects above the broth in the bottle. The instrument withdraws this CO_2-containing atmosphere and measures the amount of radioactivity present. Bottles that yield a radioactive index, called a *growth index,* greater than or equal to 10 are considered positive
Septi-Chek AFB System (Becton Dickinson Diagnostic Systems)	Biphasic culture system made up of a modified Middlebrook 7H9 broth with a three-sided paddle containing chocolate, egg-based, and modified 7H11 solid agars. The bottle is inverted regularly to inoculate the solid media. Growth is detected by observing the three-sided paddle
Mycobacteria Growth Indicator Tube (MGIT) (Becton Dickinson)	Culture tube contains Middlebrook 7H9 broth and a fluorescent compound embedded in a silicone sensor. Growth is detected visually using an ultraviolet light. Oxygen (O_2) diminishes the fluorescent output of the sensor; therefore, O_2 consumption by organisms present in the medium are detected as an increase in fluorescence. This system is conducive to possible automation
Continuous Growth Monitoring Systems	
ESP Culture System II Accumed International, Inc., Chicago, Ill.	Organisms are cultured in a modified Middlebrook 7H9 broth with enrichment and a cellulose sponge to increase the culture's surface area. The instrument detects growth by monitoring pressure changes that occur as a result of O_2 consumption or gas production by the organisms as they grow
BACTEC 9000 MB (Becton Dickinson)	Organisms are cultured in a modified Middlebrook 7H9 broth. The instrument detects growth by monitoring O_2 consumption by means of a fluorescent sensor

GROWTH RATE   The rate of growth is an important criterion for determining the initial category of an isolate. Rapid-growers will usually produce colonies within 3 to 4 days after subculture. Even a rapid-grower, however, may take longer than 7 days to initially produce colonies because of inhibition by a harsh decontaminating procedure. Therefore, the growth rate (and pigment production) must be determined by subculture; the method is described in Procedure 60-7. The dilution of the organism used to assess growth rate is critical. Even slow-growing mycobacteria will appear to produce colonies in less than 7 days if the inoculum is too heavy. One organism particularly likely to exhibit false-positive rapid growth is *M. flavescens.* This species therefore serves as an excellent quality-control organism for this procedure.

PIGMENT PRODUCTION As previously discussed, mycobacteria may be categorized into three groups on the basis of pigment production. Procedure 60-7

describes how to determine pigment production. To achieve optimum photochromogenicity, colonies should be young, actively metabolizing, isolated, and well-aerated. Although some species, such as *M. kansasii,* turn yellow after a few hours of light exposure, others, such as *M. simiae,* may take a prolonged exposure to light. Scotochromogens produce pigmented colonies even in the absence of light, and colonies often become darker with prolonged exposure to light (Figure 60-4). One member of this group, *M. szulgai,* is peculiar in that it is a scotochromogen at 35° C and nonpigmented when grown at 25° to 30° C. For this reason, all pigmented colonies should be subcultured to test for photoactivated pigment at both 35° C and 25° to 30° C. Nonchromogens are not affected by light.

Biochemical testing
Once placed into a preliminary subgroup based on its growth characteristics, an organism must be definitively identified to species or complex level.

TABLE 60-9 CONTROLS AND MEDIA USED FOR THE BIOCHEMICAL IDENTIFICATION OF MYCOBACTERIA

BIOCHEMICAL TEST	CONTROL ORGANISMS POSITIVE	CONTROL ORGANISMS NEGATIVE	RESULT POSITIVE	RESULT NEGATIVE	MEDIUM USED AND AMOUNT	DURATION	INCUBATION CONDITIONS
Niacin	*M. tuberculosis*	*M. intracellulare*	Yellow	No color change	0.5 mL DH$_2$O*	15-30 min	Room temperature
Nitrate	*M. tuberculosis*	*M. intracellulare*	Pink or red	No color change	0.3 mL DH$_2$O	2 hrs	37° C bath
Urease	*M. fortuitum*	*M. avium*	Pink or red	No color change	Urea broth for AFB	1, 3, and 5 days	37° C incubator without CO$_2$
68° C Catalase	*M. fortuitum* or *M. gordonae*	*M. tuberculosis*	Bubbles	No bubbles	0.5 mL phosphate buffer [pH 7]	20 min	68° C bath
SQ Catalase†	*M. kansasii* or *M. gordonae*	*M. avium*	>45 mm	<45 mm	Commercial medium	14 days	37° C incubator (with CO$_2$)
Tween 80	*M. kansasii*	*M. intracellulare*	Pink or red	No color change	1 mL DH$_2$O	5 or 10 days	37° C incubator (in the dark, without CO$_2$)
Tellurite	*M. avium*	*M. tuberculosis*	Smooth, fine black precipitate (smoke-like action)	Gray clumps (no smokelike action)	Middlebrook, 7H9 broth	7, then 3 additional days	37° C incubator (with CO$_2$)
Arylsulfatase	*M. fortuitum*	*M. intracellulare*	Pink or red	No color change	Wayne's arylsulfatase medium	3 days	37° C incubator (without CO$_2$)
5% NaCl	*M. fortuitum*	*M. gordonae*	Substantial growth	Little or no growth	Commercial slant with and without 5% NaCl	28 days	37° C incubator (with CO$_2$)
TCH‡	*M. bovis*	*M. tuberculosis*	No growth (i.e., susceptible)	Growth (i.e., resistant or ≥1% of colonies are resistant)	TCH slant	3 weeks	37° C incubator (with CO$_2$)

*DH$_2$O, distilled water.
†SQ, Semiquantitative.
‡TCH, Thiophene-2-carboxylic acid hydrazide.

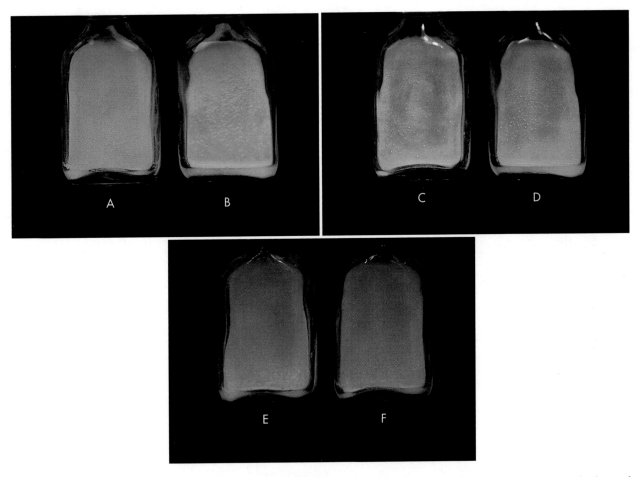

FIGURE 60-4 Initial grouping of mycobacteria based on pigment production before and after exposure to light. In one test system, subcultures of each isolate are grown on two agar slants. One tube is wrapped in aluminum foil to prevent exposure of the organism to light, and the other tube is allowed light exposure. After sufficient growth is present, the wrapped tube is unwrapped, and the tubes are examined together. **Photochromogens** are unpigmented when grown in the dark (*tube A*) and develop pigment after light exposure (*tube B*). **Scotochromogens** are pigmented in the dark (*tube C*); the color does not intensify after exposure to light (*tube D*). **Nonphotochromogens** are nonpigmented when grown in the dark (*tube E*) and remain so even after light exposure (*tube F*).

Conventional biochemical tests can be used for this purpose, although new methods (discussed later in this section) have replaced biochemical tests for the identification of many mycobacterial species. Table 60-10 summarizes distinctive properties of the more commonly cultivable mycobacteria isolated from clinical specimens; key biochemical tests for each of the major mycobacterial groupings, including *M. tuberculosis* complex, are listed in Table 60-11. Key biochemical tests are discussed below; detailed procedures are given in other texts.[1,11]

NIACIN Niacin (nicotinic acid) plays an important role in the oxidation-reduction reactions that occur during mycobacterial metabolism. Although all species produce nicotinic acid, *M. tuberculosis* accumulates the largest amount. (*M. simiae* and some strains of *M. chelonae* also produce niacin.) Niacin therefore accumulates in the medium in which these

organisms are growing. A positive niacin test is preliminary evidence that an organism that exhibits a buff-colored, slow-growing, rough colony may be *M. tuberculosis* (Figure 60-5). The method is delineated in Procedure 60-8. This test is not sufficient, however, for confirmation of the identification. If sufficient growth is present on an initial L-J slant (the egg-base medium enhances accumulation of free niacin), a niacin test can be performed immediately. If growth on the initial culture is scanty, the subculture used for growth rate determination can be used. If this culture yields only rare colonies, the colonies should be spread around with a sterile cotton swab (after the growth rate has been determined) to distribute the inoculum over the entire slant. The slant is then reincubated until light growth over the surface of the medium is visible. For reliable results, the niacin test should be performed only from cultures on L-J that are at least 3 weeks old and show at least 50 colonies;

TABLE 60-10 DISTINCTIVE PROPERTIES OF COMMONLY CULTIVABLE MYCOBACTERIA ENCOUNTERED IN CLINICAL SPECIMENS

GROUP/COMPLEX	SPECIES	OPTIMAL TEMP (° C)	USUAL COLONY MORPHOLOGY[a]	NIACIN	GROWTH ON TCH (10 µg/mL)[b]	NITRATE REDUCTION	SEMIQUANTITATIVE CATALASE (>45 mm)
TB	M. tuberculosis	37	R	+	+	+	−
	M. bovis	37	Rt	−	−	−	−
	M. africanum	37	R	−	V	−	−
Photochromogens	M. marinum	30	S/SR	∓	+	−	−
	M. kansasii	37	SR/S	−	+	+	+
	M. simiae	37	S	+	+	−	+
	M. asiaticum	37		−	+	−	+
Scotochromogens	M. scrofulaceum	37	S	−	+	−	+
	M. szulgai	37	S or R	−	+	+	+
	M. gordonae	37	S	−	+	−	+
Nonphotochromogens	M. avium complex	37	St/R	−	+	−	−
	M. genavense[e]	37	St/R	−	+	−	+
	M. gastri	37	S/SR/R	−	+	−	−
	M. malmoense	37	S	−	+	−	−
	M. haemophilum[f]	30	R	−	+	−	−
	M. shimoidei	37	R	−	+	−	−
	M. ulcerans	30	R	−	+	−	−
	M. flavescens[g]	37	S	−	+	+	+
	M. xenopi[h]	42	Sf	−	+	−	−
	M. terrae complex (M. terrae, M. triviale,[i] M. nonchromogenicum)	37	SR	−	−	+	+

68° C CATALASE	TWEEN HYDRO-LYSIS, 5 DAYS	TELLURITE REDUC-TION	TOLER-ANCE TO 5% NaCl	IRON UPTAKE	ARYLSUL-FATASE, 3 DAYS	GROWTH ON MAC-CONKEY AGAR	UREASE	PYRAZIN-AMIDASE, 4 DAYS
−	−c	∓	−	−	−	−	+	+
−	−	∓	−	−	−	−	+	−
−	−	−	−	−	−	−	+	−
−	+	∓	−	−	∓d	−	+	+
+	+	∓	−	−	−	−	+	−
+	−	+	−	−	−	−	+	+
+	+	−	−	−	−	−	−	−
+	−	∓	−	−	V	−	V	±
+	∓c	±	−	−	V	−	+	+
+	+	−	−	−	V	−	V	∓
±	−	+	−	−	−	∓	−	+
+	+			−	−		+	+
−	+	∓	−	−	−	−	∓	−
±	+	+	−	−	−	−	−	+
−	−	−	−	−	−	−	−	+
−	+	−	−	−	−	−	−	+
+	−	−	−	−	−	−	−	−
+	+	∓	+	−	−	−	+	+
+	−	∓	−	−	±	−	−	V
+	+	−	−	−	−	V	−	V

Continued

TABLE 60-10 DISTINCTIVE PROPERTIES OF COMMONLY CULTIVABLE MYCOBACTERIA ENCOUNTERED IN CLINICAL SPECIMENS—CONT'D

GROUP/COMPLEX	SPECIES	OPTI-MAL TEMP (° C)	USUAL COLONY MORPH-OLOGY[a]	NIACIN	GROWTH ON TCH (10 μg/mL)[b]	NITRATE REDUC-TION	SEMIQUAN-TITATIVE CATALASE (>45 mm)
Rapidly Growing	*M. fortuitum* group		Sf/Rf	∓	−	+	+
	M. chelonae		S/R	V	−	−	+
	M. abscessus		S/R	V		−	+
	M. smegmatis		R/S			+	+

Plus and minus signs indicate the presence or absence, respectively, of the feature; blank spaces indicate either that the information is not currently available or that the property is unimportant. *V*, Variable; ±, usually present; ∓, usually absent.

[a]*R*, Rough; *S*, smooth; *SR*, intermediate in roughness; *t*, thin or transparent; *f*, filamentous extensions.

[b]*TCH*, Thiophene-2-carboxylic acid hydrazide.

[c]Tween hydrolysis may be positive at 10 days.

[d]Arylsulfatase, 14 days, is positive.

[e]Requires mycobactin for growth on solid media.

[f]Requires hemin as a growth factor.

[g]Young cultures may be nonchromogenic or possess only pale pigment that may intensify with age.

[h]Strains of *M. xenopi* can be nonphotochromogenic or scotochromogenic.

[i]*M. triviale* is tolerant to 5% NaCl, and a rare isolate may grow on MacConkey agar.

*See American Society for Microbiology. 1992. Clinical microbiology procedures handbook, vol 1. The Society, Washington, ed 6. American Society for Microbiology, Washington, D.C. for other mycobacterial species' biochemical reactions and addi-

TABLE 60-11 KEY BIOCHEMICAL REACTIONS TO HELP DISTINGUISH MYCOBACTERIA BELONGING TO THE SAME MYCOBACTERIAL GROUP

MYCOBACTERIAL GROUP	KEY BIOCHEMICAL TESTS
M. tuberculosis complex	Niacin, nitrate reduction, susceptibility to TCH if *M. bovis* is suspected
Photochromogens	Tween 80 hydrolysis, nitrate reduction, pyrazinamidase, 14-day arylsulfatase, urease, niacin
Scotochromogens	Permissive growth temperature, Tween 80 hydrolysis, nitrate reduction, semi-quantitative catalase, urease, 14-day arylsulfatase
Nonphotochromogens	Heat-resistant and semiquantitative catalase activity, nitrate reduction, Tween 80 hydrolysis, urease, 14-day arylsulfatase, tellurite reduction, acid phosphatase activity
Rapidly growing	Growth on MacConkey agar, nitrate reduction, Tween 80 hydrolysis, 3-day arylsulfatase, iron uptake

otherwise, enough niacin might not have been produced to be detected.

NITRATE REDUCTION This test is valuable for the identification of *M. tuberculosis, M. kansasii, M. szulgai,* and *M. fortuitum*. The ability of acid-fast bacilli to reduce nitrate is influenced by age of the colonies, temperature, pH, and enzyme inhibitors. Although rapid-growers can be tested within 2 weeks, slow-growers should be tested after 3 to 4 weeks of luxuriant growth. Commercially available nitrate strips yield acceptable results only with strongly nitrate-

68° C CATALASE	TWEEN HYDRO-LYSIS, 5 DAYS	TELLURITE REDUC-TION	TOLER-ANCE TO 5% NaCl	IRON UPTAKE	ARYLSUL-FATASE, 3 DAYS	GROWTH ON MAC-CONKEY AGAR	UREASE	PYRAZIN-AMIDASE, 4 DAYS
+	V	+	+	+	+	+	+	+
V	V	+	−	−	+	+	+	+
V	V	+	+	−	+	+		
+	+	+	+	+	−	−		

D.C. and Nolte, F.S. and Metchock, B. 1995. In Murray, P.R., Baron, E.J., Pfaller, M.A., et al. Manual of clinical microbiology, tional biochemical reactions on the mycobacteria included in this table.

positive organisms, such as *M. tuberculosis*. This test (Procedure 60-9) may be tried first because of its ease of performance. The *M. tuberculosis*–positive control must be strongly positive in the strip test or the test results will be unreliable. If the paper strip test is negative or if the control test result is not strongly positive, the chemical procedure (Procedure 60-10) must be carried out using strong and weakly positive controls.

CATALASE Most species of mycobacteria, except for certain strains of *M. tuberculosis* complex (some isoniazid-resistant strains) and *M. gastri*, produce the intracellular enzyme catalase, which splits hydrogen peroxide into water and oxygen. Catalase is assessed in the following two ways:

1. By the relative activity of the enzyme, as determined by the height of a column of bubbles of oxygen (Figure 60-6) formed by the action of untreated enzyme produced by the organism (semiquantitative catalase test). On the basis of the semiquantitative catalase test, mycobacteria are divided into two groups: those producing <45 mm of bubbles and those producing >45 mm of bubbles.

2. By the ability of the catalase enzyme to remain active after heating, a measure of the heat stability of the enzyme (heat-stable catalase test). When heated to 68° C for 20 minutes, the catalase of *M. tuberculosis*, *M. bovis*, *M. gastri*, and *M. haemophilum* becomes inactivated.

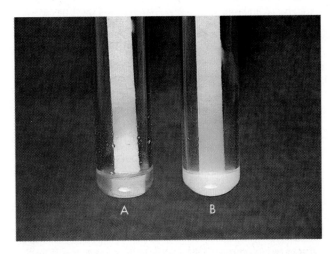

FIGURE 60-5 Niacin test performed with filter paper strips. The positive test (**A**) displays a yellow color. The negative result (**B**) remains milky white or clear.

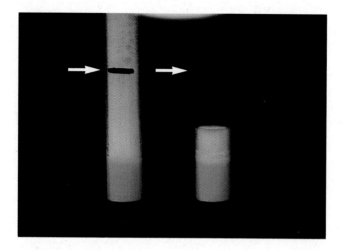

FIGURE 60-6 Semiquantitative catalase test. The tube on the left contains a column of bubbles that has risen past the line (*arrow*) indicating 45-mm height (a positive test). The tube on the right is the negative control.

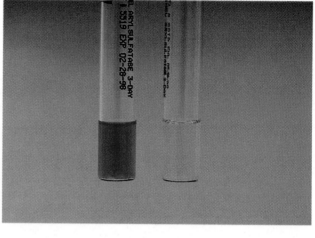

FIGURE 60-7 A positive arylsulfatase test is shown on the left; the tube containing the negative control is on the right.

TWEEN 80 HYDROLYSIS The commonly nonpathogenic, slow-growing scotochromogens and nonphotochromogens produce a lipase that is able to hydrolyze Tween 80 (the detergent polyoxyethylene sorbitan monooleate) into oleic acid and polyoxyethylated sorbitol, whereas pathogenic species do not. Tween 80 hydrolysis is useful for separating species of photochromogens, nonchromogens, and scotochromogens. Because laboratory-prepared media have a very short shelf life, the CDC recommends the use of a commercial Tween 80 hydrolysis substrate (Difco Laboratories or Remel Laboratories) that is stable for up to 1 year.

TELLURITE REDUCTION Some species of mycobacteria reduce potassium tellurite at variable rates. The ability to reduce tellurite in 3 to 4 days distinguishes members of *M. avium* complex from most other nonchromogenic species. All rapid-growers reduce tellurite in 3 days.

ARYLSULFATASE The enzyme arylsulfatase is present in most mycobacteria. Test conditions can be varied to differentiate different forms of the enzyme. The rate by which this enzyme breaks down phenolphthalein disulfate into phenolphthalein (which forms a red color in the presence of sodium bicarbonate) and other salts helps to differentiate certain strains of mycobacteria. The 3-day test is particularly useful for identifying the potentially pathogenic rapid-growers, *M. fortuitum* and *M. chelonae.* Slow-growing *M. marinum* and *M. szulgai* are positive in the 14-day test (Figure 60-7).

GROWTH INHIBITION BY THIOPHENE-2-CARBOXYLIC ACID HYDRAZIDE (TCH) This test is used to distinguish *M. bovis* from *M. tuberculosis,* because only *M. bovis* is unable to grow in the presence of 10 mg per mL TCH.

OTHER TESTS Other tests (Table 60-11) are often performed to make more subtle distinctions between species. It is not cost effective for routine clinical microbiology laboratories to be able to perform all the procedures necessary for definitive identification of mycobacteria, because excellent reference laboratories are available in every state. With a minimal number of basic procedures, however, the majority of strains isolated can be presumptively identified, and those that require further testing can be forwarded to regional laboratories.

Nucleic acid

DNA hybridization is used to identify some of the more common mycobacterial species isolated on solid culture media or from broth culture. Of importance, these tests can be performed with sufficient growth from primary cultures. Nonisotopically labeled (i.e., acridinium ester–labeled) DNA probes specific for mycobacterial ribosomal RNA (rRNA) sequences are commercially available (GenProbe, San Diego, Calif.) and are listed in Box 60-4. rRNA is released from the mycobacteria by means of a lysing agent, sonication, and heat. The specific DNA probe is allowed to react with the extracted rRNA to form a stable DNA-RNA hybrid. Any unhybridized DNA-acridinium ester probes are chemically degraded. When an alkaline hydrogen peroxide solution is added to elicit chemiluminescence, only the hybrid-bound acridinium ester is available to emit light; the amount of light emitted is directly related to the amount of hybridized probe. The light produced is measured on a chemiluminometer. Numerous laboratories have incorporated these tests into their routine procedures. By modifying the probe assay protocol, some laboratories have successfully combined the BACTEC 460 TB system with

M. tuberculosis complex
M. avium complex
M. avium
M. intracellulare
M. kansasii
M. gordonae

probes to detect and identify mycobacteria directly from BACTEC 12B culture vials. Rapid detection and identification of *M. tuberculosis* complex in the BACTEC 460 TB has also been accomplished using polymerase chain reaction.[21,46]

Chromatographic analysis

The analysis of mycobacterial lipids by chromatographic methods, including thin-layer chromatography, gas-liquid chromatography (GLC), capillary gas chromographic methods, and reverse-phase high-performance liquid chromatography (HPLC), has been used to identify mycobacteria. HPLC of extracted mycobacteria is a very specific and rapid method for identification of species.[51] Many state health departments and the CDC now use this method routinely. The long-chain mycolic acids are separated better by HPLC than by GLC, because they do not withstand the high temperatures needed for GLC. The patterns produced by different species are very reproducible, and a typical identification requires only a few hours.

Serodiagnosis

Immunodiagnostic methods based on the detection of antibodies to various mycobacterial antigens have not come into widespread clinical use compared with other infectious diseases.

ANTIMICROBIAL SUSCEPTIBILITY TESTING AND THERAPY

M. TUBERCULOSIS COMPLEX

Because of the increased incidence of tuberculosis in the United States in the early 1990s, new recommendations by the Centers for Disease Control and Prevention and the American Thoracic Society require in vitro drug susceptibility testing of *M. tuberculosis* isolates from all patients.[15] Susceptibility testing of *M. tuberculosis* requires meticulous care in the preparation of the medium, selection of adequate samples of colonies, standardization of the inoculum, use of appropriate controls, and interpretation of results. Laboratories that see very few positive cultures should consider sending isolates to a reference labora-

tory for testing. Isolates must be saved in sterile 10% skim milk in distilled water at −70° C for possible future additional studies (such as susceptibilities if the patient does not respond well to treatment).

Direct vs. indirect susceptibility testing

Susceptibilities may be performed by either the direct or the indirect method. The direct method uses as the inoculum a smear-positive concentrate containing more than 50 acid-fast bacilli per 100 oil immersion fields; the indirect method uses a culture as the inoculum source. Although direct testing provides more rapid results, this method is less standardized and contamination may occur.

Conventional methods

Development of primary drug resistance in tuberculosis represents an increase in the proportion of resistant organisms. This increase in resistant organisms results from a spontaneous mutation and subsequent selection to predominance of these drug-resistant mutants by the action of a single or ineffective drug therapy. A poor clinical outcome is predicted with an agent when more than 1% of bacilli in the test population are resistant.[31] If an isolate is reported as resistant to a drug, treatment failure will most likely occur if this drug is used for therapy. Drug resistance is defined for *M. tuberculosis* complex in terms of the critical concentration of the drug. The **critical concentration** of a drug is the amount of drug required to prevent growth above the 1% threshold of the test population of tubercle bacilli.

Four general methods are used throughout the world for determining susceptibility of isolates of *M. tuberculosis* to various antituberculous agents; these methods are summarized in Table 60-12. The proportion and BACTEC radiometric methods are most commonly used in the United States. Usually the initial isolate of *M. tuberculosis* is tested against five antimicrobials; these antimicrobials are referred to as **primary drugs** and are listed in Box 60-5. If any drug resistance is detected to any of the primary drugs, a second battery of agents are tested; these agents are also listed in Box 60-5.

New approaches

Several new technologies recently have been introduced that are promising in terms of being faster, more reliable, and/or easier to perform than most conventional methods for susceptibility testing. For example, the MGIT system has been used to perform in vitro susceptibility testing of *M. tuberculosis,* and mutations leading to rifampin resistance have been detected using molecular methods. In addition, Jacobs et al.[30] used an innovative approach to perform susceptibility testing by using a luciferase-reporter

TABLE 60-12 OVERVIEW OF CONVENTIONAL METHODS TO DETERMINE SUSCEPTIBILITY OF *M. TUBERCULOSIS* ISOLATES TO ANTIMYCOBACTERIAL AGENTS

METHOD	PRINCIPLE
Absolute Concentration	For each drug tested, a standardized inoculum is inoculated to control (drug-free) media and media containing several appropriate graded drug concentrations. Resistance is expressed as the lowest concentration of drug that inhibits all or almost all of the growth, that is, the minimum inhibitory concentration (MIC)
Resistance Ratio	The resistance of the test organism is compared with that of a standard laboratory strain. Both strains are tested in parallel by inoculating a standard inoculum to media containing two-fold serial dilutions of the drug. Resistance is expressed as the ratio of the MIC of the test strain divided by the MIC for the standard strain for each drug
Proportion	For each drug tested, several dilutions of standardized inoculum are inoculated onto control and drug-containing agar medium. The extent of growth in the absence or presence of drug is compared and expressed as a percentage. If growth at the critical concentration of a drug is >1%, the isolate is considered clinically resistant[27]
Radiometric	Employing the principles of the proportion method, this rapid method uses liquid medium containing ^{14}C-labeled growth substrate. Growth is indicated by the amount of ^{14}C-labeled–carbon dioxide (CO_2) released, as measured by the BACTEC 460 instrument. For each drug tested, a standardized inoculum is inoculated into a drug-free and drug-containing vial. The rate and amount of CO_2 produced in the absence or presence of drug is then compared

BOX 60-5

ANTITUBERCULOUS AGENTS COMMONLY TESTED AGAINST *M. TUBERCULOSIS*

Primary Drugs
Streptomycin
Isoniazid
Rifampin
Ethambutol
Pyrazinamide

Secondary Drugs
Ethionamide
Capreomycin
Ciprofloxacin
Ofloxacin
Kanamycin
Cycloserine
Rifabutin

mycobacteriophage (bacterial viruses). The basis for this assay is simple—only viable mycobacteria can become infected with and replicate the mycobacteriophage; dead tubercle bacilli cannot. The mycobacteriophage was constructed to have the firefly luciferase gene next to a mycobacterial promoter; therefore, the presence and growth of the mycobacteriophage is detected by chemiluminescence. In brief, the isolate of *M. tuberculosis* to be tested is grown in the presence

and absence of drug and the specially constructed mycobacteriophage is added. Following infection, a substrate of luciferase, luciferin is added. If organisms are viable (i.e., thereby allowing infection of the bacteriophage and subsequent transcription and translation of the luciferase gene), the luciferin is broken down and light is emitted that can be measured; the amount of light emitted is directly proportional to the number of viable *M. tuberculosis*. Therefore if an organism is resistant to the drug, light will be emitted, whereas organisms susceptible to the drug will not emit any light. In addition to these molecular approaches, the E-test (see Chapter 18) has been successfully applied to susceptibility testing of *M. tuberculosis.*

Therapy

Therapy directed against *M. tuberculosis* is dependent on the susceptibility of the isolate to various antimicrobial agents. To prevent the selection of resistant mutants, treatment of tuberculosis requires two or three drugs. The most common two-drug regimen is INH and rifampin administered for 9 months in cases of uncomplicated tuberculosis; if pyrazinamide is added to this regimen during the first 2 months, the total length of therapy can be shortened to 6 months.[40] Finally, INH prophylaxis is recommended for those individuals with a recent skin-test conversion who are disease free.

NONTUBERCULOUS MYCOBACTERIA (NTM)

In vitro susceptibility testing of most NTM, including isolates of *M. avium* complex is not routinely performed. Studies are lacking in which in vitro susceptibility test results have been correlated with clinical outcome and, as yet, there are no standardized methods for performing these tests. However, susceptibility testing of rapidly growing mycobacteria is one exception and should be performed, particularly if these organisms are isolated from wounds. The E-test, broth microdilution, and agar disk elution are three methods that can be employed.[28] Drugs usually tested against the rapidly growing mycobacteria include aminoglycosides, quinolones, macrolides, cephalosporins, sulfonamides, tetracycline, and a carbapenem.

Many infections caused by NTMs are not amenable to antimicrobial therapy, and most clinically significant NTMs are resistant to primary antituberculous agents.[40] In some instances, such as infections caused by *M. avium* complex, the organisms are multidrug resistant and treatment requires at least four or more agents.

PREVENTION

As previously mentioned, prophylactic chemotherapy with INH is used in situations in which known or suspected primary tuberculous infection causes a risk for clinical disease. At present, the BCG vaccine (named after Calmette and Guérin) is the only available vaccine against tuberculosis. The effectiveness of this live vaccine is controversial, because studies have demonstrated ineffectiveness to 80% protection. The greatest potential value for this vaccine is in developing countries that have high prevalence rates for tuberculosis.

Controlled clinical trials are under way to determine the best therapeutic approach for disseminated MAC infection in patients infected with HIV. Regardless, azithromycin or clarithromycin (new macrolide antibiotics) should be included along with ethambutol, rifabutin, rifampin, ciprofloxacin, or amikacin.

Procedures

60-1 *N*-ACETYL-L-CYSTEINE-SODIUM HYDROXIDE METHOD FOR LIQUEFACTION AND DECONTAMINATION OF SPECIMENS

PRINCIPLE

Sodium hydroxide (NaOH), a decontaminating agent, also acts as an emulsifier. Because of its potential toxicity, NaOH should be used at the lowest concentration that effectively digests and decontaminates the specimen. The addition of a mucolytic agent, *N*-acetyl-L-cysteine (NALC), reduces the concentration of NaOH required and also shortens the time required for decontamination, thus aiding the optimal recovery of acid-fast bacilli.

METHOD

1. Reagent preparation:
   A. NALC-NaOH preparation:
   For each day's cultures, add up the total volume of specimens to be treated and prepare an equal volume of the digestant-decontamination mixture, as follows:

   1 N (4%) NaOH (50 mL)
   0.1 M (2.94%) trisodium citrate · 3H$_2$O (50 mL)
   NALC powder (0.5 g)

   Use sterile distilled water for preparation of solutions to minimize chances of inadvertently adding acid-fast tap water contaminants to the specimens. Mix, sterilize, and store the NaOH and the citrate in sterile, screw-capped flasks for later use. This solution should be used within 24 hours after the NALC is added.
   B. 0.67 M phosphate buffer, pH 6.8 preparation
   Make of the following solutions:
   Solution A (0.067 M disodium phosphate):
   Sodium monohydrogen phosphate (anhydrous) 9.47 g
   Distilled water 1000 mL
   Solution B (0.067 M monopotassium phosphate):
   Potassium dihydrophosphate 9.07 g
   Distilled water 1000 mL
   Add 50 mL of solution B to 50 mL of solution A and adjust pH to 6.8.

60-1 N-ACETYL-L-CYSTEINE-SODIUM HYDROXIDE METHOD FOR LIQUEFACTION AND DECONTAMINATION OF SPECIMENS—cont'd

2. Work within a biological safety cabinet and wear protective clothing, gloves, and mask. Transfer a maximum of 10 mL of sputum, urine, or other fluid to be processed to a sterile, disposable, plastic 50-mL conical centrifuge tube with a leakproof and aerosol-free plastic screw-cap. Tubes with easily visible volume indicator marks are best.

3. Add an equal volume of freshly prepared digestant to the tube, being very careful when pouring digestant not to touch the lip of the specimen container, which might inadvertently transfer positive material to a negative specimen. Tighten the cap completely.

4. Vortex the specimen for approximately 15 seconds or for a maximum of 30 seconds, being certain to create a vortex in the liquid and not to merely agitate the material. Check for homogeneity by inverting the tube. If clumps remain, vortex the specimen intermittently while the rest of the specimens are being digested. An extra pinch of NALC crystals may be necessary to liquefy mucoid sputa.

5. Start a 15-minute timer when the first specimen is finished being vortexed. Continue digesting the other specimens, noting the amount of time that the entire run takes. The digestant should remain on the specimens for a maximum exposure of 20 minutes.

6. After 15 minutes of digestion, add enough phosphate buffer to reach within 1 cm of the top, screw the cap tightly closed, and invert the tube to mix the solutions and stop the digestion process. Addition of this solution also reduces the specific gravity of the specimen, aiding sedimentation of the bacilli during centrifugation.

7. Centrifuge all tubes at 3600× *g* for 15 minutes, using aerosol-free sealed centrifuge cups.

8. Carefully pour off the supernatant into a splash-proof container. To ensure that the specimen does not run down the outside of the tube after pouring, the lip of the tube may be wiped with an amphyl- or phenol-soaked gauze to absorb drips. Be careful not to touch the lip of any tube to another container. It is helpful to watch the sediment carefully as the supernatant is being decanted, because a very mucoid sediment may be loose and may pour out with the supernatant. If the sediment begins to slip, stop decanting and use a sterile capillary pipette to remove the supernatant without losing the sediment.

9. Resuspend the sediment in 1 to 2 mL phosphate buffer, pH 6.8 buffer (with bovine serum albumin [BSA]).

10. Inoculate the sediment to culture media and prepare slides.

60-2 PREPARATION OF SMEARS FOR ACID-FAST STAIN FROM DIRECT OR CONCENTRATED SPECIMENS

METHOD

1. Vortex concentrated sediment, unconcentrated sputum, other purulent material, or stool. Aspirate 0.1 to 0.2 mL into a Pasteur pipette and place two to three drops on the slide. Place the end of the pipette or a sterile applicator stick parallel to the slide and slowly spread the liquid uniformly to make a thin smear.

2. For cerebrospinal fluid sediment, vortex thoroughly and apply to the slide in heaped drops. A heaped drop is allowed to air dry, and a second application of sediment is placed on the same spot and allowed to dry. A minimum of three layers, applied to the same 1-cm diameter circle, should facilitate detection of small numbers of bacilli. (Note: some laboratories have stopped performing acid-fast stains on CSF because positive stains are extremely rare.)

3. Fix the smear at 80° C for 15 minutes or for 2 hours at 65° to 70° C on an electric hot plate.

Note: Survival of mycobacteria at this temperature has been reported; handle all specimens with proper precautions.

4. Stain slides by Ziehl-Neelsen or fluorochrome stain.

60-3 AURAMINE-RHODAMINE FLUOROCHROME STAIN

PRINCIPLE
The fluorochrome dyes used in this stain complex to the mycolic acids in acid-fast cell walls. Detection of fluorescing cells is enhanced by the brightness against a dark background.

METHOD
1. Heat-fix slides at 80° C for at least 15 minutes or for 2 hours at 65° to 70° C.

2. Flood slides with auramine-rhodamine reagent and allow to stain for 15 to 20 minutes at room temperature.

3. Rinse with deionized water and tilt slide to drain.

4. Decolorize with 0.5% acid-alcohol (70% ethanol and 0.5% hydrochloric acid) for 2 to 3 minutes.

5. Rinse with deionized water and tilt slide to drain.

6. Flood slides with 0.5% potassium permanganate for 2 to 4 minutes.

7. Rinse with deionized water and air dry.

8. Examine under low power (250×) for fluorescence.

EXPECTED RESULTS
Mycobacterium spp. will fluoresce yellow to orange depending on the filter system used.

60-4 ZIEHL-NEELSEN ACID-FAST STAIN

PRINCIPLE
Heating the slide allows greater penetration of carbolfuchsin into the cell wall. Mycolic acids and waxes complex the basic dye, which then fails to wash out with mild acid decolorization.

METHOD
1. Heat-fix slides as previously described.

2. Flood smear with carbolfuchsin stain reagent and steam the slides gently for 1 minute by flaming from below the rack with a gas burner or by staining the slides directly on a special hot plate. Do not permit the slides to boil or dry out.

3. Allow the stain to remain on the slides for an additional 4 to 5 minutes without heat.

4. Rinse with deionized water and tilt slides to drain.

5. Decolorize with 3.0% acid-alcohol (95% ethanol and 3.0% hydrochloric acid) for 2 minutes. Rinse slides with deionized water and tilt to drain.

6. Flood slides with methylene blue reagent for 1 minute.

7. Rinse with deionized water and allow to air dry.

8. Examine under oil immersion (1000×) for presence of acid-fast bacilli.

EXPECTED RESULTS
Mycobacterium spp. will appear red or have a red-blue, beaded appearance, whereas nonmycobacteria will appear blue.

60-5 KINYOUN STAIN

PRINCIPLE
By increasing the concentration of basic fuchsin and phenol, the need for heating the slide is avoided.

METHOD
1. Heat-fix slides as previously described.

2. Flood slides with Kinyoun carbolfuchsin reagent and allow to stain for 5 minutes at room temperature.

3. Rinse with deionized water and tilt slide to drain.

4. Decolorize with 3.0% acid-alcohol (70% ethanol and 0.5% hydrochloric acid) for 2 minutes.

5. Rinse with deionized water and drain standing water from slide surface by tipping slide.

6. Flood slide with methylene blue counterstain and allow to stain for 1 to 3 minutes.

7. Rinse with deionized water and allow to air dry.

8. Examine under oil immersion (1000×).

EXPECTED RESULTS
Mycobacterium spp. will appear red or have a red-blue beaded appearance, whereas nonmycobacteria will appear blue.

60-6 QUALITY CONTROL FOR MYCOBACTERIOLOGY

REAGENTS

1. Media: routine media used for cultivation of mycobacteria

2. Quality control organism: a recent isolate of *M. tuberculosis* or *M. tuberculosis* strain H37Rv

3. Other materials:
   Autoclaved sputum (AFB-negative)
   7H9 liquid medium containing 15% glycerol
   Sterile buffer, pH 7.0
   50-mL plastic, conical centrifuge tubes

METHOD

1. Suspend several colonies of H37Rv in a tube containing 3 mL of 7H9 liquid medium and several plastic or glass beads. Mix vigorously on a test tube mixer; then allow large particles to settle for 15 minutes.

2. Prepare a dilution of approximately 10^6 organisms per mL by adding the above cell suspension drop-by-drop to 1 mL of buffer until a barely turbid suspension occurs. Transfer 0.5 mL of the 10^6 cells per mL dilution to 4.5 mL of glycerol broth to give a suspension of 10^5 cells per mL. Repeat the procedure to make a 10^4 per mL and a 10^3 per mL suspension.

3. Label 15 3-dram vials for each suspension (10^5, 10^4, and 10^3). Transfer 0.3 mL of the appropriate suspension to each vial. Store the vials at $-70°$ C to use for future quality control testing.

4. Thaw one vial of each of the three dilutions each time the quality control procedure is performed.

5. Add 2.7 mL autoclaved sputum to each cell suspension to effect a tenfold dilution, and inoculate three sets of the media used for primary isolation with each of the three dilutions of sputum. Inoculate 0.1 mL of sputum per bottle.

6. Decontaminate and concentrate the remainder as with sputum specimens. Reconstitute the sediments with sterile buffer to 2.6 mL, resuspend vigorously, and inoculate a second set of media with 0.1 mL of each of the concentrated and resuspended samples.

7. Incubate at 35° C in 5% to 10% CO_2 for 21 days.

INTERPRETING AND RECORDING RESULTS

Egg media should have been inoculated with approximately 10^4, 10^3, and 10^2 organisms, respectively. The first dilution should produce semiconfluent growth, and the second and third dilutions should produce countable colonies in each bottle. Because of the retrospective nature of these determinations, close comparisons must be made between current and previous results to note trends or developing deficiencies. Failures may be the result of faulty media, lethal effects of decontamination and concentration procedures, improperly prepared reagents, or overexposure of specimens to these reagents. Should deficiencies become evident, techniques should be reviewed and attempts made to determine the source of the problem. New batches of media must be substituted for deficient media, and the latter rechecked to verify deficiencies. Personnel should be included in all discussions of problems and corrective measures. All deficiencies and corrective actions should be recorded in the appropriate section of the quality control records.

EXAMPLE OF INTERPRETING QUALITY CONTROL TEST RESULTS OF DECONTAMINATION AND CONCENTRATION PROCEDURE

SPUTUM SAMPLE	SPUTUM UNPROCESSED 10^4	10^3	10^2	PROCESSED 10^4	10^3	10^2	INTERPRETATION
1	3+	2+	50-100 colonies	2+	1+ 2+	Approximately 10 colonies	Media and decontamination procedures are acceptable
2	3+	2+	50-100 colonies	1+	0	0	Media acceptable; procedure too toxic
3	2+ or 1+	2+ or 1+	0	1+ or 0	1+ or 0	0	One or more of the media is not supporting growth of AFB adequately

60-7 DETERMINATION OF PIGMENT PRODUCTION AND GROWTH RATE

PRINCIPLE
Certain mycobacteria produce carotenoids, either dependently or independently of exposure to light. This characteristic, in addition to their doubling time under standard conditions, is useful for initial identification.

METHOD
1. After the broth culture has incubated for 5 to 7 days, adjust the turbidity to that of a McFarland 0.5 standard.

2. Dilute the broth (McFarland 0.5 turbidity) 10^4.

3. Inoculate 0.1 mL of the diluted broth to each of three tubes of Löwenstein-Jensen agar. Completely wrap two of the tubes in aluminum foil to block all light. If the isolate was obtained from a skin lesion or the initial colony was yellow-pigmented (possible *M. szulgai*), six tubes should be inoculated. The second set of tubes, two of them also wrapped with aluminum foil, is incubated at 30° C, or at room temperature if a 30° C incubator is not available.

4. Examine the cultures after 5 and 7 days for the appearance of grossly visible colonies. Examine again at intervals of 3 days. Interpretation: rapid-growers produce visible colonies in less than 7 days; slow-growers require more than 7 days.

5. When colonies are mature, expose the growth from a foil-wrapped tube to a bright light, such as a desk lamp, for 2 hours. The cap must be loose during exposure, because pigment production is an oxygen-dependent reaction.[54] The tube is rewrapped and returned to the incubator, and the cap is left loose.

6. The three tubes are examined 24 and 48 hours after light exposure. For tubes incubating at 30° C, pigment may require 72 hours for development.

EXPECTED RESULTS
Interpret as shown in Figure 60-4.

60-8 NIACIN TEST WITH COMMERCIALLY AVAILABLE PAPER STRIPS*

PRINCIPLE
The accumulation of niacin in the medium caused by lack of an enzyme that converts niacin to another metabolite in the coenzyme pathway is characteristic for *M. tuberculosis* and a few other species. Niacin is measured by a colored end product.

METHOD
1. Add 1 mL sterile distilled water to the surface of the egg-based medium on which the colonies to be tested are growing.

2. Lay the tube horizontally, so that the fluid is in contact with the entire surface. Using a pipette, scratch or lightly poke through the surface of the agar; this allows niacin in the medium to dissolve in the water.

3. Allow the tube to sit for 15 to 30 minutes at room temperature. It can incubate longer to achieve a stronger reaction.

4. Remove 0.6 mL of the distilled water (which appears cloudy at this point) to a clean, 12×75-mm screw-cap or snap-top test tube. Insert a niacin test strip with the arrow down, following manufacturer's instructions.

5. Cap the tube tightly and incubate at room temperature, occasionally shaking the tube to mix the fluid with the reagent on the bottom of the strip.

6. After 20 minutes, observe the color of the liquid against a white background (see Figure 60-5).

EXPECTED RESULTS
Yellow liquid indicates a positive test. The color of the strip should not be considered when evaluating results. If the liquid is clear, the test is negative. Discard the strip into alkaline disinfectant (10% NaOH) to neutralize the cyanogen bromide.

*Difco Laboratories, Detroit, Mich.

60-9 NITRATE REDUCTION TEST WITH COMMERCIALLY AVAILABLE PAPER STRIPS*

PRINCIPLE

The presence of the enzyme nitroreductase can be detected by the ability of a suspension of organisms to produce a colored end product from substrates that combine with nitrite, the product of nitroreductase. *M. tuberculosis* and several other species of mycobacteria possess this enzyme.

METHOD

1. Add 1 mL sterile saline to a sterile 13×100-mm screw-cap test tube.

2. Emulsify two very large clumps of growth from a 4-week-old culture in the saline. The solution should be very turbid (milky).

3. Using sterile forceps, carefully insert a nitrite test strip according to the package insert instructions. The strip should touch only the fluid at the bottom of the tube, not the sides.

4. Cap the tube tightly and incubate upright for 2 hours at 35° C. Incubation in a water bath will ensure maintenance of adequate temperature.

5. After the first hour, shake the tube gently without tilting.

6. After the 2-hour incubation, tilt the tube 6 times, wetting the entire strip.

7. Place the tube in a slanted position for 10 minutes with the liquid covering the strip.

EXPECTED RESULTS

Observe the top portion of the strip for any blue color change, indicating a positive reaction. A negative reaction (lack of nitroreductase) yields no color change.

*Difco Laboratories, Detroit, Mich.

60-10 NITRATE REDUCTION TEST USING CHEMICAL REAGENTS

PRINCIPLE

As in the conventional nitrate test, the presence of nitrite (product of the nitroreductase enzyme) is detected by production of a red-colored product on the addition of several reagents. If the enzyme has reduced nitrate past nitrite to gas, then addition of zinc dust (which converts nitrate to nitrite) will detect the lack of nitrate in the reaction medium.

METHOD

1. Prepare the dry crystalline reagent as follows:
   Sulfanilic acid (Sigma Chemical Co., St. Louis, Mo.)   1 part
   N-(1-Naphthyl) ethylenediamine dihydrochloride (Eastman Chemical Co., Rochester, N.Y.)   1 part
   1-Tartaric acid (Sigma Chemical Co.)   10 parts
   These crystals can be measured with any small scoop or tiny spoon, because the proportions are by volume, not weight. The mixture should be ground in a mortar and pestle to ensure adequate mixing, because the crystals are of different textures. The reagent can be stored in a dark glass bottle at room temperature for at least 6 months.

2. Add 0.2 mL sterile distilled water to a 16×125 mm screw-cap tube. Emulsify two very large clumps of growth from a 4-week culture on Löwenstein-Jensen agar in the water. The suspension should be milky.

3. Add 2 mL nitrate substrate broth (Difco Laboratories or Remel, Lenexa, Kan.) to the suspension and cap tightly. Shake gently and incubate upright for 2 hours in a 35° C water bath.

4. Remove from water bath and add a small amount of the crystalline reagent. A wooden stick or a small spatula can be used to add crystals; the amount is not critical. Examine immediately.

EXPECTED RESULTS

Development of a pink to red color indicates the presence of nitrite, demonstrating the ability of the organism to reduce nitrate to nitrite. If no color results, the organisms may have reduced nitrate beyond nitrite (as in the conventional nitrate test). Add a small amount of powdered zinc to the negative tube. If a red color develops, that indicates that unreduced nitrate was present in the tube and the organism was nitroreductase-negative.

References

1. American Society for Microbiology. 1992. Clinical Microbiology Procedures Handbook, vol 1. Washington, D.C.

2. Anargyros, P., et al. 1990. Comparison of improved BACTEC and Löwenstein-Jensen media for culture of mycobacteria from clinical specimens. J. Clin. Microbiol. 28:1288.

3. Arachi, A. 1991. The global tuberculosis situation and the new control strategy of the World Health Organization. Tubercle 72:1.

4. Badak, F.Z., et al. 1996. Comparison of mycobacteria growth indicator tube with BACTEC 460 for detection and recovery of mycobacteria from clinical specimens. J. Clin. Microbiol. 34:2236.

5. Banales, J.L., et al. 1991. Adenosine deaminase in the diagnosis of tuberculous pleural effusions. Chest 99:355.

6. Bass, J.B. Jr., et al. 1990. Diagnostic standards and classification of tuberculosis. Am. Rev. Respir. Dis. 142:725.

7. Bates, J.H. and Stead, W.W. 1993. The history of tuberculosis as a global epidemic. Med. Clin. North Am. 77:1205.

8. Boggs, D.S. 1995. The changing spectrum of pulmonary infections due to nontuberculous mycobacteria. J. Okla. State Med. Assoc. 88:373.

9. Boyd, J.C. and Marr, M.J. 1988. Decreasing reliability of acid-fast smear techniques for detection of tuberculosis. Ann. Intern. Med. 82:489.

10. Bull, T.J. et al. 1995. A new group (type 3) of *Mycobacterium celatum* isolated from AIDS patients in the London area. Int. J. Syst. Bacteriol. 45:861.

11. Centers for Disease Control. U.S. Department of Health and Human Services, Public Health. In Kent, P.T. and Kubica, G.P., editors. 1995. Public health mycobacteriology: a guide for the level III laboratory, Centers for Disease Control, Atlanta.

12. Centers for Disease Control. U.S. Department of Health and Human Services, Public Health. Core curriculum on tuberculosis, ed 2. Centers for Disease Control, Atlanta.

13. Centers for Disease Control. Morb. Mortal. Wkly. Rep. 1991. Nosocomial transmission of multi-drug-resistant tuberculosis among HIV-infected persons: Florida and New York, 1988-1991, Centers for Disease Control, Atlanta. M.M.W.R. 40:585.

14. Centers for Disease Control. Morb. Mortal. Wkly. Rep. CDC Survival Summary. 1991. Tuberculosis morbidity in the United States: final data, Centers for Disease Control, Atlanta. M.M.W.R. 40:23.

15. Centers for Disease Control. 1993. Initial therapy for tuberculosis in the era of multidrug resistance: recommendations of the advisory council for the elimination of tuberculosis. Morb. Mortal. Wkly. Rep. 42(RR-7):1.

16. Colston, M.J. 1993. The microbiology of *Mycobacterium leprae*; progress in the last 30 years. Trans. Royal Soc. Trop. Med. Hyg. 87:508.

17. Dahl, D.M., Klein, D., and Morgentaler, A. 1996. Penile mass caused by the newly described organism *Mycobacterium celatum*. Urology 47:266.

18. Daniel, T.M. 1989. Rapid diagnosis of tuberculosis: laboratory techniques applicable in developing countries. Rev. Infect. Dis. 11(suppl. 2):S471.

19. Debrunner, M., et al. 1992. Epidemiology and clinical significance of nontuberculous mycobacteria in patients negative for human immunodeficiency virus in Switzerland. Clin. Infect. Dis. 15:330.

20. Forbes, B.A. 1995. Current and future applications of mycobacterial amplification assays. Clin. Microbiol. Newsletter 17:145.

21. Forbes, B.A. and Hicks, K.E.S. 1994. Ability of PCR assay to identify *Mycobacterium tuberculosis* in BACTEC 12B vials. J. Clin. Microbiol. 32:1725.

22. Good, R.C. 1985. Opportunistic pathogens in the genus *Mycobacterium*. Annu. Rev. Microbiol. 39:347.

23. Griffith, D.E., Girard, W.M., and Wallace, R.J. 1993. Clinical features of pulmonary disease caused by rapidly growing mycobacteria. Am. Rev. Respir. Dis. 147:1271.

24. Havlik, J.A., et al. 1992. Disseminated *Mycobacterium avium* complex infection: clinical identification and epidemiologic trends. J. Infect. Dis. 165:577.

25. Havlik, J.A., et al. 1993. A prospective evaluation of *Mycobacterium avium* complex colonization of the respiratory and gastrointestinal tracts of persons with human immunodeficiency virus infection. J. Infect. Dis. 168:1045.

26. Hoop, R.K., Böttger, E.C., and Pfyffer, G.E. 1996. Etiologic agents of mycobacteriosis in pet birds between 1986 and 1995. J. Clin. Microbiol. 34:991.

27. Horsburgh, C., et al. 1992. Clinical implications of recovery of *Mycobacterium avium* complex from the stool or respiratory tract of HIV-infected individuals. AIDS 6:512.

28. Inderlied, C.B. 1994. Antimycobacteriology susceptibility testing: present practices and future trends. Eur. J. Clin. Microbiol. Infect. Dis. 13:980.

29. Isenberg, H.D., et al. 1991. Collaborative feasibility study of a biphasic system (Roche Septi-Check AFB) for rapid detection and isolation of mycobacteria. J. Clin. Microbiol. 29:1719.

30. Jacobs, W.R., et al. 1993. Rapid assessment of drug susceptibilities of *Mycobacterium tuberculosis* by means of luciferase reporter phages. Science 260:819.

31. Kent, P.T. and Kubica, G.P. 1985. Public health mycobacteriology: a guide for the level III laboratory. US Department of Health and Human Service, Public Health Service, Centers for Disease Control, Atlanta.

32. Kiehn, T.E., et al. 1996. *Mycobacterium genavense* infections in pet animals. J. Clin. Microbiol. 34:1840.

33. Lichtenstein, I.H. and MacGregor, R.R. 1983. Mycobacterial infections in renal transplant recipients: report on 5 cases and review of the literature. Rev. Infect. Dis. 5:216.

34. Lipsky, B.J., et al. 1984. Factors affecting the clinical value for acid-fast bacilli. Rev. Infect. Dis. 6:214.

35. Meissner, G. and Anz, W. 1977. Sources of *Mycobacterium avium* complex infection resulting in human diseases. Am. Rev. Respir. Dis. 116:1057.

36. Morris, A., et al. 1993. Mycobacteria in stool specimens: the nonvalue of smears for predicting culture results. J. Clin. Microbiol. 31:1385.

37. Nolte, F.S. and Metchock, B. *Mycobacterium.* 1995. In Murray, P.R., Baron, E.J., Pfaller, M.A., et al., editors. Manual of clinical microbiology, American Society for Microbiology, Washington, D.C.

38. Noordeen, S.K. 1993. Epidemiology and control of leprosy: a review of progress over the last 30 years. Trans. Royal Soc. Trop. Med. Hyg. 87:515.

39. O'Reilly, L.M. and Daborn, C.J. 1995. The epidemiology of *Mycobacterium bovis* infections in animals and man: a review. Tubercle Lung Dis. 76(suppl. 1):1.

40. Pfaller, M.F. 1994. Application of new technology to the detection, identification, and antimicrobial susceptibility testing of mycobacteria. Am. J. Clin. Pathol. 101:329.

41. Picardeau, M., et al. 1997. Genotypic characterization of five species of *Mycobacterium kansasii.* J. Clin. Microbiol. 35:25.

42. Portaels, F. 1995. Epidemiology of mycobacterial diseases. Clin. Dermatol. 13:207.

43. Runyon, E.H. 1959. Anonymous bacteria in pulmonary disease. Med. Clin. North Am. 43:273.

44. Salo, W.L., et al. 1994. Identification of *Mycobacterium tuberculosis* DNA in a pre-Columbian Peruvian mummy. Proc. Natl. Acad. Sci. USA, 91:2091.

45. Shinnick, T.M. and Good, R.C. 1994. Mycobacterial taxonomy. Eur. J. Clin. Microbiol. Infect. Dis. 13:884.

46. Smith, M.B., Bergman, J.S., Woods, G.L. 1997. Detection of *Mycobacterium tuberculosis* in BACTEC 12B broth cultures by Roche Amplicor PCR assay. J. Clin. Microbiol. 35:900.

47. Snider, D.E. and Roper, W.L. 1992. The new tuberculosis. N. Engl. J. Med. 326:703.

48. Springer, B., et al. 1995. *Mycobacterium conspicuum* sp nov., a new species isolated from patients with disseminated infections. J. Clin. Microbiol. 33:2805.

49. Stager, C.E., et al. 1991. Role of solid media when used in conjunction with the BACTEC system for mycobacterial isolation and identification. J. Clin. Microbiol. 29:154.

50. Steele, J.H. and Ranney, A.F. 1958. Animal tuberculosis. Am. Rev. Tuberculosis 77:908.

51. Thibert, L. and Lapierre, S. 1993. Routine application of high-performance liquid chromatography for identification of mycobacteria. J. Clin. Microbiol. 31:1759.

52. Thorel, M.F., Krichevsky, M., and Levy-Frebault, V.V. 1990. Numerical taxonomy of mycobactin-dependent mycobacteria: emended description of *Mycobacterium avium,* and description of *Mycobacterium avium* subsp *avium* subsp nov., *Mycobacterium avium* subsp *paratuberculosis* subsp nov., and *Mycobacterium avium* subsp *silvaticum* subsp nov. Int. J. Syst. Bacteriol. 40:254.

53. van Griethuysen, A.J., Jansz, A.R., and Buiting, A.G.M. 1996. Comparison of fluorescent BACTEC 9000 MB System, Septi-Check AFB System, and Löwenstein-Jensen medium for detection of mycobacteria. J. Clin. Microbiol. 43:2391.

54. Wallace, J.M. and Hannah, J.B. 1988. *Mycobacterium avium-complex* infection in patients with the acquired immunodeficiency syndrome. Chest 93:926.

55. Wallace, R.J. 1994. Recent changes in taxonomy and disease manifestations of the rapidly growing mycobacteria. Eur. J. Clin. Microbiol. Infect. Dis. 13:953.

56. Wallace, R.J., et al. 1990. Diagnosis and treatment of disease caused by nontuberculous mycobacteria. Am. Rev. Respir. Dis. 142:940.

57. Wayne, L.G. 1964. The role of air in the photochromogenic behavior of *Mycobacterium kansasii.* Am. J. Clin. Pathol. 42:431.

58. Wayne, L.G. and Sramek, H.A. 1992. Agents of newly recognized or infrequently encountered mycobacterial disease. Clin. Microbiol. Rev. 5:1.

59. Welch, D.F., et al. 1993. Timely culture for mycobacteria which utilizes a microcolony method. J. Clin. Microbiol. 31:2178.

60. Wolinsky, E. 1979. Nontuberculous mycobacteria and associated diseases. Am. Rev. Respir. Dis. 119:107.

61. Wolinsky, E. 1992. Mycobacterial diseases other than tuberculosis. Clin. Infect. Dis. 15:1.

62. Woods, G.L. and Washington, J.A., II. 1987. Mycobacteria other than *Mycobacterium tuberculosis:* review of microbiologic and clinical aspects. Rev. Infect. Dis. 9:275.

63. Yajko, D.M., et al. 1993. Comparison of four decontamination methods for recovery of *Mycobacterium avium* complex from stools. J. Clin. Microbiol. 31:302.

Bibliography

Böttger, E.C. 1994. *Mycobacterium genavense:* an emerging pathogen. Eur. J. Clin. Microbiol. Infect. Dis. 13:932.

Goutzamanis, J.J. and Gilbert, G.L. *Mycobacterium ulcerans* infection in Australian children: report of eight cases and review. Clin. Infect. Dis. 21:1186.

Henriques, B., et al. 1994. Infection with *Mycobacterium malmoense* in Sweden: report of 221 cases. Clin. Infect. Dis. 18:596.

Hoffner, S.E. Pulmonary infections caused by less frequently encountered slowly growing mycobacteria. Eur. J. Clin. Microbiol. Infect. Dis. 13:925.

Kiehn, T.E. and White, M. *Mycobacterium haemophilum:* an emerging pathogen. Eur. J. Clin. Microbiol. Infect. Dis. 13:932.

Inderlied, C.B. 1996. Antimycobacterial agents: *in vitro* susceptibility testing, spectra of activity, mechanisms of action and resistance, and assays for activity in biologic fluids. In Lorian, V., editor. Antibiotics in laboratory medicine. Williams & Wilkins, Baltimore.

Wayne, L.G. and Sramek, H.A. 1992. Agents of newly recognized or infrequently encountered mycobacterial diseases. Clin. Microbiol. Rev. 5:1.

Zaugg, M., et al. 1993. Extrapulmonary and disseminated infections due to *Mycobacterium malmoense:* case report and review. Clin. Infect. Dis. 16:540.

61 OBLIGATE INTRACELLULAR AND NONCULTURABLE BACTERIAL AGENTS

Genera and Species to be Considered

- *Chlamydia trachomatis*
- *Chlamydia psittaci*
- *Chlamydia pneumoniae*
- *Rickettsia akari*
- *Rickettsia conorii*
- *Rickettsia rickettsii*

- *Rickettsia prowazekii*
- *Rickettsia typhi*
- *Rickettsia tsutsugamushi*
- *Ehrlichia chaffeensis*
- *Ehrlichia equi*-like organism
- *Ehrlichia sennetsu*

- *Coxiella burnetii*
- *Tropheryma whippelii*
- *Calymmatobacterium granulomatis*

The organisms addressed in this chapter are obligate intracellular bacteria or are considered either extremely difficult to culture or unable to be cultured. Organisms of the genera *Chlamydia*, *Rickettsia*, and *Ehrlichia* are prokaryotes that differ from most other bacteria with respect to their very small size and obligate intracellular parasitism. Three other organisms, *Coxiella*, *Calymmatobacterium granulomatis*, and *Tropheryma whippelli*, are also reviewed here because they are difficult to cultivate or are noncultivable.

CHLAMYDIA

The genus *Chlamydia* comprises four species: *C. trachomatis*, *C. psittaci*, *C. pneumoniae*, and *C. pecorum*, with less than 33% DNA-DNA homology among the species.[4] Chlamydiae possess a heat-stable, genus-specific antigen that is an essential component of the cell membrane lipopolysaccharide[3]; species and type-specific protein antigens also exist. Members of the family *Chlamydiaceae* are obligate intracellular bacteria that were once regarded as viruses. These organisms require the biochemical resources of the eukaryotic host cell's environment to fuel their metabolism for growth and replication because they are unable to produce high-energy compounds such as ATP. Chlamydiae have a unique developmental life cycle, with an intracellular growth, or replicative, form, the reticulate body (RB) and an extracellular, metabolically inert, infective form, the elementary body (EB).[28] Structurally, the chlamydial EB closely resembles a gram-negative bacillus, however, its cell wall lacks a peptidoglycan layer. The life cycle is illus-

trated in Figure 61-1. *C. trachomatis*, *C. pneumoniae*, and *C. psittaci* are important causes of human infection; *C. psittaci* and *C. pecorum* are common pathogens among animals. The three species that infect humans differ with respect to their antigens, host cell preference, antibiotic susceptibility, EB morphology, and inclusion morphology (Table 61-1).

CHLAMYDIA TRACHOMATIS

Over the past few decades, the importance of infections caused by this organism has been recognized. Of significance, the association of infertility and ectopic pregnancy with *C. trachomatis* infections and the realization that the majority of *C. trachomatis* infections are asymptomatic has recently been recognized.

General characteristics

C. trachomatis infects humans almost exclusively and is responsible for various clinical syndromes. Based on major outer membrane protein (MOMP) antigenic differences, *C. trachomatis* is divided into 17 different serovars that are associated with different primary clinical syndromes.[23]

Epidemiology and pathogenesis

C. trachomatis causes significant infection and disease worldwide. In the United States *C. trachomatis* is the most common sexually transmitted bacterial pathogen and a major cause of **pelvic inflammatory disease (PID)** (see Chapter 26 for more information on PID); an estimated 4 million *C. trachomatis* infections occur annually in the United States.[40] Another disease caused by this organism, **ocular trachoma**,

TABLE 61-1 DIFFERENTIAL CHARACTERISTICS AMONG *CHLAMYDIA* SPP. THAT CAUSE HUMAN DISEASE

PROPERTY	C. TRACHOMATIS	C. PSITTACI	C. PNEUMONIAE
Host range	Humans (except one biovar that causes mouse pneumonitis)	Birds, lower mammals, humans (rare)	Humans
Elementary body morphology	Round	Round	Pear-shaped
Inclusion morphology	Round, vacuolar	Variable, dense	Round, dense
Glycogen-containing inclusions	Yes	No	No
Plasmid DNA	Yes	Yes	No
Susceptibility to sulfonamides	Yes	No	No

affects 500 million people, with 7 to 9 million of those infected becoming blind.[39]

C. trachomatis infections are primarily spread from human to human by sexual transmission. Some infections, such as neonatal pneumonia or inclusion conjunctivitis, are transmitted from mother to infant during birth. The various routes of transmission for *C. trachomatis* infection are summarized in Table 61-2.

The natural habitat of *C. trachomatis* is humans. The mechanisms by which *C. trachomatis* causes inflammation and tissue destruction are poorly understood. Although the natural history of *C. trachomatis* infections is unclear, it appears that chronic asymptomatic or persistent infections might play an important role.[2,5] For example, the importance of multiple, recurrent infections with *C. trachomatis* is recognized for the development of ocular trachoma; some data suggest that the host's immune response may account for most of the inflammation and tissue destruction. Of importance, immunity provides little protection from reinfection and appears to be short-lived, following infection with *C. trachomatis*.

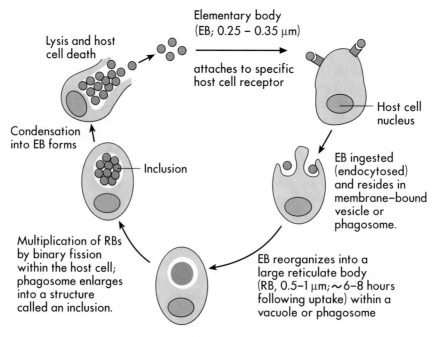

FIGURE 61-1 The life cycle of chlamydiae. The entire cycle takes approximately 48 to 72 hours.

—————   TABLE 61-2   PRIMARY SYNDROMES CAUSED BY *C. TRACHOMATIS*   —————

SUBTYPES	CLINICAL SYNDROME	ROUTE(S) OF TRANSMISSION
A, B, Ba, C	Endemic trachoma (multiple or persistent infections that ultimately lead to blindness)	Hand to eye from fomites, flies
L1, L2, L3	Lymphogranuloma venereum	Sexual
D-K	Urethritis, cervicitis, pelvic inflammatory disease, epididymitis, infant pneumonia and conjunctivitis (does not lead to blindness)	Sexual, hand to eye by autoinoculation of genital secretions; eye to eye by infected secretions; neonatal

Spectrum of disease

As previously mentioned, infection with *C. trachomatis* can lead to several different clinical syndromes. These infections are summarized in Table 61-2.

TRACHOMA **Trachoma** is manifested by a chronic inflammation of the conjunctiva and remains a major cause of preventable blindness worldwide. As the infection progresses, the conjunctiva becomes scarred, which causes distortion of the eyelids such that the eyelashes become misdirected and turn in. The eyelashes then mechanically damage the cornea, resulting in ulceration, scarring, and visual loss.

LYMPHOGRANULOMA VENEREUM **Lymphogranuloma venereum (LGV)** is a sexually transmitted disease that is unusual in Europe and North America but relatively frequent in Africa, Asia, and South America. The disease is characterized by a primary genital lesion at the initial infection site, which lasts a short time. This lesion is often small and may be unrecognized, especially by female patients. The second stage, acute lymphadenitis, often involves the inguinal lymph nodes, causing them to enlarge and become matted together, forming a large area of groin swelling, or bubo. During this stage, infection may become systemic and cause fever or may spread locally, causing granulomatous proctitis. In a few patients (more women than men) the disease progresses to a chronic third stage, causing the development of genital hyperplasia, rectal fistulas, rectal stricture, draining sinuses, and other manifestations.

Diagnosis of LGV is established by the isolation of an LGV strain from a bubo or other infected site. However, organism recovery rates of only 24% to 30% are reported. An intradermal skin test of LGV antigen, Frei's test, lacks sensitivity in early LGV and lacks specificity later, because Frei's antigen is only a genus-specific antigen. Moreover, the Frei's test can remain positive for many years, limiting its usefulness.

OCULOGENITAL INFECTIONS *C. trachomatis* can cause an acute inclusion conjunctivitis in adults. These infections are associated with a purulent discharge. In contrast to trachoma, inclusion conjunctivitis does not lead to blindness in adults (or newborns).

C. trachomatis infections have surpassed gonococcal infections as a cause of sexually transmitted disease in the United States. Similar to gonococci, *C. trachomatis* is a cause of urethritis, cervicitis, bartholinitis, proctitis, salpingitis, epididymitis, and acute urethral syndrome (see Chapter 26 for more information on genital tract infections) in women. In the United States, 60% of cases of nongonococcal urethritis are caused by chlamydiae. Both chlamydiae and gonococci are major causes of PID, contributing significantly to the rising rate of infertility and ectopic pregnancies in young women. After only one episode of PID, as many as 10% of women may become infertile because of tubal occlusion. The risk increases dramatically with each additional episode.

Many genital chlamydial infections in both sexes are asymptomatic or not easily recognized by clinical criteria; asymptomatic carriage in both men and women may persist, often for months.[24] As many as 25% of men[38] and 70% to 80% of women[36] identified as having chlamydial genital tract infections have no symptoms. Of significance, these asymptomatic infected individuals serve as a large reservoir to sustain transmission within a community.

PERINATAL INFECTIONS Approximately one quarter to half of infants that are born to women infected with *C. trachomatis* will develop inclusion conjunctivitis. Usually, the incubation period is 5 to 12 days from birth, but may be as late as 6 weeks. Although most develop inclusion conjunctivitis, about 10% to 20% of infants develop pneumonia.[35] Of significance, perinatally acquired *C. trachomatis* infection may persist in the nasopharynx, urogenital tract, or rectum for more than 2 years.[2]

Laboratory diagnosis

Diagnosis of *C. trachomatis* can be achieved by cytology, culture, direct detection of antigen or nucleic acid, and serologic testing.

SPECIMEN COLLECTION AND TRANSPORT The organism can be recovered from or detected in infected cells of the urethra, cervix, conjunctiva, nasopharynx, and rectum and from material aspirated from the fallopian tubes and epididymis. For collecting specimens from the endocervix (preferred anatomic site to collect screening specimens from women), the specimen for *C. trachomatis* culture should be obtained after all other specimens (e.g., those for Gram-stained smear, *Neisseria gonorrhoeae* culture, or Pap smear). A large swab should first be used to remove all secretions from the cervix. The appropriate swab (for nonculture tests, use the swab supplied or specified by the manufacturer) or endocervical brush is inserted 1 to 2 cm into the endocervical canal, rotated against the wall for 10 to 30 seconds, withdrawn without touching any vaginal surfaces, and then placed in the appropriate transport medium or swabbed onto a slide prepared for direct fluorescent antibody (DFA) testing.[6]

Urethral specimens should not be collected until 2 hours after the patient has voided. A urogenital swab (or one provided or specified by the manufacturer) is gently inserted into the urethra (females, 1 to 2 cm; males, 2 to 4 cm), rotated at least once for 5 seconds; and then withdrawn. Again, swabs should be placed into the appropriate transport medium or onto a slide prepared for DFA testing.

CULTIVATION Cultivation of *C. trachomatis* is discussed before methods for direct detection and serodiagnosis because all nonculture methods for the diagnosis of *C. trachomatis* are compared with culture.

Several different cell lines have been used to isolate *C. trachomatis* in cell culture, including McCoy, Hela, and monkey kidney cells; cycloheximide-treated McCoy cells are commonly used. Centrifugation of the specimen onto the cell monolayer (usually growing on a coverslip in the bottom of a vial, the "shell vial") presumably facilitates adherence of elementary bodies. After 48 to 72 hours of incubation, monolayers are stained with iodine (Figure 61-2) or an immunofluorescent stain and examined microscopically for inclusions. Figure 61-3 is an overview of the basic steps in processing of specimens for chlamydial culture. Procedure 61-1 describes a method for isolation of chlamydiae. Although its specificity approaches 100%, the sensitivity of culture has been estimated at between 70% and 90% in experienced laboratories.[31] Limitations of *Chlamydia* culture that contribute to this lack of sensitivity include prerequisites to maintain viability of patient specimens by either rapid or frozen transport and to ensure the quality of the specimen submitted for testing (i.e., endocervical specimens devoid of mucus and containing endocervical epithelial or metaplastic cells and/or urethral epithelial cells).[22] In addition to these issues is the requirement for a sensitive cell culture system and the minimum delay of at least 2 days between specimen receipt and the availability of results. Despite these limitations, culture is still recommended as the test of choice for some situations (Table 61-3). In particular, only chlamydia cultures should be used in situations with legal implications (e.g., sexual abuse) when the possibility of a false-positive test is unacceptable.[6]

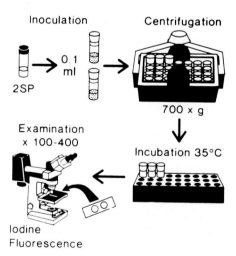

FIGURE 61-3 Processing of specimens for the cultivation of *C. trachomatis.* (From Smith, T.F. 1982. Role of the diagnostic virology laboratory in clinical microbiology: tests for *Chlamydia trachomatis* and enteric toxins in cell culture. In de la Maza, L.M. and Peterson, E.M. editors. Medical virology, vol. 1 Elsevier Biomedical, New York.)

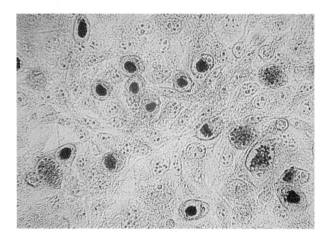

FIGURE 61-2 Iodine-stained inclusions in McCoy cell monolayer infected with *Chlamydia trachomatis.* (Courtesy Nadine Bartholoma, SUNY Health Science Center, Syracuse, N.Y.)

DIRECT DETECTION METHODS

Cytologic Examination. Cytologic examination of cell scrapings from the conjunctiva of newborns or persons with ocular trachoma can be used to detect *C. trachomatis* inclusions, usually after Giemsa staining. Cytology has also been used to evaluate endocervical and urethral scrapings, including those obtained for Pap smears. However, this method is insensitive compared with culture or other methods discussed below.[9,37]

Antigen Detection and Nucleic Acid Hybridization. To circumvent the shortcomings of cell culture, antigen detection methods are commercially available.

Direct fluorescent antibody (DFA) staining methods employ fluorescein-isothiocyanate-conjugated monoclonal antibodies to either outer membrane proteins or lipopolysaccharides of *C. trachomatis* to detect elementary bodies in smears of clinical material

(Figure 61-4). The MicroTrak DFA (Syva Co., San Jose, Calif.) is a good commercial system. DFAs achieve at least a 90% sensitivity compared with culture in many evaluations of symptomatic men and high-risk women (i.e., sex partners of *Chlamydia*-positive men or women attending STD clinics).

Chlamydial antigen can also be detected by enzyme-linked immunosorbent assays (ELISA).[20,31] Numerous FDA-approved kits are commercially available. These assays employ polyclonal or monoclonal antibodies that detect chlamydial LPS. These tests are not species-specific for *C. trachomatis* and may cross-react with LPS of other bacterial species present in the vagina or urinary tract and thereby produce a false-positive result. Also available are nucleic acid hybridization tests that use a chemiluminescent type of DNA probe (PACE, Gen-Probe, San Diego, Calif.) that is complimentary to a sequence of ribosomal RNA (rRNA) in the chlamydial genome. Once

TABLE 61-3 USE OF DIFFERENT LABORATORY TESTS TO DIAGNOSE *C. TRACHOMATIS* INFECTIONS

PATIENT POPULATION	SPECIMEN TYPE	PREVALENCE*	ACCEPTABLE DIAGNOSTIC TEST†
Prepubertal girls	Vaginal	-	Culture
Neonates and infants	Nasopharyngeal	-	Culture
	Rectal	-	Culture
	Conjunctiva	-	Culture, DFA, EIA, NAH
Women	Cervical	High	Culture, DFA, EIA, NAH, AMPA
		Low	Culture, DFA‡, EIA‡, NAH‡, AMPA
Women and men	Urethral	High (or symptomatic)	Culture, DFA, EIA, NAH, AMPA
		Low (or asymptomatic)	Culture, AMPA
	Rectal	-	Culture
	Urine	-	AMPA§

*High prevalence ≥ 5%.

†*NA*, nonapplicable; *DFA*, direct fluorescent antibody staining; *EIA*, enzyme-linked immunosorbentassay; *NAH*, nucleic acid hybridization; *AMPA*, nucleic acid amplification assay.

‡Must be confirmed in a population with a low prevalence (<5%) of *C. trachomatis* infection.

§EIA can be used on urine from symptomatic men but not on urine from older men. Also, a positive result must be confirmed in a population with a low prevalence of *C. trachomatis* infection.

Modified from Centers for Disease Control and Prevention. 1993. Recommendations for the prevention and management of *Chlamydia trachomatis* infections. Centers for Disease Control and Prevention, Atlanta; Weinstock, H., Dean, D., and Bolan, C. 1994. Infect. Dis. Clin. North Am. 8:797.

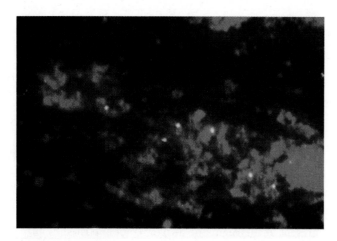

FIGURE 61-4 Appearance of fluorescein-conjugated, monoclonal antibody-stained elementary bodies in direct smear of urethral cell scraping from a patient with chlamydial urethritis. (Courtesy Syva, Co., San Jose, Calif.)

formed, the DNA-rRNA complex is absorbed onto a magnetic bead and detected by a luminometer. This assay is species-specific for *C. trachomatis*.

Based on numerous studies, these nonculture tests are more reliable in patients who are symptomatic and shedding large numbers of organisms than in those who are asymptomatic and most likely shedding fewer organisms. For the most part, these assays have sensitivities of greater than 70% and specificities of 97% to 99% in populations with prevalences of *C. trachomatis* infection of 5% or more.[6] In a low-prevalence population, that is, less than 5%, a significant proportion of positive tests will be falsely positive (see Chapter 5 for more information). Therefore, a positive result in a low-prevalence population should be handled with care, and it is desirable to verify a positive result.[6,41] Positive results can be validated by the following methods:

- Culture
- Performing a second nonculture test that identifies a *C. trachomatis* antigen or nucleic acid sequence that is different from that used in the screening test
- Using a blocking antibody or competitive probe that verifies a positive test result by preventing attachment of a labeled antibody or probe used in the standard assay[6]

Amplification Assays. At the time of this writing, three commercial assays using nucleic acid amplification are FDA-approved for the laboratory diagnosis of *C. trachomatis* infection. These assays use three different formats: polymerase chain reaction, ligase chain reac-

tion, and transcription-mediated amplification (see Chapter 14 for more information). The first two assay formats amplify or target DNA sequences present in the cryptic plasmid that is present in 7 to 10 copies in the chlamydial EB, while the last format amplifies ribosomal RNA sequences. Studies indicate that these tests are more sensitive than culture and as specific.[29,34] Because of the increased sensitivity of detection, promising results have been obtained using these amplification assays to detect *C. trachomatis* in first-voided urine specimens from symptomatic and asymptomatic men and women,[7,29] thereby affording a noninvasive means of chlamydia testing. Table 61-3 summarizes the various uses of the different chlamydia laboratory diagnostic tests.

SERODIAGNOSIS   Serologic testing has limited value for diagnosis of urogenital infections in adults. Most adults with chlamydial infection have had a previous exposure to *C. trachomatis* and are therefore seropositive. Serology can be used to diagnose LGV. Antibodies to a genus-specific antigen can be detected by complement fixation, and a single-point titer greater than 1:64 is indicative of LGV. This test is not useful in diagnosing trachoma, inclusion conjunctivitis, or neonatal infections. The microimmunofluorescence assay (micro-IF), a tedious and difficult test, is used for type-specific antibodies of *C. trachomatis* and can also be used to diagnose LGV. A high titer of IgM (1:32) suggests a recent infection; however, not all patients produce IgM; in contrast to CF, micro-IF may be used to diagnose trachoma and inclusion conjunctivitis using acute and convalescent phase sera. Detection of *C. trachomatis*-specific IgM is useful in diagnosis of neonatal infections. Negative serology can reliably exclude chlamydial infection.

Antibiotic susceptibility testing and therapy
Because *C. trachomatis* is an obligate intracellular bacteria, susceptibility testing is not practical in the routine clinical microbiology laboratory setting and is performed in only a few laboratories. Antibiotics that have activity against *C. trachomatis* include erythromycin and other macrolide antibiotics, tetracyclines, and fluoroquinolones.

Prevention
Because no effective vaccines are available, strategies to prevent chlamydial urogenital infections focus on trying to manifest behavioral changes. By identifying and treating persons with genital chlamydia before infection is transmitted to sexual partners or in the case of pregnant women to babies, the risk of acquiring or transmitting infection may be significantly decreased.

CHLAMYDIA PSITTACI

Although members of this chlamydial species are common in birds and domestic animals, infections in humans are relatively uncommon.

General characteristics

C. psittaci differs from *C. trachomatis* in that it is not sulfonamide-sensitive and in the morphology of its EB and inclusion bodies (Table 61-1).

Epidemiology and pathogenesis

C. psittaci is an endemic pathogen of all bird species. Psittacine birds (e.g., parrots, parakeets) are a major reservoir for human disease, but outbreaks have occurred among turkey-processing workers and pigeon aficionados. The birds may show diarrheal illness or may be asymptomatic. Humans acquire the disease by inhalation of aerosols. The organisms are deposited in the alveoli; some are ingested by alveolar macrophages and then carried to regional lymph nodes. From there they are disseminated systemically, growing within cells of the reticuloendothelial system. Human-to-human transmission is rare, thus obviating the need for isolating patients if admitted to the hospital.

Spectrum of disease

Disease usually begins after an incubation period of 5 to 15 days. Onset may be insidious or abrupt. Clinical findings associated with this infection are diverse and include pneumonia, severe headache, mental status changes, and hepatosplenomegaly. The severity of infection ranges from inapparent or mild disease to a life-threatening systemic illness with significant respiratory problems.

Laboratory diagnosis

Diagnosis of psittacosis is almost always by serologic means. Because of hazards associated with working with the agent, only laboratories with Biosafety Level 3 biohazard containment facilities can culture *C. psittaci* safely. State health departments take an active role in consulting with clinicians about possible cases. Complement fixation has been the most frequently used serologic test to detect psittacosis infection. A more recent test, indirect microimmunofluorescence, is more sensitive but difficult to perform. For this new test, different strains of *C. psittaci* are grown in hens' egg yolk sac cultures. The cultures, rich with elementary bodies, are diluted and suspended in buffer. Next, a dot of each antigen suspension is placed in a geometric array on the surface of a glass slide. Several identical arrays, each containing one dot of every antigen, are prepared on a single slide. These slides are fixed and frozen until use. At the time of the test, serial dilutions of a patient's serum are placed over the antigen arrays. After incubation and washing steps, fluorescein-conjugated antihuman immunoglobulin (either IgG or IgM) is overlaid on the slide. The slide is finally read for fluorescence of particular antigen dots using a microscope. Either a fourfold rise in titer between acute and convalescent serum samples or a single IgM titer of 1:32 or greater in a patient with an appropriate illness is considered diagnostic of an infection.

Antibiotic susceptibility testing and therapy

Because *C. psittaci* is an obligate intracellular pathogen and its incidence of infection is rare, susceptibility testing is not practical in the routine clinical microbiology laboratory. Tetracycline is the drug of choice for psittacosis. If left untreated, the fatality rate is about 20%.

Prevention

Prevention of disease is accomplished by treatment of infected birds and/or quarantining of imported birds for a month.

CHLAMYDIA PNEUMONIAE (TWAR)

The TWAR strain of *Chlamydia* was first isolated from the conjunctiva of a child in Taiwan. It was initially considered to be a psittacosis strain, because the inclusions produced in cell culture resembled those of *C. psittaci*. The Taiwan isolate (TW-183) was shown to be serologically related to a pharyngeal isolate (AR-39) isolated from a college student in the United States, and thus the new strain was called "TWAR," an acronym for TW and AR (acute respiratory). To date, only this one serotype of the new species, *C. pneumoniae,* has been identified.

General characteristics

C. pneumoniae is considered more homogeneous than either *C. trachomatis* or *C. psittaci*, because all isolates tested are immunologically similar. One significant difference between *C. pneumoniae* and the other chlamydiae is the pear-shaped appearance of its EB (Figure 61-5).

Epidemiology and pathogenesis

C. pneumoniae appears to be a human pathogen; no bird or animal reservoirs have been identified. The mode of transmission from person to person is by aerosolized droplets via the respiratory route.[13] The spread of infection is low. Antibody prevalence to *C. pneumoniae* starts to rise in school-age children and reaches 30% to 45% in adolescents; more than half of adults in the United States and in other countries have *C. pneumoniae* antibody. Of interest, *C. pneumoniae* infections are both endemic and epidemic.

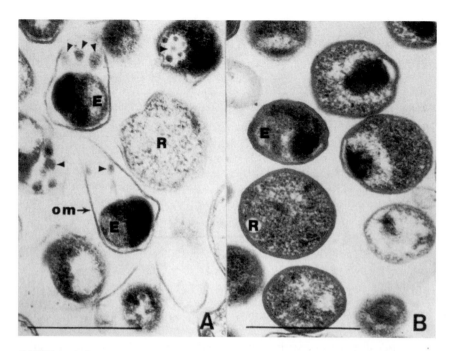

FIGURE 61-5 Electron micrograph of *C. pneumoniae* **(A)** and *C. trachomatis* **(B)** (bar = 0.5 μm). *E*, elementary body; *R*, reticulate body; *OM*, outer membrane; *arrowhead*, small electron-dense bodies of undetermined junction. (From Grayston, J.T. et al. 1989. *Chlamydia trachomatis* spp. Nov. for *Chlamydia* sp. strain TWAR. Int. J. Syst. Bacteriol. 39:88.)

Unfortunately, little is known about the pathogenesis of *C. pneumoniae* infections.

Spectrum of disease

C. pneumoniae has been associated with pneumonia, bronchitis, pharyngitis, sinusitis, and a flulike illness. Infection in young adults is usually mild to moderate; the microbiologic differential diagnosis primarily includes *Mycoplasma pneumoniae*. Severe pneumonia may occur in elderly or respiratory-compromised patients. Of note, asymptomatic infection or unrecognized, mildly symptomatic illnesses caused by *C. pneumoniae* is common. In addition, an association exists between *C. pneumoniae* infection and the development of asthmatic symptoms.[18] Finally, an association between coronary artery disease and other atherosclerotic syndromes and *C. pneumoniae* infection has been suggested by seroepidemiologic studies and the demonstration of the organism in atheromatous plaques (yellow deposits within arteries containing cholesterol and other lipid material).

Laboratory diagnosis

The laboratory diagnosis of *C. pneumoniae* infections is accomplished by cell culture, serology, or amplification of *C. pneumoniae*-specific DNA.

DIRECT DETECTION METHODS   To date, assays to directly detect *C. pneumoniae* antigens have poor sensitivity.

However, several *C. pneumoniae*-specific primers have been used in PCR assays to detect organisms and appear to be more sensitive than cell culture.[15] The organism has been detected in throat swabs and other specimens, such as nasopharyngeal, bronchoalveolar lavage fluids, and sputum.

CULTIVATION   Specimens for isolation are usually swabs of the oropharynx; techniques for isolation from sputum are unsatisfactory. Swabs should be placed into chlamydial transport media, transported on ice, and stored at 4° C; organisms are rapidly inactivated at room temperature or by rapid freezing or thawing. A cell culture procedure similar to that used for *C. trachomatis* but using the more sensitive HL or Hep-2 cell lines must be substituted for McCoy cells. *C. pneumoniae* species-specific monoclonal antibodies can detect the organism in cell culture.

SERODIAGNOSIS   Diagnosis of *C. pneumoniae* infection can also be achieved by serology. Complement fixation using a genus-specific antigen has been used, but it is not specific for *C. pneumoniae*. A microimmunofluorescence test using *C. pneumoniae* elementary bodies as antigen is more reliable but available only in specialized laboratories. A fourfold rise in either IgG or IgM is diagnostic, and a single IgM titer of 16 or greater or an IgG titer of 512 or greater is suggestive of recent infection. Commercial kits for diagnosing *C.*

pneumoniae are not available yet because of the difficulty in developing species-specific detection methods.

Antibiotic susceptibility testing and therapy

Methods for susceptibility testing of *C. pneumoniae* have been largely adapted from those used for *C. trachomatis*.[19] Similar to *C. trachomatis*, susceptibility testing is not practical for the clinical microbiology laboratory, and, because methods are not yet standardized, the results can be influenced by several variables. Treatment with tetracycline and erythromycin has been successful.

Prevention

Little is known regarding effective ways to prevent *C. pneumoniae* infections.

RICKETTSIA AND EHRLICHIA

Agents that are responsible for human disease within the family *Rickettsiaceae* belong to two genera, *Rickettsia* and *Ehrlichia*. Until recently, two other genera causing human disease, *Coxiella* and *Bartonella*, were included in this family. However, based on phylogenetic differences, it has been proposed that these two genera be removed from the *Rickettsiaceae* family. *Bartonella* spp. can be cultured on standard bacteriologic media; therefore, this group of organisms is addressed in Chapter 44. Because *Coxiella burnetii* can survive extracellularly, unlike the rickettsiae, yet requires cultivation in cell culture similar to the rickettsiae, this organism is discussed separately in this chapter.

GENERAL CHARACTERISTICS

Rickettsiae are fastidious bacteria that are obligate, intracellular parasites. These bacterial agents survive only briefly outside of a host (reservoir or vector) and multiply only intracellularly. Organisms are small (0.3 μm × 1.0 to 2.0 μm), pleomorphic, gram-negative bacilli that multiply by binary fission in the cytoplasm of host cells; the release of mature rickettsiae results in the lysis of the host cell.

EPIDEMIOLOGY AND PATHOGENESIS

This group of organisms infects wild animals, with humans acting as accidental hosts in most cases. Most of these organisms are passed between animals by an insect vector. Similarly, humans become infected following the bite of an infected arthropod vector. Characteristics, including the respective arthropod vector of the prominent species of *Rickettsia* and *Ehrlichia*, are summarized in Table 61-4.

Organisms belonging to the genus *Rickettsia* do not undergo any type of intracellular developmental cycle.[17] Different species of *Rickettsia* not only share some antigenic properties and are genetically similar, but also share a similar mechanism of pathogenesis. After being deposited directly into the bloodstream through the bite of an arthropod vector, these organisms induce the host's endothelial cells of blood vessels to engulf them and are carried into the cell's cytoplasm within a vacuole. Following infection, organisms escape the vacuole, becoming free in the cytoplasm. *Rickettsia* spp. then multiply, causing cell injury and death. Subsequent vascular lesions caused by *Rickettsia*-induced damage to endothelial cells account for the changes that occur throughout the body, particularly in the skin, heart, brain, lung, and muscle.

In contrast to *Rickettsia* spp., organisms belonging to the genus *Ehrlichia* undergo an intracellular developmental cycle following infection of circulating leukocytes. Similar to chlamydiae, *Ehrlichia* spp. go through the three developmental stages of elementary bodies, initial bodies, and morulae (Figure 61-6). Two species have been recognized as causes of disease in humans following tick bites.[10,14] *E. chaffeensis* that primarily infects monocytes and causes human monocytic ehrlichiosis (HME); an *Ehrlichia* species that is similar to *E. phagocytophila* and *E. equi* primarily infects neutrophils, causing human granulocytic ehrlichiosis (HGE). The pathogenesis of ehrlichiosis is unknown.

SPECTRUM OF DISEASE

The genus *Rickettsia* encompasses the following three groups of bacteria: the spotted fever group, the typhus group, and the scrub typhus group, based on the arthropod mode of transmission, clinical manifestations, rate of intracellular growth, rate of intracellular burden, and extent of intracellular growth (Table 61-4).[10] The triad of fever, headache, and rash is the primary clinical manifestation in patients with an exposure to insect vectors. Infections cause by *Rickettsia* spp. may be severe and are sometimes fatal.

Although distinct infections, the clinical findings associated with HME or HGE are similar. In general, patients with ehrlichial infections present with nonspecific symptoms such as fever, headache, and myalgias; rashes occur only rarely. The severity of illness can range from asymptomatic to mild to severe. Of note, the majority of patients infected with *E. chaffeensis* have required hospitalization and 2% to 3% have died.[14]

LABORATORY DIAGNOSIS

Because rickettsial and ehrlichial infections can be severe or even fatal, a timely diagnosis is essential.

TABLE 61-4 CHARACTERISTICS OF PROMINENT *RICKETTSIA* AND *EHRLICHIA* SPP.

AGENT	DISEASE	VECTOR	DISTRIBUTION	DIAGNOSTIC TESTS
Spotted fever group				
Rickettsia akari	Rickettsialpox	Mites	Worldwide	Serology, immunohistology
R. conorii	Mediterranean spotted fever	Ticks	Southern Europe, Mideast, Africa	Serology, immunohistology
R. rickettsii	Rocky Mountain spotted fever	Ticks	North and South America; particularly in southeastern states and Oklahoma in the United States	Serology, immunohistology, PCR
Typhus group				
R. prowazekii	Epidemic typhus	Lice	Worldwide	Serology, PCR
	Brill-Zinsser	None	Worldwide	Serology, PCR
R. typhi	Murine typhus	Fleas	Worldwide	Serology, PCR
Scrub typhus group				
R. tsutsuga-mushi	Scrub typhus	Chiggers	South and Southeast Asia, South Pacific	Serology, PCR
Ehrlichia				
Ehrlichia chaffeensis	Human monocytic ehrlichiosis	Ticks	Southeast and South Central United States	Serology, PCR, immunohistology
E. equi-like organism	Human granulocytic ehrlichiosis	Ticks	United States	Serology, PCR, immunohistology, peripheral blood smear
E. sennetsu	Sennetsu fever	Ticks	Southeast Asia (primarily Japan)	Serology

Modified from Drancourt, M. and Raoult, D. 1994. FEMS Microbiol. Rev. 13:13; Hackstadt, T. 1996. Infect. Agents Dis. 5:127.

Direct detection methods

Immunohistology and PCR have been used to diagnose rickettsial and ehrlichial infections, but these methods are available in only a limited number of laboratories. Biopsy of skin tissue from the rash caused by the spotted fever group rickettsiae is the preferred specimen to demonstrate their presence. Organisms are identified using polyclonal antibodies; detection is achieved by fluorescein-labeled antibodies or enzyme-labeled indirect procedures. The sensitivity of these techniques is about 70% and is dependent on correct tissue sampling, examination of multiple tissue levels, and biopsying before or during the first 24 hours of therapy.[11] In addition, PCR has been used to detect several rickettsial and ehrlichial specimens (see Table 61-4). Direct microscopic examination of Giemsa- or Diff-Quik–stained peripheral blood buffy-coat smears can detect morulae during the febrile stage of infection in ehrlichiosis; morula-like structures also can be observed in cerebrospinal fluid cells and tissues.[12]

Cultivation

Although the rickettsiae can be cultured in embryonated eggs and in tissue culture, the risk of laboratory-acquired infection is extremely high, limiting availabil-

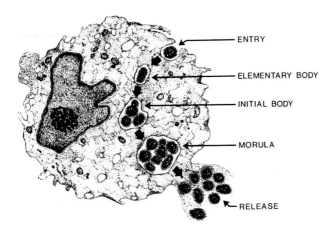

FIGURE 61-6 Schematic representation of the growth cycle of ehrlichiae in an infected cell. Elementary bodies (EB; individual ehrlichiae) enter the leukocyte by phagocytosis and multiply. After 3 to 5 days, small numbers of tightly packed EBs are observable and are called *initial bodies*. During the next 7 to 12 days, additional growth and replication occurs and the initial bodies develop into mature inclusions, which appear by light microscopy as mulberry or morula forms. This morula configuration is a hallmark of ehrlichiae infection. (From McDade, J. 1990. Ehrlichiosis: a disease of animals and humans. J. Infect. Dis. 161:609. The University of Chicago Press, Chicago, Ill.)

ity of culture to a few specialized laboratories. To date, culture of *Ehrlichia* is limited, however, the agent of HGE has been successfully cultivated in cell cultures.

Serodiagnosis

Diagnosis of rickettsial disease and ehrlichiosis is primarily accomplished serologically. The least specific but most widely used test in the United States is the Weil-Felix reaction, the fortuitous agglutination of certain strains of *Proteus vulgaris* by serum from patients with rickettsial disease. Although this test has been clinically useful, false-positive tests are a continuing problem. Table 61-5 outlines the positive reactions associated with each disease. Commercial suppliers produce kits that contain all necessary antigens and human antirickettsial control antisera for performance of the Weil-Felix test. This test is only presumptive, however, and a more specific serologic test, such as CF, micro-IF or direct immunofluorescence, must be used for confirmation of disease. These tests are performed primarily by reference laboratories, and kits can be obtained from the Centers for Disease Control and Prevention in Atlanta for performing tests for IgG antibody.

The microimmunofluorescent dot test, described previously for diagnosis of psittacosis, has shown excellent sensitivity for detecting antibodies to rickettsiae. By overlaying the antigen dots on the slide with fluorescein-conjugated antihuman IgM after the patient's serum has been allowed to form complexes, an early diagnosis of Rocky Mountain spotted fever

(RMSF) can be achieved within 7 to 10 days after onset of symptoms. Both immunofluorescent and a recently developed latex agglutination test method have shown good sensitivity for diagnosis of RMSF within the first week. Although Q fever, ehrlichiosis, and rickettsialpox do not induce Weil-Felix antibody in infected patients, specific immunofluorescent tests are valuable.

Except for latex agglutination, immunofluorescence, and direct fluorescent antibody testing for diagnosing RMSF, none of the serologic tests is useful for diagnosing disease in time to influence therapy. This lack of utility for serology is because antibodies to rickettsiae other than *R. rickettsii* cannot be reliably detected until at least 2 weeks after the patient has become ill. With newer immunologic reagents under development, the potential exists for new tests for all the rickettsial diseases.

To date, the sensitivity and specificity of serologic assays for ehrlichiosis is unknown; an indirect immunofluorescence assay is available using cells infected with *E. chaffeensis*, as well as by enzyme-linked immunosorbent assay. A fourfold or greater rise in antibody titer during the course of disease is considered significant.

ANTIBIOTIC SUSCEPTIBILITY TESTING AND THERAPY

Tetracyclines are the primary drugs of choice for treatment of most infections caused by *Rickettsia* spp. Depending on the specific species of *Rickettsia*, some fluoroquinolones may be used, as well as chloramphenicol. Likewise, tetracyclines are effective for treating ehrlichiosis.

PREVENTION

The best means of prevention for rickettsial and ehrlichial infection is to avoid contact with the respective vectors.

COXIELLA

Coxiella burnetii is the causative agent of **Q fever**, an acute systemic infection that primarily affects the lung.

GENERAL CHARACTERISTICS

C. burnetti is smaller than *Rickettsia* spp. and is more resistant to various chemical and physical agents. Recent phylogenetic studies of this gram-negative coccobacillus have demonstrated that it is far removed from the rickettsiae. In contrast to the rickettsiae, *C. burnetii* can survive extracellularly; however, it can be grown only in lung cells. The organism has a spore-like life cycle and can exist in two antigenic states.[30] When isolated from animals, *C. burnetii* is in a phase

TABLE 61-5 REACTION OF *PROTEUS* STRAINS IN WEIL-FELIX TEST

DISEASE	OX-19	OX-2	OX-K
Brill-Zinsser	V	V	−
Epidemic typhus	+	V	−
Murine typhus	+	V	−
Rickettsialpox	−	−	−
Rocky Mountain spotted fever	+	+	−
Scrub typhus	−	−	V
Q fever	−	−	−
Ehrlichiosis	−	−	−

+, >90% positive agglutination; −, >90% negative agglutination; *V*, variable results.

I form and is highly infectious. In its phase II form, *C. burnetii* has been grown in cultured cell lines and is not infectious.

EPIDEMIOLOGY AND PATHOGENESIS

The most common animal reservoirs for the zoonotic disease caused by *C. burnetti* are cattle, sheep, and goats. In infected animals, organisms are shed in urine, feces, milk, and birth products. Usually, the infected animals are asymptomatic. Humans are infected by the inhalation of contaminated aerosols. Of significance, because of its resistance to desiccation and sunlight, *C. burnetii* is able to withstand harsh environmental conditions. Q fever is endemic worldwide except in Scandinavia.

Following infection, *C. burnetii* is passively phagocytized by host cells and multiplies within vacuoles. The incubation period is about 2 weeks to 1 month. After infection and proliferation in the lungs, organisms are picked up by macrophages and carried to the lymph nodes, from which they then reach the bloodstream.

SPECTRUM OF DISEASE

After the incubation period, initial clinical manifestations of *C. burnetii* infections are systemic and non-specific: headache, fever, chills, myalgias. In contrast to rickettsial infections, a rash does not develop. Both acute and chronic forms of the disease are recognized. Possible clinical manifestations are listed in Box 61-1.

LABORATORY DIAGNOSIS

Because laboratory-acquired infections caused by *C. burnetii* have occurred, cultivation of the organism has been discouraged. However, the use of a shell vial

BOX 61-1   CLINICAL MANIFESTATIONS OF *C. BURNETII* INFECTION[30]

Febrile, self-limited illness
Atypical pneumonia
Granulomatous hepatitis
Endocarditis
Neurologic manifestations (e.g., encephalitis, meningoencephalitis)
Osteomyelitis

assay with human lung fibroblasts to isolate the organism from buffy coat and biopsy specimens has not resulted in any laboratory-acquired infections.[30] Once inoculated, cultures are incubated for 6 to 14 days at 37° C in carbon dioxide. The organism is detected using a direct immunofluorescent assay.[27]

Although organisms can be detected by nucleic acid amplification assays, serology is the most convenient and commonly used diagnostic tool. Three serologic techniques are available: indirect immunofluorescent antibody (IFA), complement fixation, and EIA. IFA is considered the reference method that is both highly specific and sensitive and is recommended for its reliability, cost effectiveness, and ease of performance.[30]

ANTIBIOTIC SUSCEPTIBILITY TESTING AND THERAPY

Because *C. burnetii* does not multiply in bacteriologic culture media, susceptibility testing has been performed in only a very limited number of laboratories. Tetracyclines are recommended for the treatment of acute Q fever.

PREVENTION

The best way to prevent infection with *C. burnetii* is to avoid contact with infected animals. A vaccine is commercially available in Australia and Eastern European countries; a vaccine is being developed in the United States.[42]

TROPHERYMA WHIPPELII

Although observed in diseased tissue, some organisms are nonculturable yet associated with specific disease processes, making the development of "traditional" diagnostic assays difficult (e.g., serology or antigen detection). However, with the ability to detect and classify bacteria using molecular techniques without culture, some of these agents have been detected and characterized. This strategy uses PCR to amplify ribosomal DNA sequences by PCR of the unknown agent followed by sequencing and phylogenetic analysis.[31] Using this approach, *Tropheryma whippelii* has been identified as the causative agent of Whipple's disease.[32]

GENERAL CHARACTERISTICS

Phylogenetic analysis shows that this organism is a gram-positive actinomycete that is not closely related to any other genus known to cause infection.

EPIDEMIOLOGY, PATHOGENESIS, AND SPECTRUM OF DISEASE

Whipple's disease, found primarily in middle-age men, is characterized by the presence of PAS-staining macrophages (indicating mucopolysaccharide or glycoprotein) in almost every organ system. The bacillus is seen in macrophages and affected tissues but it has never been cultured. Patients develop diarrhea, weight loss, arthralgia, lymphadenopathy, hyperpigmentation, often a long history of joint pain, and a distended and tender abdomen. Neurologic and sensory changes often occur. It has been suggested that a cellular immune defect is involved in pathogenesis of this disease.

LABORATORY DIAGNOSIS

Detection of *T. whippelii* is performed by only a few laboratories using PCR.

ANTIBIOTIC SUSCEPTIBILITY TESTING AND THERAPY

No susceptibility testing is performed, because this organism is nonculturable. Patients usually respond well to long-term therapy with antibacterial agents, including trimethoprim/sulfamethoxazole, tetracycline, and penicillin; tetracycline has been associated with serious relapses, however. Colchicine therapy appears to control symptoms.[26] Without treatment the disease is uniformly fatal.

PREVENTION

Little is known about the prevention of this disease.

CALYMMATOBACTERIUM GRANULOMATIS

Calymmatobacterium granulomatis is the etiologic agent of **granuloma inguinale,** or **donovanosis,** a sexually transmitted disease.

GENERAL CHARACTERISTICS

C. granulomatis is an encapsulated, pleomorphic gram-negative bacillus that is usually observed in vacuoles in the cells of large mononuclear cells.[33]

EPIDEMIOLOGY AND PATHOGENESIS

Granuloma inguinale is uncommon in the United States but is recognized as a major cause of genital ulcers in India, Papua New Guinea, the Caribbean, Australia, and parts of South America. The causative agent is sexually transmitted, although there is a possibility that it may be nonsexually transmitted as well. Infectivity of this bacillus must be low, because sexual partners of infected patients often do not become infected.

SPECTRUM OF DISEASE

Granuloma inguinale is characterized by subcutaneous nodules that enlarge and evolve to form beefy, erythematous, granulomatous, painless lesions that bleed easily. The lesions, which usually occur on the genitalia, have been mistaken for neoplasms. Patients often have inguinal lymphadenopathy.

LABORATORY DIAGNOSIS

The organism can be visualized in scrapings of lesions stained with Wright's or Giemsa stain. Subsurface infected cells must be present; surface epithelium is not an adequate specimen. Groups of organisms are seen within mononuclear endothelial cells; this pathognomonic entity is known as a *Donovan body,* named after the physician who first visualized the organism in such a lesion. The organism stains as a blue rod with prominent polar granules, giving rise to a "safety pin" appearance, surrounded by a large, pink capsule. The capsular polysaccharide shares several cross-reactive antigens with *Klebsiella* species, and there also appears to be a high degree of molecular homology, fueling speculation that *Calymmatobacterium* is closely related to *Klebsiella.*[1]

Cultivation in vitro is very difficult, but it can be done using media containing some of the growth factors found in egg yolk. A medium described by Dienst[8] has been used to culture *Calymmatobacterium* from aspirated bubo material.[16] More recently, this agent was cultured in human monocytes from biopsies of genital ulcers of patients with donovanosis.[21]

ANTIBIOTIC SUSCEPTIBILITY TESTING AND THERAPY

No antibiotic susceptibility testing is performed. Although gentamicin and chloramphenicol are the most effective drugs for the therapy of granuloma inguinale, tetracycline or ampicillin are the drugs of first choice. Trimethoprim/sulfamethoxazole or erythromycin (in pregnancy) also provide effective treatment for granuloma inguinale.

Procedure

61-1 CELL CULTURE METHOD FOR ISOLATION OF CHLAMYDIAE

METHOD

1. Collect a swab or tissue in sucrose transport medium containing antibiotics. Keep specimens refrigerated at all times before transport to the laboratory. If the culture is not inoculated within 24 hours of collection, freeze the specimen at −70° C.

2. Vortex the specimen vigorously and remove the swab. (Mince the tissue and grind to make a cell suspension at a 1:10 dilution using transport medium.)

3. Use aseptic technique and aspirate the cell culture medium above the McCoy cell monolayer in the shell vial (commercially available), and add 0.2 mL of the patient specimen to the vial.

4. Centrifuge the vial for at least 1 hour at 2500 to 3000 × g at ambient temperature in a temperature-controlled centrifuge. Do not allow the temperature to increase past 40° C.

5. Aseptically remove the inoculum and add 1 mL fresh maintenance medium containing 1 μg/mL cycloheximide.

6. Recap the vials tightly, and incubate at 35° C for 48 to 72 hours.

7. Remove medium from the vials, fix (according to manufacturer's instructions), and stain with iodine or fluorescein-conjugated chlamydial antibody.

8. Examine slides at 200 to 400 × magnification for dark-brown (iodine stain; Figure 61-4) or apple-green (fluorescent stain) intracytoplasmic inclusions (Figure 61-4).

References

1. Bastian, I. and Bowden, F.J. 1996. Amplification of *Klebsiella*-like sequences from biopsy samples from patients with donovanosis. Clin. Infect. Dis. 23:1328.

2. Bell, T.A., et al. 1992. Chronic *Chlamydia trachomatis* infections in infants. J.A.M.A. 267:400.

3. Brade, H., Brade, L. and Nano, F.E. 1987. Chemical and serological investigations on the genus-specific lipopolysaccharide epitope of *Chlamydia*. Proc. Natl. Acad. Sci. (USA). 84:2508.

4. Brunham, R.C., Pelling, R.W. 1994. *Chlamydia trachomatis* antigens: role in immunity and pathogenesis. Infect. Agents Dis. 3:218.

5. Campbell, L.A., et al. 1993. Detection of *Chlamydia trachomatis* deoxyribonucleic acid in women with tubal infertility. Fertil. Steril. 59:45.

6. Centers for Disease Control and Prevention. 1993. Recommendations for the prevention and management of *Chlamydia trachomatis* infections, Centers for Disease Control, Atlanta.

7. Deguchi, T., et al. 1996. Comparison among performances of a ligase chain reaction-based assay and two enzyme immunoassays in detecting *Chlamydia trachomatis* in urine specimens from men with nongonococcal urethritis. J. Clin. Microbiol. 34:1708.

8. Dienst, R.B. and Brownell, G.H. 1984. Genus *Calymmatobacterium Aragao* and *Vianna 1913*. In Krieg, N.R., and Holt, J.G., editors. Bergey's manual of systematic bacteriology, vol 1. Williams & Wilkins, Baltimore.

9. Dorman, S.A., et al. 1983. Detection of chlamydial cervicitis by Papanicolaou stained smears and culture. Am. J. Clin. Pathol. 79:421.

10. Drancourt, M. and Raoult, D. 1994. Taxonomic position of the Richettsiae: current knowledge, FEMS Microbiol. Rev. 13:13.

11. Dumler, J.S. and Walker, D.H. 1994. Diagnostic tests for Rocky Mountain spotted fever and other rickettsial diseases. Dermatol. Clin. 12:25.

12. Dumler, J.S. and Bakken, J.S. 1995. Ehrlichial diseases of human: emerging tick-borne infections. Clin. Infect. Dis. 20:1102.

13. Falsey, A.R. and Walsh, E.E. 1993. Transmission of *Chlamydia pneumoniae*. J. Infect. Dis. 168:493.

14. Fishbein, D.B., Dawson, J.E., and Robinson, L.E. 1994. Human ehrlichiosis in the United States, 1985-1990. Ann. Intern. Med. 120:736.

15. Gaydos, C.A., et al. 1994. Diagnostic utility of PCR-enzyme immunoassay, culture, and serology for detection of *Chlamydia pneumoniae* in symptomatic and asymptomatic patients. J. Clin. Microbiol. 32:903.

16. Goldberg, J. 1959. Studies on granuloma inguinale. IV. Growth requirements of *Donovania granulomatis* and its relationship to the natural habitat of the organism. Br. J. Ven. Dis. 35:266.

17. Hackstadt, T. 1996. The biology of rickettsiae, Infect. Agents Dis. 5:127.

18. Hahn, D.L., Dodge, R.W., and Golubjatnikov, R. 1991. Association of *Chlamydia pneumoniae* (strain TWAR) infection with wheezing, asthmatic bronchitis, and adult-onset asthma, J.A.M.A. 266:225.

19. Hammerschlag, M.R. 1994. Antimicrobial susceptibility and therapy of infections caused by *Chlamydia pneumoniae.* Antimicrob. Agents Chemother. 38:1873.

20. Hipp, S.S., Yangsook, H., and Murphy, D. 1987. Assessment of enzyme immunoassaay and immunofluorescence test for detection of *Chlamydia trachomatis.* J. Clin. Microbiol. 25:1938.

21. Kharsany, A.B.M., et al. 1996. Culture of *Calymatobacterium granulomatis.* Clin. Infect. Dis. 22:391.

22. Kellogg, J. 1995. Impact of variation in endocervical specimen collection and testing techniques on frequency of false-positive and false-negative chlamydia detection results. Am. J. Clin. Pathol. 104:554.

23. Martin, D.H. 1990. Chlamydial infections. Med. Clin. North Amer. 74:1367.

24. McCormack, W.M., et al. 1979. Fifteen-month follow up study of women infected with *Chlamydia trachomatis.* N. Engl. J. Med. 300:123.

25. McDade, J.E. 1990. Ehrlichiosis: a disease of animals and humans. J. Infect. Dis. 161:609.

26. McMenemy, A. 1992. Whipple's disease, a familial Mediterranean fever, adult-onset Still's disease, and enteropathic arthritis. Curr. Opin. Rheumatol. 4:479.

27. Musso, D. and Raoult, D. 1995. *Coxiella burnetii* blood cultures from acute and chronic Q-fever patients. J. Clin. Microbiol. 33:3129.

28. Nurminen, M., et al. 1983. The genus-specific antigen of *Chlamydia:* resemblance to the lipopolysaccharide of enteric bacteria. Science 220:1279.

29. Pasternack, R., et al. 1996. Detection of *Chlamydia trachomatis* infections in women by Amplicor PCR: comparison of diagnostic performance with urine and cervical specimens. J. Clin. Microbiol. 34:995.

30. Raoult, D. and Marrie, T. 1995. Q fever. Clin. Infect. Dis. 20:489.

31. Relman, D.A. 1983. The identification of uncultured microbial pathogens. J. Infect. Dis. 168:1.

32. Relman, D.A., et al. 1992. Identification of the uncultured bacillus of Whipple's disease. N. Engl. J. Med. 327:293.

33. Richens, J. 1991. The diagnosis and treatment of donovanosis (granuloma inguinale). Genitourin. Med. 67:441.

34. Schachter, J., et al. 1992. Nonculture tests for genital tract chlamydial infection. Sex. Transm. Dis. 19:243.

35. Schachter, J., et al. 1986. Prospective study of perinatal transmission of *Chlamydia trachomatis.* J.A.M.A. 255:3374.

36. Schachter, J., et al. 1983. Screening for chlamydial infections in women attending family planning clinics. West. J. Med. 138:375.

37. Smith, J.W., et al. 1987. Diagnosis of chlamydial infection in women attending antenatal and gynecologic clinics. J. Clin. Microbiol. 25:868.

38. Stamm, W.E., et al. 1984. *Chlamydia trachomatis* urethral infections in men: prevalence, risk factors, and clinical manifestations. Ann. Intern. Med. 100:47.

39. Thylefors, B. 1985. Development of trachoma control programs and the involvement of natural resources. Rev. Infect. Dis. 7:774.

40. Washington, A.E., Johnson, R.E. and Sander, L.L. 1987. *Chlamydia trachomatis* infections in the United States: what are they costing us? J.A.M.A. 257:2070.

41. Weinstock, H., Dean, D., and Bolan, C. 1994. *Chlamydia trachomatis* infections. Infect. Dis. Clin. North Am. 8:797.

42. Williams, J.C. and Waag, D. 1991. Antigens, virulence factors, and biological response modifiers of *Coxiella burnetii:* strategies for vaccine development. In Williams, J.C. and Thompson, H.A., editors. Q fever: the biology of *Coxiella burnetii.* CRC Press, Boca Raton, Fla.

Bibliography

Black, C.M. Current methods of laboratory diagnosis of *Chlamydia trachomatis* infections. Clin. Microbiol. Rev. 10:160.

Dumler, J.S. 1996. Laboratory diagnosis of human rickettsial and ehrlichial infections. Clin. Microbiol. Newsletter 18:57.

Evermann, J.F. 1987. *Chlamydia psittaci:* zoonotic potential worthy of concern. Clin. Microbiol. Newsletter 9:1.

Gaydos, C.A. 1995. *Chlamydia pneumoniae:* a review and evidence for a role in coronary artery disease. Clin. Microbiol. Newsletter 17:49.

Jones, R.B. 1991. New treatments for *Chlamydia trachomatis.* Am. J. Obstet. Gynecol. 164:1789.

Kuo, C-C., et al. 1995. *Chlamydia pneumoniae* (TWAR). Clin. Microbiol. Rev. 8:451.

Rikihisa, Y. 1991. The tribe *Ehrlichieae* and ehrlichial diseases. Clin. Microbiol. Rev. 4:286.

Schachter, J. 1988. The intracellular life of *Chlamydia.* Curr. Top. Microbiol. Immunol. 138:109.

Weinstock, H., Dean, D., and Bolan, G. 1994. *Chlamydia trachomatis* infections. Infect. Dis. Clin. North Am. 8:797.

62 | CELL WALL–DEFICIENT BACTERIA: *MYCOPLASMA* AND *UREAPLASMA*

Genera and Species to be Considered

■ *Mycoplasma pneumoniae*　　　　■ *Mycoplasma hominis*　　　　■ *Ureaplasma urealyticum*

This chapter addresses a group of bacteria, the mycoplasmas, that are the smallest known free-living forms; unlike all other bacteria, these prokaryotes do not have a cell wall. Although mycoplasmas are ubiquitous in the plant and animal kingdoms (more than 150 different species exist within this class), this chapter addresses only the most prominent varieties, that is, *Mycoplasma* spp. and *Ureaplasma urealyticum*, that colonize or infect humans and are not of animal origin.

GENERAL CHARACTERISTICS

Mycoplasmas belong to the class *Mollicutes* (Latin, meaning *soft skin*). This class comprises four orders, which, in turn, contain five families (Figure 62-1). The mycoplasmas that colonize and/or infect humans belong to the family *Mycoplasmataceae*; this family comprises two genera, *Mycoplasma* and *Ureaplasma*. Besides lacking a cell wall, these agents with their small cell size (0.3 × 0.8 μm) and small genome size, require sterols for membrane function and growth. *Mollicutes* appear to have evolved from clostridial-like cells of the gram-positive branch of the eubacterial tree. With few exceptions, most mycoplasmas are aerobic and have fastidious growth requirements.

EPIDEMIOLOGY AND PATHOGENESIS

Mycoplasmas are part of the microbial flora of humans and are found mainly in the oropharynx, upper respiratory tract, and genitourinary tract. Besides those that are considered primarily as commensals, considerable evidence indicates the pathogenicity of some mycoplasmas; for others, a role in a particular disease is less clearly delineated.

EPIDEMIOLOGY

The mycoplasmas usually considered as commensals are listed along with their respective sites of colonization in Table 62-1. One species of acholeplasma (these organisms are widely disseminated in animals), *Acholeplasma laidlawii*, has been isolated from the oral cavity of humans a limited number of times; however, the significance of these mycoplasmas and their colonization of humans remains uncertain.

Of the other mycoplasmas that have been isolated from humans, the possible role(s) that *M. genitalium*, *M. pirum*, *M. penetrans*, and *M. fermentans* might play in human disease is uncertain at this time. *M. fermentans*, *M. pirum*, and *M. penetrans* have been isolated from patients with HIV. The distribution, habitat, and transmission of these mycoplasmas is unknown.

Finally, the remaining three species of mycoplasmas that have been isolated from humans, *M. pneumoniae*, *U. urealyticum*, and *M. hominis*, have well-established roles in human infections. Both *U. urealyticum* and *M. hominis* have been isolated from the genitourinary tract of humans, and *M. pneumoniae* has been isolated form the respiratory tract.

Infants are commonly colonized with *U. urealyticum* and *M. hominis*. In general, this colonization does not persist beyond age 2.[9] Once an individual reaches puberty, colonization with these mycoplasmas can occur primarily as a result of sexual contact.[13,14] In situations in which these agents cause disease in neonates (see below), organisms are transmitted from a colonized mother to her newborn infant by an ascending route from colonization of the mother's urogenital tract, by crossing the placenta from the mother's blood, by delivery through a colonized birth canal, or postnatally from mother to infant.[17]

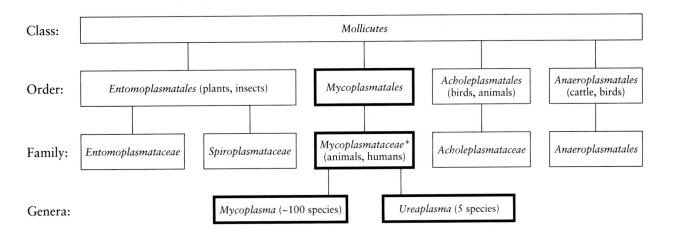

*Mollicutes that have been detected in humans.

FIGURE 62-1 Taxonomy of the class *Mollicutes*.

M. *pneumoniae* is a cause of community-acquired pneumonia (see Chapter 21); infections caused by this agent are distributed worldwide, with an estimated 2 million cases per year in the United States.[4,8] Infections can occur singly or as outbreaks in closed populations such as families and military recruit camps. Besides respiratory infection, M. *pneumoniae* can cause extrapulmonary manifestations. This organism infects children and adults and is spread by respiratory droplets produced by coughing.

PATHOGENESIS

In general, mycoplasmas colonize mucosal surfaces of the respiratory and urogenital tracts. Except for those mycoplasmas noted, most rarely produce invasive disease except in immunocompromised hosts. Of the mycoplasmas that are established as causes of human infections, these agents reside extracellularly, attaching with great affinity to ciliated and nonciliated epithelial cells. With respect to the mycoplasmas that are clearly able to cause disease, many of the disease processes are thought to be immunologically mediated.[3] Subsequent events following attachment remain unclear as of this writing. In addition to adherence properties and possibly immune-mediated injury, the ability to cause localized cell injury appears to contribute to their pathogenicity.

Of interest, the mycoplasmas associated with patients with HIV (M. *fermentans*, M. *penetrans*, and M. *pirum*) are all capable of invading humans cells and modulate the immune system.[16] Based on these findings, some investigators have proposed that these mycoplas-

TABLE 62-1 MYCOPLASMAS THAT ARE CONSIDERED NORMAL FLORA OF THE OROPHARYNX OR GENITAL TRACT

ORGANISM	SITE OF COLONIZATION
M. *orale*	Oropharynx
M. *salivarium*	Oropharynx
M. *buccale*	Oropharynx
M. *faucium*	Oropharynx
M. *lipohilum*	Oropharynx
M. *primatum*	Genital tract
M. *spermatophilum*	Genital tract

mas might play a role in certain disease processes in these patients.[1,2,12,16] Similarly, the possible pathogenic role for *M. genitalium* in humans remains unclear. This mycoplasma has been implicated as playing a possible role in nongonococcal urethritis (see Chapter 26).[7,10]

SPECTRUM OF DISEASE

The clinical manifestations of infections caused by *M. pneumoniae* and the pathogenic genital mycoplasmas, *U. urealyticum* and *M. hominis,* are summarized in Table 62-2.

LABORATORY DIAGNOSIS

The laboratory diagnosis of mycoplasma infections is extremely challenging because of complex and time-consuming culture requirements and the lack of reliable, widely available rapid diagnostic tests. Accurate, rapid diagnosis for *M. pneumoniae* is highly desired, because penicillin and other beta-lactam agents are ineffective treatment. The laboratory diagnosis of the mycoplasmas well-recognized as able to cause human disease (i.e., *M. pneumoniae, U. urealyticum,* and *M. hominis*) is addressed.

SPECIMEN COLLECTION, TRANSPORT, AND PROCESSING

Various specimens are appropriate for the diagnosis of mycoplasma infections by culture or other means of detection. Acceptable specimens include body fluids (e.g., blood, joint fluid, amniotic fluid, urine, prostatic secretions, semen, pleural secretions, bronchoalveolar lavage specimens), tissues, and swabs of the throat, nasopharynx, urethra, cervix, or vagina. Blood for culture of genital mycoplasmas should be collected without anticoagulants and immediately inoculated into an appropriate broth culture medium (see below). Swab specimens should be obtained without the application of any disinfectants, analgesics, or lubricant[19]; Dacron, polyester, or calcium alginate swabs on aluminum or plastic shafts should be used.

Because mycoplasmas have no cell wall, they are highly susceptible to drying and therefore transport media are necessary, particularly when specimens are collected on swabs. Transport and storage conditions of various types of specimens are summarized in Table 62-3.

DIRECT DETECTION METHODS

At present, no direct methods for identifying *M. pneumoniae, U. ureaplasma,* or *M. hominis* in clinical samples are recommended, although some methods have been described, such as PCR, immunoblotting, and indirect immunofluorescence.[11,22] PCR appears to hold the greatest promise for being a highly sensitive and specific means to detect *M. pneumoniae* directly in clinical specimens, however, the assays must be further evaluated.

TABLE 62-2 CLINICAL MANIFESTATIONS OF MYCOPLASMA INFECTIONS CAUSED BY *MYCOPLASMA PNEUMONIAE, UREAPLASMA UREALYTICUM,* AND *M. HOMINIS*

ORGANISM	CLINICAL MANIFESTATIONS
Mycoplasma pneumoniae	Asymptomatic infection
	Upper respiratory tract infection in young children: Mild, nonspecific symptoms including runny nose, coryza, and cough; most without fever
	Lower respiratory tract infection in adults: Typically mild illness with nonproductive cough, fever, malaise, pharyngitis, myalgias; 3%-13% of patients will develop pneumonia; complications include rash, arthritis, encephalitis, myocarditis, pericarditis, and hemolytic anemia
Genital mycoplasmas: *U. urealyticum* and *M. hominis*	Systemic infections in neonates: Meningitis, abscess, and pneumonia. *U. urealyticum* is also associated with the development of chronic lung disease
	Invasive disease in imunosuppressed patients: Bacteremia, arthritis (particularly in patients with agammaglobulinemia), abscesses and other wound infections, pneumonia, peritonitis
	Urogenital tract infections: Prostatitis, pelvic inflammatory disease, bacterial vaginosis, amnionitis, nongonococcal urethritis

CULTIVATION

In general, the media for mycoplasma isolation contains a beef or soybean protein with serum, fresh yeast extract, and other factors. Of importance, because these organisms grow more slowly than most other bacteria, the media must be selective to prevent overgrowth of faster-growing organisms that may be present in a clinical sample. Culture media and incubation conditions for these organisms are summarized in Table 62-4. Culture methods for *M. pneumoniae*, *U. urealyticum*, and *M. hominis* are provided at the

end of this chapter in procedures 62-1, 62-2, and 62-3, respectively. The quality control of the growth media with a fastidious isolate is of great importance.

For the most part, the different metabolic activity of the mycoplasmas for different substrates is used to detect their growth. Glucose (dextrose) is incorporated into media selective for *M. pneumoniae*, because this mycoplasma ferments glucose to lactic acid; the resulting pH change is then detected by a color change in a dye indicator. Similarly, urea and/or arginine can be incorporated into media to detect *U.*

TABLE 62-3   TRANSPORT AND STORAGE CONDITIONS FOR *MYCOPLASMA PNEUMONIAE*, *UREAPLASMA UREALYTICUM*, AND *M. HOMINIS*

SPECIMEN TYPE	TRANSPORT CONDITIONS	TRANSPORT MEDIA (EXAMPLES)	STORAGE	PROCESSING
Body fluid or liquid specimens*	Within 1 hr of collection on ice or at 4° C	Not required	4° C up to 24 hrs[†]	Concentrate by high-speed centrifugation and dilute (1:10 and 1:100) in broth culture media to remove inhibitory substances and contaminating bacteria
Swabs	Place immediately into transport media	0.5% albumin in trypicase soy broth	4° C up to 24 hrs[†]	None
		Modified Stuart's		
		2SP (sugar-phosphate medium with 10% heat-inactivated fetal calf serum)		
		Shepard's 10B broth for ureaplasmas[18]		
		SP-4 broth[20] for other mycoplasmas and *M. pneumoniae*[‡]		
		Mycoplasma transport medium (trypticase phosphate broth, 10% bovine serum albumin, 100,000 U of penicillin/milliliter)[5]		
Tissue	Within 1 hr of collection on ice or at 4° C	Not required as long as prevented from drying out	4° C up to 24 hrs[†]	Mince and dilute (1:10 and 1:100) in transport media

*Except blood (see text).

[†]Can be stored indefinitely at −70° C if diluted in transport media following centrifugation.

[‡]SP-4 broth: sucrose phosphate buffer, 20% horse serum, *Mycoplasma* base, and neutral red.

TABLE 62-4 CULTIVATION OF MYCOPLASMA *PNEUMONIAE*, *UREAPLASMA UREALYTICUM*, AND *M. HOMINIS*

ORGANISM	MEDIA (EXAMPLES)	INCUBATION CONDITIONS
M. pneumoniae	Biphasic SP-4 (pH 7.4)	Agar and broths: 37° C, ambient air for up to 4 wks
	Triphasic system (Mycotrim RS, Irvine Scientific, Irvine, Calif.)	
	PPLO broth or agar with yeast extract and horse serum	
	Modified New York City medium	
U. urealyticum[*]/ *M. hominis*[†]	A7 or A8 agar medium[‡]	Broths: 37° C, ambient air for up to 7 days
	New York City medium	Agars: 37° C in 5% CO_2 or anaerobically for 2 to 5 days
	Modified New York City medium	
	SP-4 broth with arginine[§]	
	SP-4 broth with urea[‖]	
	Triphasic system (Mycotrim GU, Irvine Scientific)	
	Shepard's 10B broth (or Ureaplasma 10C broth)[‖]	

[*]Utilizes urea and requires acidic medium.
[†]Converts arginine to ornithine and grows over a broad pH range.
[‡]Commercially available.
[§]For *M. hominis* isolation.
[‖]For *U. urealyticum* isolation.

ureaplasma and *M. hominis*, respectively. If a color change, that is, a pH change, is detected, a 0.1 to 0.2-mL aliquot is immediately subcultured to fresh broth and agar media.

In some clinical situations, it may be necessary to provide quantitative information regarding the numbers of genital mycoplasmas in a clinical specimen. For example, quantitation of specimens taken at different stages during urination or after prostatic massage can help determine the location of mycoplasmal infection in the genitourinary tract.[15]

APPROACH TO IDENTIFICATION

On agar, *M. pneumoniae* will appear as spherical, grainy, yellowish forms that are embedded in the agar, with a thin outer layer similar to those shown in Figure 62-2. The agar surface is examined under 40× magnification every 24 to 72 hours. Because only *M. pneumoniae* and one serovar of *U. urealyticum* hemadsorb, definitive identification of *M. pneumoniae* is accomplished by overlaying suspicious colonies with 0.5% guinea pig erythrocytes in phosphate-buffered saline instead of water. After 20 to 30 minutes at room temperature, colonies are observed for adherence of red blood cells.

Cultures for the genital mycoplasmas are handled in a similar fashion, including culture examination and the requirement for subculturing. Colonies may be definitely identified on A8 agar as *U. urealyticum* by urease production in the presence of a calcium chloride indicator.[16] *U. urealyticum* colonies (15 to 60 μm in diameter) will appear as dark brownish clumps. Colonies that are typical in appearance for *U. urealyticum* are shown in Figure 62-3. *M. hominis* are large (about 20 to 300 μm in diameter) and will be urease-negative (Figure 62-3), with a characteristic "fried egg" appearance. On conventional blood agar, strains of *M. hominis*, but not *U. urealyticum*, produce nonhemolytic, pinpoint colonies that do not stain with Gram stain. These colonies can be stained with the Dienes or acridine orange stains.

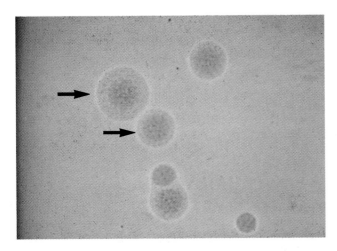

FIGURE 62-2 Colonies of *Mycoplasma pneumoniae* visualized under 100× magnification. Note the variation in the size of the colonies (*arrows*). (Courtesy Clinical Microbiology Laboratory, SUNY Health Science Center, Syracuse, N.Y.)

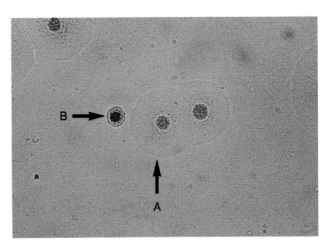

FIGURE 62-3 Isolation of *Mycoplasma hominis* and *Ureaplasma urealyticum* (100× magnification). Note the "fried egg" appearance of the large *M. hominis* colony (*arrow A*) and the relatively small size of the *U. urealyticum* colony (*arrow B*). (Courtesy Clinical Microbiology Laboratory, SUNY Health Science Center, Syracuse, N.Y.)

SERODIAGNOSIS

Laboratory diagnosis of *M. pneumoniae* is usually made serologically. Nonspecific production of cold agglutinins occurs in approximately half of patients with atypical pneumonia caused by this organism. The most widely used serologic tests today are complement fixation (CF) and enzyme-linked immunosorbent assay (ELISA), although newly developed indirect fluorescent antibody tests are being used with some success. Although a fourfold rise in titer from acute to convalescent serum is better evidence, CF titers of 1:128 or greater in single serum specimens suggest recent infection.

Although serologic tests such as indirect hemagglutination and metabolism inhibition for genital mycoplasmas are available, they are rarely used. Because of the antigenic complexity of the mycoplasmas, the development of a specific and useful serologic assay is a challenge.

SUSCEPTIBILITY TESTING AND THERAPY

Although an agar dilution method and another method called the MI method may be used to deter-

mine antibiotic susceptibilities, the complex growth requirements of mycoplasmas have restricted their performance to only a few laboratories. Further complicating matters, the lack of standardized methods and interpretive criteria have also contributed to problems with susceptibility tests for the mycoplasmas. Most mycoplasmal infections are treated empirically.

The majority of *M. pneumoniae* infections are self-limited and usually do not require treatment. However, treatment can markedly shorten the illness, although complete eradication of the organism takes a long time, even after therapy. *M. pneumoniae* is usually susceptible to the macrolides and tetracycline antibiotics. Of the quinolones, sparfloxacin appears to have activity against *M. pneumoniae*.[21] Unfortunately, the susceptibility of *M. hominis* and *U. urealyticum* to various agents is not as predictable. For the most part, the tetracyclines are the drugs of choice for these agents.

PREVENTION

As of this writing, no vaccines have been developed for the mycoplasmas.

Procedures

62-1 ISOLATION OF *MYCOPLASMA PNEUMONIAE*

PRINCIPLE

Mycoplasma pneumoniae is primarily a respiratory tract pathogen that is extremely fastidious. Because this bacterial agent lacks a cell wall and has strict nutritional requirements, the recovery of this organism from

clinical specimens requires special conditions both in terms of culture media and incubation conditions.

METHOD

1. Prepare biphasic SP-4 culture media as follows:

A. Combine the ingredients for the media base.
   Mycoplasma broth base (BBL), 3.5 g
   Tryptone (Difco), 10 g
   Bacto-Peptone (Difco), 5 g
   (50% aqueous solution,

   Continued

62-1 ISOLATION OF *MYCOPLASMA PNEUMONIAE*—cont'd

filter sterilized), 10 mL
Distilled water, 615 mL

B. Stir the solids into boiling water to dissolve them and adjust the pH to 7.5. Autoclave according to the manufacturer's instructions. Cool to 56° C before adding the following supplements for each 625 mL of base to make a final volume of 1 L:

CMRL 1066 tissue culture medium with glutamine, 10× (GIBCO), 50 mL
Aqueous yeast extract (prepared as described in New York City Medium), 35 mL
Yeastolate (Difco; 2% solution), 100 mL
Fetal bovine serum (heat inactivated at 56° C for 30 minutes), 170 mL
Penicillin G sodium, 1 million U, or ampicillin, 1 mg/mL
Amphotericin B, 0.5 g
Polymyxin B, 500,000 U

C. To prepare the agar necessary for making biphasic media, add 8.5 g Noble agar (Difco) to the basal medium ingredients before adding the supplements.

D. Dispense 1 mL of SP-4 agar aseptically into the bottom of sterile 4-mL screw-capped vials. Allow the agar to set and dispense 2 mL of SP-4 broth above the agar layer in each vial. Seal caps tightly and store at −20° C.

2. Place 0.1 to 0.2 mL of liquid specimen or dip and twirl a specimen received on a swab in a vial of biphasic SP-4 culture medium. After expressing as much fluid as possible from the swab, remove the swab to prevent contamination.

3. Seal the vial tightly and incubate in air at 35° C for up to 3 weeks.

4. Inspect the vial daily. During the first 5 days, a change in pH, indicated by a color shift from orange to yellow or violet, or increased turbidity is a sign that the culture is contaminated and should be discarded.

5. If either a slight acid pH shift (yellow color) with no increase in turbidity or no change occurs after 7 days' incubation, subculture several drops of the broth culture to agar. Continue to incubate the original broth.

6. If broth that exhibited no changes at 7 days shows a slight acid pH shift at any time, subculture to agar as above. Broths that show no change at 3 weeks are blindly subcultured to agar.

7. Incubate the agar plates in a very moist atmosphere with 5% CO_2 at 35° C for 7 days.

8. Observe the agar surface under 40× magnification after 5 days

for colonies, which appear as spherical, grainy, yellowish forms, embedded in the agar, with a thin outer layer (Figure 62-2).

9. Definitive identification of *M. pneumoniae* is accomplished by overlaying agar plates showing suspicious colonies with 5% sheep or guinea pig erythrocytes in 1% agar prepared in physiologic saline (0.85% NaCl) instead of water. The 1% agar is melted and cooled to 50° C, the blood cells are added, and a thin layer is poured over the original agar surface.

10. Reincubate the plate for 24 hours, and observe for beta hemolysis around colonies of *M. pneumoniae* caused by production of hydrogen peroxide. Additional incubation at room temperature overnight enhances the hemolysis. No other species of *Mycoplasma* produces this reaction.

Modified from Clyde, W.A., Jr., Kenny G.E., and Schachter, J. 1984. Cumitech 19: laboratory diagnosis of chlamydial and mycoplasmal infections. Drew, M.L., coordinating editor. American Society for Microbiology, Washington D.C.

62-2 ISOLATION OF *UREAPLASMA UREALYTICUM*

PRINCIPLE
Similar to other mycoplasmas, *Ureaplasma urealyticum* requires special cultural conditions to allow recovery of this bacterial agent from clinical specimens.

METHOD
1. Inoculate one *Ureaplasma* agar plate and one *Ureaplasma* broth each with 0.1 mL specimen from transport medium.

2. Incubate broth in tightly sealed test tubes for 5 days. Observe twice daily for a color change in

the broth to red, with no increase in turbidity. If color change occurs, immediately transfer one loopful to a *Ureaplasma* agar plate and streak for isolated colonies.

3. Agar plates are incubated in a candle jar or, optimally, in an anaerobic jar at 35° C. Colonies appear on agar within 48 hours. Plates are inspected in the same way as described for *M. pneumoniae* (see Procedure 62-1). *Ureaplasma* colonies appear as small, granular, yellowish spheres.

4. To identify definitively colonies on *Ureaplasma* agar after 48 hours' incubation, pour a solution of 1% urea and 0.8% $MnCl_2$ in distilled water over the agar surface. *U. urealyticum* stains dark brown because of production of urease (Figure 62-3).

62-3 ISOLATION OF *MYCOPLASMA HOMINIS*

PRINCIPLE
Similar to the other mycoplasmas, the successful recovery of *M. hominis* by culture from clinical specimens requires special conditions.

METHOD A
1. Inoculate one *M. hominis* agar plate and two *M. hominis* broth tubes, one broth containing phenol red indicator and one without the possibly inhibitory phenol red, each with 0.1-mL specimen from transport media.

2. Incubate broths in tightly sealed test tubes for 5 days. If the phenol red–containing broth changes color to red or violet, both broths are subcultured to *M. hominis* agar. After 48 hours of incubation, transfer 0.1 mL or a loopful of broth from tubes that exhibited no change or only a slight increase in turbidity to *M. hominis* agar and streak for isolated colonies.

3. *M. hominis* agar plates are incubated in the same manner as *Ureaplasma* cultures. Plates should be observed daily for up to 5 days for colonies.

METHOD B (ALTERNATIVE METHOD)
1. Inoculate specimen onto a prereduced colistin-nalidixic acid (CNA) sheep blood agar or anaerobic blood plate and incubate anaerobically for 48 to 72 hours.

2. Examine for pinpoint colonies that show no bacteria on Gram stain.

3. Streak suspicious colonies to *M. hominis* agar and incubate as in step 3 above.

References

1. Ainsworth, J.G., et al. 1994. *Mycoplasma fermentans* from urine of HIV-positive patients with AIDS. Arch. Pathol. Lab. Med. 117:511.

2. Bauer, F.A., et al. 1991. *Mycoplasma fermentans* (incognitus strain) infection in the kidneys of patients with acquired immunodeficiency syndrome and associated nephropathy: a light microscopic, immunohistochemical and ultrastructural study. Human Pathol. 22:63.

3. Cassell, G.H. and Cole, B.C. 1981. Mycoplasma as agents of human disease. N. Engl. J. Med. 304:80.

4. Centers for Disease Control and Prevention: 1993. Outbreaks of *Mycoplasma pneumoniae* respiratory infection: Ohio, Texas, and New York. Centers for Disease Control, Atlanta.

5. Clegg, A., et al. 1997. High rates of genital mycoplasma infection in highland Papua New Guinea determined both by culture and by a commercial detection kit. J. Clin. Microbiol. 35:197.

6. Clyde, W.A., Kenny, G.E., and Schachter, J. 1984. Cumitech 19: laboratory diagnosis of chlamydial and mycoplasmal infections. Drew, M.L., coordinating editor: American Society for Microbiology, Washington D.C.

7. Deguchi, T., et al. 1995. *Mycoplasma genitalium* in nongonococcal urethritis. J. STD AIDS 6:144.

8. Foy, H.M. 1993. Infections caused by *Mycoplasma pneumoniae* and possible carrier state in a different population of patients. Clin. Infect. Dis. 17(suppl. 1):S37.

9. Foy, H.M., et al. 1970. Acquisition of mycoplasmata and T-strains during infancy. J. Infect. Dis. 121:579.

10. Lanier, M., et al. 1995. Male urethritis with and without discharge: a clinical and microbiological study. Sex. Transm. Dis. 22:244.

11. Lo, L-C. 1995. New understandings of mycoplasmal infections and disease. Clin. Microbiol. Newsletter 17:169.

12. Lo, S-C., et al. 1989. Identification of *Mycoplasma incognitus* infection in patients with AIDS: an immunohistochemical *in situ* hybridization and ultrastructural. Am. J. Trop. Med. Hyg. 41:6-1.

13. McCormack, W.M., et al. 1973. Sexual experience and urethral colonization with genital mycoplasma: a study in normal men. Ann. Intern. Med. 78:696.

14. McCormack, W.M., et al. 1972. Sexual activity and vaginal colonization with genital mycoplasmas. J.A.M.A. 221:1375.

15. Meares, E.M. and Stamey, T.A. 1968. Bacteriologic localization patterns in bacterial prostatitis and urethritis. Invest. Urol. 5:492.

16. Montagnier, L. and Blanchard, A. 1993. Mycoplasmas as cofactors in infection due to the human immunodeficiency virus. Clin. Infect. Dis. 17(suppl. 1):S309.

17. Sanchez, P. 1993. Perinatal transmission of *Ureaplasma urealyticum:* current concepts based on review of the literature. Clin. Infect. Dis. 17(suppl.):S107.

18. Shepard, M.C. and Lunceford, C.D. 1978. Serological typing of *Ureaplasma urealyticum* isolates from urethritis patients by an agar growth inhibition method. J. Clin. Microbiol. 8:566.

19. Taylor-Robinson, D. and Furr, P.M. 1981. Recovery and identification of human genital tract mycoplasmas. Isr. J. Med. Sci. 17:648.

20. Tully, J.G., et al. 1977. Pathogenic mycoplasmas: cultivation and vertebrate pathogenicity of a new spiroplasma. Science 195:892.

21. Waites, K.B., et al. 1991. *In vitro* susceptibilities of *Mycoplasma pneumoniae, Mycoplasma hominis,* and *Ureaplasma urealyticum* to sparfloxacin and PD 127391. Antimicrob. Agents Chemother. 35:1181.

22. Waites, K.B., et al. 1996. Laboratory diagnosis of mycoplasmal and ureaplasmal infections. Clin. Microbiol. Newsletter 18:105.

Bibliography

Ali, N.J., et al. 1986. The clinical spectrum and diagnosis of *Mycoplasma pneumoniae* infection. Quarter. J. Med. 58:241.

Clyde, W.A. 1993. Clinical overview of typical *Mycoplasma pneumoniae* infections. Clin. Infect. Dis. 17(suppl. 1):S32.

Lo, S-C. 1995. New understandings of mycoplasmal infections and disease. Clin. Microbiol. Newsletter. 17:169.

Madoff, S. and Hooper, D.C. 1988. Nongenitourinary infections caused by *Mycoplasma hominis* in adults. Rev. Infect. Dis. 10:602.

Taylor-Robinson, D. 1996. Infections due to species of *Mycoplasma* and *Ureaplasma:* an update. Clin. Infect. Dis. 23:671.

Taylor-Robinson, D. 1995. *Mycoplasma* and *Ureaplasma.* In Murray, P.R., et al., editors. Manual of clinical microbiology, ed 6. American Society for Microbiology, Washington, D.C.

Tully, J.G. 1993. Current status of the mollicute flora of humans. Clin. Infect. Dis. 17(suppl. 1):S2.

63 | THE SPIROCHETES

Genera and Species to be Considered

- *Treponema pallidum* subsp. *pallidum*
- *Treponema pallidum* subsp. *pertenue*

- *Treponema pallidum* subsp. *endemicum*
- *Treponema carateum*
- *Borrelia recurrentis*

- *Borrelia burgdorferi*
- *Leptospira interrogans*

This chapter addresses the bacteria that belong in the order *Spirochaetales*. Although there are at least eight genera in this family, only the genera *Treponema*, *Borrelia*, and *Leptospira*, which contain organisms pathogenic for humans, are addressed in this chapter. Although one other human intestinal spirochete, *Brachyspira aalborgi*, has been isolated from biopsy material from patients with intestinal disease, a clear association between its presence and intestinal disease has not been established.

The spirochetes are all long, slender, helically curved, gram-negative bacilli, with the unusual morphologic features of axial fibrils and an outer sheath. These fibrils, or axial filaments, are flagella-like organelles that wrap around the bacteria's cell wall, are enclosed within the outer sheath, and facilitate motility of the organisms. The fibrils are attached within the cell wall by platelike structures, called *insertion disks,* located near the ends of the cells. The protoplasmic cylinder gyrates around the fibrils, causing bacterial movement to appear as a corkscrew-like winding. Differentiation of genera within the order *Spirochaetaceae* is based on the number of axial fibrils, the number of insertion disks present (Table 63-1), and biochemical and metabolic features. The spirochetes also fall into genera based loosely on their morphology (Figure 63-1): *Treponema* are slender with tight coils; *Borrelia* are somewhat thicker with fewer and looser coils; and *Leptospira* resemble *Borrelia* except for their hooked ends.

TREPONEMA

GENERAL CHARACTERISTICS

The major pathogens in the genus *Treponema*— *T. pallidum* subsp. *pallidum*, *T. pallidum* subsp. *pertenue*, and *T. pallidum* subsp. *endemicum*—infect only humans and have not been cultivated for more than one passage in vitro. Most species stain poorly with Gram's or Giemsa's methods and are best observed with the use of dark-field or phase-contrast microscopy. These organisms are considered to be microaerophilic.

Other treponemes are normal inhabitants of the oral cavity or the human genital tract; these include *T. vincentii*, *T. denticola*, *T. refringens*, *T. macrodentium*, and *T. oralis*. These organisms are cultivable anaerobically on artificial media. Acute necrotizing ulcerative gingivitis, also known as **Vincent's disease,** is a destructive lesion of the gums. Methylene blue-stained material from the lesions of patients with Vincent's disease show certain morphologic types of bacteria. Observed morphologies include spirochetes and fusiforms; oral spirochetes, particularly an unusually large one, may be important in this disease, along with other anaerobes.

EPIDEMIOLOGY AND PATHOGENESIS

Key features of the epidemiology of diseases caused by the pathogenic treponemes are summarized in Table 63-2. In general, these organisms enter the host by either penetrating intact mucous membranes (as is the case for *T. pallidum* subsp. *pallidum*—hereafter

TABLE 63-1 SPIROCHETES PATHOGENIC FOR HUMANS

GENUS	AXIAL FILAMENTS	INSERTION DISKS
Treponema	6-10	1
Borrelia	30-40	2
Leptospira	2	3-5

referred to as *T. pallidum*) or entering through breaks in the skin. After penetration *T. pallidum* subsequently invades the bloodstream and spreads to other body sites. Although the mechanisms by which damage is done to the host are unclear, *T. pallidum* has a remarkable tropism (attraction) to arterioles; infection ultimately leads to endarteritis (inflammation of the lining of arteries) and subsequent progressive tissue destruction.

SPECTRUM OF DISEASE

Treponema pallidum causes venereal (transmitted through sexual contact) syphilis. The clinical presentation of venereal syphilis is varied and complex, often mimicking many other diseases. This disease is divided into stages: incubating, primary, secondary, early latent, latent, and tertiary.[12] **Primary syphilis** is characterized by the appearance of a **chancre** (a painless ulcer) usually at the site of inoculation, most commonly the genitalia. Within 3 to 6 weeks, the chancre heals. Dissemination of the organism occurs during this primary stage; once the organism has reached a sufficient number (usually 2 to 24 weeks are required), clinical manifestations of **secondary syphilis** become apparent. During this phase the patient is ill and seeks medical attention. Systemic symptoms such as fever, weight loss, malaise, and loss of appetite are present in about half of the patients. The skin is the organ most commonly affected in secondary syphilis, with patients having a widespread

rash. After the secondary phase, the disease becomes subclinical but not necessarily dormant (inactive); during this latent period diagnosis can be made only by serologic tests. **Late or tertiary syphilis** is the tissue-destructive phase that appears 10 to 25 years after the initial infection in up to 35% of untreated patients.[12] Complications of syphilis at this stage include central nervous disease, cardiovascular abnormalities, eye disease, and granuloma-like lesions, called *gummas,* in any organ.

The other pathogenic treponemes are major health concerns in developing countries. Although morphologically and antigenically very similar, these agents differ epidemiologically and with respect to their clinical presentation from *T. pallidum*. The diseases caused by these treponemes are summarized in Table 63-2.

LABORATORY DIAGNOSIS

Direct detection

Treponemes can be detected in material taken from skin lesions by dark-field examination or fluorescent antibody staining and microscopic examination. Material for microscopic examination is collected from suspicious lesions by first cleansing the area around the lesion with a sterile gauze pad moistened in saline. The surface of the ulcer is then abraded until some blood is expressed. After blotting the lesion until there is no further bleeding, the area is squeezed until serous fluid is expressed. The surface of a clean glass slide is touched to the exudate, allowed to air-dry, and transported in a dust-free container for fluorescent antibody staining. A *T. pallidum* fluorescein-labeled antibody is commercially available for staining (Becton Dickinson Diagnostics, Sparks, Md.) as well as from the Centers for Disease Control and Prevention (CDC) in Atlanta. For dark-field examination, the expressed fluid is aspirated using a sterile pipette, dropped onto a clean glass slide, and coverslipped. The slide containing material for dark-field examination must be transported to the laboratory immediately. Because positive lesions may be teeming with viable spirochetes that are highly infec-

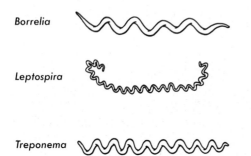

FIGURE 63-1 Species designation of spirochetes based on morphology.

TABLE 63-2   EPIDEMIOLOGY AND SPECTRUM OF DISEASE OF THE TREPONEMES PATHOGENIC FOR HUMANS

AGENT	TRANSMISSION	GEOGRAPHIC LOCATION	DISEASE	CLINICAL MANIFESTATIONS*	AGE GROUP
T. pallidum subsp. *pallidum*	Sexual contact or congenital (mother to fetus)	Worldwide	Venereal syphilis[†]	Refer to p. 776	All ages
T. pallidum subsp. *pertenue*	Traumatized skin comes in contact with an infected lesion	Humid, warm climates: Africa, South and Central America, Pacific Islands	Yaws	Skin-papules,[†] nodules, ulcers	Children
T. pallidum subsp. *endemicum*	Mouth to mouth by utensils	Arid, warm climates: North Africa, Southeast Asia, Middle East	Endemic non-venereal syphilis	Skin/mucous patches, papules, macules, ulcers, scars[†]	Children
T. carateum	Traumatized skin comes in contact with an infected lesion	Semi-arid, warm climates: Central and South America, Mexico	Pinta	Skin papules, macules	All ages

*All diseases have a relapsing clinical course and prominent cutaneous manifestations.[7]

[†]If untreated, organisms can disseminate to other parts of the body such as bone.

tious, all supplies and patient specimens must be handled with extreme caution and carefully discarded as required for contaminated materials. Gloves should always be worn.

Material for dark-field examination (see Chapter 11) is examined immediately under 400×, high-dry magnification for the presence of motile spirochetes. Treponemes are long (8 to 10 μm, slightly larger than a red blood cell) and consist of 8 to 14 tightly coiled, even spirals (Figure 63-2). Once seen, characteristic

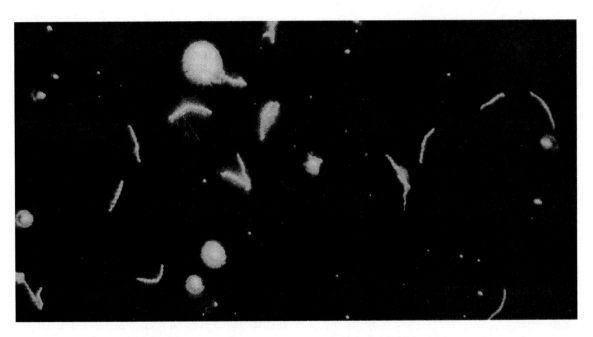

FIGURE 63-2 Appearance of *Treponema pallidum* in dark-field preparation.

forms should be verified by examination under oil immersion magnification (1000×). Although the dark-field examination depends greatly on technical expertise and the numbers of organisms in the lesion, it can be highly specific when performed on genital lesions.

Direct detection of *T. pallidum* in clinical material, such as serous exudate, has been accomplished using polymerase chain reaction technology and appears to hold some promise.

Serodiagnosis

Class serologic tests for treponematosis measure the presence of two types of antibodies: treponemal and nontreponemal. **Treponemal antibodies** are produced against antigens of the organisms themselves, whereas **nontreponemal antibodies,** often call **reaginic antibodies,** are produced by infected patients against components of mammalian cells. Reaginic antibodies, although almost always produced by patients with syphilis, are also produced by patients with other infectious diseases such as leprosy, tuberculosis, chancroid, leptospirosis, malaria, rickettsial disease, trypanosomiasis, lymphogranuloma venereum (LGV), measles, chickenpox, hepatitis, and infectious mononucleosis; noninfectious conditions such as drug addiction; autoimmune disorders, including rheumatoid disease; and factors such as old age, pregnancy, and recent immunization.

The two most widely used nontreponemal serologic tests are the **VDRL** (Venereal Disease Research Laboratory) tests and **RPR** (rapid plasma reagin) test. Each of these tests is a flocculation (or agglutination)

test, in which soluble antigen particles are coalesced to form larger particles that are visible as clumps when they are aggregated by antibody (see Chapter 16 for more information).

Specific treponemal serologic tests include the **FTA-ABS** (fluorescent treponemal antibody absorption) test and the **MHA-TP** (microhemagglutination) test. Once positive, their usefulness is limited because these tests tend to yield positive results throughout the patient's life. The FTA-ABS test is performed by overlaying whole treponemes that are fixed to a slide with serum from patients suspected of having syphilis because of a previous positive VDRL or RPR test. The patient's serum is first absorbed with non-*T. pallidum* treponemal antigens (sorbent) to reduce nonspecific cross-reactivity. Fluorescein-conjugated antihuman antibody reagent is then applied as a marker for specific antitreponemal antibodies in the patient's serum. This test should not be used as a primary screening procedure. The MHA-TP tests utilize erythrocytes from a turkey or other animal that are coated with treponemal antigens. The presence of specific antibody causes the red cells to agglutinate and form a flat mat across the bottom of the microdilution well in which the test is performed. The nontreponemal serologic tests for syphilis can be used to determine quantitative titers of antibody, which are useful for following response to therapy. The relative sensitivity of each test is shown in Table 63-3 to confirm that a positive nontreponemal test result is due to syphilis rather than to one of the other infections or biologic false-positive conditions previously mentioned.

TABLE 63-3 SENSITIVITY OF COMMONLY USED SEROLOGIC TESTS FOR SYPHILIS

METHOD	STAGE* 1°	STAGE* 2°	LATE
Nontreponemal (Reaginic Tests) — Screening			
Venereal Disease Research Laboratory (reaginic) test (VDRL)	70%	99%	1%[†]
Rapid plasma-reagin (RPR) card test and automated reagin test (ART)	80%	99%	0-1%[†]
Specific Treponemal Tests — Confirmatory			
Fluorescent treponemal antibody absorption test (FTA-ABS)	85%	100%	98%
T. pallidum hemagglutination assay (TPHA-TP)	70%	100%	95%

*Percentage of patients with positive serologic tests in treated or untreated primary or secondary syphilis.
[†]Treated late syphilis.

Modified from Tramont, E.C. 1995. Syphilis in adults: from Christopher Columbus to Sir Alexander Fleming to AIDS. Clin. Infect. Dis. 21:1361. The University of Chicago Press, Chicago, Ill.

ANTIMICROBIAL SUSCEPTIBILITY TESTING AND THERAPY

Because the treponemes are noncultivable, susceptibility testing is not performed. For all treponemal infections, penicillin G is the drug of choice.

PREVENTION

There are no vaccines available for the treponematoses. Prevention is best accomplished by early and appropriate treatment, thereby preventing person-to-person spread.

BORRELIA

GENERAL CHARACTERISTICS

Organisms belonging to the genus *Borrelia* are composed of 3 to 10 loose coils (Figure 63-1) and are actively motile. In contrast to the treponemes, *Borrelia* spp. stain well with Giemsa's stain. Species that have been grown in vitro are microaerophilic.

EPIDEMIOLOGY AND PATHOGENESIS

Although pathogens for mammals and birds, borrelias are the causative agents of tickborne and louseborne relapsing fever and tickborne Lyme disease in humans.

Relapsing fever

Human relapsing fever is caused by more than 15 species of *Borrelia* and is transmitted to humans by the bite of a louse or tick. *B. recurrentis* is responsible for louseborne or epidemic relapsing fever. This spirochete is transmitted by the louse *Pediculus humanus* subsp. *humanus* and disease is found worldwide; humans are the only reservoir for *B. recurrentis*. All other borreliae that cause disease in the United States are transmitted via tick bites and are named after the species of tick, usually of the genus *Ornithodoros*, from which they are recovered. Common species in the United States include *B. hermsii*, *B. turicatae*, *B. parkeri*, and *B. bergmanni*. Depending on the organisms and the disease, their reservoir is either humans or rodents in most cases. Although their pathogenic mechanisms are unclear, these spirochetes exhibit antigenic variability that may account for the cyclic fever patterns associated with this disease.

Lyme disease

Borrelia burgdorferi is the agent of **Lyme disease** and is transmitted by the bite of *Ixodes* ticks. Based on the CDC's case-definition criteria for reporting Lyme disease, the overall incidence in the United States was 5.2/100,000 for 1994.[3] Several genera of ticks act as vectors in the United States, including *Ixodes pacificus* in California and *I. dammini* in other areas. The ticks' natural hosts are deer and rodents, although they will attach to pets as well as humans; all stages of ticks—larva, nymph, and adult—can harbor the spirochete and transmit disease. The nymphal form of the tick is most likely to transmit disease because it is active in the spring and summer when people are dressed lightly and in the woods. Because at this stage the tick is the size of a pinhead, the initial tick bite may be overlooked. Ticks require a period of attachment of at least 24 hours before they transmit disease. Endemic areas of disease have been identified in many states, including Massachusetts, Connecticut, Maryland, Minnesota, Oregon, and California, as well as in Europe, Russia, Japan, and Australia. Direct invasion of tissues by the organism is responsible for the clinical manifestations. However, IgM antibodies are produced continually months to years after initial infection as the spirochete changes its antigens. *B. burgdorferi's* potential ability to induce an autoimmune process in the host because of cross-reactive antigens may contribute to the pathology associated with Lyme disease.[1] The pathologic findings associated with Lyme disease are also believed to be due to the release of host cytokines initiated by the presence of the organism.

SPECTRUM OF DISEASE

Relapsing fever

Two to 15 days after infection, patients have an abrupt onset of fever, headache, and myalgia that lasts for 4 to 10 days. As the host produces a specific antibody in response to the agent, organisms disappear from the bloodstream, becoming sequestered (hidden) in different body organs during the afebrile period. Subsequently, organisms reemerge with newly modified antigens and multiply, resulting in another febrile period. Subsequent relapses are usually milder and of shorter duration. Generally there are more relapses associated with cases of untreated tickborne relapsing fever, but louseborne relapsing fevers tend to be more severe.

Lyme disease

Lyme disease is characterized by three stages, not all of which occur in any given patent. The first stage, erythema migrans (EM), is the characteristic red, ring-shaped skin lesion with a central clearing that first appears at the site of the tick bite but may develop at distant sites as well (Figure 63-3). Patients may experience headache, fever, muscle and joint pain, and malaise during this stage. The second stage, beginning weeks to months after infection, may include arthritis, but the most important features are neurologic disorders (i.e., meningitis, neurologic deficits) and carditis. The third stage is usually characterized by chronic arthritis and may continue for years.

FIGURE 63-3 Appearance of the classic erythema migrans lesion of acute Lyme disease. This particular lesion has a central "target" of erythema not seen in all cases. (Courtesy Richard Dykoski, P.A., Laboratory Service, VA Medical Center, Minneapolis, Minn.)

LABORATORY DIAGNOSIS

Specimen collection, transport, and processing
Peripheral blood is the specimen of choice for direct detection of borreliae that cause relapsing fever (see below). *Borrelia burgdorferi* can be visualized and cultured, although serology is the best means to date to diagnose Lyme disease. Specimens submitted for stain and/or culture include blood, biopsy specimens, and cerebrospinal fluid; there are no special requirements for collection, transport, or processing of these specimens.

Direct detection methods
RELAPSING FEVER   Clinical laboratories rely on direct observation of the organism in peripheral blood from patients for diagnosis. Organisms can be found in 70% of cases when blood specimens from febrile patients are examined. The organisms can be seen directly in wet preparations of peripheral blood (mixed with equal parts of sterile, nonbacteriostatic saline) under dark-field illumination, in which the spirochetes move rapidly, often pushing the red blood cells around. They can also be seen by staining thick and thin films with Wright's or Giemsa stains using procedures similar to those used to detect malaria.

LYME DISEASE   *B. burgdorferi* may be visualized in tissue sections stained with Warthin-Starry silver stain and in blood and cerebrospinal fluid stained with acridine orange, a specific fluorescent stain, or Giemsa.[2] Polymerase chain reaction (PCR) has become important in diagnosing Lyme disease. PCR has detected *B. burgdorferi* DNA in clinical specimens from patients with early and late clinical manifestations; these specimens include urine, serum, and skin biopsies from patients with EM.[8,10,11]

Cultivation
Although the organisms that cause relapsing fever can be cultured in nutritionally rich media under microaerobic conditions, the procedures are cumbersome and unreliable and are used primarily as research tools. Similarly, the culture of *B. burgdorferi* may be attempted, although the yield is low. The periphery of the ring lesion of EM, blood, and cerebrospinal fluid provide the best specimens for culture. The resuspended plasma from blood, spinal fluid sediment, or macerated tissue biopsy is inoculated into a tube of modified Kelly's medium (BSK II), tightly capped, and incubated at 33° C for 6 weeks or longer.[2] Blind subcultures (0.1 mL) are performed weekly from the lower portion of the broth to fresh media, and the cultures are examined by dark-field microscopy for the presence of spirochetes.

Serodiagnosis
RELAPSING FEVER   Serologic tests for relapsing fever have not proved reliable for diagnosis because of the many antigenic shifts *Borrelia* organisms undergo during the course of disease. Patients may exhibit increased titers to *Proteus* OX K antigens (up to 1:80), but other cross-reacting antibodies are rare. Certain reference laboratories, such as those at the CDC, may perform special serologic procedures on sera from selected patients.

LYME DISEASE   Despite its inadequacies, serology continues to be the standard for the diagnosis of Lyme disease. Numerous serologic tests are commercially available; however, standardization of these tests is not yet accomplished and there is great variability in their performance characteristics. The most common of these are the indirect immunofluorescence assay (IFA), the enzyme-linked immunosorbent assay (ELISA), and Western blot. Measuring antibody by ELISA is the primary screening method because it is quick, reproducible, and relatively inexpensive. However, false-positive rates are high, mainly as a result of cross-reactivity. Patients with syphilis, HIV infection, leptospirosis, mononucleosis, parvovirus infection, rheumatoid arthritis and other autoimmune diseases commonly show positive results.[1,9] In 1994 the CDC and the Association of State and Public Health Laboratory Directors recommended a two-step approach to the serologic diagnosis of Lyme disease.[4] The first step is to use a sensitive screening test such as an ELISA or IFA; if this test is positive or equivocal, the result must be confirmed by immunoblotting. In certain clinical situations, results of serologic tests must be interpreted with caution. For example, patients with Lyme arthritis frequently remain antibody-positive despite treatment but do not necessarily have persistent infection. Conversely, patients with only a localized EM may be seronegative.

ANTIBIOTIC SUSCEPTIBILITY TESTING AND THERAPY

Because there are no standardized methods and borreliae are difficult to culture, antimicrobial susceptibility testing is not routinely performed.

Several antibiotics, including tetracycline, are effective in treating relapsing fever. Doxycycline, amoxicillin, or erythromycin are drugs of choice during the first stage of Lyme disease. Broad-spectrum cephalosporins, particularly ceftriaxone or cefotaxime, have been successful with patients who either fail initial treatment or present in later stages of the disease. Symptomatic treatment failures, particularly in patients with chronic Lyme disease, have been reported.

PREVENTION

There are no vaccines against infections caused by *Borrelia* spp. although single-protein recombinant vaccines are undergoing clinical trials. Infection is best prevented by avoiding tick-infested areas; wearing protective clothing; checking your clothing, body, and pets for ticks; and removing them promptly.

LEPTOSPIRA

GENERAL CHARACTERISTICS

The leptospires include both free-living and parasitic forms. Pathogenic species are called *Leptospira interrogans,* and most saprophytic leptospires are called *L. biflexa.* The pathogens include more than 180 serologically defined types that were formerly designated as species and are now known as serovars, or serotypes, of *L. interrogans.*

EPIDEMIOLOGY AND PATHOGENESIS

Leptospirosis, a zoonosis, has a worldwide distribution but is most common in developing countries and warm climates where contact with infected animals or water contaminated with their urine is likely to occur. *L. interrogans* can infect most mammals throughout the world, as well as reptiles, amphibians, fish, birds, and invertebrates.[5] Humans become infected through direct or indirect contact with the urine or blood of infected animals. Leptospires enter the human host through breaks in the skin, mucous membranes, or conjunctivae. Infection can be acquired in home and recreational settings (e.g., swimming, hunting, canoeing) or in certain occupational settings (e.g., farmers, ranchers, abattoir workers, trappers, veterinarians).[5]

Pathogenic leptospires rapidly invade the bloodstream after entry and spread throughout all sites in the body such as the central nervous system and kidneys. Precisely how *L. interrogans* causes disease is not completely understood, but it appears that the presence of endotoxin and other toxins may play a role.

SPECTRUM OF DISEASE

Symptoms begin abruptly 2 to 20 days after infection and include fever, headache, and myalgia. Two clinically recognizable syndromes are associated with leptospirosis.[5] The most common syndrome is **anicteric leptospirosis,** which is a self-limiting illness consisting of a septicemic stage, with high fever and severe headache that lasts 3 to 7 days, followed by the immune stage. Symptoms associated with the immune stage (onset coincides with the appearance of IgM) are varied, but in general are milder than the septicemic stage. The hallmark of the immune stage is aseptic meningitis. The second clinical syndrome is Weil's syndrome, or **icteric leptospirosis.** This illness is generally more severe, with symptoms caused by liver, kidney, and/or vascular dysfunction; death can occur in up to 10% of cases.

LABORATORY DIAGNOSIS

Specimen collection, transport, and processing

During the first 10 days of illness, leptospires are present in the blood and cerebrospinal fluid (CSF). Urine specimens can be obtained beginning in the second week of illness and up to 30 days after the onset of symptoms. Other than the time of collection of an appropriate specimen, there are no other special requirements for specimen collection, transport, or processing.

Direct detection

Blood, CSF, and urine may be examined directly by dark-field microscopy examination. Detection of motile leptospires in these specimens is optimized by centrifuging at 1500× *g* for 30 minutes; sodium oxalate or heparin-treated blood is initially spun at 500× *g* for 15 minutes to remove blood cells. Other techniques, such as fluorescent antibody staining, hybridization techniques employing leptospira-specific DNA probes, and PCR assays have also detected leptospires in clinical specimens, but only limited studies have been performed.

Cultivations

Currently, the most reliable method for laboratory diagnosis of leptospirosis is to culture the organisms from blood, CSF, or urine. A few drops of heparinized or sodium oxalate-anticoagulated blood are inoculated into tubes of semisolid media enriched with rabbit serum (Fletcher's or Stuart's) or bovine serum albumin. Urine should be inoculated soon after collection, because acidity (diluted out in the broth medium) may harm the spirochetes. One or two drops of undiluted urine and a 1:10 dilution of urine are added to 5 mL of medium. The addition of 200 μg/mL of 5-fluorouracil (an anticancer drug) may prevent contamination by other bacteria without harming the leptospires. Tissue specimens, especially from the liver and kidney, may be aseptically macerated and inoculated in dilutions of 1:1, 1:10, and 1:100 as for urine cultures.

All cultures are incubated at room temperature or 30°C in the dark for up to 6 weeks. Because organisms grow below the surface, material collected from a few centimeters below the surface of broth cultures should be examined weekly for the presence of growth using a direct wet preparation under dark-field illumination. Leptospires exhibit corkscrew-like motility.

Approach to identification

Based on the number of coils and hooked ends, leptospires can be distinguished from other spirochetes. Physiologically, the saprophytes can be differentiated from pathogens by their ability to grow to 10° C and lower, or at least 5° C lower than the growth temperature of pathogenic leptospires.[6]

Serodiagnosis

Serodiagnosis of leptospirosis requires a fourfold or greater rise in titer of agglutinating antibodies. The microscopic agglutination (MA) test using live cells is the standard serologic procedure used. Serologic diagnosis of leptospirosis is best performed using pools of bacterial antigens containing many serotypes in each pool. Positive results of the MA test are visualized by examining for the presence of agglutination under dark-field examination. However, a macroscopic agglutination procedure is more readily accessible to routine clinical laboratories. Reagents are available commercially. Indirect hemagglutination and an enzyme-linked immunosorbent assay (ELISA) test for IgM antibody are also available.

ANTIBIOTIC SUSCEPTIBILITY AND THERAPY

Standard procedures have not been developed for testing leptospires against various drugs. Penicillin, amoxicillin, doxycycline, and tetracycline are recommended for treatment of leptospirosis.

PREVENTION

General preventive measures include the vaccination of domestic livestock and pet dogs. In addition, protective clothing, rodent control measures, and preventing recreational exposures, such as avoiding freshwater ponds, are indicated in preventing leptospirosis.

References

1. Athreya, B.H. and Rose, C.D. 1996. Lyme disease. Curr. Probl. Pediatr. 26:189.
2. Barbour, A.G. 1988. Laboratory aspects of Lyme borreliosis. Clin. Microbiol. Rev. 1:399.
3. Centers for Disease Control and Prevention. 1994. Lyme disease—United States. Morb. Mortal. Wkly. Rep. 44:459.
4. Centers for Disease Control and Prevention. 1995. Recommendations for test performance and interpretation from the second national conference on serologic diagnosis of Lyme disease. Morb. Mortal. Wkly. Rep. 44:590.
5. Farr, R.W. 1995. Leptospirosis. Clin. Infect. Dis. 21:1.
6. Johnson, R.C. and Harris, V.G. 1967. Differentiation of pathogenic and saprophytic leptospires. I. Growth at low temperature. J. Bacteriol. 94:27.
7. Koff, A.B. and Rosen, T. 1993. Nonvenereal treponematoses: yaws, endemic syphilis, and pinta. J. Am. Acad. Dermatol. 29:519.
8. Liebling, M.R., et al. 1993. The polymerase chain reaction for the detection of *Borrelia burgdorferi* in human body fluids. Arthritis Rheum. 36:665.
9. Raoult, D., Hechemy, K.E., and Baranton, G. 1989. Cross-reaction with *Borrelia burgdorferi* antigen of sera from patients with human immunodeficiency virus infection, syphilis, and leptospirosis. J. Clin. Microbiol. 27:2152.
10. Schmidt, B., et al. Detection of *Borrelia burgdorferi*-specific DNA in urine specimens from patients with erythema migrans before and after antibiotic therapy. J. Clin. Microbiol. 34:1359.
11. Schwartz, I., et al. Diagnosis of early Lyme disease by polymerase chain reaction amplification and culture of skin biopsies from erythema migrans lesions. J. Clin. Microbiol. 30:3082.
12. Tramont, E.C. 1995. Syphilis in adults: from Christopher Columbus to Sir Alexander Fleming to AIDS. Clin. Infect. Dis. 21:1361.

Part Five

5

Parasitology

64 LABORATORY METHODS FOR DIAGNOSIS OF PARASITIC INFECTIONS

The field of parasitology is often associated with tropical areas; however, many parasitic organisms that infect humans are worldwide in distribution and occur with some frequency in the temperate zones. Many organisms endemic elsewhere are seen in the United States in persons who have lived or traveled in those areas. The influx of refugee populations into the United States has also served as a source of both infected patients and possible transmission of certain parasites. Another consideration is the increase in numbers of compromised patients, particularly those who are immunodeficient or immunosuppressed. Those persons are greatly at risk for certain parasitic infections. Consequently, clinicians and laboratory personnel should be aware of the possibility that these organisms may be present and should be trained in the ordering and performance of appropriate procedures for their recovery and identification. The clinician must be able to recognize and interpret the relevance of the laboratory data, and the clinical laboratory scientist must be able to review the pros and cons of each procedure, selecting those that will provide the most accurate diagnostic test results.

Although common names are frequently used to describe parasitic organisms, these names may represent different parasites in different parts of the world. To eliminate these problems, a binomial system of nomenclature is used in which the scientific name consists of the genus and species. Based on life histories and morphology, systems of classification have been developed to indicate the relationship among the various parasite species. With increased emphasis on organism differences at the molecular level, classification may dramatically change during the coming years.

Parasites of humans are classified into five major subdivisions: (1) the Protozoa (amebae, flagellates, ciliates, sporozoa, coccidia, microsporidians); (2) the Platyhelminthes, or flatworms, (cestodes, trematodes); (3) the Acanthocephala, or thorny-headed worms; (4) the Nematoda, or roundworms, and (5) the Arthropoda (insects, spiders, mites, ticks). The main groups presented here include Protozoa, Nematoda, Platyhelminthes, and Arthropoda. This classification scheme is designed to provide some order and meaning to a widely divergent group of organisms. No attempt has been made to include every possible organism but only those that are considered to be clinically relevant in the context of human parasitology are listed in Box 64-1. It is hoped this list provides some insight into the parasite groupings, thus leading to a better understanding of organism morphology, parasitic infections, and the appropriate clinical diagnostic approach.

The identification of parasitic organisms depends on morphologic criteria; these criteria, in turn, depend on correct specimen collection and adequate fixation. Improperly submitted specimens may result in failure to find the organisms or in their misidentification. Tables 64-1 and 64-2 contain information on the possible parasites recovered from different body sites and the most frequently used specimen collection approaches.

The information presented in this chapter should provide the reader with appropriate laboratory techniques and examples of morphologic criteria to permit the correct identification of the more common human parasites. Several excellent resource texts are also available.[5,22,48,49,52]

DIAGNOSTIC TECHNIQUES

FECAL SPECIMENS

Collection

The ability to detect and identify intestinal parasites, particularly protozoa, is directly related to the quality of the specimen submitted to the laboratory. Certain guidelines (see following discussion) are recommended to ensure proper collection and accurate examination of specimens.

Collection of fecal specimens for intestinal parasites should always be performed *before* radiologic studies involving barium sulfate. Because of the excess crystalline material in the stool specimen, the intestinal protozoa may be impossible to detect for at least 1 week after the use of barium. Certain medications may also prevent the detection of intestinal protozoa; these include mineral oil, bismuth, nonabsorbable

BOX 64-1

CLASSIFICATION OF HUMAN PARASITES, VECTORS, AND SIMILAR ORGANISMS

Protozoa
Amebae (Intestinal)
Entamoeba histolytica
*Entamoeba dispar**
Entamoeba hartmanni
Entamoeba coli
Entamoeba polecki
Endolimax nana
Iodamoeba bütschlii
Blastocystis hominis
Flagellates (Intestinal)
Giardia lamblia[†]
Chilomastix mesnili
Dientamoeba fragilis
Trichomonas hominis
Enteromonas hominis
Retortamonas intestinalis
Ciliates (Intestinal)
Balantidium coli
Coccidia, Microsporidia (Intestinal)
Cryptosporidium parvum
Cyclospora cayetanensis
Isospora belli
Sarcocystis hominis
Sarcocystis suihominis
Sarcocystis "lindemanni"
Microsporidia
   Enterocytozoon bieneusi
   Encephalitozoon intestinalis
Sporozoa, Flagellates (Blood, Tissue)
Sporozoa (Malaria and babesiosis)
Plasmodium vivax
Plasmodium ovale
Plasmodium malariae
Plasmodium falciparum
Babesia spp.
Flagellates (Leishmaniae, Trypanosomes)
Leishmania tropica complex
Leishmania mexicana complex
Leishmania braziliensis complex
Leishmania donovani complex
Leishmania peruviana
Trypanosoma brucei gambiense
Trypanosoma brucei rhodesiense
Trypanosoma cruzi
Trypanosoma rangeli
Amebae, Flagellates (Other Body Sites)
Amebae
Naegleria fowleri
Acanthamoeba spp.
Hartmanella spp.
Balamuthia mandrillaris (Leptomyxid ameba)
Entamoeba gingivalis
Flagellates
Trichomonas vaginalis
Trichomonas tenax
Coccidia, Sporozoa, Microsporidia (Other Body Sites)
Coccidia
Toxoplasma gondii

Sporozoa
Pneumocystis carinii[‡]
Microsporidia
Nosema
Pleistophora
Encephalitozoon
"Microsporidium"
Nematodes (Roundworms)
Intestinal
Ascaris lumbricoides
Enterobius vermicularis
Ancylostoma duodenale
Necator americanus
Strongyloides stercoralis
Trichostrongylus spp.
Trichuris trichiura
Capillaria philippinensis
Tissue
Trichinella spiralis
Visceral larva migrans (*Toxocara canis* or *Toxocara cati*)
Ocular larva migrans (*Toxocara canis* or *Toxocara cati*)
Cutaneous larva migrans (*Ancylostoma braziliense* or *Ancylostoma caninum*)
Dracunculus medinensis
Angiostrongylus cantonensis
Angiostrongylus costaricensis
Gnathostoma spinigerum
Anisakis spp. (larvae from saltwater fish)
Phocanema spp. (larvae from saltwater fish)
Contracaecum spp. (larvae from saltwater fish)
Capillaria hepatica
Thelazia spp.
Blood and Tissues (Filarial Worms)
Wuchereria bancrofti
Brugia malayi
Brugia timori
Loa loa
Onchocerca volvulus
Mansonella ozzardi
Mansonella streptocerca
Mansonella perstans
Dirofilaria immitis (usually lung lesion; in dogs, heartworm)
Dirofilaria spp. (may be found in subcutaneous nodules)
Cestodes (Tapeworms)
Intestinal
Diphyllobothrium latum
Dipylidium caninum
Hymenolepis nana
Hymenolepis diminuta
Taenia solium
Taenia saginata
Tissue (Larval Forms)
Taenia solium
Echinococcus granulosus
Echinococcus multilocularis
Multiceps multiceps
Spirometra mansonoides
Diphyllobothrium spp.

Continued

CLASSIFICATION OF HUMAN PARASITES, VECTORS, AND SIMILAR ORGANISMS—cont'd

Trematodes (Flukes)

Intestinal
Fasciolopsis buski
Echinostoma ilocanum
Heterophyes heterophyes
Metagonimus yokogawai

Liver/Lung
Clonorchis (Opisthorchis) sinensis
Opisthorchis viverrini
Fasciola hepatica
Paragonimus westermani
Paragonimus mexicanus

Blood
Schistosoma mansoni
Schistosoma haematobium
Schistosoma japonicum
Schistosoma intercalatum
Schistosoma mekongi

Arthropods

Arachnida
Scorpions
Spiders (black widow, brown recluse)
Ticks *(Dermacentor, Ixodes, Argas, Ornithodoros)*
Mites *(Sarcoptes)*

Crustacea
Copepods *(Cyclops)*

Crayfish, lobsters, crabs

Pentastomida	**Diplopoda**	**Chilopoda**
Tongue worms	Millipedes	Centipedes

Arthropods—cont'd

Insecta
Anoplura: sucking lice *(Pediculus, Phthirus)*
Dictyoptera: cockroaches
Hemiptera: true bugs *(Triatoma)*
Coleoptera: beetles
Hymenoptera: bees, wasps, and so on
Lepidoptera: butterflies, caterpillars, moths, and so on
Diptera: flies, mosquitoes, gnats, midges (e.g., *Phlebotomus, Aedes, Anopheles, Glossina, Simulium*)
Siphonaptera: fleas (e.g., *Pulex, Xenopsylla*)

*Entamoeba histolytica is being used to designate pathogenic zymodemes, while *E. dispar* is now being used to designate nonpathogenic zymodemes. However, unless trophozoites containing ingested red blood cells (*E. histolytica*) are seen, the two organisms cannot be differentiated based on morphology.
†Although some individuals have changed the species designation for the genus *Giardia* to *G. intestinalis* or *G. duodenalis*, there is no general agreement. Therefore, for this listing, we will retain the name *Giardia lamblia*.
‡*Pneumocystis carinii* has now been reclassified with the fungi.
From Garcia, L.S. and Bruckner, D.A. 1993. Diagnostic medical parasitology, ed 2. American Society for Microbiology, Washington, D.C.

---

antidiarrheal preparations, antimalarials, and some antibiotics (e.g., tetracyclines). The organisms may be difficult to detect for several weeks after the medication is discontinued. Fecal specimens should be collected in clean, wide-mouthed containers; most laboratories use a waxed, cardboard half-pint container with a tight-fitting lid. The specimen should not be contaminated with water that may contain free-living organisms. Contamination with urine should also be avoided to prevent destruction of motile organisms in the specimen. All specimens should be identified with the patient's name, physician's name, hospital number if applicable, and the time and date collected. Every fecal specimen represents a potential source of infectious material (e.g., bacteria, viruses, parasites) and should be handled accordingly.

The number of specimens required to demonstrate intestinal parasites will vary depending on the quality of the specimen submitted, the accuracy of the examination performed, and the severity of the infection. The following recommendation has been used for many years.[54] For a routine examination for parasites before treatment, a minimum of three fecal specimens is recommended: two specimens collected from normal movements and one specimen collected after a cathartic, such as magnesium sulfate or Fleet Phospho-Soda. A cathartic with an oil base should not be used, and all laxatives are contraindicated if the patient has diarrhea or significant abdominal pain. Stool softeners are inadequate for producing a purged specimen. The examination of at least six specimens ensures detection of 90% of infections; six may be recommended when amebiasis is suspected.

Although some recommend collection of only one or two specimens, opinions differ regarding this approach. It has also been suggested that three specimens be pooled and examined as a single specimen; again, this approach is somewhat controversial. It is important for any laboratory to thoroughly understand the pros and cons of each specimen collection option.[33]

Many organisms do not appear in fecal specimens daily in consistent numbers; thus collection of specimens on *alternate days* tends to yield a higher percentage of positive findings. The series of three specimens should be collected within no more than 10 days and a series of six within no more than 14 days.

The number of specimens to be examined after therapy will vary depending on the diagnosis; however, a series of three specimens collected as previously

TABLE 64-1   PARASITES RECOVERED FROM VARIOUS BODY SITES*

SITE	PARASITES	SITE	PARASITES
Blood		**Liver, Spleen**	*Echinococcus* spp.
			Entamoeba histolytica
Red cells	*Plasmodium* spp.		*Leishmania donovani*
	Babesia spp.		*Opisthorchis sinensis*
			Fasciola hepatica
White cells	*Leishmania donovani*		
	Toxoplasma gondii	**Lung**	*Pneumocystis carinii*
			(can disseminate)
Whole blood,	*Trypanosoma* spp.		*Echinococcus* spp.
Plasma	Microfilariae		*Paragonimus westermani*
			Cryptosporidium parvum
Bone marrow	*Leishmania donovani*		*Ascaris lumbricoides* larvae
			Hookworm larvae
CNS	*Taenia solium* (Cysticercosis)		
	Echinococcus spp.	**Muscle**	*Taenia solium* (Cysticerci)
	Naegleria fowleri		*Trichinella spiralis*
	Acanthamoeba spp.		*Onchocerca volvulus* (Nodules)
	Hartmanella spp.		*Trypanosoma cruzi*
	Balamuthia mandrillaris		Microsporidia
	Toxoplasma gondii		
	Trypanosoma spp.	**Skin**	*Leishmania* spp.
			Onchocerca volvulus
Cutaneous Ulcers	*Leishmania* spp.		Microfilariae
Eye	*Acanthamoeba* spp.	**Urogenital**	*Trichomonas vaginalis*
	Naegleria spp.	**System**	*Schistosoma* spp.
	Taenia solium (Cysticerci)		Microsporidia
	Loa loa		
	Microsporidia		
Intestinal Tract	*Entamoeba histolytica*	**Intestinal**	*Enterobius vermicularis*
	Entamoeba coli	**Tract**	Hookworm
	Entamoeba hartmanni	**(continued)**	*Strongyloides stercoralis*
	Endolimax nana		*Trichuris trichiura*
	Iodamoeba bütschlii		*Hymenolepis nana*
	Blastocystis hominis		*Hymenolepis diminuta*
	Giardia lamblia		*Taenia saginata*
	Chilomastix mesnili		*Taenia solium*
	Dientamoeba fragilis		*Diphyllobothrium latum*
	Trichomonas hominis		*Opisthorchis sinensis* (Clonorchis)
	Balantidium coli		*Paragonimus westermani*
	Cryptosporidium parvum		*Schistosoma* spp.
	Isospora belli		*Heterophyes* sp.
	Microsporidia		*Metagonimus* sp.
	Ascaris lumbricoides		

*This table does not include every possible parasite that could be found in a particular body site; only the most likely organisms have been listed.

Modified from Garcia, L.S. and Bruckner, D.A. 1997. Diagnostic medical parasitology, ed 3. American Society for Microbiology, Washington, D.C.

outlined is usually recommended. A patient who has received treatment for a protozoan infection should be checked 3 to 4 weeks after therapy. Patients treated for helminth infections may be checked 1 to 2 weeks after therapy and those treated for *Taenia* infections 5 to 6 weeks after therapy. Because the age of the speci-

men directly influences the recovery of protozoan organisms, the *time the specimen was collected* should be recorded on the laboratory request form. Freshly passed specimens are mandatory to detect trophic amebae or flagellates. *Liquid specimens should be examined within 30 minutes of passage* (not 30 minutes

————— **TABLE 64-2** BODY SITES AND SPECIMEN COLLECTION —————

SITE	SPECIMEN OPTIONS	COLLECTION METHOD
Blood	Smears of whole blood	Thick and thin films*
		Fresh (first choice)
	Anticoagulated blood	Anticoagulant (second choice)
		EDTA (first choice)
		Heparin (second choice)
Bone marrow	Aspirate	Sterile
CNS	Spinal fluid	Sterile
Cutaneous ulcers	Aspirates from below surface or punch biopsy (best)	Sterile plus air-dried smears
	Biopsy	Sterile, nonsterile histopathology
Eye	Biopsy	Sterile (in saline)
	Scrapings	Sterile (in saline)
	Contact lens	Sterile (in saline)
	Lens solution	Sterile
Intestinal tract	Fresh stool	½ pint waxed container
	Preserved stool (best)	5% or 10% formalin, MIF, SAF, Schaudinn's, PVA, other single-vial collection systems†
	Sigmoidoscopy material	Fresh, PVA, or Schaudinn's smears
	Duodenal contents	Entero-Test or aspirates
	Anal impression smear	Cellulose tape (pinworm examination)
	Adult worm/worm segments	Saline, 70% alcohol
Liver, spleen	Aspirates	Sterile, collected in four separate aliquots (liver)
	Biopsy	Sterile, nonsterile to histopathology
Lung	Sputum	True sputum (not saliva)
	Induced sputum	No preservative (5% or 10% formalin if time delay)

from the time they reach the laboratory), or the specimen should be placed in polyvinyl alcohol (PVA) fixative or another suitable preservative (see following section on preservation of specimens). *Semiformed or soft specimens should be examined within 1 hour of passage;* if this is not possible, the stool material should be preserved. Although the time limits are not as critical for the examination of a formed specimen, it is recommended that the material be examined on the day of passage. If these time limits cannot be met, portions of the sample should be preserved. Stool specimens should not be held at room temperature but should be refrigerated at 3° to 5° C and stored in closed containers to prevent desiccation. At this temperature, eggs, larvae, and protozoan cysts remain viable for several days. Fecal specimens should never be incubated or frozen before examination. When the proper criteria for collection of fecal specimens are not met, the laboratory should request additional samples.

COLLECTION KIT FOR OUTPATIENT USE Various commercial collection systems are available, and one can select which vials are appropriate for a particular institution. Most manufacturers will package the kit according to individual specifications and will include complete multilanguage instructions (see following section on PVA fixative for representative suppliers). PVA solution is a combination of Schaudinn's fixative plus plastic PVA resin, which serves as an adhesive used to glue the stool onto the slide. Because PVA may contain the mercuric chloride base in the Schaudinn's fixative, most manufacturers have available collection vials with child-proof caps for institutions that select that option. All commercially available vials have appropriate labels related to toxicity.

A PVA-preserved portion of the specimen may be used for the complete examination, although many laboratories prefer to use the sample in 5% or 10% buffered formalin for the concentration procedure.

TABLE 64-2   BODY SITES AND SPECIMEN COLLECTION—CONT'D

SITE	SPECIMEN OPTIONS	COLLECTION METHOD
	Bronchoalveolar lavage (BAL)	Sterile
	Transbronchial aspirate	Air-dried smears
	Tracheobronchial aspirate	Same as above
	Brush biopsy	Same as above
	Open lung biopsy	Same as above
	Aspirate	Sterile
Muscle	Biopsy	Fresh-squash preparation (Nonsterile to histopathology)
Skin	Scrapings	Aseptic, smear or vial
	Skin snip	No preservative
	Biopsy	Sterile (in saline)
		Nonsterile to histopathology
Urogenital system	Vaginal discharge	Saline swab, culturette (no charcoal), culture medium‡ air-dried smear for FA
	Urethral discharge	Same as above
	Prostatic secretions	Same as above
	Urine	Unpreserved spot specimen or 4-hr unpreserved specimen
		Mid-day urine

*Although this approach is recommended, most laboratories will receive purple top (EDTA) vacutainer tubes. It is important to make sure the tubes are filled completely (correct ratio of preservative and blood). Blood films should be prepared from this specimen as soon as possible (within an hour, if possible); prolonged standing of the blood will interfere with organism morphology (parasites distorted, stippling in *Plasmodium vivax* and *P. ovale* may not be visible).

†Newer single-vial collection systems are available. These often contain zinc sulfate; however, the formulas are proprietary. Both the concentration and permanent-stained smears can be prepared from these single vials. It is not yet known whether immunoassay procedures can also be performed from the same vial. (Apparently, some preservatives are acceptable for immunoassay and some interfere with these procedures. This information can be obtained from the individual manufacturers.)

‡Culture systems are available for *Trichomonas vaginalis*; these systems can also be used for transport. Once the specimen is collected, examination of the culture system over time does not require reentry into the pouch containing the media.

Modified from Garcia, L.S. and Bruckner, D.A. 1997. Diagnostic medical parasitology, ed 3. American Society for Microbiology, Washington, D.C.

Other laboratories may prefer to use Schaudinn's fixative for collection. The standard two-vial collection kit contains one vial of 5% or 10% buffered formalin and one vial of PVA. The benefit of the specimen vial collection system is a reduction in the lag time between when the specimen is passed and when it is fixed, thus providing better organism morphology used for identification. The unpreserved portion of the specimen (many laboratories do not request this sample unless an occult blood procedure is requested) may be examined to determine the specimen type (e.g., liquid, soft, or formed).

Preservation
Depending on specimen-to-laboratory transportation time, the laboratory workload, and the availability of trained personnel, it may often be impossible to examine the specimen within specified time limits. To maintain protozoan morphology and prevent further development of certain helminth eggs and larvae, the fecal specimen should be placed in an appropriate preservative for examination at a later time. Several preservatives are available; four of these methods— PVA (mercury- and nonmercury-based), formalin, Merthiolate (thimerosal)-iodine-formalin (MIF), and sodium acetate-acetic acid-formalin (SAF)—are discussed. When selecting an appropriate fixative, it is important to realize the limitations of each. PVA and SAF are the most commonly used fixatives from which a permanent stained smear can be easily prepared. The stained smear is extremely important in providing a complete and accurate examination for intestinal protozoa. Examination of the specimen as a wet mount only is much less accurate than the stained smear for the identification of protozoa.

With the introduction of regulatory requirements related to waste disposal, great interest has arisen in developing preservatives without the use of mercury compounds. However, studies indicate that substitute compounds generally do not provide the same quality of preservation necessary for good protozoan morphology on the trichrome or iron hematoxylin permanent-stained smear. Copper sulfate has been tried as a substitute in Schaudinn's fixative, but it does not provide results equal to those seen with mercuric chloride. The use of a zinc base compound as a substitute has been evaluated and provides better results than organism fixation seen with copper sulfate.[25] If you must eliminate mercury compounds from the laboratory, use SAF coupled with iron hematoxylin stain or PVA (nonmercury-based) and trichrome stain, although none of the substitutes provides the overall consistent quality seen with mercuric chloride.

Some of the new fixative options are now packaged as single-vial systems; they contain no mercury or formalin and both the concentration and permanent-stained smear can be performed from the fecal specimen preserved in the single vial. *Ensure that laboratory personnel realize the differences in organism morphology that will be seen with nonmercury-based fixatives; less precise morphologic detail does not indicate the laboratory has performed the procedure incorrectly!*

PVA FIXATIVE   PVA fixative solution is highly recommended as a means of preserving protozoan cysts and trophozoites for examination at a later time.[22,54] The use of PVA also permits specimens to be shipped by regular mail service to a laboratory for subsequent examination.[51] PVA, which is a combination of modified Schaudinn's fixative and a water-soluble resin, should be used in the ratio of three parts PVA to one part fecal material. Perhaps the greatest advantage in the use of PVA is that permanent-stained slides can be prepared from PVA-preserved material.[54] This is not the case with many other preservatives; some permit the specimen to be examined as a wet preparation only, a technique that may not be adequate for the correct identification of protozoan organisms. PVA vials or collection kits are generally purchased commercially.[22] This fixative remains stable for long periods (months to years) when kept in *sealed containers* at room temperature. Commercial packaging and distribution of these vials have essentially eliminated the problem of moisture loss and shelf life.

Prepared liquid PVA (ready for use) can be obtained from a number of companies, including: ALPHA-TEC Systems, Bio-Spec, Evergreen Scientific, Hardy Diagnostics, Medical Chemical Corporation, Meridian Diagnostics, MML Diagnostic Packaging, PML Microbiologicals, and Remel. Many of these companies also provide nonmercury-based fixatives and single-vial collection systems. These fixatives generally contain zinc sulfate and other compounds; the specific formulas are proprietary. Certainly not having to handle mercury disposal problems makes some of these alternatives very popular. However, overall organism morphology will not be equal to that seen with organisms preserved in fixatives with a mercuric chloride base.[25] Most laboratories will purchase collection vial kits ready for distribution to patients.

FORMALIN   Protozoan cysts, helminth eggs, and larvae are well preserved for long periods in *5% or 10% buffered formalin* (100 mL formaldehyde in 900 mL 0.85% NaCl solution). Although it is impractical for most laboratories, hot formalin (60° C) can be used to prevent further development of helminth eggs to the infective stage; this is important when bulk specimens are being saved for teaching purposes, particularly when using 5% formalin. Formalin should be used in the ratio of at least three parts formalin to one part fecal material; thorough mixing of the fresh specimen and fixative is necessary to ensure good preservation.

MIF SOLUTION   MIF solution can be used as a stain preservative for most types and stages of intestinal parasites and may be helpful in field surveys.[22,54] Helminth eggs, larvae, and certain protozoa can be identified without further staining in wet mounts, which can be prepared immediately after fixation or several weeks later. This type of wet preparation may not be adequate for the diagnosis of all intestinal protozoa. Although some workers prepare a permanent stained smear from MIF-preserved material, the majority of laboratories prefer to use PVA, SAF, or one of the single vial systems.

SAF FIXATIVE   SAF fixative contains formalin combined with sodium acetate, which acts as a buffer. It is a liquid fixative very similar to 10% aqueous formalin. When the sediment is used to prepare permanent-stained smears, one may have some difficulty in getting material to adhere to the slide. Mayer's albumin has been recommended as an adhesive.[22,54,57] This fixative generally tends to give better results when used with iron hematoxylin rather than trichrome stain.

Collection and testing options

The various options for the laboratory regarding collection and processing of fecal specimens for parasitic examination are summarized in Table 64-3.

Macroscopic examination

The consistency of the stool (formed, semiformed, soft, or liquid) may give an indication of the protozoan stages present. When the moisture content of the fecal material is decreased during *normal passage* through the intestinal tract, the trophozoite stages of

TABLE 64-3   FECAL SPECIMENS FOR PARASITES: COLLECTION AND PROCESSING OPTIONS*

OPTION	PROS	CONS
Rejection of stools from inpatients who have been in-house for >3 days	Data suggest that patients who begin to have diarrhea after they have been inpatients for a few days are not symptomatic from parasitic infections but generally other causes	There is always the chance that the problem is related to a nosocomial parasitic infection (rare), but *Cryptosporidium* and microsporidia may be possible options
Examination of a single stool	Some reports indicate that 90% of organisms can be recovered in a single stool	Depends on experience of microscopist, proper collection, and parasite load in specimen
Examination of a second stool only after the first is negative and the patient is still symptomatic	With additional examinations, yield of protozoa increased (*Entamoeba histolytica,* 22.7%; *Giardia lamblia,* 11.3%; and *Dientamoeba fragilis,* 31.1%)	Assumes the second (or third) stool is collected within the recommended 10-day time frame for a series of stools; protozoa are shed periodically. May be inconvenient for patient
Examination of a single stool and a screen (EIA, FA)	If the examinations are negative and the patient's symptoms subside, then no further testing is required	Patients may exhibit symptoms (off and on), so it may be difficult to rule out parasitic infections with only a single stool and screen
Pool three specimens for examination; perform one concentrate and one permanent stain	Three are collected over 7-10 days and may save time and expense	Some organisms may be missed; however, this may be a viable option
Pool three specimens for examination; perform a single concentrate and three permanent-stained smears.	Three are collected over 7-10 days; would maximize recovery of protozoa in areas of the country where these organisms are most common	Might miss light helminth infection (eggs, larvae) due to "numbers game"; would probably be the best option in lieu of the regular approach (routine ova and parasite examination)
Collect three stools but put sample of stool from all three into a single vial (patient given a single vial only)	Pooling of the specimens would require only a single vial	Chances very good that the vial would contain too much stool and ratio of stool to preservative would be inappropriate
Screen selected patients with FA or EIA methods for *Giardia lamblia* and/or *Cryptosporidium parvum* (Screening every stool is not cost effective and positive rate will be low unless outbreak situation)	More cost effective than screening all specimens	Laboratories rarely receive information that allows placing a patient in a particular risk group: children ≤5 yrs old, children from day care centers (may or may not be symptomatic), patients with immunodeficiencies, and patients from outbreaks
Perform FA or EIA screening on request for *Giardia lamblia* and/or *Cryptosporidium parvum*	Limits the number of stools screened for parasites using this approach	Will require education of the physician clients regarding appropriate times and patients for which screens should be ordered

It is difficult to know when you may be in an early outbreak situation where screening of all specimens for either *Giardia lamblia, Cryptosporidium parvum,* or both, may be relevant. Extensive efforts are under way to finalize guidelines for handling potential or actual outbreaks and identification and use of appropriate communications channels. Various publications from the CDC and the American Water Works Association will probably appear in 1997 or 1998.

*See references 22, 33, 36, 53.

the protozoa encyst to survive. Trophozoites (motile forms) of the intestinal protozoa are usually found in soft or liquid specimens and occasionally in a semiformed specimen; the cyst stages are normally found in formed or semiformed specimens, rarely in liquid stools. Helminth eggs or larvae may be found in any type of specimen, although the chances of finding any parasitic organism in a liquid specimen are reduced because of the dilution factor.

Occasionally, adult helminths, such as *Ascaris lumbricoides* or *Enterobius vermicularis* (pinworm), may be seen in or on the surface of the stool. Tapeworm proglottids may also be seen on the surface, or they may actually crawl under the specimen and be found on the bottom of the container. Other adult helminths, such as *Trichuris trichiura* (whipworm), hookworms, or perhaps *Hymenolepis nana* (dwarf tapeworm), may be found in the stool, but usually this occurs only after medication. The presence of blood in the specimen may indicate several factors and should always be reported. Dark stools may indicate bleeding high in the gastrointestinal tract, whereas fresh (bright-red) blood most often is the result of bleeding at a lower level. In certain parasitic infections, blood and mucus may be present; a soft or liquid stool may be highly suggestive of an amebic infection. These areas of blood and mucus should be carefully examined for the presence of trophic amebae. Occult blood in the stool may or may not be related to a parasitic infection and can result from several different conditions. Ingestion of various compounds may give a distinctive color to the stool (iron, black; barium, light tan to white).

Microscopic examination

The identification of intestinal protozoa and helminth eggs is based on recognition of specific morphologic characteristics. These studies require a good binocular microscope, good light source, and a calibrated ocular micrometer.

The microscope should have 5× and 10× oculars (wide-field oculars are often recommended) and three objectives: low power (10×), high dry (40×), and oil immersion (100×). Some laboratories are using a 50× or 60× oil immersion lens in combination with the standard 100× oil immersion lens to screen all permanent-stained fecal and blood smears. This approach prevents the contamination of the high-dry lens (40×) with oil when switching back and forth between high-dry (400×) and oil immersion (1000×) magnifications.

CALIBRATION OF MICROSCOPE Parasite identification depends on several parameters, including size; any laboratory doing diagnostic work in parasitology should have a calibrated microscope available for precise measurements. Measurements are made by means of a micrometer disk placed in the ocular of the microscope; the disk is usually calibrated as a line divided into 50 units. Because the divisions in the disk represent different measurements, depending on the objective magnification used, the ocular disk divisions must be compared with a known calibrated scale, usually a stage micrometer with a scale of 0.1- and 0.01-mm divisions. Specific directions may be found in the work of Garcia and Bruckner.[22] Although not everyone recommends recalibration each year, if your microscopes are moved periodically, accidentally bumped, or do not receive adequate maintenance, it is a good idea to recalibrate yearly.

NOTE: After each objective power has been calibrated on the microscope, *the oculars containing the disk or these objectives cannot be interchanged with corresponding objectives or oculars on another microscope.* Each microscope that will be used to measure organisms must be calibrated as a unit; *the original oculars and objectives that were used to calibrate the microscope must also be used when an organism is measured.*

Diagnostic procedures

A combination of techniques yields a greater number of positive specimens than does any one technique alone. Procedures recommended for a complete ova and parasite examination are discussed in this section.

DIRECT SMEARS Normal mixing in the intestinal tract usually ensures even distribution of helminth eggs or larvae and protozoa. However, examination of the fecal material as a direct smear may or may not reveal organisms, depending on the parasite density. The direct smear is prepared by mixing approximately 2 mg of fecal material with a drop of physiologic saline; this mixture provides a uniform suspension under a 22 × 22 mm coverslip. Some workers prefer a 1½ × 3-inch (4 × 7 cm) glass slide for the wet preparations, rather than the standard 1 × 3-inch (2.5 × 7 cm) slide most often used for the permanent-stained smear. A 2-mg sample of fecal material forms a low cone on the end of a wooden applicator stick. If more material is used for the direct mount, the suspension is usually too thick for an accurate examination; less than 2 mg results in the examination of too thin a suspension, thus decreasing the chances of finding any organisms. If present, blood and mucus should always be examined as a direct mount. The entire 22 × 22 mm coverslip should be systematically examined using the low-power objective (10×) and low light intensity; any suspect objects may then be examined under high-dry power (40×). The use of the oil immersion objective on mounts of this type is not recommended unless the coverslip (No. 1 thickness coverslip is recommended when the oil immersion objective is used) is sealed to the slide with a cotton-tipped applicator stick dipped in equal parts of heated paraffin and petroleum jelly. Most workers believe the use of oil immersion on this type of preparation is impractical, especially since morphologic detail is most easily seen and the diagnosis confirmed with oil immersion examination of the permanent-stained smear.

The direct wet mount is used primarily to detect motile trophozoite stages of the protozoa. These organisms are very pale and transparent, two characteristics that require the use of low light intensity. Protozoan organisms in a saline preparation usually appear as refractile objects. If suspect objects are seen on high-dry power, one should allow at least 15 seconds to detect motility of slow-moving protozoa. Heat applied by placing a hot penny on the edge of a slide may enhance the motility of trophic protozoa. Helminth eggs or larvae and protozoan cysts may also be seen on the wet film, although these forms are more often detected after fecal concentration procedures.

NOTE: *With few exceptions, protozoan organisms should not be identified based on a wet mount alone. Permanent-stained smears should be examined to confirm the specific identification of suspected organisms.*

After the wet preparation has been thoroughly checked for trophic amebae, a drop of iodine may be placed at the edge of the coverslip, or a new wet mount can be prepared with iodine alone. A weak iodine solution is recommended; too strong a solution may obscure the organisms. Several types of iodine solutions are available: Dobell and O'Connor's, Lugol's, and D'Antoni's.[22,23,54] *Gram's iodine used in bacteriologic work is not recommended for staining parasitic organisms.*

Many laboratories believe that the benefits gained in organism recovery and identification from receipt of preserved stool specimens far outweigh the disadvantages of not receiving a fresh specimen that can be examined for motile organisms. This is particularly true with the intestinal protozoa. No need exists to perform the direct wet smear if the specimen has been submitted to the laboratory in preservatives; any organisms present would be killed, and thus no motility would be seen. *For this reason, many laboratories that receive fecal specimens in fixative have eliminated the direct wet smear from their protocol; the fecal examination contains the concentration and permanent-stained smear. The direct wet smear would be performed on fresh specimens only.*

Protozoan cysts correctly stained with iodine contain yellow-gold cytoplasm, brown glycogen material, and paler refractile nuclei. The chromatoidal bodies may not be as clearly visible as they were in the saline mount.

Several staining solutions are available that may be used to reveal nuclear detail in the trophozoite stages. Nair's buffered methylene blue stain is effective in showing nuclear detail when used at a low pH.[54] Methylene blue (0.06%) in an acetate buffer at pH 3.6 usually gives satisfactory results; the mount should be examined within 30 minutes.[54]

CONCENTRATION PROCEDURES   Often a direct mount of fecal material fails to reveal the presence of parasitic organisms in the gastrointestinal tract. Fecal concentration procedures should be included for a complete examination for parasites; these procedures allow the detection of small numbers of organisms that may be missed using only the direct mount. Helminth eggs can usually be identified when recovered from a concentration procedure. Generally, the identification of intestinal protozoa should be considered tentative and confirmed with the permanent-stained smear. Those protozoa that might be routinely identified from a concentration procedure include cysts of *Giardia lamblia*, *Entamoeba histolytica*, *Entamoeba coli*, and *Iodamoeba bütschlii* (trophozoites are rarely identified from a concentrate). Commercially available concentrating devices may assist a laboratory in standardizing the methodology. The same companies that provide parasitology collection vial kits could be contacted for availability of concentration systems.

Various concentration procedures are available; they are either *flotation* or *sedimentation techniques* designed to separate the parasitic components from excess fecal debris through differences in specific gravity. A flotation procedure (modified zinc sulfate flotation) permits the separation of protozoan cysts and certain helminth eggs through the use of a liquid with a high specific gravity. The parasitic elements are recovered in the surface film, whereas the debris is found in the bottom of the tube. This technique yields a cleaner preparation than does the sedimentation procedure; however, some helminth eggs (operculated eggs or very dense eggs, such as unfertilized *Ascaris* eggs) and some protozoa do not concentrate well with the flotation method. The specific gravity may be increased, although this may produce more distortion in the eggs and protozoa. Any laboratory that uses only a flotation procedure may fail to recover all the parasites present; to ensure detection of all organisms in the sample, both the surface film and the sediment should be carefully examined. Of importance, directions for any flotation technique must be followed exactly to produce reliable results.

Sedimentation procedures (using gravity or centrifugation) allow the recovery of all protozoa, eggs, and larvae present; however, the sediment preparation contains more fecal debris. If a single technique is selected for routine use, the *sedimentation procedure* (Procedure 64-1) is recommended as the easiest to perform and least subject to technical error.

PERMANENT STAINED SMEARS The detection and correct identification of intestinal protozoa frequently depend on the examination of the permanent-stained smear. These slides not only provide the microscopist with a permanent record of the protozoan organisms identified but also may be used for consultations with specialists when unusual morphologic characteristics are found. In view of the number of morphologic variations possible, organisms may be found that are very difficult to identify and do not fit the pattern for any one species.

Most identifications should be considered tentative until confirmed by the permanent-stained slide. For these reasons, the permanent stain is recommended for every stool sample submitted for a routine examination for parasites.

Many staining techniques are available; those included here generally tend to give the best and most reliable results with both fresh and preserved specimens.

Preparation of Fresh Material. When the specimen arrives, use an applicator stick or brush to smear a small amount of stool on two clean slides and immediately immerse them in Schaudinn's fixative (mercury- or nonmercury-based). If the slides are prepared correctly, you should be able to read newsprint through the fecal smear. The smears should fix for a minimum of 30 minutes; fixation time may be decreased to 5 minutes if the Schaudinn's solution is heated to 60° C. However, this approach is not that practical for most clinical laboratories and is rarely used.

If a liquid specimen is received, mix 3 or 4 drops of PVA with 1 or 2 drops of fecal material directly on a slide, spread the mixture, and allow the slides to dry for several hours at 35° C or overnight at room temperature.[22]

Preparation of PVA-Preserved Material. Stool specimens preserved in PVA should be allowed to fix at least 30 minutes. *After fixation, the sample should be thoroughly mixed and a small amount of the material poured onto a paper towel to absorb excess PVA.* This is an important step in the procedure; allow the PVA to soak into the paper towel for 2 to 3 minutes before preparing the slides. With an applicator stick, apply some of the stool material from the paper towel to two slides and let them dry for several hours at 37° C or overnight at room temperature. The PVA-stool mixture should be spread to the edges of the glass slide; this causes the film to adhere to the slide during staining. It is also important to dry the slides thor-

oughly to prevent the material from washing off during staining.

Trichrome Stain. The trichrome stain was originally developed for tissue differentiation and was adapted for intestinal protozoa (Procedure 64-2).[22,54] It is an uncomplicated procedure that produces well-stained smears from both fresh and PVA-preserved material.

The trichrome stain can be used repeatedly, and stock solution may be added to the jar when the volume is decreased. Periodically, the staining strength can be restored by removing the lid and allowing the 70% alcohol carried over from the preceding jar to evaporate. Each lot number or batch of stain (either purchased commercially or prepared in the laboratory) should be checked to determine the optimal staining time, which is usually a few minutes longer for PVA-preserved material.

The 90% acidified alcohol is used as a destaining agent that will provide good differentiation; however, prolonged destaining (more than 3 seconds) may result in a poor stain. To prevent continued destaining, the slides should be quickly rinsed in 100% alcohol and then dehydrated through two additional changes of 100% alcohol. For any alcohol solution in the staining protocol other than absolute alcohol, isopropyl alcohol can be used.

Interpretation of Stained Smears. Many problems in interpretation may arise when poorly stained smears are examined; these smears are usually the result of inadequate fixation or incorrect specimen collection and submission. An old specimen or inadequate fixation may result in organisms that fail to stain or that appear as pale-pink or red objects with very little internal definition. This type of staining reaction may occur with *Entamoeba coli* cysts, which require a longer fixation time; mature cysts in general need additional fixation time and therefore may not be as well stained as immature cysts. Degenerate forms or those that have been understained or destained too much may stain pale green.

When the smear is well fixed and correctly stained, the background debris is green, and the protozoa have a blue-green to purple cytoplasm with red or purple-red nuclei and inclusions. The differences in colors between the background and organisms provide more contrast than in hematoxylin-stained smears.

Helminth eggs and larvae usually stain dark red or purple; they are usually distorted and difficult to identify. White blood cells, macrophages, tissue cells, yeast cells, and other artifacts still present diagnostic problems, since their color range on the stained smear approximates that of the parasitic organisms (Figures 64-1 to 64-3).

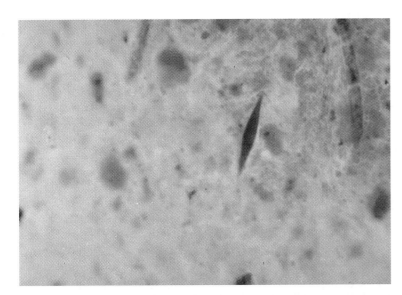

FIGURE 64-1 Charcot-Leyden crystals.

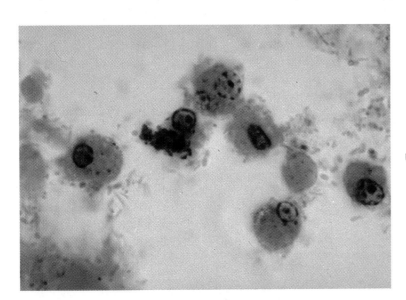

FIGURE 64-2 Polymorphonuclear leukocytes.

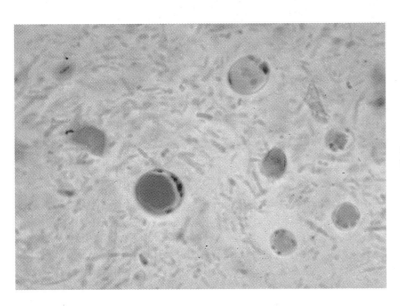

FIGURE 64-3 *Blastocystis hominis* (larger objects) and yeast cells (smaller, more homogeneous objects).

Iron Hematoxylin Stain. Although the original method produces excellent results, most laboratories that use an iron hematoxylin stain select one of the shorter methods. Several procedures are available; both those presented here can be used with either fresh or PVA-preserved (mercury- or nonmercury-based) material. This iron hematoxylin stain is being used with SAF-fixed stool specimens as an alternative to using PVA-preserved (mercury-based) specimens (Procedure 64-3). Any fecal specimen submitted in SAF fixative can be used. Fresh fecal specimens after fixation in SAF for 30 minutes can also be used. Both background debris and the organisms stain gray-blue to black, with the cellular inclusions and nuclei appearing darker than the cytoplasm.

The method described by Spencer and Monroe[54] is slightly longer than the trichrome procedure. Although the slides do not require destaining, decolorizing in 0.5% hydrochloric acid after a longer initial staining time may provide better differentiation.

Another iron hematoxylin method, described by Tompkins and Miller[54] includes the use of phosphotungstic acid as a destaining agent. This procedure also gives good, reproducible results. Another modification of the iron hematoxylin staining method incorporates a carbolfuchsin step into the protocol (see Procedure 64-3); in addition to the intestinal protozoa, acid-fast organisms could also be seen on the permanent-stained smear.[57]

General information

The most important step in preparing a well-stained fecal smear is adequate fixation of a specimen that has been submitted within specified time limits. To ensure best results, the acetic acid component of Schaudinn's fixative should be added just before use; fixation time (room temperature) may be extended overnight with no adverse effects on the smears.

After fixation, it is very important to remove completely the mercuric chloride residue from the smears. The 70% alcohol-iodine mixture removes the mercury complex; the iodine solution should be changed often enough (at least once a week) to maintain a dark, reddish brown color. If the mercuric chloride is not completely removed, the stained smear may contain varying amounts of highly refractive granules, which may prevent finding or identifying any organisms present.

When using nonmercury-based fixatives or one of the single-vial options (usually zinc-based, proprietary formula), the iodine-alcohol step can be eliminated. After drying, the smears can be placed directly into the stain (trichrome). However, remember that when material arrives for proficiency testing (e.g., College of American Pathologists), these fecal speci- *mens have been preserved using mercury-based fixatives. Therefore, you must include the iodine-alcohol step and subsequent rinse for mercury removal in the routine staining protocol. Some laboratories have decided to leave the staining protocol as is; the use of the iodine-alcohol step will not harm smears preserved using nonmercury-based fixatives.*

Good results in the final stages of dehydration (100% alcohol) and clearing (xylene or xylene substitute) depend on the use of *fresh reagents.* It is recommended that solutions be changed at least weekly and more often if large numbers of slides (10 to 50 per day) are being stained. Stock containers and staining dishes should have well-fitting lids to prevent evaporation and absorption of moisture from the air. If the clearing agent turns cloudy on addition of the slides from 100% alcohol, then water is in the solution. When clouding occurs, immediately return the slides to 100% alcohol, replace all dehydrating and clearing agents with fresh stock, and continue with the dehydration process.

Modified Acid-Fast Stain for *Cryptosporidium parvum, Isospora belli,* and *Cyclospora cayetanensis* (CLBS)

Although different procedures have been tried for the recovery and identification of *Cryptosporidium parvum* in humans, the best staining approach seems to be hot or cold modified acid-fast stains (Procedure 64-4). The hot modified acid-fast method presented here consistently provides excellent results.[22,23,38,54] In patients with few oocysts in the stool, the highly specific and sensitive fluorescent method using a monoclonal antibody reagent is more likely to reveal the organisms.[2,26] Both immunoassay methods (FA, EIA) are now available for *Cryptosporidium* and are being used as routine procedures in the diagnostic laboratory.[2,26,61]

OTHER GASTROINTESTINAL TRACT SPECIMENS

Sigmoidoscopy material

When repeated fecal examinations fail to reveal the presence of *Entamoeba histolytica*, material obtained during sigmoidoscopy may be valuable in the diagnosis of amebiasis. However, this procedure does not take the place of routine fecal examinations; a series of at least three (six is preferable) fecal specimens should be submitted for each patient having a sigmoidoscopy examination. Material from the mucosal surface should be obtained by aspiration or scraping, not with a cotton-tipped swab. If swabs must be used, most of the cotton should be removed (leave just enough to safely cover the end) and should be tightly wound to prevent absorption of the material to be examined.

The specimen should be processed and examined immediately; the number of techniques used will depend on the amount of material obtained. If the specimen is sufficient for both wet preparations and permanent-stained smears, proceed as follows. The direct mount should be examined immediately for the presence of moving trophozoites; it may take time for the organisms to become acclimated to this type of preparation, so motility may not be obvious for several minutes. Care should be taken not to confuse protozoan organisms with macrophages or other tissue cells; any suspect cells should be confirmed with the use of the permanent-stained slide. The smears for permanent staining should be prepared at the same time the direct mount is made by gently smearing some of the material onto several slides and immediately placing them into Schaudinn's fixative (mercury- or nonmercury-based). The slides can then be stained by any of the techniques mentioned for routine fecal smears. If the material is bloody, contains a large amount of mucus, or is a "wet" specimen, 1 or 2 drops of the sample can be mixed with 3 or 4 drops of PVA directly on the slide. Allow the smears to dry (overnight if possible) before staining.

Duodenal contents

In some instances, repeated fecal examinations may fail to confirm a diagnosis of *Giardia lamblia* and *Strongyloides stercoralis* infections. Because these two parasites are normally found in the duodenum, the physician may submit duodenal drainage fluid to the laboratory for examination. The specimen should be submitted without preservatives and should be received and examined within 1 hour after being taken. The amount of fluid may vary. It should be centrifuged and the sediment examined as wet mounts for the detection of motile organisms. Several mounts should be prepared and examined; because of the dilution factor, the organisms may be difficult to recover with this technique.

Another convenient method of sampling duodenal contents, which eliminates the necessity for intubation, is the use of the Entero-Test. This device consists of a gelatin capsule containing a weighted, coiled length of nylon yarn. The end of the line protrudes through the top of the capsule and is taped to the side of the patient's face. The capsule is then swallowed, the gelatin dissolves, and the weighted string is carried by peristalsis into the duodenum. After approximately 4 hours the string is recovered and the bile-stained mucus attached to the string is examined as a wet mount for the presence of organisms. This type of specimen should also be examined immediately after the string is recovered. When the mucus is examined from either duodenal drainage or the Entero-Test capsule, typical "falling leaf" motility of *G. lamblia* trophozoites is usually not visible. The flagella usually are visible as a rapid flutter, with the organism remaining trapped in the mucus. It is important to keep the light intensity low; if the light is too bright, the organisms may not be seen.

Estimation of worm burdens

Although these procedures are not routinely performed, circumstances may arise when it is helpful to know the degree of infection in a patient or perhaps to follow the effectiveness of therapy.[22,54] In certain helminth infections that have little clinical significance, the patient may not be given treatment if the numbers of parasites are small. The parasite burden may be estimated by counting the number of eggs passed in the stool.[22]

Recovery of larval-stage nematodes

Nematode infections that give rise to larval stages, which hatch either in the soil or in tissues, may be diagnosed by using culture techniques designed to concentrate the larvae.[22,54] These procedures are used in hookworm, *Strongyloides,* and *Trichostrongylus* infections. Some of these techniques, which permit the recovery of infective-stage larvae, may be helpful because the eggs of many species are identical, and specific identifications are based on larval morphology.

BAERMANN TECHNIQUE   When the stools from a patient suspected of having strongyloidiasis are repeatedly negative, the Baermann technique may be helpful in recovering larvae. The apparatus is designed to allow the larvae to migrate from the fecal material through several layers of damp gauze into water, which is centrifuged, thus concentrating the larvae in the bottom of the tube.[22] Specimens for this technique should be collected after a mild saline cathartic, not a stool softener.

AGAR PLATE CULTURE FOR *STRONGYLOIDES*   Agar plate cultures to detect *Strongyloides stercoralis* larvae are available as another method for the diagnosis of strongyloidiasis. Stool or duodenal contents are placed onto agar plates, and the plates are sealed to prevent accidental infections and held for 2 days at room temperature. As the larvae crawl over the agar, they carry bacteria with them, thus creating visible tracks over the agar. The plates are examined under the microscope for confirmation of larvae, the surface of the agar is then washed with 10% formalin, and final confirmation of larval identification made via wet examination of the sediment from the formalin washings. The agar consists of 1.5% agar, 0.5% meat extract, 1.0% peptone, and 0.5% NaCl. This approach appears to be more sensitive and is becoming more widely used.[3,22,40]

Hatching procedure for schistosome eggs

When schistosome eggs are recovered from either urine or stool, they should be carefully examined to determine viability. The presence of living miracidia within the eggs indicates an active infection, which may require therapy. The viability of the miracidium larvae can be determined in two ways:

1. The cilia on the flame cells (primitive excretory cells) may be seen on high-dry power and are usually actively moving.
2. The larvae may be released from the eggs with the use of a hatching procedure.[22]

The eggs usually hatch within several hours when placed in 10 volumes of dechlorinated or spring water. The eggs, which are recovered in the urine, are easily obtained from the sediment and can be examined under the microscope to determine viability.

Cellophane tape preparations

Enterobius vermicularis is a roundworm that is worldwide in distribution. It is very common in children and is known as the *pinworm* or the *seatworm*. The adult female migrates from the anus during the night and deposits her eggs on the perianal area. Because the eggs are deposited outside the gastrointestinal tract, examination of a stool specimen may produce negative results. Although some laboratories use the anal swab technique, pinworm infections are most frequently diagnosed by using the cellophane tape method (Procedure 64-5) for egg recovery. Another collection procedure is illustrated in Figure 64-4. Occasionally the adult female may be found on the surface of a formed stool or on the cellophane tape. Specimens should be taken in the morning *before bathing* or going to the bathroom. For small children, it is best to collect the specimen early in the morning, just before the child awakens.

A series of at least four to six consecutive negative tapes should be obtained before ruling out infection with pinworms. However, most laboratories rarely receive more than one or two tapes, and the clinician may decide to treat the patient based on symptoms or recurrent symptoms after the initial diagnosis.

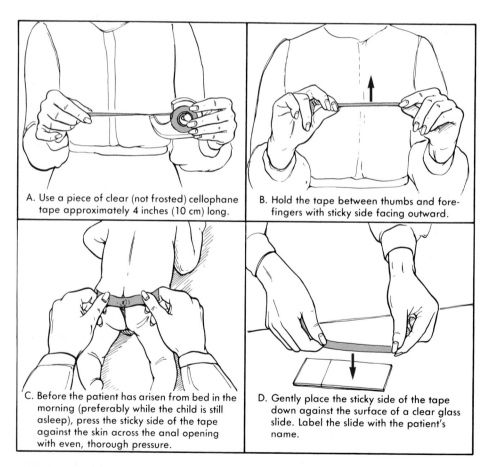

A. Use a piece of clear (not frosted) cellophane tape approximately 4 inches (10 cm) long.

B. Hold the tape between thumbs and forefingers with sticky side facing outward.

C. Before the patient has arisen from bed in the morning (preferably while the child is still asleep), press the sticky side of the tape against the skin across the anal opening with even, thorough pressure.

D. Gently place the sticky side of the tape down against the surface of a clear glass slide. Label the slide with the patient's name.

FIGURE 64-4 Method for collection of a cellulose (Scotch) tape preparation for pinworm diagnosis. This method dispenses with the tongue depressor (see Procedure 64-5), requiring only tape and a glass microscope slide. The tape must be pressed deep into the anal crack.

Identification of adult worms

Most adult worms or portions of worms that are submitted to the laboratory for identification are *Ascaris lumbricoides, Enterobius vermicularis,* or segments of tapeworms. The adult worms present no particular problems in identification (see later discussion on intestinal helminths); however, identification of the *Taenia* spp. tapeworms depends on the gravid proglottids, which contain the fully developed uterine branches. Identification as to species is based on the number of lateral uterine branches that arise from the main uterine stem in the gravid proglottids. Often the uterine branches are not clearly visible; one technique that can be used is injecting the branches with India ink (Procedure 64-6), allowing them to be easily seen and counted.

UROGENITAL SPECIMENS (TRICHOMONIASIS)

The identification of *Trichomonas vaginalis* is usually based on the examination of wet preparations of vaginal and urethral discharges and prostatic secretions. These specimens are diluted with a drop of saline and examined under low power with reduced illumination for the presence of actively motile organisms; urine sediment can be examined in the same way. As the jerky motility of the organisms begins to diminish, it may be possible to observe the undulating membrane, particularly under high-dry power (Figure 64-5). Stained smears are usually not necessary for the identification of this organism; often the number of false-positive and false-negative results reported based on stained smears would strongly suggest the value of confirmation (i.e., observation of the motile organisms).

Some studies indicate that the most sensitive method of detecting *T. vaginalis* is culture. This technique may not be the most practical, and expense may limit its availability. To improve the approach to culture, plastic envelope methods (PEM and In-Pouch TV) have been developed.[4] The medium for the PEM

pouch consists of dry ingredients that are reconstituted with water before use, with a shelf life of at least 1 year. The PEM pouch is composed of clear, transparent plastic that minimizes oxygen and water vapor transmission. The oxygen content within the medium is reduced further through the reducing action of the components, ascorbic acid and cysteine. Distilled water is added to the upper chamber of the pouch to reconstitute the medium; the medium is then pushed into the lower chamber. The patient specimen is inoculated into the small amount of fluid left in the upper chamber, then pushed into the lower chamber. The pouch comes with a small, rigid plastic frame that holds the pouch in a horizontal position for viewing directly under the microscope. Some advantages of this system are that (1) the pouch containing dry medium can be held at room temperature for at least a year before use; (2) holding medium is not necessary for specimen transport to the laboratory; the pouch can be inoculated with the specimen at the time of collection; and (3) the system is inexpensive to manufacture and purchase. Another excellent system, In Pouch TV test, containing liquid medium, is also available and has proved to be an excellent approach within the diagnostic laboratory.[6,7] In this system, the medium is liquid and ready for immediate use. When using any culture options, positive control strains should be used each time patient material is cultured.

Although more sensitive and specific methods using monoclonal antibodies have been developed, the fluorescent procedure may or may not be available commercially. The specimen is smeared onto a glass slide, allowed to air dry, and then transported to the laboratory before testing by fluorescent methodology (Figure 64-6). It is hoped this reagent may again become available for purchase.

SPUTUM (PARAGONIMIASIS)

When sputum is submitted for examination, it should be deep sputum from the lower respiratory passages, not a specimen that is mainly saliva. The specimen should be collected early in the morning (before eating or brushing teeth) and immediately delivered to the laboratory. Sputum is usually examined as a saline or iodine wet mount under low and high-dry microscope power. If the quantity is sufficient, the formalin-ether sedimentation technique can be used. A very mucoid or thick sputum can be centrifuged after the addition of an equal volume of 3% sodium hydroxide. With any technique, the sediment should be carefully examined for the presence of brownish spots or "iron filings," which may be *Paragonimus* eggs.

NOTE: Care should be taken not to confuse *Entamoeba gingivalis,* which may be found in the mouth and might be seen in the sputum, with *E.*

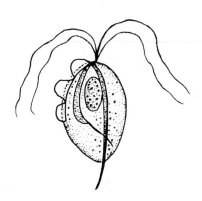

FIGURE 64-5 *Trichomonas vaginalis* trophozoite. (Illustration by Nobuko Kitamura.)

histolytica from a pulmonary abscess. *E. gingivalis* contains ingested polymorphonuclear neutrophils (PMNs); *E. histolytica* does not.

ASPIRATES

The diagnosis of certain parasitic infections may be based on procedures using aspirated material. These techniques include microscopic examination, animal inoculation, and culture.

Examination of aspirated material from lung or liver abscesses may reveal the presence of *Entamoeba histolytica;* however, the demonstration of these parasites is often extremely difficult for several reasons. Hepatic abscess material taken from the peripheral area, rather than the necrotic center, may reveal organisms, although they may be trapped in the thick pus and not exhibit any motility. The Amoebiasis Research Unit, Durban, South Africa, has recommended using proteolytic enzymes, such as 10 units of streptodornase per milliliter of thick pus, to free the organisms from the aspirate material.[22] The enzyme is incubated with the specimen for 30 minutes at 37° C with repeated shaking, the suspension is centrifuged, and the sediment is examined as for other direct examination methods.

Aspiration of cyst material, usually liver or lung, for the diagnosis of hydatid disease is usually performed when open surgical techniques are used for cyst removal. The aspirated fluid is submitted to the laboratory and examined for the presence of hydatid sand (scolices) or hooklets; the absence of this material does not rule out the possibility of hydatid disease, since some cysts are sterile (Figure 64-7).

Material from lymph nodes, spleen, liver, bone marrow, or spinal fluid may be examined for the presence of trypanosomes or leishmanial forms. Part of the specimen should be examined as a wet preparation to demonstrate motile organisms. Impression smears can also be prepared and stained with Giemsa stain (see later section on detection of blood parasites). This type of material can also be cultured (see later section on culture techniques).

Specimens obtained from cutaneous ulcers should be aspirated from below the ulcer bed rather than the surface; this type of sample is more likely to contain the intracellular leishmanial organisms and is free of bacterial contamination. A few drops of sterile saline may be introduced under the ulcer bed (through uninvolved tissue) by needle (25 gauge) and syringe (1 or 2 mL). The aspirated fluid should be examined as stained smears and should be inoculated into appropriate media (see culture techniques).

Nasopharyngeal aspirates and material from the sinus cavities can be examined for the presence of microsporidia; these specimens can be stained with the modified trichrome stain and examined for the presence of spores. The material also can be examined using one of the optical brightening agents such as calcofluor white.

SPINAL FLUID

Cases of primary meningoencephalitis are infrequently seen, but the examination of spinal fluid may reveal the causative agent, *Naegleria fowleri* (Figure 64-8), if present. The spinal fluid may range from cloudy to purulent, with or without red blood cells. The cell count ranges from a few hundred to more than 20,000 white blood cells per milliliter, primarily

FIGURE 64-6 Fluorescein-conjugated, monoclonal antibody–stained *Trichomonas vaginalis* in vaginal discharge. (Courtesy Meridian Diagnostics, Cincinnati.)

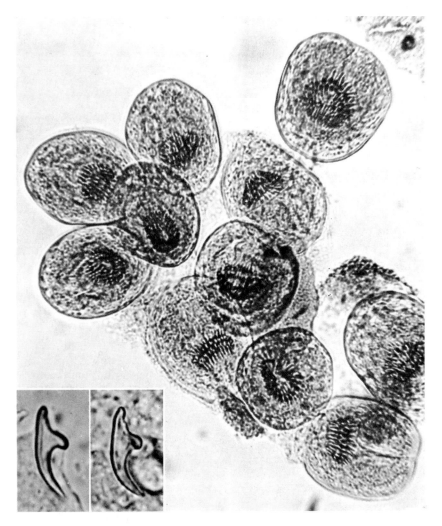

FIGURE 64-7 *Echinococcus granulosus,* hydatid sand (300×). *Inset.* Two individual hooklets (1000×).

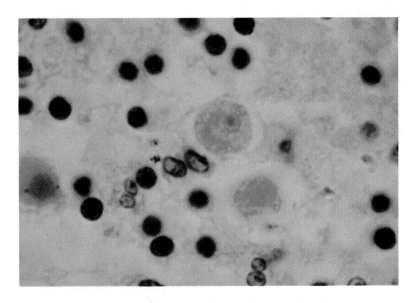

FIGURE 64-8 *Naegleria fowleri* in brain tissue. Hematoxylin and eosin stain.

neutrophils; the failure to find bacteria in this type of spinal fluid should alert one to the possibility of primary meningoencephalitis. Motile amebae may be found in unstained spinal fluid; however, one should be very careful not to confuse organisms with various blood and tissue cells that may also be motile, particularly if the spinal fluid is in a counting chamber.[22,76]

The classification, transmission, virulence, and disease pathogenesis of the free-living amebic genera *Naegleria, Acanthamoeba,* and *Balamuthia mandrillaris* (Leptomyxid ameba) are receiving considerable attention at this time.* The question of increasing infections has not been answered, although studies of thermally polluted or enriched waters indicate the presence of such amebae in these situations. Studies have also shown that these organisms have been recovered from asymptomatic human carriers (nasal passages, nasopharyngeal secretions) and have been implicated in chronic or subacute meningoencephalitis, acute primary amebic meningoencephalitis, and granulomatous amebic encephalitis. These infections have definitely been linked to the compromised patient.

BIOPSY MATERIAL

In some patients, biopsy material may be used to confirm the diagnosis of certain parasitic infections. Most of these specimens are sent for routine tissue processing (fixation, embedding, sectioning, staining). However, fresh material may be sent directly to the laboratory for examination; it is imperative that these specimens be received immediately to prevent deterioration of any organisms present.

Pneumocystis carinii is recognized as an important cause of pulmonary infection in patients who are immunosuppressed as a result of therapy or in patients with congenital or acquired immunologic disorders, including acquired immunodeficiency syndrome (AIDS).[80] It is also becoming more widely recognized that *Pneumocystis* can present as disseminated disease found in various tissues, particularly in severely compromised patients.[22] The organisms can be demonstrated in stained impression smears of lung material obtained by open, transbronchial, or brush biopsy. *Pneumocystis* can be seen in stained smears of tracheobronchial aspirates, although preparations of lung tissue are more likely to reveal the organisms. Bronchoalveolar lavage has a lower yield than transbronchial biopsy, but it is helpful when biopsy is contraindicated. The usefulness of induced sputum has been evaluated, and it is becoming more widely used, particularly when obtained from patients with AIDS and when coupled with the monoclonal antibody diagnostic procedure (Figure 64-9).[74,84] Sputum speci-

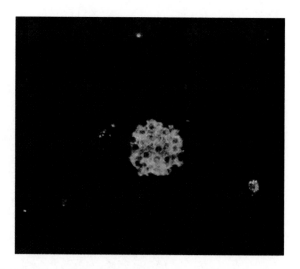

FIGURE 64-9 Cluster of *Pneumocystis carinii* cysts and trophozoites from a bronchoalveolar lavage specimen, stained with a monoclonal antibody–fluorescent stain (Merifluor *Pneumocystis,* Meridian Diagnostics). Note the "honeycomb" appearance caused by peripheral staining of cysts. (Courtesy Meridian Diagnostics, Cincinnati.)

mens are generally considered unacceptable for the detection of *Pneumocystis*.

Regardless of the technique used, multiple specimens may have to be examined to confirm the presence of *P. carinii* organisms. Although several different stains are available, the stain most often recommended is Gomori's methenamine silver (GMS), which clearly stains *Pneumocystis* organisms in dark brown or black (Figure 64-10, *A*). Preparation of the silver stain is relatively complex, and positive control slides should be included with each specimen to ensure accurate identification of the smears. Although the silver stain yields the most definitive results, toluidine blue O stains are much easier and faster to perform and yield reliable results in the hands of skilled technologists (Figure 64-10, *B*). Although not as sensitive as the monoclonal antibody procedures, a rapid method for the detection of *Pneumocystis* cysts is the use of calcofluor white (CFW). This is a chemofluorescent dye with an affinity for the polysaccharide polymers of amebic cysts. PCR has also been used to diagnose this infection.[10,62] *Pneumocystis* is now classified with the fungi; additional information can be found in the section on mycology.

Cryptosporidia are coccidial protozoa usually found in the intestinal tract. Respiratory cryptosporidiosis has also been reported with infection of the gallbladder and biliary tree. Human cryptosporidiosis has been confirmed from gastrointestinal biopsies and stool from patients whose immune responses were diminished, including AIDS patients, as well as those with normal immune systems. Cryptosporidia differ from other coccidia found in the intestine in that they are limited to the microvillus layer of gastrointestinal

---

*References 44, 58, 59, 70, 72, 76, 79.

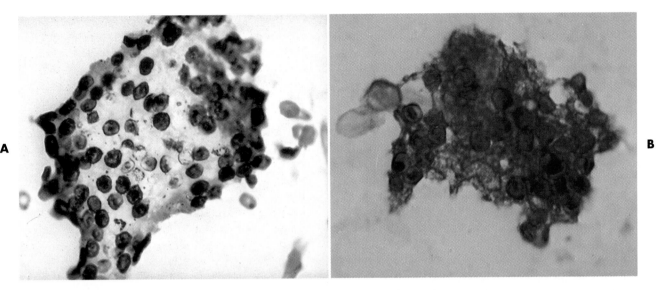

FIGURE 64-10 **A,** Cysts of *Pneumocystis carinii* in bronchial lavage, stained with modified Gomori methenamine silver stain (1000×). **B,** Toluidine blue O–stained cysts of *P. carinii* in smear made from centrifuged sediment from bronchial lavage specimen (1000×).

cells. Tissue diagnosis has been based on electron microscopy (EM) studies and Giemsa-stained smears from jejunal biopsy material and routine tissue processing (paraffin embedding and sectioning). Currently, more than 1500 published papers appear in the literature related to cryptosporidiosis, including information on nosocomial transmission, day care center outbreaks, and waterborne outbreaks.* This is remarkable, given that before 1980 less than 30 publications were available. This organism has definitely become more widely recognized as a potential pathogen in both immunosuppressed and immunocompetent hosts, and other diagnostic procedures have been developed to facilitate organism recovery and identification, that is, modified acid-fast staining methods of stool (modified Kinyoun stain; see Procedure 64-4 and later section on coccidia). The organisms may also be visible as refractile bodies in smears made directly or from concentrated feces.

Biopsy specimens from the intestinal tract, particularly in AIDS patients, may reveal **microsporidian** organisms, although EM may be necessary for confirmation.[9] Techniques that do not require tissue embedding or EM are now becoming more popular. Touch preparations of fresh biopsy material, which are air dried, methanol fixed, and Giemsa stained, have been used; however, screening must be performed at 1000× magnification. Cytocentrifugation followed by Giemsa staining has also been used; however, all results were confirmed using EM. Other more recent procedures use a modified trichrome stain in which the chromotrope 2R component added to the stain is

10 times the concentration normally used in the routine trichrome stain for stool; the background counterstain also varies, depending on the procedure.* These stains are excellent and can be used within the clinical laboratory. It is important to remember that the stool preparations must be very thin, the staining time is 90 minutes, and the slide must be examined at 1000× magnification (or higher). The staining time may be reduced when using the "hot" method.

Unfortunately, many objects within stool material are oval, stain pinkish with trichrome, and measure approximately 1.5 to 3 μm. If this stain is used to identify microsporidia in stool, positive control material should be available for comparison (see Procedures 64-7 to 64-9). Although the individual procedures are not difficult to perform, interpretation of the results, particularly in stool, can be difficult. Both false-negatives and false-positives have been reported related to the amount of artifact material normally found in stool. The specimen can be fresh stool or stool that has been preserved in 5% or 10% formalin, SAF, or some of the newer single-vial system fixatives. Actually, any specimen other than tissue thought to contain microsporidia could be stained by this method.

The use of optical brightening agents (calcofluor white, Fungi-Fluor, Uvitex 2B) has also been recommended. However, these reagents provide nonspecific staining and a number of false-positive results have been seen.[45] Polymerase chain reaction (PCR) has also been used.[1,20]

The newest approach for the identification of

---

*References 12, 13, 18, 35, 46.

*References 14, 15, 24, 41, 63, 81, 82.

microsporidial spores in clinical specimens uses anti-serum in an indirect fluorescent antibody (IFA) procedure.[8] Fluorescing microsporidial spores were distinguished by a darker cell wall and internal visualization of the polar tubule as diagonal lines or cross-lines within the cell. In another study using this same antiserum, 9/27 (30%) patients who had already been diagnosed as having cryptosporidiosis (eight with AIDS, one without AIDS but immunodeficient), were found to have *Enterocytozoon bieneusi* or *Septata intestinalis* in the stool.[24] Although some cross-reactivity occurs with bacteria, this technique offers a more sensitive approach than routine staining methods currently available for the examination of stool specimens. Trials related to new diagnostic immunoassay kits for the diagnosis of microsporidiosis are ongoing. As clinicians begin to suspect these infections and laboratorians become more familiar with the diagnostic methods, the number of positive patients may increase dramatically, particularly those who are immunocompromised.

NOTE: It is important to remember that microsporidial spores in specimens from body sites with less artifact material (i.e., tissues, urine) will be much easier to detect using either modified trichrome stains or optical brightening agents. *The use of multiple methods is recommended, particularly when examining fecal specimens.*

Corneal scrapings or biopsy specimens can be examined by routine histology, CFW stain, or culture for the presence of *Acanthamoeba*. Although these infections are relatively rare, the medical community has become much more aware of *Acanthamoeba* keratitis, especially among contact lens wearers, during the past few years.[58,59,76]

Skin biopsy specimens for the diagnosis of cutaneous amebiasis or cutaneous leishmaniasis should be submitted for tissue processing. A portion of the tissue for the diagnosis of leishmaniasis can be teased apart with sterile needles and inoculated into appropriate culture media (see later section on culture techniques).

The diagnosis of onchocerciasis *(Onchocerca volvulus)* may be confirmed by the examination of "skin snips," very thin slices of skin, which are teased apart in saline to release the microfilariae.

Biopsy specimens taken from lymph nodes are submitted for routine tissue processing; impression smears can also be prepared and stained with Giemsa stain (see later section on detection of blood parasites). The diagnosis of trichinosis is usually based on clinical findings; however, confirmation may be obtained by the examination of a muscle biopsy (Figure 64-11). The encapsulated larvae can be seen in small pieces of fresh tissue, which are pressed between two slides and examined under low power of the microscope. At necropsy the larvae are most abundant in the diaphragm, masseter muscle, or tongue. Larvae can also be recovered from tissue that has undergone digestion in artificial digestive fluid at 37° C. Figure 64-12 shows the life cycle of this organism.

Tapeworm larvae occasionally may be recovered from a muscle specimen and should be carefully dissected from the capsules. They should then be pressed between two slides and examined under low power for the presence of a scolex with four suckers and a circle of hooks. If no hooks are present, it may be a species other than *Taenia solium*.

In some patients with schistosomiasis, the eggs may not be recovered in the stool or urine; however, examination of the rectal or bladder mucosa may reveal eggs of the appropriate species. The mucosal tissue should be compressed between two slides and examined under low power and decreased illumination. The eggs should be carefully examined to determine viability (see earlier discussion of schistosome eggs). Small pieces of tissue may be digested with 4% sodium hydroxide for 2 to 3 hours at 60° to 80° C. The eggs, which are recovered by sedimentation or centrifugation, can be examined under the microscope.

CULTURE TECHNIQUES

Most clinical laboratories do not provide culture techniques for the diagnosis of parasitic organisms; however, the lack of culture procedures should not prevent the correct identification of most parasites. If culture methods are used, they are probably for the isolation and identification of *Trichomonas vaginalis*. Many different media have been developed for the culture of protozoan organisms (some of which are available commercially), and specific directions for their preparation are available in the literature.[4,6,7,75]

ANIMAL INOCULATION

Most laboratories have neither the time nor the facilities for animal care to provide animal inoculation procedures for the diagnosis of parasitic infections. Host specificity for many parasites also limits the types of laboratory animals available for these procedures. Occasionally, animal studies may be requested; included here are several procedures that can be used.

The hamster is the animal of choice for inoculation procedures designed to recover leishmanial organisms. After intraperitoneal or intratesticular inoculation, the infection may develop very slowly over several months; in some cases a generalized infection develops more quickly, and the animal may die in several days. Splenic and testicular aspirates should be examined for the presence of intracellular

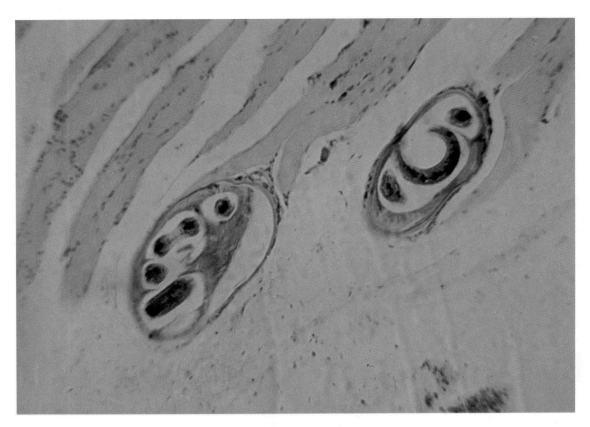

FIGURE 64-11 *Trichinella spiralis* larvae encysted in muscle.

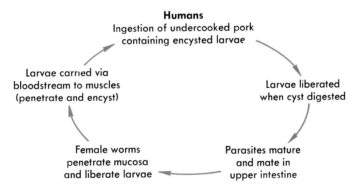

FIGURE 64-12 Life cycle of *Trichinella spiralis*.

organisms; stained smears should be prepared and carefully examined with the oil immersion lens.

Mice are generally used for the isolation of *Toxoplasma gondii*, although most cases are diagnosed on clinical and serologic findings.[8,17,50] Mice inoculated through the peritoneum develop a fulminating infection that leads to death within a few days. Organisms can be recovered easily from the ascitic fluid and should be examined as stained smears. Giemsa stain is recommended for both intraperitoneal and intratesticular inoculation studies (specific staining techniques are discussed later in this chapter).

SERODIAGNOSIS AND IMMUNODIAGNOSTIC PROCEDURES

Although serologic procedures (antibody detection) for the diagnosis of parasitic diseases have been available for many years, they are generally not performed by most clinical laboratories. The procedures vary in both sensitivity and specificity and at times may be difficult to interpret. The Centers for Disease Control and Prevention (CDC) offers a wide variety of serologic procedures for diagnostic purposes, some of which are still in the experimental stages and not available elsewhere. (For a discussion of specific procedures, see Chapter 16).

At present, immunodiagnostic procedures (antigen detection) are most widely used for amebiasis, giardiasis, cryptosporidiosis, toxoplasmosis, and trichomoniasis. Numerous diagnostic kits are available commercially and include EIA, DFA, and IFA formats.[22,84]

DETECTION OF BLOOD PARASITES

Laboratory diagnosis of malaria

Malaria is one of the few acute parasitic infections that can be life-threatening. For this reason, any laboratory that offers this type of diagnostic service must be willing to provide technical expertise on a 24-hour basis, 7 days per week.[55]

The definitive diagnosis of malaria is based on the demonstration of the parasites in the blood. Two types of blood films are used. The "thick smear" (Figure 64-13) allows the examination of a larger amount of blood and is used as a screening procedure; the "thin film" allows species identification of the parasite.

Blood films are usually prepared when the patient is admitted; samples should be taken at intervals of 6 to 18 hours for at least 3 successive days. From 200 to 300 microscopic fields should be examined before a film is signed out as negative. If possible, the smears should be prepared from blood obtained from the finger or earlobe; the blood should flow freely. If patient contact is not possible and the quality of the submitted slides may be poor, a tube of fresh blood should be requested (ethylenediaminetetraacetic acid [EDTA] anticoagulant is recommended), and smears should be prepared immediately after the blood is received. It is even better to use the blood remaining in the needle from a venipuncture for smear preparation because this blood has not been in contact with any anticoagulant.

To prepare the thick film, place 2 or 3 small drops of fresh blood on an alcohol-cleaned slide. With the corner of another slide, and using a circular motion, mix the drops and spread the blood over an area about 2 cm in diameter. Continue stirring for about 30 seconds to prevent formation of fibrin strands, which may obscure the parasites after staining. If the blood is too thick or any grease remains on the slide, the blood will flake off during staining. Allow the film to air dry (room temperature) in a dust-free area. Never apply heat to a thick film because heat will fix the blood, causing the red blood cells to remain intact during staining; the result is stain retention and subsequent inability to identify any parasites present.

The thin blood film is used primarily for specific parasite identification, although the number of organisms per field is much reduced compared with the thick film. The thin film is prepared in exactly the same manner used for the differential blood count. After the film has air dried (do not apply heat), it may be stained. The necessity for fixation before staining depends on the stain selected.

Staining blood films

For accurate identification of blood parasites, it is very important that a laboratory develop proficiency in the use of at least one good staining method. As a general rule, blood films should be stained as soon as possible, since prolonged storage results in stain retention. The stains used are generally of two types. One has the fixative in combination with the staining solution so that both fixation and staining occur at the same time. Wright stain is an example of this type of staining solution. Giemsa stain represents the other type of staining solution, in which the fixative and stain are separate; thus the thin film must be fixed before staining (Procedures 64-10 and 64-11).

Methods are also available for the identification of malarial parasites with the use of acridine orange and other deoxyribonucleic acid (DNA)–binding dyes, with smears examined using fluorescence microscopy. The QBC tube is available from Becton Dickinson; this is a qualitative screening method for rapidly detecting the presence of malaria and other blood parasites in centrifuged capillary or venous blood. Although parasites are detected by observing fluorescence of acridine orange within the tube, speciation requires examination of routine stained blood films.[19]

A new *P. falciparum* antigen detection system, the ParaSight F test (PFT) (Becton Dickinson) has been found to be very effective in field trials.[16,65] This procedure is based on an antigen capture approach and has been incorporated in a dipstick format; the entire test takes approximately 10 minutes. However, to date, it has not been approved for use within the United States. As with other parasites, PCR has also been used.[37]

When removed from either type of staining solution, slides should be dried in a vertical position. After being air dried, they may be examined under oil immersion by placing the oil directly on the uncovered blood film.

Specimens may be submitted from patients with *Plasmodium falciparum* infections who do not yet have gametocytes in the blood. Consequently, a low-level parasitemia with delicate ring forms might be missed without extensive oil immersion examination of the blood films (at least 200 to 300 oil immersion fields). NOTE: Remember that automated differential

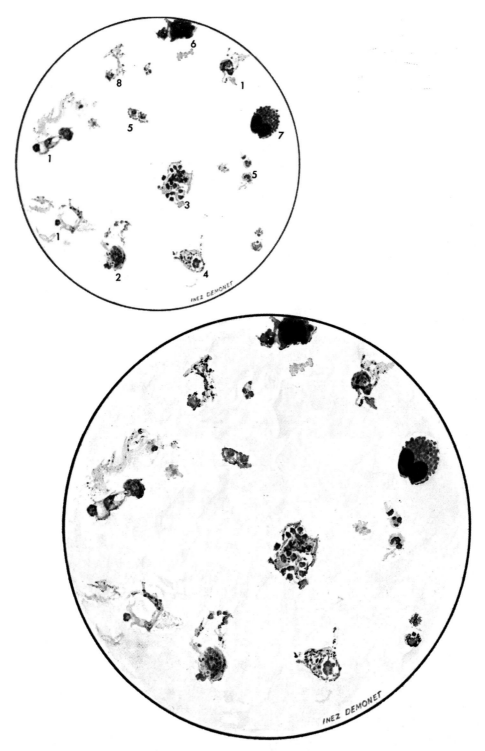

FIGURE 64-13 *Plasmodium vivax* in thick smear. **1,** Ameboid trophozoites. **2,** Schizont, two divisions of chromatin. **3,** Mature schizont. **4,** Microgametocyte. **5,** Blood platelets. **6,** Nucleus of neutrophil. **7,** Eosinophil. **8,** Blood platelet associated with cellular remains of young erythrocytes. (From Wilcox, A. 1960. Manual for the microscopical diagnosis of malaria in man. U.S. Public Health Service, Washington, D.C.)

instruments used in hematology laboratories are not designed to recognize intracellular (red blood cell) parasites. Any suspected parasitic infection or presumptive diagnosis of fever of unknown origin mandates a manual blood smear examination.[22,27]

Many studies have used immunodiagnostic procedures for the diagnosis of malaria; however, these procedures are not routinely performed in most laboratories. The use of thick-smear antigens prepared from washed parasitized blood cells has been reported. This type of antigen is used in the IFA procedure, which has a 95% sensitivity and a false-positive rate of 1% at a titer of 1:16.[27] The indirect hemagglutination assay (IHA) has also been used and evaluated by various workers. The newly developed PCR has even been used for amplification and subsequent detection of malarial DNA in blood.

IDENTIFICATION OF ANIMAL PARASITES

INTESTINAL PROTOZOA

The protozoa are unicellular organisms, most of which are microscopic. They possess a number of specialized organelles, which are responsible for life functions and which allow further division of the group into classes.

The class Sarcodina contains the organisms that move by means of cytoplasmic protrusions called *pseudopodia*. Included in this group are free-living organisms, as well as nonpathogenic and pathogenic organisms found in the intestinal tract and other areas of the body.

The Mastigophora, or flagellates, contain specialized locomotor organelles called *flagella*: long, thin cytoplasmic extensions that may vary in number and position depending on the species. Different genera may live in the intestinal tract, the bloodstream, or various tissues. Detection of the blood- and tissue-dwelling flagellates is discussed in the previous section.

The class Ciliata contains species that move by means of cilia, short extensions of cytoplasm that cover the surface of the organism. This group contains only one organism that infects humans: *Balantidium coli* infects the intestinal tract and may produce severe symptoms.

Members of the class Sporozoa are found in the blood and other tissues and have a complex life cycle that involves both sexual and asexual generations. The four species of *Plasmodium*, the cause of malaria, are found in this group; their diagnosis is discussed in the previous section. Members of the genera *Isospora, Cryptosporidium, Cyclospora,* and the microsporidia can be found in the intestinal mucosa and other

tissues. These organisms have been seen with increasing frequency in specimens from immunosuppressed patients, particularly those with AIDS.[56]

Isospora, Cryptosporidium, and *Cyclospora* sp. (formerly called CLBs) are passed in the stool as oocysts; the other members of the protozoa exist in the intestinal tract in the trophozoite or cyst stages. With *Enterocytozoon* and *Encephalitozoon,* two of the genera of microsporidia that can infect humans, spore stages in the life cycle are passed in the stool. Within the past few years, diagnostic methods have become available for the identification of microsporidial spores in stool specimens.[41,63,81] However, considering the spore size (approximately 1 to 4 μm), the clinical laboratory may have to wait until monoclonal antibodies are available commercially for detection of spores.[24,86] The coccidian parasite *Toxoplasma gondii* is acquired by humans via ingestion, although its life cycle includes stages in several animal hosts, particularly cats (Figure 64-14). The trophozoites may be seen in squash preparations from lymph nodes or brain tissue, usually examined by pathologists.

The important characteristics of the intestinal protozoa are found in Tables 64-4 to 64-9. The clinically important intestinal protozoa are generally considered to be *Entamoeba histolytica, Dientamoeba fragilis, Giardia lamblia, Isospora belli, Cryptosporidium, Cyclospora,* and *Balantidium coli. E. histolytica* is the most important species and may invade other tissues of the body, resulting in severe symptoms and possible death. *D. fragilis* has been associated with diarrhea, nausea, vomiting, and other nonspecific abdominal complaints. *G. lamblia* is probably the most common protozoan organism found in persons in the United States and is known to cause symptoms ranging from mild diarrhea, flatulence, and vague abdominal pains to acute, severe diarrhea to steatorrhea and a typical malabsorption syndrome. Various documented waterborne and

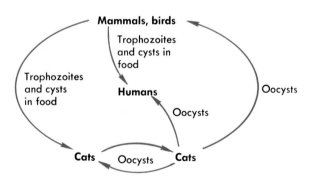

FIGURE 64-14 Life cycle of *Toxoplasma gondii.*

foodborne outbreaks have occurred during the past several years, and the beaver has been implicated as an animal reservoir host for *G. lamblia*. It is speculated that other animals may be involved as well. With present improved culture techniques and the ability to harvest material for antigen, immunoassay tests (FA, EIA) for giardiasis have been developed and are commercially available.

Another organism, *Blastocystis hominis* (Figures 64-3 and 64-15), is also now considered to be potentially pathogenic for humans. It has been reclassified as a protozoan, and there are reports of *B. hominis* as an agent of enteric disease in humans.* However, another report indicated that *B. hominis* is an incidental finding where other proven pathogens may be present in low numbers.[47] We recommend that it be reported and quantitated; this information may be valuable in helping assess the pathogenicity of the organism.[22]

Sarcocystis sp. appear in Table 64-9 but are not discussed in detail. According to the literature, extraintestinal human sarcocystosis is rare, with a much lower incidence than that seen with the intestinal infection.

The identification of intestinal protozoan parasites is difficult at best, and the importance of the permanent-stained slide should be reemphasized (Procedure 64-2). It is important to remember that many artifacts (vegetable material, debris, cells of human origin) may mimic protozoan organisms on a wet mount (Figure 64-15). The important diagnostic characteristics are visible on the stained smear, and the *final identification of protozoan parasites should be confirmed with the permanent stain.*

Amebae

Occasionally, when fresh stool material is examined as a direct wet mount, motile trophozoites may be observed. *Entamoeba histolytica* is described as having directional and progressive motility, whereas the other amebae tend to move more slowly and at random. The cytoplasm usually appears finely granular, less frequently coarsely granular, or vacuolated. Bacteria, yeast cells, or debris may be present in the cytoplasm. The presence of red blood cells in the cytoplasm is usually considered to be diagnostic for *E. histolytica* (Figures 64-16 and 64-17).

Although many people throughout the world are infected with this organism, only a small percentage will develop clinical symptoms. Morbidity and mortality as a result of *E. histolytica* vary, depending on geographic area, organism strain, and the patient's

immune status. For many years, the issue of pathogenicity has been very controversial with, essentially, two points of view. Some felt that what was called *E. histolytica* was really two separate species of *Entamoeba,* one being pathogenic, causing invasive disease, and the other being nonpathogenic, causing mild or asymptomatic infections. Others felt that all organisms designated *E. histolytica* were potentially pathogenic, symptoms depending on the result of host or environmental factors, including intestinal flora.

Recent evidence supports the differentiation of the pathogenic *E. histolytica* from the nonpathogenic *E. dispar* as two distinct species. Based on current knowledge, pathogenic *E. histolytica* is considered to be the etiologic agent of amebic colitis and extraintestinal abscesses, while nonpathogenic *E. dispar* produces no intestinal symptoms and is not invasive in humans.[29,30,31,60]

Nuclear morphology is one of the most important criteria used for identification; nuclei of the genus *Entamoeba* contain a relatively small karyosome and have chromatin material arranged on the nuclear membrane (Figures 64-16 to 64-25). Unfortunately, unless the trophozoite contains ingested red blood cells, it is impossible to identify *Entamoeba* to the species level *(E. histolytica* vrs. *E. dispar).* The nuclei of the other two genera, *Endolimax* and *Iodamoeba,* tend to have very large karyosomes with no peripheral chromatin on the nuclear membrane (Figures 64-26 to 64-30).

The trophozoite stages may often be pleomorphic and asymmetrical, whereas the cysts are usually less variable in shape, with more rigid cyst walls. The number of nuclei in the cysts may vary, but their general morphology is similar to that found in the trophozoite stage. There are various inclusions in the cysts, such as chromatoidal bars or glycogen material, that may be helpful in identification.

Members of the *Acanthamoeba* and *Naegleria* genera may be identified from stained smears of culture material when specific organisms are suspected based on the original clinical specimen. Key characteristics are the typical pseudopods, *Naegleria* having lobed pseudopods and *Acanthamoeba* having the spiky acanthapodia. Also, when organisms from the culture plates are placed in distilled water, *Naegleria fowleri* undergoes transformation within a few hours to a pear-shaped flagellate, usually with two flagella, occasionally with three or four flagella. The flagellate stage is a temporary, nonfeeding stage and usually reverts back to the trophozoite stage.[76] Both genera of free-living amebae are also characterized by having the typical hexagonal, double-walled cyst that can be seen using calcofluor white stain.[76]

---

*References 21, 64, 66, 69, 85.

Text continued on p. 828

TABLE 64-4   INTESTINAL PROTOZOA: TROPHOZOITES OF COMMON AMEBAE

CHARACTERISTIC	ENTAMOEBA HISTOLYTICA	ENTAMOEBA HARTMANNI
Size* (diameter or length)	12-60 μm; usual range: 15-20 μm; invasive forms may be larger than 20 μm	5-12 μm; usual range: 8-10 μm
Motility	Progressive, with hyaline, finger-like pseudopodia; motility may be rapid	Usually nonprogressive
Nucleus: number and visibility	Difficult to see in unstained preparations; one nucleus	Usually not seen in unstained preparations; one nucleus
Peripheral chromatin (stained)	Fine granules, uniform in size and usually evenly distributed; may have beaded appearance	Nucleus may stain more darkly than E. histolytica although morphology is similar; chromatin may appear as solid ring rather than beaded (trichrome)
Karyosome (stained)	Small, usually compact; centrally located but may also be eccentric	Usually small and compact; may be centrally located or eccentric
Cytoplasm appearance (stained)	Finely granular, "ground glass" appearance; clear differentiation of ectoplasm and endoplasm; if present, vacuoles are usually small	Finely granular
Inclusions (stained)	Noninvasive organism may contain bacteria; presence of red blood cells diagnostic	May contain bacteria; no red blood cells

*Sizes refer to wet-preparation measurements. Organisms on a permanent-stained smear may be 1 to 1.5 μm smaller because of artificial shrinkage.

ENTAMOEBA COLI	*ENDOLIMAX NANA*	*IODAMOEBA BÜTSCHLII*
15-50 μm; usual range: 20-25 μm	6-12 μm; usual range: 8-10 μm	8-20 μm; usual range: 12-15 μm
Sluggish; nondirectional, with blunt, granular pseudopodia	Sluggish; usually nonprogressive	Sluggish, usually nonprogressive
Often visible in unstained preparations; one nucleus	Occasionally visible in unstained preparations; one nucleus	Usually not visible in unstained preparations; one nucleus
May be clumped and unevenly arranged on the membrane; may also appear as solid, dark ring with no beads or clumps	Usually no peripheral chromatin; nuclear chromatin may be quite variable	Usually no peripheral chromatin
Large, not compact; may or may not be eccentric; may be diffuse and darkly stained	Large, irregularly shaped; may appear "blotlike"; many nuclear variations are common; may mimic *E. hartmanni* or *Dientamoeba fragilis*	Large; may be surrounded by refractile granules that are difficult to see ("basket nucleus")
Granular, with little differentiation into ectoplasm and endoplasm; usually vacuolated	Granular, vacuolated	
Bacteria, yeast, other debris	Bacteria	

TABLE 64-5   INTESTINAL PROTOZOA: CYSTS OF COMMON AMEBAE

CHARACTERISTIC	ENTAMOEBA HISTOLYTICA	ENTAMOEBA HARTMANNI
Size* (diameter or length)	10-20 μm; usual range: 12-15 μm	5-10 μm; usual range: 6-8 μm
Shape	Usually spherical	Usually spherical
Nucleus: number and visibility	Mature cyst: four; immature cyst: one or two nuclei; nuclear characteristics difficult to see on wet preparation	Mature cyst: four; immature cyst: one or two nuclei; two nucleated cysts very common
Peripheral chromatin (stained)	Peripheral chromatin present; fine, uniform granules, evenly distributed; nuclear characteristics may not be as clearly visible as in trophozoite	Fine granules evenly distributed on the membrane; nuclear characteristics may be difficult to see
Karyosome (stained)	Small, compact, usually centrally located but occasionally may be eccentric	Small, compact, usually centrally located
Cytoplasm, chromatoidal bodies (stained)	May be present; bodies usually elongate, with blunt, rounded, smooth edges; may be round or oval	Usually present; bodies usually elongate with blunt, rounded, smooth edges; may be round or oval
Glycogen (stained with iodine)	May be diffuse or absent in mature cyst; clumped chromatin mass may be present in early cysts (stains reddish brown with iodine)	May or may not be present as in E. histolytica

*Sizes refer to wet-preparation measurements. Organisms on a permanent-stained smear may be 1 to 1.5 μm smaller because of artificial shrinkage.

ENTAMOEBA COLI	ENDOLIMAX NANA	IODAMOEBA BÜTSCHLII
10-35 μm; usual range: 15-25 μm	5-10 μm; usual range: 6-8 μm	5-20 μm; usual range: 10-12 μm
Usually spherical; may be oval, triangular, or other shapes; may be distorted on permanent-stained slide because of inadequate fixative penetration	Usually oval, may be round	May vary from oval to round; cyst may collapse because of large glycogen vacuole space
Mature cyst: eight; occasionally 16 or more nuclei may be seen; immature cysts with two or more nuclei are occasionally seen	Mature cyst: four; immature cyst: two; very rarely seen and may resemble cysts of *Enteromonas hominis*	Mature cyst: one
Coarsely granular; may be clumped and unevenly arranged on membrane; nuclear characteristics not as clearly defined as in trophozoite; may resemble *E. histolytica*	No peripheral chromatin	No peripheral chromatin
Large, may or may not be compact and/or eccentric; occasionally may be centrally located	Smaller than karyosome seen in trophozoites, but generally larger than those of the genus *Entamoeba*	Larger, usually eccentric refractile granules may be on one side of karyosome ("basket nucleus")
May be present (less frequently than *E. histolytica*); splinter-shaped with rough, pointed ends	Rare chromatoidal bodies present; occasionally, small granules or inclusions seen; fine linear chromatoidals may be faintly visible on well-stained smears	No chromatoidal bodies present; occasionally small granules may be present
May be diffuse or absent in mature cyst; clumped mass occasionally seen in mature cysts (stains reddish brown with iodine)	Usually diffuse if present (will stain reddish brown with iodine)	Large, compact, well-defined mass (will stain reddish brown with iodine)

TABLE 64-6   INTESTINAL PROTOZOA: TROPHOZOITES OF FLAGELLATES

ORGANISM	CHARACTERISTIC SHAPE AND SIZE	MOTILITY
Dientamoeba fragilis	Shaped like amebae. 5-15 μm; usual range: 9-12 μm	Usually nonprogressive; pseudopodia are angular, serrated, or broad-lobed and almost transparent
Giardia lamblia	Pear-shaped, 10-20 μm; width, 5-15 μm	"Falling leaf" motility may be difficult to see if organism in mucus
Chilomastix mesnili	Pear-shaped, 6-24 μm; usual range: 10-15 μm; width, 4-8 μm	Stiff, rotary
Trichomonas hominis	Pear-shaped, 5-15 μm; usual range: 7-9 μm; width 7-10 μm	Jerky, rapid
Trichomonas tenax	Pear-shaped, 5-12 μm; average of 6.5-7.5 μm width, 7-9 μm	Jerky, rapid
Enteromonas hominis	Oval, 4-10 μm; usual range: 8-9 μm; width, 5-6 μm	Jerky
Retortamonas intestinalis	Pear-shaped or oval, 4-9 μm; usual range: 6-7 μm; width, 3-4 μm	Jerky

NUMBER OF NUCLEI AND VISIBILITY	NUMBER OF FLAGELLA (USUALLY DIFFICULT TO SEE)	OTHER FEATURES
Percentage may vary, but 40% of organisms have one nucleus and 60% have two nuclei; not visible in unstained preparations; no peripheral chromatin; karysome is composed of a cluster of 4-8 granules	No visible flagella	Cytoplasm finely granular and may be vacuolated with ingested bacteria, yeasts, and other debris; may be great variation in size and shape on a single smear
Two; not visible in unstained mounts	Four lateral, two ventral, two caudal	Sucking disk occupying $\frac{1}{2}$-$\frac{3}{4}$ of ventral surface pear-shaped front view; spoon-shaped side view
One; not visible in unstained mounts	Three anterior, one in cytostome	Prominent cytostome extending $\frac{1}{3}$-$\frac{1}{2}$ length of body; spiral groove across ventral surface
One; not visible in unstained mounts	Three to five anterior, one posterior	Undulating membrane extends length of body; posterior flagellum extends free beyond end of body
One; not visible in unstained mounts	Four anterior, one posterior	Seen only in preparations from mouth; axostyle (slender rod) protrudes beyond posterior end and may be visible; posterior flagellum extends only halfway down body and there is no free end
One; not visible in unstained mounts	Three anterior, one posterior	One side of body flattened; posterior flagellum extends free posteriorly or laterally
One; not visible in unstained mount	One anterior, one posterior	Prominent cytostome extending approximately $\frac{1}{2}$ length of body

TABLE 64-7 INTESTINAL PROTOZOA: CYSTS OF FLAGELLATES

SPECIES	SIZE	SHAPE	NUMBER OF NUCLEI	OTHER FEATURES
Dientamoeba fragilis, Trichomonas hominis, Trichomonas tenax	No cyst stage			
Giardia lamblia	No cyst stage; 8-19 μm; usual range: 11-14 μm; width, 7-10 μm	Oval, ellipsoidal, or may appear round	Four; not distinct in unstained preparations; usually located at one end	Longitudinal fibers in cysts may be visible in unstained preparations; deep-staining median bodies usually lie across the longitudinal fibers; there is often shrinkage and the cytoplasm pulls away from the cyst wall; there may also be a halo effect around the outside of the cyst wall because of shrinkage caused by dehydrating reagents
Chilomastix mesnili	No cyst stage; 6-10 μm; usual range: 7-9 μm; width, 4-6 μm	Lemonshaped with anterior hyaline knob	One; not distinct in unstained preparations	Cytostome with supporting fibrils, usually visible in stained preparation; curved fibril along side of cytostome usually referred to as "shepherd's crook"
Enteromonas hominis	No cyst stage; 4-10 μm; usual range: 6-8 μm; width, 4-6 μm	Elongate or oval	One to four; usually two lying at opposite ends of cyst; not visible in unstained mounts	Resembles *E. nana* cyst; fibrils or flagella usually not seen
Retortamonas intestinalis	4-9 μm; usual range: 4-7 μm; width, 5 μm	Pear-shaped or slightly lemonshaped	One; not visible in unstained mounts	Resembles *Chilomastix* cyst; shadow outline of cytostome with supporting fibrils extends above nucleus; bird beak fibril arrangement

TABLE 64-8   INTESTINAL PROTOZOA: CILIATES — *BALANTIDIUM COLI*

SPECIES	SHAPE AND SIZE	MOTILITY	NUMBER OF NUCLEI	OTHER FEATURES
Balantidium coli Trophozoite	Ovoid with tapering anterior end; 50-100 μm in length; 40-70 μm in width; usual range: 40-50 μm	Ciliates: rotary, boring; may be rapid	One large, kidney-shaped macronucleus; one small, round micronucleus, which is difficult to see even in stained smear; macronucleus may be visible in unstained preparation	Body covered with cilia, which tend to be longer near cytostome; cytoplasm may be vacuolated
Cyst	Spherical or oval, 50-70 μm; usual range: 50-55 μm		One large macronucleus visible in unstained preparation; micronucleus difficult to see	Macronucleus and contractile vacuole are visible in young cysts; in older cysts, internal structure appears granular; cilia difficult to see within cyst wall

TABLE 64-9 Morphologic Criteria Used to Identify Intestinal Protozoa (Coccidia, Microsporidia, *Blastocystis hominis*)

SPECIES	SHAPE/SIZE	OTHER FEATURES
Cryptosporidium parvum	Oocyst generally round, 4-6 μm, each mature oocyst containing four sporozoites	Oocyst, usual diagnostic stage in stool, sporozoites occasionally visible within oocyst wall; acid-fast positive using modified acid-fast stains; various other stages in life cycle can be seen in biopsy specimens taken from gastrointestinal tract (brush border of epithelial cells) and other tissues; disseminated infection well documented in compromised host
Cyclospora cayetanensis (formerly called CLBs)	Oocyst generally round, 8-10 μm, oocysts are not mature, no visible internal structure; oocysts may appear wrinkled	Oocyst, usual diagnostic stage in stool; acid-fast variable using modified acid-fast stains; color range from clear to deep purple (tremendous variation); oocysts may appear wrinkled (like crumpled cellophane); mimic *Cryptosporidium* oocysts but are twice as large
Isospora belli	Ellipsoidal oocyst; range 20-30 μm in length, 10-19 μm in width; sporocysts rarely seen broken out of oocysts but measure 9-11 μm	Mature oocyst contains two sporocysts with four sporozoites each; usual diagnostic stage in feces is immature oocyst containing spherical mass of protoplasm (intestinal tract); oocysts are modified acid-fast positive; whole oocyst may stain pink, not just the internal sporocyst
Sarcocystis hominis S. *suihominis* S. *bovihominis*	Oocyst thin-walled and contains two mature sporocysts, each containing four sporozoites; frequently thin oocyst wall ruptures; ovoidal sporocysts each measure 10-16 μm in length and 7.5-12 μm in width	Thin-walled oocyst or ovoid sporocysts occur in stool (intestinal tract)
S. *"lindemanni"*	Shapes and sizes of skeletal and cardiac muscle sarcocysts vary considerably	Sarcocysts contain from several hundred to several thousand trophozoites, each of which measures from 4-9 μm in width and 12-16 μm in length. The sarcocysts may also be divided into compartments by septa, not seen in *Toxoplasma* cysts (tissue/muscle)
Microsporidia	Small, oval spores (1-4 μm) can be found in routine histologic sections; however, electron microscopy has been used most successfully; spores in fecal specimens can also appear round; diagonal or horizontal stripe in modified trichrome stains often visible in some spores; some will also contain a terminal vacuole; the spore wall will fluoresce with optical brightening agents	Spores shed from enterocytes can now be identified in stool specimens. In addition to the modified trichrome procedure, preliminary results indicate monoclonal antibodies may provide a more sensitive detection method; optical brightening agents (Calcofluor, Fungi-Fluor, Uvitex 2B) are nonspecific staining agents for microsporidial spores; specimens from other body sites are also acceptable and have been found to be positive, depending on the genus involved
Blastocystis hominis	Organisms are generally round, measure approximately 6-40 μm, and are usually characterized by a large, central body (looks like a large vacuole); this stage has been called the *central body form*	The more amebic form can be seen in diarrheal fluid but will be difficult to identify; the central body forms vary tremendously in size, even on a single fecal smear; this is the most common form seen; routine fecal examinations may indicate a positive rate much higher than other protozoa; some laboratories report figures of 20% and higher

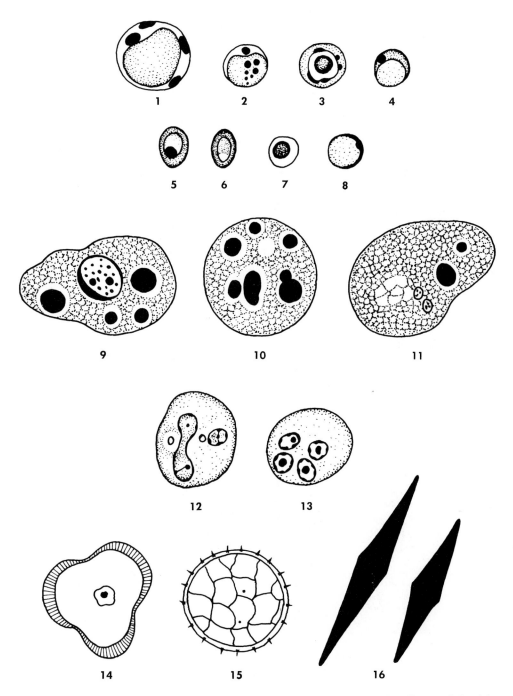

FIGURE 64-15 Various structures that may be seen in stool preparations. **1, 2,** and **4,** *Blastocystis hominis.* **3,** and **5** to **8,** Various yeast cells. **9,** Macrophage with nucleus. **10** and **11,** Deteriorated macrophage without nucleus. **12** and **13,** Polymorphonuclear leukocytes. **14** and **15,** Pollen grains. **16,** Charcot-Leyden crystals. (Modified from Markell, E.K. and Voge, M. 1981. Medical parasitology, ed 5. W.B. Saunders, Philadelphia. Illustration by Nobuko Kitamura.)

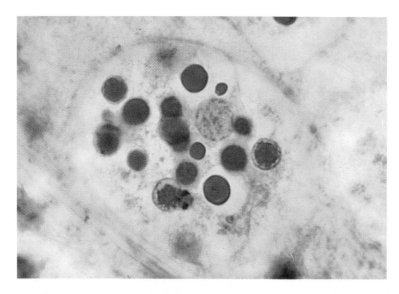

FIGURE 64-16 *Entamoeba histolytica* trophozoite containing ingested red blood cells.

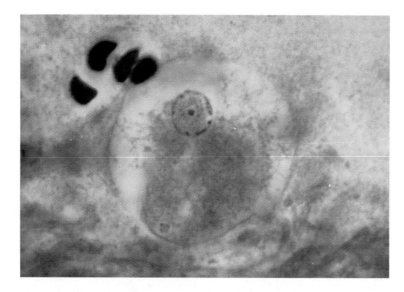

FIGURE 64-17 *Entamoeba histolytica* trophozoite.

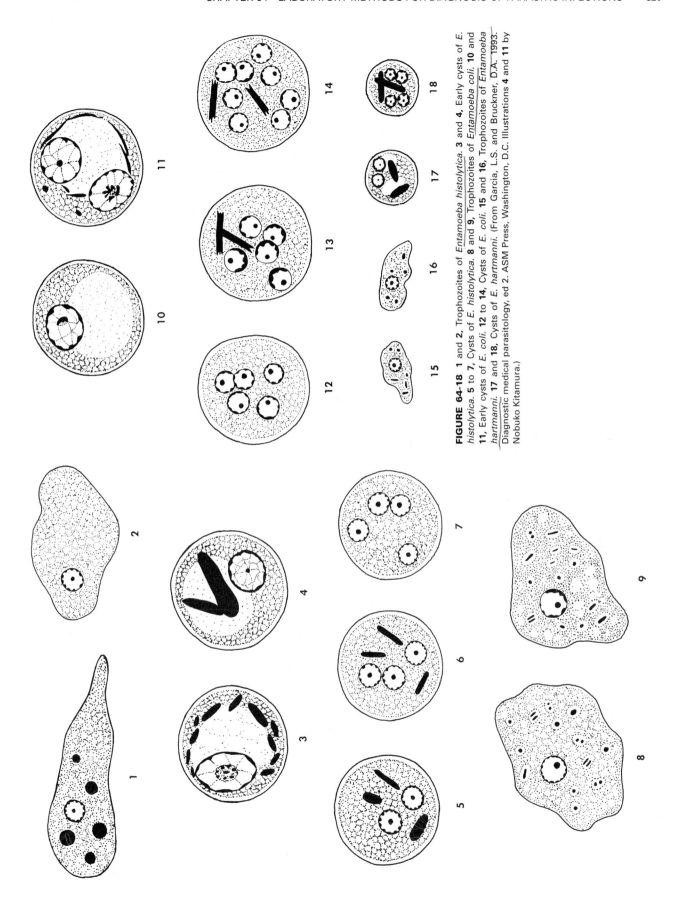

FIGURE 64-18 **1** and **2**, Trophozoites of *Entamoeba histolytica*. **3** and **4**, Early cysts of *E. histolytica*. **5** to **7**, Cysts of *E. histolytica*. **8** and **9**, Trophozoites of *Entamoeba coli*. **10** and **11**, Early cysts of *E. coli*. **12** to **14**, Cysts of *E. coli*. **15** and **16**, Trophozoites of *Entamoeba hartmanni*. **17** and **18**, Cysts of *E. hartmanni*. (From Garcia, L.S. and Bruckner, D.A. 1993. *Diagnostic medical parasitology*, ed 2. ASM Press, Washington, D.C. Illustrations **4** and **11** by Nobuko Kitamura.)

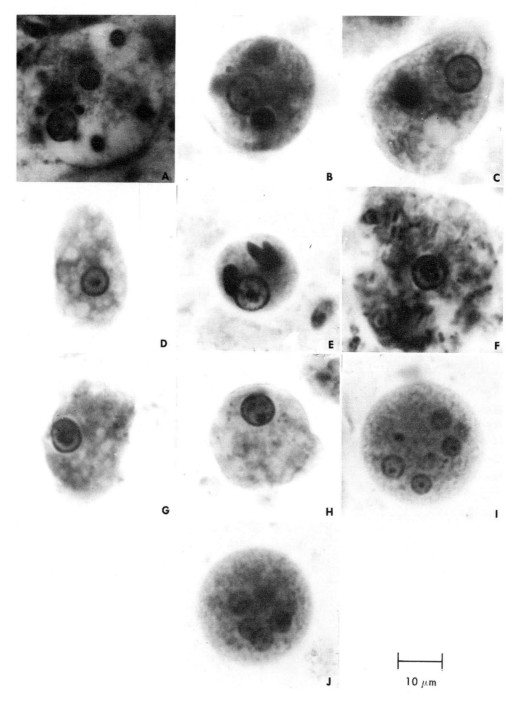

FIGURE 64-19 **A** to **D,** Trophozoites of *Entamoeba histolytica.* **E,** Early cysts of *E. histolytica.* **F** to **H,** Trophozoites of *Entamoeba coli.* **I** and **J,** Cysts of *E. coli.*

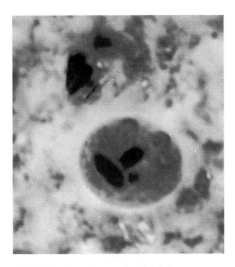

FIGURE 64-20   *Entamoeba histolytica* cyst.

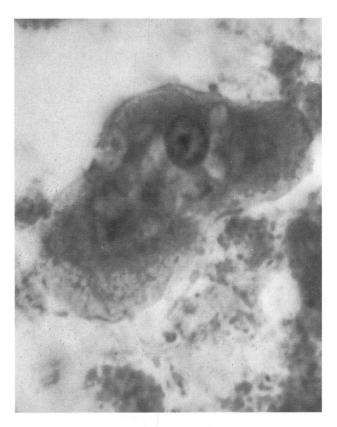

FIGURE 64-21   *Entamoeba coli* trophozoite.

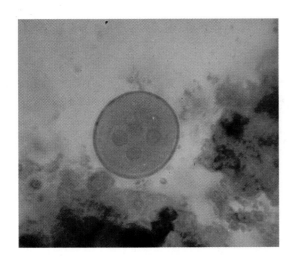

FIGURE 64-22   *Entamoeba coli* cyst, iodine stain.

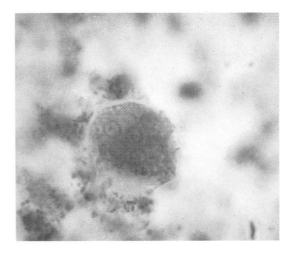

FIGURE 64-23   *Entamoeba coli* cyst, trichrome stain (poor preservation).

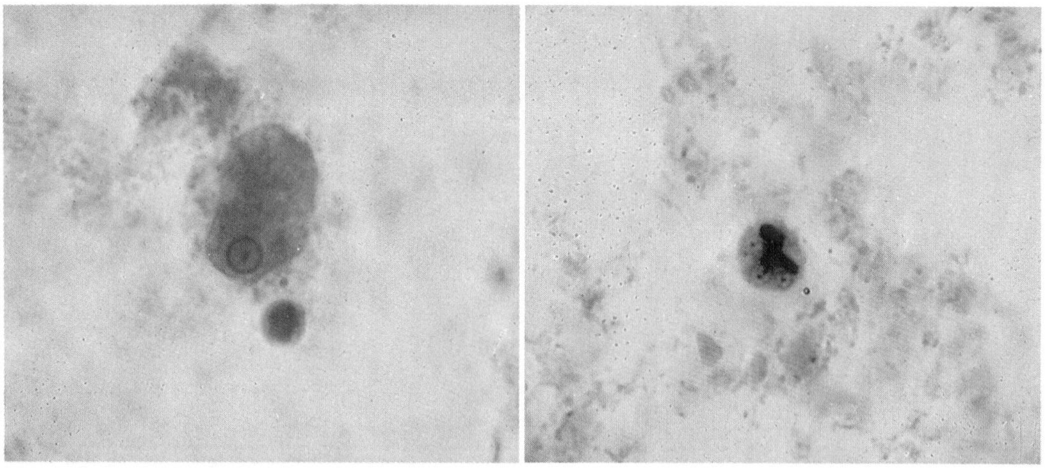

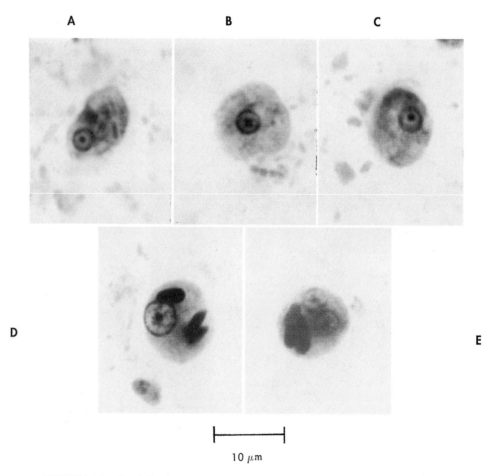

FIGURE 64-24 *Left, Entamoeba hartmanni* trophozoite. *Right, E. hartmanni* cyst.

FIGURE 64-25 A to **C,** Trophozoites of *Entamoeba hartmanni.* **D** and **E,** Cysts of *E. hartmanni.*

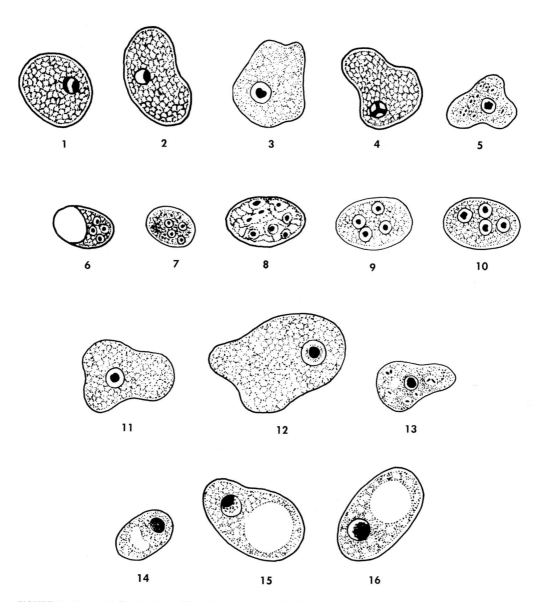

FIGURE 64-26 **1** to **5,** Trophozoites of *Endolimax nana.* **6** to **10,** Cysts of *E. nana.* **11** to **13,** Trophozoites of *Iodamoeba bütschlii.* **14** to **16,** Cysts of *I. bütschlii.* (From Garcia, L.S. and Bruckner, D.A. 1993. Diagnostic medical parasitology, ed 2. ASM Press, Washington, D.C.)

A   B   C

D   E

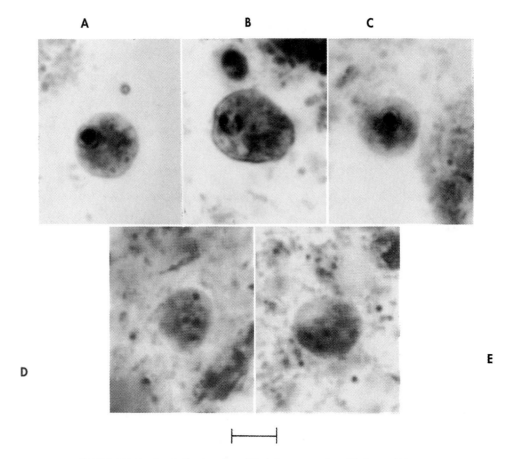

FIGURE 64-27 **A** to **C**, Trophozoites of *Endolimax nana.* **D** and **E,** Cysts of *E. nana.*

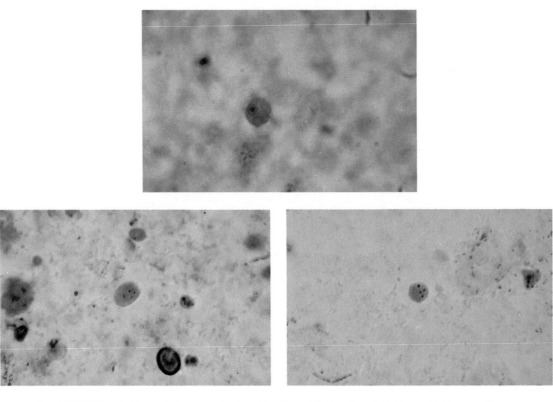

FIGURE 64-28 *Top, Endolimax nana* trophozoite. *Bottom left, E. nana* cyst. *Bottom right, E. nana* cyst.

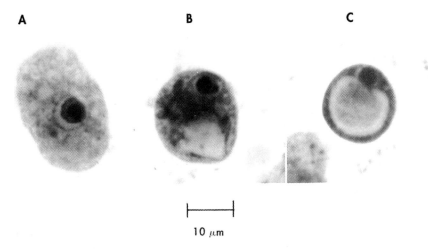

10 μm

FIGURE 64-29   **A,** Trophozoites of *Iodamoeba bütschlii.* **B** and **C,** *Cysts of I.* bütschlii.

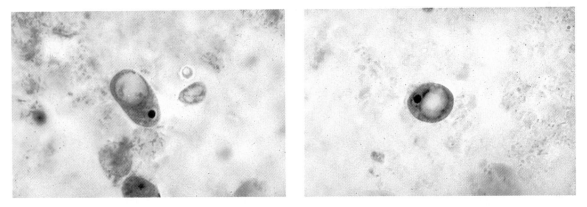

FIGURE 64-30   *Top, Iodamoeba bütschlii* trophozoites. *Bottom left, I. bütschlii* cyst. *Bottom right, I. bütschlii* cyst.

Both trophozoites and cysts of *B. mandrillaris* (Leptomyxid ameba) are found in many of the same central nervous system tissues as are *Acanthamoeba*.[79] The trophozoites are characterized by extensive branching and a single nucleus (occasionally, binucleate forms are seen) with a central karyosome, and they measure 15 to 60 μm. Occasionally a few elongated forms with several contractile vacuoles are seen. The cysts have a single nucleus (occasionally, binucleate forms are seen), have the typical double wall with the outer wall thick and irregular, and measure 15 to 30 μm. Leptomyxid amebae have also been shown not to grow well on *E. coli*-seeded nonnutrient agar plates.

Flagellates

Four common species of flagellates are found in the intestinal tract: *Giardia lamblia, Chilomastix mesnili, Trichomonas hominis,* and *Dientamoeba fragilis* (Figures 64-31 to 64-38). Several other smaller flagel-

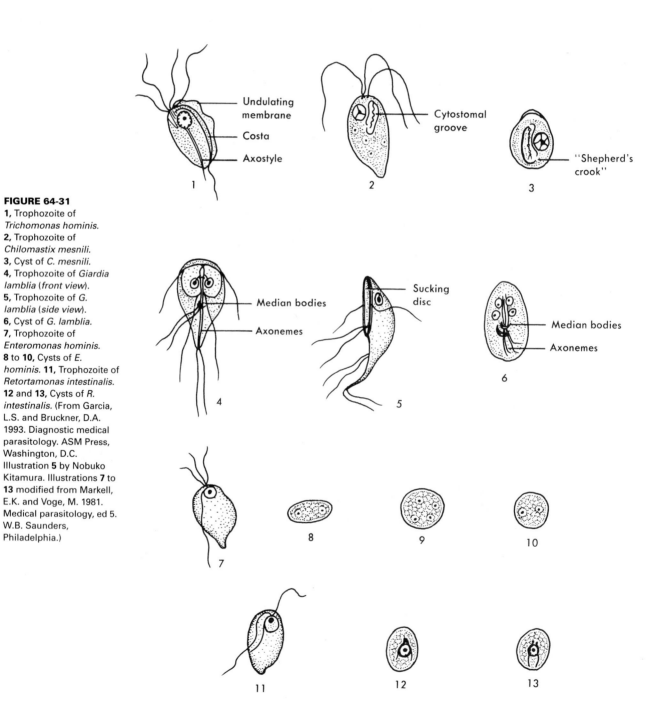

FIGURE 64-31
1, Trophozoite of *Trichomonas hominis.* **2,** Trophozoite of *Chilomastix mesnili.* **3,** Cyst of *C. mesnili.* **4,** Trophozoite of *Giardia lamblia* (front view). **5,** Trophozoite of *G. lamblia* (side view). **6,** Cyst of *G. lamblia.* **7,** Trophozoite of *Enteromonas hominis.* **8 to 10,** Cysts of *E. hominis.* **11,** Trophozoite of *Retortamonas intestinalis.* **12 and 13,** Cysts of *R. intestinalis.* (From Garcia, L.S. and Bruckner, D.A. 1993. Diagnostic medical parasitology. ASM Press, Washington, D.C. Illustration **5** by Nobuko Kitamura. Illustrations **7** to **13** modified from Markell, E.K. and Voge, M. 1981. Medical parasitology, ed 5. W.B. Saunders, Philadelphia.)

lates, such as *Enteromonas hominis* and *Retortamonas intestinalis* (Figure 64-31), are rarely seen, and none of the flagellates in the intestinal tract, except for *G. lamblia* and *D. fragilis,* is considered pathogenic. *Trichomonas vaginalis* is pathogenic but occurs in the urogenital tract. *Trichomonas tenax* is occasionally found in the mouth and may be associated with poor oral hygiene.

Except for *Dientamoeba*, the flagellates can be recognized by their characteristic rapid motility, which has been described as a "falling leaf" motion for *Giardia* and a jerky motion for the other species. Most flagellates have a characteristic pear shape and possess different numbers and arrangements of flagella, depending on the species. The sucking disk and axonemes of *Giardia*, the cytostome and spiral groove of *Chilomastix,* and the undulating membrane of *Trichomonas* are all distinctive criteria for identification (Figures 64-31 to 64-35).

Until recently, *Dientamoeba* was grouped with the amebae; however, electron microscopy (EM) studies have confirmed its correct classification with the flagellates, specifically the trichomonads. *Dientamoeba* has no known cyst stage and is characterized by having one or two nuclei, which have no peripheral chromatin and which have four to eight chromatin granules in a central mass. This organism varies in size and shape and may contain many ingested bacteria and other debris. *Dientamoeba* is inconspicuous in the wet mount and is consistently overlooked without the use of the stained smear (Figures 64-36 to 64-38).

Organisms can be recovered in fecal specimens from asymptomatic persons, but reports in the literature describe a wide range of symptoms, which include intermittent diarrhea, abdominal pain, nausea, anorexia, malaise, fatigue, poor weight gain, and unexplained eosinophilia.

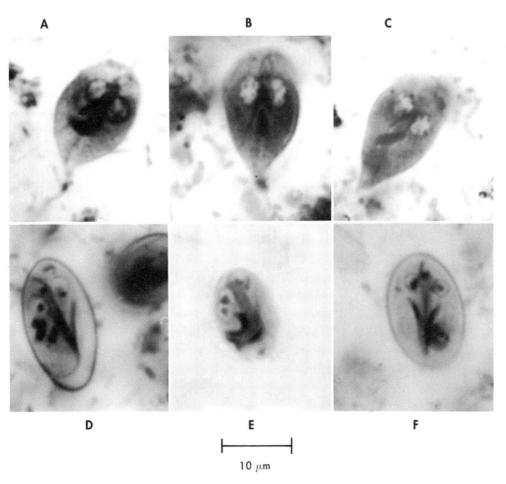

FIGURE 64-32 **A** to **C**, Trophozoites of *Giardia lamblia.* **D** to **F**, Cysts of *G. lamblia.*

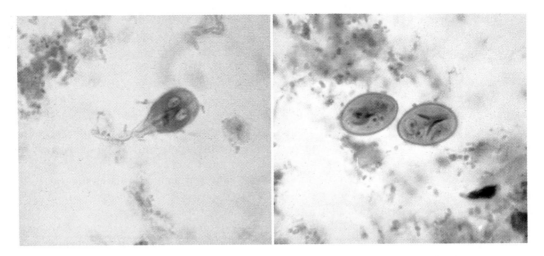

FIGURE 64-33 *Left, Giardia lamblia* trophozoite. *Right, G. lamblia* cysts.

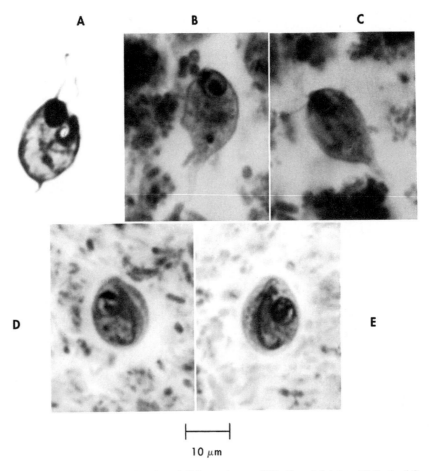

FIGURE 64-34 **A** to **C,** Trophozoites of *Chilomastix mesnili* (**A,** silver stain). **D** and **E,** Cysts of *C. mesnili.*

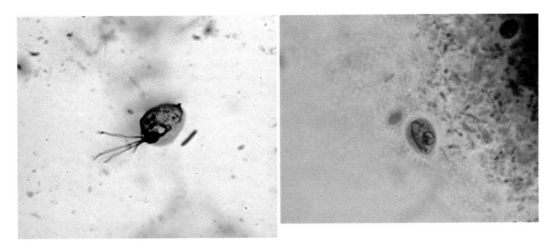

FIGURE 64-35   *Left, Chilomastix mesnili* trophozoite, silver stain. *Right, C. mesnili* cyst.

FIGURE 64-36   Trophozoites of *Dientamoeba fragilis.* (From Garcia, L.S. and Bruckner, D.A. 1993. Diagnostic medical parasitology, ed 2. ASM Press, Washington, D.C.)

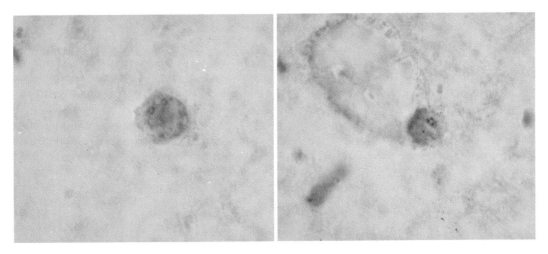

FIGURE 64-37   *Left, Dientamoeba fragilis,* two nuclei. *Right, D. fragilis,* one nucleus.

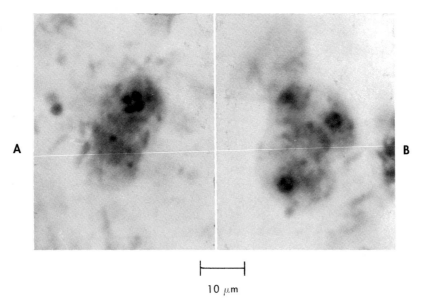

10 μm

FIGURE 64-38 **A** and **B**, Trophozoites of *Dientamoeba fragilis*.

Ciliates

Balantidium coli is the largest protozoan and the only ciliate that infects humans (Figure 64-39). The living trophozoites have a rotatory, boring motion, which is usually rapid. The organism's surface is covered by cilia, and the cytoplasm contains both a kidney-shaped macronucleus and a smaller, round micronucleus that is often difficult to see. The number of nuclei in the cyst remains the same as that in the trophozoite. *B. coli* infections are rarely seen in the United States; however, the organisms are easily recognized (Figure 64-40).

Coccidia

Isospora belli is now considered to be the only valid species of the genus *Isospora* that infects humans, since *Isospora hominis* has been placed in the genus *Sarcocystis*. *I. belli* is released from the intestinal wall as immature oocysts, so all stages from the immature oocyst containing a mass of undifferentiated protoplasm to those oocysts containing fully developed sporocysts and sporozoites are found in the stool (Figure 64-41). If passed in the immature condition, they mature within 4 or 5 days to form sporozoites. *I. belli* infections are becoming increasingly important as a cause of diarrhea in immunodeficient and immunosuppressed patients. Many recently reported infections have occurred in patients with AIDS. The diagnosis of *Isospora* infection is usually accomplished by finding oocysts in stool concentrates or rarely by direct wet-mount examination during routine ova and parasite examinations. Multiple stool examinations are recommended, since the oocysts are not continually shed in the stool during infections. Oocysts can be stained with rhodamine-auramine, modified acid-fast procedures, or Giemsa. The oocyst wall and sporocyst fluoresce bright yellow when stained with rhodamine-auramine. With the modified acid-fast stain (see Procedure 64-4), the oocyst appears pink with bright-red sporocysts. Giemsa (see Procedure 64-10) stains both the oocyst and the sporocyst pale blue; trichrome stain is not recommended because it stains the oocyst poorly or not at all. Multiple biopsies of the jejunal-duodenal brush border can also lead to diagnosis, since in many patients apparently only the asexual stages are seen.

Cryptosporidium sp. is another coccidian parasite that has been implicated in intestinal disease, primarily in immunosuppressed patients and particularly those with AIDS. These organisms may not be host specific, are probably transmitted by the fecal-oral route, and may also infect persons with a competent immune system. The developmental stages occur within a vacuole of host cell origin, which is located at the microvillous surface of the host epithelial cell (Figure 64-42). Although the infection is generally self-limiting in the immunocompetent person, the presence of autoinfective oocysts may explain why a small inoculum can lead to an overwhelming infection in compromised patients and why they may have persistent, life-threatening infections in the absence of documentation of repeated exposure to oocysts. In the immunodeficient individual, *Cryptosporidium* is not always confined to the gastrointestinal tract and

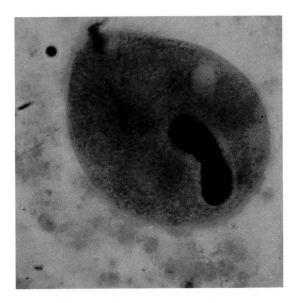

FIGURE 64-39 *Balantidium coli* trophozoite, iodine stain.

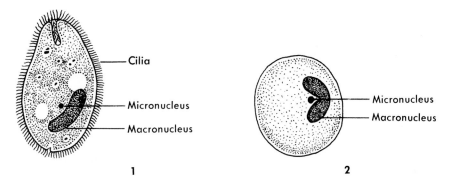

FIGURE 64-40 1, Trophozoite of *Balantidium coli.* **2,** Cyst of *B. coli.* (From Garcia, L.S. and Bruckner, D.A. 1993. Diagnostic medical parasitology, ed 2. ASM Press, Washington, D.C.)

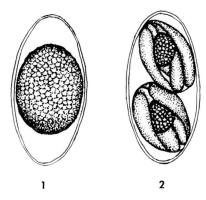

FIGURE 64-41 1, Immature oocyst of *Isospora belli.* **2,** Mature oocyst of *I. belli.* (Illustration by Nobuko Kitamura.)

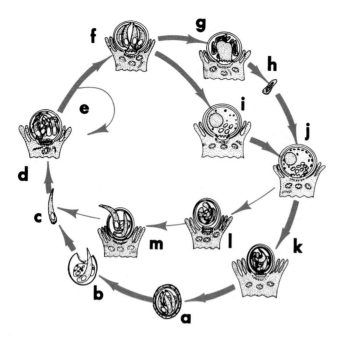

FIGURE 64-42 Proposed life cycle of *Cryptosporidium*. *(a)* Sporulated oocyst in feces. *(b)* Excystation in intestine. *(c)* Free sporozoite in intestine. *(d)* Type I meront (six or eight merozoites). *(e)* Recycling of Type I merozoite. *(f)* Type II meront (four merozoites). *(g)* Microgametocyte with approximately 16 microgametes. *(h)* Microgamete fertilizes macrogamete *(i)* to form zygote *(j)*. Approximately 80% of the zygotes form thick-walled oocysts *(k)*, which sporulate within the host cell. About 20% of the zygotes do not form an oocyst wall; their sporozoites are surrounded only by a unit membrane *(l)*. Sporozoites within "autoinfective," thin-walled oocysts *(l)* are released into the intestinal lumen *(m)* and reinitiate the endogenous cycle (at *c*). (Life cycle from William L. Current, Lilly Research Laboratories, Greenfield, Ind.)

has been associated with cholecystitis (biliary tree and gallbladder epithelium) and respiratory tract infections.

Previously, most patients were diagnosed by light or electron microscopic examination of small or large bowel biopsy material. However, cases are now diagnosed routinely from stool specimens and the use of specific stains and concentration techniques are often more sensitive than intestinal biopsies in diagnosing *Cryptosporidium*. The oocysts range from 4 to 6 μm and can be confused with yeast or overlooked without special techniques; various modified acid-fast stains are one of the best approaches.[22] In wet preparations when iodine is added, yeast cells stain with the iodine, but *Cryptosporidium* does not take up the stain. However, in light infections with few oocysts and many artifacts in the stool, this difference is very difficult to visualize. Sheather's sugar flotation is no longer routinely used by most laboratories and has been replaced by the routine ova and parasite concentration (Figure 64-43).[22,54] In addition to routine modified acid-fast stains, several immunoassay kits (EIA, FA) are now available commercially.

NOTE: The number of oocysts per stool specimen varies considerably, both from patient to patient and from specimen to specimen during an infection in a single patient. Because therapy for this infection is less than optimal, even with the use of spiramycin, one should check multiple specimens before assuming the patient is cured of the infection. With the development of very specific monoclonal reagents directed against components of the oocyst wall (Figure 64-44), the ability to screen large numbers of specimens is now available.[26]

During the past few years, several outbreaks of diarrhea have been associated with a spherical organism measuring 8 to 10 μm in diameter; the distribution of case reports is worldwide. The organisms were subsequently thought to be a new pathogen, possibly an oocyst, a flagellate, an unsporulated coccidian, a large *Cryptosporidium*, a blue-green alga (cyanobacterium-like body), or a coccidian-like body.[28,39,43,67] An epidemiologic report of one outbreak implicated exposure to a contaminated water source.[34] These organisms are now identified as coccidia in the genus *Cyclospora*. The name *Cyclospora cayetanensis* has been proposed for this newly described, disease-producing coccidian organism from humans.[56] In clean wet mounts the organisms are seen as nonrefractile spheres, stain orange with safranin, and are acid-fast variable with the modified acid-fast stain; those that are unstained appear as glassy, wrinkled spheres. Modified acid-fast methods stain the oocysts from light pink to deep red, some of which contain granules or a bubbly appearance (Figure 64-45). It is very important to be aware of these organisms when using the modified acid-fast stain for *Cryptosporidium parvum*. If other similar but larger structures (approximately twice the size of *Cryptosporidium* oocysts) are seen in the stained smear, laboratories should measure them. Also, the *Cyclospora* oocysts autofluoresce (strong green or intense blue) under ultraviolet (UV) epifluorescence (Figure 64-46).

Microsporidia

The microsporidia are obligate intracellular parasites that can infect both animals and humans, probably through the gastrointestinal tract, and have become associated with infections in the compromised host.* To date, six genera have been recognized in humans: *Encephalitozoon, Nosema, Vittaforma, Pleistophora, Enterocytozoon,* and "*Microsporidium*," a catchall genus for those organisms not yet classified. Classification criteria include spore size, configuration of the nuclei within the spores and developing forms, the number of polar tubule coils within the spore, and the

---

*References 20, 68, 71, 77, 78.

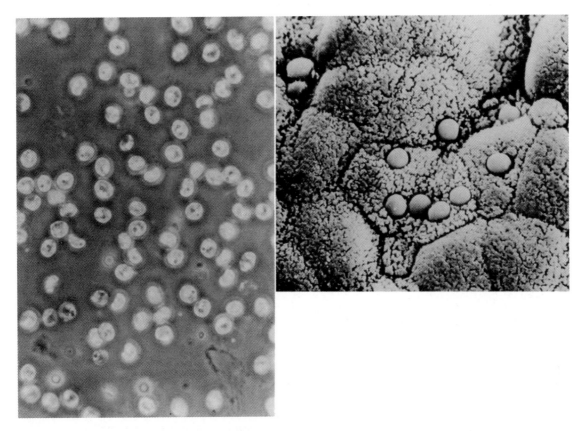

FIGURE 64-43 *Cryptosporidium.* **A,** Oocysts recovered from a Sheather's sugar flotation; organisms measure 4 to 6 μm. **B,** Scanning electron microscopy view of organisms at brush border of epithelial cells. (From Garcia, L.S. and Bruckner, D.A. 1993. Diagnostic medical parasitology, ed 2. ASM Press, Washington D.C.)

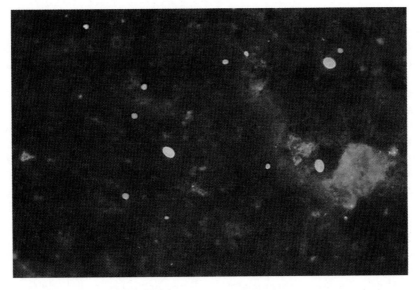

FIGURE 64-44 *Cryptosporidium* oocysts and *Giardia* cysts stained with monoclonal antibody–conjugated fluorescent reagent. (Courtesy Merifluor, Meridian Diagnostics, Cincinnati.)

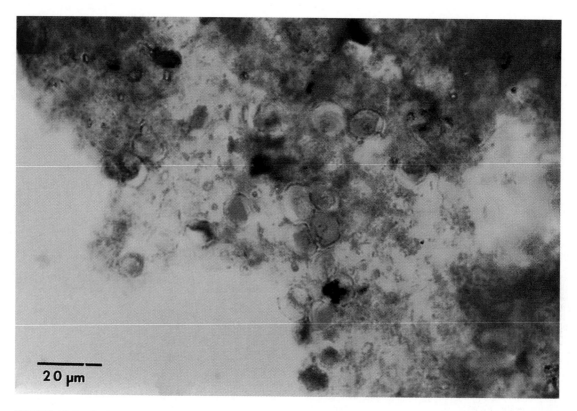

FIGURE 64-45 *Cyclospora* sp. (formerly called CLBs) oocysts after modified acid-fast staining. Note the variability in the intensity of stain. These oocysts are approximately twice the size of *Cryptosporidium* sp. (Courtesy Charles R. Sterling, University of Arizona.)

relationship between the organism and host cell. Infection occurs with the introduction of infective sporoplasm through the polar tubule into the host cell. The microsporidia multiply extensively within the host cell cytoplasm; the life cycle includes repeated divisions by binary fission (merogony) or multiple fission (schizogony) and spore production (sporogony) (Figures 64-47 and 64-48). Both merogony and sporogony can occur in the same cell at the same time. During sporogony, a thick spore wall is formed, thus providing environmental protection for this infectious stage of the parasite. An example of infection potential is illustrated by *Enterocytozoon bieneusi*, an intestinal pathogen. The spores are released into the intestinal lumen and are passed in the stool. These spores are environmentally resistant and can then be ingested by other hosts. There is also evidence for inhalation of spores and evidence in animals that suggests human microsporidiosis may also be transmitted via the rectal route.

The number of documented cases is dramatically increasing, with the majority being reported in patients with AIDS; infection with *E. bieneusi* and *Encephalitozoon intestinalis* tend to be the most com-

mon microsporidia currently being reported in these patients. Chronic intractable diarrhea accompanied by fever, malaise, and weight loss are symptoms of *E. bieneusi* and *E. intestinalis* infections, similar to those seen with cryptosporidiosis or isosporiasis. Patients diagnosed with AIDS tend to have four to eight watery, nonbloody stools daily, which can be accompanied by nausea and anorexia. *E. intestinalis* can disseminate to other body sites, primarily the kidneys; in these patients, microsporidia spores can also be found in urine specimens.

The diagnosis can be made by identifying the spores in biopsy or autopsy specimens; all body organs have been involved, including the eye (Figure 64-49 and 64-50). The resistant spores measure approximately 1 to 4 μm and do not stain well with hematoxylin-eosin. They are occasionally acid-fast and have a periodic acid-Schiff (PAS)–positive polar granule at the anterior end. Definitive diagnosis can also be made using electron microscopy (Figure 64-51). (See earlier section on biopsy material for information on the concentrated trichrome stain and IFA procedures.) Several modified trichrome stains are being routinely used on concentrated, formalin-fixed stool.[41,63,81]

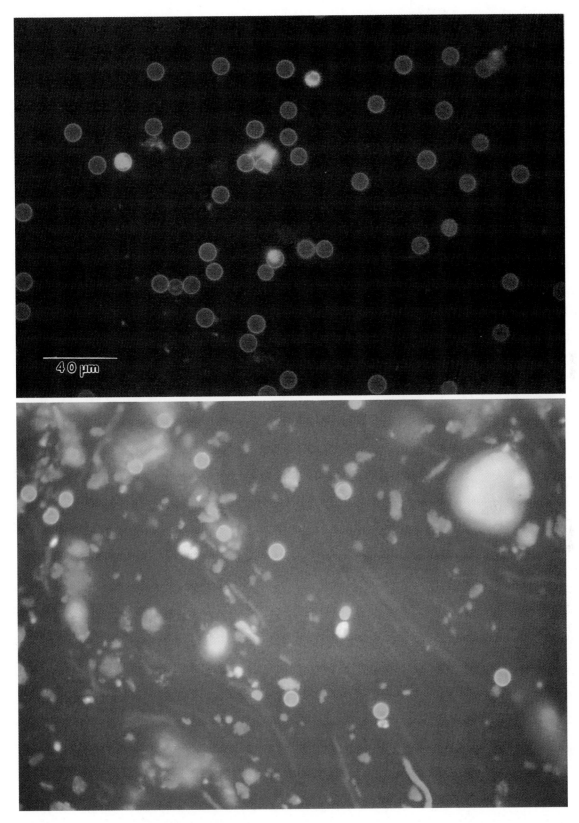

FIGURE 64-46 *Cyclospora* sp. (formerly called CLBs) oocysts exhibiting autofluorescence. (Top photograph courtesy Charles R. Sterling, University of Arizona; bottom photograph courtesy G. Long, Centers for Disease Control.)

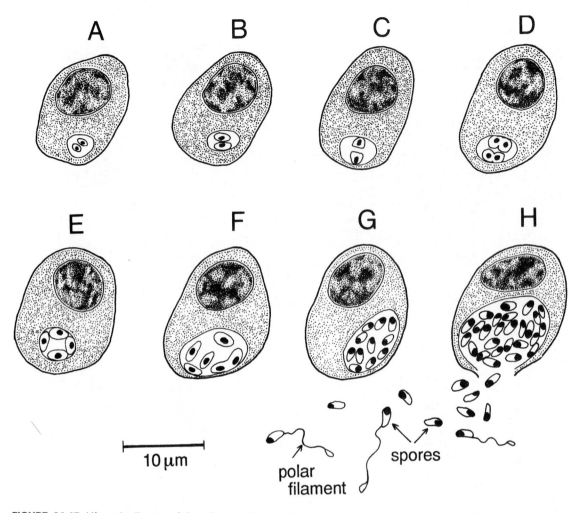

FIGURE 64-47 Life cycle diagram of the microsporidia. **A** to **G,** Asexual development of sporoblasts. **H,** Release of spores. (Modified from Gardiner, C.H., Fayer, R., and Dubey, J.P. 1988. An atlas of protozoan parasites in animal tissues. U.S. Dept of Agriculture, Agriculture Handbook No. 651. Illustration by Sharon Belkin.)

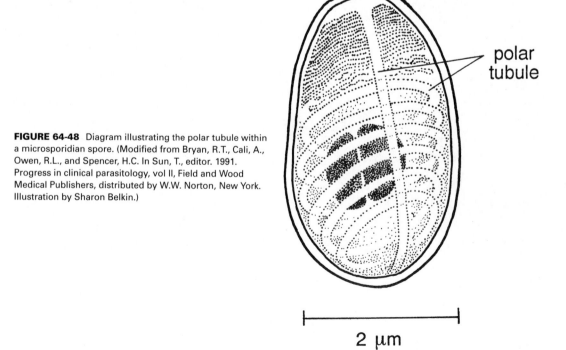

FIGURE 64-48 Diagram illustrating the polar tubule within a microsporidian spore. (Modified from Bryan, R.T., Cali, A., Owen, R.L., and Spencer, H.C. In Sun, T., editor. 1991. Progress in clinical parasitology, vol II, Field and Wood Medical Publishers, distributed by W.W. Norton, New York. Illustration by Sharon Belkin.)

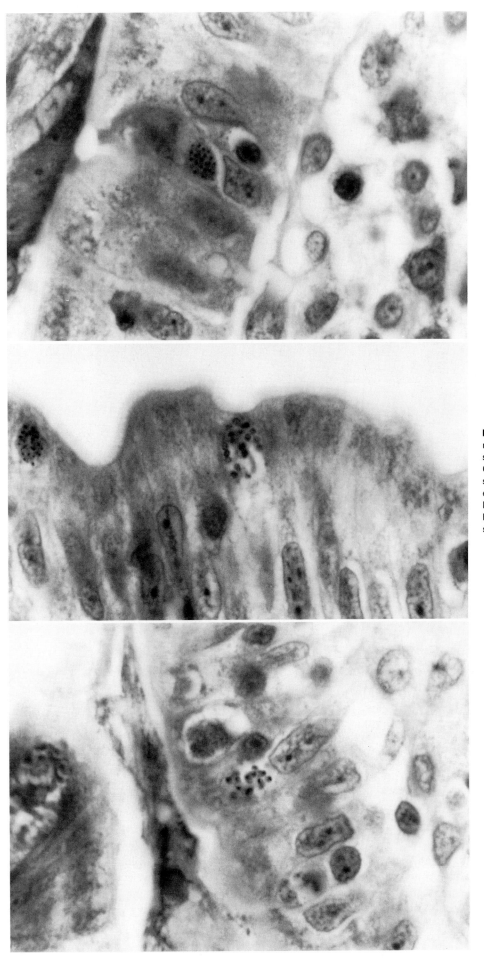

FIGURE 64-49 Routine histology micrograph of microsporidian spores in enterocytes, stained with Giemsa stain. *Upper,* note the small size. *Middle,* the spores are more easily seen; note the position between the nucleus and the brush border of the cell. *Lower,* in these spores, the granule is easily seen.

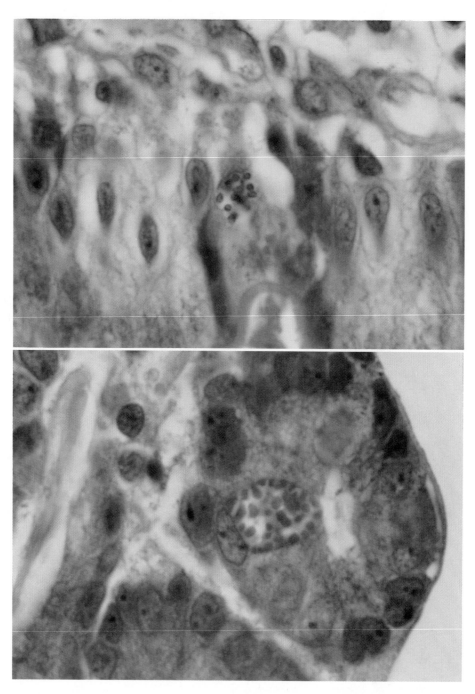

FIGURE 64-50 Routine histology micrograph of microsporidian spores in enterocytes, stained with Giemsa stain. *Upper,* note the fully formed spores. *Lower,* these spores are not fully mature.

BLOOD PROTOZOA

Malaria

Malaria is caused by four species of the protozoan genus *Plasmodium: P. vivax, P. falciparum* (Figure 64-52), *P. malariae,* and *P. ovale* (Table 64-10). Humans become infected when the sporozoites are introduced into the blood from the salivary secretion of the infected mosquito when the mosquito vector takes a blood meal. These sporozoites then leave the blood and enter the parenchymal cells of the liver, where they undergo asexual multiplication. This development in the liver before red blood cell invasion is called the **preerythrocytic cycle;** if further liver development takes place after red blood cell invasion, it is called the **exoerythrocytic cycle.** The length of time for the preerythrocytic cycle and the number of asexual generations vary depending on the species; however, the schizonts eventually rupture, releasing thou-

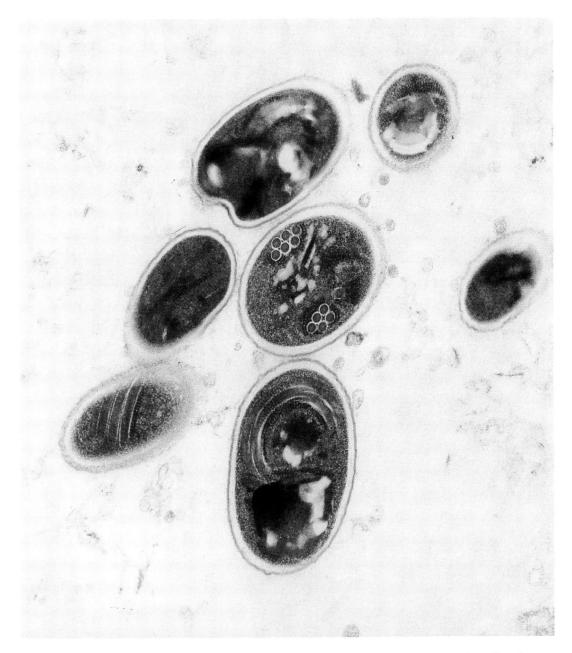

FIGURE 64-51 Transmission electron micrograph of six *Enterocytozoon bieneusi* spores in a stool specimen. Note the cross sections of the polar tubule. (From Orenstein, J.M. 1991. J. Parasitol. 77:843.)

sands of merozoites into the bloodstream, where they invade the erythrocytes. The early forms in the red blood cells are called **ring forms,** or **young trophozoites** (Figure 64-53). As the parasites continue to grow and feed, they become actively ameboid within the red blood cell. They feed on hemoglobin, which is incompletely metabolized; the residue that is left is called **malarial pigment** and is a compound of hematin and protein.

During the next phase of the cycle, the chromatin (nuclear material) becomes fragmented through-

out the microorganism, and the cytoplasm begins to divide, each portion being arranged with a fragment of nuclear material. These forms are called **mature schizonts** and are composed of individual merozoites. The infected red blood cell then ruptures, releasing the merozoites and also metabolic products into the bloodstream. If many red blood cells rupture simultaneously, a malarial paroxysm may result from the amount of toxic materials released into the bloodstream. In the early stages of infection or in a mixed infection with two species, rupture of the red blood

TABLE 64-10   Microscopic Identification of Plasmodia of Humans in Giemsa-Stained Thin Blood Films

	PLASMODIUM VIVAX	PLASMODIUM MALARIAE	PLASMODIUM FALCIPARUM	PLASMODIUM OVALE
Persistence of exoerythrocytic cycle	Yes	No	No	Yes
Relapses	Yes	No, but long-term recrudescences are recognized	No long-term relapses	Possible, but usually spontaneous recovery
Time of cycle	44-48 hrs	72 hrs	36-48 hrs	48 hrs
Appearance of parasitized red blood cells; size and shape	1½ to 2 times larger than normal; oval to normal; may be normal size until ring fills ½ of cell	Normal shape; size may be normal or slightly smaller	Both normal	60% of cells larger than normal and oval; 20% have irregular, frayed edges
Schüffner's dots (eosinophilic stippling)	Usually present in all cells except early ring forms	None	None; occasionally, commalike red dots are present (Maurer's dots)	Present in all stages, including early ring forms; dots may be larger and darker than in P. vivax
Color of cytoplasm	Decolorized, pale	Normal	Normal, bluish tinge at times	Decolorized, pale
Multiple rings/cell	Occasional	Rare	Common	Occasional
Developmental stages present in peripheral blood	All stages present	Ring forms few, as ring stage brief; mostly growing and mature trophozoites and schizonts	Young ring forms and no older stages; few gametocytes	All stages present
Appearance of parasite; young trophozoite (early ring form)	Ring is ⅓ diameter of cell, cytoplasmic circle around vacuole; heavy chromatin dot	Ring often smaller than in P. vivax, occupying ⅛ of cell; heavy chromatin dot; vacuole at times "filled in"; pigment forms early	Delicate, small ring with small chromatin dot (frequently two); scanty cytoplasm around small vacuoles; sometimes at edge of red cell (appliqué form) or filamentous slender form; may have multiple rings per cell	Ring is larger and more ameboid than in P. vivax, otherwise similar to P. vivax

Growing trophozoite	Multishaped, irregular ameboid parasite; streamers of cytoplasm close to large chromatin dot; vacuole retained until close to maturity; increasing amounts of brown pigment	Nonameboid rounded or band-shaped solid forms; chromatin may be hidden by coarse, dark brown pigment	Heavy ring forms; fine pigment grains	Ring shape maintained until late in development; nonameboid compared with *P. vivax*
Mature trophozoite	Irregular ameboid mass; one or more small vacuoles retained until schizont stage; fills almost entire cell; fine, brown pigment	Vacuoles disappear early; cytoplasm compact, oval, band-shaped, or nearly round almost fills cell; chromatin may be hidden by peripheral, coarse, dark brown pigment	Not seen in peripheral blood (except in severe infections); development of all phases following ring form occurs in capillaries of viscera	Compact; vacuoles disappear; pigment dark brown, less than in *P. malariae*
Schizont (presegmenter)	Progressive chromatin division; cytoplastic bands containing clumps of brown pigment	Similar to *P. vivax* except smaller; darker, larger pigment granules peripheral or central	Not seen in peripheral blood (see above)	Smaller and more compact than *P. vivax*
Mature schizont	Merozoites, 16 (12-24) each with chromatin and cytoplasm, filling entire red cell, which can hardly be seen	8 (6-12) merozoites in rosettes or irregular clusters filling normal-sized cells, which can hardly be seen; central arrangement of brown-green pigment	Not seen in peripheral blood	$\frac{3}{4}$ of cells occupied by 8 (8-12) merozoites in rosettes or irregular clusters
Macrogametocyte	Rounded or oval homogeneous cytoplasm; diffuse, delicate, light brown pigment throughout parasite; eccentric compact chromatin	Similar to *P. vivax*, but fewer in number; pigment darker and more coarse	Sex differentiation difficult; "crescent" or "sausage" shapes characteristic; may appear in "showers," black pigment near chromatin dot, which is often central	Smaller than *P. vivax*
Microgametocyte	Large pink to purple chromatin mass surrounded by pale or colorless halo; evenly distributed pigment	Similar to *P. vivax* but fewer in number; pigment darker and more coarse	Same as macrogametocyte (described above)	Smaller than *P. vivax*
Main criteria	Large, pale red cell; trophozoite irregular; pigment usually present; Schüffner's dots not always present; several phases of growth seen in one smear; gametocytes appear as early as third day	Red cell normal in size and color; trophozoites compact, stain usually intense, band forms not always seen; no stippling of red cells; coarse pigment; gametocytes appear after a few weeks	Development following ring stage takes place in blood vessels of internal organs; delicate ring forms and crescent-shaped gametocytes are only forms normally seen in peripheral blood; gametocytes appear after 7-10 days	Red cell enlarged, oval, with fimbriated edges; Schüffner's dots seen in all stages; gametocytes appear after 4 days or as late as 18 days

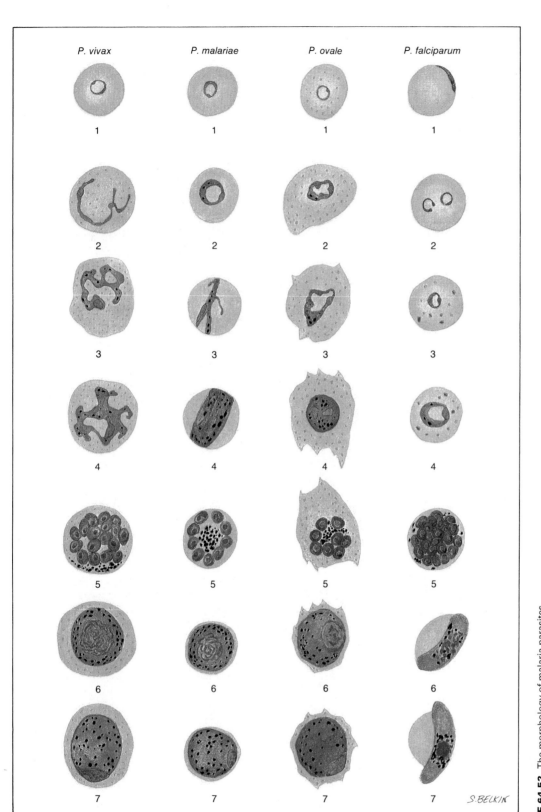

FIGURE 64-52 The morphology of malaria parasites.

Plasmodium vivax
1, Early trophozoite (ring form) **2,** Late trophozoite with Schüffner's dots (note enlarged red blood cell) **3,** Late trophozoite with ameboid cytoplasm **5,** Mature schizont with merozoites (18) and clumped pigment **6,** Microgametocyte with dispersed chromatin **7,** Macrogametocyte with compact chromatin

Plasmodium malariae
1, Early trophozoite (ring form) **2,** Early trophozoite with thick cytoplasm **3,** Early trophozoite (band form) **4,** Late trophozoite (band form) with heavy pigment **5,** Mature schizont with merozoites (9) arranged in rosette **6,** Microgametocyte with dispersed chromatin **7,** Macrogametocyte with compact chromatin

Plasmodium ovale
1, Early trophozoite (ring form) with Schüffner's dots **2,** Early trophozoite (note enlarged red blood cell) **3,** Late trophozoite in red blood cell with fimbriated edges **4,** Developing schizont with irregular-shaped red blood cell **5,** Mature schizont with merozoites (8) arranged irregularly **6,** Microgametocyte with dispersed chromatin **7,** Macrogametocyte with compact chromatin

Plasmodium falciparum
1, Early trophozoite (accolé or appliqué form) **2,** Early trophozoite (one ring is in headphone configuration/double chromatin dots) **3,** Early trophozoite with Maurer's dots **4,** Late trophozoite with larger ring and Maurer's dots **5,** Mature schizont with merozoites (24) **6,** Microgametocyte with dispersed chromatin **7,** Macrogametocyte with compact chromatin

NOTE: Without the appliqué form, Schüffner's dots, multiple rings/cell, and other developing stages, differentiation among the species can be very difficult. It is obvious that the early rings of all four species can mimic one another very easily. *Remember: one set of negative blood films cannot rule out a malaria infection.* (Reprinted by permission of the publisher from Garcia, L.S. and Bruckner, D.A. 1993. Diagnostic medical parasitology, ed 2. p. 126. Copyright by American Society for Microbiology, Washington, D.C.)

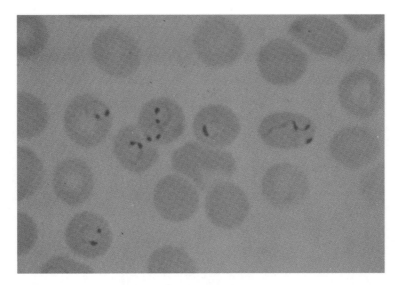

FIGURE 64-53 *Plasmodium falciparum,* early ring forms.

cells is usually not synchronous; consequently, the fever may be continuous or daily rather than intermittent. After several days, a 48- or 72-hour periodicity is usually established.

After several generations of erythrocytic schizogony, the production of gametocytes begins. These forms are derived from merozoites, which do not undergo schizogony but continue to grow and form the male and female gametocytes that circulate in the bloodstream. When the mature gametocytes are ingested by the appropriate mosquito vector, the sexual cycle is initiated within the mosquito with the eventual production of the sporozoites, the infective stage for humans (Figure 64-54).[27]

The asexual and sexual forms just described circulate in the human bloodstream in three species of *Plasmodium.* However, in *P. falciparum* infections, as the parasite continues to grow, the red blood cell membrane becomes sticky, and the cells tend to adhere to the endothelial lining of the capillaries of the internal organs. Thus, only the ring forms and crescent-shaped gametocytes occur in the peripheral blood.[22,49,83] Interference with normal blood flow in these vessels gives rise to additional problems, which are responsible for the different clinical manifestations of this type of malaria.

In some areas of the world where *P. falciparum* is endemic, many people are carriers of hemoglobin S (HbS), thalassemia, and glucose-6-phosphate dehydrogenase (G6PD) deficiency. These factors are associated with increased resistance to falciparum malaria. It has been suggested that the resistance of many West Africans and approximately 70% of American blacks to *P. vivax* may be related to the high incidence of Duffy-negative erythrocytes in these groups.[49]

There continue to be imported cases of malaria into the United States each year.[11,42,73] It is also well known that appropriate vectors for mosquito-borne transmission are present within the United States. *The request for examination of blood smears for parasites is always considered a STAT request.*

Babesiosis

Babesia are tickborne (they also can be transmitted via a blood transfusion) sporozoan parasites that have generally been considered parasites of animals (Texas cattle fever) rather than of humans. However, there are now many well-documented cases in humans, with some infections occurring in splenectomized patients and others in patients with intact spleens. *Babesia* organisms infect the red blood cells and appear as pleomorphic, ringlike structures when stained with any of the recommended stains used for blood films (Figure 64-55). They may be confused with the ring forms in *Plasmodium* infections; however, in a *Babesia* infection, four or five rings often are seen per red blood cell, and the individual rings are small compared with those found in malaria infections.[22,32]

Hemoflagellates

Hemoflagellates are blood and tissue flagellates, two genera of which are medically important for humans: *Leishmania* and *Trypanosoma.* Some species may circulate in the bloodstream or at times may be present in lymph nodes or muscle. Other species tend to parasitize the reticuloendothelial cells of the hematopoietic organs. The hemoflagellates of human beings have four morphologic types (Figure 64-56): amastigote (leishmanial form, or Leishman-Donovan [L-D] body), pro-

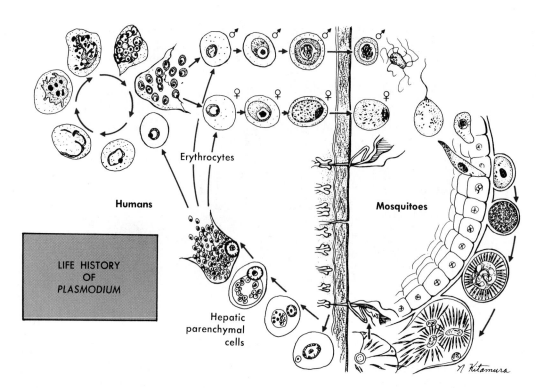

FIGURE 64-54 Life cycle of *Plasmodium*. (Modified from Wilcox, A. 1960. Manual for the microscopical diagnosis of malaria in man. U.S. Public Health Service, Washington, D.C. Illustration by Nobuko Kitamura.)

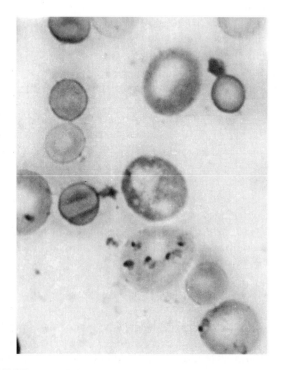

FIGURE 64-55 *Babesia* in red blood cells. (Photomicrograph by Zane Price. From Markell, E.K., and Voge, M. 1981. Medical parasitology, ed 5. W.B. Saunders, Philadelphia.)

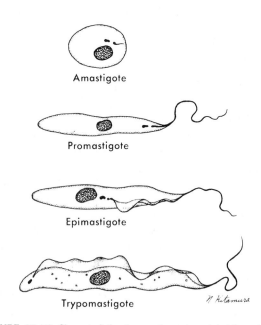

FIGURE 64-56 Characteristic stages of species of *Leishmania* and *Trypanosoma* in human and insect hosts. (Illustration by Nobuko Kitamura.)

mastigote (leptomonal form), epimastigote (crithidial form), and trypomastigote (trypanosomal form).

The amastigote form is an intracellular parasite in the cells of the reticuloendothelial system and is oval, measuring approximately 1.5 to 5 μm, and contains a nucleus and kinetoplast. *Leishmania* spp. usually exist as the amastigote form in humans and in the promastigote form in the insect host. The life cycle is essentially the same for all three species, and the clinical manifestations vary depending on the species involved. As the vector takes a blood meal, the promastigote form is introduced into a human, thus initiating the infection. Depending on the species, the parasites then move from the site of the bite to the organs of the reticuloendothelial system (liver, spleen, bone marrow) or to the macrophages of the skin (Figure 64-57).

Species based on clinical grounds are morphologically the same; however, differences exist in serologies and in growth requirements for culture. Great biologic variation occurs among the many strains that make up these groups. *Leishmania tropica* causes oriental sore or cutaneous leishmaniasis of the Old World, *L. braziliensis* causes mucocutaneous leishmaniasis of the New World, and *L. donovani* causes visceral leishmaniasis (Dumdum fever, or kala-azar) (Figure 64-58). Additional species have been delineated based on the buoyant density of kinetoplast DNA, the isoenzyme patterns, and serologic testing.

In tissue impression smears or sections, *Histoplasma capsulatum* must be differentiated from the L-D bodies. *H. capsulatum* does not have a kinetoplast and stains with both PAS and GMS stains, neither of which stains L-D bodies. Diagnosis of leishmanial organisms is based on the demonstration of the L-D bodies or the recovery of the promastigote in culture.

Three species of trypanosomes are pathogenic for humans: *Trypanosoma brucei gambiense* (Figure 64-59) causes West African sleeping sickness, *Trypanosoma brucei rhodesiense* causes East African sleeping sickness, and *Trypanosoma cruzi* causes South American trypanosomiasis, or Chagas' disease (Figure 64-60). The first two species are morphologically similar and produce African sleeping sickness, an illness characterized by both acute and chronic stages. In the acute stage of the disease, the organisms can usually be found in the peripheral blood or lymph node aspirates. As the disease progresses to the chronic stage, the organisms can be found in the cerebrospinal fluid (CSF) (comatose stage: "sleeping sickness"). *T. rhodesiense* produces a more severe infection, usually resulting in death within 1 year.

In the early stages of infection with *T. cruzi,* the trypomastigote forms appear in the blood but do not multiply (Figure 64-61, *A*). They then invade the endothelial or tissue cells and begin to divide, producing many L-D bodies, which are most often found in cardiac muscle (Figure 64-61, *B*). When these forms are liberated into the blood, they transform into the trypomastigote forms, which are then carried to other sites, where tissue invasion again occurs.

Diagnosis of the infection is based on demonstration of the parasites, most often on wet unstained or stained blood films. Both thick and thin films should be examined; these can be prepared from peripheral blood or buffy coat. The sediment recovered from CSF can also be examined for the presence of trypomastigotes. Specific techniques for culture, animal inoculation, handling of aspirate and biopsy material, and serologic procedures are presented in earlier sections of this chapter.

Another technique often used in endemic areas for the diagnosis of Chagas' disease is xenodiagnosis. Triatomids, the insect vector, are raised in the laboratory and are determined to be free from infection with *T. cruzi.* These insects are allowed to feed on the blood of an individual suspected of having Chagas' disease, and after 2 weeks the intestinal contents are checked for the presence of the epimastigote forms.[22]

INTESTINAL HELMINTHS

The intestinal helminths that infect humans belong to two phyla: the Nematoda, or roundworms, and the platyhelminths, or flatworms. The platyhelminths, most of which are hermaphroditic, have a flat, bilaterally symmetric body. The two classes, Trematoda and Cestoda, contain organisms that are parasitic for human beings.

The trematodes (flukes) are leaf-shaped or elongate, slender organisms (blood flukes: *Schistosoma* spp.) that possess hooks or suckers for attachment. Members of this group, which parasitize humans, are found in the intestinal tract, liver, blood vessels, and lungs.

The cestodes (tapeworms) typically have a long, segmented, ribbonlike body, which has a special attachment portion, or scolex, at the anterior end. Adult forms inhabit the small intestine; however, humans may be host to either the adult or the larval forms, depending on the species. The cestodes, as well as the trematodes, require (with few exceptions) one or more intermediate hosts for the completion of their life cycle (Figures 64-62 and 64-63).

The nematodes, or roundworms, are elongate, cylindric worms with a well-developed digestive tract. The sexes are separate, with the male usually smaller than the female. Intermediate hosts are required for larval development in certain species; many species parasitize the intestinal tract and certain tissues of humans (Figures 64-64 to 64-66).

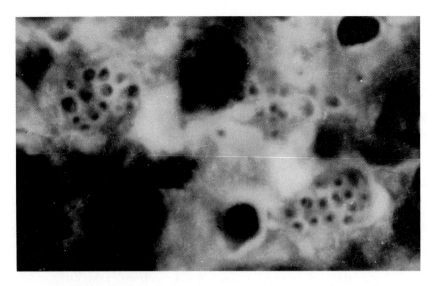

FIGURE 64-57 *Leishmania donovani* parasites in Küpffer cells of liver (2000×).

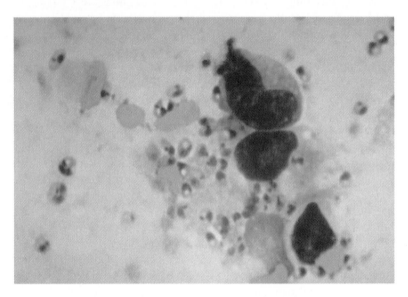

FIGURE 64-58 *Leishmania donovani* amastigotes.

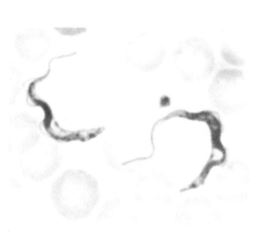

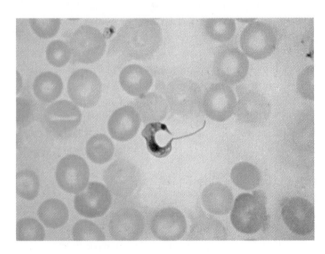

FIGURE 64-59 *Trypanosoma gambiense* in blood film (1600×).

FIGURE 64-60 *Trypanosoma cruzi* trypomastigote.

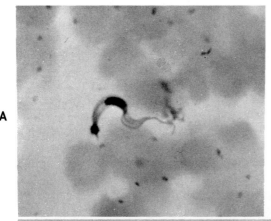

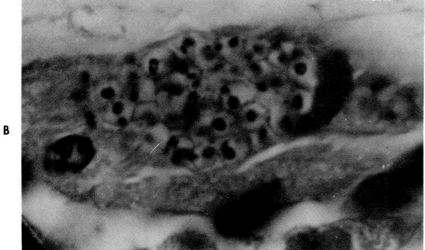

FIGURE 64-61 A, *Trypanosoma cruzi* in blood film (1600×). **B,** *Trypanosoma cruzi* parasites in cardiac muscle (2500×). (From Markell, E.K. and Voge, M. 1981. Medical parasitology, ed 5. W.B. Saunders, Philadelphia.)

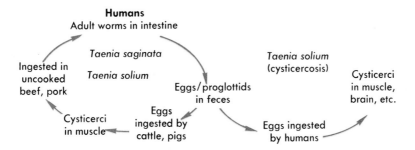

FIGURE 64-62 Life cycle of *Taenia saginata* and *Taenia solium.*

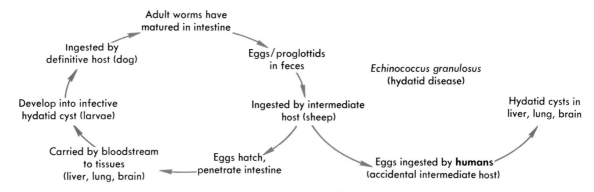

FIGURE 64-63 Life cycle of *Echinococcus granulosus* (hydatid disease).

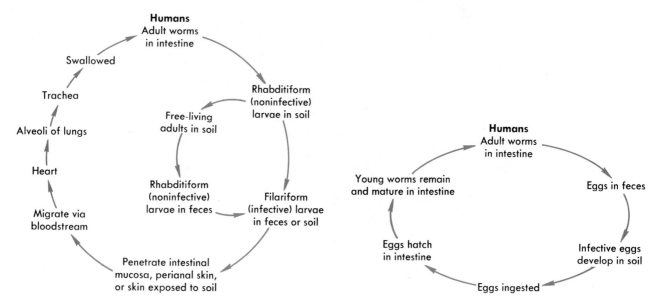

FIGURE 64-64 Life cycle of *Strongyloides stercoralis*.

FIGURE 64-65 Life cycle of *Enterobius vermicularis* and *Trichuris trichiura* (direct type of cycle).

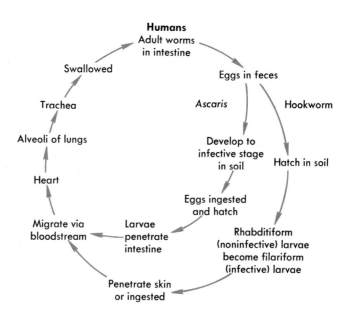

FIGURE 64-66 Life cycle of *Ascaris lumbricoides* and hookworms (indirect type of cycle).

Diagnosis of most intestinal helminth infections is based on the detection of the characteristic eggs and larvae in the stool; occasionally, adult worms or portions of worms may also be found. No permanent stains are required, and most diagnostic features can easily be seen on direct wet mounts or in mounts of the concentrated stool material.

Nematodes

Most nematodes are diagnosed by finding the characteristic eggs in the stool (Figures 64-67 and 64-68). The eggs of *Ancylostoma duodenale* and *Necator americanus* are essentially identical, so an infection with either species is reported as "hookworm eggs present." *Trichostrongylus* eggs may easily be mistaken for those of hookworms; however, the eggs of *Trichostrongylus* are somewhat larger, and one end tends to be more pointed.

Strongyloides stercoralis is passed in the feces as the noninfective rhabditiform larva (Figure 64-69). Although hookworm eggs are normally passed in the stool, these eggs may continue to develop and hatch if the stool is left at room temperature for several days. These larvae may be mistaken for those of *Strongyloides*. Figure 64-70 shows the morphologic differences between the rhabditiform larvae of hookworm and *Strongyloides*. Recovery of *Strongyloides* larvae in duodenal contents is mentioned earlier in this chapter.

The appropriate techniques for recovery of *Enterobius vermicularis* (pinworm) eggs are given earlier in the section on cellophane tape preparations (see Procedure 64-5). Other worms do not deposit eggs in the same site; the characteristic morphology of pinworm eggs is shown in Figure 64-71. Eggs of the other nematodes are relatively easy to find and differentiate from one another.

Cestodes

Except for *Diphyllobothrium latum*, tapeworm eggs are embryonated and contain a six-hooked oncosphere (Figure 64-72 and Table 64-11). *Taenia saginata* and *T. solium* cannot be speciated based on egg morphology; gravid proglottids (Figure 64-73) or the scolices must be examined. *T. saginata* (beef tapeworm) proglottids have approximately 15 to 30 main lateral branches, and the scolex has no hooks. The proglottids of *T. solium* have seven to 12 main lateral branches, and the scolex has a circle of hooks.

Hymenolepis nana has an unusual life cycle in that the ingestion of the egg can lead to the adult worm in humans (Figure 64-74). The eggs of *H. nana* (more common) and *Hymenolepis diminuta* are very similar; however, *H. nana* eggs are smaller and have polar filaments, which are present in the space between the oncosphere and the eggshell (Figure 64-72).

Eggs of *Dipylidium caninum* are occasionally found in humans, particularly children, and are passed in the feces in packets of five to 15 eggs each (Figures 64-72 and 64-75). The proglottids may also be found; they may resemble cucumber seeds or, when dry, rice grains (white).

The fish tapeworm, *Diphyllobothrium latum*, does not have embryonated eggs. These eggs have a somewhat thicker shell and are operculated, similar to the trematode eggs (see Figure 64-72). Humans acquire the parasite by eating undercooked or raw freshwater fish (Figure 64-76).

Trematodes

Humans acquire most fluke infestations by ingesting the encysted metacercariae (Figure 64-77). Most trematodes have operculated eggs, which are best recovered by the sedimentation concentration technique rather than the flotation method. Many of these eggs are very similar, both in size and morphology, and often careful measurements must be taken to speciate the eggs (Figure 64-78). The eggs of *Clonorchis (Opisthorchis), Heterophyes,* and *Metagonimus* are very similar and small; they are easily missed if the concentration sediment is examined with the 10× objective only. The eggs of *Fasciola hepatica* and *Fasciolopsis buski* are also very similar but much larger than those just mentioned.

Paragonimus westermani eggs are not only found in the stool but also may be found in sputum (Figure 64-78). These eggs are very similar in size and shape to the egg of the fish tapeworm, *Diphyllobothrium latum*. Schistosomiasis, a great source of morbidity in much of the developing world, is acquired during activities (e.g., bathing, swimming, doing laundry, planting rice, fishing) that involve contact with water infested with the intermediate snail host of the parasite. The free-swimming cercarial form attaches to a vertebrate host and penetrates through intact skin (Figure 64-79). In temperate zones, such as in the areas near the Great Lakes, cercariae of other trematodes that normally infect birds or other animals can penetrate human skin and cause a strong local reaction (swimmer's itch), although they cannot go on to develop into adult worms in the unnatural human host. Probably the easiest trematode eggs to identify are those of the schistosomes: *Schistosoma mansoni* eggs are characterized by having a very prominent lateral spine, *S. haematobium* a terminal spine, and *S. japonicum* a small lateral spine that may be difficult to see (Figure 64-78). These eggs are nonoperculated. Schistosome eggs stain acid fast. Specific procedures for their recovery and identification are found in the earlier section on the hatching procedure.

Text continued on p. 858

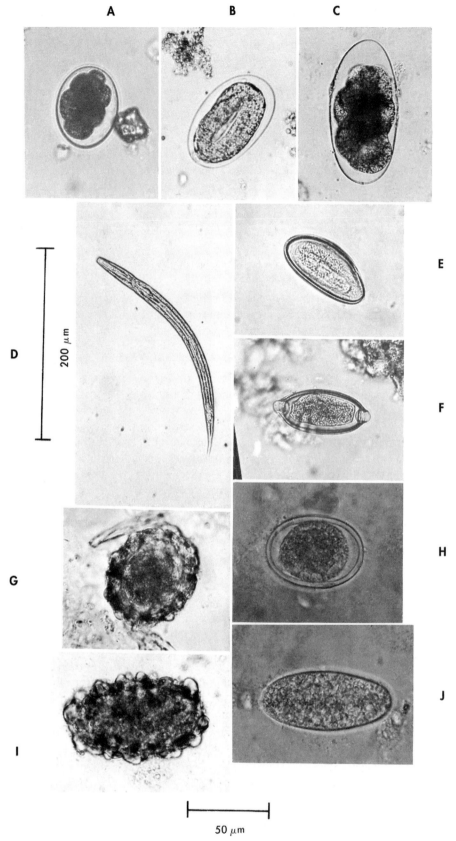

FIGURE 64-67 **A,** Immature hookworm egg. **B,** Embryonated hookworm egg. **C,** *Trichostrongylus orientalis,* immature egg. **D,** *Strongyloides stercoralis,* rhabditiform larva (200 μm). **E,** *Enterobius vermicularis* egg. **F,** *Trichuris trichiura* egg. **G,** *Ascaris lumbricoides,* fertilized egg. **H,** *A. lumbricoides,* fertilized egg, decorticate. **I,** *A. lumbricoides,* unfertilized egg. **J,** *A. lumbricoides,* unfertilized egg, decorticate.

200 μm

50 μm

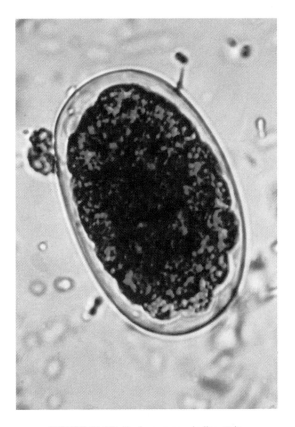

FIGURE 64-68 Hookworm egg, iodine stain.

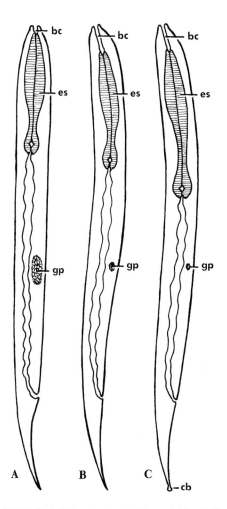

FIGURE 64-70 Rhabditiform larvae. **A,** *Strongyloides.* **B,** Hookworm. **C,** *Trichostrongylus. bc,* Buccal cavity; *es,* esophagus; *gp,* genital primordia; *cb,* beadlike swelling of caudal tip. (Illustration by Nobuko Kitamura.)

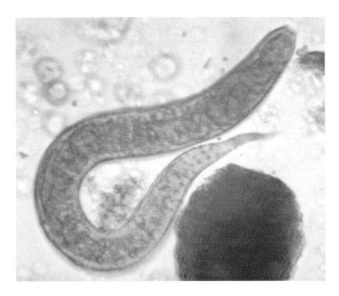

FIGURE 64-69 *Strongyloides stercoralis* larva, iodine stain.

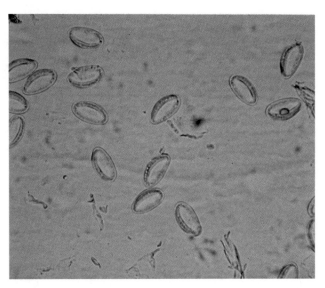

FIGURE 64-71 *Enterobius vermicularis* eggs (cellophane [Scotch] tape preparation).

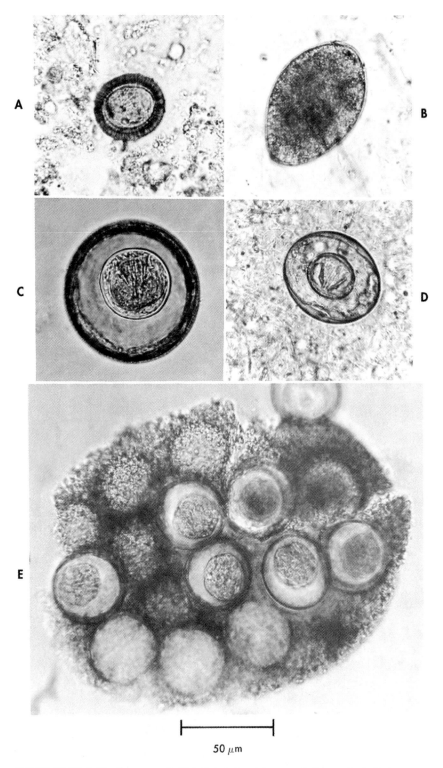

50 μm

FIGURE 64-72 **A,** *Taenia* spp. egg. **B,** *Diphyllobothrium latum* egg. **C,** *Hymenolepis diminuta* egg. **D,** *Hymenolepis nana* egg. **E,** *Dipylidium caninum* egg packet.

TABLE 64-11   DIFFERENTIAL CHARACTERISTICS OF SOME IMPORTANT TAPEWORMS OF HUMANS

	TAENIA SAGINATA	*TAENIA SOLIUM*	*HYMENOLEPIS NANA*	*DIPHYLLOBOTHRIUM LATUM*
Length	4-8 m	3-5 m	2.5-4 cm	4-10 m
Scolex				
Shape	Quadrilateral	Globular	Usually not seen	Almondlike
Size	1 × 1.5 mm	1 × 1 mm		3 × 1 mm
Rostellum and hooklets	No	Yes		No
Suckers	Four	Four		Two (grooves)
Terminal proglottids (gravid)				
Size	19 × 7 mm, longer than wide	11 × 5 mm	Usually not seen	3 × 11 mm, wider than long
Primary lateral uterine branches	15-30 on each side	6-12 on each side		Rosette-shaped
Color	Milky white	Milky white		Ivory
Appearance in feces	Usually appear singly	Five or six segments		Varies from a few inches to a few feet in length
Ova				
Shape	Spheroid	Spheroid	Broadly oval	Oval
Size	35 μm	35 μm	30 × 47 μm	45 × 70 μm
Color	Rusty brown	Rusty brown	Pale	Yellow-brown
Embryo with hooklets	Yes	Yes	Yes	No
Operculum	No	No	No	Yes (difficult to see)

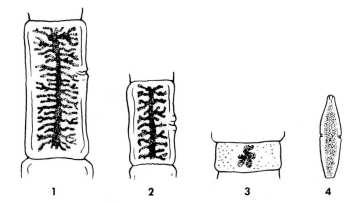

FIGURE 64-73 Gravid proglottids. **1,** *Taenia saginata.* **2,** *Taenia solium.* **3,** *Diphyllobothrium latum.* **4,** *Dipylidium caninum.* (From Garcia, L.S. and Bruckner, D.A. 1993. Diagnostic medical parasitology, ed 2. ASM Press, Washington D.C.)

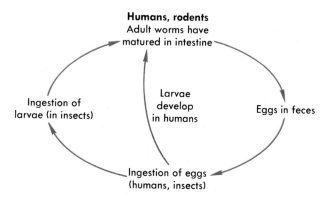

FIGURE 64-74 Life cycle of *Hymenolepis nana.*

FIGURE 64-75 *Dipylidium caninum* egg packet. (Illustration by Nobuko Kitamura.)

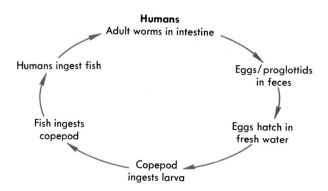

FIGURE 64-76 Life cycle of *Diphyllobothrium latum.*

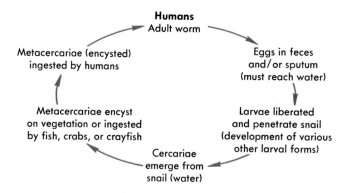

FIGURE 64-77 Life cycle of trematodes acquired by humans through ingestion of raw fish, crabs, or crayfish and vegetation.

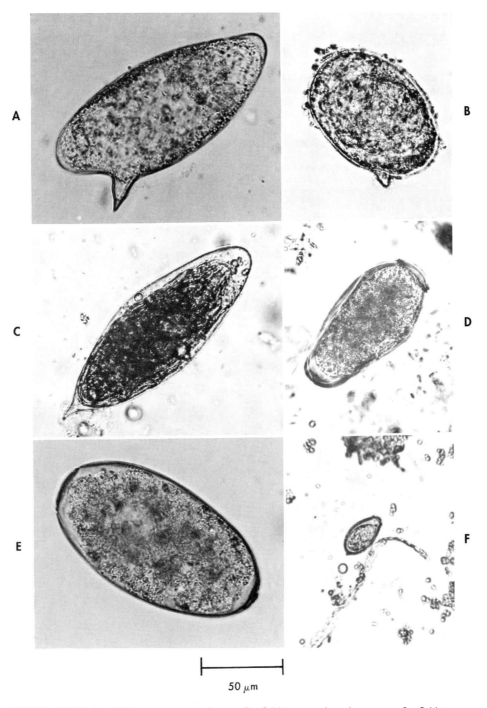

50 μm

FIGURE 64-78 **A,** *Schistosoma mansoni* egg. **B,** *Schistosoma japonicum* egg. **C,** *Schistosoma haematobium* egg. **D,** *Paragonimus westermani* egg. **E,** *Fasciola hepatica* egg. **F,** *Clonorchis (Opisthorchis) sinensis* egg.

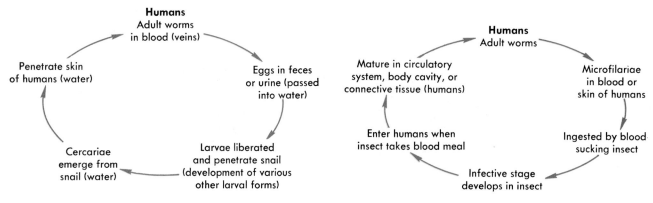

FIGURE 64-79 Life cycle of human schistosomes.

FIGURE 64-80 Life cycle of human filarial worms.

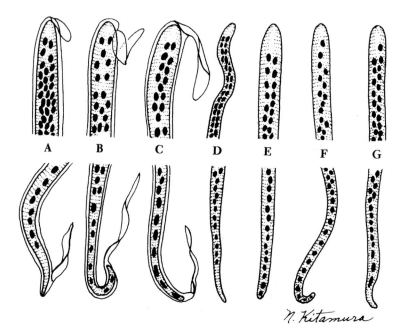

FIGURE 64-81 Anterior and posterior ends of microfilariae found in humans.
A, *Wuchereria bancrofti.* **B,** *Brugia malayi.* **C,** *Loa loa.* **D,** *Onchocerca volvulus.*
E, *Mansonella perstans.* **F,** *Mansonella streptocerca.* **G,** *Mansonella ozzardi.*

BLOOD HELMINTHS

Filariae

The filarial worms are long, thin nematodes that inhabit parts of the lymphatic system and the subcutaneous and deep connective tissues. A life cycle has been included here to show the various stages present in the human host (Figure 64-80). Most species produce microfilariae, which can be found in the peripheral blood; two species, *Onchocerca volvulus* and *Dipetalonema streptocerca,* produce microfilariae found in the subcutaneous tissues and dermis.

Diagnosis of filarial infections is often based on clinical grounds, but demonstration of the parasite is the only accurate means of confirming the diagnosis (Figure 64-81). Fresh blood films may be prepared; actively moving microfilariae can be observed in a preparation of this type. If the patient has a light infection, thick blood films can be prepared and stained. The Knott concentration procedure and the membrane filtration technique may also be helpful in recovering the organisms.[22] Microfilariae of some strains tend to exhibit nocturnal periodicity; thus the time the blood is drawn may be critical in demonstrating the parasite. The microfilariae of *O. volvulus* and *D. streptocerca* are found in "skin snips," very thin slices of skin, which are teased apart in normal saline to release the organisms. Differentiation of the species depends on (1) the presence or absence of the sheath and (2) the distribution of nuclei in the tail region of the microfilaria (Figure 64-82).

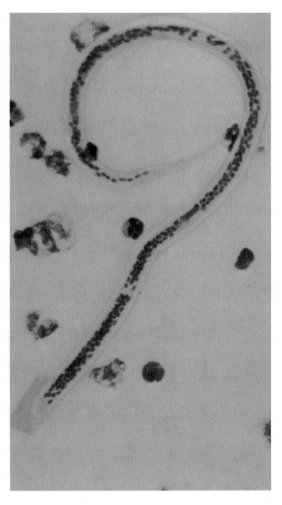

FIGURE 64-82 Microfilaria of *Wuchereria bancrofti* in thick blood film. (From Markell, E.K. and Voge, M. 1981. Medical parasitology, ed 5. W.B. Saunders, Philadelphia.)

Procedures

64-1 FORMALIN-ETHER (FORMALIN-ETHYL ACETATE) SEDIMENTATION TECHNIQUES

PRINCIPLE

Formalin fixes the eggs, larvae, and cysts, so that they are no longer infectious, as well as preserves their morphology. Fecal debris is extracted into the ethyl acetate phase of the solution, freeing the sedimented parasitic elements from at least some of the artifact material in stool. Numerous ether substitutes are available; the term *ethyl acetate* used throughout this chapter is used in the general sense (ether substitute).

METHOD

1. Transfer ¼ to ½ teaspoon of fresh stool into 10 mL of 5% or 10% formalin in a 15-mL shell vial, unwaxed paper cup, or 16 × 125 mm tube (container may vary depending on individual preferences) and comminute thoroughly. Let stand 30 minutes for adequate fixation.

2. Filter this material (funnel or pointed paper cup with end cut off) through two layers of gauze into a 15-mL centrifuge tube.

3. Add physiologic saline or 5% or 10% formalin to within ½ inch (1.5 cm) of the top and centrifuge for 10 minutes at 500× g.

4. Decant, resuspend the sediment (should have 0.5 to 1 mL sediment) in saline to within ½ inch (1.5 cm) of the top, and centrifuge again for 10 minutes at 500× g. This second wash may be eliminated if the supernatant fluid after the first wash is light tan or clear.

5. Decant and resuspend the sediment in 5% or 10% formalin (fill the tube only half full). If the amount of sediment left in the bottom of the tube is very small, do not add ethyl acetate in step 6; merely add the formalin, then spin, decant, and examine the remaining sediment.

6. Add approximately 3 mL of ethyl acetate, stopper, and shake vigor-

Continued

64-1 FORMALIN-ETHER (FORMALIN-ETHYL ACETATE) SEDIMENTATION TECHNIQUES—cont'd

ously for 30 seconds. Hold the tube so that the stopper is directed away from your face; remove stopper carefully to prevent spraying of material caused by pressure within the tube.

7. Centrifuge for 10 minutes at 500 $\times$ g. Four layers should result: a small amount of sediment in the bottom of the tube, containing the parasites; a layer of formalin; a plug of fecal debris on top of the formalin layer; and a layer of ether substitute at the top.

8. Free the plug of debris by ringing with an applicator stick, and decant all the fluid. After proper decanting, a drop or two of fluid remaining on the side of the tube will drain down to the sediment. Mix the fluid with the sediment and prepare a wet mount for examination.

The formalin-ethyl acetate sedimentation procedure may be used on PVA-preserved material. Steps 1 and 2 differ as follows:

1. Fixation time with PVA should be at least 30 minutes. Mix contents of PVA bottle (stool-PVA mixture: 1 part stool to 2 or 3 parts PVA) with applicator sticks. Immediately after mixing, pour approxi-mately 2 to 5 mL (amount will vary depending on the viscosity and density of the mixture) of the stool-PVA mixture into a 15-mL shell vial, 16- $\times$ 125-mm tube, or such, and add approximately 10 mL physiologic saline or 5% or 10% formalin.

2. Filter this material (funnel or paper cup with pointed end cut off) through two layers of gauze into a 15-mL centrifuge tube.

Steps 3 through 8 will be the same for both fresh and PVA-preserved material.

NOTE: Tap water may be substituted for physiologic saline throughout this procedure; however, saline is recommended. Some workers prefer to use 5% or 10% formalin for all the rinses (steps 3 and 4).

NOTE: The introduction of ethyl acetate (or Hemo-De, or other substitute) as a substitute for diethyl ether (ether) in the formalin-ether sedi-mentation concentration procedure provides a much safer chemical for the clinical laboratory. Tests comparing the use of these two compounds on formalin-preserved and PVA-preserved material indicate that the differences in organism recovery and identifi-cation are minimal and probably do not reflect clinically relevant differences.

When examining the sediment in the bottom of the tube:

1. Prepare a saline mount (1 drop of sediment and 1 drop of saline solution mixed together), and scan the whole 22 $\times$ 22 mm coverslip under low power for helminth eggs or larvae.

2. Iodine may then be added to aid in the detection of protozoan cysts and should be examined under high-dry power. If iodine is added before low-power scanning, be certain that the iodine is not too strong; otherwise, some of the helminth eggs will stain so darkly that they will be mistaken for debris.

3. Occasionally a precipitate is formed when iodine is added to the sediment obtained from a con-centration procedure with PVA-preserved material. The precipitate is formed from the reaction between the iodine and excess mercuric chloride that has not been thoroughly rinsed from the PVA-preserved material. The sediment can be rinsed again to remove any remaining mercuric chloride, or the sediment can be examined as a saline mount without the addition of iodine.[54]

64-2 TRICHROME STAIN

PRINCIPLE
The internal elements that distinguish among cysts and trophozoites can best be visualized with a stain that enhances the morphologic features. In addition, such a stained smear provides a permanent record of the results.

REAGENTS
A. FORMULA

Chromotrope 2R	0.6 g
Light green SF	0.3 g
Phosphotungstic acid	0.7 g
Acetic acid (glacial)	1 mL
Distilled water	100 mL

B. STAIN PREPARATION
1. Add 1 mL of glacial acetic acid to the dry components.

2. Allow the mixture to stand for 15 to 30 minutes to "ripen"; then add 100 mL of distilled water. This preparation gives a highly uniform and reproducible stain; the stain should be purple. Store in Coplin jars.

METHOD
1. Prepare fresh fecal smears or PVA smears as described.

2. Place in 70% ethanol for 5 minutes. (This step may be eliminated for PVA smears.)

3. Place in 70% ethanol plus D'Antoni's iodine (dark reddish brown) for 2 to 5 minutes.

4. Place in two changes of 70% ethanol—one for 5 minutes* and one for 2 to 5 minutes.

64-2 TRICHROME STAIN—cont'd

5. Place in trichrome stain solution for 10 minutes.

6. Place in 90% ethanol, acidified (1% acetic acid) for up to 3 seconds (*do not leave the slides in this solution any longer*).

7. Dip once in 100% ethanol.

8. Place in two changes of 100% ethanol for 2 to 5 minutes each.*

9. Place in two changes of xylene or toluene for 2 to 5 minutes each.*

10. Mount in Permount or some other mounting medium; use a No. 1 thickness coverglass.

EXPECTED RESULTS

Background debris will be green and protozoa will show blue-green to purple cytoplasm. The nuclei and inclusions will be red or purple-red and sharply delineated from background.

*At this stage, slides can be held several hours or overnight.

64-3 MODIFIED IRON HEMATOXYLIN STAIN

PRINCIPLE

The internal elements that distinguish among cysts and trophozoites can best be visualized with a stain that enhances the morphologic features. Also, the stained smear provides a permanent record of the results.

REAGENTS

A. MAYER'S ALBUMIN
Add an equal quantity of glycerine to a fresh egg white. Mix gently and thoroughly. Store at 4° C and indicate an expiration date of 3 months. Mayer's albumin from commercial suppliers can normally be stored at 25° C for 1 year (example: Product #756, E.M. Diagnostic Systems, Inc., 480 Democrat Road, Gibbstown, N.J. 08027; 800-443-3637).

B. STOCK SOLUTION OF HEMATOXYLIN STAIN
Hematoxylin powder   10 g
Ethanol (95% or 100%) 1000 mL

1. Mix well until dissolved.

2. Store in a clear glass bottle, in a light area. Allow to ripen for 14 days before use.

3. Store at room temp with an expiration date of 1 year.

C. MORDANT
Ferrous ammonium
  sulfate   10 g
  [Fe(NH$_4$)$_2$(SO$_4$)$_2$ • 6H$_2$O]
Ferric ammonium sulfate 10 g
  [FeNH$_4$(SO$_4$)$_2$ • 12 H$_2$O]

Hydrochloric acid (HCl) 10 mL
  (concentrated)
Add distilled water to   1000 mL

D. WORKING SOLUTION OF HEMATOXYLIN STAIN
1. Mix equal quantities of stock solution of stain and mordant.

2. Allow mixture to cool thoroughly before use (prepare at least 2 hours before use). The working solution should be made fresh every week.

E. PICRIC ACID
Mix equal quantities of distilled water and an aqueous saturated solution of picric acid to make a 50% saturated solution.

F. ACID-ALCOHOL DECOLORIZER
Hydrochloric acid (HCl) 30 mL
  (concentrated)
Alcohol to   1000 mL

G. 70% ALCOHOL AND AMMONIA
70% alcohol   50 mL
Ammonia   0.5-1 mL
Add enough ammonia to bring the pH to approximately 8.0.

H. CARBOLFUCHSIN
Basic fuchsin (solution A): Dissolve 0.3 g basic fuchsin in 10 mL of 95% ethanol.

Phenol (solution B): Dissolve 5 g of phenol crystals in 100 mL distilled water. (Gentle heat may be needed.)

1. Mix solution A with solution B.

2. Store at room temperature. Stable for 1 year.

METHOD

1. Slide preparation
   a. Place 1 drop of Mayer's albumin on a labeled slide.
   b. Mix the sediment from the SAF concentration well with an applicator stick.
   c. Add approximately 1 drop of the fecal concentrate to the albumin and spread the mixture over the slide.

2. Allow slide to air dry at room temperature (smear will appear opaque when dry).

3. Place slide in 70% alcohol for 5 minutes.

4. Wash in container (not running water) of tap water for 2 minutes.

5. Place slide in Kinyoun stain for 5 minutes.

6. Wash slide in running tap water (constant stream of water into container) for 1 minute.

7. Place slide in acid/alcohol decolorizer for 4 minutes.*

8. Wash slide in running tap water (constant stream of water into container) for 1 minute.

64-3 MODIFIED IRON HEMATOXYLIN STAIN—cont'd

9. Place slide in iron hematoxylin working solution for 8 minutes.

10. Wash slide in distilled water (in container) for 1 minute.

11. Place slide in picric acid solution for 3 to 5 minutes.

12. Wash slide in running tap water (constant stream of water into container) for 10 minutes.

13. Place slide in 70% alcohol plus ammonia for 3 minutes.

14. Place slide in 95% alcohol for 5 minutes.

15. Place slide in 100% alcohol for 5 minutes.

16. Place slide in two changes of xylene for 5 minutes.

*This step can also be performed as follows:
a. Place slide in acid/alcohol decolorizer for 2 minutes.
b. Wash slide in running tap water (constant stream of water into container) for 1 minute.
c. Place slide in acid/alcohol decolorizer for 2 minutes.
e. Wash slide in running tap water (constant stream of water into container) for 1 minute.
f. Continue staining sequence with step 9 above (iron hematoxylin working solution).

PROCEDURE NOTES

1. The first 70% alcohol step acts with the Mayer's albumin to "glue" the specimen to the glass slide. The specimen may wash off if insufficient albumin is used or if the slides are not completely dry before staining.

2. The working hematoxylin stain should be checked each day of use by adding a drop of stain to alkaline tap water. If a blue color does not develop, prepare fresh working stain solution.

3. The picric acid differentiates the hematoxylin stain by removing more stain from fecal debris than from the protozoa and removing more stain from the organism cytoplasm than the nucleus. When properly stained, the background should be various shades of gray-blue and protozoa should be easily seen with medium-blue cytoplasm and dark blue-black nuclei.

64-4 MODIFIED ACID-FAST STAIN FOR *CRYPTOSPORIDIUM PARVUM*

REAGENTS

A. CARBOLFUCHSIN
Basic fuchsin   4 g
Phenol   8 mL
Alcohol (95%)   20 mL
Distilled water 100 mL
Dissolve the basic fuchsin in the alcohol, and add the water slowly while shaking. Melt the phenol in a 56° C water bath, and add 8 mL to the stain, using a pipette with a rubber bulb.
B. DECOLORIZER
Ethanol (95%)   97 mL
Concentrated HCl   3 mL
Add the hydrochloric acid to the alcohol slowly, working under a chemical fume hood.
C. COUNTERSTAIN
Methylene blue 0.3 g
Distilled water 100 mL

"COLD" MODIFIED ACID-FAST STAIN METHOD (KINYOUN)

1. Spin an aliquot of 10% formalinized stool for 10 minutes at $500\times$ g.

2. Remove upper layer of sediment with pipette and place a thin layer onto a microscope slide.
   NOTE: If the stool specimen contains a lot of mucus, 10 drops of 10% KOH can be added to the sediment (step 2), vortexed, rinsed with 10% formalin, and respun before smear preparation. Some laboratories use this approach routinely before smear preparation.

3. Heat fix the smear at 70° C for 10 minutes.

4. Stain the fixed smear for 3 to 5 minutes (no heat necessary).

5. Wash in distilled, filtered water and shake off excess water.

6. Flood with decolorizer for approximately 1 minute. Check to see that no more red color runs when the slide is tipped. Add a bit more decolorizer for very thick slides or those that continue to bleed red dye.

7. Wash thoroughly with filtered water as above and shake off excess.

8. Flood with counterstain for approximately 1 minute.

9. Wash with distilled water and drain by standing slides upright. Do not blot dry.

By the addition of a detergent or wetting agent the staining of acid-fast organisms may be accelerated. Tergitol No. 7 (Sigma Chemical Co.) may be used. Add 1 drop of Tergitol No. 7 to every 30 to 40 mL of the Kinyoun carbolfuchsin stain.

Acid-fast bacteria stain red with carbolfuchsin stains. The background color is dependent on the counterstain; methylene blue imparts a blue color to nonacid-fast material, whereas brilliant green results in green background, and picric acid results in yellow.

HOT MODIFIED ACID-FAST STAIN METHOD

1. Spin an aliquot of 10% formalinized stool for 10 minutes at $500\times$ g.

2. Remove upper layer of sediment with pipette and place a thin layer onto a microscope slide.
   NOTE: If the stool specimen contains a lot of mucus, 10 drops of 10% KOH can be added to the sediment (step 2), vortexed, rinsed with 10% formalin, and respun before smear preparation. Some laboratories use this approach routinely before smear preparation.

64-4 MODIFIED ACID-FAST STAIN—cont'd

3. Heat fix the smear at 70° C for 10 minutes.

4. Place slide on staining rack and flood with carbolfuchsin.

5. Heat to steaming and allow to stain for 5 minutes. If the slide begins to dry, more stain is added without additional heating.

6. Rinse the smear with tap or distilled water.

7. Decolorize with 5% aqueous sulfuric acid for 30 seconds (thicker smears may require a longer time).

8. Rinse smear with tap or distilled water, drain, and flood smear with methylene blue counterstain for 1 minute.

9. Rinse with tap or distilled water, drain, and air dry.

Kinyoun. 1915. Am. J. Pub. Health 5:867.

64-5 EXAMINATION OF CELLOPHANE* TAPE PREPARATIONS

PRINCIPLE

The female adult pinworm deposits her eggs during the night on the surface of the skin surrounding the anus. The eggs will adhere to the sticky surface of clear cellophane tape, on which they can be visualized microscopically.

METHOD

1. Place a strip of cellophane tape on a microscope slide, starting ½ inch (1.5 cm) from one end and, running toward the same end, continuing around this end across the slide; tear off the strip even with the other end. Place a strip of paper, ½ × 1 inch (1.5 × 2.5 cm), between the slide and the tape at the end where the tape is torn flush.

2. To obtain the sample from the perianal area, peel back the tape by gripping the label, and, with the tape looped adhesive side outward over a wooden tongue depressor held against the slide and extended about 1 inch (2.5 cm) beyond it, press the tape firmly against the right and left perianal folds.

3. Spread the tape back on the slide, adhesive side down.

4. Place name and date on the label. NOTE: *Do not use Magic transparent tape, but use regular clear cellophane tape.*

5. Lift one side of the tape and apply 1 small drop of toluene or xylene; press the tape down onto the glass slide.

6. The tape is now cleared; examine under low power and low illumination. The eggs should be visible if present; they are described as football-shaped with one slightly flattened side.

*Refers to cellulose (Scotch) tape.

64-6 IDENTIFICATION OF ADULT TAPEWORMS

PRINCIPLE

The uterus of gravid tapeworm proglottids (*Taenia* spp.) can be injected with India ink for visualization of the branches; the number of branches determines the species of tapeworm.

METHOD

1. Using a 1-mL syringe and 25- to 26-gauge needle, inject India ink into the central stem or into the uterine pore, filling the uterine branches with ink.

2. Press the proglottid between two slides, hold it up to the light, and count the branches (see Figure 64-73). (Besides Permount, Euparol is another mounting medium that can be used for tapeworm proglottids.)

NOTE: *Caution should be used in handling proglottids of* Taenia *spp. because the eggs of* T. solium *are infective for humans.*

64-7 MODIFIED TRICHROME STAIN FOR THE MICROSPORIDIA (WEBER-GREEN)

PRINCIPLE

The oval shape, spore wall, and diagonal or horizontal "stripe" that distinguish microsporidia spores can best be visualized with a stain that enhances the morphologic features.[81] In addition, such a stained smear provides a permanent record of the results.

REAGENTS

A. MODIFIED TRICHROME STAIN

Chromotrope 2R	6 g*
Fast green	0.15 g
Phosphotungstic acid	0.7 g
Acetic acid (glacial)	3 mL
Distilled water	100 mL

1. Prepare the stain by adding 3 mL of acetic acid to the dry ingredients. Allow the mixture to stand (ripen) for 30 minutes at room temperature.

2. Add 100 mL of distilled water. Properly prepared stain will be dark purple.

3. Store in a glass or plastic bottle at room temperature. The shelf life is at least 24 months.

B. ACID-ALCOHOL

90% ethyl alcohol	995.5 mL
Acetic acid (glacial)	4.5 mL

Prepare by combining the two solutions.

METHOD

1. Using a 10-μL aliquot of unconcentrated, preserved liquid stool (5% or 10% formalin or SAF), prepare the smear by spreading the material over an area of 45 × 25 mm.

2. Allow the smear to air dry.

3. Place the smear in absolute methanol for 5 minutes.

4. Allow the smear to air dry.

5. Place in trichrome stain for 90 minutes.

6. Rinse in acid-alcohol for no more than 10 seconds.

7. Dip slides several times in 95% alcohol. Use this step as a rinse.

8. Place in 95% alcohol for 5 minutes.

9. Place in 100% alcohol for 10 minutes.

10. Place in xylene substitute for 10 minutes.

11. Mount with coverslip (No. 1 thickness), using mounting medium. Check the specimen for ad-herence to the slide.

12. Examine smears under oil immersion (1000×) and read at least 100 fields; the examination time will probably be at least 10 minutes per slide.

EXPECTED RESULTS

Known parasites should be detected readily. When the smear is thoroughly fixed and the stain is performed correctly, the spores will be ovoid and refractile, with the spore wall being bright pinkish red. Occasionally, the polar tube can be seen either as a stripe or as a diagonal line across the spore. The majority of the bacteria and other debris will tend to stain green. However, there will still be some bacteria and debris that will stain red.

*10 times the normal trichrome stain formula.

64-8 MODIFIED TRICHROME STAIN FOR THE MICROSPORIDIA (RYAN-BLUE)

PRINCIPLE

The oval shape, spore wall, and diagonal or horizontal "stripe" that distinguish microsporidia spores can best be visualized with a stain that enhances the morphologic features. In addition, such a stained smear provides a permanent record of the results.

Numerous variations to the modified trichrome (Weber-Green) were tried in an attempt to improve the contrast between the color of the spores and the background staining. Optimal staining was achieved by modifying the composition of the trichrome solution.[63] This stain is also available commercially from several suppliers. The specimen can be fresh stool or stool that has been preserved in 5% or 10% formalin,

SAF, or some of the newer single-vial system fixatives. Actually, any specimen other than tissue thought to contain microsporidia can be stained by this method.

REAGENTS

A. TRICHROME STAIN (MODIFIED FOR MICROSPORIDIA)[63] (RYAN-BLUE)

Chromotrope 2R	6 g*
Aniline blue	0.5 g
Phosphotungstic acid	0.25 g
Acetic acid (glacial)	3 mL
Distilled water	100 mL

1. Prepare the stain by adding 3 mL of acetic acid to the dry ingredients. Allow the mixture to stand (ripen) for 30 minutes at room temperature.

2. Add 100 mL of distilled water and adjust the pH to 2.5 with 1M HCl. Properly prepared stain will be dark purple. The staining solution should be protected from light.

3. Store in a glass or plastic bottle at room temperature. The shelf life is at least 24 months.

B. ACID-ALCOHOL (SEE PROCEDURE 64-7)

METHOD

1. Using a 10-μL aliquot of unconcentrated, preserved liquid stool (5% or 10% formalin or SAF), prepare the smear by spreading the material over an area of 45 × 25 mm.

64-8   MODIFIED TRICHROME STAIN FOR THE MICROSPORIDIA (RYAN-BLUE)—cont'd

2. Allow the smear to air dry.

3. Place the smear in absolute methanol for 5 or 10 minutes.

4. Allow the smear to air dry.

5. Place in trichrome stain for 90 minutes.

6. Rinse in acid-alcohol for no more than 10 seconds.

7. Dip slides several times in 95% alcohol. Use this step as a rinse (no more than 10 seconds).

8. Place in 95% alcohol for 5 minutes.

9. Place in 95% alcohol for 5 minutes

10. Place in 100% alcohol for 10 minutes.

11. Place in xylene substitute for 10 minutes.

12. Mount with coverslip (No. 1 thickness), using mounting medium.

13. Examine smears under oil immersion (1000×) and read at least 100 fields; the examination time will probably be at least 10 minutes per slide.

EXPECTED RESULTS

Known parasites should be detected readily. When the smear is thoroughly fixed and the stain is performed correctly, the spores will be ovoid and refractile, with the spore wall being bright pinkish red. Occasionally the polar tube can be seen either as a stripe or as a diagonal line across the spore. The majority of the bacteria and other debris will tend to stain blue. However, there will still be some bacteria and debris that will stain red.

PROCEDURE NOTES FOR MODIFIED TRICHROME STAINING METHODS (WEBER OR RYAN)

1. It is mandatory that positive control smears be stained and examined each time patient specimens are stained and examined.

2. Because of the difficulty in getting stain penetration through the spore wall, prepare thin smears and do not reduce the staining time in trichrome. Also, make sure the slides are not left too long in the decolorizing agent (acid-alcohol). If the control organisms are too light, leave them in the trichrome longer and shorten the time to two dips in the acid-alcohol solution. Also, remember that the 95% alcohol rinse after the acid-alcohol should be performed quickly to prevent additional destaining from the acid-alcohol reagent.

3. When you purchase the chromotrope 2R, obtain the highest dye content available. Two sources are Harleco (Gibbstown, N.J.) and Sigma Chemical Co. (St. Louis, Mo.) (Dye content among the highest [85%]). Fast green and aniline blue can be obtained from Allied Chemical and Dye, New York.

4. In the final stages of dehydration, the 100% ethanol and the xylenes (or xylene substitutes) should be kept as free from water as possible. Coplin jars must have tight-fitting caps to prevent both evaporation of reagents and absorption of moisture. If the xylene becomes cloudy after addition of slides from 100% alcohol, return the slides to 100% alcohol and replace the xylene with fresh stock.

PROCEDURE LIMITATIONS FOR MODIFIED TRICHROME STAINING METHODS (WEBER OR RYAN)

1. Although this staining method will stain the microsporidia, the range of stain intensity and the small size of the spores will cause some difficulty in identifying these organisms. Because this procedure will result in many other organisms or objects staining in stool specimens, differentiation of the microsporidia from surrounding material will still be very difficult. There also tends to be some slight size variation among the spores.

2. If the patient has severe watery diarrhea, there will be less artifact material in the stool to confuse with the microsporidial spores; however, if the stool is semiformed or formed, the amount of artifact material will be much greater; thus, the spores will be much harder to detect and identify. Also, remember that the number of spores will vary according to the stool consistency (the more diarrhetic, the more spores that will be present).

3. Those who developed some of these procedures feel that concentration procedures result in an actual loss of microsporidial spores; thus there is a strong recommendation to use unconcentrated, formalinized stool. However, there are no data indicating what centrifugation speeds, etc., were used in the study.

4. In the UCLA Clinical Microbiology Laboratory, we have generated data (unpublished) to indicate that centrifugation at 500 3 g for 10 minutes increases dramatically the number of microsporidial spores available for staining (from the concentrate sediment). This is the same protocol we use for centrifugation of all stool specimens, regardless of the suspected organism.

5. Avoid the use of wet gauze filtration (an old, standardized method of filtering stool before centrifugation) with too many layers of gauze that may trap organisms and allow them to flow into the fluid to be concentrated. It is recommended that no more than two layers of gauze be used. Another option is to use the commercially available concentration systems that use metal or plastic screens for filtration.

*10 times the normal trichrome stain formula.

64-9 MODIFIED TRICHROME STAIN FOR THE MICROSPORIDIA[41] (KOKOSKIN-HOT METHOD)

PRINCIPLE
The oval shape, spore wall, and diagonal or horizontal "stripe" that distinguish microsporidia spores can best be visualized with a stain that enhances the morphologic features. Changes in temperature from room temperature to 50° C and the staining time from 10 to 90 min have been recommended as improvements for the modified trichrome staining methods. In addition, such a stained smear provides a permanent record of the results.

METHOD
1. Using a 10-µL aliquot of unconcentrated, preserved liquid stool (5% or 10% formalin or SAF), prepare the smear by spreading the material over an area of 45 × 25 mm.

2. Allow the smear to air dry.

3. Place the smear in absolute methanol for 5 minutes.

4. Allow the smear to air dry.

5. Place in trichrome stain for 10 minutes at a temperature of 50° C.

6. Rinse in acid-alcohol for no more than 10 seconds.

7. Dip slides several times in 95% alcohol. Use this step as a rinse (no more than 10 seconds).

8. Place in 95% alcohol for 5 minutes.

9. Place in 100% alcohol for 10 minutes.

10. Place in xylene substitute for 10 minutes.

11. Mount with coverslip (No. 1 thickness), using mounting medium.

12. Examine smears under oil immersion (1000×) and read at least 100 fields; the examination time will probably be at least 10 minutes per slide.

64-10 STAINING THIN FILMS: GIEMSA STAIN

PRINCIPLE
By spreading the blood cells in a thin layer, the size of red cells, inclusions, and extracellular forms can be more easily visualized.

METHOD
1. Fix blood films in absolute methanol (acetone-free) for 30 seconds.

2. Allow slides to air dry.

3. Immerse slides in a solution of 1 part Giemsa stock (commercial liquid stain or stock prepared from powder) to 10 to 50 parts of Triton-buffered water (pH 7.0 to 7.2). Stain 10 to 60 minutes (see note below). Fresh working stain should be prepared from stock solution each day.

4. Dip slides briefly in Triton X-100 buffered water.

5. Drain thoroughly in vertical position and allow to air dry.

    NOTE: *A good general rule for stain dilution vs. staining time is that if dilution is 1:20, stain for 20 minutes; if 1:30, stain for 30 minutes; and so forth.*
    However, a series of stain dilutions and staining times should be tried to determine the best dilution/time for each batch of stock stain.

EXPECTED RESULTS
Giemsa stain colors the components of blood as follows: erythrocytes, pale gray-blue; nuclei of white blood cells, purple and pale purple cytoplasm; eosinophilic granules, bright purple-red; neutrophilic granules, deep pink-purple. Parasitic forms are blue to purple, with reddish nuclei. Their characteristics morphologies are used for differentiation. Inexperienced workers may confuse platelets with parasites.

64-11 STAINING THICK FILMS: GIEMSA STAIN

PRINCIPLE
A large amount of blood can be examined for parasitic forms by lysing the red blood cells and staining for parasites. The lack of methanol fixation allows lysis of red blood cells by the aqueous stain solution. Although parasites can be found in the larger volume of blood, definitive morphologic criteria necessary for specific organism identification may be more difficult to see.

METHOD
The procedure to be followed for thick films is the same as for thin films, except that the first two steps are omitted. If the slide has a thick film at one end and a thin film at the other, fix only the thin portion, and then stain both parts of the film simultaneously.

References

1. Aldras, A.M., Orenstein, J.M., Kotler, D.P., et al. 1994. Detection of microsporidia by indirect immunofluorescence antibody test using polyclonal and monoclonal antibodies. J. Clin. Microbiol. 32:608.

2. Alles, A.J., Waldron, M.A., Sierra, L.S., and Mattia, A.R. 1995. Prospective comparison of direct immunofluorescence and conventional staining methods for detection of *Giardia* and *Cryptosporidium* spp. in human fecal specimens. J. Clin. Microbiol. 33:1632.

3. Arakaki, T., Iwanaga, M., Kinjo, F., et al. 1990. Efficacy of agar-plate culture in detection of *Strongyloides stercoralis* infection. J. Parasitol. 76:425.

4. Beal, C., Goldsmith, R., Kotby, M., et al. 1992. The plastic envelope method: a simplified technique for culture diagnosis of trichomoniasis. J. Clin. Microbiol. 30:2265.

5. Beaver, P.C., Jung, R.C., and Cupp, E.W. 1984. Clinical parasitology, ed 9. Lea & Febiger, Philadelphia.

6. Borchardt, K.A. 1994. Trichomoniasis: its clinical significance and diagnostic challenges. American Clinical Laboratory, Sept. 20, p. 1.

7. Borchardt, K.A. and Smith, R.F. 1991. An evaluation of an InPouch TV culture method for diagnosing *Trichomonas vaginalis* infection. Genitourin. Med. 67:149.

8. Bretagne, S., Costa, J.M., Fleury-Feith, J., et al. 1995. Quantitative competitive PCR with bronchoalveolar lavage fluid for diagnosis of toxoplasmosis in AIDS patients. J. Clin. Microbiol. 33:1662.

9. Bryan, R.T. 1995. Microsporidiosis as an AIDS-related opportunistic infection. Clin. Infect. Dis. 21:S62.

10. Cartwright, C.P., Nelson, N.A., and Gill, V.J. 1994. Development and evaluation of a rapid and simple procedure for detection of *Pneumocystis carinii* by PCR. J. Clin. Microbiol. 32:1634.

11. Centers for Disease Control. 1995. Local transmission of *Plasmodium vivax* malaria—Houston, Texas, 1994. Morb. Mortal. Wkly. Rep. 44:295.

12. Centers for Disease Control. 1994. *Cryptosporidium* infections associated with swimming pools—Dane County, Wisconsin, 1993. Morb. Mortal. Wkly. Rep. 43:561.

13. Current, W.L. and Garcia, L.S. 1991. Cryptosporidiosis. Clin. Microbiol. Rev. 3:325.

14. DeGirolami, P.C., Ezratty, C.R., Desai, G. et al. 1995. Diagnosis of intestinal microsporidiosis by examination of stool and duodenal aspirate with Weber's modified trichrome and Uvitex 2B stains. J. Clin. Microbiol. 33:805.

15. Didier, E.S., Orenstein, J.M., Aldras, A., et al. 1995. Comparison of three staining methods for detecting microsporidia in fluids. J. Clin. Microbiol. 33:3138.

16. Dietze, R., Perkins, M., Boulos, M., et al. 1995. The diagnosis of *Plasmodium falciparum* infection using a new antigen detection system. Am. J. Trop. Med. Hyg. 52:45.

17. Dupon, M., Cazenave, J., Pellegrin, J.L., et al. 1995. Detection of *Toxoplasma gondii* by PCR and tissue culture in cerebrospinal fluid and blood of human immunodeficiency virus-seropositive patients. J. Clin. Microbiol. 33:2421.

18. DuPont, H.L., Chappell, C.L., Sterling, C.R., et al. 1995. The infectivity of *Cryptosporidium parvum* in healthy volunteers. N. Engl. J. Med. 332:855.

19. Clendennen, T.E., Long, G.W., and Baird, J.K. 1995. QBC™ and Giemsa-stained thick blood films: diagnostic performance of laboratory technologists. Trans. Royal Soc. Trop. Med. Hyg. 89:183.

20. Franzen, C., Müller, A., Hegener, P., 1995. Detection of microsporidia (*Enterocytozoon bieneusi*) in intestinal biopsy specimens from human immunodeficiency virus-infected patients by PCR. J. Clin. Microbiol. 33:2294.

21. Garavelli, P.L., Zierdt, C.H., Fleisher, T.A., et al. 1995. Serum antibody detected by fluorescent antibody test in patients with symptomatic *Blastocystis hominis* infection. Recenti Prog. Med. 86:398.

22. Garcia, L.S. and Bruckner, D.A. 1997. Diagnostic medical parasitology, ed 3. American Society for Microbiology, Washington, D.C.

23. Garcia, L.S., Bullock-lacullo, S., Palmer, J., and Shimizu, R.Y. 1995. Diagnosis of parasitic infections: collection, processing and examination of specimens. In Murray, P.R., Baron, E.J., Pfaller, M.A., et al., editors. Manual of clinical microbiology, ed 6. American Society for Microbiology, Washington, D.C.

24. Garcia, L.S., Shimizu, R.Y., and Bruckner, D.A. 1994. Detection of microsporidial spores in fecal specimens from patients diagnosed with cryptosporidiosis. J. Clin. Microbiol. 32:1739.

25. Garcia, L.S., Shimizu, R.Y., Shum, A.C., and Bruckner, D.A. 1993. Evaluation of intestinal protozoan morphology in polyvinyl alcohol preservative: comparison of zinc-based and mercuric chloride-based compounds for use in Schaudinn's fixative. J. Clin. Microbiol. 31:307.

26. Garcia, L.S., Shum, A.C., and Bruckner, D.A. 1992. Evaluation of a new monoclonal antibody combination reagent for the direct fluorescent detection of *Giardia* cysts and *Cryptosporidium* oocysts in human fecal specimens. J. Clin. Microbiol. 30:3255.

27. Garcia, L.S., Sulzer, A.J., Healy, G.R. et al. 1995. Murray, P.R., Baron, E.J., Pfaller, M.A., et al., editors. Blood and tissue protozoa. In Manual of clinical microbiology, ed 6. American Society for Microbiology, Washington, D.C.

28. Gascon, J., Corachan, M., Bombi, J.A., et al. 1995. *Cyclospora* in patients with traveller's diarrhea. Scan. J. Infect. Dis. 27:511.

29. Gonzalez-Ruiz, A., Haque, R., Rehman, T., et al. 1994. Diagnosis of amebic dysentery by detection of *Entamoeba histolytica* fecal antigen by an invasive strain-specific, monoclonal antibody-based enzyme-linked immunosorbent assay. J. Clin. Microbiol. 32:964.

30. Haque, R., Neville, L.M., Hahn, P., and Petri, W.A. 1995. Rapid diagnosis of *Entamoeba* infection by using *Entamoeba* and *Entamoeba histolytica* stool antigen detection kits. J. Clin. Microbiol. 33:2558.

31. Healy, G.R. and Garcia, L.S. 1995. Intestinal and urogenital protozoa. In Murray, P.R., Baron, E.J., Pfaller, M.A., et al., editors. Manual of clinical microbiology, ed 6. American Society for Microbiology, Washington, D.C.

32. Herwaldt, B.L., Springs, F.E., Roberts, P.P., et al. 1995. Babesiosis in Wisconsin: a potentially fatal disease. Am. J. Trop. Med. Hyg. 53:146.

33. Hiatt, R.A., Markell, E.K., and Ng, E. 1995. How many stool examinations are necessary to detect pathogenic intestinal protozoa? Am. J. Trop. Med. Hyg. 53:36.

34. Huang, P., Weber, J.T., Sosin, D.M., et al. 1995. The first reported outbreak of diarrheal illness associated with *Cyclospora* in the United States. Ann. Int. Med. 123:409.

35. Juranek, D.D. 1995. Cryptosporidiosis: sources of infection and guidelines for prevention. Clin. Infect. Dis. 21:S57.

36. Kehl, K.S.C. 1996. Screening stools for *Giardia* and *Cryptosporidium*: are antigen tests enough? Clin. Microbiol. Newsletter, 18:133.

37. Kimura, M., Miyake, H., Kim, H.S., et al. 1995. Species-specific PCR detection of malaria parasites by microtiter plate hybridization: clinical study with malaria patients. J. Clin. Microbiol. 33:2342.

38. Kinyoun. 1915. A note on Uhlenhuth's method for sputum examination for tubercle bacilli. Am. J. Pub. Health 5:867.

39. Kocka, F., Peters, C., Dacumos, E., et al. 1991. Outbreaks of diarrheal illness associated with Cyanobacteria (blue-green algae)-like bodies—Chicago and Nepal, 1989 and 1990. Morb. Mortal. Wkly. Rep. 40:325.

40. Koga, K., Kasuya, S., Khamboonruang, C., et al. 1991. A modified agar plate method for detection of *Strongyloides stercoralis*. Am. J. Trop. Med. Hyg. 45:518.

41. Kokoskin, E., Gyorkos, T.W., Camus, A., et al. 1994. Modified technique for efficient detection of microsporidia. J. Clin. Microbiol. 32:1074.

42. Layton, M., Parise, M.E., Campbell, C.C., 1995. Mosquito-transmitted malaria in New York City, 1993. Lancet 346:729.

43. Long, E.G., White, E.H., Carmichael, W.W., et al. 1991. Morphologic and staining characteristics of a *Cyanobacterium*-like organism associated with diarrhea. J. Infect. Dis. 164:199.

44. Lowichik, A., Rollins, N., Delgado, R., et al. 1995. Leptomyxid amebic meningoencephalitis mimicking brain stem glioma. Am. J. Neuroradiol. 16:926.

45. Luna, V.A., Stewart, B.K., Bergeron, D.L., et al. 1995. Use of the fluorochrome calcofluor white in the screening of stool specimens for spores of microsporidia. Am. J. Clin. Pathol. 103:656.

46. Mackenzie, W.R., Schell, W.L., Blair, K.A., et al. 1995. Massive outbreak of waterborne cryptosporidium infection in Milwaukee, Wisconsin: recurrence of illness and risk of secondary transmission. Clin. Infect. Dis. 21:57.

47. Markell, E.K. and Udkow, M.P. 1986. *Blastocystis hominis*: pathogen or fellow traveler? Am. J. Trop. Med. Hyg. 35:1023.

48. Markell, E.K. and Voge, M. 1981. Medical parasitology, ed 5. W.B. Saunders, Philadelphia.

49. Markell, E.K., Voge, M., and John, D.T. 1992. Medical parasitology, ed 7. W.B. Saunders, Philadelphia.

50. Mayes, J.T., O'Connor, B.J., Avery, R., et al. 1995. Transmission of *Toxoplasma gondii* infection by liver transplantation. Clin. Infect. Dis. 21:511.

51. McVicar, J.W. and Suen, J. 1994. Packaging and shipping biological materials. In Laboratory safety: principles and practices, ed 2. American Society for Microbiology, Washington, D.C.

52. Melvin, D.M. and Brooke, M.M. 1974. Laboratory procedures for the diagnosis of intestinal parasites. DHEW Pub. No. (CDC) 75-8282, U.S. Government Printing Office, Washington, D.C.

53. Morris, A.J., Wilson, M.L., and Reller, L.B. 1992. Application of rejection criteria for stool ovum and parasite examinations. J. Clin. Microbiol. 30:3213.

54. National Committee for Clinical Laboratory Standards. 1993. Procedures for the recovery and identification of parasites from the intestinal tract. Proposed guideline M28-P. National Committee for Clinical Laboratory Standards, Villanova, Pa.

55. National Committee for Clinical Laboratory Standards. 1990. Use of Blood film examination of parasites. Tentative guideline M15-T. National Committee for Clinical Laboratory Standards, Villanova, Pa.

56. Ortega, Y., Sterling, C.R., Gilman, R.H., et al. 1993. *Cyclospora cayetanensis*: a new protozoan pathogen of humans. N. Eng. J. Med. 328:1308.

57. Palmer, J. 1991. Modified iron hematoxylin/kinyoun stain. Clin. Microbiol. Newsletter 13:39 (letter).

58. Perry, H.D., Donnenfeld, E.D., Foulks, G.N., et al. 1995. Decreased corneal sensation as an initial feature of *Acanthamoeba* keratitis. Ophthalmology 102:1565.

59. Radford, C.F., Bacon, A.S., Dart, J.K.G., and Minassian, D.C., 1995. Risk factors for *Acanthamoeba* keratitis in contact lens users: a case-control study. Br. Med. J. 310:1567.

60. Reed, S.L. 1995. New concepts regarding the pathogenesis of amebiasis. Clin. Infect. Dis. 21:S182.

61. Rosenblatt, J.E. and Sloan, L.M. 1993. Evaluation of an enzyme-linked immunosorbent assay for detection of *Cryptosporidium* spp. in stool specimens. J. Clin. Microbiol. 31:1468.

62. Roux, P., Lavrard, I., Poirot, J.L., et al. 1994. Usefulness of PCR for detection of *Pneumocystis carinii* DNA. J. Clin. Microbiol. 32:2324.

63. Ryan, N.J., Sutherland, G., Coughlan, K., et al. 1993. A new trichrome-blue stain for detection of microsporidial species in urine, stool, and nasopharyngeal specimens. J. Clin. Microbiol. 31:3264.

64. Sheehan, D.J., Raucher, B.G., and McKitrick, J.C. 1986. Association of *Blastocystis hominis* with signs and symptoms of human disease. J. Clin. Microbiol. 24:548.

65. Shiff, C.J., Minjas, J., and Premji, Z. 1994. The ParaSight™F Test: a simple, rapid manual dipstick test to detect *Plasmodium falciparum* infection. Parasito. Today 10:494.

66. Shlim, D.R., Hoge, C.W., Rajah, R., et al. 1995. Is *Blastocystis hominis* a cause of diarrhea in travellers? a prospective controlled study in Nepal. Clin. Infect. Dis. 21:97.

67. Sifuentesosornio, J., Porrascortes, G., Bendall, R.P., et al. 1995. *Cyclospora cayetanensis* infection in patients with and without AIDS: biliary disease as another clinical manifestation. Clin. Infect. Dis. 21:1092.

68. Silveira, H. and Canning, E.U., 1995. *Vittaforma cornae* n. comb. for the human microsporidium *Nosema corneum* Shadduck, Meccoli, Davis and Font, 1990 based on its ultrastrucure in the liver of experimentally infected athymic mice. J. Eukaryo. Microbiol. 42:158.

69. Singh, M., Suresh, K., Ho, L.C., et al. 1995. Elucidation of the life cycle of the intestinal protozoan *Blastocystis hominis.* Parasitol. Res. 81:446.

70. Sison, J.P., Kemper, C.A., Loveless, M., et al. 1995. Disseminated *Acanthamoeba* infection in patients with AIDS: case reports and review. Clin. Infect. Dis. 20:1207.

71. Sobottka, I., Albrecht, H., Schafer, H., et al. 1995. Disseminated *Encephalitozoon (Septata) intestinalis* infection in a patient with AIDS: novel diagnostic approaches and autopsy-confirmed parasitological cure following treatment with albendazole. J. Clin. Microbiol. 33:2948.

72. Spargano, O., Drouet, E., Brebant, R., et al. 1995. Use of monoclonal antibodies to distinguish pathogenic *Naegleria fowleri* (cysts, trophozoites, or flagellate forms) from other *Naegleria* species. J. Clin. Microbiol. 31:2758.

73. Svenson, J.E., Maclean, J.D., Gyorkos, T.W., and Keystone, J. 1995. Imported malaria: clinical presentation and examination of symptomatic travelers. Arch. Int. Med. 155:861.

74. Tiley, S.M., Marriott, D.J.E., and Harkness, J.L., 1995. An evaluation of four methods for the detection of *Pneumocystis carinii* in clinical specimens. Pathology 26:325.

75. Visvesvara, G.S. 1992. In Isenberg, H.D., editor. Clinical microbiology procedures handbook (parasitology section). American Society for Microbiology, Washington, D.C.

76. Visvesvara, G.S. 1995. Pathogenic and opportunistic free-living amebae. In Murray, P.R., Baron, E.J., Pfaller, M.A., et al., editors. Manual of clinical microbiology, ed 6. American Society for Microbiology, Washington, D.C.

77. Visvesvara, G.S., Da Silva, A.J., Croppo, G.P., et al. 1995. In vitro culture and serologic and molecular identification of *Septata intestinalis* isolated from urine of a patient with AIDS. J. Clin. Microbiol. 33:930.

78. Visvesvara, G.S., Leitch, G.J., Da Silva, A.J., et al. 1994. Polyclonal and monoclonal antibody and PCR-amplified small-subunit rRNA identification of a microsporidian, *Encephalitozoon hellem,* isolated from an AIDS patient with disseminated infection. J. Clin. Microbiol. 32:2760.

79. Visvesvara, G.S., Schuster, F.L. and Martinez. A.J., 1993. *Balamuthia mandrillaris,* new genus, new species, agent of amebic meningoencephalitis in humans and animals. J. Eukaryot. Microbiol. 40:504.

80. Walzer, P.D. 1994. *Pneumocystis carinii* pneumonia, ed 2. Marcel Dekker, New York.

81. Weber, R., Bryan, R.T., Owen, R.L. et al. and the Enteric Opportunistic Infections Working Group. 1992. Improved light-microscopical detection of microsporidia spores in stool and duodenal aspirates. N. Engl. J. Med. 326:161.

82. Weber, R., Sauer, B., Spycher, M.A., et al. 1994. Detection of *Septata intestinalis* in stool specimens and coprodiagnostic monitoring of successful treatment with albendazole. Clin. Infect. Dis. 19:342.

83. Wilcox, A. 1960. Manual for the microscopical diagnosis of malaria in man. U.S. Public Health Service, Washington, D.C.

84. Wilson, M., Schantz, P., and Pieniazek, N. 1995. Diagnosis of parasitic infections: immunologic and molecular methods. In Murray, P.R., Baron, E.J., Pfaller, M.A., et al., editors. Manual of clinical microbiology, ed 6. American Society for Microbiology, Washington, D.C.

85. Zierdt, C.H. 1988. *Blastocystis hominis:* a long-misunderstood intestinal parasite. Parasitol. Today 4:15.

86. Zierdt, C.H., Gill, V.J., and Zierdt, W.S. 1993. Detection of microsporidian spores in clinical samples by indirect fluorescent-antibody assay using whole-cell antisera to *Encephalitozoon cuniculi* and *Encephalitozoon hellem.* J. Clin. Microbiol. 31:3071.

Part Six

6

 Mycology

65 LABORATORY METHODS IN BASIC MYCOLOGY

Historically, the fungi were regarded as relatively insignificant causes of infection and only until recently did clinical laboratories offer little more than passing interest in providing mycologic services. It is now well documented that the fungi are common causes of infection, particularly in immunocompromised patients. Some clinical microbiology laboratories have kept pace with changing times and have offered somewhat complete mycologic services; however, because of the current economic and health care environments, many laboratories are unable to offer these services. In most instances, diagnostic clinical mycology is performed by a few reference laboratories that have varying degrees of experience. The lack of experience in clinical mycology has been influenced by a shortage of trained individuals, lack of quality educational programs, and the inability of clinical laboratories to support the cost of sending personnel to training courses. Commonly those individuals with experience who retire or leave their position are rarely replaced by someone with experience. Despite these limitations, some clinical microbiology laboratories are able to offer quality mycologic services. The concern is that the changing health care environment and implementation of cost containment measures will prevent future generations from being well trained in diagnostic clinical mycology.

This chapter is designed to assist technologists or microbiologists with the basics of diagnostic clinical mycology with the hope that this information will allow some laboratories to offer clinical mycology services.

OVERVIEW OF CLINICAL MYCOLOGY

There are more than 50,000 valid species of fungi, but only 100 to 150 species are generally recognized as a cause of disease in humans. Many of these organisms normally live a **saprophytic** (living on dead or decayed organic matter) existence in nature, enriched by the nitrogenous matter, but they are capable of maintaining a separate existence as a parasite in humans or animals. Fungal infections are not communicable in the usual sense of person-to-person or animal-to-person transfer; humans become an accidental host by inhalation of spores or by their introduction into tissue by trauma. Except for the dimorphic fungi, humans are relatively resistant to infections caused by the fungi. The major factor responsible for the increase in the number of fungal infections has been an alteration in the immune system. Whether caused by the introduction of immunosuppressive agents or serious underlying diseases, this alteration may lead to infection by organisms that are normally considered to be nonpathogenic. Such infections may occur in patients with debilitating diseases such as infection with the human immunodeficiency virus (HIV), diabetes, or impaired immunologic function resulting from corticosteroid or antimetabolite therapy. Common predisposing factors also include long-term intravenous cannulation, gastrointestinal surgical procedures, and long-term antimicrobial chemotherapy. It is because of the need to provide effective health care to immunocompromised patients that laboratories now must be able to identify and report all fungi recovered from clinical specimens. Because of the vast array of fungal species known to cause disease in humans, it is impossible for the laboratory to determine the significance of an organism recovered in a clinical specimen.

GENERAL FEATURES OF THE FUNGI

Fungi seen in the clinical laboratory can generally be separated into two groups based on the appearance of the colonies formed. The **yeasts** produce moist, creamy, opaque, or pasty colonies on media, whereas the **filamentous fungi** or **molds** produce fluffy, cottony, woolly, or powdery colonies. Several pathogenic species of fungi that exhibit either a yeast and filamentous form are referred to as being **dimorphic**. When dimorphism is temperature dependent, the fungi are designated as **thermally dimorphic**. In general, these fungi produce a mold form at 25° to 30° C and a yeast or spherule form at 35° to 37° C under certain circumstances. The medically important dimorphic fungi include *Histoplasma capsulatum, Blastomyces dermatitidis, Coccidioides immitis, Paracoccidioides* *brasiliensis, Sporothrix schenckii,* and *Penicillium marneffei.* Additionally, some of the medically important yeasts, particularly the *Candida* spp., may produce yeast forms, pseudohyphae, and/or true hyphae. The polymorphic features of this group of organisms are not temperature dependent.

YEASTS

Yeasts are unicellular organisms that are round to oval and range in size from 2 to 60 μm. The microscopic morphologic features usually appear similar among genera and are not particularly helpful in their differentiation. In some cases, this also makes the direct microscopic and histopathologic examination of clinical specimens difficult. However, the size, presence or absence of a capsule, or "narrow-necked" budding are features that can be helpful in differentiating *Cryptococcus* spp. from *Candida* spp. Most of the important yeasts and yeastlike organisms may belong to the Ascomycota, Basidiomycota, or Deuteromycota.

In general, the yeasts reproduce asexually by **blastoconidia** formation (**budding**) (Figure 65-1) and sexually by the production of ascospores or basidiospores. The process of budding begins with a weakening and subsequent out-pouching of the yeast cell wall. This process continues until the bud, or daughter cell, is completely formed. The cytoplasm of the bud is contiguous with the cytoplasm for the original cell. A cell wall septum is created between a mother and daughter yeast cell. The daughter cell usually detaches from the mother cell and a residual defect occurs at the site of the budding; this is known as a **bud scar.**

If a delineating septum is not formed after bud formation and an elongation process begins, a **germ tube** is formed (Figure 65-2). The germ tube is the initial phase of pseudohyphae formation. **Pseudohyphae** represent elongated buds connected to one another to create a "links of sausage" appearance (Figure 65-3). Pseudohyphae do not contain true septae but have cell wall constrictions, rather than true intracellular septations delineating the borders.

Yeasts divide by binary fission, undergoing cell wall expansion followed by mitosis. Following mitosis, a new cell wall septum is created between the two newly formed nuclei. The two cells may either separate after the septum is formed or remain attached through subsequent rounds of mitosis. Binary fission is a characteristic of a parasitic form of *Penicillium marneffei* (Figure 65-4).

MOLDS

The basic structural units of the molds are tubelike projections known as **hyphae.** As the hyphae grow,

they become intertwined to form a loose network, the **mycelium,** which penetrates the substrate from which the mold obtains the necessary nutrients for growth. **Vegetative hyphae** comprise the body of the fungus, in contrast to specialized **reproductive hyphae.** The nutrient-absorbing and water-exchanging portion of the fungus is called a **vegetative mycelium.** The portion extending above the substrate surface is known as **aerial mycelium;** aerial mycelia often give rise to fruiting bodies from which asexual spores are borne.

Three types of hyphae exist in the medically important fungi. These include the **coenocytic (sparsely septate) hyphae** of the Zygomycetes (Figure 65-5), the dark and pigmented septate hyphae of the dematiaceous fungi (Figure 65-6) and the septate, nonpigmented hyphae of the **hyaline** molds (Figure 65-7). The terms **dematiaceous** and **hyaline** describe the presence or absence, respectively, of pigmentation within the hyphae or the spores. Hyphal pigmentation is a useful feature to differentiate some fungi and will be discussed later in this chapter.

The hyphae of Zygomycetes are wider in diameter when compared to the those of the fungi producing septate hyaline or dematiaceous hyphae. The branching of the Zygomycetes often occurs at angles greater than 90 degrees in contrast to the acutely branching hyphae of the septate dematiaceous and hyaline molds. Hyphae of the Zygomycetes are most often nonseptate; however, a few septae may be present depending on the organism and the age of the culture. Those septae that are present are often located near specialized reproductive hyphae. In a nonseptate hyphae, nuclei and cytoplasm are free to flow throughout the length of the hyphal element.

When a commonly employed technique of tissue grinding is used to process clinical specimens, the hyphae of Zygomycetes may be destroyed. Nonseptate hyphae are not compartmentalized by septae; disruption of any area of the hyphae causes cytoplasmic leakage and eventually death of the organism. The septate hyphae of dematiaceous and hyaline molds will survive tissue processing because of compartmentalization of the hyphae by septae.

The vegetative hyphae of the dematiaceous and hyaline molds are morphologically similar. Each have true septate hyphae that branch at acute angles. The major differentiating feature is the presence or absence of pigmentation in the hyphae. The cell wall of the dematiaceous fungi contain melanin-related compounds, which make the hyphae darkly pigmented when observed by light microscopy. In histopathologic sections, these pigments may be accentuated by the use of special stains (e.g., Fontana-Masson stain).

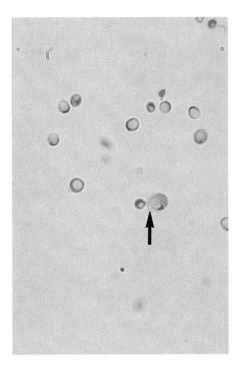

FIGURE 65-1 Blastoconidia (budding cells [*arrow*]) characteristic of the yeasts (430×).

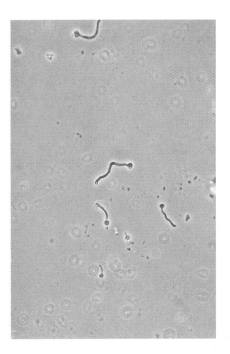

FIGURE 65-2 *C. albicans,* germ tube test, showing yeast cells with germ tubes present (430×).

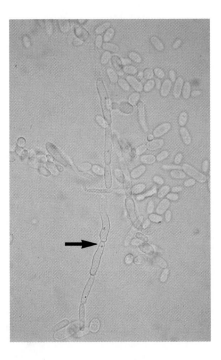

FIGURE 65-3 Pseudohyphae consisting of elongated cells (*arrow*) with constrictions where attached (430×).

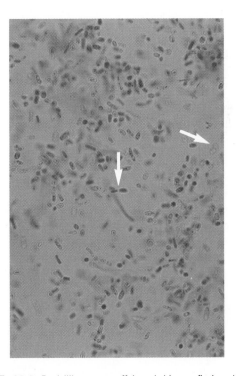

FIGURE 65-4 *Penicillium marneffei* and binary fission (*arrows*) (500×).

TAXONOMY OF THE FUNGI

Fungi comprise a vast array of organisms that are unique compared with plants and animals. Included are the mushrooms; rusts and smuts; molds and mildews, and the yeasts. Despite the great variation in morphologic features of the fungi, all share similar characteristics. These include:

- The presence of chitin in the cell wall
- The presence of ergosterol in the cell membrane
- Reproduction by means of spores, either asexually or sexually
- The lack of chlorophyll
- Lack of susceptibility to antibacterial antibiotics
- Their heterotopic nature (i.e., they derive nutrition from organic materials)

Traditionally, the fungi have been categorized into four well-established phyla: the Zygomycota, Ascomycota, Basidiomycota, and Deutromycota. The phylum Zygomycota includes those organisms that produce nonseptate or sparsely septate hyphae and exhibit asexual reproduction by sporangiospores and sexual reproduction by zygospores. A few of the clinically important genera in this phylum include *Rhizopus*, *Mucor*, *Absidia*, *Cunninghamella*, *Saksenaea*, and several other genera uncommonly seen in the clinical microbiology laboratory.

The Ascomycota include many fungi that reproduce asexually by the formation of **conidia** (asexual spores) and sexually by the production of ascospores. The filamentous ascomycetes are ubiquitous in nature and all produce true septate hyphae. All exhibit a sexual form (**teleomorph**) but also exist in an asexual form (**anamorph**). In general, the anamorphic form correlates well with the teleomorphic classification.

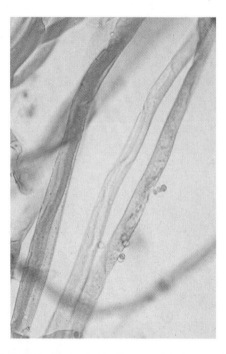

FIGURE 65-5 Hyaline hyphae lacking septations (aseptate) (430×).

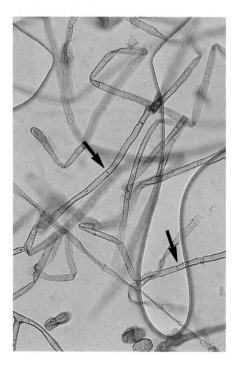

FIGURE 65-6 Dematiaceous hyphae showing pigmentation and septations (*arrows*) (430×).

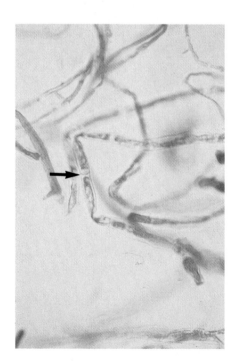

FIGURE 65-7 Hyaline hyphae showing septations (*arrow*) (430×).

However, different teleomorphic forms may have an identical anamorph. *Pseudallescheria boydii* (Figure 65-8), in addition to having the *Scedosporium apiospermum* anamorph (Figure 65-9), may exhibit a *Graphium* anamorph (Figure 65-10). This anamorph may be seen with several other fungi. Examples of other clinically important fungi that belong to the phylum Ascomycota include *Histoplasma capsulatum* and *Blastomyces dermatitidis,* which have a teleomorph designated as *Ajellomyces.* Some species of *Aspergillus* have a teleomorph, *Eurotium.* Additionally, numerous yeast species belong to the Ascomycota and include *Saccharomyces* and some species of *Candida.*

The phylum Basidiomycota includes those fungi that reproduce sexually by the formation of basidiospores on specialized structures called the basidia. The basidiomycetes are generally plant pathogens or environmental organisms that rarely cause disease in humans. Included are the smuts, rusts, mushrooms, and the sexual form of *Cryptococcus neoformans (Filobasidiella neoformans).*

The phylum Deutromycota includes those fungi that lack a sexual reproductive cycle and are characterized by their asexual reproductive structures, primarily conidia. It is possible that the sexual form of organisms in this group exist but have not been described.

PRACTICAL CLASSIFICATION OF THE FUNGI

The botanic taxonomic schema for grouping the fungi has little value in a clinical microbiology laboratory. Table 65-1 is a simplified taxonomic schema illustrating the major groups of fungi; these have been previously described within this chapter.

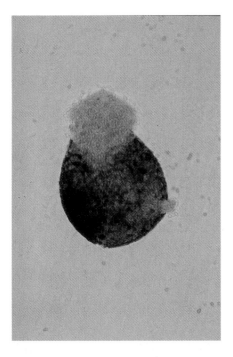

FIGURE 65-8 *P. boydii,* showing cleistothecia and numerous ascospores (750×).

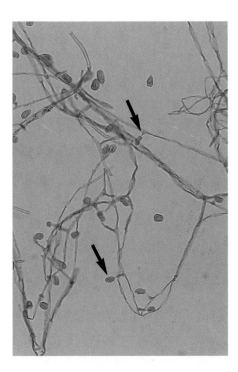

FIGURE 65-9 *S. apiospermum,* showing asexually produced conidia borne singly on conidiophores (annellophores [*arrows*]) (430×).

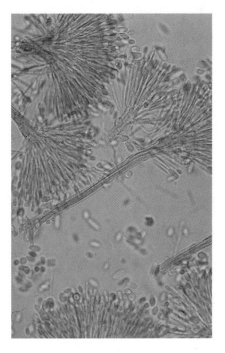

FIGURE 65-10 Graphium anamorph of *P. boydii* (500×).

— TABLE 65-1 PHYLOGENETIC POSITION OF MEDICALLY SIGNIFICANT FUNGI —

CLASS	ORDER	GENUS/SPECIES
Phylum Zygomycota Zygomycetes	Entomophthorales Mucorales	*Basidiobolus*[‡]
		Absidia[‡]
		Cunninghamella[‡]
		Mucor[†]
		Rhizopus[†]
		Syncephalastrum
Phylum Ascomycota Hemiascomycetes	Endomycetales	*Endomyces (Geotrichum* sp.)*
		*Kluyveromyces (Candida pseudotropicalis)**
Loculoascomycetes	Myriangiales Microascales	*Piedraia hortae*
		Pseudallescheria boydii[†]
		Scedosporium prolificans[‡]
Plectomycetes	Eurotiales Onygerales	*Emericella (Aspergillus nidulans)**
		Ajellomyces (Histoplasma capsulatum,[†] *Blastomyces dermatitidis*[†])*
		Arthroderma (Trichophyton sp.[†] and *Microsporum* sp.[†])*
Phylum Basidiomycota Teliomycetes	Filobasidiales	*Filobasidiella (Cryptococcus neoformans*[†])*
Phylum Deutromycota Blastomycetes Hyphomycetes	Moniliales	*Candida*[†]
		Acremonium[‡]
		Aspergillus[†]
		Blastomyces[†]
		Chrysosporium
		Coccidioides[†]
		Epidermophyton[†]
		Geotrichum
		Gliocladium
		Histoplasma[†]
		Microsporum[†]
		Paecilomyces[‡]
		Paracoccidioides[†]
		Penicillium[‡]

TABLE 65-1 PHYLOGENETIC POSITION OF MEDICALLY SIGNIFICANT FUNGI—CONT'D		
CLASS	ORDER	GENUS/SPECIES

Phylum Deutromycota—cont'd
  Blastomycetes
  Hyphomycetes
		Sepedonium

		Scopulariopsis
		Sporothrix[†]
		Trichoderma
		Trichophyton[†]
		Alternaria
		Aureobasidium
		Bipolaris[‡]
		Cladosporium
		Curvularia[‡]
		Drechslera
		Exophiala[‡]
		Exserohilum
		Fonsecaea[‡]
		Helminthosporium
		Madurella[‡]
		Nigrospora
		Phaeoannellomyces[‡]
		Phialophora[‡]
		Rhinocladiella
		Stemphylium
		Ulocladium
		Xylohypha[‡]
		Epicoccum
		Fusarium[†]
		Phoma

[*]When the sexual form is known.
[†]Most commonly encountered as causes of infection.
[‡]Infrequently cause of infection.

Modified from Kwon-Chung, K.J. and Bennett, J.E. 1992. Medical mycology, Lea & Febiger, Philadelphia.

Clinicians find value in placing the fungi into four categories of mycoses:

- The superficial or cutaneous mycoses
- Subcutaneous mycoses
- Systemic mycoses
- Opportunistic mycoses

The **superficial** or **cutaneous mycoses** are fungal infections that involve the hair, skin, or nails without direct invasion of the deeper tissue. The fungi in this category include the dermatophytes (i.e., agents of rignworm, athlete's foot) and agents of infections such as tinea, tinea nigra, and piedra; all infect keratinized tissues.

Some fungi produce infections that are confined to the subcutaneous tissue without dissemination to distant sites; examples of subcutaneous infections include chromoblastomycosis, mycetoma, and phaeohyphomycosis; these will be discussed under subcutaneous mycosa later in this chapter.

As traditionally defined, agents of systemic fungal infections contain the genera *Blastomyces*, *Coccidioides*, *Histoplasma,* and *Paracoccidioides*. Infections caused by these organisms primarily involve the lungs but also may become widely disseminated and involve any organ system. Recently, *Penicillium marneffei* has been added to this group.

The fungi causing opportunistic infections include an ever-expanding list of organisms. The infections they cause are found primarily in patients who are immunocompromised, either by an underlying disease process or immunosuppressive agent(s). Fungi previously thought to be nonpathogenic may be the cause of infections. The commonly encountered organisms include *Aspergillus*, *Zygomycetes*, *Candida,* and *Cryptococcus*, all of which may cause disseminated (systemic) disease.

This type of classification allows the clinician to attempt to categorize organisms in a logical fashion into groups having clinical relevance. Table 65-2 presents an example of a clinical classification of infections and their etiologic agents useful to clinicians.

PRACTICAL WORKING SCHEMA

To assist persons working in clinical microbiology laboratories with the identification of clinically important fungi, Koneman and Roberts[59] have suggested the use of a practical working schema designed to:

- Assist in the recognition of fungi most commonly encountered in clinical specimens
- Assist with the recognition of fungi recovered on culture media that belong to the strictly pathogenic fungi
- Provide a pathway that is easy to follow and that allows an identification to be made based on a few colonial and microscopic features.

Table 65-3 presents these features; however, it must be recognized that it includes only those organisms commonly seen in the clinical laboratory. With practice, most laboratorians should be able to recognize them on a day-to-day basis. The use of various texts containing photomicrographs is helpful in the identification of other less commonly encountered fungi.

The use of this schema requires that one first examine the culture for the presence or absence of septae. If the hyphae appear to be predominantly nonseptate, Zygomycetes should be considered. If the hyphae are septate, they must be examined further for the presence or absence of pigmentation. If a dark pigment is present in the conidia or the hyphae, the organism is considered to be dematiaceous and the conidia are then examined for their morphologic features and arrangement on the hyphae. If the hyphae are nonpigmented, they are considered to be hyaline. They are then examined for the type and arrangement of the conidia produced. The molds are identified by recognition of their characteristic microscopic features (Table 65-3). Murray has developed an expanded morphologic classification of medically important fungi based on general microscopic features and colonial morphology. (Box 65-1).[74]

VIRULENCE FACTORS OF THE MEDICALLY IMPORTANT FUNGI

The increase in the number of opportunistic fungal infections in humans is due to the immunocompromised nature of the host. It is speculated that fungi produce certain virulence factors that make it easier for them to invade tissues and cause disease. Some virulence factors have been known for years and include:

- The size of an organism (it must be small enough for it to reach the alveoli of the lungs before infection can be established)
- The ability of an organism to grow at 37° C at a neutral pH
- Conversion of the dimorphic fungi from the mycelial form to the corresponding yeast or spherule form within the host
- Perhaps toxin production

Most of the fungi have a place in the environment where they live out a saphrophytic existence (Table 65-4). Perhaps the fungi that cause disease in humans have developed various mechanisms over time that have allowed them to establish disease in a somewhat foreign setting (the human host). A recent review by Hogan, et al.[46] describes the known or speculative virulence factors for certain of the fungi known to be pathogenic for humans. Table 65-5 presents a summary of those virulence factors.

— **TABLE 65-2   GENERAL CLINICAL CLASSIFICATION OF PATHOGENIC FUNGI** —

CUTANEOUS	SUBCUTANEOUS	OPPORTUNISTIC	SYSTEMIC
Superficial mycoses	Chromoblastomycosis	Aspergillosis	Aspergillosis
Tinea	Sporotrichosis	Candidosis	Blastomycosis
Piedra	Mycetoma (eumycotic)	Cryptococcosis	Candidosis
Candidosis	Phaeohyphomycosis	Geotrichosis	Coccidioidomycosis
Dermatophytosis		Zygomycosis	Histoplasmosis
		Fusariosis	Cryptococcosis
		Trichosporonosis	Geotrichosis
			Paracoccidioidomycosis
			Zygomycosis
			Fusariosis
			Trichosporonosis

ASPERGILLUS

It is well documented that there is an increased risk of aspergillosis in individuals who have received high doses of corticosteroids; this is thought to be the result of the suppression of macrophage and perhaps T-cell function. In addition, the association of neutropenia and neutrophil dysfunction has long been suspected, particularly in those patients having chronic granulomatous disease of childhood. Because the ability of *Aspergillus* to establish disease in humans is directly related to the immunocompromised host, it is difficult to assess whether specific fungal virulence factors play a role in the development of the disease.

Aspergillus fumigatus is known to produce two elastases, including a serum proteinase and a metalloproteinase,[58,70] that may act on elastin, which compromises approximately 30% of the lung tissue. However, no definitive evidence has been shown that elastases are related to invasive pulmonary infection. Aspergilli also are known to produce catalase that may be a virulence factor contributing to aspergillosis associated with chronic granulomatous disease of childhood.[14] The association of catalase production with any other clinical presentation of aspergillosis is uncertain. Fungi ordinarily do not produce toxins that affect humans. However, *Aspergillus flavus* is a notable exception. The aflatoxin produced by this fungus is a known carcinogenic hepatotoxin.

BLASTOMYCES DERMATITIDIS

Little is known about the pathogenesis or virulence mechanisms of *Blastomyces dermatitidis*. It is assumed that either conidia or hyphal fragments are the infectious particles responsible for establishing disease. Thermal dimorphism plays a role in the conversion from the mold to the yeast form that occurs in patients with active disease. It has been shown that decreased virulence is associated with an increase in the amount of α-1,3-glucan content present in *Blastomyces dermatitidis*.[15] It appears that the increased ability of *Blastomyces dermatitidis* to produce a granulomatous response is directly related to an increased phospholipid component found in an alkali-soluble fraction of the organism.

CANDIDA SPP.

The pathogenesis of *Candida* is extremely complex and probably varies with each species. Adhesion of *Candida* to the epithelium of the gastrointestinal or urinary tract is critically important. *Candida* commonly colonizes the mucosal surfaces, and its ability to invade and cause infection is first dependent on binding.[49] Fibronectin, a component of the host extracellular matrix, may play a role in the initiation and dissemination of *Candida albicans* infection.[123] Three distinct aspartyl proteinases have been described in *C. albicans*; those strains having high levels of proteinases had an increased ability to cause disease in experimental

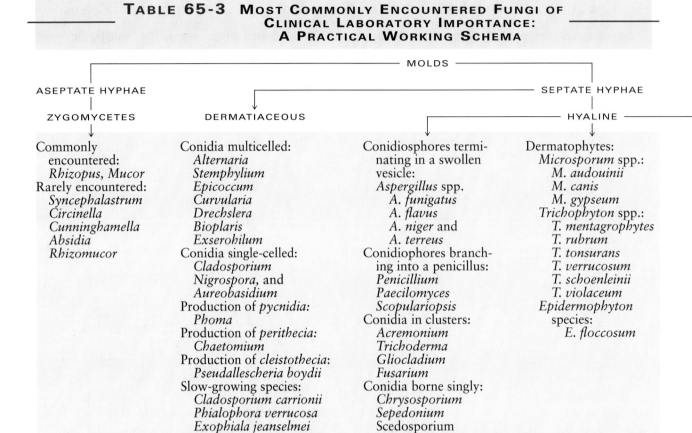

TABLE 65-3   MOST COMMONLY ENCOUNTERED FUNGI OF
CLINICAL LABORATORY IMPORTANCE:
A PRACTICAL WORKING SCHEMA

MOLDS

ASEPTATE HYPHAE   SEPTATE HYPHAE

ZYGOMYCETES   DERMATIACEOUS   HYALINE

Commonly encountered:	Conidia multicelled:	Conidiosphores terminating in a swollen vesicle:	Dermatophytes:
Rhizopus, Mucor	*Alternaria*	*Aspergillus* spp.	*Microsporum* spp.:
Rarely encountered:	*Stemphylium*	A. *funigatus*	M. *audouinii*
Syncephalastrum	*Epicoccum*	A. *flavus*	M. *canis*
Circinella	*Curvularia*	A. *niger* and	M. *gypseum*
Cunninghamella	*Drechslera*	A. *terreus*	*Trichophyton* spp.:
Absidia	*Bioplaris*	Conidiophores branching into a penicillus:	T. *mentagrophytes*
Rhizomucor	*Exserohilum*	*Penicillium*	T. *rubrum*
	Conidia single-celled:	*Paecilomyces*	T. *tonsurans*
	Cladosporium	*Scopulariopsis*	T. *verrucosum*
	Nigrospora, and	Conidia in clusters:	T. *schoenleinii*
	Aureobasidium	*Acremonium*	T. *violaceum*
	Production of *pycnidia:*	*Trichoderma*	*Epidermophyton* species:
	Phoma	*Gliocladium*	E. *floccosum*
	Production of *perithecia:*	*Fusarium*	
	Chaetomium	Conidia borne singly:	
	Production of *cleistothecia:*	*Chrysosporium*	
	Pseudallescheria boydii	*Sepedonium*	
	Slow-growing species:	*Scedosporium* state of	
	Cladosporium carrionii	*Pseudallescheria boydii*	
	Phialophora verrucosa	*Scedosporium inflatum*	
	Exophiala jeanselmei		
	Fonsecaea pedrosoi and		
	Xylohypha bantiana		

*Rudimentary hyphae may be present.

From Koneman, E.W. and Roberts, G.D. 1985. Practical laboratory mycology, ed 3. Williams & Wilkins, Baltimore.

animal models.[77] Hydrophobic molecules present on the surface of *Candida* spp. also appear to be important in the pathogenesis, and a strong correlation has been shown between adhesion and surface hydrophobicity.[29] Also, an increased level of phospholipase found in strains of *C. albicans* has been shown to correlate well with a higher mortality rate for experimental animals compared with a lower rate for strains having a lower level of phopholipase.[52] Furthermore, the dimorphism seen in *Candida albicans* and phenotypic switching may also play a role in the pathogenesis.[115] Finally, the immunosuppressive activity of yeast mannans also may contribute to virulence of these yeasts.[17]

COCCIDIOIDES IMMITIS

The arthroconidia of *Coccidioides immitis* measure 2 to 5 μm and are easily deposited within the alveoli of the lungs. From the arthroconidia, spherules develop and ulti-

mately segment into numerous endospores that are extruded through the spherule cell wall as it ruptures; this provides for local and hematologic dissemination of the organism. It has been noted that spherules are coated with an extracellular matrix that appears to inhibit polymorphonuclear leukocyte (PMN) access[32] and may be responsible for the resistance of spherules to killing by PMNs. Characteristically, PMNs aggregate around endospores as they are released from the spherule.

The dissemination of *Coccidioides immitis* from the lungs and the marked tissue damage caused by the organism has led to speculation that an extracellular proteinase may be responsible. It has been shown that the proteinase, primarily found in hyphae and spherules, promotes dissolution of the wall of the cell matrix as the endospores become mature. Concurrently, the action of the proteinase on immunoglobulins and against elastin in the lungs may be a virulence

YEAST

Dimorphic molds:	Hyphae formed on corn-meal-Tween 80 agar:	Hyphae not formed on cornmeal-Tween 80 agar:	Arthroconidia:

Dimorphic molds:
Blastomyces
  dermatitidis
Histoplasma
  capsulatum
Coccidioides immitis
Paracoccidioides
  brasiliensis
Sporothrix schenckii
Branching filamentous
  bacteria:
Actinomyces
Nocardia
Streptomyces

Hyphae formed on corn-meal-Tween 80 agar:
Pseudohyphae:
Candida spp.:
  C. albicans
  C. tropicalis
  C. parapsilosis
  C. kefyr
    (pseudotropicalis)
  C. krusei
  C. guilliermondii

Hyphae not formed on cornmeal-Tween 80 agar:
Cryptococcus
Rhodotorula
Candida glabrata
*Saccharomyces**

Arthroconidia:
Geotrichum
Trichosporon

factor that promotes tissue damage when the ruptured spherule releases endospores and any residual proteinase.[124]

Men are several times more likely than women to develop disseminated infection. Pregnancy reverses the resistance of women to dissemination and with each increasing term, the risk of dissemination increases. In fact, numerous hormones during pregnancy have been studied for their effects on *C. immitis*. Several were shown to stimulate growth in vitro by accelerating the rate of spherule development and endospore release. It is speculated that estrogen-binding proteins found in the cytosols of *C. immitis* may play a role in controlling its growth in vitro.[25]

CRYPTOCOCCUS NEOFORMANS

Cryptococcosis is most often seen in immunocompromised patients with impaired cell-mediated immunity.[21] Some common risk factors include acquired immunodeficiency syndrome (AIDS), corticosteroid therapy, lymphoma, and T-cell dysfunction.

Cryptococcus neoformans is the only pathogenic yeast known to have a polysaccharide capsule. Under drying conditions,[13] the capsule collapses and protects the yeast from desiccation. The cell size reduction, caused by capsular collapse, places the yeast in the ideal size range for alveolar deposition.[78] In addition, the polysaccharide capsule contains compounds that are not recognized by phagocytes.[61] The acapsular strains of *C. neoformans* are more easily phagocytosed. In most instances, *C. neoformans* elicits no tissue response in infected individuals.

Phenoloxidase, an enzyme found in *C. neoformans*, is responsible for melanin production. It is speculated that melanin might act as a virulence factor by making the organism resistant to leukocyte

Text continued on p. 886

BOX 65-1	TAXONOMIC CLASSIFICATION OF MEDICALLY IMPORTANT FUNGI

Subdivision: *Zygomycotina*
  Class: *Zygomycetes*
    Order: *Mucorales*
      Genus: *Absidia*
      Genus: *Cunninghamella*
      Genus: *Mucor*
      Genus: *Rhizomucor*
      Genus: *Saksenaea*
    Order: *Entomophthorales*
      Genus: *Basidiobolus*
      Genus: *Conidiobolus*
Subdivision: *Ascomycotina*
  Class: *Ascomycetes*
    Order: *Endomycetales*
      Genus: *Pichia* (teleomorph [sexual] stage of some *Candida* spp.)
      Genus: *Saccharomyces*
    Order: *Eurotidales*
      Family: *Trichocomaceae*
      Genus: *Emericella* (*Aspergillus* teleomorph stage)
      Genus: *Eurotium* (*Aspergillus* teleomorph stage)
      Genus: *Neosartorya* (*Aspergillus* teleomorph stage)
    Order: *Onygenales*
      Genus: *Ajellomyces* (*Histoplasma* and *Blastomyces* teleomorph stages)
      Genus: *Arthroderma* (*Trichophyton* and *Microsporum* teleomorph stages)
Subdivision: *Basidiomycotina*
  Class: *Basidiomycetes*
    Order: *Agaricales*
      Genus: *Amanita*
    Order: *Filobasidiales*
      Genus: *Filobasidiella* (*Cryptococcus* teleomorph stage)
Subdivision: *Deuteromycotina*
  Class: *Blastomycetes*
    Order: *Cryptococcales*
      Genus: *Candida*
      Genus: *Cryptococcus*
      Genus: *Hansenula*
      Genus: *Malassezia*
      Genus: *Rhodotorula*
      Genus: *Torulopsis*
      Genus: *Trichosporon*
  Class: *Hyphomycetes*
    Order: *Moniliales*
      Family: *Moniliaceae*
      Genus: *Acremonium*
      Genus: *Aspergillus*
      Genus: *Chrysosporium*
      Genus: *Coccidioides*
      Genus: *Epidermophyton*
      Genus: *Fusarium*
      Genus: *Paecilomyces*
      Genus: *Paracoccidioides*
      Genus: *Pseudallescheria*
      Genus: *Scedosporium*
      Genus: *Scopulariopsis*
      Genus: *Sporothrix*
      Family: *Dematiaceae*
      Genus: *Alternaria*
      Genus: *Bipolaris*

Genus: *Cladosporium*
Genus: *Curvularia*
Genus: *Exophiala*
Genus: *Exserohilum*
Genus: *Fonsecaea*
Genus: *Phialophora*
Genus: *Wangiella*
Genus: *Xylohypha* (*Cladophialophora*)
  Class: *Coelomycetes*
    Order: *Sphaeropsidales*
      Genus: *Phoma*

Morphologic Classification of Medically Important Fungi, Monomorphic Yeasts and Yeastlike Organisms

1. Pseudohyphae with blastoconidia
   Candida spp.
   Hansenula
   Saccharomyces
2. Yeastlike cells only (usually no hyphae or pseudohyphae)
   Cryptococcus
   Hansenula
   Malassezia
   Prototheca
   Rhodotorula
   Saccharomyces
   Sporobolomyces
   Torulopsis
   Ustilago
3. Hyphae and arthroconidia or annelloconidia
   Blastoschizomyces
   Geotrichum
   Trichosporon

Thermally Dimorphic Fungi

1. *Blastomyces dermatitidis*
2. *Histoplasma capsulatum*
3. *Paracoccidioides brasiliensis*
4. *Penicillium marneffei*
5. *Sporothrix schenckii*

Thermally Monomorphic Molds

1. White, cream, or light gray surface; nonpigmented reverse
   a. With microconidia or macroconidia
      Acremonium
      Beauveria
      Chrysosporium
      Emmonsia
      Fusarium
      Graphium
      Microsporum
      Pseudallescheria
      Sepedonium
      Stachybotrys
      Trichophyton
      Verticillium
   b. Having sporangia or sporangiola
      Absidia
      Apophysomyces
      Basidiobolus
      Conidiobolus
      Cunninghamella
      Mucor
      Rhizomucor
      Rhizopus

Saksenaea
  c. Having arthroconidia
    Coccidioides
    Geotrichum
  d. Having only hyphae with chlamydoconidia
    Microsporum
    Trichophyton
2. White, cream, beige, or light gray surface; yellow, orange, or reddish reverse
    Acremonium
    Chaetomium
    Microsporum
    Trichophyton
3. White, cream, beige, or light gray surface; red to purple reverse
    Microsporum
    Penicillium
    Trichophyton
4. White, cream, beige, or light gray surface; brown reverse
    Chaetomium
    Chrysosporium
    Cokeromyces
    Emmonsia
    Madurella
    Microsporum
    Scopulariopsis
    Sporotrichum
    Trichophyton
5. White, cream, beige, or light gray surface; black reverse
    Chaetomium
    Graphium
    Nigrospora
    Phoma
    Pseudallescheria
    Scedosporium
    Trichophyton
6. Tan to brown surface
  a. Having small conidia
    Aspergillus
    Botrytis
    Chrysosporium
    Cladosporium
    Dactylaria
    Emmonsia
    Paecilomyces
    Phialophora
    Pseudallescheria
    Scopulariopsis
    Sporotrichum
    Trichophyton
    Verticillium
  b. Having large conidia or sporangia
    Alternaria
    Apophysomyces
    Basidiobolus
    Bipolaris
    Botrytis
    Cokeromyces
    Conidiobolus
    Curvularia
    Epicoccum

Epidermophyton
    Fusarium
    Microsporum
    Rhizomucor
    Rhizopus
    Stemphylium
    Trichophyton
    Ulocladium
  c. Having miscellaneous microscopic morphology
    Chaetomium
    Coccidioides
    Madurella
    Phoma
    Ustilago
7. Yellow to orange surface
    Aspergillus
    Chrysosporium
    Epicoccum
    Epidermophyton
    Microsporum
    Monilia
    Penicillium
    Sepedonium
    Sporotrichum
    Trichophyton
    Trichothecium
    Verticillium
8. Pink to violet surface
    Acremonium
    Aspergillus
    Beauveria
    Chrysosporium
    Fusarium
    Gliochadium
    Microsporum
    Monilia
    Paecilomyces
    Sporotrichum
    Trichophyton
    Trichothecium
    Verticillium
9. Green surface; light reverse
    Aspergillus
    Epidermophyton
    Gliocladium
    Penicillium
    Trichoderma
    Verticillium
10. Dark gray or black surface; light reverse
    Aspergillus
    Syncephalastrum
11. Green, dark gray, or black surface; dark reverse
  a. Having small conidia
    Aureobasidium
    Botrytis
    Cladosporium
    Exophiala
    Fonsecaea
    Phaeoannellomyces
    Phialophora
    Pseudallescheria

Continued

BOX 65-1

TAXONOMIC CLASSIFICATION OF MEDICALLY IMPORTANT FUNGI—CONT'D

11. Green, dark gray, or black surface; dark reverse
    a. Having small conidia—cont'd
       Scedosporium
       Wangiella
       Xylohypha
    b. Having large conidia
       Alternaria
       Bipolaris
       Curvularia
       Dactylaria
       Epicoccum
       Helminthosporium
       Nigrospora
       Pithomyces

      Stachybotrys
      Stemphylium
      Ulocladium
  c. Having only hyphae (with or without chlamydoconidia)
      Madurella
      Piedraia
  d. Having large fruiting bodies
      Chaetomium
      Phoma

From Murray, P.R. 1996. ASM pocket guide to clinical microbiology. ASM Press, Washington, D.C.

TABLE 65-4 SUMMARY OF COMMON PATHOGENS

ORGANISM	NATURAL HABITAT	INFECTIOUS FORM	MODE OF TRANSMISSION	COMMON SITES OF INFECTION	CLINICAL FORM
Aspergillus spp.	Ubiquitous, plants	Conidia	Inhalation	Lung, eye, skin, nails	Hyphae
Blastomyces dermatitidis	Unknown ?soil/wood	Probably conidia	Probably inhalation	Lungs, skin, long bones	Yeast
Candida spp.	Human flora	Yeast, pseudo- and true hyphae	Direct Invasion/ dissemination	GI and GU tract, nails, viscera, blood	Yeast, Pseudo- and true hyphae
Coccidioides immitis	Soil of many arid regions	Arthroconidia	Inhalation	Lungs, skin, meninges	Spherules, endospores
Cryptococcus neoformans	Bird feces, soil	Yeast	Inhalation	Lungs, skin, meninges	Yeast
Histoplasma capsulatum	Bat and bird feces	Conidia	Inhalation	Lungs, bone marrow, blood	Yeast
Paracoccidiodes brasilensis	?Soil, plants	Conidia	Inhalation/ trauma	Lungs, skin, mucous membranes	Yeast
Sporothrix schenckii	Soil, plants	Conidia/ hyphae	Trauma/rarely inhalation	Skin and lymphatics, lungs, meninges	Yeast
Dermatophytes	Human disease, animals, soil	Conidia/ hyphae	Contact	Skin, hair, or nails	Hyphae

—— **TABLE 65-5 VIRULENCE FACTORS OF MEDICALLY IMPORTANT FUNGI** ——

FUNGAL PATHOGEN	PUTATIVE VIRULENCE FACTOR

Aspergillus spp.	Elastase-serine protease
	  Proteases
	  Toxins
	  Other
	Elastase-metalloprotease
	Aspartic acid proteinase
	Aflatoxin
	Catalase
	Lysine biosynthesis
	p-aminobenzoic acid synthesis

Blastomyces dermatitidis	Cell wall α-1,3-glucan
Coccidioides immitis	Extracellular proteinases EBP
Cryptococcus neoformans	Capsule Phenoloxidase melanin synthesis Varietal differences
Dematiaceous fungi	Phenoloxidase melanin synthesis
Histoplasma capsulatum	Cell wall α-1,3-glucan Intracellular growth Thermotolerance

Paracoccidioides brasiliensis	Estrogen-binding proteins
	Cell wall components
	  β-glucan
	  α-1,3-glucan

Sporothrix schenckii	Thermotolerance Extracellular enzymes

Modified from Hogan, L.H., Klein, B.S., and Levitz, S.M. 1996. Virulence factors of medically important fungi. Clin. Microbiol. Rev. 9:(4)469.

attack.[117] Evidence has also been presented to show that increased melanin production can decrease lymphocyte proliferation and tumor necrosis factor production.[50] Whether phenoloxidase is truly a virulence factor has yet to be determined. It is interesting to note that the brain is rich with substrates that can react with phenoloxidase such as L-dopamine, which might account for the affinity of *C. neoformans* to invade the central nervous system.

HISTOPLASMA CAPSULATUM

Histoplasma capsulatum is another of the thermally dimorphic fungi. The microconidia and/or small hyphal fragments measure 2 to 5 μm and are within the range for alveolar deposition. Conversion of the mold to the yeast form is a prerequisite for the organism to cause infection. *Histoplasma capsulatum* is an intracellular parasite and can survive within macrophages for extended periods.[122] Its ability to survive within the macrophages is perhaps a key virulence factor. It is speculated that the organism modulates the pH within the phagolysosomes.[46]

PARACOCCIDIOIDES BRASILIENSIS

Paracoccidioides brasiliensis is another of the thermally dimorphic fungi in which virulence is influenced by the conversion of the mold to the yeast form. Infection is initiated via inhalation of conidia,[71] then rapid convertion to the yeast form within the lungs. As with *Coccidioides immitis*, *Paracoccidioides brasiliensis* is 13 to 80 times more common in men than in women.[102] It is speculated that an estrogen-binding protein inhibits conversion to the yeast form.[9] A beta-glucan has been implicated as an important immunomodulator in initiation of the host inflammatory response.[7]

SPOROTHRIX SCHENCKII

Conversion of *Sporothrix schenckii* to the yeast form is necessary before infection with this mold can occur. Conidia and/or small hyphal fragments are capable of deposition within the alveoli when inhaled. The organism most often causes infection after it is acquired by traumatic implantation into the skin and subcutaneous tissues. Kwon-Chung reported that several clinical isolates of *Sporothrix schenckii* failed to grow at 37° C and that they caused infection only in the skin and testes of infected rats, suggesting that it requires a lower temperature.[63] It is not documented exactly how thermotolerance affects the virulence of the organism. Extracellular enzymes, such as acid phosphatases, have been speculated to play a role in the interaction of *Sporothrix schenckii* yeast forms within macrophages. However, confirmation of these effects has not been documented.

DERMATOPHYTES

Dermatophytes, which involve skin, hair, or nails, secrete extracellular enzymes that may aid in the colonization of keratinous tissues.[8] Extracellular enzymes that may be associated with virulence include keratinase, elastase, and lipase.[66]

GENERAL CONSIDERATIONS FOR THE LABORATORY DIAGNOSIS OF FUNGAL INFECTIONS

COLLECTION, TRANSPORT, AND CULTURING OF CLINICAL SPECIMENS

The diagnosis of fungal infections is dependent entirely on the selection and collection of an appropriate clinical specimen for culture. Many fungal infections are similar clinically to mycobacterial infections, and often the same specimen is cultured for both fungi and mycobacteria. Most infections have a primary focus in the lungs; respiratory tract secretions are almost always included among the specimens selected for culture. It should be emphasized that dissemination to distant body sites often occurs, and fungi may be commonly recovered from nonrespiratory sites. The proper collection of specimens and their rapid transport to the clinical laboratory are of major importance for the recovery of fungi. In many instances, specimens not only contain the etiologic agents but also contain contaminating bacteria or fungi that will rapidly overgrow some of the slower growing pathogenic fungi. It is common for many yeasts, for example, *Candida* spp., to be recovered on routine bacteriology media and fungal culture media. A few specific comments concerning specimen collection and culturing are included in this chapter.

Respiratory tract secretions

Respiratory tract secretions (sputum, induced sputum, bronchial washings, bronchoalveolar lavage, and tracheal aspirations) are perhaps the most commonly submitted specimens for fungal culture. To ensure the optimal recovery of fungi and prevent overgrowth by contaminants, antibacterial antibiotics should be included in the battery of media to be used. Cycloheximide, an antifungal agent that prevents overgrowth by rapidly growing molds, should be included in at least one of the culture media used. As much specimen as possible (0.5 mL) should be used to inoculate each medium.

Cerebrospinal fluid

Cerebrospinal fluid (CSF) collected for culture should be filtered through a 0.45-μm membrane filter attached to a sterile syringe. After filtration, the filter is removed and is placed onto the surface of an appropriate culture medium with the microorganism side down. Cultures should be examined daily and the filter moved to another location on an every-other-day basis. If less than 1 mL of specimen is submitted for culture, it should be centrifuged and 1-drop aliquots of the sediment should be placed onto several areas on the agar surface. Media used for the recovery of fungi from CSF should contain no antibacterial or antifungal agents. Once submitted to the laboratory, CSF specimens should be processed promptly. If prompt processing is not possible, samples should be kept at room temperature or placed in a 30° C incubator, since most organisms will continue to replicate in this environment.

Blood
Disseminated fungal infections are more prevalent than previously recognized, and blood cultures provide an accurate method for determining their etiology. Only a few manual fungal blood culture systems have been available over past years, and most are not used routinely by clinical microbiology laboratories. Currently, several automated blood culture systems, including the BACTEC (Becton Dickinson Diagnostic Instrument Systems, Towson, Md.), BACT/ALERT (Organon-Tekinika Corporation, Durham, N.C.) and ESP (Difco Laboratories, Detroit, Mich.), all are adequate systems for the recovery of yeasts. However, those laboratories having a high incidence of dimorphic fungi recovered from blood are encouraged to use the lysis-centrifugation system, the Isolator (Wampole Laboratories, Cranbury, N.J.). It has been shown to be optimal for the recovery of *H. capsulatum* and other filamentous fungi.[5,6] Using this system, red blood cells and white blood cells that may contain the microorganisms are lysed, and centrifugation serves to concentrate the organisms before culturing. The concentrate is inoculated onto the surface of appropriate culture media, and most fungi are detected within the first 4 days of incubation. However, occasional isolates of *Histoplasma capsulatum* may require approximately 10 to 14 days for recovery. The optimal temperature for fungal blood cultures is 30° C, and the suggested incubation time is 21 days.

Hair, skin, and nail scrapings
These specimens are usually submitted for dermatophyte culture and are contaminated with bacteria and/or rapidly growing fungi. Samples collected from lesions may be obtained by scraping the skin or nails with a scalpel blade or microscope slide; infected hairs are removed by plucking them with forceps. These specimens should be placed in a sterile petri dish or paper envelope before culturing; they should not be refrigerated. Mycosel agar, which contains

chloramphenicol and cycloheximide, is satisfactory for the recovery of dermatophytes. Cultures should be incubated for a minimum of 30 days at 30° C before being reported as negative.

Urine
Urine samples collected for fungal culture should be processed as soon after collection as possible. Twenty-four–hour urine samples are unacceptable for culture. The usefulness of quantitative cultures is undetermined and it should not be done. All urine samples should be centrifuged and the sediment cultured using a loop to provide adequate isolation of colonies. Because urine is often contaminated with gram-negative bacteria, it is necessary to use media containing antibacterial agents to ensure the recovery of fungi.

Tissue, bone marrow, and sterile body fluids
All tissues should be processed before culturing by mincing or grinding or placement in a Stomacher (Tekmar, Cincinnati, Ohio). The Stomacher, expresses the cytoplasmic contents of cells by pressure exerted from the action of rapidly moving metal paddles against the tissue in a broth suspension. After processing, at least 1 mL of specimen should be spread onto the surface of appropriate culture media and incubation should be at 30° C for 30 days.

Bone marrow may be placed directly onto the surface of appropriate culture media and incubated in the manner previously mentioned. Sterile body fluids should be concentrated by centrifugation before culturing, and at least 1 mL of specimen should be placed onto the surface of appropriate culture media. An alternative is to place bone marrow and other body fluids in an Isolator tube and process it as a blood culture. All specimens should be cultured as soon as they are received by the laboratory to ensure the recovery of fungi from these important sources.

CULTURE MEDIA AND INCUBATION REQUIREMENTS
Any of a number of fungal culture media are satisfactory for use in the clinical microbiology laboratory. Most are adequate for the recovery of fungi, and selection is usually left up to each individual laboratory director. Table 65-6 lists various media and the indications for their use. For optimal recovery a battery of media should be used, and the following are recommended:

1. Media with and without blood enrichment
2. Media with and without cycloheximide
3. All media should contain antibacterial agents; specimens from sterile sites may be cultured onto media lacking antimicrobials

TABLE 65-6 FUNGAL CULTURE MEDIA: INDICATIONS FOR USE

MEDIA	INDICATIONS FOR USE
Primary Recovery Media	
Brain-heart infusion agar	Primary recovery of saprobic and pathogenic fungi
Brain-heart infusion agar with antibiotics	Primary recovery of pathogenic fungi exclusive of dermatophytes
Brain-heart infusion biphasic blood culture bottles	Recovery of fungi from blood
Dermatophyte test medium	Primary recovery of dermatophytes, recommended as screening medium only
Inhibitory mold agar	Primary recovery of pathogenic fungi exclusive of dermatophytes
Potato flake agar	Primary recovery of saprobic and pathogenic fungi
Mycosel or mycobiotic agar	Primary recovery of dermatophytes
SABHI agar	Primary recovery of saprobic and pathogenic fungi
Yeast-extract phosphate agar	Primary recovery of pathogenic fungi exclusive of dermatophytes
Differential Test Media	
Ascospore agar	Detection of ascospores in ascosporogenous yeasts such as *Saccharomyces* spp.
Cornmeal agar with Tween 80 and trypan blue	Identification of *C. albicans* by chlamydospore production; identification of *Candida* by microscopic morphology
Cottonseed conversion agar	Conversion of dimorphic fungus *B. dermatitidis* from mold to yeast form
Czapek's agar	Recovery and differential identification of *Aspergillus* spp.
Niger seed agar	Identification of *C. neoformans*
Nitrate reduction medium	Detection of nitrate reduction in confirmation of *Cryptococcus* spp.
Potato dextrose agar	Demonstration of pigment production by *T. rubrum;* preparation of microslide cultures
Rice medium	Identification of *M. audouinii*
Trichophyton agars 1–7	Identification of members of *Trichophyton* genus
Urea agar	Detection of *Cryptococcus* spp.; differentiate *T. mentagrophytes* from *T. rubrum;* detection of *Trichosporon* spp.
Yeast fermentation broth	Identification of yeasts by determining fermentation
Yeast nitrogen base agar	Identification of yeasts by determining carbohydrate assimilation

From Koneman, E.W. and Roberts, G.D. 1985. Practical laboratory mycology, ed 3. Williams & Wilkins, Baltimore.

Agar plates or screw-capped agar tubes are satisfactory for the recovery of fungi, but plates are preferred, since they provide better aeration of cultures, a large surface area for better isolation of colonies, and greater ease of handling by technologists making microscopic preparations for examination. Agar has a tendency to dehydrate during the extended incubation period required for fungal recovery; however, this problem can be minimized by using culture dishes containing at least 40 mL of agar and placing them in a humid field incubator. Dishes should be opened and examined only within a certified biological safety cabinet. Many laboratories discourage the use of culture dishes because of safety considerations; however, the advantages of using them outweigh the disadvantages.

Compared with agar plates, screw-capped culture tubes are more easily stored, require less space for incubation, and are more easily handled. In addition, they have a lower dehydration rate, and laboratory workers feel that cultures are less hazardous to handle when in tubes. However, disadvantages, such as relatively poor isolation of colonies, a reduced surface area for culturing, and a tendency to promote anaerobiosis, discourage their routine use in most clinical microbiology laboratories. If culture tubes are used, the tube should be as large as possible to provide an adequate surface area for isolation. After inoculation, tubes should be placed in a horizontal position for at least 1 to 2 hours to allow the specimen to absorb to the agar surface and avoid settling at the bottom of the tube. Cotton-plugged tubes are unsatisfactory for fungal cultures.

Cultures should be incubated at room temperature, or preferably at 30° C, for 30 days before reporting as negative. A relative humidity in the range of 40% to 50% can be achieved by placing an open pan of water in the incubator. Cultures should be examined at least three times weekly during incubation.

As previously mentioned, most clinical specimens are contaminated with bacteria and/or rapidly growing fungi. The need for antifungal and antibacterial agents is obvious. The addition of 0.5 mg/mL of cycloheximide and 16 μg/mL of chloramphenicol to media has been traditionally advocated to inhibit the growth of contaminating molds and bacteria, respectively. Better results have been achieved using a combination of 5 μg/mL of gentamicin and 16 μg/mL of chloramphenicol as antibacterial agents. Also, ciprofloxacin, 5 μg/mL, may be used.

Certain of the fungi seem to have a requirement for blood, and enrichment with 5% to 10% sheep blood is recommended for at least one of the culture media used. Cycloheximide may be added to any of the media that contain or lack antibacterial antibiotics. However, if cycloheximide is included as a part of the battery of culture media used, a medium lacking this ingredient should also be included. Certain of the pathogenic fungi (*Cryptococcus neoformans, Candida krusei,* other *Candida* spp., *Trichosporon beigelii, Pseudallescheria boydii,* and *Aspergillus* spp.) are partially or completely inhibited by cycloheximide.

Although the use of antibiotics in fungal culture media is necessary for the optimal recovery of organisms, the use of decontamination and concentration methods advocated for the recovery of mycobacteria is not appropriate as many fungi are killed by sodium hydroxide treatment.[90]

DIRECT MICROSCOPIC EXAMINATION OF CLINICAL SPECIMENS

Direct microscopic examination of clinical specimens has been used for many years; however, its usefulness should be reemphasized.[73] Because the mission of a clinical microbiology laboratory is to provide a rapid and accurate diagnosis, the mycology laboratory can provide this service in many instances by direct examination (particularly the Gram stain) of the clinical specimen submitted for culture. Microbiologists are encouraged to become familiar with the diagnostic features of fungi commonly encountered in clinical specimens and to recognize them when stained by dyes. This very important procedure can often provide the first microbiologic proof of etiology in patients with fungal infection. Currently, this is the most rapid method available.

Tables 65-7 and 65-8 present the methods available for the direct microscopic detection of fungi in clinical specimens and a summary of the characteristic microscopic features of each. Figures 65-11 to 65-28 present photomicrographs of some of the fungi commonly seen in clinical specimens.

Traditionally, the potassium hydroxide preparation has been the recommended method for the direct microscopic examination of specimens.[88] However, calcofluor white stain is now viewed as superior.[40] Slides prepared by this method may be observed using fluorescent or bright-field microscopy as used for the potassium hydroxide preparation; the former is optimal because fungal cells will fluoresce.

Molecular detection methods are becoming popular in all areas of clinical microbiology; however, none has been accepted as a routine diagnostic tool in clinical mycology. Ideally, a panel of primers specific for the detection of fungi in clinical specimens would include the most common organisms known to cause disease in immunocompromised patients (including the dimorphic fungi and *Pneumocystis carinii*). However, currently, no commercially methods are available to the clinical laboratory and reports in the literature deal only with selected organisms. The large number

Text continued on p. 898

TABLE 65-7 SUMMARY OF METHODS AVAILABLE FOR THE DIRECT MICROSCOPIC DETECTION OF FUNGI IN CLINICAL SPECIMENS

METHOD	USE	TIME REQUIRED	ADVANTAGES	DISADVANTAGES
Acid-fast stain	Detection of mycobacteria and *Nocardia*	12 min	Detects *Nocardia** and *B. dermatitidis*	Tissue homogenates are difficult to observe because of background staining
Calcofluor white	Detection of fungi	1 min	Can be mixed with KOH: detects fungi rapidly because of bright fluorescence	Requires use of a fluorescence microscope; background fluorescence prominent, but fungi exhibit more intense fluorescence; vaginal secretions are difficult to interpret
Gram stain	Detection of bacteria	3 min	Is commonly performed on most clinical specimens submitted for bacteriology and will detect most fungi, if present	Some fungi stain well; however, others (e.g., *Cryptococcus* spp.) show only stippling and stain weakly in some instances; some isolates of *Nocardia* fail to stain or stain weakly
India ink	Detection of *C. neoformans* in CSF	1 min	When positive in CSF, is diagnostic of meningitis	Positive in less than 50% of cases of meningitis; not reliable in non-HIV-infected patients
Potassium hydroxide (KOH)	Clearing of specimen to make fungi more readily visible	5 min; if clearing is not complete, an additional 5-10 min is necessary	Rapid detection of fungal elements	Experience required since background artifacts are often confusing, clearing of some specimens may require an extended time
Methenamine silver stain	Detection of fungi in histologic section	1 hr	Best stain to detect fungal elements	Requires a specialized staining method that is not usually readily available to microbiology laboratories
Papanicolaou stain	Examination of secretions for presence of malignant cells	30 min	Cytotechnologist can detect fungal elements	
Periodic acid-Schiff (PAS) stain	Detection of fungi	20 min; 5 min additional if counterstain is employed	Stains fungal elements well; hyphae of molds and yeasts can be readily distinguished	*Nocardia* spp. do not stain well; *B. dermatitidis* appears pleomorphic
Wright stain	Examination of bone marrow or peripheral blood smears	7 min	Detects *H. capsulatum* and *C. neoformans*	Detection is limited to *H. capsulatum* and *C. neoformans*

*Acid-fast bacterium.

From Merz, W.G., and Roberts, G.D. 1995. Detection and recovery of fungi from clinical specimens. In Murray, P.R., Baron, E.J., Pfaller, M.A., et al., editors. Manual of clinical microbiology, ed 6. ASM Press, Washington, D.C.

TABLE 65-8   SUMMARY OF CHARACTERISTIC FEATURES OF FUNGI SEEN IN DIRECT EXAMINATION OF CLINICAL SPECIMENS

MORPHOLOGIC FORM FOUND IN SPECIMENS	ORGANISM(S)	SIZE RANGE (DIAMETER, μm)	CHARACTERISTIC FEATURES
Yeastlike	*Histoplasma capsulatum*	2-5	Small; oval to round budding cells; often found clustered within histiocytes; difficult to detect when present in small numbers
	Sporothrix schenckii	2-6	Small; oval to round to cigar-shaped; single or multiple buds present; uncommonly seen in clinical specimens
	Cryptococcus neoformans	2-15	Cells exhibit great variation in size; usually spherical but may be football-shaped; buds single or multiple and "pinched off"; capsule may or may not be evident; occasionally, pseudohyphal forms with or without a capsule may be seen in exudates of CSF
	Malasezzia furfur (in fungemia)	1.5-4.5	Small; bottle-shaped cells, buds separated from parent cell by a septum; emerge from a small collar
	Blastomyces dermatitidis	8-15	Cells are usually large, double refractile when present; buds usually single; however, several may remain attached to parent cells; buds connected by a broad base
	Paracoccidiodes brasiliensis	5-60	Cells are usually large and are surrounded by smaller buds around the periphery ("mariner's wheel appearance"); smaller cells may be present (2-5 μm) and resemble *H. capsulatum*; buds have "pinched-off" appearance
Spherules	*Coccidiodes immitis*	10-200	Spherules vary in size; some may contain endospores, others may be empty; adjacent spherules may resemble *B. dermatitidis*; endospores may resemble *H. capsulatum* but show no evidence of budding; spherules may produce multiple germ tubes if a direct preparation is kept in a moist chamber for ≥24 hrs
	Rhinosporidium seeberi	6-300	Large, thick-walled sporangia containing sporangiospores are present; mature sporangia are larger than spherules of *C. immitis*; hyphae may be found in cavitary lesions
	Pseudallescheria boydii (cases other than mycetoma)		Hyphae are septate and are impossible to distinguish from those of other hyaline molds
Yeast and pseudohyphae or hyphae	*Candida* spp.	3-4 (yeast) 5-10 (pseudohyphae)	Cells usually exhibit single budding; pseudohyphae, when present, are constricted at the ends and remain attached like links of sausage; hyphae, when present, are septate
	M. furfur (in tinea versicolor)	3-8 (yeast) 2.5-4 (hyphae)	Short, curved hyphal elements are usually present along with round yeast cells that retain their spherical shape in compacted clusters

Continued

TABLE 65-8 SUMMARY OF CHARACTERISTIC FEATURES OF FUNGI SEEN IN DIRECT EXAMINATION OF CLINICAL SPECIMENS—CONT'D

MORPHOLOGIC FORM FOUND IN SPECIMENS	ORGANISM(S)	SIZE RANGE (DIAMETER, μm)	CHARACTERISTIC FEATURES
Nonseptate hyphae	Zygomycetes; *Mucor, Rhizopus,* and other genera	10-30	Hyphae are large, ribbonlike, often fractured or twisted; occasional septa may be present; smaller hyphae are confused with those of *Aspergillus* spp., particularly *A. flavus*
Hyaline septate hyphae	Dermatophytes Skin and nails	3-15	Hyaline, septate hyphae are commonly seen; chains of arthroconidia may be present
	Hair	3-15	Arthroconidia on periphery of hair shaft producing a sheath are indicative of ectothrix infection; arthroconidia formed by fragmentation of hyphae within the hair shaft are indicative of endothrix infection
		3-15	Long hyphal filaments or channels within the hair shaft are indicative of favus hair infection
	Aspergillus spp.	3-12	Hyphae are septate and exhibit dichotomous, 45-degree angle branching; larger hyphae, often disturbed, may resemble those of Zygomycetes
	Geotrichum spp.	4-12	Hyphae and rectangular arthroconidia are present and are sometimes rounded; irregular forms may be present
	Trichosporon spp.	2-4 by 8	Hyphae and rectangular arthroconidia are present and sometimes rounded; occasionally blastoconidia may be present
Dematiaceous septate hyphae	*Bipolaris* spp. *Cladosporium* spp. *Curvularia* spp. *Drechslera* spp. *Exophiala* spp. *Exserohilum* spp. *Phialophora* spp. *Wangiella dermatitidis*	2-6	Dematiaceous polymorphous hyphae are seen; budding cells with single septa and chains of swollen rounded cells are often present; occasionally aggregates may be present in infection caused by *Phialophora* and *Exophiala* spp.
	Phaeoannellomyces (Exophiala) werneckii	1.5-5	Usually large numbers of frequently branched hyphae are present along with budding cells
Sclerotic bodies	*Cladosporium carrionii* *Fonsecaea compacta* *Fonsecaea pedrosoi* *Phialophora verrucosa* *Rhinocladiella aquaspersa*	5-20	Brown, round to pleomorphic, thick-walled cells with transverse septations; commonly, cells contain two fission planes that form a tetrad of cells; occasionally, branched septate hyphae may be found along with sclerotic bodies

TABLE 65-8 SUMMARY OF CHARACTERISTIC FEATURES OF FUNGI SEEN IN DIRECT EXAMINATION OF CLINICAL SPECIMENS — CONT'D

MORPHOLOGIC FORM FOUND IN SPECIMENS	ORGANISM(S)	SIZE RANGE (DIAMETER, μm)	CHARACTERISTIC FEATURES
Granules	*Acremonium* A. falciforme A. kiliense A. recifei	200-300	White, soft granules without a cementlike matrix
	Curvularia C. geniculata C. lunata	500-1,000	Black, hard grains with a cementlike matrix at periphery
	Aspergillus A. nidulans	65-160	White, soft granule without a cementlike matrix
	Exophiala E. jeanselmei	200-300	Black, soft granules, vacuolated, without a cementlike matrix, made of dark hyphae and swollen cells
	Fusarium F. moniliforme	200-500	White, soft granules without a cementlike matrix
	F. solani	300-600	
	Leptosphaeria L. senegalensis	400-600	Black, hard granules with cementlike matrix present
	L. tompkinsii	500-1,000	Periphery composed of polygonal swollen cells and center of a hyphal network
	Madurella M. grisea	350-500	Black, soft granules without a cementlike matrix, periphery composed of polygonal swollen cells and center of a hyphal network
	M. mycetomatis	200-900	Black to brown, hard granules of two types: (1) Rust brown, compact, and filled with cementlike matrix (2) Deep brown, filled with numerous vesicles, 6-14 μm in diameter, cementlike matrix in periphery, and central area of light-colored hyphae
	Neotestudina N. rosatti	300-600	White, soft granules with cementlike matrix present at periphery
	Pseudallescheria P. boydii	200-300	White, soft granules composed of hyphae and swollen cells at periphery in cementlike matrix
	Pyrenochaeta P. romeri	300-600	Black, soft granules composed of polygonal swollen cells at periphery, center is network of hyphae, no cementlike matrix present

From Merz, W.G. and Roberts, G.D. 1995. Detection and recovery of fungi in clinical specimens. In Murray, P.R., Baron, E.J., Pfaller, M.A., et al., editors. Manual of clinical microbiology, ed 6. ASM Press, Washington, D.C.

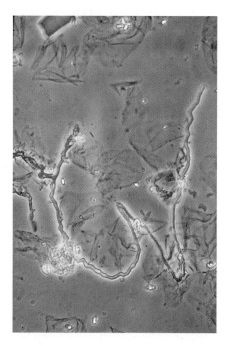

FIGURE 65-11 Potassium hydroxide preparation of skin, phase-contrast, dermatophyte, showing septate hyphae intertwined among epithelial cells (500×).

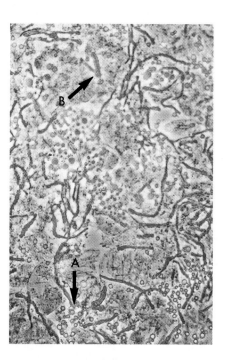

FIGURE 65-12 Potassium hydroxide preparation of skin, phase-contrast, *M. furfur,* showing spherical yeast cells (**A**) and short hyphal fragments (**B**) (500×).

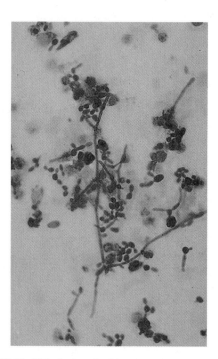

FIGURE 65-13 PAS of urine, *C. albicans,* showing blastoconidia and pseudohyphae.

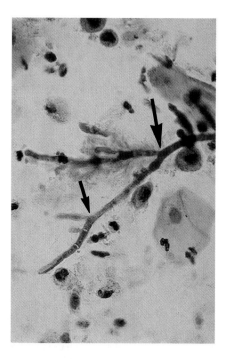

FIGURE 65-14 Papanicolaou stain of sputum, *A. fumigatus,* showing dichotomously branching septate hyphae (*arrows*).

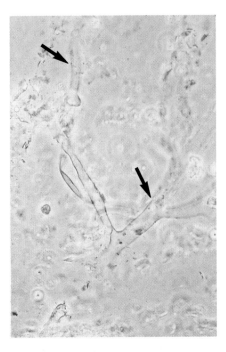

FIGURE 65-15 Potassium hydroxide preparation of sputum, phase contrast, *Rhizopus* spp., showing fragmented portions (*arrows*) of nonseptate hyphae of varying size.

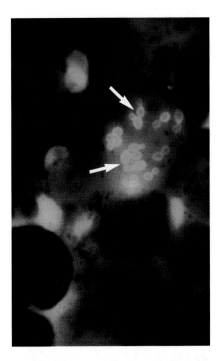

FIGURE 65-16 Calcofluor white stain of sputum, *H. capsulatum,* showing intracellular yeast cells (*arrows*) 2 to 5 μm in diameter.

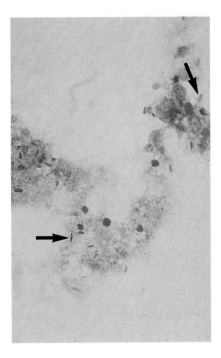

FIGURE 65-17 Periodic acid-Schiff stain of exudate. *S. schenckii,* showing cigar- to oval-shaped yeast cells (*arrows*).

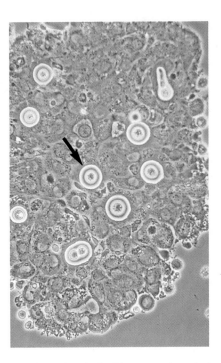

FIGURE 65-18 Potassium hydroxide preparation of pleural fluid, phase contrast, *C. neoformans,* showing encapsulated, spherical yeast cells (*arrow*) of varying sizes.

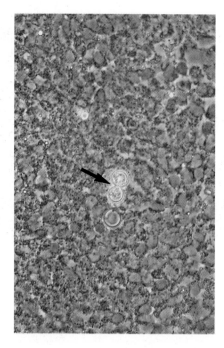

FIGURE 65-19 Potassium hydroxide preparation of exudate, phase contrast, *B. dermatitidis,* showing large budding yeast cell with a distinct broad base (*arrow*) between the cells.

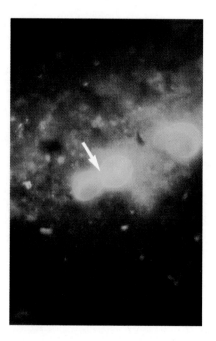

FIGURE 65-20 Auramine-rhodamine preparation of bone lesion. *Blastomyces dermatitidis* showing characteristic broad-based budding yeast cell (*arrow*).

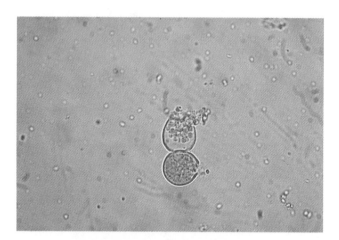

FIGURE 65-21 Potassium hydroxide preparation of sputum, bright field, *C. immitis,* showing two spherules, filled with endospores, lying adjacent to each other that resemble *B. dermatitidis.*

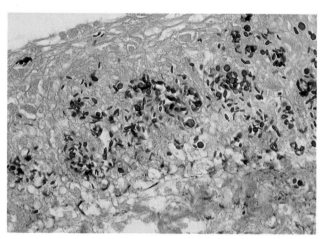

FIGURE 65-22 *Sporothrix schenkii (deeply stained bodies)* in mouse testis.

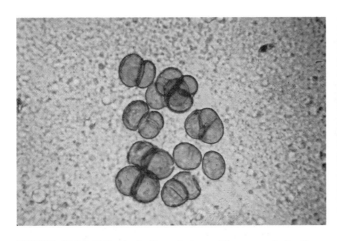

FIGURE 65-23 Sclerotic bodies of chromoblastomycosis in tissue (400×) (Courtesy Upjohn Co.).

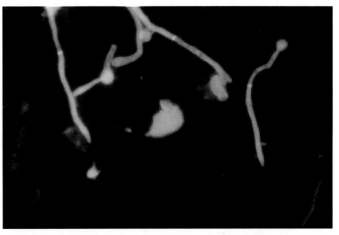

FIGURE 65-24 *Candida albicans* in urine, calcofluor white stain.

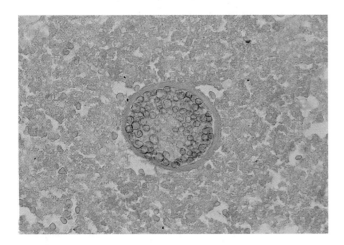

FIGURE 65-25 *Coccidioides immitis.* Spherule with endospores.

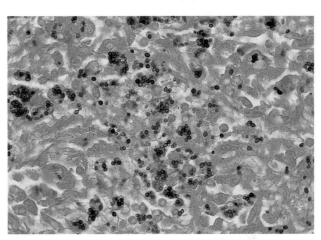

FIGURE 65-26 *Histoplasma capsulatum* (deeply stained bodies) in lung. Methenamine silver stain (430×).

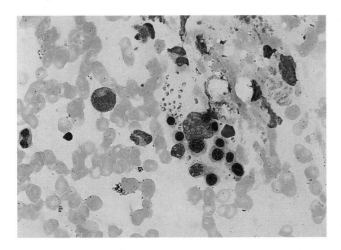

FIGURE 65-27 *Histoplasma capsulatum* in peripheral blood smear (1000×).

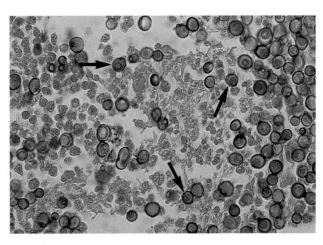

FIGURE 65-28 *Blastomyces dermatitidis* (*arrows*) in tissue. Methenamine silver stain (430×).

of fungi may limit the development of a screening method that is cost effective. Studies by Makimura,[68] Hopfer,[47,48] Sandhu,[92] and Lu[67] present what has been done with molecular methods in mycology for the detection of fungi in clinical specimens.

EXTENT OF IDENTIFICATION OF FUNGI RECOVERED FROM CLINICAL SPECIMENS

The question of when and how far to go with identification of fungi recovered from clinical specimens presents an interesting challenge. The current emphasis on cost containment and the ever-increasing number of opportunistic fungi causing infection in compromised patients causes one to consider whether all fungi recovered from clinical specimens should be thoroughly identified and reported. Murray et al.[75] in 1977 were concerned with the time and expense associated with the identification of yeasts from respiratory tract specimens. Because these are the specimens most commonly submitted for fungal culture, they questioned whether it was important to provide an identification for every organism recovered. After evaluating the clinical usefulness of information provided by the identification of yeast recovered from respiratory tract specimens, they suggested the following:

- The routine identification of yeasts recovered in culture from respiratory secretions is not warranted, but that all yeasts should be screened for the presence of *Cryptococcus neoformans*
- All respiratory secretions submitted for fungal culture, regardless of the presence or absence of oropharyngeal contamination, should be cultured since common pathogens such as *Histoplasma capsulatum, Blastomyces dermatitidis, Coccidioides immitis,* and *Sporothrix schenckii* may be recovered
- The routine identification of yeasts in respiratory secretions is of little or no value to the clinician and probably represents normal flora except for *C. neoformans*

The extent of identification of yeasts from other specimen sources is discussed under yeast identification later in this chapter.

When and how far to proceed with an identification of the molds is a much more difficult question to answer. All commonly encountered molds should be identified and reported regardless of the clinical source. Those organisms that fail to sporulate after a reasonable time should be reported as being present but the identification need not be attempted if the dimorphic fungi have been ruled out or if the clinician feels that the organism is not clinically significant. Ideally, all laboratories should identify all fungi recov-

ered from clinical specimens; however, the limits of practicality and economic considerations play a definite role in deciding how far to go with their identification. Each individual laboratory director, in consultation with the clinicians being served, will have to make this decision after considering the patient population, laboratory practice, and economic impact.

As shown in Table 65-9, an increasing number of fungi are seen in the clinical microbiology laboratory that have not been discussed thus far. They are considered to be environmental flora, but in reality they must be regarded as potential pathogens since infections with a number of these organisms have been reported, including *Pseudallescheria boydii,*[4] *Bipolaris,*[1,24,72] *Exserohilum,*[1,24,72] *Trichosporon,*[107] *Aureobasidium,*[11] and others.[76] It is necessary for the laboratory to identify and report all organisms recovered from clinical specimens so that their clinical significance can be determined. In many instances, the presence of environmental fungi is unimportant; however, that is not always the case. The following section will present the molds most commonly recovered from clinical specimens and a brief description of their colonial and morphologic features. In addition, Tables 65-10 and 65-11 present the molds and yeasts implicated in causing infection in humans, the time required for their identification, most likely site for their recovery, and the clinical implications of each.

GENERAL CONSIDERATIONS FOR THE IDENTIFICATION OF MOLDS

The identification of molds is made using a combination of:

- The growth rate
- Colonial morphologic features
- Microscopic morphologic features

In most instances the latter provides the most definitive means for identification. The determination of the growth rate can be the most helpful when examining a mold culture. However, it may be of limited value since the growth rate of certain fungi is variable, depending on the amount of inoculum present in a clinical specimen. In general, the growth rate for the dimorphic fungi, including *B. dermatitidis, C. immitis, H. capsulatum,* and *P. brasiliensis,* is slow; 1 to 4 weeks are usually required before colonies become visible. In some instances, however, cultures of *B. dermatitidis, H. capsulatum,* and *C. immitis* may be detected within 3 to 5 days. This is a somewhat uncommon circumstance and is encountered only when many of the organism are present in the specimen. In contrast, colonies of the Zygomycetes may appear within 24 hours, while the other hyaline and

Text continued on p. 904

TABLE 65-9   FUNGI MOST COMMONLY RECOVERED FROM CLINICAL SPECIMENS AT MAYO CLINIC

BLOOD	CEREBROSPINAL FLUID	GENITOURINARY TRACT	RESPIRATORY TRACT	SKIN
Candida albicans	Cryptococcus neoformans	Candida albicans	Yeast, not Cryptococcus	Trichophyton rubrum
Candida tropicalis	Candida albicans	Candida glabrata	Penicillium spp.	Trichophyton mentagrophytes
Candida parapsilosis	Candida parapsilosis	Candida tropicalis	Aspergillus spp.	Alternaria spp.
Cryptococcus neoformans	Candida tropicalis	Candida parapsilosis	Aspergillus fumigatus	Candida albicans
Histoplasma capsulatum	Coccidioides immitis	Penicillium spp.	Cladosporium spp.	Penicillium spp.
Candida lusitaniae	Histoplasma capsulatum	Candida krusei	Alternaria spp.	Scopulariopsis spp.
Candida krusei		Cryptococcus neoformans	Aspergillus niger	Epidermophyton floccosum
Saccharomyces spp.		Saccharomyces spp.	Geotrichum candidum	Candida parapsilosis
Candida pseudotropicalis		Histoplasma capsulatum	Fusarium spp.	Aspergillus spp.
Candida zeylanoides		Cladosporium spp.	Aspergillus versicolor	Acremonium spp.
Trichosporon beigelii		Aspergillus spp.	Aspergillus flavus	Aspergillus versicolor
Coccidioides immitis		Trichosporon beigelii	Acremonium spp.	Cladosporium spp.
Candida guilliermondii		Alternaria spp.	Scopulariopsis spp.	Fusarium spp.
Malasezzia furfur			Beauveria spp.	Trichosporon beigelii
			Trichosporon beigelii	Phialophora spp.

TABLE 65-10 COMMON FILAMENTOUS FUNGI IMPLICATED IN HUMAN MYCOTIC INFECTIONS

ETIOLOGIC AGENT	TIME REQUIRED FOR IDENTIFICATION	PROBABLE RECOVERY SITES	CLINICAL IMPLICATION
Acremonium	2-6 days	Skin, nails, respiratory secretions, cornea, vagina, gastric washings	Skin and nail infections, mycotic keratitis, mycetoma
Alternaria spp.	2-6 days	Skin, nails, conjunctiva, respiratory secretions	Skin and nail infections, sinusitis, conjunctivitis, hypersensitivity pneumonitis, skin abscess
Aspergillus flavus	1-4 days	Skin, respiratory secretions, gastric washings, nasal sinuses	Skin infections, allergic bronchopulmonary infection, sinusitis, myocarditis, disseminated infection, renal infection, subcutaneous mycetoma
Aspergillus fumigatus	2-6 days	Respiratory secretions, skin, ear, cornea, gastric washings, nasal sinuses	Allergic bronchopulmonary infection, fungus ball, invasive pulmonary infection, skin and nail infections, external otomycosis, mycotic keratitis, sinusitis, myocarditis, renal infection
Aspergillus niger	1-4 days	Respiratory secretions, gastric washings, ear, skin	Fungus ball, pulmonary infection, external otomycosis, mycotic keratitis
Aspergillus terreus	2-6 days	Respiratory secretions, skin, gastric washings, nails	Pulmonary infection, disseminated infection, endocarditis, onychomycosis, allergic bronchopulmonary infection
Bipolaris spp.	2-6 days	Respiratory secretions, skin, nose, bone	Sinusitis, brain abscess, peritonitis, subcutaneous abscess, pulmonary infection, osteomyelitis, encephalitis
Blastomyces dermatitidis	6-21 days (recovery time) (additional 3-14 days required for confirmatory identification)	Respiratory secretions, skin, oropharyngeal ulcer, bone, prostate	Pulmonary infection, skin infection, oropharyngeal ulceration, osteomyelitis, prostatitis, arthritis, CNS infection, disseminated infection
Cladosporium spp.	6-10 days	Respiratory secretions, skin, nails, nose, cornea	Skin and nail infections, mycotic keratitis, chromoblastomycosis caused by *Cladosporium carrionii*
Coccidioides immitis	3-21 days	Respiratory secretions, skin, bone, cerebrospinal fluid, synovial fluid, urine, gastric washings	Pulmonary infection, skin infection, osteomyelitis, meningitis, arthritis, disseminated infection
Curvularia spp.	2-6 days	Respiratory secretions, cornea, brain, skin, nasal sinuses	Pulmonary infection, disseminated infection, mycotic keratitis, brain abscess, mycetoma, endocarditis
Drechslera spp.	2-6 days	Respiratory secretions, skin, peritoneal fluid (following dialysis)	Pulmonary infection (rare)
Epidermophyton floccosum	7-10 days	Skin, nails	Tinea cruris, tinea pedis, tinea corporis, onychomycosis

ETIOLOGIC AGENT	TIME REQUIRED FOR IDENTIFICATION	PROBABLE RECOVERY SITES	CLINICAL IMPLICATION
Exserohilum spp.	2-6 days	Eye, skin, nose, bone	Keratitis, subcutaneous abscess, sinusitis, endocarditis, osteomyelitis
Fusarium spp.	2-6 days	Skin, respiratory secretions, cornea	Mycotic keratitis, skin infection (in burn patients), disseminated infection, endophthalmitis
Geotrichum spp.	2-6 days	Respiratory secretions, urine, skin, stool, vagina, conjunctiva, gastric washings, throat	Bronchitis, skin infection, colitis, conjunctivitis, thrush, wound infection
Histoplasma capsulatum	10-45 days (recovery time) (additional 7-21 days required for confirmatory identification)	Respiratory secretions, bone marrow, blood, urine, adrenals, skin, cerebrospinal fluid, eye, pleural fluid, liver, spleen, oropharyngeal lesions, vagina, gastric washings, larynx	Pulmonary infection, oropharyngeal lesions, CNS infection, skin infection (rare), uveitis, peritonitis, endocarditis, brain abscess, disseminated infection
Microsporum audouinii	10-14 days (recovery time) (additional 14-21 days required for confirmatory identification)	Hair	Tinea capitis
Microsporum canis	5-7 days	Hair, skin	Tinea corporis, tinea capitis, tinea barbae, tinea manuum
Microsporum gypseum	3-6 days	Hair, skin	Tinea capitis, tinea corporis
Mucor spp.	1-5 days	Respiratory secretions, skin, nose, brain, stool, orbit, cornea, vitreous humor, gastric washings, wounds, ear	Rhinocerebral infection, pulmonary infection, gastrointestinal infection, mycotic keratitis, intraocular infection, external otomycosis, orbital cellulitis, disseminated infection
Penicillium spp.	2-6 days	Respiratory secretions, gastric washings, skin, urine, ear, cornea	Pulmonary infection, skin infection, external otomycosis, mycotic keratitis, endocarditis
Phialophora spp.	6-21 days	Respiratory secretions, gastric washings, skin, cornea, conjunctiva	Some species produce chromoblastomycosis or mycetoma; mycotic keratitis, conjunctivitis, intraocular infection

Continued

TABLE 65-10 COMMON FILAMENTOUS FUNGI IMPLICATED IN HUMAN MYCOTIC INFECTIONS—CONT'D

ETIOLOGIC AGENT	TIME REQUIRED FOR IDENTIFICATION	PROBABLE RECOVERY SITES	CLINICAL IMPLICATION
Pseudallescheria boydii	2-6 days	Respiratory secretions, gastric washings, skin, cornea	Pulmonary fungus ball, mycetoma, mycotic keratitis, endocarditis, disseminated infection, brain abscess
Rhizopus species	1-5 days	Respiratory secretions, skin, nose, brain, stool, orbit, cornea, vitreous humor, gastric washings, wounds, ear	Rhinocerebral infection, pulmonary infection, mycotic keratitis, intraocular infection, orbital cellulitis, external otomycosis, disseminated infection
Scedosporium prolificans	2-6 days	Respiratory secretions, skin, nasal sinuses, bone	Arthritis, osteomyelitis, sinusitis, endocarditis
Scopulariopsis spp.	2-6 days	Respiratory secretions, gastric washings, nails, skin, vitreous humor, ear	Pulmonary infection, nail infection, skin infection, intraocular infection, external otomycosis
Sporothrix schenckii	3-12 days (recovery time) (additional 2-10 days required for confirmatory identification)	Respiratory secretions, skin, subcutaneous tissue, maxillary sinuses, synovial fluid, bone marrow, bone, cerebrospinal fluid, ear, conjunctiva	Pulmonary infection, lymphocutaneous infection, sinusitis, arthritis, osteomyelitis, meningitis, external otomycosis, conjunctivitis, disseminated infection
Trichophyton mentagrophytes	7-10 days	Hair, skin, nails	Tinea barbae, tinea capitis, tinea corporis, tinea cruris, tinea pedis, onychomycosis
Trichophyton rubrum	10-14 days	Hair, skin, nails	Tinea pedis, onychomycosis, tinea corporis, tinea cruris
Trichophyton tonsurans	10-14 days	Hair, skin, nails	Tinea capitis, tinea corporis, onychomycosis, tinea pedis
Trichophyton verrucosum	10-18 days	Hair, skin, nails	Tinea capitis, tinea corporis, tinea barbae
Trichophyton violaceum	14-18 days	Hair, skin, nails	Tinea capitis, tinea corporis, onychomycosis
Wangiella dermatitidis	5-21 days	Respiratory secretions, skin, eye	Phaeohyphomycosis, endophthalmitis, pneumonia

From Koneman, E.W. and Roberts, G.D. 1991. Mycotic disease. In Henry, J.B., editor. Clinical diagnosis and management by laboratory methods, ed 18. W.B. Saunders, Philadelphia.

_____ **TABLE 65-11   COMMON YEASTLIKE ORGANISMS IMPLICATED** _____
IN HUMAN INFECTION*

ETIOLOGIC AGENT	PROBABLE RECOVERY SITES	CLINICAL IMPLICATION
Candida albicans	Respiratory secretions, vagina, urine, skin, oropharynx, gastric washings, blood, stool, transtracheal aspiration, cornea, nails, cerebrospinal fluid, bone, peritoneal fluid	Pulmonary infection, vaginitis, urinary tract infection, dermatitis, fungemia, mycotic keratitis, onychomycosis, meningitis, osteomyelitis, peritonitis, myocarditis, endocarditis, endophthalmitis, disseminated infection, thrush, arthritis
Candida glabrata	Respiratory secretions, urine, vagina, gastric washings, blood, skin, oropharynx, transtracheal aspiration, stool, bone marrow, skin (rare)	Pulmonary infection, urinary tract infection, vaginitis, fungemia, disseminated infection, endocarditis
Candida tropicalis	Respiratory secretions, urine, gastric washings, vagina, blood, skin, oropharynx, transtracheal aspiration, stool, pleural fluid, peritoneal fluid, cornea	Pulmonary infection, vaginitis, thrush, endophthalmitis, endocarditis, arthritis, peritonitis, mycotic keratitis, fungemia
Candida parapsilosis	Respiratory secretions, urine, gastric washings, blood, vagina, oropharynx, skin, transtracheal aspiration, stool, pleural fluid, ear, nails	Endophthalmitis, endocarditis, vaginitis, mycotic keratitis, external otomycosis, paronychia, fungemia
Saccharomyces spp.	Respiratory secretions, urine, gastric washings, vagina, skin, oropharynx, transtracheal aspiration, stool	Pulmonary infection (rare), endocarditis
Candida krusei	Respiratory secretions, urine, gastric washings, vagina, skin, oropharynx, blood, transtracheal aspiration, stool, cornea	Endocarditis, vaginitis, urinary tract infection, mycotic keratitis
Candida guilliermondii	Respiratory secretions, gastric washings, vagina, skin, nails, oropharynx, blood, cornea, bone, urine	Endocarditis, fungemia, dermatitis, onychomycosis, mycotic keratitis, osteomyelitis, urinary tract infection
Rhodotorula spp.	Respiratory secretions, urine, gastric washings, blood, vagina, skin, oropharynx, stool, cerebrospinal fluid, cornea	Fungemia, endocarditis, mycotic keratitis
Trichosporon spp.	Respiratory secretions, blood, skin, oropharynx, stool	Pulmonary infection, brain abscess, disseminated infection, piedra
Cryptococcus neoformans	Respiratory secretions, cerebrospinal fluid, bone, blood, bone marrow, urine, skin, pleural fluid, gastric washings, transtracheal aspiration, cornea, orbit, vitreous humor	Pulmonary infection, meningitis, osteomyelitis, fungemia, disseminated infection, endocarditis, skin infection, mycotic keratitis, orbital cellulitis, endophthalmic infection
Cryptococcus albidus subsp. _albidus_	Respiratory secretions, skin, gastric washings, urine, cornea	Meningitis, pulmonary infection

Continued

TABLE 65-11   COMMON YEASTLIKE ORGANISMS IMPLICATED IN HUMAN INFECTION* —CONT'D

ETIOLOGIC AGENT	PROBABLE RECOVERY SITES	CLINICAL IMPLICATION
Candida kefyr (pseudotropicalis)	Respiratory secretions, vagina, urine, gastric washings, oropharynx	Vaginitis, urinary tract infection
Cryptococcus luteolus	Respiratory secretions, skin, nose	Not commonly implicated in human infection
Cryptococcus laurentii	Respiratory secretions, cerebrospinal fluid, skin, oropharynx, stool	Not commonly implicated in human infection
Cryptococcus albidus subsp. *diffluens*	Respiratory secretions, urine, cerebrospinal fluid, gastric washings, skin	Not commonly implicated in human infection
Cryptococcus terreus	Respiratory secretions, skin, nose	Not commonly implicated in human infection

*Arranged in order of occurrence in the clinical laboratory.

From Koneman, E.W. and Roberts, G.D. 1991. Mycotic disease. In Henry, J.B., editor. Clinical diagnosis and management by laboratory methods, ed 18. W.B. Saunders, Philadelphia.

dematiaceous fungi often exhibit growth within 1 to 5 days. The growth rate of an organism is important, but it must be used in combination with other features before a definitive identification can be made.

The colonial morphologic features may be of limited value in identifying the molds, because of natural variation among isolates and colonies grown on different culture media. Although it may be possible to recognize common organisms recovered repeatedly in the laboratory, colonial morphology is an unreliable criterion and should be used only to supplement the microscopic morphologic features of the organism. Incubation conditions and culture media must be considered. For example, *H. capsulatum,* which appears as a white to tan fluffy mold on brain-heart infusion agar, may appear yeastlike in appearance when grown on the same medium containing blood enrichment.

In general, the microscopic morphologic features of the molds are stable and exhibit minimal variation. The definitive identification is based on the characteristic shape, method of production, and arrangement of spores; however, the size of the hyphae also provides helpful information. The large ribbonlike hyphae of the Zygomycetes are easily recognized, while small hyphae, 1 to 2 μm in diameter, may suggest the presence of one of the dimorphic fungi.

The fungi may be prepared for microscopic observation using several techniques. The procedure traditionally used by most laboratories is the adhesive tape preparation (Procedure 65-1, Figure 65-29). It can be prepared easily and quickly and often is sufficient to make the identification for most of the fungi. However, the wet mount (Procedure 65-2, Figure 65-30) is preferred by some laboratories, and a microslide culture method (Procedure 65-3, Figure 65-31) may be used when greater detail of the morphologic features is required for identification.

LABORATORY SAFETY CONSIDERATIONS

Although there are risks associated with the handling of fungi recovered from clinical specimens, a common sense approach concerning their handling will protect the laboratory from contamination and workers from acquiring an infection.

It is necessary that all mold cultures and clinical specimens be handled in a class II, B3 biological safety cabinet (BSC), with no exceptions. Some laboratory directors feel that it is necessary to use gloves in handling mold cultures within an enclosed BSC; however, this is not necessary if a laminar flow BSC is used. It is permissible, however, to handle yeast cultures on the bench-top; they must be treated as infectious agents. The use of an electric incinerator or a gas flame is suitable for the decontamination of a loop used for transfer of yeast cultures. Cultures of organisms suspected of being pathogens should be sealed with tape to prevent laboratory contamination and should be autoclaved as soon as the definitive identification is made. Few problems with laboratory contamination or acquired infection by laboratory personnel will develop if common safety precautions are used.

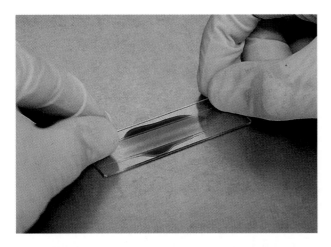

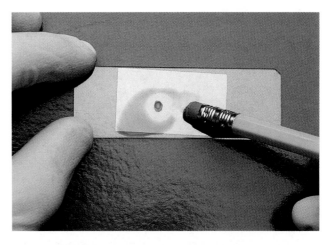

FIGURE 65-29 Scotch tape preparation, showing placement of tape onto slide containing lactophenol cotton or aniline blue.

FIGURE 65-30 Performance of a wet mount, showing agar positioned under coverslip before using pressure to disperse growth.

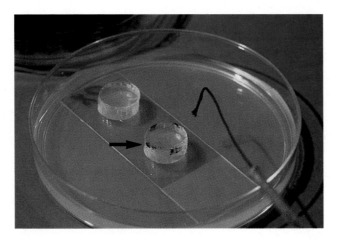

FIGURE 65-31 Microslide culture, showing inoculation of agar plug (*arrow*).

GENERAL MORPHOLOGIC FEATURES OF THE MOLDS

Specialized types of vegetative hyphae may be helpful in placing an organism into a certain group. For example, dermatophytes often produce several types of hyphae, including antler hyphae that are curved, freely branching, and antlerlike in appearance (Figure 65-32). Racquet hyphae are enlarged, club-shaped structures (Figure 65-33). In addition, certain of the dermatophytes produce spiral hyphae that are coiled or exhibit corkscrew-like turns seen within the hyphal strand (Figure 65-34). These structures are not characteristic for any certain group; however, they are found most commonly in dermatophytes.

Some species of fungi produce sexual spores in a large, saclike structure called an **ascocarp** (Figure 65-35). The ascocarp, in turn, contains smaller sacs called **asci**, each of which contain four to eight **ascospores**. This type of sexual reproduction is uncom-

monly seen in the fungi recovered in the clinical microbiology laboratory; most exhibit only asexual reproduction. It is possible that all fungi possess a sexual form, but this form may not have been observed yet on artificial culture media. Conidia are produced by most fungi and represent the asexual reproductive cycle (Figure 65-36). The type of conidia, their morphology, and arrangement are important criteria for establishing the definitive identification of an organism.

The simplest type of sporulation is the development of a spore directly from the vegetative hyphae. **Arthroconidia** are formed directly from the hyphae by fragmentation through the points of septation (Figure 65-37). When mature, they appear as square, rectangular, or barrel-shaped, thick-walled cells. These result from the simple fragmentation of the hyphae into spores, which are easily dislodged and disseminated into the environment. **Chlamydospores (chlamydoconidia)** are round, thick-walled spores

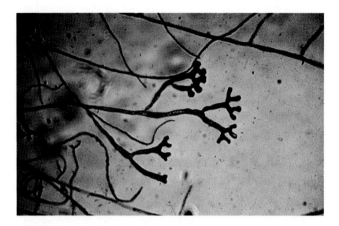

FIGURE 65-32 Antler hyphae showing swollen hyphal tips resembling antlers, with lateral and terminal branching (favic chandeliers) (500×).

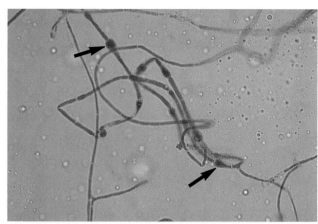

FIGURE 65-33 Racquet hyphae showing swollen areas (*arrows*) resembling a tennis racquet.

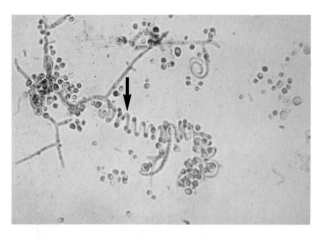

FIGURE 65-34 Spiral hyphae (*arrow*) exhibiting corkscrewlike turns (430×).

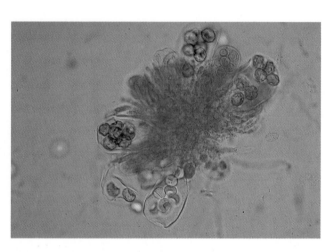

FIGURE 65-35 Ascocarp showing dark-appearing ascospores (430×).

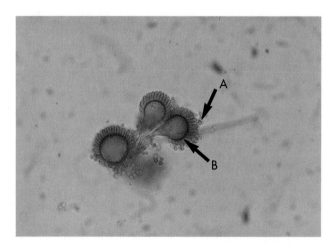

FIGURE 65-36 Conidia (asexual spores [**A**]) produced on specialized structures (conidiophores [**B**]) of *Aspergillus* (430×).

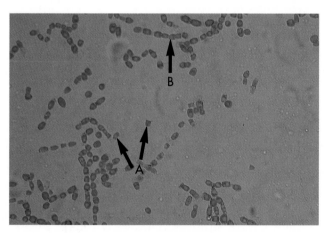

FIGURE 65-37 Arthroconidia formation (**A**) produced by the breaking down of a hyphal strand (**B**) into individual rectangular units (430×).

formed directly from the differentiation of hyphae in which there is a concentration of protoplasm and nutrient material (Figure 65-38). These appear to be resistant resting spores produced by the rounding up and enlargement of the terminal cells of the hyphae. Chlamydospores may be **intercalary** (within the hyphae) or **terminal** (end of hyphae).

Various other types of spores occur with many species of fungi. Conidia are asexual spores produced singly or in groups by specialized hyphal strands called **conidiophores**. In some instances, the conidia are freed from their point of attachment by pinching off, or abstriction. Some conidiophores terminate in a swollen vesicle. Secondary, small, flask-shaped phialides are formed from the surface of the vesicle. These, in turn, give rise to long chains of conidia. This type of fruiting structure is characteristic of the

aspergilli. A single, simple, slender, tubular conidiophore (**phialide**) that supports a cluster of conidia, held together as a gelatinous mass, is characteristic of certain fungi, including the genus *Acremonium* (Figure 65-39). In other instances, conidiophores form a branching structure termed a **penicillus**. Each branch terminates in secondary branches (**metulae**) and phialides from which chains of conidia are borne (Figure 65-40). Species of *Penicillium* and *Paecilomyces* are representative of this type of sporulation. In other instances, fungi may produce conidia of two sizes: **microconidia** that are small, unicellular, round, elliptic, or pyriform in shape, or **macroconidia** that are large, usually multiseptate, and club- or spindle-shaped (Figure 65-41). Microconidia may be borne directly on the side of a hyphal strand or at the end of a conidiophore. Macroconidia are usually borne on a

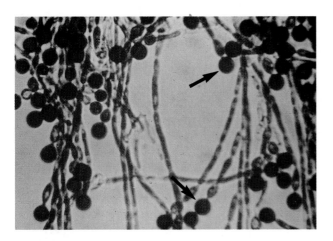

FIGURE 65-38 Chlamydospores composed of thick-walled spherical cells (*arrows*) (430×).

FIGURE 65-39 Simple tubular phialide with a cluster of conidia at its tip (*arrow*), characteristic of *Acremonium* (430×).

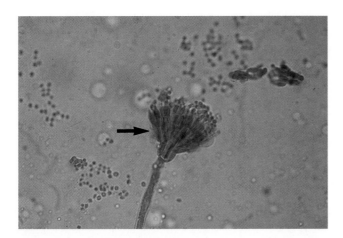

FIGURE 65-40 Complex method of sporulation in which conidia are borne on phialides produced on secondary branches (metulae [*arrow*]), characteristic of *Penicillium* (430×).

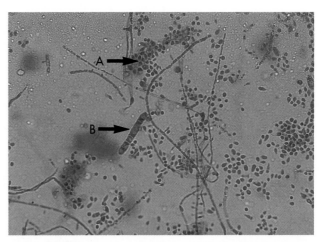

FIGURE 65-41 Numerous small, spherical microconidia (**A**) contrasted with a large, elongated macroconidium (**B**) (430×).

short to long conidiophore and may be smooth or rough-walled. Microconidia and macroconidia are seen in some fungal species and are not specific, except as they are used to differentiate a limited number of genera.

The hyphae of the Zygomycetes are sparsely septate. Sporulation takes place by progressive cleavage during maturation within the sporangium, a saclike structure produced at the tip of a long stalk called a **sporangiophore**. **Sporangiospores**, spores produced within the sporangium, are released by the rupture of the sporangial wall (Figure 65-42). Rarely, some isolates may produce **zygospores**, rough-walled spores produced by the union of two matching types of a zygomycite; this is an example of sexual reproduction.

HYALINE, ASEPTATE, OR SPARSELY SEPTATE MOLDS

ZYGOMYCETES

Although the Zygomycetes (*Rhizopus, Mucor, Absidia, Syncephalastrum, Cunninghamella,* and others) are a less common cause of infection compared with the aspergilli, they may be a significant cause of morbidity and mortality in immunocompromised patients. These organisms are distributed worldwide and commonly are found on decaying vegetable matter or in soil. The infection is generally acquired by inhalation of spores followed by subsequent development of infection. Immunocompromised patients, particularly those having uncontrolled diabetes mellitus and those receiving prolonged corticosteroid, antibiotic, or cytotoxic therapy, are at greatest risk. The organisms involved in causing **zygomycosis** (infection caused by a Zygomycete) have a marked propensity for vascular invasion and rapidly produce thrombosis and necrosis of tissue. One of the most common presentations is the rhinocerebral form wherein the nasal mucosa, palate, sinuses, orbit, face, and brain are involved; each shows massive necrosis with vascular invasion and infarction. Other types of infection involve the lungs and gastrointestinal tract; some patients develop disseminated infection. Organs, including the liver, spleen, pancreas, and kidneys, may be involved. The Zygomycetes have also been implicated in causing skin infections among patients with burns and in infections of subcutaneous tissue of patients undergoing surgery.

Direct detection

The rapid diagnosis of zygomycosis may be made by examination of tissue specimens or exudate from infected lesions in a calcofluor white or potassium hydroxide preparation. Branching, predominantly nonseptate hyphae are observed (Figure 65-15). It is very important that the laboratory notify the clinician of these findings since Zygomycetes grow very rapidly and vascular invasion occurs at a rapid rate.

Culture and identification

The colonial morphologic features of the Zygomycetes allow one to immediately suspect organisms belonging to this group. Colonies characteristically produce a fluffy, white to gray or brown hyphal growth that diffusely covers the surface of the agar within 24 to 96 hours (Figure 65-43). The hyphae appear to be coarse and fill the entire culture dish or tube rapidly with loose, grayish hyphae dotted with brown or black sporangia. It is impossible to distinguish between the different genera and species of Zygomycetes based on their colonial morphologic features since most are identical in appearance.

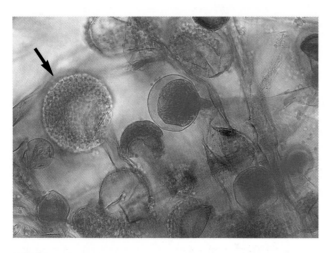

FIGURE 65-42 Large, saclike sporangia that contains sporangiospores (*arrow*), characteristic of the zygomycetes (250×).

FIGURE 65-43 *Rhizopus* colony.

The Zygomycetes characteristically produce large, ribbonlike hyphae that are irregular in diameter and nonseptate; however, occasional septa may exist in older cultures. The specific identification of these organisms is confirmed by observing the characteristic saclike fruiting structures (**sporangia**), which produce internally spherical, yellow or brown spores (**sporangiospores**) (Figure 65-44). Each sporangium is formed at the tip of a supporting structure (**sporangiophore**). During maturation, the sporangium becomes fractured and sporangiospores are released into the environment. Sporangiophores are usually connected to each other by occasionally septate hyphae called **stolons**, which attach at contact points where rootlike structures (**rhizoids**) anchor the organism to the agar surface. The identification of the three most common Zygomycetes—*Mucor, Rhizopus*, and *Absidia*—is based on the presence or absence of rhizoids and the position of the rhizoids in relation to the sporangiophores.

Mucor is characterized by sporangiophores that are singularly produced or branched and have at their tip a round sporangium filled with sporangiospores. It does not have rhizoids or stolons; this distinguishes it from the other genera of Zygomycetes (Figure 65-45).

Rhizopus has unbranched sporangiophores with rhizoids that appear at the point at which the stolon arises (Figure 65-44).

Absidia, an uncommon isolate in the clinical laboratory, is characterized by the presence of rhizoids that originate between sporangiophores (Figure 65-46). The sporangia are pyriform and have a funnel-shaped area (**apophysis**) at the junction of the sporangium. Usually a septum is formed in the sporangiophore just below the sporangium.

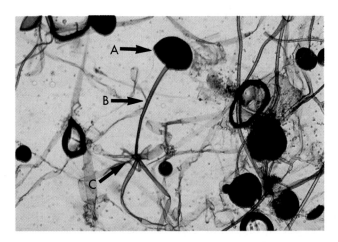

FIGURE 65-44 *Rhizopus* spp., showing sporangium (**A**) on long sporangiophore (**B**) arising from nonseptate hyphae. Note presence of characteristic rhizoids (**C**) at the base of the sporangiophore (250×).

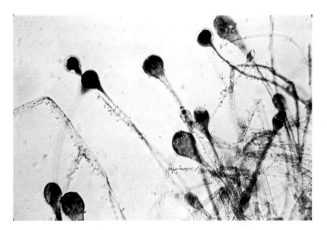

FIGURE 65-45 *Mucor* spp., showing numerous sporangia in the absence of rhizoids (430×).

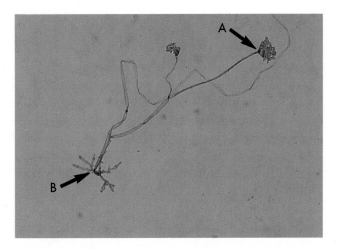

FIGURE 65-46 *Absidia* spp. (**A**), showing sporangia on long sporangiophores arising from nonseptate hyphae (**B**). Note that rhizoids are produced between sporangiophores and not at their bases (250×).

Other genera of Zygomycetes that are encountered much less frequently in the clinical laboratory include *Saksenaea, Cunninghamella, Conidiobolus,* and *Basidiobolus*.[86]

HYALINE, SEPTATE, MONOMORPHIC MOLDS

CUTANEOUS INFECTIONS (DERMATOMYCOSES)

The dermatophytes produce infections that involve the superficial areas of the body, including the hair, skin, and nails. The genera *Trichophyton, Microsporum,* and *Epidermophyton* are the principle etiologic agents of the dermatomycoses. Such cutaneous mycoses are perhaps the most common fungal infections of humans and are usually referred to as **tinea** (Latin for "worm" or "ringworm"). The gross appearance of the lesion is that of an outer ring of an active, progressing infection with central healing within the ring. These infections may be characterized by another Latin noun to designate the area of the body involved. For example, tinea corporis (body), tinea cruris (groin), tinea capitis (scalp and hair), tinea barbae (beard), and tinea unguium (nail) are used to designate the type of infection produced. The dermatophytes break down and utilize keratin as a source of nitrogen but are usually incapable of penetrating subcutaneous tissue. The genus *Trichophyton* is capable of invading the hair, skin, and nails, whereas the genus *Microsporum* involves only the hair and skin; the genus *Epidermophyton* involves the skin and nails. Common species of dermatophytes recovered from clinical specimens, in order of frequency, include *T. rubrum, T. mentagrophytes, E. floccosum, T. tonsurans, M. canis,* and *T. verrucosum*.[120] Other geographically limited species are described elsewhere.

Direct detection

Calcofluor white or potassium hydroxide preparations will reveal the presence of hyaline septate hyphae and/or arthroconidia (Figure 65-11). The direct microscopic examination of infected hairs may reveal the hair shaft to be filled with masses of large (4 to 7 μm) arthroconidia in chains, characteristic of an **endothrix** type of invasion. In other instances, the hair will show external masses of spores that ensheath the hair shaft; this is characteristic of the **ectothrix** type of hair invasion. Hairs infected with *Trichophyton schoenleinii* reveal hyphae and airspaces within the shaft.

Culture and identification

Because the dermatophytes generally present a similar microscopic appearance within infected hair, skin, or nails, the final identification can be made only by culture. A summary of the colonial and microscopic morphologic features of these fungi is presented in Table 65-12. Figure 65-47 presents an identification schema that will be of use to the clinical laboratory for the identification of commonly encountered dermatophytes. Figure 65-47 begins with the microscopic features of the dermatophytes as they might be observed in an initial examination of the culture. In many instances the primary recovery medium fails to function well as a sporulation medium. It is commonly necessary to subculture the initial growth onto cornmeal agar or potato dextrose agar so that sporulation will occur.

TRICHOPHYTON

Members of the genus *Trichophyton* are widely distributed and are the most important and common causes of infections of the feet and nails; they may be responsible for tinea corporis, tinea capitis, and tinea barbae. They are most commonly seen in adult infections, which vary in their clinical manifestations. Most cosmopolitan species are **anthropophilic,** or "human-loving"; a few are **zoophilic,** primarily infecting animals.

Generally, hairs infected with members of the genus *Trichophyton* do not fluoresce under the ultraviolet light of a Wood's lamp; the demonstration of fungal elements inside, surrounding, and penetrating the hair shaft, or within a skin scraping is necessary to make a diagnosis of a dermatophyte infection by direct examination. The recovery and identification of the causative organism are necessary for confirmation.

Microscopically, the genus *Trichophyton* is characterized by smooth, club-shaped, thin-walled macroconidia with 8 to 10 septa ranging in size from 4 × 8 μm to 8 × 15 μm. The macroconidia are borne singly at the terminal ends of hyphae or on short conidiophores; the microconidia predominate and are usually spherical, pyriform (teardrop-shaped), or clavate (club-shaped), and 2 to 4 μm in size (Figure 65-48). Only the common species of *Trichophyton* are described here.

Trichophyton rubrum and *T. mentagrophytes* are the most common species recovered in the clinical laboratory. *T. rubrum* is a slow-growing organism that produces a flat or heaped-up colony that is generally white to reddish with a cottony or velvety surface. The characteristic cherry-red color is best observed on the reverse side of the colony; however, this is produced only after 3 to 4 weeks of incubation. Occasional strains may lack the deep red pigmentation on primary isolation. Colonies may be either fluffy or granular. Microconidia are uncommon in most of the fluffy strains but are more common in the

TABLE 65-12   CHARACTERISTICS OF DERMATOPHYTES COMMONLY RECOVERED IN THE CLINICAL LABORATORY

DERMATOPHYTE	COLONIAL MORPHOLOGY	GROWTH RATE	MICROSCOPIC IDENTIFICATION
*Microsporum audouinii**	Downy white to salmon-pink colony: reverse tan to salmon-pink	2 wks	Sterile hyphae: terminal chlamydospores, favic chandeliers, and pectinate bodies; macroconidia rarely seen—bizarre-shaped if seen; microconidia rare or absent
Microsporum canis	Colony usually membranous with feathery periphery; center of colony white to buff over orange-yellow; lemon-yellow or yellow-orange apron and reverse	1 wk	Thick-walled, spindle-shaped, multiseptate, rough-walled macroconidia, some with a curved tip; microconidia rarely seen
Microsporum gypseum	Cinnamon-colored, powdery colony; reverse light tan	1 wk	Thick-walled, rough, elliptic, multiseptate macroconidia; microconidia few or absent
Epidermophyton floccosum	Center of colony tends to be folded and is khaki green; periphery is yellow; reverse yellowish brown with observable folds	1 wk	Macroconidia large, smooth-walled, multiseptate, clavate, and borne singly or in clusters of two or three; microconidia not formed by this species
Trichophyton mentagrophytes	Different colonial types; white to pinkish, granular and fluffy varieties; occasional light yellow periphery in younger cultures; reverse buff to reddish brown	7-10 days	Many round to globose microconidia most commonly borne in grapelike clusters or laterally along the hyphae; spiral hyphae in 30% of isolates; macroconidia are thin-walled, smooth, club-shaped, and multiseptate; numerous or rare depending upon strain
Trichophyton rubrum	Colonial types vary from white downy to pink granular; rugal folds are common; reverse yellow when colony is young; however, wine red color commonly develops with age	2 wks	Microconidia usually teardrop, most commonly borne along sides of the hyphae; macroconidia usually absent, but when present are smooth, thin-walled, and pencil-shaped
Trichophyton tonsurans	White, tan to yellow or rust, suedelike to powdery; wrinkled with heaped or sunken center; reverse yellow to tan to rust red	7-14 days	Microconidia are teardrop or club-shaped with flat bottoms; vary in size but usually larger than other dermatophytes; macroconidia rare and balloon forms found when present
*Trichophyton schoenleinii**	Irregularly heaped, smooth white to cream colony with radiating grooves; reverse white	2-3 wks	Hyphae usually sterile; many antler-type hyphae seen (favic chandeliers)
*Trichophyton violaceum**	Port wine to deep violet colony, may be heaped or flat with waxy-glabrous surface; pigment may be lost on subculture	2-3 wks	Branched, tortuous hyphae that are sterile; chlamydospores commonly aligned in chains

Continued

TABLE 65-12 CHARACTERISTICS OF DERMATOPHYTES COMMONLY RECOVERED IN THE CLINICAL LABORATORY—CONT'D

DERMATOPHYTE	COLONIAL MORPHOLOGY	GROWTH RATE	MICROSCOPIC IDENTIFICATION
Trichophyton verrucosum	Glabrous to velvety white colonies; rare strains produce yellow-brown color; rugal folds with tendency to sink into agar surface	2-3 wks	Microconidia rare; large and teardrop when seen; macroconidia extremely rare, but form characteristic "rat-tail" types when seen; many chlamydospores seen in chains, particularly when colony is incubated at 37° C

*These organisms are rarely recovered in the United States.

From Koneman, E.W. and Roberts, G.D. 1985. Practical laboratory mycology, ed 3. Williams & Wilkins, Baltimore.

granular strains and occur as small, teardrop-shaped conidia often borne laterally along the sides of the hyphae (Figure 65-48). Macroconidia are seen uncommonly, although they are sometimes found in the granular strains, where they appear as thin-walled, smooth-walled, multicelled, pencil-shaped conidia with three to eight septa. *T. rubrum* has no specific nutritional requirements. It does not perforate hair in vitro (Procedure 65-4) or produce urease.

T. mentagrophytes produces two distinct colonial forms: the downy variety recovered from cases of tinea pedis and the granular variety recovered from lesions acquired by contact with animals. The rapidly growing colonies may appear as white, cottony or downy colonies to cream-colored or yellow colonies that are coarsely granular to powdery. Granular colonies may show evidence of red pigmentation. The reverse side of the colony is usually rose-brown, occasionally orange to deep red, and may be confused with *T. rubrum*. The white, downy colonies produce only a few spherical microconidia; the granular colonies sporulate freely, with numerous small, spherical microconidia produced in grapelike clusters and thin-walled, smooth-walled, cigar-shaped macroconidia measuring 6 × 20 μm to 8 × 50 μm in size, with two to five septa (Figure 65-49). Macroconidia characteristically exhibit a definite narrow attachment to their base. Spiral hyphae may be found in one third of the isolates recovered.

T. mentagrophytes produces urease within 2 to 3 days after inoculation onto Christensen's urea agar. In contrast to *T. rubrum*, *T. mentagrophytes* perforates hair (Figure 65-50). This latter criterion may be used it is difficult to distinguish between these two species.[108]

Trichophyton tonsurans is responsible for an epidemic form of tinea capitis that occurs most commonly in children but also occurs occasionally in adults. It has displaced *M. audouinii* as a primary cause of tinea capitis in most of the United States. The fungus causes a low-grade superficial lesion of varying severity and produces circular, scaly patches of alopecia (loss of hair). The stubs of hair remain in the epidermis of the scalp after the brittle hairs have broken off and may give the typical "black dot" ringworm appearance. Since the infected hairs do not fluoresce under a Wood's lamp, a careful search for the embedded stub should be carried out by the physician with the use of a bright light.

Cultures of *T. tonsurans* develop slowly and are typically buff to brown, wrinkled and suedelike in appearance. The colony surface shows radial folds and often develops a craterlike depression in the center with deep fissures. The reverse side of the colony is yellowish to reddish brown. Microscopically, numerous microconidia with flat bases are produced on the sides of hyphae. With age, the microconidia tend to become pleomorphic, are swollen to elongated, and are referred to as *balloon forms* (Figure 65-51). Chlamydospores are abundant in old cultures; swollen and fragmented hyphal cells resembling arthroconidia may be seen. *T. tonsurans* grows poorly on media lacking enrichments (casein agar); however, growth is greatly enhanced by the presence of thiamine. or inositol in casein agar.

Trichophyton verrucosum causes various lesions in cattle and in humans; it is most often seen in farmers who acquire their infection from cattle. The lesions are found chiefly on the beard, neck, wrist, and back of the hands; they are deep, pustular, and inflammatory. With pressure, short stubs of hair may be recovered from the purulent lesion. Direct examination of the hair shaft reveals sheaths of isolated chains of large (5 to 10 μm) spores (**ectothrix**) and hyphae within the hair (**endothrix**). Masses of these conidia may also be seen in exudate from the lesions.

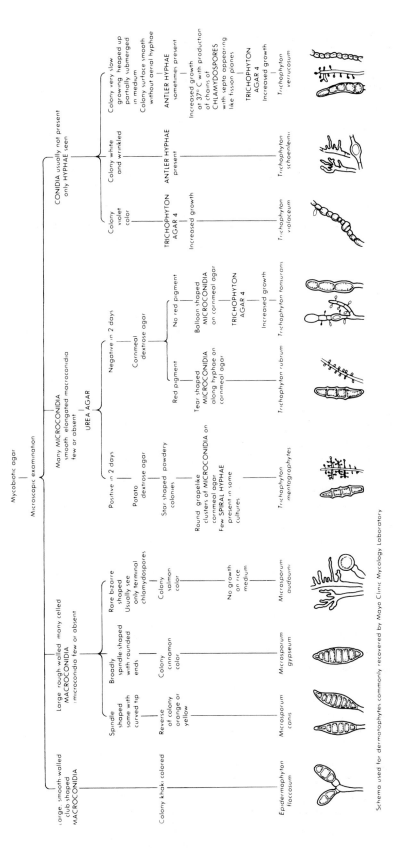

FIGURE 65-47 Dermatophyte identification schema. (From Koneman, E.W. and Roberts, G.D. 1985. Practical laboratory mycology, ed 3. Williams & Wilkins, Baltimore.)

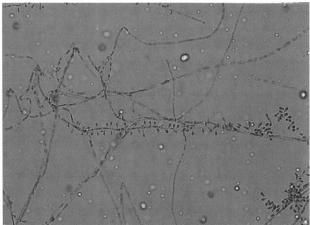

FIGURE 65-48 *T. rubrum,* showing numerous pyriform microconidia borne singly on hyphae (750×).

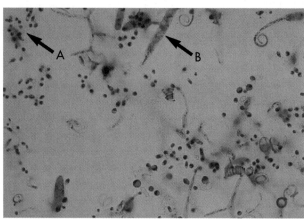

FIGURE 65-49 *T. mentagrophytes,* showing numerous microconidia in grapelike clusters (**A**). Also shown are several thin-walled macroconidia (**B**) (500×).

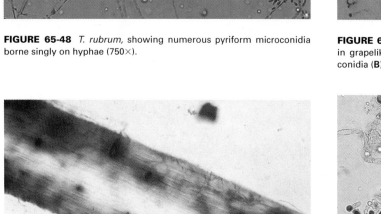

FIGURE 65-50 Hair perforation by *T. mentagrophytes.* Wedge-shaped areas (*arrow*) illustrate hair perforation (100×).

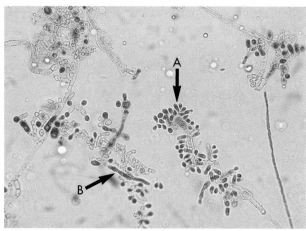

FIGURE 65-51 *T. tonsurans,* showing numerous microconidia (**A**) borne singly or in clusters. A single macroconidium (**B**) (rare) is also present (600×).

T. *verrucosum* grows slowly (14 to 30 days); growth is enhanced at 35° to 37° C and also on media enriched with thiamine and inositol. T. *verrucosum* may be suspected when slow-growing colonies appear to embed themselves into the agar surface. Kane and Smitka[53] described a medium of 4% casein and 0.5% yeast extract for the early detection and identification of T. *verrucosum*. The organism is recognized by its early hydrolysis of casein and very slow growth rate. Chains of chlamydospores are formed regularly at 37° C. The early detection of hydrolysis, formation of characteristic chains of chlamydospores, and the restrictive slow growth rate of T. *verrucosum* differentiate it from T. *schoenleinii,* another slow-growing organism. Colonies are small, heaped, and folded, occasionally flat and disk-shaped. At first they are glabrous and waxy, with a short aerial mycelium. Colonies range from gray and waxlike to a bright ochre. The reverse of the colony is most often nonpigmented but may be yellow.

Chlamydospores produced in chains and antler hyphae may be the only structures observed microscopically (Figures 65-38 and 65-32). Chlamydo-spores may be abundant at 35 to 37° C. Microconidia may be produced by some cultures if the medium is enriched with yeast extract or a vitamin (Figure 65-52). Conidia, when present, are borne laterally from the hyphae and are large and clavate. Macroconidia are rarely formed, vary considerably in size and shape, and are referred to as having a rat-tail appearance.

Trichophyton schoenleinii causes a severe type of infection called **favus**. It is characterized by the formation of yellowish, cup-shaped crusts, or scutulae; considerable scarring of the scalp; and sometimes permanent alopecia. Infections are common among members of the same family. A distinctive invasion of the infected hair, the favic type, is demonstrated by the presence of large inverted cones of hyphae and arthroconidia at the base of the hair follicle and branching hyphae throughout the length of the hair shaft. Longitudinal tunnels or empty spaces appear in the hair shaft where the hyphae have disintegrated. In calcofluor white or potassium hydroxide preparations, these tunnels are readily filled with fluid; air bubbles may also be seen in these tunnels.

T. schoenleinii is a slow-growing organism (30 days or longer) and produces a white to light gray colony that has a waxy surface. Colonies have an irregular border that consists mostly of submerged hyphae and that tends to crack the agar. The surface of the colony is usually nonpigmented or tan, furrowed, and irregularly folded. The reverse side of the colony is usually tan or nonpigmented. Microscopically, conidia are not formed commonly. The hyphae tend to become knobby and club-shaped at the terminal ends, with the production of many short lateral and terminal branches (Figure 65-53). Chlamydospores are generally numerous. All strains of *T. schoenleinii* may be grown in a vitamin-free medium and grow equally well at room temperature or at 35° to 37° C.

Trichophyton violaceum produces an infection of the scalp and body and is seen primarily in persons living in the Mediterranean region, the Middle and Far East, and Africa. Hair invasion is of the endothrix type; the typical black dot type of tinea capitis is observed clinically. Direct microscopic examination of the calcofluor white or potassium hydroxide preparation of the nonfluorescing hairs shows dark, thick hairs filled with masses of arthroconidia arranged in chains, similar to those seen in *T. tonsurans* infections.

Colonies of *T. violaceum* are very slow-growing, beginning as cream-colored, glabrous, and cone-shaped, and later becoming heaped up, verrucous (warty), violet to purple, and waxy in consistency. Colonies may often be described as being port wine in color. The reverse side of the colony is purple or nonpigmented. Older cultures may develop a velvety area of mycelium and sometimes lose their pigmentation. Microscopically, microconidia and macroconidia are generally not present; only sterile, distorted hyphae and chlamydospores are found. In some instances, however, swollen hyphae containing cytoplasmic granules may be seen. The growth of *T. violaceum* is enhanced on media containing thiamine.

MICROSPORUM

The genus *Microsporum* is immediately recognized by the presence of large (8 to 15 × 35 to 150 μm) spindle-shaped macroconidia with (thick rough walls up to 4 μm) that contain 4 to 15 septa (Figure 65-54). The exception is *M. nanum,* which characteristically produces macroconidia having two cells. Microconidia, when present, are 3 to 7 μm and club-shaped and are borne on hyphae, either laterally or on short conidiophores. Cultures of *Microsporum* develop either

rapidly or slowly (5 to 14 days) and produce aerial hyphae that may be velvety, powdery, glabrous, or cottony, varying in color from whitish, buff, to a cinnamon brown, with varying shades on the reverse side of the colony.

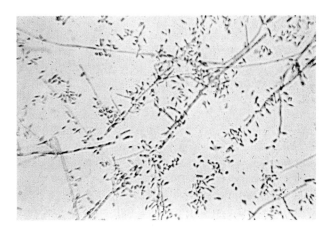

FIGURE 65-52 *T. verrucosum,* showing microconidia, which are rarely seen (500×).

FIGURE 65-53 *T. schoenleinii,* showing swollen hyphal tips with lateral and terminal branching (favic chandeliers). Microconidia and macroconidia are absent (500×).

FIGURE 65-54 Large, rough-walled macronidia of *M. canis* (430×).

Microsporum audouinii was, in past years, the most important cause of epidemic tinea capitis among schoolchildren in the United States. This organism is anthropophilic and is spread directly by means of infected hairs on hats, caps, upholstery, combs, or barber clippers. The majority of infections are chronic; some heal spontaneously, whereas others may persist for several years. Infected hair shafts fluoresce yellow-green using a Wood's lamp. Colonies of *M. audouinii* generally grow more slowly (10 to 21 days), than other members of the genus *Microsporum* and they produce a velvety aerial mycelium that is colorless to light gray to tan. The reverse side often appears salmon-pink to reddish-brown. Colonies of *M. audouinii* do not usually sporulate in culture. The addition of yeast extract may stimulate growth and the production of macroconidia in some instances. Most commonly, atypical vegetative forms, such as terminal chlamydospores and antler and racquet hyphae, are the only clues to the identification of this organism. It is common to identify *M. audouinii* by exclusion of all the other dermatophytes as a cause of infection.

Microsporum canis is primarily a pathogen of animals (zoophilic); it is the most common cause of ringworm infection in dogs and cats in the United States. Children and adults acquire the disease through contact with infected animals, particularly puppies and kittens, although human-to-human transfer has been reported. Hairs infected with *M. canis* fluoresce a bright yellow green using a Wood's lamp, which is a useful tool for screening pets as possible sources of human infection. Direct examination of a calcofluor white or potassium hydroxide preparation of infected hairs reveals small spores (2 to 3 μm) outside the hair. Culture must be performed to provide the specific identification.

Colonies of *M. canis* grow rapidly, are granular or fluffy with a feathery border, are white to buff, and characteristically have a lemon-yellow or yellow-orange fringe at the periphery. On aging, the colony becomes dense and cottony and a deeper brownish-yellow or orange and frequently shows an area of heavy growth in the center. The reverse side of the colony is bright yellow, becoming orange or reddish-brown with age. Rarely, strains are recovered that show no reverse side pigment. Microscopically, *M. canis* shows an abundance of large (15 to 20 × 60 to 125 μm), spindle-shaped, multisegmented (four to eight) macroconidia with curved ends (Figure 65-55). These are thick-walled with warty (**echinulate**) projections on their surfaces. Microconidia are usually few in number; however, large numbers may occasionally be seen.

Microsporum gypseum, a free-living organism of the soil (**geophilic**) that only rarely causes human or animal infection, may be occasionally seen in the clinical laboratory. Infected hairs generally do not fluoresce using a Wood's lamp. However, microscopic examination of the infected hairs shows them to be irregularly covered with clusters of spores (5 to 8 μm), some in chains. These arthroconidia of the ectothrix type are considerably larger than those of other *Microsporum* spp.

M. gypseum grows rapidly as a flat, irregularly fringed colony with a coarse powdery surface that appears to be a buff or cinnamon color. The underside of the colony is conspicuously orange to brownish. Microscopically, macroconidia are seen in large numbers and are characteristically large, ellipsoidal, have rounded ends, and are multisegmented (three to nine) with echinulated surfaces (Figure 65-56). Although they are spindle-shaped, these macroconidia are not as pointed at the distal ends as those of *M. canis.* The appearance of the colonial and microscopic

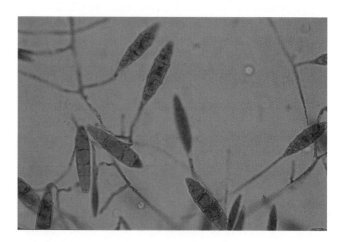

FIGURE 65-55 *M. canis,* showing several spindle-shaped, thick-walled, multicelled macroconidia (500×).

FIGURE 65-56 *M. gypseum,* showing ellipsoidal, multicelled macroconidia (750×).

morphologic features is sufficient to make the distinction between *M. gypseum* and *M. canis*.

EPIDERMOPHYTON

Epidermophyton floccosum, the only member of the genus *Epidermophyton*, is a common cause of tinea cruris and tinea pedis. In direct examination of skin scrapings using the calcofluor white or potassium hydroxide preparation, the fungus is seen as fine branching hyphae. *E. floccosum* grows slowly, and growth appears as an olive-green to khaki color, with the periphery surrounded by a dull, orange-brown color. After several weeks, colonies develop a cottony white aerial mycelium that completely overgrows the colony and is sterile, and remains so, even after subculture.

Microscopically, numerous smooth, thin-walled, club-shaped, multiseptate (2 to 4 μm) macroconidia are seen (Figure 65-57). They are rounded at the tip and are borne singly on a conidiophore or in groups of two or three. Microconidia are absent, spiral hyphae are rare, and chlamydospores are usually numerous.

This organism is susceptible to cold temperatures and for this reason it is recommended that specimens submitted for dermatophyte culture not be refrigerated before culture and that cultures not be stored at 4° C.

HYALINE, SEPTATE, MONOMORPHIC MOLDS (OPPORTUNISTIC MYCOSES)

The opportunistic mycoses are a group of fungal infections that occur almost exclusively in immuno-compromised patients. The type of patient who acquires an opportunistic fungal infection is one who is compromised by some underlying disease process such as HIV infection, lymphoma, leukemia, diabetes mellitus, or another defect of the immune system. Many patients are often placed on treatment with corticosteroids, cytotoxic drugs, or other immunosuppressive agents. Many fungi previously thought to be nonpathogenic are now recognized as etiologic agents of opportunistic fungal infections. Because most of the organisms known to cause infection in this group of patients are commonly encountered in the clinical laboratory as **saprobes** (saprophytic fungi), it is impossible for the laboratory to determine the significance of these isolates recovered from clinical specimens. It is necessary for the laboratory to identify and report completely the presence of all fungi recovered, since each is a potential pathogen. Many of the organisms associated with opportunistic infections are frequently acquired during construction, demolition, or remodeling of buildings, or are hospital-acquired. Other information regarding the specific clinical aspects of the opportunistic fungal infections will be discussed with each individual organism.

DIRECT DETECTION

The direct examination of clinical specimens, including tissues, reveals the presence of septate hyaline hyphae, which exhibit dichotomous branching (Figure 65-14). Although often considered to represent an *Aspergillus* species, these cannot be reliably distinguished from hyphae of *Fusarium* spp., *Pseudallescheria boydii*, or other hyaline molds.

ASPERGILLUS

Several species of aspergilli are among the most frequently encountered organisms in the clinical laboratory; some are pathogenic, whereas others are infrequently associated with infection or do not cause infection at all. The aspergilli are widespread in the environment, where they colonize grain, leaves, soil, and living plants. Conidia of the aspergilli are easily dispersed into the environment, and humans become infected by inhaling them. These organisms are capable not only of causing disseminated infection, as is seen in immunocompromised patients, but also of causing various other infections, including invasive lung infection, pulmonary fungus ball, allergic bronchopulmonary aspergillosis, external otomycosis, mycotic keratitis, **onychomycosis** (infection of nail and surrounding tissue), sinusitis, endocarditis, and central nervous system infection. Most often, immunocompromised patients acquire a primary pulmonary infection that rapidly disseminates and causes infection in virtually every organ.

It is very difficult to assess the significance of members of the genus *Aspergillus* in a clinical specimen. They are found frequently in cultures of respiratory secretions, skin scrapings, and other specimens. Table 65-13 presents the Mayo Clinic experience with the recovery of species of *Aspergillus* from clinical

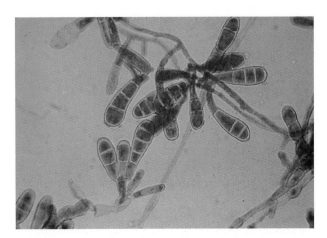

FIGURE 65-57 *E. floccosum*, showing numerous smooth, multiseptate, thin-walled macroconidia that appear club-shaped (1000×).

specimens. It has been reported that *Aspergillus* is significant in only 10% of cases[104]; however, this depends upon the hospital setting and type of patient population seen. Table 65-13 also illustrates the diversity of aspergilli seen in clinical specimens.

Because aspergilli are recovered frequently, it is imperative that the organism be demonstrated in the direct microscopic examination of fresh clinical specimens and that it be recovered repeatedly from patients having a compatible clinical picture to ensure that the organism is clinically significant. Most species of *Aspergillus* are susceptible to cycloheximide; therefore, specimens submitted for its recovery or subcultures should not be inoculated onto media containing this compound, that is, mycobiotic agar.

Aspergillus fumigatus is the most common species recovered from immunocompromised patients; moreover, it is the species most often seen in the clinical laboratory. In addition, *Aspergillus flavus* is sometimes recovered from immunocompromised patients and represents a frequent isolate in the clinical microbiology laboratory. The recovery of *A. fumigatus* or *A. flavus* from surveillance (nasal) cultures is strongly correlated with subsequent invasive aspergillosis.[113] The absence of a positive nasal culture does not preclude infection, however. *Aspergillus niger* is seen commonly in the clinical laboratory, but its association with clinical disease is somewhat limited. *Aspergillus terreus* is a significant cause of infection in immunocompromised patients, but its frequency of recovery is much lower than the previously mentioned species.

Specimens submitted for direct microscopic examination may contain septate hyphae that usually show evidence of **dichotomous branching** (45° angle, Figure 65-14). In addition, some hyphae may show the presence of rounded, thick-walled cells.

A. fumigatus is a rapidly growing mold (2 to 6 days) that produces a fluffy to granular, white to blue-green colony. Mature sporulating colonies most often exhibit the blue-green powdery appearance. Microscopically, *A. fumigatus* is characterized by the presence of septate hyphae and short or long conidiophores having a characteristic "foot cell" at their base. The tip of the conidiophore expands into a large, dome-shaped vesicle that has bottle-shaped phialides covering the upper half or two thirds of its surface. Long chains of small (2 to 3 μm in diameter), spherical, rough-walled, green conidia form a columnar mass on the vesicle (Figure 65-58). Cultures of *A. fumigatus* are thermotolerant and are able to withstand temperatures up to 45° C.

A. flavus, a somewhat more rapidly growing species (1 to 5 days), produces a yellow-green colony. Microscopically, vesicles are globose and phialides are produced directly from the vesicle surface (**uniserate**) or from a primary row of branches (metulae; **biserate**). The phialides give rise to short chains of yellow-orange elliptic or spherical conidia that become roughened on the surface with age (Figure 65-59).

A. niger, sometimes referred to as being dematiaceous, produces mature colonies within 2 to 6 days. Growth begins initially as a yellow colony that soon develops a black, dotted surface as conidia are produced. With age, the colony becomes jet black and powdery while the reverse remains a buff or cream color; this occurs on any culture medium.

Microscopically, *A. niger* exhibits septate hyphae, long conidiophores that support spherical vesicles that give rise to large metulae and smaller phialides from which long chains of brown to black, rough-walled conidia are produced (Figure 65-60). The entire surface of the vesicle is involved in sporulation.

Aspergillus terreus is less commonly seen in the clinical laboratory; it produces colonies that are tan and resemble cinnamon. Vesicles are hemispheric, as seen microscopically, and phialides cover the entire surface and are produced from a primary row of branches (metulae; biserate). Phialides produce globose to elliptic conidia arranged in chains. This species produces larger cells, **aleuriospores,** which are found on submerged hyphae (Figure 65-61). Refer to the chapter by Kennedy et al.[56] for further information regarding other species.

FUSARIUM

Infection caused by species of *Fusarium* is becoming more common, particularly in immunocompromised patients.[34] The organisms are common environmental flora and have long been known to cause mycotic keratitis after traumatic implantation into the cornea. However, cases of disseminated disease accompanied by fungemia are becoming more common. Necrotic skin lesions are common. Other types of infection caused by *Fusarium* include sinusitis, wound (burn) infection, allergic fungal sinusitis, and endophthalmitis. *Fusarium* spp. are commonly recovered from respiratory tract secretions, skin, and other specimens of patients having no evidence of infection. Interpretation of culture results rests with the clinician. Colonies of *Fusarium* grow rapidly, within 2 to 5 days, and are fluffy to cottony and may appear pink, purple, yellow, green, or other shades, depending on the species. Microscopically, the hyphae are small and septate and give rise to phialides that produce either single-celled microconidia, usually borne in gelatinous heads similar to those seen in *Acremonium* (Figure 65-39), or large macroconidia (single to multicelled) that are sickle- or boat-shaped and contain numerous septations (Figure 65-62). It is common to find numerous chlamydospores produced by some cultures

TABLE 65-13   SPECIES OF *ASPERGILLUS* RECOVERED FROM CLINICAL SPECIMENS DURING A 10-YEAR PERIOD AT MAYO CLINIC

| ORGANISM | CLINICAL SPECIMEN SOURCE | | | | | |
	RESPIRATORY SECRETIONS	GASTROINTESTINAL	GENITOURINARY	SKIN, SUBCUTANEOUS TISSUE	BLOOD, BONE, CNS, ETC.
A. clavatus	97/93*	1/1	—	1/1	—
A. flavus	1298/740	10/10	11/11	177/131	2/2
A. fumigatus	3247/2656	11/9	14/14	175/137	8/8
A. glaucus	503/307	1/1	—	8/8	1/1
A. nidulans	52/48	—	—	5/3	—
A. niger	1484/1376	18/18	17/17	151/124	11/11
A. terreus	164/146	—	—	23/21	3/3
A. versicolor	1237/1202	6/6	24/22	226/224	16/16
Other species of Aspergillus	3463/3418	18/14	32/32	319/314	16/16

*Numerator, number of cultures; denominator, number of patients.

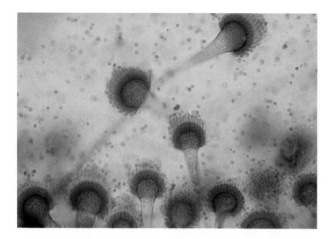

FIGURE 65-58 *A. fumigatus* conidiophore and conidia (400×).

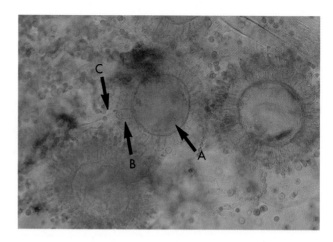

FIGURE 65-59 *A. flavus,* showing spherical vesicles (**A**) that give rise to metulae (**B**) and phialides (**C**) that produce chains of conidia (750×).

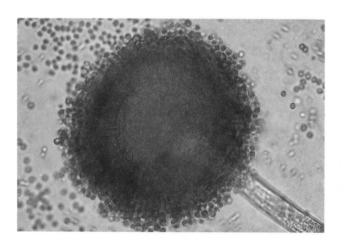

FIGURE 65-60 *A. niger,* showing larger spherical vesicle that gives rise to metulae, phialides, and conidia (750×).

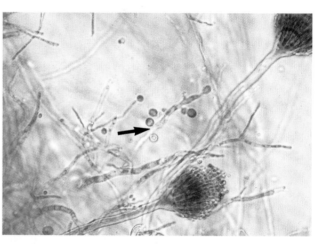

FIGURE 65-61 *A. terreus,* showing typical head of aspergillus and aleuriospores (*arrow*) found on submerged hyphae of this species (500×).

of *Fusarium.* The most common medium used to induce sporulation is cornmeal agar. Keys for identification of species by *Fusarium* are based on growth on potato dextrose agar.

GEOTRICHUM

Geotrichum is an uncommon cause of infection but has been implicated in causing wound infections and oral thrush.[45] *Geotrichum* often initially appears as white to cream, yeastlike colonies; some isolates may appear as white, powdery molds. Hyphae are septate and produce numerous rectangular to cylindric to barrel-shaped arthroconidia (Figure 65-63). Arthroconidia do not alternate, unlike *Coccidioides immitis* (Figure 65-64). Blastoconidia are not produced.

ACREMONIUM

Acremonium has been associated with disseminated infection, fungemia, subcutaneous lesions, and esoph-

agitis.[93] Colonies of *Acremonium* are rapidly growing and often appear yeastlike when initial growth is observed. Mature colonies become white to gray to rose or reddish-orange. Microscopically, small septate hyphae that produce single, unbranched, tubelike phialides are observed. Phialides give rise to clusters of elliptic, single-celled conidia contained in a gelatinous cluster at the tip of the phialide (Figure 65-39).

PENICILLIUM

Penicillium is one of the most common organisms recovered by the clinical laboratory. When clinically significant, some of the clinical manifestations include allergic bronchopulmonary penicillinosis, endocarditis, and cutaneous ulcers of the extremities. Colonies of *Penicillium* may be shades of green, blue-green, pink, white, or other colors. The surface of the colonies may be velvety to powdery as a result of the presence of conidia. Microscopically, hyphae are hya-

FIGURE 65-62 *Fusarium* spp., showing characteristic multicelled, sickle-shaped macroconidia (500×).

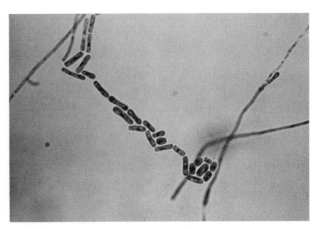

FIGURE 65-63 *G. candidum,* showing numerous arthroconidia. Note that arthroconidia do not alternate with a clear cell as in the case of *C. immitis* (430×).

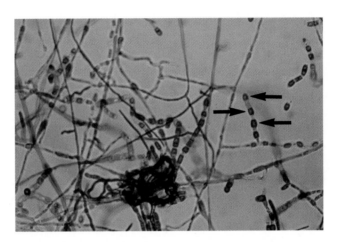

FIGURE 65-64 *C. immitis,* mycelial form, showing numerous thick-walled, rectangular or barrel-shaped (*arrows*) alternate arthroconidia (500×).

line and septate and produce brushlike conidiophores (penicillus). Conidiophores produce metulae from which phialides producing chains of conidia arise (Figure 65-65). *Penicillium marneffei,* a particularly virulent species in that genus, will be discussed in the section on hyaline, septate, dimorphic molds.

PAECILOMYCES

Of the *Paecilomyces* spp., *P. lilacinus* appears to be the most pathogenic and has been associated with endopthalmitis, cutaneous infections, and arthritis. Other species of *Paecilomyces* have been implicated in endocarditis and fungemia. Colonies of *Paecilomyces* spp. are often velvety, tan to olive brown, and somewhat powdery. Colonies of *P. lilacinus* exhibit shades of lavender to pink. Microscopically, *Paecilomyces* resembles a small *Penicillium;* however, phialides are long, delicate, and tapering (Figure 65-66). The penicillus produces numerous chains of small, oval conidia that

are easily dislodged. Single phialides producing chains of conidia may also be present.

SCOPULARIOPSIS

Scopulariopsis has been associated with onychomycosis, pulmonary infection, and fungus ball. Colonies of *Scopulariopsis* initially appear white but later become light brown and powdery. Colonies often resemble those of *M. gypseum.* Microscopically, *Scopulariopsis* resembles a large *Penicillium* at first glance. The hyaline and septate conidiophores (annellophores) are branched and produce a penicillus. Conidia are produced from annellophores and may also be produced singly directly from the hyphae. Conidia are large, have a flat base, and are rough walled (Figure 65-67). *Scopulariopsis brumptii* is a dematiaceous species and is occasionally recovered in the clinical laboratory. This species was reported to cause brain abscess in a patient who received a liver transplant.[82]

FIGURE 65-65 *Penicillium* spp., showing typical brushlike conidiophores (penicillus) (430×).

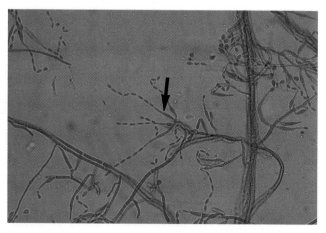

FIGURE 65-66 *Paecilomyces* spp. showing long, tapering, delicate phialides (*arrow*).

Various saprobic fungi not discussed here may be encountered in the clinical laboratory but are seen less commonly. Other references are recommended for further information regarding the identification of these organisms.[18,56,94,98]

HYALINE, SEPTATE, DIMORPHIC MOLDS

The dimorphic fungi produce systemic fungal infections that may involve any of the internal organs of the body, as well as lymph nodes, bone, subcutaneous tissue, and skin. Asymptomatic infection is common, may go unrecognized clinically, and may be detected only by serologic testing; in some cases roentgenographic examination may reveal healed lesions. Symptomatic infections may present signs of only a mild or more severe, but self-limited disease, with positive supportive evidence from cultural or immunologic findings. Disseminated or progressive infection may reveal severe symptoms, with spread of the initial disease to several organs. Some cases of disseminated infection may exhibit little in the way of signs or symptoms of disease for long periods, only to exacerbate later. Immunocompromised patients, particularly those infected with HIV, or those receiving long-term corticosteroid therapy, most often present with disseminated infection.[63]

DIRECT DETECTION

The microscopic morphologic features of the parasitic forms of the dimorphic fungi vary with the genus. *Blastomyces dermatitidis* produces broad-based budding yeast cells 8 to 15 µm in size (Figure 65-19). *Coccidioides immitis* produces spherules 20 to 200 µm in diameter, which may contain nonbudding endospores 2 to 5 µm in diameter (Figure 65-25). *Histoplasma capsulatum* produces small, budding yeast cells, 2 to 5 µm in diameter, which are generally found within macrophages (Figure 65-16). *Paracoccidioides brasiliensis* produces multiply budding yeast cells 8 to 40 µm in size (Figure 65-68). *Sporothrix schenkii* produces small round to oval to cigar-shaped yeast cells, 2 to 5 µm in size (Figure 65-22). *Penicillium marneffei* produces small, yeastlike cells, that have internal crosswalls; no budding cells are produced (Figure 65-4).

CULTURE AND IDENTIFICATION

Traditionally, the systemic mycoses have included only blastomycosis, coccidioidomycosis, histoplasmosis, and paracoccidioidomycosis. The fungi responsible for these infections, although unrelated genetically and morphologically dissimilar, both show thermal dimorphism. The dimorphic fungi involved exist in nature as a mold form, which is distinct from the parasitic, or invasive form, sometimes called the *tissue form*. Distinct morphologic differences may be observed with the dimorphic fungi both in vivo and in vitro as discussed later in this chapter. Temperature (35° to 37° C), certain nutritional factors, and stimulation of growth in tissue independent of temperature are among the factors necessary to initiate the transformation of the mold form to the parasitic form.[38,109] In general, 25° to 30° C is the optimal temperature for the recovery and identification of dimorphic fungi from clinical specimens.

GENERAL FEATURES OF DIMORPHIC FUNGI

Commonly, dimorphic fungi are regarded as slow-growing organisms that require 7 to 21 days for visible growth to appear at 25° to 30° C. However, exceptions to this rule occur with some frequency. Cultures of *B. dermatitidis* and *H. capsulatum* are recovered in as short a time as 2 to 5 days when large numbers of organisms are present in the clinical specimen. In contrast, when small numbers of colonies of *B. dermati-*

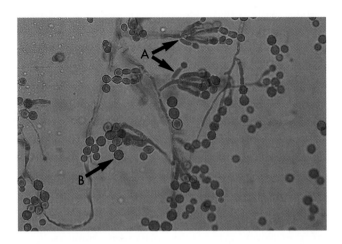

FIGURE 65-67 *Scopulariopsis* spp., showing a large penicillus (**A**) with echinulate conidia (**B**) (430×).

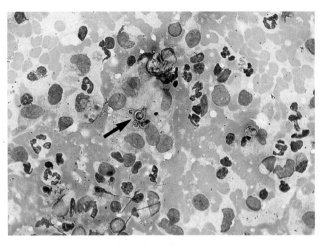

FIGURE 65-68 *P. brasiliensis* in bone marrow showing yeast cell having multiple buds (*arrow*).

tidis and *H. capsulatum* are present, sometimes 21 to 30 days of incubation are required before they are detected. *C. immitis* is consistently recovered within 3 to 5 days of incubation, but when large numbers of organisms are present, colonies may be detected within 48 hours. Single colonies may require 14 to 21 days before visible growth can be detected. Cultures of *P. brasiliensis* are commonly recovered within 5 to 25 days, with a usual incubation period of 10 to 15 days. As can be seen, the growth rate, if slow, might lead one to suspect the presence of a dimorphic fungus; however, considerable variation in the time for recovery exists. The exceptions to this are *C. immitis* and *P. marneffei*, which may be recovered within 3 to 5 days.

Textbooks present descriptions for the dimorphic fungi that readers assume are typical for each organism. As is true in other areas of microbiology, variation in the colonial morphologic features also occurs depending on the strain and type of medium used. One must be aware of this variation and not rely heavily on colonial morphologic features for the identification of members of this group of fungi.

The pigmentation of colonies is sometimes helpful but also varies widely; colonies of *B. dermatitidis* and *H. capsulatum* are described as being fluffy white with a change to tan or buff with age. Some isolates initially appear darkly pigmented with colors ranging from gray or dark brown to red.[51] On media containing blood enrichment, these organisms appear heaped, wrinkled, glabrous, and neutral in color, and yeastlike in appearance; often tufts of aerial hyphae project from the top of the colony. Some colonies may appear pink to red because of the adsorption of hemoglobin from the blood in the medium. *C. immitis* is described as being fluffy white with scattered areas of hyphae that are adherent to the agar surface so as to give an overall cobweb appearance to the colony. However, numerous morphologic forms,

including textures ranging from woolly to powdery and pigmentation ranging from pink-lavender or yellow to brown or buff, have been reported.

At 25° C, *Penicillium marneffei* grows rapidly and produces blue-green to yellowish colonies on Sabouraud agar. A soluble, red to maroon pigment that diffuses into the agar and is often best observed by viewing the reverse of the colony is highly suggestive of *P. marneffei*. Although the growth rate and colonial morphologic features may help one to recognize the presence of a dimorphic fungus, they should be used in combination with the microscopic morphologic features to make a tentative identification.

The definitive identification of a dimorphic fungus has traditionally been made by observing both the mold and parasitic forms of the organism. Cultures of the mold form are easily recovered at 25° to 30° C; however, the tissue form is usually not recovered from the clinical specimen. Previously the definitive identification was made by the in vitro conversion of a mold form to the corresponding yeast or spherule form by animal inoculation or by in vitro conversion on a blood-enriched medium incubated at 35° to 37° C. The conversion of dimorphic molds to the yeast form (except for *C. immitis*) can be accomplished with some difficulty, as outlined in Procedure 65-5. The exoantigen test (Procedure 65-6), is used by some laboratories for the identification of the dimorphic pathogens. However, it requires an extended incubation before cultures may be identified.

Perhaps the most significant advance in clinical mycology is the development of specific nucleic acid probes for the identification of some of the dimorphic fungi (Procedure 65-7). Probes are commercially available from Gen-Probe Inc., San Diego, Calif., and are DNA probes that are complimentary to species-specific ribosomal RNA. Fungal cells are disrupted by a lysing agent and sonication, heat-killed, and the nucleic acid is exposed to a species-specific DNA probe that has been labeled with a chemiluminescent tag (acridinium ester). The labeled DNA probe combines with an organism's ribosomal RNA to form a stable DNA/RNA hybrid. All unbound DNA probe is quenched and DNA/RNA hybrids are measured in a luminometer. Total time for testing is less than 1 hour, and young colonies may be tested.[103]

Nucleic acid probe identification is sensitive, specific, and rapid. Contaminated colonies may be tested; however, results from colonies recovered on a blood-enriched medium must be interpreted with caution because hemin may cause false-positive chemiluminescence. The major disadvantage of nucleic acid probe identification relates to the cost per test. It is recommended that nucleic acid probes be used whenever possible to confirm the identification of an organism suspected of being *H. capsulatum*, *B. dermatitidis*, or *C. immitis*.

The inclusion of only the dimorphic fungi in the group of systemic mycoses is somewhat misleading, since other fungi, including *C. neoformans*, *Candida* and *Aspergillus*, spp. and others also cause disseminated infection. They are not discussed in this portion of the chapter, because they are not dimorphic, as previously defined, and because they usually cause infection only in immunocompromised patients. They are discussed in a subsequent section on the opportunistic mycoses.

BLASTOMYCES

Blastomyces dermatitidis commonly produces a chronic suppurative and granulomatous infection. The disease (**blastomycosis**) is most commonly found on the continent of North America and extends southward from Canada to the Mississippi, Ohio, and Missouri River valleys, Mexico, and Central America. Some isolated cases have also been reported from Africa. Most cases occur in the Mississippi, Ohio, and Missouri River valley regions. Blastomycosis begins as a respiratory infection and is probably acquired by inhalation of the conidia or hyphal fragments of the organism. The exact ecologic niche for this organism has not been determined; however, patients with a history of exposure to soil or wood have the highest incidence of infection. The infection may spread and involve the lungs, long bones, soft tissue, and skin. Several outbreaks have been reported and have been related to a common exposure. Blastomycosis is more common in men and seems to be associated with outdoor occupations.

Direct detection

The diagnosis of blastomycosis may easily be made when a clinical specimen is observed by direct microscopy. *B. dermatitidis* appears as large, spherical, thick-walled yeast cells 8 to 15 μm in diameter, usually with a single bud that is connected to the parent cell by a broad base (Figures 65-19 and 65-20).

Culture and identification

Blastomyces dermatitidis commonly requires 5 days to 4 weeks or longer for growth to be detected but may be detected in as short a time as 2 to 3 days. On enriched culture media, the mold form develops initially as a glabrous- (or waxy) appearing colony that may become off-white to white in color. With age, the aerial hyphae often turn gray to brown; however, on media enriched with blood, colonies have a more waxy, yeastlike appearance. Tufts of hyphae often project upward from the colonies and are referred to as the *prickly state* of the organism. Some isolates appear fluffy on primary recovery and remain so throughout the incubation period.

Microscopically, hyphae of the mold form are septate, delicate, and measure 1 to 2 μm in diameter. Commonly, ropelike strands of hyphae are seen; however, these are found with most of the dimorphic fungi. The characteristic microscopic morphologic features are single, pyriform conidia produced on long to short conidiophores that resemble lollipops (Figure 65-69). The production of conidia in some isolates is minimal or absent, particularly on a medium containing blood enrichment.

When incubated at 37° C, colonies of the yeast form develop within 7 days and appear waxy and wrinkled, and cream to tan in color. Microscopically, large, thick-walled yeast cells (8 to 13 μm) with buds attached by a broad base are seen (Figure 65-70). Some strains may produce yeast cells as small as 2 to 5 μm and may resemble *C. neoformans*. During the conversion process, swollen hyphal forms and immature cells with rudimentary buds may be present. Because the conversion of *B. dermatitidis* is easily accomplished, this is feasible in the clinical laboratory; however, this is the only instance where mold-to-yeast conversion should be attempted. Some isolates may produce small yeast cells 2 to 5 μm in size, and these strains are being seen more often.

B. dermatitidis may also be identified by the presence of a specific (i.e., A) band in the exoantigen test[101] or by nucleic acid probe testing. In some instances, *H. capsulatum*, *P. boydii*, or *T. rubrum* might be confused microscopically with *B. dermatitidis*. The relatively slow growth rate of *B. dermatitidis* and careful examination of the microscopic morphologic features will usually differentiate it from these fungi.

COCCIDIOIDES IMMITIS

Coccidioides immitis is found primarily in the desert southwestern portion of the United States, as well as semiarid regions of Mexico and Central and South

America. Although the geographic distribution of the organism is well defined, infection may be seen in any part of the world because of the ease of travel. The infection (**coccidioidomycosis**) is acquired by inhalation of the infective arthroconidia of *C. immitis*. Approximately 60% of the cases are asymptomatic and self-limited respiratory tract infections. The infection, however, may become disseminated, with extension to visceral organs, meninges, bone, skin, lymph nodes, and subcutaneous tissue. Fewer than 1% of individuals who acquire coccidioidomycosis ever become seriously ill; dissemination does, however, occur most frequently in persons with dark skin. Pregnancy also appears to predispose women to disseminated infection. This infection has been known to occur in epidemic proportions and persons who visit endemic areas and return to a distant location will present to their local physician. All laboratories should be prepared to deal with the diagnosis of coc-

cidioidomycosis. In 1992 an epidemic occurred in northern California in which more than 4000 cases were seen in Kern County near Bakersfield.

Direct detection

In direct microscopic examinations of sputum or other body fluids, *C. immitis* appears as a nonbudding, thick-walled spherule, 20 to 200 μm in diameter, containing either granular material or numerous small (2 to 5 μm in diameter), nonbudding endospores (Figure 65-71). The endospores are freed by rupture of the spherule wall; therefore, empty and collapsed "ghost" spherules may be present. Small, immature spherules measuring 10 to 20 μm may be confused with *B. dermatitidis* when two are lying adjacent. In instances in which the identification of *C. immitis* is questionable, a wet preparation of the clinical specimen may be made using sterile saline, and the edges of the coverglass may be sealed with petrolatum

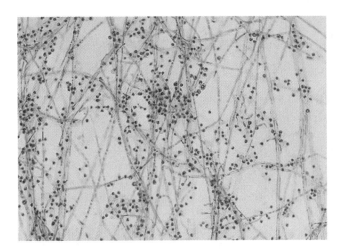

FIGURE 65-69 *B. dermatitidis,* mycelial form, showing oval conidia borne laterally on branching hyphae (1000×).

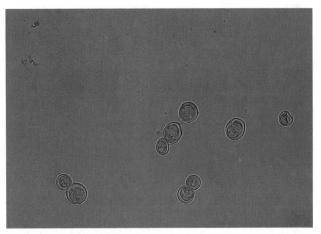

FIGURE 65-70 *B. dermatitidis,* yeast form, showing thick-walled, oval to round, single-budding, yeastlike cells (500×).

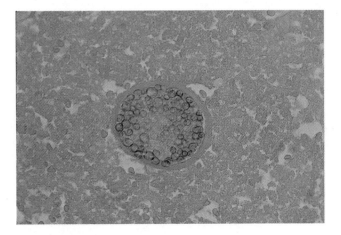

FIGURE 65-71 *C. immitis,* tissue form, showing spherule containing numerous spherical endospores (1000×).

and incubated overnight. When spherules are present, multiple hyphal strands will be produced from the endospores.

Culture and identification

Cultures of *C. immitis* represent a biohazard to laboratory workers, and strict safety precautions must be followed when examining cultures. Mature colonies may appear within 3 to 5 days of incubation and may be present on most media, including those used in bacteriology. *Laboratory workers are cautioned not to open cultures of fluffy white molds unless they are placed inside a BSC.* Colonies of *C. immitis* often appear as delicate, cobweblike growth after 3 to 21 days of incubation. Some portions of the colony will exhibit aerial hyphae, whereas others will have the hyphae adherent to the agar surface. Most isolates appear fluffy white; however, colonies of varying colors ranging from pink to yellow to purple and black have been reported.[51] On blood agar, some colonies exhibit a greenish discoloration, whereas others appear yeastlike, smooth, wrinkled, and tan.

Microscopically, some cultures show small septate hyphae that often exhibit right-angle branches and racquet forms. With age, the hyphae form arthroconidia that are characteristically rectangular or barrel-shaped in appearance. The arthroconidia are larger than the hyphae from which they are produced and stain darkly with lactophenol cotton or aniline blue. The arthroconidia are separated from one another by clear or lighter staining nonviable cells and are referred to as **alternate arthroconidia** (Figure 65-64). Arthroconidia have been reported to range from 1.5 to 7.5 μm in width and 1.5 to 30.0 μm in length, whereas most are 3.0 to 4.5 μm in width and 3.0 μm in length. Variation has been reported in the shape of arthroconidia and ranges from rounded to square or rectangular to curved; however, most are barrel-shaped. Even if alternate arthroconidia are observed microscopically, the definitive identification should be made using nucleic acid probe testing.

If a culture is suspected of being C. immitis, it should be sealed with tape to prevent chances of laboratory-acquired infection. Because *C. immitis* is the most infectious of all the fungi, extreme caution should be used when handling cultures of this organism.

Safety precautions include:

1. If culture dishes are used, they should be handled only in a biological safety cabinet. Cultures should be sealed with tape if the specimen is suspected of containing *C. immitis*.
2. The use of cotton plug test tubes is discouraged,

and screw-capped tubes should be used if culture tubes are preferred. All handling of cultures of *C. immitis* in screw-capped tubes should be performed inside a biological safety cabinet.
3. All microscopic preparations for examination should be prepared in a biological safety cabinet.
4. Cultures should be autoclaved as soon as the final identification of *C. immitis* is made.

Other, usually nonvirulent fungi that resemble *C. immitis* microscopically may be found in the environment. Some species produce alternate arthroconidia that tend to be more rectangular, and it is necessary to consider them when making an identification. *Geotrichum candidum* and species of *Trichosporon* produce hyphae that disassociate into arthroconidia (Figures 65-63 and 65-72). The colonial morphologic features of older cultures may resemble *C. immitis*. However, in contrast to *C. immitis*, the arthroconidia do not appear to be alternate and are even-staining. It is also important to remember that if confusion in identification does arise, or when occasional strains of *C. immitis* that fail to sporulate are encountered, identification by exoantigen or nucleic acid probe testing may be performed.

HISTOPLASMA

Histoplasma capsulatum produces a chronic, granulomatous infection (**histoplasmosis**) that is primary and begins in the lung and eventually invades the reticuloendothelial system. Approximately 95% of cases are asymptomatic and self-limited. Chronic pulmonary infections occur, and dissemination to the lymphatic tissue, liver, spleen, kidneys, meninges, and heart is becoming more common, especially in

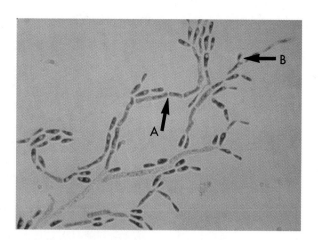

FIGURE 65-72 *Trichosporon* spp., showing arthroconidia (**A**) and an occasional blastoconidium (**B**).

immunocompromised patients (e.g., people with AIDS). Ulcerative lesions of the upper respiratory tract may occur. Histoplasmosis is prevalent in the Ohio, Mississippi, and Missouri River valleys, where conditions are optimal for growth of the organism in soil enriched with bird manure or bat guano.

Outbreaks of histoplasmosis have been associated with activities that disperse aerosolized microconidia or hyphal fragments. Infection is acquired via inhalation of these infective structures from the environment. The severity of the disease is generally related directly to the inoculum size. Numerous cases of histoplasmosis have been reported in persons cleaning an old chicken house or barn that has been undisturbed for long periods or from working in or cleaning those areas that have served as roosting places for starlings and similar birds. Spelunkers are commonly exposed to the organism when it is aerosolized from bat guano in caves.

It is estimated that 500,000 persons are infected annually by *H. capsulatum*. Although it is perhaps one of the most common fungal infections seen in the midwestern and southern parts of the United States, history of exposure is often impossible to document.

Direct detection
The direct microscopic examination of respiratory tract specimens and other similar specimens often fails to reveal the presence of *H. capsulatum*. The organism, however, may be detected by an astute laboratorian when examining Wright- or Giemsa-stained specimens of bone marrow and rarely in peripheral blood. *H. capsulatum* is found intracellularly within mononuclear cells as small, round to oval yeast cells 2 to 5 μm in diameter (Figure 65-73).

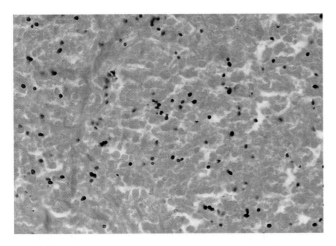

FIGURE 65-73 *H. capsulatum,* yeast form, showing intracellular, oval yeast cells, deeply stained (2000×).

Culture and identification
H. capsulatum is easily cultured from clinical specimens; however, it may be overgrown by bacteria and/or rapidly growing molds. A procedure useful for the recovery of *H. capsulatum, B. dermatitidis,* and *C. immitis* from contaminated specimens (e.g., sputa) utilizes a yeast extract phosphate medium[7] and a drop of concentrated ammonium hydroxide (NH₄OH) placed on one side of the inoculated plate of medium.[95] In the past, it has been recommended that specimens not be kept at room temperature before culture, since *H. capsulatum* will not survive. The organism will survive transit in the mail for as long as 16 days.[42] It is, however, recommended that specimens be cultured as soon as possible to ensure the optimal recovery of *H. capsulatum* and other dimorphic fungi. The method just described works well with specimens shipped via mail.

H. capsulatum is usually considered to be a slow-growing mold at 25° to 30° C and commonly requires 2 to 4 weeks or more for colonies to appear. The organism commonly can be recovered in 5 days or less if large numbers of cells are present in the clinical specimen. Isolates of *H. capsulatum* have been reported to be recovered from blood cultures with the Isolator within a mean time of 8 days.[6] Textbooks describe the colonial morphology of *H. capsulatum* as being a white, fluffy mold that turns brown to buff with age. Some isolates ranging from gray to red have also been reported. The organism commonly produces wrinkled, moist, heaped, yeastlike colonies that are soft and cream, tan, or pink. Tufts of hyphae often project upward from the colonies as described with *B. dermatitidis*. It is often impossible to differentiate *H. capsulatum* and *B. dermatitidis* from each other using colonial morphologic features.

Microscopically, the hyphae of *H. capsulatum* are small (1 to 2 μm in diameter) and are often intertwined to form ropelike strands. Commonly, large (8 to 14 μm in diameter) spherical or pyriform, smooth-walled macroconidia are seen in young cultures. With age, the macroconidia become roughened or tuberculate and provide enough evidence to make a tentative identification (Figure 65-74). The macroconidia are produced either on short or long lateral branches of the hyphae. Some isolates produce round to pyriform, smooth microconidia (2 to 4 μm in diameter) in addition to the characteristic tuberculate macroconidia. Some isolates of *H. capsulatum* fail to sporulate despite numerous attempts to induce sporulation.

Conversion of the mold to the yeast form is usually difficult and is not recommended. Microscopically, a mixture of swollen hyphae and small budding yeast cells 2 to 5 μm in size should be observed.

These are similar to the intracellular yeast cells seen in mononuclear cells in infected tissue. The yeast form of *H. capsulatum* cannot be recognized unless the corresponding mold form is present on another culture or unless the yeast is converted directly to the mold form by incubation at 25° to 30° C after yeast cells have been observed. The exoantigen test can be used for identification, but nucleic acid probe testing is recommended as the most definitive means of providing a rapid identification of this organism.

Sepedonium, an environmental organism found growing on mushrooms, is always mentioned as being confused with *H. capsulatum*, since it produces similar tuberculate macroconidia. This organism is almost never recovered from clinical specimens, does not have a yeast form, fails to produce characteristic bands in the exoantigen test with *H. capsulatum* antiserum, and does not react in nucleic acid probe tests.

PARACOCCIDIOIDES

Paracoccidioides brasiliensis produces a chronic granulomatous infection (**paracoccidioidomycosis**) that begins as a primary pulmonary infection. It is often asymptomatic and then disseminates to produce ulcerative lesions of the mucous membranes. Ulcerative lesions are commonly present in the nasal and oral mucosa, gingivae, and less commonly in the conjunctivae. Lesions occur most commonly on the face in association with oral mucous membrane infection. The lesions are characteristically ulcerative, with a **serpiginous** (snakelike) active border and a crusted surface. Lymph node involvement in the cervical area is common. Pulmonary infection is seen most often, and progressive chronic pulmonary infection is found in approximately 50% of cases. Dissemination to other anatomic sites, including the lymphatic system, spleen, intestines, liver, brain, meninges, and adrenal glands, occurs in some patients.

The infection is most commonly found in South America, with the highest prevalence in Brazil, Venezuela, and Colombia. It also has been seen in many other areas, including Mexico, Central America, and Africa. Occasional imported cases are seen in the United States and Europe.

The exact mechanism by which paracoccidioidomycosis is acquired is unclear; however, it is speculated that its origin is pulmonary and that it is acquired by inhalation of the organism from the environment. Because mucosal lesions are an integral part of the disease process, it is also speculated that the infection may be acquired through trauma to the oropharynx caused by vegetation that is commonly chewed by residents of the endemic areas. The specific ecologic niche of the organism in nature is undetermined.

Direct detection

Specimens submitted for direct microscopic examinations are important for the diagnosis of paracoccidioidomycosis. Large, round or oval, multiply budding yeast cells (8 to 40 μm in diameter) are usually recognized in sputum, mucosal biopsy, and other exudates. Characteristic multiply budding yeast forms resemble a mariner's wheel (Figure 65-68). The yeast cells surrounding the periphery of the parent cell range from 8 to 15 μm in diameter. Some cells may be as small as 2 to 5 μm, but still exhibit multiple buds.

Culture and identification

Colonies of *P. brasiliensis* grow very slowly (21 to 28 days) and are heaped, wrinkled, moist, and yeastlike. With age, colonies may become covered with a short aerial mycelium and turn tan to brown. The surface of colonies is often heaped with crater formations.

Microscopically, the mold form is similar to that seen with *B. dermatitidis*. Small hyphae (1 to 2 μm in diameter) are seen, along with numerous chlamydospores. Small, delicate (3 to 4 μm) globose or pyriform conidia may be seen arising from the sides of the hyphae or on very short conidiophores (Figure 65-75). Most often, cultures reveal only fine septate hyphae and numerous chlamydospores.

After temperature-based conversion on a blood-enriched medium, the colonial morphology of the yeast form is characterized by smooth, soft-wrinkled, yeastlike colonies that are cream to tan. Microscopically, the colonies are composed of yeast cells 10 to 40 μm in diameter that are surrounded by narrow-necked yeast cells around the periphery, as previously described (Figure 65-68).

If in vitro conversion to the yeast form is unsuccessful, the exoantigen test[100] (Procedure 65-6) should

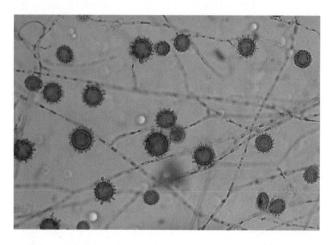

FIGURE 65-74 *H. capsulatum*, mycelial form, showing characteristic tuberculate macroconidia (1000×).

be used to make the definitive identification of *P. brasiliensis*. Nucleic acid probe testing is not available for this organism.

SPOROTHRIX

Sporothrix schenkii, also a dimorphic fungus, is most often associated with chronic subcutaneous infection. The organism is found worldwide, and its natural habitat is living or dead vegetation. Humans acquire the infection (**sporotrichosis**) through trauma (e.g., thorn or splinter puncture wounds), usually to the hand, arm, or leg. The primary lesion begins as a small, nonhealing ulcer, commonly of the index finger or the back of the hand. With time, the infection is characterized by the development of nodular lesions of the skin or subcutaneous tissues at the point of contact and later involves the lymphatic channels and lymph nodes draining the region. The subcutaneous nodules break down and ulcerate to form an infection that becomes chronic. Only rarely is the disease disseminated; pulmonary infection may be seen. The infection is an occupational hazard for farmers, nursery workers, gardeners, florists, and miners. It is commonly known as *rose gardener's disease.*

Direct detection

Exudate aspirated from unopened subcutaneous nodules or from open draining lesions is often submitted for culture and direct microscopic examination. Direct examination of this material is usually of little diagnostic value because it is difficult to demonstrate the characteristic yeast forms, even with special stains; reasons for this are unclear. If identified, *S. schenkii* usually appears as small (2 to 5μm), round to oval to cigar-shaped yeast cells (Figure 65-22). If stained with a PAS stain, an amorphous pink material may be seen surrounding the yeast cells and resemble a star-shaped inclusion (asteroid body) (Figure 65-17).

Culture and identification

Colonies of *S. schenckii* grow rapidly (3 to 5 days) and are initially usually small, moist, and white to cream-colored. On further incubation, these become membranous, wrinkled, and coarsely matted, with the color becoming irregularly dark brown or black and the colony becoming leathery in consistency. Commonly the clinical microbiology laboratory mistakes a young culture of *S. schenckii* for that of a yeast until the microscopic features are observed. One of the commercially available yeast identification systems, API-20C AUX (bioMérieux, St. Louis, Mo.), has a characteristic biochemical profile for the identification of *S. schenckii.*

Microscopically, hyphae are delicate (1 to 2 μm thick), septate, exhibit branching, and bear one-celled conidia that are 2 to 5 μm in diameter. These are borne in clusters from the tips of single conidiophores. Each conidium is attached to the conidiophore by an individual, delicate, threadlike structure (**denticle**) that may require examination under oil immersion to be visible. As the culture ages, single-celled, thick-walled, black-pigmented conidia may also be produced along the sides of the hyphae, simulating the arrangement of microconidia produced by *T. rubrum*. The common designations for these types of sporulation are the "flowerette" and "sleeve" arrangements, respectively (Figure 65-76).

Because of similar morphologic features, saprophytic species of the genus *Sporotrichum* may be confused with *S. schenckii,* and it is necessary to distinguish between them. During incubation of a culture at 37° C, the colony of *S. schenckii* transforms to a soft,

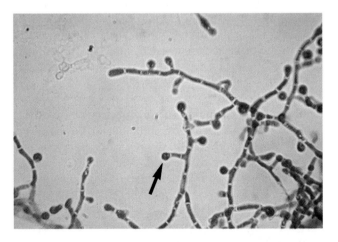

FIGURE 65-75 *Paracoccidioides brasiliensis,* mycelial form, showing septate hyphae and pyriform conidia singly borne (*arrow*) (430×).

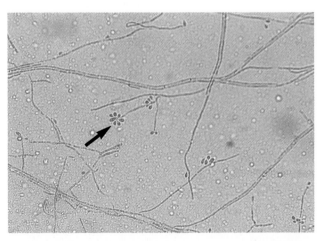

FIGURE 65-76 *S. schenckii,* mycelial form, showing pyriform to ovoid microconidia borne in a flowerette at the tip of the conidiophore (*arrow*) (750×).

cream-colored to white, yeastlike colony. Microscopically, singly or multiply budding, spherical, oval, or cigar-shaped yeast cells are observed without difficulty (Figure 65-77). Conversion from the mold form to the yeast form is easily accomplished and usually occurs within 1 to 5 days after transfer of the culture to a medium containing blood enrichment; most isolates of *S. schenckii* are converted to the yeast form within 12 to 48 hours at 37° C; *Sporotrichum* does not produce a yeast form.

PENICILLIUM MARNEFFEI

Penicillium marneffei is a pathogenic dimorphic fungus endemic to Southeast Asia, particularly in the Guangxi Zhuang Autonomous Region of the Peoples Republic of China.[64] It is an emerging pathogen that may infect either immunocompetent or immunosuppressed individuals. *P. marneffei* has been associated with the bamboo rat *(Rhizomys pruinosus)*. In one study, the internal organs from 18 of 19 bamboo rats were culture postive for *P. marneffei*.[20] This fungus has also been associated with the Vietnamese bamboo rat *(Rhizomys sinensis)*.

P. marneffei causes either a focal cutaneous or mucocutaneous infection, or may produce a progressive disseminated and often fatal infection. Granulomatous, suppurative, and necrotizing inflammatory responses have been demonstrated.[19] The mode of transmission and the primary source in the environment is unknown.

Direct detection

The direct examination of infected tissues and exudates that contain *P. marneffei* disclose yeastlike organisms that measure 2 to 6 μm. After careful examination, crosswalls or internal septae and the absence of budding is evident (see Figure 65-4).

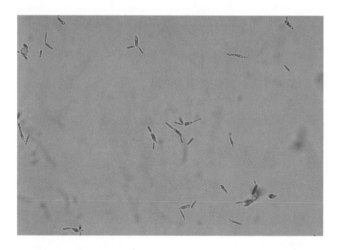

FIGURE 65-77 *S. schenckii*, yeast form, showing cigar-shaped and oval budding cells (500×).

Culture and identification

At 25° C, *Penicillium marneffei* grows rapidly and produces blue-green to yellowish colonies. A soluble red to maroon pigment, which diffuses into the agar, is highly suggestive of *P. marneffei*. At 37° C, the mycelial-to-yeast conversion occurs in approximately 2 weeks. Oval, yeastlike cells (2 to 6 μm) with septae, abortive, extensively branched, and highly septate hyphae may be seen (see Figure 65-4).

Table 65-14 presents a summary of the colonial and microscopic morphologic features of the dimorphic fungi in addition to other organisms previously discussed.

SEPTATE, DEMATIACEOUS MOLDS (SLOW-GROWING SPECIES [7 TO 10 DAYS])

The dematiaceous fungi are known agents of superficial and subcutaneous mycoses that involve the skin and subcutaneous tissues, generally without dissemination to other organs of the body (see Table 65-A on p. 944). The etiologic agents are found in several unrelated fungal genera, all of which exist in nature. Humans and animals serve as accidental hosts after traumatic inoculation of the organism into cutaneous and subcutaneous tissues. Several superficial and subcutaneous mycoses are considered here and include:

- Tinea nigra
- Black piedra
- Chromoblastomycosis
- Mycetoma
- Phaeohyphomycosis

DIRECT DETECTION

Dematiaceous fungal elements, including septate hyphae, sclerotic bodies, or ascospores, are seen in clinical specimens collected from the diverse group of infections caused by these fungi (Figures 65-6, 65-23 and 65-35, respectively). Individual characteristics will be discussed for each specific infection presented.

SUPERFICIAL INFECTIONS

Tinea nigra

Tinea nigra is an infection manifested by blackish-brown macular patches on the palm of the hand or sole of the foot. Lesions have been compared with silver nitrate staining of the skin. The etiologic agent is *Exophiala werneckii*, a dematiaceous fungus. Initial colonies of *E. werneckii* may be black, shiny, and yeastlike in appearance (Figure 65-78). With age, colonies become filamentous with velvety-gray aerial hyphae. Microscopically, the yeastlike growth consists of olive-colored budding cells that are one- or two-celled. Older colonies exhibit one- or two-celled conidia that are produced by annellophores that bear successive rings (**annellides**) that are difficult to see microscopically. Direct microscopic examination of a clinical specimen may show dematiaceous hyphae and small budding yeast cells and/or hyphal fragments.

Black piedra

Black piedra is a fungal infection of the hair of the scalp and rarely of axillary and pubic hair. The etiologic agent is the dematiaceous fungus *Piedraia hortae*. The disease occurs primarily in tropical areas of the world; cases have been reported in Africa, Asia, and in Latin America. Portions of hair are examined (in wet mounts using potassium hydroxide that is gently heated) for the presence of nodules composed of cemented mycelium. When mature nodules are crushed, oval asci, containing two to eight aseptate ascospores 19 to 55 μm long by 4 to 8 μm in diameter are seen. The asci are spindle-shaped and have a filament at each pole. The organism is easily cultured on any fungal culture medium lacking cycloheximide.

Chromoblastomycosis

Chromoblastomycosis is a chronic fungal infection acquired via traumatic inoculation of an organism, primarily into the skin and subcutaneous tissue. The infection is characterized by the development of a papule at the site of the traumatic insult that slowly spreads to form warty or tumorlike lesions characterized as cauliflower-like. There may be secondary infection and ulceration. The lesions are usually confined to the feet and legs but may involve the head, face, neck, and other body surfaces. Histologic examination of lesion tissue reveals characteristic **sclerotic bodies** (copper pennies), copper-colored, septate cells that appear to be dividing (Figure 65-23). Brain abscess (cerebral chromoblastomycosis) caused by the dematiaceous fungi has been reported with some frequency and will be discussed with phaeohyphomycosis.[80]

The disease is widely distributed, but most cases occur in tropical and subtropical areas of the world. Occasional cases are reported from temperate zones, including the United States. The infection is seen most often in areas where agricultural workers fail to wear protective clothing and suffer thorn or splinter puncture wounds through which an organism enters from the soil.

CULTURE AND IDENTIFICATION

The group of fungi known to cause chromoblastomycosis are dematiaceous. All grow slowly and produce heaped-up and slightly folded, darkly pigmented colonies with a gray to olive to black velvety appearance. The reverse side of the colonies is jet black.

The taxonomy of the organisms that cause chromoblastomycosis is complex.[94] Their identification is based on somewhat distinct microscopic morphologic features. Three genera, *Cladosporium*, *Phialophora*, and *Fonsecaea*, are included.

The genus *Cladosporium* includes those species that produce long chains of budding conidia (blastoconidia) that have a dark septal scar. In rare instances some isolates may also produce phialides. The genus has been renamed *Cladophialophora*[112]; however, the genus *Cladosporium* is retained in this chapter.

The genus *Phialophora* includes those species that produce short, flask-shaped to tubular phialides, each having a well-developed collarette. Clusters of conidia are produced by the phialides through an apical pore and often remain aggregated in a gelatinous mass, when microscopic mounts are observed.

The genus *Fonsecaea* includes those organisms that exhibit a mixed type of sporulation, which uniquely includes one-celled primary conidia on denticles that are produced on either side of conidiophores (**sympodially**). The primary conidia give rise to secondary conidia that appear to occur in loose heads. This is known as the *rhinocladiella* type of sporulation and predominates, depending on the strain isolated. A mixture of the rhinocladiella and *Cladosporium* (long chains of budding cells) types may occur; moreover, phialides with collarettes may also be present. Two species, *Fonsecaea pedrosoi* and *Fonsecaea compacta*, are etiologic agents of chromoblastomycosis. Both are morphologically distinct: *F. pedrosoi* is differentiated from *F. compacta* by the production of loose heads, in contrast to the more compact heads produced by *F. compacta*.

The diagnostic features of the three genera are summarized as follows:

1. *Cladosporium (Cladosporium carrionii):* *Cladosporium* type of sporulation with long chains of elliptic conidia (2 to 3 μm × 4 to 5 μm) borne from erect, tall, branching conidiophores (Figure 65-79).
2. *Phialophora (Phialophora verrucosa):* Tubelike or flask-shaped phialides, each with a distinct collarette. Conidia are produced endogenously and occur in clusters at the tip of the phialide (Figure 65-80).
3. *Fonsecaea (F. pedrosoi and F. compacta):* Conidial heads with sympodial arrangement of conidia, with primary conidia giving rise to secondary conidia (Figure 65-81). Cladosporium type of sporulation may occur, and phialides with collarettes may also be present.

The laboratory diagnosis of chromoblastomycosis is made easily. Scrapings from crusted lesions added to 10% potassium hydroxide show the presence

TABLE 65-14 SUMMARY OF THE CHARACTERISTIC FEATURES OF FUNGI KNOWN TO BE COMMON CAUSES OF FUNGAL INFECTION IN HUMANS

INFECTION	ETIOLOGIC AGENT	GROWTH RATE (DAYS)	CULTURAL CHARACTERISTICS AT 30° C	
			BLOOD-ENRICHED MEDIUM	MEDIUM LACKING BLOOD ENRICHMENT
Blastomycosis	*Blastomyces dermatitidis*	2-30	Colonies are cream to tan, soft, moist, wrinkled, waxy, flat to heaped, and yeastlike; "tufts" of hyphae often project upward from colonies	Colonies are white to cream to tan, some with drops of exudate present, fluffy to glabrous, and adherent to the agar surface
Histoplasmosis	*Histoplasma capsulatum*	3-45	Colonies are heaped, moist, wrinkled, yeastlike, soft, and cream, tan, or pink in color; "tufts" of hyphae often project upward from colonies	Colonies are white, cream, tan, or gray, fluffy to glabrous; some colonies appear yeastlike and adherent to the agar surface; many variations in colonial morphology occur
Paracoccidioidomycosis	*Paracoccidioides brasiliensis*	21-28	Colonies are heaped, wrinkled, moist, and yeastlike; with age, colonies may become covered with short aerial mycelium and may turn brown	
Zygomycosis	*Rhizopus* spp., *Mucor* spp., and other *Zygomycetes*	1-3	Colonies are extremely fast growing, woolly, and gray to brown to gray-black	

MICROSCOPIC MORPHOLOGIC FEATURES		RECOMMENDED SCREENING TESTS	MICROSCOPIC MORPHOLOGIC FEATURES OF TISSUE FORM	RECOMMENDED CONFIRMATORY TESTS FOR IDENTIFICATION
BLOOD-ENRICHED MEDIUM	NONBLOOD-ENRICHED MEDIUM			
Hyphae 1-2 μm in diameter are present; some are aggregated in ropelike clusters; sporulation is rare	Hyphae 1-2 μm in diameter are present; single pyriform conidia are produced on short to long conidiophores; some cultures produce few conidia	Not available	8-15 μm, broad-based budding cells with double-contoured walls are seen; cytoplasmic granulation is often obvious	1. Specific nucleic acid probe 2. Broad-based budding cells may be seen after in vitro conversion on cottonseed agar 3. Exoantigen test
Hyphae 1-2 μm in diameter are present; some are aggregated in ropelike clusters; sporulation is rare	Young cultures usually have a predominance of smooth-walled macroconidia that become tuberculate with age; macroconidia may be pyriform or spherical; some isolates produce small pyriform microconidia in the presence or absence of macroconidia	Not available	2-5 μm, small, oval to spherical budding cells often seen inside of mononuclear cells	1. Specific nucleic acid probe 2. Exoantigen test
Hyphae 1-2 μm in diameter are present; some isolates produce conidia similar to those of *B. dermatitidis;* chlamydospores may be numerous, and multiple budding yeast cells 10-25 μm in diameter may be present		Not available	10-25 μm, multiple budding cells (buds 1-2 μm) resembling a mariner's wheel may be present; buds are attached to the parent cell by a narrow neck	Exoantigen test
1. *Rhizopus* spp.—rhizoids are produced at the base of sporangiophore 2. *Mucor* spp.—no rhizoids are produced		Not available	Large, ribbonlike (10-30 μm), twisted, often distorted pieces of aseptate hyphae may be present; septa occasionally may be seen	Identification is based on characteristic morphologic features

Continued

TABLE 65-14   SUMMARY OF THE CHARACTERISTIC FEATURES OF FUNGI KNOWN TO BE COMMON CAUSES OF FUNGAL INFECTION IN HUMANS—CONT'D

INFECTION	ETIOLOGIC AGENT	GROWTH RATE (DAYS)	CULTURAL CHARACTERISTICS AT 30° C	
			BLOOD-ENRICHED MEDIUM	MEDIUM LACKING BLOOD ENRICHMENT
Aspergillosis	*Aspergillus fumigatus* *Aspergillus flavus* *Aspergillus niger* *Aspergillus terreus* *Aspergillus* spp.	3-5	Colonies of *A. fumigatus* are usually blue-green to gray-green, whereas those of *A. flavus* and *A. niger* are yellow-green and black, respectively; colonies of *A. terreus* resemble powdered cinnamon; other species of *Aspergillus* exhibit a wide range of colors; blood-enriched media usually have little effect on the colonial morphologic features	
Coccidioidomycosis	*Coccidioides immitis*	2-21	Colonies may be white and fluffy to greenish on blood-enriched media; some isolates are yeastlike, heaped, wrinkled, and membranous	Colonies usually are fluffy white but may be pigmented gray, orange, brown, or yellow; mycelium is adherent to the agar surface in some portions of the colony

From Thomson, R.B. and Roberts, G.D. 1982. A practical approach to the diagnosis of fungal infections of the respiratory tract. Clin. Lab. Med. 2:321.

of the sclerotic bodies, which appear rounded, brown, and 4 to 10 μm in diameter and resemble copper pennies having fission planes (Figure 65-23).

MYCETOMA

Mycetoma is a chronic granulomatous infection that usually involves the lower extremities but may occur in any part of the body. The infection is characterized by swelling, purplish discoloration, tumorlike deformities of the subcutaneous tissue, and multiple sinus tracts that drain pus containing yellow, white, red, or black granules. The infection gradually progresses to involve the bone, muscle, or other contiguous tissue and ultimately requires amputation in most cases.

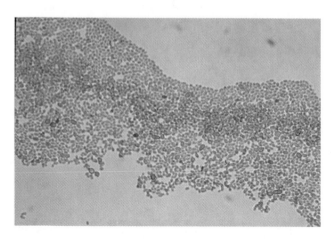

FIGURE 65-78 *Exophiala werneckii* yeast.

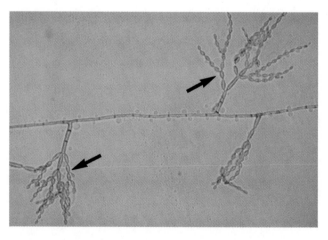

FIGURE 65-79 *Cladosporium* spp., showing *Cladosporium* type of sporulation (*arrows*) with chains of elliptic conidia (430×).

MICROSCOPIC MORPHOLOGIC FEATURES		RECOMMENDED SCREENING TESTS	MICROSCOPIC MORPHOLOGIC FEATURES OF TISSUE FORM	RECOMMENDED CONFIRMATORY TESTS FOR IDENTIFICATION
BLOOD-ENRICHED MEDIUM	NONBLOOD-ENRICHED MEDIUM			
1. *A. fumigatus*—uniserate heads with phialides covering the upper half to two thirds of the vesicle 2. *A. flavus*—uniserate or biserate or both with phialides covering the entire surface of a spherical vesicle 3. *A. niger*—biserate with phialides covering the entire surface of a spherical vesicle; conidia are black 4. *A. terreus*—biserate with phialides covering the entire surface of a hemispherical vesicle; aleuriospores are formed on submerged hyphae		Not available	Septate hyphae 5-10 μm in diameter that exhibit dichotomous branching	Identification is based on microscopic morphologic features and colonial morphology; *A. fumigatus* can tolerate elevated temperatures ≥45° C
Chains of alternate, barrel-shaped arthroconidia are characteristic; some arthroconidia may be elongated; hyphae are small and often arranged in ropelike strands, and racquet forms are seen in young cultures		Not available	Round spherules 30-60 μm in diameter containing 2-5 μm endospores are characteristic Empty spherules are commonly seen	1. Specific nucleic acid probe 2. Exoantigen test

Occasionally there may be dissemination to other organs, including the brain; however, this type of infection is relatively uncommon.

Mycetoma is common among persons who live in tropical and subtropical regions of the world and who work outdoors without protective clothing, predisposing them to trauma.

Two types of mycetomas are described: actinomycotic (bacterial) mycetoma, which is caused by species of the aerobic actinomycetes, including *Nocardia, Actinomadura,* and *Streptomyces;* and eumycotic (fungal) mycetoma, caused by a heterogeneous group of fungi having septate hyphae.[79] Etiologic agents of eumycotic mycetoma include: *Pseudallescheria boydii,*

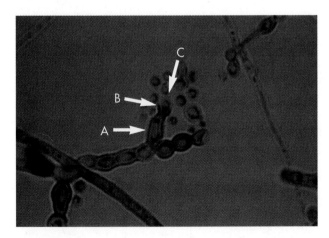

FIGURE 65-80 *P. verrucosa,* showing flask-shaped phialide (**A**) with distinct collarette (**B**) and conidia (**C**) near its tip (750×).

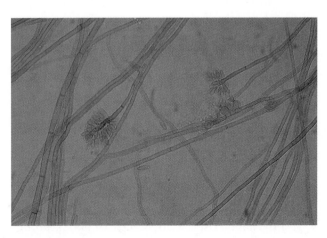

FIGURE 65-81 *F. pedrosoi,* showing conidial heads with sympodial arrangement of conidia (430×).

Exophiala jeanselmei, Acremonium, Curvularia, Madurella mycetomatis, and numerous other genera and species. The most common etiologic agent of mycetoma in the United States is *P. boydii,* a member of the Ascomycota. The organism is commonly found in soil, standing water, and sewage; humans acquire the infection by traumatic implantation of the organism into the skin and subcutaneous tissues.

Direct examination

Macroscopic examination of granules from lesions of mycetoma caused by *P. boydii* reveal them to be white to yellow, and 0.2-2.0 mm in diameter. Microscopically, the granules of *P. boydii* consist of loosely arranged, intertwined, septate hyphae, cemented together.

Culture and identification

P. boydii grows rapidly (5 to 10 days) on common laboratory media. Initial growth begins as a white, fluffy colony that changes in several weeks to a brownish-gray (mousey) mycelium; the reverse of the colony is black. *P. boydii* has undergone several name changes in the past; it exhibits both asexual and sexual reproduction. The asexual form (anamorph) is called *Scedosporium apiospermum* and microscopically produces golden brown elliptic (sperm-shaped), single-celled conidia borne singly from the tips of long or short conidiophores (annellophores) (Figure 65-9). This anamorph predominates in cultures from clinical specimens. Clusters of conidiophores with conidia produced at the ends sometimes occur and are referred to as *coremia*. This anamorphic form is often referred to as the *Graphium* stage of *P. boydii* (Figure 65-10).

P. boydii is the teleomorphic (sexual form) form of the organism. The sexual form produces brown cleistothecia, pseudoparenychatous saclike structures that contain asci and ascospores. When the latter are fully developed, the large (50 to 200 μm), thick-walled cleistothecia rupture and liberate the asci and ascospores (Figure 65-8). The ascospores are oval and delicately pointed at each end and resemble the conidia of the asexual form. Isolates of *P. boydii* may be induced to form cleistothecia by culturing on plain water agar; however, they are seldom found on primary recovery of a culture from a clinical specimen.

P. boydii is also involved in causing infections elsewhere in the body. Included are infections of the nasal sinuses and septum, meningitis, arthritis, endocarditis, mycotic keratitis, external otomycosis, and brain abscess. Most of these more serious infections occur primarily in immunocompromised patients. Recently, another species of *Scedosporium, S. prolifi-*

cans, has been associated with infections other than mycetoma, such as arthritis.[97] It differs from *S. apiospermum* by producing inflated annellophores that appear flask-shaped.

It should be mentioned that it is impossible to predict the specific etiologic agent of mycetoma. Culture media containing antibiotics should not be used alone for culturing clinical specimens, because species of the aerobic actinomycetes are susceptible to antibacterial antibiotics and may be inhibited by these agents incorporated into routine culture media. For further information on other fungi involved in mycetoma, refer to references listed at the end of this chapter.[79]

PHAEOHYPHOMYCOSIS

Phaeohyphomycosis is a general term used to describe any infection caused by a dematiaceous organism. The strict definition does not include chromoblastomycosis or mycetoma. However, in the context of this chapter, it refers to subcutaneous, localized, or systemic infection caused by any of a number of dematiaceous fungi. Infections include subcutaneous abscess, brain abscess, sinusitis, endocarditis, mycotic keratitis, pulmonary infection, and systemic infection.[1,24,72,121]

In general, dematiaceous fungal hyphae are seen in clinical specimens by direct microscopic examination or by histopathologic examination of tissue obtained at surgery or autopsy. In some instances, yeastlike elements may also be observed.

Common etiologic agents of subcutaneous infection include:

- *Exophiala jeanselmei*
- *Phialophora richardsiae*
- *Exophiala dermatitidis*
- *Alternaria*
- *Bipolaris*
- *Exserohilum*

Common agents associated with sinusitis include:

- *Bipolaris*
- *Exserohilum*
- *Curvularia*

Common etiologic agents of brain abscess include:

- *Xylohypha (Cladophialophora) bantiana*
- *Exophiala dermatitidis*
- *Bipolaris*

Mycotic keratitis and endophthalmitis have been caused by:

- *Exophiala dermatitis*
- *Curvularia*

Direct examination

The laboratory diagnosis is made by observing yellowish-brown, septate, moniliform hyphae with or without budding cells present. The methenamine silver stain of tissue often overstains fungal elements, and the observation of pigmented hyphae in hematoxylin-eosin or unstained histopathologic sections will provide the diagnosis of this clinical entity. Culture of the specific etiologic agent is necessary for final confirmation.

Culture and identification

Exophiala jeanselmei is a dematiaceous organism that grows slowly (7 to 21 days) and initially produces shiny black, yeastlike colonies. With age, colonies become filamentous, velvety, and gray to olive to black. The microscopic features of young colonies exhibit dematiaceous budding yeast cells with annella-

tions (difficult to observe). Filamentous colonies produce dematiaceous hyphae and conidiogenous cells that are cylindric and have a tapered tip. Annellations may be visible at the tip, and clusters of oval to round conidia are apparent (Figure 65-82). Cultures can grow at 37° C and not at 40° C and will grow on a medium containing KNO_3 as the only NO_3 substrate.

Exophiala dermatitidis, a dematiaceous organism, grows slowly (7 to 21 days) and produces shiny, often mucoid, black to brown, yeastlike colonies initially (Figure 65-83). With age, colonies become filamentous, velvety, and gray to olive to black. The microscopic features of young colonies exhibit dematiaceous budding yeast cells. Filamentous colonies produce dematiaceous hyphae and long tubular conidiophores (annellophores). Annellophores produce single-celled dematiaceous conidia in clusters (Figure 65-84). Balls of conidia may appear to slide down the

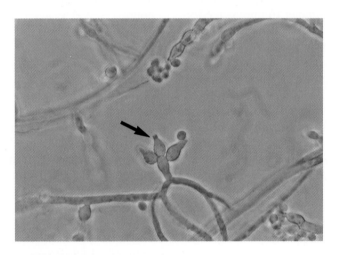

FIGURE   65-82 *E. jeanselmei,* showing elongated conidiophore (annellophore) (*arrow*) with a narrow, taped tip (500×).

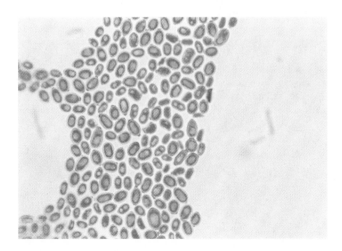

FIGURE 65-83 *E. dermatitidis,* showing dematiaceous budding yeast cells from a young culture (500×).

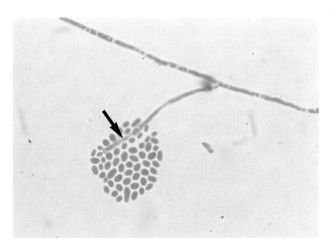

FIGURE 65-84 *E. dermatitidis,* showing elongated tubular annellophores (*arrow*) lacking a distinct collarette (500×).

hyphae. Cultures can grow at 40° C and cannot utilize KNO₃ in contrast to *E. jeanselmei*.

Phialophora richardsiae is a dematiaceous organism that grows rapidly and produces colonies that are wooly, olive-brown to brownish-gray; some may appear to have concentric zones of color. Microscopically, hyphae are dematiaceous and sporulation is common. Phialides with distinct saucerlike collarettes are readily observed (Figure 65-85). Pleomorphic phialides may also be seen; however, all produce either or both hyaline elliptic conidia or brown elliptic conidia produced within the phialides.

DEMATIACEOUS, SEPTATE MOLDS (RAPID-GROWING MOLDS [< 7 TO 10 DAYS])

The rapid-growing dematiaceous molds, like those that are slow-growing, are ubiquitous and exist as saprophytes and plant pathogens. When nonsterile body sites are cultured, it is difficult to impossible for the laboratorian to determine the significance of these molds. They may represent true pathogens; however, more often, they represent transient flora, inhaled spores, or contaminants. Cultures from sterile body sites, if aseptically obtained, should not contain these molds. To more effectively establish the diagnosis of mycotic infection caused by these organisms, cultures should be interpreted in conjunction with the results of the direct examination for fungal elements and corresponding histopathology. If colonies of common saprophytic molds occur near the edge of the plate *and* are clearly away from the innoculum, then they should be considered contaminants.

ALTERNARIA

Colonies of *Alternaria* grow rapidly and appear to be fluffy and gray to gray-brown or dark to gray-green in color. Microscopically, hyphae are septate and golden-brown–pigmented, and conidiophores are simple but sometimes branched. Conidiophores bear a chain of large brown conidia resembling a drumstick and contain both horizontal and longitudinal septa (Figure 65-86). It is sometimes difficult to observe chains of conidia since they are easily dislodged as the culture mount is prepared.

BIPOLARIS

Colonies of *Bipolaris* are gray-green to dark brown and slightly powdery. Hyphae are dematiaceous and septate, as observed microscopically. Conidiophores are twisted at the ends where conidia are attached; conidia are arranged sympodially, are oblong to fusoid, and the hilum protrudes only slightly (Figure 65-87). Germ tubes are formed at one or both ends, parallel to the long axis of the conidium, when the fungus is incubated in water at 25° C for up to 24 hours.

CLADOSPORIUM

Colonies of *Cladosporium* most commonly appear as velvety or suedelike, heaped, and folded. Microscopically, hyphae are septate and brown. Conidiophores are long and branched and give rise to chains of darkly pigmented budding conidia. Conidia are usually single-celled and exhibit prominent attachment scars (disjunctors) that may resemble "shield" cells (Figure 65-88). This organism also fails to reveal chains of conidia on wet mounts because the conidia dislodge easily.

CURVULARIA

Colonies of *Curvularia* are rapid-growing and appear to resemble those of *Alternaria*. Most are fluffy and gray to gray-brown or black to gray-green in color.

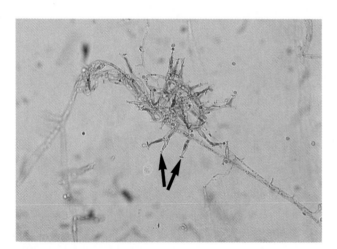

FIGURE 65-85 *P. richardsiae,* showing phialides having prominent saucerlike collarette (*arrows*) (500×).

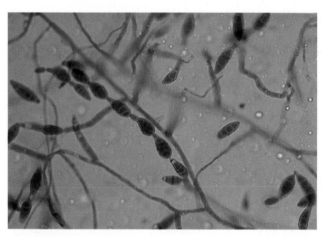

FIGURE 65-86 *Alternaria* spp., showing muriform, dematiaceous conidia with horizontal and longitudinal septa.

Hyphae are dematiaceous and septate, as observed microscopically. Conidiophores are twisted at the ends where conidia are attached; conidia are arranged sympodially, are golden-brown, multicelled, and curved with a central swollen cell (Figure 65-89). The end cells are lighter in color.

DRECHSLERA

Drechslera is a dematiaceous organism that has been described erroneously as *Helminthosporium* or *Bipolaris* in many textbooks. Colonies of *Drechslera* are fluffy to velvety and gray to brown or black in color. Microscopically, the hyphae are septate and darkly pigmented, and conidiophores are **geniculate** (series of bent-knee structures). Conidia are produced sympodially (Figure 65-90). However, sporulation is generally sparse with this organism, and it is not commonly seen. The conidia of *Drechslera* is impossible to differentiate from those of *Bipolaris* based on morphologic criteria alone. The germ tube test may be used to differentiate these organisms. Conidia are placed in a drop of water, coverslipped, and observed after an incubation period of at least 24 hours. Any of the cells of the conidia belonging to *Drechslera* spp. may germinate and these may grow perpendicular to the long axis of the conidium. Only the end cells of the conidia belonging to *Bipolaris* germinate and these grow predominantly parallel to the longitudinal axis of the conidium.

EXSEROHILUM

Colonies of *Exserohilum* resemble those of *Bipolaris*. Hyphae are septate and dematiaceous. Conidiophores are geniculate, and conidia are produced sympodially.

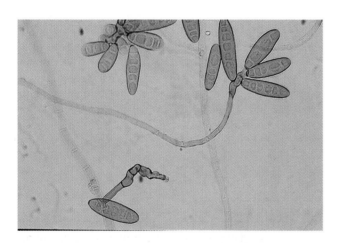

FIGURE 65-87 *Bipolaris* spp., showing dematiaceous, multicelled conidia produced sympodially (430×).

FIGURE 65-88 *Cladosporium* spp., showing branching chains of dematiaceous blastoconidia that are easily dislodged during the preparation of a microscopic mount (430×).

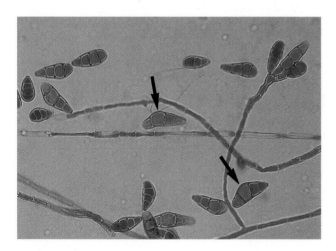

FIGURE 65-89 *Curvularia* spp., showing twisted conidiophore and curved conidia having a swollen central cell (*arrows*) (500×).

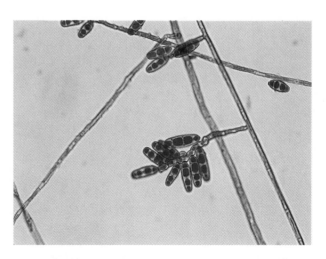

FIGURE 65-90 *Drechslera* spp., showing dematiaceous, multicelled conidia. Most isolates produce only a few conidia.

Conidia are elongate, ellipsoid to fusoid, and exhibit a prominent hilum that is truncated and protruding (Figure 65-91). The conidia are multicellular, with only perpendicular septae, and usually contain 5 to 9 septae.

PNEUMOCYSTIS CARINII (AN ATYPICAL FUNGUS)

BACKGROUND AND DISEASE

Pneumocystis carinii is an opportunistic, atypical fungus that infects immunocompromised hosts.[116] Originally thought to be a trypanosome, the precise taxonomic categorization of *P. carinii* has remained a challenge.[12] The clinical response to antiprotozoal drugs and the lack of response to antifungal drugs in patients with pneumocystosis further supported the notion that *P. carinii* was a protozoan parasite.[105] The inability to successfully maintain and propagate this organism in culture has further limited its characterization.

P. carinii differs from other fungi in various aspects. The cell membrane of *P. carinii* contains cholesterol rather than ergosterol.[54] A flexible-walled trophozoite form exists that is susceptible to osmotic disturbances. Additionally, this organism contains only one to two copies of the small ribosomal subunit gene. Most other fungi contain numerous copies of this gene.[37]

DNA sequence analysis of the small ribosomal subunit gene has been performed on *P. carinii*. These comparative studies have disclosed a greater sequence homology with the fungi than with the protozoa.[26,27,106] Two independent analyses that compared DNA sequenes of *P. carinii* with those of other fungi have been performed. These studies confirm the position of *P. carinii* within the fungal kingdom and place it somewhere between the ascomyctes and the basidiomycetes.[10,114]

P. carinii most commonly presents as pneumonia in an immunocompromised host. After inhalation, the organism is thought to adhere to type I pneumocytes.[116] The organisms are thought to exist and replicate extracellularly, while bathed in alveolar lining fluid. Successful replication of the organism results in alveolar spaces being filled with foamy material. These changes result in impaired oxygen diffusing capacity and hypoxemia. A predominantly interstitial, mononuclear inflammatory response is associated with this type of pnuemonia.

DIRECT DETECTION AND IDENTIFICATION

Currently, the diagnosis of *P. carinii* pneumonia is based on the clinical presentation, radiographic studies, and direct and/or pathologic examination of bronchoalveolar lavage fluid or biopsy material. The flexible-walled trophozoite forms are the predominant form of the organism.[105] These forms are visible on Giemsa-stained material; however, their pleomorphic appearance makes this form of the organism difficult to identify. A firmer-walled cystic form also exists; however, cysts are outnumbered by trophozoites 1:10.[105] The cystic forms are more easily recognizable than the trophozoites and may be definitively identified using calcofluor white (Procedure 65-14) or methenamine silver stains (Figure 65-92). Commercial kits employing monoclonal antibodies may be slightly more sensitive than microscopy; however, these tests are also more expensive.

INTRODUCTION TO YEASTS

During recent years, a significant increase in the number of fungal infections caused by yeasts and yeastlike fungi has been reported in the literature. Numerous species of *Candida* and other yeasts have been impli-

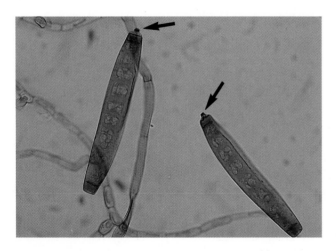

FIGURE 65-91 *Exserohilum* spp., showing elongated, multicelled conidia with prominent hila (*arrows*).

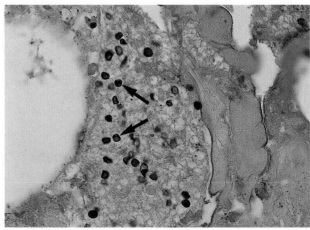

FIGURE 65-92 *P. carinii* cystic forms (*arrows*), methenamine silver (500×).

cated. These infections are primarily seen in immuno-compromised patients (including those with AIDS), and some of the yeasts have proved to be resistant to antifungal therapy.[22,39,65,110] Most infections are seen in patients having surgery and patients on long-term intravenous therapy without adequate catheter care. Microbiologists should be aware that any of the genera and species of yeast are potential pathogens in this group of patients.

Just how far the laboratory should go with a complete identification of all yeast species is of question. It is recommended that:

- All yeasts recovered from sterile body fluids, including cerebrospinal fluid, blood, urine, paracentesis, and other fluids, should be identified.
- Yeasts from all seriously ill or immunocompromised patients or in whom a mycotic infection is suspected should be identified.
- Yeasts from respiratory secretions should not be identified routinely; however, they should be screened for the presence of *Cryptococcus neoformans*.
- Yeasts recovered in large amounts from any clinical source should be identified.
- Yeasts recovered from several successive specimens, except respiratory secretions, should be identified. Each laboratory director will have to decide how much time, effort, and expense is to be spent on the identification of yeasts in the laboratory.

The development of commercially available yeast identification systems provides laboratories of all sizes with the capability of using accurate and standardized methods. Some have extensive computer databases that include biochemical profiles of large numbers of yeasts. Variations in the reactions of carbohydrates and other substrates utilized are considered in the identification of yeasts provided by these systems. Commercially available systems are recommended for all laboratories; however, they may be used in conjunction with some less expensive and rapid screening tests that will provide the presumptive identification of *C. neoformans* and a definitive identification of *C. albicans*. Some laboratories might prefer the use of conventional systems; therefore, the information presented within this section discusses rapid screening methods for the presumptive identification of yeasts, commercially available systems, and a conventional schema that will provide for the identification of commonly encountered species of yeast.

CANDIDA

Candida spp. are responsible for the most frequently encountered opportunistic fungal infections. *Candida* infections are caused by various species, but *Candida albicans* is the most frequent etiologic agent, followed by *Candida tropicalis*, *Candida parapsilosis*, and *Candida glabrata*. Numerous other species have been involved in infection in immunocompromised patients; however, their incidence is not as high. *C. albicans* and others are a part of the normal endogenous microbial flora, and infections are believed to be endogenous in origin. The organisms may be recovered from the oropharynx, gastrointestinal tract, genitourinary tract, and skin. *Candida* spp. are responsible for several types of infections in normal and immunocompromised patients. Included are intertriginous candidosis, in which skin folds are involved, paronychia, onychomycosis, perleche, vulvovaginitis, thrush, pulmonary infection, eye infection, endocarditis, meningitis, fungemia, and disseminated infection. The latter two types of infection are most commonly seen in immunocompromised patients, while the others mentioned may occur in normal hosts. Onychomycosis and esophagitis produced by *C. albicans* are very common in patients with AIDS.

The clinical significance of *Candida* recovered from respiratory tract secretions is difficult to determine, since it is considered to be part of the normal flora of humans. A study at Mayo Clinic evaluated the clinical significance of yeasts recovered from respiratory secretions, except for *Cryptococcus neoformans*, and concluded that they are part of the normal flora and that their routine identification is unnecessary.[75] The repeated recovery of different species of yeasts from multiple specimens from the same patient usually indicates colonization. The simultaneous recovery of the same species of yeast from several body sites, including urine, is a good indicator of disseminated infection and the subsequent development of fungemia.

Direct examination

The direct microscopic examination of clinical specimens containing *Candida* will reveal budding yeast cells (**blastoconidia**) 2 to 4 μm in diameter and/or pseudohyphae (see Figure 65-13) showing regular points of constriction, resembling lengths of sausages, or true septate hyphae. The blastoconidia, hyphae, and pseudohyphae are strongly gram-positive. It is advisable to report the approximate number of such forms since the presence of large numbers in a fresh clinical specimen may be diagnostically significant.

Culture and identification

The colonial and microscopic morphologic features of *Candida* spp. are of little value in making a definitive identification. *C. albicans*, however, may be identified by the production of germ tubes or chlamydospores

(Figures 65-2 and 65-93). Other species of *Candida* must be identified by the utilization of, and sometimes fermentation of, specific substrates. The morphologic features of yeast on cornmeal agar containing Tween 80 often allow for the tenative identification of selected species (Table 65-15).

CRYPTOCOCCUS

Cryptococcosis, specifically caused by *Cryptococcus neoformans,* is a subacute or chronic fungal infection that has several manifestations. In the immunocompromised patient, it is common to see disseminated disease with or without meningitis; meningitis occurs in approximately two thirds of patients with disseminated infection. Disseminated cryptococcosis is becoming a very common clinical entity in patients with AIDS.[16] Occasionally, patients with disseminated infection will exhibit painless skin lesions that may ulcerate. Other uncommon manifestations of cryptococcosis include endocarditis, hepatitis, renal infection, and pleural effusion. It is of interest to note that

a review of patient records at the Mayo Clinic revealed more than 100 cases of colonization of the respiratory tract with *C. neoformans* without subsequent development of infection. Follow-up on these immunocompetent patients was as long as 6 years; none in this group was considered to be immunocompromised. This makes the clinical significance of *C. neoformans* somewhat difficult to assess; however, its presence in clinical specimens from immunocompromised patients should be considered to be significant. In many instances, the clinical symptoms are suppressed by corticosteroid therapy, and culture or serologic evidence (detection of crytococcal antigens) provides the earliest proof of infection. There is a strong association of cryptococcal infection with debilitating diseases such as leukemia, lymphoma, and the immunosuppressive therapy that may be required for these and other underlying diseases. The presence of *C. neoformans* in clinical specimens in some instances precedes the symptoms of an underlying disease. The infection is probably more frequent than is commonly

---

TABLE 65-15   CHARACTERISTIC MICROSCOPIC FEATURES OF COMMONLY ENCOUNTERED YEASTS ON CORNMEAL TWEEN 80 AGAR

ORGANISM	ARTHROCONIDIA	BLASTOCONIDIA	PSEUDOHYPHAE OR HYPHAE
Candida albicans	—	Spherical clusters at regular intervals on pseudohyphae	Chlamydospores present on pseudohyphae
C. glabrata	—	Small, spherical, and tightly compacted	—
C. krusei	—	Elongated; clusters occur at septa of pseudohyphae	Branched pseudohyphae
C. parapsilosis	—	Present but not characteristic	Sagebrush appearance; large (giant) hyphae present
C. kefyr (pseudotropicalis)	—	Elongated, lie parallel to pseudohyphae	Pseudohyphae present but not characteristic
C. tropicalis	—	Produced randomly along hyphae or pseudohyphae	Pseudohyphae present but not characteristic
Cryptococcus	—	Round to oval, vary in size, separated by a capsule	Rare
Saccharomyces	—	Large and spherical	Rudimentary hyphae sometimes present
Trichosporon	Numerous; resemble *Geotrichum*	May be present but difficult to find	Septate hyphae present

recognized; it is estimated that there may be 300 new cases each year in the United States.[63]

Four serotypes of *C. neoformans* have been described (A, B, C, and D) with somewhat different geographic distribution.[63] The organism has been proved to be a basidiomycete by the discovery of its teleomorph, *Filobasidiella*, which has two varieties: *F. neoformans* var. *neoformans* and *F. neoformans* var. *gattii*. The former includes the serotypes A and D; the latter includes serotypes B and C.[63] *Cryptococcus* will continue to be used as the genus name since it is well-established by tradition. Despite the discovery of two varieties of *Cryptococcus*, there is no difference in disease produced or in the response to chemotherapy between the two species.

C. neoformans exists as a saprobe in nature. It is most often found associated with the excreta of pigeons. The hypothesis that pigeon habitats serve as reservoirs for infection in humans is substantiated by numerous reports; the pigeon manure apparently serves as an enrichment for *C. neoformans* because of its chemical makeup. It is believed that *C. neoformans* is widely distributed in nature, becomes aerosolized, and is inhaled before infection.

Direct examination

Traditionally, the India ink preparation has been the most widely used method for the rapid detection of *C. neoformans* in clinical specimens. An evaluation of 39 consecutive non-AIDS patients with cryptococcal meningitis seen at the Mayo Clinic showed that only 40% gave positive India ink preparations of the cerebrospinal fluid. The purpose of the method is to delineate the large capsule of *C. neoformans*, since the ink particles cannot penetrate the capsular polysaccharide material. Because of the low positivity rate of the India ink preparation, it is not recommended as a routine tool in the clinical microbiology laboratory. It should be replaced with the cryptococcal latex test for antigen, which will be described in a subsequent section. However, the India ink preparation is commonly positive in specimens from patients with AIDS. Laboratories examining large numbers of specimens from these patients may wish to retain the use of this procedure in combination with the cryptococcal latex test for antigen.

The microscopic examination of other clinical specimens, including respiratory secretions, can also be of value in making a diagnosis of cryptococcosis. *C. neoformans* appears as a spherical, single or multiple budding, thick-walled, yeastlike organism 2 to 15 μm in diameter, usually surrounded by a wide, refractile polysaccharide capsule (Figure 65-18). Perhaps the most important characteristic of *C. neoformans* is the extreme variation in the size of the yeast cells; this is unrelated to the amount of polysaccharide capsule present. It is important to remember that not all isolates of *C. neoformans* exhibit a capsule.

Culture and identification

C. neoformans is easily cultured on routine fungal culture media without cycloheximide. The organism is inhibited by the presence of cycloheximide at 25° to 30° C. For the optimal recovery of *C. neoformans* from cerebrospinal fluid, it is recommended that a 0.45-μm membrane filter be used with a sterile syringe. The filter is placed on the surface of the culture medium and is removed at daily intervals so that growth under the filter can be visualized. An alternative to the membrane filter technique is the use of centrifugation.

Colonies of *C. neoformans* usually appear on culture media within 1 to 5 days and begin as a smooth, white to tan colony that may become mucoid and cream to brown in color (Figure 65-94). It is

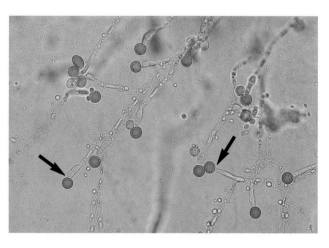

FIGURE 65-93 *C. albicans.* Chlamydospores (*arrows*).

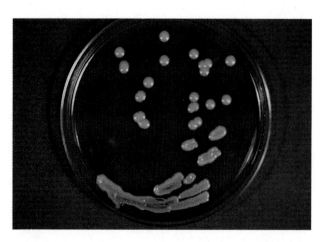

FIGURE 65-94 *C. neoformans,* showing colonies that appear shiny and mucoid because of the presence of a polysaccharide capsule.

--- **TABLE 65-A** DEMATIACEOUS (Dm) FUNGI ---

ORGANISM	DISEASE	SITE	TISSUE FORM
Slow-Growing Species			
Cladosporium spp.	Chromoblastomycosis	Subcutaneous	Sclerotic bodies
	Phaeohyphomycosis	Subcutaneous, brain	Septate, dematiaceous hyphae
Dactylaria gallopava	Phaeohyphomycosis	Brain, subcutaneous, lungs	Septate, dematiaceous hyphae
Exophiala dermatitidis	Phaeohyphomycosis	Brain, eye, subcutaneous, and dissemination	Dematiaceous hyphal fragments and budding yeasts
	Pneumonia	Lungs	
Exophiala jeanselmei	Mycetoma	Subcutaneous	Dematiaceous hyphal fragments and budding yeasts
	Phaeomycotic cyst	Subcutaneous	
Exophiala werneckii	Tinea nigra	Skin	Dematiaceous hyphal fragments and budding yeast
Fonsecaea spp.	Chromoblastomycosis	Subcutaneous	Sclerotic bodies
	Phaeohyphomycosis	Brain	Septate, dematiaceous hyphae
	Cavitary lung disease	Lungs	Septate, dematiaceous hyphae
Phialophora spp.	Chromoblastomycosis	Subcutaneous	Sclerotic bodies
	Phaeohyphomycosis	Subcutaneous	Septate, dematiaceous hyphae
	Septic arthritis	Joints	Septate, dematiaceous hyphae
Piedraia hortae	Black piedra	Hair	Asci containing nodules cemented to hair shafts
Madurella mycetomatis	Mycetoma	Subcutaneous	Dematiaceous hyphal fragments

TABLE 65-A   DEMATIACEOUS (Dm) FUNGI—CONT'D

ORGANISM	DISEASE	SITE	TISSUE FORM
Rapid-Growing Species			
Alternaria spp.	Phaeohyphomycosis	Subcutaneous	Septate, dematiaceous hyphae
	Sinusitis	Sinuses	Septate, dematiaceous hyphae; possible fungus-ball
	Nasal septal erosion	Nasal septum	Septate, dematiaceous hyphae
	Ulcers and onychomycosis	Skin, Nails	Septate, dematiaceous hyphae
Bipolaris spp.	Phaeohyphomycosis	Subcutaneous, brain, eye, bones	Septate, dematiaceous hyphae
	Sinusitis, fungus-ball	Sinuses	Septate, dematiaceous hyphae; possible fungus-ball
Curvularia spp.	Sinusitis	Sinuses	Septate, dematiaceous hyphae; possible fungus-ball
	Phaeohyphomycosis	Subcutaneous, heart valves, eye, and lungs	Septate, dematiaceous hyphae
Drechslera spp.	Phaeohyphomycosis	Subcutaneous and brain	Septate, dematiaceous hyphae
	Sinusitis	Sinuses	Septate, dematiaceous hyphae
Exoserohilum spp.	Phaeohyphomycosis	Subcutaneous	Septate, dematiaceous hyphae
Pseudallescheria boydii	Mycetoma	Subcutaneous	Granules of hyaline hyphae
	Phaeohyphomycosis	Subcutaneous, skin, joints, bones, brain, and lungs	Septate, hyaline hyphae
Xylohypha bantiana	Phaeohyphomycosis	Brain	Septate, dematiaceous hyphae

important to recognize the colonial morphology on different culture media since variation does occur; for example, on inhibitory mold agar, *C. neoformans* appears as a golden yellow, nonmucoid colony. Textbooks typically characterize the colonial morphology as being *Klebsiella*-like because of the large amount of polysaccharide capsule material present (see Figure 65-94). In reality, most isolates of *C. neoformans* do not have large capsules and may not have the typical mucoid appearance.

The microscopic examination of colonies of *C. neoformans* may be of help in providing a tentative identification of *C. neoformans* since the cells will be spherical and exhibit a wide variation in their size. A presumptive identification of *C. neoformans* may be based on urease production and failure to utilize an inorganic nitrate substrate. The final identification of *C. neoformans* is usually based on typical substrate utilization patterns and pigment production on niger seed agar (Figure 65-95). A discussion of tests useful for the identification of *C. neoformans* and other species of cryptococci will be presented in a subsequent section.

TRICHOSPORON

Trichosporonosis, caused by *Trichosporon beigelii,* a yeastlike organism, occurs almost exclusively in immunocompromised patients, particularly those with leukemia.[2] Disseminated trichosporonosis is the most common clinical manifestation. Skin lesions accompanied by fungemia are frequently seen. Endocarditis, endophthalmitis, and brain abscess have been reported. *Trichosporon beigelli* is commonly recovered from respiratory tract secretions, skin, and the oropharynx and stool of patients who have no evidence of infection and may represent colonization of those individuals.

White piedra is an uncommon fungal infection of immunocompetent patients and is found in both tropical and temperate regions of the world. It is characterized by the development of soft, yellow or pale brown aggregations around hair shafts in the axillary, facial, genital, and scalp regions of the body. The etiologic agent is *T. beigelii*. It frequently invades the cortex of the hair and causes damage.

Direct detection

The microscopic examination of clinical specimens reveals hyaline hyphae, numerous round to rectangular arthroconidia and, occasionally, a few blastoconidia. Usually hyphae and arthroconidia predominate.

In white piedra, white nodules are removed and observed using the potassium hydroxide preparation after applying light pressure to the coverslip so that crushing of the nodule occurs. Hyaline hyphae, 2 to 4 μm in width, and arthroconidia are found in the preparation of the cementlike material binding the hyphae together. The organism may be identified in culture by blastoconidia and arthroconidia formation (Figure 65-72). *T. beigelii* may be distinguished from the other species in the genus by its inability to ferment carbohydrates and ability to aerobically utilize certain substrates.

Culture and identification

Colonies of *T. beigelli* vary in their morphology; however, most are cream-colored, heaped, dry to moist, and wrinkled. Some may appear white, dry, powdery, and wrinkled.

Arthroconidia, which are rectangular, often with rounded ends, predominate along with septate hyaline hyphae. Blastoconidia are sometimes present but are not seen in all cultures. Urease is produced and is often helpful in making a tentative identification. The final identification is based on characteristic substrate utilization profiles; fermentation does not occur with *Trichosporon*.

Trichosporon shares an antigen that is similar to that produced by *C. neoformans*. Sera from patients having trichosporonosis often yield false-positive cryptococcal antigen tests when latex agglutination methods are used and this is sometimes used as tentative evidence of trichosporonosis.

MALASEZZIA

Malasezzia furfur causes **tinea versicolor,** a skin infection that is characterized by superficial, brownish scaly areas on light-skinned persons and lighter areas on dark-skinned persons. The lesions occur on the smooth surfaces of the body, namely, the trunk, arms, shoulders, and face. It is distributed worldwide. *Malasezzia furfur* is currently being seen as the cause of disseminated infection in infants and young children and even adults given lipid replacement ther-

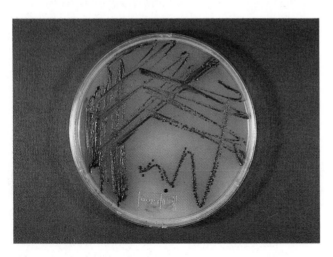

FIGURE 65-95 Niger seed agar. *C. neoformans.*

apy.[69] *Malassezia pachydermatis,* another species, may be recovered from skin lesions of patients; it requires no lipid additives. It is now seen as the cause of fungemia in immunocompromised patients.

Direct detection

Most often, detection of *M. furfur* is made by direct microscopic examination of skin scales. Here the organism will easily be recognized as oval-or bottle-shaped cells that exhibit monopolar budding in the presence of a cell wall with a septum at the site of the bud scar. In addition, small hyphal fragments are observed (Figure 65-12). In cases of fungemia, the morphologic form seen in the direct examination of the broth of blood cultures are small yeasts without the presence of pseudohyphae.

Culture and identification

M. furfur is not usually cultured in the clinical laboratory. Recovery of the organism is not required to establish a diagnosis (in skin infections), and it is seldom attempted; if one wishes to culture the organism, an agar medium overlaid with a fatty acid (olive oil) is used.

M. furfur may be recovered from the blood of patients with fungemia. In most instances, the residual lipid (from lipid replacement therapy) is adequate to support the primary growth of the organism in culture. Subculture onto additional media requires overlayment of the inoculum with olive oil.

RAPID SCREENING TESTS FOR THE IDENTIFICATION OF YEASTS

RAPID UREASE TEST

The rapid urease test (Procedure 65-8) is a most useful tool for screening for urease-producing yeasts recovered from respiratory secretions and other clinical specimens. Alternatives to this method include the heavy inoculum of the tip of a slant of Christensen's urea agar and subsequent incubation at 35° to 37° C. In many instances, a positive reaction will occur within several hours; however, 1 to 2 days of incubation may be required. It is of interest to note that strains of *Rhodotorula, Candida,* and *Trichosporon* occasionally hydrolyze urea. Therefore, the microscopic morphologic features will be of help in interpreting the usefulness of the urease test for the detection of *Cryptococcus.* Another alternative method is the rapid selective urease test (Procedure 65-9). This method appears to be useful in detecting *C. neoformans* from specimens cultured onto Sabouraud dextrose agar. It has not been evaluated using isolates tested from other media.

All of the tests mentioned provide a tentative identification of *C. neoformans;* however, they must be supplemented with additional tests before a preliminary identification can be reported. These include the rapid nitrate reduction test and the detection of phenoloxidase production.

It is helpful to use these screening tests to make a presumptive identification of *C. neoformans.* In instances in which inoculum is limited, the laboratorian must use tests that can be performed and then prepare a subculture so that additional tests can be performed later. It is often just as fast to inoculate the organism onto the surface of a plate of niger seed agar; results may be obtained during the same day of incubation at 25° C. This will be discussed further within this section.

GERM TUBE TEST

The germ tube test (Procedure 65-10) is the most generally accepted and economical method used in the clinical laboratory for the identification of yeasts. Approximately 75% of the yeasts recovered from clinical specimens are *C. albicans,* and the germ tube test usually provides a definitive identification of this organism within 3 hours.

Germ tubes appear as hyphal-like extensions of yeast cells, produced usually without a constriction at the point of origin from the cell (see Figure 65-2). In the past, it has been emphasized that *Candida stellatoidea* also produces germ tubes, and the distinction between *C. albicans* should be made. *The Yeasts: A Taxonomic Study,* written by Kreger-Van Rij,[62] indicates that *C. stellatoidea* is no longer a valid species and has been combined with *C. albicans.* The germ tube test is specific for the identification of *C. albicans* with the exception of an occasional isolate of *C. tropicalis* that may rarely produce germ tubes.

Another method of identification of *C. albicans* is based on the presence of chlamydospores (see Figure 65-93) on cornmeal agar containing 1% Tween 80 and trypan blue incubated at room temperature for 24 to 48 hours. In addition, the appearance of spiderlike colonies on eosin methylene blue agar is characteristic of *C. albicans* and may be used to make a final identification by persons experienced with this method.

Because the germ tube test requires 3 hours of incubation before results are available, a search for a more rapid alternative is under way. *Candida albicans* produces β-glactose aminidase and L-proline aminopeptidase. Other *Candida* spp. may produce one enzyme, but not both. Assays including the BactiCard Candida (Remel Laboratories, Lexena, Kan.), Murex-C.albicans-50 (Murex Diagnostics, Norcross, Ga.), and Albicans-sure (Clinical Standards Laboratories, Rancho Dominguez, Calif.) were designed to detect these enzymes.

Heelan et al.[44] compared the germ tube test to

BactiCard, Murex-C.albicans-50, Albicans-sure, and the API 20C AUX yeast identification system. All rapid enzymatic screening methods were sensitive and specific for the rapid identification of *C. albicans*. Compared with the germ tube test, all required less time (5 to 30 minutes), were more expensive, and required some additional equipment. Overall, all methods provided rapid and objective alternatives to the germ tube test. Laboratories that wish to use these methods would need to prepare a financial impact analysis.

COMMERCIALLY AVAILABLE YEAST IDENTIFICATION SYSTEMS

As previously mentioned, commercially available yeast identification systems have provided laboratories of all sizes with standardized identification methods. The methods for the most part are rapid, and results are available within 72 hours. The major advantage is that the systems provide an identification based on a database of thousands of yeast biotypes that considers a number of variations and substrate utilization patterns. Another advantage is that manufacturers of these products provide computer consultation services to help the laboratorian with the identification of isolates that give an atypical result.

API-20C AUX YEAST SYSTEM

The API-20C AUX yeast identification system (bio-Mérieux, St Louis, Mo.) has perhaps the most extensive computer-based data set of all commercial systems available. The system consists of a strip that contains 20 microcupules, 19 of which contain dehydrated substrates for determining utilization profiles of yeasts. Reactions are compared with growth in the first cupule, which lacks a carbohydrate substrate. Reactions are read, and results are converted to a seven-digit biotype profile number. Most of the yeasts are identified within 48 hours; however, species of *Cryptococcus* and *Trichosporon* may require up to 72 hours. It should be mentioned that the API-20C AUX yeast identification system, as well as all the other commercially available products, requires that the microscopic morphologic features of yeast grown on cornmeal agar containing 1% Tween 80 and trypan blue be used in conjunction with the substrate utilization patterns. This is particularly helpful when more than one possibility for an identification is provided and the microscopic morphologic features can be used to distinguish between the possibilities given by the profile register. Several evaluations of the API-20C AUX yeast identification system have been made and results have all been favorable. This system is limited in that it cannot identify unusual species; however,

most of those seen in the clinical laboratory are accurately identified to the species level.

A new system, API Candida (bioMérieux, St Louis, Mo.), was developed for the identification of yeasts. The sensitivity with and without additional tests was high while it was unsatisfactory without additional testing. This system proved to be easy to use and generated results within 18 to 24 hours. This system appears to have promise for use by clinical microbiology laboratories.[33]

UNI-YEAST TEK SYSTEM

The Uni-Yeast Tek yeast identification system (Remel Laboratories, Lexena, Kan.) consists of a sealed, multicompartment plate containing media used to indicate carbohydrate utilization, nitrate utilization, urease production, and cornmeal agar morphology. Past evaluations of this product have shown that it is generally satisfactory for the identification of commonly encountered yeasts.[118] A major disadvantage of this system is that it requires up to 7 days for the complete identification of some yeasts. It is now being redeveloped.

MICROSCAN YEAST IDENTIFICATION PANEL

The MicroScan Yeast Identification Panel (Baxter-MicroScan, West Sacramento, Calif.) is a 96-well microtiter plate containing 27 dehydrated substrates and was introduced as an alternative to the API-20C AUX yeast identification system. It utilizes chromogenic substrates to assess specific enzyme activity detected within 4 hours. Specific enzyme profiles have been generated for many of the yeasts commonly encountered in the clinical microbiology laboratory. The most recent evaluation of the method showed that it was moderately accurate within 4 hours using no supplementary tests. When supplementary tests were used, the sensitivity was excellent compared with the API-20C AUX yeast identification system. The accuracy for the identification of common yeasts was high while uncommon yeasts were identified in most instances.[99] An evaluation of the MicroScan Yeast Identification Panel used with the Baxter MicroScan WalkAway was reported by Riddle et al.[87] and was less satisfactory when the panel was used with morphology on cornmeal agar. It is recommended that Cornmeal Tween-80 agar be used in conjunction with the Yeast Identification Panel.

VITEK BIOCHEMICAL CARD

The Yeast Biochemical Card (Vitek Systems, Inc., Hazelwood, Mo.) is a 30-well disposable plastic card that contains conventional biochemical tests and negative controls. The Yeast Biochemical Card is used with the automated AutoMicrobic System, which is utilized for bacterial identification and susceptibility testing in many laboratories. The most recent evaluation of this system showed an overall accuracy of identification near 100% compared with API-20C AUX. Less than one fourth of the yeasts required supplemental biochemical or morphologic features to confirm their identification. Of all correctly identified yeasts, over half were reported after 24 hours of incubation.[28] The accuracy of identification of common and uncommon species was satisfactory. It is not necessary to identify germ-tube–positive yeasts using this system. For laboratories already utilizing this system, accurate and reliable identification of most commonly encountered yeasts can be accomplished. Another evaluation by Dooley et al.[23] showed that the Vitek YBC system correctly identified 93% of common yeasts; however, only 55% of uncommon yeasts were correctly identified. The latter included isolates such as *Candida guilliermondii, Candida krusei, Candida lambica, Candida lusitaniae, Candida rugosa,* species of *Cryptococcus* (non-neoformans), *Geotrichum candidum, Rhodotorula, Saccharomyces,* and *Trichosporon.* It appears that the databases need an update with the inclusion of these organisms.

Interest in commercially available yeast identification systems has taken precedence over the more cumbersome and labor-intensive conventional yeast identification methods. Currently the rapid identification methods are affordable and enable laboratories of all sizes to identify yeasts. Commercially available systems are recommended for all laboratories and provide accurate and rapid identification of yeasts and yeastlike organisms. In general, they are easy to use and interpret and are relatively inexpensive compared with conventional methods. In most instances, they are faster than conventional systems, provide more standardized results, and require less technical skill to perform.

CHROMAGAR

CHROMagar (CHROMagar Co., Paris, France) is a differential medium useful for the recovery, isolation of colonies, and differentiation of species of *Candida* found in clinical specimens. Each species reacts with a chromogenic substrate to yield a characteristic colony color. When used with colonial morphologic features, the medium can provide a presumptive identification. Sand-Millan et al.[91] reported an evaluation of 1537 isolates of yeast, which after 48 hours of incubation at 37° C, showed CHROMagar had a sensitivity and specificity near 100% for *C. albicans, C. tropicalis,* and *Candida krusei.*

Another evaluation by Pfaller et al.[83] showed that more than 95% of stock and clinical isolates of *C. albicans, C. tropicalis,* and *C. krusei* were correctly identified. Further, a similar sensitivity was observed

for *Candida glabrata.* In addition, CHROMagar was evaluated as a recovery medium and was found to detect mixed cultures of *Candida* spp. Considering that the previously mentioned species account for approximately 90% of the yeast recovered in the clinical laboratory, CHROMagar appears to be a suitable alternative to the other yeast identification systems.

RAPIDID YEAST PLUS SYSTEM

The RapidID Yeast Plus System (Innovative Diagnostic Systems, Norcross, Ga.) is a qualitative micromethod employing conventional and chromogenic substrates to identify the medically important yeasts. Kitch et al.[57] showed that the system correctly identified more than 90% of 286 strains tested within 5 hours, without the need for additional tests. It appears that the RapidID Yeast Plus System can be a useful alterative to other commercial systems.

CONVENTIONAL YEAST IDENTIFICATION METHODS

A few laboratories still prefer to use conventional methods to identify yeasts.[43,118] Regardless of the type of identification system used, the germ tube test is a beginning step in screening a large number of isolates. As previously mentioned, approximately 75% of yeasts recovered in the clinical laboratory can be identified using the germ tube test.

CORNMEAL AGAR MORPHOLOGY

The second major step using this practical identification schema is to use cornmeal agar morphology as a means to determine if the yeast produces blastoconidia, arthroconidia, pseudohyphae, true hyphae, and/or chlamydospores (Procedure 65-11). In the past, cornmeal agar morphology was used successfully to detect characteristic chlamydospores produced by *C. albicans.* This method is currently satisfactory for the definitive identification of *C. albicans* when the germ tube test is negative. In other instances, microscopic morphologic features on cornmeal agar differentiate the genera *Cryptococcus, Saccharomyces, Candida, Geotrichum,* and *Trichosporon.* Previously, it was believed that the morphologic features of the common species of *Candida* were distinct enough to provide a presumptive identification. This can be accomplished for *C. albicans, C. (Torulopsis) glabrata, C. krusei, C. parapsilosis, C. tropicalis,* and *C. pseudotropicalis* if one keeps in mind that there are numerous other species, uncommonly recovered in the clinical laboratory, that might resemble microscopically any of the previously mentioned species. In general, this method performs well since the previously mentioned genera and species are more commonly seen in clinical laboratories. For the uncommonly encountered isolates, cornmeal agar morphology will have less value. It is, however, recommended for use with most commercially available yeast identification systems and plays a major role in the differentiation between genera that yield similar biochemical profiles.

CARBOHYDRATE UTILIZATION

Carbohydrate utilization patterns are the most commonly used conventional methods for the definitive identification of yeast recovered in a clinical laboratory. Various methods have been advocated for use in determining carbohydrate utilization patterns by clinically important yeast, and all work equally well. Procedure 65-12 outlines the method previously found to be most useful by the Mayo Clinic Mycology Laboratory.

Once the carbohydrate utilization profile is obtained, reactions may be compared with those listed in tables in most mycology laboratory manuals.[62] In most instances, carbohydrate utilization tests provide the definitive identification of an organism, and additional tests are unnecessary. Carbohydrate fermentation tests are preferred by some laboratories and are simply performed using purple broth containing different carbohydrate substrates. In general, carbohydrate fermentation tests are unnecessary and are not recommended for routine use.

PHENOLOXIDASE DETECTION USING NIGER SEED AGAR

A simplified *Guizotia abyssinica* medium (niger seed medium) is a definitive method for detection of phenoloxidase production on yeasts.[55,81] Most isolates of *C. neoformans* readily produce phenoloxidase; however, some do not (Procedure 65-13). In addition, in some instances cultures of *C. neoformans* have been shown to contain both phenoloxidase-producing and phenoloxidase-deficient colonies within the same culture. If conventional methods are used, it is necessary to use all the criteria, including urease production, carbohydrate utilization, and the phenoloxidase test, before making a final identification of *C. neoformans*.

ANTIFUNGAL AGENTS

Numerous antifungal agents have been developed, and newer agents are on the horizon. The growing numbers of immunosuppressed patients and the increase in antimicrobial drug resistance makes the development and appropriate use of antimicrobial agents one of the most important areas in microbiology and infectious diseases. This section is meant only to introduce readers to the more commonly used antifungal agents and is by no means comprehensive. This section is not to be used as a guide for therapy. Therapeutic guidelines may be found in the referenced texts.

POLYENE MACROLIDE ANTIFUNGALS

Polyene macrolide antifungal agents consist of a group of complex organic molecules, most of which contain multiple, conjugated double bonds and one to three ring structures. This group includes many of the most commonly used antifungal agents, such as amphotericin B, nystatin, 5-fluorocytosine, and griseofulvin.

AMPHOTERICIN B

Amphotericin B is produced by the actinomycete *Streptomyces nodosus*. It is commonly infused intravenously to treat deep-seated fungal infections caused by the dimorphic pathogens, as well as those caused by *Candida* spp., *Crytococcus*, and members of the Zygomycetes. Amphotericin B binds the ergosterol component of the fungal cell membrane and alters the selective permeability of this membrane. However, other sterols, including those present in mammalian cell membranes, are also bound. The most significant adverse reaction associated with amphotericin B therapy is renal insufficiency. A newer agent, liposomal amphotericin B, reportedly diminishes this adverse reaction.

NYSTATIN

Nystatin, an antifungal antibiotic produced by *Streptomyces noursei*, is not absorbed by the gastrointestinal tract. It is principally used locally to treat oral or vulvovaginal candidiasis. Toxicity of this drug is prohibitive to parenteral use.

5-FLUOROCYTOSINE (FLUCYTOSINE)

Flucytosine is a pyrimidine base, which is fluorinated in the fifth position. Flucytosine is metabolized to 5-fluorouracil, which is incorporated into fungal RNA. This subsequently inhibits protein synthesis. Flucytosine is also metabolized into fluorodeoxyuridine monophosphate, which is a potent inhibitor of DNA synthesis. Flucytosine and amphotericin B act synergistically and are useful as combination therapy for treating infections by *Candida* spp. and *C. neoformans*.

GRISEOFULVIN

Griseofulvin is an antifungal antibiotic produced by a species of *Penicillium*. Its mechanism of action consists of binding microtubular proteins, which are required for mitosis. Griseofulvin is an oral agent used to treat dermatophytoses, which are not responsive to azole antifungal therapy. Headache, gastrointestinal disturbances, and photosensitivity are a few of the adverse reactions that limit the usefulness of this drug.

AZOLE ANTIFUNGAL DRUGS

The azole group of antifungal agents consist of the imidazoles and the triazoles. These compounds contain six carbon-ring structures with conjugated double bonds, chloride residues, and five carbon ring structures that contain at least two nitrogen molecules. Commonly used agents included in this group include clotrimazole, miconazole, fluconazole, itraconazole, and ketoconazole. These antifungal agents disrupt the integrity of the fungal cell membrane by interfering with the synthesis of ergosterol.[3]

CLOTRIMAZOLE AND MICONAZOLE

Clotrimazole and miconazole, which are both synthetic imidazoles, are covered together because of their many similarities. These agents are available for topical or intravaginal applications. They are useful in mild cases of dermatophytosis, including tinea versicolor. Adverse reactions are generally limited to burning, itching, and/or skin irritation.

FLUCONAZOLE

Fluconazole is a triazole, which is exceptionally soluble in water. The solubility of this compound allows either oral or intravenous administration. Fluconazole has excellent activity against both *Candida* spp. and *Cryptococcus neoformans;* therapeutic levels are easily reached in the central nervous system. Side effects of fluconazole therapy are usually minimal.

ITRACONAZOLE

Itraconazole is a triazole with a spectrum of activity that encompasses that of ketoconazole. Additionally, itraconazole is effective in cases of aspergillosis, sporotrichosis, cryptococcosis, and onchymycosis. Adverse reactions principally include gastrointestinal disturbances; however, vestibular disturbances, edema, and skin irritations have been reported.

KETOCONAZOLE

Ketoconazole is an imidazole, which is either taken orally or applied topically. It is useful in mild cases of paracocidioidomycosis and is an alternative to amphotericin B for infections caused by *Blastomyces* or *Histoplasma*. Ketoconazole is the drug of choice for prolonged oral therapy for chronic mucocutaneous candidiasis. One group has reported some success in the treatment of *P. boydii* infections with ketoconazole.[36] In vivo, ketoconazole is fungistatic,

because fungicidal levels are not achievable with therapeutic concentrations. Adverse reactions include transient elevations in liver enzymes, nausea, and dose-related gynecomastia, decreased libido, and oligospermia in males.

OTHER ANTIFUNGAL AGENTS

Selenium sulfide

Selenium sulfide shampoos, available commercially, disclose antifungal activity against *M. furfur,* the causative agent of tinea versicolor. Additionally, selenium sulfide is sporocidal for *T. tonsurans* and therefore may be used as an adjuvant to griseofulvin therapy.

Potassium iodide

Potassium iodide is the therapy of choice for cutaneous/lymphatic sporotrichosis. Localized heat therapy may also be use adjunctively. Some individuals are allergic to potassium iodide. Adverse reactions include a bitter taste, allergic rash, and anorexia.

ANTIFUNGAL SUSCEPTIBILITY TESTING

Antifungal susceptibility tests[30,31,35] are designed to provide information that will allow the physician to select the appropriate antifungal agent useful for treating a specific infection. Unfortunately, antifungal susceptibility testing has not progressed as far as tests used for determining susceptibility of bacteria to antimicrobial agents. However, much effort has gone into the standardization of a method that is reproducible between laboratories. All variables have been standardized, and efforts are under way to develop interpretative guidelines for different antifungal agents. The National Committee for Clinical Laboratory Standards (NCCLS) has tentatively accepted the M27-A document,[35] which consists of current guidelines for antifungal susceptibility testing for *Candida* and *Cryptococcus*.

It must be emphasized that the methodology and interpretation of antifungal susceptibility tests are still evolving. These costly and time-consuming tests may be of value in the following circumstances:

- When determining antibiograms for isolates within an institution
- To aid in the management of patients with refractory oropharyngeal candidiasis
- To aid in the management of patients having invasive candidiasis when the use of the azoles is questioned in infections caused by non-*C. albicans*

The interpretative breakpoints for fluconazole, itraconazole, and flucytosine are based on experiences drawn from treating patients with mucosal infections but appear to be consistent with information assembled for invasive infections. Problems that complicate the interpretative guidelines include:

- Physical condition of the patient, i.e., immunologic status
- Type of infection and ability of the drug to penetrate a closed space (in the case of an abscess)
- Dose of the drug and its pharmacokinetics
- Susceptibility testing method used and serum level of drug that is administered

Many isolates of the same species exhibit a different minimum inhibitory concentration (MIC). Some isolates of *Candida glabrata* exhibit susceptibility to fluconazole while others do not; the trend is that most isolates are resistant.

Anecdotal experience provides the best experience with antifungal chemotherapy. For example, it is known that patients infected with *Candida krusei* do not respond to therapy with fluconazole. Patients infected with *Candida lusitaniae* do not respond to amphotericin B treatment. The interpretative guidelines as defined by the NCCLS are as follows:

ANTIFUNGAL AGENT	SUSCEPTIBLE CONCENTRATION	RESISTANT CONCENTRATION
Fluconazole	≤ 8.0 µg/ml	≥ 64.0 µg/ml
Itraconazole	≤ 0.125 µg/ml	≥ 1.0 µg/ml
Flucytosine	≤ 4.0 µg/ml	≥ 32.0 µg/ml

Amphotericin B susceptibility or resistance cannot be distinguished using the NCCLS method. It is suggested that an MIC ≥ 1 µg/mL is considered as resistant; however, this information is tentative. Ketoconazole susceptibility testing has suggested that isolates having an MIC of between 0.313 and 16 µg/mL are considered as susceptible.

Despite the problems associated with antifungal susceptibility testing, many physicians believe that these tests are important for the selection of an appropriate antifungal agent and as a method to detect the development of resistance of certain organisms during chemotherapy. The method described by the NCCLS (M27-A)[36] is advocated in the United States. A laboratory not equipped for such studies should be prepared to send an important isolate to a reference laboratory for testing. Amphotericin B, 5-fluorocytosine, ketoconazole, itraconazole, and fluconazole are common antifungal agents currently available. Just how one can use the data generated by antifungal susceptibility testing remains to be determined. Progress is being made; a method has been standardized and now newer methods can be compared. Unfortunately, it will require some time before interpretative guidelines will be definite. Simply stated, most patients are clinically too complex to allow for an objective evaluation of therapeutic failure or success.

Demand for susceptibility testing of itraconazole and fluconazole continues to increase despite problems with the interpretation of results. References that may be helpful to the reader to understand the usefulness of these tests are included at the end of this chapter.[31,36,84,85]

Procedures

65-1 ADHESIVE (SCOTCH) TAPE PREPARATION

METHOD

1. Touch the adhesive side of a small length of transparent tape to the surface of the colony.

2. Adhere the length of tape to the surface of a microscope slide to which a drop of lactophenol cotton or aniline blue has been added (see Figure 65-29).

3. Observe microscopically for the characteristic shape and arrangement of the spores.

The transparent adhesive tape preparation allows one to observe the organism microscopically approximately the way it sporulates in culture. The spores are usually intact, and the microscopic identification of an organism can be made easily. If the tape is not pressed firmly enough to the surface of the colony, the sample may not be adequate for an identification. In instances in which spores are not observed, a wet mount should be made as a backup step. There have been situations where the macroconidia of *H. capsulatum* were seen in wet mount preparations when the adhesive tape preparation revealed only hyphal fragments. On the contrary, in some instances cultures have sporulated heavily and revealed only the presence of conidia when the adhesive tape preparation was observed. In this type of situation, a second adhesive tape preparation should be made from the periphery of the colony where sporulation is not as heavy.

Some laboratories prefer to use the microslide culture (Procedure 65-3) for making the microscopic identification of an organism. This method might appear to be the most suitable because it allows one to observe microscopically the fungus growing directly underneath the coverslip. Microscopic features should be easily discerned, structures should be intact, and many representative areas of growth are available for observation.

65-2 WET MOUNT

METHOD

1. With a wire bent at a 90-degree angle, cut out a small portion of an isolated colony. The portion should be removed from a point intermediate between the center and the periphery. The portion removed should contain a small amount of the supporting agar.

2. Place the portion onto a slide to which a drop of lactophenol cotton or aniline blue has been added (see Figure 65-30).

3. Place a coverslip into position and apply gentle pressure with a pencil eraser or other suitable object to disperse the growth and the agar. Examine microscopically.

The major disadvantage of the wet mount is that the characteristic arrangement of spores is disrupted when pressure is applied to the coverslip. This method is suitable in many instances since characteristic spores are often seen but their arrangement cannot be determined; it is not adequate to make a definitive identification.

65-3 MICROSLIDE CULTURE

METHOD

1. Cut a small block of a suitable agar medium that has been previously poured into a culture dish to a depth of approximately 2 mm. The block may be cut using a sterile scalpel blade or with a sterile test tube that has no lip (which produces a round block).

2. Place a sterile microscope slide onto the surface of a culture dish containing sterile 2% agar. Alternatively, place a round piece of filter paper or paper towel into a sterile culture dish, add two applicator sticks, and position the microscope slide on top.

3. Add the agar block to the surface of the sterile microscope slide.

4. With a right-angle wire, inoculate the four quadrants of the agar plug with the organisms (see Figure 65-31).

5. Apply a sterile coverslip onto the surface of the agar plug.

6. If the filter paper applicator stick method is used, add a small amount of sterile water to the bottom of the culture dish. Replace the lid of the culture and allow it to incubate at 30° C.

7. After a suitable incubation period, remove the coverslip (working inside of a biological safety cabinet) and place it on a microscope slide containing a drop of lactophenol cotton or aniline blue. It is often helpful to place the coverslip near the opening of an incinerator-burner to allow rapid drying of the organism on the coverslip to occur before adding it to the stain.

8. Observe microscopically for the characteristic shape and arrangement of spores.

9. The remaining agar block may be used later (if the microslide culture is unsatisfactory for the microscopic identification) if it is allowed to incubate further. The agar plug is then removed and discarded, and a drop of lactophenol cotton or aniline blue is placed on the area of growth and a coverslip is positioned into place. Many laboratorians like to make two cultures on the same slide so that if characteristic microscopic features are not observed on examination of the first culture, the second will be available after an additional incubation period.

Although this method is ideal for making a definitive identification of an organism, it is the least practical of all the methods described. It should be reserved for those instances in which an identification cannot be made based on an adhesive tape preparation or wet mount. CAUTION: *Do not make slide cultures of slow-growing organisms suspected to be dimorphic pathogens such as* H. capsulatum, B. dermatitidis, C. immitis, P. brasiliensis, *or* S. schenckii. Microslide cultures must be observed only after a coverslip has been removed from the agar plug and not while it is in position on top of the agar plug. The latter method of observation is very dangerous and is not to be used by employees in the clinical laboratory to reduce the possibility of a laboratory-acquired infection.

65-4 HAIR PERFORATION TEST

METHOD

1. Place a filter paper disk into the bottom of a sterile culture dish.

2. Cover the surface of the paper disk with sterile distilled water.

3. Add a small portion of sterilized hair into the water.

4. Inoculate a portion of the colony to be studied directly onto the hair.

5. Incubate at 25° C for 10 to 14 days.

6. Observe hairs regularly by placing them into a drop of water on a microscope slide. Position a coverslip and examine microscopically for the presence of conical perforations of the hair shaft (see Figure 65-50).

65-5 IN VITRO CONVERSION OF DIMORPHIC MOLDS[119]

PRINCIPLE

Dimorphic molds exist in the yeast or spherule form in infected tissue. Proof that a mold is actually one of the systemic dimorphic fungi can be achieved by simulating the environment of the host and converting the mold to the yeast form. This method is recommended only for the identification of *B. dermatitidis*, *P. marneffei*, or *S. schenckii*. Conversion of *C. immitis* to its spherule form requires special media and is not recommended for clinical laboratories. *Histoplasma capsulatum* most often will not convert from the mold to the yeast form in vitro, and for this reason, it should not be attempted.

METHOD

1. Transfer a large inoculum of the mold form of the culture onto the surface of a fresh, moist slant of brain-heart infusion agar containing 5% to 10% sheep blood. If *B. dermatitidis* is suspected, a tube of cottonseed conversion medium should be inoculated.

2. Add a few drops of sterile distilled water to provide moisture if the surface of the culture medium appears to be dry.

3. Leave the cap of the screw-capped tube slightly loose to allow the culture to have adequate oxygen exchange.

4. Incubate cultures at 35° to 37° C for several days; observe for the appearance of yeastlike portions of the colony. It may be necessary to make several subcultures of any growth that appears, since several transfers are often required to accomplish the conversion of many isolates. Cultures of *B. dermatitidis*, however, are usually easily converted and require 24 to 48 hours on cottonseed agar medium. *C. immitis* may be converted in vitro to the spherule form using a chemically defined medium; however, this method is of little use to the clinical laboratory and it should not be attempted.

QUALITY CONTROL

Because of their hazardous nature, it is not recommended that stock cultures be tested routinely. An extract of control strains can be used as a positive control. The exoantigen test (Procedure 65-6) can serve as one method to identify the dimorphic fungi.

Because the conversion of the dimorphic molds to the corresponding yeast or spherule forms is technically cumbersome to perform and long delays are often experienced, the effort to convert the dimorphic fungi is not recommended and organisms can be identified by **exoantigen testing** (Procedure 65-7). This technique is used in many laboratories to make a definitive identification of *B. dermatitidis*, *C. immitis*, *H. capsulatum*, and *P. brasiliensis*. The exoantigen test relies on the principle that soluble antigens are produced and can be extracted from fungi; they are concentrated and subsequently reacted with serum known to contain antibodies directed against the specific antigenic components of the organism being tested. Reagents and materials for the exoantigen test are available commercially.

65-6 EXOANTIGEN TEST[100,101]

PRINCIPLE

Specific antibodies developed against particular mycelial antigens will react in a gel immunodiffusion precipitin test. The mold forms of the dimorphic fungi can be identified definitively by an antigen-antibody reaction, negating the need for conversion to the yeast form.

METHOD

1. A mature fungus culture on a Sabouraud dextrose agar slant is covered with an aqueous solution of merthiolate (1:5000 final concentration), which is allowed to remain in contact with the culture for 24 hours at 25° C. It is necessary that the entire surface of the colony be covered so that effective killing of the organism is ensured and solubilization of the exoantigen is maximized.

2. Filter the aqueous solution, which overlays the culture, through a 0.45 μm-size membrane filter. This should be performed inside a biological safety cabinet.

3. Five milliliters of this solution is concentrated using a Minicon Macrosolute B-15 Concentrator (Amicon Corp., Danvers, Mass.). The solution is concentrated 50× when testing with *H. capsulatum* and *B. dermatitidis* antiserum and 5× and 25× for reaction with *C. immitis* antiserum.

4. The concentrated supernatant is used in the microdiffusion test. The supernatant is placed into wells punched into a plate of buffered, phenolized agar adjacent to the control antigen well and is tested against positive control antiserum obtained from commer-

cial sources (Immunomycologics, Washington, Okla.; Meridian Diagnostics, Cincinnati, Ohio; and Gibson Laboratories, Lexington, Ky.)

5. The immunodiffusion test is allowed to react for 24 hours at 25° C, and the plate is observed for the presence of precipitin bands of identity with the reference reagents. The sensitivity of the exoantigen test for the identification of *B. dermatitidis* may be increased by incubating the immunodiffusion plates at 37° C for 48 hours; however, bands appear sharper at 25° C after 24 hours. It is recommended that any culture suspected of being *B. dermatitidis* be incubated at both temperatures.

6. *C. immitis* may be identified for the presence of the CF, TP, or HL

65-6 EXOANTIGEN TEST[100,101]—cont'd

antigens, while *H. capsulatum* and *B. dermatitidis* may be identified by the presence of H or M bands, or both, and the A band, respectively. Detailed instructions for the performance and interpretation of the tests are included with the manufacturers' package insert.

QUALITY CONTROL

Extracts from known fungi are tested each time the test is per-formed. Lines of identity with the unknown strain are necessary for identification.

The exoantigen test is rapidly being replaced by specific nucleic acid probes that provide for the rapid and specific identification of *H. capsulatum*, *B. dermatitidis*, and *C. immitis*.

65-7 NUCLEIC ACID PROBE TESTING[41,103]

PRINCIPLE

The Gen-Probe AccuProbe System (San Diego, Calif.) for culture identification of fungi is based on DNA probes that are complimentary to species-specific ribosomal RNA. The fungal cells are lysed by sonication, heat killed, and exposed to DNA that has been labeled with a chemiluminscent tag. The labeled DNA probe combines with the organism's ribosomal RNA to form a stable DNA/RNA hybrid. A selection reagent "kills" the signal on all the unbound DNA probe. The labeled DNA/RNA hybrids are measured in a luminometer. Probes are available for the culture identification of *Blastomyces dermatitidis*, *Coccidioides immitis*, and *Histoplasma capsulatum*.

SPECIMEN

Use actively growing culture less than 1 month old taken from solid medium such as Sabouraud, inhibitory mold, brain-heart infusion, cornmeal or potato dextrose agar. Inocula taken from a medium containing blood is not recommended for direct probe testing.

METHOD

1. De-gas the water for 5 minutes. Turn on 95° C water bath (or heat block) and 60° C water bath (or heat block).

2. Add 100 μL of Reagent 1 (lysis reagent) and 100 μL Reagent 2 (hybridization buffer) to each lysing tube.

3. Working in a biological safety cabinet, transfer a full 10-μL loop of organism to be tested into the lysing reagent tube. Twirl the loop against the side of the tube to remove the inoculum from the loop.

4. Place the lysing reagent tubes in the sonicator. Be sure the water level is high enough so the contents of the tube are below the water level. Do not allow tubes to touch the sides of the sonicator. Sonicate at room temperature for 15 minutes.

5. Place the lysing tubes in the 95° C water bath (or heat block) for 15 minutes.

6. Allow tubes to cool at room temperature for 5 minutes.

7. Pipette 100 mL of the cell lysate into the probe reagent tube.

8. Incubate the tubes for 15 minutes at 60° C in the water bath (or heat block).

9. Pipette 300 μL of Reagent 3 (Selection Reagent) into each tube. Vortex, recap, and place the tube immediately back into the 60° C water bath (or heat block). Incubate 5 minutes.

10. Prepare the luminometer for operation by completing two wash cycles. Using a damp tissue, wipe each tube before inserting into the luminometer (to prevent static build-up). Read each tube and controls. The luminometer records relative light units (RLU).

FREQUENCY AND TOLERANCE OF CONTROLS

Run a positive and negative control with each batch of organisms tested.

EXPECTED VALUES

A positive result is >50,000 relative light units (RLU). Signals below 50,000 RLU are considered negative.

LIMITATIONS

This method cannot be used on fresh clinical specimens. It is for culture identification only. The use of nucleic acid probes with cultures grown on a blood-containing medium is unsatisfactory for testing, since hemin will chemiluminesce and yield a false-positive result.

65-8 RAPID UREASE TEST[89]

PRINCIPLE

The hydrolysis of urea by the enzyme urease produces ammonia and carbon dioxide. The ammonia produces alkaline conditions in the medium and changes the indicator (phenol red) from yellow to pink.

METHOD

1. Reconstitute a vial of dehydrated urea broth (Remel Laboratories, Lexena, Kan.) with 3 mL of sterile distilled water on the day of use.

2. Dispense 3 to 4 drops into each well of a microdilution plate. Determine the exact number of wells to be used for the day and use only the number necessary.

3. Transfer a heavy inoculum of a yeast colony (not including pink yeasts) to a well containing the urea broth. Colonies tested should be no older than 7 days and should be free of bacterial contamination. In some instances it may be necessary to make a subculture to obtain enough growth to provide the inoculum for the test.

4. Include positive and negative controls using *C. neoformans* and *C. albicans*, respectively.

5. Seal the microdilution wells with plastic tape and incubate 4 hours at 37° C.

6. Observe for the production of a pink to purple color, which is indicative of urease production.

QUALITY CONTROL

C. neoformans and *C. albicans* are used as the positive and negative controls, respectively.

EXPECTED RESULTS

C. neoformans should produce urease; *C. albicans* will not.

PERFORMANCE SCHEDULE

Controls should be performed with each test run.

65-9 RAPID SELECTIVE UREASE TEST[125]

PRINCIPLE

Benzalkonium chloride (1%) is added to the test medium to disassociate the cell wall of yeasts to allow the endogenous urease to be released into the test medium. The hydrolysis of urea is detected by the presence of ammonia, which changes the indicator to pink because of the alkaline conditions produced.

METHOD

1. Sweep a cotton-tipped applicator impregnated with the dehydrated urea substrate over the surface of two or three colonies so that the tip is well covered with the organism.

2. Place the applicator containing the yeast into a tube containing 3 drops of 1% benzalkonium chloride (pH 4.86 ±0.01) and whirl firmly against the bottom of the tube to place the organisms into contact with the cotton fibers.

3. Add a cotton plug to the tube and incubate at 45° C for up to 30 minutes.

4. Examine after 10, 15, 20, and 30 minutes for the presence of a color change from yellow to purple. A red or purple color indicates urease production by *C. neoformans*.

QUALITY CONTROL

Same as for Procedure 65-8.

65-10 GERM TUBE TEST

PRINCIPLE

Strains of *C. albicans* produce germ tubes from their yeast cells when placed in a liquid nutrient environment and incubated at 35° C for 3 hours (similar to the in vivo state).

METHOD

1. Suspend a very small inoculum of yeast cells obtained from an isolated colony in 0.5 mL of sheep serum (or rabbit plasma).

2. Incubate the tubes at 35° to 37° C for no longer than 3 hours.

3. After incubation, remove a drop of the suspension and place on a microscope slide. Examine under low-power magnification for the presence of germ tubes. A germ tube is defined as an appendage that is half the width and 3 to 4 times the length of the yeast cell from which it arises (Figure 65-2). In most instances, there is no point of constriction at the origin of the germ tube from the cell.

QUALITY CONTROL

C. albicans and *C. tropicalis* are used as positive and negative controls, respectively.

EXPECTED RESULTS

C. albicans will produce germ tubes, usually within 2 hours, while *C. tropicalis* will not.

PERFORMANCE SCHEDULE

Controls should be performed with each test run.

65-11 CORNMEAL AGAR MORPHOLOGY[60]

PRINCIPLE
Polysorbate (Tween) 80 is added to cornmeal agar to reduce the surface tension to allow for development of pseudohyphal, hyphal, and blastoconidial growth of yeasts. Certain species of yeasts develop characteristic morphologic features on this medium.

METHOD
1. Obtain an isolated colony from the primary culture medium.

2. Inoculate a plate of cornmeal agar containing 1% Tween 80 and trypan blue by making three parallel cuts about ½ inch apart at a 45° angle to the culture medium. A sterile coverslip may be added to one area.

3. Incubate the cornmeal agar plate at 30° C for 48 hours.

4. After 48 hours, remove and examine the areas where the cuts into the agar were made for the presence of blastoconidia, arthroconidia, pseudohyphae, hyphae, or chlamydospores. Table 65-15 presents the microscopic morphologic features of the commonly encountered yeasts on cornmeal Tween 80 agar. This method is required for use with commercial systems for yeast identification.

QUALITY CONTROL
C. albicans is tested for production of characteristic features.

EXPECTED RESULTS
C. albicans will produce chlamydospores and clusters of blastoconidia arranged at regular intervals along the pseudohyphae.

PERFORMANCE SCHEDULE
Test a strain of C. albicans each time new media are received in the laboratory or produced, and monthly thereafter.

65-12 CARBOHYDRATE UTILIZATION TESTS[118]

PRINCIPLE
Yeasts and yeastlike fungi utilize specific carbohydrate substrates. Organisms are inoculated onto a carbohydrate-free medium. Carbohydrate-containing filter paper disks are added, and utilization is determined by the presence of growth around the disk. Characteristic carbohydrate utilization profiles are used to identify species of yeasts.

METHOD
1. Prepare a suspension of the yeast in saline or distilled water to a density equivalent to a McFarland No. 4 standard.

2. Cover the surface of a yeast nitrogen base agar plate containing bromcresol purple with the suspension of the yeast cells.

3. Remove the excess inoculum and allow the surface of the agar medium to dry.

4. With sterile forceps, place selected carbohydrate disks onto the surface of the agar. These should be spaced approximately 30 mm apart from each other.

5. Incubate the carbohydrate utilization plate at 30° C for 24 to 48 hours.

6. Remove the plate and observe for the presence of a color change around the carbohydrate-containing disks or the presence of growth surrounding them.

QUALITY CONTROL
Choose a combination of control yeasts that utilize each of the carbohydrates being tested. Often two or three organisms may be necessary to control each carbohydrate (use published tables). Test control yeasts with each new lot number of carbohydrate disks and monthly thereafter.

65-13 PHENOLOXIDASE DETECTION USING NIGER SEED AGAR[81]

PRINCIPLE
C. neoformans is the only species that produces 3,4-dihydroxyphenylalanine-phenoloxidase. When reacted with L-β-3,4-dihydroxyphenylalanine and an iron compound (ferric citrate), C. neoformans oxidizes ó-diphenol to melanin, which produces a brown to black color.

METHOD
1. Place a heavy inoculum onto the surface of a plate of niger seed agar medium.

2. Incubate at 25° C for up to 7 days.

3. Observe daily for the presence of a dark brown or black pigment, which is indicative of phenoloxidase specifically produced by C. neoformans (see Figure 65-95). Many isolates of C. neoformans will produce phenoloxidase on this medium within 2 to 24 hours.

QUALITY CONTROL
C. neoformans and C. albidus are used as positive and negative controls, respectively.

EXPECTED RESULTS
C. neoformans should exhibit a positive test (brown color development on the disk surface) while C. albidus will not.

PERFORMANCE SCHEDULE
Controls should be performed with each run.

65-14 CALCOFLUOR WHITE-POTASSIUM HYDROXIDE PREPARATION

METHOD

1. Place a drop of calcofluor white (CW) reagent and a drop of 10% potassium hydroxide (KOH)-glycerin in the center of a microscope slide.

2. Add a portion of the clinical specimen to the CW-KOH solution and apply a coverslip.

3. If necessary, dissociate particles by applying gentle pressure to the coverslip with a pencil eraser. Allow to stand for 5 minutes. If particles do not dissociate, repeat step 3.

4. Examination for fungal elements is performed using a fluorescent microscope with a 400 to 500-nm exciter filter and a 500 to 520-nm barrier filter. Slides are scanned at 10× magnification for fluorescent fungal elements. The presence and nature of fungal elements are discerned using 40× magnification.

Reagents: KOH reagent: (10 gm KOH, 10 mL glycerin, and 80 mL distilled water)
CW reagent: (0.05 gm CW, 0.02 gm Evans blue, and 50 mL distilled water)

Calcofluor white is an industrial textile brightener, which nonspecifically binds to chitin and other elements in the fungal cell wall. CW fluorescence occurs maximally at an excitation wavelength of 440 nm. Under these conditions, fungal elements will fluoresce blue-white. Because CW is a nonspecific stain, an appreciation for fungal element morphology on direct examination is crucial for adequate specimen interpretation. The presence of KOH in the solution dissolves human cellular elements and debris; this allows for easier visualization of fungal elements.

References

1. Adam, R.D., Paquin, M.L., Petersen, E.A. et al. 1986. Phaeohyphomycosis caused by the fungal genera *Bipolaris* and *Exserohilum*: a report of 9 cases and review of the literature. Medicine 65:203.

2. Anaissie, E., Gokaslan, A., Hachem, R. et al. 1992. Azole therapy for trichosporonosis: clinical evaluation of eight patients, experimental therapy for murine infection, and review. Clin. Infect. Dis. 15(5):781.

3. Beggs, W.H., Andrews, F.A., and Sarosoi, G.A. 1981. Actions of imidazole-containing antifungal drugs. Life Sciences 28:111.

4. Bernstein, E.F., Schuster, M.G., and Stieritz, D.D. 1995. Disseminated cutaneous *Pseudallescheria boydii*. Br. J. Dermatol. 132(3):456.

5. Bille, J., Edson, R.S., and Roberts, G.D. 1984. Clinical evaluation of the lysis-centrifugation blood culture system for detection of fungemia and comparison with a conventional biphasic broth blood culture system. J. Clin. Microbiol. 19:126.

6. Bille, J., Stockman, L., Roberts, G.D., et al. 1983. Evaluation of a lysis-centrifugation system for recovery of yeasts and filamentous fungi from blood. J. Clin. Microbiol. 18:469.

7. Bradsher, R.W. and McDonnell, R.W. 1994. *Blastomyces dermatitidis* and *Paracoccidioides brasiliensis*. In Chmel. H., Bendinelli, M., and Friedman I., editors. Pulmonary infections and immunity. Plenum Press, N.Y.

8. Brasch, J. 1990. Pathogens and pathogenesis of dermatophytes. Hautarzt 41(1):9.

9. Brummer, E., Castaneda, E., and Restrepo, A. 1993. Paracoccidioidomycosis: an update. Clin. Microbiol. Rev. 6:89.

10. Bruns, T.D., Vigalys, R., Barns, S.M., et al. 1992. Evolutionary relationships within the fungi: analyses of nuclear small subunit rRNA sequences. Mol. Phylogenet. Evol. 1:231.

11. Caporale, N.E., Calegori, L., Perez, D., et al. 1996. Peritoneal catheter colonization and peritonitis with *Aureobasidium pullulans*. Perit. Dial. Int. 16(1):97.

12. Chagas, C. 1909. Nova trypanosomiaze humana. Mem. Inst. Oswaldo Cruz Rio J. 1:159.

13. Cherniak, R. and Sundstrom, J.B. 1994. Polysaccharide antigen of the capsule of *Cryptococcus neoformans*. Infect. Immun. 62:1507.

14. Cohen, M.S., Isturiz, R.E., Malech, H.L., et al. 1981. Fungal infection in chronic granulomatous disease: the importance of the phagocyte in defense against fungi. Am. J. Med. 71:59.

15. Cox, R.A. and Best, G.K. 1972. Cell wall composition of two strains of *Blastomyces dermatitidis* exhibiting differences in virulence for mice. Infect. Immun. 5:449.

16. Currie, B.P. and Casadevall, A. 1994. Estimation of the prevalence of cryptococcal infection among patients infected with the human immunodeficiency virus in New York City. Clin. Infect. Dis. 19:1029.

17. Cutler, J.E. 1991. Putative virulence factors of *Candida albicans*. Annu. Rev. Microbiol. 45:187.

18. deHoog, G.S. and Guarro, J. 1995. Atlas of clinical fungi. Centraal bureau voor Schimell cultures/Universita Rovira i vergilli. Baarn and Delft, The Netherlands and Reus, Spain.

19. Deng, Z., Ribas, J.L., Gibson, D.W., et al. 1988. Infections caused by *Penicillim marneffei* in China and Southeast Asia: review of eighteen published cases and report of four more Chinese cases. Rev. Infect. Dis. 10(3):640.

20. Denz, Z.L., Yun, M., and Ajello, L. 1986. Human penicilliosis *marneffei* and its relation to the bamboo rat *(Rhizomys x pruinosus)*. J. Med. Vet. Mycol. 24(5):383.

21. Diamond, R.D. 1995. *Cryptococcus neoformans*. In Mandell, G.L., Bennett, J.E., and Dolin, R., editors. Mandell, Douglas and Bennett's principles and practice of infectious diseases. Churchill Livingstone, N.Y.

22. Dick, J.D., Merz, W.G., and Saral, R. 1980. Incidence of polyene-resistant yeasts recovered from clinical specimens. Antimicrob. Agents Chemother. 18:158.

23. Dooley, D.P., Beckius, M.L., and Jeffrey, B.S. 1994. Misiden-

tification of clinical yeast isolates by using updated Vitek Yeast Biochemical Card. J. Clin. Microbiol. 32:2889.

24. Douer, D., Goldschmied-Reouven, A., Segea, S., et al. 1987. Human *Exserohilum* and *Bipolaris* infections: report of *Exserohilum* nasal infection in a neutropenic patient with acute leukemia and review of the literature. J. Med. Vet. Mycol. 25:235.

25. Drutz, D.J., Huppert, M., Sun, S.H., et al. 1981. Human sex hormones stimulate the growth and maturation of *Coccidioides immitis*. Infect. Immun. 32:897.

26. Edman, J.C., Kovacs, J.A., Masur, H., et al. 1988. Ribosomal RNA sequence shows *Pneumocystis* to be a member of the fungi. Nature 334:519.

27. Edman, J.C., Kovacs, J.A., Masur, H., et al. 1989. Ribosomal RNA genes of *Pneumocystis carinii*. J. Protozool. 36:18S.

28. El-Zaatari, M., Pasarell, L., McGinnis, M.R., et al. 1990. Evaluation of the updated Vitek Yeast Identification data base. J. Clin. Microbiol. 28:1938.

29. Ener, B. and Douglas L.J. 1992. Correlation between cell-surface hydrophobicity of *Candida albicans* and adhesion to buccal epithelial cells. FEMS Microbio. Letters 78(1):37.

30. Espinel-Ingroff, A. 1996. Antifungal susceptibility testing. Clin. Microbiol. Newsletter 184(21):161.

31. Espinel-Ingroff, A. and Pfaller, M. 1995. Antifungal agents and susceptibility testing. In Murray, P.R., Baron, E.J., Pfaller, M.A., et al., editors. Manual of clinical microbiology, ed 6. American Society for Microbiology Press, Washington, D.C.

32. Frey, C.L., and Drutz, D.J. 1986. Influence of fungal surface components on the interaction of *Coccidioides immitis* with polymorphonuclear neutrophils. J. Infect. Dis. 153:933.

33. Fricker-Hidalago, H., Lebeau, H., Kervroedan, P., et al. 1995. Auxacolor, a new commercial system for yeast identification: evaluation of 182 strains comparatively with ID32C. Annale de Biologie Clinique. 53(4):221.

34. Friedank, H. 1995. Hyalohyphomycoses due to *Fusarium* spp.: two case reports and review of the literature. Mycoses 38(1-2):69.

35. Galgiani, J.N., Bartlett, M., Ghannaoum, M.A., et al. 1996. Reference method for broth dilution antifungal susceptibility testing of yeasts, approved standard. M27-A. NCCLS, Wayne, Pa.

36. Galgiani, J.N., Stevens, D.A., Graybill, S.R., et al. 1984. *Pseudallescheria boydii* infections treated with ketoconazole: clinical evaluations of seven patients and *in vitro* susceptibility results. Chest 86:219.

37. Giuintuli, D., Stringer, S.L., and Stringer, J.R. 1994. Extraordinary low number of ribosomal RNA genes in *P. carinii*. J. Eukaryot. Microbiol. 41:88S.

38. Gray, L.D. and Roberts, G.D. 1988. Laboratory diagnosis of systemic fungal diseases. Infect. Dis. Clin. North Am. 2:779.

39. Guinet, R.J., Chanas, J., Goullier, A., et al. 1983. Fatal septicemia due to amphotericin B-resistant *Candida lusitaniae*. J. Clin. Microbiol. 18:443.

40. Hageage, G.J. and Harrington, B.J. 1984. Use of calcofluor white in clinical mycology. Lab. Med. 15:109.

41. Hall, G.S., Pratt-Rippin, K., and Washington, J.A. 1992. Evaluation of a chemiluminescent probe assay for identification of *Histoplasma capsulatum* isolates. J. Clin. Microbiol. 30:3003.

42. Harari, A.R., Hempel, H.O., Kimberlin, C.L., and Goodman, N.L. 1982. Effects of time lapse between sputum collection and culturing on isolation of clinically significant fungi. J. Clin. Microbiol. 15:425.

43. Hazen, K.C. 1995. New and emerging yeast pathogens. Clin. Microbiol. Rev. 8(4):462.

44. Heelan, J.S., Siliezar, D., and Coon, K. 1996. Comparison of rapid testing methods for enzyme production with the germ tube method for presumptive identification of *Candida albicans*. J. Clin. Microbiol. 34:2847.

45. Heinic, G.S., Greenspan, D., MacPhail, L.A., et al. 1992. Oral *Geotrichum candidum* infection associated with HIV infection: a case report. Oral Surg. Oral Med. Oral Pathol. 73(6):726.

46. Hogan, L.H., Klein, B.S., and Levitz, S.M. 1996. Virulence factors of medically important fungi. Clin. Microbiol. Rev. 9(4):469.

47. Hopfer, R.L. 1995. Use of molecular biological techniques in the diagnostic laboratory for detecting and differentiating fungi. Arch. Med. Res. 26:287.

48. Hopfer, R.L., Walden, P., Setterquist, S., et al. 1993. Detection and differentation of fungi in clinical specimens using polymerase chain reaction (PCR) amplification and restriction enzyme analysis. J. Med. Vet. Mycol. 31:65.

49. Hostetter, M.K. 1994. Adhesions and ligands involved in the interaction of *Candida* spp. with epithelial and endothelial surfaces. Clin. Microbiol. Rev. 7(1):29.

50. Huffnagle, G.B., Chen, G.H., Curtis, J.L., et al. 1995. Down-regulation of the afferent phase of T cell-mediated pulmonary inflammation and immunity by a high melanin-producing strain of *Cryptococcus neoformans*. J. Immunol 155:3607.

51. Huppert, M., Sun, S.H., and Bailey, J.W. 1967. Natural variability in *Coccidioides immitis*. In Ajello, L., editor. Coccidioidomycosis. University of Arizona Press, Tucson.

52. Ibrahim, A.S., Mirod, F., Filler, S.G., et al. 1995. Evidence implicating phospholipase as a virulence factor of *Candida albicans*. Infect. Immun. 63(5):1993.

53. Kane, J. and Smitka, C. 1978. Early detection and identification of *Trichophyton verrucosum*. J. Clin. Microbiol. 8:740.

54. Kaneshiro, E.S., Ellis, J.E., Jayasimhulu, K., et al. 1994. Evidence for the presence of "metabolic sterols" in *Pneumocystis*: identification and initial characterization of *Pneumocystis carinii* sterols. J Eukaryot. Microbiol. 41:78.

55. Kaufman, C.S. and Merz, W.G. 1982. Two rapid pigmentation tests for identification of *Cryptococcus neoformans*. J. Clin. Microbiol. 15:339.

56. Kennedy, M.J. and Sigler, L. 1995. *Aspergillus, Fusarium* and other moniliaceous fungi. In Murray, P.R., Baron, E.J., Pfaller, M.A. et al., editors. In Manual of Clinical Microbiology, ed. 6, American Society for Microbiology Press, Washington, DC.

57. Kitch, T.T., Jacobs, M.R., McGinnis, M.R., et al. 1996.

Ability of Rapid Yeast Plus system to identifying 304 clinically significant yeasts within 5 hours. J. Clin. Microbiol. 34:1069.

58. Kolattukudy, P.E., Lee, J.D., Rogers, L.M., et al. 1993. Evidence for possible involvement of an elastolytic serine protease in aspergillosis. Infect. Immun. 61:2357.

59. Koneman, E.W. and Roberts, G.D. 1985. Practical laboratory mycology, ed 3. Williams & Wilkins, Baltimore.

60. Koneman, E.W. and Roberts, G.D. 1991. Mycotic disease. In Henry, J.B., editor. Clinical diagnosis and management by laboratory methods, ed 18. W.B. Saunders, Philadelphia.

61. Kozel, T.R., Pfrommer, S.T., and Guerlain, A.S. 1988. Role of the capsule in phagocytosis of *Cryptococcus neoformans*. Rev. Infect. Dis. 19:S436.

62. Kreger-Van Rij, N.J.N. 1984. The yeasts: a taxonomic study. Elsevier Science Publishing, New York.

63. Kwon-Chung, K.J. and Bennet, J.E. 1992. Medical mycology. Lea & Febiger. Philadelphia.

64. Li, J.S., Pan, L.Q., Wu, S.X., et al. 1991. Disseminated *Penicillium marneffei* in China: report of 3 cases. Chin. Med. J. 104(3):247.

65. Libertin, C.R., Wilson, W.R., and Roberts, G.D. 1985. *Candida lusitaniae*—an opportunistic pathogen. Diag. Microbiol. Infect. Dis. 3:69.

66. Lopez-Martinez, R., Manzano-Gayosso, P., Meir, T. et al. 1994. Exoenzymes of dermatophytes isolated from acute and chronic tinea. Rev. Latinoam. Microbiol. 36(1):17.

67. Lu, J.J., Chen, C.H., Bartlett, M.S., et al. 1995. Comparison of six different methods for detection of *Pneumocystis carinii*. J. Clin. Microbiol. 33:2785.

68. Makimura, K., Murayama, S.Y., and Yamaguchi, H. 1994. Detection of a wide range of medically important fungi by the polymerase chain reaction. J. Med. Microbiol. 40:358.

69. Marcon, M.J. and Powell, D.A. 1987. Epidemiology, diagnosis and management of *Malassezia furfur* systemic infection. Diagn. Microbiol. Infect. Dis. 7:161.

70. Markaryan, A., Morozova, I., Yu, H., et al. 1994. Purification and characterization of an elastinolytic metalloprotease from *Aspergillus fumigatus* and immunoelectron microscopic evidence of secretion of this enzyme by the fungus invading the murine lung. Infect. Immun. 62:2149.

71. McEwen, J.G., Bedoya, V., Patino, M.M., et al. 1987. Experimental murine paracoccidioidomycosis induced by the inhalation of conidia. J. Med. Vet. Mycol. 25:165.

72. McGinnis, M.R., Rinaldi, M.G., and Winn, R.E. 1986. Emerging agents of phaeohyphomycosis: pathogenic species of *Bipolaris* and *Exserohilum*. J. Clin. Microbiol. 24:250.

73. Merz, W.G. and Roberts, G.D. 1995. Detection and recovery of fungi from clinical specimens. In Murray, P.R., Baron, E.J., Pfaller, M.A., et al., editors. Manual of clinical microbiology, ed 6. ASM Press, Washington, D.C.

74. Murray, P.R. 1996. ASM pocket guide to clinical microbiology. ASM Press, Washington, D.C.

75. Murray, P.R., Van Scoy, R.E., and Roberts, G.D. 1977. Should yeasts in respiratory secretions be identified? Mayo Clin. Proc. 52:42.

76. Musial, C.E., Cockerill, F.R. III, and Roberts, G.D. 1988. Fungal infections of the immunocompromised host: clinical and laboratory aspects. Rev. Infect. Dis. 1:349.

77. Neely, A.N., Orloff, M.N., and Holder, I.A. 1992. *Candida albicans* growth studies: a hypothesis for the pathogensis of *Candida* infections in burns. J. Burn Care Rehab. 13(3):323.

78. Neilson, J.B., Frontling, R.A., and Bulmer, G.S. 1977. *Cryptococcus neoformans:* size range of infectious particles from aerosolized soil. Infect. Immun. 17:634.

79. Padhye, A.A. 1995. Fungi causing eumycotic mycetoma. In Murray, P.R., Baron, E.J., Pfaller, M.A., et al., editors Manual of clinical microbiology, ed 6. ASM Press, Washington, D.C.

80. Palaoglu, S., Sav, A., Basak, T., et al. 1993. Cerebral phaeohyphomycosis. Neurosurgery 33(5):894.

81. Paliwal, D.K. and Randhawa, H.S. 1978. Evaluation of a simplified *Guizotia abyssinica* seed medium for differentiation of *Cryptococcus neoformans*. J. Clin. Microbiol. 7:346.

82. Patel, R., Gustaferro, C.A., Ruud, A.F., et al. 1994. Phaeohyphomycosis due to *Scopulariopsis brumptii* in a liver transplant recipient. Clin. Infect. Dis. 19:198.

83. Pfaller, M.A., Houston, A., and Coffman, S. 1996. Application of CHROMagar *Candida* for rapid screening of clinical specimens for *Candida albicans, Candida tropicalis, Candida krusei* and *Candida (Torulopsis) glabrata*. J. Clin. Microbiol. 34(1):58.

84. Rex, J.H., Pfaller, M.A., Barry, A.L., et al. 1995. Antifungal susceptibility testing of isolates from a randomized, multicenter trial of fluconazole versus amphotericin B as treatment of nonneutropenic patients with candidemia. Antimicrob. Agents Chemother. 39(1):40.

85. Rex, J.H., Pfaller, M.A., Rinaldi, M.G., et al. 1993. Antifungal susceptibility testing. Clin. Microbiol. Rev. 6(4):367.

86. Richardson, M.D. and Shankland, G.S. 1995. *Rhizopus, Rhizomucor, Absidia,* and other agents of subcutaneous zygomycetes. In Murray, P.R., Baron, E.J., Pfaller, M.A., et al., editors. Manual of clinical microbiology, ed 6. ASM Press, Washington, D.C.

87. Riddle, D.L., Giger, O., Miller, L., et al. 1994. Clinical comparison of the Baxter MicroScan Yeast Identification Panel and the Vitek Yeast Biochemical Card. Am. J. Clin. Pathol. 101:438.

88. Roberts, G.D. 1975. Detection of fungi in clinical specimens by phase-contrast microscopy. J. Clin. Microbiol. 2:261.

89. Roberts, G.D., Horstmeier, C.D., Land, G.A., et al. 1978. Rapid urea broth test for yeasts. J. Clin. Microbiol. 7:584.

90. Roberts, G.D., Karlson, A.G., and DeYoung, D.R. 1976. Recovery of pathogenic fungi from clinical specimens submitted for mycobacterial culture. J. Clin. Microbiol. 3:47.

91. Sand-Millan, R., Ribacoda, L., Ponton, J., et al. 1996. Evaluation of a commercial medium for identification of *Candida* sp. Eur. J. Clin. Microbiol. Infect. Dis. 15(2):153.

92. Sandhu, G.S., Kline, B.C., Stockman, L., et al. 1995. Molecular probes for diagnosis of fungal infections. J. Clin. Microbiol. 33:2913.

93. Schell, W.A. and Perfect, J.R. 1996. Fatal disseminated *Acre-*

monium strictum infection in a neutropenic host. J. Clin. Microbiol. 34:1333.

94. Schell, W.A., Pasarell J., Salkin, I.F., et al. 1995. *Bipolaris, Exophiala, Scedosporium, Sporothrix* and other dematiacious fungi. In Murray, P.R., Baron, E.J., Pfaller, M.A., et al., editors. Manual of clinical microbiology, ed 6. ASM Press, Washington, D.C.

95. Smith, C.D. and Goodman, N.L. 1975. Improved culture method for the isolation of *Histoplasma capsulatum* and *Blastomyces dermatitidis* from contaminated specimens. Am. J. Clin. Pathol. 68:276.

97. Spielberger, R.T., Tegtmeier, B.R., O'Donnel, M.R., et al. 1995. Fatal *Scedosporium prolificans (S. inflatum)* fungemia following allogenic bone marrow transplantation: report of a case in the United States. Clin. Infect. Dis. 21(4):1067.

98. St. Germain, G. and Summerbell, R. 1996. Identifying filamentous fungi: a clinical laboratory handbook. Star Publishing, Belmont, Calif.

99. St. Germain, G. and Beauchesne, D. 1991. Evaluation of the MicroScan Rapid Yeast Identification Panel. J. Clin. Microbiol. 29:2296.

100. Standard, P.G. and Kaufman, L. 1980. A rapid and specific method for the immunological identification of mycelial form cultures of *Paracoccidioides brasiliensis*. Curr. Microbiol. 4:297.

101. Standard, P.G. and Kaufman, L. 1982. Safety considerations in handling exoantigen extracts from pathogenic fungi. J. Clin. Microbiol. 15:663.

102. Stevens, D.A. 1989. The interface of mycology and endocrinology. J. Med. Vet. Mycol. 27:133.

103. Stockman, L., Clark, K.M., Hunt, J.M., et al. 1993. Evaluation of commercially available acridinium-ester-labeled chemiluminescent DNA probes for culture identification of *Blastomyces dermatitidis, Coccidioides immitis, Cryptococcus neoformans* and *Histoplasma capsulatum*. J. Clin. Microbiol. 31:845.

104. Strimlan, C.V., Dines, D.E., Rodgers-Sullivan, R.F., et al. 1980. Respiratory tract *Aspergillus*—clinical significance. Minn. Med. 63:25.

105. Stringer, J.R. 1996. *Pneumocystis carinii*: what is it, exactly? Clin. Microbiol. Rev. 9(4):489.

106. Stringer, S.L., Stringer, J.R., Blase, M.A., et al. 1989. *Pneumocystis carinii*: sequence from ribosomal RNA implies a close relationship with fungi. Exp. Parasitol. 68:450.

107. Sugita, T., Nishikawa, A., Shinoda, T., et al. 1995. Taxonomic position of deep-seated, mucosa-associated, and superficial isolates of *Trichosporon cutaneum* from trichosporonosis patients. J. Clin. Microbiol. 33(5):1368.

108. Summerbell, R.C., Rosenthal, S.A., and Kane, J. 1988. Rapid method for differentiation of *Trichophyton rubrum, Trichophyton mentagrophytes,* and related dermatophyte species. J. Clin. Microbiol. 26:2279.

109. Sun, S.H., Huppert, M., and Vukovich, K.R. 1976. Rapid in vitro conversion and identification of *Coccidioides immitis*. J. Clin. Microbiol. 3:186.

110. Terreni, A.A., Strohecker, J.S., and Dowa, H. Jr. 1987. *Candida lusitaniae* septicemia in a patient on extended home intravenous hyperalimentation. J. Med. Vet. Mycol. 25:63.

111. Thomson, R.B. and Roberts, G.D. 1982. A practical approach to the diagnosis of fungal infections of the respiratory tract. Clin. Lab. Med. 2:321.

112. Tintelnot, K., von Hunnius, P., de Hoog, G.S., et al. 1995. Systemic mycosis caused by a new *Cladophialophora* species. J. Med. Vet. Mycol. 33(5):349.

113. Treger, T.R., Visscher, D.W., Bartlett, M.S., et al. 1985. Diagnosis of pulmonary infection caused by *Aspergillus:* usefulness of respiratory cultures. J. Infect. Dis. 152:572.

114. Van der Peer, Y., Hendriks, L., Goris, A., et al. 1992. Evolution of basidiomycetous yeasts as deduced from small ribosomal subunit RNA sequences. Syst. Appl. Microbiol. 15:250.

115. Vartivarian, S.E. 1992. Virulence properties and nonimmune pathogenic mechanisms of fungi. Clin. Infect. Dis. 14:(suppl. 1)S30.

116. Walzer, P.D. 1995. *Pneumocystis carinii*. In Mandell, G.L., Bennett, J.E., and Dolin, R., editors. Principles and practice of infectious disease, ed 4. Churchill Livingstone, New York.

117. Wang, Y., Aisen, P., and Casadevall, A. 1995. *Cryptococcus neoformans* melanin and virulence: mechanism of action. Infect. Immun. 63:3131.

118. Warren, N.G. and Hazen, K.C. 1995. *Candida, Cryptococcus* and other yeasts of medical importance. In Murray, P.R., Baron, E.J., Pfaller, M.A., et al., editors. Manual of clinical microbiology, ed 6. ASM Press, Washington, D.C.

119. Weeks, R.J. 1964. A rapid simplified medium for converting the mycelial phase of *Blastomyces dermatitidis* to the yeast phase. Mycopathologia 22:153.

120. Weitzman, I., Kane, J., and Summerbell, R.C. 1995. *Trichophyton, Microsporon, Epidermophyton* and agents of superficial mycoses. In Murray, P.R., Baron, E.J., Pfaller, M.A., editors. Manual of clinical microbiology, ed 6. ASM Press, Washington, D.C.

121. Whittle, D.I. and Kominos, S. 1995. Use of itraconazole for treating subcutaneous phaeohyphomycosis casused by *Exophiala jeanselmei*. Clin. Infect. Dis. 21(4):1068.

122. Wu-Hsieh, B. and Howard, D.H. 1993. Histoplasmosis. In Murphy, J.W., Friedman, H., and Bendinelli, M. editors. Fungal infections and immune responses. Plenum Press, New York.

123. Yan, S., Negre, E., Cashel, J.A., et al. 1996. Specific induction of fibronectin binding activity by hemoglobin in *Candida albicans* grown in defined media. Infect. Immun. 64(8):2930.

124. Yuan, L., Cole, G.T., and Sun, S.H. 1988. Possible role of a proteinase endosporulation of *Coccidioides immitis*. Infect. Immun. 56:1551.

125. Zimmer, B.L. and Roberts, G.D. 1979. Rapid selective urease test for presumptive identification of *Cryptococcus neoformans*. J. Clin. Microbiol. 10:380.

Part Seven

7

Virology

66 LABORATORY METHODS IN BASIC VIROLOGY

GENERAL PRINCIPLES

VIRAL STRUCTURE

Viruses are composed of a nucleic acid genome surrounded by a protein coat called a **capsid**. Together the genome and capsid are referred to as the **nucleocapsid** (Figure 66-1). Genomes are either RNA or DNA. Viral capsids are composed of many individual subunits called **capsomeres**. Capsomeres assemble into an icosahedral or irregular-shaped capsid. Irregular-shaped capsids usually assume a helical form. Icosahedral-shaped capsids are cubical with 20 flat sides, whereas helical capsids are spiral-shaped. Some of the larger viruses have a lipid-containing envelope that surrounds the capsid. In addition, many viruses have glycoprotein spikes that extend from the surface of the virus, acting as attachment proteins or enzymes. The entire virus, including nucleic acid, capsid, envelope, and glycoprotein spikes, is called the **virion,** or viral particle (Figure 66-1).

Viruses that cause disease in humans range in size from approximately 20 to 300 nm. Even the largest viruses, such as the pox viruses, are not easily seen with a light microscope, because they are less than one fourth the size of a staphylococcal cell (Figure 66-2).

VIRAL REPLICATION

Viruses are strict intracellular parasites, reproducing or replicating only within a host cell. The steps in virus replication, called the *infectious cycle,* include attachment, penetration, uncoating, macromolecular synthesis, assembly, and release (Figure 66-3).

To initiate the infectious cycle, a virus must first recognize and bind to a suitable host cell, a process referred to as *attachment.* Typically, glycoprotein spikes bind to host cell carbohydrate receptors. Viruses recognize and attach to a limited number of host cell types, allowing infection of some tissues but not others. This is referred to as viral **tropism**.

The process by which viruses enter the host cell is called *penetration.* One mechanism of penetration involves fusion of the viral envelope with the host cell membrane. Not only does this internalize the virus but can also lead to fusion between this and other host cells nearby, forming multinucleated cells called syncytia. The detection of syncytia is used to identify the presence of virus in cell cultures or clinical specimens.

Uncoating occurs once the virus is internalized. Uncoating is necessary to release viral genome before the viral DNA or RNA is delivered to its intracellular site of replication in the nucleus or cytoplasm.

Macromolecular synthesis includes the production of nucleic acid and protein polymers. Viral transcription leads to the synthesis of mRNA, which encodes early and late viral proteins. Early proteins are nonstructural components, such as enzymes, and late proteins are viral structural proteins. Rapid identification of virus in cell culture can be accomplished by detecting early viral proteins in infected cells using immunofluorescence techniques. Replication of viral nucleic acid is necessary to provide genomes for progeny virions.

During viral *assembly,* structural proteins, genomes, and, in some cases, viral enzymes are assembled into virions. Viral envelopes are acquired during "budding" from a host cell membrane. Nuclear and cytoplasmic membranes are common areas for viral budding. Acquisition of an envelope is the final step in viral assembly.

Release of intact virions occurs following cell lysis or by budding from cytoplasmic membranes. Release by budding may not result in rapid host cell death as does release by lysis. Detection of virus in cell culture is facilitated by recognition of areas of cell lysis. Detection of virus released by budding is more difficult, because the cell monolayer remains intact. Influenza viruses, which are released with minimal cell destruction, can be detected in cell culture by an alternative technique called **hemadsorption**. Influenza virus–infected cells contain virally encoded glycoprotein hemagglutinins in the cytoplasmic membrane. Red blood cells (RBCs) added to the culture medium will adsorb to the outer membranes of infected cells but not to uninfected cells.

VIRAL CLASSIFICATION

Viruses can be classified according to morphology, type of genome, or means of replication. Morphology includes type of capsid, such as icosahedral or helical.

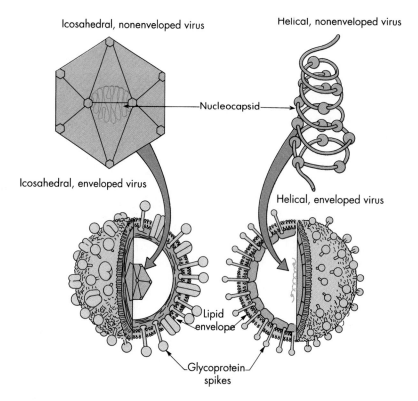

Icosahedral, nonenveloped virus

Helical, nonenveloped virus

Nucleocapsid

Icosahedral, enveloped virus

Helical, enveloped virus

Lipid envelope

Glycoprotein spikes

FIGURE 66-1 Illustration of viral particle. Enveloped and nonenveloped virions have icosahedral or irregular (helical) shape. (Modified from Murray P.R., Drew W.L., Kobayashi, G.S., et al. 1990. Medical microbiology. Mosby, St. Louis.)

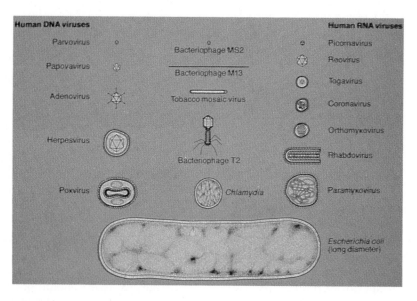

FIGURE 66-2 Relative sizes of viruses, bacteria, and chlamydia. (From Murray P.R., Drew W.L., Kobayashi, G.S., et al. 1990. Medical microbiology. Mosby, St. Louis.)

Type of genome includes RNA or DNA and whether it is single- or double-stranded. Means of replication refers to the strategy each virus uses to duplicate its genome. For example, enteroviruses have RNA genomes that synthesize additional strands of RNA directly, whereas retroviruses make RNA in a two-step process by first synthesizing DNA, which subsequently makes RNA. Viruses of medical importance to humans comprise six families of DNA viruses and 13 families of RNA viruses (Table 66-1).

—— **TABLE 66-1** **LIST OF DNA AND RNA VIRUSES OF HUMAN IMPORTANCE** ——

DNA VIRUSES		RNA VIRUSES	
FAMILY	VIRAL MEMBERS	FAMILY	VIRAL MEMBERS
Pox	Variola, vaccinia, orf. molluscum contagiosum	Orthomyxovirus	Influenza A, B, C
Herpesvirus	Herpes simplex 1 and 2, varicella-zoster, cytomegalovirus, Epstein-Barr virus, human herpesviruses 6, 7, and 8	Paramyxovirus	Parainfluenza, mumps, measles, respiratory syncytial virus
Adenovirus	Human adenoviruses include serotypes 1-48	Togavirus	Western and Eastern equine encephalitis, rubella
Papovavirus	Human papilloma virus JC and BK polyomaviruses	Flavivirus	St. Louis encephalitis, yellow fever, dengue, hepatitis C.
Hepadnavirus	Hepatitis B	Bunyavirus	California and La Crosse encephalitis, Rift Valley fever and other hemorrhagic fever agents, sin nombre and related hantaviruses
Parvovirus	B-19	Calicivirus	Calicivirus (Norwalk agent), hepatitis E
		Astrovirus	Gastroenteritis-causing astroviruses
		Coronavirus	Coronavirus
		Reovirus	Rotavirus, Colorado tick fever
		Picornavirus	Enterovirus, polio, coxsackie A, coxsackie B, echovirus, enterovirus 68-71, 72 (hepatitis A), rhinovirus
		Arenavirus	Lymphocytic choriomeningitis, Lassa and other hemorrhagic fever agents
		Rhabdovirus	Rabies
		Filovirus	Ebola and Marburg hemorrhagic fevers
		Retrovirus	Human T-lymphotropic virus (HTLV-I and II), human immunodeficiency virus

VIRAL PATHOGENESIS

Viruses are transmitted from person to person by the respiratory and fecal-oral routes, by trauma or injection with contaminated objects or needles, by tissue transplants (including blood transfusions), and by arthropod or animal bites. Once introduced into a host, the virus infects susceptible cells, frequently in the upper respiratory tract. Local infection leads to a **viremia** (viruses in the blood), which inoculates secondary target tissue distant from the primary site. Symptomatic disease ensues. Disease resolves when specific antibody- and cell-mediated immune mechanisms halt continued replication of the virus (Figure 66-4). Tissue is damaged by lysis of virus-infected cells or by immunopathologic mechanisms directed against the virus but also destructive to neighboring tissue. Some viruses, especially DNA-containing viruses, such as those in the herpes group, remain **latent** in host tissue, with no observable clinical impact. Reactivation may occur accompanying immune suppression, resulting in recurrence of clinically apparent disease.

ANTIVIRAL VACCINES

Control of many viral diseases has been accomplished by vaccination. Since Jenner developed the first vaccine against smallpox nearly 200 years ago, attenuated-live or inactivated-dead viral vaccines have been successively used to prevent yellow fever, poliomyelitis, measles, mumps, rubella, hepatitis B, and influenza. Smallpox was eliminated in 1977 by an effective vaccination program. New vaccines continue to appear. A live, attenuated varicella (chickenpox) vaccine is now recommended for all children, and an inactivated hepatitis A vaccine is available for travel-

ers or others entering areas of higher endemicity. Clinical virology laboratories occasionally detect vaccine strains of viruses, as occurs with the attenuated oral polio virus. Throat and rectal swab specimens of vaccinated children or their contacts who have become infected by the fecal-oral route may be positive for poliovirus. The vaccine strains do not cause disease in a normal host. Passive, incidental inoculation of siblings and contacts with the vaccine strain

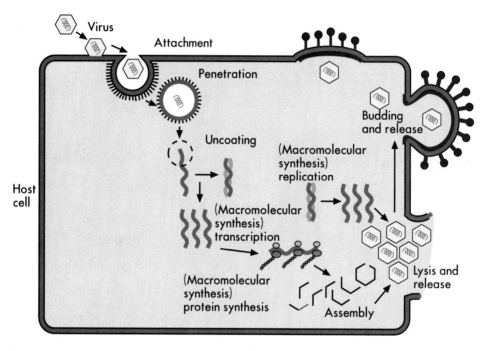

FIGURE 66-3 Illustration of viral infectious cycle. (Modified from Murray P.R., Drew W.L., Kobayashi G.S., et al. 1990. Medical microbiology. Mosby, St. Louis.)

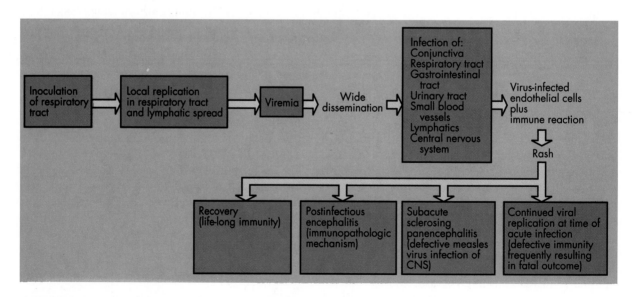

FIGURE 66-4 Viral pathogenesis illustrated by the mechanisms of the spread of measles virus within the body. (Modified from Murray P.R., Drew W.L., Kobayashi G.S., et al. 1990. Medical microbiology. Mosby, St. Louis.)

acts to vaccinate those who have not been properly immunized or boost antibody levels of those previously immunized.

ANTIVIRAL THERAPY

The availability of specific antiviral chemotherapeutic agents has added importance and urgency to the need for laboratory diagnosis of viral infections. In addition to commonly used antivirals, such as acyclovir, amantadine, azidothymidine (AZT), ribavirin, and ganciclovir, other newer antivirals are recognized as beneficial (Table 66-2). Detection of virus to establish an exact etiologic diagnosis is usually needed to justify use of these expensive and sometimes toxic agents. Although resistance does occur, routine antiviral testing, available in some reference laboratories, is not yet necessary for the management of most patients with viral disease.

OVERVIEW OF HUMAN VIRAL DISEASES

Specific disease syndromes may be caused by several unrelated viruses. For example, a partial list of viruses that can cause encephalitis includes herpes simplex virus, arboviruses, rabies virus, human immunodeficiency virus, and measles virus. Organ systems and the viruses that are important pathogens in each are summarized in Table 66-3. Conversely, individual viruses may cause many different diseases. Herpes simplex virus can cause pharyngitis, genital infection, conjunctivitis, and encephalitis. Retroviruses and most DNA viruses establish a latent state following primary infection. During the latent state, viral genome is integrated into host cell chromosome and no viral replication occurs. **Latent viruses** can silently reactivate, resulting in viral replication and shedding but no clinical symptoms, or they can reactivate, causing symptomatic or even fatal disease. Some latent viruses also are known to cause cancer. They are referred to as **oncogenic**, or cancer-causing, viruses.

ADENOVIRUSES

Adenoviruses (Table 66-4) were first isolated from human adenoid tissues. At present, 48 serotypes of human adenoviruses have been described; however, most disease is associated with only one third of these types. Adenoviruses cause less than 5% of all acute respiratory disease in the general population. In addition, adenovirus serotypes 40 and 41 cause gastroenteritis in infants and young children. Other disease occurs but is less common. Laboratory diagnosis is accomplished by conventional cell culture using HEp-2 cells.

BUNYAVIRUSES

Bunyaviruses (Table 66-5) comprise a diverse group of viruses most of which are transmitted by arthropods, primarily mosquitoes. Clinical disease caused by the California encephalitis and La Crosse arboviruses includes mild, self-limited or severe, fatal encephalitis. Bunyaviruses belonging to the *Hantavirus* genus are an exception, because these viruses are rodent-borne and transmitted through aerosolized rodent excreta. Hantavirus pulmonary syndrome, originally reported in the four corners area of the Southwestern United States, is caused by a hantavirus called *sin nombre* (no name) virus and is characterized by hemorrhagic fever, kidney disease, and acute respiratory failure. Laboratory diagnosis requires the detection of specific antibody in acute and convalescent serum specimens or in cerebrospinal fluid.

CORONAVIRUSES

Many coronaviruses of humans (Table 66-6) and other animals exist. The prefix *corona-* is used because of the crownlike surface projections that are seen when the virus is examined by electron microscopy. Human respiratory coronaviruses cause colds and occasionally

---

TABLE 66-2   ANTIVIRAL CHEMOTHERAPY

ANTIVIRAL	INTENDED VIRUS
Amantadine and rimantadine	Influenza A
Acyclovir and famciclovir	Herpes simplex virus and varicella-zoster virus
Zidovudine (AZT), didanosine, and viral protease inhibitors	Human immunodeficiency virus
Ribavirin	Respiratory syncytial virus and other viruses
Ganciclovir and foscarnet	Cytomegalovirus

— **TABLE 66-3**   **LIST OF VIRAL SYNDROMES AND COMMON VIRAL PATHOGENS** —

VIRAL SYNDROME	VIRAL ETIOLOGY
Infants and children	
Upper respiratory tract infection	Rhinovirus, coronavirus, parainfluenza, adenovirus, respiratory syncytial virus, influenza
Pharyngitis	Adenovirus, coxsackie A, herpes simplex, Epstein-Barr, rhinovirus, parainfluenza, influenza
Croup	Parainfluenza, respiratory syncytial
Bronchitis	Parainfluenza, respiratory syncytial
Bronchiolitis	Respiratory syncytial, parainfluenza
Pneumonia	Respiratory syncytial, adenovirus, influenza, parainfluenza
Gastroenteritis	Rotavirus, adenoviruses 40 and 41, calicivirus, astrovirus
Adults	
Upper respiratory tract infection	Rhinovirus, coronavirus, adenovirus, influenza, parainfluenza
Pneumonia	Influenza, adenovirus, sin nombre virus (hantavirus)
Pleurodynia	Coxsackie B
Gastroenteritis	Norwalklike virus
All Patients	
Parotitis	Mumps, parainfluenza
Myocardia/pericarditis	Coxsackie and echoviruses
Keratitis/conjunctivitis	Herpes simplex, varicella-zoster, adenovirus
Pleurodynia	Coxsackie B
Herpangina	Coxsackie A
Febrile illness with rash	Echo and coxsackie viruses
Infectious mononucleosis	Epstein-Barr, cytomegalovirus
Meningitis	Echo and coxsackie viruses, mumps, lymphocytic choriomeningitis, herpes simplex virus type 2
Encephalitis	Herpes simplex, togaviruses, bunyaviruses, flaviviruses, rabies, enteroviruses, measles, HIV, JC virus
Hepatitis	Hepatitis A, B, C, D (delta agent), E, and G and non-A, B, C, D, E, F, G
Hemorrhagic cystitis	Adenovirus
Cutaneous infection with rash	Herpes simplex, varicella-zoster, enteroviruses, Epstein-Barr, measles, rubella, parvovirus, human herpesviruses 6 and 7
Hemorrhagic fever	Ebola, Marburg, Lassa, and other viruses

TABLE 66-4 ADENOVIRUSES

Family: *Adenoviridae*   Common Name: Adenovirus

Virus: Adenovirus

Characteristics: Double-stranded DNA genome; icosahedral capsid, no envelope. Approximately 48 human serotypes

Transmission: Respiratory, fecal-oral, and direct contact (eye)

Site of latency: Replication in oropharynx

Disease: Pharyngitis, pharyngoconjunctival fever, keratoconjunctivitis, pneumonia, hemorrhagic cystitis, disseminated disease, and gastroenteritis in children

Diagnosis: Cell culture (HEp-2), EIA for gastroenteritis serotypes 40 and 41

Treatment: Supportive

Prevention: Vaccine (adenovirus serotypes 4 and 7) for military recruits

TABLE 66-5 BUNYAVIRUSES

Family: *Bunyaviridae*   Common Name: Bunyavirus

Virus: Arboviruses,* including California encephalitis virus and LaCrosse virus and hantaviruses (e.g., sin nombre virus)

Characteristics: Segmented, single-stranded, RNA genome; helical capsid with envelope

Transmission: Mosquito vector, except for hantaviruses

Disease: Encephalitis for arboviruses; pneumonia or hemorrhagic fever for hantaviruses

Diagnosis: Serology and antibody detection in CSF

Treatment: Supportive

Prevention: Avoid contact with arthropod vector; vector control programs; hantaviruses: avoid rodent urine and feces

*Arthropod-borne viruses (arboviruses) were once grouped together because of their common mode of transmission. Present classification includes three families containing arboviruses: *Togaviridae*, *Flaviviridae*, and *Bunyaviridae*. The virus group within *Togaviridae*, which includes arboviruses, is the alphavirus group. Common arboviruses are referred to as *bunyaviruses*, *flaviviruses*, and *alphaviruses*.

pneumonia in adults. Coronaviruses are thought to cause diarrhea in infants, based on the presence of coronavirus-like particles in stool of symptomatic patients that is examined by electron microscopy. No practical diagnostic methods, short of electron microscopy, are available for laboratory diagnosis.

FLAVIVIRUSES

The flaviviruses (Latin *flavus*, meaning yellow) include viruses that cause arbovirus diseases, such as yellow fever, dengue, and St. Louis encephalitis (Table 66-7). Scores of other arthropod-borne flaviviruses, most transmitted by mosquitos or ticks, cause encephalitis, hemorrhagic fever, or milder disease, such as that characterized by fever, arthralgia, and rash. In addition, molecular analysis of the hepatitis C virus genome suggests a relationship with the flaviviruses. Laboratory diagnosis of flavivirus disease in humans is performed using serologic tests. The presence of hepatitis C virus can be detected by the

---

TABLE 66-6 CORONAVIRUSES

Family: *Coronaviridae*   Common Name: Coronaviruses

Virus: Coronavirus

Characteristics: Single-stranded, RNA genome; helical capsid with envelope

Transmission: Unknown; Probably direct contact or aerosol

Disease: Common cold; possibly gastroenteritis, especially in children

Diagnosis: Electron microscopy

Treatment: Supportive

Prevention: Avoid contact with virus

TABLE 66-7 FLAVIVIRUSES

Family: *Flaviviridae*   Common Name: Flavivirus

Characteristics: Single-stranded RNA genome and icosahedral capsid with envelope

Virus: Arboviruses,* including yellow fever, dengue, and St. Louis encephalitis viruses

Transmission: Arthropod vector, usually mosquito

Disease: St. Louis encephalitis, dengue, and yellow fever

Diagnosis: Serology and antibody detection in CSF

Treatment: Supportive

Prevention: Avoid contact with vector; vector control programs

Virus: Hepatitis C virus

Transmission: Parenteral or sexual

Disease: Acute and chronic hepatitis

Diagnosis: Serology and PCR

Treatment: Supportive

Prevention: Avoid contact with virus; blood supply screened for antibody to hepatitis C virus

*Arthropod-borne viruses (arboviruses) were once grouped together because of their common mode of transmission. Present classification includes three families containing arboviruses: *Togaviridae*, *Flaviviridae*, and *Bunyaviridae*. The virus group within *Togaviridae*, which includes arboviruses, is the alphavirus group. Common arboviruses are referred to as *bunyaviruses*, *flaviviruses*, and *alphaviruses*.

---

polymerase chain reaction test, in addition to specific antibody testing.

HEPADNAVIRUSES

Hepatitis B virus (HBV) (Table 66-8) is the prototype virus found in the hepadnavirus family (*hepa* from hepatitis and *dna* from the genome type). Other mammalian and avian hepadnaviruses are known to exist. Disease in humans caused by HBV is usually asymptomatic but may result in acute or chronic hepatitis with self-limited or fatal outcomes. Diagnosis is by the detection of a battery of antibodies and viral antigens.

───────────── **TABLE 66-8   HEPADNAVIRUSES** ─────────────

Family: *Hepadnaviridae*   Common Name: Hepadnavirus

Virus: Hepatitis B virus (HBV)

Characteristics: Partly double-stranded DNA genome; icosahedral capsid with envelope; virion also called *Dane particle*; surface antigen originally termed *Australia antigen*

Transmission: Humans are reservoir and vector. Spread by direct contact including exchange of body secretions, receiving contaminated blood products, and percutaneous injection of virus

Site of latency: Liver

Disease: Acute infection with resolution (90%); fulminant hepatitis (1%); chronic hepatitis (9%), followed by resolution, asymptomatic carrier state, chronic persistent, or chronic active disease

Diagnosis: Serology and viral antigen detection

Oncogenic: Liver cancer

Treatment: Liver transplant for fulminant disease

Prevention: HBV vaccine; hepatitis B immune globulin

HERPESVIRUSES

There are eight known human herpesviruses (Table 66-9). Herpesviruses also are widely disseminated among most other animal species and do not infect humans, except for herpes B virus from nonhuman primates, which causes severe, usually fatal encephalitis in humans. Recently detected herpes viruses include human herpesviruses 6 (HHV-6) and 7 (HHV-7), which are lymphotropic viruses acquired early in life, and HHV-8. HHV-6 and possibly HHV-7 are associated with the childhood disease roseola (exanthem subitum), characterized by short-lasting fever and skin rash. HHV-8 is purported to cause Kaposi's sarcoma and some B-cell lymphomas. Herpesviruses are detected by cell culture, direct antigen or nucleic acid systems, and serologic testing of acute and convalescent serum specimens.

ORTHOMYXOVIRUSES

The orthomyxovirus family (Table 66-10) includes the influenza viruses. Influenza viruses are unique because of their ability to alter antigenic composition, thus reinfecting newly susceptible hosts. Antigenic drift, caused by mutation during genome replication, results in minor antigenic change and relatively mild influenza outbreaks every 1 to 3 years. Antigenic shift, caused by reassortment or mixing of the segmented viral genome during coinfection in nonhuman animals, results in major antigenic change and periodic worldwide outbreaks (pandemics). Coinfection occurs when a human virus infects a cell at the same time as an animal influenza virus. Influenza virus disease is detected in the laboratory by cell culture, fluorescent antibody staining, or other antigen detection methods, and serologic testing.

PAPOVAVIRUSES

The papovavirus family (Table 66-11) includes the *pa*pilloma viruses (wart viruses), *po*lyoma viruses, and simian *va*cuolating viruses (e.g., SV40 virus). The human papilloma viruses (HPV) have not been cultivated in cell culture, thus preventing the production of type-specific antigens and corresponding typing antisera. HPVs have been divided into over 70 genotypes based on DNA sequences. Genotypes have differing cellular tropisms, resulting in associations of HPV types with specific clinical types of warts. For example, HPV-1 is associated with plantar warts, HPV-2 and 4 are associated with common warts of the hands, and HPV-6, 11, and others with genital warts. Of importance, HPV types 16 and 18 are associated with genital warts and cervical carcinoma. HPV infection is detected by histopathologic or cytologic examination of cutaneous biopsy or cells, respectively, or by DNA probe assays used to identify specific genotypes within infected epithelial cells.

Polyoma viruses of concern are JC and BK viruses, named with the initials of the persons from whom the viruses were first isolated. Both have latent states in the kidney and can result in symptomatic reactivation during periods of immune suppression. JC virus reactivates to cause disease in the central

TABLE 66-9 HERPESVIRUSES

Family: *Herpesviridae*   Common name: Herpesvirus

Characteristics: Double-stranded DNA genome; icosahedral capsid with envelope. At least eight human herpesviruses known: HSV 1, HSV 2, VZV, EBV, CMV, HH6, HH7, and HH8; little is known about HH8

Virus: Herpes simplex virus types 1 and 2 (HSV 1 and 2)

Transmission: Direct contact with infected secretions

Site of latency: Sensory nerve ganglia

Disease: (Predominant virus in parentheses) Gingivostomatitis (HSV 1), pharyngitis (HSV 1), herpes labialis (HSV 1), genital infection (HSV 2), conjunctivitis (HSV 1), keratitis (HSV 1), herpetic whitlow (HSV 1 and 2), encephalitis (HSV 1), disseminated disease (HSV 1 or HSV 2 in neonates)

Detection: Cell culture (HDF, others), EIA, ELVIS, FA stain

Treatment: Acyclovir

Prevention: Avoid contact

Virus: Varicella-zoster virus (VZV)

Transmission: Close personal contact, especially respiratory

Site of latency: Dorsal root ganglia

Disease: Chicken pox (varicella), shingles (zoster)

Detection: Cell culture (HDF), shell vial culture, FA stain

Treatment: Acyclovir

Prevention: Vaccine

Virus: Epstein-Barr virus (EBV)

Transmission: Close contact with infected saliva

Site of latency: B lymphocytes

Disease: Infectious mononucleosis, progressive lymphoreticular disease, oral hairy leukoplakia

Detection: Serology

Oncogenic: Burkitt's lymphoma, nasopharyngeal carcinoma

Treatment: Supportive

Prevention: Avoid contact

Virus: Cytomegalovirus (CMV)

Transmission: Close contact with infected secretions, blood transfusions (WBCs), organ transplants, transplacental

Site of latency: White blood cells, endothelial cells, cells in various organs

Disease: Asymptomatic infection, congenital disease of newborn, symptomatic disease of immunocompromised host, heterophile-negative infectious mononucleosis

Diagnosis: Cell culture (HDF), shell vial culture, CMV antigenemia FA stain

TABLE 66-9 HERPESVIRUSES—CONT'D

Treatment: Supportive; decrease immune suppression, ganciclovir and foscarnet

Prevention: Use CMV antibody-negative blood and tissue for transfusion and transplantation, respectively

Virus: Human herpesviruses 6 (HH6), 7 (HH7), and 8 (HH8)

Transmission: Most likely close contact or respiratory

Site of latency: T lymphocytes (HH6, HH7); B lymphocytes (HH8)

Disease: Exanthem subitum (roseola) (HH6, HH7); Kaposi's sarcoma, B-cell lymphoma (HH8)

Detection: Serology

Treatment: Supportive

Prevention: Avoid contact

TABLE 66-10 ORTHOMYXOVIRUSES

Family: *Orthomyxoviridae*   Common Name: Orthomyxovirus

Characteristics: Segmented, single-stranded, RNA genome; helical capsid with envelope. Three major antigenic types, influenza A, B, and C. Types A and B cause nearly all human disease

Virus: Influenza A

Transmission: Contact with respiratory secretions

Disease: Influenza (malaise, headache, myalgia, cough); Primary influenza pneumonia; in children, bronchiolitis, croup, otitis media

Detection: Cell culture (PMK), EIA, FA stain

Epidemiology: Viral subtypes based on hemagglutinin and neuraminidase glycoproteins abbreviated *H* and *N*, respectively (e.g., H1N1 or H3N2); Infects man and other animals. Antigenic drift, resulting in minor antigenic change, causes local outbreaks of influenza every 1 to 3 years. Antigenic shift, resulting in major antigenic change, causes periodic worldwide outbreaks.

Treatment: Supportive; antiviral amantadine or rimantidine for severe, life-threatening disease

Prevention: Influenza vaccine or antiviral prophylaxis

Virus: Influenza B

Transmission: Contact with respiratory secretions

Disease: Similar to "mild" influenza

Detection: Cell culture (PMK), EIA, and FA stain

Epidemiology: Antigenic drift only, resulting in local outbreaks every 1 to 3 years

Treatment: Supportive; amantadine not effective

Prevention: Influenza vaccine

TABLE 66-11 PAPOVAVIRUSES

Family: *Papovaviridae*   Common Name: Papovavirus

Characteristics: Double-stranded DNA genome; icosahedral capsid, no envelope. Includes papilloma viruses, and BK and JC polyomaviruses

Virus: Human papilloma virus (HPV)

Characteristics: Contains more than 70 DNA types

Transmission: Direct contact

Site of latency: Epithelial tissue

Disease: Skin warts, benign head and neck tumors, anogenital warts

Diagnosis: Cytology, DNA probes

Oncogenic: Cervical and penile cancer (especially HPV types 16 and 18)

Treatment: Spontaneous disappearance the rule. Surgical or chemical removal may be necessary

Prevention: Avoid contact with infected tissue

Virus: Polyomavirus (BK and JC viruses infect humans)

Transmission: Probably direct contact with infected respiratory secretions

Site of latency: Kidney

Disease: Mild or asymptomatic primary infection; virus remains dormant in kidneys. Reactivation in immunocompromised patients causes hemorrhagic cystitis (BKV) or progressive multifocal leukoencephalopathy (JCV)

Detection: JC virus by EM; BK virus by cytology

Treatment: Supportive; decrease immune suppression

Prevention: Avoid contact with virus; prevention unlikely until more is learned about mode of transmission

---

nervous system, and BK causes a hemorrhagic cystitis. JC virus is detected using electron microscopy of brain tissue, whereas BK virus is detected by cytologic examination of urine.

PARAMYXOVIRUSESES

The paramyxovirus family (Table 66-12) includes many pathogenic viruses, especially for young children. Paramyxoviruses do not have a segmented genome and therefore do not undergo antigenic shift like the orthomyxoviruses. Respiratory syncytial virus (RSV) contains a surface protein called *F* (fusion) protein. F protein mediates host cell fusion into syncytial cells, which are a hallmark of RSV infection. Except for respiratory syncytial virus, the paramyxoviruses hemadsorb guinea pig red blood cells. Laboratory detection is performed using cell culture with hemadsorption, fluorescent antibody staining, or enzyme immunoassay.

PARVOVIRUSES

Parvoviruses (Latin *parvus* means small) (Table 66-13) have wide distribution among warm-blooded animals. Parvovirus B19 represents the one human pathogen in the family. Its replication in human cells is largely restricted to erythroid progenitor cells, making adult bone marrow and fetal liver, the site of erythropoiesis during development, the major sites of viral replication. Important diseases caused by the B19 virus include fifth disease (the fifth of the childhood exanthems), aplastic crisis in patients with underlying hemoglobinopathies, and fetal infection resulting from

─────────── **TABLE 66-12   PARAMYXOVIRUSES** ───────────

Family: *Paramyxoviridae*   Common Name: Paramyxoviruses

Characteristics: Single-stranded RNA genome; helical capsid with envelope. No segmented genome like orthomyxoviruses

Virus: Measles virus

Transmission: Contact with respiratory secretions; extremely contagious

Disease: Measles, atypical measles (occurs in those with waning "vaccine" immunity), and subacute sclerosing panencephalitis

Detection: Cell culture (PMK) and serology

Treatment: Supportive; immunocompromised patients can be treated with immune serum globulin

Prevention: Measles vaccine

Virus: Mumps

Transmission: Person-to-person contact, presumably respiratory droplets

Disease: Mumps

Detection: Cell culture (PMK) and serology

Treatment: Supportive

Prevention: Mumps vaccine

Virus: Parainfluenza virus

Transmission: Contact with respiratory secretions

Disease: Adults: upper respiratory, rarely pneumonia; Children: respiratory, including croup, bronchiolitis, and pneumonia

Detection: Cell culture (PMK), shell vial culture, and FA stain

Epidemiology: Four serotypes, disease occurs year-round

Treatment: Supportive

Prevention: Avoid contact with virus

Virus: Respiratory syncytial virus (RSV)

Transmission: Person to person by hand and respiratory contact

Disease: Primarily in infants and children. Infants: bronchiolitis, pneumonia, and croup; children: upper respiratory

Detection: Cell culture (HEp-2), EIA, and FA stain

Epidemiology: Disease occurs annually, late fall through early spring. Nosocomial transmission can occur readily

Treatment: Supportive; treat severe disease in compromised infants with immune globulin or ribavirin

Prevention: Avoid contact with virus. Immune globulin for infants with underlying lung disease. Prevent nosocomial transmission with isolation and cohorting

TABLE 66-13 PARVOVIRUSES

Family: *Parvoviridae*   Common Name Parvovirus

Virus: Parvovirus B-19

Characteristics: Single-stranded DNA virus; icosahedral capsid, no envelope. Parvovirus B-19 is the only known human parvovirus

Transmission: Close contact, probably respiratory

Disease: Erythema infectiosum (fifth disease), aplastic crises in patients with chronic hemolytic anemias, and fetal infection and stillbirth

Detection: Serology, PCR, histology

Treatment: Supportive

Prevention: Avoid contact

---

transplacental infection. Laboratory diagnosis is accomplished using parvovirus-specific IgM or virus-specific IgG antibody testing with acute and convalescent sera. Parvovirus cannot be cultivated in usual cells available in clinical virology laboratories.

PICORNAVIRUSES

Picornaviruses (Table 66-14) are small (from Italian *piccolo* meaning small) RNA viruses containing many important virus groups. The enteroviruses were originally divided into poliovirus, coxsackie virus, and echovirus groups, based on similarity of diseases caused in humans. Genetic diversity of these viruses dictates that newly characterized strains be given enterovirus type of designations because of the inexactness of the old classification system. The rhinoviruses grow best or exclusively at lower temperatures (30° C), making their detection in clinical virology laboratories, where incubation is at 35° to 37° C, unlikely. Most enteroviruses grow in cell culture. Disease caused by hepatitis A virus (enterovirus 72) is diagnosed using serology tests.

REOVIRUSES

The reoviruses (Table 66-15) were first isolated from respiratory and enteric specimens from which they gained their name, *r*espiratory-*e*nteric-*o*rphan viruses (reoviruses). "Orphan" designated the absence of an associated disease when the viruses were first described. Although reoviruses infect most mammalian species and are readily detected in water contaminated with animal feces, in general, they are not agents of serious disease. In humans, rotaviruses and

the agent of Colorado tick fever are pathogens. Reoviruses are difficult to grow, necessitating enzyme immunoassays or electron microscopy for detection.

RETROVIRUSES

The retrovirus family (Table 66-16) comprises a large group of viruses that primarily infect vertebrates. Human retroviruses include the human immunodeficiency viruses (HIV-1 and HIV-2) and the human T-cell lymphoma viruses (HTLV-1 and HTLV-2). Retroviruses are unique because they possess the enzyme reverse transcriptase. Reverse transcriptase allows the viral RNA genome to be replicated into DNA, rather than the usual RNA copies. Amino acid sequencing of the reverse transcriptase protein serves to divide the retrovirus family into its many groups. The laboratory detection of HIV and monitoring of antiviral therapy is done using serology testing, antigen detection, and quantitative testing for RNA genome by a modified PCR test.

RHABDOVIRUSES

Rhabdoviruses (Table 66-17) infect plants and animals. Rabies virus is the one member causing disease in humans. The virus is present in saliva and is readily transmitted by animal bite. Local replication at the site of the bite wound is followed by invasion of the peripheral and central nervous systems. Once in the brain, rabies virus spreads to various tissues and organs, including salivary glands, kidneys, heart, hair follicles, and corneas. The bullet-shaped virion is readily identifiable by electron microscopy; however, diagnosis is best performed by fluorescent antibody staining of infected cells or tissue.

─────────────── **TABLE 66-14**   **PICORNAVIRUSES** ───────────────

Family: *Picornaviridae*   Common Name: Picornaviruses

Characteristics: Single-stranded RNA genome; icosahedral capsid with no envelope

Virus: Enteroviruses
    Poliovirus (3 types)
    Coxsackievirus, group A (23 types)
    Coxsackievirus, group B (6 types)
    Echovirus (31 types)
    Enteroviruses (5 types)

    Transmission: Fecal-oral

    Disease: (Predominant virus in parentheses) Polio (poliovirus), herpangina (coxsackie A), pleurodynia (coxsackie B), aseptic meningitis (many enterovirus types), hand-foot-mouth disease (coxsackie A), pericarditis and myocarditis (coxsackie B), acute hemorrhagic conjunctivitis (enterovirus 70), and fever, myalgia, summer "flu" (many enterovirus types)

    Detection: Cell culture (PMK and HDF), serology

    Treatment: Supportive

    Prevention: Avoid contact with virus; vaccination for polio

Virus: Hepatitis A virus (enterovirus type 72)

    Transmission: Fecal-oral

    Disease: Hepatitis with short incubation, abrupt onset, and low mortality. No carrier state

    Detection: Serology

    Treatment: Supportive

    Prevention: Vaccine; prevent clinical illness with serum immunoglobulin

Virus: Rhinovirus (common cold virus)

    Characteristics: Approximately 100 serotypes

    Transmission: Contact with respiratory secretions

    Disease: Common cold

    Detection: Cell culture (usually not clinically necessary)

    Treatment: Supportive

    Prevention: Avoid contact with virus

TABLE 66-15 REOVIRUSES

Family: *Reoviridae*   Common Name: Reovirus

Virus: Rotavirus

Characteristics: Segmented, double-stranded RNA genome; icosahedral capsid with no envelope

Transmission: Fecal-oral. Survives well on inanimate objects

Disease: Gastroenteritis in infants and children 6 months to 2 years

Detection: EIA, latex agglutination

Epidemiology: Winter-spring seasonality in temperate climates. Nosocomial transmission can occur easily

Treatment: Supportive, especially fluid replacement

Prevention: Avoid contact with virus. Vaccine trials under way

TABLE 66-16 RETROVIRUSES

Family: *Retroviridae*   Common Name: Retroviruses

Characteristics: Single-stranded RNA genome; icosahedral capsid with envelope. Reverse transcriptase converts genomic RNA into DNA

Virus: Human immunodeficiency virus types 1 and 2 (HIV 1 and HIV 2)

Characteristics: Most disease in humans caused by HIV 1. Infected cells include CD4 (helper) T lymphocytes, monocytes, and some cells of the central nervous system

Transmission: Sexual contact, blood and blood product exposure, and perinatal exposure

Site of latency: CD4 T lymphocytes

Disease: Asymptomatic infection, acute "flu-like" disease, AIDS-related complex, and AIDS

Detection: Serology, antigen detection, PCR

Epidemiology: Those at risk of infection are homosexual or bisexual males, intravenous drug abusers, sexual contacts of HIV-infected individuals, and infants of infected mothers

Treatment: AZT, didanosine, protease inhibitors. Treat infections resulting from immunosuppression

Prevention: Avoid contact with infected blood and blood products and secretions. Blood for transfusion is screened for antibody to HIV 1 and 2

Virus: Human T-lymphotropic viruses (HTLV-1 and HTLV-2)

Transmission: Known means of transmission are similar to HIV

Disease: T-cell leukemia and lymphoma, and tropical spastic paraparesis (HTLV-1)

Detection: Serology

Epidemiology: HTLV-1 present in 0.025% of volunteer blood donors in United States. Blood is screened for antibody to HTLV 1 and 2. Rates of HTLV-1 in areas of Japan and the Caribbean are considerably higher

Oncogenic: T-cell lymphoma (HTLV-1)

Treatment: Supportive

Prevention: Avoid contact with virus

TABLE 66-17   RHABDOVIRUSES

Family: *Rhabdoviridae*   Common Name: Rhabdovirus

Virus: Rabies virus

Characteristics: Single-stranded RNA genome; helical capsid with envelope

Transmission: Bite of rabid animal most common; 20% of human rabies cases have no known exposure to rabid animal

Disease: Rabies

Detection: FA staining, serology

Treatment: Supportive

Prevention: Avoid contact with rabid animals; vaccinate domestic animals. Postexposure prophylaxis with hyperimmune antirabies globulin and immunization with rabies vaccine

---

TOGAVIRUSES

Although closely related according to virion size and genome composition, the togaviruses and flaviviruses (Table 66-18) are now classified in different families. The togavirus family includes rubella virus and the alpha viruses, a large group of mosquito-borne arboviruses. Typically, mosquitos infect a vertebrate host (such as birds and rodents), the virus multiplies (amplifies) in this host and is picked up and passed by subsequent mosquito bites. Humans are incidentially infected; they are not amplifiers of the virus but rather dead-end hosts. Human disease varies from asymptomatic infection to fatal encephalitis.

Togavirus disease is diagnosed by detecting specific serum IgG and IgM antibodies. Virus isolation is not practical in clinical laboratories.

MISCELLANEOUS VIRUSES

The arenaviruses, caliciviruses, and filoviruses are families of viruses that contain important human pathogens. Diagnosis is made using electron microscopy to detect virions or serology to detect a virus-specific antibody response.

The arenaviruses include lymphocytic choriomeningitis virus, a rare cause of aseptic meningitis or meningoencephalitis in the United States, and Lassa fever virus, which causes Lassa fever in West Africa. Lassa fever is fatal in 50% of victims and is characterized by fever, coagulopathy, hemorrhage, and shock.

Caliciviruses contain the Norwalk agent viruses, which cause diarrhea in older children and adults, and hepatitis E virus, a rare cause of disease in the United States but common in developing countries. Hepatitis E, transmitted by the fecal-oral route, causes relatively mild disease, except in pregnant women.

The filovirus family contains hemorrhagic fever viruses, including Marburg and Ebola viruses. These and similar viruses are contracted in areas covered with jungle and rain forests, especially on the African continent. Electron micrographs demonstrate long, filamentous viruses. Handling specimens from patients with suspected filovirus infection should occur in special containment facilities designed for hemorrhagic fever viruses.

METHODS FOR THE LABORATORY DIAGNOSIS OF VIRAL DISEASES

SETTING UP A CLINICAL VIROLOGY LABORATORY

The demand for clinical virology laboratory services skyrocketed during the 1980s and 1990s. This growth resulted from the commercial availability of high-quality reagents, the use by clinicians of virus-specific antiviral drugs, the development of rapid diagnostic techniques, and the application of virology cell culture procedures to chlamydial culture and *Clostridium difficile* cytotoxin testing. In deciding which virology tests to offer, a clinical laboratory should determine if the test is required for the appropriate care of the patient and if techniques are available that provide an accurate and cost-effective test result. Viral diseases that commonly require laboratory diagnosis include sexually transmitted diseases, diarrhea, respiratory disease in adults and children, aseptic meningitis, congenital disease, and infections in immunocompromised hosts. Viruses detected by a community hospital virology laboratory are shown in Table 66-19.

Those working in a clinical virology laboratory

TABLE 66-18 **TOGAVIRUSES**

Family: *Togaviridae*   Common Name: Togaviruses

Characteristics: Single-stranded RNA genome; icosahedral capsid with envelope. Family contains arboviruses and nonarthropod borne viruses

Virus: Rubella virus (togavirus)

Transmission: Respiratory, transplacental

Disease: Rubella (mild exanthematous disease), congenital rubella

Detection: Serology

Treatment: Supportive

Prevention: Rubella vaccine

Virus: Arboviruses, including alphaviruses*

Transmission: Arthropod vector, usually mosquito

Disease: Eastern, Western, and Venezuelan encephalitis

Detection: Serology and antibody detection in CSF

Treatment: Supportive

Prevention: Avoid contact with vector; vector control programs

*Arthropod-borne viruses (arboviruses) were once grouped together because of their common mode of transmission. Present classification includes three families containing arboviruses: *Togaviridae*, *Flaviviridae*, and *Bunyaviridae*. The virus group within *Togaviridae*, which includes arboviruses, is the alphavirus group. Common arboviruses are referred to as *bunyaviruses*, *flaviviruses*, and *alphaviruses*.

TABLE 66-19 **VIRUSES DETECTED BY A COMMUNITY HOSPITAL VIROLOGY LABORATORY**

VIRUS	NUMBER ISOLATED FROM	
	CHILDREN	ADULTS
Adenovirus	98	19
Cytomegalovirus	68	191
Enterovirus	229	30
Herpes simplex virus	95	656
Influenza virus	94	71
Parainfluenza virus	161	33
Respiratory syncytial virus	425	2
Rotavirus	307	0
Varicella-zoster virus	5	14
Total	1482	1016

Data from Children's Hospital Medical Center of Akron, Akron, Ohio (1986-1988).

must be familiar with cell culture, enzyme immunoassay, and immunofluorescence methods, in addition to other common laboratory techniques. Large equipment needed for a full-service virology laboratory include a laminar-flow biological safety cabinet (BSC), fluorescence microscope, inverted bright-field microscope, refrigerated centrifuge, incubator, refrigerator/freezer, and roller drum for holding cell culture tubes during incubation (Figures 66-5, 66-6 and 66-7).

Universal precautions and Biosafety Level 2 conditions are needed for community and most nonretro-viral laboratories. Requirements include standard microbiologic practices, training in biosafety, protective clothing and gloves, limited access, decontamination of all infectious waste, and a class I or II BSC. A virology laboratory floorplan that includes safety equipment and work areas is shown in Figure 66-8.

SPECIMEN SELECTION AND COLLECTION

Specimen selection depends on the specific disease syndrome, viral etiologies suspected, and time of year. Selection of specimen based on disease is confusing,

FIGURE 66-5 Roller drums used to hold cell culture tubes during incubation. Slow rotation continually bathes cells in medium. (Courtesy Children's Hospital Medical Center of Akron, Akron, Ohio.)

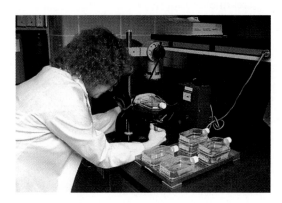

FIGURE 66-6 Inverted microscope used to examine cell monolayers growing attached to the inside surface beneath the liquid medium.

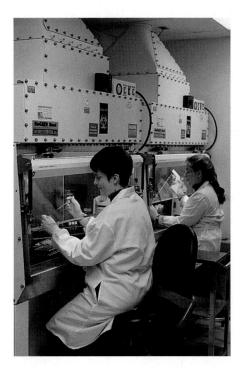

FIGURE 66-7 Class II biological safety cabinets used in clinical virology laboratory. The cabinet on the left is used for "contaminated" work, including specimen inoculation and working with positive cell cultures. The cabinet on the right is used for "clean" work, such as preparation and maintenance of uninoculated cell cultures.

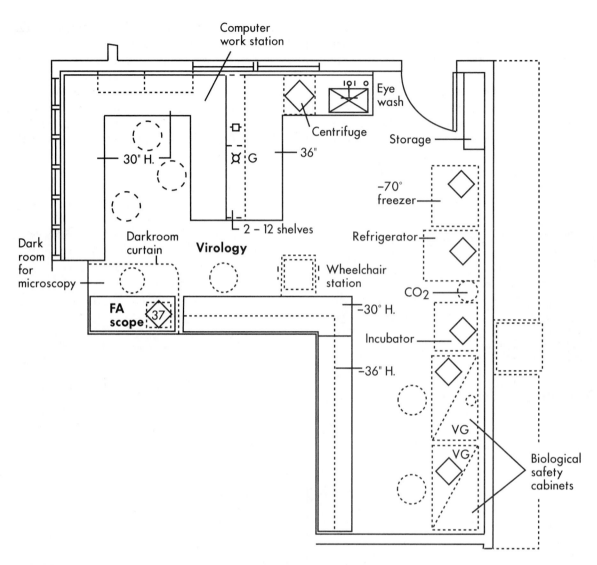

FIGURE 66-8 Floor plan of clinical virology laboratory that includes biological safety cabinet for specimen processing and cell culture handling, dark room for fluorescence microscopy, stand-up and sit-down counter space, computer station, storage areas, incubator, refrigerator, and freezers. *G*, gas; *V*, vacuum; *E*, electric outlet.

because most viruses enter via the upper respiratory tract yet infect tissues, which results in symptoms distant from the primary, oropharyngeal site. For example, aseptic meningitis, caused by many enterovirus types, is identified by detecting virus in throat, rectal swab, or CSF specimen. Pharyngitis and gastrointestinal symptoms may not be included in the patient's complaints.

Specimen selection based on virus suspected is complicated by similar clinical syndromes that are caused by many different viruses. By collection of only specimens needed to detect a specific virus, other important etiologies may be missed. For example, testing smears of nasal secretions from an infant by fluorescence staining or enzyme immunoassay for the detection of RSV does not allow for diagnosis of disease caused by influenza, parainfluenza, and adenoviruses.

Appropriate specimen selection dictates that the specimen type and viruses suspected always should be indicated on the requisition. Serum for serologic testing may be necessary, and some viral diseases need only be considered during certain months because their appearance is seasonal. Table 66-20 suggests specimens for the diagnosis of viral diseases, noting seasonality where important.

Specimens for the detection of virus should be collected as early as possible following the onset of disease. Virus may no longer be present as early as 2 days after the appearance of symptoms. Recommendations for collection of common specimens are summarized in the following section.

Throat or nasopharyngeal swab or aspirate
Nasopharyngeal aspirates are superior, in general, to swabs for recovering viruses, but swabs are consider-

ably more convenient. Throat swabs are acceptable for recovering enteroviruses, adenoviruses, and herpes simplex virus, whereas nasopharyngeal swab or aspirate specimens are preferred for the detection of respiratory syncytial, influenza, and parainfluenza viruses. Rhinovirus detection requires a nasal specimen. Throat specimens are collected by rubbing inflamed or purulent areas of the posterior pharynx with a dry, sterile swab. Avoid touching the tongue, buccal mucosa, or teeth and gums. Nasopharyngeal secretion specimens are collected by inserting a swab through the nostril to the nasopharynx or by using a bulb syringe with 3 to 7 mL of buffered saline. The saline is squirted into the nose by squeezing the bulb and aspirated back by releasing the bulb.

Bronchial and bronchoalveolar washes

Wash and lavage fluid collected during bronchoscopy are excellent specimens for the detection of viruses that infect the lower respiratory tract, especially influenza and adenoviruses.

Rectal swabs and stool specimens

Stool and rectal swabs of feces specimens are used to detect rotavirus, enteric adenoviruses (especially serotypes 40 and 41), and enteroviruses. Many agents of viral gastroenteritis do not grow in cell culture and require electron microscopy for detection (see discussion of detection of viruses in patient specimens). In general, stool specimens are preferable to rectal swabs and should be required for rotavirus and enteric adenovirus testing. Rectal swabs are acceptable for detecting enteroviruses in patients suspected of having enteroviral disease such as aseptic meningitis. Rectal swabs are collected by inserting a swab 3 to 5 cm into the rectum to obtain feces. Collection of 5 to 10 mL of freshly passed diarrheal stool is sufficient for rotavirus and enteric adenovirus detection.

Urine

Cytomegalovirus, mumps, rubella, measles, polyomaviruses, and adenoviruses can be detected in urine. Virus recovery may be increased by processing multiple (2 to 3) specimens, because virus may be shed intermittently or in low numbers. The best specimen is at least 10 mL of a clean-voided, first-morning urine.

Skin and mucous membrane lesions

Enteroviruses, herpes simplex virus, varicella-zoster virus, and, rarely, cytomegalovirus can be detected in vesicular lesions of skin and mucous membranes. Once the vesicle has ulcerated or crusted, detection of virus is very difficult.

Collection of specimen from cutaneous vesicles

for detection of HSV or VZV requires a Tzanck smear. Tzanck smears are prepared by carefully unroofing the vesicle. If a tuberculin syringe is used, the small "drop" of vesicle fluid should be aspirated first, to be used in case viral or bacterial culture is needed. Flush needle with viral transport medium and add rinse to viral transport tube. With the roof of the vesicle folded back, carefully remove excess fluid by dabbing with a sterile gauze. Press a clean glass microscope slide against the base of the ulcer. Lift, move slide slightly, and press again. Cells from the base of the ulcer will stick to the slide, making an "impression smear" of infected and noninfected cells. Additional smears can be made from other vesicles. Send slides to the laboratory for fixation and staining.

Sterile body fluids other than blood

Sterile body fluids, especially cerebrospinal, pericardial, and pleural fluids, may contain enteroviruses, herpes simplex virus, influenza viruses, or cytomegalovirus. These specimens are collected aseptically by the physician.

Blood

Viral culture of blood is used primarily to detect cytomegalovirus; however, HSV, VZV, enteroviruses, and adenovirus may occasionally be encountered. A specimen of 5 to 10 mL of anticoagulated blood collected in a Vacutainer tube is needed. Heparinized, citrated, or EDTA anticoagulated blood is acceptable for cytomegalovirus detection. Citrated blood should be used in instances in which other viruses are being sought.

Bone marrow

Bone marrow for viral culture should be added to a sterile tube with anticoagulant. Heparin, citrate, or EDTA anticoagulants are acceptable. Specimens are collected by aspiration.

Tissue

Tissue specimens are especially useful for detecting viruses that commonly infect lung (cytomegalovirus, influenza virus, adenovirus), brain (herpes simplex virus), and gastrointestinal tract (cytomegalovirus). Specimens are collected during surgical procedures.

Serum for antibody testing

Acute and convalescent serum specimens may be needed to detect antibody to specific viruses. Acute specimens should be collected as soon as possible after the appearance of symptoms. The convalescent specimen is collected a minimum of 2 to 3 weeks after the acute. In both cases, an appropriate specimen is 3 to 5 mL of serum collected by venipuncture.

TABLE 66-20 SPECIMENS FOR THE DIAGNOSIS OF VIRAL DISEASES*

DISEASE CATEGORIES AND PROBABLE VIRAL AGENTS	SEASON OF MOST COMMON OCCURRENCE	THROAT/ NASOPHARYNX	STOOL	CSF	URINE	OTHER
Respiratory syndrome						
Adenoviruses	Y	++++				
Influenza virus	W	++++				
Parainfluenza virus	Y	++++				
Respiratory syncytial virus (RSV)	W	++++				
Rhinoviruses	Y					Nasal (+++)
Dermatologic and Mucous Membrane Disease						
Vesicular						Vesicle fluid or scraping (+++)
Enterovirus	S, F	++	+++			
Herpes simplex†	Y					
Varicella-zoster†	Y	++				
Exanthematous						
Enterovirus	S, F	+++	++			
Measles	Y	++			++	Serum for antibody detection
Rubella	Y	+++			++	Serum for antibody detection
Parvovirus	Y					Serum for antibody detection, amniotic fluid (PCR)
Meningoencephalitis						
Arboviruses	S, F					CSF/serum for antibody detection
Enteroviruses	S, F	++	++	++++		
Herpes simplex	Y			++++ (PCR)		(Brain biopsy)
Lymphocytic choriomeningitis	Y					Serum for antibody detection
Mumps virus	Y					Serum for antibody detection

Virus	Season				Specimen
HIV	Y				Brain biopsy (culture)
Polyomavirus (JC virus)	Y				Brain Biopsy (EM)
Gastrointestinal					
Adenoviruses	Y	++++			Stool (EIA or EM)
Norwalk agent	S	++++			Stool (EM)
Rotavirus	W, SP	++++			Stool (EIA, Latex)
Congenital and Perinatal Disease					
Cytomegalovirus	Y		+++		Serum for antibody (IgM) detection
Enteroviruses	S, F	+++	+++	+++	
Herpes simplex virus	Y				(Vesicle fluid)
Rubella	Y		++		Serum for antibody (IgM) detection
Eye Syndrome					
Adenoviruses	Y	++			Conjunctival swab or scraping
Herpes simplex virus	Y				Conjunctival swab or scraping
Varicella-zoster virus	Y				Conjunctival swab or scraping
Posttransplantation Syndrome					
Cytomegalovirus	Y	++			Blood (++++)–shell vial, antigenemia, tissue (++++)–shell vial
Epstein-Barr virus (EBV)	Y				Serology, tissue (PCR)
Human herpesvirus 6 (HH6)	Y				Serology, blood (PCR)
Herpes simplex virus (HSV)	Y				Tissue (+++)
Myocarditis, Pericarditis, and Pleurodynia					
Coxsackie B	S, F	+++	++		Pericardial fluid (++++)

Y, year-round; SP, spring; S, summer; F, fall; W, winter; PCR, polymerase chain reaction.

*Specimens indicated beside the disease categories should be obtained in all instances; others should be obtained if the specific virus is suspected.

†Direct fluorescent antibody studies are available for herpes simplex virus and varicella-zoster virus.

SPECIMEN TRANSPORT AND STORAGE

Ideally all specimens collected for detection of virus should be processed by the laboratory immediately. Although inoculation of specimens into cell culture at the bedside has been recommended in the past, potential biohazards, sophisticated processing steps, and necessary quality control make this practice impractical today. Specimens for viral isolation should not be allowed to sit at room or higher temperature. Specimens should be placed in ice and transported to the laboratory at once. If a delay is unavoidable, the specimen should be refrigerated, *not frozen,* until processing occurs. Every attempt should be made to process the specimen within 12 to 24 hours of collection. Under unusual circumstances specimens may need to be held for days before processing. For storage up to 5 days, hold specimen at 4° C. Storage for 6 or more days should be at −20° C or preferably −70° C. Specimens for freezing should first be diluted or emulsified in viral transport medium. Significant loss of viral infectivity may occur during prolonged storage, regardless of conditions.

Many types of specimens for the detection of virus can be collected using a swab. Most types of swab material, such as cotton, rayon, and dacron, are acceptable. Once collected, it is recommended that specimens on swabs be emulsified in viral transport medium before transporting to the laboratory, especially if transport will occur at room temperature and require more than 1 hour. It has been shown, however, that transport of specimens containing herpes simplex virus can be accomplished using a Culturette (Becton Dickinson Microbiology Products, Cockeysville, Md.; modified Stuart's medium). Calcium alginate is not acceptable for the detection of herpes simplex virus because it may inactivate HSV. If calcium alginate must be used, the specimen should be transferred into viral transport medium by swirling the swab vigorously and then the swab should be discarded immediately.

Viral transport media are used to transport small volumes of fluid specimens, small tissues, scrapings, and swab specimens, especially when contamination with microbial flora is expected. Transport media contain protein, such as serum, albumin, or gelatin, to stabilize virus and antimicrobials to prevent overgrowth of bacteria and fungi. Penicillin (500 U/mL) and streptomycin (500 to 1000 μg/mL) have been used traditionally; however, a more potent mixture includes vancomycin (20 μg/mL), gentamicin (50 μg/mL), and amphotericin (10 μg/mL). If serum is added as the protein source, fetal calf serum is recommended because it is less likely to contain inhibitors, such as specific antibody. Examples of successful transport media include Stuart's medium, Amie's

medium, Leibovitz-Emory medium, Hanks balanced salt solution (HBSS), and Eagle's tissue culture medium. Respiratory and rectal/stool specimens can be maintained in modified Stuart's, modified Hanks balanced salt solution, or Leibovitz-Emory medium containing antimicrobials.

Blood for viral culture, transported in a sterile tube containing anticoagulant, must be kept at refrigeration temperature (4° C) until processing.

Blood for viral serology testing should be transported to the laboratory in the sterile tube in which it was collected. Serum should be separated from the clot as soon as possible. Serum can be stored for hours or days at 4° C or for weeks or months at −20° C or below before testing. Testing for virus-specific IgM should be done before freezing whenever possible, because IgM may form insoluble aggregates upon thawing.

SPECIMEN PROCESSING

Specimens for viral culture should be processed immediately upon receipt in the laboratory (Table 66-21). This may be best accomplished by combining bacteriology and virology processing responsibilities. Although in the past the threat of cell culture contamination dictated separation of virology procedures, the addition of broad-spectrum antimicrobials to cell cultures significantly decreased the possibility of cross-contamination with bacteria. In most laboratories, processing with other microbiology specimens allows viral cultures to be processed 7 days per week. If delays must occur, specimens should be stored in viral transport medium at 4° C as described previously. Fluid specimens that have not been added to viral transport medium and that must be stored for days before processing should be diluted in transport medium (1:2 to 1:5) before storage.

Each specimen for virus isolation should be accompanied by a requisition that provides the following information in addition to patient identification and demographics: source of specimen, clinical history or viruses suspected, and date and time of specimen collection. If this information is not available, a call should be made to the requesting physician or person caring for the patient for additional details.

Processing viral specimens should occur in a biological safety cabinet (BSC) whenever possible (Figure 66-7). This protects specimens from contamination by the processing technologist and protects those in the laboratory from infectious aerosols created when specimens are manipulated. If processing cannot be performed in a BSC, it should occur behind a protective plexiglass shield on the countertop. Latex gloves and a laboratory coat should be worn during

TABLE 66-21   LABORATORY PROCESSING OF VIRAL SPECIMENS

SOURCE	SPECIMEN	PROCESSING*	CELLS FOR COMPREHENSIVE CULTURE†
Blood	Anticoagulated blood	Separate leukocytes (see Procedure 66-1)	PMK, HDF, HEp-2
Cerebrospinal fluid (CSF)	1 mL CSF	Inoculate directly	PMK, HDF, HEp-2
Stool or rectal swab	Pea-sized aliquot of feces	Place in 2 mL of viral transport medium. Vortex. Centrifuge at $1000 \times g$ for 15 min. Use supernatant fluid for inoculum	PMK, HDF, HEp-2
Genital, skin	Vesicle fluid or scraping	Emulsify in viral transport medium	HDF
Miscellaneous	Swab, fluids	Swab: emulsify in viral transport medium Fluid: inoculate directly	PMK, HDF, HEp-2
Respiratory tract	Nasopharyngeal or throat swab or washings	Dilute with viral transport medium	PMK, HDF, HEp-2
Tissue	Tissue in sterile container	Mince with sterile scalpel and scissors and gently grind. Prepare 20% suspension in viral transport medium. Centrifuge at $1000 \times g$ for 15 min. Use supernatant fluid for inoculum	PMK, HDF, HEp-2
Urine	Fresh refrigerated urine	Inoculate directly if clear. Turbid specimens should be centrifuged at $1000 \times g$ for 15 min. Use supernatant fluid for inoculum	HDF, HEp-2 (if adenovirus suspected)

PMK, Primary monkey kidney; *HDF*, human diploid fibroblast; *HEp-2*, human epidermoid.

*All inoculum into tissue culture tubes are 0.25 mL volumes.

†Comprehensive culture implies detection of all common cultivatable viruses. A more limited cost-effective approach to the use of viral diagnostic tests is described in Figures 66-28, 66-30, 66-37, and 66-38.

manipulation of all patient specimens. Vortexing, pipetting, and centrifugation can all create dangerous aerosols. Vortexing should be done in a tightly capped tube behind a shield. After vortexing, the tube should be opened in a safety cabinet. Pipetting also should be performed behind a protective shield. Pipettes must be discarded into a disinfectant fluid, so that the disinfectant reaches the inside of the pipette or into a leakproof biosafety bag for autoclaving or incineration.

Processing virology specimens is not complicated (see Table 66-21). In general, any specimen that might be contaminated with bacteria or fungi or any swab specimen should be added to viral transport medium. Presumably sterile fluid specimens can be inoculated directly. Viral transport medium or fluid specimens not in transport medium should be vortexed just before inoculation to break up virus-

containing cells and resuspend the inoculum. Sterile glass beads added to the transport medium help break up cell clumps and release virus from cells during vortexing. Grossly contaminated or potentially toxic specimens, such as minced or ground tissue, can be centrifuged ($1000 \times g$ for 15 minutes) and the virus-containing supernatant used as inoculum. Each viral cell culture tube is inoculated with 0.25 mL of specimen. If insufficient specimen is available, dilute with viral transport medium to increase volume. Excess specimen can be saved at 4° C in case the culture is contaminated. Contaminated specimens can be reprocessed with antibiotic-containing viral transport medium, if they were not originally handled in this manner, or they can be filtered using a disposable 0.22-μm filter and the filtrate recultured. In practice, most specimens requiring reprocessing because of contamination fail to have virus detected in culture.

Table 66-21 represents an overview of processing procedures. Blood for viral culture requires special processing to isolate leukocytes, which are then inoculated to cell culture tubes (Procedure 66-1). Rapid shell vial cultures are used to detect many viruses. Handling and examining cell cultures once inoculated with specimen is covered later in this chapter.

DETECTION OF VIRUSES IN PATIENT SPECIMENS

Cytology and histology

The most readily available technique for the detection of virus is cytologic or histologic examination for the presence of characteristic viral inclusions. This involves the morphologic study of cells or tissue, respectively. Viral inclusions are intracellular structures formed by aggregates of virus or viral components within an infected cell or abnormal accumulations of cellular materials resulting from viral-induced metabolic disruption. Inclusions occur in single or syncytial cells. Syncytial cells are aggregates of cells fused to form one large cell with multiple nuclei. Papanicolaou (Pap)- or Giemsa-stained cytologic smears are examined for inclusions or syncytia. Cytology is most frequently used to detect infections with varicella-zoster and herpes simplex viruses (Figure 66-9, *A*). A stained smear of cells from the base of a skin vesicle used to detect VZV or HSV inclusions is called a **Tzanck test.** Inclusions resulting from infection with CMV, adenovirus, parvovirus, papilloma virus, and molluscum contagiosum virus are detected by histologic examination of tissue stained with hematoxylin and eosin (Figure 66-9, *B* through *F*). Less commonly, inclusions characteristic of measles and rabies viruses are detected by examining stained tissues (Figure 66-9, *G* through *H*). Rabies virus inclusions in brain tissue are called **Negri bodies.** Cytology and histology are less sensitive than culture but are especially helpful for viruses that are difficult or dangerous to isolate in the laboratory, such as parvovirus and rabies virus, respectively.

Electron microscopy

Few laboratories use electron microscopy (EM) to detect viruses because it is labor intensive and relatively insensitive. EM is most helpful for the detection of viruses that do not grow readily in cell culture and works best if the titer of virus is at least 10^6 to 10^7 particles per mL. Immune EM allows visualization of virus particles present in numbers too small for easy direct detection. The addition of specific antiserum to the test suspension causes the virus particles to form antibody-bound aggregates, which are more easily detected than are single virus particles. EM is still used by some clinical laboratories for the detection of viruses causing gastroenteritis (rotavirus, enteric adenoviruses, coronaviruses, Norwalk-agent virus, and other caliciviruses) and encephalitis (herpes simplex virus, measles virus, and JC polyomavirus) (Figure 66-10, *A* through *H*). EM is no longer needed to detect common rotavirus types causing infection in humans, because relatively simple and accurate enzyme immunoassay and latex agglutination antigen assays have been developed.

Immunodiagnosis

High-quality, commercially available, viral antibody reagents have led to the development of fluorescent antibody, enzyme immunoassay, radioimmunoassay, latex agglutination, and immunoperoxidase tests that detect viral antigen in patient specimens.

Direct and indirect immunofluorescent antibody methods are used. Direct immunofluorescent antibody testing involves use of a labeled antiviral antibody. The label is usually fluorescein isothiocyanate (FITC), which is layered over specimen suspected of containing homologous virus (Figure 66-11). The indirect immunofluorescent antibody procedure is a two-step test in which unlabeled antiviral antibody is added to the slide first, followed by a labeled (FITC) antiglobulin that binds to the first-step antibody bound to virus in the specimen (Figure 66-12). Direct immunofluorescence is, in general, more rapid and specific than indirect immunofluorescence, but less sensitive. Increased sensitivity of the indirect test results from signal amplification that occurs with the addition of the second antibody. Signal amplification decreases specificity by increasing nonspecific background fluorescence. Direct immunofluorescence is best suited to situations in which large quantities of virus are suspected or high-quality, concentrated monoclonal antibodies are used, such as for the detection of RSV in a patient specimen or the identification of viruses growing in cell culture. Indirect immunofluorescence should be used when lower quantities of virus are suspected, such as detection of respiratory viruses in specimens from adult patients. High-quality monoclonal antibodies improve the sensitivity and specificity of the indirect test. Whenever possible the direct test should be used because it is significantly faster without the second antibody step. A standard procedure for the direct fluorescent antibody test is shown in Procedure 66-2.

Strict criteria for the interpretation of fluorescent patterns must be used. This includes standard interpretation of fluorescent intensity (Table 66-22) and recognition of viral inclusion morphology. Nuclear and cytoplasmic staining patterns are typical for influenza virus, adenovirus, and the herpesviruses; cytoplasmic staining only for respiratory syncytial, parainfluenza, and mumps viruses and staining within multinucleated giant cells is typical of measles virus (Figure 66-13, *A* to *H*). False-positive staining can occur with specimens containing yeasts, certain bacteria, mucus, or leukocytes. Leukocytes, which contain Fc receptors for antibody, can also cause nonspecific binding of antibody conjugates. To verify technicians' ability to interpret fluorescent antibody tests, every laboratory should perform viral culture or some alternative detection method in parallel with immunofluorescence until in-house performance has been established.

Immunofluorescent stains that are most useful in the clinical virology laboratory are those for respiratory syncytial virus, influenza and parainfluenza viruses, adenovirus, herpes simplex, varicella-zoster, and cytomegaloviruses. A pool of antibodies can be used to screen a specimen for multiple viruses. A positive screen is then tested with each individual reagent to identify the exact virus that is present. A screening pool has been used successfully for the detection of respiratory viruses in children (Antiviral screening reagent, Meridian Diagnostics, Inc., Cincinnati, Ohio). Such pools are much less sensitive when used with specimens from adults.

Enzyme immunoassay methods used most frequently in clinical virology are the solid-phase enzyme-linked immunosorbent assay (solid-phase ELISA) and the membrane-bound enzyme-linked immunosorbent assay (membrane ELISA). Solid-phase ELISA is performed in a small test tube or microtiter tray. Breakaway strips of microtiter wells are available for low-volume test runs. The remaining unused wells can be saved for future testing. A solid-phase ELISA used for the detection of rotavirus in stool specimens is illustrated in Figure 66-14. Membrane ELISA tests have been developed for low-volume testing and situations in which rapid results are needed. They can be performed by those with

minimum training and usually require less than 30 minutes to complete. The membrane method uses a handheld reaction chamber with a cellulose-like membrane. Specimen and reagents are applied to the membrane. Following a short incubation time, a chromogenic (color) reaction occurs on the surface of the membrane and is read visually. Built-in controls on the same membrane provide convenient monitoring of test procedures. Figure 66-15 illustrates a membrane ELISA used to detect respiratory syncytial virus in a nasal wash specimen. The most useful enzyme immunoassays used for antigen detection are those for respiratory syncytial virus (solid-phase and membrane), rotavirus (solid-phase and membrane), enteric adenoviruses (solid-phase), and herpes simplex virus (solid-phase).

Advantages of enzyme immunoassays are the use of nonradioactive, relatively stable reagents and results that can be interpreted qualitatively (positive or negative) or quantitatively (titer or degree of positive reaction). It is important to note that enzyme immunoassays frequently have an indeterminant or borderline interpretative category. This result implies that low levels of viral antigen or background interference has prevented a clear-cut positive or negative result. Such results usually require testing a second specimen to avoid interference or to detect a rise in antigen level. ELISAs are more sensitive than most other assays, including fluorescent immunoassays, and can be easily automated. A drawback with ELISAs is that specimen quality cannot be evaluated; that is, the number of cells cannot be assessed, as can be done microscopically with fluorescent immunoassay. Acellular specimens are potentially inferior because specimen collection could have been inadequate and cell-associated virus would not be present.

Radioimmunoassay (RIA), immunoperoxidase staining, and latex agglutination are additional techniques used to detect viral antigen. RIA has been

Text continued on p. 995

TABLE 66-22   INTERPRETATION OF FLUORESCENCE INTENSITY

INTENSITY	INTERPRETATION
Negative	No apple-green fluorescence
1+	Faint yet unequivocal apple-green fluorescence
2+	Apple-green fluorescence
3+	Bright apple-green fluorescence
4+	Brilliant apple-green fluorescence

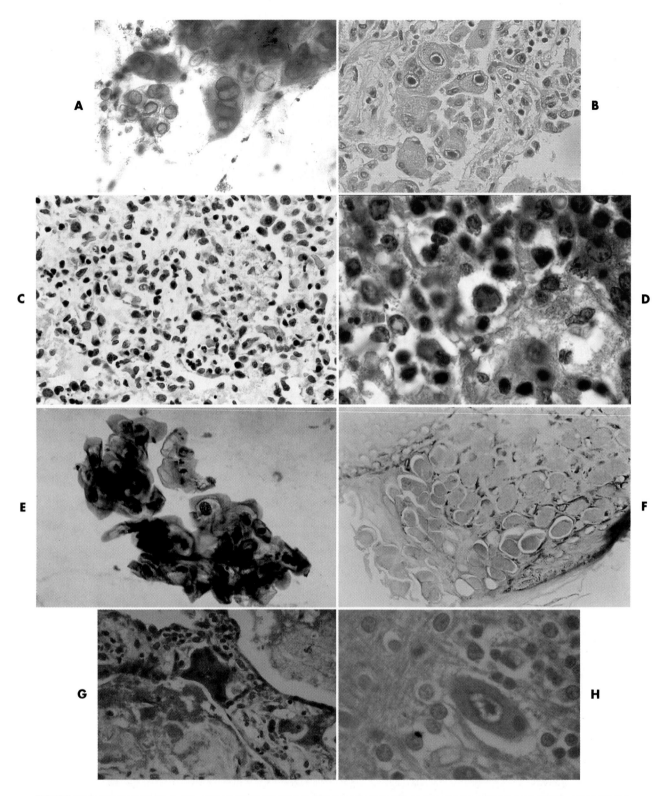

FIGURE 66-9 Viral inclusions. **A,** Pap-stained smear showing multinucleated giant cells typical of herpes simplex or varicella-zoster viruses. **B,** Hematoxylin and eosin-stained (HE) lung tissue containing intranuclear inclusion within enlarged CMV infected cells. **C,** HE-stained lung tissue containing epithelial cells with intranuclear inclusions characteristic of adenovirus. **D,** HE-stained liver from stillborn fetus showing intranuclear inclusions in erythroblasts (extramedullary hematopoiesis) resulting from parvovirus infection. **E,** Pap stain of exfoliated cervicovaginal epithelial cells showing perinuclear vacuolization and nuclear enlargement characteristic of human papillomavirus infection. **F,** HE-stained epidermis filled with molluscum bodies, which are large eosinophilic cytoplasmic inclusions resulting from infection with molluscum contagiosum virus. **G,** HE-cells infected with measles virus. **H,** HE-stained brain tissue showing oval, eosinophilic rabies cytoplasmic inclusion (Negri body). (**E** and **F** from Murray, P.R., Kobayashi, G.S., Pfaller, M.A., et al., editors. 1994. Medical microbiology, ed 2. Mosby. St. Louis.)

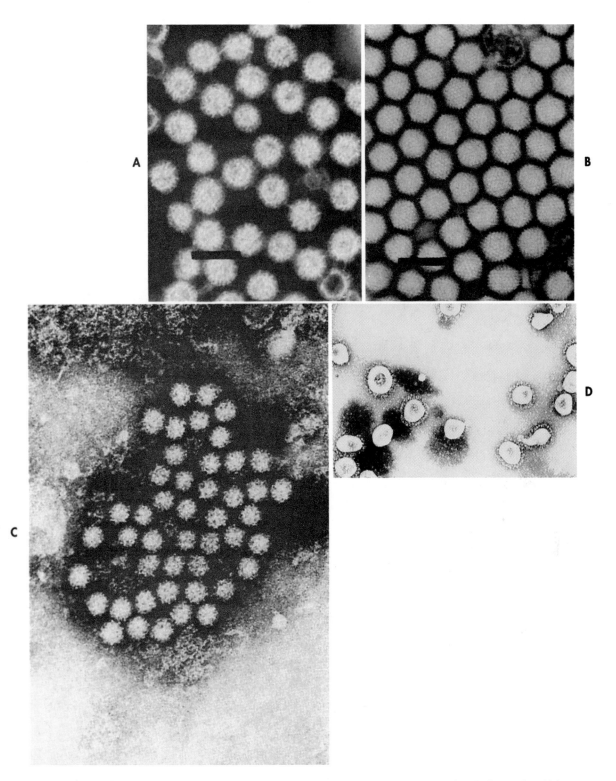

FIGURE 66-10 Electron micrographs of viruses. **A,** Rotavirus. **B,** Adenovirus. **C,** Norwalk agent virus. **D,** Coronavirus. (**C** from Howard, B.J., Klaas, J., Rubin, S.J., et al. 1987. Clinical and pathogenic microbiology. Mosby, St. Louis; **D** and **F** from U.S. Department of Health, Education, and Welfare; Public Health Service, Centers for Disease Control, Atlanta; **G** courtesy Dr. Gabriele M. ZuRhein.)

Continued

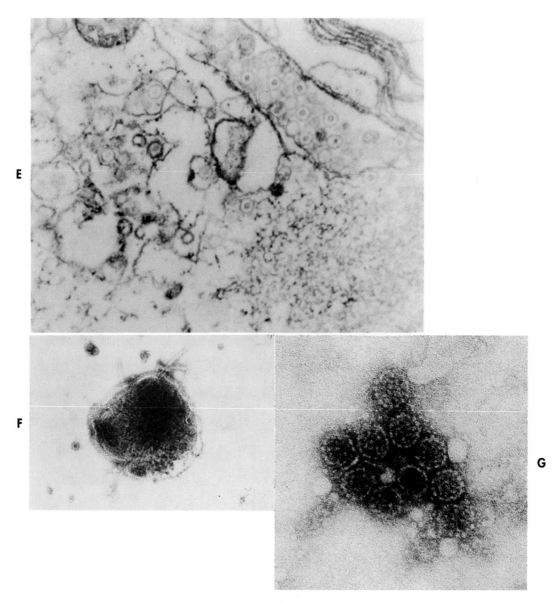

FIGURE 66-10, cont'd Electron micrographs of viruses. **E,** Herpes simplex virus. **F,** Measles virus. **G,** Negatively stained preparation of JC virus in brain tissue (Norwalk agent). (**C** from Howard, B.J., Klaas, J., Rubin, S.J., et al. 1987. Clinical and pathogenic microbiology. Mosby, St. Louis; **D** and **F** from U.S. Department of Health, Education, and Welfare; Public Health Service, Centers for Disease Control, Atlanta; **G** courtesy Dr. Gabriele M. ZuRhein.)

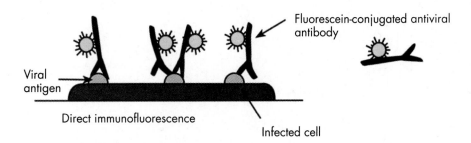

Fluorescein-conjugated antiviral antibody

Viral antigen

Direct immunofluorescence

Infected cell

FIGURE 66-11 Illustration of direct immunofluorescent staining method.

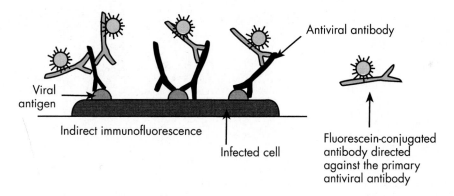

Antiviral antibody

Viral antigen

Indirect immunofluorescence

Infected cell

Fluorescein-conjugated antibody directed against the primary antiviral antibody

FIGURE 66-12 Illustration of indirect immunofluorescent staining method.

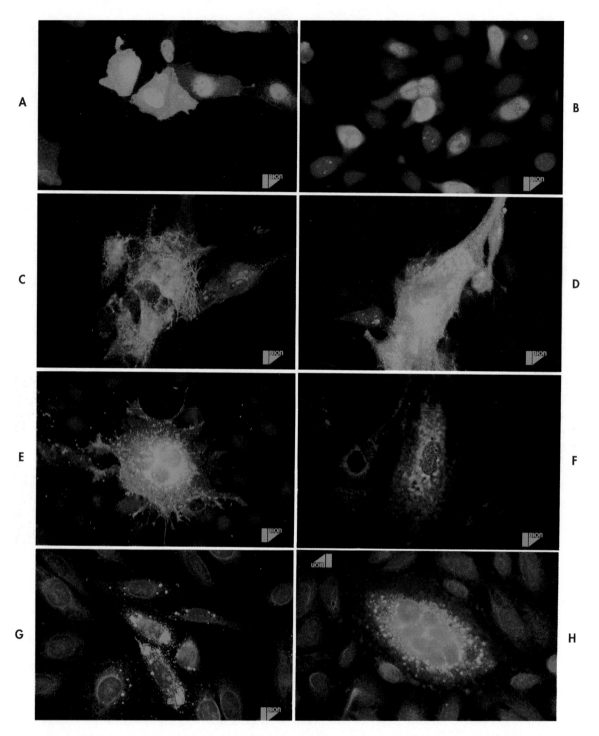

FIGURE 66-13 Fluorescent antibody staining of virus-infected cells. **A,** Influenza virus. **B,** Adenovirus. **C,** Varicella-zoster virus. **D,** Herpes simplex virus. **E,** Respiratory syncytial virus. **F,** Parainfluenza virus. **G,** Mumps virus. **H,** Measles virus. (Courtesy Bion Enterprises Ltd., Park Ridge, Ill.)

largely replaced by ELISA because of expensive equipment and disposal procedures for radioactive materials. Immunoperoxidase staining is commonly used to stain histologic sections for virus but is less popular than immunofluorescence staining in clinical virology laboratories. Latex agglutination is an easy and inexpensive method but lacks sensitivity compared with ELISA and fluorescent immunoassays.

Enzyme-linked virus-inducible system (ELVIS)

The ELVIS system uses a BHK (baby hamster kidney) cell culture system with a cloned (added) beta-galactosidase gene that is expressed only when cells are infected with a virus. In the ELVIS-HSV test system (BioWhittaker, Walkersville, Md.) the genetically engineered BHK cells are sold in multiwelled microtiter plates. Following inoculation of specimens and overnight incubation, growth of HSV results in production of the beta-galactosidase enzyme by the BHK

cells. Beta-galactosidase serves as the "reporter" molecule. When cells are fixed and stained for galactosidase activity, positive staining indicates the presence of HSV-1 or HSV-2. Wells not containing HSV show no staining.

Molecular detection using nucleic acid probes and polymerase chain reaction assays

Nucleic acid probes are short segments of DNA that are designed to hybridize with complementary viral DNA or RNA segments. The probe is labeled with a fluorescent, chromogenic, or radioactive tag that allows detection if hybridization occurs. The probe reaction can occur in situ, such as in a tissue thin section, in liquid, or on a reaction vessel surface or membrane. A DNA probe test used to detect papilloma virus DNA in a smear of cervical cells is illustrated in Figure 66-16.

Nucleic acid probes are most useful in situations

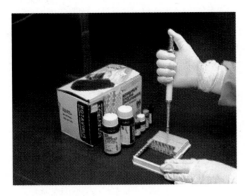

FIGURE 66-14 Solid-phase enzyme immunoassay for detection of rotavirus with breakaway strips of microtiter wells for small batch testing. (Courtesy Children's Hospital Medical Center of Akron, Akron, Ohio.)

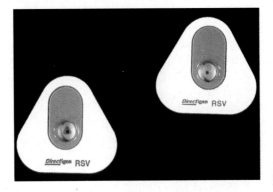

FIGURE 66-15 Positive and negative membrane ELISA for the detection of respiratory syncytial virus. Triangle or triangle and dot indicates a positive test. Dot indicates a negative test. If neither a triangle nor a dot is visible, the test is uninterpretable and must be repeated.

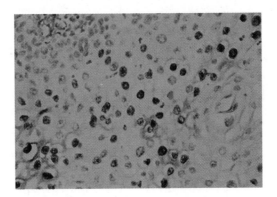

FIGURE 66-16 Smear of cervical cells stained with probe for papillomavirus DNA. Dark-staining cells contain viral DNA. (Courtesy Children's Hospital Medical Center of Akron, Akron, Ohio.)

in which the amount of virus is relatively abundant, viral culture is slow or not possible, and immunoassays lack sensitivity or specificity. Although direct probe assays have been used for detection of many viruses in research settings, there are few commercially available reagents for probe assays in clinical virology laboratories. Probe detection of virus suffers most notably from lack of sensitivity compared with tests that offer amplification of virus before detection, such as culture or the rapid shell vial assay. Probe identification following amplification by PCR, however, is very promising.

The **polymerase chain reaction (PCR)** is a method that duplicates short DNA segments thousands- to a million-fold. DNA fragments that can be identified with a specific probe, but are too few in number in the original specimen to be detected, can be duplicated (amplified) using PCR. This provides the probe with enough target to readily identify the presence of a specific virus. Probe identification of human immunodeficiency virus following PCR amplification is illustrated in Figure 66-17.

The PCR test can be used to amplify and detect RNA viruses by using the enzyme reverse transcriptase (RT). The first step in RT-PCR includes making a complementary DNA strand of the RNA segment in question. The usual PCR steps are then performed, leading to DNA amplicons whose identification signifies the presence of the original RNA sequence. The RT-PCR assay has been adopted successfully for the detection of enteroviruses in cerebrospinal fluid.

Use of PCR to identify nearly every medically important virus has been attempted in research settings. PCR is a patented technique, so commercial availability of virus detection tests has been slow to develop. Licensing arrangements of proprietary PCR procedures or the introduction of other amplification techniques, such as the ligase chain reaction, should lead soon to use of PCR or similar nucleic acid amplification techniques in the clinical laboratory.

Conventional cell culture

Viruses are strict intracellular parasites, requiring a living cell for multiplication and spread. To detect virus using living cells, suitable host cells, cell culture media, and techniques in cell culture maintenance are necessary. Host cells, referred to as *cell cultures,* originate as a few cells and grow into a monolayer on the sides of glass or plastic test tubes. Cells are kept moist and supplied with nutrients by keeping them continuously immersed in cell culture medium (Figure 66-18). Cell cultures are routinely incubated in a roller drum, which holds cell culture test tubes tilted (5 to 7 degrees) while they slowly revolve (½ to 1 RPM) at 35° to 37° C (Figure 66-5). Incubation of cell culture

tubes in a stationary rack can be used in place of a roller drum. Rapidly growing viruses, such as HSV, appear to be detected equivalently by both methods. Comparative studies are not available for most viruses.

Metabolism of growing cells in a closed tube results in the production of CO_2 and acidification of the growth liquid. To counteract the pH decrease a bicarbonate buffering system is employed in the culture medium to keep the cells at physiologic pH (pH 7.2). Phenol red, a pH indicator, that is red at physiologic, yellow at acidic, and purple at alkaline pHs is added to monitor adverse pH changes. Once inoculated with specimen, cell cultures are incubated for 1 to 4 weeks, depending on the viruses suspected. Periodically the cells are inspected microscopically for the presence of virus, indicated by areas of dead or dying cells called **cytopathic effect (CPE).**

Two kinds of media, growth medium and maintenance medium, are used for cell culture. Both are prepared with Eagle's minimum essential medium (EMEM) in Earle's balanced salt solution (EBSS) and include antimicrobials to prevent bacterial contamination. The usual antimicrobials added are vancomycin (10 μg/mL), gentamicin (20 μg/mL), and amphotericin (2.5 μg/mL). Growth medium is a serum-rich (10% fetal, newborn, or agammaglobulinemic calf serum) nutrient medium designed to support rapid cell growth. This medium is used for initiating growth of cells in a tube when cell cultures are being prepared in-house or for feeding tubes of purchased cell cultures that have incomplete cell monolayers. **Feeding** refers to the removal of old medium followed by the addition of fresh culture medium. Maintenance medium is similar to growth medium but contains less serum (0 to 2%) and is used to keep cells in a steady state of metabolism. Fetal, newborn, or agammaglobulinemic calf serum is used to avoid inhibitors, such as specific antibody, and to be free of mycoplasmas present in serum from older animals.

Several kinds of cell cultures are routinely used for isolation of viruses. A cell culture becomes a **cell line** once it has been passed or subcultured in vitro. Cell lines are classified as primary, low passage, or continuous. **Primary cell lines** are those that have been passed only once or twice since harvesting, such as primary monkey kidney cells. Further passage of primary cells results in a decreased receptivity to viral infection. **Low-passage cell lines** are those that remain virus-sensitive through 20 to 50 passages. Human diploid fibroblast cells, such as lung fibroblasts, are a commonly used low-passage cell line. **Continuous cell lines,** such as HEp-2 cells (human epidermoid carcinoma cells), can be passed and remain sensitive to some virus infections indefinitely. Unfortunately, most

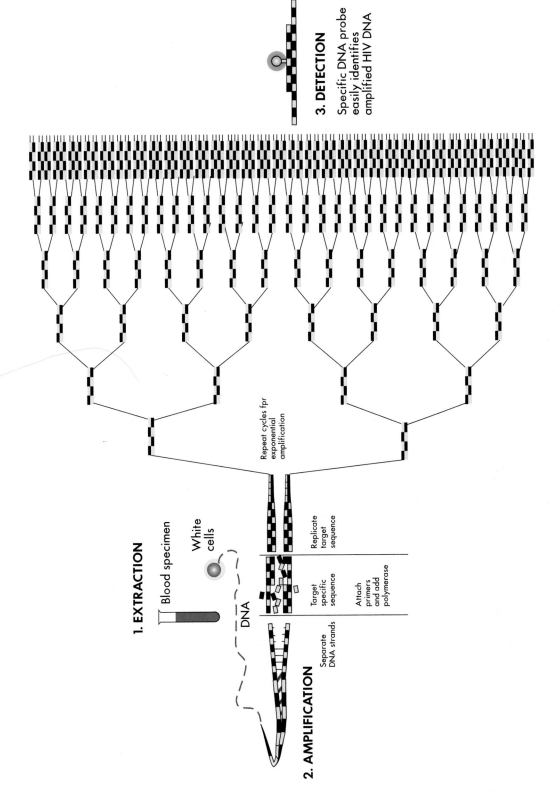

FIGURE 66-17   Probe identification of HIV DNA following PCR amplification. (Courtesy Roche Biomedical Laboratories, Research Triangle Park, N.C.)

viruses do not grow well in these cells. The majority of clinically significant viruses can be recovered using one cell culture type from each group. A combination frequently used by clinical laboratories is rhesus monkey kidney cells, MRC-5 lung fibroblast cells, and HEp-2 cells (Table 66-23).

Inoculated cell cultures should be incubated immediately at 35° C. After allowing virus to adsorb to the cell monolayer for 12 to 24 hours, it is common practice to remove the remaining inoculum and culture medium and replace it with fresh maintenance medium. This avoids most inoculum-induced cell culture toxicity and improves virus recovery. Incubation should be continued for 5 to 28 days depending on viruses suspected (see Table 66-23). Maintenance medium should be changed periodically (usually 1 to 2 times weekly) to provide fresh nutrients to the cells.

Blind passage refers to passing cells and fluid to a second cell culture tube. Blind passage is used to detect viruses that may not produce CPE in the initial culture tube but will when the "beefed-up" inoculum is passed to a second tube. Cell cultures that show nonspecific or ambiguous CPE are also passed to additional cell culture tubes. Toxicity, which causes ambiguous CPE, is diluted during passage and should not appear in the second cell culture tube. Passage, in both instances, is performed by scraping the monolayer off the sides of the tube with a pipette or disrupting the monolayer by vortexing with sterile glass beads added to the culture tube, followed by inoculating 0.25 mL of the resulting suspension into new cell cultures. Blind passage is less frequently used today, because the added time and expense do not justify detection of a few additional isolates.

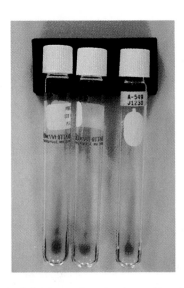

FIGURE 66-18 Cell culture tubes showing emblem indicating side that should be incubated upward. Cell monolayer forms on glass surface beneath the culture medium.

Viruses are most often detected in cell culture by the recognition of **cytopathic effects (CPE)**. Virus-infected cells change their usual morphology and eventually lyse or detach from the glass surface while dying. Viruses have distinct CPEs, just as colonies of bacteria on agar plates have unique morphologies (Figure 66-19, *A* through *K*). CPE may be quantitated as indicated in Table 66-24. Preliminary identification of a virus can sometimes be made based on the cell culture that supports viral replication, the speed at which the virus produced CPE, and a description of the CPE (see Table 66-23). Definitive identification of virus detected in cell culture is discussed later in this chapter.

Shell vial culture

The shell vial culture is a rapid modification of conventional cell culture. Virus is detected more quickly using the shell vial technique, because the infected cell monolayer is stained for viral antigens found soon after infection, before the development of CPE. Viruses that normally take days to weeks to produce CPE can be detected within 1 to 2 days. A shell vial culture tube, a 15- × 45-mm 1-dram vial, is prepared by adding a round coverslip to the bottom, covering this with growth medium, and adding appropriate cells (Figure 66-20). During incubation a cell monolayer forms on top of the coverslip. Shell vials should be used 5 to 9 days after cells have been inoculated. Shell vials with the monolayer already formed can be purchased. Specimens are inoculated onto the shell vial cell monolayer by low-speed centrifugation. This enhances viral infectivity by poorly understood mechanisms. Coverslips are stained using virus-specific immunofluorescent conjugates. Typical fluorescing inclusions confirm the presence of virus (Figure 66-21). The shell vial procedure for detecting CMV is presented in detail in Procedure 66-3.

The shell vial culture technique can be used to detect most viruses that grow in conventional cell culture. It is best used for viruses that require relatively long incubation before producing CPE, such as CMV and VZV. The advantage of shell vial is its speed; most viruses are detected within 24 hours. The disadvantage is that only one type of virus can be detected per shell vial. For example, a specimen that might contain influenza A or B or adenovirus would need to be inoculated to three separate shell vials so that each vial could be stained with a separate virus-specific conjugate.

IDENTIFICATION OF VIRUSES DETECTED IN CELL CULTURE

Viruses detected in conventional cell culture by production of CPE may be identified by noting the cell types that support viral replication, time to detection

TABLE 66-23 ISOLATION AND IDENTIFICATION OF COMMON CLINICALLY ENCOUNTERED VIRUSES

VIRUS	PMK	HEp-2	HDF	CPE DESCRIPTION	RATE OF GROWTH (DAYS)	IDENTIFICATION AND COMMENTS
Adenovirus	++*	+++	++	Rounding and aggregation of infected cells in grapelike clusters	2-10	Confirm by FA; serotype by cell culture neutralization
Cytomegalovirus (CMV)	–	–	++++	Discrete small foci of rounded cells	5-28	Distinct CPE sufficient to identify; confirm by FA
Enterovirus	++++	±	++	Characteristic refractile angular or tear-shaped CPE; progresses to involve entire monolayer	2-8	Identify by cell culture neutralization test; stable at pH3
Herpes simplex (HSV)	±	++++	++++	Rounded, swollen refractile cells. Occasional syncytia, especially with type 2; rapidly involves entire monolayer	1-3 (may take up to 7)	Distinct CPE sufficient to identify; confirm by FA
Influenza	++++	–	±	Destructive degeneration with swollen, vacuolated cells	2-10	Detect by hemadsorption or hemagglutination with guinea pig RBCs; identify by FA
Mumps	+++	±	±	CPE usually absent; occasionally, syncytia are seen	5-10	Detect by hemadsorption with guinea pig RBCs; confirm by FA
Parainfluenza	+++	–	–	CPE usually minimal or absent	4-10	Detect by hemadsorption with guinea pig RBCs; identify by FA
Respiratory syncytial virus (RSV)	+	+++	+	Syncytia in HEp-2	3-10	Distinct CPE in HEp-2 is sufficient for presumptive identification; confirm by FA
Rhinovirus	++	–	+++	Characteristic refractile rounding of cells. In PMK, CPE is identical to that produced by enteroviruses	4-10	Labile at pH 3; growth optimal at 32° to 33° C
Varicella-zoster	–	–	++	Discrete foci of rounded, swollen, refractile cells. Slowly involves entire monolayer	5-28	Confirm by FA

PMK, Primary monkey kidney; *HEp-2*, human epidermoid; *HDF*, human diploid fibroblast; *FA*, fluorescent antibody.

*Relative sensitivity of cell cultures for recovering the virus: −, none recovered; ±, rare strains recovered; +, few strains recovered; ++++, ≥ 80% of strains recovered.

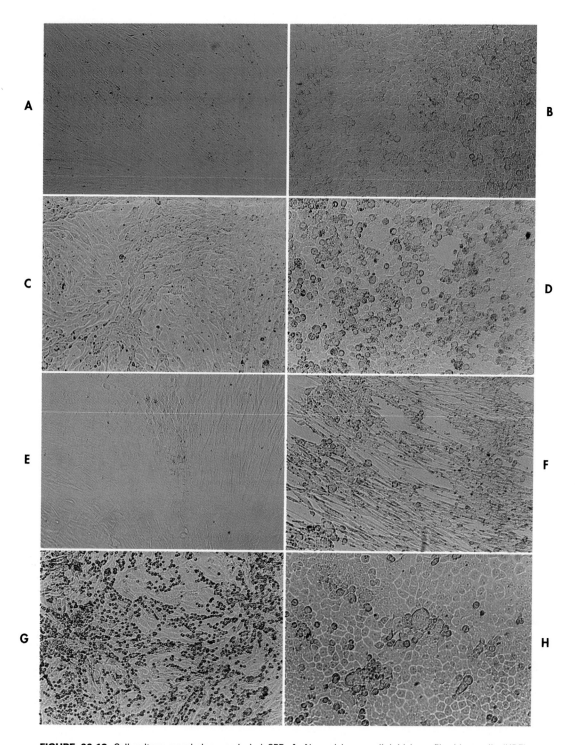

FIGURE 66-19 Cell culture morphology and viral CPE. **A,** Normal human diploid lung fibroblast cells (HDF). **B,** Normal HEp-2 cells. **C,** Normal primary monkey cells (PMK). **D,** HEp-2 cells infected with adenovirus. **E,** HDF cells infected with cytomegalovirus. **F,** HDF cells infected with herpes simplex virus. **G,** PMK cells infected with hemadsorbing virus, such as influenza, parainfluenza, or mumps, plus guinea pig erythrocytes. **H,** HEp-2 cells infected with respiratory syncytial virus.

Continued

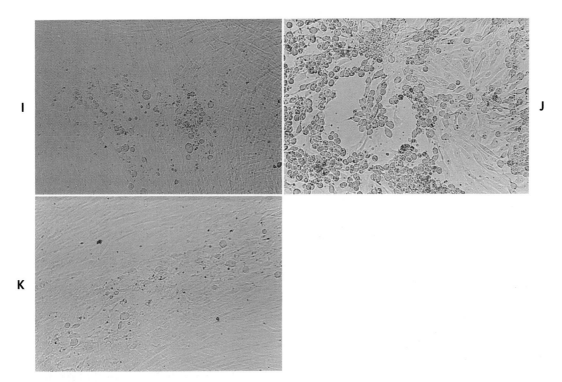

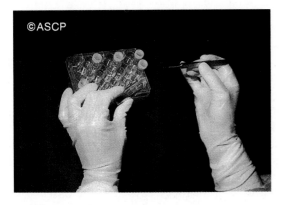

FIGURE 66-19, cont'd I, HDF cells infected with rhinovirus. J, PMK cells infected with echovirus. **K,** HDF cells infected with varicella-zoster virus. (From the U.S. Department of Health, Education, and Welfare Public Health Service, Centers for Disease Control, Atlanta.)

FIGURE 66-20 Preparation of shell vial culture tubes. (Courtesy Bostick C.C. 1992. Laboratory detection of CMV. Microbiology Tech Sample No. MB-3, p. 5.)

FIGURE 66-21 Typical fluorescing nuclei of human diploid fibroblast cells infected with cytomegalovirus as seen in the shell vial assay. (Courtesy Bostick C.C. 1992. Laboratory detection of CMV. Microbiology Tech Sample No. MB-3, p. 5.)

TABLE 66-24 QUANTITATION OF CELL CULTURE CPE

QUANTITATION	INTERPRETATION
Negative	Uninfected monolayer
Equivocal (±)	Atypical alteration of monolayer involving few cells
1+	1%-25% of monolayer exhibits CPE
2+	25%-50% of monolayer exhibits CPE
3+	50%-75% of monolayer exhibits CPE
4+	76%-100% of monolayer exhibits CPE

of CPE, and morphology of the CPE (see Figure 66-19, *A* to *K*). Experienced virologists are able to presumptively identify most viruses isolated in clinical laboratories based on these criteria. When definitive identification is required, additional testing can be performed. Fluorescent-labeled antisera, available for most viruses, are used commonly for culture confirmation. In addition, acid-lability is used to differentiate enteroviruses from rhinoviruses, and neutralization is used to identify viruses with many serotypes for which fluorescent-labeled antisera are not available. Some viruses, such as influenza, parainfluenza, and mumps, which produce little or no CPE, can be detected by **hemadsorption,** because infected cells contain viral hemadsorbing glycoproteins in their outer membranes. The addition of guinea pig red blood cells to the cell culture tube, followed by a wash to remove nonadsorbed RBCs, results in a ring of RBCs around infected cells (see Figure 66-19, *G*). Cell cultures demonstrating hemadsorption can be stained with fluorescent-labeled antisera to identify the specific hemadsorbing virus present. Detailed procedures for culture confirmation by fluorescent antibody staining, hemadsorption for the detection of influenza and parainfluenza viruses, and acid-lability test are presented in Procedures 66-4, 66-5 and 66-6.

VIRAL SEROLOGY

Serology was the primary means for the laboratory diagnosis of viral infections until the mid 1970s. At that time, culture and detection of viral antigen became more widely available because of high-quality, commercially available reagents, such as cell cultures, and a broad range of immunodiagnostic test kits for the detection of viral antigen directly in patient specimens. The use of viral serology tests has changed to complement this growing menu of diagnostic tests.

Viral serology is now used primarily to detect immune status and to make the diagnosis of infections in situations in which the virus cannot be cultivated in cell culture or detected by immunoassay.

In most viral infections, IgM is undetectable after 1 to 4 months but detectable levels of IgG remain, as a rule, for the life of the patient. If a patient is infected by an antigenically similar virus or the original strain has remained latent and reactivates at a later time, these virus-specific IgG and IgM antibody levels may again rise. The secondary IgM response may be difficult to detect; however, a significant (fourfold) IgG titer rise is readily apparent (Figure 66-22).

An immune status test measures whether a patient previously was infected by a particular virus. A positive result with a sensitive, virus-specific IgG test indicates past infection. Some immune status tests include methods that detect both IgG and IgM, to detect recent or active infections. To help diagnose active disease, two approaches are helpful. Detection of virus-specific IgM in an acute-phase specimen collected at least 10 to 14 days after the onset of infection indicates current or very recent disease. Detection of a fourfold antibody titer rise between **acute** and **convalescent sera** also indicates current or recent disease. Acute-phase serum should be collected as soon as possible after onset of symptoms. The convalescent specimen should be collected 2 to 3 weeks after the acute phase. If a single postacute serum, collected between acute and convalescent times, or convalescent specimen is all that is available for testing, an extremely high, virus-specific IgG titer may be suggestive of infection. The exact titer specific for active disease, if known at all, varies with each testing method and virus. In general, titers high enough to be diagnostic are unusual and testing of single specimens should not be performed. A reasonable policy includes using IgM tests, where available, and performing IgG testing only on paired, acute and convalescent specimens. No IgG testing need be performed on the first, acute specimen until receipt of the convalescent specimen. This eliminates useless testing of single specimens in those instances in which a second sample is never collected or submitted for analysis.

Many serologic methods are or have been routinely used to detect antiviral antibody. Prominent among these are **complement fixation (CF), enzyme-linked immunosorbent assay (ELISA),** indirect **immunofluorescence,** anticomplement immunofluorescence **(ACIF),** and **Western immunoblotting.** Complement

fixation is a labor-intensive, technically demanding method best fitted to batch testing. The major advantage of CF is the years of experience that have accumulated for interpreting results. As less demanding, easily automated techniques for batch testing, such as ELISA, are developed, the need for CF testing is disappearing. Additional advantages of ELISA are that it can be used to detect IgM-specific antibodies free of common interfering factors, in particular through use of an antibody-capture technique (Figure 66-23). Indirect immunofluorescence is best used for individual specimens or small batch testing. Immunofluorescence can also be used to detect virus-specific IgM but requires prior separation and elimination of the IgG

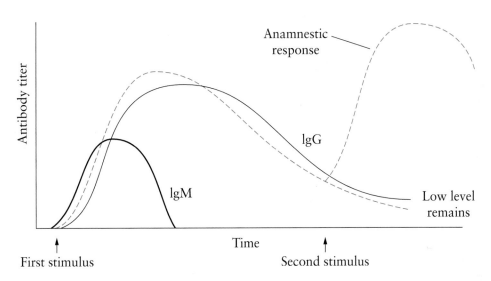

FIGURE 66-22 Relative humoral response to antigen stimulation over time.

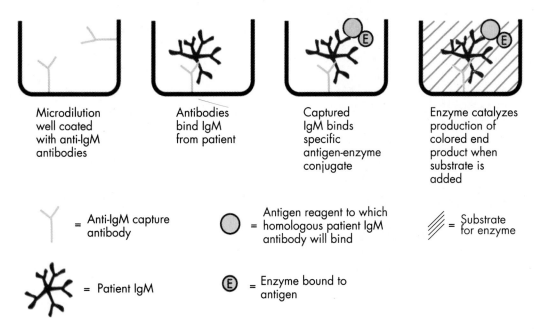

FIGURE 66-23 Illustration of antibody ELISA for IgM.

fraction which, if present, can result in both false-positive and false-negative results. IgM and IgG can be separated by either ion-exchange chromatography (Figure 66-24) or immune precipitation (Figure 66-25).

Immunoglobulin G indirect fluorescent antibody testing is subject to false-positive results because of antibody-Fc receptors that occur in cells infected with virus. IFA testing is performed using virus-infected cells fixed to a microscope slide. When the substrate cells are overlaid with patient serum, the Fc portion of the antibody molecule binds to these receptors. Fluorescent-labeled antiglobulin attaches to both homologous antibody, that bound to viral antigen, and to Fc-bound antibody. Subsequent fluorescence of Fc-bound antibody results in a false-positive or falsely elevated reading. To avoid this complication, the anti-complement immunofluorescence (ACIF) test can be used (Figure 66-26). Because fluorescent-labeled complement only binds to antigen-antibody complexes, the nonspecific antibody attached by Fc receptors, which is complement free, does not fluoresce. Western immunoblotting (see Chapter 16) is also used for viral antibody detection.

False-positive and false-negative results can occur when testing for virus-specific IgM antibodies. False-positive results occur when rheumatoid factor, an anti-IgG IgM-type of globulin, combines with homologous or virus-specific IgG present in the patient specimen. Labeled anti-IgM combines with either bound virus-specific IgM or rheumatoid factor,

causing falsely positive fluorescence. False-negative IgM tests occur when high levels of strongly binding homologous IgG antibodies prevent binding of IgM molecules, decreasing or eliminating IgM-specific fluorescence (Figure 66-27). Both problems can be eliminated by testing the IgG-free serum fraction.

PRESERVATION AND STORAGE OF VIRUSES

Clinical virology laboratories must have a method for storing and retrieving viruses. Isolates should be kept as control strains and, in rare instances, for epidemiologic investigations. Public health laboratories may use current enterovirus or influenza virus strains for typing. Viruses can be stored by freezing at $-70°$ C or in liquid nitrogen. Freezing at $-70°$ C is more practical for clinical laboratories. A method for the preservation and storage of viruses by freezing is described in Procedure 66-7.

USE OF VIRAL DIAGNOSTIC TESTS

In the past the laboratory diagnosis of viral infections was made using a comprehensive viral culture and serology batteries, both designed to detect any of the many viruses responsible for a disease syndrome. As reagents become more plentiful and expensive, comprehensive approaches to the diagnosis of disease become prohibitive. Virology laboratories should have a menu of individual culture and serology tests and an algorithm for their use, rather than one comprehensive viral culture or serology test battery for use in all situations. Tables 66-25 and 66-26 represent virus detection and serology tests that might be useful in a community clinical virology laboratory.

PROCESSING VIRUS SPECIMENS BASED ON SPECIMEN TYPE

An algorithm for the use of virus detection tests can be based on the type of specimen received or the specific virus suspected. Figure 66-28 summarizes a "specimen" algorithm. Most laboratories receive specimens with little or no clinical data and no specific virus indicated. The algorithm in Figure 66-28 is designed for use with these specimens.

All lip and genital specimens should be cultured only for herpes simplex virus. Disease by other viruses at lip and external genital sites is unusual and detection of these agents, such as VZV, should be prompted by a special request. Urine specimens require a CMV detection test. This is best done using the shell vial assay. Stool specimens from infants ($≤5$ years) in North America should be tested for rotavirus during the fall, winter, and spring. Enteric adenoviruses (serotypes 40 and 41) cause diarrhea in young children and infants year-round. Adenovirus

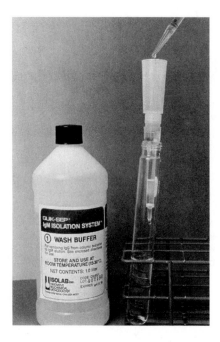

FIGURE 66-24 Separation of IgM from human serum by passing the serum through an ion-exchange column.

———— TABLE 66-25   MENU OF VIRUS DETECTION TESTS ————

TEST	FREQUENCY OF REQUEST (%)*
HSV detection (outpatient)	9
Influenza virus detection	2
Enterovirus detection	5
Rapid CMV detection (shell vial)	8
Rotavirus detection	8
RSV detection	14
Pediatric respiratory virus detection	10
Comprehensive virus detection	44
Total	100

*Frequency of request expressed as percentage of all tests.

Data from Children's Hospital Medical Center of Akron, Akron, Ohio.

———— TABLE 66-26   MENU OF VIRAL SEROLOGY TESTS ————

TEST	FREQUENCY OF REQUEST (%)*
Arbovirus antibodies	1
CMV immune status	28
CMV IgM antibodies	9
CMV IgG antibodies	9
EBV IgM antibodies	11
EBV IgG antibodies	9
HSV IgM antibodies	3
HSV IgG antibodies	2
Rubella immune status	11
Measles immune status	15
Varicella immune status	2
Total	100

*Frequency of request expressed as percent of all tests.

Data from Children's Hospital Medical Center of Akron, Akron, Ohio.

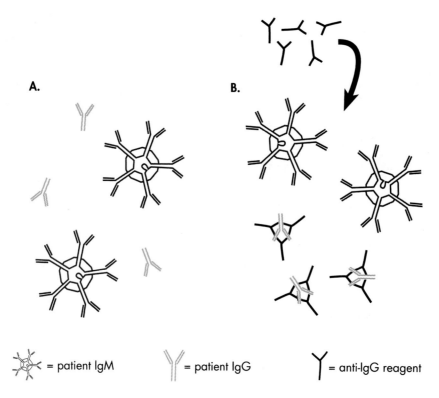

A.

B.

$\underset{\text{= patient IgM}}{}$ $\underset{\text{= patient IgG}}{}$ $\underset{\text{= anti-IgG reagent}}{}$

FIGURE 66-25 Removal of IgG antibodies from human serum by immune precipitation. **A,** Patient IgG and IgM in serum. **B,** Patient IgG inactivated by addition of anti-IgG, which binds and precipitates patient IgG. IgM-specific testing can now be performed free of IgG interference.

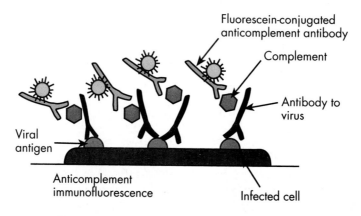

Fluorescein-conjugated anticomplement antibody

Complement

Antibody to virus

Viral antigen

Anticomplement immunofluorescence

Infected cell

FIGURE 66-26 Illustration of ACIF technique for antibody detection.

gastroenteritis appears to be more common in some geographic areas. Routine testing is only necessary in locations where disease is known to occur. Stool or rectal swabs from adults, children, and infants should be tested for enterovirus in summer and fall months as an aid in the diagnosis of aseptic meningitis. Stool for enterovirus should be tested in conjunction with throat and CSF specimens when possible. Respiratory specimens should be divided according to the patient's age and underlying medical condition. Immunocompromised patients require a comprehensive virus

detection test consisting of cell culture and shell vial for CMV or other agents if indicated by the patient's clinician. Immunocompetent adults should have an influenza virus culture during influenza season, November to April in most areas. Children (<10 years) are susceptible to serious infection caused by influenza, parainfluenza, respiratory syncytial, and adenoviruses and need a full respiratory virus culture. Infants (<2 years) are especially vulnerable to RSV bronchiolitis, which may require hospitalization and antiviral therapy with RSV-immune globulin or rib-

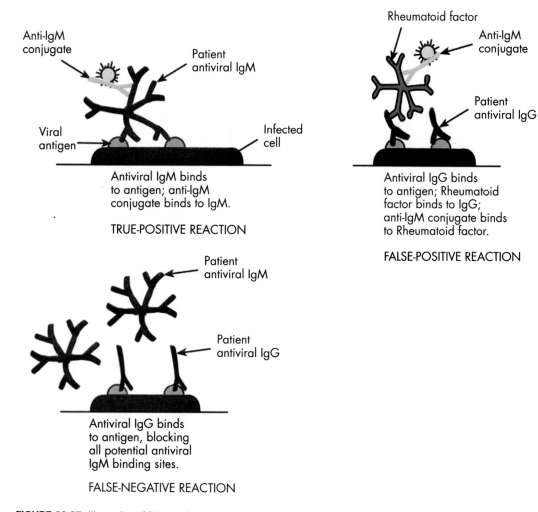

FIGURE 66-27 Illustration of false-positive and false-negative IgM test resulting from the presence of rheumatoid factor and excess IgG.

avirin. A rapid, nonculture RSV detection test, such as fluorescent antibody staining or enzyme immunoassay, is appropriate in these situations. Respiratory specimens from newborns with the possibility of congenital or perinatal viral disease should receive a comprehensive virus culture and shell vial for CMV. Blood for viral culture might contain CMV, VZV, adenovirus, or enteroviruses; however, CMV is by far the most common and is the one agent whose detection and reporting is clinically useful information. Therefore, blood for viral culture should be processed for CMV only, unless other agents are mentioned by the requesting physician. All specimens from immunocompromised hosts and tissues or fluids from presumably sterile sites should be processed for comprehensive virus detection. This algorithm represents a starting point for clinical virology laboratories. Processing of specimens should be continuously modified to match the needs of local physicians and endemic viral diseases.

PROCESSING VIRUS SPECIMENS BASED ON REQUESTS FOR SPECIFIC VIRUSES

Frequently, detecting only one virus in a specimen is advised because the physician recognizes the patient's clinical condition or that virus is the sole agent whose detection is clinically useful. Individual viruses are detected as follows.

Arboviruses

Serologic tests for **arboviruses** are offered by State Public Health Laboratories or by some commercial laboratories. Diagnosis of arbovirus encephalitis, such as Eastern, Western, Venezuelan, and St. Louis, and that caused by California encephalitis and LaCrosse viruses, requires detection of a rise in titer of antibody in acute and convalescent serum specimens. Detection of virus-specific IgM in cerebrospinal fluid is available for most agents. Culture of arboviruses for diagnostic purposes is not practical.

Cytomegalovirus

CMV can be detected in clinical specimens using conventional cell culture, shell vial assay, or antigenemia immunoassay. CMV produces CPE in diploid fibroblast cells in 3 to 28 days, averaging 7 days. Shell vial for CMV has a sensitivity equivalent to conventional cell culture but takes only 16 hours to complete. The antigenemia immunoassay uses monoclonal antibody in an indirect immunoperoxidase or indirect immunofluorescent stain to detect CMV protein (pp65) in peripheral blood leukocytes. The antigenemia assay requires only 3 to 5 hours to complete and includes separation of leukocytes by sedimentation, counting of leukocytes, smear preparation with a standard density of cells, staining, and counting of infected (fluorescing) cells (Procedure 66-8). Results are reported as number of positive leukocytes per total number of leukocytes in the smear (Figure 66-29). Interpretation of the CMV antigenemia assay is dependent on the patient popu-

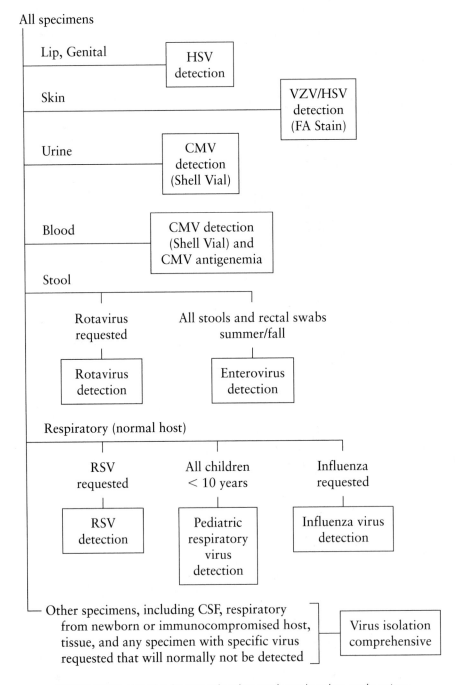

FIGURE 66-28 Algorithm for processing virus specimens based on specimen type.

lation and laboratory expertise. In general, CMV disease is accompanied by detectable virus in peripheral leukocytes. Disease severity is roughly proportional to quantity of virus, that is, number of fluorescing cells. As disease is treated and resolves, the number of positive cells decreases.

Enteroviruses

Enteroviruses can be detected by conventional cell culture and PCR. Although most enteroviruses grow in primary monkey kidney cells, some strains grow faster in diploid fibroblast, buffalo green monkey kidney, or rhabdomyosarcoma cell lines. Presumptive diagnosis is based on CPE. Confirmation or definitive diagnosis is accomplished using commercially available fluorescent antibody stains. Most enteroviruses are detected in June through December. Mean time to detection in cell culture is 4 days. An efficient cell culture procedure is summarized in Figure 66-30.

Epstein-Barr virus

Serology tests are used to help diagnose EBV associated disease, including infectious mononucleosis. Isolation of EBV (in cultured B lymphocytes) is not routinely performed in clinical laboratories. Detection of antibody to viral capsid antigen, early antigen, and Epstein-Barr nuclear antigen is used as indicated in Figure 66-31 and Table 66-27.

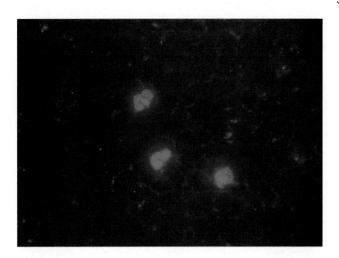

FIGURE 66-29 Typical fluorescing white blood cells containing CMV antigen as seen in the CMV antigenemia stain.

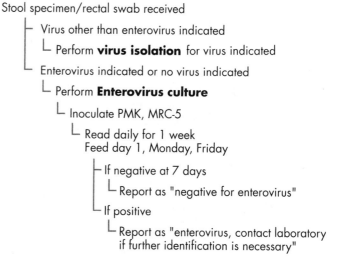

Stool specimen/rectal swab received
- Virus other than enterovirus indicated
  - Perform **virus isolation** for virus indicated
- Enterovirus indicated or no virus indicated
  - Perform **Enterovirus culture**
    - Inoculate PMK, MRC-5
      - Read daily for 1 week
        Feed day 1, Monday, Friday
        - If negative at 7 days
          - Report as "negative for enterovirus"
        - If positive
          - Report as "enterovirus, contact laboratory if further identification is necessary"

FIGURE 66-30 Flowchart for the detection and identification of enteroviruses.

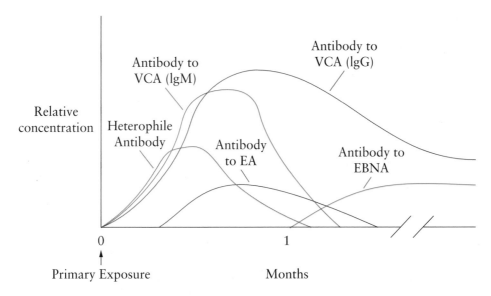

FIGURE 66-31 Illustration of time-course of immune response to EBV infection.

Hepatitis viruses

Disease or asymptomatic carriage caused by hepatitis A, B, C, D, and E viruses is detected using serology, antigen detection, or PCR tests (Table 66-28). Disease caused by hepatitis A virus is detected using serology tests specific for viral-induced IgM and IgG (Figure 66-32). Hepatitis B virus requires detection of antigen and antibody to multiple antigens to classify disease type (Figure 66-33 and Table 66-29). Diagnostic tests for HCV are still evolving. A PCR test for RNA viruses, referred to as *reverse transcriptase PCR* (RT-PCR), is used to detect early acute hepatitis C disease, because antibody detection tests may be negative. Testing by ELISA can be used to detect chronic hepatitis C disease (Figure 66-34). Antibody tests are used to detect patients infected with hepatitis E; however, disease is rare in the United States.

Other hepatitis viruses (non-A, B, C, D, and E) are being discovered by molecular technology and may be responsible for cases of acute and chronic non-A, B, C, D, and E hepatitis. In 1995 evidence of hepatitis G virus (HGV), was released. Although no serology tests have been developed, sequence analysis of the viral genome suggests HGV is a flavivirus. Although this is the virus family that includes HCV, the two are only distantly related. Hepatitis F and hepatitis GB (GBV-A and GBV-B) viruses have been reported, but further investigation is required to confirm their association with human disease.

Herpes simplex virus

HSV grows rapidly in most cell lines. MRC-5 or mink lung fibroblasts cell lines are recommended. Fifty percent of genital HSV isolates will be detected within 1 day and 100% within 3 to 5 days. Cultures should be examined daily and finalized if negative after 5 days of incubation.

Human immunodeficiency virus and other retroviruses

HIV 1 and 2 are detected by antibody, antigen, and PCR tests (Figure 66-35). HIV-1 ELISA antibody tests detect antibody to both HIV-1 and 2. Confirmation of the ELISA screening test is accomplished with an HIV-1-specific Western blot test or with an ELISA test for HIV-2 followed by an HIV-2-specific Western blot. The Western blot identifies antibody specific for several HIV antigens (Figure 66-36). Most often antibody to HIV p24 and either gp41 or gp160 confirms HIV-1 infection. HIV-1 p24 antigen testing is used to detect acutely infected patients before the appearance of antibody. PCR testing for HIV is useful for the newborn population, who may have maternal HIV antibody confounding interpretation of serology tests,

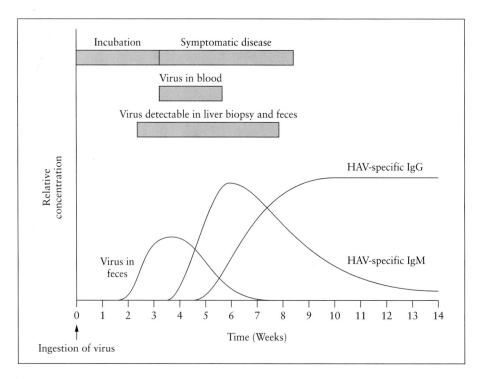

FIGURE 66-32 Illustration of time-course of disease and immune response to hepatitis A virus. (Modified from Murray, P.R., Kobayashi, G.S., Pfaller, M.A., et al., editors. 1994. Medical microbiology, ed 2. Mosby, St. Louis.)

TABLE 66-27   INTERPRETATION OF EPSTEIN-BARR VIRUS SEROLOGY TEST RESULTS

CLINICAL SITUATION	HETEROPHILE ANTIBODY	IgG-VCA*	IgM-VCA	EA	EBNA
No past infection	Usually −	−	−	−	−
Acute infection	Usually +	+	+	+	−
Convalescence phase	±	+	+ or −	+ or −	+
Past infection	Usually −	+	−	− or W+	+
Chronic or reactivation	Not useful	+	−	+	+

*VCA, viral capsid antigen; EA, early antigen; EBNA, Epstein-Barr nuclear antigen; W+, weak positive.

TABLE 66-28   SEROLOGY TESTS FOR HEPATITIS VIRUSES

DISEASE	VIRUS	DIAGNOSTIC TESTS
Hepatitis A	Enterovirus 72	Hepatitis A virus-specific IgG and IgM
Hepatitis B	Hepadnavirus	Hepatitis B surface-antigen (HBsAg) Hepatitis B e-antigen (HBeAg) Anti-HBsAg Anti-HBeAg Anti-HB core antigen
Hepatitis C	Flavivirus (?)	Antibody to hepatitis C virus
Hepatitis D	Delta agent (hepatitis D virus)	Antibody to delta agent; Delta antigen
Hepatitis E	Calicivirus-like	Antibody to hepatitis E virus
Hepatitis G	Flavivirus	Serology tests not available

TABLE 66-29   SEROLOGIC PROFILES FOLLOWING TYPICAL HEPATITIS B VIRUS (HBV) INFECTION

HBsAg	ANTI-HBsAg	HBeAg	ANTI-HBeAg	ANTI-HBc	PROBABLE INTERPRETATION
−	−	−	−	−	No (or very early) exposure to HBV
+	−	±	−	−	Early acute HBV
+	−	+	−	+	Acute or chronic HBV
+	−	−	+	+	Chronic HBV carrier state
−	−	−	+	+	Early recovery phase from acute HBV
−	+	−	+	+	Recovery from HBV with immunity
−	+	−	−	−	Distant HBV infection or HBV vaccine

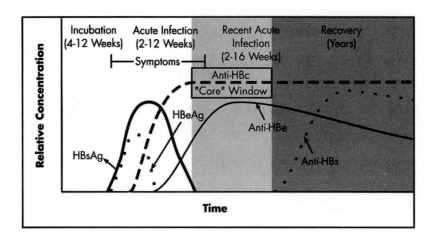

FIGURE 66-33 Illustration of time-course of antigenemia and immune response in a patient who recovers from acute hepatitis B infection.

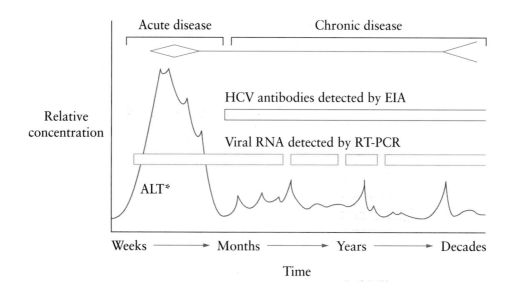

*Alanine aminotransferase.

FIGURE 66-34 Illustration of time-course of immune response and disease caused by hepatitis C virus.

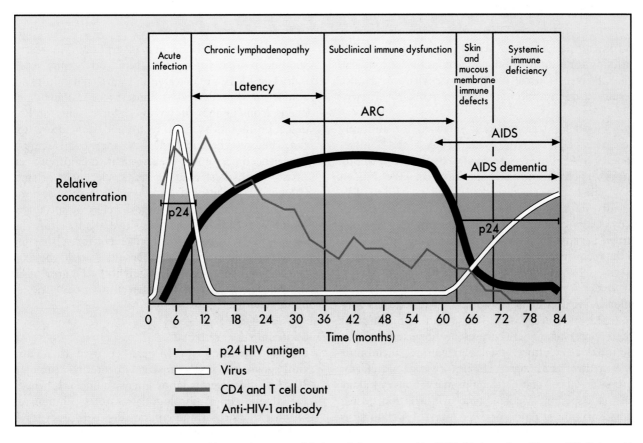

FIGURE 66-35 Illustration of time-course of immune response, viremia, and disease caused by HIV-1. (Redrawn from Murray, P.R., Kobayashi, G.S., Pfaller, M.A., et al., editors. 1994. Medical microbiology, ed 2. Mosby, St. Louis.)

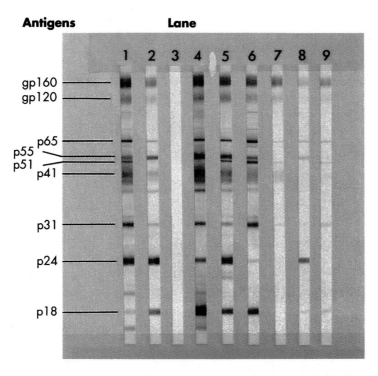

FIGURE 66-36 Western blot detecting specific HIV antibody. Lane 1 is the high-positive control, lane 2 is the low-positive control, lane 3 is the negative control, lanes 4 through 8 are positive sera, and lane 9 is an indeterminate serum. Numbers at left refer to approximate molecular weights of HIV antigens. Sera must react with specific, multiple HIV antigens before they are considered positive. (Courtesy Hodinka, R.L., Children's Hospital of Philadelphia, Philadelphia, Pa.)

and for all patients who may not produce detectable antibody for months following primary infection. The RT-PCR test is used to quantitate virus in blood. Results of quantitation are used to manage antiretroviral therapy.

Blood for transfusion is screened for antibody indicative of HIV-1 and 2 and HTLV-1 and 2 infection. HTLV-1 ELISA screening tests detect antibody to both HTLV-1 and 2. In addition, an HIV antigen (p24) test is performed to detect donors who may be recently infected and have not produced HIV antibody. Units containing antigen or antibody are discarded because of the risk of transferring latent virus to the recipient.

Influenza A and B viruses

Influenza viruses cause sporadic outbreaks each winter, epidemics every few years, and pandemics at greater intervals. These viruses are detected by conventional cell culture, shell vial culture, membrane EIA, or direct staining of respiratory tract secretions. Conventional cell culture using primary monkey kidney cells is superior. Median time to detection, using hemadsorption at day 2 or 3, is 3 days. FA staining is used for confirmation and typing of isolates as *A* or *B*. All influenza viruses will be detected after 1 week of incubation. Detection of influenza virus is outlined in Figure 66-37.

Pediatric respiratory viruses

Influenza, parainfluenza, respiratory syncytial, and adenoviruses should be sought in specimens from hospitalized infants and young children (<10 years). Influenza and parainfluenza viruses are detected in PMK cells by CPE or hemadsorption. Fluorescent staining is used for confirmation and typing. Adenovirus and RSV are detected in HEp-2 cell culture and confirmed, if necessary, by fluorescence staining. Specimens from infants or young children sent for RSV detection should be tested by a rapid, nonculture RSV test. Fluorescent antibody staining, conducted by experienced personnel, is equivalent to culture in sensitivity and is best used for single specimens or small batches. Conventional ELISA is also very accurate and is recommended for large batches of specimens. Membrane ELISA is less sensitive than culture but very rapid (<1 hour). It is convenient for stat testing. Negative membrane ELISA results should be confirmed by an alternative method. Figure 66-38 describes procedures for the detection of pediatric respiratory viruses.

Gastroenteritis viruses

Electron microscopy (EM) can be used to identify viral agents known to cause gastroenteritis (Table 66-30). EM, however, is labor intensive and not broadly available in clinical virology laboratories. Immunoassays for rotaviruses and enteric adenoviruses types 40 and 41 are commercially available. Their use is discussed earlier in this chapter.

TORCH testing

TORCH is an acronym for *T*oxoplasma, *r*ubella, *c*ytomegalovirus, and *h*erpes simplex virus. Testing for these agents is appropriate during pregnancy, because transplacental infection followed by congenital defects can occur, and postnatally, because some congenital disease of newborns can be treated to pre-

---

TABLE 66-30 HUMAN GASTROENTERITIS VIRUSES

VIRUS	MEDICAL IMPORTANCE DEMONSTRATED	EPIDEMIOLOGY	DIAGNOSTIC TESTS
Rotavirus	++++	Major cause of diarrhea in infants	EIA, latex agglutination
Enteric adenoviruses (especially types 40 and 41)	++	Diarrhea in infants and young children	EIA, EM
Norwalk viruses	++	Epidemics in children and adults	EM
Norwalklike viruses	+	Similar to Norwalk	EM
Caliciviruses	+	Diarrhea in children, adults occasionally	EM
Astroviruses	+	Diarrhea in children	EM

vent serious, permanent damage to the infant. A blanket request for TORCH tests should be avoided whenever possible, especially with specimens from newborns. Clinical presentation usually is characteristic for one or two of the TORCH agents, and tests for these etiologies only should be pursued. Table 66-31 suggests laboratory tests for the diagnosis of the viral constituents of TORCH.

Varicella-zoster virus

Varicella-zoster virus (VZV) causes chicken pox (varicella) and shingles (zoster). Varicella (a vesicular eruption) represents the clinical presentation of a primary VZV infection. VZV, a DNA-containing virus, establishes latency in the dorsal nerve root ganglion. Months to years later, during periods of relative immune suppression, VZV reactivates to cause zoster. Zoster is a modified, or limited, form of varicella, localized to a specific dermatome, the cutaneous area served by the infected nerve ganglion. Virus is present in the vesicular fluid and in the cells at the base of the vesicle. Material for virus detection should be collected from newly formed vesicles. Once the vesicle has opened and crusted over, detection is unlikely.

Virus can be detected by staining cells from the base of the vesicles or by culturing cells and vesicular fluid. The Tzanck test, which is a smear of cells from the base of the vesicle stained by Giemsa, Pap, or other suitable method, detects typical multinucleated giant

Respiratory tract specimen received
- Other viruses in addition to influenza virus indicated
  - Perform **virus isolation** for virus indicated
- Influenza virus indicated
  - Inoculate PMK
    - Read daily for 1 week
      Hemadsorb day 3 and day 7
      Feed day 3
      - If negative at 1 week
        - Report as "negative for influenza virus"
      - If positive
        - Type by FA and report as "influenza A (or B) virus detected"

FIGURE 66-37 Flowchart for the detection and identification of influenza viruses.

Respiratory tract specimen received
- RSV indicated (RSV rapid test plus cell culture-PMK and HEp-2)
  - RSV detection positive
    - Report RSV
      Discard culture tubes
      <1% have second virus
  - RSV detection negative
    - Incubate cell culture tubes
      - Read days 1-7, 10
        Hemadsorb day 3 and 10
        Feed twice/week
        - If negative at 10 days
          - Report as negative for RSV, adenovirus, influenza, and parainfluenza viruses.
        - If positive
          - Report as RSV, adenovirus, influenza, or parainfluenza. Type Influenza and parainfluenza if appropriate.
- No RSV request
  - Virus other than pediatric respiratory viruses requested
    - Perform **virus isolation** for virus indicated
  - No virus other than pediatric respiratory virus requested
    - Perform virus culture as above for "RSV detection negative"

FIGURE 66-38 Flowchart for the detection and identification of pediatric respiratory viruses.

TABLE 66-31 LABORATORY DIAGNOSIS OF THE VIRAL AGENTS OF *TORCH* INFECTIONS*

VIRUS	CULTURE	IgM	SEROLOGY IgG	PCR
CMV	**X***	X	Hold†	
HSV	**X***	**X***	Hold	X (CSF)
Rubella	X	**X***	Hold	

*Bold indicates preferred test.

†Hold until acute and convalescent can be tested together.

cells and inclusions (see Figure 66-9, *A*). The FA stain also can be used to detect VZV in Tzanck smears. Traditionally, diploid fibroblast cell culture (e.g, MRC-5) has been used to detect VZV, which requires up to 28 days before producing visible CPE. The shell vial assay reduces detection time to 48 hours and significantly increases sensitivity, identifying virus that fails to produce CPE in conventional cell culture. A comparison of FA of vesicle cells, conventional cell culture, and shell vial shows FA to be the most sensitive method for diagnosis.

Immune status testing

Immune status tests are many and varied (Table 66-32). Rubella antibody testing is used with women of childbearing age. A positive result (presence of antibody) indicates past infection or immunization and implies that congenital infection will not occur during subsequent pregnancies. Absence of antibody implies susceptibility to infection and should prompt rubella vaccination if the woman is not pregnant. Varicella and measles immune status assays are used most commonly to test health care workers. Those with no antibody must avoid diseased patients. CMV immune status is most useful for organ transplant donors and recipients and premature babies hospitalized in newborn intensive care nurseries who are likely to receive blood transfusions. Transplant recipients are susceptible to life-threatening CMV infection. Knowing the CMV status of the donor and recipient enables the physician to better diagnose and treat this disease. Newborns whose mothers were never infected with CMV are susceptible to serious, primary CMV infection that can be transmitted in white blood cells during blood transfusion. CMV-negative babies should receive only CMV-negative blood.

Serology panels

Special situations exist in which testing for antibody to individual viruses is less helpful than using a battery of antigens to test for antibody to many viruses.

The use of a combination of serologic tests to diagnose a clinical syndrome (Table 66-32) may be useful in situations in which the viruses under consideration cannot be cultured, specimens of infected tissue are not available (brain tissue), antiviral agents have been administered, or the patient is convalescing and isolation of virus is unlikely.

INTERPRETATION OF LABORATORY TEST RESULTS

Interpretation of laboratory test results must be based on knowledge of the normal viral flora in the site sampled, the clinical findings, and the epidemiology of viruses. Serologic testing, in addition to virus detection tests, may be needed to support or refute the association of a virus isolate with a disease.

VIRUSES IN TISSUE AND BODY FLUIDS

In general, the detection of any virus in host tissues, CSF, blood, or vesicular fluid is significant. Recovery of adenovirus or mumps virus in urine is usually diagnostic of disease, as opposed to the culture of CMV, which may merely reflect asymptomatic reactivation. Occasionally enteroviruses are detected in urine as a result of fecal contamination, or HSV from symptomatic or asymptomatic infection of the external urogenital tract is recovered in urine. Interpretation of these culture results requires correlation with clinical data. CMV viruria during the first 2 weeks of life establishes a diagnosis of congenital CMV infection, whereas detection at 4 weeks or after suggests intrapartum or postpartum acquisition.

VIRUSES IN THE RESPIRATORY TRACT

Detection of measles, mumps, influenza, parainfluenza, and respiratory syncytial viruses is very significant because asymptomatic carriage and prolonged shedding is unusual. Conversely, HSV, CMV, and adenoviruses can be shed in the absence of symptoms for periods ranging from a few days to many

TABLE 66-32   SEROLOGY PANELS AND IMMUNE STATUS TESTS FOR COMMON VIRAL SYNDROMES

SITUATION	VIRUSES UNDER CONSIDERATION
AIDS	Human immunodeficiency virus (HIV)
CNS infections	Western equine encephalitis Eastern equine encephalitis California encephalitis virus St. Louis encephalitis virus Lymphocytic choriomeningitis virus Measles Mumps Epstein-Barr virus Rabies
Exanthems	Measles Rubella Parvovirus
Vesicular	Herpes simplex virus Varicella-zoster virus
Hepatitis A	Hepatitis A virus
Hepatitis B	Hepatitis B virus
Heterophile-negative infectious mononucleosis syndromes	Cytomegalovirus Epstein-Barr virus
Myocarditis-pericarditis	Group B coxsackievirus, types 1-5 Influenza A, B Cytomegalovirus
Respiratory	Influenza A, B Respiratory syncytial virus Parainfluenza 1-3 Adenovirus

Serology needed to determine immune status (single serum)
  Rubella
  Hepatitis B
  Varicella-zoster
  CMV
  Measles	

months. Adenoviruses are isolated commonly from asymptomatic infants and young children. Simultaneous isolation of these agents from both throat and feces in febrile and respiratory syndromes increases the probability of association with illness. Isolation from throat but not feces has a lesser probability of association, and isolation from feces alone has the least diagnostic significance.

VIRUSES IN THE EYE

Isolation of adenoviruses, HSV, VZV, and some enteroviruses from diseased cornea and conjunctiva usually establishes the etiology of the infection.

Enterovirus type 70 is known to cause a particularly contagious form of viral conjunctivitis.

DETECTION OF ENTEROVIRUSES

Enteroviruses are most commonly found in asymptomatic infants and children, particularly during the late summer and early fall. A knowledge of the relative frequency of virus shedding is extremely helpful in assessing significance of results of throat or stool cultures. Prevalence of enteroviruses in the stools of infants and toddlers may approach 30% during peak periods. Shedding of enteroviruses in the throat is relatively transient, usually 1 to 2 weeks, whereas fecal shedding may last 4 to 16 weeks. Thus isolation of an enterovirus from the throat supports etiology of a clinically compatible illness more than isolation from the feces alone.

DETECTION OF VZV AND HSV

The detection of VZV is always significant. Asymptomatic shedding does not appear to occur with this virus as it does with other herpes viruses. The detection of HSV from cutaneous or mucocutaneous vesicles is also significant, implying primary or reactivation disease. HSV may be detected in respiratory secretions during asymptomatic "stress" reactivation, unless typical vesicles or ulcers are present. HSV in stool usually represents either severe disseminated infection or infection of the anus or perianal areas. Isolation from any newborn infant specimen suggests potentially severe infection.

DETECTION OF CMV

Interpretation of results of specimens containing CMV is most difficult. Primary CMV infection is usually asymptomatic and is commonly followed by silent reactivation of the latent virus throughout the patient's remaining life. CMV disease in immunocompromised patients can be life threatening, and ganciclovir antiviral therapy may be warranted. Detection of CMV in urine or respiratory secretions, however, is not diagnostic of significant disease. Detection of CMV in uncontaminated tissue, such as lung tissue or blood collected by venipuncture, suggests an active role in disease. The presence of virus-specific IgM or a fourfold increase in IgG antibodies may indicate disease. However, positive serology results must be interpreted with caution. False-positive IgM results have been attributed to infections by other viruses, such as EBV, and rises of both IgM and IgG may result from transfusions or the use of immune globulin therapy.

ACKNOWLEDGMENT

The author wishes to acknowledge the extensive contributions made by those with whom he has worked at both Evanston Hospital in Evanston, Illinois, and Children's Hospital Medical Center of Akron in Akron, Ohio. In particular, the help and knowledge gained from Marcia Benedict, Sandy McLachlan, Cindy Bostick, Madeline Mainzer, and Dede Pierry are appreciated.

Procedures

66-1 PROCESSING BLOOD FOR VIRAL CULTURE: LEUKOCYTE SEPARATION BY POLYMORPHPREP

PRINCIPLE

Detection of most viruses in blood specimens is best accomplished by separating and culturing leukocytes. An exception is enterovirus, which, when infecting newborns, may be found free in serum. Leukocytes were originally added to cell culture by harvesting the buffy coat following concentration by centrifugation ($100\times$ *g* for 15 minutes). This method was found to be inferior to white blood cell concentration using density gradient centrifugation and sedimentation with Ficoll-Hypaque/Macrodex. Other density gradient methods have been employed, including those using Plasmagel, Leuco-PREP, Sepracell-MN, Mono-Poly Resolving Medium, and Polymorphprep. Leukocyte preparation using Polymorphprep, a mixture of sodium metrizoate and dextran, is one of the best, because mononuclear and polymorphonuclear cells are isolated from red blood cells in a one-step procedure.

SPECIMEN

A sample of 3 to 5 mL of anticoagulated blood is optimal. EDTA, citrate, or heparin anticoagulant-containing tubes can be used. As few as 2 mL of specimen is acceptable for pediatric patients. Clotted specimens are unacceptable. Specimens should be processed immediately. When processing delay is unavoidable, hold specimen at 2° to 8° C. Processing must occur within 12 to 24 hours. Loss of virus may occur during any processing delay.

MATERIALS

1. Polymorphprep (Robbins Scientific Corp., Sunnyvale, Calif.). Store at or below 20° C.

2. Phosphate-buffered saline (PBS)— 10 ×. (Biowhittaker, Walkersville, Md.). Store at room temperature.

3. Sterile phosphate-buffered saline— 0.5 ×. Dilute 10 × PBS 1:20 with distilled water. Filter sterilize through 0.22-μm filter. Store at 2° to 8° C.

4. Eagle's minimal essential medium (2% fetal bovine serum-FBS). Store at 2° to 8° C.

5. Sterile sodium chloride (0.9%). Store at 2° to 8° C.

6. Sterile conical centrifuge tubes.

7. Sterile pipettes.

8. Centrifuge.

METHOD

1. Wear latex gloves and laboratory coat when handling patient specimens. Use biological safety cabinet whenever tube containing specimen is opened. Use biocontainment safety covers during centrifugation procedures.

2. Allow blood specimen and all reagents to come to room temperature.

66-1 PROCESSING BLOOD FOR VIRAL CULTURE: LEUKOCYTE SEPARATION BY POLYMORPHPREP—cont'd

3. Mix blood specimen well by inverting 5 times. Using a 5- or 10-mL pipette, transfer specimen to a sterile conical centrifuge tube.

4. Centrifuge at $500\times g$ for 10 minutes at room temperature ($18°$ to $22°$ C).

5. Remove plasma with a pipette.

6. Measure approximate volume of remaining cells. Using a 5- or 10-mL pipette, dilute cells with the same approximate volume of 0.9% sodium chloride. Mix well.

7. Add 3.5 mL of Polymorphprep to a 15-mL conical centrifuge tube with a 5-mL pipette.

8. Carefully layer the entire volume of diluted blood cells over the 3.5 mL of Polymorphprep in the centrifuge tube. Do not mix the blood with the Polymorphprep.

9. Centrifuge the tube containing blood and Polymorphprep at 450 to $500\times g$ for 30 minutes at room temperature ($18°$ to $22°$ C).

10. Two leukocyte bands should be visible after centrifugation. The top band at the blood sample/Polymorphprep interface consists of mononuclear cells. The lower band contains polymorphonuclear cells. Erythrocytes are pelleted at the bottom of the tube.

11. The mononuclear and polymorphonuclear bands are collected and transferred to a 15-mL conical centrifuge tube with a pasteur pipette.

12. Add 5 mL of $0.5\times$ sterile PBS to the centrifuge tube containing the harvested cells. Mix well by aspirating and expelling repeatedly with a pipette.

13. Centrifuge cells at $400\times g$ for 10 minutes at room temperature ($18°$ to $22°$ C).

14. Aspirate supernatant PBS.

15. Add 5 mL of EMEM with 2% FBS to pelleted cells. Mix well by aspirating and expelling repeatedly with a pipette.

16. Centrifuge as in step 13.

17. Aspirate supernatant EMEM and resuspend in 2 mL of fresh EMEM with 2% FBS.

18. Specimen is ready for inoculation to cell culture.

TROUBLESHOOTING

1. If all cells are at the top of the Polymorphprep after centrifugation, remix and overlay on top of a new Polymorphprep layer. It may be necessary to increase the volume of Polymorphprep so proportion is still approximately 1:1 to 1:1.5. Centrifuge at either increased force of gravity for the same length of time, or at the same force of gravity for a longer time.

2. If buffy coat is on top of the erythrocyte layer, take buffy coat and erythrocytes and relayer over a new gradient. Centrifuge for a shorter time at the same force of gravity or at a lesser force of gravity for the same time.

66-2 DETECTION OF RESPIRATORY SYNCYTIAL VIRUS BY FLUORESCENT ANTIBODY STAINING OF RESPIRATORY SECRETIONS

PRINCIPLE

Respiratory syncytial virus is recognized as a major cause of acute lower respiratory tract disease in infants and young children, commonly presenting as bronchiolitis or pneumonia. Hospitalized children with RSV disease serve as a reservoir for nosocomial transmission of the virus. Isolation or cohorting of infected patients decreases transmission. Conventional culture for the identification of virus requires 4 to 14 days and may be insensitive because of the lability of the virus. Rapid detection of RSV can be accomplished by fluorescent antibody or enzyme immunoassay detection of viral antigen. Direct fluorescent antibody staining for RSV antigen can be performed on upper respiratory wash or swab specimens that contain infected epithelial cells.

SPECIMEN

Nasal washes are the specimen of choice. Three to 5 mL of saline is quickly squirted into and aspirated back through the nostril. The nasopharyngeal lavage is sent immediately to the laboratory in a sterile container.

Nasopharyngeal swab material can be used for DFA stain, but is less likely to contain an abundance of epithelial cells. Swab material should be vigorously extracted in viral transport medium and sent to the laboratory immediately.

MATERIALS

1. Phosphate-buffered saline.

2. N-acetyl-L-cysteine (NALC). Prepare working stock by diluting 50 mg of NALC in 10 mL of 1.5% sodium citrate. Working stock may be stored for 1 month at $2°$ to $8°$ C.

3. Anti-RSV monoclonal, FITC-conjugated antibody with Evans blue counterstain (Meridian Diagnostics, Inc., Cincinnati, Ohio).

4. Buffered glycerol mounting medium (pH 8 to 8.4).

5. Centrifuge tubes.

6. Pipettes.

7. Microscope slides and coverslips.

8. Humidified incubation chamber.

9. Centrifuge.

10. Fluorescence microscope.

11. RSV control slides, each with

Continued

66-2 DETECTION OF RESPIRATORY SYNCYTIAL VIRUS BY FLUORESCENT ANTIBODY STAINING OF RESPIRATORY SECRETIONS—cont'd

two wells. One well containing RSV infected HEp-2 cells and one containing uninfected HEp-2 cells.

METHOD

1. Wear latex gloves when handling all patient specimens. Perform all specimen-handling procedures in laminar flow hood.

2. Examine specimen for the presence of mucus. If sufficient mucous is present to prevent mixing and distribution of specimen to slide, digest with mucolytic agent (NALC) as follows:
   a. Treat specimen with equal volume of stock NALC. Vortex for 20 seconds. Allow to sit at room temperature for 20 minutes.
   b. Vortex for 20 seconds. Add 5 mL PBS to cells, mix, and centrifuge at $380 \times g$ for 15 minutes.
   c. Aspirate and discard the supernatant, being careful not to disturb the cell pellet.
   d. Wash cell pellet with 5 mL PBS.
   e. Resuspend cell pellet in 0.2 mL PBS. Cell suspension should be hazy but not opaque. Dilute cell suspension with PBS if necessary.
   f. Place one drop of the suspension into the well of the microscope slide, one patient specimen per slide.

3. If specimen is free of mucus, process as follows:
   a. Centrifuge 1 mL of specimen for 2 minutes in the microcentrifuge.
   b. Carefully aspirate supernatant and wash twice with PBS.
   c. Aspirate supernatant and resuspend cell pellet in 0.2 mL PBS.
   d. Prepare slide as described in step 2, f.

4. Air dry all slides and then fix for 10 minutes in acetone at 2° to 8° C. Slides may be held for 24 hours at 2° to 8° C or for longer periods at −70° C.

5. Stain slides by adding 20 to 30 μL/well of RSV antibody conjugate to each patient slide and to both wells of the control slide. Spread conjugate to cover slide completely.

6. Place slides in humidified chamber at 36° C and incubate for 30 minutes.

7. Rinse antiserum from each slide with a gentle stream of PBS and place slides in Coplin jar filled with PBS for 5 minutes.

8. Gently rinse twice in distilled water for 1 to 2 minutes.

9. Drain excess distilled water and air dry. Do not blot.

10. Add coverslip using buffered glycerol as mounting medium.

INTERPRETATION AND QUALITY CONTROL

1. Examine using 200× magnification (10× eyepiece, 20× objective).

2. Examine control slide first. The well containing RSV-infected HEp-2 cells should show ≥ 2+ fluorescing inclusions (see Figure 66-13, *E*). The well containing uninfected HEp-2 cells should show no fluorescence. If the control slide shows other results, the batch of slides should be discarded and restained with a new control slide.

3. Examine patient slides. Fewer than 3 to 5 cells per 200× field should be considered insufficient cells for an adequate examination, and a new specimen should be requested. If sufficient cells are seen, examine for typical green-fluorescing RSV inclusions. A positive stain will show fluorescing inclusions. A negative stain will have no fluorescing inclusions.

66-3 SHELL VIAL CULTURE FOR CYTOMEGALOVIRUS

PRINCIPLE
The shell vial culture for CMV is used to rapidly detect this virus in clinical specimens. Use of the shell vial culture with urine, blood, and bronchoalveolar lavage fluid specimens has demonstrated equivalent or superior sensitivity compared with conventional cell culture.

SPECIMEN
Acceptable specimens include urine, blood, respiratory secretions and washes, tissue, and cerebrospinal fluid. Specimens are collected, transported, and stored according to standard protocols.

MATERIALS
1. Shell vials, 15 × 45 mm 1-dram vial with plastic closure (Allegiance Healthcare Corp., McGaw Park, Ill.).

2. Circular coverslips, no.1 12-mm diameter (Bellco Glass, Inc., Vineland, N.J.).

3. MRC-5 cells growing in shell vial (5 to 9 days old).

4. In place of steps 1 to 3, use MRC-5 shell vials purchased from commercial source (Biowhittaker Inc., Walkersville, Md.; Viromed Laboratories, Inc., Minneapolis, Minn.).

5. Centrifuge.

6. Sterile 15-mL conical centrifuge tubes.

7. Sterile 1- and 5-mL pipettes and cotton-plugged sterile Pasteur pipettes.

8. pH paper (3 to 9 pH).

9. Maintenance medium.

10. CMV positive control (strain AD 169).

11. Negative control (use maintenance medium with no virus).

12. Anti-CMV early nuclear protein monoclonal antibody.

13. FITC-labeled antimouse IgG conjugate.

14. Phosphate-buffered saline (PBS).

15. Methanol, forceps, buffered glycerol mounting medium, and glass microscope slides.

METHOD
1. Wear latex gloves and laboratory coat when handling patient specimens. Use biological safety cabinet whenever tube containing specimen is open. Use biocontainment safety covers during centrifugation procedures.

2. Specimen preparation
   a. Urines
      (1) Using a 5-mL pipette, add 2 mL of urine into sterile centrifuge tube.
      (2) Using a 1-mL pipette, add 0.5 mL of viral transport medium containing antimicrobials.
      (3) Centrifuge at 1000× *g* for 10 minutes at 2° to 8° C.
      (4) Aspirate supernatant with cotton-plugged sterile Pasteur pipette and transfer to a sterile test tube.
      (5) Adjust pH to 7 with sterile cold 0.1 N HCl or 0.1 N NaOH. Determine pH with pH paper.
      (6) Store at 2° to 8° C until inoculation into shell vial cultures.
   b. Blood
      (1) Refer to leukocyte separation by Polymorphprep (Procedure 66-1). Store leukocytes at 2° to 8° C until inoculation into shell vial cultures.
   c. Respiratory secretions and washes
      (1) Process as for urines above. pH adjustment is not necessary.
      (2) Specimens received in viral transport medium do not require further preparation.
      (3) Store all specimens at 2° to 8° C until inoculation into shell vial culture.
   d. Tissue
      (1) Mince and grind 5-mm³ piece of tissue gently in approximately 5 mL of viral transport medium. This results in a 10% suspension.
      (2) Centrifuge at 1000× *g* for 10 minutes at 2° to 8° C.
      (3) Aspirate supernatant with cotton-plugged sterile Pasteur pipette and transfer to small sterile test tube.
      (4) Store at 2° to 8° C until inoculation into shell vial culture.
   e. Cerebrospinal fluid
      (1) CSF and other body fluids are inoculated into shell vial culture undiluted.
      (2) Specimens with less than sufficient volume for proper inoculation (≤0.5 mL) should be diluted with enough viral transport medium to inoculate shell vial cultures (0.5 to 0.6 mL).
      (3) Specimens received in viral transport medium are ready to inoculate.
      (4) Store at 2° to 8° C until inoculation into shell vial culture.

3. Inoculation and incubation of shell vial cultures
   a. Aspirate maintenance medium from sufficient shell vials for all specimens, including one positive and one negative control.
   b. Using a new sterile 1-mL pipette for each specimen, inoculate shell vials as follows:

Specimen	No. Vials	Inoculum/ Vial
Urine	1	0.25 mL
Respiratory	1	0.1 mL
Blood	3	All 0.25 mL
All other specimens	2	0.1 and 0.25 mL
Negative control	1	0.1 mL
Positive control	1	0.1 mL

   Number of vials and volume of inoculum may vary among laboratories depending on local experience. Inoculate all specimens before inoculating controls. A negative and positive control are used with each batch of shell vial cultures. Inoculate negative control before positive.
   c. Cap vials with sterile plastic lids and label.
   d. Centrifuge vials at 700× *g* for 40 minutes at 35° C.

Continued

66-3 SHELL VIAL CULTURE FOR CYTOMEGALOVIRUS—cont'd

e. Using a 5-mL pipette, add 2 mL of maintenance medium to each shell vial.

f. Recap vials and incubate at 36° C ± 1° C for 16 to 24 hours.

4. Staining shell vial coverslips

a. Wear latex gloves and perform test in laminar flow hood.

b. Remove lids and aspirate media from shell vials.

c. Wash coverslips with 1 mL of PBS. Aspirate. Wash again with 1 mL PBS and let stand for 5 minutes. Aspirate. Additional PBS washes may be necessary for blood cultures, to remove blood cells from cell sheet. If necessary, check cell monolayer using the the microscope.

d. Fix coverslips in 1 mL cold methanol for 10 minutes.

e. Aspirate methanol. Coverslips should be allowed to dry, but not overdry. A white clearing across coverslips is apparent at the right amount of dryness. Overdrying causes nonspecific fluorescence around the edge. Underdrying causes uneven staining.

f. Add 150 µL monoclonal antibody, at current working dilution, to each vial by dropping antibody into the middle of the coverslip. Replace shell vial lids and gently rotate to spread stain over entire coverslip. Incubate at 35° to 37° C for 30 minutes.

g. After staining add 1 mL PBS to each vial (do not remove antibody first). Mix. Aspirate entire volume of fluid from vial.

h. Wash twice with 1 mL PBS. Let the second wash remain on coverslip for 5 minutes. Aspirate PBS, do not allow coverslips to dry.

i. Add 150 µL of antiimmunoglobulin conjugate (FITC), at current working dilution, to each vial by dropping antibody onto the middle of the coverslip. Spread as in step f. Incubate at 35° to 37° C for 30 minutes.

j. Repeat wash steps g and h. Wash coverslips an additional time with 1 mL distilled/deionized water for 2 minutes. A brief rinse may be sufficient for monoclonal antibody conjugates. The more thorough washing is needed for polyclonal reagents.

k. Using small forceps, remove coverslips from vials and place cell side down onto drop of mounting medium. Wipe tip of forceps with alcohol-soaked gauze between each coverslip.

5. Reading and interpretation of shell vial coverslips

a. Read entire coverslip using the 20× fluorescence microscope objective. The nuclei of CMV infected cells exhibit an apple-green fluorescence that may vary in intensity (see Figure 66-21). Infected nuclei may be distinguished from artifacts by their regular oval shape and evenly stained, slightly granular appearance. Occasionally the nuclei may appear rounded. The background should stain a dull red or a pale green (dull red if Evans blue counterstain is used with conjugate, pale green if no counterstain is used).

6. Quality control

a. The negative control cover slip should have no fluorescing nuclei typical of a CMV-infected cell.

b. The positive control cover slip should have multiple fluorescing nuclei typical of CMV-infected cells.

66-4 CULTURE CONFIRMATION BY FLUORESCENT ANTIBODY STAINING

PRINCIPLE

Many viruses can be identified presumptively in cell culture by observing CPE. Definitive identification or in some cases typing of a virus in cell culture can be accomplished by staining a smear of infected cells with a virus-specific antibody conjugate. Results are easier to interpret if there is at least 2+ CPE before collecting cells for staining.

MATERIALS

1. Virus-specific antibody conjugate.

2. Phosphate-buffered saline (PBS).

3. Glass slides with circled staining areas and coverslips.

4. 25-μL pipettes and tips.

5. Control slides with homologous virus for positive staining and nonhomologous virus for negative staining.

6. Mounting fluid (25% buffered glycerol in PBS).

7. Sterile glass beads.

8. Centrifuge.

9. Pasteur pipettes.

SPECIMEN

Cells can be collected from the surface of the glass test tube by scraping with a 1-mL pipette, adding 3 to 4 sterile glass beads and disrupting by vortexing, or detaching cells with trypsin. Scraping is best to use if there is less than 2+ CPE. Areas of CPE can be circled on the outside of the culture tube with a marker. If only these areas are scraped on the inside, the relative number of virus-infected cells will be greater and detection will be easier in the stained smear. Although trypsin treatment is the traditional method for releasing cells, use of glass beads during vortexing is simple and reliable. The glass bead method will be described.

1. Add 3 to 4 sterile glass beads to the cell culture tube. Vortex vigorously for 15 to 30 seconds. Examine microscopically to ensure that the cells are detached from the glass. Vortex a second time if cells remain on glass.

2. Pellet cells by centrifugation at $900\times g$ for 10 minutes. Carefully remove supernatant.

3. Wash cells by resuspending the pellet in PBS and repeat centrifugation.

4. Remove supernatant and resuspend in 0.05 to 0.1 mL of PBS.

5. Place 25-μm aliquot of the cell suspension onto two wells on clean labeled slides. Prepare enough smears to test with antisera for all viruses under consideration.

6. Air dry at temperatures no higher than 30° C and fix in acetone for 10 minutes. Slides may be stored at −70° C.

Determine working dilution of fluorescent conjugate if not supplied by manufacturer at proper working dilution.

1. Prepare serial doubling dilutions of the stock conjugate in PBS.

2. Test each conjugate dilution for its ability to stain smears of homologous and nonhomologous virus-infected cells, as well as smears of normal uninfected cells.

3. The working dilution of the antiserum is the highest dilution producing 3 to 4+ fluorescence of homologous virus, ≤1+ fluorescence of nonhomologous virus, and no fluorescence of uninfected cells.

METHOD

1. Apply working dilution of conjugate to circled areas on slide. Cover entire circled area (approximately 25 μL of conjugate per circle).

2. Place slides in a moist chamber and cover the chamber to minimize evaporation. Incubate slides at 35° C for 20 minutes. Be careful not to tilt slides, because conjugates will run together. Do not allow conjugate to dry on the slide. This will cause nonspecific staining.

3. Tip slides to drain conjugate from smear. Remove excess reagent with running PBS, followed by a 5-minute PBS wash in a Coplin jar. Agitate slide occasionally during wash. A brief rinse with no agitation may be sufficient for monoclonal antibody conjugates. The more thorough washing is needed for polyclonal reagents.

4. Rinse slides in distilled water to remove PBS salt crystals.

5. Allow slides to air dry at room temperature. Protect slides from bright light.

6. Apply mounting fluid and coverslip.

7. Examine with a fluorescence microscope containing appropriate filters for FITC staining (490 nm UV needed). Use 20× and 40× objectives.

INTERPRETATION OF RESULTS

1. Positive-staining cells are characterized by ≥1+ apple-green fluorescent intranuclear and/or cytoplasmic granules (see Figure 66-13).

2. Negative-staining cells are characterized by no or <1+ fluorescence. Uninfected cells may stain red if Evans blue counterstain is used.

QUALITY CONTROL

1. Positive and negative controls should be used each time the stain is performed.

2. The positive control (homologous virus) should exhibit ≥2+ staining.

3. The negative control (nonhomologous virus) should exhibit <1+ staining.

4. If control results are different than expected, stain should be repeated.

66-5 HEMADSORPTION OF PRIMARY MONKEY KIDNEY MONOLAYERS TO DETECT INFLUENZA, PARAINFLUENZA, AND MUMPS VIRUSES

PRINCIPLE

Hemadsorption of cell cultures is used to detect or confirm the presence of orthomyxoviruses (influenza A, B, and C) and some paramyxoviruses (parainfluenza 1 through 4 and mumps). Respiratory syncytial virus is not detected by this method. Cell cultures infected with these viruses may not show CPE; however, infected cells do acquire the ability to hemadsorb guinea pig red blood cells to their outer cell membranes. In addition, the presence of viral hemagglutinin in the culture medium, after release from the cell, can also cause agglutination of the red blood cells (RBCs) in the fluid medium.

Hemadsorption is performed on various days during culture incubation depending on the time of year and the viruses suspected. Influenza virus detection is performed on specimens from adult and pediatric patients during the influenza season (November through March). Early hemadsorption (day 2 or 3) decreases time to detection of positive influenza cultures. Detection of parainfluenza viruses from pediatric patient specimens is performed year-round. Hemadsorption produced by parainfluenza viruses generally occurs later than that produced by influenza. Hemadsorption to detect mumps virus is performed in most clinical settings only if this virus is suspected by the clinician.

SPECIMEN

A culture exhibiting CPE should be hemadsorbed as soon as possible to expedite reporting of a positive result. Perform hemadsorption on CPE-negative cell cultures of respiratory tract specimens or of specimens for mumps virus detection according to the following schedule:

Culture type	Perform hemadsorption on days:
Influenza virus suspected	2 and 7
Parainfluenza virus suspected	3 and 10
Mumps virus suspected	3 and 10

MATERIALS

1. Guinea pig red blood cells. Store in buffer in which cells were received at 2° to 8° C. Cells expire 10 days after date drawn from animal.

2. Sterile, 10× phosphate-buffered saline. Prepare working 1× phosphate-buffered saline (PBS) by diluting with distilled water. Filter (0.22 μm) sterilize. Store at 2° to 8° C.

3. Viral culture maintenance medium with antibiotics.

4. Positive control virus (parainfluenza virus) in tube of current lot of PMK cells

5. Negative control in uninoculated tube of current lot of PMK cells.

6. Sterile, conical centrifuge tubes.

7. Sterile pipettes.

8. Repeater Eppendorf pipettes.

9. Centrifuge.

METHOD

1. Preparation of guinea pig RBCs
   a. Transfer guinea pig cells to 15-mL conical centrifuge tube. Centrifuge for 5 minutes at 500× *g* (2° to 8° C). Aspirate supernatant and buffy coat.
   b. Resuspend RBCs in cold (2° to 8° C) PBS. Wash in cold PBS until supernate is clear (2 or 3 times).
   c. Aspirate PBS after last wash. Measure remaining volume of cells using graduations on centrifuge tube. Add a sufficient amount of PBS to make a 10% cell solution.
   d. Hemadsorption test is performed using a 0.4% guinea pig cell solution prepared with maintenance medium. Depending on the number of tubes to be tested (0.2-mL/tube), prepare as follows:

      0.2 mL 10% RBCs + 4.8 mL medium (25 tubes) or
      0.4 mL 10% RBCs + 9.6 mL medium (50 tubes)

2. Hemadsorption
   a. Wear latex gloves and laboratory coat during procedure. Perform in biological safety cabinet.
   b. Aspirate maintenance medium from all patient and control tubes to be tested.
   c. Add 0.2 mL of the 0.4% RBC suspension to each tube. Be careful not to cross-contaminate tubes. Positive and negative controls should be tested last.
   d. Gently tilt stationary rack from side to side to ensure RBC contact with cell monolayer. Place stationary rack in refrigerator (2° to 8° C) for 30 minutes. Tubes should be placed horizontally so that the erythrocyte suspension covers the monolayer.
   e. Invert tubes quickly, following incubation, to dislodge RBCs lying on the cell sheet. Examine cell culture tubes microscopically with 4× objective for RBCs that adhere to the monolayer (see Figure 66-19, *G*). Culture fluids should be inspected for hemagglutination.

INTERPRETATION

1. Negative:
   No hemadsorption of guinea pig cells and no hemagglutination of cells in fluid medium. RBCs float freely in medium.

2. Positive:
   Hemadsorption of guinea pig cells to infected PMK cells or hemagglutination of RBCs in fluid medium. Hemadsorption is graded on a scale of 1 to 4+.

   1+ = <25% of cell sheet showing hemadsorption.
   2+ = 25% to 50% of cell sheet showing hemadsorption.
   3+ = 51% to 75% of cell sheet showing hemadsorption.
   4+ = 76% to 100% of cell sheet showing hemadsorption.

3. Viruses that both hemadsorb and hemagglutinate guinea pig erythrocytes are usually influenza virus. Mumps and parainfluenza viruses generally produce only hemadsorption. Identification

66-5 HEMADSORPTION OF PRIMARY MONKEY KIDNEY MONOLAYERS TO DETECT INFLUENZA, PARAINFLUENZA, AND MUMPS VIRUSES—cont'd

of virus-causing hemadsorption should be confirmed using virus specific antibody conjugates. Certain simian viruses that may contaminate the PMK cells hemadsorb at 4° C (see quality control below). Examine all tubes as soon as possible after removal from the refrigerator, because the neuraminidase of the myxoviruses is active at room temperature and will destroy hemagglutinins if tubes are left for extended periods at this temperature.

QUALITY CONTROL

1. Positive control:
Parainfluenza virus should exhibit ≥2+ hemadsorption.

2. Negative control:
Uninoculated PMK cells should exhibit no hemadsorption. If hemadsorption attributed to an endogenous simian virus occurs, the patient must exhibit more hemadsorption than the negative control to be considered positive.

66-6 ACID LIABILITY TEST TO DIFFERENTIATE RHINOVIRUS FROM ENTEROVIRUS

PRINCIPLE

Rhinoviruses and enteroviruses may have similar patterns of CPE in cell culture. Both can be detected in upper respiratory tract specimens. Sensitivity to an acidic environment differentiates these two groups of viruses. Rhinoviruses are labile at pH 3, whereas enteroviruses are stabile at this pH.

SPECIMEN

1. 0.2-mL suspension of test virus.

2. 0.2-mL suspension of stock rhinovirus.

3. 0.2-mL suspension of stock enterovirus.

MATERIALS

1. EMEM prepared without sodium bicarbonate. pH should be approximately 3. Test with pH meter and adjust with 2 N HCl if necessary.

2. EMEM prepared the usual way with sodium bicarbonate. pH should be 7.2 to 7.4. Check with pH meter and adjust if necessary.

3. Sterile pipettes.

4. Human diploid fibroblast cells.

METHOD

1. The neutralization procedure is performed on each unknown virus and the rhinovirus and enterovirus controls.

2. For each virus to be tested, use a 1-mL pipette to add 0.1 mL of the suspension of virus to 0.9 mL of EMEM pH 3.

3. For each virus to be tested, use a 1-mL pipette to add 0.1 mL of the suspension of virus to 0.9 mL of EMEM pH 7.

4. Using a 1-mL pipette, inoculate 0.1 mL of each of the EMEM pH 3 virus suspensions to a separate human diploid fibroblast cell culture tube.

5. Inoculate a second set of culture tubes with 0.1 mL of each of the EMEM pH 7 virus suspensions.

6. Incubate cell culture tubes at 35° to 37° C. Incubate an uninocu-

lated tube of the same lot of fibroblast cells as a cell culture control. Observe daily for 7 to 10 days for CPE.

INTERPRETATION

1. Control viruses must show appropriate growth patterns. See quality control below.

2. Presence of CPE in cells inoculated with EMEM pH 3 approximately equal to that in EMEM pH 7 indicates enterovirus.

3. Absence of CPE in EMEM pH 3 and presence of CPE in EMEM pH 7 indicates rhinovirus.

QUALITY CONTROL

1. Rhinovirus control:
Presence of CPE in EMEM pH 7. Absence of CPE in EMEM pH 3.

2. Enterovirus control:
Presence of CPE in EMEM pH 7 and EMEM pH 3.

66-7 PRESERVATION AND STORAGE OF VIRUSES BY FREEZING

PRINCIPLE
Viral isolates should be saved for use as quality control strains or for epidemiologic investigation. Maximal recovery of virus is accomplished by collecting cells and fluid from infected culture tubes.

MATERIALS
1. Sterile glass beads.

2. Dimethyl sulfoxide (DMSO).

3. Dry ice and acetone.

4. Freezer tubes.

METHOD
1. Viruses other than CMV and VZV.
   a. Select tubes showing ≥ 2+ CPE or hemadsorption. RBCs may be eluted following hemadsorp-tion by placing tubes in a 37° C water bath for 15 minutes. Wash to remove RBCs.
   b. Add three sterile glass beads to cell culture tube. Tighten cap. Vortex vigorously behind shield for 30 seconds. Examine microscopically to ensure that cells have been disrupted and removed from glass surface.
   c. Add cells and medium to freezer tube. Snap freeze by swirling the tube in a slurry of dry ice and acetone.
   d. Store at −70° C.

2. CMV and VZV
   a. Use tubes showing ≥ 3+ CPE. Harvest as above except add 10% DMSO and glass beads. For example, add 0.2 mL of DMSO to 2 mL of medium in cell culture tube. Mix and proceed as described above.

3. Thaw in a 37° C water bath. Inoculate to appropriate cell culture.

66-8 CYTOMEGALOVIRUS ANTIGENEMIA STAIN

PRINCIPLE
The CMV antigenemia stain is used for the immunologic detection of cytomegalovirus lower matrix pp65 antigen in peripheral blood leukocytes. The assay is semiquantitative because white blood cells are counted and a standard number are placed on the slide for staining. Experience with the CMV antigenemia stain suggests that positive results can serve as an early indicator of active CMV disease.

SPECIMEN
Five to 7 mL of anticoagulated blood is optimal. Heparin or EDTA anticoagulant–containing tubes can be used. Specimen should be transported to the laboratory at room temperature immediately after collection. Clotted specimens are unacceptable. Specimens should be processed immediately. When processing delay is unavoidable, hold specimen at 2° to 8° C. Processing should occur within 8 hours.

MATERIALS
1. CMV Brite Kit (Biotest Diagnostic Corp., Denville, N.J.), which includes dextran solution, RBC lysing solution, fixative solution, permeabilization solution, fetal calf serum, monoclonal antibodies, and control slides. Store at 2° to 8° C.

2. Sterile pipettes and micropipettes.

3. Refrigerated centrifuge and 15-mL conical bottom centrifuge tubes.

4. Cytocentrifuge (Shandon, Inc., Pittsburgh, Pa.).

5. Phosphate-buffered saline (PBS) and calcium- and magnesium-free deionized water (dH$_2$O).

6. Coplin staining jars, moisture chamber, 37° C incubator, microscope slides for cytospin, coverslips, and mounting medium.

7. Coulter counter for WBC count (hematology laboratory).

8. Fluorescent microscope.

METHOD
1. Wear latex gloves and laboratory coat when handling patient specimens. Use biological safety cabinet whenever tube containing specimen is open. Use biocontainment safety covers during centrifugation procedures.

2. Leukocyte isolation from peripheral blood
   a. Allow kit reagents to come to room temperature (20° to 25° C).
   b. Mix 5 to 7 mL of anticoagulated blood with 1.5 mL of dextran reagent (reagent A) in a 15-mL conical centrifuge tube. If a greater volume of blood is used, the dextran-to-blood ratio of 1:4 must be maintained.
   c. Incubate tube containing blood-dextran mixture for 20 minutes at 37° C at a 45-degree angle. Leave cap loose.
   d. Remove the top leukocyte-rich layer to a 15-mL conical centrifuge tube using a pasteur pipette. Centrifuge for 10 minutes at 300× g.
   e. Dilute RBC lysing reagent (reagent B) 1:10 in dH$_2$O and allow to cool to 4° C.
   f. Discard supernatant following 10-minute centrifugation. Resuspend cell pellet in 5 mL of diluted reagent B. Vortex and incubate for 5 minutes at 4° C.
   g. Add 5 mL of PBS and centrifuge for 10 minutes at 300× g.
   h. Wash cells in 5 mL PBS, centrifuge as above and discard supernatant.

3. Counting cells and preparing cell dilution
   a. Resuspend cell pellet in 1.5 mL PBS.
   b. Aliquot 0.4 mL into a 1.5-mL microcentrifuge tube. Take this

66-8 CYTOMEGALOVIRUS ANTIGENEMIA STAIN—cont'd

aliquot to the hematology laboratory for a WBC count.

c. Dilute cell suspension with PBS to make 1 mL of 2×10^6 cell/mL in a 1.5-mL microcentrifuge tube.

Calculate the amount of cell suspension to use as follows:

$$\frac{\text{Desired number cells/mL}}{\text{Actual number cells/mL}} = \frac{\text{Amount of suspension required}}{}$$

For example:

$$\frac{2 \times 10^6 \text{ cells/mL}}{5 \times 10^6 \text{ cells/mL}} = \frac{2}{5} = 0.4 \text{ mL cell suspension}$$

Add enough PBS to increase the volume to 1 mL.

For example: If the amount of cell suspension needed is 0.4 mL, add 0.6 mL of PBS to make 1 mL.

4. Cytospin preparation of slides
   a. Prepare two cytospin slides for testing.
   b. Resuspend 1 mL cell suspension by vortexing tube.
   c. Add 100 μL of the cell suspension in cytospin assembly.
   d. Centrifuge at 800 rpm for 4 minutes.
   e. Remove slides and air dry.

5. Fixation and permeabilization
   a. Dilute the fixation reagent (reagent C) 1:5 in dH$_2$O in a fume hood (5 mL reagent C, 20 mL dH$_2$O).
   b. Flood slides with diluted reagent C and incubate at room temperature in a fume hood for 10 minutes.
   c. Prepare 100 mL of 1% washing solution by diluting reagent E 1:100 in PBS (1 mL reagent E, 99 mL PBS). Fill one Coplin jar with 1% washing solution.
   d. Dip slides 3 times in 1% washing solution and place slides in same Coplin jar for 5 minutes.
   e. Dilute permeabilization solution (reagent D) 1:5 in dH$_2$O (5 mL reagent D, 20 mL dH$_2$O).
   f. Immerse slides in dilute permeabilization solution for 5 minutes.
   g. Fill Coplin jar with fresh 1% wash solution. Dip slides 3 times in 1% wash solution and place slides in same Coplin jar for 5 minutes.
   h. Rinse slides in dH$_2$O for 15 seconds, then let slides dry. When dry, slides are ready for immediate staining. If not stained immediately store at 2° to 8° C for 24 hours or up to 24 hours at −20° C. Wrap slides in foil and store in a desiccated environment for extended storage.

6. Immunofluorescent staining
   a. Allow slides, including control slide, to come to room temperature. Mark a small circle around the cell area (cytospin area) on the slide to contain the antibody solution during staining.
   b. Immerse slides in Coplin jar containing PBS for 5 minutes. Once slides have been immersed, do not allow specimen area to dry through remainder of staining procedure.
   c. Working with one slide at a time, remove slide from Coplin jar and carefully dry the area around the specimen with a sterile cotton swab. Place slide in moisture chamber and apply 35 μL of monoclonal antibody (reagent F).
   d. Incubate slides for 30 minutes at 37° C.
   e. Remove slides from staining chambers and rinse in a Coplin jar with fresh PBS for 3 minutes.
   f. Again, working with one slide at a time, remove slide from Coplin jar and dry the area around the specimen with a sterile cotton swab. Place slide in moisture chamber and apply 35 μL of FITC-conjugated antibody to the slide.
   g. Incubate slides for 30 minutes at 37° C.
   h. Wash slides in PBS and rinse with dH$_2$O. Drain excess water and air dry. Do not blot.
   i. Add mounting medium and coverslip.

7. Reading and reporting
   a. Examine using fluorescent microscope and 200× magnification (10× eyepiece, 20× objective). Read soon after staining (same day), because fluorescent intensity fades. Scan the whole cell spot on both slides.
   b. Positive cells show homogenous apple-green nuclear staining as demonstrated in the positive control. Negative cells show no apple-green staining. Nonspecific staining, which appears as cytoplasmic or dull yellowish staining, should be reported as negative. Atypical greenish staining of cells on the periphery of the cell inoculum is due to drying artifact and should be reported as negative.
   c. 200,000 cells should be present on each slide, 400,000 on two slides.

   Report positive stains as, "Positive: (number) CMV-positive cells seen per 400,000 cells."

   Report negative stains as, "No CMV-positive cells seen per 400,000 cells analyzed."

8. Quality control
   a. The negative control should have no CMV-infected cells showing fluorescence.
   b. The positive control should have CMV-infected cells showing homogenous apple-green nuclear staining.
   c. If controls are not as above, results are unacceptable and test must be repeated.

Bibliography

Bean, B., 1992. Antiviral therapy: current concepts and practices. Clin. Microbiol. Rev. 5:146.

Chernesky, M.A., Ray, C.G., and Smith, T.F. 1982. Cumitech 15: laboratory diagnosis of viral infections. American Society for Microbiology, Washington, D.C.

Chonmaitree, T., Baldwin, C.D., and Lucia, H.L. 1989. Role of the virology laboratory in diagnosis and management of patients with central nervous system disease. Clin. Microbiol. Rev. 2:1.

Fields, B.N., Knipe, D.M., and Howley, P.M., editors. 1996. Virology, ed 2. Lippincott-Raven, Philadelphia.

Greenberg, S.B. and Krilov, L.R. 1986. Cumitech 21: laboratory diagnosis of viral respiratory disease. American Society for Microbiology, Washington, D.C.

Isenberg, H.D., editor. 1992. Clinical microbiology procedures handbook. American Society for Microbiology, Washington, D.C.

Johnson, F.B. 1990. Transport of viral specimens. Clin. Microbiol. Rev. 3:120.

Leland, D.S. 1996. Clinical virology. WB Saunders, Philadelphia.

Lennette, E.H. and Schmidt, N.J., editors. 1979. Diagnostic procedures for viral, rickettsial, and chlamydial infections, ed 5. American Public Health Association, Washington, D.C.

Levine, A.J. 1991. Viruses. Scientific American Library, New York.

Minnich, L.L., Smith, T.F., and Ray, C.G. 1988. Cumitech 24: rapid detection of viruses by immunofluorescence. American Society for Microbiology, Washington, D.C.

Murray, P.R., Baron, E.J., Pfaller, M.A., et al., editors. 1995. Manual of clinical microbiology, ed 6. American Society for Microbiology, Washington, D.C.

Murray, P.R., Kobayashi, G.S., Pfaller, M.A., et al., editors. 1994. Medical microbiology, ed 2. Mosby, St. Louis.

Okano, M., Thiele, G.M., Davis, J.R., et al. 1988. Epstein-Barr virus and human diseases: recent advances in diagnosis. Clin. Microbiol. Rev. 1:300.

Persing, D.H. 1991. Polymerase chain reaction: trenches to benches. J. Clin. Microbiol. 29:1281.

Sherlock, C.H., Brandt, C.J., Middleton, P.J., et al. 1989. Cumitech 26: laboratory diagnosis of viral infections producing enteritis. American Society for Microbiology, Washington, D.C.

Specter, S. and Lancz, G.J. 1986. Clinical virology manual. Elsevier, New York.

White, D.O. and Fenner, F.J. 1994. Medical virology, ed 4. Academic Press, San Diego.

Wolcott, M.J. 1992. Advances in nucleic acid-based detection methods. Clin. Microbiol. Rev. 5:370.

*Page numbers in italics indicate illustrations; t indicates tables.